Metabolism and Pathophysiology of Bariatric Surgery

Metabolism and Pathophysiology of Bariatric Surgery

Nutrition, Procedures, Outcomes, and Adverse Effects

Rajkumar Rajendram
Stoke Mandeville Hospital, Aylesbury, United Kingdom;
King's College London, London, United Kingdom;
King Abdulaziz Medical City, Ministry of National Guard Hospital Affairs,
Riyadh, Saudi Arabia

Colin R. Martin
Buckinghamshire New University, Middlesex, United Kingdom

Victor R. Preedy
King's College London, London, United Kingdom

AMSTERDAM • BOSTON • HEIDELBERG • LONDON • NEW YORK • OXFORD • PARIS
SAN DIEGO • SAN FRANCISCO • SINGAPORE • SYDNEY • TOKYO

Academic Press is an imprint of Elsevier

Academic Press is an imprint of Elsevier
125 London Wall, London EC2Y 5AS, United Kingdom
525 B Street, Suite 1800, San Diego, CA 92101-4495, United States
50 Hampshire Street, 5th Floor, Cambridge, MA 02139, United States
The Boulevard, Langford Lane, Kidlington, Oxford OX5 1GB, United Kingdom

Notices

Knowledge and best practice in this field are constantly changing. As new research and experience broaden our understanding, changes in research methods, professional practices, or medical treatment may become necessary.

Practitioners and researchers must always rely on their own experience and knowledge in evaluating and using any information, methods, compounds, or experiments described herein. In using such information or methods they should be mindful of their own safety and the safety of others, including parties for whom they have a professional responsibility.

To the fullest extent of the law, neither the Publisher nor the authors, contributors, or editors, assume any liability for any injury and/or damage to persons or property as a matter of products liability, negligence or otherwise, or from any use or operation of any methods, products, instructions, or ideas contained in the material herein.

British Library Cataloguing-in-Publication Data
A catalogue record for this book is available from the British Library

Library of Congress Cataloging-in-Publication Data
A catalog record for this book is available from the Library of Congress

ISBN: 978-0-12-804011-9

For Information on all Academic Press publications
visit our website at https://www.elsevier.com

Working together
to grow libraries in
developing countries

www.elsevier.com • www.bookaid.org

Publisher: Mica Haley
Acquisition Editor: Stacy Masucci
Editorial Project Manager: Sam Young
Production Project Manager: Karen East and Kirsty Halterman
Designer: Victorial Pearson

Typeset by MPS Limited, Chennai, India

Contents

5. Percutaneous Electrical Neurostimulation of Dermatome T6 to Reduce Appetite

J. Ruiz-Tovar and C. Llavero

6. The Management of Obesity: An Overview

M. Camilleri, B. Abu Dayyeh and A. Acosta

Section II
Surgical and Postsurgical Procedures

7. Why Patients Select Weight Loss Bariatric Surgery

W. Guan and P.J. Brantley

8. Best Practices for Bariatric Procedures in an Accredited Surgical Center

T. Javier Birriel and M. El Chaar

Section III
Safety and Outcomes

Section IV
Metabolism, Endocrinology and Organ Systems

32. Endoscopic Treatments for Obesity-Related Metabolic Diseases

G. Lopez-Nava, M. Galvao Neto and J.W.M. Greve

33. Sleeve Gastrectomy: Mechanisms of Weight Loss and Diabetes Improvements

P.K. Chelikani and D. Sekhar

34. Postprandial Hyperinsulinemic Hypoglycemia in Bariatric Surgery

L.J.M. de Heide, M. Emous and A.P. van Beek

35. Bariatric Surgery Improves Type 2 Diabetes Mellitus

J.W.M. Greve

41. Thyroid Hormone Homeostasis in Weight Loss and Implications for Bariatric Surgery

L.B. Sweeney, G.M. Campos and F.S. Celi

42. The Ghrelin—Cannabinoid 1 Receptor Axis After Sleeve Gastrectomy

C. Fedonidis, D. Mangoura and N. Alexakis

43. *PNPLA3* Variant p.I148M and Bariatric Surgery

R. Liebe, F. Lammert and M. Krawczyk

Section V
Nutritional Aspects

44. Dietary Reference Values

M.Y. Price and V.R. Preedy

Section VI
Cardiovascular, Body Composition, and Physiological Aspects

Section VIII
Resources

List of Contributors

Barham Abu Dayyeh, Mayo Clinic, Rochester, MN, United States

Andres Acosta, Mayo Clinic, Rochester, MN, United States

Sanjay Agrawal, Homerton University Hospital NHS Foundation Trust & Honorary, London, United Kingdom

Varun Agrawal, University of Vermont College of Medicine, Burlington, VT, United States

Nicholas Alexakis, University of Athens, Athens, Greece

Martin A. Alpert, University of Medicine-Columbia School of Medicine, Columbia, MO, United States

Abdallah Al-Salameh, Hôpital Bicêtre, Le Kremlin-Bicêtre, France; Assistance Publique-Hôpitaux de Paris, Paris, France

Konstantinos G. Apostolou, Laiko General Hospital, Athens, Greece

Ali Ardestani, Harvard Medical School, Boston, MA, United States; Brigham and Women's Hospital, Boston, MA, United States

Ibrahim Aslan, Research and Education Hospital, Antalya, Turkey

Mutay Aslan, Akdeniz University Medical Faculty, Antalya, Turkey.

Brenno Astiarraga, University of Pisa, Pisa, Italy

Molly Atwood, Ryerson University, Toronto, ON, Canada

Alfonso Barbarisi, Second University of Naples, Naples, Italy

Mohamed Bekheit, El Kabbary general Hospital, Alexandria, Egypt

Josiah Billing, Eviva Bariatrics, Edmonds, WA, United States

Peter Billing, Eviva Bariatrics, Edmonds, WA, United States

Giuseppe Boldrini, Catholic University Medical School, Roma, Italy

Phillip J. Brantley, Pennington Biomedical Research Center, Baton Rouge, LA, United States

Luca Busetto, Padova University Hospital, Padova, Italy

Virginia F. Byron, Ben-Gurion University of the Negev, Be'er-sheva, Israel

Stefania Camastra, University of Pisa, Pisa, Italy

Michael Camilleri, Mayo Clinic, Rochester, MN, United States

Guilherme M. Campos, Virginia Commonwealth University, Richmond, VA, United States

Marilia Carabotti, University "Sapienza", Rome, Italy

Joseph Caruana, ECMC, Buffalo, NY, United States

Daniela S. Casagrande, Hospital de Clinicas de Porto Alegre, Porto Alegre, RS, Brazil

Stephanie Cassin, Ryerson University, Toronto, ON, Canada

Everton Cazzo, State University of Campinas (UNICAMP), Campinas, Brazil

Francesco S. Celi, Virginia Commonwealth University, Richmond, VA, United States

Elinton A. Chaim, State University of Campinas (UNICAMP), Campinas, Brazil

Ron Charach, Ben-Gurion University of the Negev, Be'er-sheva, Israel

Gabriela V. Chaves, Brazilian National Cancer Institute, Rio de Janeiro, RJ, Brazil

Prasanth K. Chelikani, University of Calgary, Calgary, AB, Canada

Mingyi Chen, Davis Medical Center, Sacramento, CA, United States

Eva M. Conceição, Universidade do Minho, Braga, Portugal

Louise Crovesy, Federal University of Rio de Janeiro, Rio de Janeiro, RJ, Brazil

Paresh Dandona, State University of New York at Buffalo, Buffalo, NY, United States

Nickolas Dasher, University of Michigan, Ann Arbor, MI, United States

Lauren David, Ryerson University, Toronto, ON, Canada

Dafydd A. Davies, The IWK Children's Health Centre, Halifax, NS, Canada

Mariane de Almeida Cardeal, Gastrocirurgia de Brasília, Brasilia, DF, Brazil

Loek J.M. de Heide, Medical Centre Leeuwarden, Leeuwarden, The Netherlands

Marcela F. de Novais, Gastrocirurgia de Brasília, Brasilia, DF, Brazil

Maria R.M. de Oliveira, Bioscience Institute of the University Júlio de Mesquita Filho, UNESP, Botucatu, SP, Brazil

Chiara De Panfilis, University of Parma, Parma, Italy

Elizabeth Dettmer, Hospital for Sick Children, Toronto, ON, Canada

Parag Dhumane, Lilavati Hospital and Research Center, Mumbai, India

Fernando Dip, Cleveland Clinic Florida, Weston, FL, United States

Gianfranco Donatelli, Hôpital Privé des Peupliers, Paris, France

Dana L. Duren, University of Missouri, Columbia, MO, United States

Shenelle Edwards-Hampton, Wake Forest Baptist Medical Center, Winston-Salem, NC, United States

Maher El Chaar, St. Luke's University Hospital and Health Network, Allentown, PA, United States

Athar S. Elward, Cairo University, Giza, Egypt

Marloes Emous, Medical Centre Leeuwarden, Leeuwarden, The Netherlands

Roberto Fabris, Padova University Hospital, Padova, Italy

Melissa G. Farb, Boston University School of Medicine, Boston, MA, United States

Gil Faria, Hospital de Santo António, Porto, Portugal; Abel Salazar Biomedical Sciences Institute, Porto, Portugal

Orlando P. Faria, Gastrocirurgia de Brasília, Brasilia, DF, Brazil

Silvia L. Faria, University of Brasilia, Brasilia, DF, Brazil

Constantinos Fedonidis, BRFAA, Athens, Greece

Andrew W. Froehle, Wright State University Boonshoft School of Medicine, Kettering, OH, United States

Michael S. Furman, Warren Alpert Medical School of Brown University, Providence, RI, United States; Rhode Island Hospital, Providence, RI, United States

Marinos Fysekidis, Assistance Publique-Hôpitaux de Paris, Paris, France; Hôpital Avicenne, Bobigny, France

Alfredo Genco, La Sapienza University, Rome, Italy

Irene Generali, University of Parma, Parma, Italy

Ina Gesquiere, University Hospitals Leuven/KU Leuven, Leuven, Belgium

Husam Ghanim, State University of New York at Buffalo, Buffalo, NY, United States

Noyan Gokce, Boston University School of Medicine, Boston, MA, United States

Nicolas Goossens, University Hospital Geneva, Geneva, Switzerland; Tisch Cancer Institute, New York, NY, United States

Ralph Green, Davis Medical Center, Sacramento, CA, United States

Jan Willem M. Greve, Medical Director Obesity Clinics South, Heerlen, The Netherlands; Zuyderland Medical Center, Heerlen, The Netherlands

Win Guan, Pennington Biomedical Research Center, Baton Rouge, LA, United States

Jill Hamilton, Hospital for Sick Children, Toronto, ON, Canada

Luzia J. Hintze, University of Ottawa, Ottawa, ON, Canada

Tomas Javier Birriel, St. Luke's University Hospital and Health Network, Allentown, PA, United States

Amanda M. Johner, The Oregon Clinic and Legacy Weight & Diabetes Institute, Portland, OR, United States

Minoa Jung, University Hospital Geneva, Geneva, Switzerland

Jan P. Kamiński, University of Illinois Metropolitan Groups Hospitals, Chicago, IL, United States

Natraj Katta, University of Medicine-Columbia School of Medicine, Columbia, MO, United States

Jedediah Kaufman, Eviva Bariatrics, Edmonds, WA, United States

Silvana M.B. Kelles, Hospital das Clínicas da Universidade Federal de Minas Gerais, Belo Horizonte, Minas Gerais, Brazil

Muhammad Faisal Khan, Aga Khan University Hospital, Karachi, Pakistan

Timothy R. Koch, MedStar-Washington Hospital Center, Washington, DC, United States; Center for Advanced Laparoscopic General & Bariatric Surgery, Washington, DC, United States

Katerina Kotzampassi, Aristotle's University of Thessaloniki School of Medicine, Thessaloniki, Greece

Marcin Krawczyk, Saarland University Medical Center, Homburg, Germany; Medical University of Warsaw, Warsaw, Poland

Blandine Laferrère, Columbia University College of Physicians and Surgeons, New York, NY, United States

Frank Lammert, Saarland University Medical Center, Homburg, Germany

Jacob C. Langer, Hospital for Sick Children, Toronto, ON, Canada

Derek Larkin, Edge Hill University, Ormskirk, United Kingdom

Richard T. Laughlin, Wright State University Boonshoft School of Medicine, Dayton, OH, United States

Belinda Lennerz, Boston Children's Hospital, Boston, MA, United States

Roman Liebe, Saarland University Medical Center, Homburg, Germany

Carolina Llavero, Electrostimulation Unit, Garcilaso Clinic, Madrid, Spain

Emanuele Lo Menzo, Cleveland Clinic Florida, Weston, FL, United States

G. Lopez-Nava, Madrid Sanchinarro University Hospital, Madrid, Spain; San Pablo University of Madrid (CEU), Madrid, Spain

Michele Lorenzo, ASL NA 3 SUD, Torre Annunziata, Italy

William Lynn, Homerton University Hospital NHS Foundation Trust, London, United Kingdom

Fernanda C.C.M. Magno, Federal University of Rio de Janeiro, Rio de Janeiro, RJ, Brazil

Ajay V. Maker, University of Illinois Metropolitan Groups Hospitals, Chicago, IL, United States; University of Illinois at Chicago, Chicago, IL, United States

Vijay K. Maker, University of Illinois Metropolitan Groups Hospitals, Chicago, IL, United States; University of Illinois at Chicago, Chicago, IL, United States

Dimitra Mangoura, BRFAA, Athens, Greece

Colin R. Martin, Buckinghamshire New University, Middlesex, United Kingdom

Christophe Matthys, University Hospitals Leuven/KU Leuven, Leuven, Belgium

Milene Moehlecke, Hospital de Clinicas de Porto Alegre, Porto Alegre, RS, Brazil

Violeta Moizé, Hospital Clinic, Barcelona, Spain

Scott Monte, School of Pharmacy & Pharmaceutical Sciences, Amherst, NY, United States

Claudio Cora Mottin, Hospital de Clinicas de Porto Alegre, Porto Alegre, RS, Brazil

Giuseppe Nanni, Catholic University Medical School, Roma, Italy

Nelson Nardo, State University of Maringá, Maringá, PR, Brazil

Anand Nath, MedStar-Washington Hospital Center, Washington, DC, United States

Manoel G. Neto, Gastro Obeso Center and Mario Covas Hospital, São Paulo, SP, Brazil; Florida International University, Miami, FL, United States; ABC Medical School, Santo Andre, Brazil

Alex Ordonez, Previty Clinic, Beaumont, TX, United States

José C. Pareja, State University of Campinas (UNICAMP), Campinas, SP, Brazil

Alfons Pomp, New York Presbyterian–Weill Cornell Medical College, New York, NY, United States

Victor R. Preedy, King's College London, London, United Kingdom

Mina Y. Price, Royal Free London NHS Foundation Trust, London, United Kingdom

Hong Qiu, Davis Medical Center, Sacramento, CA, United States

Karina Quesada, University of Marília, Marília, SP, Brazil

Rajkumar Rajendram, Stoke Mandeville Hospital, Aylesbury, United Kingdom; King's College London, London, United Kingdom; King Abdulaziz Medical City, Ministry of National Guard Hospital Affairs, Riyadh, Saudi Arabia

Yudi P.G. Ramirez, School of Pharmacological Sciences of the University Júlio de Mesquita Filho, UNESP, Araraquara, SP, Brazil

Irineu Rasera, Center of Excellence in Bariatric Surgery of Piracicaba, Piracicaba, SP, Brazil

Michele N. Ravelli, School of Pharmacological Sciences of the University Júlio de Mesquita Filho, UNESP, Araraquara, SP, Brazil

Kevin M. Reavis, The Oregon Clinic and Legacy Weight & Diabetes Institute, Portland, OR, United States

D.D. Rosa, Hospital de Clinicas de Porto Alegre, Porto Alegre, RS, Brazil

Eliane L. Rosado, Federal University of Rio de Janeiro, Rio de Janeiro, RJ, Brazil

Raul J. Rosenthal, Cleveland Clinic Florida, Weston, FL, United States

Jaime Ruiz-Tovar, Electrostimulation Unit, Garcilaso Clinic, Madrid, Spain

Giuseppe Scalera, Second University of Naples, Naples, Italy

Beatriz D. Schaan, Hospital de Clinicas de Porto Alegre, Porto Alegre, RS, Brazil

Luigi Schiavo, Second University of Naples, Naples, Italy

Deepa Sekhar, University of Calgary, Calgary, AB, Canada

Roberto Serra, Padova University Hospital, Padova, Italy

Carola Severi, University "Sapienza", Rome, Italy

Eyal Sheiner, Ben-Gurion University of the Negev, Be'er-sheva, Israel

Richard J. Sherwood, University of Missouri, Columbia, MO, United States

Eric G. Sheu, Harvard Medical School, Boston, MA, United States; Brigham and Women's Hospital, Boston, MA, United States

Timothy R. Shope, MedStar-Washington Hospital Center, Washington, DC, United States; Georgetown University School of Medicine, Washington, DC, United States

Anne D. Shrewsbury, West Sussex, United Kingdom

Jacqueline S. Silva, Federal University of Rio de Janeiro, Rio de Janeiro, RJ, Brazil

George Stavrou, Thessaloniki, Greece

Lee L. Swanstrom, Oregon Health and Sciences University, Portland, OR, United States; Institut Hopitalo Universitaire, Strasbourg, France

Lori B. Sweeney, Virginia Commonwealth University, Richmond, VA, United States

David W. Swenson, Warren Alpert Medical School of Brown University, Providence, RI, United States; Rhode Island Hospital, Providence, RI, United States

Samuel Szomstein, Cleveland Clinic Florida, Weston, FL, United States

Ali Tavakkoli, Harvard Medical School, Boston, MA, United States; Brigham and Women's Hospital, Boston, MA, United States

Renee M. Tholey, New York Presbyterian—Weill Cornell Medical College, New York, NY, United States

Andre P. van Beek, University Medical Centre Groningen, Groningen, The Netherlands

Bart Van der Schueren, University Hospitals Leuven/KU Leuven, Leuven, Belgium

Roberto Vettor, Padova University Hospital, Padova, Italy

Josep Vidal, Hospital Clinic, Barcelona, Spain

Kristen L. Votruba, University of Michigan, Ann Arbor, MI, United States

Martin Wabitsch, Ulm University Medical Center, Ulm, Germany

Sharlene Wedin, Medical University of South Carolina, Charleston, SC, United States

Foreword

Having trained and worked in two of the largest bariatric centers in the United Kingdom (King's College Hospital NHS Foundation Trust and Imperial Healthcare NHS Trust, London), within both clinical and academic settings, I fully appreciate the enormity of the challenge obesity poses to health care systems.

The World Health Organization (WHO) estimated that globally, in 2014, more than 1.9 billion adults were overweight and more than half a billion were obese. The prevalence more than doubled between 1980 and 2014. Obesity, which was once associated only with high-income countries, is now also prevalent in low- and middle-income countries as well. Furthermore, overweight children are likely to become obese adults, thus, childhood obesity and overweight is one of the most serious public health challenges of the 21st century. Obesity and overweight can have a variety of adverse health consequences and metabolic effects associated with a high rate of death, such as type 2 diabetes mellitus, hypertension, dyslipidemia, obstructive sleep apnea, steatohepatitis, and certain types of cancer. It is estimated that around 65% of the world's population lives in a country where overweight and obesity kills more people than underweight.

The huge economic burden to the health systems across the world highlights the importance of collectively addressing this global epidemic. Currently, bariatric (weight loss) surgery has become the only long-term effective treatment for severe (morbid) obesity, as calorie-restricted diets and drug therapy have had disappointing results for weight loss. Remarkably, bariatric surgery not only helps to achieve significant and sustained weight loss, but also leads to multiple metabolic benefits. Thus, the consensus among experts in the field is to refer to bariatric surgery as "metabolic surgery." Over the past decade, there has been a significant global rise in basic science and translational research studies looking into the pathophysiology of bariatric surgery, which potentially holds the key to the long-term management of this epidemic. With all its merits in contributing to metabolic improvements, bariatric surgery is also associated with complications that can be devastating if not managed appropriately in well-established bariatric centers of excellence.

This book, "Pathophysiology of Bariatric Surgery: Metabolism, Nutrition Procedures, Outcomes, and Adverse Effects," takes the reader on a journey through the complex world of obesity, highlighting that there is more to obesity than excess weight or an expanding waistline. The authors provide an in-depth review of how weight contributes to obesity-related comorbidities, and the rationale behind the different surgical procedures used. There is also comprehensive insight into the nutritional and metabolic complications, as well as the much-neglected psychological and behavioral aspects, of obesity before and after bariatric surgery. Ideally, the intended audience will appreciate and relate to the multiple challenges faced by obese patients in dealing with their physical and mental health issues as they weigh the favorable results and adverse outcomes of bariatric surgery. Ultimately, it is a life-changing surgical intervention that leads to either euphoria or despair.

The content of this book is relevant and current to my own clinical practice and obesity-related research, which makes it a timely publication. It is intended to be both informative and easy to read as an important reference for clinicians, allied health professionals, students, and all those interested in the complexity and management of obesity.

Dr. Royce P. Vincent, *MBBS, MSc, EuSpLM, FRCPath, MD*
Consultant Chemical Pathologist and Clinical Lead, Department of Clinical Biochemistry,
King's College Hospital NHS Foundation Trust, London, UK
Honorary Senior Lecturer, Department of Nutrition and Dietetics, King's College London, UK

Preface

In the United States, about one-third of the population are obese. While no state in the United States has an incidence less than 20%, in some communities, 50% of the adult population may be obese. The adverse effects of obesity include increased cancer rates, metabolic syndrome and diabetes, heart disease, stroke, sleep apnea, liver disease, musculoskeletal problems, and various psychological changes. Thus, obesity can be considered a disease that affects not only the individual, but the family unit, the local community, and the nation as a whole. For example, the medical cost of obesity in the United States alone is $150 billion annually.

There are various strategies to reduce obesity. These include dietary changes, behavioral modifications (including exercise), and drugs (that can cause malabsorption or alterations in satiety−appetite signaling). However, they are not always effective, and at that stage bariatric surgery becomes a viable alternative.

Bariatric surgery has been shown to improve numerous psychological, metabolic, physiological, and functional parameters. These include quality of life measures, diabetes, hypertension, hyperlipidemia, and sleep apnea. However, there are different types of bariatric surgery, including Roux-en-Y gastric bypass (RYGB), gastric banding, sleeve gastrectomy (SG), biliopancreatic diversion (BPD), and other variations of these procedures. The various weight loss procedures have different levels of popularity, outcomes, and success rates. Their effects on reducing obesity and comorbidities are dissimilar as well. Dissimilar bariatric procedures also have variable cellular and tissue effects, as well as nutritional complications. It is therefore clear that there are complex interrelationships between obesity and metabolic profiles before and after bariatric surgery. However, understanding these relationships has been difficult as the information relating to nutrition, surgical procedures, outcomes, and side effects have never been marshaled into a single text. *Pathophysiology of Bariatric Surgery* addresses this in a comprehensive way.

The book has eight sections:
1. Features of Obesity and Strategies for Weight Loss
2. Surgical and Postsurgical Procedures
3. Safety and Outcomes
4. Metabolism, Endocrinology, and Organ Systems
5. Nutritional Aspects
6. Cardiovascular, Body Composition, and Physiological Aspects
7. Psychological and Behavioral Aspects
8. Resources

The Editors recognize the difficulties in assigning some chapters to different sections. Very often chapters cover different scientific domains, and they could equally fit into one of several sections of the book. However, this is resolved by the excellent indexing system compiled by Elsevier.

Novel features in each chapter include a *Mini-Dictionary of Terms, Key Facts, and Summary Points.*

Contributors are authors of international and national standing, leaders in the field, and trendsetters. The emerging fields of obesity and bariatric surgery, as well as important discoveries relating to diet and nutritional health, are also incorporated in *Pathophysiology of Bariatric Surgery*. This represents essential reading for nutritionists, dietitians, surgeons, health care professionals, research scientists, molecular and cellular biochemists, physicians, general practitioners, public health workers, and anyone interested in well-being in general.

Editors
Rajkumar Rajendram, Colin R. Martin & Victor R. Preedy

Section I

Features of Obesity and Strategies for Weight Loss

Chapter 1

Obesity and Cardiac Failure: Pathophysiology, Epidemiology, Clinical Manifestations, and Management

M.A. Alpert and N. Katta

University of Medicine-Columbia School of Medicine, Columbia, MO, United States

LIST OF ABBREVIATIONS

BMI   body mass index
CO   cardiac output
HF   heart failure
HTN   systemic hypertension
LV   left ventricular
LVEF left ventricular ejection fraction
LVH   left ventricular hypertrophy
RAAS renin-angiotensin-aldosterone system

INTRODUCTION

The relation of obesity to heart disease has been a subject of interest since ancient times, but has been studied most extensively during the past half century [1−7]. Obesity affects the heart in multiple ways. However, the most intense focus, as it relates to bariatric surgery, has been on the impact of obesity on cardiac performance and morphology and its relation to heart failure (HF) [3−7]. The purpose of this chapter is to explore these issues.

CLASSIFICATION OF OBESITY

The World Health Organization classifies body weight on the basis of body mass index (BMI). Table 1.1 summarizes the current and proposed body weight classifications. For purposes of this review, the term severe obesity will refer to BMI $\geq 40.0 \text{ kg/m}^2$. Central obesity is commonly defined as waist circumference >102 cm in men and >88 cm in women, or waist−hip ratio >1.0 in men and ≥ 0.8 in women.

CARDIAC PERFORMANCE AND MORPHOLOGY IN OBESITY

Obesity causes changes in cardiac performance that may produce alterations in cardiac morphology and impairment of ventricular function in adults. Such maladaptation is most pronounced in severely obese persons, but may occur to a lesser extent in overweight, class I or class II obese patients [1,3−45]. These alterations in cardiac structure and function have also been reported in obese children and adolescents [1,3−5]. The following comments on the effects of obesity on cardiac performance and morphology apply primarily to severe obesity.

TABLE 1.1 Body Weight Classification

Classification	BMI (kg/m^2)
Underweight[a]	<18.5
Normal weight[a]	18.5−24.9
Overweight[a]	25.0−29.9
Class I obesity[a]	30.0−34.9
Class II obesity[a]	35.0−39.9
Class III obesity[b,c]	≥40.0
Class IV obesity[b,d]	≥50.0
Class V obesity[b]	≥60.0

[a]World Health Organization classification.
[b]Proposed classification by American Heart Association.
[c]Also referred to as severe, extreme, or morbid obesity.
[d]Sometimes referred to as super-obesity.
Source: Adapted from Poirier P, Giles TD, Bray GA, et al. Obesity and cardiovascular disease: pathophysiology, evaluation, and effect of weight loss: an update of the 1997 American Heart Association Scientific Statement on obesity and Heart Disease from the Obesity Committee of the Council on Nutrition, Physical Activity, and Metabolism. Circulation 2006;113:898−918; Bastien M, Poirier P, Lemieux I, Despres JP. Overview of epidemiology and contribution of obesity to cardiovascular disease. Prog Cardiovas Dis 2014;56:369−81.

Altered Hemodynamics Associated With Obesity

Obesity, particularly severe obesity, is a high cardiac output (CO) state [1,3−15]. It was originally thought that elevated CO in obese individuals resulted exclusively from excess adipose accumulation. However, increased fat mass alone does not completely account for the increase in cardiac CO. Recent studies indicate that fat-free mass contributes to augmentation of CO, possibly to a greater extent than fat mass [1,4,5,7]. The rise in CO is accompanied by a decrease in systemic vascular resistance [3−5,7−10]. Heart rate in obese persons is reportedly similar to or slightly higher than that of normal weight individuals. Thus, increased left ventricular (LV) stroke volume is the predominant cause of increased CO [3−10]. LV dP/dt was normal and LV V$_{max}$ was lower than predicted for lean patients in one study of class II−III obese subjects [11]. Myocardial oxygen consumption was greater than predicted for normal weight patients in the same study [11]. LV end-diastolic pressure and pulmonary capillary wedge pressure at rest are commonly, but not invariably, elevated at rest in severely obese patients [1−5,7,8,10−15]. Pulmonary artery pressure may be elevated in severely obese persons due to left HF, obstructive sleep apnea, and obesity hypoventilation [1−5,8−13]. In such individuals, right ventricular end-diastolic pressure and mean right atrial pressure may be elevated [1−5,8−13]. Pulmonary vascular resistance is often increased in severely obese persons due to pulmonary arterial hypertension from sleep apnea/obesity hypoventilation and/or from pulmonary arterial vasoconstriction in those with LV failure [1,3−5,7,8]. It is not uncommon to encounter a transpulmonic pressure gradient in severely obese patients [8].

Exercise substantially increases central blood volume and LV dP/dt in severely obese patients [1,3−5,8,14,15]. In a study by Kaltman and Goldring, LV end-diastolic pressure increased from 21 to 31 mmHg with aerobic exercise in severely obese patients [14,15]. At a workload three times that of the resting level, augmentation of CO is blunted in severely obese patients [8−10]. Arteriovenous oxygen difference, usually normal at rest, may become substantially elevated during exercise [8−10]. LV end-diastolic pressure rises out of proportion to stroke work in such individuals, indicating reduced LV compliance [8−10].

Changes in Cardiac Morphology Associated With Obesity

Smith and Willius reported autopsy findings in 135 obese individuals [15]. In most, heart weight was greater than that predicted for normal body weight. In nine normotensive patients who died of HF, there was no evidence of primary myocardial disease. The authors attributed increased heart weight to excess epicardial fat, an observation that was later disproved. Subsequently, three studies of postmortem findings in severely obese subjects, comprising a total of 33 patients, reported LV hypertrophy (LVH) in all patients, and right ventricular hypertrophy in six patients [17−19]. Excess epicardial fat was present in 21 patients [17−19]. These studies included patients with systemic hypertension (HTN) and coronary artery disease. Thus, it is uncertain to what extent the pathology described is attributable to obesity.

In 1992, Kasper et al. published a study of 43 obese patients and 409 lean patients with HF [12]. Of those who underwent myocardial biopsy, a specific cause of HF was identified in 64.5% of lean subjects, but only 23.3% of obese subjects. The most common histologic abnormality in obese subjects was LVH. These findings lend credence to the existence of a "cardiomyopathy of obesity," one that is characterized primarily by LVH.

A large number of studies employing noninvasive cardiac diagnostic techniques have compared cardiac morphology in obese and normal weight patients [3−5,7,20]. Nearly all of these studies have shown that LV mass is significantly greater in obese than in normal weight patients. This is certainly true for severely obese subjects, but has also been reported in patients with class I and class II obesity. LV wall thickness is commonly, but not always, increased in obesity. Several studies have shown a strong positive correlation between body weight indices and LV mass [3,4,20]. Some studies have shown a positive correlation between body weight indices and LV diastolic chamber dimension, but this observation is less consistent than the relation between body weight indices and LV mass [3,4,20].

A variety of factors may contribute to LVH development in obese persons [3−7,22−26]. HTN is perhaps the most common risk factor [3−7]. Elevated LV end-systolic wall stress has also been shown to contribute to LVH, even in the absence of HTN [22]. Duration of obesity is also an important predictor of LVH [24]. Volume overload due to obesity increases LV preload, which also contributes to the development of LVH [3,4,22]. Fat-free mass contributes to increased LV mass to a greater extent than fat mass [25]. Multiple neurohormonal and metabolic alterations commonly present with obesity have been associated with the development of LVH [3−5,26−28]. These include activation of the renin-angiotensin-aldosterone system (RAAS) and the sympathetic nervous system, hyperleptinemia associated with leptin resistance, and insulin-resistance with hyperinsulinemia [3−5,26−28]. Insulin-related growth factors have been associated with LVH in human and animal studies [3−5,7,27]. In addition, several studies of murine models of lipotoxicity have reported the presence of LVH [3−5,28].

Based on the hemodynamic alterations that occur with obesity (especially severe obesity), it was predicted that uncomplicated (normotensive) obesity would predispose a patient to eccentric LVH. Early studies appeared to support this hypothesis [20]. However, multiple studies have shown that concentric LV remodeling or concentric LVH occurs as frequently, and in some cases more frequently, than eccentric LVH in obese subjects whose LV geometry is abnormal [3−5,23,27−31]. Some studies showed a predominance of eccentric LVH in uncomplicated obesity [3,4,20,22,32,33]. Some of the studies showing a predominance of concentric LVH or concentric LV remodeling did not exclude patients with HTN, although one study adjusted for it [27]. None of these studies considered the relative duration or severity of obesity and hypertension. However, scrutiny of older studies suggests that concentric LVH or concentric LV remodeling was present to some degree in obese patients [3,4,20]. Possible explanations for concentric LVH or concentric LV remodeling in obese patients include the presence of HTN or pre-HTN, activation of the RAAS, increased sympathetic nervous system tone, the effects of growth factors associated with insulin-resistance and hyperinsulinemia, and failure to consider the relative duration and severity of obesity and HTN [3,4]. In addition, the definition of concentric LVH has changed. What used to be called "eccentric−concentric LVH" (commonly present in hypertensive obese patients) is currently classified as concentric LVH (Fig. 1.1).

LV Diastolic Function in Obesity

Hemodynamic studies of obese subjects have shown that LV filling pressure is frequently high in obese subjects [8,10−15]. Moreover, LV filling pressure may increase substantially during exercise due to reduced LV compliance [14,15]. This is particularly true in severely obese persons [14,15].

LV diastolic function in obese subjects has been extensively studied using the full array of noninvasive cardiac diagnostic techniques [3−5,7,35−38]. Multiple studies comparing LV diastolic function in normal weight and obese subjects have consistently shown LV diastolic filling or relaxation to be impaired in obese subjects relative to normal weight patients, regardless of the severity of obesity and regardless of the diagnostic technique used [3,4,35]. LV diastolic function becomes more impaired as the severity of obesity worsens. Pascual et al. noted abnormal LV diastolic filling in 12% of class I, 35% of class II, and 45% of class III obese patients using Doppler echocardiographic techniques [36]. Some studies have reported greater impairment of LV diastolic function to progressive increases in LV mass [3,4,34,35]. One study of severely obese subjects using Doppler echocardiography showed that impaired LV diastolic filling was present only in those with increased LV mass [34]. Others have described diastolic dysfunction in the absence of LVH [3,4,33]. LV diastolic filling is more severely impaired with a longer duration of obesity [24]. Tissue Doppler imaging studies have shown reduced diastolic mitral annular velocities in obese subjects [37,38].

When considered in total, these studies suggest that LV diastolic dysfunction is common in obesity, and is particularly common in severe obesity. Most, but not all, studies relate this to increased LV mass. The results of tissue Doppler imaging suggest that diastolic dysfunction in obesity may in part be load independent. The degree to which neurohormonal and metabolic factors influence LV diastolic function in obesity apart from their effect on LV mass is uncertain. Several studies of murine models of lipotoxicity have described the presence of LV diastolic dysfunction [3–5,26].

LV Systolic Function in Obesity

LV systolic function, determined using LV ejection phase indices such as LV fractional shortening or LV ejection fraction (LVEF), has been less extensively studied than LV mass and LV diastolic function in obese patients [3–5,7,37–41]. In most obese subjects LV systolic function is normal or supranormal [3–5,7]. Comparison of LV ejection phase indices using noninvasive cardiac techniques have not consistently shown significant differences in obese and normal weight patients [3,4]. When LV systolic function is diminished in such patients, it is usually mildly reduced. The presence of moderate to severe LV systolic dysfunction should elicit an evaluation for comorbidities such as coronary heart disease that may produce such changes. Duration of obesity, severity of obesity, systolic blood pressure, and LV end-systolic wall stress have correlated negatively with LV ejection phase indices [3,4,24,39]. Recent studies using tissue Doppler imaging and LV speckle track imaging have shown decreased mitral annular velocities in systole and abnormal radial LV strain in obese (often asymptomatic) subjects with a normal LVEF [3,4,40,41]. This may indicate that LV systolic dysfunction is more common in obesity than was previously thought. Abnormal tissue Doppler imaging raises the question of whether there exists an intrinsic abnormality of LV systolic function in obese persons. Recently, adipokines including leptin, adiponectin, and resistin, and gut hormones such as glucagon-like peptide-1 and glucose-dependent insulintopic polypeptide, have been identified as possible mediators of LV systolic dysfunction in obesity [42]. In addition, some murine models of lipotoxicity have reported reduced LV fractional shortening [3–5,28].

Right Ventricular Function in Obesity

Right ventricular function has not been extensively studied in obese patients [41,43]. In the MESA-Right Ventricle Study, Chahal et al. reported a larger mean right ventricular end-diastolic volume, a larger mean right ventricular stroke volume, and a significantly lower mean right ventricular ejection fraction after adjustment for LV parameters [41]. Abnormal lateral tricuspid annular systolic and diastolic velocities on tissue Doppler imaging and abnormal circumferential and radial strain on speckle track imaging have also been described in obese subjects, possibly suggesting subclinical right ventricular dysfunction [43].

Obesity and HTN: Effects on the Heart

HTN is present in nearly 50% of class I and class II obese persons, and occurs in up to 60% of severely obese individuals [1–4]. The coexistence of obesity and HTN produces alterations in cardiac structure and function that differ somewhat from those associated with obesity and HTN alone [3,4,44–46]. To fully appreciate the relative contributions of obesity and HTN to cardiac structure and function in any individual, it is important to know the duration and severity of both.

With long-standing severe obesity and poorly controlled HTN, CO and stroke volume remain increased, but less so than in normotensive obese patients. LV stroke work is greater in obese hypertensives than in normotensive obese persons [8,44,45]. Systemic vascular resistance is higher in obese hypertensives than in normotensive obese subjects [3,4,8,44,45]. LV end-diastolic pressure is often elevated in obese hypertensives [3,4,8,44,45]. A hybrid form of LVH may occur in which LV wall thickness is greater and LV diastolic chamber size is less dilated than in normotensive obese patients [3,4,45]. Once called eccentric–concentric LVH, it is now classified as a form of concentric LVH. Left atrial enlargement occurs commonly in obesity HTN, and LV diastolic dysfunction occurs with high frequency [3,4,44]. LV systolic function usually remains normal [3,4,44].

HF AND OBESITY

Obesity is a risk factor for HF, and in severely obese individuals may serve as a primary cause of HF [1,3−5]. The changes in cardiac performance and morphology described previously predispose patients to HF in all classes of obesity, and in severely obese persons may be sufficient to serve as the primary pathophysiologic basis for HF [1,3−5].

Epidemiology

Kenchaiah et al. studied 5881 patients enrolled in the Framingham Heart Study and reported that 8.4% of those with class I and II obesity developed HF during a mean follow-up period of 14 years [47]. For every 1 kg/m^2 increase in BMI, there was an increase in risk of HF of 7% in women and 5% in men [47]. The risk of HF was significantly greater in overweight patients than in obese subjects [47]. Similarly, the risk of HF was significantly greater in obese than in overweight patients [47]. Alpert and colleagues reported that 24 of 74 class III obese subjects had clinical evidence of HF [48]. The prevalence of HF rose to 90% in those who were severely obese for more than 20 years [48]. A retrospective analysis of data from the National Health and Nutrition Examination Survey (NHANES-1) study also suggested that obesity serves as a risk factor for HF [49]. In a study of a low-risk Mediterranean outpatient population, obesity was identified as an independent risk factor for HF [50]. Obesity is also a risk factor for HF in hospitalized patients. In a study of more than 6000 inpatients discharged with a diagnosis of HF, Owan et al. reported an incidence of obesity of 41.4% in patients with HF with a preserved LVEF and 35.5% in subjects with HF with a reduced LVEF [51].

The Obesity Paradox in Patients With HF

Although there is no question that HF is associated with reduced survival, there is increasing evidence of the presence of an obesity paradox in such patients with respect to mortality [3,4,6,52]. The paradox is that overweight and obese patients with HF live longer than normal weight patients with similar degrees of severity of HF. In a meta-analysis of 28,209 patients with HF reported by Oreopoulos et al., all-cause mortality was 16% lower in overweight patients and 33% lower in obese subjects with HF compared to normal weight patients [52]. A variety of studies have shown that underweight patients have higher mortality rates than normal weight or class I obese patients [3,4,6]. In some studies, mortality in class II obese patients is lower than in normal weight patients, but in other studies there is a trend toward higher mortality risk [3,4,6]. Mortality in class III obese patients has not been extensively studied or compared to other obesity classes. Limited data suggest that mortality risk in such patients is higher than that of normal weight, overweight, and class I and II obese patients, and is more comparable to that of underweight subjects [3,4,6]. Thus, it is likely that mortality risk in HF patients forms a "U" curve with respect to weight with the highest mortality rates in underweight and severely obese patients. The obesity paradox as it relates to HF appears to be applicable to a variety of populations including males and females, elderly and nonelderly patients, those with acute or chronic HF, patients with HF with preserved or reduced LVEF, and those with central and peripheral obesity. Possible explanations for the obesity paradox in patients with HF include greater metabolic reserves and less cachexia, greater muscle mass, better cardiorespiratory fitness, attenuation of the RAAS response, earlier diagnosis and more aggressive medical therapy in overweight and class I and II obese patients than in normal weight subjects [3,4,6].

OBESITY CARDIOMYOPATHY

Obesity cardiomyopathy can be defined as HF that is due predominantly or entirely to obesity [3−5,53]. Obesity cardiomyopathy occurs almost exclusively in severely obese patients [3,4,53]. To date, no unique cardiac structural or histologic abnormality has been consistently described in humans with obesity cardiomyopathy [3,4,54]. LVH is the most common structural abnormality encountered [3,4,54]. The hemodynamic and structural abnormalities described previously are noted in patients with obesity cardiomyopathy, except that alterations in cardiac performance and morphology are more pronounced in this syndrome than in asymptomatic obese persons [3,4,48,53]. The pathophysiology of obesity cardiomyopathy is summarized in Fig. 1.1.

Clinical Manifestations

Obesity cardiomyopathy is associated with symptoms and signs of HF, some of which are identical to those of other causes of HF and some of which are unique to severe obesity [3,4,53]. General symptoms and signs include dyspnea on exertion, paroxysmal nocturnal dyspnea, lower extremity edema (often brawny) weight gain, increased abdominal girth, jugular venous distension, pulmonary crackles, and gallop rhythm. Symptoms and signs that are more specific to severe

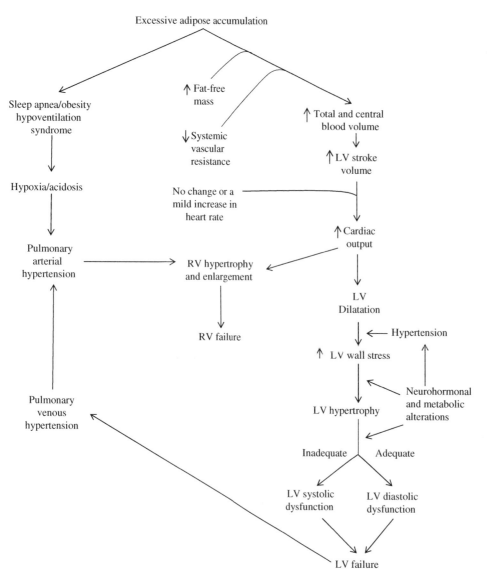

FIGURE 1.1 Pathophysiology of obesity cardiomyopathy. *Adapted from Lavie CJ, Milani RV, Ventura HO. Obesity and cardiovascular disease: risk factor, paradox, and impact of weight loss. J Am Coll Cardiol 2009;53:1925−32.*

obesity include mental confusion and disorientation, somnolence, cyanosis, periodic breathing, subconjunctival suffusion, retinal venous congestion and papilledema, and in some cases sudden death. Cardiac murmurs are frequently absent. HF tends to be episodic and follows recent weight gain. Most patients have been severely obese for at least 10 years. Sleep apnea is present in up to 50% of patients, and obesity hypoventilation occurs in 10−20% [1,3,4,53].

Plasma natriuretic peptide levels are lower in obese patients than in lean patients with comparable degrees of severity of HF. In severely obese patients they may be up to 50% lower [3,4].

Management of Obesity Cardiomyopathy: General Measures

Exacerbations of HF are treated with sodium restriction, low-flow inspired oxygen, and loop diuretics [3,4]. With biventricular failure associated with bowel edema, intravenous loop diuretics are commonly used as initial therapy [3,4]. Drugs such as torsemide or bumetanide may be preferable to furosemide in such patients. If moderate to severe LV systolic dysfunction is present, then RAAS blockers should be considered [3,4]. Appropriate treatment of HTN should be provided. There are no specific antihypertensive drug regimens that are preferred, but angiotensin-converting enzyme inhibitors and angiotensin receptor blockers are commonly used as initial therapy [3,4,44]. Digoxin is used to help control the ventricular rate in patients with atrial fibrillation and in those with severe LV systolic dysfunction who remain symptomatic despite diuresis, beta-blockade, and RAAS blockade [3,4]. The role of beta-blockers, calcium channel blockers, direct-acting vasodilators, and endothelin antagonists in obese hypertensives with normal LV systolic function without pulmonary hypertension is uncertain.

Effect of Weight Loss on Cardiac Performance, Cardiac Morphology, and HF

The most effective method for reversing alterations in cardiac performance and morphology in patients with obesity cardiomyopathy is substantial weight loss [3,4,7,42,54–61]. Bariatric surgery has been more effective than diet, exercise, and pharmacotherapy in achieving hemodynamic and structural changes, presumably due to the greater degree of weight reduction achieved using bariatric surgery techniques [1,3,5,42,54–61]

Multiple studies have demonstrated that substantial weight loss is capable of reducing total and central blood volume, CO, and stroke volume [3,4,13,54–57]. Systemic vascular resistance increases. LV end-diastolic pressure and pulmonary capillary wedge pressure do not consistently decrease following weight loss [3,4,14,54–57]. Whether right heart pressures decrease following weight loss depends on whether LV filling pressure decreases and whether sleep-disordered breathing improves [3,4]. Most studies assessing LV diastolic filling or relaxation after weight loss have shown an improvement, possibly related to regression of LVH [56,58]. The response of LV systolic function to substantial weight loss is variable. LV systolic function increased significantly in subjects with obesity cardiomyopathy and LV systolic dysfunction in one study [39]. When LV systolic function is due to comorbidities such as prior myocardial infarction, the response of LV function to weight reduction may be blunted [42].

Relatively few studies have assessed the effect of weight loss on HF symptoms and signs, functional capacity, and quality of life. Four small case series have demonstrated reversal of signs of HF, improvement in New York Heart Association functional class, reduction of dyspnea and edema, and improvement of quality of life with weight loss (predominantly from bariatric surgery) in severely obese patients with obesity cardiomyopathy [48,59–61]. A more extensive discussion of the effects of bariatric surgery on cardiac performance, morphology, and HF is presented in a later chapter.

MINI-DICTIONARY OF TERMS

- *Body mass index*: Body weight index calculated by dividing weight in kilograms by height in meters squared.
- *Heart failure*: Clinical syndrome resulting from any structural or functional impairment of ventricular filling or ejection of blood (American Heart Association).
- *Cardiac output*: Product of LV stroke volume and heart rate.
- *Systemic vascular resistance*: Physiologic relationship of systemic pressure and flow calculated by dividing the difference between mean blood pressure and mean right atrial pressure by CO.
- *Systemic hypertension*: Blood pressure >140 mmHg systolic and/or >90 mmHg diastolic.
- *Severe obesity*: Defined herein as BMI ≥ 40 kg/m^2.
- *Left ventricular hypertrophy*: Increased LV mass.
- *Eccentric left ventricular hypertrophy*: Form of LVH characterized by a high LV radius-to-thickness or volume-to-mass ratio; usually caused by LV volume overload states.
- *Concentric left ventricular hypertrophy*: Form of LVH characterized by increased wall thickness and a normal or reduced radius-to-thickness or volume-to-mass ratio; usually caused by LV pressure overload states.
- *Concentric left ventricular remodeling*: LV geometry similar to concentric LVH, but with LV mass insufficient to fulfill criteria for LVH.

KEY FACTS

- Obesity is a risk factor for the development of cardiac (heart) failure
- Obesity, particularly severe obesity, produces alterations in cardiac structure and function that predispose to heart failure
- Heart failure may occur in severely-obese individuals in the absence of other causes of heart disease
- Obesity cardiomyopathy is a term used to describe heart failure due predominantly or entirely to severe chronic obesity
- Overweight and mildly-obese persons with heart failure live longer than normal weight or underweight individuals with heart failure of comparable severity. This is known as the obesity paradox
- High blood pressure and sleep apnea occur commonly in obese persons and may cause changes in cardiac structure and function that make the development of heart failure more likely in obese individuals
- The most effective treatment for obesity cardiomyopathy is voluntary weight loss which may be accomplished by diet and exercise or bariatric surgery. Substantial weight loss is capable of reversing many of the abnormalities of cardiac structure and function as well as many of the clinical manifestations of obesity cardiomyopathy

SUMMARY POINTS

- Severe obesity produces hemodynamic alterations that predispose patients to changes in cardiac morphology, which may lead to impairment of ventricular function and subsequent HF.
- HF due predominantly or entirely to obesity is known as obesity cardiomyopathy.
- Various neurohormonal and metabolic factors may contribute to obesity cardiomyopathy.
- Obesity cardiomyopathy is predominantly characterized by LV failure.
- Pulmonary arterial hypertension due to left HF, sleep apnea, and obesity hypoventilation may lead to right HF in severely obese patients.
- Many of the pathophysiological and clinical alterations associated with obesity cardiomyopathy are reversible following substantial weight loss.

REFERENCES

[1]   Poirier P, Giles TD, Bray GA, et al. Obesity and cardiovascular disease: pathophysiology, evaluation, and effect of weight loss: an update of the 1997 American Heart Association Scientific Statement on obesity and Heart Disease from the Obesity Committee of the Council on Nutrition, Physical Activity, and Metabolism. Circulation 2006;113:898−918.

[2]   Bastien M, Poirier P, Lemieux I, Despres JP. Overview of epidemiology and contribution of obesity to cardiovascular disease. Prog Cardiovas Dis 2014;56:369−81.

[3]   Alpert MA, Omran J, Mehra A, Ardhanari S. Impact of obesity and weight loss on cardiac performance and morphology in adults. Prog Cardiovasc Dis 2014;56:391−400.

[4]   Alpert MA. Obesity and cardiac disease. In: Ahima RS, editor. Metabolic syndrome: a. comprehensive textbook. New York, NY: Springer Meteor; 2016. p. 619−36.

[5]   Wong C, Marwick TH. Obesity cardiomyopathy. Pathogenesis and pathophysiology. Nat Clin Pract Cardiovasc Med 2007;4:436−43.

[6]   Lavie CJ, Milani RV, Ventura HO. Obesity and cardiovascular disease: risk factor, paradox, and impact of weight loss. J Am Coll Cardiol 2009;53:1925−32.

[7]   Abel ED, Litwin SE, Sweeney G. Cardiac remodeling in obesity. Physiol Rev 2008;88:389−419.

[8]   Alexander JK, Alpert MA. Hemodynamic alterations with obesity in man. In: Alpert MA, Alexander JK, editors. The heart and lung in obesity. Armonk, NY: Futura Publishing Co; 1998. p. 45−56.

[9]   Alexander JK, Dennis EW, Smith WG, et al. Blood volume, cardiac output and distribution of systemic blood flow in extreme obesity. Cardiovasc Res Center Bull 1962;1:39−44.

[10] Alexander JK. Obesity and cardiac performance. Am J Cardiol 1964;14:860−5.

[11] DeDivitiis O, Fazio S, Petitto M, et al. Obesity and cardiac function. Circulation 1981;64:477−82.

[12] Kasper EK, Hruban RH, Baughman KL. Cardiomyopathy of obesity: a clinical pathological evaluation of 43 obese patients with heart failure. Am J Cardiol 1992;70:921−4.

[13] Alaud-din A, Meteressian S, Lisbona R, et al. Assessment of cardiac function in patients who were morbidly obese. Surgery 1982;92:226−35.

[14] Kaltman AJ, Goldring RM. Role of circulatory congestion in the cardiorespiratory failure of obesity. Am J Med 1976;60:645−53.

[15] Backman C, Freyschuss U, Halberg D, et al. Cardiovascular function in extreme obesity. Acta Med 1973;193:437−46.

[16] Smith HL, Willius FA. Adiposity of the heart. Arch Intern Med 1933;52:911−31.

[17] Amad RH, Brennan JC, Alexander JK. The cardiac pathology of obesity. Circulation 1965;32:740−5.

[18] Alexander JK, Pettigrove JR. Obesity and congestive heart failure. Geriatrics 1967;22:101−8.

[19] Warnes CA, Roberts WC. The heart in massive (more than 300 pounds or 136 kilograms) obesity. Analysis of 12 patients studies at necropsy. Am J Cardiol 1989;54:1087−91.

[20] Alpert MA, Alexander JK. Cardiac morphology and obesity in man. In: Alpert MA, Alexander JK, editors. The heart and lung in obesity. Armonk, NY: Futura Publishing Co; 1998. p. 25−49.

[21] Alpert MA, Terry BE, Kelly DL. Effect of weight loss on cardiac chamber size, wall thickness and left ventricular function in morbid obesity. Am J Cardiol 1985;55:783−6.

[22] Alpert MA, Lambert CR, Terry BE, et al. Effect of weight loss on left ventricular mass in non-hypertensive morbidity obese patients. Am J Cardiol 1994;73:918−21.

[23] Lauer MS, Anderson KM, Kannel WB, et al. The impact of obesity on left ventricular mass and geometry. J Am Med Assoc 1991;266:231−6.

[24] Alpert MA, Lambert CR, Panayiotou H, et al. Relation of duration of morbid obesity to left ventricular mass, systolic function and diastolic filling, and effect of weight loss. Am J Cardiol 1996;76:1194−7.

[25] Bella NJ, Devereutx RB, Roman MJ, et al. Relations of left ventricular mass to fat-free and adipose body mass: The Strong Heart Study. Circulation 1998;98:2538−44.

[26] Amador N, de Jesus Encarnacion J, Rodriguez L, et al. Relationship between left ventricular mass and heart sympathetic activity in male obese subjects. Arch Med Res 2004;34:411−15.

[27] Iacobellis G, DiGuardo MC, Zapaterrenno A, et al. Relationship of insulin sensitivity and left ventricular mass in uncomplicated obesity. Obes Res 2003;11:578−84.

[28] McGavock JM, Victor R, Unger RH, Szczepaniak LS. Adiposity of the heart revisited. Ann Intern Med 2006;144:515–24.

[29] Woodwiss AJ, Libhaber CD, Majane OHI, et al. Obesity promotes left ventricular concentric rather than eccentric geometric remodeling and hypertrophy independent of blood pressure. Am J Hypertens 2008;21:1149–53.

[30] Aurigemma GP, de Simone G, Fitzgibbons TP. Cardiac remodeling in obesity. Circ Cardiovasc Imaging 2013;6:442–52.

[31] Peterson LR, Waggoner AD, Schectman KB, et al. Alterations in left ventricular structure and function in young healthy obese women. J Am Coll Cardiol 2004;43:1388–404.

[32] Okpura IC, Adediran OS, Odia OJ, et al. Left ventricular geometric patterns in obese Nigerian adults: an echocardiographic study. Internet J Intern Med 2010;9:1–7.

[33] Iacobellis G, Ribaudo MC, Zappaterreno A, et al. Adapted changes in left ventricular structure and function in severe uncomplicated obesity. Obes Res 2004;12:1616–21.

[34] Alpert MA, Lambert CR, Terry BE, et al. Effect of weight loss on left ventricular diastolic filling in morbid obesity. Am J Cardiol 1997;80:736–40.

[35] Chakko S, Alpert MA, et al. Abnormal left ventricular diastolic filling in eccentric left ventricular hypertrophy of obesity. Am J Cardiol 1991;68:95–8.

[36] Pascual M, Pascual DA, Soria F, et al. Effects of isolated obesity on systolic and diastolic left ventricular function. Heart 2003;89:1152–6.

[37] Kossaify A, Nicolais N. Impact of overweight and obesity on left ventricular diastolic function and value of tissue Doppler echocardiography. Clin Med Insights Cardiol 2013;7:43–50.

[38] Barbosa MM, Beleigoli AM, de Fatima Diniz M, et al. Strain imaging and morbid obesity: insight into subclinical ventricular dysfunction. Clin Cardiol 2011;34:288–93.

[39] Alpert MA, Terry BE, Lambert CR, et al. Factors infusing left ventricular systolic function in non-hypertensive morbidly obese patients and effect of weight loss induced by gastroplasty. Am J Cardiol 1993;75:773–7.

[40] Tumuklu MM, Etikan I, Kucasik B, et al. Effect of obesity on left ventricular structure and myocardial systolic function: assessment by tissue Doppler imaging and strain/strain rate imaging. Echocardiology 2007;24:802–9.

[41] Urhan AL, Uslu N, Davi SU, et al. Effects of isolated obesity on left and right ventricular function: a tissue Doppler and strain rate imaging study. Echocardiography 2010;22:236–43.

[42] Grapsa J, Tan TL, Paschou SA, et al. The effect of bariatric surgery on echocardiographic indices: a review of the literature. Eur J Clin Invest 2013;42:1224–30.

[43] Chahal H, McCelland RL, Tandai H, et al. Obesity and right ventricular structure and function: the MESA-Right Ventricular Study. Chest 2012;141:388–95.

[44] Thakur V, Richards R, Reisen E. Obesity, hypertension and the heart. Am J Med Sci 2001;321:242–8.

[45] Fazio S, Ferraro S, De Simone G, et al. Hemodynamic adaptation in severe obesity with or without arterial hypertension. Cardiologia 1989;34:967–72.

[46] Messerli FH, Sundgaard-Riise K, Reisin ED, Dreslinski GR, et al. Dimorphic cardiac adaptation to obesity and arterial hypertension. Ann Intern Med 1983;94:757–61.

[47] Kenchaiah S, Evans J, Levy D, et al. Obesity and the risk of heart failure. N Eng J Med 2002;347:305–13.

[48] Alpert MA, Terry BE, Mulekar M, et al. Cardiac morphology and left ventricular function in morbidly obese patients with and without congestive heart failure. Am J Cardiol 1997;80:736–40.

[49] He J, Ogden L, Bazzano L, et al. Risk factors for congestive heart failure in US men and women. NHANES I epidemiologic follow-up study. Arch Intern Med 2001;161:996–1002.

[50] Baena-Diez JM, Bynam AO, Grau M, et al. Obesity as an independent risk factor for heart failure: Zona Franco Cohort Study. Clin Cardiol 2010;33:760–4.

[51] Owan TE, Hodge DO, Herges RM, et al. Trends in prevalence and outcome of heart failure and preserved ejection proction. N Engl J Med 2006;355:251–9.

[52] Oreopoulos A, Padwal R, Kalantar-Zadeh K, Fonarow GC, Norris CM, McCallister FA. Body mass index in heart failure: a meta-analysis. Am Heart J 2008;156:13–22.

[53] Alexander JK, Alpert MA. Pathogenesis and clinical manifestations of obesity cardiomyopathy. In: Alpert MA, Alexander JK, editors. The heart and lung in obesity. Armonk NY: Futura Publishing Co; 1998. p. 133–46.

[54] Backman L, Freyschuss U, Hallberg D, et al. Reversibility of cardiopulmonary changes in extreme obesity. Act Med Scand 1979;205:367–73.

[55] Alexander JK, Petersen KL. Cardiovascular effects of weight reduction. Circulation 1972;45:310–18.

[56] Ashrafian H, le Roux CW, Darzi A, et al. Effects of bariatric surgery on cardiovascular function. Circulation 2008;118:2091–102.

[57] Rider OJ, Franco JM, Ali MK, et al. Beneficial cardiovascular effects on bariatric surgical and dietary weight loss in obesity. J Am Coll Cardiol 2009;54:718–26.

[58] Luaces M, Cachofeiro V, Garcia-Munoz-Najar A, et al. Anatomical and functional alterations of the heart in morbid obesity. Changes after bariatric surgery. Rev Esp Cardiol 2012;65:14–21.

[59] Estes EH, Sieker HO, McIntosh HD, et al. Reversible cardiopulmonary syndrome with extreme obesity. Circulation 1957;41:179–87.

[60] Ramani GY, McClaskey C, Ramanathan RC, Mathies MA. Safety and efficacy of bariatric surgery in morbidly obese patients with severe systolic heart failure. Clin Cardiol 2008;31:516–20.

[61] Miranda WR, Batsis JA, Sarr MG, et al. Impact of bariatric surgery on quality of life, function capacity and symptoms in patients with heart failure. Obes Surg 2013;23:1101–5.

Chapter 2

Obesity and Adipose Tissue Microvascular Dysfunction

M.G. Farb and N. Gokce

Boston University School of Medicine, Boston, MA, United States

LIST OF ABBREVIATIONS

BMI   body mass index
CAD   coronary artery disease
CLS   crown-like structures
CRP   C-reactive protein
CT   computed tomography
EAT   epicardial adipose tissue
eNOS   endothelial nitric oxide synthase
FFA   free fatty acids
FMD   flow-mediated dilation
IL   interleukin
NO   nitric oxide
PVAT   perivascular adipose tissue
TNF-α   tumor necrosis factor-alpha
VEGF   vascular endothelial growth factor

INTRODUCTION

Obesity has emerged as one of the most critical healthcare problems worldwide. Affecting both high- and low-income countries, nearly 2.2 billion people worldwide are currently overweight (body mass index [BMI] ≥ 25 kg/m^2) or obese (BMI ≥ 30 kg/m^2) [1]. Obesity prevalence in adults and children is continuing to rise, with significant short- and long-term health, social, and economic consequences [2]. Obesity is a strong predictor of all-cause mortality, and is closely linked to the development of several common medical conditions, including insulin resistance, type 2 diabetes mellitus, cancer, and cardiovascular disease. Premature heart disease and stroke are currently the major causes of death in this population [3]; thus, elucidating mechanisms of obesity-related vascular dysfunction are critical. Obesity is associated with a state of chronic systemic inflammation, and increasing evidence suggests that cardiovascular disease may be a consequence of adipose tissue dysregulation driven by an imbalance of pro- and anti-inflammatory adipokines and cytokines released from dysfunctional adipose tissues [4]. It is postulated that the adipose microenvironment may influence whole-body metabolic and vascular function through systemic release of adipocytokines that mediate pathophysiology in distant organs. A number of abnormalities can already be detected in the microvasculature within fat depots. The current chapter will describe these findings in the adipose milieu of obese humans that may have connections to systemic disease.

ADIPOSOPATHY IN OBESITY

Adiposopathy, or "sick fat," is described as pathogenic adipose tissue changes that are associated with metabolic dysregulation and activation of proinflammatory pathways. While cardiometabolic risk generally increases as a function of adipose tissue volume (quantity), qualitative alterations that develop within fat tissues in obesity have also been linked to endocrine, metabolic, and immune perturbations that promote cardiometabolic disease [5]. Inflammation in fat is largely driven by nonadipocyte cell populations that infiltrate and reside in the stromal—vascular fraction of adipose tissue compartments. While numbers of T cells, B cells, neutrophils, and mast cells increase, macrophages are the most abundant immune cell in the adipose tissue of obese individuals [6,7]. Both animal models and clinical data have linked the extent of adipose inflammation to metabolic dysfunction such as insulin resistance. Adipose macrophages appear to exist in at least two different activated states characterized as M1 classically activated macrophages that produce proinflammatory cytokines linked to insulin resistance and atherosclerosis, and alternative M2 macrophages that are generally involved in immunosuppressive functions [4]. Both M1 and M2 macrophage populations have been described in human fat [8] and tend to aggregate around dying adipocytes forming distinct "crown-like structures" (CLS, Fig. 2.1). In clinical studies, obese subjects lacking CLS in their fat stores tend to exhibit reduced proinflammatory adipose gene expression and more favorable systemic cardiometabolic profiles compared to age- and BMI-matched individuals with evidence of CLS [9—11]. Cardiovascular disease may thus be the "collateral damage" of cytokine imbalances in adipose tissues that shape systemic phenotypes.

Adipose tissue dysfunction likely occurs in all fat depots under obesogenic stress, however clinical data suggest that visceral fat may be relatively more prone to dysfunction. Central adiposity and the deposition of intraabdominal visceral fat have been consistently linked with increased cardiovascular and metabolic disease risk, which may in part be related to upregulated synthesis and release of adipokines, cytokines, and lipolysis in these compartments. In contrast, the expansion of subcutaneous fat has been shown to be a lesser contributor, or in some cases even protective in the development of obesity-associated cardiometabolic dysfunction, although this latter concept remains controversial [12—14]. Individuals with a greater degree of visceral fat have higher circulating levels of free fatty acids, interleukin (IL)-6, C-reactive protein, and tumor necrosis factor (TNF)-α compared to individuals with peripheral obesity [15—18]. In addition, IL-6, vascular endothelial growth factor (VEGF), vasoconstrictor prostaglandins, plasminogen activator inhibitor-1, noncanonical wingless-related integration site (WNT)5A, and TNF-α are released in greater quantities from abdominal visceral compared to subcutaneous fat [8,15,19—21]. In contrast, levels of antiatherogenic adiponectin, omentin, and secreted frizzled-related protein (SFRP)5 are reduced in an obesogenic environment [4,19]. Mediators produced by adipose tissue that have been implicated in cardiovascular disease mechanisms are listed in Table 2.1, and also reviewed extensively elsewhere [4].

Epicardial adipose tissue (EAT) is also emerging as a potential candidate regulator of cardiovascular function given its close anatomic proximity to the coronary vasculature and myocardium, and shared microcirculation. EAT has been viewed as the "visceral fat depot" of the heart [22] and shares embryologic origin with intraabdominal fat. EAT volume has been identified as an independent predictor of cardiovascular disease risk in population-based studies [22,23]. Epicardial fat measured by different methods including echocardiography, computed tomography (CT), and magnetic

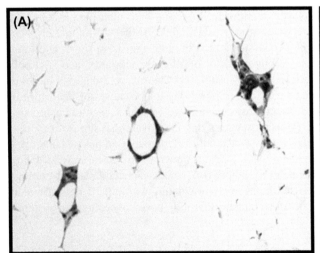 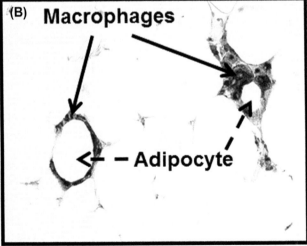

FIGURE 2.1   Histological illustration of inflamed human adipose tissue as demonstrated by light microscopy. As a hallmark of local chronic inflammation in human adipose tissue, CD68 + macrophages organize into "crown-like structures" (CLS, brown color [dark gray in print versions]) that encircle necrotic adipocytes. (A) 10 × power; (B) 20 × power, dotted arrows identify adipocytes and solid arrows indicate CD68 + macrophages.

TABLE 2.1 Mediators Produced From Dysfunctional Adipose Tissue Implicated in Cardiometabolic Disease

Angiopoietin-like 4 (ANGPTL-4)

Angiotensinogen

Apelin

C-reactive protein (CRP)

Chemokine (C–C motif) ligand-5 (CCL-5)

Free fatty acids (FFA)

Intercellular adhesion molecule-1 (ICAM-1)

Interleukin-1β (IL-1β)

Interleukin-6 (IL-6)

Interleukin-18 (IL-18)

JNK

Leptin

Matrix metalloproteinase

Monocyte chemotactic protein-1 (MCP-1)

Nuclear factor kappa B (NF-κB)

Plasminogen activator inhibitor-1 (PAI-1)

Prostaglandins

P-selectin

Retinol binding protein 4 (RBP-4)

Resistin

Serum amyloid A (SAA)

Toll-like receptor-4 (TLR-4)

Tumor necrosis factor-alpha (TNF-α)

Vascular cell adhesion molecule-1 (VCAM-1)

Vascular endothelial growth factor- $A_{165}b$ (VEGF-A_{165}b)

Visfatin

WNT5A

Adipose-derived mediators implicated in cardiovascular disease are listed in alphabetical order.

resonance imaging correlates with degree of intraabdominal visceral adiposity [24]. Epicardial fat in coronary artery disease (CAD) patients displays greater infiltration of macrophages with M1 polarization and cytokine production consistent with a proinflammatory phenotype compared to non-CAD patients [25,26]. While the literature suggests that EAT has capacity for production of proinflammatory mediators with paracrine effects that may mediate cardiovascular disease [22], the relation of epicardial fat to atherosclerosis remains associational and future studies may usher in causal relationships.

ADIPOSE TISSUE MICROVASCULAR DYSFUNCTION

Blood vessels are lined by an endothelial layer in direct contact with circulating blood that is essential in maintaining and regulating arterial tone, local blood flow, inflammation, and thrombosis. It also plays a key role in nutrient exchange in metabolic tissues such as fat, and can grow to meet the blood supply demands of expanding adipose tissue. The term "endothelial dysfunction" describes a pathophysiological state that is characterized by a loss of vascular homeostatic properties that promotes a vasoconstrictive, prothrombotic, proatherogenic, and antiangiogenic environment [27]. Endothelial dysfunction represents the earliest stage of the atherosclerotic process and severity of dysfunction in both coronary and peripheral vessels has been shown to independently predict future cardiovascular events [27]. As will be discussed, endothelial dysfunction can be detected in adipose tissue microvessels likely as a consequence of proatherogenic mediators released from dysfunctional fat. While degree of obesity has been linked to endothelial dysfunction such as impaired brachial artery flow-mediated dilation (FMD) [27,28], emerging clinical data make a compelling case that qualitative features of adipose tissue are also instrumental in shaping cardiovascular risk independent of total adiposity burden [29,30].

Adipose tissue qualitative features can be assessed noninvasively using CT attenuation, which utilizes a quantitative radiodensity scale measured as Hounsfield units. Radiodensity attenuation of fat is linked to insulin resistance, cardiac risk factors, and all-cause mortality, independent of total fat volume [29−33]. Animal studies suggest that lower radiodensity may be associated with higher lipid content, fibrosis, and inflammation, which have been linked to insulin resistance and endothelial dysfunction [31]. However, direct clinical interpretation of CT findings is limited because pathological tissue validation is largely lacking in humans. As such, histological adipose tissue inflammation has been shown to be significantly associated with vascular dysfunction in severely obese subjects [9,11]. While noninvasive imaging provides some perspective regarding adipose tissue quantity and "quality," their relationship to vascular dysfunction tends to be primarily associational, and limited information is available regarding causal mechanisms with regard to how adipose tissue may directly cause vascular disease.

To address this issue, experimental approaches utilizing videomicroscopy or myography have been developed for studying adipose tissue arterioles and directly probe pathophysiology in intact segments of human blood vessels that can be removed from living subjects during bariatric surgery. This methodology can be utilized to gain insight into pathways that are differentially altered in disease conditions [8,20,34−46]. Given the systemic nature of endothelial dysfunction, the technique represents a pragmatic approach to studying vascular pathophysiology with potential translation to mechanisms relevant to systemic vessels. The method involves removal of fresh adipose tissue from visceral or subcutaneous fat depots that can be harvested intraoperatively by the surgeon during bariatric surgery or via minimally invasive percutaneous needle biopsy of subcutaneous fat. Adipose arterioles (75−250 μm internal diameter) are carefully isolated from surrounding fat, cannulated between two glass capillary pipettes in a heated organ bath, and perfused under physiological conditions. The organ chamber is attached to a video microscope that allows for the quantification of the adipose vascular diameter in response to various chemical and physical stimuli. A representative image of a cannulated adipose microvessel is displayed in Fig. 2.2.

Utilization of *ex vivo* microvascular studies has provided insight into potential pathologic connections between adiposopathy and the local microvasculature. In experiments that examined paired subcutaneous and intraabdominal visceral adipose tissue samples collected from severely obese (BMI ≥ 40 kg/m^2) subjects during bariatric surgery, endothelium-dependent, acetylcholine-mediated vasodilation was significantly impaired in visceral compared to subcutaneous adipose tissue arterioles [8,45]. Studying paired depots from the same person removes the confounding effect of patient differences in systemic metabolic parameters and yields depot-specific signatures with evidence of profound dysfunction in the visceral milieu. The degree of vasomotor impairment is consistent across several endothelium-dependent vasodilators, including bradykinin, shear stress, and insulin [8,45,46]. Additionally, impairment is specific to the state of obesity since arterioles isolated from visceral fat of lean subjects display preserved endothelium-dependent vasodilation [38,43,44]. Responses to endothelium-independent vasodilators such as sodium nitroprusside and papaverine are generally preserved, which suggest intact vascular smooth muscle cell responses and selective impairment primarily at the level of the endothelium in adipose microvessels [8,20,44,45]. Complementary studies in endothelial cells isolated from visceral fat demonstrate impairment in endothelial nitric oxide synthase (eNOS) phosphorylation at the activating site serine 1177, suggesting abnormalities in nitric oxide (NO) bioactivity as a significant contributing factor to obesity-related vascular dysfunction [20]. A significant correlation between phosphorylated-eNOS expression in visceral adipose endothelial cells and brachial arterial flow-mediated vasodilation has been reported, suggesting parallel abnormalities in adipose and systemic circulations [47].

While profound derangements in visceral fat have been emphasized, perturbations are also evident in the subcutaneous depot of obese subjects which display blunted endothelium-dependent vasodilation compared to subcutaneous arterioles in lean subjects. In fact, there appears to be a disease gradient with extreme microenvironmental abnormalities in the visceral fat of obese subjects, with lesser, yet still prominent, perturbations in their subcutaneous depots compared to lean subjects [37,38]. Even moderate obesity adversely impacts subcutaneous adipose microvascular endothelial

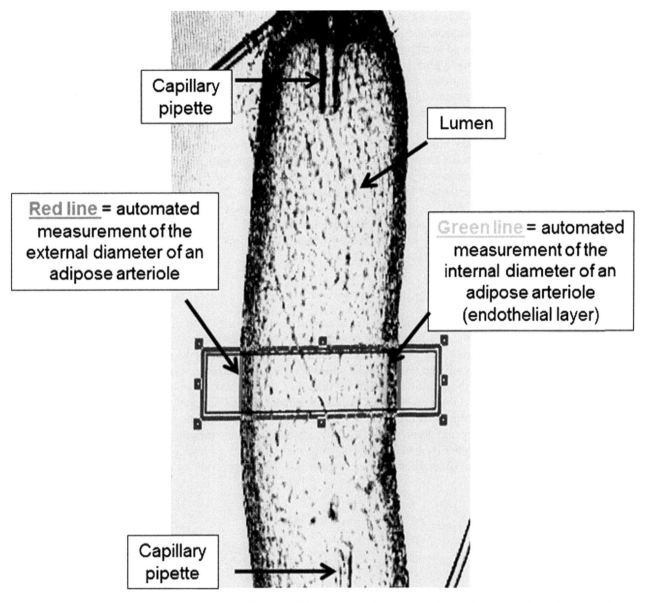

FIGURE 2.2 Illustration of adipose microvascular dilation using videomicroscopy. Representative image of a human adipose tissue arteriole suspended in a videomicroscopy organ chamber set-up that allows for assessment of vasodilation using an automated edge-detection tracking system.

function, particularly in women [42]. The degree of vasomotor impairment is worsened when obesity is associated with diabetes, metabolic syndrome, or hypertension, and linked to systemic inflammation and decreased eNOS activity [36–38,42,48]. There are likely multiple mechanisms that lead to microvascular dysfunction in diseased fat. Adipose proinflammatory gene expression correlates negatively with acetylcholine-mediated arteriolar vasodilation [8], suggesting inflammation as an important factor. Arterioles and isolated endothelial cells isolated from the visceral depot of obese subjects display enhanced expression of proinflammatory mediators such as CCL-5, IL-6, JNK, TNF-α, and toll-like receptor-4 [8,43,44]. Moreover, vasomotor dysfunction is reversed following treatment with IL-6 and TNF-α antagonists [39,44]. Other pathogenic pathways involving oxidative stress, mitochondrial dysfunction, and endoplasmic reticulum stress are also likely to contribute to adiposopathy and vascular diathesis. Adipose arterioles of type 2 diabetic obese subjects display impaired NO-dependent vasodilation along with abnormalities in mitochondrial structure and function [40]. Additionally, increasing telomerase activity in adipose arterioles of obese patients with CAD restores endothelial function by limiting vascular inflammation and mitochondrial reactive oxygen species production [34]. Systemic medications have direct effects on the adipose tissue microvasculature as treatment with a direct renin inhibitor aliskiren or angiotensin-conversing enzyme inhibitor ramipril are associated with the correction of microvascular structural alterations [35]. Glucagon-like peptide 1 receptor agonists, in addition to improving glucose utilization, increase eNOS-mediated vasodilation in adipose arterioles via AMP-activated protein kinase activity in diabetic patients [41]. Lastly, the eicosanoid/cyclooxygenase pathway may also be important in obesity-linked vascular disease, as cyclooxygenase-mediated vasoconstrictor prostanoids appear to contribute to adipose microvascular dysfunction [20]. Collectively, as clinical data consistently link adiposity to cardiovascular risk, the ability to directly access and examine dysfunctional human blood vessels provides major opportunities to discover novel mechanisms that may have whole-body cardiometabolic implications.

Videomicroscopy and myograph culture methodologies are also used to examine the influence of perivascular adipose tissue (PVAT) on the microvasculature. Defined as fat that directly surrounds blood vessels, PVAT has the potential to elicit pathogenic signals upon the local microvasculature, analogous to epicardial fat. In healthy conditions, PVAT modulates vascular contractile tone by release of vasodilatory mediators, including NO and adiponectin, having an "anticontractile" effect [49]. Under obesogenic stress, however, PVAT loses its protective anticontractile phenotype, becoming pathogenic to the local vasculature most likely due to macrophage activation, oxidative stress, and inflammation [39,50]. Moreover, removal of presumably dysfunctional PVAT from arterioles of obese subjects restores endothelium-dependent vasodilation [43].

ANGIOGENIC DYSFUNCTION IN OBESITY

Generation of new blood vessels, termed angiogenesis, as adipose tissue expands with progressive obesity is important in maintaining metabolic and oxygen exchange, and consequently whole-body homeostasis. Experimental studies suggest that expanding adipocytes may outgrow their blood supply due to deficient tissue angiogenesis that may initiate localized ischemia, hypoxia, necrosis, and inflammation within the adipose microenvironment, which may lead to metabolic dysfunction [51]. As with the vasodilator functions described above, there is growing evidence that angiogenic properties of microvessels can exhibit abnormalities in human adipose depots. Clinical studies have examined angiogenic capacity in human adipose tissue using an *ex vivo* sprout assay [52,53]. As displayed in Fig. 2.3, different fat depots can exhibit varying angiogenic profiles quantified by capillary growth emanating from fat pads, confirmed by immunofluorescence. In human obesity, subcutaneous adipose tissue appears to display higher capillary density and angiogenic capacity compared to visceral fat despite paradoxical higher expression of proangiogenic factors such as

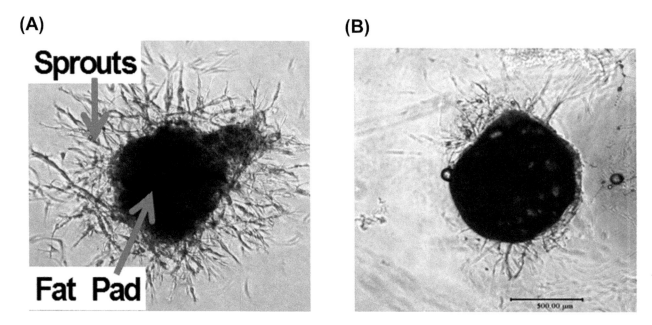

(A)

(B)

FIGURE 2.3 Assessment of adipose tissue angiogenic capacity *ex vivo*. Representative images of (A) normal and (B) blunted capillary growth from human fat pad explants after 7 days of culture.

VEGF-A [52,53] in the visceral depot. Among several mediators, proangiogenic angiopoietin-like 4 expression is down-regulated [52], while expression of antiangiogenic isoform VEGF-A$_{165}$b is increased in visceral fat and associated with impaired adipose tissue angiogenesis, which can be reversed upon targeted inhibition [53]. Capillary density and angiogenic potential of subcutaneous adipose tissue is blunted in severely obese compared to overweight individuals, which may have global cardiometabolic implications [52]. While vascularity and angiogenic features of fat may impact its metabolic functions, whether modulation of angiogenesis could influence clinical sequelae of obesity remains unknown.

WEIGHT LOSS AND ADIPOSE MICROVASCULAR FUNCTION

Bariatric surgery is currently the most effective and durable weight-loss intervention for obesity. The operation improves cardiac risk factors, remits diabetes, and to date represents the only clinical weight-loss intervention shown to improve long-term (>10 year) total and cardiovascular mortality by up to 50%, mainly from reduced myocardial infarction risk [54−58]. Meta-analysis of weight loss intervention studies show that systemic arterial endothelial function assessed by brachial artery FMD improves significantly following weight decline [59], and degree of vascular recovery may depend on subject characteristics or type of weight loss treatment. While mechanisms of benefit are incompletely understood, clinical data suggest that improved insulin sensitivity may be a dominant factor in cardiovascular risk reduction [58,60,61], and reversing systemic insulin resistance and endothelial dysfunction may be important clinical targets. At the adipose tissue level, weight loss has been shown to reduce ectopic fat burden and favorably remodel adipose tissue by attenuating macrophage-mediated inflammation [62−64]. Additionally, anticontractile functions of PVAT are restored following bariatric intervention and attributed to reduced inflammation and oxidative stress, and increased adiponectin and NO bioavailability [49]. Moreover, obesity-induced changes to the adipose microvascular structure also improve following bariatric surgical weight loss [65]. While bariatric surgery saves lives, it is obviously not an option for everyone, and ≤1% of eligible individuals undergo this procedure. However, by studying how the human vasculature favorably remodels following surgery, valuable physiological information can be learned and potentially translated for therapeutic applications.

CONCLUSIONS

With obesity rates on the rise worldwide, it will remain one of the most important global healthcare challenges for decades to come. A summary concept schematic illustrating local and systemic effects of obesity-induced adipose tissue dysfunction in promoting cardiometabolic disease is provided in Fig. 2.4. Endothelial dysfunction can be detected in adipose tissue arterioles of human subjects, and clinical studies have identified the dysfunctional adipose milieu and cytokine imbalance as contributors to vascular disease mechanisms. With clinical data consistently linking obesity to cardiovascular risk, examination of dysfunctional human blood vessels in adipose tissue domains may provide us with opportunities to discover novel translational clues to vascular disease mechanisms in human obesity.

MINI-DICTIONARY OF TERMS

- *Adipokines*: Adipose-derived mediators produced by fat cells.
- *Adiposopathy*: Pathogenic abnormalities that develop within fat tissue that lead to functional endocrine, metabolic, and immune changes that promote obesity-associated cardiometabolic disease.
- *Angiogenesis*: The generation of new blood vessels.
- *Central adiposity*: Preferential abdominal (vs peripheral) deposition of adipose tissue.
- *Endothelial dysfunction*: A pathophysiological state characterized by the loss of normal homeostatic properties of the vasculature that support a vasoconstrictive, prothrombotic, and proatherogenic environment leading to atherosclerosis.
- *Endothelium*: A single cell layer that lines the inside walls of blood vessels, important in maintaining and regulating arterial tone, blood flow, inflammation, and thrombosis.
- *Epicardial adipose tissue*: Fat that surrounds the heart.
- *Perivascular adipose tissue*: Fat that surrounds blood vessels.
- *Videomicroscopy*: An *ex vivo* method that measures the microvascular vasodilatory function of live arterioles.

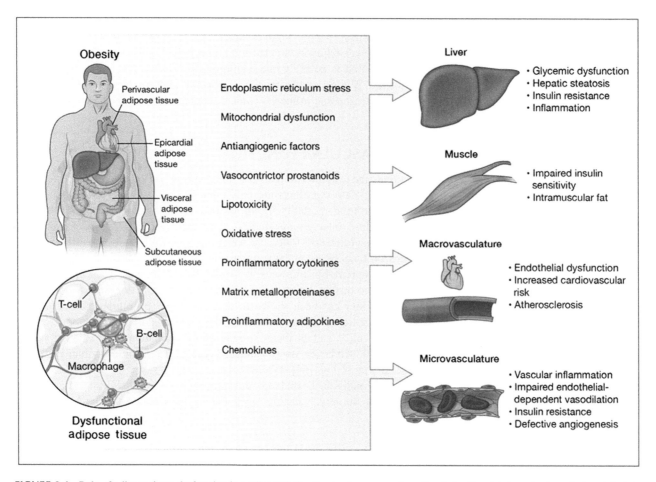

FIGURE 2.4 Role of adipose tissue dysfunction in cardiometabolic disease. Summary schematic of obesity-related mechanisms that contribute to cardiometabolic disease.

KEY FACTS

- The inner lining of blood vessels termed the endothelium plays a key role in regulating vascular function, including vasodilation and blood flow.
- NO is the major endothelium-derived vasoactive mediator that promotes vasodilation, and is produced by the enzyme eNOS.
- Impairment in endothelial function is a key event in the early stages of atherosclerosis that occurs, in part, due to an imbalance of vasodilator and vasoconstrictor mediators.
- Obesity and metabolic disease cause endothelium-dependent vasodilator dysfunction.
- The severity of impairment in the coronary and peripheral vasculature independently predicts future risk of cardio-vascular events.

SUMMARY POINTS

- Obesity is a strong predictor of all-cause mortality; premature heart disease and stroke are the leading causes of death in this population.
- In most forms of obesity, adipose tissue develops pathogenic changes that lead to cytokine imbalance that may promote cardiometabolic disease.
- The inflammatory response in adipose tissue originates predominately from infiltrating macrophages.
- Adipose tissue arterioles from obese subjects exhibit blunted endothelium-dependent vasodilation compared to arterioles from healthy, lean individuals.

- Visceral adipose microvessels have profoundly impaired endothelium-mediated vasodilation compared to subcutaneous arterioles in obese humans.
- PVAT loses its anticontractile properties in obesity and promotes endothelial dysfunction.
- Proinflammatory cytokines derived from fat tissue have been implicated in mechanisms of endothelial dysfunction.
- Angiogenic capacity is reduced in obesity and may be linked to mechanisms of adiposopathy.
- Bariatric weight loss improves endothelial function, reduces adipose tissue inflammation, and restores anticontractile properties of perivascular fat.
- Assessment of adipose arteriolar function correlates with *in vivo* measures of systemic endothelial function and cerebrovascular responses within an individual, and associate with cardiovascular risk factors including hypertension, smoking, diabetes, and inflammation.

REFERENCES

[1] Ng M, Fleming T, Robinson M, Thomson B, et al. Global, regional, and national prevalence of overweight and obesity in children and adults during 1980-2013: a systematic analysis for the Global Burden of Disease Study 2013. Lancet 2014;384:766.

[2] Gortmaker SL, Swinburn BA, Levy D, Carter R, Mabry PL, Finegood DT, et al. Changing the future of obesity: science, policy, and action. Lancet 2011;378:838.

[3] Prospective Studies C, Whitlock G, Lewington S, Sherliker P, et al. Body-mass index and cause-specific mortality in 900 000 adults: collaborative analyses of 57 prospective studies. Lancet 2009;373:1083.

[4] Fuster JJ, Ouchi N, Gokce N, Walsh K. Obesity-induced changes in adipose tissue microenvironment and their impact on cardiovascular disease. Circ Res 2016;118:1786.

[5] Bays HE. Adiposopathy is "sick fat" a cardiovascular disease? J Am Coll Cardiol 2011;57:2461.

[6] Cancello R, Tordjman J, Poitou C, Guilhem G, et al. Increased infiltration of macrophages in omental adipose tissue is associated with marked hepatic lesions in morbid human obesity. Diabetes 2006;55:1554.

[7] Curat CA, Wegner V, Sengenes C, Miranville A, Tonus C, Busse R, et al. Macrophages in human visceral adipose tissue: increased accumulation in obesity and a source of resistin and visfatin. Diabetologia 2006;49:744.

[8] Farb MG, Ganley-Leal L, Mott M, Liang Y, et al. Arteriolar function in visceral adipose tissue is impaired in human obesity. Arterioscler Thromb Vasc Biol 2012;32:467.

[9] Apovian CM, Bigornia S, Mott M, Meyers MR, et al. Adipose macrophage infiltration is associated with insulin resistance and vascular endothelial dysfunction in obese subjects. Arterioscler Thromb Vasc Biol 2008;28:1654.

[10] Bigornia SJ, Farb MG, Mott MM, Hess DT, Carmine B, Fiscale A, et al. Relation of depot-specific adipose inflammation to insulin resistance in human obesity. Nutr Diabetes 2012;2:e30.

[11] Farb MG, Bigornia S, Mott M, Tanriverdi K, et al. Reduced adipose tissue inflammation represents an intermediate cardiometabolic phenotype in obesity. J Am Coll Cardiol 2011;58:232.

[12] McLaughlin T, Lamendola C, Liu A, Abbasi F. Preferential fat deposition in subcutaneous versus visceral depots is associated with insulin sensitivity. J Clin Endocrinol Metab 2011;96:E1756.

[13] Neeland IJ, Turer AT, Ayers CR, Powell-Wiley TM, et al. Dysfunctional adiposity and the risk of prediabetes and type 2 diabetes in obese adults. JAMA 2012;308:1150.

[14] Porter SA, Massaro JM, Hoffmann U, Vasan RS, O'Donnel CJ, Fox CS. Abdominal subcutaneous adipose tissue: a protective fat depot? Diabetes Care 2009;32:1068.

[15] Fain JN, Madan AK, Hiler ML, Cheema P, Bahouth SW. Comparison of the release of adipokines by adipose tissue, adipose tissue matrix, and adipocytes from visceral and subcutaneous abdominal adipose tissues of obese humans. Endocrinology 2004;145:2273.

[16] Fontana L, Eagon JC, Trujillo ME, Scherer PE, Klein S. Visceral fat adipokine secretion is associated with systemic inflammation in obese humans. Diabetes 2007;56:1010.

[17] Park HS, Park JY, Yu R. Relationship of obesity and visceral adiposity with serum concentrations of CRP, TNF-alpha and IL-6. Diabetes Res Clin Pract 2005;69:29.

[18] Tsigos C, Kyrou I, Chala E, Tsapogas P, Stavridis JC, Raptis SA, et al. Circulating tumor necrosis factor alpha concentrations are higher in abdominal versus peripheral obesity. Metabolism 1999;48:1332.

[19] Catalan V, Gomez-Ambrosi J, Rodriguez A, Perez-Hernandez AI, et al. Activation of noncanonical Wnt signaling through WNT5A in visceral adipose tissue of obese subjects is related to inflammation. J Clin Endocrinol Metab 2014;99:E1407.

[20] Farb MG, Tiwari S, Karki S, Ngo DT, et al. Cyclooxygenase inhibition improves endothelial vasomotor dysfunction of visceral adipose arterioles in human obesity. Obesity 2014;22:349.

[21] Fuster JJ, Zuriaga MA, Ngo DT, Farb MG, Aprahamian T, Yamaguchi TP, et al. Noncanonical wnt signaling promotes obesity-induced adipose tissue inflammation and metabolic dysfunction independent of adipose tissue expansion. Diabetes 2015;64:1235.

[22] Ngo DT, Gokce N. Epicardial adipose tissue: a benign consequence of obesity? Circ Cardiovasc Imaging 2015;8:e003156.

[23] Shimabukuro M, Hirata Y, Tabata M, Dagvasumberel M, et al. Epicardial adipose tissue volume and adipocytokine imbalance are strongly linked to human coronary atherosclerosis. Arterioscler Thromb Vasc Biol 2013;33:1077.

[24] Fitzgibbons TP, Czech MP. Epicardial and perivascular adipose tissues and their influence on cardiovascular disease: basic mechanisms and clinical associations. J Am Heart Assoc 2014;3:e000582.

[25] Hirata Y, Tabata M, Kurobe H, Motoki T, et al. Coronary atherosclerosis is associated with macrophage polarization in epicardial adipose tissue. J Am Coll Cardiol 2011;58:248.

[26] Mazurek T, Zhang L, Zalewski A, Mannion JD, et al. Human epicardial adipose tissue is a source of inflammatory mediators. Circulation 2003;108:2460.

[27] Gokce N. Clinical assessment of endothelial function: ready for prime time? Circ Cardiovasc Imaging 2011;4:348.

[28] Parikh NI, Keyes MJ, Larson MG, Pou KM, et al. Visceral and subcutaneous adiposity and brachial artery vasodilator function. Obesity 2009;17:2054.

[29] Rosenquist KJ, Massaro JM, Pedley A, Long MT, Kreger BE, Vasan RS, et al. Fat quality and incident cardiovascular disease, all-cause mortality, and cancer mortality. J Clin Endocrinol Metab 2015;100:227.

[30] Rosenquist KJ, Pedley A, Massaro JM, Therkelsen KE, Murabito JM, Hoffmann U, et al. Visceral and subcutaneous fat quality and cardiometabolic risk. JACC Cardiovasc Imaging 2013;6:762.

[31] Baba S, Jacene HA, Engles JM, Honda H, Wahl RL. CT hounsfield units of brown adipose tissue increase with activation: preclinical and clinical studies. J Nucl Med 2010;51:246.

[32] Abraham TM, Pedley A, Massaro JM, Hoffmann U, Fox CS. Association between visceral and subcutaneous adipose depots and incident cardiovascular disease risk factors. Circulation 2015;132:1639.

[33] Hu HH, Chung SA, Nayak KS, Jackson HA, Gilsanz V. Differential computed tomographic attenuation of metabolically active and inactive adipose tissues: preliminary findings. J Comput Assist Tomogr 2011;35:65.

[34] Beyer AM, Freed JK, Durand MJ, Riedel M, et al. Critical role for telomerase in the mechanism of flow-mediated dilation in the human microcirculation. Circ Res 2016;118:856.

[35] De Ciuceis C, Savoia C, Arrabito E, Porteri E, et al. Effects of a long-term treatment with aliskiren or ramipril on structural alterations of subcutaneous small-resistance arteries of diabetic hypertensive patients. Hypertension 2014;64:717.

[36] Dharmashankar K, Welsh A, Wang J, Kizhakekuttu TJ, Ying R, Gutterman DD, et al. Nitric oxide synthase-dependent vasodilation of human subcutaneous arterioles correlates with noninvasive measurements of endothelial function. Am J Hypertens 2012;25:528.

[37] Georgescu A, Popov D, Constantin A, Nemecz M, Alexandru N, Cochior D, et al. Dysfunction of human subcutaneous fat arterioles in obesity alone or obesity associated with Type 2 diabetes. Clin Sci 2011;120:463.

[38] Grassi G, Seravalle G, Scopelliti F, Dell'Oro R, Fattori L, Quarti-Trevano F, et al. Structural and functional alterations of subcutaneous small resistance arteries in severe human obesity. Obesity 2010;18:92.

[39] Greenstein AS, Khavandi K, Withers SB, Sonoyama K, et al. Local inflammation and hypoxia abolish the protective anticontractile properties of perivascular fat in obese patients. Circulation 2009;119:1661.

[40] Kizhakekuttu TJ, Wang J, Dharmashankar K, Ying R, Gutterman DD, Vita JA, et al. Adverse alterations in mitochondrial function contribute to type 2 diabetes mellitus-related endothelial dysfunction in humans. Arterioscler Thromb Vasc Biol 2012;32:2531.

[41] Koska J, Sands M, Burciu C, D'Souza KM, et al. Exenatide protects against glucose- and lipid-induced endothelial dysfunction: evidence for direct vasodilation effect of GLP-1 receptor agonists in humans. Diabetes 2015;64:2624.

[42] Suboc TM, Dharmashankar K, Wang J, Ying R, Couillard A, Tanner MJ, et al. Moderate obesity and endothelial dysfunction in humans: influence of gender and systemic inflammation. Physiol Rep 2013;1:e00058.

[43] Virdis A, Duranti E, Rossi C, Dell'agnello U, Santini E, Anselmino M, et al. Tumour necrosis factor-alpha participates on the endothelin-1/nitric oxide imbalance in small arteries from obese patients: role of perivascular adipose tissue. Eur Heart J 2015;13:784.

[44] Virdis A, Santini F, Colucci R, Duranti E, et al. Vascular generation of tumor necrosis factor-alpha reduces nitric oxide availability in small arteries from visceral fat of obese patients. J Am Coll Cardiol 2011;58:238.

[45] Grizelj I, Cavka A, Bian JT, Szczurek M, et al. Reduced flow-and acetylcholine-induced dilations in visceral compared to subcutaneous adipose arterioles in human morbid obesity. Microcirculation 2015;22:44.

[46] Farb MG, Gokce N. Visceral adiposopathy: a vascular perspective. Horm Mol Biol Clin Investig 2015;21:125.

[47] Karki S, Farb MG, Ngo DT, Myers S, Puri V, Hamburg NM, et al. Forkhead box o-1 modulation improves endothelial insulin resistance in human obesity. Arterioscler Thromb Vasc Biol 2015;35:1498.

[48] Grassi G, Seravalle G, Brambilla G, Facchetti R, Bolla G, Mozzi E, et al. Impact of the metabolic syndrome on subcutaneous microcirculation in obese patients. J Hypertens 2010;28:1708.

[49] Aghamohammadzadeh R, Greenstein AS, Yadav R, Jeziorska M, et al. Effects of bariatric surgery on human small artery function: evidence for reduction in perivascular adipocyte inflammation, and the restoration of normal anticontractile activity despite persistent obesity. J Am Coll Cardiol 2013;62:128.

[50] Withers SB, Agabiti-Rosei C, Livingstone DM, Little MC, Aslam R, Malik RA, et al. Macrophage activation is responsible for loss of anticontractile function in inflamed perivascular fat. Arterioscler Thromb Vasc Biol 2011;31:908.

[51] Corvera S, Gealekman O. Adipose tissue angiogenesis: impact on obesity and type-2 diabetes. Biochim Biophys Acta 2014;1842:463.

[52] Gealekman O, Guseva N, Hartigan C, Apotheker S, et al. Depot-specific differences and insufficient subcutaneous adipose tissue angiogenesis in human obesity. Circulation 2011;123:186.

[53] Ngo DT, Farb MG, Kikuchi R, Karki S, et al. Antiangiogenic actions of vascular endothelial growth factor-A165b, an inhibitory isoform of vascular endothelial growth factor-A, in human obesity. Circulation 2014;130:1072.

[54] Ashrafian H, le Roux CW, Darzi A, Athanasiou T. Effects of bariatric surgery on cardiovascular function. Circulation 2008;118:2091.

[55] Kwok CS, Pradhan A, Khan MA, Anderson SG, Keavney BD, Myint PK, et al. Bariatric surgery and its impact on cardiovascular disease and mortality: a systematic review and meta-analysis. Int J Cardiol 2014;173:20.

[56] Romeo S, Maglio C, Burza MA, Pirazzi C, et al. Cardiovascular events after bariatric surgery in obese subjects with type 2 diabetes. Diabetes Care 2012;35:2613.

[57] Sjostrom L, Narbro K, Sjostrom CD, Karason K, et al. Effects of bariatric surgery on mortality in Swedish obese subjects. N Engl J Med 2007;357:741.

[58] Sjostrom L, Peltonen M, Jacobson P, Sjostrom CD, et al. Bariatric surgery and long-term cardiovascular events. JAMA 2012;307:56.

[59] Joris PJ, Zeegers MP, Mensink RP. Weight loss improves fasting flow-mediated vasodilation in adults: a meta-analysis of intervention studies. Atherosclerosis 2015;239:21.

[60] Bigornia SJ, Farb MG, Tiwari S, Karki S, et al. Insulin status and vascular responses to weight loss in obesity. J Am Coll Cardiol 2013;62:2297.

[61] Lupoli R, Di Minno MN, Guidone C, Cefalo C, Capaldo B, Riccardi G, et al. Effects of bariatric surgery on markers of subclinical atherosclerosis and endothelial function: a meta-analysis of literature studies. Int J Obes 2016;40:395.

[62] Cancello R, Henegar C, Viguerie N, Taleb S, et al. Reduction of macrophage infiltration and chemoattractant gene expression changes in white adipose tissue of morbidly obese subjects after surgery-induced weight loss. Diabetes 2005;54:2277.

[63] Clement K, Viguerie N, Poitou C, Carette C, et al. Weight loss regulates inflammation-related genes in white adipose tissue of obese subjects. FASEB J 2004;18:1657.

[64] Gaborit B, Jacquier A, Kober F, Abdesselam I, et al. Effects of bariatric surgery on cardiac ectopic fat: lesser decrease in epicardial fat compared to visceral fat loss and no change in myocardial triglyceride content. J Am Coll Cardiol 2012;60:1381.

[65] De CC, Porteri E, Rizzoni D, Corbellini C, et al. Effects of weight loss on structural and functional alterations of subcutaneous small arteries in obese patients. Hypertension 2011;58:29.

Chapter 3

Ghrelin-Producing Cells in Stomachs: Implications for Weight Reduction Surgery

R. Rosenthal, F. Dip, E. Lo Menzo and S. Szomstein
Cleveland Clinic Florida, Weston, FL, United States

LIST OF ABBREVIATIONS

BMI body mass index
CIN cinnamon
GC ghrelin cell
GH ghrelin hormone
HP Helicobacter pylori
LAGB laparoscopic adjustable gastric band
RYGB Roux-en-Y gastric bypass
SG sleeve gastrectomy

INTRODUCTION

In the last several years the rising incidence of metabolic syndrome in the obese population has been a concern both for the medical community and researchers. Obesity is associated with complex mechanisms regulated by different cells and hormones secondary to an imbalance of excess energy and fat tissue deposition [1].

According to estimations from 2015, approximately 1.9 billion people in the world are overweight. Metabolic syndrome is strongly associated with obesity and diabetes. This condition increases a patient's morbidity, hospital stay, and medication intake, and also increases the number of deaths and overall healthcare costs [2].

In recent years the discovery of two hormones, leptin and ghrelin, has improved surgeons' understanding of obesity and patients' appetite control. Leptin is produced in the adipose tissue and is responsible for decreasing both body weight and blood sugar via activation of the hypothalamus.

Ghrelin hormone (GH) was discovered by Kojima and Kangawa [3] in 1999. The presence of the hormone in different species like fish, amphibians, and many mammals has allowed researchers to investigate and understand its role in obesity and the outcomes of different surgical treatments. It is produced by approximately 1% of the epithelial cells of the stomach by endocrine cells, the "ghrelin cells." Ghrelin cells deliver two different forms of hormones to the blood, one that is the active GH and one inactive form, desacyl—GH. It is produced mostly in the stomach and demonstrates action in the brain, pancreas, small bowel, liver, and lung. After a bariatric procedure, GH production may be increased in other sites. That is why after a gastrectomy the reduction of GH reaches only 65% [4].

GH-producing cells, "X/A-like cells," can be characterized as the second most populous endocrine cells in the stomach. They are round or ovoid in shape, with compact and electrodense granules, and are located in the oxinic mucosa. There are two different kinds of cells. The classification has been created according to the relationship they have with the lumen of the organ. "Close type" cells have no physical connection with the lumen, while "open type" cells are connected with the lumen of the glands.

The main mechanism of action of GH is to release growth hormone from the pituitary gland. This has been described as a direct mechanism through an activation of a GH receptor at the pituitary gland. The second mechanism is most common and is produced after an activation of a ghrelin receptor in the arcuate neurons of the hypothalamus. Growth hormone release produces a chain of effects: insulin's action is antagonized, lypolysis is activated, and omental fat mass is reduced.

The positive and negative feedback mechanisms for the activation of the pituitary gland receptors may be transmitted by an unmyelinated segment of the vagus nerve situated in the subdiaphragmatic area. These fibers are able to sense different levels of GH between the mucosa and the submucosa. The relationship between the neuro- and endocrine systems has been proven by the increased marker of neural activation "FOS protein" after intravenous administration of GH.

GH also has different effects on stomach acid secretion. When GH is intravenously administered, the secretion of the acid and the motility of the stomach is increased. This effect is dose dependent. On the other hand, when a vagotomy is performed, the effect is canceled [5]. The latter effect is opposite of leptin hormone. Overall, GH is considered an orexigenic hormone with the purpose of increasing body weight. The impact of GH on carbohydrate metabolism is well known. In fact, when GH is administrated intravenously in humans, glycemia increases considerably. Its mechanism of action on insulin secretion, however, remains unclear, as reductions of insulin levels have been reported even in the presence of high levels of glucose. In animal studies the ablation of ghrelin receptors induces gluconeogenesis and glycogenolysis, while insulin sensitivity is improved [6].

A Complex Circuit

GH is not the only component responsible for the improvement of comorbidities such as diabetes in obese patients, but it is one of the main factors that may contribute to a better quality of life for patients.

Hypothalamic changes in the feeding center have been demonstrated in animal models after diet modifications and bariatric procedures. Levels and behavior of GH have been analyzed. In the central nervous system, ghrelin acts directly by activating NPY/Agp receptors and different neurotransmissions. Cowley reported a decrease of hypothalamic melanocortine tone with a consequent increase in energy intake after GH administration.

Dopamine is also involved in the GH circuit. The concept was supported after the discovery of an elevation of extracellular dopamine content in the nucleus accumbens after intravenous GH administration [7,8].

Interestingly, there is a reverse relationship between obesity and GH plasma levels. GH levels are lower in obese patients and those with type II diabetes, and are higher in patients with anorexia, bulimia nervosa, and cachexia [4].

Grinspoon et al. reported the relationship between ghrelin and recombinant human insulin-like growth factor-1 (IGF-1). It is known that after the administration of IGF-1 the secretion of GH decreases. This mechanism produces an indirect reduction of GH levels [9].

The immediate improvement of metabolic syndrome after bariatric procedures has been commonly reported. Granata et al. demonstrated that even pancreatic cells have GH receptors, and as a result, the calcium pathway may be activated in the pancreas [11]. When GH is elevated the release of insulin is lower. Otherwise, if an immunoneutralization of endogenous ghrelin is performed, insulin is released.

A similar relationship has been found between glucagon and GH. In fact, Date et al. [10] described an increase of glucagon secretion after GH administration in diabetic animals. Moreover, glucagon may stimulate gene transcription of ghrelin as well [12].

Ghrelin Hormone and Diet

Food intake affects the regulation of GH. In fact, increased glucose blood levels block ghrelin cells, determining a negative feedback. Fatty acid and lactate seem to work in the same way. Both can be linked to specific receptors encoding specific channels and decreasing the release of GH [13].

The phenomenon of negative feedback that is observed in the nonobese population seems to be different than in patients with metabolic syndrome. In the former, ghrelin cell concentration is elevated in the stomach and in the small bowel. Food intake cannot suppress the neuroendocrine axis, and GH levels are higher after ingestion [14].

Even though the mechanism is complex and not totally elucidated, different authors support the idea of the existence of a hypothalamic resistance to GH in the obese population [15]. Moreover, the resistance may be due not to obesity per se, but to an increase of leptin hormone that disrupts the hypothalamic response. This phenomenon may be responsible for weight maintenance in obese patients.

Ghrelin Hormone and Drugs

There are substances that have antidiabetic and antiobesity properties in nature such as cinnamon (CIN). CIN is a spice that interacts with different cells in the gut, and improves glucose blood levels and modifies food intake. It has been used in traditional medicine for years without knowing the mechanism of action. When CIN is administrated to mice induced to obesity, the animals experience a reduction of weight and glucose sensitivity.

Studies have demonstrated the presence of TRPA1 mRNA expression at ghrelin cells' surfaces. Ingestion of CIN activates a chemical sequence in the ghrelin cells and may help in the deactivation of GH. After TRPA1 is activated by CIN, the secretion of GH decreases and insulin receptors are upregulated. Additionally, two hours after the administration of CIN, both food intake and gut motility decrease [16].

When CIN was periodically administrated to mice, a reduction of cumulative body weight and body fat mass gain was demonstrated when compared to the control group. Metabolic disorders have been evaluated after CIN administration. Researchers reported a statistically significant improvement of glucose blood levels. CIN may be used in the future as an effective drug to combat obesity and as a treatment for dyslipidemic syndrome.

Surgical treatment for obesity remains the gold standard option for weight loss when other alternatives have failed. Restrictive and malabsorptive procedures are performed on patients with different changes in body mass index (BMI) and variable improvement of metabolic syndrome.

The effects on GH levels appear to be different depending on the bariatric procedures performed. The variation of GH levels may be due to the direct contact between food and the gastric mucosa, or because of the different neurohormonal changes determined by the surgeries.

Gastric Banding and Ghrelin Hormone Levels

Laparoscopic adjustable gastric band (LAGB) placement was introduced in 1970 into clinical practice. During this procedure an adjustable silicon ring is placed in the upper portion of the stomach in order to create a smaller virtual gastric pouch at the bottom of the esophagus. As a consequence, the volume of each ingested meal is reduced.

Initially banding was classified as a restrictive procedure. Recent investigations attribute a neuroendocrine mechanism to the technique, resulting in changes in the peripheral satiety mechanism. As a consequence, these patients experience changes in their eating behaviors [17]. Clinical data supports a direct correlation between the pressure of the band and decreased food interest and appetite. This could be partially due to a mechanical effect, but also to modifications in blood hormones. Ghrelin and leptin have been measured after gastric band placement.

The specific activation pathway of satiety has not been definitely validated. In the literature there is no consensus about the role of GH after LAGB. Different studies report a decrease of GH, while others associate this technique with no changes, or even an increase in the hormone [18,19].

Some authors hypothesize that the band impairs the production of ghrelin cells or alters the mechanism of neuroendocrine transmission of the GH. This theory is supported by a reduction of peptinogen II, observed six months after the procedure. Shak reported that the median percent weight loss (% excess weight loss, %EWL) over 12 months was 45.7%, with median BMI decreasing from 43.2 at baseline to 33.8 at 12 months postsurgery in 24 patients; however, no significant reduction of ghrelin plasma levels was found [20].

Knerr et al. studied the expression of leptin and ghrelin in adipose tissue after LAGB. They described an increase of leptin receptors in subcutaneous fat. Ghrelin serum values were instead decreased after LAGB when compared to a control group [21].

Krieger analyzed sleep and metabolism modifications in 30 patients after LAGB placement. He found no changes in ghrelin values at 12 months after surgery [22].

Alternatively, Hady et al. evaluated concentration of ghrelin, insulin, glucose, triglycerides, total and high-density lipoprotein (HDL)-cholesterol, as well as aspartate transaminase and alanine transaminase levels in plasma in more than 100 patients who underwent an LAGB or sleeve gastrectomy (SG). They found that patients with SG experienced a reduction of all the variables, while LAGB increased the GH levels [23]. The same results were published by Gelisgen et al., who showed that fundic production of ghrelin was significantly increased after six months of LAGB placement [24]. Moreover, the same authors concluded that a high level of ghrelin after an LAGB does not work as a predictor of decreased weight loss. Busetto evaluated LAGB outcomes in patients with high and normal preoperative ghrelin plasma levels. High plasma ghrelin concentrations at surgery did not modify the %EWL after LAGB surgery in 113 patients [25].

If any endocrine mechanism is altered after LAGB, the effect is not permanent. Changes in metabolic syndrome and weight loss usually decrease one year after surgery.

Sleeve Gastrectomy

SG consists of the creation of a long and thin gastric pouch made by stapling the stomach longitudinally. Initially the SG was conceived as the first step for super morbid obese patients too high-risk to undergo biliopancreatic diversion with duodenal switch. Once the patient had achieved improvement in comorbid conditions and achieved substantial weight loss, the second part of the malabsorptive surgery was performed.

Surprisingly the majority of these patients continued to experience significant weight loss over time and never required a second intervention. More interesting was the improvement of comorbidities before the weight loss, probably due to the relationship between SG and hormonal effects [26]. Modifications of ghrelin cells and SG have been studied and reported by different authors.

In order to confirm the cause—effect phenomenon, localization and patterns of ghrelin-producing cells in the stomach in obese populations have been studied. The concentration of ghrelin cells in the stomach has been observed in gastric specimens. Abdemur et al. reported a statistically significant difference in the amount of ghrelin cells between the fundus of the stomach when compared with the body and antrum. Age, sex, BMI, and diabetic status seem not to modify the quantity of cells. On the contrary, when ethnicity was evaluated as a variable, ghrelin cells were more abundant in the Hispanic population [26].

Goiten et al. described that the average counts of ghrelin cells declined from 60 ± 40 to 45 ± 20 and 39 ± 13 cells/ high power fields in the fundus, body, and preantral region, respectively. He emphasized the importance of a meticulous fundus resection during SG [27]. Some authors support the idea that by removing ghrelin cells in the stomach there is a consequent decrease of GH and a negative feedback in the neuro—hormonal axis. Fedonidis evaluated the modulations of the ghrelin receptors in the brain after weight loss in rats after SG. Downregulation of CB1R and an increase in MC4R receptors were described after 90 days following the procedure. This may explain one of the body weight loss mechanisms [28]. GH blood level reduction was reported after SG. This phenomenon may be due to different mechanisms of action. Probably the principal hormonal effect of SG is the decrease of cells after fundus resection. A second mechanism may be due to a paracrine effect after glucagon-like peptide-1 cells are activated and the remaining ghrelin cells are blocked [29]. Most studies attribute a GH reduction following SG after 3, 6, and 12 months [30].

Miyazaki et al. evaluated ghrelin expression and its clinical significance in 52 obese patients who underwent SG. The author described two groups determined by the number of ghrelin positive cells (GPC) in the mucosa of the fundus. Those with high numbers of GPC presented a higher percent of body weight loss after the bariatric procedure [31].

Mans et al. evaluated the effect of SG on hunger suppression, gastric and gallbladder motility, and gastrointestinal hormone response in nonobese patients, morbidly obese patients, and morbidly obese patients who had SG. Interestingly, patients after SG showed lower ghrelin concentration and an improved insulin resistance when compared with the other participants [32]. Plasma GH changes may be involved in the improvements of metabolic syndrome after SG. Buzca et al., after following 37 patients who underwent SG, concluded that fasting glucose, leptin, and ghrelin significantly decrease after the procedure [33]. SG may have an antidiabetic effect that can last more than 5 years in some cases [34].

Roux-en-Y Gastric Bypass

Roux-en-Y gastric bypass (RYGB) is one of the most effective treatments for obesity. It consists of the creation of a small gastric pouch, which empties directly into the jejunum. Bile and pancreatic juices are mixed with food after the anastomosis of the alimentary limb. Consequently, the procedure has a restrictive mechanism that decreases the storage food space in the stomach, and a malabsorptive action. Gastric bypass has a clear relationship with hormonal changes. Decreased appetite after RYGB has been associated with different changes in gut hormones. Elevated peptide YY and glucagon-like peptide 1 are involved in the mechanism of weight loss after gastric bypass [35].

There is some controversy regarding the role of ghrelin in post-RYGB weight loss; some studies have reported ghrelin reduction following RYGB surgery, but most have found either no significant change or a slight increase in ghrelin secretion in the long term. While the mechanism is not yet understood, authors support the idea that GH may be inhibited because of the lack of natural contact between the nutrients and some intestinal areas. Food, in fact, bypasses the ghrelin-activating center in the duodenum through the gastrojejunal anastomosis [36]. Geloneze et al. found lower levels of GH one year after RYGB compared to preoperative values [37]. Patients that achieved more than 36% weight loss had a lower concentration of ghrelin circulating hormone compared to controls. Roth et al. measured ghrelin and obestatin plasma levels in 18 patients after RYGB. They report decreased levels of ghrelin, while obestatin levels remained stable after massive weight loss in long-term follow-up studies [38].

Couce et al. evaluated the changes of obestatin plasma levels in 49 morbidly obese subjects who underwent laparoscopic Roux-Y-gastric bypass. His study shows a significant reduction of ghrelin and adiponectin levels 2 hours after the procedure. However, once glucose and insulin plasma levels were normalized, GH returned to normal values. They concluded that weight loss occurred without significant changes in ghrelin levels [39].

Other researchers have described a different relationship between weight loss and GH after RYGB. Faraj et al. evaluated ghrelin plasma levels in morbidly obese patients after RYGB. They bypassed 95% of the stomach and isolated the fundus from contact with nutrients. GH increased in those patients who actively lost weight. Those patients who did not achieve a change in BMI did not exhibit a decrease in GH levels [36].

Interestingly this modification in the GH regulations does not change in diabetic and nondiabetic subjects. It seems that after a 38% body weight loss, there are still no changes in GH secretion.

Different behavior of GH levels during RYGB may be due to different surgical approaches to the vagus nerve. Those surgeons who transect the vagus nerve may observe a decrease in GH, while the surgeons who preserve the vagus axis may witness constant levels of GH [40].

Helicobacter pylori *and Ghrelin Levels*

Helicobacter pylori (HP) is a gram-negative bacterium that infects between 8.7% and 85.5% of the population. This variability seems to have an ethnic geographical predisposition, with the German population expressing the lowest incidence and the Saudi bariatric population expression the highest incidence [41]. The presence of the bacterium in the stomach may induce changes in the gastric mucosa, like chronic gastritis, gastric and duodenal ulcer, and in some cases, cancer [42]. There is an intense humoral and cellular immune response concomitant with the infection that makes HP eradication difficult. Destruction of the gastric glands consequent to HP infection may alter GH production. Whether HP is a protective factor for obesity in infected patients, or its cure even a factor in weight loss failure, or whether it is responsible for weight regain after bariatric surgery, is still controversial.

The correlation between BMI and GH was analyzed in HP-infected and noninfected patients. Patients with HP have a weak correlation, suggesting influence of the bacterium on GH release [43]. In the last few years, more than 70% of researchers agree that GH is lower when HP is positive [44,46]. Also, the quantity of ghrelin cells in the stomach was measured by Date et al. [10] after comparing HP-infected and noninfected individuals. The authors concluded that the number of ghrelin cells is statistically higher in non-HP-infected patients.

Gastric diseases can modify levels of GH in fasting blood independently from HP infection. Atrophic gastritis with negative HP may decrease the levels of GH more than gastritis and duodenal ulcer with positive HP [45]. Eun Bae et al. [47] studied 154 patients with atrophic gastritis. Eighty-five percent of the population was infected with HP, and atrophic gastritis was reported in 24% of patients. GH was significantly decreased in patients with atrophic chronic gastritis compared to patients without (170.4 pg/mL, vs 201.1 pg/mL in patients without atrophy; $p < 0.001$) [46].

Future Horizons in Obesity Treatment

Vaccine

Authors have stated the possibility of disrupting the endogenous signal that activates the center of satiety. Different types of vaccines have been created to induce a decrease of GH activity.

One of the first examples that was tested was a monoclonal antighrelin antibody. The main objective was to inhibit GH production. Initially, an acute orexigenic effect was demonstrated; however, the vaccine failed to produce long-term results [47].

Later, Zakhari et al. combined different monoclonal antibodies targeting different haptens. They could demonstrate a decrease of food intake and an increase of energy expenditure.

A virus-like particles vaccine that blocks ghrelin has also been studied [48]. The vaccine generates antibodies against the hormone. Data has shown that vaccinated DIO mice have high levels of anti-GH, and these mice presented a reduced weight gain of up to 50%. They present significantly decreased food intake and an increase in energy expenditure [49].

A ghrelin receptor antagonist was demonstrated to be useful to reduce deprivation of food intake and improve glucose tolerance. Unfortunately, these vaccines are associated with an important inflammatory response.

In order to develop a safer and more effective antighrelin vaccine that could be used for human use, novel small molecules are now being characterized. They will be used as a ghrelin inverse agonist [50].

Near Infrared Guided Surgery

In the last few years a new type of technology has been developed in order to improve surgical outcomes. It uses near infrared (NIR) light for structure and tissue identification. Specific bindings between fluorescent dyes and antibodies

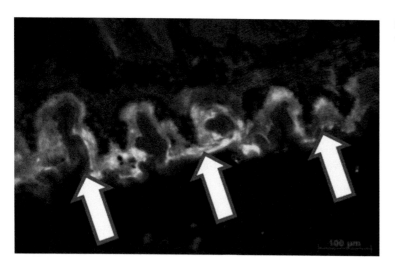

FIGURE 3.1   Gastric fundus mucosa. Ghrelin cell tagged with antighrelin dye (arrow).

can be obtained in order to illuminate cells. GH has been tagged to ghrelin antibody (H-40) and indocyanine green, allowing researchers to understand GH activation and its pathways. Fluorescent identification of GH-producing cells was achievable in the mouse model.

Abdemur et al. [26] reported that antibody-conjugated antibody was detected in 99.61% of GH-producing cells. Flow cytometry demonstrated ghrelin expression in gastric fundus, brain, and lung cells, while no significant expression was detected in liver cells ($p < 0.0001$). Similar results were detected by immunofluorescence, and different dyes have been investigated. (Fig. 3.1). In terms of efficacy in each fluorophore, ghrelin cells were highly sensitive to fluorescein when compared to indocyanine green. Further studies are needed to demonstrate the clinical implication of NIR-guided surgery in the treatment of obesity.

MINI-DICTIONARY OF TERMS

- *Metabolic syndrome*: Group of disorders such as obesity, hypertension, and insulin resistance that increase the rate of morbidity and mortality of patients.
- *Bariatric surgery*: Surgical procedure that is performed in order to decrease the weight of patients and improve metabolic syndrome.
- *Ghrelin hormone*: Substance produced by cells localized mainly in the fundus of the stomach that are involved in the regulation of other hormones involved in the control of hypertension, energy, and food intake.
- *Sleeve gastrectomy*: Surgical procedure that removes the fundus of the stomach and decreases the level of ghrelin cells.
- *Helicobacter pylori*: Gram-negative bacteria that infects the stomach and is able to modify the microenvironment of the stomach mucosa.
- *Near infrared guided surgery*: Novel technique that uses fluorescent light to enhance a cell or structure during a surgical procedure.

KEY FACTS

- Ghrelin cells are located in the stomach and in the gastrointestinal tract.
- Ghrelin cells produce a peptide hormone (GH) that interacts with the central nervous system.
- GH regulates the distribution and rate of energy.
- GH levels increase when the stomach is empty and decrease when it is full.
- A higher concentration of ghrelin cells are localized in the fundus.
- After a sleeve gastrectomy, GH reduces significantly.
- GH is involved in weight loss and improvement of metabolic syndrome.

SUMMARY POINTS

- Bariatric surgical procedures modify the neuro–hormonal axis.
- Hormonal changes induce long-term weight-loss changes after surgery.

- GH changes may have an impact on metabolic syndrome improvement after SG and gastric bypass.
- GH interacts with other gut hormones modifying BMI and glucose levels.
- Future bariatric procedures involve direct action on neuro−hormonal axis.

REFERENCES

[1] Naser KA, Gruber A, Thomson GA. The emerging pandemic of obesity and diabetes: are we doing enough to prevent a disaster? Int J Clin Pract 2006;60:1093−7.

[2] World Health Organization. Fact sheet 311: obesity and overweight [Internet]. Geneva: World Health Organization; 2013.

[3] Kojima M, Kangawa K. Ghrelin: structure and function. Physiol Rev 2005;85(2):495−522.

[4] Cummings DE, Weigle DS, Frayo RS, Breen PA, Ma MK, Dellinger EP, et al. Plasma ghrelin levels after dietinduced weight loss or gastric bypass surgery. N Engl J Med 2002;346:1623−30.

[5] Masuda Y, Tanaka T, Inomata N, Ohnuma N, Tanaka S, Itoh Z, et al. Ghrelin stimulates gastric acid secretion and motility in rats. Biochem Biophys Res Commun 2000;276:905−8.

[6] Li Z, Mulholland M, Zhang W. Ghrelin O-acyltransferase (GOAT) and energy metabolism. Sci China Life Sci 2016;59(3):281−91.

[7] Andrews ZB, Erion D, Beiler R, Liu ZW, Abizaid A, Zigman J, et al. Ghrelin promotes and protects nigrostriatal dopamine function via a UCP2-dependent mitochondrial mechanism. J Neurosci 2009;29(45):14057−65.

[8] Perelló M, Zigman JM. The role of ghrelin in reward-based eating. Biol Psychiatry 2012;72(5):347−53.

[9] Grinspoon S, Miller KK, Herzog DB, Grieco KA, Klibanski A. Effects of estrogen and recombinant human insulin-like growth factor-I on ghrelin secretion in severe undernutrition. J Clin Endocrinol Metab 2004;89(8):3988−93.

[10] Date Y, Kangawa K. Ghrelin as a starvation signal. Obes Res Clin Pract 2012 Oct-Dec;6(4):e263−346. Available from: http://dx.doi.org/10.1016/j.orcp.2012.08.195.

[11] Granata R. Peripheral activities of growth hormone-releasing hormone. J Endocrinol Invest 2016;39(7):721−7 Feb 18. [Epub ahead of print]

[12] Wei W, Wang G, Qi X, Englander EW, Greeley Jr. GH. Characterization and regulation of the rat and human ghrelin promoters. Endocrinology 2005;146(3):1611−25 Epub 2004 Dec 16.

[13] Cigdem Arica P, Kocael A, Tabak O, Taskin M, Zengin K, Uzun H. Plasma ghrelin, leptin, and orexin-A levels and insulin resistance after laparoscopic gastric band applications in morbidly obese patients. Minerva Med 2013;104(3):309−16.

[14] English PJ, et al. Food fails to suppress ghrelin levels in obese humans. J Clin Endocrinol Metab 2002;87:2984.

[15] Briggs DI, et al. Diet-induced obesity causes ghrelin resistance in arcuate NPY/AgRP neurons. Endocrinology 2010;151:4745−55.

[16] Camacho S, Michlig S, de Senarclens-Bezençon C, Meylan J, Meystre J, Pezzoli M, et al. Anti-obesity and anti-hyperglycemic effects of cinnamaldehyde via altered ghrelin secretion and functional impact on food intake and gastric emptying. Sci Rep 2015;5:7919. Available from: http://dx.doi.org/10.1038/srep07919.

[17] de Jong JR, van Ramshorst B, Gooszen HG, Smout AJ, Tiel-VanBuul MM. Weight loss after laparoscopic adjustable gastric banding is not caused by altered gastric emptying. Obes Surg 2009;19:287−92.

[18] Fruhbeck G, Rotellar F, Hernandez-Lizoain JL, et al. Fasting plasma ghrelin concentrations 6 months after gastric bypass are not determined by weight loss or changes in insulinemia. Obes Surg 2004;14:1208−15.

[19] Leonetti F, Silecchia G, Iacobellis G, et al. Different plasma ghrelin levels after laparoscopic gastric bypass and adjustable gastric banding in morbid obese subjects. J Clin Endocrinol Metab 2003;88:4227−31.

[20] Shak JR, Roper J, Perez-Perez GI, Tseng CH, Francois F, Gamagaris Z, et al. The effect of laparoscopic gastric banding surgery on plasma levels of appetite-control, insulinotropic, and digestive hormones. Obes Surg 2008;18(9):1089−96. Available from: http://dx.doi.org/10.1007/s11695-008-9454-6. Epub 2008 Apr 12.

[21] Knerr I, Herzog D, Rauh M, Rascher W, Horbach T. Leptin and ghrelin expression in adipose tissues and serum levels in gastric banding patients. Eur J Clin Invest 2006;36(6):389−94 Jun.

[22] Krieger AC, Youn H, Modersitzki F, Chiu YL, Gerber LM, et al. Effects of laparoscopic adjustable gastric banding on sleep and metabolism: a 12-month follow-up study. Int J Gen Med 2012;5:975−81.

[23] Hady HR, Dadan J, Gołaszewski P. 100 obese patients after laparoscopic adjustable gastric banding − the influence on BMI, gherlin and insulin concentration, parameters of lipid balance and co-morbidities. Adv Med Sci 2012;57(1):58−64.

[24] Gelisgen R, Zengin K, Kocael A, Baysal B, Kocael P, Erman H, et al. Effects of laparoscopic gastric band applications on plasma and fundic acylated ghrelin levels in morbidly obese patients. Obes Surg 2012;22(2):299−305. Available from: http://dx.doi.org/10.1007/s11695-011-0498-7.

[25] Busetto L, Segato G, De Luca M, Foletto M, Pigozzo S, Favretti F, et al. High ghrelin concentration is not a predictor of less weight loss in morbidly obese women treated with laparoscopic adjustable gastric banding. Obes Surg 2006;16(8):1068−74.

[26] Abdemur A, Slone J, Berho M, Gianos M, Szomstein S, Rosenthal RJ. Morphology, localization, and patterns of ghrelin-producing cells in stomachs of a morbidly obese population. Surg Laparosc Endosc Percutan Tech 2014;24(2):122−6.

[27] Goitein D, Lederfein D, Tzioni R, Berkenstadt H, Venturero M, Rubin M. Mapping of ghrelin gene expression and cell distribution in the stomach of morbidly obese patients--a possible guide for efficient sleeve gastrectomy construction. Obes Surg 2012;22(4):617−22.

[28] Fedonidis C, Alexakis N, Koliou X, Asimaki O, Tsirimonaki E, Mangoura D. Long-term changes in the ghrelin-CB1R axis associated with the maintenance of lower body weight after sleeve gastrectomy. Nutr Diabetes 2014;4:e127.

[29] Frezza EE, Chiriva-Internati M, Wachtel MS. Analysis of the results of sleeve gastrectomy for morbid obesity and the role of ghrelin. Surg Today 2008;38(6):481–3.

[30] Anderson B, Switzer NJ, Almamar A, Shi X, Birch DW, Karmali S. The impact of laparoscopic sleeve gastrectomy on plasma ghrelin levels: a systematic review. Obes Surg 2013;23(9):1476–80. Available from: http://dx.doi.org/10.1007/s11695-013-0999-7.

[31] Miyazaki Y, Takiguchi S, Seki Y, Kasama K, Takahashi T, Kurokawa Y, et al. Clinical significance of ghrelin expression in the gastric mucosa of morbidly obese patients. World J Surg 2013;37(12):2883–90.

[32] Mans E, Serra-Prat M, Palomera E, Suñol X, Clavé P. Sleeve gastrectomy effects on hunger, satiation, and gastrointestinal hormone and motility responses after a liquid meal test. Am J Clin Nutr 2015;102(3):540–7.

[33] Bužga M, Zavadilová V, Holéczy P, Švagera Z, Švorc P, Foltys A, et al. Dietary intake and ghrelin and leptin changes after sleeve gastrectomy. Wideochir Inne Tech Maloinwazyjne 2014;9(4):554–61.

[34] Basso N, Capoccia D, Rizzello M, Abbatini F, Mariani P, Maglio C, et al. First-phase insulin secretion, insulin sensitivity, ghrelin, GLP-1, and PYY changes 72 h after sleeve gastrectomy in obese diabetic patients: the gastric hypothesis. Surg Endosc 2011;25(11):3540–50.

[35] le Roux CW, Welbourn R, Werling M, Osborne A, Kokkinos A, Laurenius A, et al. Gut hormones as mediators of appetite and weight loss after Roux-en-Y gastric bypass. Ann Surg 2007;246(5):780–5.

[36] Faraj M, Havel PJ, Phelis S, Blank D, Sniderman AD, Cianflone K. Plasma acylation-stimulating protein, adiponectin, leptin, and ghrelin before and after weight loss induced by gastric bypass surgery in morbidly obese subjects. J Clin Endocrinol Metab 2003;88:1594–602.

[37] Geloneze B, Tambascia MA, Pilla VF, Geloneze SR, Repetto EM, Pareja JC. Ghrelin: a gut-brain hormone: effect of gastric bypass surgery. Obes Surg 2003;13:17–22.

[38] Roth CL, Reinehr T, Schernthaner GH, Kopp HP, Kriwanek S, Schernthaner G. Ghrelin and obestatin levels in severely obese women before and after weight loss after Roux-en-Y gastric bypass surgery. Obes Surg 2009;19(1):29–35.

[39] Couce ME, Cottam D, Esplen J, Schauer P, Burguera B. Is ghrelin the culprit for weight loss after gastric bypass surgery? A negative answer. Obes Surg 2006;16(7):870–8 Jul.

[40] Williams DL, Kaplan JM, Cummings DE, Grill HJ. Vagotomy dissociates consumption- and deprivation-related controls of endogenous ghrelin. In: Proc 11th Annual Meeting of Society for the Study of Ingestive Behavior, Groningen, The Netherlands; 2003.

[41] Küper MA, Kratt T, Kramer KM, Zdichavsky M, Schneider JH, Glatzle J, et al. Effort, safety, and findings of routine preoperative endoscopic evaluation of morbidly obese patients undergoing bariatric surgery. Surg Endosc 2010;24:1996–2001.

[42] Goodwin CS, Warren JR, Murray R, Blincow ED, Blackbourn SJ, Phillips M, et al. Prospective double-blind trial of duodenal ulcer relapse after eradication of Campylobacter pylori. Lancet 1988;2(8626–8627):1437–42.

[43] Tatsuguchi A, Miyake K, Gudis K, Futagami S, Tsukui T, Wada K, et al. Effect of *Helicobacter pylori* infection on ghrelin expression in human gastric mucosa. Am J Gastroenterol 2004;99:2121–7.

[44] Nwokolo CU, Freshwater DA, O'Hare P, Randeva HS. Plasma ghrelin following cure of *Helicobacter pylori*. Gut 2003;52:637–40.

[45] Nweneka CV, Prentice AM. *Helicobacter pylori* infection and circulating ghrelin levels - a systematic review. BMC Gastroenterol 2011;11:7 Jan 26.

[46] Lu SC, Xu J, Chinookoswong N, et al. An acyl-ghrelin-specific neutralizing antibody inhibits the acute ghrelin-mediated orexigenic effects in mice. Mol Pharmacol 2009;75:901–7.

[47] Eun Bae S, Hoon Lee J, Soo Park Y, Ok Kim S, Young Choi J, Yong Ahn J, et al. Decrease of serum total ghrelin in extensive atrophic gastritis: comparison with pepsinogens in histological reference. Scand J Gastroenterol 2016;51(2):137–44. Available from: http://dx.doi.org/10.3109/00365521.2015.1083049.

[48] Andrade S, Pinho F, Ribeiro AM, Carreira M, Casanueva FF, Roy P, et al. Immunization against active ghrelin using virus-like particles for obesity treatment. Curr Pharm Des 2013;19(36):6551–8.

[49] Zakhari JS, Zorrilla EP, Zhou B, Mayorov AV, Janda KD. Oligoclonal antibody targeting ghrelin increases energy expenditure and reduces food intake in fasted mice. Mol Pharm 2012;9:281–9.

[50] Kong J, Chuddy J, Stock IA, Loria PM, Straub SV, Vage C, et al. Pharmacological characterization of the first in class clinical candidate PF-05190457: a selective ghrelin receptor competitive antagonist with inverse agonism that increases vagal afferent firing and glucose-dependent insulin secretion ex vivo. Br J Pharmacol 2016;173(9):1452–64.

Chapter 4

Asthma in Obesity and Diabetes: Novel Mechanisms and Effects of Bariatric Surgery

P. Dandona[1], H. Ghanim[1], S. Monte[2] and J. Caruana[3]

[1]State University of New York at Buffalo, Buffalo, NY, United States, [2]School of Pharmacy & Pharmaceutical Sciences, Amherst, NY, United States, [3]Synergy Bariatrics ECMC, Buffalo, NY, United States

LIST OF ABBREVIATIONS

ADAM-33	a disintegrin and metalloproteinase
BMI	body mass index
CCR	chemokine receptor
CRP	C-reactive protein
FEV1	forced expiratory volume in one second
FVC	Forced vital capacity
FRC	Functional residual capacity
HMG-1	High-mobility group protein-1
HOMA	homeostatic model assessment
ICAM-1	intercellular adhesion molecule 1
iNOS	inducible nitric oxide synthase
IRS-1	insulin receptor substrate-1
IL-1-β	interleukin-1
IL-6	interleukin-6
LPS	lipopolysaccharides
LTBR	lymphotoxin-β receptor
MCP-1	monocyte chemoattractant protein-1
MMP-9	matrix metalloprotease-9
MNC	mononuclear cells
NOM	nitric oxide metabolites
NFκB	nuclear factor-κB
PTP-1B	phosphotyrosine phosphatase 1B
PMN	polymorph nuclear cells
ROS	reactive oxygen species
SOCS-3	suppressor of cytokine signaling
TSLP	thymic stromal lymphopoietin
TLR	toll-like receptor
TNF-α	tumor necrosis factor
VC	vital capacity

OBESITY, ASTHMA, AND BARIATRIC SURGERY

Obesity and diabetes are known to be associated with bronchial asthma [1,2]. Obese persons and type 2 diabetics have a prevalence of asthma of 2–2.5 times greater than that in the normal population. Marked obesity and increase in abdominal fat leads to an elevation of the diaphragm and a reduction in lung volume. This could potentially lead to impairment of pulmonary function. In addition, there is a bronchoconstrictive element. Several factors contribute to bronchoconstriction. Most of these are proinflammatory. Since both obesity and type 2 diabetes are chronic inflammatory states, the association of asthma with these two disease states is an obvious one.

PROINFLAMMATORY STATE OF OBESITY, INSULIN RESISTANCE, AND THE ANTIINFLAMMATORY ACTION OF INSULIN

Previous work on obesity has shown that it is associated with chronic low-grade inflammation [3–6]. The original observation that the expression of tumor necrosis factor (TNF)-α is increased in the adipose tissue of the ob/ob mouse, and that the infusion of soluble TNF-α receptor while neutralizing TNF-α also reverses its insulin resistance, was the initial observation that led to the discovery of this concept and the subsequent explosion of investigations in this area [3]. Thus, plasma concentration of TNF-α was shown to be elevated in obese humans; it fell after caloric restriction and weight loss [4]. Other proinflammatory cytokines like interleukin-6 (IL-6) and interleukin-1 (IL-1)-β were also shown to be increased in obese persons, and it was shown that TNF-α interfered with insulin signal transduction at the level of insulin receptor substrate (IRS)-1 [7]. Subsequently, several other proinflammatory serine kinases were also shown to impair insulin signaling by inducing serine phosphorylation of IRS-1 [8].

Obesity has also been shown to be associated with an increase in oxidative stress [9]. This diminishes rapidly after restriction of caloric intake. Since oxidative stress also interferes with insulin signaling, and since insulin exerts a rapid and potent reactive oxygen species (ROS) suppressive and antiinflammatory effect [10], insulin-resistant states are likely to be proinflammatory. In addition, since macronutrient intake as high-fat and high-calorie diets induces oxidative stress and inflammation, obese persons have a dual problem: insulin resistance and poor diets. It is of interest that a single high-fat meal of 900 calories induces not only comprehensive oxidative stress and inflammation, but also induces several factors that interfere with insulin signal transduction, including suppressor of cytokine signaling, several proinflammatory serine kinases, and phosphotyrosine phosphatase (PTP)-1B [11]. Consistent with these observations, the adipose tissue and the circulating mononuclear cells (MNC) in obese persons have been shown to be in a proinflammatory state [5] with a concomitant biochemical state consistent with insulin resistance [12].

In addition, leptin, secreted by adipocytes, is also proinflammatory and may therefore contribute to increased airway reactivity. Leptin bears a homology with IL-6, a key proinflammatory cytokine. It is also relevant that adipose tissue is infiltrated by inflammatory cells, including macrophages that participate in creating the inflammatory state associated with obesity [13].

ASTHMA-RELATED GENES IN OBESITY

The relationship of IL-4 and IgE concentrations to asthma is long-established [14–16]. While the binding of IgE to its receptor activates mast cells and basophils to induce the release of histamine and prostaglandins to trigger bronchoconstriction, IL-4 stimulates the production of IgE. Recent work has shown that IL-4 expression in MNC is increased in nonasthmatic obese persons [17]. In addition, there is an increase in the expression of matrix metalloprotease (MMP)-9, tumor necrosis factor superfamily member 14 (LIGHT); chemokine receptor (CCR)-2 is also increased. MMP-9 is a matrix proteinase that helps to spread inflammation in the interstitial tissue and mediates the remodeling of bronchial architecture. LIGHT, through its binding to its receptor, lymphotoxin-β receptor (LTBR), also mediates bronchial remodeling in asthma [17,18]. CCR-2 is the receptor for the chemokine monocyte chemoattractant protein-1 (MCP-1), and thus mediates the chemotaxis of inflammatory cells into the bronchi [19–21]. Finally, the plasma concentration of nitric oxide metabolites (NOM) is also increased in obese persons. Since circulating NOM is mainly the product of NO generated by inducible nitric oxide synthase, this is also a reflection of the inflammatory state of obesity. Asthma is known to be associated with increased NO in exhaled air from inflammatory cells in the bronchi [22].

It is noteworthy that the expression and concentrations of several of these factors are significantly related to body mass index (BMI) and homeostatic model assessment of insulin resistance (HOMA-IR) [17].

REVERSAL OF THE PROINFLAMMATORY STATE FOLLOWING BARIATRIC SURGERY

There is clear evidence that weight loss in obese persons leads to a reversal of inflammatory and oxidative stress. This trend is even more intense following bariatric surgery since the magnitude of caloric restriction and weight loss is marked. Thus, there is a reduction in plasma concentration of endotoxin, C-reactive protein, MMP-9, and MCP-1 [23].

Intranuclear binding of nuclear factor-κB (NFκB), the key proinflammatory transcription factor, also falls simultaneously. There is a concomitant reduction in the expression of toll-like receptor (TLR)-4, the receptor for lipopolysaccharides (LPS) and CD-14, which facilitates the binding of LPS to TLR-4 [23]. The expression of TLR-2, which binds to peptidoglycans (the products of Gram-positive bacteria), also fell. Bacterial inflammation is important in the pathogenesis of asthma. In parallel, there is also a reduction in the levels/expression of asthma-related genes [17]. In addition, since proinflammatory factors contribute to interference with insulin signal transduction, their reduction leads to an improvement in insulin sensitivity, and frequently, a resolution of the diabetic state. An increase in insulin sensitivity is important since insulin exerts an antiinflammatory action of its own, including the suppression of asthma-related inflammatory factors, as described below [24].

As far as asthma-related factors are concerned, there was a significant reduction in the expression of IL-4 by 49%, LIGHT by 29%, LTBR by 33%, A Disintegrin and Metalloproteinase-(ADAM)-33 by 20%, MMP-9 by 59%, MCP-1 by 23%, and CCR-2 by 27% following bariatric surgery [17]. These findings emphasize the point that while the reversal of general inflammation is impressive following bariatric surgery, there is also a significant reduction in the expression and concentrations of asthma-related factors [17,23].

Published data are limited as far as the clinical outcomes for the indices of pulmonary function following bariatric surgery. However, in morbidly obese patients with asthma, marked weight loss is associated with a reversal of the abnormalities in lung volumes, including forced expiratory volume in one second (FEV1), forced vital capacity, vital capacity, and functional residual capacity. There is also a decrease in responsiveness to methacholine, a known bronchoconstrictor used for assessing bronchial reactivity. Parallel improvements in asthma symptoms are also associated with the reversal of these abnormalities. A majority of patients were able to go drug-free. Those who had to continue with drugs were shown to be atopic [25]. In another study, the frequency of emergency room visits for acute exacerbations of asthma by morbidly obese patients was shown to be markedly reduced after bariatric surgery and weight loss [26].

EFFECT OF INTRAVENOUS INSULIN INFUSION ON INFLAMMATION IN GENERAL AND IN RELATION TO ASTHMA-RELATED FACTORS

A low-dose insulin infusion (2u/h) has been shown to exert a rapid and potent antioxidant and antiinflammatory effect in obese subjects [10]. Thus, this infusion results in a reduction in ROS generation, the expression of the $p47^{phox}$ subunit of NADPH oxidase, and intranuclear NFκB binding, as well as increases in NFκBα (IκBα) inhibitor expression, and plasma concentrations of MCP-1 and intercellular adhesion molecule 1 (ICAM-1). In addition, it reduces the plasma concentrations of tissue factor, plasminogen activator inhibitor (PAI)-1, MMP-2, and MMP-9. It also reduces CRP and SAA [27], and reduces the size of the infarct in patients with acute stance phase elevation myocardial infarction [28]. It has also been shown that insulin reduces the expression and plasma concentrations of chemokines and the expression of CCR. This includes eotaxin and one of its receptors, CCR-5 [29].

More recently, a low-dose insulin infusion has been shown to reduce the concentration/expression of asthma-related factors [24]. These include IL-4, ADAM-33, LIGHT, and LTBR. IL-4 expression falls dramatically: 25% within 2 hours, 34% at 4 hours, and 43% at 6 hours. Thus, IL-4 expression continues to decrease even after the cessation of insulin infusion at 4 hours, while the suppression of other genes ceases with the end of the insulin infusion. In addition, the plasma concentrations of NOM, LIGHT, TGFβ, MCP-1, and MMP-9 fall significantly during 4 hours of infusion. In addition, eotaxin and CCR-5 expression also fall after insulin infusion [29]. The reduction in ROS generation is also important since increased isoprostane levels in exhaled breath are reflections of oxidative stress and increased lipid peroxidation in asthma. Thymic stromal lymphopoietin, a recently described factor related to the pathogenesis of asthma, does not change after insulin infusion.

While these actions of insulin comprehensively link insulin resistance to inflammation and asthma, they also provide a potential novel function for insulin in the treatment of acute asthma. Since high doses of corticosteroids exert certain proinflammatory factors [30], and since these proinflammatory factors are suppressed by insulin, it is possible that a combination of insulin and corticosteroids may provide a more potent and safer mode of antiinflammatory therapy. The use of such a combination may also allow the use of lower doses of corticosteroids. For example, a single injection of 300 mg of hydrocortisone (=60 mg prednisolone) induces increases in the expression of MMP-9 and high-mobility group protein-1 (HMG-1) [30]. HMG Box 1 (HMGB-1) is associated with mortality in mice given endotoxin. The same dose of hydrocortisone is also known to suppress ROS generation by polymorph nuclear cells and MNC by >90% [31,32], which reduces the ability of these leukocytes to kill bacteria. Thus, an opportunity to reduce the dose of corticosteroids while enhancing the antiinflammatory effect would be worthwhile.

CONCLUSIONS

The data shown above are consistent with the concept that the proinflammatory state of obesity is associated with an increased expression of asthma-related factors, and that this increase is related to BMI and insulin resistance. In addition, insulin is not only suppressive of oxidative stress and inflammation in general, but also of asthma-related factors. Bariatric surgery leading to weight loss results in the suppression of these factors in parallel with the restoration of insulin sensitivity and the resolution of clinical symptoms (Fig. 4.1).

MINI-DICTIONARY OF TERMS

- *Insulin resistance*: Less than optimal response to insulin.
- *Inflammation (metabolic)*: Subclinical acute or chronic innate response to harmful stimuli.
- *Mononuclear cells*: White blood cells, including monocytes and lymphocytes.

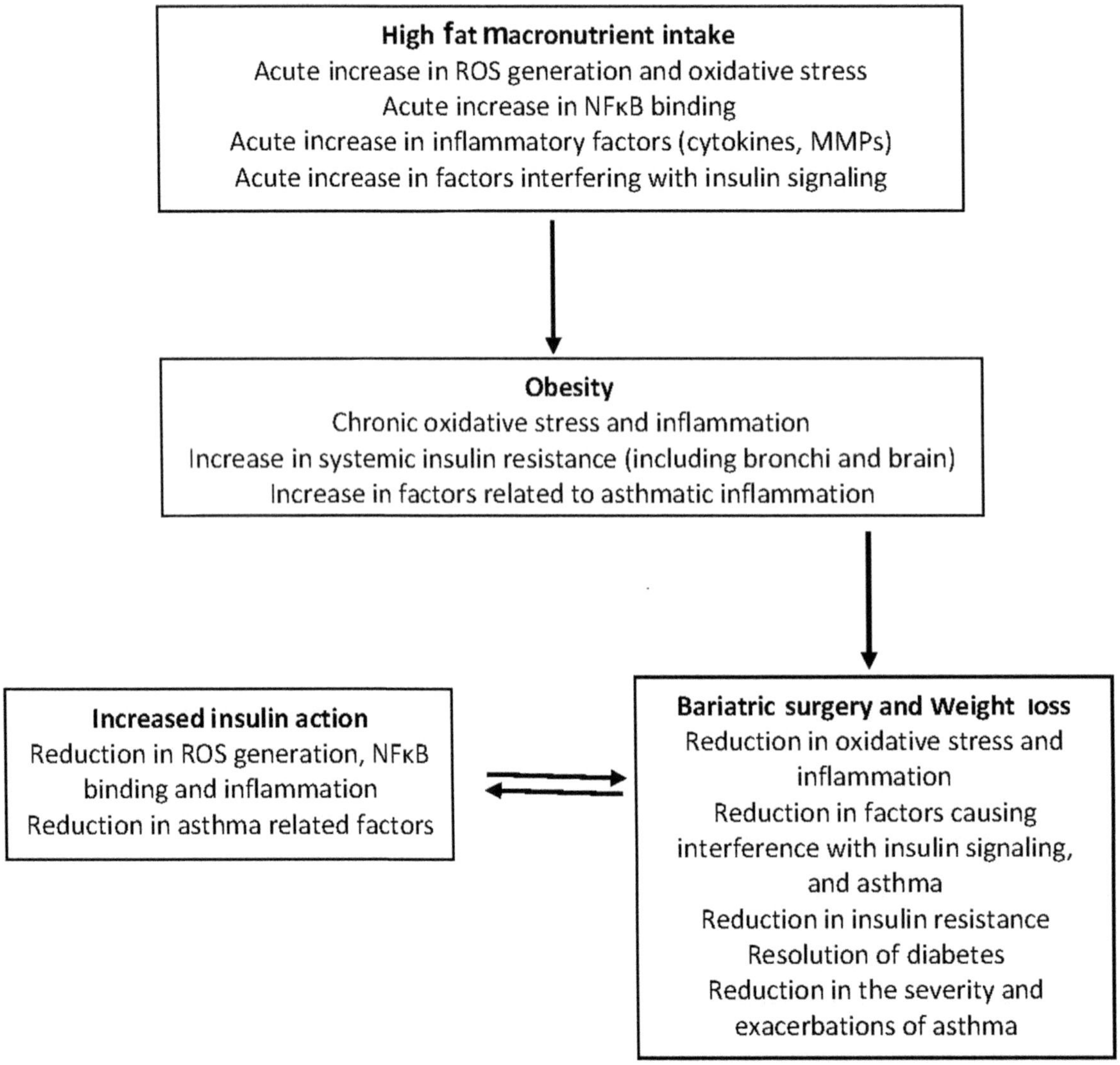

High fat macronutrient intake
Acute increase in ROS generation and oxidative stress
Acute increase in NFκB binding
Acute increase in inflammatory factors (cytokines, MMPs)
Acute increase in factors interfering with insulin signaling

Obesity
Chronic oxidative stress and inflammation
Increase in systemic insulin resistance (including bronchi and brain)
Increase in factors related to asthmatic inflammation

Increased insulin action
Reduction in ROS generation, NFκB binding and inflammation
Reduction in asthma related factors

Bariatric surgery and Weight loss
Reduction in oxidative stress and inflammation
Reduction in factors causing interference with insulin signaling, and asthma
Reduction in insulin resistance
Resolution of diabetes
Reduction in the severity and exacerbations of asthma

FIGURE 4.1 Link between food intake and obesity with asthma-related mediators, and its reversal with bariatric surgery and insulin sensitization.

- *FEV1*: volume of air that can be forcibly blown out in 1 second.
- *Interleukins*: A group of cytokines (inflammatory mediators) secreted mainly by the immune cells; part of the inflammatory response.

KEY FACTS

- Insulin resistance is a state of deficient response of cells to insulin.
- Insulin resistance is considered to be one of the main pathophysiological causes of type 2 diabetes.
- Increase in inflammation, oxidative stress, and other stresses contribute to insulin resistance.
- Patients suffering from metabolic syndrome, obesity, and type 2 diabetes are characterized by increased inflammatory and oxidative stress, and therefore are also in insulin-resistance states.
- Insulin resistance can be evaluated using HOMA calculated as (Insulin $\times$ glucose (mMm))/22.5, with normal values being around 1. However, the gold standard for quantifying insulin resistance is the hyperinsulinemic euglycemic clamp, which measures the amount of glucose necessary to compensate for the increased insulin level required to maintain normal glycemic levels.

SUMMARY POINTS

- Obesity and type 2 diabetes are associated with bronchial asthma.
- Obesity and type 2 diabetes are also associated with chronic inflammation in general and an increase in the expression/concentrations of factors related to asthma.
- There is a concomitant increase in insulin resistance and interference with insulin signal transduction.
- Bariatric surgery resulting in marked weight loss reduces inflammation and insulin resistance.
- Bariatric surgery also suppresses the expression/concentration of asthma-related factors and the clinical severity of asthma.

REFERENCES

[1] Mokdad AH, Ford ES, Bowman BA, Dietz WH, Vinicor F, Bales VS, et al. Prevalence of obesity, diabetes, and obesity-related health risk factors, 2001. JAMA 2003;289(1):76−9.

[2] Beuther DA, Sutherland ER. Overweight, obesity, and incident asthma: a meta-analysis of prospective epidemiologic studies. Am J Respir Crit Care Med 2007;175(7):661−6.

[3] Hotamisligil GS, Shargill NS, Spiegelman BM. Adipose expression of tumor necrosis factor-alpha: direct role in obesity-linked insulin resistance. Science 1993;259(5091):87−91.

[4] Dandona P, Weinstock R, Thusu K, Abdel-Rahman E, Aljada A, Wadden T. Tumor necrosis factor-alpha in sera of obese patients: fall with weight loss. J Clin Endocrinol Metab 1998;83(8):2907−10.

[5] Ghanim H, Aljada A, Hofmeyer D, Syed T, Mohanty P, Dandona P. Circulating mononuclear cells in the obese are in a proinflammatory state. Circulation 2004;110(12):1564−71.

[6] Nathan C. Epidemic inflammation: pondering obesity. Mol Med 2008;14(7−8):485−92.

[7] Hotamisligil GS, Peraldi P, Budavari A, Ellis R, White MF, Spiegelman BM. IRS-1-mediated inhibition of insulin receptor tyrosine kinase activity in TNF-alpha- and obesity-induced insulin resistance. Science 1996;271(5249):665−8.

[8] Hotamisligil GS. Inflammation and metabolic disorders. Nature 2006;444(7121):860−7.

[9] Dandona P, Mohanty P, Ghanim H, Aljada A, Browne R, Hamouda W, et al. The suppressive effect of dietary restriction and weight loss in the obese on the generation of reactive oxygen species by leukocytes, lipid peroxidation, and protein carbonylation. J Clin Endocrinol Metab 2001;86(1):355−62.

[10] Dandona P, Aljada A, Mohanty P, Ghanim H, Hamouda W, Assian E, et al. Insulin inhibits intranuclear nuclear factor kappaB and stimulates IkappaB in mononuclear cells in obese subjects: evidence for an anti-inflammatory effect? J Clin Endocrinol Metab 2001;86(7):3257−65.

[11] Ghanim H, Abuaysheh S, Sia CL, Korzeniewski K, Chaudhuri A, Fernandez-Real JM, et al. Increase in plasma endotoxin concentrations and the expression of Toll-like receptors and suppressor of cytokine signaling-3 in mononuclear cells after a high-fat, high-carbohydrate meal: implications for insulin resistance. Diabetes Care 2009;32(12):2281−7.

[12] Ghanim H, Aljada A, Daoud N, Deopurkar R, Chaudhuri A, Dandona P. Role of inflammatory mediators in the suppression of insulin receptor phosphorylation in circulating mononuclear cells of obese subjects. Diabetologia 2007;50(2):278−85.

[13] Weisberg SP, McCann D, Desai M, Rosenbaum M, Leibel RL, Ferrante Jr. AW. Obesity is associated with macrophage accumulation in adipose tissue. J Clin Invest 2003;112(12):1796−808.

[14] Munitz A, Brandt EB, Mingler M, Finkelman FD, Rothenberg ME. Distinct roles for IL-13 and IL-4 via IL-13 receptor alpha1 and the type II IL-4 receptor in asthma pathogenesis. Proc Natl Acad Sci U S A 2008;105(20):7240−5.

[15] Suarez CJ, Parker NJ, Finn PW. Innate immune mechanism in allergic asthma. Curr Allergy Asthma Rep 2008;8(5):451−9.

[16] Galli SJ, Tsai M, Piliponsky AM. The development of allergic inflammation. Nature 2008;454(7203):445−54.

[17] Dandona P, Ghanim H, Monte SV, Caruana JA, Green K, Abuaysheh S, et al. Increase in the mediators of asthma in obesity and obesity with type 2 diabetes: reduction with weight loss. Obesity (Silver Spring) 2014;22(2):356−62.

[18] Doherty TA, Soroosh P, Khorram N, Fukuyama S, Rosenthal P, Cho JY, et al. The tumor necrosis factor family member LIGHT is a target for asthmatic airway remodeling. Nat Med 2011;17(5):596−603.

[19] de Faria IC, de Faria EJ, Toro AA, Ribeiro JD, Bertuzzo CS. Association of TGF-beta1, CD14, IL-4, IL-4R and ADAM33 gene polymorphisms with asthma severity in children and adolescents. J Pediatr (Rio J) 2008;84(3):203−10.

[20] Romagnani S. Cytokines and chemoattractants in allergic inflammation. Mol Immunol 2002;38(12−13):881−5.

[21] Weiss ST, Raby BA, Rogers A. Asthma genetics and genomics 2009. Curr Opin Genet Dev 2009;19(3):279−82.

[22] Kharitonov SA, Barnes PJ. Effects of corticosteroids on noninvasive biomarkers of inflammation in asthma and chronic obstructive pulmonary disease. Proc Am Thorac Soc 2004;1(3):191−9.

[23] Monte SV, Caruana JA, Ghanim H, Sia CL, Korzeniewski K, Schentag JJ, et al. Reduction in endotoxemia, oxidative and inflammatory stress, and insulin resistance after Roux-en-Y gastric bypass surgery in patients with morbid obesity and type 2 diabetes mellitus. Surgery 2012;151(4): 587−93.

[24] Ghanim H, Green K, Abuaysheh S, Batra M, Kuhadiya ND, Patel R, et al. Suppressive effect of insulin on the gene expression and plasma concentrations of mediators of asthmatic inflammation. J Diabetes Res 2015;2015:202406.

[25] Boulet LP, Turcotte H, Martin J, Poirier P. Effect of bariatric surgery on airway response and lung function in obese subjects with asthma. Respir Med 2012;106(5):651−60.

[26] Hasegawa K, Tsugawa Y, Chang Y, Camargo Jr. CA. Risk of an asthma exacerbation after bariatric surgery in adults. J Allergy Clin Immunol 2015;136(2) 288-94.e8.

[27] Chaudhuri A, Janicke D, Wilson MF, Tripathy D, Garg R, Bandyopadhyay A, et al. Anti-inflammatory and profibrinolytic effect of insulin in acute ST-segment-elevation myocardial infarction. Circulation 2004;109(7):849−54.

[28] Selker HP, Udelson JE, Massaro JM, Ruthazer R, D'Agostino RB, Griffith JL, et al. One-year outcomes of out-of-hospital administration of intravenous glucose, insulin, and potassium (GIK) in patients with suspected acute coronary syndromes (from the IMMEDIATE [Immediate Myocardial Metabolic Enhancement During Initial Assessment and Treatment in Emergency Care] Trial). Am J Cardiol 2014;113(10): 1599−605.

[29] Ghanim H, Korzeniewski K, Sia CL, Abuaysheh S, Lohano T, Chaudhuri A, et al. Suppressive effect of insulin infusion on chemokines and chemokine receptors. Diabetes Care 2010;33(5):1103−8.

[30] Dandona P, Ghanim H, Sia CL, Green K, Abuaysheh S, Dhindsa S, et al. A mixed anti-inflammatory and pro-inflammatory response associated with a high dose of corticosteroids. Curr Mol Med 2014;14(6):793−801.

[31] Dandona P, Mohanty P, Hamouda W, Aljada A, Kumbkarni Y, Garg R. Effect of dexamethasone on reactive oxygen species generation by leukocytes and plasma interleukin-10 concentrations: a pharmacodynamic study. Clin Pharmacol Ther 1999;66(1):58−65.

[32] Dandona P, Thusu K, Hafeez R, Abdel-Rahman E, Chaudhuri A. Effect of hydrocortisone on oxygen free radical generation by mononuclear cells. Metabolism 1998;47(7):788−91.

Chapter 5

Percutaneous Electrical Neurostimulation of Dermatome T6 to Reduce Appetite

J. Ruiz-Tovar and C. Llavero

Electrostimulation Unit, Garcilaso Clinic, Madrid, Spain

LIST OF ABBREVIATIONS

BMI   body mass index
PENS   percutaneous electrical neurostimulation

INTRODUCTION

About one-third of the population in developed countries is obese to some degree. Obesity itself is a health risk factor that influences the development and progression of various diseases, such as dyslipidemia, ischemic heart disease, hypertension, type 2 diabetes mellitus, and sleep apnea—hypopnea syndrome, thereby worsening the quality of life of patients, limiting their activities, and causing psychosocial problems. There is a direct relationship between body mass index (BMI) and morbidity and mortality risks in obese patients; it is derived from associated pathologies, and elevates obesity itself to a disease [1—3].

Physical exercise is the first therapeutic step in the dietary treatment of obesity, and patient motivation is essential (though often lacking) for effectiveness. Obese patients often tire of following a low-calorie diet for long periods of time. A continuous feeling of hunger is the major cause of dietary treatment failure. Another problem associated with diets is that once abandoned, the patient often regains the weight previously lost [1,4]. Second-line therapeutic weapons include pharmacological drugs and endoscopic techniques. All these issues will be discussed later.

The final treatment option for obesity is bariatric surgery. Surgery is the best method to obtain a significant and maintained weight loss. However, bariatric surgery has specific indications (patients with $BMI > 40 \, kg/m^2$ or $BMI > 35 \, kg/m^2$ associated with obesity-related comorbidities). Though most techniques involve a laparoscopic approach, which lowers morbidity and mortality rates, this therapy remains the most risky option and is therefore reserved for those cases where dietary treatment has failed [5—7].

THEORETICAL BASIS FOR THE DEVELOPMENT OF PERCUTANEOUS ELECTRICAL NEUROSTIMULATION OF DERMATOME T6 TO REDUCE APPETITE

The initial development of this technique was based on four key points: (1) effect of the gastric pacemaker, (2) percutaneous electrical neurostimulation (PENS) of the posterior tibial nerve, (3) observation of weight loss after central or peripherical neural stimulation, and (4) anatomical background.

1. *Gastric pacemaker:*

   An implantable gastric stimulator (gastric pacemaker) has been used to treat obesity, and has exhibited promising results. It applies cyclic electric pulses of 40 Hz every 4—12 minutes to the gastric wall. The stimulator induces gastric distention in the fasting state and inhibits postprandial antral contractions, thereby impairing stomach emptying, which may lead to early satiety and reduced food intake. The induction of gastric distension in the fasting state results in the activation of stretch receptors, causing satiety [7]. It has been observed that this technique achieves excess weight loss of up to 40% after 1 year. This stimulator can be placed laparoscopically or endoscopically; both techniques present a small risk to the patient, but are still invasive [8,9]. The modulation of neuronal activities and release of certain hormones with an implantable gastric stimulator may also explain the reduction of appetite and the increase of satiety. A decrease in ghrelin levels could be one mechanism that explains weight loss and appetite reduction after implantable gastric stimulation [10]. Chen [8] reported that electric gastric stimulation with a gastric pacemaker may affect the central nervous system by segregating hormones in the stomach and regulating satiety and/or appetite, with ghrelin in particular involved in this process.

2. *Percutaneous electrical neurostimulation of the posterior tibial nerve:*

   PENS was originally developed to treat urinary and fecal incontinence by stimulation of the posterior tibial nerve. The mechanism of action involves the creation of a somato−somatic reflex: the posterior tibial nerve is the afferent pathway that leads the electrical impulse to root S3, and the efferent pathway is the pudendal nerve, which is responsible for innervation of the anal sphincter [11,12].

3. *Weight loss after central or peripheral neural stimulation:*

   Pereira and Foster [13] observed an excess weight loss of 20% associated with decreased appetite in two morbidly obese patients in whom spinal cord stimulators were set up at the T6 and T7 levels to control intractable lumbar pain and lumbosacral radiculitis secondary to lumbar disc herniation. These patients did not increase their physical activity or follow any type of diet, but still experienced significant appetite reduction. These authors were the first to hypothesize that spinal cord stimulation could affect the stomach. Other authors have reported that transcutaneous electrical gastric stimulation may alter gastric motility, delay gastric emptying, and lead to postprandial satiety [14−16]. They believed that electrical stimulation was transmitted to the stomach through the abdominal wall when the electrode was placed in the left upper quadrant of the abdomen. However, we think it is more likely that the effect is produced by the creation of a somato−autonomic reflex rather than by transcutaneous transmission of the electrical stimuli, similar to the transcutaneous electrical stimulation of the posterior tibial nerve in incontinence treatment [17]. Moreover, it is difficult to believe that electrical stimulation could have some effect when traversing a thick abdominal wall (as in morbidly obese patients), particularly considering the presence of adipose tissue, which is not a good electrical conductor. The same authors also postulated that the effect of gastric stimulation, which is associated with the delay of gastric emptying, might also decrease ghrelin segregation in the gastric fundus and inhibit appetite through the central nervous system [14−16].

4. *Anatomical background:*

   The parasympathetic fibers of the vagal nerve that specifically stimulate the stomach arise from the T6 root of the spinal cord. These fibers mainly innervate the gastric body and fundus [18].

Considering these four issues, we initially hypothesized that based on the creation of a somato−autonomic reflex, the stimulation of sensory nerve terminals located in dermatome T6 may cause a reflex for which the efferent pathways end in vagal nerve branches stimulating the gastric wall, similar to a gastric pacemaker.

PENS OF DERMATOME T6 METHODOLOGY

In this methodology, the Urgent PC 200 Neuromodulation System (Uroplasty, Minnetonka, MN, USA) was used, a device originally developed to treat fecal and urinary incontinence. The participants underwent one 30-minute session every week for 12 consecutive weeks. Each patient was placed in a supine position without anesthesia, and PENS was delivered by a needle electrode inserted in the left upper quadrant along the medioclavicular line, 2 cm below the ribcage at a 90° angle toward the abdominal wall at a depth of approximately 0.5−1 cm. Successful placement was confirmed by the feeling of electrical sensation at least 5 cm beyond the dermatome territory. PENS was undertaken at a frequency of 20 Hz at the highest amplification (0−20 mA) without causing pain (Fig. 5.1).

PRELIMINARY RESULTS

A pilot study was performed by our group in 2013 [19] that compared 45 morbidly obese patients undergoing PENS of dermatome T6 on a 1200 Kcal/day diet with 45 patients following only the 1200 Kcal/day diet. In order to eliminate a possible placebo effect associated with the intervention, 15 obese patients undergoing PENS of posterior tibial nerve for fecal incontinence and following the same 1200 Kcal/day diet were also included.

The patients undergoing PENS of dermatome T6 on the hypocaloric diet presented a mean weight loss of 7.1 kg and an excess weight loss of 10.7% versus 2 kg weight loss and 3.2% excess weight loss in the patients following only the

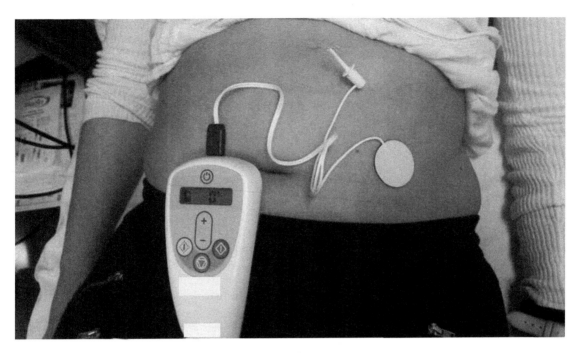

FIGURE 5.1   Placement of the needle in dermatome T6 and the neurostimulator.

diet. Referencing appetite perception, as measured by the Visual Analogic Scale (ranging from 0 to 10), the patients undergoing PENS of T6 experienced an appetite reduction from 6 to 1.5; there were no differences in the patients following only the diet and those experiencing the diet with PENS of the posterior tibial nerve. This demonstrated that appetite reduction was not just a placebo effect related to the intervention. Dietary compliance after 12 weeks was 93.3% in patients undergoing PENS of dermatome T6, while 50% of the patients following only the diet said they had abandoned it because the perception of hunger was unbearable. The second week (range 1st−6th) was the median treatment week, in which decreased appetite was reported by the patients.

From these preliminary results, we observed that appetite was associated with diet compliance, which, logically, was related to weight loss. We did not analyze the isolated effect of PENS of dermatome T6 without diet, but this therapy itself does not justify a relevant weight loss. As shown in this study, its main effect is appetite reduction; all of the patients presented with mild feelings of hunger or even absence of hunger after PENS of dermatome T6. Though some weight loss could be obtained based simply on lower food intake (secondary to appetite reduction), a specific low-calorie diet must be established in order to obtain maximum benefits from appetite reduction.

In terms of bariatric surgery aims, an excess weight loss over 50% with a final BMI $<35\ kg/m^2$ is considered a satisfactory result [20]. Only one morbidly obese patient undergoing PENS of dermatome T6 achieved these goals before deciding to abandon the bariatric surgery program. Typically, the therapy was intended to achieve a weight reduction before bariatric surgery in order to reduce surgical risk. Therefore, after finishing the treatment with PENS of dermatome T6, the patients underwent a bariatric technique. Given these results, PENS of dermatome T6 cannot be considered a bariatric approach because a mean weight loss of 7.1 kg and a mean excess weight loss of 10.7% are not enough to achieve the intended aims. From the results of this first study, it remains unknown whether prolonging the therapy would have had additional effects and how long the effect of this therapy would have lasted. It has been determined that in the PENS of posterior tibial nerve for treating fecal or urinary incontinence, a secondary treatment period every 2 weeks over a 3-month period adds some benefits to the initial treatment [21,22].

LONG-TERM EFFECT OF PENS OF DERMATOME T6

All of the patients undergoing PENS of dermatome T6 in this first study presented with BMIs $>35\ kg/m^2$, and as already mentioned, the weight reduction obtained in these patients was not enough to alleviate their morbid obesity. However, in overweight patients or those presenting with mild obesity, this therapy would most likely help them lose their excess weight and return them to a normal weight status.

Thus, looking for an answer to all these questions, we recently performed a prospective study that included 150 consecutive obese patients with BMIs between 30 and 40 kg/m^2 and previous dietary treatment failure [23]. A significant reduction of weight and BMI was observed between pretreatment values and values after 12 weeks of treatment. This significant reduction was maintained during the long-term follow-up (3 and 9 months after finishing the therapy). Mean weight loss after 12 weeks of treatment was 11.8 kg, 3 months after finishing the treatment mean weight loss was 14.6 kg, and 9 months after finishing it was 14.5 kg. Excess weight lost was 66%. Nine months after finishing treatment (i.e., 1 year after beginning), 42% of all patients presented a BMI within the normal range (BMI: 20−25 kg/m^2), and the remaining patients were within the range considered overweight (BMI: 25−30 kg/m^2) (Fig. 5.2).

Significant appetite reduction was observed after completing the 12 weeks of therapy, and was maintained during the following 3 months; later, however, appetite sensation gradually recovered. Ninety-six percent of the patients presented a decrease in their appetite to some degree. Dietary compliance after 12 weeks of treatment was 90%, 3 months after finishing the treatment the compliance was 84%, and 9 months after finishing it was still 62% (Fig. 5.3).

It is widely known that the main reason for dietary treatment failure is the weight regain once the diet is abandoned. This phenomenon is called the "yo-yo effect" [1,4]. The T6 method allows excellent dietary compliance during therapy, which results in a mean weight loss of 11.8 kg after 12 weeks. The main advantage of this technique, however, is that after finishing therapy, the satiation effect lasts for at least 3 months more, which allows greater weight loss during this period. Most surprising of all is that 12 months after beginning treatment (i.e., 9 months after finishing it), appetite is partially restored, but a weight regain is not observed. Here we want to remark on one fact: the prescribed diet associated with the T6 method is based on common foods eaten by the patients, but with a caloric restriction achieved by reducing the quantity of food and limiting hypercaloric elements. Many diets are based on dysbalances between carbohydrates, proteins, and fats (e.g., hyperproteic diets), or on the ingestion of commercial nutritional formulas, some of which mimic "normal" food; the main drawback of these diets is that patients often regain the lost weight once they abandon these types of diets and return to their normal, balanced diet [24,25]. In our opinion, apart from the satiating effect obtained with PENS of dermatome T6, which is essential for initial dietary compliance, the key factor in achieving significant weight loss is intense monitoring of the diet by a dietitian, adapting the diet to the individual features of

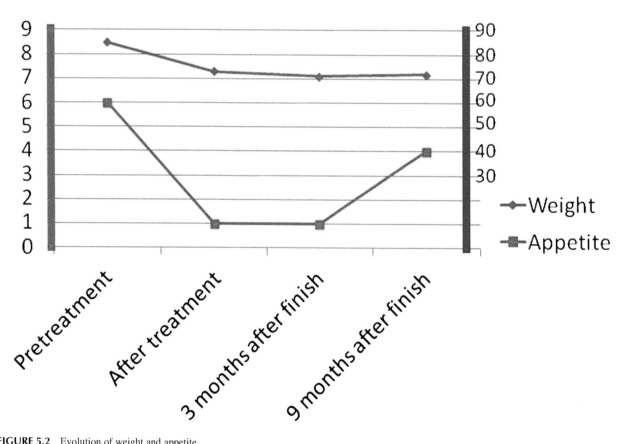

FIGURE 5.2 Evolution of weight and appetite.

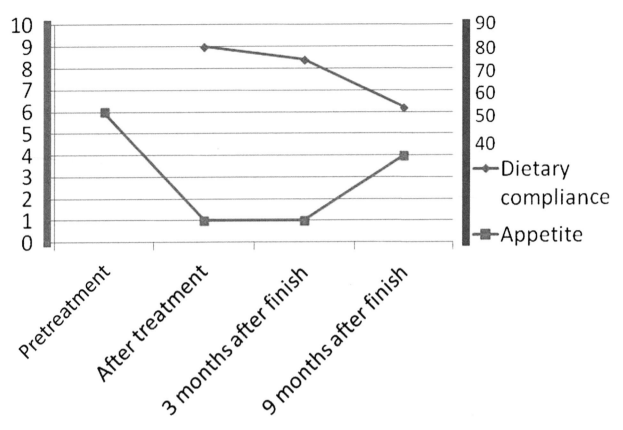

FIGURE 5.3   Evolution of dietary compliance and appetite.

each subject, and suggesting "tricks" to ease dietary compliance. Patients can tire of following a low-calorie diet for long periods of time, even in the absence of a hunger sensation, and guidance by a dietitian is essential to maintaining motivation. However, close monitoring by the dietitian cannot be maintained indefinitely. Therefore, the key factor in maintaining weight loss once the satiating effect of the therapy ends and close monitoring by the dietitian is also complete, is that the patient has learned how to maintain a healthy diet and adapt it to personal conditions.

HORMONAL EFFECT OF PENS OF DERMATOME T6

We have already presented the clinical results of dermatome T6, but little is known about the neurohormonal effect of this therapy. The initial hypothesis postulated that, based on the creation of a somato−autonomic reflex, the stimulation of sensory nerve terminals located in dermatome T6 might cause a reflex, for which the efferent pathways would end in vagal nerve branches stimulating the gastric wall [19]. Similar to the effect of the gastric pacemaker, once stimulated, the stomach's gastric emptying would be impaired, leading to early satiety and reduced food intake. Moreover, the induction of gastric distension in the fasting state could result in the activation of stretch receptors, causing satiety [8].

Ghrelin is one of the main hormones involved in this satiety mechanism. When gastric emptying is delayed, stretch receptors located in the gastric fundus inhibit the segregation of ghrelin to the bloodstream. Ghrelin is an orexygen hormone acting in the appetite center in the hypothalamus, increasing appetite sensation. In cases of decreased ghrelin segregation, the appetite sensation is also reduced, which is the mechanism that explains weight loss and appetite reduction after PENS of dermatome T6 [10,14−16].

Our group is also performing a prospective study of 20 consecutive morbidly obese patients undergoing PENS of dermatome T6, investigating the ghrelin changes at five different time points: (1) basal, (2) after finishing the first stimulation (initial peak point), (3) before undergoing the last stimulation (valley point), (4) after the last stimulation (accumulative peak point), and (5) 30 days after the last stimulation (residual point). Preliminary data suggest a ghrelin decrease at the valley point and the accumulative peak point; these decreased levels are maintained in the residual point.

These results suggest, at least partially, that the clinical effect of appetite reduction is based on a reduction in the orexygenic hormone ghrelin.

MINI-DICTIONARY OF TERMS

- *Ghrelin*: The main orexygenic hormone in humans.
- *Percutaneous electrical neurostimulation*: This technique was initially developed as posterior tibial nerve stimulation for the treatment of urinary and fecal incontinence. It consists of the stimulation of a targeted organ by means of the creation of an artificial reflex.
- *Dermatome*: An area of skin that is mainly supplied by a single spinal nerve. There are 8 cervical nerves, 12 thoracic nerves, 5 lumbar nerves, and 5 sacral nerves. Each of these nerves relays sensation (including pain) from a particular region of the skin to the brain.
- *Reflex*: A reflex action is an involuntary and nearly instantaneous movement in response to a stimulus. Scientific use of the term "reflex" refers to a behavior that is mediated via the reflex arc. Human reflexes (stretched reflexes) are ancestral defense mechanisms. They are used to provide information on the integrity of the central nervous system and peripheral nervous system.
- *Diet compliance*: The main failure of diets is that patients abandon them because of the continuous feeling of hunger. Good diet compliance is essential for achieving significant and maintained weight loss. Diet compliance is defined as the ability of the patient to exactly follow the prescribed diet in terms of quantities and types of food.
- *Gastric pacemaker*: Similar to the heart pacemaker, the gastric pacemaker applies cyclic electrical pulses to the gastric wall. This inhibits postprandial antral contractions, thereby impairing stomach emptying, which may lead to early satiety and reduced food intake.

KEY FACTS

- *Percutaneous electrical neurostimulation*
  - It was first described for the treatment of fecal and urinary incontinence as posterior tibial nerve stimulation.
  - It develops an artificial reflex whose efferent pathway stimulates the target organ.
  - PENS of dermatome T6 was first described by our group in 2014.
  - It reduces appetite and allows better diet compliance.
  - It was initially applied in morbidly obese patients to reduce weight prior to bariatric surgery.
- *Ghrelin*
  - It is a peptide hormone.
  - It is produced in the gastric fundus when the stomach is empty.
  - It has orexigenic properties (stimulates appetite).
  - In obesity its production is increased because gastric emptying is accelerated.
  - PENS of dermatome T6 reduces the release of ghrelin.

SUMMARY POINTS

- PENS of dermatome T6 is a promising new therapy for weight loss.
- Though its effect as a bariatric technique in morbidly obese patients has not yet been demonstrated, it is widely useful as a means to significantly reduce the weight of mildly to moderately obese patients, and for many, to achieve their normal weight.
- The effect of this therapy does not end after completing the last stimulation; its anorexygenic effect continues for at least three more months.
- PENS of dermatome T6 must always be combined with a hypocaloric diet, and monitoring by a dietitian is important so that the patient acquires healthy alimentary habits, thus avoiding weight regain once the effect tapers off.
- Preliminary results suggest that the effect of appetite reduction is mediated by a reduction in ghrelin plasma levels.

REFERENCES

[1] Bray GA. Medical consequences of obesity. J Clin Endocrinol Metab 2004;2583–9.
[2] Vest AR, Heneghan HM, Agarwal S, et al. Bariatric surgery and cardiovascular outcomes: a systematic review. Heart 2012;98:1763–77.

[3] Nguyen NT, Magno CP, Lane KT, et al. Association of hypertension, diabetes, dyslipidemia and metabolic syndrome with obesity: findings from the National Health and Nutrition Examination Survey 1999 to 2004. J Am Coll Surg 2008;207:928−34.

[4] Martin Duce A, Diez del Val I. Cirugía de la obesidad mórbida. Guías Clínicas de la Asociación Española de Cirujanos. Madrid, Aran; 2007.

[5] Sullivan PW, Ghushchyan VH, Ben-Joseph R. The impact of obesity on diabetes, hyperlipidemia and hypertension in the United States. Qual Life Res 2008;17:1063−71.

[6] Kaul A, Sharma J. Impact of bariatric surgery on comorbidities. Surg Clin North Am 2011;91:1295−312.

[7] Buchwald H, Avidor Y, Braunwald E, et al. Bariatric surgery: a systematic review and meta-analysis. JAMA 2004;292:1724−37.

[8] Chen J. Mechanisms of action of the implantable gastric stimulator for obesity. Obes Surg 2004;28−32.

[9] Yao SK, Ke MY, Wang ZF, et al. Visceral response to acute retrograde gastric electrical stimulation in healthy human. World J Gastroenterol 2005;11:4541−6.

[10] De Luca M, Segato G, Busetto L, et al. Progress in implantable gastric stimulation: summary of results of the European multi-center study. Obes Surg 2004;14:33−9.

[11] Van der Pal F, Van Balken MR, Heesakkers JP, et al. Percutaneous tibial nerve stimulation in the treatment of overactive bladder syndrome: is maintenance treatment a necessity? BJU Int 2006;97:547−50.

[12] Boyle DJ, Prosser K, Allison ME, et al. Percutaneous tibial nerve stimulation for the treatment of urge fecal incontinence. Dis Colon Rectum 2010;53:432−7.

[13] Pereira E, Foster A. Appetite suppression and weight loss incidental to spinal cord stimulation for pain relief. Obes Surg 2007;17:1272−4.

[14] Wang J, Song J, Hou X, et al. Effects of cutaneous gastric electrical stimulation on gastric emptying and postprandial satiety and fullness in lean and obese subjects. J Clin Gastroenterol 2010;44:335−9.

[15] Yin J, Ouyang H, Wang Z, et al. Cutaneous gastric electrical stimulation alters gastric motility in dogs: new option for gastric electrical stimulation? J Gastroenterol Hepatol 2009;24:149−54.

[16] Abell TL, Minocha A, Abidi N. Looking to the future: electrical stimulation for obesity. Am J Med Sci 2006;331:226−32.

[17] Vitton V, Damon H, Roman S, et al. Transcutaneous posterior tibial nerve stimulation for fecal incontinence in inflammatory bowel disease patients: a therapeutic option? Inflamm Bowel Dis 2009;15:402−5.

[18] Moore KL, Dalley II AF. Clinically oriented anatomy. Philadelphia: Lippincott Wilkins & Williams; 2006. p. 321−5.

[19] Ruiz-Tovar J, Oller I, Diez M, et al. Percutaneous electrical neurostimulation of dermatome T6 for appetite reduction and weight loss in morbidly obese patients. Obes Surg 2014;24:205−11.

[20] Lemanu DP, Srinivasa S, Singh PP, et al. Laparoscopic sleeve gastrectomy: its place in bariatric surgery for the severely obese patient. N Z Med J 2012;125:41−9.

[21] Monga AK, Tracey MR, Subbaroyan J. A systematic review of clinical studies of electrical stimulation for treatment of lower urinary tract dysfunction. Int Urogynecol J 2012;23:993−1005.

[22] Findlay JM, Maxwell-Armstrong C. Posterior tibial nerve stimulation and faecal incontinence: a review. Int J Colorectal Dis 2011;26:265−73.

[23] Ruiz-Tovar J, Llavero C. Long − term effect of percutaneous electrical neurostimulation of dermatome T6 for appetite reduction and weight loss in obese patients. J Laparoendosc Adv Surg Tech 2016;26(3):212−15. Epub ahead of print.

[24] Langeveld M, De Vries JH. The long-term effect of energy restricted diets for treating obesity. Obesity (Silver Spring) 2015;23:1529−38.

[25] Leeds AR. Formula food-reducing diets: a new evidence-based addition to the weight management tool box. Nutr Bull 2014;39:238−46.

Chapter 6

The Management of Obesity: An Overview

M. Camilleri, B. Abu Dayyeh and A. Acosta

Mayo Clinic, Rochester, MN, United States

LIST OF ABBREVIATIONS

BMI   body mass index
ER   extended release
EWL   excess weight loss
GABA γ-aminobutyric acid
GERD   gastroesophageal reflux disease
GLP-1   glucagon-like peptide 1
HbA1c   hemoglobin glycosylated test
PYY   peptide tyrosine-tyrosine
RYGB   Roux-en-Y gastric bypass
SR   sustained release
T2DM   type 2 diabetes mellitus

INTRODUCTION

Obesity results from weight gain due to calorie intake in excess of energy expenditure over a prolonged period. This chapter reviews the most common forms of obesity management, but it does not address the hedonic aspects of energy intake or expenditure, or the strictly behavioral approaches to obesity therapy. The chapter instead focuses on hypocaloric diets, pharmacologic agents (approved for long-term treatment of at least 3 months), and endoscopic and surgical approaches to the management of obesity.

The National Health And Nutrition Examination Survey (NHANES) 2007−12 in the United States led to guidelines in 2013 whereby the number of adults recommended for weight loss treatment increased by 20.9%, from 116.0 million to 140.2 million; thus, 64.5% of nonpregnant, noninstitutionalized US adults are candidates for weight loss treatment [1,2], including overweight people having only one risk factor (obesity comorbidity) and those having a large waist circumference. With these recommendations, up to 53.4% of adults could be considered for pharmacological therapy in addition to lifestyle therapy, and up to 14.7% could be considered for bariatric surgery.

To understand the current approaches to the management of obesity, it is essential to review the therapeutic targets in the brain−gut axis, including the mechanisms involved in satiation and appetite centrally and in the gastrointestinal tract.

FOOD INTAKE: THE BRAIN−GUT AXIS

Food intake is regulated by a balance between appetite, satiation, and satiety, which is controlled by the brain−gut axis. Thus, gastrointestinal signals inform brain centers about energy intake status (reviewed elsewhere [3]). The gastrointestinal signals involved in food intake regulation include the motor and sensory functions of the stomach, such as the rate of emptying and gastric volume and accommodation. The latter signals hunger and increased appetite in response to

Metabolism and Pathophysiology of Bariatric Surgery.

fasting, while calorie and volume ingestion signal satiation and fullness. In addition, the peripherally released peptides and hormones (such as ghrelin, motilin, cholecystokinin, glucagon-like peptide 1 (GLP-1), peptide YY, and oxyntomodulin) provide feedback from the arrival of nutrients in different regions of the gut from where they are released to signal satiation or regulate metabolism through their incretin effects. Ultimately, the highly organized hypothalamic and vagal centers influence energy intake during meal ingestion and the return of appetite and hunger during fasting [4].

Hypothalamic centers also control food intake through the actions of several peptides and their receptors. The arcuate nucleus receives input from the brainstem (e.g., vagal) nuclei and from circulating hormones. Neurons in the arcuate nucleus are either orexigenic (e.g., contain neuropeptide Y via Y1 receptors or agouti-related peptide (AgRP)) or anorexigenic (e.g., contain proopiomelanocortin (POMC), and cocaine- and amphetamine-related transcript). POMC is a precursor of α-melanocyte-stimulating hormone. Ultimately, other regions of the hypothalamus (the paraventricular nucleus and lateral hypothalamus) and higher centers (such as amygdala, limbic system, and cerebral cortex) are stimulated to change feeding behavior through actions mediated by hypothalamic nuclei.

GASTRIC DETERMINANTS OF POSTPRANDIAL SYMPTOMS AND SATIATION

Satiation is the sense of feeling full during a meal, which induces meal termination; satiety is the degree of fullness that persists until the consumption of a subsequent meal after a period of fasting [5,6], and regulates meal frequency. Satiation is appraised in practice by the volume ingested to reach fullness and maximum tolerated volume of Ensure nutrient drink (1 kcal/mL) ingested at a rate of 30 kcal/minute [7]. Satiety is appraised by the total calorie intake at an *ad libitum* buffet meal [8] after a standard period of fasting (e.g., 4 hours) and a standard prior meal (e.g., a 300 kcal liquid breakfast meal).

Studies based on hundreds of patients with dyspepsia convincingly showed that gastric motor functions (such as emptying and accommodation), intragastric pressure [9], and gastric sensation are important determinants of intraprandial and postprandial symptoms [10–12].

Other studies have shown increased gastric volume in patients with binge eating disorder [13].

TRAITS ASSOCIATED WITH FOOD INTAKE, POSTPRANDIAL SYMPTOMS, AND OBESITY

Prospective studies [13] identified the following quantitative traits in the brain–gut axis that represent latent dimensions that contribute to the obesity trait, based on a principal components analysis: increased appetite or higher fasting plasma ghrelin levels; larger fasting gastric volume (>300 mL); accelerated gastric emptying of solids (GE $T_{1/2} < 85$ minutes); decreased satiation (or increased calorie intake to sensation of fullness (>800 kcal) during ingestion of a liquid nutrient at a constant rate of 30 mL/minute); increased calorie intake at an *ad libitum* meal (>800 kcal ingested at buffet meal); decreased peak postprandial GLP-1; and abnormal eating behaviors and affect based on validated questionnaires.

In addition, among 169 patients with obesity, almost two-thirds had single, or a combination of two, alterations in quantitative traits: 22 patients with exclusive acceleration of gastric emptying, 29 patients with large fasting gastric volume, 24 patients with abnormal satiation, and 30 with abnormal satiation/satiety associated with either large gastric volume or accelerated gastric emptying. These gastrointestinal traits were largely independent of the behavioral traits [13].

APPROACHES AVAILABLE FOR THE MANAGEMENT OF OBESITY

Weight Loss Programs

A recent review of 45 studies involving commercial weight loss programs based on diet and behavioral modification (which included 39 randomized, controlled trials) showed that at 12 months the commercial diets achieved greater weight loss than control/education and counseling: Weight Watchers by $>2.6\%$, Jenny Craig by $>4.9\%$, and very-low-calorie programs (e.g., Medifast and OPTIFAST) by $>4.0\%$ [14]. These commercial programs incorporate group sessions (e.g., Weight Watchers) or more expensive one-on-one counseling (e.g., Jenny Craig). Table 6.1 shows a recent analysis of weight loss programs [14].

Approved Pharmacological Therapies

Mechanism of Action and Efficacy Relative to Placebo

A brief description of the pharmacological actions and efficacy relative to placebo of the currently approved medications or combinations for the long-term treatment of obesity is provided in Table 6.2 [15–17]. There is moderate

TABLE 6.1 Efficacy of Commercial Weight Loss Programs: Weight Change Between Commercial Programs (Weight Watchers, Jenny Craig, and Nutrisystem) and Comparators

Program	Intensity	Nutrition	Physical Activity	Behavioral Strategies	Support
Weight Watchers	High	Low-calorie conventional foods Points tracking	Activity tracking	Self-monitoring	Group sessions Online coaching Online community forum
Jenny Craig	High	Low-calorie meal replacements	Encourages increased activity	Goal setting Self-monitoring	One-on-one counseling
Nutrisystem	High	Low-calorie meal replacements	Exercise plans	Self-monitoring	One-on-one counseling Online community forum
HMR	High	Very-low-calorie or low-calorie meal replacements	Encourages increased activity	Goal setting	Group sessions Telephone coaching Medical supervision
Medifast	High	Very-low-calorie or low-calorie meal replacements	Encourages increased activity	Self-monitoring	One-on-one counseling Online coaching
OPTIFAST	High	Very-low-calorie or low-calorie meal replacements	Encourages increased activity	Problem solving	One-on-one counseling Group support Medical supervision

Source: From Gudzune KA, Doshi RS, Mehta AK, et al. Efficacy of commercial weight-loss programs: an updated systematic review. Ann Intern Med 2015;162:501−12.

overall average weight loss of 3−8 kg after at least 12 weeks of treatment with novel pharmacological agents: lorcaserin, GLP-1 agonists, phentermine−topiramate extended release (ER), and bupropion−naltrexone sustained release (SR) [18−21].

Market Penetration

Despite the approval of several novel pharmacological therapies that suppress appetite, increase satiation/satiety, and result in weight loss, the public and physicians have not embraced use of these medications for obesity for several reasons, including safety issues with past diet drugs (e.g., phentermine−fenfluramine, sibutramine, and rimonabant), significant costs or copays, and physician propensity to wait a year or longer after approval before prescribing new weight loss drugs in order to allow for unforeseen safety issues to emerge [22]. Current therapy is still based on the assumption that one-treatment-fits-all in obesity; this approach may explain the highly variable response to treatment with current pharmacological approaches.

Recently, we identified quantitative traits [23] that are regulated by the brain−gut axis that can be measured reliably in humans and provide "actionable" approaches to control food intake. In two proof of concept, small, randomized, controlled trials, we demonstrated that phentermine−topiramate ER resulted in significantly greater weight loss in patients who ingested >900 kcal at an *ad libitum* buffet meal [23], and the short-acting GLP-1 receptor agonist, exenatide, retarded gastric emptying and produced numerically (not significantly) greater weight loss compared to placebo in obese patients with accelerated gastric emptying [24]. Thus, there is the opportunity to individualize pharmacotherapy for obesity based on quantifiable gastrointestinal physiological and behavioral traits [25]. The combination of phentermine and topiramate ER caused significant weight loss, slowed gastric emptying, and decreased calorie intake. Importantly, a baseline abnormal satiety test in the form of an *ad libitum* meal significantly predicted weight loss response to phentermine and topiramate ER [23]. However, it still remains to be proven that the physiological traits can

TABLE 6.2 Pharmacological Actions and Efficacy of Currently Approved Medications for Long-Term Treatment of Obesity

Drug	Mechanism of Action	Dose	Weight Loss (kg) Versus Placebo	Adverse Effects
Orlistat 60 mg (Alli) or 120 mg (Xenical; 3 × within 1 h of a fat-containing meal)	Lipase inhibitor causing excretion of approximately 30% of ingested triglycerides in stool	60 mg, or 120 mg, with 3 meals	60 mg dose: −2.5 (−1.5 to −3.5)	Oily spotting, flatus, fecal urgency, fatty oily stool, increased defecation, fecal incontinence
			120 mg dose: −3.4 (−3.2 to −3.6)	
Lorcaserin 10 mg (Belviq; 2 ×)	Highly selective serotonergic 5-HT2C receptor agonist causing appetite suppression	10 mg twice daily	−3.2 (−2.7 to −3.8)	Headache, dizziness, fatigue, nausea, dry mouth, cough, and constipation; and in T2DM back pain, cough, and hypoglycemia
Phentermine plus topiramate ER (Qsymia; 3.75 mg/ 23 mg for 2 weeks, increased to 7.5 mg/ 46 mg)	Noradrenergic + GABA-receptor activator, kainite/ AMPA glutamate receptor inhibitor causing appetite suppression	3.75 mg/23 mg to 7.5 mg/ 46 mg; once daily	7.5 mg/46 mg: −6.7 (−5.9 to −7.5)	Paresthesiae, dizziness, taste alterations, insomnia, constipation, dry mouth, elevation in heart rate, memory or cognitive changes
			15 mg/92 mg: −8.9 (−8.3 to −9.4)	
Liraglutide (1.2 mg escalating over 4 weeks to 3.0 mg SQ)	Glucagon-like peptide 1 agonist causing appetite suppression and possibly delaying gastric emptying	1.2 mg escalating to 3.0 mg SQ once daily	1.2 mg dose: −2.1 (−3.6 to −0.6)	Nausea and vomiting
			1.8 mg dose: −2.8 (−4.3 to −1.3)	
			2.4 mg dose: −3.5 (−5.0 to −2.0)	
			3 mg dose: −4.4 (−6.0 to −2.9)	
Bupropion SR/ Naltrexone SR	Dopamine and norepinephrine reuptake inhibitor and opioid receptor antagonist	360/32 mg p.o. daily	Naltrexone 16 mg dose: −4.9 (−4.6 to −5.2)	Headache, nausea, insomnia, constipation tremor
			Naltrexone 32 mg dose: −6.1 (−5.8 to −6.4)	

ER = extended release; SR = sustained release; T2DM = type 2 diabetes mellitus; − = weight loss
Source: Summarized from Yanovski SZ, Yanovski JA. Long-term drug treatment for obesity: a systematic and clinical review. JAMA 2014;311:74−86; Astrup A, Rössner S, Van Gaal L, Rissanen A, Niskanen L, Al Hakim M, et al. Effects of liraglutide in the treatment of obesity: a randomised, double-blind, placebo-controlled study. Lancet 2009;374:1606−16; Pi-Sunyer X, Astrup A, Fujioka K, Greenway F, Halpern A, Krempf M, et al. A randomized, controlled trial of 3.0 mg of liraglutide in weight management. NEJM 2015;373:11−22.

impact the response to therapy. Overall, there is an imbalance between the perceived clinical need for weight loss at the individual patient and population levels and the average efficacy (and long-term costs) of commercial weight loss programs and drug therapy. Hence, clinically significant weight loss therapy is still dominated by bariatric surgery and, to a lesser extent, endoscopic therapies.

Bariatric Surgery

In the last decade, the number of bariatric procedures performed in the United States has virtually doubled to ∼190,000 operations per year (Table 6.3). The most frequently performed operations are Roux-en-Y gastric bypass (RYGB) and vertical sleeve gastrectomy; these have dwarfed laparoscopic banding in the last 4 years [26].

TABLE 6.3 Summary From the American Society for Metabolic and Bariatric Surgery of Estimated Number of Bariatric Surgical Cases Performed per Year

Year	2011	2012	2013	2014
Total	158,000	173,000	179,000	193,000
Roux-en-Y gastric bypass	36.7%	37.5%	34.2%	26.8%
Laparoscopic adjustable gastric banding	35.4%	20.2%	14%	9.5%
Sleeve gastrectomy	17.8%	33%	42.1%	51.7%
Biliopancreatic diversion/duodenal switch	0.9%	1%	1%	0.4%
Revisions	6%	6%	6%	11.5%
Other	3.2%	2.3%	2.7%	0.1%

Source: From Ponce J, Nguyen NT, Hutter M, Sudan R, Morton JM. American Society for Metabolic and Bariatric Surgery estimation of bariatric surgery procedures in the United States, 2011-2014. Surg Obes Relat Dis 2015;11(6):1199–200. pii: S1550-7289(15)00803-5. http://dx.doi.org/10.1016/j.soard.2015.08.496. [Epub ahead of print].

The efficacy of bariatric surgery is based on the effects of gastrointestinal physiology. Specifically, with partial gastrectomy in the RYGB procedure or vertical sleeve gastrectomy, there is reduction in the volume of the stomach, induction of early satiation, increased postprandial GLP-1 response (which enhances glycemic control), increased early postprandial peptide tyrosine-tyrosine (PYY) (a satiety hormone), and decreased postprandial ghrelin (orexigen). In addition, some bariatric procedures divert nutrients and result in malabsorption, such as RYGB and biliopancreatic diversion with duodenal switch. These bariatric procedures are also established as remedies for metabolic diseases, including correction of glycemia, which occurs even before the benefits of weight loss; these effects are in part attributed to the earlier and higher postprandial incretin responses [27], to activation of farnesoid X receptor-mediated mechanisms [28], and to changes in bile acids and microbiome [29].

Long-term effects of bariatric surgery on health status and morbidity related to obesity include long-term remission of type 2 diabetes mellitus (T2DM) with micro- and macro-vascular complications [30], which is more efficient than the usual care in the prevention of T2DM in obese persons [31], better diabetes control than medical care (with RYGB) [32], better 3-year outcomes (such as those of hemoglobin glycosylated test (HbA1c), body mass index, and a number of diabetes medications) with RYGB and vertical sleeve gastrectomy compared to intensive medical therapy for diabetes [33], lower incidence of fatal and total cardiovascular events [34], improvement of cardiovascular risk factors with biliopancreatic diversion [35], reduced cancer incidence (particularly in women, diabetics, and smokers [36]), and reduced transaminases proportionate to weight loss in patients with nonalcoholic steatohepatitis [37].

Novel Devices Proposed or Approved in Obesity Treatment

Fig. 6.1 illustrates the three categories of devices proposed or in development for treatment of obesity: those that divert nutrients, those that occupy space in the stomach, and those that alter gastric emptying or capacity. The efficacy and mechanisms of action have been extensively reviewed elsewhere [38].

Devices That Divert Nutrients

AspireAssist Aspiration Therapy System (Aspire Bariatrics, King of Prussia, PA) diverts food out of the stomach after a meal, preventing absorption in the small bowel and resulting in mean weight loss of 15 kg at 6 months, with improved blood glucose and HbA1C [39]. Adverse effects arise from electrolyte losses (e.g., sodium and potassium). A major benefit [39] of this device is its positive impact on patients whose calorie consumption cannot be reduced in any other way.

DuodenoJejunal Bypass Liner (GI Dynamics, Lexington, MA) mimics RYGB, inducing weight loss and improving metabolic control through malabsorption, and improving glycemic control through release of incretin hormones (GLP-1, PYY) [40–42]. The DuodenoJejunal Bypass Liner also delays gastric emptying and prevents proper mixing of food with pancreaticobiliary juices [40]. Three open-label studies of the EndoBarrier DuodenoJejunal Bypass System resulted in 105 patients achieving 35.3% excess weight loss (EWL; 95% CI, 24.6–46.1) at 12 months; however, four

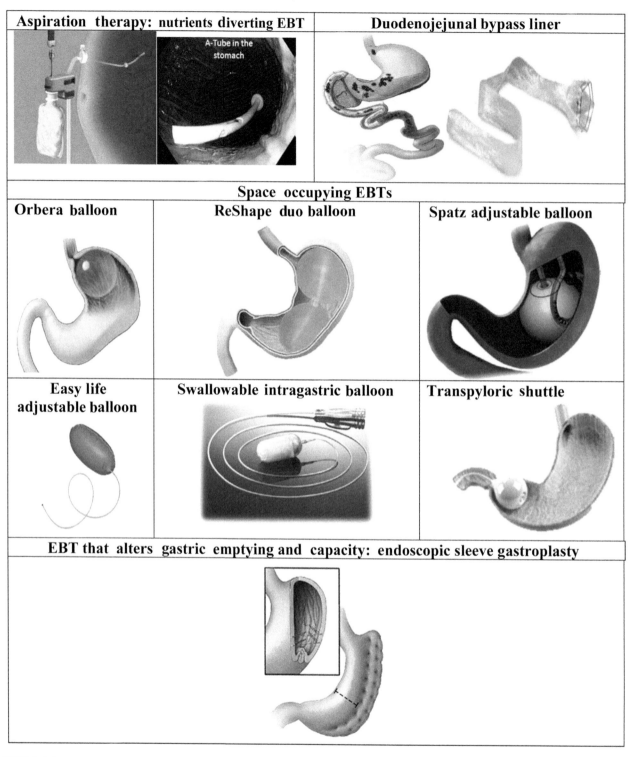

FIGURE 6.1 Endoscopic bariatric therapies (EBTs).

randomized controlled trials of 12−24 weeks' treatment with the EndoBarrier DuodenoJejunal Bypass System (90 subjects) compared to a sham or control arm (84 subjects) revealed a mean %EWL of only 9.4% (95% CI, 8.26−10.65). Nevertheless, improvement in HgA1c was statistically significant compared with a sham or control treatment in the group of patients with diabetes, as the EndoBarrier DuodenoJejunal Bypass System resulted in an additional 1% reduction (95% CI, −1.67 to −0.4; $p = 0.001$) in HBA1c compared to the control treatment group [43].

Devices That Occupy Stomach Space

The proximal stomach is not the only part of the organ that has the capacity to store food; the antrum also accommodates food [44]. Hence, interventions that reduce gastric reservoir capacity must include more than the fundus to be effective in the long term. This may explain the disappointing results with surgical vertical banded gastroplasty, gastric banding, or the endoscopic transoral gastroplasty [45].

These intragastric space-occupying devices limit the portion of the stomach available to accommodate food, and result in satiation with lower volume intake. There is some evidence that balloons may delay gastric emptying and reduce plasma ghrelin [46], though the effect on ghrelin is not consistently demonstrated at different times after insertion of the device [47,48].

Two types of balloon devices have been approved by the FDA: Orbera (previously known as Bioenteric Balloon, Apollo Endosurgery, Austin, TX) and ReShape Intragastric Duo-Balloon (ReShape Medical, San Clemente, CA). Orbera intragastric balloon was recently shown, based on a meta-analysis of 17 studies that included 1638 patients, to induce 25.44% EWL at 12 months (95% CI, 21.47−29.4), though there was a high degree of heterogeneity (I^2 Z 97.4%) [43].

ReShape Intragastric Duo-Balloon attempts to occupy space in the proximal and distal stomach and results in weight loss with shorter duration of symptoms (<7 days) such as nausea, vomiting, and retching [49].

Multiple other intragastric balloons with unique design features that might enhance tolerability, weight loss, and durability are currently available outside the United States [50]. The Spatz Adjustable Balloon (Spatz Medical, Great Neck, NY) and the Easy Life Intragastric Balloon (Life Partners, Bagnolet, France) allow for volume adjustment while the balloon remains in the stomach for a longer duration of therapy (12 months). The Obalon Gastric Balloon (Obalon Therapeutics Inc., Carlsbad, CA) and the Elipse Intragastric Balloon (Allurion Technologies, Wellesley, MA) are placed under fluoroscopic guidance without the need for endoscopy. The latter balloon is spontaneously excreted from the gastrointestinal tract after 4 months.

Other space-occupying endoscopic bariatric devices include the TransPyloric Shuttle (BAROnova Inc., Goleta, CA). This device consists of a large spherical bulb connected to a smaller cylindrical bulb by a flexible tether that passes freely into the duodenum to position the device across the pylorus. This may delay gastric emptying, which reduces caloric intake and enables weight loss. It is associated with nausea, vomiting, pain, and gastroesophageal reflux disease, especially in the first 30 days after deployment [51].

Sequential use of intragastric balloons, or in conjunction with pharmacotherapy, have been reported to enhance the magnitude and durability of weight loss [52,53]. A randomized study indicated that the Orbera balloon was more efficacious than pharmacotherapy in accomplishing weight reduction. In that study, 50 patients with obesity were randomized to either lifestyle modifications combined with the Orbera balloon for 6 months (n = 30), or to lifestyle modifications combined with sibutramine (pharmacotherapy group, n = 20) for 6 months [53]. After Orbera balloon removal, patients were randomly assigned to lifestyle (Orbera/lifestyle) or lifestyle plus pharmacotherapy (Orbera/pharmacotherapy) for an additional 6 months. Patients in the Orbera/pharmacotherapy arm had more significant weight loss, suggesting a potential synergistic effect of the device and pharmacotherapy.

Devices That Alter Gastric Emptying or Capacity

Delay in gastric emptying results in increased satiation and reduced food intake. Examples include reversible vagal block (VBLOC), gastric electrical stimulation, and endoscopic sleeve gastroplasty.

VBLOC was efficacious in retarding gastric emptying, inducing earlier satiation, and reducing hunger, which resulted in weight loss, especially in patients in whom there was effective vagal inhibition [54]. VBLOC was associated with significantly greater weight loss compared to sham block [55]. Other beneficial effects that may be independent of weight loss were 1% mean reduction in HbA1c at 12 months of treatment compared to baseline, mean 28 mg/dL reduction in fasting blood glucose in obese patients with T2DM, and reduced blood pressure among those with hypertension [56]. The beneficial effects on glycemia have been postulated to result from interruption of hepatic afferent vagal pathways, which reduces hepatic glucose production experimentally [57].

Gastric electrical stimulation may induce antegrade or retrograde propulsion. Overall, results with several devices are associated with variable degrees of weight loss, even with the same device [58]. The mechanisms of action include changes in gastric emptying or gastric accommodation and induction of satiety [59,60].

Endoscopic sleeve gastroplasty results in significant weight loss [61], which is thought to result from reduced gastric capacity, inducing fullness (early satiation) and, possibly, altered gastric emptying [62]. Results of 1-year follow-up and effects on satiation and gastric emptying show >50% reduction in maximum tolerated volume in a satiation drink test and marked retardation in gastric emptying, possibly because of the creation of a proximal gastric pouch with reduced size of the outflow from this pouch to the rest of the stomach [62,63].

WHAT PHARMACOTHERAPIES MIGHT BE DIRECTED AT DIFFERENT "ACTIONABLE" OBESITY TRAITS?

Different "actionable" obesity traits can be targets in therapeutic trials for obesity.

1. *Patients with accelerated gastric emptying*: The class of amylin agonists (e.g., pramlintide [64]) or GLP-1 agonists (e.g., liraglutide and exenatide) might be preferred for this group of patients, as they retard gastric emptying [65–67].
2. *Patients with enlarged gastric volume*: Ghrelin or motilin agonists may be preferred as they reduce gastric accommodation [68,69]; however, ghrelin agonists may not be the preferred class of medications since they increase appetite [70]. The opioid antagonist naltrexone reduces gastric accommodation [71]; however, it is unclear if SR of the combination bupropion/naltrexone has any effect on gastric volume and accommodation. In contrast, GLP-1 agonists, which enhance gastric volume and postprandial accommodation [65], may not be the preferred medications for patients with enlarged gastric volume.
3. *Patients with prominent psychological/depression scores* may be amenable to treatment with centrally acting medications that may also impact affect, such as lorcaserin (a selective serotonin 5-HT$_{2C}$ agonist that activates hypothalamic centers to reduce food intake) or the combination of bupropion and naltrexone. Bupropion appears to reduce food intake by acting on adrenergic and dopaminergic receptors in the hypothalamus, while naltrexone is an opioid receptor antagonist that might block inhibitory influences of opioid receptors activated by β-endorphin that is released in the hypothalamus that stimulates feeding [72]. In addition, effects of lorcaserin and bupropion/naltrexone may be mediated through higher brain centers by regulating feeding behavior and cravings.
4. *Patients with low levels of satiation or satiety* may respond to centrally acting medications that act on the hypothalamic nuclei of satiation, such as phentermine/topiramate ER. Phentermine reduces appetite through increasing norepinephrine in the hypothalamus, and topiramate may reduce appetite through its effect on γ-aminobutyric acid receptors.

WHAT BARIATRIC INTERVENTIONS ARE DIRECTED AT DIFFERENT "ACTIONABLE" OBESITY TRAITS?

1. Reduced reservoir in stomach and reduced ghrelin; examples include RYGB, sleeve gastrectomy, intragastric space-occupying devices, and endoscopic sleeve gastroplasty.
2. Malabsorption: nutrient diversion from stomach, RYGB with long Roux limb, and duodenal switch.
3. Enhanced incretins and PYY: RYGB, sleeve gastrectomy, and duodenojejunal bypass liner.
4. Reduced gastric emptying: sleeve gastroplasty and intragastric space-occupying devices.

CONCLUSION

While bariatric surgery remains the most efficacious treatment for obesity, and the use of sleeve gastrectomy has surpassed that of RYGB in the last 2 years, surgery is not a feasible obesity treatment for a large population of candidates. Therefore, it appears that interventional therapy by endoscopy and personalized drug therapy based on gastrointestinal and behavioral traits constitute alternative approaches that have the potential to impact the obesity epidemic.

MINI-DICTIONARY OF TERMS

- *Obesity*: The amount of excess body fat at which health risks to individuals begin to increase.
- *Appetite*: The desire to eat food, sometimes due to hunger.
- *Satiation*: The sense of feeling full during a meal, which induces meal termination.
- *Satiety*: The degree of fullness or satiation before the consumption of the next meal.
- *Endoscopy*: An examination or intervention by means of an endoscope, usually without any skin rupture and using natural orifices (e.g., oral cavity).
- *Bariatric Surgery*: Surgical intervention with the purpose of restricting or reducing absorption of calories.

KEY FACTS

- Obesity is the epidemic of the century, i.e., nonpregnant, noninstitutionalized US adults are candidates for weight loss treatment.
- Disorders of appetite, gastric function, incretin hormones, and psychological/behavioral factors contribute to the development of obesity and constitute disorders of the brain−gut axis.
- There are several medications and devices approved by regulatory agencies for the treatment of obesity or overweight with comorbidity.
- Bariatric surgery aims to reduce the reservoir capacity of the stomach or to divert nutrients away from digestive mechanisms to induce malabsorption.
- Laparoscopic sleeve gastrectomy and RYGB are the two most commonly performed bariatric surgery procedures, and achieve approximately the same effects on weight, glycemic control, and blood pressure.
- The different mechanisms of action of approved medications provide opportunity to personalize drug therapy in obesity.

SUMMARY POINTS

- Obesity is the epidemic of the century, and results from weight gain due to greater calorie intake than energy expenditure over a prolonged period.
- Interventions for obesity target the brain−gut axis, including the mechanisms involved in satiation and appetite centrally and in the gastrointestinal tract.
- Bariatric surgery remains the most efficacious treatment for obesity.
- Novel endoscopic procedures result in significant weight loss and are less invasive than bariatric surgery.
- Personalized drug or device therapy, based on gastrointestinal and psychological or behavioral traits identified in individual patients, have the potential for greater impact on obesity than nonselective use of these therapies.

REFERENCES

[1] Stevens J, Oakkar EE, Cui Z, et al. US adults recommended for weight reduction by 1998 and 2013 obesity guidelines, NHANES 2007-2012. Obesity (Silver Spring) 2015;23:527−31.

[2] Jensen MD, Ryan DH, Donato KA, et al. Guidelines (2013) for managing overweight and obesity in adults. Obesity 2014;22(Suppl. 2):S1−410.

[3] Acosta A, Abu Dayyeh BK, Port JD, et al. Recent advances in clinical practice challenges and opportunities in the management of obesity. Gut 2014;63:687−95.

[4] Camilleri M. Peripheral mechanisms in appetite regulation. Gastroenterology 2015;148:1219−33.

[5] Blundell J, Rogers P, Hill A. Evaluating the satiating power of foods: implications for acceptance and consumption. In: Solms J, Booth DA, Pangborn RM, Raunhardt O, editors. Food acceptance and nutrition. London: Academic Press; 1987. p. 205−19.

[6] Gibbons C, Finlayson G, Dalton M, et al. Metabolic Phenotyping Guidelines: studying eating behaviour in humans. J Endocrinol 2014;222: G1−12.

[7] Chial HJ, Camilleri C, Delgado-Aros S, et al. A nutrient drink test to assess maximum tolerated volume and postprandial symptoms: effects of gender, body mass index and age in health. Neurogastroenterol Motil 2002;14:249−53.

[8] Vazquez Roque MI, Camilleri M, Stephens DA, et al. Gastric sensorimotor functions and hormone profile in normal weight, overweight and obese people. Gastroenterology 2006;131:1717−24.

[9] Janssen P, Verschueren S, Tack J. Intragastric pressure as a determinant of food intake. Neurogastroenterol Motil 2012;24:612−15.

[10] Tack J, Bisschops R. Mechanisms underlying meal-induced symptoms in functional dyspepsia. Gastroenterology 2004;127:1844−7.

[11] Delgado-Aros S, Camilleri M, Cremonini F, et al. Contributions of gastric volumes and gastric emptying to meal size and post-meal symptoms in functional dyspepsia. Gastroenterology 2004;127:1685−94.

[12] Delgado-Aros S, Camilleri M, Castillo EJ, et al. Effect of gastric volume or emptying on meal-related symptoms after liquid nutrients in obesity: a pharmacological study. Clin Gastroenterol Hepatol 2005;3:997−1006.

[13] Geliebter A, Yahav EK, Gluck ME, et al. Gastric capacity, test meal intake, and appetitive hormones in binge eating disorder. Physiol Behav 2004;81:735−40.

[14] Gudzune KA, Doshi RS, Mehta AK, et al. Efficacy of commercial weight-loss programs: an updated systematic review. Ann Intern Med 2015;162:501−12.

[15] Yanovski SZ, Yanovski JA. Long-term drug treatment for obesity: a systematic and clinical review. JAMA 2014;311:74−86.

[16] Astrup A, Rössner S, Van Gaal L, Rissanen A, Niskanen L, Al Hakim M, et al. Effects of liraglutide in the treatment of obesity: a randomised, double-blind, placebo-controlled study. Lancet 2009;374:1606−16.

[17] Pi-Sunyer X, Astrup A, Fujioka K, Greenway F, Halpern A, Krempf M, et al. A randomized, controlled trial of 3.0 mg of liraglutide in weight management. NEJM 2015;373:11–22.

[18] Bray GA. Medical treatment of obesity: the past, the present and the future. Best Pract Res Clin Gastroenterol 2014;28:665–84.

[19] Yanovski SZ, Yanovski JA. Naltrexone extended-release plus bupropion extended-release for treatment of obesity. JAMA 2015;313:1213–14.

[20] Vilsbøll T, Christensen M, Junker AE, et al. Effects of glucagon-like peptide-1 receptor agonists on weight loss: systematic review and meta-analyses of randomised controlled trials. BMJ 2012;344:d7771.

[21] Wadden TA, Hollander P, Klein S, et al. Weight maintenance and additional weight loss with liraglutide after low-calorie-diet-induced weight loss: the SCALE maintenance randomized study. Int J Obes (Lond) 2013;37:1443–51.

[22] http://www.wsj.com/articles/weight-loss-drugs-seek-acceptance-from-patients-and-physicians-1426526252.

[23] Acosta A, Camilleri M, Shin A, et al. Quantitative gastrointestinal and psychological traits associated with obesity and response to weight-loss therapy. Gastroenterology 2015;148:537–46. e4.

[24] Acosta A, Camilleri M, Burton D, O'Neill J, Eckert D, Carlson P, et al. Exenatide in obesity with accelerated gastric emptying: a randomized, pharmacodynamics study. Physiol Rep 2015;3(10):e12610. Available from: http://dx.doi.org/10.14814/phy2.12610.

[25] Camilleri M, Acosta A. Gastrointestinal traits: individualizing therapy for obesity with drugs and devices. Gastrointest Endosc 2015;83 (1):48–56. Aug 10. pii: S0016-5107(15)02741-8. Available from: http://dx.doi.org/10.1016/j.gie.2015.08.007. [Epub ahead of print].

[26] Ponce J, Nguyen NT, Hutter M, Sudan R, Morton JM. American Society for Metabolic and Bariatric Surgery estimation of bariatric surgery procedures in the United States, 2011-2014. Surg Obes Relat Dis 2015;11(6):1199–200. pii: S1550-7289(15)00803-5. Available from: http://dx.doi.org/10.1016/j.soard.2015.08.496. [Epub ahead of print].

[27] le Roux CW, Welbourn R, Werling M, et al. Gut hormones as mediators of appetite and weight loss after Roux-en-Y gastric bypass. Ann Surg 2007;246:780–5.

[28] Ryan KK, Tremaroli V, Clemmensen C, et al. FXR is a molecular target for the effects of vertical sleeve gastrectomy. Nature 2014;509:183–8.

[29] Raghow R. Ménage-à-trois of bariatric surgery, bile acids and the gut microbiome. World J Diabetes 2015;6:367–70.

[30] Sjöström L, Peltonen M, Jacobson P, et al. Association of bariatric surgery with long-term remission of type 2 diabetes and with microvascular and macrovascular complications. JAMA 2014;311:2297–304.

[31] Carlsson LM, Peltonen M, Ahlin S, et al. Bariatric surgery and prevention of type 2 diabetes in Swedish obese subjects. N Engl J Med 2012;367:695–704.

[32] Ikramuddin S, Korner J, Lee WJ, et al. Roux-en-Y gastric bypass vs intensive medical management for the control of type 2 diabetes, hypertension, and hyperlipidemia: the Diabetes Surgery Study randomized clinical trial. JAMA 2013;309:2240–9.

[33] Schauer PR, Bhatt DL, Kirwan JP, et al. Bariatric surgery versus intensive medical therapy for diabetes--3-year outcomes. N Engl J Med 2014;370:2002–13.

[34] Sjöström L, Peltonen M, Jacobson P, et al. Bariatric surgery and long-term cardiovascular events. JAMA 2012;307:56–65.

[35] Piché MÈ, Martin J, Cianflone K, et al. Changes in predicted cardiovascular disease risk after biliopancreatic diversion surgery in severely obese patients. Metabolism 2014;63:79–86.

[36] Sjöström L, Gummesson A, Sjöström CD, et al. Effects of bariatric surgery on cancer incidence in obese patients in Sweden (Swedish Obese Subjects Study): a prospective, controlled intervention trial. Lancet Oncol 2009;10:653–62.

[37] Burza MA, Romeo S, Kotronen A, et al. Long-term effect of bariatric surgery on liver enzymes in the Swedish Obese Subjects (SOS) study. PLoS One 2013;8(3):e60495.

[38] ASGE Bariatric Endoscopy Task Force and ASGE Technology Committee, Abu Dayyeh BK, Edmundowicz SA, et al. Endoscopic bariatric therapies. Gastrointest Endosc 2015;81:1073–86.

[39] Forssell H, Norén E. A novel endoscopic weight loss therapy using gastric aspiration: results after 6 months. Endoscopy 2015;47:68–71.

[40] de Moura EG, Lopes GS, da Costa Martins B, et al. Effects of duodenal-jejunal bypass liner (EndoBarrier®) on gastric emptying in obese and type 2 diabetic patients. Obes Surg 2015;25(9):1618–25. Feb 18. [Epub ahead of print].

[41] de Jonge C, Rensen SS, Verdam FJ, et al. Endoscopic duodenal-jejunal bypass liner rapidly improves type 2 diabetes. Obes Surg 2013;23:1354–60.

[42] Koehestanie P, de Jonge C, Berends FJ, et al. The effect of the endoscopic duodenal-jejunal bypass liner on obesity and type 2 diabetes mellitus, a multicenter randomized controlled trial. Ann Surg 2014;260:984–92.

[43] ASGE Bariatric Endoscopy Task Force and ASGE Technology Committee, Abu Dayyeh BK, Kumar N, Edmundowicz SA, Jonnalagadda S, Larsen M, Sullivan S, et al. ASGE Bariatric Endoscopy Task Force systematic review and meta-analysis assessing the ASGE PIVI thresholds for adopting endoscopic bariatric therapies. Gastrointest Endosc 2015;82(3): 425–38. e5. Available from: http://dx.doi.org/10.1016/j.gie.2015.03.1964. Epub 2015 Jul 29.

[44] Bouras EP, Delgado-Aros S, Camilleri M, et al. SPECT imaging of the stomach: comparison with barostat and effects of sex, age, body mass index, and fundoplication. Gut 2002;51:781–6.

[45] Brethauer SA, Chand B, Schauer PR, et al. Transoral gastric volume reduction as intervention for weight management: 12-month follow-up of TRIM trial. Surg Obes Relat Dis 2012;8:296–303.

[46] Mion F, Napoléon B, Roman S, et al. Effects of intragastric balloon on gastric emptying and plasma ghrelin levels in non-morbid obese patients. Obes Surg 2005;15:510–16.

[47] Martinez-Brocca MA, Belda O, Parejo J, et al. Intragastric balloon-induced satiety is not mediated by modification in fasting or postprandial plasma ghrelin levels in morbid obesity. Obes Surg 2007;17:649–57.

[48] Mion F, Gincul R, Roman S, et al. Tolerance and efficacy of an air-filled balloon in non-morbidly obese patients: results of a prospective multi-center study. Obes Surg 2007;17:764−9.

[49] Ponce J, Quebbemann BB, Patterson EJ. Prospective, randomized, multicenter study evaluating safety and efficacy of intragastric dual-balloon in obesity. Surg Obes Related Dis 2013;9:290−5.

[50] Abu-Dayyeh BK, Sarmiento R, Rajan E, et al. Endoscopic treatments of obesity and metabolic disease: are we there yet? Rev Esp Enferm Dig 2014;106:467−76.

[51] Marinos G, Eliades C, Raman Muthusamy V, et al. Weight loss and improved quality of life with a nonsurgical endoscopic treatment for obesity: clinical results from a 3- and 6-month study. Surg Obes Relat Dis 2014;10:929−34.

[52] Genco A, Maselli R, Cipriano M, et al. Long-term multiple intragastric balloon treatment--a new strategy to treat morbid obese patients refusing surgery: prospective 6-year follow-up study. Surg Obes Relat Dis 2014;10:307−11.

[53] Farina MG, Baratta R, Nigro A, et al. Intragastric balloon in association with lifestyle and/or pharmacotherapy in the long-term management of obesity. Obes Surg 2012;22:565−71.

[54] Camilleri M, Toouli J, Herrera MF, et al. Intra-abdominal vagal blocking (VBLOC therapy): clinical results with a new implantable medical device. Surgery 2008;143:723−31.

[55] Ikramuddin S, Blackstone RP, Brancatisano A, et al. Effect of reversible intermittent intra-abdominal vagal nerve blockade on morbid obesity: the ReCharge randomized clinical trial. JAMA 2014;312:915−22.

[56] Shikora S, Toouli J, Herrera MF, et al. Vagal blocking improves glycemic control and elevated blood pressure in obese subjects with type 2 diabetes mellitus. J Obes 2013;2013:245683.

[57] Bernal-Mizrachi C, Xiaozhong L, Yin L, et al. An afferent vagal nerve pathway links hepatic PPARalpha activation to glucocorticoid-induced insulin resistance and hypertension. Cell Metab 2007;5:91−102.

[58] Cha R, Marescaux J, Diana M. Updates on gastric electrical stimulation to treat obesity: Systematic review and future perspectives. World J Gastrointest Endosc 2014;6:419−31.

[59] Yao S, Ke M, Wang Z, et al. Retrograde gastric pacing reduces food intake and delays gastric emptying in humans: a potential therapy for obesity? Dig Dis Sci 2005;50:1569−75.

[60] Zhang Y, Du S, Fang L, et al. Retrograde gastric electrical stimulation suppresses calorie intake in obese subjects. Obesity (Silver Spring) 2014;22:1447−51.

[61] Sharaiha RZ, Kedia P, Kumta N, et al. Initial experience with endoscopic sleeve gastroplasty: technical success and reproducibility in the bariatric population. Endoscopy 2015;47:164−6.

[62] Abu Dayyeh BK, Acosta Cardenas AJ, Camilleri M, et al. Endoscopic sleeve gastroplasty alters gastric physiology and induces loss of body weight in obese individuals. Clin Gastroenterol Hepatol 2015; pii: S1542-3565(15)01714-0. Available from: http://dx.doi.org/10.1016/j.cgh.2015.12.030. [Epub ahead of print].

[63] Abu Dayyeh B, Acosta A, Topazian M, et al. One-year follow-up and physiological alterations following endoscopic sleeve gastroplasty for treatment of obesity. Gastroenterology 2015;148(Suppl. 1):S11−12.

[64] Samsom M, Szarka LA, Camilleri M, et al. Pramlintide, an amylin analog, selectively delays gastric emptying: potential role of vagal inhibition. Am J Physiol 2000;278:G946−51.

[65] Delgado-Aros S, Kim DY, Burton DD, et al. Effect of GLP-1 on gastric volume, emptying, maximum volume ingested and postprandial symptoms in humans. Am J Physiol Gastrointest Liver Physiol 2002;282:G424−31.

[66] Cervera A, Wajcberg E, Sriwijitkamol A, et al. Mechanism of action of exenatide to reduce postprandial hyperglycemia in type 2 diabetes. Am J Physiol Endocrinol Metab 2008;294:E846−52.

[67] van Can J, Sloth B, Jensen CB, et al. Effects of the once-daily GLP-1 analog liraglutide on gastric emptying, glycemic parameters, appetite and energy metabolism in obese, non-diabetic adults. Int J Obes (Lond) 2014;38:784−93.

[68] Tack J, Depoortere I, Bisschops R, et al. Influence of ghrelin on interdigestive gastrointestinal motility in humans. Gut 2006;55:327−33.

[69] Cuomo R, Vandaele P, Coulie B, et al. Influence of motilin on gastric fundus tone and on meal-induced satiety in man: role of cholinergic pathways. Am J Gastroenterol 2006;101:804−11.

[70] Wren AM, Seal LJ, Cohen MA, et al. Ghrelin enhances appetite and increases food intake in humans. J Clin Endocrinol Metab 2001;86:5992−5.

[71] Janssen P, Pottel H, Vos R, et al. Endogenously released opioids mediate meal-induced gastric relaxation via peripheral mu-opioid receptors. Aliment Pharmacol Ther 2011;33:607−14.

[72] Bray GA, Ryan DH. Update on obesity pharmacotherapy. Ann NY Acad Sci 2014;1311:1−13.

Surgical and Postsurgical Procedures

Chapter 7

Why Patients Select Weight Loss Bariatric Surgery

W. Guan and P.J. Brantley

Pennington Biomedical Research Center, Baton Rouge, LA, United States

LIST OF ABBREVIATIONS

BMI body mass index
CTG conservative treatment group
LAGB laparoscopic adjustable gastric banding
QOL quality of life
SG surgical intervention group

INTRODUCTION

Patients seeking bariatric surgery are often motivated by factors related to minimizing the presence of comorbidities related to obesity, the potential for future comorbidities to develop, and the psychosocial effects of obesity, including, but not limited to, mental health and quality of life (QOL) impairments. There are several reasons why the motivations of patients seeking bariatric surgery are important. First, bariatric surgery is elective and often described as a "behavioral surgery" given its requirement of pre- and postoperative care and maintenance [1]. This means that patient motivations for surgical intervention can have significant effects on whether the intervention itself is successful.

Second, the definition of "success" of bariatric surgery can vary widely from patient to patient. This can depend on both the motivating factors for choosing surgical intervention and subsequent patient expectations for postsurgery. Patients motivated by health concerns may expect tangible results such as improvement of existing comorbidities (e.g., improved glycemic control) and thereby define success as it relates to the manifestation of this result [2]. Other patients are more motivated by appearance and body image concerns. As a result, these patients define success as it relates to weight loss and subsequent self-defined appearance outcomes [3]. Moreover, success is not necessarily tangible since many patients seeking bariatric surgery are doing so in order to diminish the future potential for developing comorbidities of obesity.

Additionally, understanding *why* patients choose bariatric surgery can further the existing knowledge of *who* chooses bariatric surgery. This knowledge has implications at both the micro and macro levels. At the micro level, this can help health practitioners anticipate who will seek bariatric surgery. Perhaps more importantly, however, knowing and understanding the reasons for choosing bariatric surgery can aid researchers in resolving macro-level sociodemographic gaps in the bariatric patient population.

As a result, this chapter focuses on emerging themes from both quantitative and qualitative research on patients who have undergone bariatric surgery, and prospective patients showing interest in surgical intervention. The chapter is organized into three primary sections. The first section reviews findings regarding sociodemographic and psychosocial characteristics of patients choosing bariatric surgery. The second section asks why patients choose surgery, including a look at patients' common experiences and motivations. The final section discusses the potential implications of these findings.

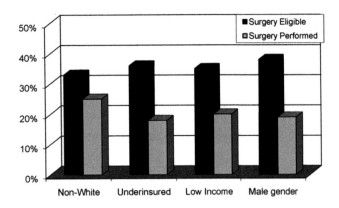

FIGURE 7.1 Sociodemographic effects on the likelihood of undergoing bariatric surgery. *With permission from Martin M, Beekley A, Kjorstad R, Sebesta J. Socioeconomic disparities in eligibility and access to bariatric surgery: a national population-based analysis, Surg Obes Relat Dis 2010;6:8–15*, this figure shows that patients who are male, low income, underinsured, and of minority status often do not choose to undergo bariatric surgery despite being eligible.

TABLE 7.1 Demographic of Physiological Differences Between Patients Choosing Bariatric Surgery and Patients Choosing Conservative Treatment Options

	SG (*n* = 249)	CTG (*n* = 256)	*p* value
Female	180 (72%)	173 (68%)	0.304
Age (years)	41 (11)	44 (13)	0.002
BMI (kg/m²)	46.5 (6.2)	43.2 (5.5)	<0.001
Body weight (kg)	136 (23.2; range 90–226)	127 (20.5; range 82–200)	<0.001
Waist circumference (cm)	136 (15)	131 (14)	<0.001
Hip circumference (cm)	137 (12)	133 (15)	<0.001
Waist-to-hip ratio	1.00 (0.10)	1.00 (0.10)	0.696
Maximum weight (kg)	143 (24)	134 (23)	<0.001
Max weight, age (years)	38 (11)	41 (14)	0.057
Previous bariatric surgery	11 (4%)	5 (2%)	0.135

This table compares means and percentages between patients of the surgical group (SG) and the conservative treatment group (CTG). *p*-values of less than 0.05 indicate a statistically significant difference between the two groups.
Source: With permission from Jakobsen GS, Hofsø D, Røislien J, Sandbu R, Hjelmesæth J, Morbidly obese patients—who undergoes bariatric surgery? Obes Surg 2010;20:1142–8.

WHO CHOOSES BARIATRIC SURGERY?

Sociodemographic Patterns

Patients who choose surgery are predominantly white, female, and in their early 40s [4–6]. These individuals are also often in the latter quartiles of income [7]. As a result, the demographics of patients undergoing bariatric surgery often do not reflect the general population of individuals suffering from obesity. For instance, although individuals with low income and of minority status are much more likely to be obese, they are much less likely to undergo bariatric surgery [8].

Fig. 7.1 presents results of a study [7] examining the major gap between those who are eligible for bariatric surgery and those who actually undergo the surgical procedure. As shown in the figure, over 30% of nonwhite minorities are eligible for surgery but only about 23% choose to have surgery. This gap is even greater for those individuals who are underinsured and low income. The largest gap is found in gender differences. Although almost 40% of men are characterized as "surgery eligible," less than half of these individuals choose to have surgery.

Table 7.1 shows results from another study [9] examining the same question through a different method and perspective. The patients in this study are given the choice between a surgical treatment group (SG) and conservative treatment group (CTG). As a result, the researchers in this study are able to compare sociodemographic characteristics between patients in both groups. As the table shows, patients who chose surgical intervention were younger and had greater obesity measures such as body mass index (BMI), body weight, and various body measurements. Interestingly, gender did not have a significant effect on whether patients chose surgery or conservative behavioral treatment.

Psychosocial Variations

In addition to demographic and medical patterns, there can also be significant differences in psychosocial health among the patient population. This is due to the potential for patients suffering from obesity to have significant issues with stress, anxiety, depression, self-esteem, and QOL as a result of their obesity [10,11]. For instance, Kolotkin et al. [10] found that those individuals seeking surgical intervention were more likely to report lower QOL measures. Castellini et al. [12] found that patients seeking bariatric surgery have higher levels of depression and higher rates of binge eating.

However, the findings are not entirely consistent from study to study. More specifically, although a recent study [13] echoes prior findings suggesting QOL differences [10], the researchers in this study found no significant differences in several psychosocial measures, including impulsivity, anxiety, depression, or eating behavior between surgical and nonsurgical patients. Moreover, the results in this study [13] showed significant psychosocial differences among individuals choosing gastric banding, vertical sleeve gastrectomy, and gastric bypass, implying that the type of surgery examined in each study must be taken into consideration.

There are two potential reasons for inconsistent findings. First, it can be a result of the self-selection of patients into bariatric surgery as influenced by health insurance and other socioeconomic factors [11]. A recent study controls for these measures by using unique data gathered from a large cohort of patients in Louisiana (Heads Up Project). Patients in this study were given the option of a surgical or nonsurgical intervention in which expenses were covered entirely by an insurance company. The results in this study revealed that surgical patients have more previously failed weight loss attempts, higher food cravings, more public distress, and a poorer-quality sexual life [11]. However, the results also indicated that surgical patients are less likely to experience loss of control regarding their eating habits. Additionally, no significant differences in depression were observed in this study. As a result, it appears that controlling for socioeconomic factors may not entirely account for prior inconsistent findings.

The second potential reason is that prior to a surgical procedure, physicians use a variety of selection criteria including history of psychiatric problems to exclude those individuals for which the surgery is viewed as unlikely to be successful. As a result, individuals with extreme psychosocial problems are often selected out of surgical intervention prior to observation. Therefore, this selection process may underestimate psychosocial differences or render them entirely indistinguishable.

WHY DO PATIENTS CHOOSE SURGERY?

The subsequent question following the previous section then is why are individuals with certain characteristics more likely to choose bariatric surgery? First, it may be entirely economic. Bariatric surgery costs money, and therefore low-income populations and racial minorities simply may not have the economic resources necessary for bariatric surgery. However, this still does not help us understand why there are differences in other characteristics such as gender and psychosocial measures. To better understand these differences requires examining reasons that are not purely economic. The following section reviews literature identifying the experiences and motivations of bariatric patients in order to better understand why patients choose bariatric surgery. Again, understanding patient motivations is important in understanding why patients who are eligible for surgery do not elect for surgery despite the proven benefits.

Patient Experiences

Several experiences are common for patients who choose to undergo bariatric surgery. For many, bariatric surgery is at the end of the list of treatment options given its potential for complications and fatality as compared to other more conservative interventions. As a result, many patients who choose to undergo surgery are doing so after a long history of unsuccessful lifestyle and pharmacological interventions [14]. This can often mean that past treatments have been unsuccessful in that no significant weight loss was achieved. It can also mean that patients continue to experience the "yo-yo effect," in which treatments result in weight loss but the weight is quickly regained. As a result, many individuals suffering from obesity cite a sense of hopelessness and a loss of control over their weight issues [14]. For these individuals, bariatric surgery represents a "last resort" in a desperate attempt to regain control and to improve medical and psychosocial problems related to their obesity.

Moreover, although the decision to undergo bariatric surgery is often a result of a long process, there can often be specific "trigger" moments. For instance, patients can be triggered by an onset of symptoms related to an existing medical condition related to their obesity. Other triggers can include illness in a family member or reaching some symbolic landmark such as a clothing size, age, or BMI [15].

Motivations

A significant segment of the existing literature looks at motivational factors as one way to answer why patients choose bariatric surgery over more conservative behavioral interventions. Motivations for bariatric surgery can be looked at in two ways: the degree to which the patient is motivated and what motivational factors are important for the patient. The former is relatively straightforward in that it gauges how motivated each patient is to successfully lose weight through pre- and postoperative measures. The second requires a more methodical approach in determining what kind of motivations are common. Because obese individuals have to deal with many medical, physical, and psychological effects of their obesity, there can be a multitude of factors that play a significant part in the decision-making process leading up to choosing bariatric surgery.

One method that has been used in research is to survey bariatric surgery patients, asking them to rank a list of motivation factors by importance. Several studies have utilized this method [9,16–18]. For instance, Libeton et al. [17] allowed survey respondents to rank six motivating factors by importance: appearance, medical condition, physical fitness, health concerns, embarrassment, and physical limitation. Their results are shown in Fig. 7.3. Another study [18] utilized open responses to the question, "Why are you seeking weight loss surgery?" The responses were subsequently coded into one of seven categories: body image, eating control, self-esteem, physical activity, social activity, medical pain, current health, and health prevention.

Although this method has provided information on what factors are viewed by patients as most important, it has limitations. For instance, researchers utilizing this method often focus on the primary motivating factors expressed by patients, which are health and medical conditions. This is not surprising given that those individuals choosing bariatric surgery are often coping with extreme levels of obesity and comorbidities such as diabetes and hypertension. Moreover, this method is unable to examine subsidiary questions such as why these motivational factors are important, what current medical condition is particularly challenging for the patient, and with which potential health issue the patient is particularly concerned? Similarly, in what way is appearance and body image important for the patient?

As a result, qualitative research can also present valuable information that can answer these questions and supplement the findings offered by quantitative survey methods. For example, one finding that is often overlooked is the everyday physical challenges of being obese. This issue has also been shown to be an important motivator for choosing bariatric surgery. Another example of the importance of qualitative methods is examining the psychological and social understandings of body image and appearance. These types of in-depth conceptualizations are largely impossible in quantitative research. The following sections present several themes commonly found in the existing literature on motivational factors for patients choosing bariatric surgery.

Health and Medical Issues

Health and medical concerns are consistently cited as the primary motivation for a patient's decision to undergo bariatric surgery [16–19]. Citing current health conditions as the primary motivator is not surprising given that many obese patients experience significant comorbidities such as diabetes, hypertension, sleep apnea, and joint pain [18]. The results from two example studies are shown in Figs. 7.2 and 7.3. According to the first study [16] shown in Fig. 7.2, 52% of patients

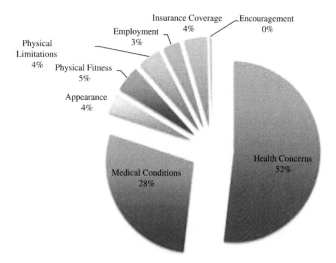

FIGURE 7.2 Primary reasons for choosing bariatric surgery over conservative behavioral interventions. *With permission from Brantley PJ, et al. Why patients seek bariatric surgery: does insurance coverage matter? Obes Surg 2014;24:961–4,* this figure shows the percentage of patients (*N* = 360) that chose each type of motivation for electing for bariatric surgery. As shown, the majority of patients are concerned with potential health effects and current medical conditions.

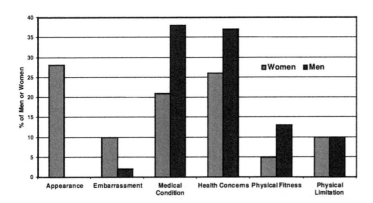

FIGURE 7.3 Gender differences in motivating factors for bariatric surgery. *With permission from Libeton M, Dixon JB, Laurie C, O'Brien PE. Patient motivation for bariatric surgery: characteristics and impact on outcomes, Obes Surg 2004;14:392−8,* this figure compares the different types of motivations indicated by male and female patients.

choosing surgical intervention were motivated by health concerns. Health concerns were defined in this study as future medical and health implications of obesity. The second leading factor (28%) was concern with current medical conditions.

Results in another study [17] indicated that only about half of patients (52%) were motivated by either current or future medical concerns. Although health is still the overwhelming motivator, this finding is not as dramatic as the results of Brantley et al. [16]. Additionally, Libeton et al. [17] found significant gender differences. These results are shown in Fig. 7.3. As illustrated, males are more likely to cite health and medical concerns as their primary motivating factor for bariatric surgery, while females report physical appearance as a key factor in their decision.

Although health and medical conditions are the primary motivation among those patients opting for surgical intervention, it does not necessarily mean that the presence of comorbidities predict surgical intervention compared to conservative treatments. In fact, a recent study [9] of 505 patients found that there is no significant difference in comorbidities between those choosing surgery and those choosing conservative treatments. There are two potential explanations.

First, regardless of the increasingly safe surgical procedures, bariatric surgery remains an intervention involving significant risk and invasiveness. Moreover, bariatric surgery is not effective for all patients undergoing the procedure [19]. As a result, even among those with serious comorbidities, patients may be more attracted to conservative treatment choices [9]. Second, bariatric surgery may be becoming more accepted as a procedure not strictly for improving comorbidities of obesity. Rather, individuals may be increasingly attracted to seeking out surgical intervention as a result of dealing with social and psychological stigma associated with obesity such as body image and appearance issues, or the physical challenges of obesity in everyday life. These motivations are discussed further in the next two sections.

Appearance and Body Image

As with other illnesses and diseases, the experience of obesity incorporates social and psychological effects in addition to health and medical issues. Aside from medical concerns, the most commonly cited motivation to seek surgical intervention for obesity is body image and physical appearance [17,19]. However, the importance of appearance for seeking bariatric surgery may often be understated. Patients surveyed in many quantitative studies represent a selective population dealing with extreme levels of obesity and therefore significant health and medical issues related to their obesity. Moreover, the selection criteria for bariatric surgery oftentimes include a BMI cut-off of >40 BMI or >35 BMI with Type II diabetes.

This potential for inconsistent findings is shown in the contrasting results between Libeton et al. [17] and Brantley et al. [16]. In the former study, 24% of patients selected appearance as a motivating factor compared to 4% in the latter study. Moreover, Libeton et al. [17] found that appearance was more important among females than males. In fact, no male in the study selected appearance as his motivating factor. Additionally, females are more likely to cite embarrassment over their obesity and males are more likely to cite physical and fitness concerns.

The type of surgical procedure offered to patients can also have a significant effect on whether appearance is shown to be an important motivator. For instance, one study [17] found that 32% of patients seeking laparoscopic adjustable gastric banding (LAGB) cited appearance as the primary motivator. However, another study [19] found that only 11% of patients were interested in surgical intervention (duodenal switch and Roux-en-Y gastric bypass) due to body image concerns. This discrepancy can be explained by the benefits of LAGB (minimal invasiveness, adjustable, and shorter recovery time). As a result, this procedure may attract patients interested in surgical intervention for nonmedically related reasons [19].

Regardless of whether appearance is the primary motivator, it remains a significant factor in choosing bariatric surgery. The self-concepts of body image and appearance-related concerns are products of internalizing the negative stigma that is attached to the social image of obesity [3]. The fat body is perceived as physically unattractive and directly under the individual's control. As a result, being overweight or obese represents individual failure. Therefore, not only are obese individuals suffering from medical comorbidities, but they also experience psychological repercussions of being obese within the social environment. Thus, many individuals seeking surgical intervention are likely experiencing stigma in the form of moral judgments and staring in public spaces [3]. The psychological effects of internalizing this stigma may include depressed mood, anger, and embarrassment.

Obesity as a Chronic Disease

In addition to medical and appearance concerns, many patients cite the difficulties of living with obesity as a significant motivator for undergoing bariatric surgery. This includes the physical challenges of a large body and the general experience of being "unfit" [15]. For instance, many patients recall difficulties in doing everyday activities such as climbing stairs and putting on socks. For many, these issues represent small factors that when put together represent a major motivation for choosing surgical intervention. Moreover, as these difficulties grew, patients became more aware of the medical issues related to obesity such as arthritis, diabetes, and hypertension.

IMPLICATIONS

Defining Success

As shown in this chapter, the majority of individuals who undergo bariatric surgery are motivated by health and medical concerns. This has significant implications for understanding the definition of "success" of bariatric surgery from the patients' perspectives. As a result, this section focuses on patients' medical expectations following bariatric surgery.

For many patients, the most valuable indicator of surgical success is weight loss [19]. According to Foster et al.'s [20] study of 397 patients, individuals undergoing surgical weight-loss treatment often expect to lose more than 30% of their current body weight. More specifically, patients referred to achieving their "dream weight" as losing on average 38% of their current weight, and "acceptable" as at least 25%.

In another study of 44 patients showing interest in bariatric surgery [19], patients reported that they would be disappointed with a weight loss of less than 24% of their current body weight. Moreover, many stated that they would not undergo bariatric surgery despite a guarantee of 20% weight loss. These findings are concerning given that 20% weight loss is considered a successful outcome of bariatric surgery according to prior studies [21,22].

A potential explanation for patients' unrealistic weight loss expectations is the gap in knowledge regarding the amount of weight loss necessary for improving medical comorbidities. As shown in this chapter, the primary reason for undergoing bariatric surgery is overwhelmingly current and future medical and health issues. As such, improvement in medical comorbidities are of primary concern. According to a study of 45 patients [2], approximately 80%, 90%, and 93% expected at least an improvement in diabetes, hypertension, and sleep apnea, respectively. As a result, patients may often be misguided as to how much weight loss is necessary for resolution or improvement of medical comorbidities.

Medical Outcomes of Surgical Procedure

Motivations for undergoing bariatric surgery can potentially have an impact on the likelihood of success of the surgical intervention. Unfortunately, few studies have examined this relationship. Research investigating patients' levels of motivation show inconclusive results for whether being more motivated for weight loss can significantly alter the outcome of the surgical intervention [17]. Even less research has been completed on differing factors of motivations. To date, Libeton et al. [17] is the only study that examined the association of motivational factors on bariatric surgery outcome. This study found no significant relationships.

Despite this finding, this direction of research is worthy of further investigation. For instance, are patients who are motivated by health and medical concerns more or less likely to achieve and sustain weight loss following surgery than patients motivated by body image concerns? Other outcomes aside from weight loss can also be important: eating behavior, QOL, physical activity, perceptions of health, and other psychosocial measures.

Healthcare

Understanding why patients undergo bariatric surgery can have significant implications on the importance of healthcare and insurance coverage for bariatric surgery. As shown in this chapter, there are major gaps between the population suffering from obesity and the population who choose to undergo bariatric surgery. According to one study [6], a large proportion of this discrepancy is due to the economic resources necessary for pursuing bariatric surgery. As a result, a side effect of the increasing obesity epidemic is the also increasing systemic inequality in access to bariatric surgical interventions. Insurance coverage can play an important role in mitigating this health inequality.

For instance, patients with private insurance are more likely to have bariatric surgery [4]. According to one study of over 80,000 patients who underwent bariatric surgery, 82% of these patients had private insurance [7]. Less than 1% of these patients were uninsured. In another study [23], 60% of bariatric surgical patients in 2002 lived in ZIP code locations with mean household incomes greater than $44,999. This is contrasted with only 5% of patients living in ZIP code locations with household incomes of less than $25,000. As a result, an important barrier or enabler of surgical intervention comes from socioeconomic factors specifically related to insurance coverage.

CONCLUSION

There are a multitude of factors that can determine whether a patient chooses to undergo bariatric surgery. This chapter highlights how the experiences, motivations, and expectations of bariatric surgery all play a major role in the decision-making process prior to choosing surgical intervention. First, the two primary reasons that are often mentioned by patients when asked why they elect to undergo surgery are concerns for health and medical issues, and appearance concerns. Second, when discussing their experiences leading up to finally deciding on bariatric surgery, many patients view bariatric surgery as a last resort following many unsuccessful attempts at sustained weight loss. Surgery for these patients is viewed as a tool for gaining control over their weight and the subsequent negative effects of obesity. These include effects such as comorbid medical conditions, difficulty in completing everyday tasks, dysfunctional eating behavior, and premature death. Finally, patients have high expectations for bariatric surgery. In regards to weight loss, many patients expect to lose much more weight than is typically achieved in what is considered a successful bariatric surgery. However, patients underestimate the positive impact that bariatric surgery has on medical comorbidities such as diabetes, hypertension, and sleep apnea.

Further research aimed at improving our understanding of how demographic, economic, and psychosocial factors influence the decision to undergo bariatric surgery may prove useful to surgical practitioners, insurance providers, and public health researchers. Learning more about patient motivations may aid health practitioners in patient selection. Second, patients' motivations and experiences can help shed light on why there are such major demographic incongruences between those individuals eligible for bariatric surgery and those who actually seek and undergo surgery. Finally, as shown in this chapter, patients undergoing bariatric surgery are primarily motivated by health and medical reasons. This finding has significant implications for healthcare policy regarding whether or not bariatric surgery should be covered through health insurance as an effective procedure for the morbidly obese.

MINI-DICTIONARY OF TERMS

- *Bariatric Surgery*: Includes all of the different types of surgical procedures that are performed on individuals in order to treat obesity.
- *Body Mass Index (BMI)*: An individual's weight (kg) divided by the square of his/her height (m). BMI is the most commonly used measure to indicate whether an individual is overweight or obese.
- *Comorbidity*: This refers to each additional disease that occurs simultaneously with the original disease.
- *Epidemic*: The spread of disease throughout a population within a short amount of time.
- *Gastric Banding*: A surgical operation used to treat obesity by placing an adjustable band around the top portion of the stomach in order to limit food consumption.
- *Gastric Bypass*: A surgical operation used to treat obesity by rerouting the digestive system in order to reduce the functional size of the stomach.
- *Obesity*: A condition in which an individual weighs more than what is considered healthy for their height. This is most commonly defined as having a BMI of >30.
- *Psychosocial*: Factors relating to individuals' mental, emotional, and social well-being.
- *Quality of Life (QOL)*: This is a concept often measured by multiple subjective indicators, including but not limited to perceived physical and mental well-being, living conditions, physical limitations, job satisfaction, housing, and education.
- *Socioeconomic Status (SES)*: This is a combined measure used to account for predominantly income, education, and occupational prestige.

- *Vertical Sleeve Gastrectomy*: A surgical procedure used to treat obesity by reducing the size of the stomach by removal of the larger portion of the stomach.

KEY FACTS

- The Heads Up Study was a collaborative project between Pennington Biomedical Research Center and Louisiana State Office of Group Benefits (OGB).
- The Heads Up Study was a project to develop management programs for severely obese adults insured by OGB.
- The Heads Up Study provided a surgical and nonsurgical (intensive medical management) intervention for program participants.
- Participants in the Heads Up Study were allowed to choose which intervention program they want to be considered for.
- Patients electing for bariatric surgery often undergo a screening process in order to evaluate whether the surgical procedure is appropriate for their specific needs.
- Decisions for whether a patient should be operated on include factors such as BMI, comorbid conditions, likelihood of surgical success, and various other mental and physical risks.
- Motivations for choosing bariatric surgery can also be a significant selection criteria for bariatric surgery. This includes both the degree of motivation and the type of motivation for electing for bariatric surgery.
- Another important motivating factor includes nutritional assessment that takes into account prior behavioral attempts at weight loss, nutritional knowledge, etc.
- Prior psychopathology is an important element to take into account when screening bariatric surgery patients.

SUMMARY POINTS

- Patients who choose surgery are predominantly white, female, and in their early 40s.
- Patients choosing bariatric surgery are often triggered by a specific experience (e.g. reaching a certain weight, developing diabetes or other comorbidities).
- Patients choosing bariatric surgery are most commonly motivated by current and future health and medical issues.
- The second leading motivation regards concerns for appearance and everyday physical limitations.
- Females are more likely to cite appearance as a motivating factor for surgical intervention.
- Bariatric surgery patients often have unrealistic weight-loss expectations following surgery.

REFERENCES

[1] Santry H, Alverdy J, Prachand V. Patient selection for bariatric surgery. In: Buchwald H, Cowan GSM, Pories WJ, editors. Surgical management of obesity. Pennsylvania: Saunders Elsevier, Philadelphia; 2007. p. 94–100.

[2] Karmali S, Kadikoy H, Brandt ML, Sherman V. What is my goal? Expected weight loss and comorbidity outcomes among bariatric surgery patients. Obes Surg 2011;21:595–603.

[3] Park J. The meanings of physical appearance in patients seeking bariatric surgery. Health Sociol Rev 2015;24:242–55.

[4] Wallace AE, Young-Xu Y, Hartley D, Weeks WB. Racial, socioeconomic, and rural–urban disparities in obesity-related bariatric surgery. Obes Surg 2010;20:1354–60.

[5] Sudan R, Winegar D, Thomas S, Morton J. Influence of ethnicity on the efficacy and utilization of bariatric surgery in the USA. J Gastrointest Surg 2013;18:130–6.

[6] Stanford FC, et al. Patient race and the likelihood of undergoing bariatric surgery among patients seeking surgery. Surg Endosc 2014;29:2794–9.

[7] Martin M, Beekley A, Kjorstad R, Sebesta J. Socioeconomic disparities in eligibility and access to bariatric surgery: a national population-based analysis. Surg Obes Relat Dis 2010;6:8–15.

[8] Flum DR, Khan TV, Dellinger E. Toward the rational and equitable use of bariatric surgery. JAMA 2007;298:1442–4.

[9] Jakobsen GS, Hofsø D, Røislien J, Sandbu R, Hjelmesæth J. Morbidly obese patients—who undergoes bariatric surgery? Obes Surg 2010;20:1142–8.

[10] Kolotkin RL, et al. Health-related quality of life in patients seeking gastric bypass surgery vs non-treatment-seeking controls. Obes Surg 2003;13:371–7.

[11] Matthews-Ewald MR, et al. Predictors for selection of insurance-funded weight loss approaches in individuals with severe obesity. Obesity 2015;23:1151–8.

[12] Castellini G, et al. Psychopathological similarities and differences between obese patients seeking surgical and non-surgical overweight treatments. Eat Weight Disord 2013;19:95–102.

[13] Miras AD, et al. Psychological characteristics, eating behavior, and quality of life assessment of obese patients undergoing weight loss interventions. Scand J Surg 2015;104:10–17.

[14] Wysoker A. The lived experience of choosing bariatric surgery to lose weight. J Am Psychiatr Nurses Assoc 2005;11:26–34.

[15] Pfeil M, Pulford A, Mahon D, Ferguson Y, Lewis MP. The patient journey to gastric band surgery: a qualitative exploration. Bariatr Surg Pract Patient Care 2013;8:69–76.

[16] Brantley PJ, et al. Why patients seek bariatric surgery: does insurance coverage matter? Obes Surg 2014;24:961–4.

[17] Libeton M, Dixon JB, Laurie C, O'Brien PE. Patient motivation for bariatric surgery: characteristics and impact on outcomes. Obes Surg 2004;14:392–8.

[18] Munoz DJ, et al. Why patients seek bariatric surgery: a qualitative and quantitative analysis of patient motivation. Obes Surg 2007;17:1487–91.

[19] Wee CC, Jones DB, Davis RB, Bourland AC, Hamel MB. Understanding patients' value of weight loss and expectations for bariatric surgery. Obes Surg 2006;16:496–500.

[20] Foster GD, Wadden TA, Phelan S, Sarwer DB, Sanderson R. Obese patients' perceptions of treatment outcomes and the factors that influence them. Arch Intern Med 2001;161:2133–9.

[21] Maggard MA, et al. Meta-analysis: surgical treatment of obesity. Ann Intern Med 2005;142:547–59.

[22] Sjöström L, et al. Lifestyle, diabetes, and cardiovascular risk factors 10 years after bariatric surgery. N Engl J Med 2004;351:2683–93.

[23] Santry HP, Gillen DL, Lauderdale DS. Trends in bariatric surgical procedures. JAMA 2005;294:1909–17.

Chapter 8

Best Practices for Bariatric Procedures in an Accredited Surgical Center

T. Javier Birriel and M. El Chaar

St. Luke's University Hospital and Health Network, Allentown, PA, United States

LIST OF ABBREVIATIONS

ACS	American College of Surgeons
ASMBS	American Society for Metabolic and Bariatric Surgery
CMS	Centers for Medicare and Medicaid Services
COE	Centres of Excellence
MBSC	Michigan Bariatric Surgery Collaborative
SAGES	Society of American Gastrointestinal and Endoscopic Surgeons
MBSAQIP	Metabolic Bariatric Surgery and Quality Improvement Program

INTRODUCTION

Bariatric surgery has evolved from an unsafe last-resort surgical option for obese and debilitated patients to a safe and effective intervention [1]. It has also been recognized as the only effective treatment to produce sustained weight loss and resolution of obesity-associated comorbidities [2–6]. Worldwide, an estimated 502 million adults are obese [7], with 78.6 million adults in the United States alone [8].

The inherent rise in obesity around the world has led to an exponential increase in the number of procedures being performed. According to data from the Nationwide Inpatient Sample examining trends in bariatric surgery in the United States from 1990 to 1997, procedures during this period more than doubled from 4925 to 12,541. This amounts from 2.7 to 6.3 bariatric surgeries per 100,000 adults [2]. In a similar study from Sweden, surgery rates from 1990 to 1996 increased from 7.7 to 13.6 per 100,000 adults [9]. Even greater increases in the number of procedures occurred between 1998 and 2002, from a reported 12,775 cases to 70,256 cases [10]. More recent estimates show an increase from 158,000 to 193,000 cases per year from 2011 to 2014 in the United States [11]. As the number of cases being performed continues to rise, the quality and safety of bariatric surgery have improved as well. This can be attributed to multiple factors, such as the use of laparoscopy, development of fellowship training programs, increased surgeon experience, accreditation amongst bariatric surgical centers, and the establishment of standards of care for bariatric patients.

ADOPTION OF LAPAROSCOPY IN BARIATRIC SURGERY

Although the laparoscopic approach to bariatric surgery was introduced in the early 1990s, it was not widely accepted or adopted until years later. Since 1998, a major shift from open to laparoscopic bariatric surgery procedures occurred, resulting in a 450% increase in the number of bariatric procedures from 1998 to 2002 [10]. Compared to the open approach, laparoscopy uses smaller incisions and produces significantly less perioperative complications [12]. Specifically, studies have revealed that the laparoscopic approach results in reduced postoperative pain, shorter hospital stays, less operative blood loss, fewer surgical site infections, deceased hernia rates, and a more rapid improvement in quality of life compared to the open approach [13–15]. The introduction of laparoscopy has encouraged many otherwise hesitant patients to consider bariatric surgery as a safe alternative to nonsurgical options. In addition, many

Metabolism and Pathophysiology of Bariatric Surgery.
© 2017 Elsevier Inc. All rights reserved.

71

surgeons have been able to acquire greater experience performing different types of laparoscopic bariatric surgery, including more complicated and invasive stapling procedures as well as nonstapling approaches [14–17].

With the immense rise in bariatric surgery and adoption of laparoscopy came an increase in the number of bariatric surgeons and in the number of institutions performing these procedures [10]. Bariatric procedures performed laparoscopically continued to grow from 20.1% in 2003 to 90.2% in 2008, with an increase in surgeon membership in the American Society for Metabolic and Bariatric Surgery (ASMBS) from 931 to 1819 members [18]. In addition to the adoption of the laparoscopic approach, the advanced training that was required of surgeons was an important factor in the evolution of weight loss surgery.

THE EFFECT OF FELLOWSHIP TRAINING

In the wake of the exponential growth in bariatric surgery, and following widespread conversion to the laparoscopic approach, there was an unfortunate increase in the number of reported surgical complications due to the early learning curve [19,20], which resulted in the need for more formal training; many studies have confirmed this [16,19]. For example, the learning curve for laparoscopic Roux-en-y gastric bypass is around 50–100 cases [16,19,20].

Efforts to address surgeons' learning curves yielded an increase in the number of training courses, mini-fellowships, and proctorships, all of which helped to a certain extent [21,22]. However, ASMBS leadership eventually determined that there was a need for more formal and standardized training, which led to the development of a year-long fellowship in laparoscopic bariatric surgery. This program, which focused on technically challenging stapling procedures and the perioperative management of bariatric patients, allowed trainees to overcome their learning curves without compromising outcomes, as well as enabled them to improve their knowledge, proficiency, and management of postoperative complications [21].

The number of fellowship programs offering advanced laparoscopy and bariatric surgery training to general surgeons continues to grow. Currently, the Fellowship Council lists 58 such programs offering 1-year bariatric surgery training in the United States and Canada [23]. The ASMBS awards a certificate of acknowledgment of satisfactory training to metabolic and bariatric surgery fellows who have completed their training and meet the educational requirements needed to practice safe and effective bariatric surgery. Case volume requirements of at least 100 weight loss operations, including a minimum of 50 intestinal bypass operations, a combined total of 10 restrictive operations, 5 revisional procedures, and exposure to and/or extensive teaching of bariatric-specific emergency procedures, are listed in the ASMBS core curriculum [24].

The rigorous nature of ASMBS-sanctioned fellowship training has yielded increasingly favorable outcomes for bariatric patients. Morbidity and mortality rates have been shown to decrease with advanced training as well as increased surgeon experience [25–30]. Furthermore, clinical practice outcomes after receiving 1-year fellowship training or extensive mentoring are of higher quality after case load requirements have been met and learning curves have been eliminated [22,31]. Fellowship training also allows for improved ability to track outcomes, manage short- and intermediate-term complications, and encourage a multidisciplinary approach to bariatric patient care [32]. Finally, following fellowship program implementation as well as greater surgeon training and experience, the overall safety of bariatric surgery has improved significantly, with a decrease in 30-day postoperative morbidity and mortality rates of 4.3% and 0.3%, respectively [33–36].

THE IMPACT OF ACCREDITATION AND STANDARDIZATION

As may be expected, the rise in the number of bariatric procedures being performed by multiple surgeons in different hospitals with divergent standards and patient volumes has led to significant variability in safety and clinical outcomes [37,38]. Therefore, the ASMBS determined the need for greater standardization, which led the Surgical Review Corporation to develop criteria for centers of excellence (COE) in 2003 [37]. The COE program was originally created to allow bariatric patients and the general public to identify centers that are committed to patient care with the establishment of standardized protocols, data collection, and quality improvement [37]. Around this time, the American College of Surgeons (ACS) also recognized the need for accreditation and the formation of a centralized database for bariatric surgery, thereby forming the Bariatric Surgery Centers Network in 2005.

In 2006, the Michigan Bariatric Surgery Collaborative (MBSC) was created as the result of a collective effort among hospitals and surgeons to improve statewide outcomes in bariatric surgery [39]. MBSC is a payer-driven endeavor requiring 29 hospitals and 75 surgeons to provide data submission to a registry in order to standardize procedures related to patient safety and clinical outcomes. As a result, MBSC hospitals and surgeons have demonstrated the ability to achieve low complication rates and decreased mortality, which highlights the importance of unified and comprehensive data collection in order to achieve best practice goals.

Following the establishment of accredited centers by the ASMBS and ACS, formal accreditation of bariatric surgery centers was required by the Centers for Medicare and Medicaid Services (CMS) in the 2006 National Coverage Determination. Many private insurers followed suit by compelling their subscribers to undergo bariatric surgery in accredited centers. More recently, certain private insurers in the United States, such as Blue Cross/Blue Shield, have developed their own COE [40,41]. By 2012, the ASMBS and ACS combined their separate accreditation programs to form the Metabolic Bariatric Surgery and Quality Improvement Program (MBSAQIP) [42]. In aiming to enhance quality improvement in all areas of bariatric surgery patient care, the MBSAQIP "accredits inpatient and outpatient metabolic and bariatric surgery centers in the United States and Canada that have undergone an independent, voluntary, and rigorous peer evaluation in accordance with nationally recognized metabolic and bariatric surgical standards." [42] The accredited institutions undergo site visits evaluating surgeon volumes, ensuring appropriate multidisciplinary staff and committees, confirming suitable equipment and structural needs for obese patients, and verifying that 30-day and long-term follow-up outcomes are reported to a national registry. There are currently more than 700 accredited centers in the United States and Canada, with greater than 150,000 bariatric cases entered annually into the MBSAQIP registry [42].

Despite these efforts, accreditation of bariatric surgery centers was challenged in 2013 when a physician from Michigan sent a letter to CMS requesting removal of the certification requirement for fear of compromising patients' access to care. The CMS decision for requirement of facility certification was subsequently reversed, citing research that found no clear advantage or improvement in outcomes for Medicare beneficiaries [34,35,43,44]. This ruling was opposed by the ASMBS, ACS, Obesity Society, Society of American Gastrointestinal and Endoscopic Surgeons (SAGES), and American Society of Bariatric Physicians due to the large number of studies demonstrating improved outcomes and lower morbidity and mortality rates in accredited versus nonaccredited centers [45]. Many studies have also demonstrated that accreditation is associated with shorter length of stay and lower overall costs, as well as providing a foundation for enhanced measurement and accountability [38,46−50] compared to nonaccredited centers [38,46−48,50,51]. As a secondary benefit, bariatric surgery center accreditation and the associated resources may improve outcomes in general laparoscopic operations for the obese patient population [52].

An important component of accreditation is to establish criteria for surgeons to safely perform bariatric surgery. To this end, a joint task force was developed in 2013 consisting of members of the ASMBS, ACS, SAGES, and the Society for Surgery of the Alimentary Tract. This group combined previous independent society principles for hospital credentialing for bariatric surgery into a unified set of guidelines that provided recommendations to local credentialing committees regarding surgeon training and experience. Although these recommendations apply to surgeons who have completed advanced fellowship training as well as those with limited or no experience in bariatric or advanced laparoscopic surgery, the importance of data maintenance and outcomes monitoring is applicable to all practitioners, as bariatric surgeons must actively participate with the MBSAQIP program and renew credentialing every 2 years [53].

It is clear that routine maintenance of a database of outcomes and establishment of standards of practice based on bariatric surgical patients at accredited programs provides ample opportunity to evaluate and verify risks and benefits of obesity surgery [37,38]. In addition, such practices allow bariatric surgery centers to objectively assess their performance and identify areas in need of quality improvement measures [42], and also facilitate the recognition and achievement of best practices in bariatric surgery.

CONCLUSION

The epidemic of morbid obesity and the subsequent recognition that it constitutes a disease process have produced an exponential rise in the need for weight loss surgery, with an increase in the number of surgeries performed. The field of bariatric surgery has evolved significantly from the early era of open abdominal cases to the development of minimally invasive laparoscopic procedures. The rapid growth of bariatric surgery has resulted in increased complications and mortality rates due to inconsistent practice standards, which has led major surgical societies to call for advanced training of practicing surgeons, including post-residency bariatric surgical fellowships. These fellowships have helped revolutionize the safety of bariatric surgery as well as propel necessary changes in the field. Furthermore, the creation of governing bodies for the accreditation of bariatric surgery programs and the credentialing of trained bariatric surgeons has yielded vast improvements in perioperative outcomes. These results have come about through a collaborative effort amongst accredited centers and surgeons in creating a national clinical database of outcomes and best practice standards.

ACKNOWLEDGMENTS

The authors would like to acknowledge Dr. Jill Stoltzfus, Ph.D., for her time, expertise, and assistance during the editing of our chapter.

MINI-DICTIONARY OF TERMS

- *Laparoscopy*: The use of a fiber optic instrument introduced through a small incision in the abdominal wall for inspection of the abdomen and/or to perform a surgical procedure.
- *Accreditation*: Certification of an organization or institution that meets and maintains required standards.

KEY FACTS

- The ASMBS was founded in 1983.
- Edward E. Mason, M.D., was the founding president of the society.
- It is the largest society in the United States for metabolic and bariatric surgery.
- Nearly 4000 general surgeons and integrated healthcare professionals constitute the society's membership.
- Surgery for Obesity and Related Diseases is the official ASMBS journal.
- The ASMBS and The Obesity Society combine their annual meetings for "ObesityWeek."

SUMMARY POINTS

- Bariatric surgery has increased in popularity and practice over the last two decades.
- The number of procedures performed and the number of bariatric surgeons have grown exponentially since the late 1990s.
- The laparoscopic approach and advanced fellowship training have been instrumental in improving the safety profile of bariatric surgery.
- National surgical societies united to develop standards and accreditation requirements for bariatric surgical programs.
- The credentialing of surgeons and accreditation of centers by the MBSAQIP has helped to produce best practices and standards of care for bariatric patients.

REFERENCES

[1] Robinson MK. Surgical treatment of obesity — weighing the facts. N Engl J Med 2009;361(5):520–1. Available from: http://dx.doi.org/10.1056/nejme0904837.

[2] Pope GD, Birkmeyer JD, Finlayson SRG. National trends in utilization and in-hospital outcomes of bariatric surgery. J Gastrointest Surg 2002;6 (6):855–60, discussion 861. Available from: http://dx.doi.org/10.1016/s1091-255x(02)00085-9.

[3] Chang S-H, Stoll CRT, Song J, Varela JE, Eagon CJ, Colditz GA. The effectiveness and risks of bariatric surgery. JAMA Surg 2014;149 (3):275–313. Available from: http://dx.doi.org/10.1001/jamasurg.2013.3654.

[4] Buchwald H, Avidor Y, Braunwald E, et al. Bariatric surgery: a systematic review and meta-analysis. JAMA 2004;292(14):1724–37. Available from: http://dx.doi.org/10.1001/jama.292.14.1724.

[5] Arterburn DE, Olsen MK, Smith VA, et al. Association between bariatric surgery and long-term survival. JAMA 2015;313(1):62–70. Available from: http://dx.doi.org/10.1001/jama.2014.16968.

[6] Schauer PR, Bhatt DL, Kirwan JP, et al. Bariatric surgery versus intensive medical therapy for diabetes--3-year outcomes. N Engl J Med 2014;370(21):2002–13. Available from: http://dx.doi.org/10.1056/NEJMoa1401329.

[7] World Health Organization Global Health Observatory Data|Obesity. WHO.

[8] Adult Obesity Facts|Data|Adult|Obesity|DNPAO|CDC. cdc.gov, <http://www.cdc.gov/obesity/data/adult.html> [accessed 12.04.16].

[9] Leffler E, Gustavsson S, Karlson BM. Time trends in obesity surgery 1987 through 1996 in Sweden--a population-based study. Obes Surg 2000;10(6):543–8. Available from: http://dx.doi.org/10.1381/096089200321593760.

[10] Nguyen NT, Root J, Zainabadi K, et al. Accelerated growth of bariatric surgery with the introduction of minimally invasive surgery. Arch Surg 2005;140(12):1198–202, discussion 1203. Available from: http://dx.doi.org/10.1001/archsurg.140.12.1198.

[11] Ponce J, Nguyen NT, Hutter M, Sudan R, Morton JM. American Society for Metabolic and Bariatric Surgery estimation of bariatric surgery procedures in the United States, 2011–2014. Surg Obes Relat Dis 2015;11(6):1199–200. Available from: http://dx.doi.org/10.1016/j.soard.2015.08.496.

[12] Schauer PR, Ikramuddin S, Gourash W, Ramanathan R, Luketich J. Outcomes after laparoscopic Roux-en-Y gastric bypass for morbid obesity. Ann Surg 2000;232(4):515–29.

[13] Westling A, Gustavsson S. Laparoscopic vs open Roux-en-Y gastric bypass: a prospective, randomized trial. Obes Surg 2001;11(3):284−92. Available from: http://dx.doi.org/10.1381/096089201321336610.

[14] Luján JA, Frutos MD, Hernández Q, et al. Laparoscopic versus open gastric bypass in the treatment of morbid obesity. Ann Surg 2004;239(4): 433−7. Available from: http://dx.doi.org/10.1097/01.sla.0000120071.75691.1f.

[15] Nguyen NT, Goldman C, Rosenquist CJ, et al. Laparoscopic versus open gastric bypass: a randomized study of outcomes, quality of life, and costs. Ann Surg 2001;234(3) 279−89, discussion 289−91.

[16] Wittgrove AC, Clark GW. Laparoscopic gastric bypass, Roux-en-Y- 500 patients: technique and results, with 3-60 month follow-up. Obes Surg 2000;10(3):233−9. Available from: http://dx.doi.org/10.1381/096089200321643511.

[17] Ren CJ, Patterson E, Gagner M. Early results of laparoscopic biliopancreatic diversion with duodenal switch: a case series of 40 consecutive patients. Obes Surg 2000;10(6):514−23, discussion 524. Available from: http://dx.doi.org/10.1381/096089200321593715.

[18] Nguyen NT, Masoomi H, Magno CP, Nguyen X-MT, Laugenour K, Lane J. Trends in use of bariatric surgery, 2003 − 2008. ACS 2011;213 (2):261−6. Available from: http://dx.doi.org/10.1016/j.jamcollsurg.2011.04.030.

[19] Schauer P, Ikramuddin S, Hamad G, Gourash W. The learning curve for laparoscopic Roux-en-Y gastric bypass is 100 cases. Surg Endosc. 2003;17(2):212−15. Available from: http://dx.doi.org/10.1007/s00464-002-8857-z.

[20] Oliak D, Owens M, Schmidt HJ. Impact of fellowship training on the learning curve for laparoscopic gastric bypass. Obes Surg 2004;14(2): 197−200. Available from: http://dx.doi.org/10.1381/096089204322857555.

[21] Kothari SN, Boyd WC, Larson CA, Gustafson HL, Lambert PJ, Mathiason MA. Training of a minimally invasive bariatric surgeon: are laparo-scopic fellowships the answer? Obes Surg 2005;15(3):323−9. Available from: http://dx.doi.org/10.1381/0960892053576640.

[22] Ali MR, Tichansky DS, Kothari SN, et al. Validation that a 1-year fellowship in minimally invasive and bariatric surgery can eliminate the learning curve for laparoscopic gastric bypass. Surg Endosc 2009;24(1):138−44. Available from: http://dx.doi.org/10.1007/s00464-009-0550-z.

[23] Directory of Fellowships. fellowshipcouncil.org, <https://fellowshipcouncil.org/directory-of-fellowships/?match = 1> [accessed 1.04.16].

[24] ASMBS, ed. Core Curriculum for American Society for Metabolic and Bariatric Surgery Fellowship Training Requirements, <https://asmbs. org/professional-education/fellowship> [accessed 1.04.16].

[25] Kelles SMB, Barreto SM, Guerra HL. Mortality and hospital stay after bariatric surgery in 2,167 patients: influence of the surgeon expertise. Obes Surg 2009;19(9):1228−35. Available from: http://dx.doi.org/10.1007/s11695-009-9894-7.

[26] Zevin B, Aggarwal R, Grantcharov TP. Volume-outcome association in bariatric surgery. Ann Surg 2012;256(1):60−71. Available from: http:// dx.doi.org/10.1097/SLA.0b013e3182554c62.

[27] Markar SR, Penna M, Karthikesalingam A, Hashemi M. The impact of hospital and surgeon volume on clinical outcome following bariatric sur-gery. Obes Surg 2012;22(7):1126−34. Available from: http://dx.doi.org/10.1007/s11695-012-0639-7.

[28] Nguyen NT, Paya M, Stevens CM, Mavandadi S, Zainabadi K, Wilson SE. The relationship between hospital volume and outcome in bariatric surgery at academic medical centers. Ann Surg 2004;240(4):586−94. Available from: http://dx.doi.org/10.1097/01.sla.0000140752.74893.24.

[29] Flum DR, Salem L, Elrod JAB, Dellinger EP, Cheadle A, Chan L. Early mortality among Medicare beneficiaries undergoing bariatric surgical procedures. JAMA 2005;294(15):1903−8. Available from: http://dx.doi.org/10.1001/jama.294.15.1903.

[30] Courcoulas A, Schuchert M, Gatti G, Luketich J. The relationship of surgeon and hospital volume to outcome after gastric bypass surgery in Pennsylvania: a 3-year summary. Surgery 2003;134(4):613−21. Available from: http://dx.doi.org/10.1016/S0039-6060(03)00306-4.

[31] Sánchez-Santos R, Estévez S, Tomé C, et al. Training programs influence in the learning curve of laparoscopic gastric bypass for morbid obe-sity: a systematic review. Obes Surg 2011;22(1):34−41. Available from: http://dx.doi.org/10.1007/s11695-011-0398-x.

[32] Agrawal S. Impact of bariatric fellowship training on perioperative outcomes for laparoscopic Roux-en-Y gastric bypass in the first year as con-sultant surgeon. Obes Surg 2011;21(12):1817−21. Available from: http://dx.doi.org/10.1007/s11695-011-0482-2.

[33] Longitudinal Assessment of Bariatric Surgery (LABS) Consortium, Flum DR, Belle SH, et al. Perioperative safety in the longitudinal assess-ment of bariatric surgery. N Engl J Med 2009;361(5):445−54. Available from: http://dx.doi.org/10.1056/NEJMoa0901836.

[34] Flum DR, Kwon S, MacLeod K, et al. The use, safety and cost of bariatric surgery before and after Medicare's National Coverage Decision. Ann Surg 2011;254(6):860−5. Available from: http://dx.doi.org/10.1097/SLA.0b013e31822f2101.

[35] Dimick JB, Nicholas LH, Ryan AM, Thumma JR, Birkmeyer JD. Bariatric surgery complications before vs after implementation of a national policy restricting coverage to centers of excellence. JAMA 2013;309(8):792−9. Available from: http://dx.doi.org/10.1001/jama.2013.755.

[36] Encinosa WE, Bernard DM, Du D, Steiner CA. Recent improvements in bariatric surgery outcomes. Med Care 2009;47(5):531−5. Available from: http://dx.doi.org/10.1097/MLR.0b013e31819434c6.

[37] Champion JK, Pories WJ. Centers of excellence for bariatric surgery. Surg Obes Relat Dis 2005;1(2):148−51. Available from: http://dx.doi.org/ 10.1016/j.soard.2005.02.002.

[38] Telem DA, Talamini M, Altieri M, Yang J, Zhang Q, Pryor AD. The effect of national hospital accreditation in bariatric surgery on periopera-tive outcomes and long-term mortality. Surg Obes Relat Dis 2015;11(4):749−57. Available from: http://dx.doi.org/10.1016/j.soard.2014.05.012.

[39] Share DA, Campbell DA, Birkmeyer N, et al. How a regional collaborative of hospitals and physicians in Michigan cut costs and improved the quality of care. Health Aff (Millwood) 2011;30(4):636−45. Available from: http://dx.doi.org/10.1377/hlthaff.2010.0526.

[40] Schirmer B, Jones DB. The American College of Surgeons Bariatric Surgery Center Network: establishing standards. Bull Am Coll Surg 2007;92(8):21−7.

[41] Bradley DW, Sharma BK. Centers of excellence in bariatric surgery: design, implementation, and one-year outcomes. Surg Obes Relat Dis 2006;2(5):513−17. Available from: http://dx.doi.org/10.1016/j.soard.2006.06.005.

[42] MBSAQIP Resources for Optimal Care of the Metabolic and Bariatric Surgery Patient 2016 - Standards Manual V2.0. April 2016:1−63.

[43] Birkmeyer NJO, Dimick JB, Share D, et al. Hospital complication rates with bariatric surgery in Michigan. JAMA 2010;304(4):435—42. Available from: http://dx.doi.org/10.1001/jama.2010.1034.

[44] Livingston EH. Bariatric surgery outcomes at designated centers of excellence vs nondesignated programs. Arch Surg 2009;144(4):319—25, discussion 325. Available from: http://dx.doi.org/10.1001/archsurg.2009.23.

[45] Ponce J, Hoyt D, Grill H, Fried G, Bryman D. ASMBS response to CMS 7.26.2013. July 2013:1—8.

[46] Kohn GP, Galanko JA, Overby DW, Farrell TM. High case volumes and surgical fellowships are associated with improved outcomes for bariatric surgery patients: a justification of current credentialing initiatives for practice and training. ACS 2010;210(6):909—18. Available from: http://dx.doi.org/10.1016/j.jamcollsurg.2010.03.005.

[47] Nguyen NT, Nguyen B, Nguyen VQ, Ziogas A, Hohmann S, Stamos MJ. Outcomes of bariatric surgery performed at accredited vs nonaccredited centers. ACS 2012;215(4):467—74. Available from: http://dx.doi.org/10.1016/j.jamcollsurg.2012.05.032.

[48] Kwon S, Wang B, Wong E, Alfonso-Cristancho R, Sullivan SD, Flum DR. The impact of accreditation on safety and cost of bariatric surgery. Surg Obes Relat Dis 2013;9(5):617—22. Available from: http://dx.doi.org/10.1016/j.soard.2012.11.002.

[49] El Chaar M, Claros L, Ezeji GC, Miletics M, Stoltzfus J. Improving outcome of bariatric surgery: best practices in an accredited surgical center. Obes Surg. 2014;1—7. Available from: http://dx.doi.org/10.1007/s11695-014-1209-y.

[50] Morton JM, Garg T, Nguyen N. Does hospital accreditation impact bariatric surgery safety? Ann Surg 2014;260(3):504—9. Available from: http://dx.doi.org/10.1097/SLA.0000000000000891.

[51] Jafari MD, Jafari F, Young MT, Smith BR, Phalen MJ, Nguyen NT. Volume and outcome relationship in bariatric surgery in the laparoscopic era. Surg Endosc. 2013;27(12):4539—46. Available from: http://dx.doi.org/10.1007/s00464-013-3112-3.

[52] Gebhart A, Young M, Phelan M, Nguyen NT. Impact of accreditation in bariatric surgery. Surg Obes Relat Dis 2014;10(5):767—73. Available from: http://dx.doi.org/10.1016/j.soard.2014.03.009.

[53] Inabnet WB, Bour E, Carlin AM, et al. Joint task force recommendations for credentialing of bariatric surgeons. SOARD 2013;9(5):595—7. Available from: http://dx.doi.org/10.1016/j.soard.2013.06.014.

Chapter 9

Anesthesia for Bariatric Surgery

R. Rajendram[1,2,3], M.F. Khan[4] and V.R. Preedy[1]

[1]King's College London, London, United Kingdom, [2]Stoke Mandeville Hospital, Aylesbury, United Kingdom, [3]King Abdulaziz Medical City, Ministry of National Guard Hospital Affairs, Riyadh, Saudi Arabia, [4]Aga Khan University Hospital, Karachi, Pakistan

LIST OF ABBREVIATIONS

BP   blood pressure
CPAP continuous positive airway pressure
ERV   expiratory reserve volume
FRC   functional residual capacity
GA   general anesthesia
IBW   ideal body weight
LBW   lean body weight
NIV   noninvasive ventilation
OSA   obstructive sleep apnea
PEEP positive end-expiratory pressure
RM   recruitment maneuvers
TBW   total body weight
VD   volume of distribution

INTRODUCTION

Bariatric surgery patients are a major subgroup of the morbidly obese. Despite the significant comorbidities in this cohort, the overall 30-day mortality rate after elective bariatric surgery is surprisingly low (0.3%) [1]. As significant reductions in morbidity and mortality are associated with weight loss in this cohort, this perceived safety has led to a massive increase in bariatric surgery. However, the risk of morbidity and mortality of any individual patient may be significantly higher.

An estimation of an individual patient's risk of postoperative complications is made by the patient's anesthetist and surgeon during their preoperative assessments. This assessment determines the anesthetic management plan.

There is a global epidemic of obesity, so the challenges of bariatric patients are not isolated to specialist centers where bariatric surgery is performed. As the prevalence of obesity increases worldwide, an increasing number of obese surgical patients will require anesthesia. Obesity is typically defined by body mass index (BMI), which is the ratio of weight (in kilograms) to the square of height (in meters). In adults, obesity is defined as BMI ≥ 30 kg/m^2.

The increasing prevalence of obesity means that anesthetists are more frequently confronted with obese patients in their daily practice. This chapter reviews the changes in anatomy and physiology in obese patients that affect anesthetic management, anesthetic drug dosing, planning the anesthetic, equipment, appropriate monitoring, and analgesic plans in obese patients undergoing bariatric surgery, as they differ from patients with normal BMIs.

The principles of anesthesia for the patient having bariatric surgery described in this chapter can be applied to any obese patient having any elective or emergency surgery. Although this chapter focuses on anesthesia for the patient having bariatric surgery, it includes data on the management of obese patients after nonbariatric surgery (which adds important insights).

TABLE 9.1 Respiratory and Cardiovascular Changes Associated With Obesity

Respiratory Changes	Cardiovascular Changes
Restriction of lung volumes and chest movement	Increased circulating blood volume
Respiratory rate increase	Decreased systemic vascular resistance
Functional residual capacity (FRC) decrease	Increased cardiac output
Expiratory reserve volume (ERV) decrease	Left ventricular hypertrophy
Ventilation–perfusion mismatch	Left ventricular failure
Intrapulmonary right-to-left shunt	Right heart failure

This table summarizes the physiological and pathophysiological changes in the respiratory and cardiovascular systems associated with obesity.

TABLE 9.2 Relevance to Anesthetists of the Changes in Respiratory Physiology in Obese Patients

1. Decreased time to desaturation during apnea.
2. Increased O_2 requirements.
3. Hypoventilation during spontaneous ventilation in the supine position.

This table summarizes the relevance to anesthetists of the changes in respiratory physiology in obese patients.

PHYSIOLOGICAL CHANGES

Increasing obesity leads to respiratory and cardiovascular changes that influence anesthesia and perioperative analgesia. These changes are summarized in Table 9.1, but are discussed in more detail below.

Respiratory Physiology

The respiratory effects of obesity result from physical restrictions of chest movement and lung volume, as well as the increased metabolism by excess tissue. Respiratory rates increase, and functional residual capacity (FRC) and expiratory reserve volume decrease [2]. If FRC is reduced sufficiently, the small airways and alveoli remain closed during spontaneous ventilation. This causes ventilation–perfusion mismatch and right-to-left shunt. Lung volumes fall and the intrapulmonary shunt increases under general anesthesia (GA) in all patients. However, this effect is much more significant in obese patients [3]. The supine position and obstructive sleep apnea (OSA) further exacerbate these effects [4].

These changes increase the work of breathing, increase oxygen (O_2) consumption, and impair ventilation–perfusion matching [5]. The relevance to anesthetists of these respiratory changes is summarized in Table 9.2.

Cardiovascular Physiology

Cardiovascular physiological and pathophysiological changes in obesity include the following: [6]

1. Increased circulating blood volume. However, circulating blood volume is a lower proportion of total body weight (TBW) (50 mL/kg as compared with 75 mL/kg) than in patients with normal BMI.
2. Decreased systemic vascular resistance.
3. Increased cardiac output. Cardiac output increases by 20–30 mL per kilogram of excess body fat. Stroke volume increases, but stroke index, cardiac index, and heart rate do not increase.
4. Left ventricular hypertrophy.
5. Increased cardiac output can cause left ventricular failure (especially when associated with hypertension), right heart failure (especially if associated with the hypoxia and hypercapnia of OSA), or both.
6. Hypertension and ischemic heart disease are also more common in obese patients [6].

DOSING ANESTHETIC DRUGS

Drug dosing in obese patients may be based on TBW, lean body weight (LBW), or ideal body weight (IBW), depending upon the pharmacokinetic properties of the drug [7]. When there is no reliable data to guide the dosing for a specific drug, it is reasonable to base doses on LBW, unless the drug is highly lipophilic when TBW should be used [7].

Doses should be modified because obesity increases TBW, LBW, cardiac output, and blood volume, and also affects regional blood flow. This affects the peak plasma concentrations, clearance, and elimination half-lifes of many drugs [7].

The loading dose of a drug is mainly dependent on the volume of distribution (Vd). The physiochemical attributes of a drug predominantly determine Vd, which varies with plasma protein binding and tissue blood flow. The Vd of lipophilic drugs is increased by obesity. However, as perfusion of adipose tissue is less than vessel-rich or lean tissue, there is little change in the Vd of hydrophilic drugs in obese patients [7]. Changes are not consistent for all drugs within a class. The precise pharmacokinetics and pharmacodynamics of many medications remain undetermined [7].

Clearance of drugs is generally higher in the obese than the nonobese [7]. This is largely due to hepatic and renal physiology. Obesity affects hepatic metabolic pathways in different ways. Some are significantly enhanced in obese patients [7]. Renal elimination includes glomerular filtration, tubular secretion, and tubular reabsorption. Changes do occur in obese patients, but vary by drug, and the available data are limited.

The elimination half-life of a drug determines dosing interval and dosing of continuous infusions. The half-life varies with Vd and inversely with clearance. Both are altered by obesity.

Pharmacodynamic effects also change. For example, therapeutic windows narrow and side effects increase in severity.

INTRAOPERATIVE ANESTHETIC MANAGEMENT

Intraoperative Monitoring

Intraoperative monitoring devices should be used as recommended by guidelines such as those provided by the Association of Anaesthetists of Great Britain and Ireland [8]. However, additional equipment may be required to safely monitor obese patients. For example, accurate noninvasive blood pressure (BP) readings require appropriately sized BP cuffs. Obese patients often have upper arms that are conical in shape. In these patients placement of a standard BP cuff around the upper arm is difficult. Alternate sites for cuff placement such as the forearm or calf can also be used. However, there are no data to confirm the accuracy of this practice [9]. If there are any concerns about the accuracy of noninvasive monitoring or beat-to-beat monitoring of BP, and cardiac output is required, then invasive arterial BP monitoring should be considered.

Choice of Anesthetic for Bariatric Surgery

GA, regional anesthetic, and sedation can all be used safely in obese patients. However, bariatric surgery requires GA. Adaptation of GA for obese patients is required because of the changes in cardiovascular and respiratory physiology described above. The incidence of respiratory complications is higher in obese patients. As obese patients desaturate more quickly when apneic, anticipation, prevention, and prompt treatment of respiratory complications is crucial.

Premedication should provide anxiolysis without abolishing airway reflexes prior to induction of GA. The prevalence of sleep apnea and sensitivity to sedatives is high in obese patients. Therefore, long-acting respiratory depressants should not be used.

When patients undergo open bariatric surgery, the addition of neuroaxial analgesia to GA can improve postoperative pain control, reduce postoperative use of opioid analgesia, and decrease potential for drug-induced respiratory depression. Postoperative pain management with epidural infusions mitigates respiratory dysfunction in obese individuals, compared with systemic opioids. However, neuroaxial analgesia does not improve outcomes [10], and redundant tissue makes neuraxial anesthesia technically difficult to perform. These challenges may be overcome with appropriate equipment (e.g., long needles) and ultrasound guidance [11].

However, as laparoscopic bariatric surgery has more or less superseded open surgery, the use of neuraxial analgesia is rarely required.

Preparation for Induction

Preoxygenation should be performed using a tight-fitting facemask using 100% oxygen (O_2) at a high enough flow to prevent rebreathing (10−12 L/min), aiming for an end-tidal O_2 concentration of greater than 90% to maximize safe apnea time. Three minutes of tidal volume breathing or eight vital-capacity breaths over 60 seconds should be sufficient to achieve this [12]. Manually applied positive end-expiratory pressure (PEEP), or the use of noninvasive ventilation (NIV), for preoxygenation will further increase the nonhypoxic apneic period in obese patients who can tolerate it [13].

Passive apneic oxygenation via nasal cannula can further prolong the time to desaturation during laryngoscopy in high-risk patients [14]. However, absorption atelectasis occurs when high concentration oxygen is administered [15]. Use of a recruitment maneuver and PEEP after intubation may reverse this.

The supine position and induction of anesthesia reduces lung volumes [16], so preoxygenation is best with the patient sitting or head-up (reverse Trendelenburg). "Ramping" with a stack of blankets or preformed ramp that elevates the patient's torso and aligns the external auditory meatus with the sternal notch in the horizontal plane facilitates mask ventilation and improves laryngoscopic view [17].

Induction

The choice of induction agent depends on patient-specific factors rather than the presence of obesity. However, it is best to use a rapidly acting neuromuscular blocker (e.g., succinylcholine or rocuronium) to minimize the interval between induction and tracheal intubation.

Airway Management

The type of surgery, length of surgery, patient position, and risk of aspiration determine airway management. Bariatric surgery requires endotracheal intubation, relaxation of abdominal muscles, and therefore controlled ventilation.

Patients should be mask ventilated between induction and intubation unless rapid sequence intubation is required. If mask ventilation is difficult, then a supraglottic airway may be used to ventilate the patient prior to tracheal intubation.

Obese patients desaturate rapidly during apnea. Thus, there is less time to rescue a difficult airway. Ideally, awake tracheal intubation should be considered when there are risk factors for difficult intubation and difficult mask ventilation.

Devices designed to facilitate difficult intubation, medications and equipment to topically anesthetize the airway, and expert assistance must be immediately available for any obese patient having GA.

Patient Positioning

Inappropriate positioning impairs organ function (e.g., restricts ventilation) and damages tissues (e.g., neuropraxia and rhabdomyolysis). Additional staff and special equipment must therefore be available to safely and correctly position obese patients. Various devices are available to facilitate the movement of patients between operating tables, stretchers, and beds. These range from simple boards and sheets to hover mattresses. These devices improve patient safety and prevent injury to health care professionals during manual handling.

Beds and equipment used to support obese patients must also support the additional girth and mass and provide sufficient space to avoid pressure from side rails. Large beds and operating tables are required. Additional arm supports and a second operating table should be available in case the patient is too wide to fit on a standard operating table. Padding pressure points will help to prevent neuropraxia, and a bean bag can be used to provide additional support.

Maximum weight limits are only valid if the patient is positioned exactly as described in the literature provided by the manufacturer. If the patient is shifted on the table, or the table is unlocked, then these weight limits are significantly reduced.

The position of obese patients should be checked regularly during GA. Mattresses and patients can move when the operating table is tilted. Intraoperative repositioning may be required. Velcro can be used to fix the mattress to reduce slipping.

Maintenance of Anesthesia

Anesthesia can be maintained with either a volatile anesthetic (e.g., isoflurane, sevoflurane, desflurane) or an intravenous anesthetic (e.g., propofol). The limited data comparing these agents in obese patients have yielded conflicting results. Some data suggest that emergence and recovery are fastest with desflurane [18], but other trials have failed to confirm this [19,20].

Ventilation Management

When ventilation is controlled, a protective ventilation strategy should be used to maintain adequate oxygenation and ventilation while minimizing ventilator-associated lung injury [21]. This involves low tidal volumes (6−8 mL/kg IBW), adequate oxygenation ($PaO_2 > 8$ KPa; $SpO_2 > 88\%$), PEEP, and recruitment maneuvers.

If patients are allowed to breathe spontaneously (whether via a supraglottic airway or an endotracheal tube), minute ventilation and end-tidal carbon dioxide (CO_2) must be monitored closely to ensure ventilation is adequate.

Continuous positive airway pressure (CPAP) can be used to improve oxygenation during spontaneous ventilation. If patients cannot maintain adequate tidal volumes, then ventilation should be assisted or controlled. The addition of pressure support to PEEP may improve ventilation. If tidal volumes remain inadequate, then ventilation should be controlled with either pressure or volume with a lung protective strategy, as described above.

Fluid Management

There is little specific data to guide perioperative fluid therapy for obese patients. Therefore, clinical judgment must be based on assessment of volume status and tissue perfusion.

Extubation

At emergence the head-up position improves oxygenation and reduces the work of breathing. Obese patients may emerge from anesthesia slowly. Avoiding premature extubation is crucial. Laryngospasm and airway edema would exacerbate an already challenging intubation.

Emergency airway equipment and personnel to assist in difficult airway management must be available to manage potential complications. Therefore, extubation should usually be performed in the operating room prior to transfer to the postanesthesia care area.

Besides meeting standard extubation criteria, obese patients should only be extubated when fully awake after complete reversal of any neuromuscular blockade, which can be brought about by neostigmine (an acetylcholinesterase inhibitor) or sugammadex, a slightly lipophilic drug that chelates steroidal nondepolarizing neuromuscular blockers (i.e., rocuronium and vecuronium). Sugammadex results in significantly faster (approximately 3 minutes rather than 10 minutes) and better recovery from neuromuscular blockade than neostigmine [22]. However, sugammadex cannot be used to reverse the effects of atracurium or pancuronium.

POSTANESTHESIA CARE UNIT MANAGEMENT

Monitoring

Patients require continuous pulse oximetry until they can maintain adequate oxygenation when unstimulated. Patients who cannot maintain adequate oxygenation when unstimulated cannot be discharged from the hospital.

Obese patients are relatively hypoxic in comparison to nonobese patients due to postoperative changes in physiology. After extubation oxygen should be administered to keep O_2 saturation greater than 90%. If standard oxygen therapy is not sufficient, then incentive spirometry or chest physiotherapy can improve pulmonary function and reduce postoperative complications [23]. The use of NIV after abdominal surgery also reduces the incidence of reintubation and severe complications [24]. While there are no specific data for obese patients, they are also likely to benefit.

Despite concerns that forced aspiration of air during CPAP could damage recent intestinal anastomoses, administration of CPAP after gastric bypass did not affect risk of anastomotic leak [25]. After gastrointestinal surgery, a decision on the use of CPAP should involve the surgeon, anesthetist, and physiotherapist/respiratory technician.

Ventilation

Hypoventilation should be suspected in patients who remain drowsy or desaturate despite administration of oxygen. Arterial blood gas analysis is the best investigation for suspected hypoventilation. Oversedation should be excluded; pharmacologic reversal of benzodiazepines or opioids may be considered. If the upper airway is obstructed, simply waking a somnolent patient with a reminder to breathe deeply is often sufficient. However, this reminder may need to be repeated frequently. If the patient tolerates a jaw thrust without waking, an oropharyngeal airway, a nasopharyngeal airway, or both, may keep the airway patent. If this is not sufficient, the use of NIV may prevent reintubation.

Management of Pain and Anxiety

The risk and complications of respiratory depression are high in obese patients. Therefore, opioids, sedatives, and anxiolytics (e.g., benzodiazepines) should be used cautiously. Although titration to effect with small doses can reduce side effects, it is best to minimize their use through a multimodal approach to analgesia. However, an optimal analgesic regimen has not yet been found for obese patients [26].

The use of paracetamol is almost universal. Although this reduces postoperative opioid use [27], there are no specific data on obese patients. An evidence-based technique involves use of potent nonsteroidal antiinflammatory analgesics (NSAIDs) and local anesthetic wound infiltration [28].

Other agents that may be used to augment analgesia include ketamine, alpha-2 agonists (e.g., clonidine and dexmedetomidine), and antiepileptic drugs (pregabalin and gabapentin). These may reduce the need for intraoperative and postoperative opioids. Obese patients treated with alpha-2 adrenergic receptor agonists (e.g., preoperative oral clonidine, or intraoperative intravenous dexmedetomidine) used less opioid, and in some cases less antiemetic, and had shorter postanesthesia care unit stays [29].

Criteria for Step Down From the Postanesthesia Care Area

Specific data on duration of postoperative monitoring in morbidly obese patients are limited. However, there is generic guidance for the step down of surgical patients from the postanesthesia care area to an unmonitored environment (e.g., Association of Anaesthetists of Great Britain and Ireland guidelines on immediate postanesthesia recovery) [30], which can obviously be applied to obese patients. However, a low threshold for prolonged monitoring must be based on each individual patient's comorbidities and perioperative course. In some cases it may not be appropriate to discharge the patient to an unmonitored environment. Admission to a high-dependency area or intensive care unit may be required. Management of patients in intensive care after bariatric surgery is described in detail in this book [31].

MINI-DICTIONARY OF TERMS

- *Mechanical or Invasive Ventilation*: Invasive ventilation is positive pressure delivered to the patient's lungs via an endotracheal tube or a tracheostomy tube.
- *Noninvasive ventilation*: NIV is ventilatory support that is provided via a face mask to the patient's upper airway.
- *Obstructive sleep apnea*: This is a serious condition that occurs during sleep and has consequences for the subject's quality of life. Apnea means "to stop breathing." There is a narrowing of the throat, which can block the airway for the passage of air (breathing). Sleep is interrupted.
- *Hypertrophy*: This is an enlargement of the cell size. Collectively, an entire organ may be enlarged, and this too can be termed hypertrophy. Hypertrophy is usually an adaptive response that can become pathological (dangerous). Weightlifters can have hypertrophy of their skeletal muscles. When the heart is involved (as in cardiac hypertrophy), there is concern as cardiac enlargement is a risk factor for sudden death. The term hyperplasia is related to hypertrophy; it means increase in the number of cells. Organ enlargement may often be accompanied by both hyperplasia to some degree, as well as hypertrophy of the cells.
- *Pharmacokinetics*: This is the behavior of drugs within the body. It covers aspects of absorption, distribution, metabolism, and finally, excretion. Related to pharmacokinetics is the term "pharmacodynamics."
- *Pharmacodynamics*: This is the effect a drug has on the body. Both pharmacokinetics and pharmacodynamics are interrelated in a two-way process.

KEY FACTS

- Elective bariatric surgery generally has a low risk of complications.
- Estimation of an individual patient's risk of postoperative complications is made by the patient's anesthetist and surgeon during their preoperative assessments.
- The preoperative assessment guides the perioperative management of patients having bariatric surgery.
- No single specific anesthetic technique is superior to another with respect to important patient outcomes.
- Management of bariatric patients requires institutional investment in specialized equipment and training of staff.

SUMMARY POINTS

- The effect of obesity on respiratory function leads to rapid desaturation during apneic periods.
- Increased blood volume, decreased systemic vascular resistance, and increased cardiac output may lead to either left or right heart failure or both.
- The pharmacokinetic and pharmacodynamic parameters of a specific drug determine dosing in the obese.
- Use LBW to guide drug dosing unless there is specific guidance.
- GA is required for bariatric surgery as relaxation of the abdominal wall is required.
- No specific induction or maintenance agent has been shown to be superior.
- Monitoring in the postanesthesia care area should continue until patients can maintain an unobstructed airway and adequate oxygenation when left alone.

REFERENCES

[1]   Flum DR, Belle SH, King WC, et al. Perioperative safety in the longitudinal assessment of bariatric surgery. N Engl J Med 2009;361:445—54.

[2]   Jones RL, Nzekwu MM. The effects of body mass index on lung volumes. Chest 2006;130:827.

[3]   Söderberg M, Thomson D, White T. Respiration, circulation and anaesthetic management in obesity. Investigation before and after jejunoileal bypass. Acta Anaesthesiol Scand 1977;21:55.

[4]   Lee MY, Lin CC, Shen SY, et al. Work of breathing in eucapnic and hypercapnic sleep apnoea syndrome. Respiration 2009;77:146.

[5]   Littleton SW. Impact of obesity on respiratory function. Respirology 2012;17:43.

[6]   Alpert MA, Hashimi MW. Obesity and the heart. Am J Med Sci 1993;306:117.

[7]   Hanley MJ, Abernethy DR, Greenblatt DJ. Effect of obesity on the pharmacokinetics of drugs in humans. Clin Pharmacokinet 2010;49:71.

[8]   Association of Anaesthetists of Great Britain and Ireland. Draft recommendations for standards of monitoring during anaesthesia and recovery. 5th ed. <https://www.aagbi.org/sites/default/files/Standards%20of%20monitoring%2020150812.pdf>; 2015 [accessed 25.08.16].

[9]   Schumann R, Jones SB, Cooper B, et al. Update on best practice recommendations for anaesthetic perioperative care and pain management in weight loss surgery, 2004—2007. Obesity (Silver Spring) 2009;17:889.

[10] von Ungern-Sternberg BS, Regli A, Reber A, Schneider MC. Effect of obesity and thoracic epidural analgesia on perioperative spirometry. Br J Anaesth 2005;94:121.

[11] de Filho GR, Gomes HP, da Fonseca MH, et al. Predictors of successful neuraxial block: a prospective study. Eur J Anaesthesiol 2002;19:447.

[12] Baraka AS, Taha SK, Aouad MT, et al. Preoxygenation: comparison of maximal breathing and tidal volume breathing techniques. Anesthesiology 1999;91:612.

[13] Carron M, Zarantonello F, Tellaroli P, Ori C. Perioperative noninvasive ventilation in obese patients: a qualitative review and meta-analysis. Surg Obes Relat Dis 2016;12:681.

[14] Weingart SD, Levitan RM. Preoxygenation and prevention of desaturation during emergency airway management. Ann Emerg Med 2012;59:165.

[15] Edmark L, Kostova-Aherdan K, Enlund M, Hedenstierna G. Optimal oxygen concentration during induction of general anaesthesia. Anesthesiology 2003;98:28.

[16] Altermatt FR, Muñoz HR, Delfino AE, Cortínez LI. Pre-oxygenation in the obese patient: effects of position on tolerance to apnoea. Br J Anaesth 2005;95:706.

[17] El-Orbany M, Woehlck H, Salem MR. Head and neck position for direct laryngoscopy. Anesth Analg 2011;113:103.

[18] Juvin P, Vadam C, Malek L, et al. Postoperative recovery after desflurane, propofol, or isoflurane anaesthesia among morbidly obese patients: a prospective, randomized study. Anesth Analg 2000;91:714.

[19] Leykin Y, Pellis T, Del Mestro E, et al. Anaesthetic management of morbidly obese and super-morbidly obese patients undergoing bariatric operations: hospital course and outcomes. Obes Surg 2006;16:1563.

[20] De Baerdemaeker LE, Jacobs S, Den Blauwen NM, et al. Postoperative results after desflurane or sevoflurane combined with remifentanil in morbidly obese patients. Obes Surg 2006;16:728.

[21] Schumann R. Pulmonary physiology of the morbidly obese and the effects of anaesthesia. Int Anesthesiol Clin 2013;51:41.

[22] Gaszynski T, Szewczyk T, Gaszynski W. Randomized comparison of sugammadex and neostigmine for reversal of rocuronium-induced muscle relaxation in morbidly obese undergoing general anaesthesia. Br J Anaesth 2012;108:236.

[23] Thomas JA, McIntosh JM. Are incentive spirometry, intermittent positive pressure breathing, and deep breathing exercises effective in the prevention of postoperative pulmonary complications after upper abdominal surgery? A systematic overview and meta-analysis. Phys Ther 1994;74:3.

[24] Squadrone V, Coha M, Cerutti E, et al. Continuous positive airway pressure for treatment of postoperative hypoxemia: a randomized controlled trial. JAMA 2005;293:589.

[25] Ramirez A, Lalor PF, Szomstein S, Rosenthal RJ. Continuous positive airway pressure in immediate postoperative period after laparoscopic Roux-en-Y gastric bypass: is it safe? Surg Obes Relat Dis 2009;5:544.

[26] Sollazzi L, Modesti C, Vitale F, et al. Preinductive use of clonidine and ketamine improves recovery and reduces postoperative pain after bariatric surgery. Surg Obes Relat Dis 2009;5:67.

[27] Apfel CC, Turan A, Souza K, et al. Intravenous acetaminophen reduces postoperative nausea and vomiting: a systematic review and meta-analysis. Pain 2013;154:677.

[28] Govindarajan R, Ghosh B, Sathyamoorthy MK, et al. Efficacy of ketorolac in lieu of narcotics in the operative management of laparoscopic surgery for morbid obesity. Surg Obes Relat Dis 2005;1:530.

[29] Tufanogullari B, White PF, Peixoto MP, et al. Dexmedetomidine infusion during laparoscopic bariatric surgery: the effect on recovery outcome variables. Anesth Analg 2008;106:1741.

[30] Association of Anaesthetists of Great Britain and Ireland. AAGBI SAFETY GUIDELINE. Immediate Post-anaesthesia Recovery. Association of Anaesthetists of Great Britain and Ireland, UK, 2013.

[31] Rajendram R, Martin C, Preedy VR. Management of patients on intensive care after bariatric surgery. In: Rajendram R, Martin C, Preedy VR, editors. Metabolism and pathophysiology of bariatric surgery. Academic Press; In press.

Chapter 10

Laparoscopic Roux-ᴇɴ-Y Gastric Bypass

W. Lynn[1] and S. Agrawal[2]

[1]*Specialist Registrar Homerton University Hospital NHS Foundation Trust, London, United Kingdom,* [2]*Consultant Surgeon Homerton University Hospital NHS Foundation Trust & Honorary Senior Lecturer, Queen Mary University of London, London, United Kingdom*

LIST OF ABBREVIATIONS

BMI   Body mass index
DJ   Duodeno-jejunal flexure
EGD   Eosphagogastroduodenoscopy
GJ   Gastro-jejunal
JJ   Jejuno-jejunal
NSAID   Non-steroidal anti-inflammatory drug
RYGB   Roux-en-Y gastric bypass

ROUX-en-Y GASTRIC BYPASS

Roux-en-Y gastric bypass (RYGB) has been proven to be a successful operative treatment for morbid obesity [1]. RYGB has been shown to outperform even vigorous best medical therapy for both weight loss and type 2 diabetes remission [2,3]. The RYGB operation was first described by Mason et al. in 1966 [4], and the laparoscopic approach first reported by Wittgrove et al. in 1994 [5]. RYGB may be undertaken via either an open or laparoscopic technique. The vast majority of operations are undertaken laparoscopically. Compared to an open approach, laparoscopic surgery has been shown to be safe, with an associated decrease in postoperative wound complications, reduced inpatient stays and blood loss, and improved quality of life postoperatively [6,7]. The rate of incisional hernia is significantly reduced in laparoscopic surgery [8], as is the 30-day complication rate after surgery [9,10]. However, it has been suggested that the initial cost of laparoscopic surgery is higher compared to open surgery [11]. The lower rate of adhesions associated with laparoscopic surgery may also lead to an increase in the rate of internal herniation [12].

RYGB is associated with a significant learning curve of between 75 and 100 cases [13−15]. The importance of fellowship training for surgeons undertaking RYGB has been widely reported [16].

There are multiple different techniques for performing laparoscopic gastric bypass surgery, and there is significant variation in technique between surgeons [17]. The three main areas of difference with regard to the technique employed in RYGB are the following:

1. Anastomotic Technique
   a. Linear stapler
   b. Circular stapler
   c. Hand sewn
2. Alimentary limb configuration
   a. Antecolic or Retrocolic
   b. Antegastric or Retrogastric
3. Limb length of bilio-pancreatic (BP) limb

Of 3817 gastric bypasses performed in the United Kingdom in 2010, 22.4% of the operations used the circular stapling technique, 36.2% the linear stapling technique, and 33.4% a hand sewn technique [18] The authors preference is

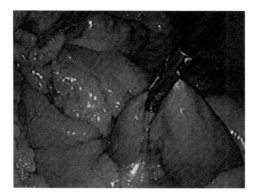

FIGURE 10.1   Identification of the DJ flexure.

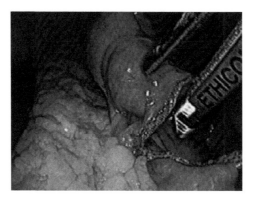

FIGURE 10.2   Jejunal division prior to JJ anastomosis.

for an antegastric antecolic anastomosis using the linear stapler technique. With regard to limb configuration, an antecolic approach reduces the risk of internal herniation, as there is no mesocolonic window created. However, the distance that the small bowel needs to travel to reach the gastric pouch is increased when compared to the retrocolic technique.

Linear Stapling Technique

There are two variations of this technique. The first variation is to perform the jejuno-jejunal (JJ) anastomosis first, and then the gastro-jejunal (GJ) anastomosis. In the second variation this process is reversed.

Jejuno-Jejunal First Technique

Jejuno-Jejunal Anastomosis

The first step is the displacement of the omentum cephalad to facilitate identification of the duodenal-jejunal (DJ) flexure and the ligament of Treitz (Fig. 10.1). The BP limb is measured to 25 cm from the ligament of Treitz and divided with a laparoscopic stapling device, and the mesentery divided with the use of a surgical energy device (Fig. 10.2). The Roux limb is then measured from the distal stapled end to the length of choice (100 cm if body mass index (BMI) < 40, 150 cm if BMI > 40). The Roux limb and BP limb are then approximated with stay sutures and enterotomies made on both limbs. The JJ anastomosis is then performed with a single firing of a laparoscopic stapling device and the enterotomy closed with a continuous absorbable suture (Fig. 10.3). The mesenteric defect is closed with the use of a continuous nonabsorbable braided suture in a purse string fashion to reduce the risk of internal herniation in the postoperative period.

Gastric Pouch Formation

The next step is the formation of the gastric pouch (Figs. 10.4 and 10.5). With the aid of liver retraction, a window is created on the lesser curve of the stomach to the lesser sac at the perigastric border between the second and third vessel of the lesser curve. The pouch is then created with the aid of a laparoscopic stapling device. The orogastric tube is

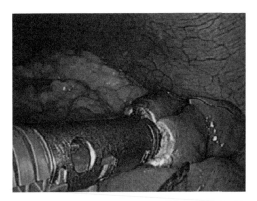

FIGURE 10.3 Creation of the JJ anastomosis.

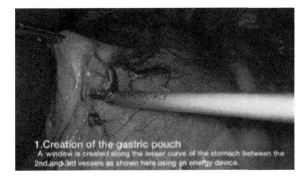

FIGURE 10.4 Creation of the gastric pouch—lesser curve dissection.

FIGURE 10.5 Creation of the gastric pouch—stapling with boogie guide.

removed prior to the first firing of the stapler in a horizontal direction. The orogastric tube is then replaced, and the pouch created over the tube with further firings of the stapler in a cephalad direction toward the angle of His, which may require mobilization to facilitate complete division of the stomach.

Gastro-Jejunal Anastomosis

The next step is the formation of a gastrostomy at the angle of the first and second firings of the stapling device (the area of the pouch with the poorest vascular supply). The previously divided alimentary limb is then delivered to the pouch in an antecolic manner. An enterotomy is made in the jejunum and an anastomosis created between the gastric pouch and alimentary limb with a single firing of a laparoscopic stapling device. The anastomotic defect is then closed with two continuous absorbable sutures as per the JJ anastomosis. A leak test is performed with 50 mL of diluted methylene blue dye to ensure anastomotic integrity.

FIGURE 10.6 Completed omega loop bypass prior to division.

Omega Loop Technique

In this technique (Fig. 10.6), the first step is the creation of the gastric pouch as described above. Following this, the omentum is divided. The BP limb is measured from the DJ flexure to the desired length and brought toward the gastric pouch. An enterotomy is made at the desired point and a stapled anastomosis is performed in the same manner as described above. The enterotomy is closed with a single-layer continuous suture. A further enterotomy is made on the proximal BP limb at this stage. The next step is the measurement of the Roux limb to the desired length. This is then brought toward the enterotomy on the BP limb and a further enterotomy made on the Roux limb at the desired length. The BP and Roux limbs are then anastomosed with a linear stapling device and the enterotomy closed with a continuous absorbable suture. Mesenteric defects (JJ and Petersen's) are then closed in the same fashion as described above. The BP limb is then divided from the GJ anastomosis with a laparoscopic stapling device. Leak testing is performed as described above.

Circular Stapling Technique

The circular stapling technique may be performed either with transoral placement of the anvil as first described by Wittgrove et al. [5] or via transabdominal placement of the anvil. Smaller anvil sizes have been associated with increased risk of postoperative complications [19]. Transoral placement has also been associated with increased wound and intraabdominal infection rates as well as esophageal trauma [20].

Transoral Technique

The first step in this technique is formation of the jejuno-jejunostomy. The greater omentum is divided and the jejunum measured from the DJ flexure to the desired length. This is then divided and a further length measured and brought to lie next to the divided bowel. A side-to-side anastomosis is created with a linear stapler and the enterotomy closed with a further firing of the device (tristapler technique). The distal end of the bowel is then brought up to the pouch to form the GJ anastomosis. The pouch is then created as described above. In this technique the anvil of the stapling device is then attached to an orogastric tube and the tube passed into the gastric pouch by the anesthetist. A small gastrotomy is made to allow passage of the orogastric tube through the pouch. The tube is then passed through the pouch until the anvil of the stapling device is seen through the gastrotomy. At this stage the orogastric tube is removed via one of the port sites. The staple gun is then entered into the blind ending jejunum and the spike advanced through the bowel wall approximately 5 cm from the end of the jejunum and then attached to the anvil in the pouch. The anastomosis is then created and the blind ending loop resected with the aid of a laparoscopic stapling device.

Transabdominal Technique

The jejuno-jejunostomy is created in the same manner as described above, however, the gastric pouch is not completed before the anvil is inserted into it. An initial firing of a linear stapler is performed, and a small gastrotomy made above this. A further larger gastrotomy is made in the remnant stomach and the anvil of the stapler introduced transabdominally. The anvil is placed into the gastrotomy in the remnant stomach and manipulated so that the spike of the anvil protrudes from the gastrotomy in the pouch. The gastric pouch is then completed with the anvil in place with the same technique as described for the linear stapling technique, and the anastomosis completed as described for the transoral approach. The remnant stomach gastrotomy may be closed via suturing or a further firing of the linear stapling device used to create the pouch.

Hand Sewn Technique

In this technique the first step is to enter the lesser sac by dividing the gastrocolic omentum. The DJ flexure is then identified and a window is made in the mesentery of the transverse colon to facilitate the transfer of the jejunum into the supracolic compartment. The BP limb and Roux limb are then created to the desired length. The jejuno-jejunostomy is created in the same way as described in the linear stapler technique, and the mesenteric defects are closed with a non-absorbable suture. The gastric pouch is created using the linear stapling technique. The gastro-jejunostomy is then formed using a hand sewn technique with a bougie acting as a sizer to ensure anastomotic patency. A two-layer continuous suturing technique is used to create an end-to-side anastomosis. Leak testing may then be performed as previously described.

Fully Stapled Technique

The fully stapled technique has been popularized by Dillemans et al. [21]. The gastric pouch creation is performed as per the linear stapling technique. The gastro-jejunostomy is then formed as follows. A gastrotomy is made in the pouch in the same area as per linear stapling and stretched. A purse string suture is then placed around this gastrotomy and a circular stapler anvil is introduced into the abdomen and manipulated into the pouch. The purse string is then tied. Splitting of the greater omentum is then undertaken, if required. The required length of BP limb is measured out and brought toward the pouch in an antecolic fashion. Five centimeters (5 cm) before the desired site of anastomosis, an enterotomy is formed, and the circular stapling device introduced into the jejunum and the stapler connected to the anvil as per the circular stapling technique. The blind loop is resected to close the introduction site of the stapling device. Formation of the jejuno-jejunostomy begins by measurement of the Roux limb, and an enterotomy is formed at the desired point as well as in the BP limb. Two firings of a laparoscopic stapling device are then used to form the jejuno-jejunostomy. The enterotomy from the JJ is then lifted with three stay sutures and the defect stapled closed with a further firing of a laparoscopic stapling device. Excess blind BP limb is then resected and the mesenteric defects closed.

COMPLICATIONS OF ROUX-en-Y GASTRIC BYPASS

Gastrointestinal Bleeding

Bleeding after gastric bypass may be either intraabdominal or intraluminal in nature. Any of the structures may bleed. However, the most common is the GJ anastomosis. Other sources of bleeding (hepatic, splenic, mesenteric) should also be considered. Management is centered on prompt resuscitation, identification of the bleeding source, and measures to control the hemorrhage. This may include diagnostic laparoscopy or esophagogastroduodenoscopy (EGD). A protocol for the management of postoperative bleeding was suggested by Mehran et al. [22].

Anastomotic Leak

The rate of anastomotic leak has been reported in the region of 2% [23,24]. Common symptoms of anastomotic leak include tachycardia, pain, pyrexia, and raised respiratory rate. Tachycardia (> 120) and elevated respiratory rate are the most sensitive indicators of leak [25]. The most common site for anastomotic leak is the gastro-jejunostomy. In a series of 59 leaks from 1200 patients undergoing RYGB, the leak site was noted to be at the gastro-jejunostomy in 68% of cases, 10% in the gastric pouch, and the JJ anastomosis in 5% of cases [26].

A leak may be diagnosed intraoperatively (via the methylene blue test described above) or postoperatively. Intraoperative diagnosis can facilitate immediate repair of any anastomotic defect and prevent postoperative complications.

A recent systematic review indicated no benefit toward the routine use of postoperative upper gastrointestinal imaging studies to exclude anastomotic leak over the selective use in patients displaying clinical signs (tachycardia, raised respiratory rate) [27]. However, the accuracy of reliance upon postoperative imaging studies to diagnose leak [28] is limited and cannot image the JJ anastomosis or remnant stomach accurately [29]. Therefore, high index of clinical suspicion combined with early relook diagnostic laparoscopy should be considered the gold standard technique for the assessment of postoperative anastomotic leak. The use of prophylactic drains in both the detection and treatment of anastomotic leak has not been shown to be of benefit [30,31].

The operative technique involves identification of the leak site and attempt at primary repair in the acute phase if the tissues are suitable. Jacobsen et al. [32] reported a success rate of 90% of primary repair in early leaks (<5 days postoperatively). Otherwise, prompt, accurate placement of drains to divert intestinal connect, and establishment of appropriate enteral or paraenteral feeding, are the cornerstones of management. Early surgical intervention (within 24 hours of symptoms) in suspected leaks has been shown to reduce morbidity and hospital stay [25]. The use of endoscopic stenting may have a role in the management of delayed leaks [33]. Nonoperative management may be appropriate in patients with contained leaks.

Anastomotic Ulceration

The rate of anastomotic ulceration has been reported to be between 1% and 16%, however, the underlying etiology is unclear [34]. Ying et al. [35] reported a significant risk reduction in marginal ulcer formation with the use of prophylactic proton pump inhibitors after RYGB. Patients may present with pain, dysphagia, or vomiting.

Potential causative factors include *Helicobacter pylori* infection [29], diabetes, steroid usage [36], smoking [37], and long pouch formation [38]. Antecolic placement of the Roux limb has also been suggested as a potential risk factor for anastomotic ulceration [39]. Benedwald et al. [40] found no difference in anastomotic ulcer rates between hand sewn, linear stapled, or circular stapled anastomoses, however, the use of nonabsorbable suture material has been shown to increase the risk of marginal ulcer formation [41].

Anastomotic ulcers are initially treated conservatively with risk factor modification (smoking/NSAID cessation), and acid suppression therapy with proton pump inhibitors potentially combined with sucralfate usage. Reported rates of revisional surgery vary from 20% [36] to 44% [42]. Revisional surgery involves revision of the GJ anastomosis, pouch resizing if required, and mobilization of the Roux limb if required.

Anastomotic Stricture

The rate of anastomotic stricture has been reported to be in the region of 3−5% [43,44]. The risk of anastomotic stricture has been shown to be significantly increased in patients undergoing circular stapled anastomoses when smaller (21 mm) staplers are used [19]. Common presenting symptoms include nausea, vomiting, or dysphagia. Balloon dilatation is the treatment of choice, with good results [19,45]. Perforation rates of less than 5% have been reported, however, this risk may be ameliorated if sequential dilatation is performed [21].

Internal Herniation Following Roux-en-Y Gastric Bypass

The mesenteric defects created by performing a RYGB may either be closed or left open. There are three potential spaces created during surgery that may lead to postoperative internal herniation: Petersen's space, mesocolon space, and jejunal mesentery space [46]. The large amount of weight lost after surgery and the use of laparoscopy rather than open surgery also increase the risk of internal herniation. The rate of symptomatic internal hernia formation has been reported to be in the range of 3% [1,47].

A small randomized trial conducted by Rosas et al. [48] showed no difference in the rate of symptomatic internal herniation whether or not the mesenteric defects were closed. However, other studies have suggested that the closure of the mesenteric defects is associated with a significant decrease in the rate of internal herniation. Stenberg et al. [49] randomized 2500 patients to undergo either mesenteric closure or leave the potential spaces open. They reported a significant decrease in the rate of postoperative small bowel obstruction in the group in whom mesenteric defects were closed at the expense of a higher rate of postoperative complications (kinking of the jejuno-jejunostomy due to defect closure). Chowbey et al. [50] reported a retrospective cohort of patients finding a significant decrease in internal herniation following the closure of the defects. However, it is important to note that the closure of the spaces, while significantly decreasing the risk of internal hernia formation, does not eliminate it entirely [51].

The Roux limb may have either an antecolic or retrocolic orientation as previously described. The advantage of the retrocolic technique is that it provides the shortest route for reconstruction and minimizes any potential tension on the anastomosis [52]. However, to facilitate this a window must be made in the transverse mesocolon, a potential site for internal herniation. This need to create a window is obviated by the antecolic approach. The antecolic approach has been shown to reduce the risk of development of internal hernias post-RYGB when compared to the retrocolic technique [53]. The rate of internal herniation with the use of the retrocolic technique has been associated with a four to eight times increased risk of internal hernia formation [54−56].

Other techniques employed to reduce the risk of internal hernias include omental division, not dividing the mesentery of the jejuno-jejunostomy, and ensuring right-sided orientation of the Roux limb. Quebbemann et al. reported a significant decrease in the rate of internal formation when a switch from left-sided orientation of the Roux limb was changed to right-sided [57].

Dumping and Hypoglycemia

Dumping represents a constellation of symptoms that may be seen after any time of gastric surgery. Up to 50% of patients undergoing gastric bypass surgery can experience dumping if a high carbohydrate load meal is ingested [58]. It may be divided into two clinical subtypes:

- Early Dumping. This usually begins within 15 minutes of eating. The high osmolar value of food transported directly to the small bowel leads to fluid shift stimulating the sympathetic nervous system. Sweating, tachycardia, and abdominal pain may be the clinical features.
- Late Dumping. This occurs some hours after eating as a result of the hyperinsulinemic effect of the high-calorie food delivered to the small bowel [59]. Patients present with dizziness, fatigue, and sweating.

Dumping may in fact be an advantageous side effect of RYGB as it leads to modulation of food behaviors and the avoidance of high-calorie foods [60,61] that may promote dumping. However, this has been questioned in other studies [23], and indeed, persistence of dumping in the later postoperative period may indicate failure to follow dietary advice and subsequent failure of maximal weight loss [62].

Persistent symptomatic postoperative hypoglycemia following RYGB is rare—the incidence of RYGB-related hyperinsulinemic hypoglycemia is reported to be less than 1% [63]. Younger patients, or those with a lower BMI at surgery, may be at higher risk [64]. Patients who experience symptoms suggestive of potential hypoglycemia (abnormalities of cognition, speech, weakness, lethargy) in the presence of low plasma glucose should be considered as potentially suffering from RYGB-related hyperinsulinemic hypoglycemia [65]. Potential mechanisms of hyperinsulinemic hypoglycemia postbypass include changes to insulin secretion, increases in islet cell function coupled with an increase in the sensitivity of the body tissues to the actions of insulin postoperatively [66]. Most patients can be managed with dietary modifications, and revisional surgery is not commonly required. Medical therapy in the form of acarbose or octreotide may be effective in certain subgroups. In those patients in whom medical therapy is not successful, conversion of the bypass to either normal anatomy or sleeve gastrectomy should be considered [67].

Outcomes Following Roux-en-Y Gastric Bypass

RYGB is a safe operation. Mortality rates in the early postoperative period have been reported to be in the range of 0.2% [68]. However, in the early period rates of reattendences vary in the literature from <5% [69] to 20−24% of cases [70,71].

Weight Loss

Weight loss outcomes after RYGB are good. The second National Bariatric Surgery Register (NBSR) report indicated excess weight loss of 68.7% after gastric bypass 1 year after surgery with excess weight loss rates at 3 years after surgery of 65.4%. The results from the Swedish Obese Subjects trial [72] showed maximal weight loss outcome was achieved at 2 years, with a tendency of slight weight regain in the following years.

Diabetes Remission

RYGB outperforms best medical therapy for remission of diabetes [2], with results of 60% with RYGB compared to 6% with best medical therapy in one randomized control trial. The meta-analysis published by Buchwald et al. [1] reported remission rates of 80% in patients undergoing RYGB. Even in those patients in whom remission of diabetes is not achieved, RYGB leads to a significant reduction in the use of oral hypoglycemics and insulin [73]. Schauer et al. [74] compared medical therapy alone or in combination with RYGB or sleeve gastrectomy. Thirty-eight percent of patients achieved the primary endpoint (HBa1c <6.0) in the RYGB group compared to 5% of patients in the medical therapy alone group. Mingrone et al. [75] also compared best medical therapy with RYGB. No patient in the medical group achieved diabetes remission, compared with 75% of patients undergoing RYGB. Five-year data was published by the same group, and showed remission rates of 37% in patients with at least a 5-year history of diabetes [76]. In has been suggested that younger patients, without associated complications of diabetes, and those not using insulin, are more likely to achieve diabetes remission [77].

Cardiovascular Disease

Sjostrom et al. [78] reported a significant decrease in cardiovascular mortality in patients undergoing bariatric surgery compared to matched controls (Hazard ratio = 0.47; 95% CI, 0.29−0.76; $p = 0.002$). In a study of 500 patients by Torquati et al. [79], gastric bypass decreased absolute risk of cardiac events by a mean of 63% in diabetic patients. The absolute risk of cardiac events was 5.4% at baseline compared to 2.7% at 1 year postoperatively. Eliasson et al. [80] compared 6312 patients undergoing RYGB and 6312 controls. Myocardial infarction rates were 49% lower in the RYGB group (HR 0.51, 0.29−0.91; $p = 0.021$). The rate of cardiovascular death was 59% lower in the Roux-en-Y group (0.41, 0.19−0.90; $p = 0.026$).

CONCLUSION

RYGB is a safe operation and leads to effective and sustained weight loss compared to other surgical procedures and best medical therapy. Laparoscopic gastric bypass has superseded open surgery. Rates of resolution of comorbidities are good, with bypass outperforming gastric banding and sleeve gastrectomy, and lower rates of significant nutritional deficiencies seen with duodenal switch or biliary−pancreatic diversion operations. Multiple different operative techniques for laparoscopic gastric bypass have been reported, with no one technique offering superior results over another at present.

MINI-DICTIONARY OF TERMS

- *Antecolic Roux Limb*: The Roux limb is brought to the gastric pouch in front of the colon.
- *Biliary−Pancreatic Limb*: Afferent limb acting as conduit for pancreatic and biliary secretion intestinal continuity.
- *Enterotomy*: Surgically created opening in the small bowel.
- *Gastrotomy*: Surgically created opening in the stomach.
- *Jejuno-Jejunal (JJ) Anastomosis*: Anastomosis between the Roux and biliary−pancreatic limb to restore function.
- *Retrocolic Roux Limb*: The Roux limb is brought to the gastric pouch through a surgically created opening in the mesentery of the transverse colon.
- *Roux Limb*: Efferent limb acting as primary conduit for ingested material.

SUMMARY POINTS

- Operation may be performed either laparoscopically or open.
- Laparoscopic gastric bypass was first performed in 1994.
- RYGB outperforms best medical therapy with regards to weight loss.
- Multiple surgical techniques have been reported with equivalent results.
- Limb lengths vary according to surgeon preference.
- Antecolic limb orientation reduces internal herniation.
- Complications of RYGB include anastomotic leak, stenosis, and internal herniation.
- Diabetes remission in patients undergoing RYGB is superior to those treated with best medical and lifestyle therapy.
- RYGB is associated with a significant reduction in cardiovascular mortality.

REFERENCES

[1] Buchwald H, Avidor Y, Braunwald E, Jensen MD, Pories W, Fahrbach K, et al. Bariatric surgery: a systematic review and meta-analysis. JAMA 2004;292(14):1724−37.

[2] Cummings DE, Arterburn DE, Westbrook EO, Kuzma JN, Stewart SD, Chan CP, et al. Gastric bypass surgery vs intensive lifestyle and medical intervention for type 2 diabetes: the CROSSROADS randomised controlled trial. Diabetologia 2016;59(5):945−53.

[3] Buchwald H, Estok R, Fahrbach K, Banel D, Jensen MD, Pories WJ, et al. Weight and type 2 diabetes after bariatric surgery: systematic review and meta-analysis. Am J Med 2009;122:248−56.

[4] Mason EE. Gastric bypass. Ann Surg 1969;170:329−39 - post gastrectomy patients experienced significant sustained weight-loss.

[5] Wittgrove AC, Clark GW, Tremblay LJ. Laparoscopic gastric bypass, Roux-en-Y: preliminary report of five cases. Obes Surg 1994;4:4353−7.

[6] Luján JA, Frutos MD, Hernández Q, Liron R, Cuenca JR, Valero G, et al. Laparoscopic versus open gastric bypass in the treatment of morbid obesity: a randomized prospective study. Ann Surg 2004;239(4):433−7.

[7] Nguyen NT, Goldman C, Rosenquist CJ, Arango A, Cole CJ, Lee SJ, et al. Laparoscopic versus open gastric bypass: a randomized study of outcomes, quality of life, and costs. Ann Surg 2001;234(3):279−91.

[8] Puzziferri N, Austrheim-Smith IT, Wolfe BM, Wilson SE, Nguyen NT. Three-year follow-up of a prospective randomized trial comparing laparoscopic versus open gastric bypass. Ann Surg 2006;243(2):181−8.

[9] Hutter MM, Randall S, Khuri SF, Henderson WG, Abbott WM, Warshaw AL. Laparoscopic versus open gastric bypass for morbid obesity: a multicenter, prospective, risk-adjusted analysis from the National Surgical Quality Improvement Program. Ann Surg 2006;243(5):657−66.

[10] Paxton JH, Matthews JB. The cost effectiveness of laparoscopic versus open gastric bypass surgery. Obes Surg 2005;15(1):24−34.

[11] Jones Jr KB, Afram JD, Benotti PN, Capella RF, Cooper CG, Flanagan L, et al. Open versus laparoscopic Roux-en-Y gastric bypass: a comparative study of over 25,000 open cases and the major laparoscopic bariatric reported series. Obes Surg 2006;16(6):721−7.

[12] Champion JK, Williams M. Small bowel obstruction and internal hernias after laparoscopic Roux-en-Y gastric bypass. Obes Surg 2003;13:596−600.

[13] Schauer P, Ikramuddin S, Hamad G, Gourash W. The learning curve for laparoscopic Roux-en-Y gastric bypass is 100 cases. Surg Endosc 2003;17(2):212−15.

[14] Oliak D, Owens M, Schmidt HJ. Impact of fellowship training on the learning curve for laparoscopic gastric bypass. Obes Surg 2004;14(2): 197−200.

[15] Aguilo R, Mukherjee S, Agrawal S. Effect of Post-CCT bariatric fellowship on the learning curve of laparoscopic Roux-en-Y gastric bypass as Consultant Surgeon. In British Journal of Surgery, vol. 99; 2012. p. 18−18.

[16] Agrawal S. Impact of bariatric fellowship training on perioperative outcomes for laparoscopic Roux-en-Y gastric bypass in the first year as consultant surgeon. Obes Surg 2011;21(12):1817−21.

[17] Madan AK, Harper JL, Tichansky DS. Techniques of laparoscopic gastric bypass: on-line survey of American Society for Bariatric Surgery practicing surgeons. Surg Obes Relat Dis 2008;4(2):166−72; discussion 172−3.

[18] Welbourn R, Fiennes A, Kinsman R, Walton P. The National Bariatric Surgery Registry: First Registry Report to March 2010. Henley-on-Thames: Dendrite Clinical Systems; 2011.

[19] Nguyen NT, Stevens CM, Wolfe BM. Incidence and outcome of anastomotic stricture after laparoscopic gastric bypass. J Gastrointest Surg 2003;7(8):997−1003; discussion 1003.

[20] Alasfar F, Sabnis A, Liu R, Chand B. Reduction of circular stapler related wound infection in patients undergoing laparoscopic Rouxen-Y gastric bypass, Cleveland clinic technique. Obes Surg 2010;20(2):168−72.

[21] Dillemans B, Sakran N, Van Cauwenberge S, Sablon T, Defoort B, Van Dessel E, et al. Standardization of the fully stapled laparoscopic Roux-en-Y gastric bypass for obesity reduces early immediate postoperative morbidity and mortality: a single center study on 2606 patients. Obes Surg 2009;19(10):1355−64.

[22] Mehran A, Szomstein S, Zundel N, Rosenthal R. Management of acute bleeding after laparoscopic Roux-en-Y gastric bypass. Obes Surg 2003;13(6):842−84.

[23] Gonzalez R, Sarr MG, Smith CD, Baghai M, Kendrick M, Szomstein S, et al. Diagnosis and contemporary management of anastomotic leaks after gastric bypass for obesity. J Am Coll Surg 2007;204(1):47−55.

[24] Chousleb E, Szomstein S, Podkameni D, Soto F, Lomenzo E, Higa G, et al. Routine abdominal drains after laparoscopic Roux-en-Ygastric bypass: a retrospective review of 593 patients. Obes Surg 2004;1203−7.

[25] Hamilton EC, Sims TL, Hamilton TT, Mullican MA, Jones DB, Provost DA. Clinical predictors of leak after laparoscopic Roux-en-Y gastric bypass for morbid obesity. Surg Endosc 2003;17(5):679−84.

[26] Ballesta C, Berindoague R, Cabrera M, Palau M, Gonzales M. Management of anastomotic leaks after laparoscopic Roux-en-Y gastric bypass. Obes Surg 2008;18(6):623−30.

[27] Quartararo G, Facchiano E, Scaringi S, Liscia G, Lucchese M. Upper gastrointestinal series after Roux-en-Y gastric bypass for morbid obesity: effectiveness in leakage detection. A systematic review of the literature. Obes Surg 2014;24(7):1096−101.

[28] Doraiswamy A, Rasmussen JJ, Pierce J, Fuller W, Ali MR. The utility of routine postoperative upper GI series following laparoscopic gastric bypass. Surg Endosc 2007;21(12):2159−62.

[29] Lee S, Carmody B, Wolfe L, DeMaria E, Kellum JM, Sugerman H, et al. Effect of location and speed of diagnosis on anastomotic leak outcomes in 3828 gastric bypass cases. J Gastrointest Surg 2007;11(6):708−13.

[30] Liscia G, Scaringi S, Facchiano E, Quartararo G, Lucchese M. The role of drainage after Roux-en-Y gastric bypass for morbid obesity: a systematic review. Surg Obes Relat Dis 2014;10(1):171−6.

[31] Dallal RM, Bailey L, Nahmias N. Back to basics−clinical diagnosis in bariatric surgery. Routine drains and upper GI series are unnecessary. Surg Endosc 2007;21(12):2268−71.

[32] Jacobsen HJ, Nergard BJ, Leifsson BG, Frederiksen SG, Agajahni E, Ekelund M, et al. Management of suspected anastomotic leak after bariatric laparoscopic Roux-en-y gastric bypass. Br J Surg 2014;101(4):417−23.

[33] Puli SR, Spofford IS, Thompson CC. Use of self-expandable stents in the treatment of bariatric surgery leaks: a systematic review and meta-analysis. Gastrointest Endosc 2012;75(2):287−93.

[34] Rasmussen JJ, Fuller W, Ali MR. Marginal ulceration after laparoscopic gastric bypass: an analysis of predisposing factors in 260 patients. Surg Endosc 2007;21(7):1090−4.

[35] Ying VWC, Kim SHH, Khan KJ, Farrokhyar F, D'Souza J, Gmora S, et al. Prophylactic PPI help reduce marginal ulcers after gastric bypass surgery: a systematic review and meta-analysis of cohort studies. Surg Endosc 2015;29(5):1018—23.

[36] Coblijn UK, Lagarde SM, de Castro SM, Kuiken SD, van Wagensveld BA. Symptomatic marginal ulcer disease after Roux-en-Y gastric bypass: incidence, risk factors and management. Obes Surg 2015;25(5):805—11.

[37] El-Hayek K, Timratana P, Shimizu H, Chand B. Marginal ulcer after Roux-en-Y gastric bypass: what have we really learned? Surg Endosc 2012;26(10):2789—96.

[38] Azagury DE, Abu DB, Greenwalt IT, Thompson CC. Marginal ulceration after Roux-en-Y gastric bypass surgery: characteristics, risk factors, treatment, and outcomes. Endoscopy 2011;43(11):950—4.

[39] Ribeiro-Parenti L, Arapis K, Chosidow D, Marmuse JP. Comparison of marginal ulcer rates between antecolic and retrocolic laparoscopic Roux-en-Y gastric bypass. Obes Surg 2015;25(2):215—21.

[40] Bendewald FP, Choi JN, Blythe LS, Selzer DJ, Ditslear JH, Mattar SG. Comparison of hand-sewn, linear-stapled, and circular-stapled gastroje-junostomy in laparoscopic Roux-en-Y gastric bypass. Obes Surg 2011;21(11):1671—5.

[41] Sacks BC, et al. Incidence of marginal ulcers and the use of absorbable anastomotic sutures in laparoscopic Roux-en-Y gastric bypass. Surg Obes Relat Dis 2006;2(1):11—16.

[42] Moon RC, Teixeira AF, Goldbach M, Jawad MA. Management and treatment outcomes of marginal ulcers after Roux-en-Y gastric bypass at a single high volume bariatric center. Surg Obes Relat Dis 2014;10(2):229—34.

[43] Gonzalez R, Lin E, Venkatesh KR, Bowers SP, Smith D. Gastrojejunostomy during laparoscopic gastric bypass: analysisof 3 techniques. Arch Surg 2003;138(2):181—4.

[44] Ahmad J, Martin J, Ikramuddin S, Schauer P, Slivka A. Endoscopic balloon dilation of gastroenteric anastomotic stricture after laparoscopic gastric bypass. Endoscopy 2003;35(9):725—8.

[45] Ukleja A, Afonso BB, Pimentel R, Szomstein S, Rosenthal R. Outcome of endoscopic balloon dilation of strictures after laparoscopic gastric bypass. Surg Endosc 2008;22(8):1746—50.

[46] Higa KD, Ho T, Boone KB. Internal hernias after laparoscopic Roux-en-Y gastric bypass: incidence, treatment and prevention. Obes Surg 2003;13(3):350—4.

[47] Blachar A, Federle MP, Pealer KM, et al. Gastrointestinal complications of laparoscopic Roux-en-Y gastric bypass surgery: clinical and imaging findings. Radiology 2002;223:625—32.

[48] Rosas U, Ahmed S, Leva N, Garg T, Rivas H, Lau J, et al. Mesenteric defect closure in laparoscopic Roux-en-Y gastric bypass: a randomized controlled trial. Surg Endosc 2015;29(9):2486—90.

[49] Stenberg E, Szabo E, Ågren G, Ottosson J, Marsk R, Lönroth H, et al. Closure of mesenteric defects in laparoscopic gastric bypass: a multicentre, randomised, parallel, open-label trial. Lancet 2016;387(10026):1397—404.

[50] Chowbey P, Baijal M, Kantharia NS, Khullar R, Sharma A, Soni V. Mesenteric defect closure decreases the incidence of internal hernias following laparoscopic Roux-En-Y gastric bypass: a retrospective cohort study. Obes Surg 2016;1—6.

[51] de la Cruz-Muñoz N, Cabrera JC, Cuesta M, Hartnett S, Rojas R. Closure of mesenteric defect can lead to decrease in internal hernias after Roux-en-Y gastric bypass. Surg Obes Relat Dis 2011;7(2):176—80.

[52] Wittgrove AC, Clark GW. Laparoscopic gastric bypass, Roux en-Y-500 patients: technique and results, with 3—60 month follow-up. Obes Surg 2000;10(3):233—9.

[53] Steele KE, Prokopowicz GP, Magnuson T, Lidor A, Schweitzer M. Laparoscopic antecolic Roux-en-Y gastric bypass with closure of internal defects leads to fewer internal hernias than the retrocolic approach. Surg Endosc 2008;22(9):2056—61.

[54] Escalona A, Devaud N, Pérez G, Crovari F, Boza C, Viviani P, et al. Antecolic versus retrocolic alimentary limb in laparoscopic Roux-en-Y gastric bypass: a comparative study. Surg Obes Relat Dis 2007;3(4):423—7.

[55] Ahmed AR, Rickards G, Husain S, Johnson J, Boss T, O'Malley W. Trends in internal hernia incidence after laparoscopic Roux-en-Y gastric bypass. Obes Surg 2007;17(12):1563—6.

[56] Hwang RF, Swartz DE, Felix EL. Causes of small bowel obstruction after laparoscopic gastric bypass. Surg Endosc 2004;18(11):1631—5.

[57] Quebbemann BB, Dallal RM. The orientation of the antecolic Roux limb markedly affects the incidence of internal hernias after laparoscopic gastric bypass. Obes Surg 2005;15(6):766—70.

[58] Banerjee A, Ding Y, Mikami DJ, Needleman BJ. The role of dumping syndrome in weight loss after gastric bypass surgery. Surg Endosc 2013;27(5):1573—8.

[59] Elliot K. Nutritional considerations after bariatric surgery. Crit Care Nurs Q 2003;26:133—8.

[60] Cummings DE, Overduin J, Foster-Schubert KE. Gastric bypass for obesity: mechanisms of weight loss and diabetes resolution. J Clin Endocrinol Metab 2004;89:2608—15.

[61] Tadross JA, le Roux CW. The mechanisms of weight loss after bariatric surgery. Int J Obes (Lond) 2009;33(Suppl. 1):S28—32.

[62] Sarwer DB, Dilks RJ, West-Smith L. Dietary intake and eating behavior after bariatric surgery: threats to weight loss maintenance and strategies for success. Surg Obes Relat Dis 2011;7(5):644—51.

[63] Kellogg TA, Bantle JP, Leslie DB, et al. Postgastric bypass hyperinsulinemic hypoglycemia syndrome: characterization and response to a modified diet. Surg Obes Relat Dis 2008;4(4):492—9.

[64] Nielsen JB, Pedersen AM, Gribsholt SB, Svensson E, Richelsen B. Prevalence, severity, and predictors of symptoms of dumping and hypoglycemia following Roux-En-Y gastric bypass. Surg Obes Relat Dis 2016.

[65] Ceppa EP, Ceppa DP, Omotosho PA, et al. Algorithm to diagnose etiology of hypoglycemia after Rouxen- Y gastric bypass for morbid obesity: case series and review of the literature. Surg Obes Relat Dis 2012;8(5):641−7.

[66] Marsk R, Jonas E, Rasmussen F, Näslund E. Nationwide cohort study of post-gastric bypass hypoglycaemia including 5,040 patients undergoing surgery for obesity in 1986−2006 in Sweden. Diabetologia 2010;53(11):2307−11.

[67] Fischer LE, Belt-Davis D, Khoraki J, Campos GM. Post-gastric bypass hypoglycemia: diagnosis and management. Bariatric surgery complications and emergencies. Springer International Publishing; 2016. p. 253−68.

[68] Flum DR, Belle SH, King WC, Wahed AS, Berk P, Chapman W, et al. Perioperative safety in the longitudinal assessment of bariatric surgery. N Engl J Med 2009;361(5):445−54.

[69] Reyes-Pérez A, Sánchez-Aguilar H, Velázquez-Fernández D, Rodríguez-Ortíz D, Mosti M, Herrera MF. Analysis of causes and risk factors for hospital readmission after Roux-en-Y gastric bypass. Obes Surg 2016;26(2):257−60.

[70] Kellogg TA, Swan T, Leslie DA, Buchwald H, Ikramuddin S. Patterns of readmission and reoperation within 90 days after Roux-en-Y gastric bypass. Surg Obes Relat Dis 2009;5(4):416−23.

[71] Saunders JK, Ballantyne GH, Belsley S, Stephens D, Trivedi A, Ewing DR, et al. 30-day readmission rates at a high volume bariatric surgery center: laparoscopic adjustable gastric banding, laparoscopic gastric bypass, and vertical banded gastroplasty-Roux-en-Y gastric bypass. Obes Surg 2007;17(9):1171−7.

[72] Sjöström L. Review of the key results from the Swedish Obese Subjects (SOS) trial—a prospective controlled intervention study of bariatric surgery. J Intern Med 2013;273(3):219−34.

[73] Ikramuddin S, Korner J, Lee WJ, Connett JE, Inabnet WB, Billington CJ, et al. Roux-en-Y gastric bypass vs intensive medical management for the control of type 2 diabetes, hypertension, and hyperlipidemia: the Diabetes Surgery Study randomized clinical trial. JAMA 2013;309(21):2240−9.

[74] Schauer PR, Bhatt DL, Kirwan JP, Wolski K, Brethauer SA, Navaneethan SD, et al. Bariatric surgery versus intensive medical therapy for diabetes—3-year outcomes. N Engl J Med 2014;370(21):2002−13.

[75] Mingrone G, Panunzi S, De Gaetano A, Guidone C, Iaconelli A, Leccesi L, et al. Bariatric surgery versus conventional medical therapy for type 2 diabetes. N Engl J Med 2012;366(17):1577−85.

[76] Mingrone G, Panunzi S, De Gaetano A, et al. Bariatric-metabolic surgery versus conventional medical treatment in obese patients with type 2 diabetes: 5 year follow-up of an open-label, singlecentre, randomised controlled trial. Lancet 2015;386:964−73.

[77] Iacobellis G, Xu C, Campo RE, Nestor F. Predictors of short-term diabetes remission after laparoscopic Roux-en-Y gastric bypass. Obes Surg 2015;25(5):782−7.

[78] Sjöström L, Peltonen M, Jacobson P, Sjöström CD, Karason K, Wedel H, et al. Bariatric surgery and long-term cardiovascular events. JAMA 2012;307(1):56−65.

[79] Torquati A, Wright K, Melvin W, Richards W. Effect of gastric bypass operation on Framingham and actual risk of cardiovascular events in class II to III obesity. J Am Coll Surg 2007;204(5):776−82.

[80] Eliasson B, Liakopoulos V, Franzén S, Näslund I, Svensson AM, Ottosson J, et al. Cardiovascular disease and mortality in patients with type 2 diabetes after bariatric surgery in Sweden: a nationwide, matched, observational cohort study. Lancet Diabetes Endocrinol 2015;3(11):847−54.

Chapter 11

Omega Loop Gastric Bypass

A. Ordonez[1], E. Lo Menzo[2] and R.J. Rosenthal[2]

[1]Previty Clinic, Beaumont, TX, United States, [2]Cleveland Clinic Florida, Weston, FL, United States

LIST OF ABBREVIATIONS

RYGBP Roux-en-Y gastric bypass
BMI Body mass index
AST Aspartate aminotransferase
ALT Alanine aminotransferase
WBC White blood cell
MCV Mean corpuscular volume

INTRODUCTION

Obesity has become a major healthcare concern in the United States, and even worldwide. This has led to a continued rise in the number of bariatric surgery procedures being performed. Laparoscopic Roux-en-Y gastric bypass (RYGBP) is one of the most effective therapeutic options for the treatment of morbid obesity. RYGBP is not only among the most common bariatric procedures, but also the gold standard to which all other operations are compared [1].

With the intent of developing procedures of lesser complexity, and possibly less morbid intervention, gastric bypass variations have been developed. Mason loop gastric bypass was an early version that was later abandoned due to multiple problems related to biliary reflux disease and esophagitis [2].

The mini-gastric bypass (MGBP), also known as omega loop gastric bypass, is a simpler version of laparoscopic RYGBP. It is a single anastomosis procedure initially introduced by Robert Rutledge in 2001 [3]. His first work included 1274 cases with a subsequent 6-year follow-up study in 2005 including 2410 patients [4].

The proponents of this operation claim that even though laparoscopic RYGBP is a safe alternative, it is technically challenging, the learning curve is steep, and it is associated with longer operative times, and even higher perioperative complication rates [5−7].

Over the past 15 years, MGBP has gained increasing attention and popularity because of its simplicity (when compared to RYGBP), as well as for its effectiveness for weight loss and improvement of obesity-related comorbidities. Some studies have found results comparable to those obtained with RYGBP and sleeve gastrectomy [3−5,8,9].

Despite recent increasing popularity and the presence of several dozens studies, it is reasonable to say that laparoscopic MGBP is still a controversial procedure given the lack of randomized controlled studies (only one has been done [5]) and long-term follow-up. The major concerns associated with this procedure are marginal ulcer, biliary reflux, and a potential risk of esophageal and gastric cancer [10−12].

METABOLIC CONSEQUENCES

A study published in 2015 evaluated the differences in glucose and insulin dynamics between RYGBP and MGBP. The data showed that after 4 years both groups (nondiabetic patients) displayed a significant enhancement of the baseline parameters reflecting Beta-cell activity evaluated by fasting plasma glucose, C-peptide, and fasting plasma insulin. This was thought to be likely secondary to the significant weight loss the participants experienced in both groups [13]. It was also concluded that because the only anatomical difference between the two procedures is the presence (or absence) of a Roux limb, the increased insulin secretion seen after MGBP could be related to the absence of the Roux limb. A potential explanation of this finding could be (1) the fact that after construction of a Roux limb, the jejunal mucosa is exposed to nondiluted nutrients at the gastro−jejunal (GJ) anastomosis and (2) the mix with the gastric contents and biliopancreatic fluids occurring 50 cm distally, as opposed to the immediate mix of nutrients that occurs in the presence of an MGBP [13]. The study, however, has some limitations, including low power, possible significant differences between groups (i.e., weight), and the lack of long-term data.

SURGICAL TECHNIQUE

Rutledge [3] initially described the technique in 2001 with five laparoscopic ports. Using a linear stapler the stomach was divided at the junction of the body and antrum, making sure that the jejunal loop was brought up comfortably to this location. A gastric evacuator ("Ewald tube") was passed into the stomach and held against the lesser curvature (size of the tube was not described). The stomach was then divided along the gastric tube with the linear stapler to form the gastric pouch. He emphasized that the division of the stomach should be parallel to the lesser curvature and directed toward the angle of His. The short gastric vessels were not divided. The next step of the procedure consisted of selecting a point on the small bowel about 200 cm from the ligament of Treitz. The intestinal (jejunal) loop was then brought up in an antecolic fashion and the linear stapler was used to create an end-to-side GJ anastomosis.

There have been some modifications proposed by different groups. For instance, the length of the biliopancreatic limb has been tailored according to the initial body mass index (BMI) of the patient [14]. Similarly, in an attempt to decrease the possibility of potential complications, alternative techniques for the GJ anastomosis were described, which aimed to reduce the exposure of the gastric mucosa to the biliopancreatic secretions given the potential carcinogenic effects associated with prolonged exposure. This technique was first described in 2004, and consisted of fixing the intestinal (jejunal) loop to the gastric pouch several centimeters cephalad to the anastomosis. As a result, the biliopancreatic secretions had a lower contact time with the gastric mucosa [9].

OUTCOMES

Weight Loss

Upon the initial report by Rutledge in 2001, an average excess weight loss (%EWL) of 77% was reported at 2 years [3]. On his follow-up series, the excess body weight loss was 80% at 1 year, with more than 95% of patients maintaining the weight loss within 25 pounds of the maximum after a 5-year follow-up period [4].

Another large study with a follow-up of up to 6 years with 1054 patients reported a mean %EWL of 84% and 85% at 1 and 6 years, respectively [15]. Conversely, other large series have also described lower rates of excess body weight: 56% at 5 years [16]. Most other series reported mean excess body weight between 50% and 85% [4,5,9,17,18].

Comorbidities

The only prospective randomized controlled trial performed in 2005 reported a 100% resolution of metabolic syndrome after 2 years; however, the result was exactly the same in the RYGBP group [5].

Most other series reported high percentages of resolution of comorbidities: sleep apnea remission in 75−100% of patients; type 2 diabetes mellitus remission in 70−100% of patients; and hyperlipidemia remission in 80−100% of patients. Follow-up time was up to 6 years [3−5,15,16,19−22].

Quality of Life

Several studies included analysis of quality of life outcomes. [5,16,17]. Overall, all studies found a positive change and improvement in gastrointestinal quality of life in patients who underwent MGBP; however, when compared head-to-head during the randomized controlled trial, no significant difference was found between laparoscopic RYGBP and MGBP. Nonetheless, both groups demonstrated a significant improvement in the quality of life index when compared to preoperative values. This difference was no longer present for either group 5 years after surgery [5,16].

Mini-Gastric (Omega Loop) Bypass Versus Other Bariatric Procedures

There have been dozens of publications about MGBP, however, only a few of those studies included large series of patients, and all of them have only mid-term follow-up results. In the only randomized controlled trial reported [5], the %EWL was 58.7% for the laparoscopic RYGBP group, whereas the MGBP group had a 64.9 %EWL after 2 years. The difference was not statistically significant; however, they pointed out that a higher percentage of MGBP patients achieved excess body weight loss greater than 50% than the RYGBP patients (95% vs 75%). Other findings achieving statistical significance ($p < 0.05$) in favor of MGBP were mean operative time (205 vs 147 minutes; <0.001), early postoperative complications (20% vs 7.5%; <0.05), analgesic use (3.4 vs 2.0 units; <0.05), and postoperative hospital stay (6.9 vs 5.5 days; <0.001). On the other hand, mortality (0 for both), conversion rate (2.5% vs 0%), intraoperative blood loss (42.5 vs 48.3 mL), postoperative flatus passage (2.4 vs 2.3 days), and late complications requiring readmission (7.5% for both) did not exhibit a significant difference.

Regarding comorbidities, except for excess weight reduction (%) at 1 year ($p = 0.025$) and hemoglobin level ($p = 0.02$), all the data about comorbidities did not reach statistical significance. The factors analyzed included: BMI at 1 and 2 years, excess weight reduction (%) at 2 years, metabolic syndrome, systolic and diastolic blood pressure, blood glucose, cholesterol, triglycerides, uric acid, aspartate aminotransferase (AST), alanine aminotransferase (ALT), albumin, white blood cell (WBC) count, and mean corpuscular volume (MCV).

A small prospective database analysis with 53 patients where MGBP was compared to sleeve gastrectomy reported a trend toward higher diabetes mellitus type 2 remission rates relative to sleeve gastrectomy; however, the authors concluded that further studies were needed to provide definite conclusions [20]. A recent report from 2016 with a higher number of patients did find a significant difference ($p < 0.01$) in diabetes type 2 remission when MGBP (85.4%) was compared to sleeve gastrectomy (60.9%) [23].

Advantages and Disadvantages

Proponents of the omega loop bypass have claimed multiple advantages over RYGBP. As previously noted, the only prospective randomized clinical trial reported shorter length of stay, shorter operative time, lower complication rates, and even less postoperative pain for MGBP. Many other reports coincide with these findings and also claim MGBP is a safer and more effective procedure [5,16,17]. They go on to theorize that omega loop MGBP has the advantage of using a lower antecolic GJ anastomosis which would be much easier to perform than a higher antecolic or retrocolic anastomosis when performing an RYGBP [5]. Given the fact there is only one anastomosis (gastro-jejunostomy), this would lower the risk for leaks overall. Furthermore, the absence of a jejuno-jejunostomy and division of the intestine would, in theory, decrease the chances of an internal hernia. The lower complexity of the operation when compared to an RYGBP would also favor surgeons by shortening the learning curve [16,15].

With the presence of a longer biliopancreatic limb created during an MGBP procedure, it was also assumed this would result in a more extensive malabsorptive surface area [3].

Most detractors of MGBP point to biliary reflux as a major concern. Furthermore, esophagitis and gastritis, with the potential for increased gastric cancer risk, have also been reported [24]. On the other hand, studies actually measuring esophageal or gastric exposure to biliary fluid have not been performed.

In terms of the potential need for a revisional procedure, the groups who favored MGBP also claimed a reduced complexity level, even if a reversal of the procedure was needed [3,4,17,21].

Complications

A retrospective report from 2015, including 2321 patients and describing the incidence, presentation, and surgical management of leaks in patients who underwent laparoscopic MGBP, reported a leak rate of 1.5% [25]. Leaks were classified into two different categories. Type 1 leaks where the most common. They were located in the gastric tube and could be explained by the presence of a long gastric tube with a decreased compliance and vascularization. Type 2 leaks arose from the gastro-jejunostomy; they were reported in 0.17% of patients. They emphasized that this low rate coincided with the leak rate reported in other series and attributed this to a tension-free anastomosis and the absence of mesenteric division.

The randomized controlled trial reported no major complications in patients undergoing MGBP compared to a 5% major complication rate for laparoscopic RYGBP patients (both related to leaks). No leaks were reported for the MGBP group. Similarly, minor complication rates were also lower for MGBP patients (7.5% vs 15%). Minor complications for the RYGBP group were described as upper gastrointestinal bleeding, ileus, and leakage from the draining tube; for the other group these included wound infection and upper gastrointestinal bleeding [5].

The incidence of marginal ulcer in this trial was higher in the omega MGBP group (5%) when compared to the RYGBP patients (3%). They did not report if this represented a significant difference.

Reports from other large series included a leak rate of 0.2−1.3%, mortality of 0−0.1%, early complications (30-day morbidity) of 2.7−6.7%, late complications of 4.7−11.9%, and marginal ulcer of 0.6−5.6%. Long-term complications included iron deficiency anemia of 4.9−7.6% and %EWL of 0.1−1.2% [3−5,15,16,18,22,23].

Revisional surgery from a large series varied between 1% and 1.7% [4,19]. The randomized controlled trial did not report revision rate initially, however, on a follow-up study several years later, the revision rate did not differ significantly between the MGBP and RYGBP groups [5,16].

Revision rates were perhaps one of the weakest areas surrounding MGBP. Most series did not report or provide valid information about this topic, possibly reflecting the lack of adequate or long-term follow-up. A study with information from multiple centers investigating revision and conversions rates noted that all revisions had taken place in different hospitals, which could certainly lead to poor reporting and awareness of late complications. The most common reason for patients requiring revision was bile reflux. Other complications leading to reoperation included marginal ulcer, leaks, malnutrition, and weight gain. Most patients required conversion to RYGBP [26].

CONCLUSIONS

The ever-increasing obese population worldwide has led to a rise in the number of bariatric surgery procedures performed every year. Laparoscopic RYGBP is not only the most common bariatric procedure, but also the gold standard to which all other operations are compared. For more than a decade, different versions of this operation have been developed.

MGBP, also known as omega loop gastric bypass, is a simpler version of laparoscopic RYGBP. There is only one anastomosis accompanied by a long biliopancreatic limb.

Proponents of the omega loop MGBP claim this is a simpler, faster, and safer procedure, with similar or even better results that its counterpart, RYGBP.

Dozens of studies from around the globe including more than 7000 patients have presented outcomes since MGBP was initially described by Rutledge in 2001. However, only one randomized controlled study has been performed to date. Most of the other series were retrospective analyses with short- to mid-term follow-up. Long-term follow-up for MGBP is simply nonexistent. Furthermore, the major concerns about this procedure (bile reflux, gastritis, potential for gastric cancer) have not been objectively evaluated. More studies with long-term follow-up, as well as more prospective, randomized studies, are certainly needed.

With all this in mind, even though most (if not all) of the series reported excellent results with MGBP, one must be careful when the data is compared with the decisive outcomes RYGBP has shown for decades.

MINI-DICTIONARY OF TERMS

- *Roux-en-Y gastric bypass*: Surgical procedure aimed at determining weight loss by reducing both the amount of food ingested and the absorption of it.
- *Laparoscopic omega loop (mini) gastric bypass*: Variation of the traditional gastric bypass in which only one anastomosis is utilized.
- *Biliary reflux disease*: The chronic inflammation of the stomach and/or esophagus caused by chronic reflux of bile.
- *Sleeve gastrectomy*: Partial gastrectomy of the greater curvature of the stomach intended to induce weight loss.
- *BMI*: Body mass index is a calculation of the degree of obesity based on weight and height.
- *AST*: Aspartate aminotransferase enzyme produced by the liver parenchyma. Its serum elevation is indicative of liver parenchymal damage or inflammation.
- *ALT*: Alanine aminotransferase enzyme produced by the liver parenchyma. Its serum elevation is indicative of liver parenchymal damage or inflammation.
- *WBC*: White blood cell count blood.
- *MVC*: Microtic cell volume. An indicator of the size of red blood cells that helps distinguish different types of anemia.
- *Biliopancreatic limb*: The intestinal limb extending from the duodenum that contains the bile and pancreatic secretions.

KEY FACTS

- Obesity is an expanding pandemic phenomenon.
- Obesity is the direct cause of several comorbid conditions (hypertension, diabetes, dyslipidemia, coronary artery disease, cancer, among many others) responsible for decreased life expectancy.
- Weight loss has been proven to reverse, and often resolve, some of these comorbid conditions.
- Surgical procedures aimed to induce weight loss are far more effective than medical and dietary interventions.
- Gastric bypass is considered the gold standard of surgical weight loss operations.

- The MGBP is an easier technical variation of the gastric bypass. However the safety and efficacy of this procedure has not been fully established.
- It is also known as omega loop gastric bypass.
- It is a simpler version of laparoscopic RYGBP, as it only has one anastomosis accompanied by a long biliopancreatic limb.
- Proponents of the omega loop MGBP claim this is a simpler, faster, and safer procedure, with similar or even better results than its counterpart, RYGBP.
- Only one randomized controlled study has been performed to date.
- Long-term follow-up for MGBP is nonexistent.
- The major concerns about this procedure (bile reflux, gastritis, potential for gastric cancer) have not been objectively evaluated.
- More studies with long-term follow-up, as well as more prospective, randomized studies, are needed.

SUMMARY POINTS

- Laparoscopic RYGBP is currently the gold standard of bariatric surgery procedures. With the intent to create an easier, faster, and more effective operation, a variation of RYGBP was proposed 15 years ago.
- The laparoscopic omega loop MGBP has gained popularity in many centers around the globe.
- Multiple large series have suggested MGBP as an equal or even better option than RYGBP, claiming better outcomes and fewer complications.
- Only one randomized prospective trial has been performed.
- Long-term data is lacking, which could suggest insufficient follow-up and consequently inaccurate information, especially when it comes to late complications.
- More prospective, randomized studies are needed, as well as long-term follow-up reports.
- Until then, caution should be exercised when considering a laparoscopic MGBP over RYGBP for the surgical treatment of morbid obesity.

REFERENCES

[1] Cho M, Carrodeguas L, Pinto D, et al. Diagnosis and management of partial small bowel obstruction after laparoscopic antecolic antegastric Roux-en-Y gastric bypass for morbid obesity. J Am Coll Surg 2006;202:262–9.
[2] Mason R, Taylor P, Filipe M, et al. Pancreaticoduodenal secretions and the genesis of gastric stump carcinoma in the rat. Gut 1988;29:830–4.
[3] Rutledge R. The mini-gastric bypass: experience with the first 1,274 cases. Obes Surg 2001;11:279–80.
[4] Rutledge R, Walsh T. Continued excellent results with the mini-gastric bypass: six-year study in 2,410 patients. Obes Surg 2005;15:1304–8.
[5] Lee W, Yu P, Wand W, et al. Laparoscopic Roux-en-Y versus mini-gastric bypass for the treatment of morbid obesity. Ann Surg 2005;242:20–8.
[6] Westling A, Gustavsson S. Laparoscopic vs open Roux-en-Y gastric bypass: a prospective, randomized trial. Obes Surg 2001;11:284–92.
[7] Reddy R, Riker A, Marra D, et al. Open Roux-en-Y gastric bypass for the morbidly obese in the era of laparoscopy. Am J Surg 2002;184:611–16.
[8] Lee W, Chong K, Lin Y, et al. Laparoscopic sleeve gastrectomy versus single anastomosis (mini) gastric bypass for the treatment of type 2 diabetes mellitus: 5-year results of a randomized trial and study of incretin effect. Obes Surg 2014;24:1552–62.
[9] Garcia-Caballero M, Carbajo M. One anastomosis gastric bypass: a simple, safe, and efficient surgical procedure for treating morbid obesity. Nutr Hosp 2004;19:372–5.
[10] Fisher B, Buchwald H, Clark W, et al. Mini-gastric bypass controversy. Obes Surg 2001;11:773–7.
[11] Mahawar K, Jennings N, Brown J, et al. Mini gastric bypass: systematic review of a controversial procedure. Obes Surg 2013;23:1980–8.
[12] Georgiadou D, Sergentanis T, Nixon A, et al. Efficacy and safety of laparoscopic mini gastric bypass. A systematic review. Surg Obes Relat Dis 2014;10:984–91.
[13] Himpens J, Villalonga R, Cadiere G, et al. Metabolic consequences of the incorporation of a Roux limb in an omega loop (mini) gastric bypass: evaluation by a glucose tolerance test at mid-term follow-up. Surg Endosc 2015;1–11.
[14] Lee W, Wang W, Lee Y, et al. Laparoscopic mini-gastric bypass: experience with tailored bypass according to body weight. Obes Surg 2008;18:294–9.
[15] Kular K, Manchanda N, Rutledge R. A 6-year experience with 1,054 mini-gastric bypasses - First study from the Indian subcontinent. Obes Surg 2014;24:1430–5.
[16] Lee W, Ser K, Lee Y, et al. Laparoscopic Roux-en-Y vs. Mini-gastric bypass for the treatment of morbid obesity: a 10-year experience. Obes Surg 2012;22:1827–34.

[17] Carbajo M, Garcia-Caballero M, Toledano M, et al. One-anastomosis gastric bypass by laparoscopy: results of the first 209 patients. Obes Surg 2005;15:398−404.

[18] Chakhtoura G, Zinzindohoue F, Ghanem Y, et al. Primary results of laparoscopic mini-gastric bypass in a French obesity surgery specialized university hospital. Obes Surg 2008;18:1130−3.

[19] Wang W, Wei P, Lee Y, et al. Short-term results of laparoscopic mini-gastric bypass. Obes Surg 2005;15:648−54.

[20] Milnone M, Di Minno M, Leongito M, et al. Bariatric surgery and diabetes remission: sleeve gastrectomy or mini gastric bypass? World J Gastroenterol 2013;19:6590−7.

[21] Noun R, Skaff J, Riachi E, et al. One-thousand consecutive mini-gastric bypass: short- and long- term outcome. Obes Surg 2012;22:697−703.

[22] Kim Z, Hur K. Laparoscopic mini-gastric bypass for type 2 diabetes: the preliminary report. World J Surg 2011;35:631−6.

[23] Mussella M, Apers J, Rheinwalt K, et al. Efficacy of Bariatric surgery in type 2 diabetes mellitus remission: the role of mini gastric bypass/one anastomosis gastric bypass and sleeve gastrectomy at 1 year of follow-up. A European survey. Obes Surg 2016;26:933−40.

[24] Mahawar K, Carr W, Balupuri S, et al. Controversy surrounding "mini" gastric bypass. Obes Surg 2014;28:156−63.

[25] Genser A, Carandina S, Tabbara M, et al. Presentation and surgical management of leaks after mini-gastric bypass for morbid obesity. Surg Obes Rel Dis 2015;12:305−12.

[26] Johnson W, Fernandez A, Farrell T, et al. Surgical revision of loop ("mini") gastric bypass procedure: multicenter review of complications and conversions to Roux-en-Y gastric bypass. Obes Surg 2007;3:37−41.

[27] Griffith P, Birch D, Sharma A, Karmali S. Managing complications associated with laparoscopic Roux-en-Y gastric bypass for morbid obesity. Can J Surg 2012;55(5):329−36.

Chapter 12

Laparoscopic Sleeve Gastrectomy

J. Kaufman, J. Billing and P. Billing
Eviva Bariatrics, Edmonds, WA, United States

LIST OF ABBREVIATIONS

BE	Barrett's esophagus
BiPAP	bilevel positive airway pressure
BMI	body mass index
BPL	biliopancreatic limb
BS	bariatric surgery
CC	common channel
CPAP	continuous positive airway pressure
DS	duodenal switch
DVT	deep venous thrombosis
EWL	excess weight loss
GEJ	gastroesophageal junction
GERD	gastroesophageal reflux disease
HH	hiatal hernia
HP	*Helicobacter pylori*
JIB	jejunal-ileal bypass
LAGB	laparoscopic adjustable gastric banding
NASH	nonalcoholic steatohepatitis
PE	pulmonary embolism
PVT	portal vein thrombosis
RYGBP	Roux-en-Y gastric bypass
SG	sleeve gastrectomy
SLR	staple line reinforcement
SMV	superior mesenteric vein
STOP-BANG	snoring, tiredness, observed apnea, blood pressure, body mass index, age, neck circumference, gender
VBG	vertical banded gastroplasty
VTE	venous thromboembolism

INTRODUCTION

Obesity is a chronic, progressive disease affecting every human organ system. According to the Centers for Disease Control (CDC), in 2012 an estimated 36% of the US adult population was obese, with levels expected to reach 44% by 2030 [1]. Costs are estimated to surpass $210 billion/year. Morbid obesity increases risks to every organ system and increases surgical complications in cardiac, thoracic, abdominal, gynecological, breast, orthopedic, vascular, esthetic, transplant, gastrointestinal, and colorectal surgery compared to patients of lower body mass index (BMI). Healthcare costs for morbidly obese people are higher than their nonobese counterparts. Wage and employment discrimination are commonly reported against morbidly obese people. A growing body of literature convincingly proves bariatric surgery (BS) is the only durable means of achieving sustained weight loss [2–8]. Sleeve gastrectomy (SG) is the most commonly performed bariatric operation in the United States, with outcomes now approaching those of Roux-en-Y Gastric Bypass (RYGBP).

HISTORY

RYGBP, historically heralded as the gold standard of bariatric operations, is now challenged by SG. Complications of RYGBP, such as internal hernias, intestinal volvulus, marginal ulcers, nutrition malabsorption, and anastomotic complications, drive the need for less anatomically complex and safer operations. LAGB, while lowest in operative and 30-day risk, has seen use plummet as long-term complications have increased and poor weight loss have become the norm. SG began as the first of the two-stage duodenal switch (DS) operations. Patients undergoing the first stage of DS often lost significant weight, thus avoiding the second stage. SG is now the most performed bariatric procedure in the United States [7].

SG creates a narrow tubular structure from the large stomach reservoir by removing 70−80% of gastric volume. SG accelerates gastric emptying, improves insulin production and sensitivity, and reduces several gut peptides responsible for hunger [9,10]. Other theorized effects include cell-signaling and changes in gut microbes in yet undetermined ways. After SG, significant alterations occur for ghrelin, leptin, polypeptide-YY, glucagon-like hormone 1, and glucagon-dependent insulinotropic peptide, as well as significant immediate improvements in insulin resistance. [10,11] Long-term results are also likely affected by changes in gastric emptying, gut microbiota, inflammatory mediators, and other unknown mechanisms [12,13].

PREOPERATIVE EVALUATION

To qualify for BS, patients must have a BMI over 35 kg/m^2 and one or more obesity-related complications, or a BMI over 40 kg/m^2 with or without related complications. These BMI-related definitions of obesity and morbid obesity have been accepted since the 1991 Consensus Conference Statement on Gastrointestinal Surgery for Severe Obesity. Obesity was further classified by the National Institutes of Health in 1998 as Class I (BMI = 30.0−34.9), Class II (BMI = 35.0−39.9), and Class III (BMI ≥ 40 kg m^2) [14]. Mounting evidence supports SG in low-BMI patients or Class I obesity, especially for control or treatment of diabetes [5]. The American Society for Metabolic and Bariatric Surgery (ASMBS) states high-level data exists supporting BS for Class I obesity. Since the LAGB imparts no weight-loss independent variables effecting metabolic and endocrine improvements, the ASMBS now recommends SG or RYGBP for patients with Class I obesity and diabetes.

We regularly evaluate for cardiopulmonary disease, gastroesophageal reflux disease (GERD), nicotine/drug abuse, and mental health disorders. We focus on a patient's ability to comply with an aftercare program. Financial consultation is completed before the patient advances into the program. Specialty consultations are considered based on the patient's medical history. This often includes cardiac, sleep medicine, and hematologic consults if the patient has had unexplained deep venous thrombosis (DVT). Initial studies always include an EKG and baseline laboratory studies. Upper endoscopy is often recommended for those with GERD symptoms, age greater than 50, history of smoking, or high index of suspicion for other foregut problems. We find great value in the bariatric surgeon performing the upper endoscopy, especially when patients require conversion from LAGB to SG. The diagnosis of Barrett's esophagus (BE) needs to be carefully discussed with the patient before proceeding with SG because the 2015 Sleeve Consensus Statement considers it a contraindication. Screening colonoscopy and mammography are recommended per standard healthcare maintenance. We utilize the sleep apnea screening tool called the STOP BANG (Snoring, Tiredness, Observed apnea, Blood pressure, BMI, Age, Neck circumference, Gender) questionnaire. A score of 3 or more indicates need for a sleep study [15]. Patients with continuous positive airway pressure or bilevel positive airway pressure are required to stay overnight for monitoring after surgery in our outpatient center. In addition, patients with a BMI of 50−54.9 require chart review with a bariatric anesthesiologist, and patients with a BMI over 55 require a face-to-face consultation.

Patients are seen 2−4 weeks before the operation date for clearance by the bariatric surgeon with a review of consultations and studies. Patients typically begin their low-carb diet 2 weeks preoperatively to decrease liver volume and improve laparoscopic access and visualization. Patients have two presurgical appointments with the nutritionist, one to perform an assessment for candidacy for BS, and the other to discuss the preoperative and postoperative dietary requirements. Psychological concerns are addressed prior to this appointment. A support person comes to the appointment.

PREOPERATIVE EVALUATIONS

Routine Endoscopy

Routine preoperative use of esophagogastroduodenoscopy (EGD) is debated amongst bariatric surgeons. Those in favor quote high postoperative morbidity of untreated *Helicobacter pylori* [16]; many asymptomatic endoscopic conditions

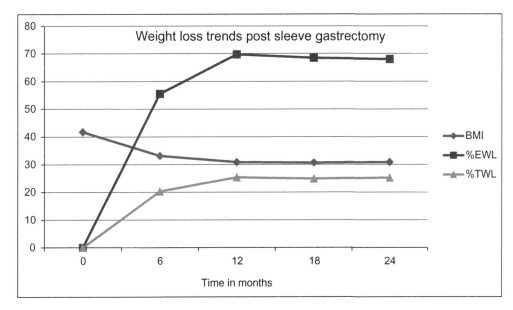

FIGURE 12.1 Weight loss trends after laparoscopic sleeve gastrectomy. All procedures were performed at a single freestanding ambulatory surgical center (Eviva, Seattle, WA). Data was collected from 2013 to 2015. N = 1037. Percent excess weight loss (%EWL) was calculated for the weight the patient would be with a body mass index (BMI) of 25. Revisions such as conversions from a gastric band to a sleeve were excluded from the data set.

are found in the morbidly obese [17]. Those supporting selective EGD endorse studies showing the number of patients required for screening to find one course-altering finding prohibitively high [18] and confounding literature now showing complications after SG unrelated to *H. pylori* [19]. Selective EGD is often utilized for patients with GERD symptoms or those with comorbidities, including age over 50, diabetes, history of smoking, recent weight loss, bleeding, dysphagia, or odynophagia [19]. Any patient with a prior bariatric or gastric operation who is considering revision requires an EGD [20]. Barium studies are helpful in determining anatomy for patients considering primary or revision surgery. Preoperatively identifying hiatal hernias (HHs) and BE may further define the appropriateness of SG or RYGBP (Fig. 12.1).

Gastroesophageal Reflux Disease

Debate continues over whether SG is appropriate in patients with GERD. An expanding body of literature shows significant GERD improvement in patients and minimal de novo post-SG GERD [21], while other studies show increased de novo and/or worsening of preexisting GERD after SG [22,23]. Technique variability related to aggressively identifying and repairing HHs, bougie size (32−40 French), oversewing staple lines, variable amount of retained antrum (distance from pylorus), suturing the omentum to the antrum (omentopexy), and undiagnosed esophageal motility disorders all contribute to the variability in GERD after SG. We know strictures at the incisura, tight HH repairs, pyloric stenosis, tortuosity or twisting of the gastric sleeve, and esophageal motility disorders worsen GERD after SG. Furthermore, many LAGB patients are opting to convert to SG. These gastric sleeves are more prone to strictures, gastric staple line leaks, and retained fundus, especially when previous fundoplication is not completely reduced at the time of the conversion. Prospective studies using esophageal manometry and 48-hour Bravo pH testing may help determine if some patients would do better with RYGBP for severe GERD. We don't advocate routine upper endoscopy in our center. Patients reporting GERD symptoms, bleeding, dysphagia, or new onset abdominal pain, or those on proton pump inhibitors, or who have a long history of smoking or other concerns, should have upper endoscopy (Fig. 12.2).

BARRET'S ESOPHAGUS

RYGBP is still considered the best operation for preventing GERD in morbidly obese patients and BE progression [24]. Currently, expert surgeons consider BE a contraindication to SG [25] because SG does not anatomically eliminate acid and bile exposure to esophageal mucosa, and the portion of stomach utilized to create a neo-esophagus is removed. The Fifth International Consensus Conference for Current Status of Sleeve Gastrectomy (International Federation for the

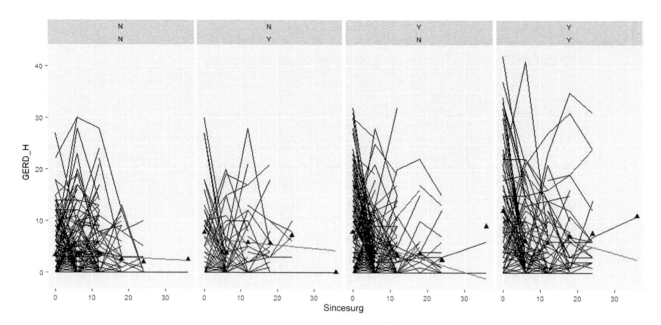

FIGURE 12.2 GERD before and after sleeve gastrectomy. Line graph illustrating the trends of patients reporting heartburn symptoms post-SG. The first row of Y/N indicates whether a hiatal hernia repair was done at the time of surgery. The second row indicates whether the patient was treated with a prescription drug for GERD preoperatively, such as a proton pump inhibitor or an H2 blocker. Patients answered nine questions on the GERD-H questionnaire regarding their heartburn symptoms pre-op and at each follow-up. Data was measured in months postop. Each line represents individual patient's GERD-H scores ($n = 707$).

Surgery of Obesity and Metabolic Disorders) in 2014 found that most experts agree that BE precludes SG; interestingly, most nonexperts disagree. Further studies are needed to better understand the role of SG and control of GERD in patients with short- versus long-segment BE [24,26]. We currently recommend patients with long-segment BE or concerning pathology undergo RYGBP. As further treatment options for BE expand, SG may become more widely accepted in patients with BE (Table 12.1).

ADDITIONAL PRE- AND POSTOPERATIVE EVALUATIONS

All patients undergo nutritional, psychological, and fitness evaluations. The exercise physiologist performs biometric testing to calculate lean body mass, percent body fat, and resting metabolic rate. All patients undergo pre- and postoperative nutrition, fitness, and psychiatric evaluations, helping to insure the appropriateness of BS while achieving reasonable weight-loss goals. Preventing patients with active substance addiction or significant eating disorders from undergoing surgery prior to treatment is imperative. Significant nutrition and fitness teaching occurs pre- and postoperatively, and many commonly held misconceptions about food, exercise, and nutrition, and the establishment of realistic weight-loss expectations, are addressed. Nutritional baseline is assessed with complete metabolic and blood counts, including studies of iron, zinc, vitamins A, B1, B12, and D, and thyroid issues. Testosterone is drawn in males as evidence points to insufficient or poorly sustained weight loss in hypogonadal men [27]. We address age-appropriate screening and healthcare maintenance with routine colonoscopy, mammograms, and other recommended screening tests. Patients are required to stop smoking 8 weeks prior to surgery, and most achieve this goal. Several studies show that smoking increases risks for serious complications after SG, including leaks, stricture, and DVT or portal vein/superior mesenteric vein (SMV) thrombosis [28,29].

Once the workup is completed, we determine if patients are suitable for an inpatient or outpatient setting for SG. Currently, 80% of surgeries at our center are performed in our freestanding outpatient ambulatory surgery center. Patients with sleep apnea are kept overnight as a 23-hour stay and discharged early the next day (Table 12.2).

SURGICAL TECHNIQUE

Patients are put under general anesthesia after receiving low-molecular-weight heparin, antibiotics, and antiemetics, and sequential compression devices are placed accordingly. Patients are secured with gel padding, multiple padded bed

TABLE 12.1 Complications After Sleeve Gastrectomy at Freestanding Outpatient Surgery Center

Complication	Occurrence
Readmission	18 (1.73%)
Bleeding	14 (1.35%)
Transfer from ASC to hospital	12 (1.16%)
Reoperation	10 (1.04%)
Wound infection	8 (0.77%)
Leak	6 (0.58%)
Stricture	6 (0.58%)
Portal vein thrombosis	6 (0.58%)
Pneumonia	4 (0.39%)
Dehydration	2 (0.19%)
Deep vein thrombosis	2 (0.19%)
Myocardial infarction	1 (0.097%)
Deaths	1 (0.097%)

Complications postsleeve gastrectomy. All data is based off results from a free standing ambulatory surgical center (Eviva, Seattle, WA).
$n = 1036$ patients. Revisions were excluded from this study.

TABLE 12.2 Estimates of Bariatric Operations in the United States

	2011	2012	2013	2014
Total	158,000	173,000	179,000	193,000
RNY	36.7%	37.5%	34.2%	26.8%
Band	35.4%	20.2%	14%	9.5%
Sleeve	17.8%	33%	42.1%	51.7%
BPD/DS	0.9%	1%	1%	0.4%
Revisions	6%	6%	6%	11.5%
Other	3.2%	2.3%	2.7%	0.1%

Estimates of the number of bariatric procedures for inpatient data in the United States from 2011 to 2014. No outpatient data is included.

straps, and a footboard. Both arms remain extended on padded arm boards with generous use of gel pads. We don't routinely place urinary catheters. A Nathanson retractor post is secured to the patient's bed on the right. Hook cautery and a vessel sealing device are utilized. Initial peritoneal access is via a transparent 5 mm trocar placed in the left supra umbilical position with a 5 mm 0-degree laparoscope. A steep reverse Trendelenburg position facilitates hiatus and Angle of His exposure. We perform a six-incision, five-trocar SG. After initial trocar placement, 2, 5 mm trocars are placed in the left subcostal region. A 12 mm trocar is placed to the right of the umbilicus at a similar level as the initial trocar. A final 5 mm trocar is placed in the R subcostal region in the midaxillary line, and a small subxiphoid incision is made for the Nathanson retractor. A 5 mm or 10 mm, 30- or 45-degree laparoscope is utilized. A suction irrigator is immediately available; however, gauze strips or absorbent patties are usually sufficient.

If HH repair is indicated, the pars flaccida is divided and a plane between the right crus and esophagus is opened and an intrathoracic circumferential esophageal dissection performed, dividing the gastrophrenic ligament while preserving the vagus nerves. The left crus is identified, and a tunnel created under the esophagus. The gastroesophageal junction is exposed with fat pad dissection from the proximal stomach to decrease retain fundus and Angle of His diaphragm attachments divided. HH is repaired with posterior cruroplasty utilizing a braided permanent suture. Short gastric and omental vessels are now divided with a Harmonic scalpel or Ligasure from approximately 3−6 cm proximal to the pylorus. Bougie (38 French) is now passed through the hiatus under direct laparoscopic guidance and placed along the lesser curvature just proximal to the pylorus. A laparoscopic stapler with a green load (2.0 mm closed staple height) is now utilized, beginning 3−6 cm proximal to the pylorus. Minimum width at incisura is approximately 3 cm. Gold (1.8 mm closed staple height) or blue loads (1.5 mm closed staple height) are used for the mid-to-upper stomach. Thicker stomachs may require an initial black load (2.3 mm closed staple height). Revisions often require black or green loads, allowing for scar and thicker tissues with the potential increase in bleeding. Great care is taken to avoid encroachment on the incisura. We then cinch the bougie up to the Angle of His. The gastroesophageal junction (GEJ) must be avoided during stapling as this creates high leak risk. A 15-round fluted drain is placed in high-risks patients or revisions, and removed approximately 1 week later. We don't currently use staple line reinforcement (SLR) as studies are mixed as to the utility in leak prevention [29−34]. Large meta-analyses and small randomized trials show decreased staple line bleeding with SLR [31,35]. Anatomic omentopexy, performed by reattaching the divided greater omentum to the posterior aspect of the antrum, may prevent reflux and torsion or twist due to adhesions. Evidence of omentopexy benefits is sparse, with small studies showing decreased GERD at the expense of slightly more antinausea medication requirements. We do not perform leak tests as they've failed to change operative outcomes or technique, and several studies have shown trends toward increased leaks when tests are utilized. Selective leak tests or intraoperative EGD is prudent when concern for staple line failure is present. The staple line is judiciously cauterized at crossing vessel and hemorrhage sites. Use of fibrin glue, SLR, and oversewing staple lines all remain topics of debate. Clinically significant postoperative bleeding is usually from the staple line.

RESULTS

Weight loss after SG now approximates results after RYGBP. Outcomes at 5 years show similar reductions in diabetes mellitus, heart disease, sleep apnea, and nonalcoholic steatohepatitis, and improvements in quality of life indices comparing RYGBP and SG. BMI reduction and percent excess weight loss (%EWL) were superior for RYGBP (76.2% + / − 21/7%) compared to SG (63.2 + / − 24.5%) [36]. More recent studies show equivalence between RYGBP and SG in both %EWL and reduction of comorbidities [8,37−39], and several now show superior micronutrient maintenance with SG [39]. Recent multicenter studies comparing SG, RYGBP, and LAGB prove LAGB weight-loss inferiority, higher long-term complication rates, and the need for frequent revision, while showing outcome equivalence between SG and RYGBP [8].

SG results in excellent control or resolution of diabetes in both insulin-resistant and insulin-dependent patients. Surgery is superior to intensive medical therapy for patients with uncontrolled type 2 diabetes. [40] Small, randomized, controlled trials comparing SG and RYGBP have not shown superiority in HGB A1-c or diabetes control in RYGBP [40−45]. Other studies show superior responses to beta-cell function, truncal fat reduction, and improved insulin sensitivity in RYGBP compared to SG [46]. SG decreases LDL, increases HDL, improves insulin sensitivity, and decreases predicted coronary heart disease risk and vascular age [47]. In addition, SG improves markers of systemic inflammation, lipid profiles, and markers of heart disease. Life expectancy is prolonged by 8−10 years with maintained weight loss.

COMPLICATIONS

Complications after SG are divided into early (1−4 days) or late (> 10 days). This is based on evaluation of staple line leaks after RYGBP, as defined by Csendes in 2005 [45]. Overall 30-day, risk-adjusted serious morbidity appears to be lower with SG than RYGBP, with equivalent mortality [48]. Risk calculators show that preoperative history of heart failure, chronic steroid use, male gender, diabetes, total serum bilirubin, BMI, and hematocrit correlate with postoperative comorbidity [48]. However, studies also show even extremely high-risk patients can safely undergo BS [49].

STAPLE LINE LEAK

Over 95% of leaks occur at the proximal staple line. Current leak rate in low-risk patients is around 1%, with most occurring 4−21 days postoperatively. Ischemia and increased pressure gradient from strictures may contribute to this potentially life-threatening complication. Leaks into the mediastinum can cause highly morbid broncho−esophageal, broncho—gastric, and rarely, esophageal—aorto fistulae. If the proximal staple line migrates into the mediastinum, HH repair should be performed with sufficient proximal esophageal dissection to bring the GEJ into the peritoneal cavity without tension. The high-pressure SG leak requires multidisciplinary care, endoscopic interventions, or reoperation. Leak rate has dropped to between 0.9% and 2.4% in large studies [50]. Few techniques such as SLR or oversewing prove to decrease leak rates in randomized controlled trials. Meta-analysis shows some benefit of SLR with bovine pericardium for prevention of leak and bleed [30], however, other studies refute these findings and show higher rates of leak with bovine pericardium [34] and improved outcomes with absorbable polymer membrane strips [29]. Proximal staple line failure is postulated as ischemic and/or pressure-related [31]. The most proximal portion of the staple line was responsible in 75—89% of leaks found in large reviews [50,51], and was more common with smaller than 40fr bougie [50]. Leak tests, either intraoperatively with air, EGD, or methylene blue, or postoperatively with UGI, are now considered superfluous by most centers [51—53]. Risk factors for leak include diabetes [54], BMI, age over 50 [50], bougie size (<40), male gender, history of venous thromboembolism (VTE), smoking history, and coronary artery disease [55].

POSTOPERATIVE HEMORRHAGE

Postoperative hemorrhage occurs in approximately 1—3% of patients, commonly from the staple line, intraluminal, and rarely, other sources such as trocar sites or solid organ injury. Symptoms usually manifest within 24 hours of surgery and include tachycardia, hemoglobin/hematocrit drop, and orthostatic hypotension. Although staple line hemorrhage is usually self-limited, laparoscopic removal of large peritoneal clot burden may improve recovery and decrease risk of abscess. Meta-analysis in eight randomized control trials did show a slight decrease in staple line bleeding with SLR [33], and may decrease operation time due to improved hemorrhage control without further maneuvers such as over sewing, cautery, or fibrin glue [34,35]. Using SLR materials in high-risk patients is cost-effective when reoperation and admission costs are factored in.

Preemptive maneuvers commonly performed to minimize postoperative staple line bleeding include increasing intraoperative systolic blood pressure to over 140 mmHg [56], use of fibrin glue, selective and complete oversewing of the staple line, cautery, and the use of local procoagulants.

Venous Thromboembolism

BS patients are considered higher risk for VTE, with some groups now anticoagulating all patients postsurgery with slightly higher risk of postoperative hemorrhage. Decreased flow through the splenic system has been postulated as causal for portal vein, SMV, and/or splenic vein thrombosis. This potentially disastrous complication occurs 8—45 days after surgery; symptoms range from vague abdominal pain to systemic collapse and death. Higher-risk populations include smokers, users of oral contraception, and those with a personal or family history of VTE. We selectively anticoagulate higher-risk patients (significant mobility restrictions, prior DVT/PE, or positive family history) for 4 weeks. Several strategies exist for postoperative anticoagulation with no consensus on superiority.

LATE COMPLICATIONS

Long-term complications and risk of reoperation after SG are relatively low compared to RYGBP and LAGB. Internal hernias, adhesions, intussusception, and nonhealing ulcers are all seen more frequently in RYGBP. Chronic fistulas to spleen, colon, pancreas, thoracic cavity, or other organs are reported. Micronutrient and potentially devastating thiamine or B12 deficiencies are rare, but certainly do occur, and physicians must be vigilant in patients lost to follow-up or those presenting with prolonged vomiting.

SG anatomy is susceptible to strictures due to angulation at the incisura, twisting of the staple line during creation of the SG, healing decreasing the lumen diameter, or other sources. Symptomatic stenosis is a complication of SG occurring in 0.1—3.9% of patients. The anatomy of the stomach creates a tendency for surgeons to relatively narrow the sleeve at the incisura, resulting in funnel physiology and relative obstruction. Studies now show the use of a variety of achalasia balloons and stents are highly successful at treating even long strictures after SG [57].

Gastric volume, anxiety, higher fat intake, and higher energy intake are all associated with weight regain after SG in some studies. Anemia, iron deficiency, vitamin B12 deficiency, calcium deficiency, and other nutritional complications have all been described after SG. RYGBP is usually considered higher risk for most vitamin and mineral deficiencies.

SG has quickly gained immense popularity in the United States due to excellent results, significant improvement in comorbidities, and fewer risks compared to RYGBP. Debates continue related to BE, SLR, bougie size, omentopexy, revisions, and GERD. Further studies are required to guide clinical decisions.

MINI-DICTIONARY OF TERMS

- *Sleeve Gastrectomy*: A bariatric operation forming a tubular-shaped stomach by removing the fundus of the stomach.
- *Roux-en-Y Gastric Bypass*: A bariatric operation forming a small proximal stomach pouch with an anastomosis to a jejunal limb. Downstream the enteral limb is reconnected to the biliary and pancreatic fluids, forming a "Y" configuration (Roux-en-Y).
- *Duodenal Switch*: A bariatric operation consisting of an SG, then an RYGBP or loop ileal bypass to the duodenum, just beyond the pylorus.
- *Dumping Syndrome*: A phenomenon where ingestion of high-carbohydrate or high-sugar content foods delivered to the gastrointestinal tract quickly induces fluid shifts, rapid changes in gut peptides, and low blood sugars, which create symptoms of nausea, flushing, abdominal cramps, sweating, weakness, rapid heart rate, and fatigue. It is divided into early (30−60 minutes) and late (1−3 hours) dumping syndromes.

KEY FACTS

- SG has become the most commonly performed bariatric operation in the United States.
- SG weight-loss results are approaching those of RYGBP in some centers.
- Mortality after SG is now less than 0.3%.
- SG and other forms of BS provide the most durable weight loss in the morbidly obese.
- Access to BS continues to be a problem for those suffering from the disease of obesity.
- SG can resolve or greatly improve diabetes, hypertension, sleep apnea, depression, hyperlipidemia, osteoarthritis, and many other comorbid conditions.

SUMMARY POINTS

- Obesity is increasing worldwide. Its etiology is poorly understood.
- BS is currently the only proven durable solution for the treatment of morbid obesity.
- SG carries lower complication risk than RYGBP, while providing similar weight-loss patterns.
- Most SG patients have improvement of GERD.
- SG achieves weight loss and comorbid conditions resolution similar to RYGBP.
- Understanding the mechanisms for weight loss after SG may lead to improved understanding of the metabolic cause of obesity.
- The SG patient requires a multidisciplinary approach to weight management.
- SG complications may require conversion to RYBGP. These complications can include chronic dysphagia, vomiting, or regurgitation, resulting from HH, kinking, or gastric stricture.
- For high BMI patients (BMI > 50), DS may be done as a second stage if there is inadequate weight loss.

REFERENCES

[1] F as in Fat: How Obesity Threatens America's Future 2012, a report released by Trust for America's Health [TFAH] and the Robert Wood Johnson Foundation [RWJF], <http://www.cdc.gov/obesity/adult/defining.html.ponce>.

[2] Cummings DE, Arterburn DE, Westbrook EO, et al. Gastric bypass surgery vs intensive lifestyle and medical intervention for type 2 diabetes: the CROSSROADS randomised controlled trial. Diabetologia 2016;59(5):945−53.

[3] Ribaric G, Buchwald JN, McGlennon TW. Diabetes and weight in comparative studies of bariatric surgery vs conventional medical therapy: a systematic review and meta-analysis. Obes Surg 2014;24(3):437−55.

[4] Sumithran P, Prendergast LA, Haywood CJ, et al. Review of 3-year outcomes of a very-low-energy diet-based outpatient obesity treatment programme. Clin Obes 2016;6(2):101−7.

[5] Schauer PR, Bhatt DL, Kirwan JP, et al. Bariatric surgery versus intensive medical therapy for diabetes--3-year outcomes. STAMPEDE investigators. N Engl J Med 2014;370(21):2002−13.

[6] Douglas IJ, Bhaskaran K, Batterham RL, et al. Bariatric surgery in the United Kingdom: a cohort study of weight loss and clinical outcomes in routine clinical care. PLoS Med 2015;12(12):e1001925.

[7] Çetinkünar S, Erdem H, Aktimur R, et al. The effect of laparoscopic sleeve gastrectomy on morbid obesity and obesity-related comorbidities: a cohort study. Ulus Cerrahi Derg 2015;31(4):202−6. Available from: http://dx.doi.org/10.5152/UCD.2015.2993. eCollection 2015 BMC Obes. 2015 Aug 26;2:30.

[8] Biter LU, Gadiot RP, Grotenhuis BA, et al. The sleeve bypass trial: a multicentre randomized controlled trial comparing the long term outcome of laparoscopic sleeve gastrectomy and gastric bypass for morbid obesity in terms of excess BMI loss percentage and quality of life. BMC Obes 2015;2:30.

[9] Svane MS, Bojsen-Møller KN, Madsbad S, Holst JJ. Updates in weight loss surgery and gastrointestinal peptides. Curr Opin Endocrinol Diabetes Obes 2015;22(1):21−8.

[10] Anderson B, Switzer NJ, Almamar A, et al. The impact of laparoscopic sleeve gastrectomy on plasma ghrelin levels: a systematic review. Obes Surg 2013;23(9):1476−80.

[11] Dimitriadis E, Daskalakis M, Kampa M, et al. Alterations in gut hormones after laparoscopic sleeve gastrectomy: a prospective clinical and laboratory investigational study. Ann Surg 2013;257(4):647−54.

[12] Damms-Machado A, Mitra S, Schollenberger AE. Effects of surgical and dietary weight loss therapy for obesity on gut microbiota composition and nutrient absorption. Biomed Res Int 2015;2015:806248.

[13] Mans E, Serra-Prat M, Palomera E, Suñol X, Clavé P. Sleeve gastrectomy effects on hunger, satiation, and gastrointestinal hormone and motility responses after a liquid meal test. Am J Clin Nutr 2015;102(3):540−7.

[14] Clinical guidelines on the identification, evaluation, and treatment of overweight and obesity in adults--the evidence report. National Institutes of Health. Obes Res 1998 (6 Suppl 2):51S−209S. [No authors listed].

[15] Chung F, Yang Y, Liao P. Predictive performance of the STOP-Bang score for identifying obstructive sleep apnea in obese patients. Obes Surg 2013;23(12):2050−7.

[16] Fernandes SR, Meireles LC, Carrilho-Ribeiro L, Velosa J. The role of routine upper gastrointestinal endoscopy before bariatric surgery. Obes Surg 2016;26(9):2105−10. [Epub ahead of print].

[17] Carabotti M, Avallone M, Cereatti F, et al. Usefulness of upper gastrointestinal symptoms as a driver to prescribe gastroscopy in obese patients candidate to bariatric surgery. A prospective study. Obes Surg 2016;26(5):1075−80.

[18] Schigt A, Coblijn U, Lagarde S, et al. Is esophagogastroduodenoscopy before Roux-en-Y gastric bypass or sleeve gastrectomy mandatory? Surg Obes Relat Dis 2014;10(3):411−17. quiz 565−6.

[19] Almazeedi S, Al-Sabah S, Alshammari D, et al. The impact of *Helicobacter pylori* on the complications of laparoscopic sleeve gastrectomy. Obes Surg 2014;24(3):412−15.

[20] Clapp B, Yu S, Sands T, Wilson E, Scarborough T. Preoperative upper endoscopy is useful before revisional bariatric surgery. JSLS 2007;11(1): 94−6.

[21] Traish AM. Testosterone and weight loss: the evidence. Curr Opin Endocrinol Diabetes Obes 2014;21(5):313−22.

[22] Rebecchi F, Allaix ME, Giaccone C, Ugliono E, Scozzari G, Morino M. Gastroesophageal reflux disease and laparoscopic sleeve gastrectomy: a physiopathologic evaluation. Ann Surg 2014;260(5):909−14.

[23] Sheppard CE, Sadowski DC, de Gara CJ, et al. Rates of reflux before and after laparoscopic sleeve gastrectomy for severe obesity. Obes Surg 2015;25(5):763−8.

[24] Kindel TL, Oleynikov D. The improvement of gastroesophageal reflux disease and Barrett's after bariatric surgery. Obes Surg 2016;26(4): 718−20.

[25] Gagner M, Hutchinson C, Rosenthal R. Fifth international consensus conference: current status of sleeve gastrectomy. Surg Obes Relat Dis 2016; Jan 25. pii: S1550-7289[16]00027-7.

[26] Braghetto I, Csendes A. Patients having bariatric surgery: surgical options in morbidly obese patients with Barrett's esophagus. Obes Surg 2016;26(7):1622−6. May 11. [Epub ahead of print].

[27] Khan A, Kim A, Sanossian C, Francois F. Impact of obesity treatment on gastroesophageal reflux disease. World J Gastroenterol 2016;22(4): 1627−38.

[28] Oor JE, Roks DJ, Ünlü Ç, Hazebroek EJ. Laparoscopic sleeve gastrectomy and gastroesophageal reflux disease: a systematic review and meta-analysis. Am J Surg 2016;211(1):250−67.

[29] Carlin AM, Zeni TM, English WJ, Hawasli AA. The comparative effectiveness of sleeve gastrectomy, gastric bypass, and adjustable gastric banding procedures for the treatment of morbid obesity. Michigan Bariatric surgery collaborative. Ann Surg 2013;257(5):791−7.

[30] Steele KE, Schweitzer MA, Prokopowicz G, et al. The long-term risk of venous thromboembolism following bariatric surgery. Obes Surg 2011;21(9):1371−6.

[31] Gagner M, Buchwald JN. Comparison of laparoscopic sleeve gastrectomy leak rates in four staple-line reinforcement options: a systematic review. Surg Obes Relat Dis 2014;10(4):713−23.

[32] Shikora SA, Mahoney CB. Clinical benefit of gastric staple line reinforcement [SLR] in gastrointestinal surgery: a meta-analysis. Obes Surg 2015;25(7):1133−41.

[33] Wang Z, Dai X, Xie H, et al. The efficacy of staple line reinforcement during laparoscopic sleeve gastrectomy: a meta-analysis of randomized controlled trials. Int J Surg 2016;25:145−52.

[34] Timucin Aydin M, Aras O, Karip B, et al. Staple line reinforcement methods in laparoscopic sleeve gastrectomy: comparison of burst pressures and leaks. Obes Surg 2015;25(9):1577−83.

[35] Sroka G, Milevski D, Shteinberg D, et al. Minimizing hemorrhagic complications in laparoscopic sleeve gastrectomy--a randomized controlled trial. Obes Surg 2015;25(3):418−22.

[36] Barreto TW, Kemmeter PR, Paletta MP, Davis AT. A comparison of a single center's experience with three staple line reinforcement techniques in 1,502 laparoscopic sleeve gastrectomy patients. Obes Surg 2014;24(12):2014−20.

[37] Shah SS, Todkar JS, Shah PS. Buttressing the staple line: a randomized comparison between staple-line reinforcement versus noreinforcement during sleeve gastrectomy. Obes Surg 2014;24(10):1617−24.

[38] Zhang Y, Zhao H, Cao Z, et al. A randomized clinical trial of laparoscopic Roux-en-Y gastric bypass and sleeve gastrectomy for the treatment of morbid obesity in China: a 5-year outcome. Acta Gastroenterol Latinoam 2015;45(2):143−54.

[39] Acquafresca PA, Palermo M, Duza GE, et al. Gastric bypass versus sleeve gastrectomy: comparison between type 2 diabetes weight loss and complications. Review of randomized control trails. Acta Gastroenterol Latinoam 2015;45(2):143−54.

[40] Dogan K, Gadiot RP, Aarts EO, et al. Effectiveness and safety of sleeve gastrectomy, gastric bypass, and adjustable gastric banding in morbidly obese patients: multicenter, retrospective, matched cohort study. Obes Surg 2015;25(7):1110−18.

[41] Lee WJ, Pok EH, Almulaifi A, et al. Medium-term results of laparoscopic sleeve gastrectomy: a matched comparison with gastric bypass. Obes Surg 2015;25(8):1431−8.

[42] Schauer PR, Bhatt DL, Kirwan JP, et al. STAMPEDE investigators. Bariatric surgery versus intensive medical therapy for diabetes--3-year outcomes. N Engl J Med 2014;370(21):2002−13.

[43] Keidar A, Hershkop KJ, Marko L, et al. Roux-en-Y gastric bypass vs sleeve gastrectomy for obese patients with type 2 diabetes: a randomised trial. Diabetologia 2013;56(9):1914−18.

[44] Benaiges D, Flores Le-Roux JA, Pedro-Botet J, et al. Sleeve gastrectomy and Roux-en-Y gastric bypass are equally effective in correcting insulin resistance. Int J Surg 2013;11(4):309−13.

[45] Kashyap SR, Bhatt DL, Wolski K, et al. Metabolic effects of bariatric surgery in patients with moderate obesity and type 2 diabetes: analysis of a randomized control trial comparing surgery with intensive medical treatment. Diabetes Care 2013;36(8):2175−82.

[46] Peterli R, Borbély Y, Kern B, Gass M, et al. Early results of the Swiss Multicentre Bypass or Sleeve Study [SM-BOSS]: a prospective randomized trial comparing laparoscopic sleeve gastrectomy and Roux-en-Y gastric bypass. Ann Surg 2013;258(5):690−4.

[47] Schauer PR, Kashyap SR, Wolski K, et al. Bariatric surgery versus intensive medical therapy in obese patients with diabetes. N Engl J Med 2012;366(17):1567−76.

[48] Abou Rached A, Basile M, El Masri H. Gastric leaks post sleeve gastrectomy: review of its prevention and management. World J Gastroenterol 2014;20(38):13904−10.

[49] Young MT, Gebhart A, Phelan MJ, Nguyen NT. Use and outcomes of laparoscopic sleeve gastrectomy vs laparoscopic gastric bypass: analysis of the American College of Surgeons NSQIP. J Am Coll Surg 2015;220(5):880−5.

[50] Aminian A, Brethauer SA, Sharafkhah M, Schauer PR. Development of a sleeve gastrectomy risk calculator. Surg Obes Relat Dis 2015;11(4):758−64.

[51] Aminian A, Jamal MH, Andalib A, et al. Is laparoscopic bariatric surgery a safe option in extremely high-risk morbidly obese patients? J Laparoendosc Adv Surg Tech A 2015;25(9):707−11.

[52] Aurora AR, Khaitan L, Saber AA. Sleeve gastrectomy and the risk of leak: a systematic analysis of 4,888 patients. Surg Endosc 2012;26(6):1509−15.

[53] Sakran N, Goitein D, Raziel A, et al. Gastric leaks after sleeve gastrectomy: a multicenter experience with 2,834 patients. Surg Endosc 2013;27(1):240−5.

[54] Brockmeyer JR, Simon TE, Jacob RK, et al. Upper gastrointestinal swallow study following bariatric surgery: institutional review and review of the literature. Obes Surg 2012;22(7):1039−43.

[55] Sethi M, Magrath M, Somoza E, et al. The utility of radiological upper gastrointestinal series and clinical indicators in detecting leaks after laparoscopic sleeve gastrectomy: a case-controlled study. Surg Endosc 2016;30(6):2266−75.

[56] Rogula T, Khorgami Z, Bazan M, et al. Comparison of reinforcement techniques using suture on staple-line in sleeve gastrectomy. Obes Surg 2015;25(11):2219−24.

[57] D'Ugo S, Gentileschi P, Benavoli D, et al. Comparative use of different techniques for leak and bleeding prevention during laparoscopic sleeve gastrectomy: a multicenter study. Surg Obes Relat Dis 2014;10(3):450−4.

Chapter 13

Laparoscopic Adjustable Gastric Banding

R. Rajendram[1,2,3] and V.R. Preedy[1]

[1]*King's College London, London, United Kingdom,* [2]*Stoke Mandeville Hospital, Aylesbury, United Kingdom,* [3]*King Abdulaziz Medical City, Ministry of National Guard Hospital Affairs, Riyadh, Saudi Arabia*

LIST OF ABBREVIATIONS

EBW   excess body weight
FDA   Food and Drug Administration
GEJ   gastroesophageal junction
ICERs   incremental cost-effectiveness ratios
LAGB   laparoscopic adjustable gastric banding
NHS   National Health Service
QALYs   quality adjusted life years
QOL   quality of life
USA   United States of America

INTRODUCTION

The management of obesity includes conservative therapies (i.e., lifestyle changes and behavioral modification), medical therapies, and surgery. Motivated patients can lose 5–10% of excess body weight with lifestyle changes (e.g., dietary restriction and increased exercise), behavioral modification, and medical therapy [1]. However, these ascetic regimens do not improve the complications of obesity. So few patients persist with these demanding yet ineffective treatments, and unfortunately any weight lost then reaccumulates [2].

Only bariatric surgery achieves sufficient sustained weight loss to reduce the burden of obesity-associated diseases. The various options for surgical management of morbid obesity are classified in Table 13.1. The restrictive techniques are the most commonly performed operations for morbid obesity; of these, gastric banding is the least invasive, most easily reversible, and one of the most commonly performed [3].

The placement of a constricting, silicone band distal to the gastroesophageal junction (GEJ) around the fundus of the stomach induces weight loss by reducing oral intake. Initially gastric bands were fixed (i.e., not adjustable), so uptake was poor. The adjustable bands that are in current use are silicone rings with an inner inflatable balloon. This balloon can be inflated or deflated by adding or removing saline via a subcutaneous access port to adjust the size of the stoma.

The adjustable gastric band is usually inserted laparoscopically. This operation is now often performed as a day-case procedure. Laparoscopic adjustable gastric banding (LAGB) is a reversible, restrictive procedure that avoids the complications of the techniques that cause malabsorption.

Following LAGB procedures, patients require frequent postoperative follow-up to adjust the tightness of the band. Patients also require advice on nutrition, such as micronutrient supplementation, until target weight loss is achieved. To ensure that this weight loss is sustained, patients require ongoing follow-up.

This intervention induces sustained weight loss by restricting calorie intake. Patients must modify their eating habits; consuming smaller amounts, more slowly. The introduction of LAGB has transformed the management of obesity because it is safe and effective. The perioperative mortality is 0–0.1% and percentage of excess weight loss (%EWL) is 50–60% [3–6]. Furthermore, the weight loss achieved by LAGB is associated with significant improvement in obesity-related comorbidities.

TABLE 13.1 The Options for Surgical Management of Morbid Obesity

Restrictive	Adjustable gastric banding
	Fixed gastric banding
	Vertical band gastroplasty
Malabsorptive	Duodenal switch
Restrictive/resective	Sleeve gastrectomy
Restrictive/malabsorptive	Roux-en-Y gastric bypass Biliopancreatic diversion with duodenal switch

This table classifies the techniques for the surgical management of morbid obesity. Many of these techniques may be performed open or laparoscopically.

EVOLUTION OF THE LAGB TECHNIQUE

Szinicz and Schnapka first proposed adjustable gastric banding in the early 1980s [7], and in 1986 Kusmak performed this as an open procedure [8]. However, the open technique did not attract much interest. Then, as new technologies allowed complex laparoscopic surgery to flourish, Belachew et al. was able to introduce the Lap-Band system into clinical practice in 1993 [9].

LAGB was attractive because the weight loss was controlled by adjustment of the band, which could be inserted laparoscopically without any bowel resection or anastomoses. As a result, LAGB rapidly became the most common bariatric procedure performed throughout the developed world. However, there were significant gaps in the understanding of the operative technique and the intensive postoperative follow-up that was required [6]. As a consequence, the incidence of complications was initially quite high.

Significant advances have been made in this millennium. Over 1000 peer-reviewed papers have contributed to the evolution of the process of LAGB. The current version of the Lap-Band is known as the Lap-Band Advanced Platform, but several adjustable gastric bands are now available. The two approved by the US Food and Drug Administration (FDA) are the Lap-Band (Apollo Endosurgery, Texas) and the Realize Band (Ethicon Endosurgery, Ohio) [6]. Others (e.g., Mid-Band, Heliogast, and Minimizer) lack sufficient efficacy data to be approved by the FDA, and so their uptake has been poor [6].

The best outcomes from LAGB are achieved when patients are treated by an experienced surgical team that uses the best technique to insert a well-designed LAGB device, and then provides intensive postoperative management [3–6]. The absence of any one of these factors results in poor outcomes [3–5].

The initially high incidence of postoperative complications after LAGB (e.g., gastric herniation or prolapse) offset its early promise. These complications were due to the initial perigastric technique for LAGB, which was reported by Belachew et al. [9]. The perigastric technique places the gastric band low on the stomach [9]. This is a complicated technique that requires perigastric dissection along the lesser curve of the stomach into the lesser sac [4,10]. The extent of perigastric dissection depends on the width of the stomach and where the surgeon starts to dissect [4,10]. Unfortunately, the resulting postoperative anatomy can allow the posterior aspect of the stomach to slip through the gastric band and cause a symptomatic gastric prolapse [10].

Although this groundbreaking procedure has advanced significantly in this millennium, it continues to evolve as data on the outcomes of the various techniques of LAGB accumulate. For example, using the pars flaccida approach for LAGB results in far fewer complications than the perigastric technique [4,10].

The dissection for the pars flaccida technique begins at the anterior aspect of the lowermost fibers of the right crus of the diaphragm and exposes the left crus. The dissection passes along the left crus to join an opening in the peritoneum at the angle of His [4,6]. This technique requires minimal dissection and places the band in a high position near the GEJ away from the body of the stomach and out of the lesser sac [4,6].

The pars flaccida technique is relatively straightforward, safe, reproducible and easy to teach. It also reduces the incidence of gastric herniation and prolapse. Placement of the band away from the lesser sac and peristalsing stomach minimizes dissection of the stomach's attachments and gastric-to-gastric suturing. Not adding fluid to the band until 6 weeks postoperatively reduces the incidence of prolapse [4,6]. The risk of erosion of the band into the stomach is markedly reduced by not suturing the gastric wall to the crura of the diaphragm or pushing it on to the buckle of the band [4,6,10].

The band is adjusted by percutaneous injection of saline into the access port. The access port should therefore be placed in a position that is easy to inject. If inserted on the anterior rectus sheath the gastric band can be adjusted by specialist nurses in outpatient clinics [4,6]. This allows frequent changes to the band, saving time and reducing cost. The rate of weight loss is better controlled and therefore increases without causing vomiting and heartburn.

THE PROFILE OF WEIGHT LOSS

The ability to adjust the band frequently allows smooth progression toward the target weight. Overtightening and excessive early weight loss can result in nutritional deficiencies and should be avoided. Removal of fluid from the band releases the restriction. This may be required during pregnancy, major illness or operation, or remote travel.

Graded increases of gastric restriction by LAGB allows slow, steady, asymptomatic weight loss [6]. The %EWL characteristically increases steadily for the first 2 years. The patient's weight should then remain stable at around 50 %EWL over the next 4 years [6,11]. The ability to adjust the band many years after insertion should result in a maintenance of weight loss that cannot be achieved with restrictive procedures [6,11].

Mechanism of Weight Loss

The gastric band is usually placed around the cardia at the top of the stomach, close to the GEJ. In this position the gastric band causes satiety (the state of not being hungry) and satiation (resolution of hunger with eating) [11,12].

The compression of the band on the wall of the stomach should be adjusted until the patient's appetite, interest in food, and concerns about not eating all decrease. Although the patient may feel hungry occasionally, they should generally feel satiated [11,12].

When the band is correctly adjusted, each morsel of food passing across the band is squeezed by oesophageal peristalsis. This increases the pressure on that part of the wall of the stomach and reduces any hunger that the patient may have felt. So satiation is induced by small amounts of food. This should ideally reduce the patient's calorific intake to 1000−1200 kcal/day [11,12].

HEALTH, QUALITY OF LIFE, AND ECONOMIC ANALYSIS

Reduction of weight is one of the most effective medical therapies currently available. Weight loss results in significant improvements in a range of obesity-related diseases (e.g., diabetes mellitus, hypercholesterolemia, and hypertension) [6,10]. However, more importantly from the patient's perspective is the effect of LAGB on quality of life (QOL).

Quality of Life

Obesity is associated with poor health-related QOL as determined by the Medical Outcomes Study Short Form-36 [6,13]. LAGB results in significant improvement in health-related QOL [6,13].

Economic Analysis

In 2006 the direct cost to the National Health Service (NHS) of overweight and obese patients (BMI ≥ 22) was estimated at £3.2 billion [14]. The Parliamentary briefing paper Obesity Statistics 2016 updated this to £4.2 billion in 2007 with projected costs in 2015, 2025, and 2050 to be £6.3, £8.3, and £9.7 billion, respectively [15]. The majority of this expenditure was for the treatment of obesity-related diseases (e.g., stroke, coronary heart disease, hypertensive disease, and diabetes mellitus). The financial cost to society at large is much higher.

The cost of the operation for LAGB was reported to be approximately £4300 [13]. Approximately £1200−2000 of this cost is for consumables (i.e., gastric bands and other single-use equipment). Preoperative assessments cost around £1100, and the postoperative follow-up costs around £1900. So the total cost of LAGB is around £8800 [13]. In comparison, the cost of laparoscopic gastric bypass was estimated at £11,500. These figures are similar to those currently reported by NHS England [16].

Management of morbid obesity (BMI > 40) with LAGB is more expensive than nonsurgical management, but results in better outcomes in terms of quality adjusted life years (QALYs). The incremental cost-effectiveness ratios (ICERs) ranged from £2000−4000 per QALY [13]. LAGB management of moderate-to-severely obese patients (BMI of 30−40) with Type 2 diabetes is more expensive than nonsurgical management, but the outcomes are better. These ICERs are cost-effective from the perspective of decision-making in the NHS [17].

FUTURE CHALLENGES

Being safe, cost-effective, and generally acceptable to patients, LAGB could have a major impact on the global pandemic of obesity. However, there are still many challenges to overcome. Good outcomes require highly skilled laparoscopic surgical teams to insert the best-designed bands into carefully selected patients using the best technique. These bands must then be adjusted frequently and correctly.

A robust network of allied healthcare professionals are required to provide the advice, motivation, and support patients require before LAGB and intensive ongoing postoperative care, with adjustment of the band after the surgery.

The technique of LAGB and the protocols for postoperative management are still evolving. Measurement of the effectiveness of current practice and technologies is required to drive further developments.

CONCLUSION

LAGB induces significant weight loss, thus improving health, reducing the comorbidities of obesity, and increasing QOL. It is a short, laparoscopic procedure that can be performed as a day case because it does not cause significant pain or limitation. The key to its long-term effectiveness is the ability to adjust the band during postoperative follow-up.

MINI-DICTIONARY OF TERMS

- *Adjustable Gastric Band*: A constricting, silicone band inserted around the stomach close to the GEJ that induces weight loss by reducing oral intake. An inner balloon can be inflated or deflated by adding or removing saline via a subcutaneous access port.
- *Incremental Cost-effectiveness Ratios (ICERs)*: This statistic is used to summarize the cost-effectiveness of medical interventions. It is the difference in cost of two possible interventions divided by the difference in their effect.
- *Laparoscopy*: This minimally invasive abdominal surgery uses ports inserted through the abdominal wall via small incisions. The abdominal cavity is inflated with carbon dioxide before a camera and instruments are passed through these ports to visualize the intraabdominal organs and perform the operation.
- *Quality Adjusted Life Years (QALYs)*: This is a generic measure of burden of disease that is used to assess the cost-effectiveness of medical interventions. It depends on both the quality as well as quantity of life. One QALY equates to 1 year of good health.
- *Quality of Life (QOL)*: These are generally self-perceived, quantifiable measures of well-being. Very often they have various domains, such as those relating to physical and emotional health. A variety of questionnaires exist for measuring and quantifying QOL in numerical terms. At the end of the questionnaire a numerical score is derived.

KEY FACTS

- QOL measures are an attempt to place numerical indices or scores on the individual's perception of their daily lives.
- There are many different types of QOL questionnaires, some of which have been specifically developed for particular diseases such as asthma, HIV, age-related macular degeneration, and even bariatric surgery (The Bariatric QOL Index for example).
- Well-known questionnaires include the World Health Organization Quality of Life (WHOQOL), Euro Quality of Life Questionnaire (EQ-5D), and Short Form Health Survey (SF-36).
- The Short Form Health Survey (SF-36) has a total of eight areas covered by 36 different questions. Questions in the SF-36 pertain to health, changes in health, relationships between health and daily activities, impact of health on work, emotional problems, social activities, pain, and other areas.
- Bariatric surgery improves QOL measures.

SUMMARY POINTS

- Elective bariatric surgery generally has a low risk of complications.
- LAGB is safe, cost-effective, and acceptable to patients.

- LAGB induces satiety and satiation, which results in loss of weight.
- The excess weight loss after LAGB is sustained, and reduces the incidence of obesity-related diseases.
- The best outcomes require treatment by an experienced surgical team who insert a well-designed LAGB device using the best surgical technique, and then provide intensive postoperative management.
- Management of bariatric patients requires institutional investment in specialized equipment and training of staff.

REFERENCES

[1] Lau DC, Douketis JD, Morrison KM. Canadian clinical practice guidelines on the management and prevention of obesity in adults and children. CMAJ 2006;2007(176):S1−13.

[2] Birmingham CL, Muller JL, Palepu A, et al. The cost of obesity in Canada. CMAJ 1999;160:483−8.

[3] Dixon JB, O'Brien PE. Health outcomes of severely obese type 2 diabetic subjects 1 year after laparoscopic adjustable gastric banding. Diabetes Care 2002;25:358−63.

[4] Fielding GA, Allen JW. A step-by-step guide to placement of the LAP-BAND adjustable gastric banding system. Am J Surg 2002;184:26S−30S.

[5] Fisher BL, Schauer P. Medical and surgical options in the treatment of severe obesity. Am J Surg 2002;184:9S−16S.

[6] O'Brien PE. Surgical treatment of obesity. Endotext, <http://www.endotext.org/section/obesity/> [accessed 29.08.16].

[7] Szinicz G, Schnapka G. A new method in the surgical treatment of disease. Acta Chirirgica Austrica 1982;(Suppl. 43):43.

[8] Kuzmak LI. A review of seven years' experience with silicone gastric banding. Obes Surg 1991;1:403−8.

[9] Belachew M, Legrand M, Vincenti V, et al. Laparoscopic placement of adjustable silicone gastric band in the treatment of morbid obesity: how to do it. Obes Surg 1995;5:66−70.

[10] O'Brien PE, Dixon JB. Laparoscopic adjustable gastric banding in the treatment of morbid obesity. Arch Surg 2003;138:376−82.

[11] Dixon AF, Dixon JB, O'Brien PE. Laparoscopic adjustable gastric banding induces prolonged satiety: a randomised blind crossover study. J Clin Endocrinol Metab 2004;90(2):813−19.

[12] Colles SL, Dixon JB, O'Brien PE. Hunger control and regular physical activity facilitate weight loss after laparoscopic adjustable gastric banding. Obes Surg 2008;18:833−40.

[13] Picot J, Jones J, Colquitt JL, Gospodarevskaya E, Loveman E, Baxter L, et al. The clinical effectiveness and cost-effectiveness of bariatric (weight loss) surgery for obesity: a systematic review and economic evaluation. Health Technol Assess 2009;13:41.

[14] House of Commons Health Committee. Obesity 2004. Third Report of Session 2003−04, vol. 1. London: UK. House of Commons.

[15] Baker C, Bate A. Obesity statistics. Briefing paper 3336, February 2016. House of Commons UK.

[16] NHS UK. Weight loss surgery, <http://www.nhs.uk/Conditions/weight-loss-surgery/Pages/introduction.aspx> [accessed 2.09.16].

[17] Allender S, Rayner M. The burden of overweight and obesity-related ill health in the UK. Obes Rev 2007;8:467−73.

Chapter 14

Biliopancreatic Diversion

G. Nanni and G. Boldrini

Catholic University Medical School, Roma, Italy

LIST OF ABBREVIATIONS

BPD   biliopancreatic diversion
BPD-DS   biliopancreatic diversion duodenal switch
ERCP   endoscopic retrograde cholangiopancreatography
EWL   excess weight loss
GLP1   glucagon-like peptide 1
GIP   glucose-dependent insulinotropic polypeptide
RYGBP   Roux-en-Y gastric bypass
SG   sleeve gastrectomy
SADI-S   single anastomosis duodenal—ileal bypass with SG

INTRODUCTION

Since the 1970s, biliopancreatic diversion (BPD) has represented a valid malabsorptive procedure for the surgical treatment of obesity. It combines the best results in terms of weight loss with an acceptable rate of nutritional and metabolic adverse effects and complications [1].

In contrast to the aggressive and uncontrolled malabsorptive effect of previously performed jejuno—ileal bypass, BPD produces at the same time a major reduction in fat absorption and an adequate absorption of carbohydrates; furthermore, due to the technical adjustments of the last 20 years, it reduces the risk of protein malnutrition.

Restrictive procedures, and all the surgical measures that hamper oral intake of food, such as gastroplasty, gastric banding, and gastric bypass (in use between the 1960s and 2000s), have proved to a large extent to be ineffective due to the progressive rise of food consumption and consequent weight regain. On the contrary, due to its well-balanced malabsorptive effect, BPD stands as the most effective procedure, which is most evident in patients with a body mass index (BMI) > 50 [2].

SURGICAL TECHNIQUE

BPD was devised by Nicola Scopinaro in 1976 [3] to reduce the absorption of ingested calories, mainly from fat, altering the contact between biliopancreatic secretions and food, and thus limiting nutrient absorption. This result is obtained by preventing the transit of ingested food through the duodenum by performing a distal resection of the stomach. A long limb Roux-en-Y reconstruction is realized, which leaves a short common loop in which nutrients and biliopancreatic secretions mix. The branch of the Y sutured to the resected stomach is the alimentary limb; the biliopancreatic loop carries digestive juices coming from the duodenum; and juices and food mix in the common loop.

At first, BPD was referred to as biliopancreatic bypass (1974—85) [4], similar to the previous jejuno—ileal bypass. In a certain way it can be considered analogous to the Roux-en-Y gastric bypass (RYGBP) because the ingested food skips part of the gastrointestinal tract. RYGBP was originally devised to reduce the role of gastric volume in the nutritional status, excluding the stomach as well as the duodenum from food transit. Due to the lengthening of the duodeno—jejuno—ileal limb of the Y reconstruction (biliopancreatic branch), with the aim of reducing the mix

Metabolism and Pathophysiology of Bariatric Surgery.
© 2017 Elsevier Inc. All rights reserved.

119

of food and digestive juices (short common limb), BPD can be defined a biliopancreatic bypass. The classic definition, "biliopancreatic diversion," is more effective in elucidating the underlying physiological principles.

From a technical point of view, BPD can be performed in two ways. In the original form, Scopinaro resected the distal part of the stomach with a Roux-en-Y reconstruction. A more recent (1990) American/Canadian version called the "duodenal switch" biliopancreatic diversion (BPD-DS) is frequently utilized: a pylorus-preserving vertical subtotal gastrectomy is performed with a postpyloric Roux-en-Y reconstruction, which prevents food transit through the duodenum.

Scopinaro's Original Procedure

The procedure is a distal gastric resection, similar to the Billroth 2 procedure for the cure of peptic ulcer, but with a Y reconstruction according to Roux. With respect to the original version, a few details in the gastric resection, as well as in the ileal reconstruction, have changed over the years.

At the beginning, the gastric resection was particularly wide in order to limit food consumption during the first postoperative weeks, thus taking advantage of the decrease in appetite and the postcibal syndrome. To obtain better results, the residual volume of the gastric pouch was selected according to preoperative weight and other features of the patient, performing the so-called "ad hoc stomach—between 200 and 500 mL—BPD" (AHS BPD) [4].

With the improvement of the surgical experience, it became clear that a greater residual volume of the gastric pouch, set between "less than a hemigastrectomy and a little more than an antrectomy," clearly reduced the incidence of malnutrition, particularly proteins, without relevant untoward effects on weight loss [5,6].

In the Roux reconstruction, Scopinaro modified the length of the limbs in the first years of his practice. At the beginning, he chose a short biliopancreatic limb (30 cm) with a long alimentary tract and 100 cm of the common one. Nowadays, a 50 cm common limb and a 200 cm alimentary tract characterize the latest and extensively utilized version. For many years surgeons were determined to counteract protein malnutrition at the expense of preserving a too-low body weight; for this reason, a slightly longer alimentary limb (between 200 and 250 cm) and common limb (between 50 and 75 cm) have been preferred [7].

The classic version of this procedure includes cholecystectomy due to the high incidence of gallstone formation after surgery [8].

The Duodenal Switch Procedure

Since the 1980s, great attention has been paid to long-term postoperative malnutrition, and thus several surgeons have chosen to build up a longer common limb (100 cm), as well as to modify the gastric resection.

Inspired by the De Meester procedure [9] for the treatment of alkaline gastritis, in 1990 Marceau [10] followed Hess's idea [11] of performing a parietal gastrectomy, the so-called "sleeve gastrectomy" (SG), in order to counteract the gastrointestinal symptoms and malabsorption caused by dumping syndrome. The aim of the so-called "duodeno—ileal switch" was to reduce gastric size, prevent peptic ulcer, and preserve an adequate digestive role for the stomach.

After a brief experience in which the duodenal switch was performed by stapling an uninterrupted duodenum, a great number of cases of weight regain were observed; these were due to reopening of the sutured duodenum. It was decided to definitely interrupt the duodenum. In the mid-1990s, the BPD-DS technique expanded. Nowadays, this procedure follows a definite standard: it requires a 100 mL SG, with an antral pouch 5 cm proximal to the pylorus, a 250 cm alimentary tract sutured to the subpyloric duodenum, and a 100 cm common limb. Transmesocolic or antecolic position of the alimentary limb is considered irrelevant, as in the Scopinaro BPD [5].

Laparoscopic versus Laparotomy Procedure

Since the early 2000s, laparoscopic techniques have been extended to bariatric surgery and received great notoriety, mainly because of the expansion of gastric banding (Adjustable Silicon Gastric Banding). The goal of the laparoscopic approach is the reduction of intra- and postoperative complications, which result more frequently after bariatric procedures than after surgery in nonobese patients. Laparoscopic techniques allow a net reduction of pulmonary complications, ventral hernias, wound infections, postoperative pain, and adhesions; furthermore, the overall hospital stay is shortened. On the contrary, there has been a rise in the incidence of bleeding, stenosis of anastomosis, bowel obstruction, and cardiorespiratory alterations due to pneumoperitoneum [12,13].

Both the standard Scopinaro BPD and the DS-BPD can be performed using the laparoscopic technique [14,15], but limitations are found in patients with the highest BMI values.

Recently, in most centers, cholecystectomy is not included in the laparoscopic approach, and is limited to open procedures [16]. Cholelitiasis is frequently found in obese patients; furthermore, after bariatric surgery there is a great chance of developing gallbladder stones (28–40% of patients) [17]. For these reasons, cholecystectomy was once routinely included in the bariatric procedure. With the expansion of laparoscopy, a definite mindset developed: restrict concomitant cholecystectomy to patients with biliary stones (ideally if symptomatic) [18]. As a matter of fact, cholecystectomy was usually considered for patients who were developing gallstones after the bariatric procedure [19,20]. It has to be pointed out that after BPD, as well after RYGBP, treatment for the complications of biliary stones, such as common duct lithiasis and acute pancreatitis, becomes impossible through endoscopic retrograde cholangiopancreatography (ERCP). In the presence of common bile duct stones, a few authors recommend percutaneous transgastric ERCP [21].

Recently, the laparoscopic single anastomosis technique versus Roux-en-Y reconstruction has gained popularity, either in patients submitted to GBP (the so called "mini-gastric bypass") [21], or in subjects treated with the BPD-DS single anastomosis duodenal–ileal bypass with SG (SADI-S technique) [22]. Notably, the single anastomosis procedure ("omega reconstruction") had already been utilized by Scopinaro in his early work [23]. Nowadays its use is under evaluation, particularly as revisional bariatric surgery after failed SG [24]. As a matter of fact, this technique still awaits standardization in terms of length of the biliopancreatic limb in relation to the alimentary/common tract.

REMARKS

Weight Loss

BPD, in any of its technical variations, stands as the most successful procedure in bariatric surgery. In long-term follow-ups, it achieves the best results in terms of weight loss, either expressed as excess weight loss (%EWL) or as the percentage of patients reaching a defined value of BMI (<40, <35, or as low as <30). In his personal experience on a series of 3000 patients with a 25-year follow-up, Scopinaro demonstrated that weight loss remains unchanged at around 70 %EWL [25]. We obtained similar results in our series of 1235 patients with a 23-year follow-up [26].

Bariatric procedures were compared in a 2004 Buchwald meta-analysis [27]. Results suggested that BPD was better in terms of weight loss.

The relevant role of BPD in the surgical treatment of morbid obesity is also paradoxically enhanced by the recent expansion of SG [28], the first step of the BPD-DS, for its successful results as an isolated procedure. The frequent failure of SG alone underscores the ability of BPD to almost always result in substantial weight loss [29].

An 18-month follow-up has recently proved that 60% of SG patients maintain a BMI value <35 versus 85% of the BPD-DS patients. [30]. Marceau demonstrated that at a 5-year follow-up in patients with BMI > 50, the %EWL was 32% for SG versus 90% for BPD-DS [31]; 86% of patients with initial BMI < 50 reached a value <35, while 78% of those with BMI > 50 reached a value of <40 [32]. No other procedure has revealed similar achievements if the results were linked to a definite final BMI value. Mean values either of BMI or of %EWL were unreliable, as their distribution usually did not fall under a Gaussian curve.

In his well-documented study on a large series of BPD-DS patients with a 20-year follow-up, Marceau [5] defined a BMI of <35 or <40 as a target that is "realistic, reasonable and coincident with patient's expectation" in the case of preoperative values respectively lower and higher than 50.

In a series from our institution, we showed that at the 2-year follow-up, the percentage of patients reaching BMI values of <35 and <30 were higher after BPD with respect to RYGB. [6]. These values were consistent with mild obesity or overweight status; thus, they underlined in a more accurate and pragmatic way the target of bariatric surgery and delineated the success of the procedure.

The growing application of the single anastomosis duodenal–ileal bypass as a revisional measure after failed SG (SADI-S technique) [33,34] further underlined the clinical relevance of BPD-DS.

Comorbidities

Comorbidities related to obesity commonly represent a strong indication to bariatric surgery, even for BMI < 40. Many papers confirm that BPD in its different versions allows the best results in terms of disappearance or improvement. The 2004 and 2009 Buchwald meta-analyses [27,35] and the 2014 Cochrane review [36] showed that, in treating associated conditions, the results of BPD outscored those of RYGBP of 15−25%. Even better results can be found by comparing BPD with all the other bariatric procedures. Greater impact can be observed on type 2 diabetes, hypertension, hyperlipidemia, hyperuricemia, and sleep apnea syndrome. Diabetes being the main determinant of cardiovascular risk in obesity, a great number of published papers deal with the treatment of type 2 diabetes, which is detected in up to 15% of bariatric surgery candidates.

The mechanism of diabetes remission after bariatric surgery was extensively investigated after evidence that the changes in gastrointestinal transit, induced by BPD and RYGBP, were effective in large measure independent of the weight loss [37,38]. Changes in gastrointestinal hormones, mostly incretins like glucagon-like peptide 1, gastric inhibitory polypeptide, peptide YY (PYY), and ghrelin, played a role in diabetes control, either if the main effect was conditioned by the lack of duodenal transit (foregut hypothesis), or if it was induced by direct access of partially digested food into the distal bowel (hindgut hypothesis) [39]. At our institution we proved that BPD has an early impact on diabetes [40], that the resolution of its comorbidities can be predicted [41], and that its complications can be improved or reversed [42]. These results induced several authors to perform bariatric surgery even in slightly obese or overweight diabetic patients; the benefits were significant [43−45]. Recently we showed with a 5-year follow-up that BPD is more effective with respect to medical treatment, as well as to RYGBP, in the control of diabetes [46]. Also, the SADI-S variation of BPD-DS [47] showed good results in terms of resolution of diabetes.

Adverse Events and Nutritional Follow-Up

Early and late surgical complications of BPD such as bleeding, intestinal obstruction, and stomal ulcer show similar patterns with respect to other bariatric procedures.

In the medium- and long-term control of bariatric patients, the nutritional and metabolic alterations represent a great concern. From this point of view, BPD requires a tight follow-up to prevent protein malnutrition, multifactorial anemia, and bone demineralization, mostly due to secondary hyperparathyroidism. To a large extent, these alterations are similar to those observed after RYGBP [48], being due to the lack of food transit through the duodenum and jejunum.

BPD implies a greater rate of intestinal malabsorption. Partial impairment of gastric function, skipping the duodenum and jejunum transit, shortening of the ileum, and the presence of steatorrhea impair the absorption of iron, vitamin B12, folic acid, calcium, vitamin D, and other fat-soluble vitamins, and to a certain extent, proteins. Moreover, in a certain number of patients, malabsorption causes meteorism and foul smelling stools, which can partially regress with adequate alimentary habits.

No differences have been observed between open, more frequently performed methods and laparoscopic procedures [8].

Revisional surgery is not infrequent after bariatric surgery. The main causes of reoperation can be failure in weight loss, major nutritional imbalances, and gastric or intestinal transit alterations. In contrast, the recent technical adjustments of BPD have made reoperative surgery sporadic, mainly consisting of elongation of the common limb to correct serious protein malnutrition, which is observed in 1−2% of patients [49].

Potential nutritional and metabolic derangements imply adequate lifelong prevention through the administration of multivitamins and minerals, and in some cases also via the parenteral route (according to European Guidelines) [50]. The greater problems concern multifactorial anemia, hyperparathyroidism, and fat-soluble vitamin deprivation states. Supportive measures require strict periodic, clinical, and laboratory monitoring.

CONCLUSIONS

BPD offers a wide therapeutic spectrum. Modulating the length of intestinal limbs and/or the volume of the gastric pouch, it is possible to tailor the procedure to specific features in the individual patient.

A physiological approach to this procedure aims to attain the best results in terms of weight loss and quality of life, thus minimizing nutritional problems and side effects.

MINI-DICTIONARY OF TERMS

- *Gastrectomy*: Partial or total surgical removal of the stomach, either to treat some diseases (cancer or peptic ulcer), or to decrease weight (reduction of gastric volume or prevention of the normal duodenal transit of food to create malabsorption).
- *Gastrointestinal Anastomosis*: Connection between two different tracts of the gastrointestinal system to reroute its contents.

- *Colecystectomy*: Surgical removal of the gallbladder, usually containing biliary stones (cholelitiasis).
- *Duodenal Switch*: Recent definition of a surgical bypass that realizes a BPD, preserving the pyloric sphincter in place.
- *Laparoscopy*: A limited-access surgical technique that allows operations to be performed through minor abdominal incisions and an optical device, with the aim of reducing postoperative pain and complications, and shortening of the hospital stay.

KEY FACTS

- *Bariatric Surgery*: Since the 1950s, it has been a surgical treatment for obesity showing, in comparison to other therapies, the best results in terms of weight loss and control of comorbidities.
- *Malabsorptive Procedures*: Procedures limiting nutrient absorption (particularly fat) are more effective in terms of durable weight loss with respect to those limiting food ingestion.
- *Biliopancreatic Diversion*: A kind of controlled malabsorption obtained through the separation of food and digestive secretions, and readily adjustable to specific patient features.
- *Diabetes Treatment*: One of the most relevant results of bariatric surgery, obtained not only through weight loss, but also to a large extent through modification of gastrointestinal hormones induced by variations in food transit.
- *Nutritional Complications*: Major weight loss (even through strict diet) exposes patients to deprivation of macro- and micronutrients; therefore, bariatric surgery requires close follow-up and adequate integrations.
- *Laparoscopy*: A limited-access surgical technique that allows operations to be performed through minor abdominal incisions, with the aim of reducing postoperative pain and complications, and shortening of hospital stays.

SUMMARY POINTS

- BPD represents a malabsorptive procedure in the surgical treatment of obesity.
- BPD was devised by Scopinaro to reduce the absorption of ingested calories, mainly from fat, thus limiting the contact between biliopancreatic secretions and food.
- Limitation of the contact between biliopancreatic secretions and food is achieved by bypassing duodenal–jejunum transit.
- BPD can be performed in two ways, the original Scopinaro way and the "duodenal switch," a pylorus-preserving gastrectomy.
- BPD combines the best results in weight loss with an acceptable rate of nutritional adverse effects.
- BPD allows resolution or improvement of comorbidities, particularly diabetes. BPD is effective in diabetes treatment, not only reducing weight, but also changing gastrointestinal transit.
- In long-term follow-up, BPD requires tight control to prevent protein malnutrition, anemia, and secondary hyperparathyroidism.
- Adequate lifelong prevention through the administration of multivitamins and minerals is necessary.
- BPD offers a wide therapeutic spectrum. Modulating the length of the intestinal limbs and/or the volume of the gastric pouch, it is possible to tailor the procedure to specific features in the individual patient.
- A physiological approach to this procedure aims to reach the best results in terms of weight loss and quality of life, thus minimizing nutritional problems.

REFERENCES

[1] Richards W, Schirmer BD. Morbid obesity. In: Townsend CM, editor. Sabiston textbook of surgery. Saunders/Elsevier Pub; 2008. p. 399–430. [chapter 17].
[2] Risstad H, Søvik TT, Engström M, Aasheim ET, Fagerland MW, Olsén MF, et al. Five-year outcomes after laparoscopic gastric bypass and laparoscopic duodenal switch in patients with body mass index of 50 to 60. A randomized clinical trial. JAMA Surg 2015;150:352–61.
[3] Scopinaro N, Gianetta E, Civalleri D, Bonalumi U, Bachi V. Bilio-pancreatic bypass for obesity: II. Initial experience in man. Br J Surg 1979;66:618–20.
[4] Scopinaro N, Adami GF, Marinari GM, Traverso E, Papadia F, Camerini G. Biliopancreatic diversion: two decades of experience. In: Deitel M, editor. Update: surgery for the morbidly obese patient. Pub. Mothersill Printing Inc; 2000. p. 227–58. [chapter 23].
[5] Biron S, Hould FS, Lebel S, Marceau S, Lescelleur O, Simard S, et al. Twenty years of biliopancreatic diversion: what is the goal of the surgery? Obes Surg 2004;14:160–4.

[6] Nanni G, Familiari P, Mor A, Iaconelli A, Perri V, Rubino F, et al. Effectiveness of the transoral endoscopic vertical gastroplasty (TOGa®): a good balance between weight loss and complications, if compared with gastric bypass and biliopancreatic diversion. Obes Surg 2012;22:1897–902.

[7] Gracia JA, Martinez M, Aguilella V, Elia M, Royo P. Postoperative morbidity of biliopancreatic diversion depending on common limb length. Obes Surg 2007;17:1306–11.

[8] Scopinaro N. Laparoscopic biliopancreatic diversion. In: Lucchese M, Scopinaro N, editors. Minimally invasive bariatric and metabolic surgery. Springer Pub; 2015. p. 209–26. [chapter 20].

[9] DeMeester TR, Fuchs KH, Ball CS, Albertucci M, Smyrk TC, Marcus JN. Experimental and clinical results with proximal end-to-end duodeno-jejunostomy for pathologic duodenogastric reflux. Ann Surg 1987;206:414–26.

[10] Marceau P, Hould FS, Potvin M, Lebel S, Biron S. Biiopancreatic diversion with duodenal switch procedure. In: Deitel M, editor. Update: surgery for morbidly obese patient. Pub. Mothersill Printing Inc; 2000. p. 259–65. [chapter 24].

[11] Hess DS, Hess DW. Biliopancreatic diversion with a duodenal switch. Obes Surg 1998;8:267–826.

[12] Lucchese M, Sturiale A, Quartararo G, Facchiano E. The role of laparoscopy in bariatric surgery. In: Lucchese M, Scopinaro N, editors. Minimally invasive bariatric and metabolic surgery. Springer Pub; 2015. p. 99–108. [chapter 10].

[13] Lim RB, Blackburn GL, Jones DB. Benchmarking best practices in weight loss surgery. Curr Probl Surg 2010;47:69–171.

[14] Scopinaro N, Marinari GM, Camerini G. Laparoscopic standard Biliopancreatic diversion: technique and preliminary results. Obes Surg 2002;12:362–5.

[15] Iannelli A, Martini F. Laparoscopic duodenal switch. In: Lucchese M, Scopinaro N, editors. Minimally invasive bariatric and metabolic surgery. Springer Pub; 2015. p. 227–36. [chapter 21].

[16] Anthone GJ. The duodenal switch operation for morbid obesity. Surg Clin N Am 2005;85:819–33.

[17] Coupaye M, Castel B, Sami O, Tuyeras G, Msika S, Ledoux S. Comparison of the incidence of cholelithiasis after sleeve gastrectomy and Roux-en-Y gastric bypass in obese patients: a prospective study. Surg Obes Relat Dis 2015;11:779–84.

[18] Raziel A, Sakran N, Szold A, Goitein D. Concomitant cholecystectomy during laparoscopic sleeve gastrectomy. Surg Endosc 2015;29:2789–93.

[19] Taylor J, Leitman IM, Horowitz M. Is routine cholecistectomy necessary at the time of Rou-and-Y gastric bypass. Obes Surg 2006;16:759–61.

[20] Moon RC, Teixeira AF, DuCoin C, Varnadore S, Jawad MA. Comparison of cholecystectomy cases after Roux-en-Y gastric bypass, sleeve gastrectomy, and gastric banding. Surg Obes Relat Dis 2014;10:64–8.

[21] Brockmeyer JR, Grover BT, Kallies KJ, Kothari SN. Management of biliary symptoms after bariatric surgery. Am J Surg 2015;210:1010–17.

[22] Rutledge R. The mini-gastric bypass: experience with the first 1,274 cases. Obes Surg 2001;11:276–80.

[23] Scopinaro N, Gianetta E, Civalleri D, Bonalumi U, Bachi V. Two years of clinical experience with biliopancreatic bypass for obesity. Am J Clin Nutr 1980;33:506–14.

[24] Grueneberger JM, Karcz-Socha I, Marjanovic G, Kuesters S, Zwirska-Korczala K, Schmidt K, et al. Pylorus preserving loop duodeno-enterostomy with sleeve gastrectomy - preliminary results. BMC Surg 2014;11(14):20.

[25] Scopinaro N. Biliopancreatic diversion: mechanisms of action and long-term results. Obes Surg 2006;16:683–9.

[26] Nanni GG, Bertoncini M, Falotti C, Balduzzi G, Demichelis P, Scansetti M, et al. Standard biliopancreatic diversion for obesity. 21 Years of clinical experience. Obes Surg 2011;21:990–1.

[27] Buchwald H, Avidor Y, Braunwald E, Jensen MD, Pories W, Fahrbach K, et al. Bariatric surgery: a systematic review and meta-analysis. JAMA 2004;292:1724–37.

[28] Nanni G. Biliopancreatic diversion is more effective than sleeve gastrectomy. N Am J Med Sci 2014;6:39–40.

[29] Abraham A, Ikramuddin S, Jahansouz C, Arafat F, Hevelone N, Leslie D. Trends in Bariatric surgery: procedure selection, revisional surgeries, and readmissions. Obes Surg 2016;26:1371–7.

[30] Sucandy I, Titano J, Bonanni F, Antanavicius G. Comparison of vertical sleeve gastrectomy versus biliopancreatic diversion. N Am J Med Sci 2014;6:35–8.

[31] Marceau P, Biron S, Marceau S, Hould FS, Lebel S, Lescelleur O, et al. Biliopancreatic diversion-duodenal switch: independent contributions of sleeve resection and duodenal exclusion. Obes Surg 2014;24:1843–9.

[32] Biron S, Hould FS, Lebel S, Marceau S, Lescelleur O, Biertho L. BPD-DS: 9000 patients years of follow-up. Obes Surg 2014;24:1158.

[33] Vilallonga R, Fort JM, Caubet E, Gonzalez O, Balibrea JM, Ciudin A, et al. Robotically assisted single anastomosis duodenoileal bypass after previous sleeve gastrectomy implementing high valuable technology for complex procedures. J Obes 2015;2015:586419.

[34] Sánchez-Pernaute A, Rubio MÁ, Conde M, Arrue E, Pérez-Aguirre E, Torres A. Single-anastomosis duodenoileal bypass as a second step after sleeve gastrectomy. Surg Obes Relat Dis 2015;11:351–5.

[35] Buchwald H, Estok R, Fahrbach K, Banel D, Jensen MD, Pories WJ, et al. Weight and type 2 diabetes after bariatric surgery. Systematic review and meta-analysis. Am J Med 2009;122:248–56.

[36] Colquitt JL, Pickett K, Loveman E, Frampton GK. Surgery for weight loss in adults. Cochrane Database Syst Rev 2014;8:8.

[37] Panunzi S, De Gaetano A, Carnicelli A, Mingrone G. Predictors of remission of diabetes mellitus in severely obese individuals undergoing bariatric surgery: do BMI or procedure choice matter? A meta-analysis. Ann Surg 2015;261:459–67.

[38] Mingrone G, Castagneto-Gissey L. Mechanisms of early improvement/resolution of type 2 diabetes after bariatric surgery. Diabetes Metab 2009;35:518–23.

[39] Kamvissi V, Salerno A, Bornstein SR, Mingrone G, Rubino F. Incretins or anti-incretins? A new model for the "entero-pancreatic axis". Horm Metab Res 2015;47:84–7.

[40] Guidone C, Manco M, Valera-Mora E, Iaconelli A, Gniuli D, Mari A, et al. Mechanisms of recovery from type 2 diabetes after malabsorptive bariatric surgery. Diabetes 2006;55:2025—31.

[41] Valera-Mora ME, Simeoni B, Gagliardi L, Scarfone A, Nanni G, Castagneto M, et al. Predictors of weight loss and reversal of comorbidities in malabsorptive bariatric surgery. Am J Clin Nutr 2005;81:1292—7.

[42] Iaconelli A, Panunzi S, De Gaetano A, Manco M, Guidone C, Leccesi L, et al. Effects of bilio-pancreatic diversion on diabetic complications: a 10-year follow-up. Diabetes Care 2011;34:561—7.

[43] Scopinaro N, Adami GF, Papadia FS, Camerini G, Carlini F, Briatore L, et al. Effects of gastric bypass on type 2 diabetes in patients with BMI 30 to 35. Obes Surg 2014;24:1036—43.

[44] Scopinaro N, Adami GF, Papadia FS, Camerini G, Carlini F, Fried M, et al. Effects of biliopancreatic diversion on type 2 diabetes with BMI 25 to 35. Ann Surg 2011;253:699—703.

[45] Chiellini C, Rubino F, Castagneto M, Nanni G, Mingrone G. The effect of bilio-pancreatic diversion on type 2 diabetes in patients with BMI <35 kg/m^2. Diabetologia 2009;52:1027—30.

[46] Mingrone G, Panunzi S, De Gaetano A, Guidone C, Iaconelli A, Nanni G, et al. Bariatric-metabolic surgery versus conventional medical treatment in obese patients with type 2 diabetes: 5 year follow-up of an open-label, single-centre, randomised controlled trial. Lancet 2015;386:964—73.

[47] Sánchez-Pernaute A, Rubio MÁ, Cabrerizo L, Ramos-Levi A, Pérez-Aguirre E, Torres A. Single-anastomosis duodenoileal bypass with sleeve gastrectomy (SADI-S) for obese diabetic patients. Surg Obes Relat Dis 2015;11:1092—8.

[48] Pajecki D, Dalcanalle L, Souza de Oliveira CP, Zilberstein B, Halpern A, Garrido AB, et al. Follow-up of Roux-en-Y gastric bypass patients at 5 or more years postoperatively. Obes Surg 2007;17:601—7.

[49] Scopinaro N, Marinari G, Camerini G, Papadia F. Biliopancreatic diversion: physiological and metabolic aspects. In: Parini U, Nebiolo PE, editors. Bariatric surgery - multidisciplinary approach and surgical technique. Musumeci Editore Aosta; 2007. p. 314—42.

[50] Fried M, Yumuk V, Oppert JM, Scopinaro N, Torres A, Weiner R, et al.International Federation for Surgery of Obesity and Metabolic Disorders-European Chapter (IFSO-EC); European Association for the Study of Obesity (EASO); European Association for the Study of Obesity Obesity Management Task Force (EASO OMTF) Interdisciplinary European Guidelines on metabolic and bariatric surgery. Obes Surg 2014;24:42—55.

[51] Sánchez-Pernaute A, Herrera MA, Pérez-Aguirre ME, Talavera P, Cabrerizo L, Matía P, et al. Single anastomosis duodeno-ileal bypass with sleeve gastrectomy (SADI-S). One to three-year follow-up. Obes Surg 2010;20:1720—6.

Chapter 15

Endoluminal Procedures for the Treatment and Management of Bariatric Patients

A.M. Johner and K.M. Reavis

The Oregon Clinic and Legacy Weight & Diabetes Institute, Portland, OR, United States

LIST OF ABBREVIATIONS

RYGB Roux-en-Y gastric bypass
LSG Laparoscopic sleeve gastrectomy
EGD Esophagogastroduodenoscopy
BPD Biliopancreatic diversion
GERD Gastroesophageal reflux disease
GI Gastrointestinal
GJ Gastrojejunal
JJ Jejuno-jejunal
POSE Primary obesity surgery endolumenal
IOP Incisionless operating platform
BMI Body mass index
EWL Excess weight loss
TBWL Total body weight loss
FDA Federal drug administration
EUS Endoscopic ultrasound

OBJECTIVES

1. To understand the role of endoscopic therapies as options in the bariatric patient.
2. To understand the physiologic principles of different bariatric endoscopic therapies.
3. To understand the role of endoscopy in the revisional bariatric surgery candidate.
4. To understand the future direction of completely flexible endoscopic approaches to index and revisional bariatric surgical procedures.

INTRODUCTION

A steady rise in the overall prevalence of obesity has occurred over the past 20 years [1]. Obesity poses a major public health challenge and a significant burden on healthcare resources, both in terms of direct and indirect costs [2]. This obesity epidemic has continued to expand despite the availability of diet and lifestyle counseling, pharmacologic therapy, and bariatric surgery. While lifestyle modification is effective in achieving initial weight loss, follow-up studies demonstrate weight regain within 1 year [3–5]. Pharmacotherapy has been disappointing, often revealing marginal weight loss with poor side-effect profiles, which preclude widespread use [6–8]. Despite the benefits of bariatric surgery for obesity and related comorbidities, only a fraction of obese patients will undergo bariatric surgery. At present,

Roux-en-Y gastric bypass (RYGB) and laparoscopic sleeve gastrectomy (LSG) constitute the majority of weight loss procedures performed in the United States. Both surgical procedures have demonstrated good efficacy in terms of weight loss and improvement in comorbidities such as diabetes [9]; however, such operations remain highly invasive and carry a significant postoperative mortality rate of 0.31%, adverse event rate of 10−17%, and a failure rate of 10−20% with weight regain reported in 20−30% of the patient population [10]. Furthermore, only approximately 1% of eligible patients consider surgical methods of weight loss, due in part to its invasiveness and availability [11].

There remains a huge unmet need for minimally invasive, safe, and effective long-term therapies for obesity. Endoscopic procedures have the ability to bridge the treatment gap between medical therapy and surgery. Recent advances made in the field of endoluminal devices and techniques have provided an increased array of nonsurgical options for the management of obesity. Endoscopic treatments have the opportunity to provide a more cost-effective, accessible, and reversible intervention that can function as a bridge or alternative therapeutic option to bariatric surgery. The use of endoscopy in the treatment of obesity may prove advantageous in potentially reducing postoperative convalescence with decreased pain, incisional hernia development, and surgical site infections compared to standard surgical techniques.

Current treatments of morbid obesity utilize a multidisciplinary approach involving medical as well as surgical intervention. Medical providers who plan to offer endoscopic bariatric therapies in their clinical practice need appropriate training in both medical and endoscopic management of obesity and its associated comorbidities. All patients should be managed by a multidisciplinary team encompassing nutritional, psychological, medical, and social support in order to achieve optimal success [12]. Endoscopists should have familiarity with their target population, and the benefits, contraindications, and adverse events for each device or procedure. A central tenet surrounding the practice of endoscopy in patients before or after bariatric surgery is the need for close consultation or coordination with the surgeon/surgical team by the endoscopist if the endoscopist is not part of the bariatric surgery team. Once selected for treatment, patients can undergo endoluminal procedures in endoscopy suites or formal operative theaters. The procedure can be performed in the supine or decubitus position and under general anesthesia or deep sedation. The specific set up is typically determined by the level of invasiveness of the procedure, the desired ergonomics of the interventionist, and the comorbid conditions of the patient.

Four roles of endoscopy in the care of the bariatric patient will be discussed here: (1) preoperative, (2) intraoperative, (3) postoperative, and (4) primary endoscopic weight loss procedures. Traditional bariatric surgical procedures cause weight loss via gastric volume restriction, malabsorption, or through a combination of the two. Similarly, current primary endoscopic bariatric therapies can be classified as restrictive, bypass, or space-occupying.

Preoperative Evaluation

Preoperative endoscopy with esophagogastroduodenoscopy (EGD) can identify patients with asymptomatic anatomic findings that may alter surgical planning. After undergoing index malabsorptive procedures such as RYGB or biliopancreatic diversion (BPD), the remnant stomach and/or duodenum are not easily accessible from a standard transoral endoscopic approach. For this reason it has been suggested that preoperative endoscopy be performed in all patients undergoing these procedures if there is any potential concern for pathology. This includes patients who have unexplained iron deficiency or suspected *Helicobacter pylori* infection [13]. That being said, since only 3−5% of asymptomatic patient's undergo a change in surgical planning when endoscopy is routinely performed, many surgeons are proponents of selective preoperative endoscopy [14,15]. Patients with symptoms of gastroesophageal reflux disease (GERD), such as heartburn, regurgitation, dysphagia, or any postprandial symptoms that suggest a foregut pathology, and/or who chronically use antisecretory medications, should have an upper gastrointestinal (GI) endoscopic evaluation before bariatric surgery [16]. Multiple published studies have demonstrated that routine EGD before surgery can identify a variety of conditions, including hiatal hernia, esophagitis, ulcers, and tumors [17,18]. Surgeons advocate hiatal hernia reduction and crural closure in patients with hiatal hernia undergoing any weight loss operation [16]. It is therefore useful for surgeons planning weight loss interventions to know the measured size of any hiatal hernia (in centimeters).

INTRAOPERATIVE ENDOSCOPY

During resectional procedures, endoscopy allows for visualization of staple lines to assess for bleeding or leakage. Clinically relevant endoluminal bleeding occurs in 1−4% of patients undergoing gastric bypass [19]. Although some bleeding occurs from sites inaccessible to upper endoscopy, thereby requiring conservative or reconstructive approaches, when the bleeding is noted at the gastrojejunal (GJ) staple line (in approximately one-third of cases), the treatment can involve homeostatic injection or clip application in addition to guidance for laparoscopic oversewing of the suspect area. Energy sources are generally avoided to prevent delayed thermal injury and perforation. In addition to bleeding, intraoperative leaks can be detected and treated endoscopically. An endoscopic air leak test with an anastomotic or sleeve staple line submerged under saline can identify leaks in many cases. This method of leak testing has been found superior to methylene blue infusion via an orogastric tube with laparoscopic visualization [13]. Haddad et al. [20] reviewed 2311 gastric bypass cases at their institution from 2002−11 in which 2308 (99.9%) successfully underwent intraoperative upper endoscopy to assess the GJ. Forty six (2%) clinically significant leaks were noted and received immediate intraoperative surgical repair, 44 required no additional intervention, and two of the 46 initial leaks recurred and required further treatment postoperatively.

POSTOPERATIVE EVALUATION

The presence of adverse GI symptoms after bariatric surgery and the management of surgical complications are the most common indications for postoperative endoscopy. Often these symptoms present in the first six postoperative months and include abdominal pain, unrelenting nausea, vomiting, dysphagia, heartburn, regurgitation, diarrhea, anemia, and weight regain. Endoscopy is recommended as a first-line diagnostic study in the evaluation of the postoperative bariatric patient with abdominal pain, nausea, or vomiting [16]. In one study, 62% of patients presenting with persistent nausea and vomiting and 30% of those presenting with abdominal pain or dyspepsia after RYGB had significant findings on upper endoscopy, including marginal ulcers, stomal stenosis, and staple line dehiscence [21]. Postoperative GERD symptoms should be managed as in nonbypass patients [16], with endoscopy considered to rule out inciting factors such as obstruction of the GJ anastomosis, increased pouch size, and distal limb obstruction, as well as for the evaluation of symptoms refractory to medical therapy. Complications of diagnostic endoscopy are rare (0.4%), and most commonly involve those related to conscious sedation [22]. When considering endoscopy in the postbariatric surgery patient, the endoscopist should review pertinent operative notes and previous imaging studies (both pre- and postoperative), and must understand the expected anatomy, including the presence of altered gastric anatomy, the extent of resection, and length of intestinal limbs. Following a careful history and physical exam, radiographs such as an upper GI swallow study can provide clues regarding current GI enteric flow, regurgitation, obstruction, and stasis. The choice of endoscope will depend on the indication for endoscopy and the need for intubation of the excluded limb or for therapeutic intervention.

The goal of short-term postoperative endoscopy is to assess for structural pathology. The expected endoscopic findings after RYGB include a normal esophagus and GE junction with a small gastric pouch, often too small to permit safe retroflexion; this maneuver is not recommended in this setting. Pouch length and width, stoma size, and the presence of visible suture material are useful information that should routinely be included in endoscopy reports. The length of the blind afferent limb segment should also be documented as excessively long blind pouch limbs at the GJ may be a cause of postprandial pain. The jejuno-jejunal (JJ) area can sometimes be reached with an upper endoscope, depending on the length of the Roux limb. The most common diagnoses found with upper endoscopy after bariatric surgery are marginal ulcer, anastomotic stricture, staple line dehiscence, band erosion or slippage, and GERD sequelae [23]. Progressive food intolerance three or more weeks after gastric bypass or BPD suggests stenosis of the proximal anastomosis. Anastomotic stricture, defined as an anastomoses that is <10 mm in diameter, is a common adverse event of RYGB, occurring in 3% − 28% of patients [24], and is typically hallmarked by dysphagia, nausea, and vomiting. Endoscopic balloon dilatation with a 12−20 mm balloon is the preferred initial treatment modality, although this can also be achieved via wire-guided bougie dilation [16]. If needed, the procedure can be repeated 2 weeks later and serially thereafter. The success rate reported is around 90% [13]. The perforation rate of balloon dilation of RYGB strictures is estimated to be 2% − 5% [25]. Marginal ulcers occur at the GJ anastomosis, usually on the intestinal side, and are thought to arise as a result of a number of factors, including local ischemia, staple line disruption, effects of acid on exposed intestinal mucosa, and the presence of exposed staples or suture material. If a marginal ulcer is present, medical therapy is recommended as long as the anastomosis is patent [16]. Stenosis after sleeve gastrectomy is less common, but usually presents acutely. Balloon dilatation and stenting are feasible, with a risk of perforation or stent migration. Conversion of the sleeve to a formal RYGB serves as definitive treatment in the case of failed, repeated endoscopic treatments.

PRIMARY ENDOSCOPIC PROCEDURES

Currently available index endoluminal treatments range from noninvasive and reversible procedures such as intragastric balloons and malabsorptive sleeves to permanently intended gastric plication with suturing and stapling devices. Obese patients who are typically at greatest risk for incision-related complications such as failure-to-heal, infection, and hernia development will potentially benefit the most from these advances.

Restrictive Techniques

The LSG is a surgical treatment for morbid obesity that combines an aspect of restriction from the narrow lesser curve-based sleeve with a metabolically based decrease in hunger sensation due to the resection of the greater curvature and gastric fundus where ghrelin and other hunger-based hormones are manufactured. Attempts to achieve similar results endoscopically have been launched, both with staplers and suture. This endoscopic gastric volume reduction potentially allows for an incisionless procedure with lower costs and shorter recovery times compared with surgery.

Primary Obesity Surgery Endolumenal Procedure

The primary obesity surgery endolumenal (POSE) procedure involves use of the incisionless operating platform (IOP; USGI Medical, San Clemente, CA) to apply suture−anchor plications in the gastric fundus (to limit gastric fundal accommodation), and in the distal gastric body near the proximal antral inlet (to delay gastric emptying). The IOP consists of the TransPort, a flexible, steerable, multilumen access device that is passed transorally under general anesthesia. The four working channels enable placement of full-thickness serosa-to-serosa transmural nitinol Snowshoe tissue anchors. The anchors are designed to distribute the compression force of tissue approximation along the larger surface area of the anchor mesh, which theoretically results in superior durability compared with suture approximation. Eight to ten plications are created in the gastric fundus (in retroflexion) in two parallel ridges until the fundic apex is brought down to the level of the GE junction. The device is then straightened so that the distal gastric body is visualized. A tissue ridge is created with three to four plications in the distal gastric body across from the incisura.

Initial experience with the POSE procedure was reported by Espinos et al. in an observational 45-patient pilot study [26]. Patients had a baseline mean body mass index (BMI) of 36.7 +/− 3.8. Mean operative time was 69.2 +/− 26.6 minutes, and a mean of 8.2 suture−anchor plications were placed in the fundus and three anchors along the distal body wall. BMI loss was 5.8 kg/m^2, percentage of excess weight loss (%EWL) was 49.4%, and percentage of total body weight loss (%TBWL) was 15.5% at 6 months. No mortality or operative morbidity was observed. Adverse events associated with the procedure included one case of low-grade fever and one case of chest pain. Lopez-Nava and colleagues recently published the largest case series to date of a POSE procedure involving 147 obese patients with a baseline BMI of 38 [27]. Follow-up at 1 year was obtained in 79% of patients who achieved a mean TBWL of 16.6 +/− 9.7 kg, percentage of total weight loss (%TWL) of 15.1 +/− 7.8, and %EWL of 44.9 +/− 24.4. No serious short- or long-term adverse events were reported. POSE is currently being studied in the US multicenter randomized sham-controlled ESSENTIAL trial.

Apollo OverStitch for Endoscopic Sleeve Gastroplasty

The Apollo OverStitch device (Apollo Endosurgery, Austin, TX) can place full-thickness stitches in a variety of interrupted or running patterns. The OverStitch includes a curved needle driver attached to the tip of the endoscope, a catheter-based suture anchor, and an actuating handle attached near the endoscopic controls. A double-channel endoscope is necessary. The device is being used to perform endoscopic sleeve gastroplasty by placing running stitches in a triangular configuration starting in the antrum and working proximally. Each suture is used to create two conjoined triangles with 8−14 sutures placed in this fashion. The procedure includes fundic reduction in retroflexion. The sleeve is reinforced with interrupted stitches.

Two independent feasibility studies in 2013 and 2014 demonstrated in nine total patients the ability to safely perform the OverStitch technique for vertical gastroplasty [28,29]. Lopez-Nava and colleagues demonstrated encouraging safety and efficacy results [30]. They demonstrated a robust 17.8 %TBWL at 6 months in the 20-patient study with a mean baseline BMI of 38.5. No major adverse events were reported except intraprocedural bleeding that required injection therapy in two patients. The investigators further reported their experience in a larger cohort of 50 patients, with 13 having reached 1 year follow-up [31]. The procedure duration averaged 66 minutes, during which an average of 6−8 sutures were placed. All patients were discharged in less than 24 hours, and there were no major intraprocedural, early, or delayed adverse events. Mean %TBWL was 19 and BMI reduced from 37.7 to 30.9 at 1 year. Oral contrast studies and endoscopy revealed a preserved sleeve gastroplasty configuration at 1 year of follow-up. The Primary Obesity Multicenter Incisionless Suturing Evaluation trial to study efficacy of endoscopic sleeve gastroplasty using OverStitch is ongoing in the United States.

EndoCinch/RESTORe for Endoscopic Gastroplasty

The EndoCinch (Bard/Davol, Warwick, RI) is a superficial-thickness endoscopic suturing system. The device uses suction to acquire tissue in a hollow capsule, and then passes a needle through the tissue. The safety and efficacy of this endoluminal vertical gastroplasty was first reported in a cohort of 64 obese patients with average BMI of 39.9 [32]. They reported %EWL after 1 year was 58.1% with 97% of patients attaining a 30% or greater EWL. No serious adverse events were reported. The device was then modified and named the RESTORe, and was capable of both full-thickness suturing and suture reloading in vivo. The device was studied in a two-site trial including 18 patients [33]. There were no significant adverse events. One-year mean weight loss was 11.0 +/− 10 kg, or 27.7 %EWL. Average weight circumference declined by 12.6 +/− 9.5 cm. Blood pressure decreased significantly (systolic −15.2 mmHg, diastolic −9.7 mmHg). However, follow-up endoscopy revealed partial or complete release of plications in 13 of 18 patients.

TransOral Gastroplasty

The TransOral Gastroplasty (Satiety Inc., Palo Alto, CA) procedure utilizes a flexible endoscopic stapler capable of full-thickness tissue apposition. An 8.6 mm endoscope is passed through the device and retroflexed to visualize the procedure. The procedure begins with dilation of the esophagus to 60 F with a Savary dilator. The device is inserted into the stomach over a guidewire. Once in position, vacuum apposes the gastric walls, acquiring tissue into the device. Firing the stapler creates a 4.5-cm sleeve around the stapler using titanium staples. The device has to be removed for reloading, and the firing process is repeated once more distally, overlapping the first sleeve. A restrictor is then inserted over the guidewire, with the endoscope adjacent to the device. Vacuum acquires tissue into the device at the distal sleeve, and firing the restrictor creates a 2.5 cm-long stapled narrowing at the outlet of the sleeve. Initial data were promising; however, Food and Drug Adminitration (FDA) studies did not reach targeted clinical outcomes [34].

Gastric Volume Reduction Using the ACE Stapler

The ACE stapler (Boston Scientific Corporation, Natick, MA) represents a full-thickness stapling device system with a head capable of both 360-degree rotation and complete retroflexion. A 5-mm endoscope enables visualization, with the device itself 16 mm in diameter. The stapler head acquires gastric tissue using vacuum suction; and firing the stapler creates a full-thickness plication using a 10-mm plastic ring with eight titanium staples. For gastric volume reduction, eight plications are typically placed in the fundus and two in the antrum. A 17-patient Netherlands safety and feasibility study (median BMI of 40.2 kg/m^2) reported a median procedure time of 123 minutes. Occasional abdominal pain, nausea, and vomiting were reported postprocedure; however, these symptoms resolved within 2 weeks. The authors noted a 34.9 %EWL at 12-month follow-up. EGD at the end of the study revealed 6−9 plications in place with preserved reduced gastric volume in all patients [35].

Transoral Endoscopic Restrictive Implant System

The transoral endoscopic restrictive implant system (TERIS; BaroSense, Redwood, CA) is an implanted gastric restriction device. A gastric pouch is created by implanting a diaphragm, with a 10-mm orifice, which is then attached to the gastric cardia. To do so, a 22-mm endogastric tube is inserted. A gastroscope with a stapling device is retroflexed, and a full-thickness plication is created in the cardia. An anchor is attached to the plication. This is repeated until five anchors have been implanted. The restrictive diaphragm is then attached to the anchors. A study of TERIS in 13 patients reported 3 adverse events: one gastric perforation and two cases of pneumoperitoneum [36]. The device has not been FDA-approved and is no longer available in the United States.

Bypass/Diversion

Animal models suggest that duodenal exclusion and accelerated arrival of partially digested meals to mid-jejunum and ileum are partially responsible for the beneficial effects of gastric bypass in diabetes and obesity. Endoscopically implanted devices have been developed to reproduce this effect.

EndoBarrier Duodenal−Jejunal Bypass Liner

The EndoBarrier device (GI dynamics, Lexington, MA USA) is a nickel−titanium implant attached to an impermeable 60-cm polymer sleeve that extends from the duodenal bulb into the jejunum. The liner prevents food from contacting the mucosa of the small intestine, but allows pancreaticobiliary secretions to move along the outside of the device to the jejunum. This precludes mixing until the jejunum, similar to an RYGB. The device is placed endoscopically with fluoroscopic guidance under general anesthesia.

A prospective, randomized sham-controlled trial including 47 patients revealed an increased 3-month %EWL in the EndoBarrier group (11.1% vs 2.7%, $p < 0.05$) [37]. Eight patients were excluded from the study due to hematemesis, nausea, and abdominal pain. A meta-analysis evaluating five EndoBarrier randomized controlled trials revealed an average %EWL of 12.5% compared to diet [38]. The 2014 ENDO trial, a large multicenter study assessing the device, was prematurely terminated during interim analysis. Trial data revealed the development of hepatic abscesses in seven patients in the EndoBarrier treatment group [39]. Initial studies illustrate therapeutic potential for the EndoBarrier device; however, device modifications with an improved safety profile will be required prior to further clinical study.

Gastro-Duodeno-Jejuno Bypass

The ValentX bypass sleeve (ValenTx, Inc., Carpinteria, CA) is another endoscopically placed removable sleeve designed for malabsorptive weight loss. It is placed not in the duodenum but at the gastroesophageal junction, thus bypassing the stomach as well. The 120-cm sleeve extends distally to the jejunum, mimicking the surgical bypass and compartmentalization of the stomach. A 2011 prospective observational study investigated 22 patients who underwent device implantation with 3-month follow-up weight assessment and subsequent device removal. Seventeen patients completed the full 12-week trial, with an average %EWL of 39.7% [40]. A follow-up study in 2015 reviewed 13 patients, with 3 of these patients withdrawing from the trial due to esophagitis and general intolerance [41]. The remaining 10 patients tolerated the device for 1 year; at follow-up 12-month EGD, 4 patients had partial device detachment. In the patients who fully maintained the bypass sleeve, the 12-month EWL was 54%.

Lumen Apposing Metal Stents

The advent of lumen apposing metal stents under the guidance of endoscopic ultrasound (EUS) has afforded an opportunity to create fistulas between various abdominal organs and cavities. It is FDA-approved, but use for obesity is off-label. EUS-guided gastrojejunostomy has recently been reported for palliation of gastric and duodenal outlet obstructions [42]. Given that some of the endoscopic devices attempt duodenal bypass for the purposes of distal nutrient exposure and altered physiology, this technique and technology is being investigated for its bariatric properties.

Magnetic Jejunoileal Bypass

Self-assembling magnets for endoscopy (GI Windows Inc., Bridgewater, MA) have been studied in a porcine model for creation of incisionless, magnetic compression gastrojejunostomy. Ryou et al. [43] recently described insertion of two magnetic devices via a catheter into the foregut and the hindgut of five pigs. The devices transform from a linear shape to an octagonal geometry, and then couple to form an anastomosis. Coupled devices are eventually expelled naturally, leaving behind an anastomosis without residual foreign material. Within a few days this method worked to create an intestinal bypass which was demonstrated to be large-caliber and leak-free. Three-month follow-up endoscopy confirmed the anastomosis to be large and patent. The incisionless anastomosis system (IAS) was further studied on 10 obese patients with mean BMI of 41 [44]. Four patients had diabetes and three were classified and prediabetic. The study found that the dual-path enteral diversion was safely created in all patients, and the IAS devices were expelled without incident. At 6 months, investigators observed that all patients experienced significant reductions in HbA1c and fasting blood glucose levels. Patients with Type 2 diabetes showed a decrease of HbA1c from a mean baseline of 7.8% − 6.0% at 6 months, with a decrease in fasting blood glucose levels from 177 mg/dL to 111 mg/dL. All patients had fasting blood glucose levels move from the diabetic or prediabetic range to the normal range at 6 months. The mean weight loss for all patients was approximately 28 pounds (12.9 kg), representing a 10.6% decrease in TWL.

Duodenal Mucosal Resurfacing

The Revita duodenal mucosal resurfacing (DMR) system (Fractyl Laboratories, Cambridge, MA) is being used to perform endoscopic DMR for the treatment of obesity and metabolic disease. The system consists of a console and single-use balloon catheter. The procedure involves superficial mucosal thermal ablation using recirculating hot water within a balloon catheter. Neto and colleagues found a two-point HbA1c reduction in a single-center small unpublished case series in Chile, and currently a definitive multicenter prospective double-blind sham-controlled trial is underway in Europe and South America [45]. The Revita system remains under clinical investigation and is not commercially available.

Aspiration Therapy

The AspireAssist Device (Aspire Bariatrics, King of Prussia, PA) is essentially a reverse gastrostomy tube (30 Fr) whereby 15–20 minutes after food ingestion, a portion of the consumed bolus is externally voided by aspiration of gastric contents through the AspireAssist siphon. This process physically removes 30% of calories in an ingested meal, leading to a reduced daily caloric intake and gradual weight loss. In a pilot study conducted in 2013, 18 patients were randomized to aspiration therapy and lifestyle changes versus lifestyle changes alone [46]. At 1-year follow-up, %EWL was higher in the aspiration and lifestyle group versus lifestyle group (49% vs 14.9%, $p < 0.04$). Seven of the 10 patients in the aspiration therapy group opted to continue for another year with a final %EWL of 54.6%. No compensation for aspirated calories with increased food intake or binge eating was observed. The US prospective 171 subject multicenter PATHWAY trial completed enrollment in 2014, and presentation of results is pending. The AspireAssist device is now FDA-approved for adults with a BMI of between 35 and 55.

Space-Occupying Devices

Space-occupying devices displace volume and induce gastric distention, but may also alter GI motility, nutrient transit, and hormone levels. Balloons have built a track record of safety and efficacy in Europe, and have found a role as a bridge to bariatric surgery in patients with high risk for anesthesia, temporary use in patients eligible for bariatric surgery but unwilling to undergo it, and temporary use in patients not eligible for bariatric surgery, as part of an integrated medical weight loss program.

Orbera Intragastric Balloon

The Orbera balloon (Apollo Endosurgery, Austin, TX) is an endoscopically implanted spherical silicone elastomer that is attached to a preloaded catheter and advanced blindly into the stomach. Under direct visualization with EGD, the balloon is inflated with 450–700 mL of saline and resides within the stomach for 6 months. The balloon is designed to act as a bezoar and move freely within the stomach while inducing weight loss via gastric distention, reducing gastric emptying and increasing baroreceptor stipulation, thereby inducing satiety [47].

The device was studied in a 2008 meta-analysis of 3608 patients in 15 studies; early device removal was required in 4.2% of patients. Reported adverse events included nausea, vomiting, bowel obstruction (0.8%), and gastric perforation (0.1%). Average weight loss after 6 months was 14.7 kg or 32.1 %EWL, with a drop in BMI of 5.7 kg/m^2 [48]. The largest included study was an Italian cohort of 2515 patients that demonstrated a 9 kg/m^2 BMI reduction at 6 months, with a %EWL of 25.6% [49]. Statistically significant improvements in blood pressure, fasting glucose, and lipid profiles were seen. A significant decrease in, or normalization of, HbA1c was reported in 87.2% of the 488 diabetic patients. There was an associated 2.8% complication rate and 0.19% gastric perforation rate.

The benefit of repeat or serial Orbera balloon placement for maintaining weight loss remains unclear. In 2010, 19 patients requesting repeat Orbera placement were nonrandomized: 8 patients underwent direct repeat balloon placement at 6 months, while 11 patients had a balloon-free interval [50]. Those patients undergoing a second balloon placement with a balloon-free interval regained 13.6 kg on average during that interval. The second balloon did result in weight loss, although its magnitude was smaller than that of the initial therapy (9 kg vs 14.6 kg, or 18.2 %EWL vs 49.3 %EWL). The effect of second balloon placement dissipated by the third year of follow-up.

The utility of Orbera as a bridge to gastric bypass was studied in 60 consecutive superobese subjects (average BMI of 66.5 kg/m^2), comparing patients who received Orbera within 3 months prior to laparoscopic gastric band to patients who solely underwent surgery [51]. The preop balloon placement group demonstrated a BMI reduction of 5.5 kg/m^2 with a percentage of excess BMI loss (%EBL) of 11.2% at the time of surgery. The mean operative time (146 vs 201 minutes, $p < 0.01$) and the composite adverse event endpoint (ICU admission, open laparotomy conversion, death) were significantly lower in the preoperative balloon group. Weight loss was similar between groups 1 year after gastric bypass. This study suggests a role for balloon therapy as a preoperative method to optimize success of bariatric surgery. Orbera balloon therapy has also been reported to improve several metabolic parameters. An Italian prospective study involving 130 obese patients showed significant improvement in glycemia, insulin resistance, triglyceridemia, and liver steatosis in addition to significant weight loss in 91 responders to the intragastric balloon [52].

The Orbera balloon gained FDA approval in 2015 for insertion for up to 6 months in obese patients with BMI between 30 and 40 after a multicenter US pivotal trial involving 215 patients. An average of 10 %TBWL was seen at the time of balloon removal, as compared to 4% in the control group at 6 months. The reported %EWL was approximately 40% versus 13%, respectively. Three months after device removal, the mean %EWL was 26.5% in the balloon group [53].

ReShape Duo Balloon System

ReShape Duo Integrated Dual Balloon System (ReShape Medical Inc., San Clemente, CA) is a dual-balloon implant that is endoscopically placed and retrieved following 6 months of treatment. The two silicone spheres are filled with a total of 900 mL of saline. The dual balloon design is thought to provide enhanced gastric space filling while potentially reducing the risk of intestinal migration [54]. However, as compared to the Orbera balloon, the ReShape Duo is a relatively new device with significantly less published clinical data regarding safety and efficacy. In a prospective sham-controlled US pivotal trial, the ReShape balloon resulted in significantly greater %EWL (25.1% intent-to-treat, 27.9% completed cases) as compared to patients managed with diet and exercise alone at 24 weeks. There were no deaths, intestinal obstructions, gastric perforations, or device migrations; balloon deflation without migration occurred in 6%; and early retrieval for nonulcer intolerance was required in 9.1%. Gastric ulceration at the incisura was observed in 39%, but was significantly reduced to 10% after a minor device change [54]. The ReShape Duo Balloon gained FDA approval in 2015 for patients with a BMI of 30 − 40 with unsuccessful attempts at weight loss, and should be placed for obese patients enrolled in a structured weight loss program.

Heliosphere Bag

The Heliosphere Bag (Helioscope, France) is another temporary intragastric balloon system; however, it differs in that it is filled with 950 mL of air rather than fluid. It has been postulated that the lightness (<30 g) of the air-filled bag may limit nausea and vomiting. In a study of 60 patients (average BMI of 46.3), the Heliosphere Bag was compared with the Orbera balloon. The Heliosphere group achieved a BMI decrease of 4.2 versus 5.7 in the Orbera group. The Heliosphere group had significantly longer extraction procedure time and significantly more discomfort during extraction [55]. Another prospective study of 91 patients compared Orbera (72 patients) to Heliosphere (18 patients). Balloons were implanted for 6 months, with 13.2% removed early due to intolerance. Average weight reduction at 6 months was 13.3 kg, and BMI reduction was 5 kg/m². Eighty-eight percent of weight reduction occurred in the first 3 months. Weight loss was similar between balloon types. Again, the Heliosphere Bag was significantly more likely to result in retrieval complications [56].

Obalon Intragastric Balloon

The Obalon intragastric balloon (Obalon Therapeutics, Carlsbad, CA) is a 250 mL gas-filled balloon that is swallowed under fluoroscopic visualization rather than inserted endoscopically. The balloon is enclosed in a capsule, and a catheter that extends through the esophagus and outside the mouth, is used to fill the balloon with gas. The device is removed endoscopically. If it is tolerated and induces weight loss, a second balloon can be swallowed at 4 weeks and a third at 8 weeks. A study with 17 patients with BMI ranging from 27 to 35 reported that 98% of balloons were swallowed successfully [57]. Abdominal pain (76%) and nausea (41%) were the most frequent adverse events. All were removed endoscopically at 12 weeks.

Transpyloric Shuttle

The Transpyloric Shuttle (Baronova Inc., Goleta, CA) consists of a large spherical bulb attached to a smaller cylindrical bulb by a flexible tether. The cylinder is small enough to enter the duodenal bulb with peristalsis and pulls the spherical bulb to the pylorus, occluding it intermittently to reduce gastric emptying, which leads to early and prolonged satiety. The device is delivered transorally via a catheter and removed endoscopically. A single-center nonblinded prospective trial of 20 patients (average BMI of 36) reported loss of 8.9 kg or 31.3 %EWL at 3 months [58]. Six-month weight loss was 14.6 kg or 50 %EWL. Two patients required early removal due to persistent ulcer. An initial US multicenter trial is currently being planned.

Satisphere

Satisphere (Endosphere, Columbus, OH) is a duodenal implantable device that consists of a nitinol backbone with multiple attached spheres that putatively function to delay duodenal motility and alter gut hormone signaling and glucose metabolism. A trial of 31 patients (average BMI of 41.3) compared 21 Satisphere patients with 10 controls. Device migration was reported in 10 of 21 implanted patients. Emergency surgery was necessary in two patients, which prompted study termination. Of patients completing the trial, 3-month weight loss was 6.7 kg in the Satisphere group versus 2.2 kg in the controls. Satisphere was associated with delayed glucose absorption, delayed insulin secretion, and altered glucagon-like peptide 1 kinetics [59].

REVISIONAL ENDOSCOPIC PROCEDURES

Bariatric surgical interventions have consistently demonstrated the best long-term outcomes regarding persistent weight loss; however, there is risk for eventual weight regain. Enlargement of the remnant gastric pouch or GJ stoma post-RYGB has been demonstrated to be an independent predictor of weight regain [60]. This patient subset would potentially benefit from stomal reduction. Endoscopic stoma reduction is feasible, thus avoiding much of the risk associated with the laparoscopic and open approaches.

Suturing Devices

Due to the complexity and risks associated with surgical revision, endoscopic suturing has been explored as a minimally invasive and safe option for stomal revisions. The 2013 RESTORe trial evaluated this vacuum-based superficial-thickness suturing system for transoral reduction in a randomized, sham-controlled study consisting of 77 post-RYGB patients (55 RESTORe vs 27 sham) [61]; at 6 months postintervention, the RESTORe group demonstrated significantly increased percentages of %EWL (3.9% vs 0.2%, $p < 0.014$) and increased rates of weight stabilization or loss (96% vs 76%, $p < 0.019$). The RESTORe system was compared to the Apollo OvertStitch in a retrospective matched cohort study in 2014 by Kumar et al., and 59 patients were compared in each arm [62]. The full-thickness stitch group had significantly better %EWL compared to the superficial-thickness stitch group (20.4% vs 8.1% at 6 months, $p < 0.01$) and (18.9% vs 9.1% at 12 months, $p < 0.03$). These results suggest that full-thickness suturing may be more durable for transoral outlet reduction.

Endoscopic Sclerotherapy and Radiofrequency Ablation

Sclerotherapy with sodium morrhuate injection at the GJ anastomosis creates submucosal blebs to reduce GJ aperture diameter. Serial radiofrequency ablation of the gastric pouch and stoma may alter pouch compliance, stomal aperture diameter, and GI physiology to promote weight loss. The largest published series of 231 patients noted that 78% of patients experienced weight-regain stabilization at 12 months postprocedure [63]. Average weight loss at 6 months was 4.5 kg.

Ovesco Over-the-Scope Clip

The Ovesco clip (Ovesco, Tubingen, Germany) is an elastic nitinol memory alloy that, following deployment, retakes an unbent shape and exerts tissue compression. The clip is loaded onto the application cap, which is placed on the end of the gastroscope. The desired end of the GJ aperture is reacquired with graspers; the tissue is then apposed via suctioning into the cap, followed by clip deployment. A 2011 trial of 94 patients post-RYGB, with an average GJ diameter of 35 mm, underwent Ovesco clipping, yielding an average postclipping diameter of 8 mm [64]. Initial average study patient BMI was 32.8, with significant improvement at 3 months (BMI of 29.7) and 1 year (BMI of 27.4).

Plication Devices

The StomaphyX device (Endogastric Solutions, Redwood City, CA) acts by pulling tissue into a suction chamber and then deploying polypropylene H-fasteners in a transmural fashion to create full-thickness plications and reduction in gastric pouch and stomal diameter. Small case series have shown modest EWL 12 months following treatment.

The EndoCinch device (C.R. Bard, Murray Hill, NJ) is an overtube device that incorporates a sewing-machine motion for the passage and retrieval of a suturing carrying needle. Target tissue is drawn into the device jay where the needle and suture are passed through the tissue. Small published case series have shown stoma diameter reduction of greater than 65% with moderate %EWL reported [61].

The Revision Obesity Surgical Endolumenal plication device creates transmural serosa-to-serosa anchoring plications that can shorten the gastric pouch and reduce GJ diameter. A 2010, 116-patient multicenter prospective study revealed a mean 50% reduction of the GJ diameter, as well as a mean 44% pouch length reduction [65]. At the 6-month follow-up, 32% of previous weight gain had been lost, and an 18 %EWL was observed.

CONCLUSIONS

To date, no medical therapy exists that matches the impact demonstrated by bariatric surgery on reduction in weight, comorbidity control, or mortality. Although the safety profile, efficacy, and comprehensive care model for the surgical treatment of morbid obesity has rapidly improved in the last several decades, we are now transitioning to even less invasive therapeutic approaches. The role of endoscopy in bariatric surgery is continually evolving. Initially used as a diagnostic tool to assess for concerning anatomy and pathologic processes in the preoperative state, it evolved to include diagnostic and therapeutic use intra- and postoperatively. Flexible endoscopy is allowing surgeons to safely and efficiently address concerning issues that are potentially hazardous to approach from a standard surgical perspective. As endoscopic instrumentation and skill sets continue to improve, we are witnessing a transition of endoscopically assisted surgical procedures to, in some cases, completely endoscopically performed surgical interventions. The recent FDA approval of the Orbera balloon, ReShape balloon, and Aspire device suggests that bariatric endoscopy is entering a commercial era. Insurance coverage and bariatric device training programs will need to be established to promote wider device accessibility. Behavioral modification, learned through counseling and education, is an important component for long-term success of any weight loss intervention, and is an important reason why endoscopic weight loss procedures should be offered only as components of a multidisciplinary weight management approach.

Looking forward, bariatric endoscopy will encompass varying roles, including primary obesity therapy, bridge to surgery, and postbariatric surgical revision. Endoscopic techniques have the potential to reduce adverse events, cost, and recovery times. Despite significant progress by select centers, the generalizability of these techniques awaits evolution of the basic endoscope as well as its effector instruments. Given the broad types of devices in development, and the increasingly diverse indications, the future for endoscopic treatment of obesity and associated diseases is exciting.

REFERENCES

[1]   Flegal KM, Carroll MD, Ogden CL, Curtin LR. Prevalence and trends in obesity among US adults, 1999–2008. JAMA 2010;303:235–41.

[2]   Must A, Spadano J, Coakley EH, et al. The disease burden associated with overweight and obesity. JAMA 1999;282(16):1523–9.

[3]   Sacks FM, Bray GA, Carey VJ, et al. Comparison of weight-loss diets with different compositions of fat, protein, and carbohydrates. N Engl J Med 2009;360(9):859–73.

[4]   Wadden TA, Berkowitz RI, Womble LG, et al. Randomized trial of lifestyle modification and pharmacotherapy for obesity. N Engl J Med 2005;353(20):211–20.

[5]   Wing RR, Hill JO. Successful weight loss maintenance. Annu Rev Nutr 2001;21:323–41.

[6]   Yanovski SZ, Yanovski JA. Long-term drug treatment for obesity: a systematic and clinical review. JAMA 2014;311(1):74–86.

[7]   Astrup A. Drug management of obesity - efficacy versus safety. N Engl J Med 2010;363(3):288–90.

[8]   Smith SR, Weissman NJ, Anderson CM, Behavioural Modification and Lorcaserin for Overweight and Obesity Management (BLOOM) Study Group, et al. Multicenter, placebo-controlled trial of lorcaserin for weight management. N Engl J Med 2010;363(3):245–56.

[9]   Buchwald H, Estok R, Fahrbach K, et al. Weight and type 2 diabetes after bariatric surgery: systematic review and meta-analysis. Am J Med 2009;122(3):248–56.

[10]   Adam TD, Gress RE, Smith SC, et al. Long-term mortality after gastric bypass surgery. N Engl J Med 2007;357(8):753.

[11]   Avidor Y, Still C, Brunner M, et al. Primary care and subspecialty management of morbid obesity: referral patterns for bariatric surgery. Surg Obes Relat Dis 2007;3:392–407.

[12]   ASGE Bariatric Endoscopy Task Force, Sullivan S, Kumar N, et al. ASGE position statement on endoscopic bariatric therapies in clinical practice. Gastrointest Endosc 2015;82(5):767–72.

[13]   Bradley DD, Reavis KM. The role of endoscopy in bariatric surgery. In: Nguyen NT, Blackstone RP, Morton JM, et al., editors. The ASMBS textbook of bariatric surgery. New York, NY: Springer; 2015. p. 391–404.

[14]   Azagury D, Domonceau JM, Morel P, et al. Preoperative work-up in asymptomatic patients undergoing Roux-en-Y gastric bypass: is endoscopy mandatory? Obes Surg 2006;16(10):1304–11.

[15]   Schirmer B, Erenoglu C, Miller A. Flexible endoscopy in the management of patients undergoing Roux-en-Y gastric bypass. Obes Surg 2002;12(5):634–8.

[16]   ASGE Standards of Practice Committee, Evan JA, Muthesamy VR, et al. The role of endoscopy in the bariatric surgery patient. Gastrointest Endosc 2015;81(5):1063–72.

[17]   Korenkov M, Sauerland S, Shah S, et al. Is routine preoperative upper endoscopy in gastric banding patients really necessary? Obes Surg 2006;16:142–6.

[18]   Zeni TM, Frantzides CT, Mahr C, et al. Value of preoperative upper endoscopy in patients undergoing laparoscopic gastric bypass. Obes Surg 2006;16:142–6.

[19]   Nguyen NT, Rivers R, Wolfe BM. Early gastrointestinal hemorrhage after laparoscopic gastric bypass. Obes Surg 2003;13(1):62–5.

[20] Haddad A, Tapazoglou N, Singh K, Averbach A. Role of intraoperative esophagogastroenteroscopy in minimizing gastrojejunostomy-related morbidity: experience with 2,311 laparoscopic gastric bypasses with linear stapler anastomosis. Obes Surg 2012;22:1928—33.

[21] Wilson JA, Romangnuolo J, Byrne TK, et al. Predictors of endoscopic findings after Roux-en-Y gastric bypass. Am J Gastroenterol 2006;101:2194—9.

[22] Schirmer BD. The role of endoscopy in bariatric surgery. In: Nguyen NT, Scott-Conner CE, editors. The SAGES manual, Advanced laparoscopy and endoscopy. New York, NY: Springer; 2012. p. 73—96.

[23] Huang C. The role of the endoscopist in a multidisciplinary obesity center. Gatrointest Endosc 2009;70(4):763—7.

[24] Carrodeguas L, Szomstein S, Zundel N, et al. Gatrojejunal anastomotic strictures following laparoscopic Roux-en-Y gastric bypass surgery: analysis of 1291 patients. Sure Obes Relat Dis 2006;2:92—7.

[25] Wetter A. Role of endoscopy after Roux-en-Y gastric bypass surgery. Gastrointest Endosc 2007;66:253—5.

[26] Espinos JC, Turro R, Mata A, et al. Early experience with the incisionless operating platform. OBEs Surg 2013;23:1375—83.

[27] Lopez-Nava G, Bautista-Castano I, Jimenez A, et al. The primary obesity surgery endoluminal (POSE) procedure: one-year patient weight loss and safety outcomes. Sure Obes Relat Dis 2015;11(4):861—5.

[28] Kumar N, Sahdala HN, Shaikh S, et al. Endoscopic sleeve gastroplasty for primary therapy of obesity: initial human cases. Gastroenterology 2014;146:S571—2.

[29] Abu Dayyeh BK, Rajan E, Gostout CJ. Endoscopic sleeve gastroplasty: a potential endoscopic alternative to surgical sleeve gastrectomy for treatment of obesity. Gastrointest Endosc 2013;78(3):530—5.

[30] Lopez-Nava G, Galvao MP, da Bautista-Castano I, et al. Endoscopic sleeve gastroplasty for the treatment of obesity. Endoscopy 2015;47 (5):449—52.

[31] Lopez-Nava G, Galvao MP, da Bautista-Castano I, et al. Endoscopic sleeve gastroplasty: how I do it? Obes Surg 2015;25(8):1534—8.

[32] Fogel R, De Fogel J, Bonilla Y, et al. Clinical experience of transoral suturing for an endoluminal vertical gastroplasty: 1-year follow-up in 64 patients. Gastrointest Endosc 2008;68(1):51—8.

[33] Brethauer SA, Chand B, Schauer PR, et al. Tranoral gastric volume reduction as intervention for weight management: 12-month follow-up of TRIM trial. Sure Obes Relat Dis 2012;8(3):296—303.

[34] Familiari P, Costamagna G, Blero D, et al. Transoral gastroplasty for morbid obesity: a multicenter trial with a 1-year outcome. Gastrointest Endosc 2011;74(6):1248—58.

[35] Verlaan T, Paulus GF, Mathus-Vliegen EM. Endoscopic gastric volume reduction with a novel articulating plication device is safe and effective in the treatment of obesity. Gastrointest Endosc 2015;81(2):312—20.

[36] de Jong K, Mathus-Vliegen EM, Veldhuyzen EA, et al. Short-term safety and efficacy of the trans-oral endoscopic restrictive implant system for the treatment of obesity. Gastrointest Endosc 2010;72(3):497—504.

[37] Gersin KS, Rothstein RI, Rosenthal RJ, et al. Open-label, sham-controlled trial of an endoscopic duodenojejunal bypass liner for preoperative weight loss in bariatric surgery candidates. Gastrointest Endosc 2010;71(6):976—82.

[38] Rohde U, Hedback N, Gluud LL, et al. Effect of the EndoBarrier Gastrointestinal Liner on obesity and type 2 diabetes: systematic review and meta-analysis. Diabetes Obes Metab 2015;5.

[39] https://clinicaltrials.gov/ct2/show/NCT01728116.

[40] Sandler BJ, Rumbaut R, Swain CP, et al. Human experience with an endoluminal endoscopic, gastrojejunal bypass sleeve. Sure Endosc 2011;25(9):3028—33.

[41] Sandler BJ, Rumbaut R, Swain CP, et al. One-year human experience with a novel endoluminal, endoscopic gastric bypass sleeve for morbid obesity. Sure Endosc 2015;29(11):3298—303.

[42] Khashab MA, Kumbhari V, Grimm IS, et al. EUS-guided gastroenterostomy: the first US clinical experience. Gastrointest Endosc 2015;82(5):932—8.

[43] Ryou M, Agoston AT, Thompson CC. Endoscopic intestinal bypass creation using self-assembling magnets in a porcine model. Gastrointest Endosc 2015;5107(14): 03039-4.

[44] http://giwindows.com/news/view/gi_windows_announces_data_from_first_ever_clinical_study.

[45] ASGE Bariatric Endoscopy Task Force; ASGE Technology Committee, Abu Dayyeh BK, Endmundowicz SA, et al. Endoscopic bariatric therapies. Gastrointest Endosc 2015;81(5):1073—86.

[46] Sullivan S, Stein R, Jonnalagadda S, et al. Aspiration therapy leads to weight loss in obese subjects: a pilot study. Gastroenterology 2013;145 (6):1245—62.

[47] ASGE Bariatric Endoscopy Task Force and ASGE Technology Committee. ASGE bariatric task force systematic review and meta-analysis assessing the ASGE PIVI thresholds for adopting endoscopic bariatric therapies. Gastrointest Endosc 2015;82(3):425—38.

[48] Imaz I, Martinez-Cervell, Garcia-Alvarez EE, et al. Safety and effectiveness of the intragastric balloon for obesity. A meta-analysis. Obes Surg 2008;18(7):841—6.

[49] Genco A, Bruni T, Doldi SB, et al. BioEnterics intragastric balloon: the Italian experience with 2,515 patients. Obes Surg 2005;15(8):1161—4.

[50] Dumonceau JM, Francois E, Hittelet A, et al. Single vs repeated treatment with the intragastric balloon: a 5-year weight loss study. Obes Surg 2010;20(6):692—7.

[51] Zerrweck C, Maunoury V, Caiazzo R, et al. Preoperative weight loss with intragastric balloon decreases the risk of significant adverse outcomes of laparoscopic gastric bypass in super-super obese patients. Obes Surg 2012;22(5):777—82.

[52] Forlano R, Ippolito AM, Iacobellis A, et al. Effect of the BioEnterics intragastric balloon on weight, insulin resistance, and liver steatosis in obese patients. Gastrointest Endosc 2010;71(6):927—33.

[53] Abu Dayyeh BK, Eaton LL, Woodman G, et al. A randomized, multicenter study to evaluate the safety and effectiveness of an intragastric balloon as an adjunct to a behavioural modification program, in comparison with a behavioural modification program alone in the weight management of obese subjects. Digestive Disease Week 2015, Washington DC.

[54] Ponce J, Woodman G, Swain J, et al. The reduce pivotal trial: a prospective, randomized controlled pivotal trial of a dual intragastric balloon for the treatment of obesity. Sure Obes Relat Dis 2015;11(4):874–81.

[55] Giardiello C, Borrelli A, Silvestri E, et al. Air-filled vs water-filled intragastric balloon: a prospective randomized study. Obes Surg 2012;22:1916–19.

[56] de Castro ML, Morales MJ, Martinez-Olmos MA, et al. Safety and effectiveness of gastric balloons associated with hypocaloric diet for the treatment of obesity. Rev Esp Enferm Dig 2013;105:529–36.

[57] Mion F, Ibrahim M, Marjoux S, et al. Swallowable obalon gastric balloons as an aid for weight loss: a pilot feasibility study. Obes Surg 2013;23:730–3.

[58] Marinos G, Eliades C, Muthusamy V, et al. First clinical experience with the TransPyloric Shuttle device, a non-surgical endoscopic treatment for obesity: Results from a 3-month and 6-month study. SAGES, 2013:abstract.

[59] Sauer N, Rosch T, Pezold J, et al. A new endoscopically implantable device (Satisphere) for treatment of obesity—efficacy, safety, and metabolic effects on glucose, insulin and GLP-1 levels. Obes Surg 2013;23:1727–33.

[60] Heneghan HM, Yimcharoen P, Brethauer SA, et al. Influence of pouch and stoma size on weight loss after gastric bypass. Sure Obes Relat Dis 2012;8:408–15.

[61] Thompson CC, Chand B, Chen YK, et al. Endoscopic suturing for transoral outlet reduction increases weight loss after roux-en-y gastric bypass surgery. Gastroenterology 2013;145(1):129–37.

[62] Kumar N, Thompson CC. Comparison of a superficial suturing device with a full-thickness suturing device for transoral outlet reduction. Gastrointest Endosc 2014;79(6):984–9.

[63] Abu Dayyeh BK, Jirapinyo P, Weitzner Z, et al. Endoscopic sclerotherapy for the treatment of weight regain after roux-en-y gastric bypass: outcomes, complications, and predictors of response in 575 procedures. Gastrointest Endosc 2012;76:275–82.

[64] Heylen AM, Jacobs A, Lybeer M, et al. The OTSC-clip in revisional endoscopy against weight gain after bariatric gastric bypass surgery. Obes Surg 2011;21(10):1629–33.

[65] Horgan S, Jacobsen G, Weiss GD, et al. Incisionless revision of post-Roux-en-Y bypass stomal and pouch dilation: multicenter registry results. Surg Obes Relat Dis 2010;6:290–5.

Chapter 16

Intragastric Balloon for the Treatment of Morbid Obesity

M. Lorenzo[1] and A. Genco[2]

[1]ASL NA 3 SUD, Torre Annunziata, NA, Italy, [2]La Sapienza University, Rome, Italy

LIST OF ABBREVIATIONS

BE	binge eating
BIB	bioenterics intragastric balloon
BMI	body mass index
EDNOS	eating disorders not otherwise specified
LSG	laparoscopic sleeve gastrectomy
NBE	not binge eating
PWS	prader willi syndrome

Bariatric surgery is generally considered the only therapeutic option for patients with morbid obesity who are affected by life-threatening comorbidities. A wide range of surgical options are available. [1] The standardization of surgical techniques and protocols is also a work in progress that differs according to country and the experience of the surgical staff. A contemporary, and less invasive, endoscopic approach is positioning of an intragastric balloon. This minimally invasive approach is potentially attractive to healthcare practitioners who have experienced poor results with dietary programs, drug treatment, and behavioral therapy [2,3]. Moreover, intragastric balloon positioning has been recommended as a weight reduction adjuvant therapy prior to bariatric surgery and all kinds of planned surgery in the morbidly obese to reduce life-threatening comorbidities and lower surgical risk [1−4].

Treatment of obesity with only intragastric balloon positioning has also recently been hypothesized.

A LITTLE HISTORY OF THE INTRAGASTRIC BALLOON

The use of intragastric devices to promote weight reduction is not novel [5−6]. Several authors over the years have tested different types of balloon devices as they were thought to be promising and less invasive than surgery for the treatment of morbid obesity [7−11]. During the 1990s several prospective and controlled studies evidenced that balloons, such as Garren−Edwards gastric bubbles and Ballobes, had no significant adjuvant effects on weight reduction in morbidly obese patients [6−11]. Reasons for this were considered to be the small volume of the balloon (220 mL for Garren−Edwards and 400 mL for Ballobes), the air filling having no weight effect on stomach walls, and the cylindrical shape of these devices. In addition, these devices had high rates of complications (Mallory−Weiss tears: 11%; gastric ulcer: 14%; gastric erosion: 26%; and complete deflations, migrations, and intestinal obstruction) [6−11].

After these unsuccessful experiences, a meeting was organized in Tarpon Springs to determine the specifications for the "ideal" intragastric balloon: spherical shape, high and variable volume capacity (400−700 mL), and a saline solution filling. [12]

INTRAGASTRIC BALLOONS: POSITIONING, INFLATIONS, ADJUSTABILITY, AND REMOVAL

For several years only the bioenterics intragastric balloon (BIB; Orbera, Allergan, Irvine, USA) was used, and improved according to physician feedback. This intragastric balloon, according to suggestions made at the Tarpon Springs meeting, has a spherical shape, high-volume capacity (400−700 mL), and uses saline solution as its filling [13]. Almost all kinds of intragastric balloons were positioned with the patient in a lateral decubitus position. They were usually inserted during conscious sedations (propofol: 2 mg/kg) and under endoscopic surveillance [14]. Intubation was eventually carried out in patients with body mass index (BMI) > 40 kg/m^2 affected by sleep apnea or Chronic Obstructive Pulmonary Disease; this was a preanesthesia protocol. After positioning, the balloon was filled with saline (500−700 cc + 10 cc of methylene blue). Prospective studies and extensive clinical experience have shown that its positioning is safe, with low complication rates and mortality [13,15−17]. The extraction after 6 months was done with a dedicated instrument or via needle catheter puncture, and with an endoscopic grasp under endoscopic vision.

More recently, other intragastric balloons have been commercialized. These new devices have different characteristics such as an air filling, adjustability, positioning by swallowing a capsule, or multiple balloon positioning.

Several articles have reported prospective and retrospective experience with a new air-filled balloon [18,19]. This device also adheres to many Tarpon Springs criteria as it is smooth and spherical-shaped, with an X-ray opaque marker (Heliosphere bag: Hescopie, Vienne, France). This high-volume capacity balloon is made with a polyurethane reservoir with an external silicone envelope. The defining characteristic of this device is the higher volume (>900 mL vs 400/700 mL) and the lower weight (30 g vs 520 g at 500 cc) as compared to BIB [19]. These basic differences suggest a different mechanism of action between these two intragastric devices, but studies on this argument are lacking. Despite these differences, the weight loss parameters were similar. Other studies underscored the lack of significant differences in terms of weight loss obtained with these intragastric balloons; the differences were with regards to tolerance and safety [18,19].

According to patient exigencies, Spatz balloons (Spatz Intragastric Balloon) were projected and manufactured to be adjustable and longstanding as compared to all other intragastric balloons [20,21]. Spatz patients were usually placed under unconscious sedation, but rarely under general anesthesia. The device was pushed through the mouth with a guidewire. Once the balloon passed the cardiac region, it was inflated by injecting saline (500 mL). The adjustment procedure for the Spatz device was performed by extracting the filling tube, exclusively by direct endoscopic surveillance, which left the balloon in place. This procedure usually took no longer than 15−20 minutes. Spatz balloons were deflated via the inflation valve or via needle catheter puncture, and then extracted with a standard polypectomy snare. Different generations of this balloon have been commercialized, but patients' tolerance was low and the rate of complications linked to the system were still higher as compared to other balloons [20,21].

The Obalon Gastric Balloon (Obalon Therapeutics, San Diego, CA, USA) is a 250 cc gas-filled, less than 6 g-weight device that is composed of strong, light, multilayer, smooth, and bioresistant nylon and polyurethane material. The balloon has a small, self-sealing, smooth, and flush valve that is radiopaque under X-ray (to check balloon position). This device was projected and manufactured to be swallowable, thus minimizing the use of endoscopy and providing an adjustable effect by positioning one, two, or three balloons. Usually two balloons were placed together in the same session, while a third balloon was ingested on demand. The patient swallowed a capsule attached to a microcatheter, with the balloon inside this gelatin capsule. Once in the stomach (verified by fluoroscopy and the inflation system gauge), the balloon was remotely inflated with gas (nitrogen) using the microcatheter (71 cm, 2 French) without endoscopy and/or anesthesia. After inflation, the microcatheter (extended from the stomach to the mouth) was detached and removed, which left the balloon in the stomach. [22] (Table 16.1).

TABLE 16.1 Main Characteristics of Most Diffuse Intragastric Balloons

Balloon	Material	Volume (mL)	Filling	Positioning duration (months)
BIB (Orbera)	Silicone	500−700	Saline	6
Heliogast bag	Polyurethane	900	Air	6
Spats balloon	Silicone	500−700	Saline	9
Obalon	Nylon + polyuretane	250	Nitrogen	3

PATIENT SELECTION AND MANAGEMENT

Patients who underwent intragastric balloon positioning were usually selected according to National Institutes of Health criteria and guidelines for obesity surgery [23], and were independently evaluated by different members of the bariatric staff (internists, dieticians, and psychologists). Exclusion criteria included all conditions that precluded safe endoscopy, such as esophagitis, a large hiatal hernia (>5 cm), chronic therapy with steroids or nonsteroidal drugs, active peptic ulcer or its previous complications, previous gastric surgery, lesions considered at risk for bleeding, or anticoagulant therapy, pregnancy, and a physical inability to maintain regular follow-ups.

On the first postoperative day, isotonic liquids with proton pump inhibitors, ondansetron (8 mg/die), and butylscopolamine bromide (20 mg $\times$ 3/die) were given to all patients. After the fifth postoperative day they began a progressive solid 1000 kcal diet (glucide, 146 g; lipids, 68 g; proteins, 1 g/kg Ideal Weight).

In those patients who underwent a second, liquid-filled, consecutive balloon positioning, after removal of the first balloon, patients were given 1 month of medical therapy. This was followed by a second balloon positioning according to the same protocol as the first positioning; it was removed after 6 months. Patients were followed weekly during the first month, then monthly until balloon removal. Patients were discharged the same day if the balloon was well-tolerated, or during the second or third day postplacement.

It was mandatory to respect alimentary rules for these patients in order to avoid life-threatening complications such as gastric perforation or ab ingestis bronchopneumonia.

INTRAGASTRIC BALLOON EFFICACY AND SAFETY

Safety and effectiveness of intragastric balloons are the most-studied aspects of BIB. In multicenter reports and meta-analyses, it has been shown that the mortality related to the devices was low, ranging between 2.5% and 4.2%. Severe complications such as gastric perforation and intestinal obstruction were considered exceptional (0.2%). Patients intolerance was considered the main reason for balloon removal during the first week. There is a lack of data for large series of patients treated with Heliosphere bags, Spatz balloons, and Obalon. In comparative studies it has been demonstrated that complications due to Heliosphere bags, Spatz balloons, and BIB were similar. Spatz balloons still have technical problems regarding structure and adjustability. Moreover, the main problem with these devices was the patients' discomfort at time of removal [19−21].

The efficacy of intragastric balloons has been reported in several investigations, and confirmed in meta-analysis [15−17]. Efficacy was usually evaluated in terms of BMI, % excess weight loss (EWL), or comorbidity improvement. In a prospective randomized sham study it was demonstrated that BIB was significantly more effective than diet in terms of weight loss. In nonrandomized studies, different metaanalyses showed 14.7 kg or 17.8 (range 4.9−28.1) kg of weight loss. BMI loss was between 4.6 and 9.0 kg/m^2, and mean %EWL was 32.1%. Moreover, the weight loss induced by intragrastric balloons reduced insulin resistance and improved liver damage as shown by alanine aminotrasferase and gamma glutamine transferase. This benefit depended on the decrease of BMI by higher than 10% [24].

Dumonceau observed that patients with a BMI between 30.0 and 39.9 kg/m^2 who have failed to lose weight via other approaches represented the best candidates for BIB therapy; superobese patients preparing for bariatric surgery were good candidates as well [16]. BIB therapy was followed by a significant improvement in comorbidities in the short term. Imaz stated that the use of BIB within a multidisciplinary weight management program was a short-term effective treatment to lose weight, but it was still not yet possible to verify its capacity to maintain the weight loss over a long period of time [15].

Intragastric balloons have also influenced the quality of life, as shown by Mui et al. in a prospective study [25].

Most authors have also emphasized that the efficiency of a single intragastric balloon was time-related in almost all patients. Several months after balloon removal, patients started to regain the weight lost, eventually rising to their initial weight or higher. This fact strongly influenced the real indications on clinical use of intragastric balloons in morbidly obese patients.

COMPARISON WITH MORE INVASIVE SURGICAL PROCEDURES

The safety and efficacy of intragastric balloons have rarely been compared to bariatric surgical procedures. Two studies considered sleeve gastrectomy when it was only the first step in a duodenal switch. Milone et al. compared two nonhomogeneous patients with BMI >50 kg/m^2, and indicated that laparoscopic sleeve gastrectomy (LSG) was superior to BIB positioning for both mean weight loss (45.5 kg vs 22.3 kg) and %EWL (35 vs 24). BMI decreased from 69 to 53 and from 59 to 51 in LSG and BIB patients, respectively. Comorbid conditions decreased in 90% of both groups of patients. LSG was a safe procedure, while intragastric balloon was tolerated by 93% of patients. They concluded that in the superobese, although the BIB procedure showed efficacy in reducing weight, LSG did so faster and to a greater degree [26]. Genco and colleagues, in a case control study, showed that the mean weight loss parameters at time of BIB removal were similar to those observed at 6 months follow up in patients treated by LSG. A similar nonsignificant pattern was observed for comorbidities. After another 6 months of follow-up, patients treated with BIB tended to regain weight, while patients with sleeve gastrectomy continued to lose weight [27].

INDICATIONS

Indications for clinical use of intragastric balloons can be summarized as follows: (1) preoperative weight loss in a bariatric surgery candidate with high anesthesiological risk; (2) temporary weight loss treatment in a patient with a BMI in the range of bariatric surgery (>35) who refuses surgery, has possible low compliance to surgery, or who is on a long waiting list for surgery; and (3) temporary weight loss treatment for a patient with no indications for surgery in the context of an integrated medical approach to obesity treatment (BMI < 35) (Table 16.2).

Most authors agree with the statement that the efficacy of intragastric balloons is time-related in almost all patients [2,3]. Some months after balloon removal patients start to regain the weight lost, eventually rising to their initial weight or higher. This fact strongly influences the indications for clinical use of intragastric balloons in morbidly obese patients. Angrisani et al. reported the need for a bariatric laparoscopic procedure to permit weight maintenance and/or continued weight loss. The intragastric balloon was mainly indicated as a way to improve laparoscopic results and reduce intra- and postoperative complications associated with more invasive procedures [28].

Another question arises from different experiences regarding the influence of intragastric balloons as a predictive tool following bariatric operations.

Predictive Role of Intragastric Balloon on Subsequent Bariatric Surgery

Several investigations have demonstrated that weight loss with a very-low-calorie diet before surgery may have some advantages such as the reduction of visceral adipose tissue, fat liver volume, and a reduction of surgery duration and blood loss [28,29]. Other authors have observed that placement of an intragastric balloon at 6 months added other advantages such as a reduction in the rate of intraoperative complications and laparotomic conversion. For these reasons, in many bariatric centers an intragastric balloon is positioned independent of the type of bariatric procedure. Moreover, it is still unclear if preoperative treatment with an intragastric balloon is predictive of success of the subsequent bariatric procedure. This problem has mainly been studied with Laparoscopic Adjustable Gastric Banding.

Many years ago Loffredo and colleagues, after observation of a weight loss at the 12-month follow-up in 18 patients who underwent gastric banding positioning after intragastric balloon positioning, suggested the use of this device as a test for patient selection for gastric restrictive procedures. They determined that in the case of BIB-test positive ($>10\%$ kg weight loss), the patient could be considered highly eligible for a restrictive gastric surgical procedure [30].

TABLE 16.2 Main Indications of Intragastric Balloons

1. Preoperative positioning:
   a. Any kind of bariatric procedure.
   b. Other surgical procedure (i.e., total knee prosthesis replacement).
2. Patients refusing surgery and entering a program of sequential treatment (two or more intragastric balloons positioning separated by 1 month balloon free at last).
3. Patients without surgical indications:
   a. Out of bariatric range (<35).
   b. Teenagers.
   c. Elderly.
   d. Morbidly obese who are programmed for chemotherapy.

In 2004, Busetto and colleagues, in a case control study with 43 patients who had undergone intragastric balloon positioning before gastric banding placement, compared these patients with those who had undergone gastric banding alone. They showed a reduction in the rate of laparotomic conversion, as well as the risk of intraoperative complications in superobese patients. Significantly better weight loss was observed in the group treated with intragastric balloons during the first 6 months of follow-up; however, at the 3-year follow-up, the mean weight loss tended to be the same [31].

De Goederen-van der Meij and colleagues, in a prospective cohort study with 40 patients with BMI $>40\,kg/m^2$, stated that intragastric balloons did not predict the success of subsequent gastric banding. In this study, after balloon removal, patients were allocated into two groups according to a cut-off of 10% loss of their initial weight. Patients who lost >10% of their initial weight had a significantly lower BMI at 12 months follow-up after gastric band positioning; however, the weight loss between the two groups was the same during this period [32].

Genco and colleagues in a multicenter study on more than 1000 patients observed that the differences in terms of mean BMI and mean %EWL observed at time of balloon removal were maintained after gastric banding after 1 ($p<0.001$), 3 ($p<0.01$), and 5 ($p<0.05$) years, respectively. Moreover, this observation was not adequate enough to affirm that positive results at the time of balloon removal had positive predictive value in the short and medium term for gastric banding outcomes. Results of this study suggested that patients with good intragastric balloon treatment (>25%EWL) gained an advantage in terms of BMI loss and %EWL, and that these results were maintained until short- and medium-term follow-up after gastric banding [33].

INTRAGASTRIC BALLOON AS SUPPORT THERAPY WITHOUT SURGERY

Intragastric balloons are also safe and effective in treating overweight patients with BMIs of 27–30. From this observation it has been suggested that it can be a useful tool for clinicians to prevent progression to obesity and reduce important comorbidities associated with excess weight, and may also have implications for the future management of overweight patients.

Indications for the prolonged use of intragastric balloons are derived from experiences in which patients refused to remove the balloon because of excellent results or longer waiting lists for treatment. Genco et al. conducted several studies prolonging the balloon treatment by inserting a second balloon 1 month after the removal of the first. Moreover, patients were clearly informed that the positioning of a second balloon should not be considered a definitive treatment, but only a bridge for a more invasive bariatric treatment. The second balloon was positioned without difficulty or adjunctive complications, and was tolerated without significant complications. The positioning of a second balloon significantly improved weight loss in terms of BMI and %EWL as compared with patients treated with only one balloon and diet adjustments [34,35].

Intragastric Balloon in Eating Disorders Not Otherwise Specified

The term "eating disorders" refers to a group of conditions characterized by abnormal eating habits that may involve either insufficient or excessive food intake with compensatory behavior (i.e., purging or restrictive). In the Diagnostic and Statistical Manual of Mental Disorders—Text Revision (DSM-IV-TR), there are three principal categories of eating disorders: anorexia nervosa, bulimia nervosa, and eating disorders not otherwise specified (EDNOS; i.e., binge eating, BE). In bariatric practice, eating behavior before surgery is still considered a great predictive factor for the postoperative results in terms of weight loss or regain. Not only bulimia, but also other eating disorders (like EDNOS), could influence the outcomes, and could be influenced by the bariatric procedure performed [36,37]. This influence on the outcome of bariatric surgery is not yet well established, but the presence of BE status is considered a major predictive factor for the postoperative course of body weight after gastric restrictive operations [36,37].

Positioning of the intragastric balloon in morbidly obese binge eaters has shown different results depending upon the experience. Puglisi and colleagues showed that intragastric balloon treatment achieved significant weight loss in BE and nonbinge eating (NBE) obese patients. However, the degree of BMI reduction in BE patients was significantly less than in the NBE patients. Moreover, the rate of complications and failure in the BE group was statistically higher than in the NBE group. These results suggested that the presence of BE behavior is a negative predictive factor for the success of the treatment [38]. On the contrary, the study by Genco et al., in which they positioned the same manufacturer balloon for treatment of morbid obesity, showed a significant weight loss in BE and NBE patients, both at 6 months and at 13 months in those patients in whom a second balloon was positioned 1 month after removal of the first balloon [39].

INTRAGASTRIC BALLOON IN TEENAGERS AND ADOLESCENTS

Experiences with intragastric balloon positioning in morbidly obese adolescents are scarce, and usually presented together with experiences in adults. Sometimes experiences with particular subgroups of obese adolescents have been published. De Peppo et al. reported their experience in young patients affected by Prader–Willi Syndrome (PWS). Obesity in PWS is progressive, severe, and resistant to dietary, pharmacological, and behavioral treatment. A body weight reduction is mandatory to reduce the risk of cardio–respiratory and metabolic complications. De Peppo and colleagues positioned 21 BIBs in 12 PWS patients. In five patients, BIB treatment was repeated more than once. They concluded that when noninvasive pharmacological therapies fail, BIB may be effective for controlling body weight in PWS patients with morbid obesity, particularly if treatment is started in early childhood [40].

Nobili et al. recently reported their experience with a new kind of intragastric balloon (Obalon Intragastric Balloon). In this study they evaluated the efficacy of this intragastric balloon on weight loss and on metabolic and cardiovascular parameters in a pediatric population with severe obesity. Ten children with severe obesity underwent Obalon positioning [41]. The researchers concluded that the Obalon had a positive effect on the decrease of weight, BMI, and percentage of excess body weight within 3 months of placement. Moreover, this safe, minimally invasive device improved the cardio–metabolic profiles of obese children. They concluded that Obalon could be a useful tool in the difficult management of pediatric patients with morbid obesity, inducing a meaningful weight loss in the short-term.

Although lifestyle interventions and intragastric balloons are effective in the short term, they are often ineffective in the long-term treatment of pediatric obesity. Data on weight recidivism after intragastric balloon insertion in adolescents are lacking, but the belief is that as in adults, the weight returns to the starting level in a couple of years.

CONCLUSIONS

Bariatric surgery seems to be the only therapeutically valid option for superobese patients at this time. In the patients who have life-threatening complications, treatment with intragastric balloons is a change intended to reduce the surgical and anesthesiological risks of both bariatric and nonbariatric surgical procedures. In overweight patients, and in those who refuse the subsequent, more invasive, bariatric step, the intragastric balloon seems to be a valid option to improve comorbidities and avoid progression of the disease.

KEY FACTS

In 1991 in prelaparoscopic era, the statement of National Institute of Health restrict the bariatric surgical procedures to adult patients with BMI ≥ 40 Kg/m^2 or in patients with BMI > 35 Kg/m^2 in the presence of life treathening comorbidities such as diabetes or hypertension, and previous experience of failed dietetics attempts.

After the worldwide diffusion of laparoscopic approach, and the increased baritric experience it has been clear that comorbidities have a major role in the surgical indications. Moreover the original indications of NIH based on BMI are still considered in the majority of baritric centers.

SUMMARY POINTS

- The intragastric balloon for temporary weight loss in morbidly obese patients is used worldwide.
- An intragastric balloon can be positioned before every type of bariatric and non bariatric procedure.
- Use of intragastric balloons lowers anesthesiological risk, intra- and postoperative complications.
- Different kind of intragastric balloon are commercially available, with similar effects on weight loss.
- Mortality rate is very low.
- The main reason of early balloon removal is the psychological patient intolerance.
- Patients undergoing intragastric balloon positioning must be carefully selected, according to NIH criteria.
- Soon after balloon removal a bariatric surgical procedure should be performed, as the majority of patients regain their weight lost in a few months.
- Recently, good results of positioning a second intragastric balloon after the first has been obtained.
- In selected patients this surgical-less therapy prevents morbid obesity, and its life-threatening complications.

REFERENCES

[1] Angrisani L, Lorenzo M. Bariatric super worldwide: overview and results. In: Foletto M, Roenthal RJ, editors. The globesity challenge to general surgery. London: Springer; 2014. p. 237—46.

[2] Weiner A, Gutberlet H, Bockhorn H. Preparation of extremely obese patients for laparoscopic gastric banding by gastric balloon therapy. Obes Surg 1999;261—4.

[3] De Waele B, Reynaert H, Urbain D, Willelms G. Intragastric balloons for preoperative weight reduction. Obes Surg 2000;10:58—60.

[4] Genco A, Cipriano M, Bacci V, et al. Bioenterics intragastric balloon (BIB): a double blind, randomised, controlled, cross-over study. Int J Obes 2006;30:129—33.

[5] Nieben OG, Harboe H. Intragatric balloon as an artificial bezoar for treatment of obesity. Lancet 1982;1:198—9.

[6] McFarland RJ, Grundy A, Gazet JC, et al. The intragastric balloon: a novel idea proved ineffective. Br J Surg 1987;74:137—9.

[7] Ramhamadany EM, Fowler J, Baird IM. Effect of the gastric balloon versus sham procedure on weight loss in obese subjects. Gut 1989;30:1054—7.

[8] Benjamin SB, Maher KA, Cattau EL, et al. Double blind controlled trial of the Garren-Edward gastric bubble: an adjunctive treatment for exogenous obesity. Gastroenterology 1988;95:581—8.

[9] Mathus-Vliegen EMH, Tytgat GNJ. Intragastric balloons for morbid obesity: results, patient tolerance and balloon life span. Br J Surg 1990;77:77—9.

[10] Hogan RB, Johnston JH, Long BW, et al. A double blind, randomised, sham controlled trial of the gastric bubble for obesity. Gastrointest Endosc 1989;35:381—5.

[11] Meshkinpour H, Hsu D, Farivar S. Effect of gastric bubble as a weight reduction device: a controlled, crossover study. Gastroenterology 1988;95:589—92.

[12] Schapiro M, Benjamin S, Blackburn G, et al. Obesity and the gastric balloon: a comprehensive workshop. Tarpon Springs, Florida, March 19-21, 1987. Gatrointest Endosc 1987;33:323—7.

[13] Genco A, Bruni T, Doldi SB, et al. BioEnterics intragastric balloon: the Italian experience with 2,515 patients. Obes Surg 2005;15:1161—4.

[14] Messina T, Genco A, Favaro R, Maselli R, et al. Intragastric balloon positioning and removal: sedation or general anesthesia? Surg Endosc 2011;25:3811—14.

[15] Imaz I, Martinez-Cervell C, Garcia-Alvarez EE, Sendra-Gutierrez JM, Gonzales-Enriquez J. Safety and effectiveness of the intragastric balloon for obesity. A meta-analysis. Obes Surg 2008;18:841—6.

[16] Dumonceau LM. Evidence based review of the Bioenterics intragastric balloon for weight loss. Obes Surg 2008;18:1611—17.

[17] Spyropoulos C, Katsakoulis E, Mead N, Vagenas K, Kalfarentsos F. Intragastric balloon for high risk supero-bese patients: a prospective analysis of efficacy. Surg Obes Relat Dis 2007;3:78—83.

[18] Caglar E, Dobrucali A, Bal K. Gastric balloon to treat obesity: filled with air or fluid? Dig Endosc 2013;25:502—7.

[19] Giardiello C, Borrelli A, Silvestri E, Antognozzi V, Iodice G, Lorenzo M. Air-filled vs water-filled intragastric balloon: a prospective randomized study. Obes Surg 2012;22:1916—19.

[20] Daniel F, Abou Fadel C, Houmani Z, Salti N. Spatz 3 adjustable intragastric balloon: long-term safety concerns. Obes Surg 2016;26:159—60.

[21] Genco A, Dellepiane D, Baglio G, et al. Adjustable intragastric balloon vs non-adjustable intragastric balloon: case-control study on complications, tolerance, and efficacy. Obes Surg 2013;23:953—8.

[22] Mion F, Ibrahim M, Marjoux S, Ponçhon T, Dugardeyn S, Roman S, et al. Swallowable Obalon® gastric balloons as an aid for weight loss: a pilot feasibility study. Obes Surg 2013;23:730—3.

[23] NIH Conference. Gastrointestinal surgery for severe obesity. Consensus development conference panel. Ann Intern Med 1991;115:956—61.

[24] Ricci G, Bersani G, Rossi A, Pigò F, De Fabritiis G, Alvisi V. Bariatric therapy with intragastric balloon improves liver dysfunction and insulin resistance in obese patients. Obes Surg 2008;18:1438—42.

[25] Mui WLM, Ng EKW, Yuk B, Tsung S, Lam H, Yung MI. Impact on obesity-related illnesses and quality of life following intragastric balloon. Obes Surg 2010;10:1571—4.

[26] Milone L, Strong V, Gagner M. Laparoscopic sleeve gastrectomy is superior to endoscopic intragastric balloon as a first stage procedure for superobese patients (BMI > or = 50). Obes Surg 2005;15:612—17.

[27] Genco A, Cipriano M, Materia A, et al. Laparoscopic sleeve gastrectomy versus intragastric balloon: a case control study. Obes Surg 2009;23:1849—53.

[28] Angrisani L, Lorenzo M, Borrelli V, Giuffré M, Fonderico C, Capece G. Is bariatric surgery necessary after intragastric balloon treatment? Obes Surg 2006;16:1135—7.

[29] Tarnoff M, Kaplan LM, Shikora S. An evidenced-based assessment of preoperative weight loss in bariatric surgery. Obes Surg 2008;18:1059—61.

[30] Loffredo A, Cappuccio M, De Luca M, de Werra C, Galloro O, Naddeo M, et al. Three years experience with the new intragastric balloon, and a preoperative test for success with restrictive surgery. Obes Surg 2001;11:330—3.

[31] Busetto L, Segato G, De Luca M, Bortolozzi E, Maccari T, Magon A, et al. Preoperative weight loss by intragastric balloon in superobese patients treated with laparoscopic gastric banding: a case control study. Obes Surg 2004;14:671—6.

[32] De Goederen-van del Meij S, Pierik RGJM, Oudkerk M, Gouma DJ, Mathus-Vkiegen LM. Six months of balloon treatment does not predict the success of gastric banding. Obes Surg 2007;17:88—94.

[33]   Genco A, Lorenzo M, Baglio G, Furbetta F, Rossi A, et al. Does the intragastric balloon have a predictive role on subsequent lap-band® surgery? Italian multicenter study results at 5-years follow-up. Surg Obes Relat Dis 2015;10:474−8.

[34]   Genco A, Cipriano M, Bacci V, et al. Intragastric balloon followed by diet vs intragastric balloon followed by another balloon: a prospective Study on 100 patients. Obes Surg 2010;20:1496−500.

[35]   Genco A, Maselli R, Cipriano M, Lorenzo M, Basso N, Redler A. Long-term multiple intragastric balloon treatment − a new strategy to treat morbid obese patients refusing surgery. Prospective 6-years follow up study. Surg Obes Relat Dis 2014;10:307−12.

[36]   Burgmer R, Grigutsch K, Zipfel S, et al. The influence of eating behavior and eating pathology on weight loss after gastric restriction operations. Obes Surg 2005;15:684−91.

[37]   de Zwaan M, Hilbert A, Swan-Kremeier L, et al. Comprehensive interview assessment of eating behavior 18-35 months after gastric bypass surgery for morbid obesity. Surg Obes Relat Dis 2010;6:79−85.

[38]   Puglisi F, Antonucci N, Capuano P, et al. Intragastric balloon and binge eating. Obes Surg 2007;17:504−50.

[39]   Genco A, Maselli R, Cipriano M, Frangella F, et al. Effects of consecutive intragastric balloon plus diet versus single BIB plus diet on eating disorders not otherwise specified. (EDNOS). Obes Surg 2013;23:2075−9.

[40]   De Peppo F, Di Giorgio G, Germani M, et al. BioEnterics Intragastric balloon for treatment of morbid obesity in Prader-Willy syndrome. Specific risks and benefits. Obes Surg 2008;18:1443−9.

[41]   Nobili V, Corte CD, Liccardo D, et al. Obalon intragastric balloon in the treatment of pediatric obesity: a pilot study. Pediatr Obes 2015;10:e1−4.

Chapter 17

Jejunoileal Bypass: Physiologic Ramifications of an Obsolete Procedure That Has Resurfaced in Oncologic Surgery

J.P. Kamiński[1], V.K. Maker[1,2] and A.V. Maker[1,2]

[1]University of Illinois Metropolitan Groups Hospitals, Chicago, IL, United States, [2]University of Illinois at Chicago, Chicago, IL, United States

LIST OF ABBREVIATION

JIB jejunoileal bypass

INTRODUCTION

The first prominent surgical procedure for morbid obesity was the jejunoileal bypass (JIB). The very first JIB was described in dogs in 1952, and later in an experimental human subject in 1954 at the University of Minnesota [1]. Subsequently, Payne and colleagues described a case series of JIB that showed weight loss [2]. This procedure became popular in the 1960s and 1970s as it was the most effective surgical intervention at that time. The basis of the effectiveness of the procedure was the malabsorptive component, since the goal of the procedure was to bypass 90% of the small intestine. However, with the rise of safer bariatric alternatives, this procedure was eventually abandoned because its postoperative course was plagued with metabolic complications, including diarrhea, acute liver failure, cirrhosis, renal lithiasis, and renal insufficiency [3].

TECHNIQUE AND VARIATIONS

The technique first described by Payne and colleagues describes transecting the jejunum 15 in. from the ligament of Treitz and performing an end-to-side anastomosis to the transverse colon [2]. However, this resulted in massive diarrhea, dehydration, and severe electrolyte imbalances that ultimately ended in reversing the procedure or converting it to a JIB. Eventually, they advised against it and recommended the JIB [4]. This consists of transecting the jejunum 15 in. from the ligament of Treitz and performing an end-to-side anastomosis approximately 10 in. proximal to the ileocecal valve (Fig. 17.1). This relatively simple maneuver came to be the classic JIB, and resulted in about 90% of patients having satisfactory weight loss.

For the 10% of patients who either regained the weight or did not have satisfactory weight loss due to reflux into the bypassed segment, Scott and colleagues developed a new technique [5]. This involved transecting the jejunum at 30 cm distal to the ligament of Treitz and creating an end-to-end anastomosis with the last 30, 20, or 15 cm of ileum [5−8]. The bypassed small bowel is then drained into the colon with an end-to-side anastomosis to the cecum, transverse colon, or sigmoid colon (Fig. 17.2) [5,7]. This depends on whether the jejunoileal anastomosis is created beneath or superior to the root of the mesentery.

Another variation of the JIB procedure is the so-called biliointestinal bypass (BIB) [9]. This procedure includes the anatomy of the classic JIB, however, the proximal end of the blind jejunal limb is anastomosed to the gallbladder. This allows the bile to flow down the blind end from the gallbladder while the patient fasts as the sphincter of Oddi is

Jejunoileal bypass

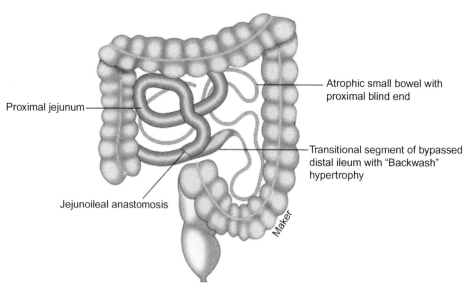

FIGURE 17.1   Anatomy of the jejunoileal bypass (JIB) and surgical considerations. Scott's end-to-end operation involved 30 cm of jejunum anastomosed to the last 30, 20, or 15 cm of ileum and the 90% bypassed small bowel was drained into the colon. The classic JIB anatomy (shown) consisted of 15 in. of jejunum anastomosed side-to-side to the distal 10 in. of ileum. During a period of several years post-JIB, the proximal jejunum and terminal ileum distal to the anastomosis become longer, thickened, and dilated due to adaptive hyperplasia and hypertrophy. The bypassed proximal bowel develops two zones: a proximal longer segment that atrophies in length and diameter, and a second zone of terminal ileum just proximal to the jejunoileal anastomosis. This second transitional zone appears similar to the distal "in-line" terminal ileum due to backwash reflux and, as a result, plays a role in the functional adaptive response of the small bowel [50]. *With permission from Elsevier.*

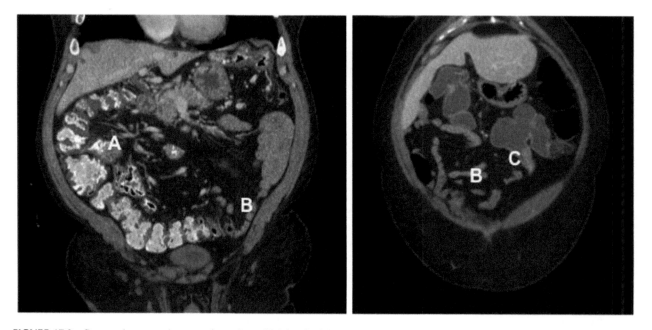

FIGURE 17.2   Computed tomography scan of a patient with jejunoileal bypass. In the classic jejunoileal bypass, oral contrast can be seen in the markedly hypertrophied distal ileum emptying into the cecum (left panel, A), adjacent to a nonoral contrast opacified and atrophied bypassed jejunal segment (B). In a different patient with a variation of the bypass (right panel), the jejunoileal anastomosis is end-to-end and the defunctionalized small bowel is anastomosed to the transverse colon. In this case, there is a tumor in the ileal lumen just proximal to the colon anastomosis (C).

closed. This method was supposed to curb diarrhea significantly compared to the classic JIB [10]. Other variations of the JIB procedure include automatic stapling, plication of the ileum [11], or formation of a valve [12].

ADVERSE EVENTS AND NUTRITIONAL FOLLOW-UP

The most common adverse event after a JIB is diarrhea, which can lead to severe electrolyte imbalances, especially hypokalemia [13]. The abnormalities to monitor in the immediate postoperative course include hypokalemia, hypomagnesemia, and hypocalcemia. Most of these imbalances can be controlled with oral supplementation until the diarrhea comes under control. A small percentage of patients have to be hospitalized for parenteral supplementation for more severe deficits. Micronutrient deficiencies, including the fat-soluble vitamins A, D, E, K, and B12, can also be encountered given the altered enterohepatic cycle. Having the patient avoid excessive intake of liquids and foods high in fat may help in the control of diarrhea. The diarrhea usually resolves within 12−18 months as the remaining bowel adapts to increase absorption of nutrients [14].

The most feared complication of JIB is hepatic failure, which occurs in 7−10% of patients, and can range from asymptomatic elevations in liver enzymes to death secondary to acute liver failure [3,15,16]. When present, the latter usually occurs within the first 2 years after JIB. Multifactorial reasons for hepatic insufficiency include increased fat deposition, protein deficiency, and exposure to toxic substances from bacterial overgrowth and bile salts [17−21].

Gallstones and kidney stones are also common in patients' status post-JIB [13]. Thirty percent of morbidly obese patients are found to have biliary calculi before bypass surgery, and afterwards the incidence increases by 5% per year [22−24]. Urinary calculi and nephrocalcinosis incidence varies between 1% and 32% within the first 2 years [25−28]. Lithogenic factors for gallstones include alterations in cholesterol and bile-salt metabolism and bacterial infection [22,24,29], while nephrolithiasis results from increased absorption of exogenous oxalates, especially in the colon, and alterations in either the hepatic or enteric oxalate pathway [25].

Bone disease is also a common sequela of the JIB, with incidence of up to 48% [30−32]. Alterations in vitamin D metabolism is attributed to malabsorption, steatorrhea, liver disease, and the blind loop syndrome [33]. Further, arthromyalgia occurs in up to 20% of JIB patients [34,35]. This is associated with dermatitis bypass syndrome, antibodies to bacteria, and a blind loop syndrome with bacterial overgrowth [36,37].

Renal insufficiency occurs in up to 9% of JIB patients, with 2% requiring permanent dialysis and <1% undergoing renal transplantation [3]. The mechanisms of renal failure in the JIB population include risk of renal stones developing, prerenal factors from decreased bowel absorptive surface area and chronic loss of fluid in diarrhea, and oxalate deposition damaging the renal parenchyma [38−40]. Clinical gout is rare in JIB patients, although transitory hyperuricemia is common. Most patients hospitalized for gout after JIB had a history of gout before the surgery [7,41].

Intestinal complications included intussusception, bypass enteropathy, intestinal pneumatosis, pseudoobstruction, transmural ileocolitis, and the blind loop syndrome [42]. Colonic pseudoobstructions can be treated with antibiotics and conservative management even in the setting of intestinal pneumatosis. Surgical resection should be considered only when bleeding or obstruction is observed clinically.

CANCER RISK

The most common source of death more than 1 year after surgery in patients undergoing bariatric surgery in the era of JIB is malignancy [43]. The JIB patient has a presumed increased risk of colorectal cancer due to several mechanisms of the unique anatomy, though clinical studies have shown conflicting results [44−46]. The colon in the JIB patient is exposed to up to 10 times the amount of bile acids, which are a known carcinogen to the colonic mucosa [47,48]. Further, increased cell proliferation has been found in the colon of JIB patients compared to non-JIB patients, and JIB animal models show increased colon cancers compared to sham-operated animals [49].

Once a JIB patient is diagnosed with a gastrointestinal luminal cancer, attention must be taken preoperatively and intraoperatively given the unique anatomy and consequences of resection (Fig. 17.2) [50]. Specifically, the previous operative report determines if the patient has an intact ileocecal valve, the length of functional small bowel, and the amount of functional colon. This should be confirmed with enteral contrast plain films and/or computed tomography (CT) enterography. Intraoperatively, the jejunal and ileal segments involved in nutrient absorption are markedly hypertrophic and should be measured to determine the length of functional small bowel. Added to this distance is the distal hypertrophic length of bypassed ileum caused by reflux at the jejunoileal anastomosis, since this portion is also engaged in nutrient absorption. Treatment for potential malignancy is performed with the appropriate oncologic resection. The decision to reverse the bypass depends on multiple factors. The bypass may not necessarily need to be reversed after resection if the combined length of functional small bowel is 50–70 cm with an intact colon; there are 50–70 cm of functional small bowel with an intact ileocecal valve; or there is 100–150 cm of functional small bowel without any colon or an intact ileocecal valve [51]. If a reversal is necessary, it is safe to perform during the same operation, but a nutritional bridge with parenteral nutrition may be necessary.

In summary, the JIB is a historical malabsorptive bypass procedure with high morbidity and mortality, resulting in reversals in up to 30% of patients. Knowledge of the anatomy of this procedure and its metabolic ramifications are important in the design of new bariatric procedures and in the surgical management of these patients who are now presenting with malignancy or requiring gastrointestinal surgery.

MINI-DICTIONARY OF TERMS

- *Jejunoileal bypass*: A malabsorptive operation where 90% of the small intestine is bypassed to help morbidly obese patients lose weight.
- *Payne bypass*: The "classic" JIB procedure where the proximal 15 in. of the jejunum are anastomosed in an end-to-side fashion to the distal terminal ileum.
- *Scott bypass*: A variation of the original JIB where the proximal 30 cm of jejunum is anastomosed in an end-to-end fashion to the distal 30, 20, or 15 cm of the terminal ileum, with the bypassed bowel drained into the colon.
- *Biliointestinal bypass*: A variation of the original JIB where the "classic" anatomy is maintained except with the proximal end of the blind jejunal limb attached to the gallbladder.
- *Functional small bowel*: The bowel in a JIB patient that is in continuity with the working gastrointestinal tract and/or involved in nutrient absorption. It is a factor in determining need for reversal of JIB after resection.

KEY FACTS

- JIB is a malabsorptive procedure that bypasses 90% of the small intestine.
- The Payne procedure, Scott procedure, and the BIB are all variations of the JIB.
- JIB was the most effective weight loss procedure at the time.
- The most common complication is diarrhea.
- The most feared complication is hepatic insufficiency.
- Other common complications are biliary lithiasis, nephrolithiasis, bone disease, renal insufficiency, and intestinal complications.
- JIB has been largely abandoned for other less debilitating procedures.

SUMMARY POINTS

- The JIB is a malabsorptive operation designed to treat morbidly obese patients.
- The classic procedure bypasses 90% of the small bowel by transecting the jejunum about 15 in. distal to the ligament of Treitz and performing an end-to-side anastomosis to the distal ileum.
- The JIB was the most effective weight loss operation at the time; however, a number of metabolic complications have been associated with this procedure, and it has largely been abandoned in favor of less debilitating procedures.
- The specific adverse events associated with the JIB procedure include electrolyte abnormalities, liver insufficiency, biliary lithiasis, nephrolithiasis, bone disease, renal insufficiency, and intestinal complications.
- A potential increased risk of colorectal cancer is associated with this procedure, and specific operative management is warranted when removing gastrointestinal tumors in these patients.

REFERENCES

[1] Kremen AJ, Linner JH, Nelson CH. An experimental evaluation of the nutritional importance of proximal and distal small intestine. Annu Surg 1954;140(3):439–48.
[2] Payne JH, Dewind LT, Commons RR. Metabolic observations in patients with jejunocolic shunts. Am J Surg 1963;106:273–89.
[3] Requarth JA, Burchard KW, Colacchio TA, Stukel TA, Mott LA, Greenberg ER, et al. Long-term morbidity following jejunoileal bypass. The continuing potential need for surgical reversal. Arch Surg 1995;130(3):318–25.
[4] Payne JH, DeWind LT. Surgical treatment of obesity. Am J Surg 1969;118(2):141–7.
[5] Scott Jr. HW, Sandstead HH, Brill AB, Burko H, Younger RK. Experience with a new technic of intestinal bypass in the treatment of morbid obesity. Annu Surg 1971;174(4):560–72.

[6] Scott HW. The surgical management of patients with morbid obesity. J R Coll Surg Edinb 1977;22(4):241–54.

[7] Scott Jr. HW, Dean R, Shull HJ, Abram HS, Webb W, Younger RK, et al. Considerations in use of jejunoileal bypass in patients with morbid obesity. Annu Surg 1973;177(6):723–35.

[8] Scott Jr HW, Dean RH, Shull HJ, Gluck F. Results of jejunoileal bypass in two hundred patients with morbid obesity. Surg Gynecol Obstet 1977;145(5):661–73.

[9] Hallberg D, Holmgren U. Bilio-intestinal shunt. A method and a pilot study for treatment of obesity. Acta chirurgica Scandinavica 1979;145 (6):405–8.

[10] Hallberg D. A survey of surgical techniques for treatment of obesity and a remark on the bilio-intestinal bypass method. Am J Clin Nutr 1980;33(suppl. 2):499–501.

[11] Starkloff GB, Stothert JC, Sundaram M. Intestinal bypass: a modification. Annu Surgery 1978;188(5):697–700.

[12] Wiklund B, Hallberg D. A reflux-preventing valve in jejunoileal bypass. Acta chirurgica Scandinavica Supplementum 1978;482:77.

[13] Scott Jr. HW, Dean RH, Shull HJ, Gluck FW. Metabolic complications of jejunoileal bypass operations for morbid obesity. Annu Rev Med 1976;27:397–405. Available from: http://dx.doi.org/10.1146/annurev.me.27.020176.002145.

[14] Wright HK, Tilson MD. The short gut syndrome: pathophysiology and treatment. Curr Probl Surg 1971;3–51.

[15] Hocking MP, Duerson MC, O'Leary JP, Woodward ER. Jejunoileal bypass for morbid obesity. Late follow-up in 100 cases. N Engl J Med 1983;308(17):995–9. Available from: http://dx.doi.org/10.1056/NEJM198304283081703.

[16] Fikri E, Cassella R. Jejunoileal bypass for massive obesity: results and complications in fifty-two patients. Annu Surg 1974;179(4):460–4.

[17] O'Leary JP, Maher JW, Hollenbeck JI, Woodward ER. Pathogenesis of hepatic failure after obesity bypass. Surg Forum 1974;25(0):356–9.

[18] Moxley III RT, Pozefsky T, Lockwood DH. Protein nutrition and liver disease after jejunoileal bypass for morbid obesity. N Engl J Med 1974;290(17):921–6. Available from: http://dx.doi.org/10.1056/NEJM197404252901701.

[19] Holzbach RT, Wieland RG, Lieber CS, DeCarli LM, Koepke KR, Green SG. Hepatic lipid in morbid obesity. Assessment at and subsequent to jejunoileal bypass. N Engl J Med 1974;290(6):296–9. Available from: http://dx.doi.org/10.1056/NEJM197402072900602.

[20] Hollenbeck JI, O'Leary JP, Maher JW, Woodward ER. An etiologic basis for fatty liver after jejunoileal bypass. J Surg Res 1975;18(2):83–9.

[21] Brown RG, O'Leary JP, Woodward ER. Hepatic effects of jejunoileal bypass for morbid obesity. Am J Surg 1974;127(1):53–8.

[22] Halverson JD, Wise L, Wazna MF, Ballinger WF. Jejunoileal bypass for morbid obesity. A critical appraisal. Am J Med 1978;64(3):461–75.

[23] Payne JH, DeWind L, Schwab CE, Kern WH. Surgical treatment of morbid obesity. Sixteen years of experience. Arch Surg 1973;106 (4):432–7.

[24] Ayub A, Falcon WW. Gallstones, obesity, and jejunoileostomy. Surg Clin North Am 1979;59(6):1095–101.

[25] Clayman RV, Williams RD. Oxalate urolithiasis following jejunoileal bypass. Surg Clin North Am 1979;59(6):1071–7.

[26] Salmon PA. The results of small intestine bypass operations for the treatment of obesity. Surg Gynecol Obstet 1971;132(6):965–79.

[27] MacLean LD, Shibata HR. The present status of bypass operations for obesity. Surg Annu 1977;9:213–30.

[28] Bray GA. Current status of intestinal bypass surgery in the treatment of obesity. Diabetes 1977;26(11):1072–9.

[29] Sorensen TI, Bruusgaard A. Lithogenic index of bile after jejunoileal bypass operation for obesity. Scand J Gastroenterol 1977;12(4):449–51.

[30] Parfitt AM, Miller MJ, Frame B, Villanueva AR, Rao DS, Oliver I, et al. Metabolic bone disease after intestinal bypass for treatment of obesity. Annu Intern Med 1978;89(2):193–9.

[31] Compston JE, Horton LW, Laker MF, Ayers AB, Woodhead JS, Bull HJ, et al. Bone disease after jejuno-ileal bypass for obesity. Lancet 1978;2(8079):1–4.

[32] Compston JE, Ayers AB, Horton LW, Tighe JR, Creamer B. Osteomalacia after small-intestinal resection. Lancet 1978;1(8054):9–12.

[33] Halverson JD, Teitelbaum SL, Haddad JG, Murphy WA. Skeletal abnormalities after jejunoileal bypass. Annu Surg 1979;189(6):785–90.

[34] Zapanta M, Aldo-Benson M, Biegel A, Madura J. Arthritis associated with jejunoileal bypass: clinical and immunologic evaluation. Arthritis Rheum 1979;22(7):711–17.

[35] Fernandez-Herlihy L. Arthritis after jejunoileostomy for intractable obesity. J Rheumatol 1977;4(2):135–8.

[36] Shagrin JW, Frame B, Duncan H. Polyarthritis in obese patients with intestinal bypass. Annu Intern Med 1971;75(3):377–80.

[37] Rose E, Espinoza LR, Osterland CK. Intestinal bypass arthritis: association with circulating immune complexes and HLA B27. J Rheumatol 1977;4(2):129–34.

[38] Pi-Sunyer F. Jejunoileal bypass surgery for obesity. Am J Clin Nutr 1976;29(4):409–16.

[39] Vainder M, Kelly J. Renal tubular dysfunction secondary to jejunoileal bypass. JAMA 1976;235(12):1257–8.

[40] Gelbart DR, Brewer LL, Fajardo LF, Weinstein AB. Oxalosis and chronic renal failure after intestinal bypass. Arch Intern Med 1977;137 (2):239–43.

[41] Dean RH, Scott Jr HW, Shull HJ, Gluck FW. Morbid obesity: problems associated with operative management. Am J Clin Nutr 1977;30 (1):90–7.

[42] Joffe SN. Surgical management of morbid obesity. Gut 1981;22(3):242–54.

[43] Marsk R, Freedman J, Tynelius P, Rasmussen F, Naslund E. Antiobesity surgery in Sweden from 1980 to 2005: a population-based study with a focus on mortality. Ann Surg 2008;248(5):777–81.

[44] McFarland RJ, Gazet JC, Pilkington TR. A 13-year review of jejunoileal bypass. Br J Surg 1985;72(2):81–7.

[45] McFarland RJ, Talbot RW, Woolf N, Gazet JC. Dysplasia of the colon after jejuno-ileal bypass. Br J Surg 1987;74(1):21–2.

[46] Sylvan A, Sjolund B, Janunger KG, Rutegard J, Stenling R, Roos G. Colorectal cancer risk after jejunoileal bypass: dysplasia and DNA content in longtime follow-up of patients operated on for morbid obesity. Dis Colon Rectum 1992;35(3):245–8.

[47] Hill M. Mechanism of colorectal carcinogenesis. Diet and human carcinogenesis. Amsterdam: Excerpta Medica; 1985.

[48] Koivisto P, Miettinen TA. Adaptation of cholesterol and bile acid metabolism and vitamin B12 absorption in the long-term follow-up after partial ileal bypass. Gastroenterology 1986;90(4):984−90.

[49] Appleton GV, Wheeler EE, Al-Mufti R, Challacombe DN, Williamson RC. Rectal hyperplasia after jejunoileal bypass for morbid obesity. Gut 1988;29(11):1544−8.

[50] Kaminski JP, Maker VK, Maker AV. Management of patients with abdominal malignancy after remote jejunoileal bypass: surgical considerations decades later. J Am Coll Surgeons 2013;217(5):929−39. Available from: http://dx.doi.org/10.1016/j.jamcollsurg.2013.05.030.

[51] Sundaram A, Koutkia P, Apovian CM. Nutritional management of short bowel syndrome in adults. J Clin Gastroenterol 2002;34(3):207−20.

Chapter 18

Pediatric Bariatric Surgery

D.A. Davies[1], J. Hamilton[2], E. Dettmer[2] and J.C. Langer[2]

[1]The IWK Children's Health Centre, Halifax, NS, Canada, [2]Hospital for Sick Children, Toronto, ON, Canada

LIST OF ABBREVIATIONS

ASMBS American Society of Metabolic and Bariatric Surgery
BMI body mass index
DM diabetes mellitus
LAGB laparoscopic adjustable gastric band
LRGB laparoscopic Roux-en-Y gastric bypass
LSG laparoscopic sleeve gastrectomy
NAFLD nonalcoholic fatty liver disease
NASH nonalcoholic steatohepatitis
OSA obstructive sleep apnea

INTRODUCTION

Overweight and obesity in children are assessed clinically by calculation of body mass index (BMI), with plotting of values on age- and sex-specific growth charts. Over the last quarter of the 20th century, the prevalence of pediatric and adolescent obesity steadily increased. While the increasing incidence of obesity has slowed slightly over the last 10 years in developed countries, the overall prevalence has not declined, and the proportion of children with severe obesity has increased (Fig. 18.1) [1].

The metabolic, cardiovascular, endocrine, hepatic, orthopedic, and pulmonary consequences of obesity are similar to those in adults. Many of these effects were rarely seen in children prior to the 1980s. Obstructive sleep apnea (OSA) has been reported in 19–59% of obese children [2,3]. Type 2 diabetes mellitus (T2DM) was rare in children in 1992, but by 1994 it represented 16% of new cases of T2DM, and in 1999 this figure was as high as 45%, depending on location [4]. More recent studies have closely examined obese adolescents for nonalcoholic fatty liver disease (NAFLD) and steatohepatitis (NASH) and found that 43% had elevated transaminases. In large US studies, the prevalence of dyslipidemia and hypertension in obese adolescents is upwards of 43% and 30%, respectively [5,6].

The psychosocial consequences of obesity in children and adolescents are well described, but are more difficult to quantitate. Social restrictions and social isolation contribute to limited personal, educational, and professional opportunities. Obese children underachieve in school and are often subjected to negative peer attitudes, poor self-image, and bullying [7,8]. Mental health disorders, including anxiety and depression, are more common in obese youth [9]. Of relevance to this chapter topic, eating disorders often evolve during this time, and bulimia nervosa and binge eating disorders have been reported more often in obese adolescents [10].

Perhaps most importantly, the evidence suggests that obesity during childhood is a strong predictor of obesity during adulthood. It is clear that the longer a person has obesity-related comorbidities, the greater the chance that they will progress to end-stage disease, which results in further disability and reductions in quality and duration of life. It is therefore imperative that clinicians continue to work to develop safe and effective strategies for the prevention and treatment of obesity in this population.

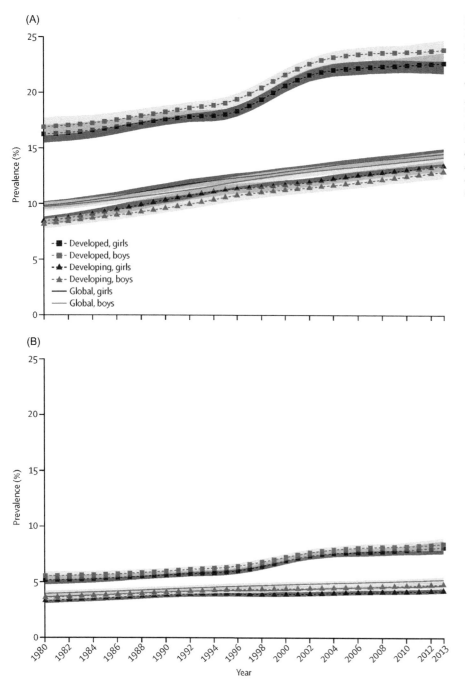

FIGURE 18.1 Age-standardized prevalence of (A) overweight status and obesity and (B) obesity alone (based on IOTF cutoffs), ages 2–19 years, by sex, 1980–2013. *IOTF*, International Obesity Task Force. *Used with permission from Elsevier Ng M, Fleming T, Robinson M, Thomson B, Graetz N, Margono C, et al. Global, regional, and national prevalence of overweight and obesity in children and adults during 1980–2013: a systematic analysis for the Global Burden of Disease Study 2013. Lancet 2014;384 (9945):766–81.*

MANAGEMENT

Great efforts are being made in various jurisdictions to prevent obesity through policy and education [11,12]. While we agree that prevention will ultimately be the best approach to childhood obesity, an immediate need is to address those who are currently affected. Most clinical programs designed to treat children and adolescents with obesity have been either designated as surgical or nonsurgical; however, the optimal arrangement may be to create a comprehensive program that offers the full range of treatment options to children and adolescents with severe obesity.

NONSURGICAL MANAGEMENT OF CHILD AND ADOLESCENT OBESITY

Nonsurgical programs generally include multidisciplinary teams designed to diagnose and monitor the consequences and comorbidities of obesity. They also attempt to stabilize weight and/or reduce obesity through supportive behavioral interventions designed to increase physical activity and improve nutrition and other factors (stress, sleep, mood) that may be interfering with positive behavior changes. The extent and duration of these interventions vary depending on the specific needs of the child/adolescent and the resources available to the team. A recent systematic review found publications describing results from 18 trials of behavior-based weight-management interventions [13]. They found that BMI reductions ranged from 0.5 to $4 \, \text{kg/m}^2$, and that these studies consistently showed reduction in adiposity, improved cardiovascular and diabetes risk factors, and increased physical fitness among participants. Program factors predicting the greatest BMI reduction included organized physical activity sessions, involvement of the parents, intervention at a younger age, and utilization of validated techniques to promote behavior change [13].

Nonsurgical programs have faced criticism that they result in only minimal or modest weight loss when compared to surgical weight loss programs. In particular, adolescents with severe obesity have low likelihood of significantly reducing BMI [14]. Some argue that these results will not have a measureable impact on health outcomes. Durability has also been questioned as many programs only demonstrate benefit during the period of intervention [15]. Despite these findings, slowing an individual's trajectory of weight gain or helping them to stop gaining weight altogether is also an important outcome. Additionally, most programs report measureable improvements in self-image and confidence, as well as increased physical activity and improved dietary intake, in their participants even without significant weight loss. It is also important to recognize that many children and adolescents are neither interested nor emotionally ready to entertain bariatric surgery.

SURGICAL MANAGEMENT OF CHILD AND ADOLESCENT OBESITY

Despite the fact that bariatric surgery has been well-validated as an effective means to manage obesity in adults, its uptake in children and adolescents has been considerably slower. Concerns about safety, growth, and micronutrient deficiencies were frequently cited as reasons not to offer bariatric surgery to children and adolescents. As the disease burden in this population grows, physicians are reconsidering the potential risks as more acceptable considering the known consequences of inaction and the potential benefits of weight loss surgery.

Surgical programs to manage obesity in children and adolescents vary greatly in the weight loss procedure utilized, the setting of the program (i.e., within a pediatric facility vs a general hospital or clinic that treats mainly adults), and the additional components of the program. The American Society of Metabolic and Bariatric Surgeons (ASMBS) recommends that weight loss surgery in children and adolescents should only be offered in the setting of a multidisciplinary team. The components of such a team are listed in Table 18.1 [16,17].

TABLE 18.1 Recommended Members of a Multidisciplinary Team Managing Children and Adolescents Considering Weight Loss Surgery [16,17]

Team Member	Description
Surgeon	Experienced in weight loss surgery
Pediatric Specialists	Experience and training in adolescent medicine, endocrinology, respirology, and gastroenterology/nutrition
Registered Dietitian	Experienced in treating obesity and working with children and families
Mental Health Specialists	Psychiatrist, psychologist, or other qualified, licensed mental health specialists with specialty training in pediatric, adolescent, and family therapy. Experienced in treating eating disorders, including obesity. They should have experience evaluating patients and families for weight loss surgery
Coordinator	Responsible for coordinating the care for each child or adolescent, and ensuring compliance and follow-up
Exercise Therapist	Exercise physiologist, physical therapist, or other individual specially trained to provide safe physical activity prescriptions to morbidly obese adolescents
Advanced Practice Nurse	Skilled and experienced in supporting and managing the multiple medical, psychosocial, and practical issues of obese children and their families

PATIENT SELECTION

The ASMBS Pediatric Committee Guidelines are often used to guide selection of patients who may benefit from weight loss surgery [16,17]:

- BMI ≥ 35 kg/m^2 and a severe comorbidity that has significant short-term effects on health (moderate to severe OSA, type-2 DM, pseudotumor cerebri, NASH).
- BMI ≥ 40 kg/m^2 with more minor comorbidities (hypertension, insulin resistance, glucose intolerance, substantially impaired quality of life of activities of daily living, dyslipidemia, OSA).
- Physical maturity (having obtained 95% of predicted adult stature based on bone age or reaching Tanner Stage IV).
- A history of sustained efforts to lose weight through changes in diet and physical activity.

Contraindications to weight loss surgery include, but are not limited to, the following [16,18]:

- An active eating disorder.
- Substance abuse (including tobacco).
- Gastrointestinal issues or anatomy precluding surgery.
- Pregnancy or planned pregnancy within the subsequent 2 years.
- An active, unstable, mental health disorder requiring treatment.
- A medically correctable cause of obesity.

PREOPERATIVE EVALUATION AND PREPARATION

The overall approach of our medical—surgical program is one of small sustainable behavioral changes to promote healthy nutrition and increased physical activity [19]. The program consists of both individual and group sessions, with increased intensity at the onset and gradual transition to community services during the final 6 months of the 2 year program. In order to promote healthy behavior change, we incorporate cognitive behavioral therapy, mindfulness techniques, motivational interviewing, and family therapy with our participants. We employ a family-centered care model. Obese adolescents interested in pursuing weight loss surgery must participate fully in this portion of the program and demonstrate commitment to healthy behavior change. Once they have been in the behavioral arm of the program for at least 3 months, interested participants undergo a presurgery mental health assessment, which screens for active mental health disorders, support, and other issues that may interfere with the patients' ability to adhere to pre- and postsurgery regimens. They then attend teaching sessions on the surgical procedures, potential complications, pre- and postsurgery diet requirements, etc. At a later date, they undergo an informed consent assessment where they are asked questions to determine if they remember and comprehend the lifelong changes in diet and health that go along with weight loss surgery. Female patients are counseled in regard to increased fertility, with recommendation of intrauterine device insertion perioperatively to provide long-term contraception.

SURGICAL OPTIONS AND OUTCOMES

At the present time, the three procedures that are widely available and have been studied most in the pediatric age range are the laparoscopic adjustable gastric band (LAGB), the laparoscopic Roux-en-Y gastric bypass (LRGB), and the laparoscopic sleeve gastrectomy (LSG). Each has its own profile of advantages and disadvantages, summarized in Table 18.2.

In early studies of weight loss surgery in obese adolescents, reported primarily on LAGB and LRGB, LAGB was favored by many centers because it is minimally invasive, reversible, and is not associated with a risk of malabsorption and micronutrient deficiency. Initial experience was encouraging, and the LAGB was shown to be significantly more effective than nonsurgical approaches in the pediatric population [20]. Unfortunately, the high rate of reoperation, which approaches 40%, and the lower achievement and maintenance of weight loss when compared to other surgical procedures, has resulted in a significant decrease in utilization of the gastric band for adolescents [21–23].

Recently, LSG has gained favor among surgeons performing weight loss surgery in adolescents. There is less risk for potential long-term nutrient deficiencies that appear to be higher with other malabsorptive procedures such as LRGB or biliopancreatic diversion [24,25]. The other theoretical advantage, which has not been well studied, is the ability to easily convert an LSG to either an LRGB or biliopancreatic diversion should weight loss either be inadequate or not sustained later in life. Weight loss and comorbidity resolution are comparable to LRGB [26]. Because of

TABLE 18.2 Advantages and Disadvantages of the Various Surgical Weight Loss Procedures Commonly Offered to Children and Adolescents

Weight Loss Procedure	Advantages	Disadvantages
Laparoscopic Adjustable Gastric Band (LAGB)	Reversible	Foreign body complications (infection, erosion, malposition)
	"Less" invasive	High reoperation rate
	Minimal micronutrient deficiencies	Less weight loss & comorbidity resolution
	Modest weight loss	Higher recidivism
Laparoscopic Roux-en-Y Gastric Bypass (LRGB)	Excellent weight loss	"More" invasive
	Maintenance of weight loss	Malabsorption & micronutrient deficiencies
	Resolution of comorbidities	High risk of bowel obstruction
		Technical challenges
Laparoscopic Sleeve Gastrectomy (LSG)	Excellent weight loss	"More" invasive
	Maintenance of weight loss	Micronutrient deficiencies
	Resolution of comorbidities	
	Less technically challenging	

the simplicity and lower risk of the procedure, we are currently recommending sleeve gastrectomy to most adolescent patients who are offered weight loss surgery.

Because it was generally considered to be the most effective operation, and because there are several decades of experience with it, LRGB continues to be employed by many centers offering weight loss surgery to adolescents. Results in this population are similar to those seen with adults showing good, sustained weight loss, and resolution of comorbidities [23]. We continue to offer this procedure to our adolescent patients who have symptomatic gastroesophageal reflux, as well as those who have a strong preference for LRGB. In many cases, such preference is based on the experience of a parent or grandparent who previously underwent a successful LRGB.

POSTOPERATIVE MANAGEMENT

Close follow-up of adolescents who have undergone weight loss surgery is of utmost importance. Unfortunately, adolescents and youth (ages 15−25) struggle with adherence across health conditions, necessitating that teams give extra attention to encouraging and rewarding attendance to follow-up visits. Micronutrient deficiencies can occur following any weight loss procedure [27−29]. Monitoring and reinforcement of healthy behaviors are key to a successful, safe program. Patients in our program are followed for 2 years after weight loss surgery for compliance, weight loss, micronutrient deficiencies, and monitoring of healthy behavior changes. This includes regular visits with each of the team members, who provide ongoing support and education. Following this, patients are transitioned to an adult bariatric surgery program for long-term follow-up, where they are assessed for micronutrient deficiencies, comorbidity management, and reinforcement of healthy behaviors.

Long-term outcomes of weight loss surgery in adolescents are limited. Two recent meta-analyses concluded that bariatric surgery in adolescents results in substantial weight loss and improvements in comorbidities with acceptable complication rates [30,31]. The Teen−Longitudinal Assessment of Bariatric Surgery (Teen−LABS), a prospective, multisite, observational study, recently completed enrollment in the United States. This study includes 242 patients undergoing weight loss surgery at five academic referral centers who are followed for 5 years. Currently, early perioperative outcomes have been published, and demonstrate early complication rates similar to those reported in adults [32]. Three-year outcomes have been reported for some of these patients, and show a 27% reduction in mean weight [33]. Results were similar regardless of whether patients had received sleeve gastrectomy or RYGB. Remission of T2DM, prediabetes, abnormal kidney function, hypertension, and dyslipidemia were 95%, 76%, 86%, 74%, and 66%, respectively. Measures of health-related quality of life also demonstrate significant improvement [34]. While we await further long-term, prospective results, the number of weight loss procedures being performed in adolescents has doubled in the United States to 1600 cases in 2009 [35].

ONGOING CONTROVERSIES

There are a number of ongoing controversies regarding bariatric surgery in the pediatric age group.

Bariatric surgery in younger children. Most surgeons continue to offer surgery to adolescents who are postpubertal, as per the recommendations mentioned previously. However, some have taken the approach that extremely obese prepubertal children who have significant morbidity should also be offered these procedures [36,37]. Proponents of this approach cite concern about long-term morbidity if the obesity is not controlled. Opponents are concerned about the potential for long-term nutritional and growth issues, and also make the point that controlling caloric intake in this population should be possible because very young children cannot independently obtain food. Ethicists raise concern about assessment of capacity to consent in younger adolescents and children.

Where should adolescent bariatric surgery be done? General recommendations are that only surgeons doing a high volume of these operations should perform weight loss surgery. Since most adolescent programs perform less than 30 procedures a year, it is difficult for a pediatric surgeon to gain the experience to perform these procedures optimally and maintain the necessary skill set. In order to optimize outcomes in our relatively low-volume bariatric surgery program, all operations involve a pediatric surgeon and an adult bariatric surgeon whose group performs over 500 bariatric operations per year. This has eliminated the "learning curve" for the program and gives us access to an experienced surgical opinion for any postoperative problems. This also facilitates the transition to the adult program for long-term follow-up [19]. No literature exists on whether adolescents can be managed exclusively in an adult weight loss surgery program. Given the complex relationships of families, medical comorbidities, and ongoing psychosocial development in adolescents, any program managing obesity in these patients should have or seek expertise in this area.

Adolescent development. Adolescents are at a stage of development when self-identity is forming, and a sense of "fitting in" with peers is critical. Unrealistic expectations regarding the outcome of surgery, or the feeling that surgery will be the "answer" to any or all of their social difficulties, may lead to unexpected negative sequelae [38]. One of the common concerns expressed by patients in our program is the "loose skin" that results with rapid weight loss, so ensuring they are realistic about outcomes is important. Adolescent sexual identity is also evolving, and unintended pregnancies have been reported following bariatric surgery [39]. This age group is more likely to engage in higher-risk behaviors (alcohol, tobacco, other illicit substances), and be less compliant with medications and supplements, so should be assessed for this on a regular basis.

Transition from pediatric to adult programs. Long-term follow-up of patients having weight loss surgery varies by program. As with many other pediatric specialties, transitioning patients to adult care is challenged by a reluctance of adult practitioners to care for medically complex young adults, and anxiety on behalf of the young adults and their families to leave the pediatric system. Many young adults are simultaneously moving away from home for school or work, putting them at increased risk for loss of follow-up. This can be improved by identifying the need for transition early, developing strong relationships with receiving teams, allowing an overlap period, and setting expectations for patients and families so that they can be prepared for the transition [40].

Specific populations of children with obesity. Long-term data on patients with specific medical conditions contributing to their obesity is limited. Prader–Willi syndrome and Trisomy 21 were initially considered contraindications to weight loss surgery due to the developmental delay associated with these conditions. Weight loss surgery has been performed in these patients, however, weight loss may not be as significant as in other adolescents, and one should anticipate challenges, such as an ongoing need to restrict food intake in these populations [41–43]. Patients with hypothalamic obesity following resection of sellar tumors, such as craniopharyngioma, have successfully been managed with weight loss surgery [44].

SUMMARY

Weight loss surgery has become a viable option for many obese adolescents provided proper screening, evaluation, and follow-up is included. Bariatric surgery should only be offered in the setting of a multidisciplinary program of practitioners experienced in managing this age group. Patients and their caregivers must fully appreciate and understand the life-altering consequences of weight loss surgery and have the maturity and ability to comprehend and consent before it is considered. Future research is required to further understand the relative efficacy of different surgical techniques, patient selection criteria, predictors of favorable outcomes, and durability of the procedures over time.

SUMMARY POINTS

- Over the last quarter of the 20th century, the prevalence of pediatric and adolescent obesity has steadily increased.
- Overweight and obesity in children is assessed clinically by calculation of BMI, with plotting of values on age- and sex-specific growth charts.
- Weight loss surgery should only be offered to children and adolescents in the setting of a multidisciplinary team dedicated to the management and long-term follow-up of these patients.
- Long-term studies show weight loss surgery can be a viable option for many obese adolescents.
- Jill Hamilton receives unrestricted research funds as the recipient of the Mead Johnson Chair in Nutritional Science.

REFERENCES

[1] Ng M, Fleming T, Robinson M, Thomson B, Graetz N, Margono C, et al. Global, regional, and national prevalence of overweight and obesity in children and adults during 1980–2013: a systematic analysis for the Global Burden of Disease Study 2013. Lancet 2014;384(9945):766–81.

[2] Marcus CL, Curtis S, Koerner CB, Joffe A, Serwint JR, Loughlin GM. Evaluation of pulmonary function and polysomnography in obese children and adolescents. Pediatr Pulmonol 1996;21(3):176–83.

[3] Verhulst SL, Van Gaal L, De Backer W, Desager K. The prevalence, anatomical correlates and treatment of sleep-disordered breathing in obese children and adolescents. Sleep Med Rev 2008;12(5):339–46.

[4] Kaufman FR. Type 2 diabetes mellitus in children and youth: a new epidemic. J Pediatr Endocrinol Metab 2002;15(suppl. 2):737–44.

[5] Centers for Disease Control and Prevention (CDC). Prevalence of abnormal lipid levels among youths --- United States, 1999–2006. MMWR Morb Mortal Wkly Rep 2010;59(2):29–33.

[6] McNiece KL, Poffenbarger TS, Turner JL, Franco KD, Sorof JM, Portman RJ. Prevalence of hypertension and pre-hypertension among adolescents. J Pediatr 2007;150(6):640–4, 644.e1.

[7] van Geel M, Vedder P, Tanilon J. Are overweight and obese youths more often bullied by their peers? A meta-analysis on the correlation between weight status and bullying. Int J Obes (Lond) 2014;38(10):1263–7.

[8] Buttitta M, Iliescu C, Rousseau A, Guerrien A. Quality of life in overweight and obese children and adolescents: a literature review. Qual Life Res 2014;23(4):1117–39.

[9] Nieman P, Leblanc CM. Canadian paediatric society, healthy active living and sports medicine committee. Psychosocial aspects of child and adolescent obesity. Paediatr Child Health 2012;17(4):205–8.

[10] Latzer Y, Stein D. A review of the psychological and familial perspectives of childhood obesity. J Eat Disord 2013;25(1), eCollection 2013.

[11] Brennan LK, Kemner AL, Donaldson K, Brownson RC. Evaluating the implementation and impact of policy, practice, and environmental changes to prevent childhood obesity in 49 diverse communities. J Public Health Manag Pract 2015;21(suppl. 3):S121–34.

[12] Waters E, de Silva-Sanigorski A, Hall BJ, Brown T, Campbell KJ, Gao Y, et al. Interventions for preventing obesity in children. Cochrane Database Syst Rev 2011;12: CD001871. doi(12):CD001871.

[13] Whitlock E, O'Connor E, Williams S, Beil T, Lutz K. Effectiveness of weight management programs in children and adolescents. Evid Rep Technol Assess (Full Rep) 2008;1–308: (170)(170).

[14] Danielsson P, Kowalski J, Ekblom O, Marcus C. Response of severely obese children and adolescents to behavioral treatment. Arch Pediatr Adolesc Med 2012;166(12):1103–8.

[15] Kelly AS, Barlow SE, Rao G, Inge TH, Hayman LL, Steinberger J, et al. Severe obesity in children and adolescents: identification, associated health risks, and treatment approaches: a scientific statement from the American Heart Association. Circulation 2013;128(15):1689–712.

[16] Pratt JS, Lenders CM, Dionne EA, Hoppin AG, Hsu GL, Inge TH, et al. Best practice updates for pediatric/adolescent weight loss surgery. Obesity (Silver Spring) 2009;17(5):901–10.

[17] Michalsky M, Reichard K, Inge T, Pratt J, Lenders C. American society for metabolic and bariatric surgery. ASMBS pediatric committee best practice guidelines. Surg Obes Relat Dis 2012;8(1):1–7.

[18] Freedman DS, Mei Z, Srinivasan SR, Berenson GS, Dietz WH. Cardiovascular risk factors and excess adiposity among overweight children and adolescents: the Bogalusa Heart Study. J Pediatr 2007;150(1):12–17.e2.

[19] Davies DA, Hamilton J, Dettmer E, Birken C, Jeffery A, Hagen J, et al. Adolescent bariatric surgery: the Canadian perspective. Semin Pediatr Surg 2014;23(1):31–6.

[20] Zitsman JL. Laparoscopic adjustable gastric banding in adolescents. Semin Pediatr Surg 2014;23(1):17–20.

[21] Tice JA, Karliner L, Walsh J, Petersen AJ, Feldman MD. Gastric banding or bypass? A systematic review comparing the two most popular bariatric procedures. Am J Med 2008;121(10):885–93.

[22] Chakravarty PD, McLaughlin E, Whittaker D, Byrne E, Cowan E, Xu K, et al. Comparison of laparoscopic adjustable gastric banding (LAGB) with other bariatric procedures; a systematic review of the randomised controlled trials. Surgeon 2012;10(3):172–82.

[23] Lennerz BS, Wabitsch M, Lippert H, Wolff S, Knoll C, Weiner R, et al. Bariatric surgery in adolescents and young adults--safety and effectiveness in a cohort of 345 patients. Int J Obes (Lond) 2014;38(3):334−40.

[24] Gehrer S, Kern B, Peters T, Christoffel-Courtin C, Peterli R. Fewer nutrient deficiencies after laparoscopic sleeve gastrectomy (LSG) than after laparoscopic Roux-Y-gastric bypass (LRYGB)-a prospective study. Obes Surg 2010;20(4):447−53.

[25] Stefater MA, Kohli R, Inge TH. Advances in the surgical treatment of morbid obesity. Mol Aspects Med 2013;34(1):84−94.

[26] Colquitt JL, Pickett K, Loveman E, Frampton GK. Surgery for weight loss in adults. Cochrane Database Syst Rev 2014;8;CD003641.

[27] Aarts EO, Janssen IM, Berends FJ. The gastric sleeve: losing weight as fast as micronutrients? Obes Surg 2011;21(2):207−11.

[28] Gletsu-Miller N, Wright BN. Mineral malnutrition following bariatric surgery. Adv Nutr 2013;4(5):506−17.

[29] Lee WJ, Pok EH, Almulaifi A, Tsou JJ, Ser KH, Lee YC. Medium-Term Results of Laparoscopic Sleeve Gastrectomy: a Matched Comparison with Gastric Bypass. Obes Surg 2015;25(8):1431−8.

[30] Paulus GF, de Vaan LE, Verdam FJ, Bouvy ND, Ambergen TA, van Heurn LW. Bariatric surgery in morbidly obese adolescents: a systematic review and meta-analysis. Obes Surg 2015;25(5):860−78.

[31] Treadwell JR, Sun F, Schoelles K. Systematic review and meta-analysis of bariatric surgery for pediatric obesity. Ann Surg 2008;248(5):763−76.

[32] Inge TH, Zeller MH, Jenkins TM, Helmrath M, Brandt ML, Michalsky MP, et al. Perioperative outcomes of adolescents undergoing bariatric surgery: the Teen-Longitudinal Assessment of Bariatric Surgery (Teen-LABS) study. JAMA Pediatr 2014;168(1):47−53.

[33] Inge TH, Courcoulas AP, Jenkins TM, Michalsky MP, Helmrath MA, Brandt ML, et al. Weight loss and health status 3 years after bariatric surgery in adolescents. N Engl J Med 2016;374(2):113−23.

[34] Zeller MH, Reiter-Purtill J, Ratcliff MB, Inge TH, Noll JG. Two-year trends in psychosocial functioning after adolescent Roux-en-Y gastric bypass. Surg Obes Relat Dis 2011;7(6):727−32.

[35] Zwintscher NP, Azarow KS, Horton JD, Newton CR, Martin MJ. The increasing incidence of adolescent bariatric surgery. J Pediatr Surg 2013;48(12):2401−7.

[36] Al-Qahtani AR. Laparoscopic adjustable gastric banding in adolescent: safety and efficacy. J Pediatr Surg 2007;42(5):894−7.

[37] Alqahtani AR, Elahmedi MO, Al Qahtani A. Co-morbidity resolution in morbidly obese children and adolescents undergoing sleeve gastrectomy. Surg Obes Relat Dis 2014;10(5):842−50.

[38] Luca P, Dettmer E, Langer J, Hamilton J. Adolescent bariatric surgery: current status in an evolving field. In: Kiess W, Wabitsch M, Maffeis C, Sharma AM, editors. Matabolic Syndrome and Obesity in Childhood and Adolescence. Basel Karger: Pediatric Adolescent Medicine; 2016. p. 179−86.

[39] Roehrig HR, Xanthakos SA, Sweeney J, Zeller MH, Inge TH. Pregnancy after gastric bypass surgery in adolescents. Obes Surg 2007;17(7):873−7.

[40] Hopper A, Dokken D, Ahmann E. Transitioning from pediatric to adult health care: the experience of patients and families. Pediatr Nurs 2014;40(5):249−52.

[41] Alqahtani AR, Elahmedi MO, Al Qahtani AR, Lee J, Butler MG. Laparoscopic sleeve gastrectomy in children and adolescents with Prader-Willi syndrome: a matched-control study. Surg Obes Relat Dis 2016;12(1):100−10.

[42] Musella M, Milone M, Leongito M, Maietta P, Bianco P, Pisapia A. The mini-gastric bypass in the management of morbid obesity in Prader-Willi syndrome: a viable option? J Invest Surg 2014;27(2):102−5.

[43] Daigle CR, Schauer PR, Heinberg LJ. Bariatric surgery in the cognitively impaired. Surg Obes Relat Dis 2015;11(3):711−14.

[44] Bretault M, Boillot A, Muzard L, Poitou C, Oppert JM, Barsamian C, et al. Clinical review: bariatric surgery following treatment for craniopharyngioma: a systematic review and individual-level data meta-analysis. J Clin Endocrinol Metab 2013;98(6):2239−46.

Chapter 19

Pregnancy After Bariatric Surgery

V.F. Byron, R. Charach and E. Sheiner

Ben-Gurion University of the Negev, Be'er-sheva, Israel

LIST OF ABBREVIATIONS

BMI body mass index
GDM gestational diabetes mellitus
IUGR intrauterine growth restriction
LAGB laparoscopic adjustable gastric banding
LGA large for gestational age
PPBS pregnancy postbariatric surgery
RYGB Roux-en-Y gastric bypass
SGA small for gestational age

PREGNANCY AND OBESITY

Obesity, defined as a body mass index (BMI) of 30 and above, has a profound health impact on reproductive-age women, affecting nearly one-third of women of childbearing age (20–39 years) [1]. The effects of obesity on fertility, pregnancy, delivery, and fetal development are far-reaching and potentially devastating. Preconception weight loss by any means has been shown to be the most effective way to increase fertility as well as decrease pregnancy complications in obese women [2].

Fertility

Obesity is well-associated with decreased fertility due to oligoovulation and anovulation. A BMI over 29 has been associated with reduced fecundity as well as poorer response to fertility treatment [3]. Studies have shown that this reduced fecundity is reversible with weight loss, and after bariatric surgery in particular. One study demonstrated that amount of weight reduction and achieved BMI were good predictors for successful conception and pregnancy. However, the absolute fertility of postsurgical patients in comparison to both the obese and general population is under further investigation. Bariatric surgery should not be considered as a primary treatment for infertility in obese women.

Antepartum Complications

Pregnancies in obese women are associated with an increased rate of complications and adverse outcomes, including spontaneous abortion, gestational diabetes mellitus (GDM), gestational hypertension, and exacerbation of existing chronic conditions in the mother, such as sleep apnea, nonalcoholic fatty liver disease, renal disease, and heart failure [4–9].

Intra- and Postpartum Complications

Deliveries of obese mothers appear to progress more slowly and are associated with an increased risk of emergent and elective cesarean delivery, as compared to normal weight mothers [10,11]. In addition, obese patients have been shown to suffer from high rates of anesthesia complications, poor wound healing, infection, and deep vein thrombosis (DVT) in the postsurgical recovery period [5,12].

Fetal Complications

There is evidence linking maternal obesity with increased rates of congenital fetal malformations, fetal and early neonatal death, and macrosomia, a risk factor for shoulder dystocia [13,14]. There is increasing support for the link between maternal obesity and long-term health effects of offspring via metabolic programming [15]. Offspring exposure to an obese maternal environment in utero is associated with increased risk for future metabolic syndrome [16].

POSTBARIATRIC SURGERY PREGNANCY

Bariatric surgery has been proven to be the most effective method for weight reduction [17]. Women of childbearing age make up a major proportion of bariatric surgery patients. Over 80% of bariatric surgeries are performed in women, with an overall mean age in the early 40s [18]. The effect of pregnancy on the efficacy of bariatric surgery in weight reduction, while still being elucidated, seems to be minimal. One study shows a slower rate of weight loss in women who become pregnant after bariatric surgery, as opposed to women who did not become pregnant postsurgery, but similar weight loss outcomes 5 years postsurgery [19].

Women who undergo pregnancy postbariatric surgery (PPBS) are best managed with a multidisciplinary approach involving a maternal fetal medicine specialist and bariatric surgeon. PPBS women are at risk for a number of complications, although these women suffer from far fewer adverse outcomes closely associated with obesity. Potential complications include delivery of small for gestational age (SGA) infants, nutrient deficiencies, GDM screening, surgical complications, and appropriate gestational weight gain. Underlying these concerns is the time from surgery to conception, although recent evidence has failed to identify a relationship between length of the interval and risk for any associated complication. Here, we explore the impact of bariatric surgery on subsequent pregnancies, presenting evidence for both maternal and fetal outcomes (Fig. 19.1 and Table 19.1).

Nutrition

Several studies have shown an increased risk for nutritional deficiencies in PPBS as compared to nonsurgical controls [22]. Malabsorptive procedures, such as Roux-en-Y gastric bypass (RYGB), pose a greater risk for nutritional deficiencies than restrictive procedures such as laparoscopic adjustable gastric banding (LAGB), although all methods are at risk for complication by iron deficiency anemia [23]. Malabsorptive procedures are associated with an increased risk

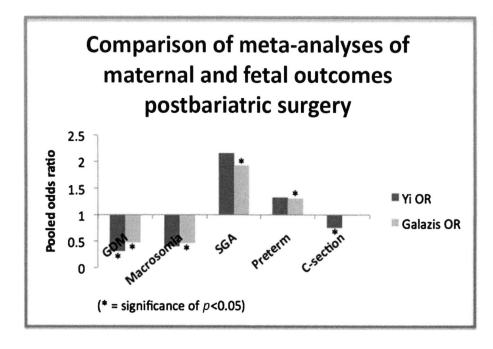

FIGURE 19.1 Meta-analysis comparison [20,21].

TABLE 19.1 Meta-analysis Comparison of Maternal Fetal Outcomes After Bariatric Surgery [25,20,21]

	Yi et al. (2015), China	Galazis et al. (2014), United Kingdom	Maggard et al. (2008), United States[a]
Objective	Meta-analysis of 11 cohort studies that reported fetal and maternal outcomes for women who underwent bariatric surgery before pregnancy to obese women who did not undergo bariatric surgery before pregnancy	Meta-analysis of 17 cohort and case-control studies that reported maternal and fetal outcomes for women postbariatric surgery as compared to (1) obese or BMI-matched controls and (2) pre- versus postsurgery pregnancies	Systematic review of 75 observational studies, including three matched cohort studies, addressing pregnancy and fertility outcomes postbariatric surgery
Findings	Women who became pregnant postbariatric surgery experienced significantly lower rates of GDM, hypertensive disorders, and macrosomia. Rates of SGA infants were increased in the postsurgical group. No differences were found for cesarean delivery, postpartum hemorrhage, or preterm delivery	Women who became pregnant postbariatric surgery experienced significantly lower rates of preeclampsia, GDM, and macrosomia. Higer rates of SGA (except after LAGB procedure), preterm birth, neonatal intensive care admission, and maternal anemia were identified	Women who became pregnant postbariatric surgery experienced significantly lower rates of GDM, low birth weight, and macrosomia. Similar rates of preterm delivery were found

[a]Maggard et al. did not report pooled quantitative results.

for calcium, vitamin D, vitamin A, and vitamin B12 deficiencies. Routine nutritional screening, as well as supplementation, is recommended for all women undergoing PPBS [24].

Gestational Diabetes Mellitus Screening

The postbariatric surgery state poses a particular challenge for routine screening for glucose tolerance, as patients who have undergone a malabsorptive procedure (i.e., RYGB) are at increased risk for "dumping syndrome," consisting of gastrointestinal (GI) complaints such as distension, cramping, nausea, and vomiting when challenged with a large glucose load [25]. It does not stand, however, for restrictive procedures. If unable to perform a glucose challenge or tolerance test, it is recommended to follow hemoglobin A1C levels as well as second trimester fasting and postprandial glucose levels in order to screen for GDM in PPBS [26].

Surgical Complications

While surgical complications of PPBS are rare, they can be life threatening when they do occur, as symptoms may be nonspecific and mimic other conditions of pregnancy. These patients should be managed by a general surgeon as well as their obstetrician. Bowel obstruction is rare, but dangerous to the lives of both the mother and fetus, as described in several case reports [27]. This complication is prevalent in up to 2% of PPBS, and is attributed to bowel ischemia and internal hernia formation post-RYGB procedures, due to the enlarging gravid uterus and elevated abdominal pressure [28]. Small bowel obstruction (SBO) must be considered in a pregnant woman post-RYGB presenting with abdominal pain due to the possibility of SBO mimicking, precipitating, or existing concurrently with an overshadowing obstetric complication [29]. The possibility of increased risk for band slippage in pregnant women who have undergone LAGB has been proposed, although a recent study demonstrated no excess risk for band revision in pregnant women as compared to nonpregnant patients post-LAGB [30]. Band adjustment may be indicated as needed to relieve nausea and vomiting throughout pregnancy and to meet gestational weight gain targets. Importantly, there is currently no consensus in the literature or in practice for band deflation during pregnancy [31]. Internal hernia repair has a good prognosis in pregnant women who have previously undergone laparoscopic RYGB [32].

Gestational Weight Gain

While there is currently no consensus on gestational weight gain postbariatric surgery, the most recent Institute of Medicine (IOM) guidelines for obese women (based on prepregnancy BMI) state a minimum weight gain of 230 g per week. A recent study showed that women who were monitored by a bariatric surgical team achieved optimal gestational weight gain in accordance with the IOM standards [33]. Caloric restriction during pregnancy is not recommended, even at the sacrifice of weight loss goals postpregnancy [26].

Surgery to Conception Interval

The first year postbariatric surgery is the period of most rapid weight loss [34]. Fertility increases as well, resulting in unintended pregnancies [35]. Concerns about malnutrition, inadequate gestational weight gain, and potential surgical complications have led to speculation about the safety of pregnancies conceived during this period. Although current ACOG guidelines (2009) recommend postponing pregnancy for 12−24 months following bariatric surgery [26], the support for this recommendation is waning, lacking conclusive evidence. There is increasing evidence demonstrating no risk associated with a shorter surgery-to-conception interval [36,37].

OBSTETRIC OUTCOMES

Gestational Diabetes Mellitus

GDM poses a great risk for both the mother and fetus, leading to complications such as macrosomia, shoulder dystocia, surgical delivery, preeclampsia, neonatal morbidity, and stillbirth. Obesity is an established risk factor for the development of GDM, opening the possibility for prevention of this condition with weight loss prior to pregnancy [38]. Several studies have shown a decreased risk of GDM in women who have previously undergone bariatric surgery, even when matched for prepregnancy BMI [39]. Although, as women postbariatric surgery are still more likely to be obese than the general population, they still experience higher rates of GDM than their normal weight counterparts [40].

Hypertensive Disorders

Obese mothers suffer from hypertensive disorders of pregnancy, including preeclampsia, at higher rates than the general population. Data suggest that women who have undergone bariatric surgery benefit from a significant reduction of this risk as compared to matched obese controls, and may even approach rates similar to the general obstetric population [9]. In a large cohort study, Adams et al. found lower rates of GDM and gestational hypertension in women post-RYGB as compared to matched nonoperated, BMI-matched controls [41].

Mode of Delivery

Both elective and emergent cesarean deliveries are more prevalent in the obese obstetric population [6]. Evidence suggests that PPBS are delivered via cesarean section at a lower rate than the obese population, but still higher than the general obstetric population [42]. Sheiner et al. suggest that some of the difference may be a result of caregiver bias, indicating physician conservatism in attempting vaginal delivery, despite a lack of evidence to support an indication to do so [43]. In a large cohort study, Parker et al. found no increased risk of failed induction of labor, cesarean delivery, or operative vaginal delivery in the postbariatric surgery group, as compared to the obese control population [44]. Roos et al. reported an increased rate of both spontaneous and medically indicated preterm deliveries in women with a BMI of less than 35 with a history of bariatric surgery [45].

FETAL OUTCOMES

Birth Weight and Fetal Growth

Fetal growth and birth weight are perhaps the most significant and well-supported complications of PPBS, with strong evidence to support increased risks of IUGR and SGA, compared to both obese and normal weight mothers. This complication is most likely due to maternal malabsorption and micronutrient deficiency [43,44,46,47].

Postsurgical mothers give birth to LGA infants at a much lower rate than their obese counterparts, reducing the risk for complications related to macrosomia, including vacuum-assisted delivery and neonatal asphyxia [23,41,48].

Congenital Anomalies

Interestingly, despite high prevalence of maternal nutrient deficiency postbariatric surgery and well-established causal effect between certain nutrient deficiencies and congenital malformation, current evidence suggests no increased prevalence of congenital anomalies in mothers who have previously undergone bariatric surgery [43,49,50]. As in the general obstetric population, it is recommended to give these mothers supplementary folic acid and to screen for congenital conditions as indicated.

MINI-DICTIONARY OF TERMS

- *Gestational diabetes mellitus (GDM)*: Impaired glucose tolerance that develops, or is first diagnosed, during pregnancy.
- *Intrauterine growth restriction (IUGR)*: Also known as fetal growth restriction (FGR), a fetus that has failed to achieve its growth potential due to maternal, placental, or fetal factors, or a combination of factors.
- *Small for gestational age (SGA)*: Below the 10th percentile of birthweight for a given gestational age.
- *Large for gestational age (LGA)*: Above the 90th percentile of birth weight for a given gestational age (exact definition variable).
- *Macrosomia*: Excessive intrauterine growth above a certain threshold, regardless of gestational age, and mostly above 4 kg.
- *Metabolic programming*: The lifelong effect of environmental factors, specifically during gestation, on metabolism and predisposition to obesity.
- *Preeclampsia*: New onset hypertension and proteinuria or end organ damage in the second half of pregnancy.
- *Gestational hypertension*: Hypertension alone that presents in the second half of pregnancy.
- *Preterm delivery*: Delivery before the 37th week of pregnancy.
- *Shoulder dystocia*: Obstruction of labor due to the failure of the anterior shoulder of the fetus to pass the pubic symphysis after delivery of the head.

SUMMARY POINTS

- Women of childbearing age comprise a large proportion of obese adults, and the vast majority of patients undergoing bariatric surgery.
- Obesity can have numerous adverse effects on pregnancy, delivery, and fetal development, many of which can be mitigated or avoided with preconception weight loss.
- Bariatric surgery is the most effective path to sustainable weight loss for women of childbearing age.
- Pregnancy does not appear to have an adverse effect on the long-term weight loss outcomes of bariatric surgery.
- PPBS are at a lower risk for GDM, hypertensive disorders, and macrosomia.
- PPBS seem to be at a higher risk for SGA fetuses, as compared to both obese and normal weight controls (Table 19.2).
- Current evidence has found no clear association between time-to-conception interval postbariatric surgery and adverse effects on pregnancy or fetal outcomes. Nevertheless, it is reasonable to delay pregnancy to more than 12 months postsurgery.
- Despite increased risk for nutritional deficiencies (Table 19.4), PPBS do not appear to be at a higher risk for developing congenital anomalies.
- Pregnant women postbariatric surgery are at a lower risk of delivering via cesarean section than obese mothers, but are still at a higher risk than the general population (Table 19.3). The failure of this rate to reach the levels of nonobese women may be attributed to caregiver bias and not medical indication.

TABLE 19.2 Summary of Selected Studies on SGA in Pregnancies After Bariatric Surgery [36,39,43,45,46,51]

Study	Study Group	Control Group	Results	Additional Findings
Johansson et al. (2015), Sweden	$n = 596$, Births to 554 mothers postbariatric surgery	$n = 2356$, Matched control births to 2278 mothers	15.6% in postsurgical group vs 7.6% in controls; odds ratio, 2.20; 95% CI 1.64–2.95; $p < 0.001$	Subgroup analysis suggested a direct relationship between length of surgery-to-conception interval and risk for SGA
Nørgaard et al. (2014), Denmark	$n = 387$, Postgastric bypass, from 3 to 1851 days surgery-to-conception interval	N/A (This study examined the effect of surgery-to-conception interval on several growth outcomes, and therefore used a continuous data set.)	18.8% SGA overall, no significant difference between pregnancies conceived before and after 18 months postsurgery	No significant correlation between surgery-to-conception interval and fetal growth index or birthweight for gestational age. Slight decreased risk with increased BMI
Roos et al. (2013), Sweden	$n = 2562$, All procedure types (gastric bypass, vertical banded gastroplasty, and gastric banding), plus a diagnosis of obesity	$n = 12338$, Fully matched (age, parity, early pregnancy BMI, smoking status, education, delivery year)	SGA in 5.2% postsurgical vs 3.0% matched controls (OR 2.0)	Decreased risk for LGA. Increased risk for preterm birth. Maternal BMI was found to be an effect modifier. No effect modification was observed with type of procedure or surgery-to-conception interval
Lesko et al. (2012), United States	$n = 70$, Postgastric bypass	$n = 140$, BMI matched	SGA in 17.4% postsurgical vs 5% matched control ($p < 0.01$)	Incidence of LGA was decreased in study group
Josefsson et al. (2011), Sweden	$n = 126$, Postbariatric surgery (gastroplasty, gastric banding, or gastric bypass)	$n = 188500$, (1) Women with no bariatric surgery and (2) Surgery after 1st pregnancy	SGA in 5.6% postsurgical vs 2.1% in both control groups, $p = 0.075$	
Sheiner et al. (2004), Israel	$n = 298$, Postsurgery (all methods)	$n = 158912$, No surgery	IUGR in 5% postsurgical vs 2% in controls ($p < 0.001$). However, this difference was nullified once confounding factors were controlled for	

Summary: Current literature supports an increased risk (around two-fold) for SGA in pregnancies postbariatric surgery. The relationship between surgery-to-conception interval and SGA is still unclear, and warrants further investigation.

TABLE 19.3 Summary of Selected Studies on Cesarean Delivery Rate in Pregnancies After Bariatric Surgery [6,49,50,52]

Study	Study Group	Control Group	Results
Burke, et al. (2010), United States	$n = 354$, Postbariatric surgery deliveries	$n = 346$, Presurgery deliveries of same mothers	Reduced risk postbariatric surgery, 28% vs 43%, adjusted OR = 0.48 (95% CI 0.32–0.73)
Weintraub et al. (2008), Israel	$n = 507$, Post-LAGB deliveries	$n = 301$, Presurgery deliveries of same mothers	Increased risk postbariatric surgery, 30.0% vs 17.9%, OR = 1.9 (95% CI 1.4–2.8, $p < 0.001$)
Ducarme et al. (2007), France	$n = 13$, Obese, post-LAGB deliveries	$n = 414$, Obese, nonpostsurgery deliveries	Reduced risk in obese women post-LAGB compared to obese women who had not undergone LAGB, 15.3% vs 34.4% ($p < 0.01$)
Sheiner et al. (2004), Israel	$n = 298$, Postbariatric surgery deliveries	$n = 158912$, Nonsurgical general obstetric population	Increased risk postbariatric surgery, 25.2% vs 12.2%, OR = 2.4 (95% CI 1.9–3.1, $p < 0.001$).

Summary: A conclusion cannot be reached in regard to the risk of cesarean section in women postbariatric surgery. Since this complication is affected in part by physician and patient preference, it is possible that the highly variable results can be attributed to cultural differences among the study groups. Further exploration into this outcome is warranted.

TABLE 19.4 Summary of Selected Studies on Micronutrient Status During Pregnancies After Bariatric Surgery [24,53–55]

Study	Study Group	Control Group	Results
Machado et al. (2015), Brazil	$n = 40$, Post-RYGB pregnancies	$n = 80$, Matched for age and prepregnancy BMI	Significantly increased rates of quantitative and functional vitamin A deficiency in the post-RYGB group
Devlieger et al. (2014), Belgium	$n = 49$, Total pregnancies, 18 postrestrictive bariatric surgery, and 31 postmalabsorptive surgery	None	Micronutrient deficiencies are common among pregnant women postbariatric surgery. No meaningful comparisons between procedure groups or with the general population can be drawn from this study
Medeiros, et al. (2015), Brazil	$n = 46$, Post-RYGB pregnancies adherent to standard supplementation protocol	None	Rates of vitamin D deficiency and insufficiency were high among the study group (17.4% and 67.4%) despite supplementation with calcium carbonate. No comparison with matched controls or the general population was made
Jans et al. (2015), Belgium	$n = 37$, Observational studies for systematic review (29 cases and 8 cohort studies)	N/A	There is insufficient evidence for a definitive analysis of micronutrient deficiencies or related fetal complications in postbariatric surgery pregnancies. Current evidence suggests global micronutrient deficiencies

Summary: Rigorous studies of nutritional status of pregnancies after bariatric surgery are lacking.

TABLE 19.5 Summary of Recommendations for Management of Pregnancy Postbariatric Surgery

- Management and follow-up by multidisciplinary team
- Appropriate micronutrient supplementation, and monitoring of compliance
- Alternate screening methods for GDM
- High index of suspicion and evaluation by general surgeon in case of severe abdominal pain
- Monitoring of gestational weight gain, and intervention if insufficient

- Special considerations in PPBS include monitoring gestational weight gain and supplementing micronutrients, and maintaining a high index of suspicion for surgical complications (Table 19.5).

REFERENCES

[1] Flegal K, Carroll MD, Kit BK, Ogden CL. Prevalence of obesity among adults: United States, 2011–2012. NCHS Data Brief 2013;131:1–8.
[2] Getahun D, Ananth CV, Peltier MR, Salihu HM, Scorza WE. Changes in prepregnancy body mass index between the first and second pregnancies and risk of large-for-gestational-age birth. Am J Obstet Gynecol 2007;196(6): pp. 530.e1–8.
[3] Law DCG, Maclehose RF, Longnecker MP. Obesity and time to pregnancy. Hum Reprod 2007;22(2):414–20.
[4] Cedergren MI. Maternal morbid obesity and the risk of adverse pregnancy outcome. Obstet Gynecol 2004;103(2):219–24.
[5] Robinson HE, O'Connell CM, Joseph KS, McLeod NL. Maternal outcomes in pregnancies complicated by obesity. Obstet Gynecol 2005;106 (6):1357–64.
[6] Sheiner E, Levy A, Menes TS, Silverberg D, Katz M, Mazor M. Maternal obesity as an independent risk factor for caesarean delivery. Paediatr Perinat Epidemiol 2004;18(3):196–201.
[7] Metwally M, Ong KJ, Ledger WL, Li TC. Does high body mass index increase the risk of miscarriage after spontaneous and assisted conception? A meta-analysis of the evidence. Fertil Steril 2008;90(3):714–26.
[8] Gaillard R, Steegers EAP, Hofman A, Jaddoe VWV. Associations of maternal obesity with blood pressure and the risks of gestational hypertensive disorders. The Generation R Study. J Hypertens 2011;29(5):937–44.

[9] Aricha-Tamir B, Weintraub AY, Levi I, Sheiner E. Downsizing pregnancy complications: a study of paired pregnancy outcomes before and after bariatric surgery. Surg Obes Relat Dis Off J Am Soc Bariatr Surg 2012;8(4):434−9.

[10] Norman SM, Tuuli MG, Odibo AO, Caughey AB, Roehl KA, Cahill AG. The effects of obesity on the first stage of labor. Obstet Gynecol 2012;120(1):130−5.

[11] Kominiarek MA, Zhang J, Vanveldhuisen P, Troendle J, Beaver J, Hibbard JU. Contemporary labor patterns: the impact of maternal body mass index. Am J Obstet Gynecol 2011;205(3):244. e1−8.

[12] Myles TD, Gooch J, Santolaya J. Obesity as an independent risk factor for infectious morbidity in patients who undergo cesarean delivery. Obstet Gynecol 2002;100(5):959−64. Pt 1.

[13] Stothard KJ, Tennant PWG, Bell R, Rankin J. Maternal overweight and obesity and the risk of congenital anomalies: a systematic review and meta-analysis. JAMA 2009;301(6):636−50.

[14] Josefsson A, Bladh M, Wiréhn A-B, Sydsjö G. Risk for congenital malformations in offspring of women who have undergone bariatric surgery. A national cohort. BJOG Int J Obstet Gynaecol 2013;120(12):1477−82.

[15] Taylor PD, Poston L. Developmental programming of obesity in mammals. Exp Physiol 2007;92(2):287−98.

[16] Boney CM, Verma A, Tucker R, Vohr BR. Metabolic syndrome in childhood: association with birth weight, maternal obesity, and gestational diabetes mellitus. Pediatrics 2005;115(3):e290−6.

[17] Colquitt JL, Picot J, Loveman E, Clegg AJ. Surgery for obesity. Cochrane Database Syst Rev 2009;2:CD003641.

[18] Nguyen NT, Masoomi H, Magno CP, Nguyen X-MT, Laugenour K, Lane J. Trends in use of bariatric surgery, 2003 − 2008. J Am Coll Surg 2011;213(2):261−6.

[19] Quyên Pham T, Pigeyre M, Caiazzo R, Verkindt H, Deruelle P, Pattou F. Does pregnancy influence long-term results of bariatric surgery? Surg Obes Relat Dis Off J Am Soc Bariatr Surg 2015;11(5):1134−9.

[20] Yi X, Li Q, Zhang J, Wang Z. A meta-analysis of maternal and fetal outcomes of pregnancy after bariatric surgery. Int J Gynaecol Obstet Off Organ Int Fed Gynaecol Obstet 2015;130(1):3−9.

[21] Galazis N, Docheva N, Simillis C, Nicolaides KH. Maternal and neonatal outcomes in women undergoing bariatric surgery: a systematic review and meta-analysis. Eur J Obstet Gynecol Reprod Biol 2014;181:45−53.

[22] Guelinckx I, Devlieger R, Vansant G. Reproductive outcome after bariatric surgery: a critical review. Hum Reprod Update 2009;15(2):189−201.

[23] Shai D, Shoham-Vardi I, Amsalem D, Silverberg D, Levi I, Sheiner E. Pregnancy outcome of patients following bariatric surgery as compared with obese women: a population-based study. J Matern Fetal Neonatal Med 2014;27(3):275−8.

[24] Machado SN, Pereira S, Saboya C, Saunders C, Ramalho A. Influence of Roux-en-Y gastric bypass on the nutritional status of vitamin A in pregnant women: a comparative study. Obes Surg 2016;26:26−31.

[25] Maggard MA, Shugarman LR, Suttorp M, Maglione M, Sugerman HJ, Livingston EH, et al. Meta-analysis: surgical treatment of obesity. Ann Intern Med 2005;142(7):547−59.

[26] A C of Obstetricians and Gynecologists. ACOG practice bulletin no. 105: bariatric surgery and pregnancy. Obstet Gynecol 2009;113(6):1405−13.

[27] Moore KA, Ouyang DW, Whang EE. Maternal and fetal deaths after gastric bypass surgery for morbid obesity. N Engl J Med 2004;351(7):721−2.

[28] Efthimiou E, Stein L, Court O, Christou N. Internal hernia after gastric bypass surgery during middle trimester pregnancy resulting in fetal loss: risk of internal hernia never ends. Surg Obes Relat Dis Off J Am Soc Bariatr Surg 2009;5(3):378−80.

[29] Wax JR, Pinette MG, Cartin A. Roux-en-Y gastric bypass-associated bowel obstruction complicating pregnancy-an obstetrician's map to the clinical minefield. Am J Obstet Gynecol 2013;208(4):265−71.

[30] Haward RN, Brown WA, O'Brien PE. Does pregnancy increase the need for revisional surgery after laparoscopic adjustable gastric banding? Obes Surg 2011;21(9):1362−9.

[31] Jefferys AE, Siassakos D, Draycott T, Akande VA, Fox R. Deflation of gastric band balloon in pregnancy for improving outcomes. Cochrane Database Syst Rev 2013;4:CD010048.

[32] Gudbrand C, Andreasen LA, Boilesen AE. Internal hernia in pregnant women after gastric bypass: a retrospective register-based cohort study,". Obes Surg 2015;25(12):2257−62.

[33] Dixon JB, Dixon ME, O'Brien PE. Birth outcomes in obese women after laparoscopic adjustable gastric banding. Obstet Gynecol 2005;106(5):965−72. Pt 1.

[34] Karmon A, Sheiner E. Timing of gestation after bariatric surgery: should women delay pregnancy for at least 1 postoperative year? Am J Perinatol 2008;25(6):331−3.

[35] Merhi ZO. Challenging oral contraception after weight loss by bariatric surgery. Gynecol Obstet Invest 2007;64(2):100−2.

[36] Nørgaard LN, Gjerris ACR, Kirkegaard I, Berlac JF, Tabor, and Danish Fetal Medicine Research Group A. Fetal growth in pregnancies conceived after gastric bypass surgery in relation to surgery-to-conception interval: a Danish national cohort study. PLoS One 2014;9(3):e90317.

[37] Ducarme G, Chesnoy V, Lemarié P, Koumaré S, Krawczykowski D. Pregnancy outcomes after laparoscopic sleeve gastrectomy among obese patients. Int J Gynecol Obstet 2015;2015(130):127−31.

[38] Burstein E, Levy A, Mazor M, Wiznitzer A, Sheiner E. Pregnancy outcome among obese women: a prospective study. Am J Perinatol 2008;25(9):561−6.

[39] Johansson K, Cnattingius S, Näslund I, Roos N, Trolle Lagerros Y, Granath F, et al. Outcomes of pregnancy after bariatric surgery. N Engl J Med 2015;372(9):814—24.

[40] Karmon A, Sheiner E. Pregnancy after bariatric surgery: a comprehensive review. Arch Gynecol Obstet 2008;277(5):381—8.

[41] Adams TD, Hammoud AO, Davidson LE, Laferrère B, Fraser A, Stanford JB, et al. Maternal and neonatal outcomes for pregnancies before and after gastric bypass surgery. Int J Obes 2005;39(4):686—94.

[42] Patel JA, Patel NA, Thomas RL, Nelms JK, Colella JJ. Pregnancy outcomes after laparoscopic Roux-en-Y gastric bypass. Surg Obes Relat Dis Off J Am Soc Bariatr Surg 2008;4(1):39—45.

[43] Sheiner E, Levy A, Silverberg D, Menes TS, Levy I, Katz M, et al. Pregnancy after bariatric surgery is not associated with adverse perinatal outcome. Am J Obstet Gynecol 2004;190(5):1335—40.

[44] Parker MH, Berghella V, Nijjar JB. Bariatric surgery and associated adverse pregnancy outcomes among obese women. J Matern-Fetal Neonatal Med Off J Eur Assoc Perinat Med Fed Asia Ocean Perinat Soc Int Soc Perinat Obstet 2015;1—4.

[45] Roos N, Neovius M, Cnattingius S, Trolle Lagerros Y, Sääf M, Granath F, et al. Perinatal outcomes after bariatric surgery: nationwide population-based matched cohort study. BMJ 2013;347:f6460.

[46] Lesko J, Peaceman A. Pregnancy outcomes in women after bariatric surgery compared with obese and morbidly obese controls. Obstet Gynecol 2012;119(3):547—54.

[47] Kjaer MM, Lauenborg J, Breum BM, Nilas L. The risk of adverse pregnancy outcome after bariatric surgery: a nationwide register-based matched cohort study. Am J Obstet Gynecol 2013;208(6):464. e1—464.e5.

[48] Berlac JF, Skovlund CW, Lidegaard O. Obstetrical and neonatal outcomes in women following gastric bypass: a Danish national cohort study. Acta Obstet Gynecol Scand 2014;93(5):447—53.

[49] Weintraub AY, Levy A, Levi I, Mazor M, Wiznitzer A, Sheiner E. Effect of bariatric surgery on pregnancy outcome. Int J Gynaecol Obstet Off Organ Int Fed Gynaecol Obstet 2008;103(3):246—51.

[50] Ducarme G, Revaux A, Rodrigues A, Aissaoui F, Pharisien I, Uzan M. Obstetric outcome following laparoscopic adjustable gastric banding. Int J Gynaecol Obstet Off Organ Int Fed Gynaecol Obstet 2007;98(3):244—7.

[51] Josefsson A, Blomberg M, Bladh M, Frederiksen SG, Sydsjo G. Bariatric surgery in a national cohort of women: sociodemographics and obstetric outcomes. Am J Obstet Gynecol 2011;205(3):206. e1—206.e8.

[52] Burke AE, Bennett WL, Jamshidi RM, Gilson MM, Clark JM, Segal JB, et al. Reduced incidence of gestational diabetes with bariatric surgery. J Am Coll Surg 2010;211(2):169—75.

[53] Medeiros M, Matos AC, Pereira SE, Saboya C, Ramalho A. Vitamin D and its relation with ionic calcium, parathyroid hormone, maternal and neonatal characteristics in pregnancy after roux-en-Y gastric bypass. Arch Gynecol Obstet 2015.

[54] Devlieger R, Guelinckx I, Jans G, Voets W, Vanholsbeke C, Vansant G. Micronutrient levels and supplement intake in pregnancy after bariatric surgery: a prospective cohort study. PloS One 2014;9(12):e114192.

[55] Jans G, Matthys C, Bogaerts A, Lannoo M, Verhaeghe J, Van der Schueren B, et al. Maternal micronutrient deficiencies and related adverse neonatal outcomes after bariatric surgery: a systematic review. Adv Nutr Bethesda Md 2015;6(4):420—9.

Chapter 20

Mortality Rate and Long-Term Outcomes After Bariatric Surgery

S.M.B. Kelles

Hospital das Clínicas da Universidade Federal de Minas Gerais, Belo Horizonte, Minas Gerais, Brazil

LIST OF ABBREVIATIONS

AGB	adjustable gastric band
BMI	body mass index
EBWL	excess body weight loss
GBP	gastric bypass
HRQoL	health-related quality of life
NNT	number needed to treat
OP	obesity-related problems scale
RGB	Roux-en-Y gastric bypass
VBG	vertical banded gastroplasty

MORTALITY RATE AND LONG-TERM OUTCOMES AFTER BARIATRIC SURGERY

Overweight and obesity comprise the fifth leading global risk for mortality in the world according to the World Health Organization (WHO), accounting for 10–13% of deaths worldwide [1]. Several studies have shown that obesity is a relevant risk factor for a number of noncommunicable diseases, increasing the risk of cardiovascular diseases, diabetes, cancer, joint disease, sleep apnea, and other conditions [2].

Bariatric surgery has been shown to be more beneficial in reducing morbidity and weight compared to more conservative approaches. Severely obese individuals undergo bariatric surgery, considered a safe procedure, for several reasons: improvement of health, social interactions, quality of life (QoL), and esthetics [3]. Thus, obese people undergo bariatric surgery to reduce the above-mentioned risk factors associated with excessive body weight. Knowledge about the effects of this procedure on risk factors, however, comes from cohort studies or case series, many of which, without control groups, may therefore be biased to a greater or lesser degree. Although it is biologically plausible to consider that surgery may reduce harm caused by excessive weight, comparative studies based on solid methodology and long-term follow-up would be required to confirm this assumption.

According to the Swedish Obese Subjects (SOS) study, weight loss after bariatric surgery has a typical pattern: weight loss reaches its nadir 1 or 2 years after surgery, followed by weight gain thereafter, irrespective of the type of procedure. Ten years later, weight was 14–25% of the preoperative weight. Fifteen years after surgery, the remaining cohort had similar percentages. Thus, over 10 years bariatric surgery results in modest weight loss (around 15% overall). Furthermore, Sjöström et al. found no associations between body mass index (BMI) and end points such as mortality, the incidence rate of cardiovascular events, cancer, or diabetes [3].

Obesity is a chronic disease; thus, it is reasonable to expect that any proposed obesity treatment should demonstrate long-term durability to be considered effective and sustainable. Therefore, this sequence of weight loss and weight regain after bariatric surgery may compromise the expected risk reduction due to weight loss. Although most patients do not fully regain their presurgical weight, the loss of weight attained during the first 2 years after surgery may not be sustainable in the long term [4].

Overall Mid- and Long-Term Mortality of Obese Individuals and Operated Subjects

Estimates of relative mortality risks associated with underweight, normal weight, overweight, and obesity may yield information about the real impact of bariatric surgery.

Weight loss has been associated with an increased mortality rate in several observational studies, suggesting that these studies were possibly unable to differentiate intentional from disease-induced weight loss [3,5]. Relative to normal weight, obesity grades 2 or 3 were associated with significantly higher all-cause mortality, with hazard ratio (HR) of 1.29 (95% CI, 1.18−1.41), in a meta-analysis of 97 studies that enrolled 2.88 million participants [6].

Data on weight loss specifically due to bariatric surgery consistently describes the associations between BMI and mortality. Pontiroli's meta-analysis [7] revealed an association between overall mortality and BMI, although there were heterogeneous differences from all-cause mortality (I-squared = 60.4%). The SOS, which was included in this metaanalysis, and was conducted by Sjöström et al., is the only prospective study that has shown an association between weight loss and decreased mortality. The authors of this study monitored obese individuals who either underwent or did not undergo bariatric surgery during a 16-year period [3]. Mortality after surgery was rare in this cohort, even though a beneficial effect on mortality could only be demonstrated after 13 years of the 16-year follow-up. The major caveat of this study was that the control group was not randomized, but was selected by pair matching [3].

Besides Sjöström's paper, eight other cohort studies showed decreased mortality after bariatric surgery; all of them, however, had pair-matched control groups that often differed from the groups of operated subjects. Adams et al. [8] (Utah, US) assessed 9949 bariatric surgery subjects with a 7.1-year follow-up period. There were 9628 community participants as controls, matched by age, sex, and self-reported weight and height recorded from their driving licenses. These authors reported a 40% decrease in mortality by any causes, as well as reductions in death due to cardiovascular diseases (56%), diabetes (92%), and cancer (60%). However, there was a 58% increase in violent deaths (not caused by disease), such as accidents and suicides. In 10,000 surgeries, 171 deaths (2.7%) caused by disease were avoided. Conversely, there were 35 violent deaths per 10,000 surgeries. The net result was 136 lives saved for 10,000 surgeries. Besides selecting controls by pair matching, weight and height were self-declared, which added a significant risk of information bias to the study.

Christou et al. [9] reported an even more significant drop in mortality (89%) compared to controls in a 5-year follow-up. But controls included subjects paired by sex and age, and were severely obese individuals who had recently been hospitalized because of obesity. Controls, therefore, had different severity characteristics compared to the study group; they were, furthermore, older and predominantly male.

Flum et al. [10] assessed the 15-year survival rate in patients undergoing bariatric surgery compared to subjects paired by sex, age, and BMI, and found that mortality decreased by 33% in the study group. Controls, however, had been hospitalized previously, rendering the comparison inadequate.

MacDonald [11] assessed only diabetic patients and pair-matched controls and reported a 32% decrease in mortality after bariatric surgery during a 6-year follow-up.

Sowemimo et al. [12] reported that the mortality rate fell by 50−82% among 908 individuals undergoing bariatric surgery, compared to 112 paired controls. Controls included subjects who were eligible for bariatric surgery but who were not operated. Gender distribution was similar in both groups, but controls had older subjects with a lower mean BMI. Reasons for not performing bariatric surgery in controls varied.

Peeters et al. [13] compared two cohorts (4- to-12-year follow-up) that either were or were not operated. The mortality rate was 72% lower in the intervention group (966 subjects) compared to a severely obese cohort (2119 subjects). Busetto et al. [14] found a 60% decrease in the mortality rate of 821 subjects with a BMI $>$ 40 kg/m^2 who underwent bariatric surgery compared to 821 individuals paired by sex, age, and BMI. Follow-up was 5 years. Guildry et al. [15] found a significant (51%) decrease in overall mortality in the bariatric cohort compared with propensity-matched controls in a median 11.9-year follow-up. Table 20.1 describes the details of the studies analyzed in this section.

Mortality decreases consistently in all of these studies in the mid- and long-term when subjects undergoing bariatric surgery are compared with severely obese nonoperated individuals. It should be pointed out that controls were often selected from higher-risk groups (hospitalized or superobese individuals), and that there were no randomized trials among available papers, i.e., there was an information bias in these results.

TABLE 20.1 Comparative Mortality Among Obese Patients Undergoing Bariatric Surgery or Not: Details of the Studies Considered

Author/ Year	Number of Patients		Follow-up (Years)	Outcome: Overall Mortality	Study Design/Groups Selection Criteria
	Surgery	Control			
Sjöström, 2013 [3]	2010	2037	16.0	Surgery was associated with an adjusted HR 0.71 (95% CI, 0.54–0.92) relative to usual care for the control subjects; 129 and 101 subjects died in the control and in the surgery group, respectively.	Prospective-controlled intervention study. Surgically treated subjects underwent AGB ($n = 376$), VBG ($n = 1369$), or GBP ($n = 265$). A control group was created using 18 matching variables.
Adams et al., 2007 [8]	9945	9628	7.1	Reduction of 40%: 37.6/10,000 person years in the surgery group and 57.1/10,000 person years in the control group.	Retrospective cohort study. Surgery: consecutive patients who had undergone GBP. Controls were community pairs matched by age, sex, and self-reported weight and height.
Christou et al, 2004 [9]	1035	5746	5.0	Reduction of 89%: 0.68% deaths among bariatric patients and 6.17% deaths among controls.	Two-cohort study. Surgery group: open GBP. Controls pair matched by age and sex with obese patients that had been hospitalized recently.
Flum et al., 2004 [10]	3328	62,781	15.0	Reduction of 33%: 16.3% nonoperated patients versus 11.8% patients who had the bariatric procedure.	Retrospective two-cohort study. Surgery group: GBP. Administrative database control: pair matched all patients hospitalized with a diagnosis of morbid obesity.
MacDonald et al, 1997 [11]	154	78	9.0 years surgical group and 6.0 years control group	Reduction of 32% in mortality: death of 28% in the control group and 9% in the surgical group.	Retrospective two-cohort study, both of morbidly obese patients with NID DM: 154 had GBP and 78 were controls who did not undergo surgery because of personal preferences or refusal by their insurance company.
Sowemimo et al., 2007 [12]	908	112	9.0	Reduction of 82% in total mortality: 2.9% in the surgical group and 14.3% in the control group.	Retrospective cohort. Surgery group: patients who met the NIH criteria for bariatric surgery. Controls were nonoperated patients eligible for bariatric surgery.
Peeters et al., 2007 [13]	966	2119	4 years surgical group 12 years community cohort controls	Reduction of 82% when the follow-up time was censored at 5 years. Deaths: 25.0/1000 person years in the community control cohort and 3.7/1000 person years in the weight loss cohort.	Two-cohort study. Severely obese population-based cohort as control (subjects were recruited via Electoral Rolls, advertisements, and community announcements). Surgery group (only LAGB).
Busetto et al. 2007 [14]	821	821	5.6 ± 1.9 years surgical cohort 7.2 ± 1.2 years reference cohort	Reduction of 60% in total mortality; RR of death in the surgical cohort 0.36 (95% CI, 0.16–0.79): Deaths: 8/821 in the surgical cohort and 36/821 in the reference cohort.	Two cohorts. Controls were pair-matched by sex, age, and BMI, selected from a sample of adults (BMI > 40 kg/m^2) seen at medical centers, nonoperated. The surgical cohort, BMI > 40 kg/m^2, was treated with LAGB.
Guildry et al, 2015 [15]	802	802	11.9	Reduction of 51%. Deaths: 6.5% in the surgery cohort and 12.7% in the control group.	Two cohorts. Controls were identified at a tertiary care center and matched by propensity scores. Surgery group received GBP.

AGB, adjustable gastric banding; VBG, vertical banded gastroplasty; GBP, gastric bypass; LAGB, laparoscopic adjustable gastric banding; NID DM, noninsulin-dependent diabetes mellitus. NIH criteria for bariatric surgery: BMI ≥ 40 kg/m^2 or BMI > 35 kg/m^2 with comorbidity.

Mid- and Long-Term Mortality Due to Specific Causes

Mortality Due to Natural Causes

Morino et al. [16] reported that 29.4% of deaths following bariatric surgery were due to complications of this procedure. Bruschi Kelles et al. [17] found that 84% of deaths in a cohort of 4344 subjects were due to natural causes, and that 16% were caused by external causes unrelated to diseases.

Descriptions of specific causes of long-term deaths in operated subjects were sparse. Data on specific causes of mortality following bariatric surgery showed that cardiovascular diseases and malignancy were the most frequent causes of death. Sjöström [3] found that malignancy and fatal acute myocardial infarction were the most common causes of death both in the control group and the surgery group, but the study lacked statistical power, and the authors were unable to estimate risk reduction according to specific causes of death. Pontiroli's meta-analysis [7] found a lower risk of cardiovascular mortality after bariatric surgery when compared to nonsurgery individuals. However, the authors pointed out that the studies in this meta-analysis had not been randomized, that they were heterogeneous in size, time, and loss of follow-up, and that determination of death was based on death certificates which, in many instances, may have distorted the death report.

Adams et al. [18] showed that patients undergoing bariatric surgery had a lower risk of death by specific causes such as stroke and cancer when compared to severely obese nonoperated controls.

Thus, after bariatric surgery, the main causes of natural death, in general, remain the same as those observed in nonoperated severely obese patients. While a decrease in overall mortality is observed in most studies, the magnitude of the effect on mortality due to cardiovascular events and cancer in individuals who lose weight after bariatric surgery is provided by observational studies (i.e., heterogeneous and lacking in statistical power). These outcomes, therefore, remain controversial.

Mortality by External or Violent Causes

Although global mortality and mortality by natural causes have decreased compared to nonoperated obese individuals, there is systematically a high rate of deaths by external causes unrelated to diseases.

Davidson et al.'s [19] cohort consisted of 7925 gastric bypass surgery patients compared with a group-matched control group consisting of 7925 severely obese nonoperated individuals identified by driver license records. The follow-up was 7.2 years. The authors sought the age groups that would benefit most from bariatric surgery in terms of survival. Except for patients aged below 35 years, survival increased in the other age groups compared to nonoperated individuals. No benefit was seen in patients below 35 years of age; the authors suggested that a high prevalence of deaths by external causes in this group accounted for this finding. The HR was 2.35 (95% CI, 1.27−5.07); among women the HR was 3.08 (95% CI, 1.4−6.7).

Omalu et al. [20] assessed the causes of death in 16,683 bariatric surgery patients from 1995 to 2004 with a 5-year follow-up. Natural causes accounted for 90% of deaths; 10% were traumatic deaths. Of 45 violent deaths, 16 were suicides, 10 were due to automobile accidents, 14 resulted from drug overdose, 3 were homicides, and 2 deaths were due to falls. The suicide rate was much higher compared to similarly aged groups in the general population.

Tindle et al. [21] reanalyzed this population and found a 6.6/10,000 suicide rate among bariatric surgery individuals. For comparison, the sex- and age-adjusted suicide rate in the United States was 2.4/10,000 in males and 0.7/10,000 in females. About 30% of suicides occurred within 2 years of surgery and 70% occurred within 3 years following surgery.

The percentage of suicides in a Brazilian cohort [17] of 4344 bariatric surgery patients with a 10-year follow-up was 10%.

Adams et al. [8] found that the risk of suicide in a bariatric surgery group after adjusting for age, BMI, and sex was double that of controls.

The causes of such a high rate of suicide required further investigation. Sansone et al. [22] found that 10% of bariatric surgery candidates reported previous suicide attempts, the most significant risk factor for fatal suicides. An association between obesity and fatal suicide or suicide attempts is controversial, but it appears that the risk is higher in bariatric surgery patients. Henegham et al. [23] found a higher suicide risk among obese individuals, and that this risk remained after bariatric surgery. Other papers have also reported a high suicide rate after bariatric surgery [24].

HOSPITAL ADMISSIONS AFTER BARIATRIC SURGERY

An increase in hospitalization is somehow expected immediately after surgery due to the complexity of this procedure. In the mid- and long-term, however, hospital admissions and use of healthcare services are expected to be less frequent compared to the time before surgery, due to decreased presurgical comorbidity rates.

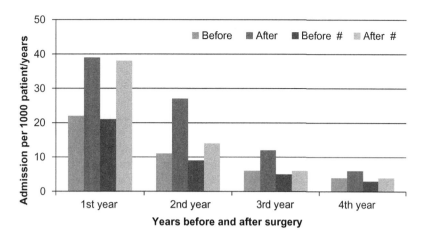

FIGURE 20.1 Inpatient hospitalizations before and after bariatric surgery, over 4 years, in the Brazilian cohort. Mean rate of admission per 1000 patients among bariatric surgery patients, before and after the procedure, over a 4-year follow-up. Each patient contributed equal follow-up time before and after surgery. For all differences: $p < 0.05$ (Wilcoxon signed-rank test). The # considered hospitalizations not including those related to pregnancy or plastic surgery. For this subgroup analysis, the $p < 0.05$ (Wilcoxon signed-rank test) for all but the fourth year. *Adapted from Bruschi Kelles SM, Machado CJ, Barreto SM. Before-and-after study: does bariatric surgery reduce healthcare utilization and related costs among operated patients? Int J Technol Assess Health Care 2015;31(6):407–13, with permission [27].*

Comorbidities such as diabetes, apnea, elevated arterial blood pressure, and the risk of cancer, among other conditions, decrease or improve after bariatric surgery.

Morgan et al. [25] assessed population data of 12,062 bariatric surgery patients in Australia, with a 41-month follow-up before and after surgery. These authors reported a significant decrease in hospitalizations for any causes and hospital admissions due to diabetes after surgery, as compared to before surgery. There was, however, a trend towards a higher hospitalization rate for sleep apnea, venous thromboembolism, gastrointestinal tract complications, sepsis, and acute kidney injury. The authors did not discriminate short- and mid-term hospital admissions after surgery, which may have confounded interpretations. Bleich et al. [26] reported consistently higher hospital costs between 2 and 6 years after bariatric surgery in a before-and-after study of diabetic patients only, which may serve as a proxy for increased healthcare use. Bruschi Kelles et al.'s [27] before-and-after study of 4004 patients during 4 years prior to and following bariatric surgery revealed a consistent increase in hospital admissions and postoperative costs compared to the period before surgery. The authors discriminated the number of hospital admissions per time period and the number of hospitalizations due to any causes. Even after excluding admissions for volitive reasons, such as pregnancy and esthetic surgery after bariatric surgery, hospital admission rates remained consistently high after the bariatric procedure (Fig. 20.1). Zingmond et al. [28] compared hospital admissions of 60,077 patients before and after bariatric surgery and reported a higher rate of hospitalizations of up to 3 years following surgery. Christou et al. compared bariatric surgery patients to matched controls and found a lower rate of hospital readmission in the operated group. Controls in this study, however, were severely obese (ICD 9) patients who had been hospitalized recently [9]. Bolen et al. [29] found consistently higher costs and use of healthcare services after surgery in an analysis of patients up to 2 years after surgery.

Therefore, although a reduction in comorbidities after surgery was expected, the need for more postoperative hospitalization suggested that new comorbid conditions acquired or developed after the bariatric procedure.

QUALITY OF LIFE

One of the goals that obese patients have when undergoing bariatric surgery is to improve QoL. Thus, assessing this outcome is relevant whether surgery yields expected results or not. Patient satisfaction and perception of QoL vary according to the moment when weight loss and regain are evaluated.

The SOS study offers one of the broadest perspectives on QoL in this area. Karlsson et al. [30] reported QoL scores of 655 patients up to 10 years after bariatric surgery. These authors concluded that improved scores were associated with weight loss, except for the item "anxiety." Gains in QoL reached a peak in the first year after surgery; scores decreased in subsequent years (1–6 years) as patients regained weight. Perceived QoL and weight remained relatively stable between 6 and 10 years after surgery. After 10 years, the QoL of operated subjects was better in the following domains: self-perception of health, social interaction, and depression. There were no changes in affect and anxiety compared to nonoperated individuals.

Raoof et al.'s cross-sectional study [31] evaluated 486 gastric bypass surgery patients 12 years after the procedure by applying the SF-36 and the obesity-related problems scale (OP). There were two sex- and age-matched control groups: one of individuals from the general population, and one comprising morbidly obese subjects waiting for bariatric surgery. These authors found improved scores in all domains of the SF-36 and OP among operated subjects compared to obese nonoperated individuals. However, the scores of operated subjects were all lower than those of the general population.

Kiewiet et al. [32] reported the QoL of operated patients, assessed by the SF-36, up to 6 years after surgery. Physical performance was the only domain that showed an improvement compared to controls. Nadaline et al. [33] monitored bariatric surgery patients for up to 4 years and found significant improvement in all dimensions of the SF-36 except mental health. QoL was perceived as better proportionally to weight loss.

Driscoll et al. published a systematic review and meta-analysis (2016) and showed that the domains of physical health and general scores were improved in operated bariatric surgery patients compared to obese nonoperated controls. The results in the domain of mental health were heterogeneous, some showing relevant gains in 5 years and others revealing no benefits. Results were heterogeneous in the domains of mental health, vitality, social functioning, and emotional role, and were due to subject heterogeneity, the analysis tools used in each study, and the follow-up of each one [34].

Andersen et al. published another systematic review in 2015 [35] to assess perceptions of QoL among bariatric surgery patients up to five or more years after surgery. Only two studies had control groups. These results showed that QoL improvements peaked within the first one or two years after surgery, followed by a decline that tended to stabilize 5 years after surgery. Changes in perceived long-term QoL were systematically higher compared to preoperative scores, but the scores were below those of the general population.

WEIGHT REGAIN TRENDS AFTER BARIATRIC SURGERY

Weight regain and its consequences are among the most disturbing outcomes after bariatric surgery. It is one of the most feared outcomes for patients undergoing major surgery that places lifelong constraints on food intake.

Sjöström's 2004 paper [36] describes the profile of weight loss and regain. There is a weight loss nadir in operated patients in the first year after surgery followed by gradual weight regain thereafter. After 10 years, bariatric surgery patients had significant weight regains compared to the weight lost within the first postoperative year. The updated data [3] showed weight regain among patients in the control group and the three surgery groups during 20 years of follow-up in the SOS study. (Fig. 20.2).

Cooper et al. [37] published the results of a cohort of 300 bariatric surgery patients Roux-en-Y gastric bypass. The mean weight loss was 39.3% compared to the preoperative weight, and the mean weight regain was 23.4% of the maximum weight lost[1] after a mean follow-up period of 6.9 years. Excessive weight regain, defined as a gain $\geq 25\%$ of the total weight lost, was seen in 37% of patients.

There is no consensus on any definition of significant weight regain and its consequences after bariatric surgery, even though weight regain has been demonstrated after all techniques in antiobesity surgery.

Several obesity-related comorbidities, such as diabetes, hyperlipidemia, arterial hypertension, and sleep apnea, may regress after surgery. But these benefits depend on long-term maintenance of weight loss. Comorbidities may reappear or, if still present, may worsen as weight increases. Other comorbidities resulting from anatomical changes of the gastrointestinal tract due to surgery may arise, such as nutrient deficiencies, osteoporosis, and other conditions. Weight regain is related to behavioral factors, although data suggest that physiological and anatomical adaptations after surgery may be possible causes [38].

Evidence on long-term sustained outcomes and safety after bariatric surgery is still scarce and lacks quality. We currently have long-term morbidity data after bariatric surgery from cohorts who have been monitored for two decades. Bariatric surgery patients are generally young adults aged around 40 years who will live significantly more than 20 years. What are the long-term consequences of surgery? We still do not know. The history of medicine has shown that extrapolating results from short-term outcomes may be flawed. Long-term information is lacking and makes it difficult to balance benefits against risks.

Most published papers have measured physical outcomes; psychosocial effects, however, may be relevant, as seen in studies that indicate significantly higher rates of postsurgery suicides.

---

1. Percentage weight regain was defined as the percentage of weight regained from the nadir weight, i.e., [100(current weight − nadir weight)]/(preoperative weight − nadir weight).

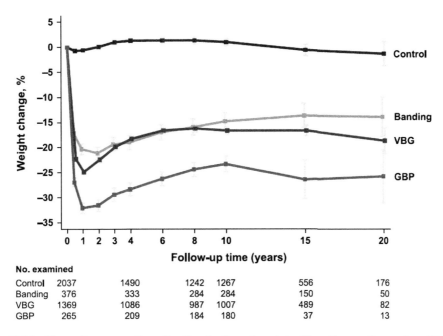

No. examined						
Control	2037	1490	1242	1267	556	176
Banding	376	333	284	284	150	50
VBG	1369	1086	987	1007	489	82
GBP	265	209	184	180	37	13

FIGURE 20.2 Mean Weight Change Percentages From Baseline for Controls and the Three Surgery Groups Over 20 Years in the Swedish Obese Subjects (SOS) Study. Mean percentage weight changes from baseline among patients in the control and the three surgery groups during 20 years of follow-up in the SOS study. Data is shown for controls receiving usual care and for surgery patients undergoing banding, vertical banded gastroplasty (VBG), or gastric bypass (GBP) at baseline. Percentage weight changes from the baseline examination are based on data available on July 1, 2011. Error bars represent 95% confidence intervals. *From Sjöström L. Review of the key results from the Swedish Obese Subjects (SOS) trial — a prospective controlled intervention study of bariatric surgery. J Inter Med 2013;273:219—341, with permission [3].*

Therefore, careful selection criteria for bariatric surgery warrants reflection, and caution is required to balance the harm and the benefits so that researchers and patients alike do not find themselves facing negative surprises in the future.

Bariatric surgery imposes lifelong changes for patients; it is only the first step in a long treatment that requires discipline and effort from patients. It certainly is an excellent opportunity for lifestyle changes. Nevertheless, the results of the procedure may compromise health and QoL if healthcare professionals and patients do not seriously invest in this concept.

MINI-DICTIONARY OF TERMS

- *SOS (Swedish Obese Subjects) Study*: A prospective, controlled intervention study of bariatric surgery.
- *SF-36*: A multipurpose, short-form health survey with 36 questions. It yields an 8-scale profile of functional health and well-being scores, as well as psychometrically based physical and mental health summary measures and a preference-based health utility index.
- *Traumatic Causes of Death*: External causes of death.
- *Postoperative Weight Gain*: Significant weight gain occurs continuously in patients after reaching the nadir weight following gastric bypass. Weight regain was higher in the superobese group.
- *Superobese*: Patients with BMI ≥ 50 kg/m^2.

KEY FACTS

- John Flanagan created the first Quality of life (QoL) scales in the 1970s. Since then, hundreds of other instruments have been developed.
- QoL can be defined as "a broad range of human experiences related to one's overall well-being. It implies value based on subjective functioning in comparison with personal expectations and is defined by subjective experiences, states and perceptions. Quality of life, by its very natures, is idiosyncratic to the individual, but intuitively meaningful and understandable to most people."

- Methods for combining clinical data with patient perceptions and goals for improvements after treatment are not standardized.
- The instruments for assessing QoL are multidimensional constructs with a number of independent domains, including physical health, psychological well-being, social relationships, functional roles, and subjective senses of life satisfaction.
- Because of the broad range of QoL instruments, it is essential to select the appropriate domains to assess the condition researchers are dealing with and to correlate the clinical findings with the results of QoL forms.
- This instrument may also yield evidence to support the reliability and validity of health-related measurements, research design, and statistical analyses.

SUMMARY POINTS

- Mortality rate in long-term assessments after bariatric procedures:
  - Death due to natural causes.
  - Death due to trauma (unrelated to the disease).
- Long-term outcomes after surgery (not directly related to the procedure):
  - Hospital admissions.
  - Quality of life.
  - Weight regain.

REFERENCES

[1] World Health Organization (WHO): Fact Sheet No.311 (updated March 2013). Available at: <www.who.int/mediacentre/factsheets/fs311/en/> [last accessed 6.04.16].

[2] Cameron AJ, Magliano DJ, Söderberg S. A systematic review of the impact of including both waist and hip circumference in risk models for cardiovascular diseases, diabetes and mortality. Obes Rev 2013;14(1):86–94.

[3] Sjöström L. Review of the key results from the Swedish Obese Subjects (SOS) trial — a prospective controlled intervention study of bariatric surgery. J Inter Med 2013;273:219–34.

[4] Revicki DA, Osoba D, Fairclough D, et al. Recommendations on Health-related quality of life research to support labeling and promotional claims in the United States. Qual Life Res 2000;9(8):887–900.

[5] Chung WS, Ho FM, Cheng NC, Lee MC, Yeh CJ. BMI and all-cause mortality among middle-aged and older adults. Public Health Nutr 2014;18(10):1839–46.

[6] Flegal KM, Kit BK, Orpana H, Graubard BI. Association of all-cause mortality with overweight and obesity using standard body mass index categories: a systematic review and meta-analysis. JAMA 2013;309(1):71–82.

[7] Pontiroli AE, Morabito A. Long-term prevention of mortality in morbid obesity through bariatric surgery. A systematic review and meta-analysis of trials performed with gastric banding and gastric bypass. Ann Surg 2011;253:484–7.

[8] Adams TD, Gress RE, Smith SC, et al. Long-term mortality following gastric bypass surgery. N Engl J Med 2007;356:753–6.

[9] Christou NV, Sampalis JS, Liberman M, et al. Surgery decrease long-term mortality, morbidity, and health care use in morbidly obese patients. Ann Surg 2004;240:416–23.

[10] Flum DR, Dellinger EP. Impact of gastric bypass operation on survival: a population-based analysis. J Am Coll Surg 2004;199:543–51.

[11] MacDonald Jr KG, Long SD, Swanson MS, et al. The gastric bypass operation reduces the progression and mortality of non-insulin-dependent diabetes mellitus. J Gastrointest Surg 1997;1:213–20.

[12] Sowemimo AO, Yood SM, Courtney J, et al. Natural history of morbid obesity without surgical intervention. Surg Obes Relat Dis 2007;3(1):73–7.

[13] Peeters A, O'Brien PE, Laurie C, et al. Substantial intentional weight loss and mortality in the severely obese. Ann Surg 2007;246(6):1028–33.

[14] Busetto L, Mirabelli D, Petroni ML, et al. Comparative long-term mortality after laparoscopic adjustable gastric banding versus nonsurgical controls. Surg Obes Relat Dis 2007;3(5):496–502.

[15] Guidry CA, Davies SW, Sawer RG, Schirmer BD, Hallowell PT. Gastric bypass improves survival compared to propensitymatched controls: a Cohort Study with over 10 year follow up. Am J Surg 2015;209(3):463–7.

[16] Morino M, Toppino M, Forestieri P, et al. Mortality after bariatric surgery: analysis of 13,871 morbidly obese patients from a national registry. Ann Surg 2007;246(6):1002–7.

[17] Bruschi Kelles SM, Machado CJ, Barreto SM. Mortality rate after open Roux-in-Y gastric bypass: a 10-year follow-up. Braz J Med Biol Res 2014;47(7):617–25.

[18] Adams TD, Mehta TS, Davidson LE. All-cause and cause-specific mortality associated with bariatric surgery: a review. Curr Atheroscler Rep 2015;17(12):74.

[19] Davidson LE, Adams TD, Kim J, et al. Association of patient age at gastric bypass surgery with long-term all-cause and cause-specific mortality JAMA Surg 2016;Published online February 10, 2016. Available from: http://dx.doi:10.1001/jamasurg.2015.5501.

[20] Omalu BI, Ives DG, Buhari AM, et al. Death rates and causes of death after bariatric surgery for Pennsylvania residents, 1995 to 2004. Arch Surg 2007;142(10):923−8.

[21] Tindle HA, Omalu B, Courcoulas A. Risk of suicide after long-term follow-up from bariatric surgery. Am J Med 2010;123:1036−42.

[22] Sansone RA, Wiederman MW, Schumacher DF, Routsong-Weichers L. The prevalence of self-harm behaviors among a sample of gastric surgery candidates. J Psychosom Res 2008;65(5):441−4.

[23] Henegham HM, Heinberg L, Windover A, et al. Weighing the evidence for an association between obesity and suicide risk. Surg Obes Relat Dis 2012;8(1):98−107.

[24] (a) Goldfeder L, Ren C, Gill J. Fatal complications of bariatric surgery. Obes Surg 2006;16(8):1050−6. (b) Mitchell J, Lancaster K, Burgard M, et al. Long-term follow-up of patients' status after gastric bypass. Obes Surg 2001;11(4):464−8.

[25] Morgan DJ, Ho KM, Armstrong J, Litton E. Long-term clinical outcomes and health care utilization after bariatric surgery: a population-based study. Ann Surg 2015;262(1):86−92.

[26] Bleich SN, Chang HY, Lau B, et al. Impact of bariatric surgery on health care utilization and costs among patients with diabetes. Med Care 2012;50:58−65.

[27] Bruschi Kelles SM, Machado CJ, Barreto SM. Before-and-after study: does bariatric surgery reduce healthcare utilization and related costs among operated patients? Int J Technol Assess Health Care 2015;31(6):407−13.

[28] Zingmond DS, McGory MI, Ko CY. Hospitalization before and after gastric bypass surgery. JAMA 2005;294:1918−24.

[29] Bolen SD, Chang HY, Weiner JP, et al. Clinical outcomes after bariatric surgery: a five-year matched cohort analysis in seven US states. Obes Surg 2012;22:749−63.

[30] Karlsson J, Taft C, Rydén A, et al. Tem-year trends in health-related quality of life after surgical and conventional treatment for severe obesity: the SOS intervention study. Int J Obes (Lond) 2007;31(8):1248−61.

[31] Raoof M, Näslund I, Rask E, et al. Health-related quality-of-life (HRQoL) on an average of 12 years after gastric bypass surgery. Obes Surg 2015;25(7):1119−27.

[32] Kiewiet RM, Durian MF, Cuijpers LP, et al. Quality of life after gastric banding im morbidly obeses Dutch patients: long-term follow-up. Obes Res Clin Pract 2008;2(3):I−I.

[33] Nadaline L, Zenti MG, Masotto L, et al. Improved quality of life after bariatric surgery in morbidly obese patients. Interdisciplinary group of bariatric surgery of Verona (GICOV). G Chir 2014;35(7):161−4.

[34] Driscoll S, Gregory DM, Fardy JM, Twells LK. Long-term health related quality of life in bariatric surgery patients: a systematic review and meta-analysis. Obesity (Silver Spring) 2016;24(1):60−70.

[35] Andersen JR, Aasprang A, Karlsen TI, et al. Health-related quality of life after bariatric surgery: a systematic review of prospective long-term studies. Surg Obes Relat Dis 2015;11(2):466−73.

[36] Sjöström L, Lindroos A-K, Peltonen M, et al. Lifestyle, diabetes, and cardiovascular risk factors 10 years after bariatric surgery. N Engl J Med 2004;351(26):2683−93.

[37] Cooper TC, Simmons EB, Webb K, et al. Trends in weight regain following Roux-en-Y gastric bypass (RYGB) bariatric surgery. Obes Surg 2015;25(8):1474−81.

[38] Yanos BR, Saules KK, Schuh LM, Sogg S. Predictors of lowest weight and long-term weight regain among Roux-em-Y gastric bypass patients. Obes Surg 2015;25(8):1364−70.

Chapter 21

Critical Care After Bariatric Surgery

R. Rajendram[1,2,3], C.R. Martin[4] and V.R. Preedy[2]

[1]Stoke Mandeville Hospital, Aylesbury, United Kingdom, [2]King's College London, London, United Kingdom, [3]King Abdulaziz Medical City, Ministry of National Guard Hospital Affairs, Riyadh, Saudi Arabia, [4]Buckinghamshire New University, Middlesex, United Kingdom

LIST OF ABBREVIATIONS

APTT	activated partial thromboplastin time
BMI	body mass index
BP	blood pressure
DM	diabetes mellitus
ICU	intensive care unit
LMWH	low-molecular-weight heparin
OHS	obesity hypoventilation syndrome
OSA	obstructive sleep apnea
TBW	total body weight
VTE	venous thromboembolism

INTRODUCTION

There is an ongoing epidemic of obesity. Patients having bariatric surgery are a major subgroup of the morbidly obese. Despite the significant comorbidities in this cohort, the 30-day mortality rate after elective bariatric surgery is surprisingly low (0.3%) [1]. As significant reduction in morbidity and mortality is associated with weight loss in this cohort, this perceived safety has led to a massive increase in bariatric surgery.

Estimation of an individual patient's risk of postoperative complications is made by the patient's anesthetist and surgeon during their preoperative assessments. This assessment determines the need for a planned postoperative admission to the intensive care unit (ICU) after elective bariatric surgery.

Most obese patients without medical comorbidities who have an uncomplicated course after bariatric surgery are managed on the standard inpatient postoperative surgical unit. Common complications that occur after bariatric surgery are related to the surgical wound, deep vein thrombosis, thromboembolism, revision surgery, or prolonged length of hospital stay. Although many of these complications can be managed on the standard inpatient surgical ward, some (e.g., emergency revision surgery) require unplanned admission to the ICU.

As there is a worldwide epidemic of obesity, the challenges of bariatric patients are not isolated to specialist centers where bariatric surgery is performed. The increasing prevalence of obesity means that intensivists are increasingly confronted with obese patients in their daily practices. The principles of postoperative management of the complicated patient after bariatric surgery described in this chapter can be applied to any obese patient developing complications after any elective surgery, and vice versa. Although this chapter focuses on the critical care of the complicated postoperative bariatric surgical patient, it includes data on the management of obese patients after nonbariatric surgery. This is where it adds important insights.

CRITERIA FOR INTENSIVE CARE UNIT ADMISSION

Despite affecting organ function (see below), obesity alone is not an independent adverse prognostic factor in surgical patients [2]. So, after elective bariatric surgery, patients do not require intensive care if they:

1. are without significant comorbidity,
2. have an uncomplicated intraoperative course,
3. have a routine recovery room admission.

If these criteria are met, patients can be transferred directly from the recovery room (postanesthesia care unit) to the inpatient surgical ward after elective bariatric surgery. However, obesity is associated with several comorbidities, and so obese patients have an increased risk of mortality [3]. If any of these criteria are not met, then close observation is required. It is sensible to consider some of these patients for planned admission to ICU after elective bariatric surgery to improve their outcome [4−7].

The selection of patients to admit to the ICU after bariatric surgery is usually protocol-based. The following criteria are important [8,9]:

1. Body mass index (BMI) $>60 \text{ kg/m}^2$.
2. Severe obstructive sleep apnea (OSA).
3. Significant cardiac disease (requiring continuous cardiac monitoring).
4. Refractory diabetes mellitus (DM).

In the event of intraoperative surgical or anesthetic complications (e.g., aspiration, bleeding, or cardiac or respiratory event) or postoperative complications (e.g., pneumonia or anastamotic leak), unplanned admission of "low risk" patients to the ICU may be required.

Intensive Care Resources Required by Obese Patients After Bariatric Surgery

The management of an obese patient with several or refractory comorbidities is challenging. Challenges specific to the postoperative management on the ICU include providing intensive medical and nursing care, hemodynamic monitoring, respiratory management, vascular access, appropriate medication dosing, and nutrition [10].

The treatment of obese patients is more physically, technically, and mentally demanding than the treatment of normal weight patients. So, obese patients often engender negative emotions in healthcare providers. The misconception that "obese patients cause problems" is widely held.

To overcome these obstacles and provide the best care to obese patients requires significant institutional investment in appropriate facilities for these patients, and training for healthcare professionals. The minimum equipment required includes specialized bariatric beds, chairs, hoists, and patient lifts [10]. This equipment must be available in the ICU as well as on the inpatient surgical ward. However, the storage and use of this equipment requires considerable space around the bedside, which is already limited by the other specialist equipment required to provide intensive care.

The amount of ICU resources consumed varies significantly between the planned and unplanned cohorts of patients. In one series of patients the median length of stay after a planned postoperative admission to the ICU after bariatric surgery was only 1 day [9]. However, after an unplanned/emergency admission the median length of stay was 8 days [9]. In this series, most unplanned ICU admissions were for patients undergoing emergency surgery for complications from a previous bariatric procedure. The patient's comorbidities also determined the amount of resources required by individual obese patients after bariatric surgery.

Associated Comorbidities

Despite the significant comorbidities in this cohort, the 30-day mortality rate after elective bariatric surgery was surprisingly low (0.3%) [1]. This low risk/benefit ratio has led to the increase of bariatric surgery.

All organs are affected by obesity. However, for management in the ICU, the cardiovascular, respiratory, and endocrine effects of obesity are particularly important. Diseases often associated with obesity include coronary artery disease, cardiac failure, venous thromboembolism (VTE) hypertension, type 2 diabetes mellitus (T2DM), and OSA [11]. Obese patients with metabolic syndrome have a seven fold increased risk of acute renal failure and postoperative stroke after coronary bypass surgery [12,13]. Approximately 70–90% of all patients with T2DM are obese. The relative risk of coronary artery disease is 3.2-fold higher in the obese than in patients with a normal weight [14].

However, many obese patients do not have any comorbidities. Obesity increases cardiac output, circulating blood volume, and sympathetic activity. Up to a third have pathological electrocardiography changes suggestive of ventricular hypertrophy, and coronary artery disease. Despite often having normal blood pressure (BP), subclinical renal and endothelial dysfunction and left ventricular hypertrophy are common in obese patients. Obese patients with cardiovascular disease are at higher risk of complications during and after surgery [15].

Absolute values of total blood volume are high, but when corrected to actual body weight may only be about 45 mL/kg (70 mL/kg in normal patients). As a result, obese patients are more hemodynamically unstable when changing position from lying to standing and during surgical stress. The use of sedatives further increases the risk of hemodynamic instability [16].

Obese patients have impaired respiratory function [17]. When adults gain weight, lung volume remains unchanged despite increasing adiposity. Compliance (lung and chest wall) is reduced by the weight of excess fat [18,19]. Abdominal and peritoneal fat cause a cranial shift of the diaphragm, impairing lung expansion. So as BMI increases, obese patients tend to have rapid, shallow breathing and increased work of breathing [20].

When lying flat the diaphragm is approximately 4 cm more cranial than when upright. So lung compliance is worse in the reclined position in bed. Careful positioning of the obese patient in the reverse Trendelenburg position may improve respiratory function, but reduction in venous return may impair cardiac output.

With sedation in intensive care, functional residual capacity (FRC) may fall well below closing capacity of the lungs, inducing a cyclic opening and closing of small airways during mechanical ventilation, and increasing the risk of ventilator-induced lung injury.

Importantly, the reduction in lung compliance and FRC increase the work of breathing, oxygen consumption, and carbon dioxide production. This considerably reduces tolerance of apnea and hypoxia [21]. A high incidence of atelectasis even 24 hours after general anesthesia predisposes obese patients to postoperative hypoxia [22].

Obesity is the main risk factor for OSA [23], which is an independent risk factor for hypertension, cardiovascular morbidity, and sudden death. Under deep sedation or general anesthesia the reduction of upper airway tone in obese patients can cause difficult mask ventilation.

INTENSIVE CARE UNIT POSTOPERATIVE MANAGEMENT

Cardiac and Hemodynamic Monitoring

Even the monitoring of BP, a basic vital sign, requires special equipment, i.e., large BP cuffs. If they are too small, the pressure generated on inflating the cuff may not be fully transmitted to the brachial artery. In this setting, the pressure in the BP cuff may be considerably higher than the intraarterial pressure, which can lead to overestimation of the systolic pressure by 10–50 mmHg in obese patients.

Complex hemodynamic monitors used in the ICU (e.g., arterial lines and central venous catheters) after bariatric surgery are typically placed in the operating theater either before or shortly after induction of anesthesia.

Placement of peripheral venous cannulae, central venous catheters, and arterial lines in the obese patient can be challenging. The usual anatomic landmarks are obscured, the distance from skin to vessel is greater than normal, and the angle of approach may be too steep to allow cannulation even after reaching the vessel. In these patients, ultrasound can be used to identify vascular structures and assist in the placement of venous and arterial catheters [24,25]. Although most theater suites in the developed world are equipped with ultrasound machines, the quality of ultrasound images decreases with depth, and so this is not a panacea.

Respiratory Management

Severely obese patients are at high risk of postoperative respiratory complications directly related to common comorbidities, such as gastroesophageal reflux, obesity hypoventilation syndrome (OHS), and OSA [26,27]. OSA and OHS can complicate management of severely obese patients because of an abnormally small upper airway and chronic hypercarbic respiratory failure, respectively. OSA is extremely common in obese patients. In one series, over 70% of patients who underwent polysomnography before bariatric surgery were found to have OSA [28]. A preoperative assessment of pulmonary function should be considered for all obese patients prior to any elective operation. This should guide perioperative management.

Extubation

Obese patients should be extubated only when they are almost completely awake, able to follow simple commands (e.g., "raise your hand"), and all neuromuscular blockade has been reversed. This applies to the obese patient having any elective surgery.

If the postoperative obese patient demonstrates either respiratory or hemodynamic instability in the immediate postoperative period in the operating theater, the patient should be transferred to the ICU and remain intubated.

If a complication occurs shortly after extubation, reintubation can be challenging. This may be because of acute swelling of the oropharyngeal tissues, the large amount of redundant pharyngeal tissue, limited mouth opening, and a thick neck with reduced flexibility [29,30]. So, obese patients may be extubated the next day when they are more likely to meet the criteria for extubation with less risk of requiring emergency reintubation.

Prolonged Invasive Ventilation

Most postoperative patients will be extubated within 24 hours of the operation. However, if there are respiratory or cardiac complications, prolonged tracheal intubation and ventilatory support may be required. Esophageal manometry and tracheostomy should be considered for obese patients who require prolonged intubation and invasive ventilation.

Role of Tracheostomy

Tracheostomy should be considered as early as possible if it appears that the patient will require prolonged mechanical ventilation. Tracheostomy can often be technically challenging, and a tracheostomy tube with a long shaft and adjustable flange may be required [31]. Percutaneous tracheostomy can be performed in morbidly obese patients, but only highly skilled clinicians with significant experience should attempt this [32].

Mechanical Ventilation

Static lung compliance is reduced in obese patients [33]. This is generally due to reduced compliance of the chest wall rather than lung parenchymal disease. If mechanical ventilation is required, initial tidal volume targets should be approximately 6−8 mL/kg of ideal body weight. Subsequent settings are determined by arterial blood gases and airway pressures [34,35].

Plateau airway pressures should usually be kept below 35 cmH$_2$O. If this does not allow adequate oxygenation or ventilation, esophageal manometry can be used to quantify the relative contribution of the chest wall to decreased compliance. Transpulmonary pressure should be kept below 35 cmH$_2$O. The addition of 10 cmH$_2$O of positive-end expiratory pressure may improve lung compliance by reopening atelectasis and increasing FRC [36].

Aggressive perioperative respiratory physiotherapy and positioning the patient head up as much as possible may reduce the incidence of aspiration- and ventilator-associated pneumonia, as well as the time required to wean off mechanical ventilation [37].

Noninvasive Positive Pressure Ventilation

Obese patients often use noninvasive positive pressure ventilation at home for treatment of sleep apnea. Those patients who do use noninvasive ventilation (NIV) at home should bring their own machine to the hospital with them. Putting patients onto their own NIV machine after extubation can reduce some postprocedural anxiety [38].

Positioning

Positioning patients in the reverse Trendelenburg position can optimize respiratory function. However, lying in one position for prolonged periods increases the risk of pressure sore formation. Pressure ulcers result from prolonged pressure on soft tissue or compression of the skin between a bony prominence or hard surface (e.g., bed sides). Pressure-induced injury ranges from nonblanching, erythematous (but intact) skin to deep ulcers down to the bone. The risk of pressure sores in critically ill patients is reduced by repositioning patients regularly [39]. When positioning patients, it is important not to occlude blood flow, which could increase the risk of VTE.

Prevention of Venous Thromboembolism and Medication Dosing

Prevention of VTE in patients with reduced mobility includes extremity compression devices (stockings and pumps), careful positioning, passive and active movement, and anticoagulant medications.

The use of anticoagulant medications highlights important aspects of pharmacokinetics that are relevant when dosing medications in obese patients. Obese patients have less lean body tissue and tissue water than those who are not obese. This results in a larger volume of distribution for hydrophobic (i.e., lipophilic) drugs and increases clearance of hydrophilic drugs [40]. In obese patients, decreased drug metabolism via the cytochrome P450 pathway, or increased clearance by glucuronidation or glomerular filtration, predisposes obese patients to both toxic and subtherapeutic effects of medication [40].

Dosing of medications requires consideration of these changes, and is therefore challenging. Appropriate weights used for dose calculation of medications used in the ICU include:

1. Ideal body weight (calculated from height).
2. Total body weight (TBW).

The volume of distribution of heparin in obese patients differs from nonobese patients. As adipose tissue has a lower blood volume than lean tissue, heparin dose does not increase linearly with body weight [41]. Most studies exclude obese patients, so the best way to prescribe unfractionated heparin and low-molecular-weight heparins (LMWHs) in obesity is unknown [41].

Weight-based dosing of parenteral anticoagulants is probably better than fixed dosing in obese patients [42]. The available data support the use of TBW to calculate the initial bolus dose and infusion rate of unfractionated heparin [43]. The initial rate should be adjusted according to the activated partial thromboplastin time (APTT) after 6 hours. The antifactor Xa activity assesses the in vivo activity of heparin and, in obese patients, may be more useful for heparin monitoring than the APTT.

LMWHs should also be adjusted according to TBW [44]. Variable absorption of LMWH by subcutaneous injection in severe obesity is a concern, and monitoring of blood levels of anti-Xa is therefore recommended [44,45].

Analgesia and Sedation

While mechanically ventilated patients should be adequately sedated. This is best achieved by starting an infusion of a short-acting sedative such as propofol. The level of sedation should be decreased to the minimum required to ensure that the patient is comfortable and alert or easily rousable. The amount of sedatives necessary to achieve this should be assessed at least daily [46]. An opiate should be provided for analgesia; remifentanil infusions have the fewest hemodynamic side effects and shortest context-sensitive half-time. After extubation, intravenous boluses of fentanyl administered via patient controlled analgesia (PCA) may be most appropriate until oral intake is established.

Severely obese patients with hypercapnea due to OHS are at risk of type 2 respiratory failure after extubation. Opiate analgesia may increase this risk. The use of epidural analgesia for the management of pain after laparotomy must therefore be considered. However, this can be technically challenging to perform. If epidural anesthesia is not possible, regional anesthesia and/or local anesthetic infiltration should be considered.

Simple non-opiate analgesia with paracetamol should be provided unless there is a contraindication. Nonsteroidal antiinflammatory drugs should be considered, but can cause nephrotoxicity, gastrointestinal tract bleeding, and cardiovascular complications.

Ketamine causes dissociative analgesia without respiratory depression. While it can be extremely useful, it is associated with vivid and disturbing dreams. Propofol and midazolam are lipophilic. When administered intravenously, these agents initially act rapidly on the central nervous system, but then redistribute to adipose tissue [40]. When titrating intravenous administration of analgesia or sedation, it is best to administer a series of small doses frequently (i.e., every 5−15 minutes) until the desired effect is achieved.

Nutrition

The estimated metabolic energy requirements of ICU patients are generally based on actual body weight using the Harris−Benedict equation [47]. However, this overestimates metabolic demand in obese patients, and leads to overfeeding [47]. Indirect calorimetry can be used to make individualized assessments of energy expenditure [47]. It should also be remembered that patients who are obese may also be malnourished as their diets, while high in calories, may be deficient in vitamins, minerals, and trace elements [47]. Measurement of levels of these micronutrients should be considered unless empirical replacement therapy is administered.

Glycemic Control

Glycemic targets for inpatients are currently being debated because tight glycemic control is associated with increased mortality [48]. Although hyperglycemia is certainly undesirable, hypoglycemia is associated with worse outcomes than hyperglycemia. The recommended blood glucose target range is 6−10 mmol/L, recognizing that hypoglycemia is also harmful [48]. Insulin and dextrose infusions may be required to achieve this while patients are not able to eat and drink [48].

Vascular Access

Peripherally inserted central catheters (PICC) should be placed for long-term venous access if ICU admission is likely to be prolonged. PICC lines, which can generally be placed under ultrasound guidance, can provide hemodynamic information [49]. Hemodynamic measurements made with an open-ended PICC line and a standard transducer system are similar to readings from standard centrally inserted catheters [49].

CONCLUSIONS

The risks associated with bariatric surgery are low, and so most obese patients who undergo bariatric surgery do not require admission to the ICU. However it is sensible to plan to admit high-risk patients to the ICU for observation after bariatric surgery. Even with intensive monitoring in a specialized environment, complications can occur. Early complications include adverse cardiac and respiratory events, anastomotic leaks, and venous thromboembolic events. Some of these complications will require unplanned admission to the ICU for observation or interventions.

MINI-DICTIONARY OF TERMS

- *Intensive care unit (ICU)*: A specialist department within a hospital where one-to-one nursing care, continuous monitoring of cardiac, respiratory, and renal function, and advanced therapies for organ support can be provided. The terms ICU, intensive therapy unit (ITU), and critical care unit (CCU) are interchangeable.
- *Mechanical or invasive ventilation*: Positive pressure delivered to the patient's lungs via an endotracheal tube or a tracheostomy tube.
- *Noninvasive ventilation (NIV)*: Ventilatory support that is provided via a face mask to the patient's upper airway.
- *Tracheostomy*: A hole made in the front of the neck so a tube can be inserted into the trachea to reduce the patient's work of breathing and reduce the sedation required by the patient.

KEY FACTS

- Elective bariatric surgery generally has a low risk of complications.
- Admission to the ICU after elective bariatric surgery is rare.
- Planned admission to the ICU after elective surgery should be protocol-based.
- Unplanned admission to the ICU after bariatric surgery is associated with a high risk of morbidity, prolonged ICU admission, and mortality.
- Management of bariatric patients requires institutional investment in specialized equipment and training of staff.

SUMMARY POINTS

- Criteria for admission to an ICU after elective bariatric surgery include "super super" morbid obesity (BMI > 60 mg/kg^2), and multiple comorbidities.
- Severely obese patients have a higher risk of developing serious complications requiring postoperative admission to the ICU for observation.
- Obese patients should be extubated as soon as possible after elective bariatric surgery.

- Obese patients should only be extubated after complete reversal of neuromuscular blockade when they are awake and can follow simple commands.
- Patients with postoperative complications that require ICU admission should have long-term venous access, a tracheostomy, and provision of nutritional support considered early.
- Medication dosing is complex in obese patients.

REFERENCES

[1] Flum DR, Belle SH, King WC, et al. Perioperative safety in the longitudinal assessment of bariatric surgery. N Engl J Med 2009;361:445−54.

[2] Mullen JT, Moorman DW, Davenport DL. The obesity paradox: body mass index and outcomes in patients undergoing nonbariatric general surgery. Ann Surg 2009;250:166−72.

[3] Guh DP, Zhang W, Bansback N, et al. The incidence of co-morbidities related to obesity and overweight: a systematic review and meta-analysis. BMC Public Health 2009;9:88.

[4] El-Solh AA. Clinical approach to the critically ill, morbidly obese patient. Am J Respir Crit Care Med 2004;169:557.

[5] Gonzalez R, Bowers SP, Venkatesh KR, et al. Preoperative factors predictive of complicated postoperative management after Roux-en-Y gastric bypass for morbid obesity. Surg Endosc 2003;17:1900.

[6] El-Solh A, Sikka P, Bozkanat E, et al. Morbid obesity in the medical ICU. Chest 2001;120:1989.

[7] Livingston EH. Procedure incidence and in-hospital complication rates of bariatric surgery in the United States. Am J Surg 2004;188:105.

[8] El Shobary H, Backman S, Christou N, Schricker T. Use of critical care resources after laparoscopic gastric bypass: effect on respiratory complications. Surg Obes Relat Dis 2008;4:698.

[9] van den Broek RJ, Buise MP, van Dielen FM, et al. Characteristics and outcome of patients admitted to the ICU following bariatric surgery. Obes Surg 2009;19:560.

[10] Abir F, Bell R. Assessment and management of the obese patient. Crit Care Med 2004;32:S87.

[11] Virdis A, Neves MF, Duranti E, et al. Microvascular endothelial dysfunction in obesity and hypertension. Curr Pharm Des 2012; undefined.

[12] Hong S, Youn YN, Yoo KJ. Metabolic syndrome as a risk factor for postoperative kidney injury after off-pump coronary artery bypass surgery. Circ J 2010;74:1121−6.

[13] Kajimoto K, Miyauchi K, Kasai T, et al. Metabolic syndrome is an independent risk factor for stroke and acute renal failure after coronary artery bypass grafting. J Thorac Cardiovasc Surg 2009;137:658−63.

[14] Nejat EJ, Polotsky AJ, Pal L. Predictors of chronic disease at midlife and beyond − the health risks of obesity. Maturitas 2010;65:106−11.

[15] Manson JE, Colditz GA, Stampfer MJ, et al. A prospective study of obesity and risk of coronary artery disease in women. N Engl J Med 1990;322:882−9.

[16] van Kralingen S, Diepstraten J, van de Garde EM, et al. Comparative evaluation of propofol 350 and 200 mg for induction of anaesthesia in morbidly obese patients: a randomized double-blind pilot study. Eur J Anaesthesiol 2010;27:572−4.

[17] Babb TG, Wyrick BL, DeLorey DS, et al. Fat distribution and end-expiratory lung volume in lean and obese men and women. Chest 2008;134:704−11.

[18] Pelosi P, Croci M, Ravagnan I, et al. Total respiratory system, lung, and chest wall mechanics in sedated-paralyzed postoperative morbidly obese patients. Chest 1996;109:144−51.

[19] Pelosi P, Croci M, Ravagnan I, et al. The effects of body mass on lung volumes, respiratory mechanics, and gas exchange during general anesthesia. Anesth Analg 1998;87:654−60.

[20] Adams JP, Murphy PG. Obesity in anaesthesia and intensive care. Br J Anaesth 2000;85:91−108.

[21] Pelosi P, Croci M, Ravagnan I, et al. Respiratory system mechanics in sedated, paralyzed, morbidly obese patients. J Appl Physiol 1997;82:811−18.

[22] Eichenberger A, Proietti S, Wicky S, et al. Morbid obesity and postoperative pulmonary atelectasis: an underestimated problem. Anesth Analg 2002;95:1788−92.

[23] Isono S. Obstructive sleep apnea of obese adults: pathophysiology and perioperative airway management. Anesthesiology 2009;110:908−21.

[24] Miller AH, Roth BA, Mills TJ, et al. Ultrasound guidance versus the landmark technique for the placement of central venous catheters in the emergency department. Acad Emerg Med 2002;9:800.

[25] Gann Jr M, Sardi A. Improved results using ultrasound guidance for central venous access. Am Surg 2003;69:1104.

[26] den Herder C, Schmeck J, Appelboom DJ, de Vries N. Risks of general anaesthesia in people with obstructive sleep apnoea. BMJ 2004;329:955.

[27] Koenig SM. Pulmonary complications of obesity. Am J Med Sci 2001;321:249.

[28] Frey WC, Pilcher J. Obstructive sleep-related breathing disorders in patients evaluated for bariatric surgery. Obes Surg 2003;13:676.

[29] Williamson JA, Webb RK, Szekely S, et al. The Australian Incident Monitoring Study. Difficult intubation: an analysis of 2000 incident reports. Anaesth Intens Care 1993;21:602.

[30] Brodsky JB, Lemmens HJ, Brock-Utne JG, et al. Morbid obesity and tracheal intubation. Anesth Analg 2002;94:732.

[31] Watters M, Waterhouse P. Percutaneous tracheostomy in morbidly obese patients. Anaesthesia 2002;57:614.

[32] Mansharamani NG, Koziel H, Garland R, et al. Safety of bedside percutaneous dilational tracheostomy in obese patients in the ICU. Chest 2000;117:1426.

[33] Ladosky W, Botelho MA, Albuquerque Jr. JP. Chest mechanics in morbidly obese non-hypoventilated patients. Respir Med 2001;95:281.

[34] Ventilation with lower tidal volumes as compared with traditional tidal volumes for acute lung injury and the acute respiratory distress syndrome. The Acute Respiratory Distress Syndrome Network. N Engl J Med 2000;342:1301.

[35] Brower RG, Ware LB, Berthiaume Y, Matthay MA. Treatment of ARDS. Chest 2001;120:1347.

[36] Pelosi P, Ravagnan I, Giurati G, et al. Positive end-expiratory pressure improves respiratory function in obese but not in normal subjects during anaesthesia and paralysis. Anesthesiology 1999;91:1221.

[37] Vaughan RW, Bauer S, Wise L. Effect of position (semirecumbent versus supine) on postoperative oxygenation in markedly obese subjects. Anesth Analg 1976;55:37.

[38] Hida W, Okabe S, Tatsumi K, et al. Nasal continuous positive airway pressure improves quality of life in obesity hypoventilation syndrome. Sleep Breath 2003;7:3.

[39] Mathison CJ. Skin and wound care challenges in the hospitalized morbidly obese patient. J Wound Ostomy Continence Nurs 2003;30:78.

[40] Brill MJ, Diepstraten J, van Rongen A, et al. Impact of obesity on drug metabolism and elimination in adults and children. Clin Pharmacokinet 2012;51:277.

[41] Nutescu EA, Spinler SA, Wittkowsky A, Dager WE. Low-molecular-weight heparins in renal impairment and obesity: available evidence and clinical practice recommendations across medical and surgical settings. Ann Pharmacother 2009;43:1064.

[42] Garcia DA, Baglin TP, Weitz JI, et al. Parenteral anticoagulants: Antithrombotic Therapy and Prevention of Thrombosis, 9th ed: American College of Chest Physicians Evidence-Based Clinical Practice Guidelines. Chest 2012;141:e24S.

[43] Myzienski AE, Lutz MF, Smythe MA. Unfractionated heparin dosing for venous thromboembolism in morbidly obese patients: case report and review of the literature. Pharmacotherapy 2010;30:324.

[44] Al-Yaseen E, Wells PS, Anderson J, et al. The safety of dosing dalteparin based on actual body weight for the treatment of acute venous thromboembolism in obese patients. J Thromb Haemost 2005;3:100.

[45] Hirsh J, Bauer KA, Donati MB, et al. Parenteral anticoagulants: American College of Chest Physicians Evidence-Based Clinical Practice Guidelines (8th Edition). Chest 2008;133:141S.

[46] Gehlbach BK, Kress JP. Sedation in the intensive care unit. Curr Opin Crit Care 2002;8:290.

[47] Elamin EM. Nutritional care of the obese intensive care unit patient. Curr Opin Crit Care 2005;11:300.

[48] Rometo D, Korytkowski M. Perioperative glycemic management of patients undergoing bariatric surgery. Curr Diab Rep 2016;16:23.

[49] Black IH, Blosser SA, Murray WB. Central venous pressure measurements: peripherally inserted catheters versus centrally inserted catheters. Crit Care Med 2000;28:3833.

REFERENCE TO A JOURNAL PUBLICATION

[1] van der Geer J, Hanraads JAJ, Lupton RA. The art of writing a scientific article. J Sci Commun 2000;163:5159.

REFERENCE TO A BOOK

[2] Strunk Jr. W, White EB. The elements of style. third ed. New York, NY: Macmillan; 1979.

REFERENCE TO A CHAPTER IN AN EDITED BOOK

[3] Mettam GR, Adams LB. How to prepare an electronic version of your article. In: Jones BS, Smith RZ, editors. Introduction to the electronic age. New York, NY: E-Publishing Inc.; 1999. p. 281–322.

Section III

Safety and Outcomes

Chapter 22

An Overview of the Safety of Bariatric Surgery

R.M. Tholey and A. Pomp

New York Presbyterian—Weill Cornell Medical College, New York, NY, United States

LIST OF ABBREVIATIONS

ACS	American College of Surgeons
ASMBS	American Society for Metabolic & Bariatric Surgeons
BMI	body mass index
CKD	chronic kidney disease
COE	center of excellence
LAGB	laparoscopic adjustable gastric band
LOS	length of stay
LRYGB	laparoscopic Roux-en-Y gastric bypass
LSG	laparoscopic sleeve gastrectomy
LVAD	left ventricular assist device
MBSAQIP	Metabolic and Bariatric Surgery Accreditation and Quality Improvement Program
NASH	nonalcoholic steatohepatitis
NSQIP	National Surgical Quality Improvement Program
SAGES	Society of American Gastrointestinal and Endoscopic Surgeons

INTRODUCTION

Bariatric surgery procedures, including laparoscopic sleeve gastrectomy (LSG), laparoscopic Roux-en-Y gastric bypass (LRYGB), and laparoscopic adjustable gastric banding (LAGB), are now commonly performed "general surgery" procedures. This chapter gives an overview of the safety of the procedures themselves, as well as the infrastructure, support, and regulating bodies required to maintain that safety. It also outlines the safety of more recent technologies such as robotic surgery, as well as the risks of operating upon special populations, including the elderly, transplant candidates, and patients with ventral hernias.

AN OVERVIEW OF MORBIDITY AND MORTALITY RATES IN BARIATRIC PROCEDURES

The safety of bariatric procedures, including morbidity, mortality, hospital readmissions, and efficacy, has been investigated in a multitude of studies. A landmark paper published in the New England Journal of Medicine in 2009 evaluated 4776 patients who underwent a first-time bariatric procedure [1]. The 30-day mortality rates for Roux-en-Y gastric bypass (RYGB) or LAGB were low at 0.3%, with a total of 4.3% of patients with one major adverse outcome; in fact, the procedures were so safe, a "composite end point" was required for analysis. A history of deep venous thrombosis, obstructive sleep apnea, impaired functional status, as well as extreme values of body mass index (BMI), were significantly associated with an increased risk of morbidity. However, this trial still included a significant number of open Roux-en-Y gastric bypasses (ORYGB), and was published before the surge in popularity of LSG. In 2011, 109 hospitals submitted data for 28,616 patients and determined that LSG has morbidity and effectiveness positioned between LAGB and LRYGB/ORYGB [2]. The study found that LSG had higher risk-adjusted morbidity, readmission, and reoperation/intervention rates compared to LAGB, but lower reoperation/intervention rates compared to LRYGB/ORYGB. The study was limited in that it only extended to 1-year outcomes. Another study evaluated the morbidity between the three procedures, including a total of 2199 patients from a single institution [3]. The leak rate was low at 0.5% for LRYGB, 0.3% for LSG, and was not applicable for LAGB. The percentages of procedures requiring reoperations due to complications or failures, however, were 14.6% in the LAGB group, 6.6% in the LRYGB group, and 1.8% in the LSG group. Finally, a recent study published in 2015 with longer-term follow-up of up to 8 years evaluated 1020 consecutive LSGs [4]. There was no mortality in this series. Early, severe, postoperative complications within 30 days of surgery were rare, but significant numbers of patients had emesis (23%), dehydration (19%), and prolonged ileus (18%). Mean length of stay (LOS) was 3.4 days with a 3.8% readmission rate. Long-term morbidity was only 3.9% of patients; the most common complication was gastroesophageal reflux disease, which occurred in 6%.

The recent shift toward LSG and away from LRYGB is due to surgeon (and patient) perception that LRYGB is more invasive, and consequently has more complications. A recent multicenter trial directly comparing LSG and LRYGB found that complications occurred in 3.7% ($n = 110$) of LSG and 4.3% ($n = 24$) of LRYGB, showing no significant difference was found between the groups regarding overall and complication-grade-specific rates [5]. Median LOS and readmission rates were not significantly different.

LRYGB and LSG have both also been proven safe and efficacious in super-super-obese patients (BMI > 60). A single institution study retrospectively evaluated 135 patients over a 5-year period and found that the excess weight loss and postoperative complications were comparable to non-super-super-obese patients [6].

THE LAPAROSCOPIC GASTRIC BAND AND THE SAFETY OF ITS CONVERSION TO OTHER PROCEDURES

LAGB has certainly become less popular over the last few years as its modest efficacy for weight loss and metabolic improvement, combined with significant incidence of complications requiring surgical revision, have become more apparent. When compared at 3 years postop, the LAGB provided 15.9% total weight loss versus 31.5% for LRYGB [7]. The same study also looked at partial remission of diabetes, and found it to be 67.5% for those patients who underwent LRYGB, and only 28.6% for those with LAGB. One long-term prospective study with a median follow-up of 95 months found that while nearly half of the patients achieved more than 50% excess weight loss, band removal was necessary for complications or insufficient weight loss in 24% of patients [8]. Another study that had 100% follow-up at a mean of 14 years showed that the band removal rate was almost 50% [9]. A modification of the band technology (EASYBAND), which allows it to be adjusted by a telemetric motor and eliminates the need for port access, has been studied; one or more complications occurred in 79.1% of patients, and reintervention was required in 15% of patients at the 2-year mark in a multicenter trial [10]. Overall, the study determined that the band was found to be an improvement over the original bands, but had significant technical problems that needed to be solved. Another multicenter study looked retrospectively at 735 bariatric patients with postoperative follow-up of 3 years [11]. LAGB showed a higher complication rate when compared to LSG and LRYGB ($p < 0.05$). Revisional surgery after LAGB was needed in 21%, while only 9% of the LSGs underwent conversion to RYGBs.

LAGB can certainly be converted to another procedure, but controversy persists as to whether this is safe to do in one operation or whether it is better to first remove the band separately before proceeding to a second bariatric operation. Conversion from LAGB to LSG in one step was analyzed using National Surgical Quality Improvement Program (NSQIP) data [12]. Data of 11,320 patients were analyzed with a 30-day composite adverse event rate (comprised of 18 postoperative adverse events) and was 6.8% in the LAGB-to-LSG group and 5.4% in the primary LSG group ($p = 0.29$). The rate of minor complications, including urinary tract infection and wound infection, were significantly higher in the revisional surgery group. Thirty-day rates of other postoperative complications, reoperation, readmission, mortality, and length of hospital stay were comparable between the two groups. Another single-center Dutch study looked at one-step conversion of LAGB to LRYGB [13]. They included 178 patients, with 8 patients (4%) needing reoperation within 30 days. Mean excess weight loss at 2 years after the conversion was 65%. They determined this reoperation rate acceptable, concluding that a one-step procedure for conversion was both feasible and safe. Another study published in Germany using a national database evaluated 240 LAGB-to-LSG conversions and found that the leak rate was statistically significant when comparing one-step operations versus two-step operations at 3.3% versus 0% [14]. This is especially important given the significant consequences of a sleeve gastrectomy leak possibly necessitating a major operation, as discussed in another chapter of this textbook. Interestingly, conversion from LAGB to LRYGB did not result in different leak rates between one- and two-step approaches (1.9% vs 2.2%). Based on the variability between studies and personal experience, it is justified to either remove the band as a separate (generally outpatient) procedure and then perform LRYGB or LSG at a later date, or to combine the band removal with the revision surgery in a single setting.

ROBOTICS IN BARIATRICS

The role of robotics in general surgery is still being defined, but several studies demonstrate equivalence to laparoscopic bariatric surgery as concerns feasibility and safety [15]. One meta-analysis included 10 studies with 2557 patients comparing robotic versus laparoscopic RYGB, and found that the rates for anastomotic leak, bleeding, stricture, and reoperation did not differ significantly [16]. An economic analysis did, however, find that the robotic RYGB was about 30% more costly at $15,447 versus the LRYGB at $11,956.

Another robotic study compared "robotic-assisted" (standard laparoscopic staplers were used) sleeve gastrectomy in morbidly obese patients versus superobese patients (BMI > 50) [17], postulating that assistance of the robot may help to overcome the operative difficulties encountered in superobese patients. When comparing the two groups there was no difference in complications, conversions, mortality, operative time, blood loss, or length of hospital stay.

CENTERS OF EXCELLENCE/QUALITY ACCREDITATION

The American Society for Metabolic & Bariatric Surgeons (ASMBS) created guidelines for hospitals to become Bariatric centers of excellence (COE) in 2004. The designated ASMBS committee designated 10 requirements a hospital must have, among which was a case volume of at least 125 cases per year, dedicated intensive care unit care if necessary, clinical pathways, patient support groups, and long-term patient follow-up submitted to the bariatric outcomes longitudinal database [18]. The database gives a goal that patient follow-up will be at least 75% at 5 years, and that outcome data be reported. In 2006 the Centers for Medicare and Medicaid Services made the COE certification mandatory and then lifted the mandate in 2013 with the thought that COE certification restricted access to care. This is contradicted by a recent study conducted from 2006 to 2013 with a total of 134,227 patients identified where access to surgery was not found to be limited [19]. A recent study published by Morton et al. in 2014 sought to evaluate whether hospital accreditation in fact impacted patient safety [20]. The study included 117,478 patients who underwent surgery at unaccredited centers and accredited centers. Compared with accredited centers, unaccredited centers had a higher mean LOS and a significantly higher incidence of complications and mortality. Regardless of the COE certification no longer being mandatory, bariatric centers should adhere to guidelines for patient safety by participating in the Metabolic and Bariatric Surgery Accreditation and Quality Improvement Program (MBSAQIP) established in 2012 as a joint venture between the ASMBS and the American College of Surgeons [21]. Additionally, the Society of American Gastrointestinal and Endoscopic Surgeons has published online guidelines and checklists pertaining to bariatric surgery and minimally invasive surgery as a whole to improve patient safety (Fig. 22.1).

EMERGENCY ROOM STAFF AND RESIDENT EDUCATION

A significant safety issue after bariatric surgery is patients presenting to nonindex hospitals after their operation. One study used a database of 38,776 patients who underwent surgery from 2010 to 2013, and found that the 30-day unplanned ED utilization rate was 11.3% and the 30-day hospital readmission rate was 5.3% [22]. The study found that almost half (46.7%) presented to a nonindex hospital and that these patients were significantly less likely to be admitted than those presenting to an index hospital. Presentation to nonindex hospitals has important implications for the accuracy of current patient safety and quality outcome measures. With the increase in the numbers of bariatric surgeries, emergency service providers need to be educated about bariatric-specific complications such as internal hernias and marginal ulcers [23,24]. Patients presenting to local hospitals where surgeons do not perform bariatric surgery are potentially at risk for missed complications and worse outcomes [25].

SPECIAL POPULATIONS: TRANSPLANT CANDIDATES, THE ELDERLY, AND VENTRAL HERNIAS PATIENTS

Bariatric surgery is being offered more frequently to those patients with severe liver, kidney, and cardiac disease. One study evaluated the effect of bariatric surgery on nonalcoholic fatty liver disease at 2 years and found that after a mean excess weight loss of 60%, steatosis disappeared in 84% and fibrosis disappeared in 50% on repeat biopsy after

MIS Safety Checklist

1. Pre-Patient Entry

A. Circulating Nurse Duties

Parameter	Actions
Surgeon Preference Card	☐ Reviewed
OR Table Position	☐ Correct orientation and weight capacity
	☐ Bean bag mattress (if indicated)
	☐ Table accessories (eg spreader bars/leg supports/foot board as indicated)
	☐ Positioned for fluoroscopy if indicated
Power sources	☐ Connected and linked to all devices
CO_2 insufflator	☐ Check CO_2 volume, pressure and flow
	☐ Backup cylinder and accessories (wrench and key) in place
	☐ Filter for CO_2 unit or tubing
Video monitors	☐ Position per procedure
	☐ Test pattern present
Suction/irrigation	☐ Canister set
	☐ Irrigation and pressure bag available
Alarms	☐ Turned on and audible
Video documentation	☐ Recording media available and operational (DVD, print, etc.)

B. Scrub Person Duties

Parameter	Actions
Reusable instruments	☐ Check movement handles and jaws, all screws present
	☐ Check sealing caps
	☐ Instrument vents closed
	☐ Check cautery insulation
Veress needle	☐ Check plunger/spring action
	☐ Flush needle and stopcock
	☐ Saline solution available
Hasson cannula	☐ Check valves, plunger, and seals
Trocars/Ports	☐ Check appropriate size/type
	☐ Close stopcocks
Laparoscope	☐ Size and type per preference
	☐ Check lens clarity
	☐ Anti-fog solution or warmed saline for lens cleaning

2. After Patient Entry

Parameter	Actions
Patient position	☐ Secured to OR table, safety strap on
	☐ Pressure sites padded
	☐ Arms out or tucked per procedure
Sequential compression device	☐ On and connected to device
Electrosurgical unit	☐ Ground pad applied
Foot controls	☐ Positioned for surgeon access
Power sources (camera, insufflator, light source, monitors, cautery, ultrasonics, bipolar)	☐ Turned on (on standby)
Miscellaneous	☐ Foley catheter (if indicated)
	☐ Naso- or orogastric tube (bougies if indicated)
Antibiotics	☐ Given as indicated

3. After Prep and Drape

Parameter	Actions
Electrosurgical unit	☐ Cautery cords connected to unit
Monopolar cautery	☐ Tip protected
Ultrasonic or bipolar device	☐ Connected to unit
	☐ Activation test performed
Line connections	☐ Camera cord
	☐ Light source (on standby)
	☐ CO_2 tubing (connected and flushed)
	☐ Suction/irrigation (suction turned on)
	☐ Smoke evacuation filter connected
Local anesthetic	☐ Syringe labeled and filled with anesthetic of choice needle connected
Fluoroscopy case	☐ Mix and dilute contrast appropriately and label
	☐ Clear tubing, syringe, catheter of air bubbles, label syringes

SAGES   This checklist has been developed by SAGES and AORN to aid operating room personnel in the preparation of equipment and other duties unique to laparoscopic surgery cases. It should not supplant the surgical time out or other hospital-specific patient safety protocols.

AORN

FIGURE 22.1 Minimally Invasive Surgery Safety Checklist. This checklist has been developed by SAGES and AORN to aid operating room personnel in the preparation of equipment and other duties unique to laparoscopic surgery cases. SAGES/AORN.

undergoing RYGB [26]. As many transplant centers require a BMI below 40 to be eligible for listing, several studies have shown quicker weight loss with bariatric intervention over diet and exercise [27]. Other studies, albeit small retrospective reviews at single institutions, compared Child A cirrhotics secondary to nonalcoholic steatohepatitis (NASH) with noncirrhotics and found no difference in complications or mortality [28], even in those with mild portal hypertension [29]. These studies may have increasing importance, as NASH leading to cirrhosis is increasingly becoming a cause for transplantation in the United States.

Kidney transplant recipients also have BMI cutoffs. In patients with reduced kidney function and chronic kidney disease (CKD) undergoing bariatric surgery, one study found that there was no statistically significant increase in 30-day postoperative complications [30]. The study included 64,589 patients, and after adjusting for patient characteristics, there was a trend toward increasing complications from stage 1 to stage 4/5 CKD, but none were statistically significant. Another study looked at patients already on dialysis and also found no increase in mortality [31]. The authors found a higher 30-day morbidity rate in dialysis patients at 5.98% compared to 2.31% in nondialysis patients, although after adjusting for confounding variables, dependence on dialysis was not found to be an independent predictor of major morbidity.

Morbid obesity also prevents patients with heart failure from becoming cardiac transplant candidates. One study evaluated the safety of LSG in patients with heart failure and left ventricular assist devices (LVADs) [32]. The study was only a small case series with six patients. At 12 months all patients had sufficient weight loss to become transplant eligible. There were no perioperative deaths, but one patient did have thrombosis of the LVAD pump at 3 weeks, which required an uneventful device exchange.

Another special population is morbidly obese patients with ventral hernias. The NSQIP database was queried from 2010 to 2011, and identified 17,117 patients who underwent LRYGB or LSG and found that a synchronous ventral hernia repair was performed in 503 (2.94%) patients [33]. The hernia repair was independently associated with surgical site infection with an odds ratio of 1.65. The study did not find a difference in infection rate between procedures (LSG vs LRYGB) when combined with hernia repair. Conversely, a smaller study collected the peritoneal washings of a total of 77 patients undergoing either LSG or LRYGB and sent the washings for culture [34]. They found no growth at 72 hours in the 51 LSG patients. In the LRYGB patients they found that 4 out of 26 (15%) were culture-positive. This led to their conclusion that the concomitant use of prosthetic mesh is acceptable for use in LSG, but not in LRYGB. Again in another small study, 23 patients underwent concomitant laparoscopic mesh placement with a bariatric procedure (22 LSG and 1 LRYGB) [35]. At 3.3 years median follow-up there were four seromas, but no mesh infections or hernia recurrences. Most surgeons still opt to repair hernias as a separate procedure, not necessarily related to the safety of using mesh, but rather with the thought that a second procedure after weight loss may prove to be more durable [36].

Finally, what is the safety of bariatric surgery in the elderly (age >60)? There are now multiple studies that have shown this group to have no increase in morbidity or mortality rate [37] with effective weight loss [38].

CONCLUSION

Bariatric surgery procedures are now some of the most commonly performed general surgery procedures. These operations are safe and extraordinarily effective in improving metabolic health problems. MBSAQIP is an example of the ability to set guidelines for quality care. It also provides maintenance of a database to record outcomes, and therefore continues to drive improvement in care. Primary care providers should not hesitate to refer suitable candidates for bariatric surgical consultation. Appropriately training all general surgeons as well as emergency room providers regarding the specific concerns of this population should be a medical education priority.

MINI-DICTIONARY OF TERMS

- *Gastric band*: A silicone device placed laparoscopically around the upper section of the stomach to restrict the amount of food that can be eaten.
- *Ileus*: A lack of peristalsis of the intestines resulting in abdominal distension and nausea.
- *Leak*: A complication after surgery where gastric contents "leak" into the intraperitoneal cavity.

KEY FACTS

- LRYGB and LSG are safe and effective procedures for morbid obesity.
- A band from LAGB is more frequently removed instead of placed due to its modest efficacy for weight loss and high complication profile.

- COE and MBSAQIP maintain standards of care for institutions performing bariatric surgery.
- Emergency room staff and residents should be required to learn about bariatric-specific complications in order to improve patient safety.
- Bariatric surgery is safe in the elderly, and may play an important role for transplant candidates.

SUMMARY POINTS

- Bariatric surgery procedures are safe and effective in improving metabolic health problems.
- Primary care providers should refer morbidly obese patients for Bariatric Surgery.
- MBSAQIP sets guidelines, and through its database drives quality care.
- Education of residents and emergency room personnel is paramount to provide adequate care for the bariatrics population.

REFERENCES

[1] Longitudinal Assessment of Bariatric Surgery, Consortium, et al. Perioperative safety in the longitudinal assessment of bariatric surgery. N Engl J Med 2009;361(5):445−54.

[2] Hutter MM, et al. First report from the American College of Surgeons Bariatric Surgery Center Network: laparoscopic sleeve gastrectomy has morbidity and effectiveness positioned between the band and the bypass. Ann Surg 2011;254(3):410−20. discussion 420−2.

[3] Fridman A, et al. Procedure-related morbidity in bariatric surgery: a retrospective short- and mid-term follow-up of a single institution of the American College of Surgeons Bariatric Surgery Centers of Excellence. J Am Coll Surg 2013;217(4):614−20.

[4] Alvarenga ES, et al. Safety and efficacy of 1020 consecutive laparoscopic sleeve gastrectomies performed as a primary treatment modality for morbid obesity. A single-center experience from the metabolic and bariatric surgical accreditation quality and improvement program. Surg Endosc 2015;30(7):2673−8.

[5] Goitein D, et al. Assessment of perioperative complications following primary bariatric surgery according to the Clavien-Dindo classification: comparison of sleeve gastrectomy and Roux-Y gastric bypass. Surg Endosc 2016;30(1):273−8.

[6] Serrano OK, et al. Weight loss outcomes and complications from bariatric surgery in the super super obese. Surg Endosc 2016;30(6):2505−11.

[7] Courcoulas AP, et al. Weight change and health outcomes at 3 years after bariatric surgery among individuals with severe obesity. JAMA 2013;310(22):2416−25.

[8] Van Nieuwenhove Y, et al. Long-term results of a prospective study on laparoscopic adjustable gastric banding for morbid obesity. Obes Surg 2011;21(5):582−7.

[9] Victorzon M, Tolonen P. Mean fourteen-year, 100% follow-up of laparoscopic adjustable gastric banding for morbid obesity. Surg Obes Relat Dis 2013;9(5):753−7.

[10] Handgraaf HJ, et al. The gastric band that is not to be: Efficacy, safety and performance of the Easyband: a multicenter experience. Obes Surg 2015;25(12):2239−44.

[11] Dogan K, et al. Effectiveness and safety of sleeve gastrectomy, gastric bypass, and adjustable gastric banding in morbidly obese patients: A Multicenter, Retrospective, Matched Cohort Study. Obes Surg 2015;25(7):1110−18.

[12] Aminian A, et al. Safety of one-step conversion of gastric band to sleeve: a comparative analysis of ACS-NSQIP data. Surg Obes Relat Dis 2015;11(2):386−91.

[13] Aarts E, et al. Revisional surgery after failed gastric banding: results of one-stage conversion to RYGB in 195 patients. Surg Obes Relat Dis 2014;10(6):1077−83.

[14] Stroh C, et al. Revisional surgery and reoperations in obesity and metabolic surgery: data analysis of the German bariatric surgery registry 2005-2012. Chirurg 2015;86(4):346−54.

[15] Tsuda S, et al. SAGES TAVAC safety and effectiveness analysis: da Vinci (R) Surgical System (Intuitive Surgical, Sunnyvale, CA). Surg Endosc 2015;29(10):2873−84.

[16] Bailey JG, et al. Robotic versus laparoscopic Roux-en-Y gastric bypass (RYGB) in obese adults ages 18 to 65 years: a systematic review and economic analysis. Surg Endosc 2014;28(2):414−26.

[17] Bhatia P, et al. Robot-assisted sleeve gastrectomy in morbidly obese versus super obese patients. JSLS 2014;18(3).

[18] Bariatric surgery center of excellence (BSCOE) Disignation Requirements. February 2016. Available from: <http://www.windbercare.org/BSCOE%20Requirements.pdf>.

[19] Bae J, et al. Effect of mandatory centers of excellence designation on demographic characteristics of patients who undergo bariatric surgery. JAMA Surg 2015;150(7):644−8.

[20] Morton JM, Garg T, Nguyen N. Does hospital accreditation impact bariatric surgery safety? Ann Surg 2014;260(3):504−8. discussion 508-9.

[21] MBSAQIP. Metabolic and bariatric surgery accreditation and quality improvement program. April 1, 2012; Available from: <https://asmbs.org/about/mbsaqip>.

[22] Telem DA, et al. Rates and risk factors for unplanned emergency department utilization and hospital readmission following bariatric surgery. Ann Surg 2016;263(5):956−60.

[23] Edwards ED, et al. Presentation and management of common post-weight loss surgery problems in the emergency department. Ann Emerg Med 2006;47(2):160–6.

[24] Mostaedi R, et al. Bariatric surgery and the changing current scope of general surgery practice: implications for general surgery residency training. JAMA Surg 2015;150(2):144–51.

[25] Bagloo MB, Pomp A. Challenging and emerging conditions in emergency medicine. The bariatric surgery patient. Hoboken, NJ: Wiley Blackwell; 2011.

[26] Furuya Jr. CK, et al. Effects of bariatric surgery on nonalcoholic fatty liver disease: preliminary findings after 2 years. J Gastroenterol Hepatol 2007;22(4):510–14.

[27] Bromberger B, et al. Weight loss interventions for morbidly obese patients with compensated cirrhosis: a Markov decision analysis model. J Gastrointest Surg 2014;18(2):321–7.

[28] Rebibo L, et al. Laparoscopic sleeve gastrectomy in patients with NASH-related cirrhosis: a case-matched study. Surg Obes Relat Dis 2014;10 (3):405–10. quiz 565.

[29] Pestana L, et al. Bariatric surgery in patients with cirrhosis with and without portal hypertension: a single-center experience. Mayo Clin Proc 2015;90(2):209–15.

[30] Saleh F, et al. Bariatric surgery in patients with reduced kidney function: an analysis of short-term outcomes. Surg Obes Relat Dis 2015;11 (4):828–35.

[31] Andalib A, et al. Safety analysis of primary bariatric surgery in patients on chronic dialysis. Surg Endosc 2016;30(6):2583–91.

[32] Chaudhry UI, et al. Laparoscopic sleeve gastrectomy in morbidly obese patients with end-stage heart failure and left ventricular assist device: medium-term results. Surg Obes Relat Dis 2015;11(1):88–93.

[33] Spaniolas K, et al. Synchronous ventral hernia repair in patients undergoing bariatric surgery. Obes Surg 2015;25(10):1864–8.

[34] Cozacov Y, et al. Is the use of prosthetic mesh recommended in severely obese patients undergoing concomitant abdominal wall hernia repair and sleeve gastrectomy? J Am Coll Surg 2014;218(3):358–62.

[35] Praveenraj P, et al. Concomitant bariatric surgery with laparoscopic intra-peritoneal onlay mesh repair for recurrent ventral hernias in morbidly obese patients: an Evolving Standard of Care. Obes Surg 2016;26(6):1191–4.

[36] Desai KA, et al. The effect of BMI on outcomes following complex abdominal wall reconstructions. Ann Plast Surg 2016;76(Suppl. 4): S295–7.

[37] Pequignot A, et al. Is sleeve gastrectomy still contraindicated for patients aged ≥ 60 years? A case-matched study with 24 months of follow-up. Surg Obes Relat Dis 2015;11(5):1008–13.

[38] Abbas M, et al. Outcomes of laparoscopic sleeve gastrectomy and Roux-en-Y gastric bypass in patients older than 60. Obes Surg 2015;25 (12):2251–6.

Chapter 23

Safety of Bariatric Surgery in Adolescents

M. Wabitsch[1] and B. Lennerz[2]

[1]Ulm University Medical Center, Ulm, Germany, [2]Boston Children's Hospital, Boston, MA, United States

LIST OF ABBREVIATIONS

AGA German Working Group on Obesity in Children and Adolescents (Arbeitsgemeinschaft Adipositas im Kindes und Jugendalter)
AGB adjustable gastric banding
ASMBS American Society for Metabolic and Bariatric Surgery
DAG expert committee commissioned by the German Association for the Study of Obesity (Deutsche Adipositas-Gesellschaft)
DGKJ German Society for Pediatrics and Adolescent Medicine (Deutsche Gesellschaft für Kinder- und Jugendmedizin)
IPEG International Pediatric Endosurgery Group
PES pediatric endocrine society
RYGB Roux-en-Y gastric bypass
VSG vertical sleeve gastrectomy
YES youth-with-extreme-obesity study

INTRODUCTION

Bariatric surgery is the only effective intervention to achieve long-term weight loss, reverse comorbidities, and thereby reduce mortality in adults with extreme obesity [1−3]. Owing to these favorable outcomes, the use of bariatric surgery has increased worldwide and has been expanded to adolescents. From 1997 to 2000 the use of adolescent bariatric surgery has increased fivefold in the United States, and has remained stable through 2009 at around 1000 cases per year [4]. Other western countries have followed suit. Similar outcomes in terms of weight reduction, improvement in somatic comorbidities, and psychosocial health have been reported in adults and adolescents alike [5−9].

Notwithstanding the clear benefits on weight and obesity comorbidities, bariatric surgery is associated with major *short-term risks*. The most important perioperative complications are cardiorespiratory events, deep venous thrombosis, pulmonary emboli, bleeding, leaks at sites of anastomoses, peritonitis, and wound-healing disorders. In addition, there are potential *long-term complications*. Most of our knowledge on long-term complications derives from studies in adults who have completed physical and psychosocial development. *General complications* relate to the drastic weight reduction and lifestyle changes. While general mortality is reduced after bariatric surgery [1], patients who have undergone bariatric surgery have a two- to threefold increased risk of suicide compared to controls [10]. Weight loss requires plastic reconstructive surgery in 15−30% of patients to remove or lift skin flaps. In the case of adolescents, the percentage might even be higher. *Specific complications* depend on the operating technique. For *restrictive procedures* such as gastric banding, complications include hernias, emesis, esophagitis, band slippage, and erosions, often requiring revisions. For *malabsorptive procedures* such as gastric bypass, complications comprise dumping syndrome with hypoglycemia, and malabsorption with resulting nutritional deficiencies in protein and vitamins A, D, E, K, and B. Nutritional deficiencies are further exacerbated by inadequate supplementation, which is more common in adolescents due to barriers to maintaining follow-up. Presenting symptoms are hair loss, dermatitis, anemia, electrolyte imbalances, osteopenia, osteoporosis, Kwashiorkor, neuropathy, parestesias, Beri-Beri, Wernicke encephalopathy, and Pellagra. While *sleeve gastrectomy* is considered a restrictive procedure, dumping syndrome and nutritional deficiencies have been described, albeit to a lesser degree compared to gastric bypass [11]. Vitamin B12 absorption is significantly impaired due to the loss of intrinsic factor, routinely requiring injections of vitamin B12 to prevent anemia and neurological complications.

CONCERNS RELATING TO ADOLESCENTS

Adolescents have unique physiological and psychosocial considerations that mandate special care [12,13]. Specific concerns relate to the inability to obtain appropriate consent, the dependence on a caregiver, the risks of major elective surgeries, long-term compliance, and unknown long-term effects on the still-developing body and mind. Due to the nature of the procedures, there are no randomized controlled trials to compare adolescent bariatric surgery with nonsurgical treatment, and we have to rely on observational data. While most studies are retrospective and/or derive from single centers, there are now three meta-analyses [6–8] and a handful of larger, prospective multicenter trials on adolescent bariatric surgery. The most commonly used procedures are adjustable gastric banding (AGB), Roux-en-Y gastric bypass (RYGB), and vertical sleeve gastrectomy (VSG). A number of other procedures have been variably performed in adolescents, and should be considered experimental.

Perioperative and Short-Term Safety

An increasing body of literature suggests that perioperative safety in adolescents is comparable to that found in adult data. Death is rare, and reported in about 0.9% after RYGB and 0% after AGB or VSG [7]. Reporting of other perioperative complications is heterogeneous, and therefore difficult to summarize.

In 2013, Lennerz et al. [14] reported data from $n = 345$ adolescents who underwent bariatric surgery between 2005 and 2010, captured in a national prospective registry in Germany. The most common surgical techniques were AGB (34%), RYGB (34%), and VSG (23%). Intraoperative complications occurred at a rate of 1–3%, general postoperative complications occurred at a rate of 3–9%, and specific postoperative complications occurred at a rate of 1–8%. Specific postoperative complications were more common after VSG. Other complication rates did not differ significantly between procedures. The most common complications included eight postoperative fevers, four urinary tract infections, four cases of peritonitis or intraabdominal abscesses, two cases of sepsis, four cases of stenosis or insufficiency of anastomoses, four wound-healing disorders, two postoperative pulmonary complications, one splenic injury, one vascular injury, and one postoperative bleeding requiring surgical revision. There was no mortality in this cohort.

In 2013, Messiah et al. reported data from a registry for excellence in bariatric surgery comprising 890 adolescents operated on at 360 centers in the United States between 2004 and 2010 [15]. Fifty-one percent of the patients underwent RYGB (99% laparoscopic) and 49% underwent AGB. Ten percent of patients required readmission after RYGB, and 2% after AGB. Reoperations were required in 6% after RYGB and 8% after AGB. Band revisions were performed in 13%. One patient died of arrhythmia 6 months after surgery. The most common complications were gastrointestinal (nausea, vomiting, bleeding, diarrhea, gallstones), surgical bleeding, general (obstruction, abscess, hernia), pulmonary, or device-related.

In 2014, Inge et al. [16] published 30-day outcomes of a prospective multicenter study, TeenLABS, conducted at five medical centers in the United States. The study comprised 242 adolescents who underwent bariatric surgery between 2007 and 2011. Main operating techniques were RYGB (67%), VSG (28%), and AGB (6%). Within 30 days of surgery 7.9% of patients experienced major, and 15% minor, complications.

Major complications included one intraoperative splenic injury; seven reoperations for intestinal obstruction, bleeding, and confirmed or suspected gastrointestinal leak; two patients with pulmonary embolus; four instances of gastrointestinal leaks; and one instance of suicidal ideation without physical harm. Minor perioperative complications included four minor injuries to solid organs at operation and six urinary tract events, abdominal/gastrointestinal complaints, and dehydration. Five patients required upper endoscopy. There was no perioperative mortality.

In summary, while mortality is low, there is a high risk of minor and major perioperative complications.

Long-Term Safety

Unlike in adults, long-term outcome data are sparse in adolescents. Most studies reporting outcome data are limited to 1–3 years of follow-up, with only a handful of small studies extending beyond that period. There are essentially no studies reporting outcomes beyond 5–10 years postoperatively. An adolescent undergoing surgery at age 17 would then be 27 years old, with many years ahead of them.

In 2016, Inge et al. reported 3-year follow-up data of 242 patients from their multicenter cohort, TeenLABS [5]. The mean percent weight loss was 27%, and did not differ significantly between AGB and VSG. Rates of obesity-related comorbidities such as hypertension and type 2 diabetes mellitus (T2DM) decreased dramatically, and quality of life improved. However, despite following one of the most rigorous postoperative care programs, these results came at a cost: 47 reoperations were performed in 30 participants (13%), all but three related to the preceding bariatric surgery. Upper endoscopic procedures (including stricture dilations) were performed in 29 participants (13%). One participant with known type 1 diabetes died 3.3 years after gastric bypass surgery from complications of a hypoglycemic event. When assessing nutritional deficiencies, 57% of patients had low ferritin levels, vitamin B12 levels declined by 35%, and 8% developed vitamin B12 deficiency. Deficiencies in vitamin A were found in 16%.

While the TeenLABS cohort reports a close to 90% attrition rate, adherence with follow-up was much lower in almost all other cohorts with loss-to-follow-up rates ranging up to 84% [7], making it impossible to accurately assess long-term outcomes. Moreover, studies in adolescents have shown that less than 20% of adolescents adhere to postoperative supplementation regimens. This is concerning because nutrient deficiencies and other complications can occur years after the surgical procedure and may remain unrecognized and untreated, especially if adherence with supplementation and follow-up are poor.

Growth and Development

Nutritional deficiencies may cause significant metabolic derangements in the still-developing adolescent. To account for this added risk, clinical practice guidelines have proposed cutoffs for developmental stage and height when considering bariatric surgery. Most experts agree that at least 95% of final adult height should be achieved and pubertal development should be at an advanced stage (Tanner stage 4). If these recommendations are followed, longitudinal growth and sexual maturation should be near completion at the time of surgery. However, other aspects of physical and psychosocial development are ongoing—e.g., peak bone mass is still building, increasing the risk for osteopenia and osteoporosis in case of interferences. In addition, there may be adverse effects on psychiatric conditions and psychosocial well-being [17], as adolescence is a particularly vulnerable period in this regard.

Ethical Considerations

Bariatric surgery in adolescents raises particular ethical challenges due to the elective nature of the procedure and the inability of the patient to provide his or her own consent—i.e., legally we must rely on a caregiver's consent. In addition, bariatric surgery is not a causal treatment, but works by causing permanent damage to a healthy organ system. The underlying genetic, physiologic, psychosocial, and environmental factors that have led to obesity persist. The procedure commits the adolescent to drastic lifestyle changes and lifelong postsurgical care. Therefore, it is mandated that the adolescent shall participate in the informed consent. Adolescents need to be evaluated for their intellectual maturity and ability to understand the scope of the procedure, associated risks and lifestyle changes, as well as the alternative treatments. However, there are no validated instruments to assess this ability. Adolescents are highly motivated to escape obesity and its associated stigma, and are likely not able to preempt the magnitude of the changes ahead of them.

Guidelines

Due to the paucity of long-term data on bariatric surgery in adolescents, it is impossible to determine a definitive risk−benefit analysis in this population. Adding to this challenge, BMI and age cutoffs are arbitrary and vary between guidelines.

In 2005, Apovian et al. suggested cutoffs of BMI ≥ 40 kg/m^2 with one serious comorbidity (e.g., T2DM, obstructive sleep apnea, severe or complicated hypertension, or pseudotumor cerebri); or BMI ≥ 50 kg/m^2 with less serious comorbidities AND failure of nonsurgical treatments for obesity [18].

The guidelines from the International Pediatric Endosurgery Group (IPEG), as well as the pediatric endocrine society (PES), state that adolescents with a BMI > 40 kg/m^2 or a BMI > 35 kg/m^2 combined with severe comorbidities should be considered for surgical intervention, if they have attained near-adult stature [19].

The American Society for Metabolic and Bariatric Surgery (ASMBS) guidelines propose a cutoff BMI of ≥ 35 kg/m^2 with major comorbidities, or a BMI of ≥ 40 kg/m^2 with other comorbidities [20].

Despite all the controversy, there is a large consensus that the situation is complex and requires a mindful, informed approach. All adolescents should be cared for by multidisciplinary teams in centers with expertise in adolescent extreme obesity and bariatric surgery. Conventional treatment approaches should be exhausted before considering surgery. Measures to improve patient understanding of the procedure and lifestyle changes, as well as compliance, need to be implemented in a preoperative treatment program. Risk−benefit ratios have to be assessed on a case-by-case basis, keeping in mind that full information on long-term risks is not available at this time.

EXPERT STATEMENT

Expert committees have formulated statements to guide the decision-making process and care associated with adolescent bariatric surgery. The recommendations below are extracted from recommendations of the German Working Group on Obesity in Children and Adolescents (Arbeitsgemeinschaft Adipositas im Kindes- und Jugendalter, AGA) expert committee commissioned by the German Association for the Study of Obesity (Deutsche Adipositas-Gesellschaft, DAG) and the German Society for Pediatrics and Adolescent Medicine (Deutsche Gesellschaft für Kinder- und Jugendmedizin, DGKJ) [21]. The statement calls for a standardized pre- and postsurgical treatment program to improve outcomes and reduce risk of adverse effects. Such a pre- and postsurgical treatment program has recently been implemented at five university centers in Germany [22]. Its outcomes are being analyzed in the prospective Youth-with-Extreme-Obesity-Study (YES) [23].

Recommendations:

1. Bariatric surgery is not a causal treatment. The underlying neurophysiological and psychological disorders that have led to low self-control and detrimental eating and exercise patterns persist and can exacerbate unforeseen complications.

2. Bariatric surgery in adolescents with extreme obesity leads to rapid weight loss due to anatomical restriction of food intake (restrictive procedures) and decreased energy uptake through diminished absorption (malabsorptive procedures). Alteration of gastric passage results in a decrease in orexigenic hormones and an increase in anorexigenic hormones. These changes lead to rapid improvements in glucose homeostasis and diabetes that cannot be attributed to weight change alone.

3. Bariatric surgery bears the risk of severe immediate and unclear long-term adverse outcomes. Due to the paucity of long-term data, it is impossible to compute a definite risk−benefit analysis.

4. Therefore, adolescent bariatric surgery remains a last-resort therapy, to be considered in selected patients who fulfill specific criteria (listed below).

5. The decision for or against surgery should be made after careful, individual assessment by an interdisciplinary team with expertise in the management of adolescents with extreme obesity. The team should comprise a pediatrician or internist, an endocrinologist, a child psychiatrist or a psychologist, a social worker, a dietician, and a bariatric surgeon. An independent ethics consultation should be obtained.

6. There is no evidence to determine a certain BMI cutoff. However, comorbidity, daily functioning, social integration, and quality of life deteriorate with increasing BMI. In our opinion, bariatric surgery can be considered if BMI is $>35 \text{ kg/m}^2$ and severe somatic and/or psychosocial comorbidities are present.

7. There is no scientific evidence supporting a certain age cutoff for bariatric surgery. However, it is likely that bariatric surgery and the associated catabolic state will impact growth and pubertal development. As such, expert groups agree that bariatric surgery should not be performed before puberty is almost complete (Tanner stage 4) and 95% of adult height is achieved.

8. All patients should be enrolled in long-term follow-up studies to close our knowledge gap regarding the lifelong sequelae of bariatric surgery in this population and ascertain optimal clinical care.

9. Bariatric surgery should only be performed if the following conditions are fulfilled:
   a. Patient conditions:
      i. Nonsurgical treatment approaches must be exhausted, including participation in a structured education and treatment program in accordance with clinical practice guidelines [24].
      ii. The patient has to participate in and adhere to a structured preparation program for the bariatric surgical intervention and the time thereafter [22].
      iii. The patient must undergo a mental health assessment to determine contraindications such as severe depression or suicidality.
      iv. The psychosocial environment of the patient needs to be supportive, competent, and stable.
      v. The adolescent and his/her family or caregivers must be aware that bariatric surgery is an ancillary treatment that cannot replace long-term lifestyle changes, including nutrition and physical activity.
      vi. A plan for long-term care should be established for the patient within his/her psychosocial environment.
      vii. The adolescent, and if possible his/her guardian, should give written consent to participate in the recommended follow-up and postsurgical care.
      viii. Compliance with long-term follow-up and recommendations must seem likely based on preoperative psychosocial assessment and preoperative compliance.

   b. Requirements of the medical center and personnel:
- **i.** Bariatric surgery has to be performed at specialized centers with a multidisciplinary team with expertise and competence in childhood obesity and bariatric surgery.
- **ii.** The multidisciplinary team has to educate the adolescent on the surgery and the associated lifestyle changes and requirements for long-term follow-up and nutritional supplementation.
- **iii.** The surgeon should have long-term experience in bariatric surgery.
- **iv.** The institution needs a multidisciplinary team that can provide long-term follow-up into adulthood. Due to patient geographic mobility, long-term follow-up in the same center is not always feasible. Therefore, a network of competence centers for postoperative care is needed.
- **v.** The centers must agree to participate in regional and national quality improvement efforts.

10. Bariatric surgery is contraindicated for the following patients:
- **a.** Adolescents with poorly controlled severe psychiatric disorders—e.g., psychosis, personality disorders with emotional instability, and severe eating disorders.
- **b.** Adolescents demonstrating noncompliance during a structured preparation program [22].
- **c.** Adolescents living in an unstable psychosocial environment.
- **d.** Adolescents with developmental disabilities who are unable to understand the implications of surgery and the associated lifestyle changes. Exceptions can be made for adolescents in secured social settings with adequate support.

MINI-DICTIONARY OF TERMS

- *Dumping syndrome:* Occurs in the setting of rapid or bypassed gastric passage. Hypertonic stomach contents enter the small intestine, leading to osmotic fluid shifts with abdominal pain, emesis, and diarrhea 15−30 minutes after the meal. Rapid inflow of glucose triggers excessive insulin secretion with reactive hypoglycemia, sweating, and dizziness 1−3 hours postprandial.
- *Osteopenia:* Decreased bone mineral density with preserved bone structure. Osteopenia constitutes a risk factor for the development of osteoporosis.
- *Osteoporosis:* Altered bone structure and strength with increased risk of fractures. Osteoporosis risk is increased later in life if normal peak bone mass is not achieved.
- *Kwashiorkor:* Severe protein deficiency. Kwashiorkor presents with edema, irritability, anorexia, ulcerating dermatitis, and fatty liver infiltrates. Kwashiorkor can be fatal.
- *Beri-Beri:* Severe form of vitamin B1 deficiency. Presentation is variable and can include neuronal damage with paralysis or ataxia, heart failure, or gastrointestinal manifestations.
- *Wernicke encephalopathy:* Acute neurological decompensation with ataxia, nystagmus, and mental status changes up to coma and death due to brain damage from vitamin B1 deficiency. It can result in permanent brain damage with altered memory function and hallucinations. The chronic form is called Korsakoff syndrome.
- *Pellagra:* Severe niacin (vitamin B3) deficiency. Pellagra presents with diarrhea, dermatitis, and mental disturbances such as depression, disorientation, and dementia. It can be fatal if untreated.
- *Intrinsic factor:* A glycoprotein produced by the parietal cells of the stomach. It binds to vitamin B12 and is required for its uptake in the small intestine. In the absence of intrinsic factor (e.g., due to removal of large parts of the stomach), vitamin B12 cannot be absorbed (Fig. 23.1).
- *Tanner stage 4:* The Tanner scale is a system to describe pubertal progression. Based on external sex characteristics, such as the size of the breasts or genitals, testicular volume, and pubic hair, a score between 1 and 5 is assigned, with 1 being prepubertal and 5 corresponding to complete sexual maturation. A Tanner stage of 4 is consistent with late (but not complete) puberty (Tables 23.1 and 23.2).

KEY FACTS

- Adolescence refers to the transition from childhood to adulthood and is commonly defined as the timespan from the onset of puberty (age 11−13 years) to completion of physical development and/or psychosocial emancipation (age 18−21 years).
- Timing of puberty varies considerably, typically starting around age 9−10 years in girls and 10−11 years in boys.

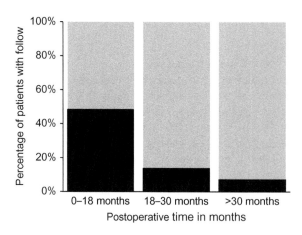

FIGURE 23.1 Follow-up declines rapidly over time. This figure shows follow-up rates in a cohort of 345 adolescents after bariatric surgery over time (in months). It is concerning that only 50% of patients have any follow-up. This number decreases rapidly to 20% after 30 months. Data was previously published in a different format [14].

TABLE 23.1 Suggested Criteria for Bariatric Surgery

- Age > 14 years.
- BMI > 40 kg/m² with serious physical or psychosocial comorbidity, or BMI > 35 kg/m² with severe physical or psychosocial comorbidity (e.g., diabetes mellitus, severe hypertension, sleep apnea syndrome, and advanced orthopedic complications).
- Completion of 95% of longitudinal growth and sexual maturation at Tanner stage 4.
- Absence of treatable etiology for obesity.
- Absence of contraindications for bariatric surgery, as established by medical and psychiatric evaluation (e.g., unstable physical or psychiatric disease, substance abuse, and pregnancy).
- Cognitive ability to provide informed consent along with parental or guardian consent.
- Stable, reliable support structures.
- Participation in a multidisciplinary treatment program over at least 6 months.
- Participation in a preoperative education program.
- A long-term follow-up concept has been established and agreed upon by the medical team, patient, and guardian.
- Indication confirmed by pediatrician, internist, endocrinologist, child psychiatrist/psychologist, bariatric surgeon, and ethics committee.

This table summarizes criteria that should to be met prior to considering bariatric surgery in an adolescent according to the YES study protocol [23].

TABLE 23.2 Ethical Considerations/Food for Thought

- Bariatric surgery in an adolescent may only be conducted if there is a positive risk–benefit ratio. As this is an individual case-by-case decision, it can only be made by the parents and child and an interdisciplinary care team with involvement of an ethics committee.
Challenges: Long-term risks and benefits are unknown. There is no systematic quality control.
- What does it mean to live with malabsorption for decades?
Challenge: Lifetime complications are unknown. It is unclear whether the necessary compliance with supplements can be maintained.
- Medical treatments must have been exhausted before considering bariatric surgery. These are costly and time consuming.
Challenge: There are insufficient treatment programs and no secure financing thereof. Moreover, only a few programs are targeted at adolescents with extreme obesity.
- The adolescent should be involved in the decision about bariatric surgery.
Challenge: Informed consent in minors. It is unclear whether adolescents have the necessary insight to make such a decision [12]. Adolescents may regret their decision later in life. There is no validated method to determine whether the adolescent is able to provide informed consent.
- In bariatric surgery, a healthy organ is treated surgically. Adolescents are unable to grasp the full scope of such a procedure.
Challenge: Is it ethically acceptable to surgically change a healthy organ when it is known that the lifestyle conditions that have contributed to the patient's obesity are only in part under the adolescent's control?
- A long-term follow-up concept has been established and agreed upon by the medical team, the patient, and their guardian.
Challenge: There are insufficient structured long-term follow-up programs.

This table lists ethical and open questions to think through when considering bariatric surgery in an adolescent. Based on these ethical considerations, iatrogenic anatomical and functional changes in a healthy organ need to be used with caution to prevent unforeseeable lifetime complications.

- Longitudinal growth peaks around age 11.5 years for girls and 13 years for boys, and slows around age 13−14 years and 15−16 years, respectively. Accordingly, girls after age 13 or boys after age 15 may be candidates for bariatric surgery if following current guidelines.
- Bone mass accrual peaks on average 1 year after the respective peaks in longitudinal growth, and is >90% complete by age 17 in girls and age 18 in boys.
- About 50% of an adult's body weight is accrued during adolescence.
- Adolescents engage in increased risk-taking behavior, likely because they hold different values and thus come to different conclusions when assessing a situation ahead.
- Adolescents give more weight to rewards, particularly social rewards, than do adults.
- Abstract thinking develops and allows adolescents to reason in a wider perspective.
- The capacity for insight and judgment that develops through experience increases later, usually between the ages of 14 and 25.
- Thoughts, ideas, and concepts developed in adolescence greatly influence one's future life, playing a major role in character and personality formation.
- Peer influence becomes more important as adolescents pull away from their core family and develop independence.
- Relationships, social surroundings, and support structures are volatile.
- Adolescents often relocate for education, vocation, or family formation.
- Opportunities available to the adolescent are likely to shape adult life in terms of vocational and personal relationships.

SUMMARY POINTS

- While mortality is low, there is a high risk of minor and major perioperative complications.
- Adolescent bariatric surgery remains controversial due to the paucity of long-term outcome data.
- All adolescents should be cared for by multidisciplinary teams in centers with expertise in adolescent extreme obesity and bariatric surgery, and included in longitudinal studies.
- Conventional treatment approaches should be exhausted before considering surgery.
- Risk−benefit ratios have to be assessed on a case-by-case basis, but full information on long-term risks is not available at this time.

REFERENCES

[1] Adams TD, et al. Long-term mortality after gastric bypass surgery. N Engl J Med 2007;357(8):753−61.
[2] Sjostrom L, et al. Lifestyle, diabetes, and cardiovascular risk factors 10 years after bariatric surgery. N Engl J Med 2004;351(26):2683−93.
[3] Sjostrom L, et al. Effects of bariatric surgery on mortality in Swedish obese subjects. N Engl J Med 2007;357(8):741−52.
[4] Kelleher DC, et al. Recent national trends in the use of adolescent inpatient bariatric surgery: 2000 through 2009. JAMA Pediatr 2013;167(2): 126−32.
[5] Inge TH, et al. Weight loss and health status 3 years after bariatric surgery in adolescents. N Engl J Med 2016;374(2):113−23.
[6] Black JA, et al. Bariatric surgery for obese children and adolescents: a systematic review and meta-analysis. Obes Rev 2013;14(8):634−44.
[7] Paulus GF, et al. Bariatric surgery in morbidly obese adolescents: a systematic review and meta-analysis. Obes Surg 2015;25(5):860−78.
[8] Treadwell JR, Sun F, Schoelles K. Systematic review and meta-analysis of bariatric surgery for pediatric obesity. Ann Surg 2008;248(5): 763−76.
[9] Zeller MH, et al. Two-year trends in psychosocial functioning after adolescent Roux-en-Y gastric bypass. Surg Obes Relat Dis 2011;7(6): 727−32.
[10] Tindle HA, et al. Risk of suicide after long-term follow-up from bariatric surgery. Am J Med 2010;123(11):1036−42.
[11] Tack J, Deloose E. Complications of bariatric surgery: dumping syndrome, reflux and vitamin deficiencies. Best Pract Res Clin Gastroenterol 2014;28(4):741−9.
[12] Inge TH, et al. Bariatric surgery for severely overweight adolescents: concerns and recommendations. Pediatrics 2004;114(1):217−23.
[13] Inge TH, et al. A critical appraisal of evidence supporting a bariatric surgical approach to weight management for adolescents. J Pediatr 2005;147(1):10−19.
[14] Lennerz BS, et al. Bariatric surgery in adolescents and young adults--safety and effectiveness in a cohort of 345 patients. Int J Obes (Lond) 2014;38(3):334−40.
[15] Messiah SE, et al. Changes in weight and co-morbidities among adolescents undergoing bariatric surgery: 1-year results from the Bariatric Outcomes Longitudinal Database. Surg Obes Relat Dis 2013;9(4):503−13.

[16] Inge TH, et al. Perioperative outcomes of adolescents undergoing bariatric surgery: the Teen-Longitudinal Assessment of Bariatric Surgery (Teen-LABS) study. JAMA Pediatr 2014;168(1):47−53.

[17] Jarvholm K, et al. Short-term psychological outcomes in severely obese adolescents after bariatric surgery. Obesity (Silver Spring) 2012;20(2): 318−23.

[18] Apovian CM, et al. Best practice guidelines in pediatric/adolescent weight loss surgery. Obes Res 2005;13(2):274−82.

[19] International Pediatric Endosurgery Group. IPEG guidelines for surgical treatment of extremely obese adolescents. J Laparoendosc Adv Surg Tech A 2009;19(suppl. 1):xiv−vi.

[20] Michalsky M, et al. ASMBS pediatric committee best practice guidelines. Surg Obes Relat Dis 2012;8(1):1−7.

[21] Wabitsch M, et al. für die Arbeitsgemeinschaft Adipositas im Kindes- und Jugendalter. Bariatrisch-chirurgische Maßnahmen bei Jugendlichen mit extremer Adipositas. Informationen und Stellungnahme der Arbeitsgemeinschaft Adipositas im Kindes- und Jugendalter (AGA). Monatsschr Kinderheilkd 2012;160:1123−8.

[22] Lennerz BWM, Geisler A, Hebebrand J, Kiess W, Moss A, Mühlig Y, et al. Manual-basiertes Vorgehen zur Vorbereitung und Nachsorge von bariatrisch-chirurgischen Eingriffen bei Jugendlichen. Adipositas 2014;8(1):5−11.

[23] Wabitsch M, et al. Medical and psychosocial implications of adolescent extreme obesity-acceptance and effects of structured care, short: Youth with Extreme Obesity Study (YES). BMC Public Health 2013;13:789.

[24] Böhler TWM, Winkler U. Konsensuspapier: Patientenschulungsprogramme für Kinder und Jugendliche mit Adipositas. Berlin: Bundesministerium für Gesundheit und Soziale Sicherung; 2004.

Chapter 24

Mortality in Bariatric Surgery: A Focus on Prediction

M. Jung[1] and N. Goossens[1,2]

[1]University Hospital Geneva, Geneva, Switzerland, [2]Icahn School of Medicine at Mount Sinai, New York, NY, United States

LIST OF ABBREVIATIONS

BMI body mass index
BPD/DS biliopancreatic diversion/duodenal switch
NAFLD nonalcoholic fatty liver disease
NASH nonalcoholic steatohepatitis
NIS National Inpatient Sample
OR odds ratio
RCT randomized controlled trial
SOS Swedish obese subjects
VA veterans affairs

INTRODUCTION

Efforts over the past several decades have worked to clarify the overall survival benefit of bariatric surgery in morbidly obese subjects across a wide range of high-risk groups [1−3]. Nevertheless, morbidly obese patients remain a clinically heterogeneous population with a high prevalence of obesity-related comorbidities such as metabolic syndrome, cardiovascular disease, hiatal hernia with gastroesophageal reflux disease, and sleep apnea, all potentially affecting perioperative and long-term outcomes after bariatric surgery. In addition, obese subjects pose unique anatomical and surgical challenges due to thick abdominal walls, increased size, and vulnerable parenchyma of the liver with underlying nonalcoholic fatty liver disease (NAFLD), as well as a high amount of visceral obesity that leads to space constraints and technical difficulties. Therefore, bariatric surgery, especially combined restrictive/malabsorptive procedures in the obese, multimorbid patient, is considered a high-risk intervention with current reports underlining significant complication rates of up to 20% and anastomotic leak rates of up to 5.1% that occasionally lead to life-threatening complications [4−6]. Early identification of high-risk subjects for increased perioperative mortality and complications is therefore important, and numerous efforts have attempted to identify key clinical variables associated with poor perioperative outcomes. In addition, in light of recent data underlining improved long-term survival after bariatric surgery, additional efforts are required to better understand baseline factors associated with long-term mortality, and possibly identify high-risk groups requiring additional follow-up or intervention.

Therefore, this chapter aims to summarize current predictive factors associated with short-term, perioperative, and long-term mortality after bariatric surgery (Fig. 24.1). We conclude with a short discussion of the current state of the field and potential future directions, including refined molecular prediction of long-term outcomes.

PERIOPERATIVE MORTALITY

Perioperative mortality is usually considered early mortality starting from the day of the operation until 30 days postoperatively. Current perioperative mortality rates of bariatric surgery are generally reported to be low, with an incidence

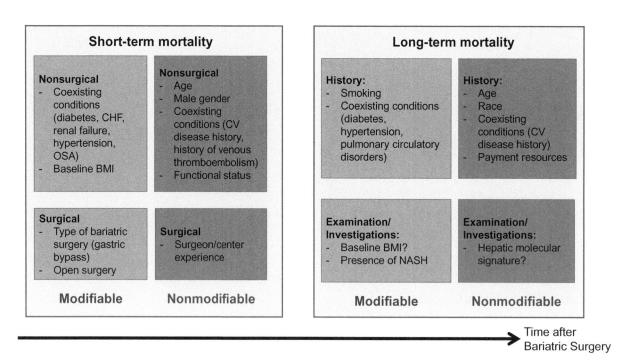

FIGURE 24.1 Summary of modifiable and nonmodifiable factors associated with short-term and long-term mortality after bariatric surgery. *BMI*, body mass index; *CHF*, congestive heart failure; *CV*, cardiovascular; *NASH*, nonalcoholic steatohepatitis; *OSA*, obstructive sleep apnea.

between 0.12% and 0.9% [7−13]. Major causes of perioperative mortality include sepsis (15−33%), cardiac causes (28%), and pulmonary embolism (17−32%) [14,15]. Factors linked with perioperative mortality after bariatric surgery can be broadly separated into surgical factors and nonsurgical factors.

Surgical Factors Linked to Perioperative Mortality

One of the first systematic reviews analyzing published mortality of bariatric surgery was written by Buchwald et al. [8]; it surveyed reports from 1990 to 2006 that included 84,931 patients with a total early mortality rate of 0.28%. Subgroup analysis by procedure showed a mortality rate of 0.30% for restrictive procedures with the open approach, and 0.07% for restrictive procedures by laparoscopy. Mortality for gastric bypass was 0.41% for open procedures and 0.16% for laparoscopic procedures. Confirming the improved outcomes of the minimal invasive approach, laparoscopy had lower mortality rates compared to the open approach, with the exception of biliopancreatic diversion/duodenal switch. The highest mortality was associated with the biliopancreatic diversion, with mortality rates of 0.76% for the open technique and 1.11% for the laparoscopic technique; the complexity of this procedure might explain why the open approach compares favorably to the laparoscopic approach. Unsurprisingly, revisional surgery also had higher mortality rates of up to 1.65%. These findings were confirmed by a recent meta-analysis by Chang et al. [9], who found the lowest perioperative mortality for adjustable gastric banding (0.07%), followed by sleeve gastrectomy (0.29%) and Roux-en-Y gastric bypass (RYGB; 0.38%). Complication rates in randomized controlled trials (RCT) for gastric bypass were highest with 21%, as compared to 13% for sleeve gastrectomy and 13% for adjustable gastric banding. This is probably not only due to decreasing complexity of the procedure, but also to the higher probability of choosing a combined malabsorptive/restrictive procedure for more morbid patients who have a higher body mass index (BMI) [16]. The subgroup analysis of Buchwald et al. revealed male patients (male-to-female ratio of 4.74 vs 0.13), the superobese (1.25% incidence), and patients 65 years and older (0.34% incidence) to be at higher risk for early mortality (≤30 days). Univariate analysis of comorbidities revealed diabetes as the most important risk factor for perioperative mortality. The number of procedures performed per year also seemed to influence mortality rates with studies with less than 50 interventions having higher mortality rates than the studies with larger patient numbers. These findings are in line with those of Nguyen et al. [17], who found a lower mortality rate in high-volume centers compared to low-volume hospitals (0.3% vs 1.2%, $p < 0.01$); the findings were later confirmed by others [18].

Improved surgical techniques and postoperative care have likely contributed to a decrease in perioperative mortality rates over the past several decades. Chang et al. [9] confirmed these findings in their meta-analysis, which included 161,756 patients and reported low mortality rates and identified RCT to have even lower perioperative ($\leq$30 days) mortality rates than observational studies (0.08% vs 0.22%), despite higher overall complication rates for RCT compared to observational studies (17% vs 10%). Reduced mortality rates over the time period of publication of selected systematic reviews by Buchwald et al. [8], Maggard et al. [19], or more recently, Chang et al. [9], were also probably due to a transition from open surgery to the minimally invasive approach with faster recovery, shorter length of stay, and reduced risk of respiratory complications. Notably, the shift from the initially lower mortality of the open technique compared with laparoscopy in earlier reports to the lower mortality of the minimally invasive approach compared to open surgery in later reports [20] is probably explained by the well-known long learning curve of the laparoscopic technique and the higher leak rates occurring in earlier publications of this technique [21].

Nonsurgical Factors Linked to Perioperative Mortality

Clinical predictors associated with short-term postoperative mortality have also been systematically studied in large cohorts. Nguyen et al. [22] analyzed a large sample size of 304,515 patients undergoing bariatric surgery using the National Inpatient Sample (NIS) database. With an in-hospital mortality rate of 0.12%, multivariate logistic regression analysis revealed male gender (odds ratio (OR) 1.7), age greater than 50 years (OR 3.8), congestive heart failure (OR 9.5), vascular disease (OR 7.4), chronic renal failure (OR 2.7), open surgery (OR 5.5), and gastric bypass (OR 1.6) as independent predictors of early, in-hospital mortality. One limitation of the NIS database, potentially explaining the low mortality rate, results from the fact that mortality rates after hospital discharge were not available in this database. In a separate analysis, Khan et al. [23] identified age greater than 45 years, male gender, BMI $\geq$ 50 kg/m^2, open surgery, diabetes, functional status, coronary intervention, dyspnea, unintended weight loss, and bleeding disorder as independent predictors for increased mortality. Other studies [24,25] confirmed advanced age and/or male sex as risk factors for adverse postoperative outcomes, although they didn't find a strong association between comorbidities and early mortality postbariatric surgery. Unlike the previous studies, the absence of association between comorbidities and early mortality was possibly due to smaller sample sizes and the low prevalence (<1.5%) of the comorbidities identified in the Nguyen et al. study in bariatric surgery patients.

In an analysis of the American College of Surgeons National Surgical Quality Improvement Program, including more than 5000 patients, Lancaster et al. [20] confirmed a significantly higher perioperative mortality after open gastric bypass compared with laparoscopic gastric bypass (0.79% vs 0.17%). BMI, age, and bleeding disorder were independent risk factors in the multivariate model for major complication, but not for mortality. Another risk factor associated with increased perioperative mortality in an analysis of the NIS was cirrhosis [26]. In particular, Mosko and Nguyen identified a disproportionately high mortality rate in patients with decompensated cirrhosis (0.3%, 0.9%, and 16.3% mortality rates in noncirrhotics, compensated cirrhotics, and decompensated cirrhotics who underwent bariatric surgery, respectively), although this analysis was limited by the absence of biopsy-proven cirrhosis.

Relatively few prospective single-center reports of predictive factors of perioperative mortality have been published. The low rates of reported early mortality in the literature complicate the detection of their predictors and risk factors, although predictors of hospital stay length after bariatric surgery are helpful as surrogate markers of complications and mortality. For instance, Ballantyne et al. [27] performed a retrospective univariate analysis in 311 patients undergoing open vertical banded gastroplasty–RYGB (159 patients) and laparoscopic RYBG (152 patients) in a 6-month period to identify predictors of increased length of hospital stay. The length of hospital stay was used as a surrogate of perioperative complications. Analysis revealed open surgery (OR 0.4), increasing BMI (60 kg/m^2; OR 0.38), increasing length of intervention (180 minutes; OR 0.48), sleep apnea (OR 2.25), asthma (OR 3.73), and hypercholesterolemia (OR 3.73) as risk factors for a prolonged hospital stay.

Demographic risk factors for perioperative mortality include male gender [22,28] as well as advanced age [25,29]. The reason why male gender represents a higher risk for early mortality might be explained by higher incidence of visceral obesity and resulting challenging space constraints, as well as the higher incidence of hypertension, diabetes, and metabolic syndrome, which seem to be correlated with visceral obesity.

Scoring Systems

Scoring systems, such as the obesity surgery mortality risk score (OS-MRS) based on multivariate analysis of preoperative mortality risk factors, have been developed [30] and validated by several groups [31–33]. This first scoring system for risk assessment integrates five preoperative factors, each of them adding 1 point to a morbidity score: BMI greater than or equal to 50 kg/m^2 (the risk factor associated with the highest OR), age greater than or equal to 45 years, male sex, hypertension, and known risk factors for pulmonary embolism (previous thromboembolism, preoperative vena cava filter, hypoventilation, pulmonary hypertension). Subjects are classified into three risk groups based on their number of risk factors: (A) low risk, with a total score between 0 and 1; (B) intermediate risk of between 2 and 3; and (C) high risk of between 4 and 5. A systematic review of the obesity surgery mortality risk score, including 9382 patients with more than 98% gastric bypass procedures, confirmed the utility of the score in mortality risk stratification. Their results, which included six studies, described a 4.3% mortality rate in class C patients, 1.3% in class B, and 0.26% in class A patients [13].

A scoring system predictive of the composite outcome of major complication and 30-day complication was developed by the Longitudinal Assessment of Bariatric Surgery (LABS) consortium. In this prospective, multicenter observational study, extreme values of BMI, history of venous thromboembolism, obstructive sleep apnea, and inability to walk 200 ft were independent predictors of the composite outcome of major morbidity and early mortality [34]. Turner et al. revealed low serum albumin, BMI, age, and functional dependence as predictive for 30-day morbidity and mortality, analyzing data of 32,426 patients from the American College of Surgeons National Security Quality Improvement Program database [35].

LONG-TERM MORTALITY AFTER BARIATRIC SURGERY

Extensive research has been performed to understand and characterize the long-term survival benefit of bariatric surgery compared to control obese subjects [1–3]. Although the literature is lacking in well-designed RCT comparing bariatric surgery to control subjects, on the basis of large, high-quality cohorts, a growing consensus has emerged that bariatric surgery is associated with reduced overall long-term mortality across a wide range of populations and risk factors. In addition, as mentioned above, a large body of research has also aimed to understand and quantify risk factors linked to perioperative mortality after bariatric surgery. In light of this, it may come as a surprise that predictive factors associated with long-term survival in patients undergoing bariatric surgery have been studied less, and that no clear scoring system has emerged to identify a "high-risk" subgroup of subjects at risk of increased long-term mortality. We therefore aim to summarize here the current state of the literature, and point toward areas of unmet needs and potential future research (Table 24.1 and Fig. 24.1).

The Swedish obese subjects (SOS) study was one of the first large, nonrandomized prospective cohorts that demonstrated a survival benefit for patients undergoing bariatric surgery compared to controls with an overall mean follow-up of nearly 11 years [2]. In multivariable analysis, predictors found to be associated with long-term mortality included the absence of bariatric surgery, male sex, age, smoking, and the presence of coexisting conditions including diabetes, a history of cardiovascular disease, high cholesterol, or psychasthenia. An important caveat, however, is that the authors combined the bariatric surgery cases and obese controls in this multivariable analysis, although the analysis was adjusted for the effect of bariatric surgery. The prognostic impact of baseline comorbidities was confirmed in an analysis of 856 bariatric surgery subjects in the veterans affairs (VA) National Surgical Quality Improvement Program, where patients with a diagnostic cost group score (a measure of predicted expenditures based on healthcare utilization in the year prior to surgery associated to patient comorbidities) of 2 or more had a greater risk of death during a median follow-up of 2.7 years [36]. Interestingly, in this cohort superobese subjects (i.e., a BMI $\geq$ 50 kg/m^2) were associated with increased mortality, but a laparoscopic surgical approach was found to be protective. A follow-up study from the same group, focusing on the benefit of bariatric surgery in 2500 high-risk VA subjects and 7462 matched controls with a mean follow-up of nearly 7 years, identified a survival benefit of bariatric surgery after 1 year of follow-up [3]. Importantly, this study found that the presence of diabetes, sex, and period of surgery did not affect the overall long-term survival benefit of bariatric surgery after 1 year of follow-up. Similarly, a recent single-practice retrospective cohort study of 7925 subjects undergoing RYGB surgery and matched obese controls confirmed the survival benefit of bariatric surgery across most age groups, except in subjects younger than 35 years where bariatric surgery was not associated with improved long-term survival benefit [37]. This was attributed by the authors to a significantly higher number of externally caused deaths in that population, particularly among women.

An interesting paper from Bruschi Kelles et al., reported results from a prospective cohort of subjects undergoing bariatric surgery under the care of a private health maintenance organization in Brazil, and identified several risk factors for mortality stratified by cause of death related or unrelated to surgery, although the authors do not clearly specify how they make this distinction [38]. Within the 4344 subjects followed for a median follow-up of 4.1 years, male gender,

TABLE 24.1 Key Studies Assessing Factors Associated With Long-Term Mortality After Bariatric Surgery

Study and Reference	Study Design and Setting	Inclusion Period	Type of Bariatric Surgery	Total Number of Subjects	Number of Deaths	Follow-up Time	Factors Associated with Long-term Mortality in Bariatric Surgery	Comments
Adams et al. [1]	Single-practice retrospective cohort, United States	1984–2002	Gastric bypass (100%)	9949 Surgical cases	Cases: 288 (2.9%)	Mean follow-up: 7.1 years		No analysis of factors associated with long-term outcome in bariatric surgery subjects
				9628 Matched controls	Controls: 425 (4.4%)			
Sjöström et al. [2]	Prospective multicenter controlled cohort, Sweden	1987–2001	Gastric bypass (13%)	2010 Surgical cases	Cases: 101 (2.9%)	Mean follow-up: 10.9 years	Male Sex / Age	Stepwise multivariable analysis of all subjects (including controls)
			Gastroplasty (68%)	2037 Matched controls	Controls: 129 (4.4%)		Smoking	
			Gastric banding (19%)				Coexisting conditions (diabetes, CVD history, dyslipidemia)	
Arterburn et al. [36]	Prospective study in 12 Veterans Affairs bariatric centers, United States	2000–2006	Gastric bypass (97%)	856 Surgical cases	54 (6.3%)	Median follow-up: 2.7 years	Baseline BMI ≥50 kg/m^2 / DCG score ≥2	Multivariable Cox model
			Gastric banding (2%)				Nonlaparoscopic procedure	Limited follow-up
			Gastroplasty (1%)					
Bruschi Kelles et al. [38]	Prospective cohort in a private HMO setting, Brazil	2001–2010	Gastric bypass (100%)	4344 Surgical cases	82 deaths (1.8%), 38 (0.9%) not related to surgery	Median follow-up: 4.1 years	Male gender / Age ≥ 50 years / Hypertension / Baseline BMI ≥ 50 kg/m^2 (only for surgery-related mortality)	Only male gender and age ≥50 years were associated with deaths unrelated to surgery

(Continued)

TABLE 24.1 (Continued)

Study and Reference	Study Design and Setting	Inclusion Period	Type of Bariatric Surgery	Total Number of Subjects	Number of Deaths	Follow-up Time	Factors Associated with Long-term Mortality in Bariatric Surgery	Comments
Arteburn et al. [3]	Retrospective cohort study in Veterans Affairs bariatric centers, United States	2000–2011	Gastric bypass (74%)	2500 Surgical cases	Cases: 263 (11%)	Mean follow-up: 6.9 years		Notably, diabetes, sex, and superobesity were not associated with different benefits from bariatric surgery in multivariate models
			Sleeve gastrectomy (15%)	7462 Matched controls	Controls: 1277 (17%)			
			Gastric banding (10%)					
			Other (1%)					
Telem et al. [39]	Database-based cohort, United States	1999–2005	Gastric bypass (57%)	7862 Surgical cases	NA	> 8-year follow-up	Age	
			Sleeve gastrectomy (8%)				Race (Spanish/Hispanic was protective)	
			Gastric banding (27%)				Payment resource	
			Gastroplasty (9%)				Comorbid conditions (congestive heart failure, diabetes, pulmonary circulatory disorders)	
Davidson et al. [37]	Single-practice retrospective cohort, United States	1984–2002	Gastric bypass (100%)	7925 Surgical cases	Cases: 213 (2.7%)	Mean follow-up: 7.2 years		Mortality benefit of bariatric surgery in all ages except in subjects <35 years.
				7925 Matched controls	Controls: 321 (4.1%)			Follow-up of study [1]
Goossens et al. [40]	Single-center, retrospective–prospective cohort, Switzerland	1997–2004	Gastric bypass (100%)	492 Surgical cases	21 (4.3%)	Median follow-up 10 years	Hypertension NASH	A 32-gene hepatic signature in patients with NASH was associated with increased mortality

BMI, body mass index; CVD, cardiovascular disease; DCG, diagnostic cost group; NASH, nonalcoholic steatohepatitis.

age greater than or equal to 50 years, hypertension, and BMI greater than or equal to $50 \, kg/m^2$ were associated with overall death; however, only male gender and age over 50 years were associated with increased mortality from death not related to surgery in the multivariable analysis, supporting previous findings that sex and age are two key predictors of long-term mortality after bariatric surgery. Similar associations were identified in another study by Telem et al., performed in the setting of a longitudinal comprehensive data reporting system aimed at capturing patients who underwent laparoscopic bariatric surgery in New York state from 1999 to 2005 with a relatively high proportion of gastric banding (27%) compared to other studies [39]. The authors confirmed the prognostic importance of age, male gender, presence of comorbidities (congestive heart failure, pulmonary circulation disorders, rheumatoid arthritis, and diabetes), as these factors were associated with increased long-term mortality; however, they also found that payment source, notably payment by Medicare or Medicaid, was a significant predictor of long-term mortality.

Our own experience in 492 consecutive patients undergoing gastric bypass surgery and followed-up for a median of 10 years, found that the presence of hypertension and histologically documented nonalcoholic steatohepatitis (NASH) were significantly associated with long-term mortality in multivariable analysis [40]. In addition, we found that a previously published liver-based 32-gene signature [41] further stratified patients with NASH into high- and low-risk groups. Although preliminary and requiring further validation, these findings my pave the way for more accurate and objective assessment of long-term outcome prediction prior to bariatric surgery.

Although discussed elsewhere in this textbook, it is useful to briefly discuss mortality from specific causes as a window into mortality prediction after bariatric surgery. For instance, an analysis of cardiovascular outcomes in the prospective SOS study with a median follow-up of nearly 15 years found that after multivariable adjustments for baseline conditions, bariatric surgery was associated with a reduced number of fatal cardiovascular deaths and a lower incidence of total cardiovascular events [42]. However, in secondary subgroup analyses, the bariatric surgery treatment benefit with respect to cardiovascular events was significantly associated with baseline plasma insulin with higher benefit in subjects with higher baseline insulin levels, suggesting that subjects with more insulin resistance benefitted more from the surgery. Another major outcome of interest is the development of cancer after bariatric surgery due to the known association between obesity and cancer. Interestingly, an analysis of cancer incidence data in the SOS study found a reduced cancer incidence in obese women undergoing bariatric surgery, but not obese men [43]. This finding was confirmed in a recent systematic review [44], potentially explaining part of the association between male gender and increased long-term mortality after bariatric surgery.

To date, predictors of resolution of comorbidities and predictors of length of hospital stay are better investigated than predictors of mortality [16,45,46]. Nevertheless, these predictors might give us indirect conclusions about the risk of mortality. When considering factors that are associated with failure of resolution of comorbidities after bariatric surgery, advanced age and preoperative insulin use seem to be associated with failure of remission of diabetes after bariatric surgery (OR 0.140; $p < 0.001$ for gastric bypass) [16]. Resolution of hypertension was, however, inversely related to the number of antihypertensive medications used: the higher the number of different therapeutic hypertension medications, the lower the chance of remission of this comorbidity ($p < 0.003$). Other data confirmed that patients who use insulin, have a higher hemoglobin A1c level, and have a longer duration of the disease are at higher risk of not resolving their diabetes [45–47]. Other studies found visceral obesity [48], preoperative BMI [46], and fasting C-peptide levels [45] as risk factors for failure of remission.

Recent efforts have attempted to better characterize prognostic factors associated with bariatric surgery using modeling tools such as Markov models [49]. However, despite an increased understanding of predictors associated with long-term outcomes after bariatric surgery, further research is needed to better stratify patients at baseline and robustly identify a subgroup of patients at higher risk of long-term outcomes who may require additional intervention.

DISCUSSION

A face-to-face comparison of predictive risk factors associated with short-term and long-term mortality after bariatric surgery reveals certain similarities (comorbidities, age, gender) and differences (surgical factors, venous thromboembolism). However, with the constant increase of bariatric surgery procedures (179,000 cases in the United States in 2013 according to the American Society for Metabolic and Bariatric Surgery), more accurate prognostic tools are needed to stratify patients more accurately, especially when considering long-term outcomes. One possibility, outlined in Fig. 24.2, is the identification of accurate clinical or molecular preoperative biomarkers associated with long-term outcomes after bariatric surgery. Patients identified as higher risk may require more intensive follow-up or focused interventions by a team of multidisciplinary healthcare professionals, hopefully leading to even better outcomes for bariatric surgery in the 21st century.

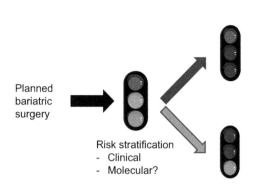

High risk
- More intensive follow-up?
- Improvement of comorbidities prior to surgery?

Low risk
- Proceed with surgery

Planned bariatric surgery

Risk stratification
- Clinical
- Molecular?

FIGURE 24.2 Proposed implementation of potential risk scores for long-term mortality after bariatric surgery; subjects are categorized into high- and low-risk groups, leading to different preparation for surgery and long-term follow-up after bariatric surgery. *Part of the images are derived from Servier Medical Art (www.servier.com), licensed under a Creative Commons Attribution 3.0 Unported License.*

MINI-DICTIONARY OF TERMS

- *Biomarker*: A measurable indicator of some biological state or condition.
- *Nonalcoholic steatohepatitis*: Liver injury, histologically characterized by the presence of hepatocyte ballooning, steatosis, and hepatic inflammation possibly accompanied with liver fibrosis.
- *Gene signature*: A group of genes whose combined expression is associated with a phenotype or outcome.

KEY FACTS

- Short- and long-term predictors of mortality after bariatric surgery are different.
- Clinical scores have been established to aid the clinician in assessing short-term perioperative risk.
- More accurate tools to assess long-term prognosis after bariatric surgery are required.

SUMMARY POINTS

- Perioperative mortality for bariatric surgery is low (0.12—0.9%).
- Major predictors of perioperative mortality include surgical factors (surgeon/center experience) and nonsurgical factors (comorbidities, gender, age, baseline BMI).
- Clinical score, including the obesity surgery mortality risk score, may help to assess risk of perioperative mortality in the individual patient.
- Predictors of long-term mortality following bariatric surgery include comorbidities, age, race, and presence of liver disease.
- Novel tools are needed to better predict long-term outcomes after bariatric surgery.

REFERENCES

[1] Adams TD, Gress RE, Smith SC, Halverson RC, Simper SC, Rosamond WD, et al. Long-term mortality after gastric bypass surgery. N Engl J Med 2007;357:753—61.

[2] Sjöström L, Narbro K, Sjöström CD, Karason K, Larsson B, Wedel H, et al. Effects of bariatric surgery on mortality in Swedish obese subjects. N Engl J Med 2007;357:741—52.

[3] Arterburn DE, Olsen MK, Smith VA, et al. Association between bariatric surgery and long-term survival. JAMA 2015;313:62—70.

[4] Lee S, Carmody B, Wolfe L, Demaria E, Kellum JM, Sugerman H, et al. Effect of location and speed of diagnosis on anastomotic leak outcomes in 3828 gastric bypass cases. J Gastrointest Surg 2007;11:708—13.

[5] DeMaria EJ, Sugerman HJ, Kellum JM, Meador JG, Wolfe LG. Results of 281 consecutive total laparoscopic Roux-en-Y gastric bypasses to treat morbid obesity. Ann Surg 2002;235:640—5. discussion 645-647.

[6] Buchwald H, Avidor Y, Braunwald E, Jensen MD, Pories W, Fahrbach K, et al. Bariatric surgery: a systematic review and meta-analysis. JAMA 2004;292:1724—37.

[7] Nguyen NT, Morton JM, Wolfe BM, Schirmer B, Ali M, Traverso LW. The SAGES bariatric surgery outcome initiative. Surg Endosc 2005;19:1429—38.

[8] Buchwald H, Estok R, Fahrbach K, Banel D, Sledge I. Trends in mortality in bariatric surgery: a systematic review and meta-analysis. Surgery 2007;142:621—32. discussion 632-625.

[9] Chang SH, Stoll CR, Song J, Varela JE, Eagon CJ, Colditz GA. The effectiveness and risks of bariatric surgery: an updated systematic review and meta-analysis, 2003-2012. JAMA Surg 2014;149:275—87.

[10] O'Rourke RW, Andrus J, Diggs BS, Scholz M, McConnell DB, Deveney CW. Perioperative morbidity associated with bariatric surgery: an academic center experience. Arch Surg 2006;141:262−8.

[11] Ballantyne GH, Belsley S, Stephens D, Saunders JK, Trivedi A, Ewing DR, et al. Bariatric surgery: low mortality at a high-volume center. Obes Surg 2008;18:660−7.

[12] Lazzati A, Audureau E, Hemery F, Schneck AS, Gugenheim J, Azoulay D, et al. Reduction in early mortality outcomes after bariatric surgery in France between 2007 and 2012: a nationwide study of 133,000 obese patients. Surgery 2016;159:467−74.

[13] Thomas H, Agrawal S. Systematic review of obesity surgery mortality risk score--preoperative risk stratification in bariatric surgery. Obes Surg 2012;22:1135−40.

[14] Smith MD, Patterson E, Wahed AS, Belle SH, Berk PD, Courcoulas AP, et al. Thirty-day mortality after bariatric surgery: independently adjudicated causes of death in the longitudinal assessment of bariatric surgery. Obes Surg 2011;21:1687−92.

[15] Mason EE, Renquist KE, Huang YH, Jamal M, Samuel I. Causes of 30-day bariatric surgery mortality: with emphasis on bypass obstruction. Obes Surg 2007;17:9−14.

[16] Hatoum IJ, Blackstone R, Hunter TD, Francis DM, Steinbuch M, Harris JL, et al. Clinical factors associated with remission of obesity-related comorbidities after bariatric surgery. JAMA Surg 2016;151:130−7.

[17] Nguyen NT, Paya M, Stevens CM, Mavandadi S, Zainabadi K, Wilson SE. The relationship between hospital volume and outcome in bariatric surgery at academic medical centers. Ann Surg 2004;240:586−93. discussion 593-584.

[18] Livingston EH. Surgical volume impacts bariatric surgery mortality: a case for bariatric surgery centers of excellence. Surgery 2010;147:751−3.

[19] Maggard MA, Shugarman LR, Suttorp M, Maglione M, Sugerman HJ, Livingston EH, et al. Meta-analysis: surgical treatment of obesity. Ann Intern Med 2005;142:547−59.

[20] Lancaster RT, Hutter MM. Bands and bypasses: 30-day morbidity and mortality of bariatric surgical procedures as assessed by prospective, multi-center, risk-adjusted ACS-NSQIP data. Surg Endosc 2008;22:2554−63.

[21] Jones Jr. KB, Afram JD, Benotti PN, Capella RF, Cooper CG, Flanagan L, et al. Open versus laparoscopic Roux-en-Y gastric bypass: a comparative study of over 25,000 open cases and the major laparoscopic bariatric reported series. Obes Surg 2006;16:721−7.

[22] Nguyen NT, Masoomi H, Laugenour K, Sanaiha Y, Reavis KM, Mills SD, et al. Predictive factors of mortality in bariatric surgery: data from the Nationwide Inpatient Sample. Surgery 2011;150:347−51.

[23] Khan MA, Grinberg R, Johnson S, Afthinos JN, Gibbs KE. Perioperative risk factors for 30-day mortality after bariatric surgery: is functional status important? Surg Endosc 2013;27:1772−7.

[24] Carbonell AM, Lincourt AE, Matthews BD, Kercher KW, Sing RF, Heniford BT. National study of the effect of patient and hospital characteristics on bariatric surgery outcomes. Am Surg 2005;71:308−14.

[25] Livingston EH, Huerta S, Arthur D, Lee S, De Shields S, Heber D. Male gender is a predictor of morbidity and age a predictor of mortality for patients undergoing gastric bypass surgery. Ann Surg 2002;236:576−82.

[26] Mosko JD, Nguyen GC. Increased perioperative mortality following bariatric surgery among patients with cirrhosis. Clin Gastroenterol Hepatol 2011;9:897−901.

[27] Ballantyne GH, Svahn J, Capella RF, Capella JF, Schmidt HJ, Wasielewski A, et al. Predictors of prolonged hospital stay following open and laparoscopic gastric bypass for morbid obesity: body mass index, length of surgery, sleep apnea, asthma, and the metabolic syndrome. Obes Surg 2004;14:1042−50.

[28] Raftopoulos Y, Gatti GG, Luketich JD, Courcoulas AP. Advanced age and sex as predictors of adverse outcomes following gastric bypass surgery. JSLS 2005;9:272−6.

[29] Poulose BK, Griffin MR, Moore DE, Zhu Y, Smalley W, Richards WO, et al. Risk factors for post-operative mortality in bariatric surgery. J Surg Res 2005;127:1−7.

[30] DeMaria EJ, Portenier D, Wolfe L. Obesity surgery mortality risk score: proposal for a clinically useful score to predict mortality risk in patients undergoing gastric bypass. Surg Obes Relat Dis 2007;3:134−40.

[31] DeMaria EJ, Murr M, Byrne TK, Blackstone R, Grant JP, Budak A, et al. Validation of the obesity surgery mortality risk score in a multicenter study proves it stratifies mortality risk in patients undergoing gastric bypass for morbid obesity. Ann Surg 2007;246:578−82. discussion 583−574.

[32] Efthimiou E, Court O, Sampalis J, Christou N. Validation of Obesity Surgery Mortality Risk Score in patients undergoing gastric bypass in a Canadian center. Surg Obes Relat Dis 2009;5:643−7.

[33] Sarela AI, Dexter SP, McMahon MJ. Use of the obesity surgery mortality risk score to predict complications of laparoscopic bariatric surgery. Obes Surg 2011;21:1698−703.

[34] Longitudinal Assessment of Bariatric Surgery C, Flum DR, Belle SH, King WC, Wahed AS, Berk P, et al. Perioperative safety in the longitudinal assessment of bariatric surgery. N Engl J Med 2009;361:445−54.

[35] Turner PL, Saager L, Dalton J, Abd-Elsayed A, Roberman D, Melara P, et al. A nomogram for predicting surgical complications in bariatric surgery patients. Obes Surg 2011;21:655−62.

[36] Arterburn D, Livingston EH, Schifftner T, Kahwati LC, Henderson WG, Maciejewski ML. Predictors of long-term mortality after bariatric surgery performed in veterans affairs medical centers. Arch Surg 2009;144:914−20.

[37] Davidson LE, Adams TD, Kim J, Jones JL, Hashibe M, Taylor D, et al. Association of patient age at gastric bypass surgery with long-term all-cause and cause-specific mortality. JAMA Surg 2016. Available from: http://dx.doi.org/10.1001/jamasurg.2015.5501.

[38] Bruschi Kelles SM, Diniz MFHS, Machado CJ, Barreto SM. Mortality rate after open Roux-in-Y gastric bypass: a 10-year follow-up. Brazilian J Med Biol Res 2014;47:617−25.

[39] Telem DA, Talamini M, Shroyer AL, Yang J, Altieri M, Zhang Q, et al. Long-term mortality rates (>8-year) improve as compared to the general and obese population following bariatric surgery. Surg Endosc 2015;29:529−36.

[40] Goossens N, Hoshida Y, Song WM, Jung M, Morel P, Nakagawa S, et al. Non-alcoholic Steatohepatitis is Associated with Increased Mortality in Obese Patients Undergoing Bariatric Surgery. Clin Gastroenterol Hepatol 2016;14(11):1619−28.

[41] King LY, Canasto-Chibuque C, Johnson KB, Yip S, Chen X, Kojima K, et al. A genomic and clinical prognostic index for hepatitis C-related early-stage cirrhosis that predicts clinical deterioration. Gut 2015;64:1296−302.

[42] Sjöström L, Peltonen M, Jacobson P, Sjöström CD, Karason K, Wedel H, et al. Bariatric surgery and long-term cardiovascular events. Jama 2012;307:56−65.

[43] Sjöström L, Gummesson A, Sjöström CD, Narbro K, Peltonen M, Wedel H, et al. Effects of bariatric surgery on cancer incidence in obese patients in Sweden (Swedish Obese Subjects Study): a prospective, controlled intervention trial. Lancet Oncol 2009;10:653−62.

[44] Tee MC, Cao Y, Warnock GL, Hu FB, Chavarro JE. Effect of bariatric surgery on oncologic outcomes: a systematic review and meta-analysis. Surg Endosc 2013;27:4449−56.

[45] Aarts EO, Janssen J, Janssen IM, Berends FJ, Telting D, de Boer H. Preoperative fasting plasma C-peptide level may help to predict diabetes outcome after gastric bypass surgery. Obes Surg 2013;23:867−73.

[46] Robert M, Ferrand-Gaillard C, Disse E, Espalieu P, Simon C, Laville M, et al. Predictive factors of type 2 diabetes remission 1 year after bariatric surgery: impact of surgical techniques. Obes Surg 2013;23:770−5.

[47] Mingrone G, Panunzi S, De Gaetano A, Guidone C, Iaconelli A, Leccesi L, et al. Bariatric surgery versus conventional medical therapy for type 2 diabetes. N Engl J Med 2012;366:1577−85.

[48] Kim MK, Lee HC, Kwon HS, Baek KH, Kim EK, Lee KW, et al. Visceral obesity is a negative predictor of remission of diabetes 1 year after bariatric surgery. Obesity (Silver Spring) 2011;19:1835−9.

[49] Schauer DP, Arterburn DE, Livingston EH, Coleman KJ, Sidney S, Fisher D, et al. Impact of bariatric surgery on life expectancy in severely obese patients with diabetes: a decision analysis. Ann Surg 2015;261:914−19.

Chapter 25

Comparing Weight Loss in Three Bariatric Procedures: Roux-en-Y Gastric Bypass, Vertical Banded Gastroplasty, and Gastric Banding

M. Bekheit[1,2] and A.S. Elward[3]

[1]El Kabbary general Hospital, Alexandria, Egypt, [2]Aberdeen Royal Infirmary, Aberdeen, United Kingdom, [3]Cairo University, Giza, Egypt

LIST OF ABBREVIATIONS

BMI   body mass index
EW   excess weight
%EWL   percentage of excess weight loss
GB   gastric banding
LOESS   locally weighted scatterplot smoothing
RYGBP   Roux-en-Y gastric bypass
VBG   vertical banded gastroplasty

INTRODUCTION

Weight loss is a primary objective of bariatric surgery, and is the most commonly reported outcome measure [1]. Therefore, it constitutes one of the largest considerations in the decision-making when choosing an intervention. The percentage of excess weight loss (%EWL) is a commonly reported measure. Patients' responses to surgery are variable, and are influenced by many factors [2]; however, little is known about most of them.

MECHANISM OF ACTION OF THE THREE INTERVENTIONS IN VIEW OF WEIGHT LOSS

To an extent, the difference in response to surgery is related to the characteristics of each intervention. Roux-en-Y gastric bypass (RYGBP) is thought to act through combined restrictive and malabsorptive components. Restriction comes from the small gastric pouch that accommodates 25−50 mL, while malabsorption reduces the amount and the spectrum of the absorbed nutrients from bypassing a certain distance from the proximal intestine. Initially, the restrictive mechanism dominates, while the malabsorption helps to maintain the weight loss after the first 1−2 years [3].

In addition to the anatomical barriers, there is also a peptide component that contributes to the mechanism of action of RYGBP [4]. Recently, an influence of RYGBP on gut microbiota was identified as being responsible for the modifications and control of body mass, but this influence was not detected in patients who underwent vertical banded gastroplasty (VBG) [5].

VBG is performed by creating a vertical partitioning of the stomach, which induces weight loss through reduction of food intake and induction of early satiety [6]. Similarly, gastric banding (GB) creates a small gastric pouch just below the gastroesophageal junction through the application of a silicon ring around the gastroesophageal junction, which reduces food intake and induces early satiety. In addition, there might be a direct stimulation of the mechanore-ceptors responsible for signaling satiety to the central nervous system [3].

Beyond the inherited potency of each intervention, the response to the same intervention is variable. We previously demonstrated that GB should not be a first choice procedure in male patients since it is less effective in this group [7]. Even the potencies of intervention of the three methods of concern in this chapter were found to be influenced by non-technical factors. Coleman [8] investigated the effect of gender and ethnicity on the outcome of weight loss after RYGBP on 860 Americans with different ethnicities: non-Hispanic White 53.6%, Hispanic 25%, and non-Hispanic Black 21.4%, with females 81.7%. He found that women were more likely to have a higher percentage of initial weight loss when compared to men, and race/ethnicity was a strong predictor of weight outcome for men, but not for women 3−5 years after RYGBP.

COMPARISON OF WEIGHT LOSS AFTER RYGBP AND GASTRIC BANDING

It is a fairly consistent finding that RYGBP is a more potent intervention compared to GB. Angrisani reported in his 5- and 10-year follow-up a greater weight loss in the bypass group over the short and long term compared to GB [9,10]. In these reports, patients with gastric bands continued to lose weight in a steady fashion; however, this loss was universally lower than the loss induced by RYGBP.

Christou [11] reported a comparison between RYGBP and GB. The %EWL during the first year and at a 5-year follow-up was higher after RYGBP than GB. However, only 40 versus 10 patients were included at 5 years, respectively. Campos and coworkers [12] reported in a matched study that %EWL after RYGBP was higher than after gastric band at 1 year, with a higher percentage of patients achieving %EWL > 50% (93% vs 31%, respectively).

Spivak [13] reported his 10-year experience with a greater %EWL after RYGBP compared to GB, with a higher reoperation rate after GB compared to RYGBP (28% vs 0%, respectively). This report included 148 GB versus 175 RYGBP patients, but they lost almost the half of these patients over the follow-up period.

Romy [14] reported a 63 %EWL after RYGBP versus 50% after GB after 6 years of follow-up in 442 patients. The slope of %EWL was steeper after RYGBP than after GB, and the majority of patients in the RYGBP group achieved % EWL > 50% within the first 6 months.

Jan [15] reported a higher %EWL following RYGBP compared to those who underwent GB. Up until 2 years, this difference was significant, but not at the study endpoint of 3 years. Patients with body mass index (BMI) > 50 kg/m^2 had significantly higher weight loss than those with BMI < 50 kg/m^2 after RYGBP early during follow-up. However, this difference was not observed after GB.

Boza [16] reported a higher %EWL, at 1 year and 5 years following RYGBP than GB. Of note, patients in this study who underwent RYGBP had a slightly higher preoperative BMI compared to those who underwent GB.

COMPARISON OF WEIGHT LOSS AFTER ROUX-ᴇɴ-Y GASTRIC BYPASS AND VERTICAL BANDED GASTROPLASTY

An interesting study by Sugerman demonstrated that the weight loss after RYGBP is higher than VBG over the whole follow-up period [17]. The efficacy of VBG, but not RYGBP, is negatively influenced by changing the patient's dietary habits into those of a sweet eater. Subsequently, he identified a better weight loss for the selective assignment of non-sweet eaters to VBG [18]. Nonetheless, the weight loss after RYGBP was still higher than that induced by VBG.

The differences in weight loss in this study were not as large as have been reported in other studies, which in part could be attributed to the change in the dietary habits of patients after surgery (i.e., conversion to sweet eaters) [6]. Olber [19] described a higher %EWL at 1 and 2 years following RYGBP surgery as compared to VBG.

Goergen [20] reported no difference between the outcome of the two interventions at 2-year follow-up. It is important not to overlook the obvious methodological shortcomings in this study, which could greatly influence the confidence in the results. However, Zimmerman [21] also reported a comparable weight loss 1 year after both RYGBP and Silastic ring VBG.

COMPARISON OF WEIGHT LOSS AFTER VERTICAL BANDED GASTROPLASTY AND GASTRIC BAND

The studies comparing VBG to GB are conflicting, and it is not clear if there is an advantage in terms of weight loss for one intervention procedure over the other.

TABLE 25.1 Data From Studies Comparing the Weight Loss After Three Bariatric Procedures

Study	RYGBP (EWL%)	GB (EWL%)	VBG (EWL%)	Follow-Up (Years)
Angrisani [9,10]	51	35		1
	67	47		3
	67	48		5
	70	46		10
Christou [11]	70	43		1
	75	61		5
Campos [12]	64	36		1
Spivac [13]	76	46		10
Romy [14]	63	59		6
Jan [15]	49	25		0.5
	64	36		1
	70	45		2
	60	57		3
Boza [16]	93	59		5
Sugerman [17]	67		37	3
Olber [19]	78		63	1
	84		60	2
Goergen [20]	82		80	2
Belachew [22]		>50	>50	0.5
Miller [23]		42	58	1
		62	59	2
Ashy [26]		50	87	0.5
Morino [28]		41	64	2
		39	59	3
Scozzari [29]		42	61	3
		32	57	5
		30	53	7

Summary of the data extracted from studies comparing the weight loss after Roux-en-Y gastric bypass (RYGBP), vertical banded gastroplasty (VBG), and gastric banding (GB).

Belachew [22] reported an effective weight loss after both procedures with %EWL > 50% at 6 months. Weight loss was similar and nearly came to plateau after the third year in one of the longest follow-up studies comparing weight loss after VBG and GB over 12 years [23]. One of the earliest studies comparing VBG and GB is the one reported by Sutor [24]. The study demonstrated that the weight loss is not different at 18−24 months follow-up, only slower in the GB group. This was similar to the initial observation by Fox [25] at 15 months. Even at long term, Miller [23] found no difference in weight loss, but a lower reoperation rate was reported after GB.

On the other hand, Ashy reported that the weight loss after VBG was larger than that after GB [26], similar to what was reported in a random controlled trial with a 2-year follow-up [27]. Morino also reported that the weight loss after VBG was larger than after GB [28]. This difference continued to exist in an update of this study at 7 years [29]. It was demonstrated that patients with GB were at a higher susceptibility rate to reoperation. Folope [30] reported that the VBG was more effective in terms of weight loss and quality of life over 5 years.

Very few studies, such as the one reported by Nilsell [31], reported a higher weight loss following GB than following VBG at 5-year follow-up. This conflict could be explained by the greatly variable outcome reported in the single-arm GB studies. It is not understood why these results are heterogonous for GB, given that no known hormonal mechanism mediates the weight loss after GB [32].

A summary of the numerical data of these studies is presented in Table 25.1, and a multiple line graph with locally weighted smoothing smoothing is shown in Fig. 25.1 for trend visualization.

STUDIES COMPARING THE THREE PROCEDURES

In a study by Hell et al. [33], there was a higher weight loss after RYGBP compared to the other two procedures. They also suggested that patients with RYGBP had a better quality of life for a long time after surgery. We previously reported in a single-institution study that the weight loss after RYGBP was comparable to that after VBG, but

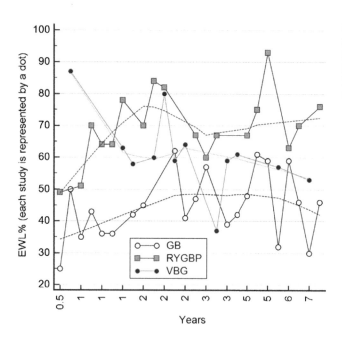

FIGURE 25.1 Multiple line graph demonstrating the pattern of weight loss in Roux-en-Y gastric bypass (RYGBP), vertical banded gastroplasty (VBG), and gastric banding (GB). The mean value of excess weight loss (%EWL) of each individual study is represented by a dot on a timeline. Dotted lines represent LOESS smoothing with 50% span trend analysis. The figure is from the corresponding author's own dataset.

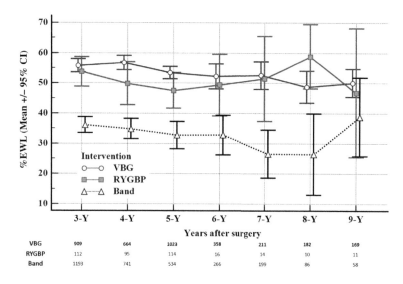

FIGURE 25.2 Multiple line graph demonstrating the pattern of weight loss after the three bariatric procedures (RYGBP, VBG, and GB) 3 years onwards following surgery. The figure is from the corresponding author's own dataset.

	3-Y	4-Y	5-Y	6-Y	7-Y	8-Y	9-Y
VBG	909	664	1023	358	211	182	169
RYGBP	112	95	114	16	14	10	11
Band	1193	741	534	266	199	86	58

significantly higher than GB [7]. In a similar, but multicenter study, we extended the data analysis to include 3515 patients [34].The study included patients with a minimum follow-up of 3 years after RYGBP, VBG, or GB.

The mean preoperative BMI and age were 44.5 ± 9.2 kg/m^2 and 38 ± 10.4 years. At the scheduled 3-year follow-up visit, only 2214 patients' records (62.98%) could be retrieved. Both RYGBP and VBG induced more weight loss than GB (Fig. 25.2). These results confirmed our previous findings where VBG and RYGBP produced an equal weight loss over a long follow-up period. They also demonstrated that GB is less potent in the long term in terms of weight loss when compared to the other two procedures.

INFLUENCE OF TECHNICAL VARIATION ON WEIGHT LOSS AFTER SURGERY

The inherited potency of each intervention is also an influential factor on the weight loss after surgery; the pattern of weight regain after each intervention is different as well. Therefore, there is a remarkable difference in response to each intervention, which could be due to intrinsic or extrinsic factors.

It appears that the restriction component of RYGBP plays an important role in weight loss. Pouch dilatation is thought to be responsible for the majority of weight regain after RYGBP [35]. In the study by Heneghan [36], positioning of the Silastic ring around the pouch led to a %EWL superior to the nonbanded groups at 2 years of follow-up (59% vs 51%, respectively). However, at 5 years of follow-up, Zarate et al. found no significant difference between the two techniques [37].

The larger the size of the gastro-jejunostomy, the greater the weight regain after RYGBP [38]. This was supported by the continued weight reduction after transoral reduction of the anastomosis size in patients with weight regain after RYGBP [39]. That is why many surgeons prefer the hand sewn fashioning of the anastomosis over the linear stapler since it gives more calibrated stoma compared to the linearly stapled one.

Likewise, the length of the different limbs might also be a factor in weight loss. In a primary RYGBP, superobese patients benefited from a long bypass, while patients with a BMI lower than 60 kg/m^2 lost an equivalent weight to those who had short RYGBP [40]. A wide range of variations was reported by different investigators regarding the length of the Roux limb (75−175 cm) and the biliopancreatic limb (35−60 cm), mainly based on the BMI at time of surgery. The efficiency of long-limb RYGBP in superobese patients was identified in a meta-analysis [41]. Furthermore, converting RYGBP to distal RYGBP was found to be an effective solution for inadequate weight loss after RYGBP [42].

Aiming to improve the durability of initial weight loss, or reducing postoperative morbidities related to RYGBP, different groups have published modifications of the RYGBP technique. The size of the gastric pouch ranges from 15−50 mL in most of the series. Hernandez-Martinez [43] used a pouch smaller than 25 mL, a fixed common limb of 230 cm, and the rest of the proximal small bowel redistributed among the Roux limb (60%) and the biliopancreatic limb (40%) in 75 patients; these proportions were reversed to 40% for the Roux limb and 60% for the biliopancreatic limb in another 75 patients. No differences in %EWL or in metabolic outcomes between the two groups of patients were found, with 69 %EWL after 8 years.

The perigastric positioning of a gastric band is obsolete, and the technical variations in modern gastric bands have little impact on weight loss. A truncal vagotomy in conjunction with GB reduced the number of adjustment requirements of the band after positioning without influence on weight loss when compared to GB alone [44]. This was attributed to the prolonged satiety in patients with truncal vagotomy. The same technique was found very useful in conjunction with VBG, where the weight loss at >5 years was twice that of gastroplasty without truncal vagotomy [45]. However, it appears that there is an important variation in weight loss after GB. While Feigel-Guiller [46] reported a modest weight loss (33, 27, and 33 %EWL at 1 year, 3 years, and 10 years, respectively), Wang [47] reported an impressive weight loss in his series in the early period following surgery with modest numbers at longer follow-up (195, 134, 114, 95, and 42 %EWL at 1−5 years, respectively).

Since the early 1980s with the introduction of VBG by Mason [48], along with the evolution of the technique by Maclean [49], and finally the adoption of laparoscopy in the 1990s, VBG results have had ups and downs, and have been abandoned in some centers [50]. The major cause of weight regain after VBG is staple line disruption [51], and there have been various stapling technique modifications to Mason's original gastroplasty methods in an effort to overcome this factor. Toppino found that the four rows of stapled gastroplasty had a lower incidence of staple line disruption [52]. Other anatomical causes of failure, such as stomal and pouch dilatation, were analyzed in the study by van Wezenbeek [53]. Suter compared three different techniques of VBG and stated that VBG with a Silastic ring induced more weight loss compared to the Marlex and adjustable sphincter [54].

WEIGHT REGAIN AND REVISION

Surgeons are frequently asked about long-lasting intervention, or more importantly, a permanent one. However, it should be clear to patients during the preoperative counseling session that such procedures do not exist. Many studies have attempted to understand the reasons behind failure of weight loss interventions, whether they are primary or secondary failures. While weight loss has been found to be associated with nonmodifiable factors, weight regain has been found to be largely associated with modifiable factors [55]. It should be clear that weight regain is common after every bariatric procedure, and that all bariatric procedures are initially effective.

Weight regain after RYGBP has been managed by various solutions. In addition to what was mentioned above, GB was also a successful option to augment the weight loss after failed RYGBP [56]. On the other hand, RYGBP remains the mainstay to correct for the failure in weight loss after GB [57]. However, it has not been concluded whether this is the best possible option. After failed VBG, conversion to RYGBP was associated with restoration of weight loss [51]. Despite that, weight regain is a common reason for revision after VBG or GB; revision is also common due to complications. Severe eating difficulties are also a common indication for revision, particularly after VBG [51]. For these reasons, VBG has been abandoned in many active bariatric centers around the world. Likewise, GB seems to be on the way out.

MINI-DICTIONARY OF TERMS

- *Mechanoreceptors*: Specialized receptors that are induced by mechanical stimulation. One example is the satiety receptors present in the upper part of the digestive tract. When they become stimulated (with stretch due to fullness after feeding), they stimulate the central nervous system with satiety.
- *Weight regain*: A term used to express (usually to quantify) the fact that after weight loss, patients regained some weight. This might be from a few grams to a weight that is higher than the initial one.
- *Malabsorptive procedure*: A procedure that relies on the reduction of the absorption of nutrients from the digestive system.
- *Restrictive procedure*: A procedure that mainly depends on the restriction of food intake as a mechanism of action.
- *Gastro-jejunostomy*: Refers to the anastomosis between the gastric pouch (i.e., the newly fashioned part of the stomach receiving food) and the part of the small intestine (jejunum) that will carry the food to the common channel where the food meets the digestive biliopancreatic secretions.
- *Gastric banding*: The positioning of a silicon ring around the uppermost part of the stomach, just below the gastro-esophageal junction, to restrict food intake.
- *Perigastric positioning of gastric band*: Refers to the positioning of the gastric band after skeletonizing the gastric wall from all of the surrounding, normally adherent, fat.

KEY FACTS

- Bariatric surgery is more effective in inducing weight loss than nonsurgical measures.
- The efficacy of a procedure is mostly measured by the amount of weight loss it induces.
- Weight loss after each one of the various procedures is influenced by many unknown or uncontrolled variables, as well as the inherited potency of the procedure.
- Initially, most of the procedures are effective.
- Weight regain is extremely common in the long term after most procedures; however, it is more significant after certain procedures than others.

SUMMARY POINTS

- Weight loss after RYGBP is generally more than after VBG or GB.
- Weight regain is common after each of the three bariatric procedures discussed herein.
- Weight loss after RYGBP is consistent among series (similarly with VBG).
- Weight loss after GB is to a large extent variable and inconsistent.
- There is an influence of the technical variation on weight loss mainly after RYGBP.
- VBG has been nearly abandoned due to significant complications and weight regain rates.

REFERENCES

[1] Madura II JA, Dibaise JK. Quick fix or long-term cure? Pros and cons of bariatric surgery. F1000 medicine reports 2012;4:19.

[2] Manning S, Carter NC, Pucci A, Jones A, Elkalaawy M, Cheung WH, et al. Age- and sex-specific effects on weight loss outcomes in a comparison of sleeve gastrectomy and Roux-en-Y gastric bypass: a retrospective cohort study. BMC Obes 2014;1:12.

[3] Kral JG, Naslund E. Surgical treatment of obesity. Nat Clin Pract Endocrinol Metabol 2007;3(8):574—83.

[4] Kashyap SR, Daud S, Kelly KR, Gastaldelli A, Win H, Brethauer S, et al. Acute effects of gastric bypass versus gastric restrictive surgery on beta-cell function and insulinotropic hormones in severely obese patients with type 2 diabetes. Int J Obes (Lond) 2010;34(3):462—71.

[5] Tremaroli V, Karlsson F, Werling M, Stahlman M, Kovatcheva-Datchary P, Olbers T, et al. Roux-en-Y gastric bypass and vertical banded gastroplasty induce long-term changes on the human gut microbiome contributing to fat mass regulation. Cell Metabol 2015;22(2):228—38.

[6] Brolin RE, Robertson LB, Kenler HA, Cody RP. Weight loss and dietary intake after vertical banded gastroplasty and Roux-en-Y gastric bypass. Ann Surg 1994;220(6):782—90.

[7] Bekheit M, Katri K, Ashour MH, Sgromo B, Abou-ElNagah G, Abdel-Salam WN, et al. Gender influence on long-term weight loss after three bariatric procedures: gastric banding is less effective in males in a retrospective analysis. Surg Endosc 2014;28(8):2406—11.

[8] Coleman KJ, Brookey J. Gender and racial/ethnic background predict weight loss after Roux-en-Y gastric bypass independent of health and lifestyle behaviors. Obes Surg 2014;24(10):1729—36.

[9] Angrisani L, Cutolo PP, Formisano G, Nosso G, Vitolo G. Laparoscopic adjustable gastric banding versus Roux-en-Y gastric bypass: 10-year results of a prospective, randomized trial. Surg Obes Relat Dis Off J Am Soc Bariatric Surg 2013;9(3):405—13.

[10] Angrisani L, Lorenzo M, Borrelli V. Laparoscopic adjustable gastric banding versus Roux-en-Y gastric bypass: 5-year results of a prospective randomized trial. Surg Obes Relat Dis Off J Am Soc Bariatric Surg 2007;3(2):127—32, discussion 32—3.

[11] Christou N, Efthimiou E. Five-year outcomes of laparoscopic adjustable gastric banding and laparoscopic Roux-en-Y gastric bypass in a comprehensive bariatric surgery program in Canada. Can J Surg 2009;52(6):E249—58.

[12] Campos GM, Rabl C, Roll GR, Peeva S, Prado K, Smith J, et al. Better weight loss, resolution of diabetes, and quality of life for laparoscopic gastric bypass vs banding: results of a 2-cohort pair-matched study. Arch Surg 2011;146(2):149—55.

[13] Spivak H, Abdelmelek MF, Beltran OR, Ng AW, Kitahama S. Long-term outcomes of laparoscopic adjustable gastric banding and laparoscopic Roux-en-Y gastric bypass in the United States. Surg Endosc 2012;26(7):1909—19.

[14] Romy S, Donadini A, Giusti V, Suter M. Roux-en-Y gastric bypass vs gastric banding for morbid obesity: a case-matched study of 442 patients. Arch Surg 2012;147(5):460—6.

[15] Jan JC, Hong D, Pereira N, Patterson EJ. Laparoscopic adjustable gastric banding versus laparoscopic gastric bypass for morbid obesity: a single-institution comparison study of early results. J Gastrointest Surg Off J Soc Surg Aliment Tract 2005;9(1):30—9, discussion 40-1.

[16] Boza C, Gamboa C, Awruch D, Perez G, Escalona A, Ibanez L. Laparoscopic Roux-en-Y gastric bypass versus laparoscopic adjustable gastric banding: five years of follow-up. Surg Obes Relat Dis Off J Am Soc Bariatric Surg 2010;6(5):470—5.

[17] Sugerman HJ, Starkey JV, Birkenhauer R. A randomized prospective trial of gastric bypass versus vertical banded gastroplasty for morbid obesity and their effects on sweets versus non-sweets eaters. Ann Surg 1987;205(6):613—24.

[18] Sugerman HJ, Londrey GL, Kellum JM, Wolf L, Liszka T, Engle KM, et al. Weight loss with vertical banded gastroplasty and Roux-Y gastric bypass for morbid obesity with selective versus random assignment. Am J Surg 1989;157(1):93—102.

[19] Olbers T, Fagevik-Olsen M, Maleckas A, Lonroth H. Randomized clinical trial of laparoscopic Roux-en-Y gastric bypass versus laparoscopic vertical banded gastroplasty for obesity. Br J Surg 2005;92(5):557—62.

[20] Goergen M, Arapis K, Limgba A, Schiltz M, Lens V, Azagra JS. Laparoscopic Roux-en-Y gastric bypass versus laparoscopic vertical banded gastroplasty: results of a 2-year follow-up study. Surg Endosc 2007;21(4):659—64.

[21] Zimmerman VV, Campos CT, Buchwald H. Weight loss comparison of gastric bypass and Silastictrade mark ring vertical gastroplasty. Obes Surg 1992;2(1):47—9.

[22] Belachew M, Jacqet P, Lardinois F, Karler C. Vertical banded gastroplasty vs adjustable silicone gastric banding in the treatment of morbid obesity: a preliminary report. Obes Surg 1993;3(3):275—8.

[23] Miller K, Pump A, Hell E. Vertical banded gastroplasty versus adjustable gastric banding: prospective long-term follow-up study. Surg Obes Relat Dis Off J Am Soc Bariatric Surg 2007;3(1):84—90.

[24] Suter M, Giusti V, Heraief E, Jayet C, Jayet A. Early results of laparoscopic gastric banding compared with open vertical banded gastroplasty. Obes Surg 1999;9(4):374—80.

[25] Fox SR, Oh KH, Fox KM. Adjustable silicone gastric banding vs vertical banded gastroplasty: a comparison of early results. Obes Surg 1993;3(2):181—4.

[26] Ashy AR, Merdad AA. A prospective study comparing vertical banded gastroplasty versus laparoscopic adjustable gastric banding in the treatment of morbid and super-obesity. Int Surg 1998;83(2):108—10.

[27] van Dielen FM, Soeters PB, de Brauw LM, Greve JW. Laparoscopic adjustable gastric banding versus open vertical banded gastroplasty: a prospective randomized trial. Obes Surg 2005;15(9):1292—8.

[28] Morino M, Toppino M, Bonnet G, del Genio G. Laparoscopic adjustable silicone gastric banding versus vertical banded gastroplasty in morbidly obese patients: a prospective randomized controlled clinical trial. Ann Surg 2003;238(6):835—41, discussion 41—2.

[29] Scozzari G, Farinella E, Bonnet G, Toppino M, Morino M. Laparoscopic adjustable silicone gastric banding vs laparoscopic vertical banded gastroplasty in morbidly obese patients: long-term results of a prospective randomized controlled clinical trial. Obes Surg 2009;19(8):1108—15.

[30] Folope V, Hellot MF, Kuhn JM, Teniere P, Scotte M, Dechelotte P. Weight loss and quality of life after bariatric surgery: a study of 200 patients after vertical gastroplasty or adjustable gastric banding. Eur J Clin Nutr 2008;62(8):1022—30.

[31] Nilsell K, Thorne A, Sjostedt S, Apelman J, Pettersson N. Prospective randomised comparison of adjustable gastric banding and vertical banded gastroplasty for morbid obesity. Eur J Surg Acta Chirurgica 2001;167(7):504—9.

[32] Pedersen SD. The role of hormonal factors in weight loss and recidivism after bariatric surgery. Gastroenterol Res Pract 2013;2013, 528450.

[33] Hell E, Miller KA, Moorehead MK, Norman S. Evaluation of health status and quality of life after bariatric surgery: comparison of standard Roux-en-Y gastric bypass, vertical banded gastroplasty and laparoscopic adjustable silicone gastric banding. Obes Surg 2000;10(3):214—19.

[34] Bekheit M. Long term outcome of bariatric surgery. University Archives: University of Alexandria; 2012.

[35] Nguyen D, Dip F, Huaco JA, Moon R, Ahmad H, LoMenzo E, et al. Outcomes of revisional treatment modalities in non-complicated Roux-en-Y gastric bypass patients with weight regain. Obes Surg 2015;25(5):928—34.

[36] Heneghan HM, Annaberdyev S, Eldar S, Rogula T, Brethauer S, Schauer P. Banded Roux-en-Y gastric bypass for the treatment of morbid obesity. Surg Obes Relat Dis Off J Am Soc Bariatric Surg 2014;10(2):210—16.

[37] Zarate X, Arceo-Olaiz R, Montalvo Hernandez J, Garcia-Garcia E, Pablo Pantoja J, Herrera MF. Long-term results of a randomized trial comparing banded versus standard laparoscopic Roux-en-Y gastric bypass. Surg Obes Relat Dis Off J Am Soc Bariatric Surg 2013;9(3):395—7.

[38] Langer FB, Prager G, Poglitsch M, Kefurt R, Shakeri-Leidenmuhler S, Ludvik B, et al. Weight loss and weight regain-5-year follow-up for circular- vs. linear-stapled gastrojejunostomy in laparoscopic Roux-en-Y gastric bypass. Obes Surg 2013;23(6):776—81.

[39] Kumar N, Thompson CC. Transoral outlet reduction for weight regain after gastric bypass: long-term follow-up. Gastrointest Endosc 2015;83(4):776—9.

[40] MacLean L, Rhode B, Nohr C. Long-or short-limb gastric bypass? J Gastrointest Surg 2001;5(5):525—30.

[41] Orci L, Chilcott M, Huber O. Short versus long Roux-limb length in Roux-en-Y gastric bypass surgery for the treatment of morbid and super obesity: a systematic review of the literature. Obes Surg 2011;21(6):797−804.

[42] Dapri G, Cadiere GB, Himpens J. Laparoscopic conversion of Roux-en-Y gastric bypass to distal gastric bypass for weight regain. J Laparoendosc Adv Surg Tech Part A 2011;21(1):19−23.

[43] Hernandez-Martinez J, Calvo-Ros MA. Gastric by-pass with fixed 230-cm-long common limb and variable alimentary and biliopancreatic limbs in morbid obesity. Obes Surg 2011;21(12):1879−86.

[44] Angrisani L, Cutolo PP, Ciciriello MB, Vitolo G, Persico F, Lorenzo M, et al. Laparoscopic adjustable gastric banding with truncal vagotomy versus laparoscopic adjustable gastric banding alone: interim results of a prospective randomized trial. Surg Obes Relat Dis Off J Am Soc Bariatric Surg 2009;5(4):435−8.

[45] Kral JG, Gortz L, Hermansson G, Wallin GS. Gastroplasty for obesity: long-term weight loss improved by vagotomy. World J Surg 1993;17(1):75−8, discussion 9.

[46] Feigel-Guiller B, Drui D, Dimet J, Zair Y, Le Bras M, Fuertes-Zamorano N, et al. Laparoscopic gastric banding in obese patients with sleep apnea: a 3-year controlled study and follow-up after 10 years. Obes Surg 2015;25(10):1886−92.

[47] Wang X, Zheng CZ, Chang XS, Zhao X, Yin K. Laparoscopic adjustable gastric banding: a report of 228 cases. Gastroenterol Rep 2013;1(2):144−8.

[48] Mason EE. Vertical banded gastroplasty for obesity. Arch Surg 1982;117(5):701−6.

[49] MacLean LD, Rhode BM, Forse RA. A gastroplasty that avoids stapling in continuity. Surgery 1993;113(4):380−8.

[50] van Wezenbeek MR, Smulders JF, de Zoete JPJGM, Luyer MD, van Montfort GSW. Long-term results of primary vertical banded gastroplasty. Obes Surg 2014;25(8):1425−30.

[51] Schouten R, van Dielen FM, van Gemert WG, Greve JW. Conversion of vertical banded gastroplasty to Roux-en-Y gastric bypass results in restoration of the positive effect on weight loss and co-morbidities: evaluation of 101 patients. Obes Surg 2007;17(5):622−30.

[52] Toppino M, Nigra II, Olivieri F, Muratore A, Bosio CA, Avagnina S, et al. Staple-line disruptions in vertical banded gastroplasty related to different stapling techniques. Obes Surg 1994;4(3):256−61.

[53] van Wezenbeek MR, Smulders JF, de Zoete JP, Luyer MD, van Montfort G, Nienhuijs SW. Long-term results of primary vertical banded gastroplasty. Obes Surg 2015;25(8):1425−30.

[54] Suter M, Jayet C, Jayet A. Vertical banded gastroplasty: long-term results comparing three different techniques. Obes Surg 2000;10(1):41−6, discussion 7.

[55] Yanos BR, Saules KK, Schuh LM, Sogg S. Predictors of lowest weight and long-term weight regain among Roux-en-Y gastric bypass patients. Obes Surg 2015;25(8):1364−70.

[56] Gobble RM, Parikh MS, Greives MR, Ren CJ, Fielding GA. Gastric banding as a salvage procedure for patients with weight loss failure after Roux-en-Y gastric bypass. Surg Endosc 2008;22(4):1019−22.

[57] Gagner M, Gumbs AA. Gastric banding: conversion to sleeve, bypass, or DS. Surg Endosc 2007;21(11):1931−5.

Chapter 26

Metabolic Predictors of Weight Loss After Bariatric Surgery

G. Faria[1,2]

[1]Hospital de Santo António, Porto, Portugal, [2]Instituto de Ciências Biomédicas de Abel Salazar, Porto, Portugal

LIST OF ABBREVIATIONS

BMI	body mass index
%EWL	percentage of excess weight loss
FBG	fasting blood glucose
GLP-1	glucagon-like peptide 1
HbA1c	glycosylated hemoglobin
POPs	persistent organic pollutants
PYY	peptide YY
RDW	red cell distribution width
RYGB	Roux-en-Y gastric bypass
TWL	total weight loss
T2DM	type 2 diabetes mellitus

INTRODUCTION

Obesity and metabolic syndrome are epidemics of the 21st century that affect more than 1.500 million people worldwide. Bariatric surgery is the only effective treatment that allows for long-term weight loss and a significant reduction in overall mortality [1]. Although the number of weight loss operations for the treatment of obesity has increased several fold all over the world during the last 2 decades, only 1% of the eligible population is being treated with surgery [2].

Following Roux-en-Y gastric bypass (RYGB), up to 18% of patients fail to achieve a body mass index (BMI) $<35 \text{ kg/m}^2$, and unsuccessful weight loss has been reported in 10−30% of patients [3]. Although excess weight loss might not be the most important outcome of surgery, it is highly correlated with the improvement in risk factors and patient satisfaction [4]. Identifying preoperative predictors of weight loss may be essential to managing patients' expectations and to identifying patients who might benefit from tailor-made surgical approaches and closer follow-up [3,5].

Given a low perioperative morbidity, weight control, resolution of comorbid conditions, and increased quality of life are the key factors to defining a successful bariatric operation. Although there is significant metabolic improvement after a modest weight loss, greater and longer-term weight control is required to achieve all of the long-term benefits associated with RYGB [6].

PREDICTORS OF WEIGHT LOSS AFTER BARIATRIC SURGERY

Several factors such as type of surgery and variations in individual techniques [7,8], gender [9,10], age [11], socioeconomic status [12,13], and psychological factors [14] have been associated with variations in weight loss. The reasons for individual differences have not been fully understood. A 2014 cohort using more than 70,000 patients from the bariatric outcomes longitudinal database concluded that most of the variability seen after bariatric surgery was due to the type of surgery, and one-third of the variation was yet to be explained [15].

One of the first major reviews about predictors of weight loss, back in 1993 [16], concluded that lower weight and younger age were related to increased success after surgery, but the high variability in weight loss was still not explained. Although preoperative weight is the strongest predictor of success after bariatric surgery, it accounts for only one-third of the variance in weight loss. While patients with higher BMIs lose more absolute weight, they still remain obese, and lose a smaller percentage of excess weight or excess BMI.

The first problem when examining weight loss success lies in the metrics selected for the evaluation. While the most-used measures are percentage of excess weight loss (%EWL) and percentage excess BMI loss (%EBMIL), these values are largely dependent upon initial weight or BMI (heavier patients have more excess weight, and despite a greater absolute weight loss, a lower %EWL). Because of this, several authors have proposed the use of percentage total weight loss (%TWL) as a better metric because it is not so dependent on the initial BMI [17]. For most of this chapter, %EWL is used since it is the most widely reported and most-studies report multivariable analysis corrected for initial weight.

Obesity has been considered a "psychological disease," and many authors have tried to identify psychological characteristics that could predict the postoperative change in weight. Several variables have been studied with different methodologies and combinations in each study. This might explain why there has been no consistent replication of associations between psychological variables and weight loss. Despite depression scores and several measures of motivation, anxiety, persistency, and social support having been linked to changes in weight loss [18−21], these results are not consistent, and most systematic reviews fail to reach any significant conclusion.

In 2001, Dixon proposed that following gastric banding, initial weight and age were the strongest predictors of excess weight loss. However, for the first time, measures of insulin resistance (higher insulin levels, HOMA-IR, and diabetes) were associated with impaired weight loss. On the contrary, psychological factors, gender, and the global burden of disease were not associated with variations in weight loss. This was the first large-volume study trying to address physiological predictors of weight loss after bariatric surgery [22]. In 2007, Chevallier [23] concluded that besides initial weight and age, the surgical team experience was also a predictive factor for weight loss following gastric banding. This could be explained by several factors, from patient selection to multidisciplinary care. For malabsorptive procedures, initial fat mass, age, and diabetes were inversely related to weight loss [24].

PREDICTORS OF WEIGHT LOSS AFTER ROUX-EN-Y GASTRIC BYPASS

Gastric bypass is the gold standard and most widely performed bariatric surgery worldwide. The first laparoscopic gastric bypass was performed by Alan Wittgrove [25] in 1994, and gave rise to the exponential growth of bariatric and metabolic surgery. It is estimated that in 2011 more than 340,000 procedures were performed worldwide [26].

In terms of technical variations, both gastric pouch size [7,27] and limb lengths [8] have been associated with differences in weight loss, although the results have not been consistent. Gastric pouch and stoma sizes contribute to the restrictive effect of the surgery. However, while pouch size was inversely correlated with weight loss [27], stoma size had no influence on medium-term weight loss [28]. The fact that pouch sizes tend to be bigger in superobese and male patients might help explain the empirical observation that male patients lose less weight [7].

Regarding patients' individual characteristics, the same results have regularly been reported: after correction for initial BMI, both age and diabetes remain predictors of reduced weight loss [3,11]. Using artificial neural network methodology [29], the following factors were associated with the 1-year %EBMIL: gender (female), race (black), preoperative BMI, high blood pressure, and the presence of diabetes.

Total body fat and lean mass have been reported as more accurate predictors of weight loss after gastric bypass surgery than BMI alone, and might predict the evolution of body weight at each given time after surgery [30]. Some authors also associated triglyceride levels with %EWL because higher triglyceride levels were related to increased success [3,12].

AGE AND WEIGHT LOSS

Several reports have linked increased age to decreased weight loss [3,11,22,23]. However, a study of older patients (>55 years) concluded that their weight loss was reasonable (64 %EWL at 12 months), and more importantly, most patients (>80%) had complete control of comorbidities [31].

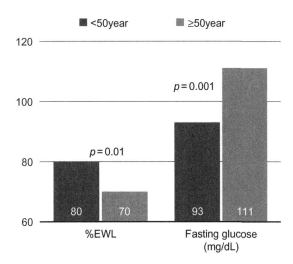

FIGURE 26.1 Association between age, weight loss, and fasting glucose levels. The bars represent the %EWL 12 months after surgery. Younger patients (*blue* (*dark gray* in print versions) bar) had both increased weight loss and lower fasting blood glucose (FBG) levels. *Data from Faria G, Pestana D, Preto J, Guimarães JT, Taveira-Gomes A, Calhau C. Age and weight loss after bariatric surgery: cause or consequence? Obes Surg 2014;24(5):824 [36].*

Several hypotheses [3,10,11] have been proposed to explain the inverse relation between age and weight loss. On one hand, aging reduces lypolitic capacity. This helps explain the increased adipose tissue deposition of older people and the decreased capacity to mobilize energy from lipids in fat stores. Thus, in order to maintain the "available energy" balance, older people tended to compensate with a larger caloric intake. This was confirmed by the fact that after RYGB young patients had a greater reduction in caloric intake than older patients, which in itself might have led to increased weight loss.

On the other hand, energy requirements usually decrease with age. After the age of 40, total energy consumption declines mainly due to a reduction in physical activity. Hatoum et al. also concluded that exercise capacity (not exercise itself or chronological age) was an independent predictor of weight loss [32]. Thus, younger patients can achieve better long-term results since they take on a more active lifestyle, have greater exercise capacity, and a greater reduction in energy intake.

Additionally, since obesity is also a psychological and social disease, some psychological traits and social determinants associated with internal motivation, social support, and desirability, might be related both to a patient's age and postoperative weight loss.

Finally, older age has been correlated with higher prevalence (up to five times) and increased severity of type 2 diabetes mellitus (T2DM) [33,34] and higher levels of glycosylated hemoglobin (HbA1c) [34,35].

Results published by our group [36–38] concluded that patients younger than 50 years achieved a greater weight loss after 12 months of follow-up (Fig. 26.1). More interestingly, in the preoperative period younger patients had significantly lower fasting glucose levels and were less contaminated with persistent organic pollutants (POPs) in visceral adipose tissue ($p < 0.0001$). Both higher fasting blood glucose (FBG) levels and higher POPs levels in adipose tissue are associated with decreased lipolysis [39,40]. This might provide the biological explanation behind the association between increased age and decreased weight loss (Fig. 26.2).

GLUCOSE METABOLISM AND WEIGHT LOSS

Several studies have associated T2DM with decreased weight loss after RYGB [3,5,9,11,27,29,32,33,41]. T2DM is related to insulin resistance and to higher circulating insulin levels. Insulin is an anabolic hormone that promotes triglyceride storage, stimulating lipogenesis and inhibiting lipolysis [42].

The UK Prospective Diabetes Study (UKPDS) concluded that intensive treatment for T2DM increased circulating levels of insulin and caused a small but significant weight gain. However, this increase in weight was associated with lower microvascular complications, thus reinforcing the notion that weight is not the sole predictor of disease [43]. Similarly, higher circulating levels of insulin might be responsible for impaired weight loss after RYGB [27,44].

These data suggest that it is not T2DM that is related to impaired weight loss, but the state of insulin resistance and poor metabolic control. In fact, several reports [3,12,35] have concluded that poor long-term glycemic control

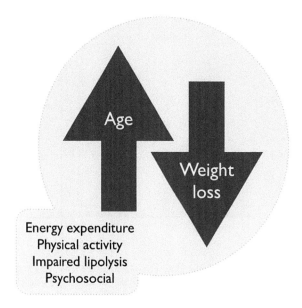

FIGURE 26.2 Putative mechanisms of association between age and weight loss.

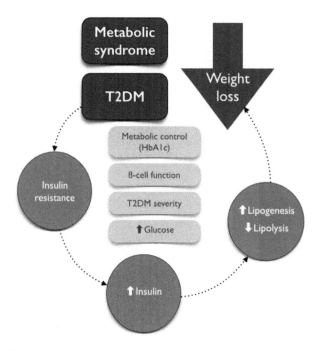

FIGURE 26.3 Putative explanation of the relation between measures of insulin resistance and impaired weight loss.

(measured by HbA1c) is related to less weight loss after RYGB. Other authors have proposed that the degree of beta-cell function and the severity of diabetes might also be related to the amount of weight loss [33,45] (Fig. 26.3).

In a 2012 study by Ortega et al., patients with diabetes had a lower %EWL (67% vs 77%; $p < 0.001$). Furthermore, when data from the oral glucose tolerance test (Fig. 26.4) was taken into account, individuals with normal glucose tolerance lost more weight than those with glucose intolerance ($p = 0.0005$) and diabetes ($p < 0.0001$). In multivariable analysis, initial BMI, age, triglyceride level, waist circumference, and HbA1c were predictors of weight loss. Whenever HbA1c and T2DM were put together in the same model, only HbA1c was independently associated with %EWL [3].

We have also reported that a simple measure of FBG could predict changes in weight loss [37]. FBG might be a surrogate marker of the "metabolic burden of disease"; higher FBG levels were associated with decreased weight loss. At the same time, FBG levels were inversely related to weight loss and directly related to increasing age and prevalence of

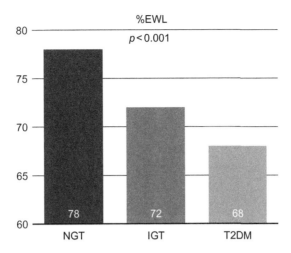

FIGURE 26.4 Percentage of excess weight loss 12 months after surgery. The bars represent %EWL 12 months after surgery (both gastric sleeve and gastric bypass combined), according to glucose tolerance class. *NGT* = normal glucose tolerance; *IGT* = impaired glucose tolerance; *T2DM* = type 2 diabetes mellitus. Data according to Ortega [3]. Our own data concluded that %EWL 12 months after gastric bypass was 84% for patients with NGT, 75% for patients with IGT, and 57% for patients with T2DM ($p < 0.05$).

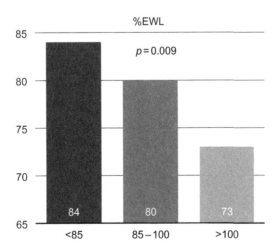

FIGURE 26.5 Percentage of excess weight loss according to fasting blood glucose (FBG) levels. Percentage of excess weight loss (%EWL) 12 months after gastric bypass decreased with increased levels of FBG. Even for normal range values (<100 mg/dL), lower values were related to increased weight loss. $p = 0.009$. *Adapted from Faria G, Preto J, Almeida AB, Guimarães JT, Calhau C, Taveira-Gomes A. Fasting glycemia: a good predictor of weight loss after RYGB. Surg Obes Relat Dis 2014;10(3):419–24.*

metabolic syndrome. Patients with T2DM were older (46 vs 36; $p < 0.001$) and had a comparable BMI at baseline, but had a lower %EWL (71.1% vs 80.4%; $p = 0.01$). However, in multivariable analysis, only initial BMI and FBG levels (and not age or diabetes status) were related to lower weight loss.

Dividing FBG into three categories (Fig. 26.5), we observed a dose–response effect, with decreasing weight loss for increasing FBG class. Worthy of mention is the fact that these FBG categories fall within the "normal range" for blood glucose measurement, thus reinforcing the idea that the metabolic disease associated with insulin resistance is a continuum and that lower glucose values lead to increased weight loss even within the normal range.

One appealing hypothesis would be that if FBG is related to weight loss, we could improve surgical results by lowering preoperative glucose levels. Most interestingly, for patients being treated with antidiabetic medications, there was no relation between FBG and weight loss at 12 months (Fig. 26.6).

Patients with higher FBG are expected to produce more insulin in order to maintain homeostasis. This environment of hyperinsulinism might be related to a resistance to weight loss, as insulin promotes lipogenesis and inhibits lipolysis [42]. This might explain why several clinical manifestations of glucose homeostasis (T2DM, insulin resistance, HbA1c, and FBG) have been associated with variations in postoperative weight loss (Table 26.1).

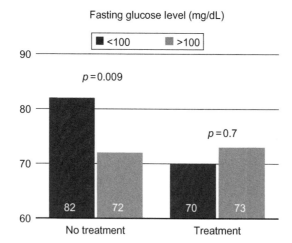

Fasting glucose level (mg/dL)

■ <100   ▪ >100

FIGURE 26.6 Percentage of excess weight loss (%EWL) according to glycemic control and hypoglycemic treatment status. The %EWL 12 months after RYGB is lower for patients under hypoglycemic treatment, and there is no difference according to the actual glucose levels. On untreated patients, lower glucose values were related with higher weight loss. *From Faria G, Preto J, Almeida AB, Guimarães JT, Calhau C, Taveira-Gomes A. Fasting glycemia: a good predictor of weight loss after RYGB. Surg Obes Relat Dis 2014;10(3):419–24.*

TABLE 26.1 Metabolic Predictors of Weight Loss

Initial BMI/Weight
Total Body Fat
T2DM & Severity
   Insulin Levels/Insulin Resistance
   HbA1c
   B-cell Function
   Fasting Glucose
Age
Race
Gender
Triglyceride
Waist Circumference

The table shows all the physiological characteristics reportedly associated with weight loss. Note that factors associated with severity of metabolic disease have been consistently reported in relation to impaired weight loss. Operating earlier in the course of metabolic disease seems to lead to improved results.

However, patients with the worst FBG had a significant improvement in metabolic control, such that 89% of patients had metabolic syndrome before RYGB and only 22% ($p < 0.001$) fulfilled the criteria 12 months after surgery [37]. Hence, it is necessary to emphasize that weight loss is not the sole (nor the most important) result of metabolic surgery. Since patients who lost less weight seemed to benefit most [3,6,46], even with modest weight loss, the final question seems to be which health indexes to use to determine which patients benefit the most from bariatric surgery.

OTHER PREDICTORS OF WEIGHT LOSS

Increased individual caloric intake capacity (ad libitum nutrient drink until reaching the sensation of fullness) before surgery is associated with increased baseline BMI and with postoperative weight loss [47]. Although individual anatomical differences can explain different intake capacities before surgery, standardization of the surgical procedure should render postoperative caloric intake more homogeneous. However, even if significantly decreased after surgery (896 kcal less during the nutrient drink test 1 year after surgery; $p < 0.001$), there was still a significant interindividual caloric intake variation. This might suggest that satiety depends on individual anatomical characteristics, but also on a complex interplay between several hormones produced by the digestive system. These hormones that lead to decreased satiation are still in effect after surgery, which leads to increased intake and impaired weight loss.

Digestive physiology begins even before any food is eaten; during this cephalic phase several enteric hormones are released [48]. Also, gastric bypass surgery can change the patterns of neural activation and reduce the appetite, food-related reward, and total food ingestion [49]. Several hormones interplay in this entero—cephalic axis, and are ultimately responsible for human appetite, food intake, energy homeostasis, and metabolism [50]. Some of the most studied have been ghrelin, PYY, and glucagon-like peptide 1 (GLP-1).

Ghrelin is the most potent orexigenic (appetite increaser) hormone, and is primarily produced in the gastric fundus [51]. Ghrelin levels increase immediately before meals and decrease rapidly after feeding [52]. Ghrelin levels are inversely related to BMI and body fat, thus, ghrelin levels are usually lower in obese patients. Also in these patients, postprandial reduction is impaired [53]. Both these factors explain why it is so difficult to lose weight by diet: on one hand, the ghrelin levels increase with decreasing BMI (thus increasing hunger), and on the other hand, there is an impaired reduction in satiety, since ghrelin does not adequately decrease with food ingestion.

Peptide YY (PYY) is an enterokine that is secreted in the bowel by L-cells (mainly on the terminal ileum) in response to the presence of nutrients in the lumen of the bowel [54]. PYY inhibits gastrointestinal motility and reduces pancreatic and intestinal secretions. These intestinal effects promote a slowing of the nutrient flow through the bowel and constitute a mechanism described as the "ileal brake" [50]. PYY is effective in suppressing appetite and food consumption, both in obese and lean subjects [55].

GLP-1 is probably the most important incretin in the physiology of obesity and weight loss. It has several functions, including stimulation of insulin secretion and inhibition of glucagon, and reduction of hunger and food intake [56]. GLP-1 is secreted by L-cells (mainly in the ileum) in response to the presence of nutrients in the bowel lumen, suppressing appetite and enhancing the "ileal brake" mechanism [57]. In addition to this effect, GLP-1 is a key factor in the rapid improvement in diabetes after RYGB: the faster arrival of nutrients to the GLP-1—producing cells of the distal bowel limits nutrient intake, increases insulin production, and improves insulin sensitivity [50,57].

The fact that faster gastric pouch emptying is related to increased weight loss [58] emphasizes the metabolic effects of RYGB: faster gastric emptying leads to increased delivery of undigested food distally and greater anorexigenic hormonal response (increased PYY and GLP-1). Thus, patients who are able to produce neurohormonal responses (younger patients and nondiabetics) might be more receptive to weight loss.

Since enteropeptide production is dynamic and highly variable according to feeding and fasting, it is difficult to study. In a study with a 400 kcal test meal, preoperative GLP-1 and PYY responses were not predictors of postoperative weight loss [59]. Similarly, the baseline ghrelin profile was also not correlated with weight loss [60]. Our own data suggests that fasting GLP-1 levels were inversely correlated with 12 months %EWL ($p = 0.02$) [61]. These results need further research and validation, but suggest that GLP-1 levels might be sensitive to the "metabolic disease burden."

A recently published report concluded that the red cell distribution width (RDW) was predictive of excess BMI lost at 1 year, and that higher RDW values were related to lower weight loss in a dose—response effect [62]. The report speculates that RDW might work as a surrogate marker of inflammation. While obesity is directly related to increased RDW, RDW has not been associated with increased obesity-related inflammation; neither inflammation has been prognostic of weight loss. However, due to the strong association reported, this needs further evaluation and validation.

If metabolic characteristics are related to postoperative weight loss, there should be a genetic determinant related to weight loss. While no such gene or genes have been found, Hatoum et al. [63] concluded that first-degree relatives had a similar response to surgery; weight loss between nonrelated cohabitants was not correlated. Also, the presence of a specific single nucleotide polymorphism (SNP, rs4771122) has been associated with increased weight loss, which underscores the existence of specific genetic mechanisms (yet to be understood) driving the metabolic changes after surgery [64].

MINI-DICTIONARY OF TERMS

- *Metabolic surgery*: Surgical procedures to treat metabolic diseases, primarily T2DM.
- *Roux-en-Y gastric bypass*: The most common metabolic surgery. Involves the creation of a small gastric pouch with a calibrated gastro-jejunal anastomosis, a biliary limb of 50—70 cm, and a Roux limb of 100—150 cm.
- *Predictor*: A variable associated with outcome, and whose variation predicts the occurrence of a defined outcome.
- *Insulin*: An anabolic hormone produced by the pancreas that promotes the absorption of glucose from the bloodstream and its conversion into triglycerides or glycogen. Insulin increases triglyceride storage, stimulating lipogenesis and inhibiting lipolysis.
- *Insulin resistance*: A pathological state in which higher concentrations of insulin are required to achieve glucose homeostasis.

- *Percentage of excess weight loss*: Reflects the percentage of weight reduction from excess weight. The formula: (Weight Lost)/(Excess Weight) × 100. There are many formulas to calculate excess weight. It can easily be calculated as the weight above the weight corresponding to a BMI of 25: (Excess Weight) = ((Current Weight) − (Weight of BMI of 25)).

KEY FACTS

- Insulin resistance is an increase in the amount of insulin required to dispose of a determined amount of glucose.
- Insulin resistance leads to an increase in circulating insulin levels.
- Insulin is a hormone that helps the body to remove glucose from the bloodstream and store it inside the cells.
- Insulin increases lipogenesis (fat storage) and decreases lipolysis (fat degradation).
- Several measures of glucose metabolism associated with insulin resistance have been shown to reduce postoperative weight loss.
- Enteropeptides are peptides produced in the digestive tract that, among other roles, are related to appetite control and energetic balance.
- Some of the most-studied enteropeptides are ghrelin, GLP-1, and PYY.
- Ghrelin is a hormone that increases hunger and tends to be inversely related to BMI, which means that a BMI decrease will increase ghrelin production, and hence hunger.
- PYY is a hormone that decreases hunger and slows the nutrient flow through the digestive tract.
- GLP-1 is mainly produced in the ileum and stimulates insulin production, reducing hunger and food intake.
- Due to their dynamic production according to food intake and energy balance, enteropeptide variations are difficult to study.

SUMMARY POINTS

- Obesity is a major epidemic of the 21st century, and bariatric surgery is the only effective treatment that reduces mortality and increases quality of life.
- The objectives of weight loss surgery are weight control and resolution of metabolic disease and comorbidities; only 1% of eligible patients are being treated with weight loss surgery.
- Initial BMI is the strongest predictor of weight loss: higher BMI patients lose more weight but a lower percentage of excess weight.
- Older patients lose less weight. Increased age is associated with reduced lipolytic activity and reduced energy expenditure, and is also associated with increased prevalence and severity of metabolic disease.
- The severity of metabolic disease and a hyperinsulinemic state are related to impaired weight loss. However, patients with metabolic disease benefit from controlling the underlying diseases.
- FBG is related to weight loss. Lower glucose values are associated with increased weight loss, even in the range of normal values.
- Appetite control and energy balance result from a complex interplay of hormones and mediators, and might be related both to severity of obesity and to surgical outcomes.
- Digestive physiology is complex, and studying the effects of surgery is helping to understand some of the fundamental mechanisms of satiety, food intake, and energy balance.

REFERENCES

[1] Sjöström L, Lindroos A-K, Peltonen M, et al. Lifestyle, diabetes, and cardiovascular risk factors 10 years after bariatric surgery. N Engl J Med 2004;351(26):2683–93.

[2] Buchwald H, Estok R, Fahrbach K, Banel D, Sledge I. Trends in mortality in bariatric surgery: a systematic review and meta-analysis. Surgery 2007;142(4):621–35.

[3] Ortega E, Morínigo R, Flores L, et al. Predictive factors of excess body weight loss 1 year after laparoscopic bariatric surgery. Surg Endosc 2012;26(6):1744–50.

[4] Dumon KR, Murayama KM. Bariatric surgery outcomes. Surg Clin North Am 2011;91(6):1313–38, −x.

[5] Júnior WS, do Amaral JL, Nonino-Borges CB. Factors related to weight loss up to 4 years after bariatric surgery. Obes Surg 2011;21 (11):1724–30.

[6] Buchwald H, Avidor Y, Braunwald E, et al. Bariatric surgery: a systematic review and meta-analysis. JAMA 2004;292(14):1724–37.

[7] Roberts K, Duffy A, Kaufman J, Burrell M, Dziura J, Bell R. Size matters: gastric pouch size correlates with weight loss after laparoscopic Roux-en-Y gastric bypass. Surg Endosc 2007;21(8):1397−402.

[8] Ciovica R, Takata M, Vittinghoff E, et al. The impact of roux limb length on weight loss after gastric bypass. Obes Surg 2008;18(1):5−10.

[9] Melton GB, Steele KE, Schweitzer MA, Lidor AO, Magnuson TH. Suboptimal weight loss after gastric bypass surgery: correlation of demographics, comorbidities, and insurance status with outcomes. J Gastrointest Surg 2007;12(2):250−5.

[10] Scozzari G, Passera R, Benvenga R, Toppino M, Morino M. Age as a long-term prognostic factor in bariatric surgery. Ann Surg 2012;256(5): 724−9.

[11] Ma Y, Pagoto SL, Olendzki BC, et al. Predictors of weight status following laparoscopic gastric bypass. Obes Surg 2006;16(9):1227−31.

[12] Lee Y-C, Lee W-J, Lee T-S, et al. Prediction of successful weight reduction after bariatric surgery by data mining technologies. Obes Surg 2007;17(9):1235−41.

[13] Júnior WS, Nonino-Borges CB. Clinical predictors of different grades of nonalcoholic fatty liver disease. Obes Surg 2012;22(2):248−52.

[14] Ray EC, Nickels MW, Sayeed S, Sax HC. Predicting success after gastric bypass: the role of psychosocial and behavioral factors. Surgery 2003;134(4):555−63.

[15] Benoit SC, Hunter TD, Francis DM, De La Cruz-Munoz N. Use of bariatric outcomes longitudinal database (BOLD) to study variability in patient success after bariatric surgery. Obes Surg 2014;24(6):936−43.

[16] Vallis M, Ross M. The role of psychological factors in bariatric surgery for morbid obesity: identification of psychological predictors of success. Obes Surg 1993;3(4):346−59.

[17] Hatoum IJ, Kaplan LM. Advantages of percent weight loss as a method of reporting weight loss after Roux-en-Y gastric bypass. Obesity 2013;21(8):1519−25.

[18] Lutfi R, Torquati A, Sekhar N, Richards WO. Predictors of success after laparoscopic gastric bypass: a multivariate analysis of socioeconomic factors. Surg Endosc 2006;20(6):864−7.

[19] van Hout GCM, Verschure SKM, van Heck GL. Psychosocial predictors of success following bariatric surgery. Obes Surg 2005;15(4):552−60.

[20] De Panfilis C, Generali I, Dall'Aglio E, Marchesi F, Ossola P, Marchesi C. Temperament and one-year outcome of gastric bypass for severe obesity. Surg Obes Relat Dis 2014;10(1):144−8.

[21] Averbukh Y, Heshka S, El-Shoreya H, et al. Depression score predicts weight loss following Roux-en-Y gastric bypass. Obes Surg 2003;13(6): 833−6.

[22] Dixon JB, Dixon ME, O'Brien PE. Pre-operative predictors of weight loss at 1-year after lap-band® surgery. Obes Surg 2001;11(2):200−7.

[23] Chevallier J-M, Paita M, Rodde-Dunet M-H, et al. Predictive factors of outcome after gastric banding: a nationwide survey on the role of center activity and patients' behavior. Ann Surg 2007;246(6):1034−9.

[24] Valera-Mora ME, Simeoni B, Gagliardi L, et al. Predictors of weight loss and reversal of comorbidities in malabsorptive bariatric surgery. Am J Clin Nutr 2005;81(6):1292−7.

[25] Wittgrove A, Clark G, Tremblay L. Laparoscopic gastric bypass, Roux-en-Y: preliminary report of five cases. Obes Surg 1994;4(4):353−7.

[26] Buchwald H, Oien DM. Metabolic/bariatric surgery worldwide 2011. Obes Surg 2013;23(4):427−36.

[27] Campos GM, Rabl C, Mulligan K, et al. Factors associated with weight loss after gastric bypass. Arch Surg 2008;143(9):877−83.

[28] Cottam DR, Fisher B, Sridhar V, Atkinson J, Dallal R. The effect of stoma size on weight loss after laparoscopic gastric bypass surgery: results of a blinded randomized controlled trial. Obes Surg 2008;19(1):13−17.

[29] Wise ES, Hocking KM, Kavic SM. Prediction of excess weight loss after laparoscopic Roux-en-Y gastric bypass: data from an artificial neural network. Surg Endosc 2016;30(2):480−8.

[30] Livingston EH, Sebastian JL, Huerta S, Yip I, Heber D. Biexponential model for predicting weight loss after gastric surgery for obesity. J Surg Res 2001;101(2):216−24.

[31] Papasavas PK, Gagné DJ, Kelly J, Caushaj PF. Laparoscopic Roux-en-Y gastric bypass is a safe and effective operation for the treatment of morbid obesity in patients older than 55 years. Obes Surg 2004;14(8):1056−61.

[32] Hatoum IJ, Stein HK, Merrifield BF, Kaplan LM. Capacity for physical activity predicts weight loss after Roux-en-Y gastric bypass. Obesity 2009;17(1):92−9.

[33] Carbonell AM, Wolfe LG, Meador JG, Sugerman HJ, Kellum JM, Maher JW. Does diabetes affect weight loss after gastric bypass? Surg Obes Relat Dis 2008;4(3):441−4.

[34] Contreras JE, Santander C, Court I, Bravo J. Correlation between age and weight loss after bariatric surgery. Obes Surg 2013;23(8):1286−9.

[35] Perna M, Romagnuolo J, Morgan K, Byrne TK, Baker M. Preoperative hemoglobin A1c and postoperative glucose control in outcomes after gastric bypass for obesity. Surg Obes Relat Dis 2011;8(6):685−90.

[36] Faria G, Pestana D, Preto J, Guimarães JT, Taveira-Gomes A, Calhau C. Age and weight loss after bariatric surgery: cause or consequence? Obes Surg 2014;24(5):824.

[37] Faria G, Preto J, Almeida AB, Guimarães JT, Calhau C, Taveira-Gomes A. Fasting glycemia: a good predictor of weight loss after RYGB. Surg Obes Relat Dis 2014;10(3):419−24.

[38] Pestana D, Faria G, Sá C, et al. Persistent organic pollutant levels in human visceral and subcutaneous adipose tissue in obese individuals— Depot differences and dysmetabolism implications. Environ Res 2014;133(C):170−7.

[39] Mclaughlin T, Yee G, Glassford A, Lamendola C, Reaven G. Use of a two-stage insulin infusion study to assess the relationship between insulin suppression of lipolysis and insulin-mediated glucose uptake in overweight/obese, nondiabetic women. Metab Clin Exp 2011;60(12):1741−7.

[40] Irigaray P, Ogier V, Jacquenet S, et al. Benzo[a]pyrene impairs beta-adrenergic stimulation of adipose tissue lipolysis and causes weight gain in mice. A novel molecular mechanism of toxicity for a common food pollutant. FEBS J 2006;273(7):1362−72.

[41] Perugini RA, Mason R, Czerniach DR, et al. Predictors of complication and suboptimal weight loss after laparoscopic Roux-en-Y gastric bypass: a series of 188 patients. Arch Surg 2003;138(5):541–5.

[42] Kahn BB, Flier JS. Obesity and insulin resistance. J Clin Invest 2000;106(4):473–81.

[43] Intensive blood-glucose control with sulphonylureas or insulin compared with conventional treatment and risk of complications in patients with type 2 diabetes (UKPDS 33). UK Prospective Diabetes Study (UKPDS) Group. Lancet 1998;352(9131):837–53.

[44] Flier J, Maratos-Flier E. Energy homeostasis and body weight. Curr Biol 2000;10(6):R215–17.

[45] Nannipieri M, Mari A, Anselmino M, et al. The role of beta-cell function and insulin sensitivity in the remission of type 2 diabetes after gastric bypass surgery. J Clin Endocrinol Metab 2011;96(9):E1372–9.

[46] Yip K, Heinberg L, Giegerich V, Schauer PR, Kashyap SR. Equivalent weight loss with marked metabolic benefit observed in a matched cohort with and without type 2 diabetes 12 months following gastric bypass surgery. Obes Surg 2012;22(11):1723–9.

[47] Gras-Miralles B, Haya JR, Moros JMR, et al. Caloric intake capacity as measured by a standard nutrient drink test helps to predict weight loss after bariatric surgery. Obes Surg 2014;24(12):2138–44.

[48] Power ML, Schulkin J. Anticipatory physiological regulation in feeding biology: cephalic phase responses. Appetite 2008;50(2–3):194–206.

[49] Ochner CN, Kwok Y, Conceição E, et al. Selective reduction in neural responses to high calorie foods following gastric bypass surgery. Ann Surg 2011;253(3):502–7.

[50] Park CW, Torquati A. Physiology of weight loss surgery. Surg Clin North Am 2011;91(6):1149–61, –vii.

[51] Higgins SC, Gueorguiev M, Korbonits M. Ghrelin, the peripheral hunger hormone. Ann Med 2007;39(2):116–36.

[52] Cummings DE. Ghrelin and the short- and long-term regulation of appetite and body weight. Physiol Behav 2006;89(1):71–84.

[53] le Roux CW, Welbourn R, Werling M, et al. Gut hormones as mediators of appetite and weight loss after Roux-en-Y gastric bypass. Ann Surg 2007;246(5):780–5.

[54] Ballantyne GH. Peptide YY(1-36) and peptide YY(3-36): Part II. Changes after gastrointestinal surgery and bariatric surgery. Obes Surg 2006;16(6):795–803.

[55] Sloth B, Holst JJ, Flint A, Gregersen NT, Astrup A. Effects of PYY1-36 and PYY3-36 on appetite, energy intake, energy expenditure, glucose and fat metabolism in obese and lean subjects. Am J Physiol Endocrinol Metab 2007;292(4):E1062–8.

[56] Efendic S, Portwood N. Overview of incretin hormones. Horm Metab Res 2004;36(11–12):742–6.

[57] Thomas S, Schauer P. Bariatric surgery and the gut hormone response. Nutr Clin Pract 2010;25(2):175–82.

[58] Akkary E, Sidani S, Boonsiri J, et al. The paradox of the pouch: prompt emptying predicts improved weight loss after laparoscopic Roux-Y gastric bypass. Surg Endosc 2009;23(4):790–4.

[59] Werling M, Fändriks L, Royce VP, Cross GF, le Roux CW, Olbers T. Preoperative assessment of gut hormones does not correlate to weight loss after Roux-en-Y gastric bypass surgery. Surg Obes Relat Dis 2014;10(5):822–8.

[60] Pellitero S, Pérez-Romero N, Martínez E, et al. Baseline circulating ghrelin does not predict weight regain neither maintenance of weight loss after gastric bypass at long term. Am J Surg 2015;210(2):340–4.

[61] Faria GR, Preto J, Guimarães J, Calhau C, Taveira-Gomes A. Fasting pre-operative Glp-1 levels predict 12 months weight loss in morbid obesity after Rygb. Obes Surg 2014;24(8): 1151–1151.

[62] Wise ES, Hocking KM, Weltz A, et al. Red cell distribution width is a novel biomarker that predicts excess body-mass index loss 1 year after laparoscopic Roux-en-Y gastric bypass. Surg Endosc 2016;1–6.

[63] Hatoum IJ, Greenawalt DM, Cotsapas C, Reitman ML, Daly MJ, Kaplan LM. Heritability of the weight loss response to gastric bypass surgery. J Clin Endocrinol Metab 2011;96(10):E1630–3.

[64] Rasmussen-Torvik LJ, Baldridge AS, Pacheco JA, et al. rs4771122 Predicts Multiple measures of long-term weight loss after bariatric surgery. Obes Surg 2015;25(11):2225–9.

Chapter 27

Long-Term Weight Loss Results After Laparoscopic Sleeve Gastrectomy

K.G. Apostolou
Laiko General Hospital, Athens, Greece

LIST OF ABBREVIATIONS

BMI	body mass index
EBMIL	excess body mass index loss
EWL	excess weight loss
F	French
LSG	laparoscopic sleeve gastrectomy
NR	not reported
TWL	total weight loss

INTRODUCTION

The safety and weight loss effectiveness of laparoscopic sleeve gastrectomy (LSG) is well-documented in both the short- and mid-term. Nowadays, LSG is often used as a stand-alone procedure. A recent meta-analysis of 2570 patients also showed a resolution of common comorbid conditions, apart from a good percentage of excess weight loss (%EWL) [1].

Considering that "durable" weight loss is one of the most important benefits of bariatric surgery, data regarding the short- and mid-term weight loss results after LSG is more than convincing. However, the load of data regarding the long-term weight loss effectiveness of LSG is rather small. Thus, a review of the existing literature would be of benefit, with the aim of evaluating the effectiveness of this popular bariatric procedure in the long-term follow-up.

A detailed search of PubMed and Medline for citations that include the term "sleeve gastrectomy," by using the keywords "sleeve gastrectomy long-term weight loss results," revealed more than 150 studies. A review of all these studies yielded a total of 44 [2–45] that fulfilled our criteria; these were included here for further analysis. The inclusion criteria were as follows: prospective and retrospective studies reporting on LSG as a primary or staged procedure that contained weight loss data at 5 or more years after LSG. Case reports, studies with fewer than five patients reaching the 5-year follow-up, as well as studies reporting on technique only, were not included. Substudies of larger series by the same group were not included in our analysis in order to avoid duplication of data.

Of the 44 studies extracted for the present review, 7 studies [2,25,35,36,38,42,44] were prospectively designed, while the remaining 37 were retrospective. Thirty-eight were from a single institution, whereas the remaining 6 studies [13,16,21,30,33,43] were multiinstitutional. Regarding the origin of the studies, 20 were from Europe, 11 were from Asia, 9 were from the United States, 2 were from Latin America, 1 was from Canada, and the remaining study originated from Asia and Africa. Of the 44 studies, 4 of them [2,8,11,19] stated clearly that LSG was used as part of a staged procedure, 37 stated that LSG was used as a stand-alone procedure, while in the remaining 3 studies [6,24,27], LSG was used either as part of a staged or as a stand-alone procedure in patients with a lower preoperative body mass index (BMI) (Table 27.1).

TABLE 27.1 Long-Term Weight Loss Outcomes of Laparoscopic Sleeve Gastrectomy

Author	Number of Patients Reaching the Follow-up (n)	Stand-Alone or Staged Procedure	Mean Preoperative BMI (kg/m²)	Bougie Size (F)	Months of Follow-Up (Follow-Up Rate %)	Mean % Excess Weight Loss (%EWL)	% Perioperative Complication Rate (30 Days)	% Postoperative Mortality (30 Days)
Weiner et al. [2]	8	Staged procedure	61.6	No calibration tube, 44 or 32	60 (NR)	40 %EBMIL	17.5%	0%
Bohdjalian et al. [3]	21	Stand-alone procedure	48.2 ± 1.3	48	60 (96.15%)	55%	NR	NR
Himpens et al. [4]	30	Stand-alone procedure	39	34	72 (78%)	53.3% ± 28.3%	12.2%	0%
D'Hondt et al. [5]	27 and 23	Stand-alone procedure	39.3	30	60 (32.5%) and 72 (27.7%)	71.3% and 55.9%	0%	0%
Strain et al. [6]	23	Stand-alone or Staged procedure	56.1 ± 14.0	40 or 60	60 (NR)	48 %EBMIL	NR	NR
Braghetto et al. [7]	60	Stand-alone procedure	38.4 ± 5.1	32 or 40	60 (11%)	57.3%	5.7%	0%
Eid et al. [8]	19, 13, and 21	Staged procedure	66 ± 7	50	72, 84, and 96 (93%)	52%, 43%, and 46%	15%	0%
Saif et al. [9]	30	Stand-alone procedure	52.2 ± 10.2	NR	60 (NR)	48 %EBMIL	NR	NR
Sarela et al. [10]	13	Stand-alone procedure	45.9	32	≥96 (65%)	69%	5%	0%
Abbatini et al. [11]	13	Staged procedure	52.1 ± 8.5	48	60 (54.1%)	56%	NR	NR
Brethauer et al. [12]	23	Stand-alone procedure	50.7 ± 10.6	NR	60 (79%)	49.5%	NR	0%
Catheline et al. [13]	45	Stand-alone procedure	49.1 ± 8.5	34	60 (82%)	50.7%	5.7%	0%
Rawlins et al. [14]	49	Stand-alone procedure	65	26.4 (endoscope used as a bougie)	60 (100%)	86%	1.9%	0%
Zachariah et al. [15]	6	Stand-alone procedure	37.4 ± 4.75	36	60 (30%)	63.7%	4.4%	0.43%
Abd Ellatif et al. [16]	859, 731, and 519	Stand-alone procedure	46 ± 9	≤36 and ≥44	60 (62%), 72 (53%), and 84 (37%)	61%, 59%, and 57%	5.1% Leaks = 0.78%	0%
Boza et al. [17]	112 and 59	Stand-alone procedure	34.9	60	60 (70%) and 72 (37%)	62.9% and 64%	3.7% Leaks = 0.6%	0%
Lim et al. [18]	14	Stand-alone procedure	40.2	32	60 (93.3%)	57.4%	NR	NR

Study	N	Procedure type	Age	Bougie size	Follow-up	Weight loss	Complications	Mortality
Marceau et al. [19]	9	Staged procedure	50.4 ± 10.8	NR	60 (21.4%)	12% ± 34%	NR		0%
Musella et al. [20]	102	Stand-alone procedure	47.9	38	60 (58.1%)	68.1%	3.3%		0.2%
Noel et al. [21]	132	Stand-alone procedure	43.8	37.5	60 (57.6%)	71.3%	4% Leaks = 1.47%	0.15%
Prevot et al. [22]	84	Stand-alone procedure	47.7 ± 7	34	60 (81%)	43% ± 25%	9.9% Leaks = 3.3%	0%
Sieber et al. [23]	54	Stand-alone procedure	43 ± 8	35	60 (91%)	57.4 %EBMIL	4.4%	0%
Van Rutte et al. [24]	19	Stand-alone or Staged procedure	44.3	34	60 (1.8%)	58.3 ± 29.1%	7.4% Leaks = 2.3%	0%
Zhang et al. [25]	26	Stand-alone procedure	38.5 ± 4.2	34	60 (81.3%)	63.2% ± 24.5%	12.5%	0%
Abelson et al. [26]	435	Stand-alone procedure	50.2 ± 10.1	40	60 and 72 (100%)	25.4 and 24.3 %TWL	11% Leaks = 1.1%	0%
Alexandrou et al. [27]	25	Stand-alone or Staged procedure	55.5 ± 1.7	29 (endoscope used as a bougie)	60 (100%)	56.4% ± 5.8%	0%	0%
Alvarenga et al. [28]	132 and 81	Stand-alone procedure	43.4 ± 5.8	38 and 42 to 48	60 (13%) and 96 (8%)	61% ± 11% and 52% ± 9.2%	63% Leaks = 0.1%	0%
Angrisani et al. [29]	99	Stand-alone procedure	47.7	40	60 (94%)	57.2%	2.9%	0%
Dogan et al. [30]	27	Stand-alone procedure	45.8 ± 6	34	60 (11%)	62.5% ± 23.8%	9% Leaks = 1.6%	NR
Golomb et al. [31]	39	Stand-alone procedure	42.6 ± 6.5	32–40	60 (69.6%)	56.1%	NR Leaks = 1.3% to 3.1%	0%
Hirth et al. [32]	14	Stand-alone procedure	43.5	32	84 (87.5%)	59.6% ± 19.9%	12.5% Leaks = 6.2%	0%
Hong et al. [33]	27	Stand-alone procedure	32.4 ± 1.6	36 and 48	60 (42.9%)	78.5% ± 30.8 %EBMIL	5.3%	0%
Lee et al. [34]	116	Stand-alone procedure	37.5 ± 6.1	36	60 (75.3%)	68.7% ± 30.3%	7.4% Leaks = 1.2%	NR
Lemanu et al. [35]	55	Stand-alone procedure	50.7	38	60 (57.3%)	40%	NR	NR
Liu et al. [36]	44	Stand-alone procedure	41 ± 7	<40: 58.6% (n = 82) ≥40: 41.4% (n = 58)	60 (84.6%)	57.2% ± 30.4%	3.6% Leaks = 1.4%	0%
Casella et al. [37]	126, 120, and 37	Stand-alone procedure	45.8 ± 7.4	48	60, 72, and 84 (85.4%,	70.2%, 67.3%, and 65.7%	5.4% Leaks = 2.7%	0%

(Continued)

TABLE 27.1 (Continued)

Author	Number of Patients Reaching the Follow-up (n)	Stand-Alone or Staged Procedure	Mean Preoperative BMI (kg/m²)	Bougie Size (F)	Months of Follow-Up (Follow-Up Rate %)	Mean % Excess Weight Loss (%EWL)	% Perioperative Complication Rate (30 Days)	% Postoperative Mortality (30 Days)
					81%, and 25%)			
Del Genio et al. [38]	54	Stand-alone procedure	51.3 ± 11.6	40	60 (100%)	56%	0%	NR
Keren et al. [39]	123	Stand-alone procedure	44.3	NR	60 (100%)	45.3% ± 19.5%	NR	NR
Lin et al. [40]	33	Stand-alone procedure	41.39 ± 6.53	NR	60 (7.8%)	67%	7.8% Leaks = 0.7%	0.2%
Pekkarinen et al. [41]	84	Stand-alone procedure	47.4	NR	60 (94.4%)	35.3%	NR	NR
Perrone et al. [42]	161	Stand-alone procedure	47.4 ± 4.2	36	60 (99.4%)	78.8% ± 23.5 %EBMIL	NR	0%
Pok et al. [43]	61, 19, and 12	Stand-alone procedure	37.3 ± 8.1	36	60, 72, and 84 (51.2%, NR, NR)	71.6% ± 32.7%, 71% ± 15.7%, and 70% ± 14.2%	8.6% Leaks = 2.1%	0.13%
Ruiz-Tovar et al. [44]	50	Stand-alone procedure	46.9 ± 8.4	50	60 (94%)	78.7% ± 7.1%	4% Leaks = 2%	1.9%
Seki et al. [45]	18	Stand-alone procedure	43.3 ± 10	36 or 45	60 (59%)	77.3% ± 36%	8.9% Leaks = 2.8%	0%
Summary: 44 studies	5136 patients	Stand-alone procedure: 37 studies; Staged procedure: 4 studies; Stand-alone or Staged procedure: 3 studies	45.8 kg/m²	60F ≥ Bougie ≥ 26.4F	64.3%	59.8 %EWL at 5 or more years	Range = 0–63% Leak Rate = 1.26% (n = 144/11,451 patients)	0.14% (n = 7 patients)

The number of patients reaching the long-term follow-up points, as well the respective weight loss results. BMI, body mass index; F, French; EWL, excess weight loss; NR, not reported; EBMIL, excess body mass index loss; TWL, total weight loss.

FOLLOWING THE WEIGHT LOSS IN DIFFERENT LAPAROSCOPIC SLEEVE GASTRECTOMY STUDIES

The total number of patients who underwent LSG and reached a follow-up of at least 60 months after surgery was 5136. The follow-up rate at 60 or more months after LSG was reported in 41 of 44 studies, and ranged from 1.8% to 100%, with a mean value of 64.3%. The mean age of patients at the time of LSG was 38 years; 71.8% of the patients were women. The mean preoperative BMI in all 44 studies was 45.8 kg/m^2. The size of the bougie used for calibration of the gastric sleeve was reported in 38 studies, and ranged from 26.4F to 60F. However, in the studies published by Rawlins et al. [14] and Alexandrou et al. [27], a 26.4F and a 29F endoscope was used as a bougie, respectively (Table 27.1).

The weight loss results at 60 or more months after LSG were expressed as %EWL in 37 studies, while in the remaining 7 studies they were expressed either as percentage of excess BMI loss (%EBMIL) [2,6,9,23,33,42] or as percentage of total weight loss (%TWL) [26]. The %EWL was derived from the following equation: %EWL = [(Initial Body Weight−Current Body Weight)/(Initial Body Weight−Ideal Body Weight)]*100. The %EBMIL was calculated as follows: %EBMIL = [(Initial BMI−Current BMI)/(Initial BMI−Ideal BMI)]*100. Finally, the %TWL was calculated as follows: %TWL = [(Initial Body Weight−Current Body Weight)/(Initial Body Weight)]*100. The mean %EWL ranged from 12% [19] in the study by Marceau et al. to 86% in the study by Rawlins et al. [14] (Table 27.1).

EXCESS WEIGHT LOSS AS A VARIABLE

Considering the studies reporting the weight loss outcome as %EWL, the 60-month follow-up group consisted of 33 studies, which enrolled 2687 patients. The mean %EWL for this group was calculated to be 60.4% (Table 27.2). The 72-month follow-up group consisted of 7 studies, with a total of 1001 patients. The mean %EWL for this group was 60.1% (Table 27.2). As for the 84-month follow-up group, it consisted of 5 studies, with a total of 595 patients. The mean %EWL for this group was 57.6% (Table 27.2). Finally, the ≥ 96-month follow-up group included 3 studies that enrolled 115 patients, with a mean %EWL of 52.8% (Table 27.2). The calculated long-term mean %EWL, defined as the mean %EWL at 60 or more months after surgery, was 59.8% (Table 27.1); its evolution is depicted in Fig. 27.1.

TABLE 27.2 Long-Term Evolution of %EWL After Laparoscopic Sleeve Gastrectomy

Author	Patients (n)	5 years %EWL	6 years %EWL	7 years %EWL	≥ 8 years %EWL
Bohdjalian et al. [3]	21	55	–	–	–
Himpens et al. [4]	30	–	53.3	–	–
D'Hondt et al. [5]	27 and 23 patients (5- and 6-year follow-up, respectively)	71.3	55.9	–	–
Braghetto et al. [7]	60	57.3	–	–	–
Eid et al. [8]	19, 13, and 21 patients (6, 7, and ≥ 8-year follow-up, respectively)	–	52	43	46
Sarela et al. [10]	13	–	–	–	69
Abbatini et al. [11]	13	56	–	–	–
Brethauer et al. [12]	23	49.5	–	–	–
Catheline et al. [13]	45	50.7	–	–	–
Rawlins et al. [14]	49	86	–	–	–
Zachariah et al. [15]	6	63.7	–	–	–
Abd Ellatif et al. [16]	859, 731, and 519 patients (5-, 6-, and 7-year follow-up, respectively)	61	59	57	–

(Continued)

TABLE 27.2 (Continued)

Author	Patients (n)	5 years %EWL	6 years %EWL	7 years %EWL	≥ 8 years %EWL
Boza et al. [17]	112 and 59 patients (5- and 6-year follow-up, respectively)	62.9	64	–	–
Lim et al. [18]	14	57.4	–	–	–
Marceau et al. [19]	9	12	–	–	–
Musella et al. [20]	102	68.1	–	–	–
Noel et al. [21]	132	71.3	–	–	–
Prevot et al. [22]	84	43	–	–	–
Van Rutte et al. [24]	19	58.3	–	–	–
Zhang et al. [25]	26	63.2	–	–	–
Alexandrou et al. [27]	25	56.4	–	–	–
Alvarenga et al. [28]	132 and 81 patients (5- and ≥ 8-year follow-up, respectively)	61	–	–	52
Angrisani et al. [29]	99	57.2	–	–	–
Dogan et al. [30]	27	62.5	–	–	–
Golomb et al. [31]	39	56.1	–	–	–
Hirth et al. [32]	14	–	–	59.6	–
Lee et al. [34]	116	68.7	–	–	–
Lemanu et al. [35]	55	40	–	–	–
Liu et al. [36]	44	57.2	–	–	–
Casella et al. [37]	126, 120, and 37 patients (5-, 6-, and 7-year follow-up, respectively)	70.2	67.3	65.7	–
Del Genio et al. [38]	54	56	–	–	–
Keren et al. [39]	123	45.3	–	–	–
Lin et al. [40]	33	67	–	–	–
Pekkarinen et al. [41]	84	35.3	–	–	–
Pok et al. [43]	61, 19, and 12 patients (5-, 6-, and 7-year follow-up, respectively)	71.6	71	70	–
Ruiz-Tovar et al. [44]	50	78.7	–	–	–
Seki et al. [45]	18	77.3	–	–	–
Summary of 37 studies	4398 patients	60.4% (33 studies, n = 2687 patients)	60.1% (7 studies, n = 1001 patients)	57.6% (5 studies, n = 595 patients)	52.8% (3 studies, n = 115 patients)

Summarizes the number of patients, as well as the respective weight loss result, at each long-term follow-up point. EWL, excess weight loss.

LSG was used both as a stand-alone and as a staged procedure. Regarding the weight loss result at 60 or more months after LSG in both groups, the stand-alone group consisted of 5053 patients, with a mean preoperative BMI of 45.6 kg/m^2. The long-term weight loss outcome for this group was 60.5%, 60.5%, 57.9%, and 54.4% at 60, 72, 84, and $\geq$ 96 months after LSG, respectively (Table 27.3). The staged-approach group included 150 patients, with a mean

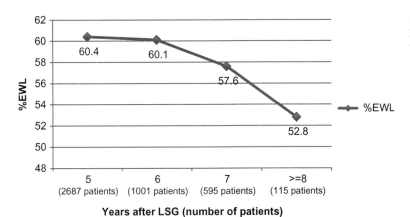

FIGURE 27.1 Mean %EWL at different follow-up points. Depicts the evolution of the weight loss outcome during the entire long-term follow-up period.

TABLE 27.3 Long-Term Weight Loss Outcomes of Laparoscopic Sleeve Gastrectomy as a Stand-Alone Operation

Author	Patients (n)	Mean Preoperative BMI (kg/m^2)	5 years %EWL	6 years %EWL	7 years %EWL	$\geq$ 8 years %EWL
Bohdjalian et al. [3]	21	48.2	55	–	–	–
Himpens et al. [4]	30	39	53.3	–	–	–
D'Hondt et al. [5]	27 and 23 patients (5- and 6-year follow-up, respectively)	39.3	71.3	55.9	–	–
Strain et al. [6]	23	56.1	48 %EBMIL	–	–	–
Braghetto et al. [7]	60	38.4	57.3	–	–	–
Saif et al. [9]	30	52.2	48 %EBMIL	–	–	–
Sarela et al. [10]	13	45.9	–	–	–	69
Brethauer et al. [12]	23	50.7	49.5	–	–	–
Catheline et al. [13]	45	49.1	50.7	–	–	–
Rawlins et al. [14]	49	65	86	–	–	–
Zachariah et al. [15]	6	37.4	63.7	–	–	–
Abd Ellatif et al. [16]	859, 731, and 519 patients (5-, 6-, and 7-year follow-up, respectively)	46	61	59	57	–
Boza et al. [17]	112 and 59 patients (5- and 6-year follow-up, respectively)	34.9	62.9	64	–	–
Lim et al. [18]	14	40.2	57.4	–	–	–

(Continued)

TABLE 27.3 (Continued)

Author	Patients (n)	Mean Preoperative BMI (kg/m²)	5 years %EWL	6 years %EWL	7 years %EWL	≥ 8 years %EWL
Musella et al. [20]	102	47.9	68.1	–	–	–
Noel et al. [21]	132	43.8	71.3	–	–	–
Prevot et al. [22]	84	47.7	43	–	–	–
Sieber et al. [23]	54	43	57.4 %EBMIL	–	–	–
Van Rutte et al. [24]	19	44.3	58.3	–	–	–
Zhang et al. [25]	26	38.5	63.2	–	–	–
Abelson et al. [26]	435	50.2	25.4 %TWL	24.3 %TWL	–	–
Alexandrou et al. [27]	25	55.5	56.4	–	–	–
Alvarenga et al. [28]	132 and 81 patients (5- and 8-year follow-up, respectively)	43.4	61	–	–	52
Angrisani et al. [29]	99	47.7	57.2	–	–	–
Dogan et al. [30]	27	45.8	62.5	–	–	–
Golomb et al. [31]	39	42.6	56.1	–	–	–
Hirth et al. [32]	14	43.5	–	–	59.6	–
Hong et al. [33]	27	32.4	78.5 %EBMIL	–	–	–
Lee et al. [34]	116	37.5	68.7	–	–	–
Lemanu et al. [35]	55	50.7	40	–	–	–
Liu et al. [36]	44	41	57.2	–	–	–
Casella et al. [37]	126, 120, and 37 (5-, 6-, and 7-year follow-up, respectively)	45.8	70.2	67.3	65.7	–
Del Genio et al. [38]	54	51.3	56	–	–	–
Keren et al. [39]	123	44.3	45.3	–	–	–
Lin et al. [40]	33	41.4	67	–	–	–
Pekkarinen et al. [41]	84	47.4	35.3	–	–	–
Perrone et al. [42]	161	47.4	78.8 %EBMIL	–	–	–
Pok et al. [43]	61, 19, and 12 patients (5-, 6-, and 7-year follow-up, respectively)	37.3	71.6	71	70	–
Ruiz-Tovar et al. [44]	50	46.9	78.7	–	–	–
Seki et al. [45]	18	43.3	77.3	–	–	–
Summary: 40 studies[a]	5053 patients	45.6 kg/m²	60.5% (n = 2695 patients)[b]	60.5% (n = 952 patients)[b]	57.9% (n = 582 patients)[b]	54.4% (n = 94 patients)[b]

[a]Including the studies by Strain et al., Van Rutte et al., and Alexandrou et al., in which LSG was performed either as a stand-alone or as part of a staged procedure.
[b]Excluding from the calculating process the studies by Strain et al., Saif et al., Sieber et al., Abelson et al., Perrone et al., and Hong et al., which report the weight loss outcome either as %EBMIL or as %TWL.
Summarizes the number of patients, as well as the respective weight loss result, at each long-term follow-up point, when LSG is used as a definitive operation. BMI, body mass index; EWL, excess weight loss; EBMIL, excess BMI loss; TWL, total weight loss.

preoperative BMI of 59.4 kg/m^2. The long-term weight loss outcome for this group was 50.8%, 52%, 43%, and 46% at 60, 72, 84, and $\geq$96 months after LSG, respectively (Table 27.4). The evolution of the weight loss result at the long-term follow-up for the stand-alone and the staged-approach group is depicted in Fig. 27.2.

Analyzing the long-term weight loss results after LSG for %EWL, 62.8%, 72.9%, 80.3%, and 74.6% of patients had %EWL > 50% at 60, 72, 84, and $\geq$96 months after LSG, respectively, which can be considered a weight loss success according to Reinhold's criteria [46].

DETERMINING THE EFFECTIVENESS OF LAPAROSCOPIC SLEEVE GASTRECTOMY ON COMORBIDITIES, AS WELL AS ITS SAFETY

Only 24 of the 44 studies contained data regarding the long-term effect of LSG on comorbidities, which are usually present in these patients. Complete or near-complete resolution was demonstrated in 61.1% of patients with type 2 diabetes mellitus (23 studies, $n = 1034$ out of 1692 patients), in 68.5% of patients with arterial hypertension (19 studies, $n = 1327$ out of 1936 patients), in 90.6% of patients with obstructive sleep apnea (12 studies, $n = 1153$ out of 1272 patients), and in 66.6% of patients with dyslipidemia (16 studies, $n = 1030$ out of 1547 patients).

Regarding the safety of LSG, the perioperative complication rate ranged from 0% to 63%, with a mean leak rate of 1.26% ($n = 144$ patients). Seven deaths were recorded in the first 30 postoperative days; thus the perioperative mortality rate was 0.14% (Table 27.1).

WEIGHT LOSS MECHANISMS OF LAPAROSCOPIC SLEEVE GASTRECTOMY

Maintenance of the weight loss outcome is the mainstay of bariatric surgery. Apart from the restrictive component of sleeve gastrectomy, one valuable contribution to its weight loss outcome might be the changes in plasma levels of ghrelin, the only known peripheral hormone so far that increases appetite and thus induces food uptake. Ghrelin is produced in the gastric fundus, thus the resection of the fundus enhances the restrictive component of LSG by decreasing appetite.

TABLE 27.4 Long-Term Weight Loss Outcomes of Laparoscopic Sleeve Gastrectomy as a Staged Procedure

Author	Patients (n)	Mean Preoperative BMI (kg/m^2)	5 years %EWL	6 years %EWL	7 years %EWL	$\geq$8 years %EWL
Weiner et al. [2]	8	61.6	40 %EBMIL	–	–	–
Strain et al. [6]	23	56.1	48 %EBMIL	–	–	–
Eid et al. [7]	19, 13, and 21 patients (6-, 7-, and $\geq$8-year follow-up, respectively)	66	–	52	43	46
Abbatini et al. [11]	13	52.1	56	–	–	–
Marceau et al. [19]	9	50.4	12	–	–	–
Van Rutte et al. [24]	19	44.3	58.3	–	–	–
Alexandrou et al. [27]	25	55.5	56.4	–	–	–
Summary: 7 studies[a]	150 patients	59.4 kg/m^2	50.8% ($n = 66$ patients)[b]	52% ($n = 19$ patients)[b]	43% ($n = 13$ patients)[b]	46% ($n = 21$ patients)[b]

[a]Including the studies by Strain et al., Van Rutte et al., and Alexandrou et al., in which LSG was performed either as a stand-alone or as part of a staged procedure.
[b]Excludes from the calculating process the studies by Weiner et al. and Strain et al., which report the weight loss outcome as %EBMIL.
Summarizes the number of patients, as well as the respective weight loss result, at each long-term follow-up point, when LSG is used as part of a staged approach. BMI, body mass index; EWL, excess weight loss; EBMIL, excess BMI loss.

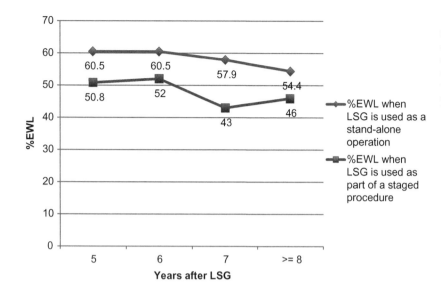

FIGURE 27.2 Mean %EWL at different follow-up points, when LSG is used stand-alone or as part of a staged procedure. Depicts the evolution of the weight loss outcome during the entire long-term follow-up period, when LSG is used as a definitive, or as part of a staged procedure.

Three studies [3,47,48] so far have demonstrated that LSG is associated with decreased plasma levels of ghrelin at different follow-up points during the postoperative period. In addition to the reduced plasma levels of ghrelin, LSG is responsible for the quick postmeal elevation of the appetite-reducing peptides peptide YY (PYY) and glucagon-like peptide 1 (GLP-1) through the quick bowel transit, which contributes to the weight loss outcome [47,49]. Despite all the aforementioned data, a recent study in mice demonstrated that neither ghrelin nor GLP-1 is essentially responsible for the weight loss result following LSG. Instead, it was demonstrated that bile acids are the key factors in determining the weight loss outcome after LSG, probably due to the increase in the levels of circulating bile acids and the associated changes to the microbial milieu of the gut [50]. Considering all these facts, it seems that the weight loss mechanism of LSG is more complex than was initially thought; future studies may elucidate this complex cascade.

WEIGHT LOSS RESULTS

The short- and mid-term weight loss results after LSG are well validated in the literature. In the systematic review published by Brethauer et al. [1], which included 24 studies with a total of 1749 patients, the mean %EWL ranged from 36% to 85%, with the %EWL being 58.8%, 77%, and 63% at the end of the first, second, and third postoperative year, respectively; the overall mean %EWL was 60.7%.

The long-term results after LSG are defined as the results at 5 or more years postoperatively, when LSG is performed without any other bariatric procedures. The present study demonstrated that the weight loss result at 5 or more years after LSG is considered successful, especially when analyzed according to Reinhold's criteria [46], although a partial weight regain is observed in some studies. At this point, however, it should be mentioned that the weight loss outcome published by Rawlins et al. [14] is far greater than the average %EWL at the 60-month follow-up mark derived by the remaining studies. The more important points here are the use of a 26.4F endoscope as a bougie, which creates a tight sleeve, the beginning of the gastric resection 3 cm from the pylorus, as well as other selection and management criteria. On the other hand, the weight loss outcomes published by Marceau et al. [19] and Pekkarinen et al. [41] are far lower than the average %EWL realized by the remaining studies. Excluding technical and anatomical factors, the significant weight regain encountered was due to patients' noncompliance to follow-up as well as to multidisciplinary team dietary guidelines.

DIFFERENCES IN THE WEIGHT LOSS RESULTS BETWEEN THE STAND-ALONE AND THE STAGED APPROACH

The well-validated weight loss results rendered LSG a valuable stand-alone operation for the treatment of severe obesity. Thus, it is of benefit to try to determine the long-term weight loss outcomes of LSG as a stand-alone operation and compare them with the respective results, when LSG is used as part of a staged approach. Comparing the two groups, it is reasonable that the patients in the stand-alone group had lower preoperative BMIs, as the patients in the staged-approach group had higher initial BMIs and thus LSG was intended as a first-step operation in an attempt to promote loss of part of their excess weight before the second-step operation. The weight loss result in the stand-alone group was nearly identical to that of the total of the included studies, and definitely better compared to the staged-approach group, at all the respective follow-up points. This fact could be attributed to the use of a smaller-size bougie in the stand-alone group, which created a tight sleeve; LSG for these patients was intended as a definitive operation with no intended second step. Thus, by using a smaller-size bougie for calibration of the gastric sleeve, a better weight loss result was expected [51,52]. At this point it should be mentioned that the weight loss results for the staged-approach group at 7 and ≥ 8 years after LSG were not satisfactory as the %EWL remained less than 50%; however, this fact should be considered with prejudice, since these data are derived from only one study [8], with a small number of patients reaching the respective follow-up points.

WEIGHT REGAIN AFTER LAPAROSCOPIC SLEEVE GASTRECTOMY

With an increasing number of LSGs being performed, the significant issue of weight regain is becoming more prevalent. Without a standardized definition, patients with weight regain, inadequate weight loss, or a procedure failure are still grouped together. A recent systematic review identified the possible mechanisms responsible for weight regain after LSG as the following: technical factors that contribute to the initial sleeve size (bougie size, gastric fundus remnant, and distance of the beginning of the gastric resection from the pylorus), the postoperative plasma levels of ghrelin, the sleeve dilatation that occurs over time, the patients' compliance with follow-up support, as well as the lifestyle and eating behaviors of the patients [53]. Moreover, homeostatic processes have been described as being in opposition to weight loss maintenance following weight loss [54,55]. Hence, large prospective studies are required to elucidate the underlying mechanisms of weight regain following LSG.

WEIGHT LOSS EVOLUTION

Weight loss results after LSG are well documented in the long-term, with sustained loss of over 50% of excess weight at 5 years postoperatively [56]. Considering Fig. 27.1, it is obvious that a decrease in the %EWL does occur after the sixth postoperative year. This reduction in the %EWL cannot arbitrarily be attributed solely to the weight regain phenomenon, due to the interfering high attrition rates in some studies. However, the %EWL still remains greater than 50% during the entire follow-up period.

At this point, it should be emphasized that the indications for LSG have broadened, with results equivalent to those reported for Roux-en-Y gastric bypass [57,58]. Thus, the long-term weight loss results after LSG are well validated, but we must wait for weight loss results at 10 or more years postoperatively before speculating on the underlying mechanisms of weight regain following LSG.

MINI-DICTIONARY OF TERMS

- *Percentage of excess weight loss (%EWL)*: The ratio of weight lost to excess weight, as the latter is calculated by the ideal BMI.
- *Body mass index (BMI)*: The ratio of body weight in kilograms (kg) to body area surface in square meters (m^2).
- *Bougie*: A thin cylinder introduced transorally into the stomach to facilitate creation of the gastric sleeve.
- *Percentage of excess BMI loss (%EBMIL)*: The ratio of BMI lost to excess BMI, as the latter is calculated by the ideal BMI.
- *Percentage of total weight loss (%TWL)*: The ratio of weight lost to the initial body weight.
- *Ghrelin*: A peptide secreted by the gastric fundus; it increases appetite.
- *Peptide YY (PYY)*: A peptide produced and released by the cells of the ileum and colon in response to food intake.
- *Glucagon-like peptide 1 (GLP-1)*: An incretin as well as a neuropeptide, produced both by intestinal L-cells and by the solitary nucleus.
- *Bile acids*: Steroid acids found in bile.
- *Staged approach*: The use of sleeve gastrectomy as a first-step operation for initial weight loss, before performance of a second-step operation.
- *Attrition rate*: The rate of patients who were lost to follow-up.
- *Gastric fundus*: The upper curvature of the stomach.
- *Gastric pylorus*: The lower section of the stomach.
- *Duodenum*: The first portion of the small intestine.

KEY FACTS

- LSG was introduced as the first step of a staged approach in patients with higher preoperative BMI.
- Nowadays it is also used as a definitive treatment of severe obesity.
- It seems that the weight loss mechanism is not only limited to the restrictive component.
- Its short- and long-term weight loss results are well validated.
- LSG also contributes to the resolution of common comorbid conditions.
- With the increase in the number of LSGs performed, the weight regain phenomenon becomes more prevalent, but without a standardized definition and explanation of the underlying mechanisms.

SUMMARY POINTS

- The calculated long-term mean percentage of %EWL was 59.8%.
- At least 6 out of 10 patients had %EWL > 50% at 5 or more years after LSG.
- The perioperative complication rate ranged from 0% to 63%, with a mean leak rate of 1.26% and a perioperative mortality rate of 0.14%.
- Not only the restrictive component of the operation is responsible for its weight loss outcome.
- A decrease in the %EWL does occur after the sixth postoperative year. However, the %EWL still remains greater than 50% during the entire follow-up period.
- Standardized definitions of weight regain are required to elucidate the underlying mechanisms of weight regain.

REFERENCES

[1] Brethauer SA, Hammel JP, Schauer PR. Systematic review of sleeve gastrectomy as staging and primary bariatric procedure. Surg Obes Relat Dis 2009;5:469—75.

[2] Weiner RA, Weiner S, Pomhoff I, Jacobi C, Makarewicz W, Weigand G. Laparoscopic sleeve gastrectomy--influence of sleeve size and resected gastric volume. Obes Surg 2007;17:1297—305.

[3] Bohdjalian A, Langer FB, Shakeri-Leidenmuhler S, Gfrerer L, Ludvik B, Zacherl J, et al. Sleeve gastrectomy as sole and definitive bariatric procedure: 5-year results for weight loss and ghrelin. Obes Surg 2010;20:535—40.

[4] Himpens J, Dobbeleir J, Peeters G. Long-term results of laparoscopic sleeve gastrectomy for obesity. Ann Surg 2010;252:319—24.

[5] D'Hondt M, Vanneste S, Pottel H, Devriendt D, Van Rooy F, Vansteenkiste F. Laparoscopic sleeve gastrectomy as a single-stage procedure for the treatment of morbid obesity and the resulting quality of life, resolution of comorbidities, food tolerance, and 6-year weight loss. Surg Endosc 2011;25:2498—504.

[6] Strain GW, Saif T, Gagner M, Rossidis M, Dakin G, Pomp A. Cross-sectional review of effects of laparoscopic sleeve gastrectomy at 1, 3, and 5 years. Surg Obes Relat Dis 2011;7:714—19.

[7] Braghetto I, Csendes A, Lanzarini E, Papapietro K, Carcamo C, Molina JC. Is laparoscopic sleeve gastrectomy an acceptable primary bariatric procedure in obese patients? Early and 5-year postoperative results. Surg Laparosc Endosc Percutan Tech 2012;22:479—86.

[8] Eid GM, Brethauer S, Mattar SG, Titchner RL, Gourash W, Schauer PR. Laparoscopic sleeve gastrectomy for super obese patients: forty-eight percent excess weight loss after 6 to 8 years with 93% follow-up. Ann Surg 2012;256:262—5.

[9] Saif T, Strain GW, Dakin G, Gagner M, Costa R, Pomp A. Evaluation of nutrient status after laparoscopic sleeve gastrectomy 1, 3, and 5 years after surgery. Surg Obes Relat Dis 2012;8:542—7.

[10] Sarela AI, Dexter SP, O'Kane M, Menon A, McMahon MJ. Long-term follow-up after laparoscopic sleeve gastrectomy: 8-9-year results. Surg Obes Relat Dis 2012;8:679—84.

[11] Abbatini F, Capoccia D, Casella G, Soricelli E, Leonetti F, Basso N. Long-term remission of type 2 diabetes in morbidly obese patients after sleeve gastrectomy. Surg Obes Relat Dis 2013;9:498—502.

[12] Brethauer SA, Aminian A, Romero-Talamas H, Batayyah E, Mackey J, Kennedy L, et al. Can diabetes be surgically cured? Long-term metabolic effects of bariatric surgery in obese patients with type 2 diabetes mellitus. Ann Surg 2013;258:628—36 discussion 636-627

[13] Catheline JM, Fysekidis M, Bachner I, Bihan H, Kassem A, Dbouk R, et al. Five-year results of sleeve gastrectomy. J Visc Surg 2013;150:307—12.

[14] Rawlins L, Rawlins MP, Brown CC, Schumacher DL. Sleeve gastrectomy: 5-year outcomes of a single institution. Surg Obes Relat Dis 2013;9:21—5.

[15] Zachariah SK, Chang PC, Ooi AS, Hsin MC, Kin Wat JY, Huang CK. Laparoscopic sleeve gastrectomy for morbid obesity: 5 years experience from an Asian center of excellence. Obes Surg 2013;23:939—46.

[16] Abd Ellatif ME, Abdallah E, Askar W, Thabet W, Aboushady M, Abbas AE, et al. Long term predictors of success after laparoscopic sleeve gastrectomy. Int J Surg 2014;12:504—8.

[17] Boza C, Daroch D, Barros D, Leon F, Funke R, Crovari F. Long-term outcomes of laparoscopic sleeve gastrectomy as a primary bariatric procedure. Surg Obes Relat Dis 2014;10:1129—33.

[18] Lim DM, Taller J, Bertucci W, Riffenburgh RH, O'Leary J, Wisbach G. Comparison of laparoscopic sleeve gastrectomy to laparoscopic Roux-en-Y gastric bypass for morbid obesity in a military institution. Surg Obes Relat Dis 2014;10:269−76.

[19] Marceau P, Biron S, Marceau S, Hould FS, Lebel S, Lescelleur O, et al. Biliopancreatic diversion-duodenal switch: independent contributions of sleeve resection and duodenal exclusion. Obes Surg 2014;24:1843−9.

[20] Musella M, Milone M, Gaudioso D, Bianco P, Palumbo R, Galloro G, et al. A decade of bariatric surgery. What have we learned? Outcome in 520 patients from a single institution. Int J Surg 2014;12(suppl. 1):S183−8.

[21] Noel P, Schneck AS, Nedelcu M, Lee JW, Gugenheim J, Gagner M, et al. Laparoscopic sleeve gastrectomy as a revisional procedure for failed gastric banding: lessons from 300 consecutive cases. Surg Obes Relat Dis 2014;10:1116−22.

[22] Prevot F, Verhaeghe P, Pequignot A, Rebibo L, Cosse C, Dhahri A, et al. Two lessons from a 5-year follow-up study of laparoscopic sleeve gastrectomy: persistent, relevant weight loss and a short surgical learning curve. Surgery 2014;155:292−9.

[23] Sieber P, Gass M, Kern B, Peters T, Slawik M, Peterli R. Five-year results of laparoscopic sleeve gastrectomy. Surg Obes Relat Dis 2014;10:243−9.

[24] van Rutte PW, Smulders JF, de Zoete JP, Nienhuijs SW. Outcome of sleeve gastrectomy as a primary bariatric procedure. Br J Surg 2014;101:661−8.

[25] Zhang Y, Zhao H, Cao Z, Sun X, Zhang C, Cai W, et al. A randomized clinical trial of laparoscopic Roux-en-Y gastric bypass and sleeve gastrectomy for the treatment of morbid obesity in China: a 5-year outcome. Obes Surg 2014;24:1617−24.

[26] Abelson JS, Afaneh C, Dolan P, Chartrand G, Dakin G, Pomp A. Laparoscopic sleeve gastrectomy: co-morbidity profiles and intermediate-term outcomes. Obes Surg 2015;26:1788.

[27] Alexandrou A, Athanasiou A, Michalinos A, Felekouras E, Tsigris C, Diamantis T. Laparoscopic sleeve gastrectomy for morbid obesity: 5-year results. Am J Surg 2015;209:230−4.

[28] Alvarenga ES, Lo Menzo E, Szomstein S, Rosenthal RJ. Safety and efficacy of 1020 consecutive laparoscopic sleeve gastrectomies performed as a primary treatment modality for morbid obesity. A single-center experience from the metabolic and bariatric surgical accreditation quality and improvement program. Surg Endosc 2016;30:2673−8.

[29] Angrisani L, Santonicola A, Hasani A, Nosso G, Capaldo B, Iovino P. Five-year results of laparoscopic sleeve gastrectomy: effects on gastroesophageal reflux disease symptoms and co-morbidities. Surg Obes Relat Dis 2016;12:960−8.

[30] Dogan K, Gadiot RP, Aarts EO, Betzel B, van Laarhoven CJ, Biter LU, et al. Effectiveness and safety of sleeve gastrectomy, gastric bypass, and adjustable gastric banding in morbidly obese patients: a multicenter, retrospective, matched cohort study. Obes Surg 2015;25:1110−18.

[31] Golomb I, Ben David M, Glass A, Kolitz T, Keidar A. Long-term metabolic effects of laparoscopic sleeve gastrectomy. JAMA Surg 2015;150:1051−7.

[32] Hirth DA, Jones EL, Rothchild KB, Mitchell BC, Schoen JA. Laparoscopic sleeve gastrectomy: long-term weight loss outcomes. Surg Obes Relat Dis 2015;11:1004−7.

[33] Hong JS, Kim WW, Han SM. Five-year results of laparoscopic sleeve gastrectomy in Korean patients with lower body mass index (30-35 kg/m(2)). Obes Surg 2015;25:824−9.

[34] Lee WJ, Pok EH, Almulaifi A, Tsou JJ, Ser KH, Lee YC. Medium-term results of laparoscopic sleeve gastrectomy: a matched comparison with gastric bypass. Obes Surg 2015;25:1431−8.

[35] Lemanu DP, Singh PP, Rahman H, Hill AG, Babor R, MacCormick AD. Five-year results after laparoscopic sleeve gastrectomy: a prospective study. Surg Obes Relat Dis 2015;11:518−24.

[36] Liu SY, Wong SK, Lam CC, Yung MY, Kong AP, Ng EK. Long-term results on weight loss and diabetes remission after laparoscopic sleeve gastrectomy for a morbidly obese Chinese population. Obes Surg 2015;25:1901−8.

[37] Casella G, Soricelli E, Giannotti D, Collalti M, Maselli R, Genco A, et al. Long-term results after laparoscopic sleeve gastrectomy in a large monocentric series. Surg Obes Relat Dis 2015.

[38] Del Genio G, Limongelli P, Del Genio F, Motta G, Docimo L, Testa D. Sleeve gastrectomy improves obstructive sleep apnea syndrome (OSAS): 5 year longitudinal study. Surg Obes Relat Dis 2016;12:70−4.

[39] Keren D, Matter I, Rainis T. Sleeve gastrectomy in different age groups: a comparative study of 5-year outcomes. Obes Surg 2016;26:289−95.

[40] Lin YH, Lee WJ, Ser KH, Chen SC, Chen JC. 15-year follow-up of vertical banded gastroplasty: comparison with other restrictive procedures. Surg Endosc 2016;30:489−94.

[41] Pekkarinen T, Mustonen H, Sane T, Jaser N, Juuti A, Leivonen M. Long-term effect of gastric bypass and sleeve gastrectomy on severe obesity: do preoperative weight loss and binge eating behavior predict the outcome of bariatric surgery? Obes Surg 2016;26:2161−7.

[42] Perrone F, Bianciardi E, Benavoli D, Tognoni V, Niolu C, Siracusano A, et al. Gender influence on long-term weight loss and comorbidities after laparoscopic sleeve gastrectomy and Roux-en-Y gastric bypass: a prospective study with a 5-year follow-up. Obes Surg 2016;26:276−81.

[43] Pok EH, Lee WJ, Ser KH, Chen JC, Chen SC, Tsou JJ, et al. Laparoscopic sleeve gastrectomy in Asia: long term outcome and revisional surgery. Asian J Surg 2016;39:21−8.

[44] Ruiz-Tovar J, Martinez R, Bonete JM, Rico JM, Zubiaga L, Diez M, et al. Long-term weight and metabolic effects of laparoscopic sleeve gastrectomy calibrated with a 50-Fr bougie. Obes Surg 2016;26:32−7.

[45] Seki Y, Kasama K, Hashimoto K. Long-term outcome of laparoscopic sleeve gastrectomy in morbidly obese japanese patients. Obes Surg 2016;26:138−45.

[46] Reinhold RB. Critical analysis of long term weight loss following gastric bypass. Surg Gynecol Obstet 1982;155:385−94.

[47] Karamanakos SN, Vagenas K, Kalfarentzos F, Alexandrides TK. Weight loss, appetite suppression, and changes in fasting and postprandial ghrelin and peptide-YY levels after Roux-en-Y gastric bypass and sleeve gastrectomy: a prospective, double blind study. Ann Surg 2008;247:401−7.

[48] Cohen R, Uzzan B, Bihan H, Khochtali I, Reach G, Catheline JM. Ghrelin levels and sleeve gastrectomy in super-super-obesity. Obes Surg 2005;15:1501–2.

[49] Melissas J, Leventi A, Klinaki I, Perisinakis K, Koukouraki S, de Bree E, et al. Alterations of global gastrointestinal motility after sleeve gastrectomy: a prospective study. Ann Surg 2013;258:976–82.

[50] Ryan KK, Tremaroli V, Clemmensen C, Kovatcheva-Datchary P, Myronovych A, Karns R, et al. FXR is a molecular target for the effects of vertical sleeve gastrectomy. Nature 2014;509:183–8.

[51] Deitel M, Gagner M, Erickson AL, Crosby RD. Third international summit: current status of sleeve gastrectomy. Surg Obes Relat Dis 2011;7:749–59.

[52] Lemanu DP, Singh PP, Berridge K, Burr M, Birch C, Babor R, et al. Randomized clinical trial of enhanced recovery versus standard care after laparoscopic sleeve gastrectomy. Br J Surg 2013;100:482–9.

[53] Lauti M, Kularatna M, Hill AG, MacCormick AD. Weight regain following sleeve gastrectomy-a systematic review. Obes Surg 2016;26:1326–34.

[54] Bobbioni-Harsch E, Morel P, Huber O, Assimacopoulos-Jeannet F, Chassot G, Lehmann T, et al. Energy economy hampers body weight loss after gastric bypass. J Clin Endocrinol Metab 2000;85:4695–700.

[55] Galtier F, Farret A, Verdier R, Barbotte E, Nocca D, Fabre JM, et al. Resting energy expenditure and fuel metabolism following laparoscopic adjustable gastric banding in severely obese women: relationships with excess weight lost. Int J Obes (Lond) 2006;30:1104–10.

[56] Diamantis T, Apostolou KG, Alexandrou A, Griniatsos J, Felekouras E, Tsigris C. Review of long-term weight loss results after laparoscopic sleeve gastrectomy. Surg Obes Relat Dis 2014;10:177–83.

[57] Abbatini F, Rizzello M, Casella G, Alessandri G, Capoccia D, Leonetti F, et al. Long-term effects of laparoscopic sleeve gastrectomy, gastric bypass, and adjustable gastric banding on type 2 diabetes. Surg Endosc 2010;24:1005–10.

[58] Kehagias I, Karamanakos SN, Argentou M, Kalfarentzos F. Randomized clinical trial of laparoscopic Roux-en-Y gastric bypass versus laparoscopic sleeve gastrectomy for the management of patients with BMI < 50 kg/m2. Obes Surg 2011;21:1650–6.

Chapter 28

Gastric Band Slippage as an Adverse Event

D.W. Swenson[1,2] and M.S. Furman[1,2]

[1]Warren Alpert Medical School of Brown University, Providence, RI, United States, [2]Rhode Island Hospital, Providence, RI, United States

LIST OF ABBREVIATIONS

AGB adjustable gastric band
LAGB laparoscopic adjustable gastric band
RYGB Roux-en-Y gastric bypass
SG sleeve gastrectomy

INTRODUCTION

First performed by Belachew in 1993 [1], laparoscopic adjustable gastric banding (AGB) became the number one preferred bariatric surgical procedure throughout the world by 2008, when over 145,000 AGB procedures were performed, accounting for 42% of bariatric surgeries that year [2]. Although by 2013 practice patterns had shifted, with most patients undergoing either Roux-en-Y gastric bypass (RYGB; 45%) or sleeve gastrectomy (SG; 37%), still over 46,000 AGBs were performed that year throughout the world [3]. In addition, there is persistent regional variation in bariatric surgical practice; e.g., in Korea, where AGB still comprised nearly 70% of bariatric surgeries in 2013 [4].

GASTRIC BAND SLIPPAGE

Multiple, large single-institution studies and meta-analyses have demonstrated that AGB is safe and effective for surgical weight loss, with comparable outcomes to RYGB and SG with regards to excess weight loss (EWL) and resolution of obesity-related comorbidities [5−9]. However, its popularity has waned largely because of concerns over long-term complications requiring reoperation. The most common serious complication of AGB is band slippage, which has been variably described in 1.1−39.9% of patients [6,10−16]. When the band slips from its correct surgical position, it generally leads to some combination of chest or abdominal pain, nausea, dysphagia, vomiting, poor diet compliance, and inadequate weight loss. Rarely, a slipped band may be associated with erosion into the gastric lumen, bleeding, infection, bowel perforation, or gastric volvulus (see Table 28.1) [10−13,17−19]. While AGB slippage may be suggested by patient symptoms, the diagnosis is generally established by radiologic studies.

In 2000, Wiesner et al. published an article on the postoperative appearance of AGBs with radiography and contrast esophograms [20], including descriptions of concentric pouch dilatation, eccentric pouch dilatation with posterior slippage, and eccentric herniation of the band. In 2006, Mehanna et al. described multimodality imaging findings of AGB complications, including a statement that one common problem was "pouch dilation" (though they acknowledged that there was no established upper limit of "normal" for pouch volume), as well as explaining band malpositioning as leading to either a medial eccentric pouch configuration, or a lateral eccentric pouch dilation with posterior band slippage [21]. A year later, Blachar et al. published a similarly themed article in the same journal, describing normal and abnormal imaging features of AGB, but in their paper they used the terms anterior and posterior slippage with eccentric upper gastric pouch dilation [19].

Metabolism and Pathophysiology of Bariatric Surgery.

TABLE 28.1 Clinical Associations With Adjustable Gastric Band Slippage

Common	Uncommon/Rare
Chest or abdominal pain	Major internal bleeding
Dysphagia	Infection
Nausea	Erosion into gastric lumen
Vomiting	Bowel perforation
Poor diet compliance	Gastric volvulus
Inadequate weight loss	

In 2011, Egan et al. [13] called attention to the lack of consensus regarding definitions for gastric band slippage, and proposed standard surgical and radiologic terminology, including straightforward terms of anterior slips, posterior slips, and concentric (or symmetric) proximal pouch dilation. Eid et al. provided a similar classification [15], while adding a category of "immediate postoperative prolapse," which they described as resulting from initial surgical placement of the band too low around the stomach. Regardless of the exact descriptions used, a familiarity with the evolution of surgical techniques for AGB is important for understanding the expected imaging findings of normal versus slipped bands [22].

The originally described surgical method for placement of AGBs is now generally referred to as the "perigastric" technique. It involved dissecting around the stomach approximately 3 cm below the diaphragm at the level of the first short gastric vessels, passing the band through the lesser sac, and intentionally creating a moderate-sized proximal gastric pouch above the band, with a volume of approximately 25–30 mL [1,23–27]. In early experience, the AGB reservoir was frequently filled with saline during the primary surgery, while calibrating with a pressure gauge at the level of the band. Alternatively, in some practices, early filling was performed in concert with a postoperative barium swallow study in order to target a presumed "optimal" luminal constriction by radiography [25,26,28,29].

Patient outcomes with the perigastric surgical technique and early practices of aggressive band tightening were suboptimal. For example, DeMaria et al. reported their experience in 2001 as part of the US Food & Drug Administration Lap-Band trial, in which only 4/37 patients achieved a targeted body mass index of less than 35 and/or EWL >50%. In addition, 41% of patients had their band removed for reasons including poor weight loss, dysphagia, progressive esophageal dilation, and band slippage [28]. The authors acknowledged that their results reflected a surgical learning curve, yet they questioned the efficacy of AGB as a primary bariatric surgical procedure. In a larger series out of Switzerland, Suter at al. reported outcomes from 317 patients with AGB over a 10-year period. While they had better primary outcomes of EWL >50% in 53.8% of patients at 2 years, and in 42.9% after 7 years, they found that 33% of patients developed late complications. Notably, they found a steady 3–4% annual increase over time, from 13% at 18 months of follow-up, to approximately 37% at 7 years. Their major complications included band erosion in 9.5%, and slippage in 6.3%, while EWL [30].

During the late 1990s and early 2000s, the surgical method for AGB placement shifted from the perigastric to a pars flaccida technique in which dissection is performed above the peritoneal reflection of the bursa omentalis, avoiding entry of the lesser sac and leaving normal gastric fixation relatively unperturbed. With this approach, the pars flaccida is dissected and the band is passed around the upper stomach at the angle of His, with ultimate band positioning within approximately 1 cm from the esophageal hiatus of the diaphragm. The resulting proximal pouch is therefore small, typically <15 mL [25,26,31]. It also became standard practice to leave the band completely deflated for several weeks to months, allowing for resolution of any perioperative mucosal edema prior to gradual band inflation, as directed by patient symptoms and weight loss trends [26,32]. In 2005, O'Brien et al. reported a prospective randomized trial comparing the perigastric and pars flaccida techniques, showing comparable effective weight loss between the two, but improved health and quality of life with the pars flaccida technique. Significantly, there was a dramatically decreased rate of band slippage in patients undergoing the pars flaccida technique (16% with perigastric, versus 4% with pars flaccida, $p < 0.01$) [31].

In numerous other studies, the pars flaccida technique has been shown to essentially eliminate the risk of what had been described as "posterior" band slips [8,22,31,32]. Now with the combination of the pars flaccida surgical technique and more gradual adjustment of band tightness, slippage is generally reported in <5% of patients, and it is nearly always "anterior" [13].

Equipped with an understanding of the expected anatomic location of the AGB following surgery with the pars flaccida technique, the diagnosis of band slippage becomes considerably easier than in years past, when variable eccentricity and dilation of the proximal pouch was considered an important subjective feature of the slipped band following perigastric placement [19—21]. In current practice, when a patient with prior AGB presents with suspicious symptoms (Table 28.1), the workup will frequently involve diagnostic imaging with radiographs, contrast esophogram, and/or computed tomography (CT), depending on the acuity and severity of symptoms. Standard treatment in this scenario includes urgent deflation of the AGB reservoir in order to minimize obstructive symptoms, followed by elective surgical revision or removal of the band if diagnostic imaging confirms slippage.

A normally positioned AGB projects very near the medial aspect of the left hemidiaphragm, roughly paralleling the diaphragm (Fig. 28.1A). Its anterior and posterior aspects typically overlap, such that the band is seen as nearly rectangular. Historically, the phi angle has been described as useful in characterizing normal band positioning. The phi angle is defined as the angle subtended by a vertical line through the spinal column and line paralleling the long axis of the band on a frontal chest or abdominal radiograph (Fig. 28.1A). A normal range of 4—58 degree has frequently been reported in the radiology literature [22]. It should be noted that the lower bound of this range has generally been significant only in the setting of posterior slips with medial eccentric pouch dilation, and again, this type of slip is almost nonexistent in the modern era of pars flaccida surgical technique [22]. Swenson et al. recently reported that using a phi angle of 58 degree as a cutoff for distinguishing normal from slipped AGBs was 91—95% sensitive, but only 52—62% specific [22]. They found a normal range of 22—84 degree (mean, 55 degree) in asymptomatic patients at routine annual follow-up visits, compared with a range of 38—145 degree (mean, 87 degree) in patients with surgically proven band slippage.

A more robust measurement for identifying AGB slippage is the shortest distance between the superolateral margin of the band and the diaphragm (Fig. 28.1). Swenson et al. described a threshold of >2.4 cm as being 95% sensitive and 97—98% specific for band slippage in a study population of 84 patients (25% of whom had surgically confirmed slippage) [22]. Given that modern surgical technique places the band within 1 cm of the diaphragm, it makes sense that a slip of the band over the fundus will necessarily move the superolateral band margin inferiorly. While not directly correlated, there is clearly an association between increasing phi angle and increasing distance of the superolateral band margin from the diaphragm. These findings both correlate with anterior slippage (which results in a lateral eccentric dilation of the proximal pouch); this is the type of slip that typically occurs with the pars flaccida surgical technique for band placement.

While the normally positioned band will typically appear nearly rectangular because its anterior and posterior aspects are superimposed, a slipped band may project as a ring on frontal radiographs of the chest or abdomen (variably oval or circular, depending on the degree of rotation in the coronal plane). This has been described as the "O" sign of AGB slippage, as reported by Pieroni et al. in 2010 [33]. Swenson et al. found this sign to be only 33—48% sensitive for band slippage, though it was 97% sensitive [22].

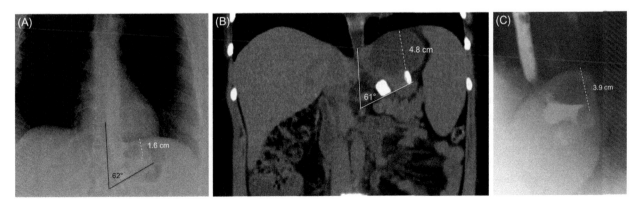

FIGURE 28.1 Radiologic studies of a 31-year-old female with AGB: Initially normal, then slipped. (A) Upright frontal chest radiograph, (B) Coronal reformatted image from CT exam performed with intravenous but not oral contrast material, and (C) Right anterior oblique fluoroscopic image from a barium swallow study.

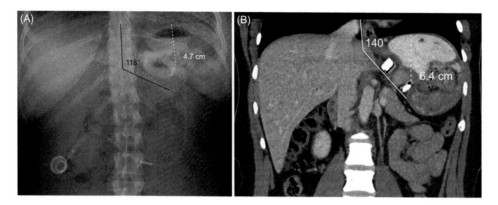

FIGURE 28.2 Radiologic studies of a 27-year-old female with slipped AGB. (A) Upright frontal abdominal radiograph, (B) Coronal reformatted image from CT exam performed with intravenous and oral contrast material.

A final sign of AGB slippage has been described as an air—fluid level within the proximal pouch above the band on a frontal upright chest or abdominal radiograph [22]. This was found to be 95% sensitive and 97—98% specific for AGB slippage. This sign can be understood in light of several facts. First, since the proximal pouch is intended to be small with the pars flaccida technique, there should not be a significant volume within the pouch to allow for layering fluid and air. Second, while the band is intended to restrict passage of solid food, it is not intended to obstruct the passage of ingested fluids. Along these lines, a study reviewing the normal findings of asymptomatic AGB patients during routine band adjustments under fluoroscopy with barium swallow exams found the mean pouch emptying time for liquid contents was only 36 seconds [22]. Therefore, the presence of an air—fluid level within the proximal pouch above the band should be interpreted as a sign of obstruction at the level of the band (Fig. 28.2).

Two illustrative examples of patients who developed AGB slippage may aid familiarity with these various signs of normal versus slippage AGB on radiologic studies. Fig. 28.1 illustrates an initial normal radiographic appearance of a gastric band in a 31-year-old female who had an AGB in place for several years, was asymptomatic, had experienced adequate weight loss, and was being seen in a surgical clinic for a routine annual follow-up visit during which a chest radiograph was obtained to confirm band positioning and orientation. Fig. 28.1A illustrates the expected normal findings of a gastric band, with phi angle of 62 degree, superolateral band margin projecting only 1.6 cm below the diaphragm, and absence of an "O" sign or air—fluid level within the proximal pouch above the band. The patient subsequently presented to the emergency department with chest and epigastric pain, described as "razor blade pain." She had developed severe dysphagia, nausea, and vomiting. Because of the severity of her symptoms, a CT scan was performed immediately (Fig. 28.1B), demonstrating that the band had shifted inferiorly in position. Now the superolateral band margin was 4.8 cm below the diaphragm, and there was a large proximal pouch (which on this single image shows fluid, but which also contained nondependent air), compatible with band slippage. Note, the phi angle remained 61 degree, which is within the range of normal variation, but in the presence of these other signs is nonspecific. The following day, a barium swallow study was performed after band deflation (Fig. 28.1C), and again demonstrated inferior displacement of the band, as well as an air—fluid/air—contrast level above the band, with no passage of contrast through the gastric lumen at the level of the band. The patient subsequently had the AGB removed surgically.

A second example is a 27-year-old female who had an AGB placed at an outside institution approximately 18 months previously, and had the reservoir inflated 2 weeks prior to presenting at our emergency department. She had developed progressive abdominal pain, nausea, and vomiting. An initial upright abdominal radiograph (Fig. 28.2A) demonstrated a phi angle of 118 degree, superolateral band margin projecting 4.7 cm below the diaphragm, a developing "O" sign, and the presence of an air—fluid level above the band, all of which were compatible with band slippage. Unfortunately, the radiographic signs were not specifically observed and interpreted as diagnostic, and therefore a CT study was also performed after the patient was asked to drink oral contrast material (Fig. 28.2B). This study confirmed findings of band slippage, with an even more distended proximal pouch, now containing mixed contrast, liquid, and air, and with a phi angle of 140 degree and inferior displacement of the superolateral band margin 6.4 cm below the diaphragm. The band was then urgently deflated, and removed the following day.

CONCLUSION

AGB remains a common, effective bariatric surgical procedure, and millions of patients have had bands placed over the last two decades throughout the world. Although the modern pars flaccida surgical technique has decreased rates of band slippage, this remains the most common major complication that leads to repeat surgery. Patients with AGB slippage generally present with pain and obstructive symptoms of varying degrees, and therefore diagnostic imaging is frequently helpful in establishing the diagnosis of slippage. Radiologic studies must be interpreted in light of the expected anatomic location of the band (resulting from the pars flaccida technique), with the phi angle, distance of the superolateral band margin from the diaphragm, "O-sign," and presence of an air—fluid level within a proximal pouch above the band being four signs that can be used in combination to make a confident diagnosis and guide surgical decision-making.

KEY FACTS

- First performed in 1993, laparoscopic AGB became the number one preferred bariatric procedure worldwide by 2008.
- The popularity of AGB has waned because of growing concerns over delayed complications, including band slippage.
- Patients with AGB slippage typically present with pain and variable obstructive symptoms.
- Over time, surgical technique evolved from a perigastric placement to a more proximal pars flaccida location, and this has effected the appearance of AGB on diagnostic imaging.
- Deflation of the AGB reservoir, followed by band removal and replacement, or conversion to an alternative bariatric procedure, is the standard treatment for slippage.

SUMMARY POINTS

- First performed in 1993, laparoscopic AGB became the number one preferred bariatric procedure worldwide by 2008.
- The popularity of AGB has waned because of growing concerns over delayed complications, including band slippage.
- Patients with AGB slippage typically present with pain and variable obstructive symptoms.
- Over time, surgical technique evolved from a perigastric placement to a more proximal pars flaccida location, and this has changed the expected appearance of AGB on diagnostic imaging.
- Familiarity with the diagnostic performance of the phi angle, the "O" sign, inferior displacement of the superolateral band margin from the diaphragm, and presence of an air—fluid level within the pouch above the band, will help in confirming band slippage with radiologic studies.
- Deflation of the AGB reservoir, followed by band removal and replacement, or conversion to an alternative bariatric procedure, is the standard treatment for slippage.

REFERENCES

[1]   Belachew M, Legrand MJ, Defechereux TH, Burtheret MP, Jacquet N. Laparoscopic adjustable silicone gastric banding in the treatment of morbid obesity. Surg Endosc 1994;8:1354—6.

[2]   Buchwald H, Oien DM. Metabolic/bariatric surgery Worldwide 2008. Obes Surg 2009;19(12):1605—11.

[3]   Angrisani L, Santonicola A, Iovino P, Formisano G, Buchwald H, Scoporinaro N. Bariatric surgery Worldwide 2013. Obes Surg 2015;25(10): 1822—32.

[4]   Lee HJ, Ahn HS, Choi YB, et al. Nationwide survey on bariatric and metabolic surgery in Korea: 2003-2013 results. Obes Surg 2016;26(3): 691—5.

[5]   Favretti F, Segato G, Ashton D, Busetto L, et al. Laparoscopic adjustable gastric banding in 1791 consecutive obese patients:12-year results. Obes Surg 2007;17:168—75.

[6]   Tice JA, Karliner L, Walsh J, Petersen AJ, Feld- man MD. Gastric banding or bypass? A systematic review comparing the two most popular bariatric procedures. Am J Med 2008;121:885—93.

[7]   O'Brien PE, MacDonald L, Anderson M, Brennan L, Brown WA. Long-term outcomes after bariatric surgery: fifteen-year follow-up of adjustable gastric banding and a systematic review of the bariatric surgical literature. Ann Surg 2013;257(1):87—94.

[8]   Victorzon M, Tolonen P. Mean fourteen-year, 100% follow-up of laparoscopic adjustable gastric banding for morbid obesity. Surg Obes Relat Dis 2013;9(5):753—7.

[9]   Colquitt JL, Pickett K, Loveman E, Frampton GK. Surgery for weight loss in adults (Review). Cochrane Database Syst Rev 2014;8:CD003641.

[10] Weber M, Muller MK, Bucher T, et al. Laparoscopic gastric bypass is superior to laparoscopic gastric banding for treatment of morbid obesity. Ann Surg 2004;240:975—83.

[11] Dumon KR, Murayama KM. Bariatric surgery outcomes. Surg Clin North Am 2011;91:1313—38.

[12]   DeMaria EJ, Jamal MK. Laparoscopic adjustable gastric banding: evolving clinical experience. Surg Clin North Am 2005;85:773−87.

[13]   Egan RJ, Monkhouse SJ, Meredith HE, Bates SE, Morgan JD, Norton SA. The reporting of gastric slip and related complications; a review of the literature. Obes Surg 2011;21:1280−8.

[14]   Snow JM, Severson PA. Complications of adjustable gastric banding. Surg Clin North Am 2011;91:1249−64.

[15]   Eid I, Birch DW, Sharma AM, Sherman V, Karmali S. Complications associated with adjustable gastric banding for morbid obesity: a surgeon's guide. Can J Surg 2011;54(1):61−6.

[16]   Shen X, Zhang X, Bi J, Yin K. Long-term complications requiring reoperations after laparoscopic adjustable gastric banding: a systematic review. Surg Obes Relat Dis 2015;11(4):956−64.

[17]   Curet MJ. Treating morbid obesity with laparoscopic adjustable gastric banding. (commentary). Am J Surg 2007;194:344−5.

[18]   Martin LF, Smits GJ, Greenstein RJ. Treating morbid obesity with laparoscopic adjustable gastric banding. Am J Surg 2007;194:333−43.

[19]   Blachar A, Blank A, Gavert N, Metzer U, Fluser G, Abu-Abeid S. Laparoscopic adjustable gastric banding surgery for morbid obesity: imaging of normal anatomic features and postoperative gastrointestinal complications. Am J Roentgenol 2007;188:472−9.

[20]   Weisner W, Schob O, Hauser RS. Adjustable laparoscopic gastric banding in patients with morbid obesity: radiographic management, results, and postoperative complications. Radiology 2000;216:389−94.

[21]   Mehanna MJ, Birjawi G, Moukaddam HA, Khoury G, Hussain M, Al-Kutoubi A. Complications of adjustable gastric banding, a radiological pictorial review. Am J Roentgenol 2006;186:522−34.

[22]   Swenson DW, Pietryga JA, Grand DJ, et al. Gastric band slippage: a case-controlled study comparing new and old radiographic signs of this important surgical complication. Am J Roentgenol 2014;203(1):10−16.

[23]   Kuzmak LI. A review of seven years' experience with silicone gastric banding. Obes Surg 1991;1:403−6.

[24]   Obrien PE, Brown WA, Smith A, McMurrick PJ, Stephens M. Prospective study of a laparoscopically placed, adjustable gastric band in treatment of morbid obesity. Brit J Surg 1999;86:113−18.

[25]   Fielding GA, Allen JW. A step-by-step guide to placement of the LAP-BAND adjustable gastric banding system. Am J Surg 2002;184(6): S26−30.

[26]   Belachew M, Zimmermann JM. Evolution of a paradigm for laparoscopic adjustable gastric banding. Amer J Surg 2002;184:21S−5S.

[27]   McBride CL, Kothari V. Evolution of laparoscopic adjustable gastric banding. Surg Clin North Am 2011;91(6):1239−47.

[28]   DeMaria EJ, Sugerman HJ, Meador JG, Doty JM, et al. High failure rate after laparoscopic adjustable silicone gastric banding for treatment of morbid obesity. Ann Surg 2001;233(6):809−18.

[29]   Swenson DW, Levine MS, Rubesin SE, Williams NN, Dumon K. Utility of routine barium studies after laparoscopically inserted gastric bands. Am J Roentgenol 2010;194:129−35.

[30]   Suter M, Calmes JM, Paroz A, Giusti V. A 10-year experience with laparoscopic gastric banding for morbid obesity: high long-term complication and failure rates. Obes Surg 2006;16:829−35.

[31]   O'Brien PE, Dixon JB, Laurie C, Anderson M. A prospective randomized trial of placement of the laparoscopic adjustable gastric band: comparison of the perigastric and pars flaccida pathways. Obes Surg 2005;15:820−6.

[32]   Favretti F, Ashton D, Busetto L, Segato G, DeLuca M. The gastric band: first choice procedure for obesity surgery World J Surg 2009;published online. Available from: http://dx.doi: 10.1007/S00268-009-0091-6.

[33]   Pieroni S, Sommer EA, Hito R, Burch M, Tkacz JN. The "O" sign, a simple and helpful tool in diagnosis of laparoscopic adjustable gastric band slippage. Am J Roentgenol 2010;195:137−41.

[34]   Singhal R, Bryant C, Kitchen M, et al. Band slippage and erosion after laparoscopic gastric banding: a meta-analysis. Surg Endosc 2010;24:2980−6.

Chapter 29

Leaks and Fistulas After Bariatric Surgery

A. Johner[1] and L.L. Swanstrom[2,3]

[1]Good Samaritan Hospital, Portland, OR, United States, [2]Oregon Health and Sciences University, Portland, OR, United States, [3]Institut Hopitalo Universitaire, Strasbourg, France

LIST OF ABBREVIATIONS

CT   computed tomography
pSEMS partially covered self-expanding metal stents
RYGB   Roux-en-Y gastric bypass
SEPS   self-expanding plastic stents
SG   sleeve gastrectomy

INTRODUCTION

Anastomotic leak is a rare complication of bariatric surgery and is an independent risk factor for mortality [1]. This complication, if not identified and treated quickly and aggressively, may lead to abdominal sepsis, which has the potential to progress either to chronic gastric fistula or to multiorgan failure and patient demise. The incidence of leak after bariatric surgery ranges from 1.7% to 2.6% after open Roux-en-Y gastric bypass (RYGB) to 2.1% to 5.2% after laparoscopic RYGB, and reaches up to 5.1% after sleeve gastrectomy (SG) [2]. In revisional bariatric surgery, the risk of anastomotic leak approaches 35% [3]. Leaks are associated with a mortality rate of 6−14.7% [4,5]. In addition to doubling the risk of mortality, leaks result in a sixfold increase in hospital stays [6]. Patients who develop a leak are at increased risk for wound infection, sepsis, respiratory failure, renal failure, thromboembolism, internal hernia, and small-bowel obstruction [6].

A gastrointestinal (GI) leak is defined as discontinuity of tissue apposition in the postoperative period. Leaks develop when intraluminal pressure exceeds tissue or suture line resistance. This can occur in the immediate postoperative period or up to months later. Clinically, leaks may range from mild microleaks as the cause of peri-sleeve abscesses and chronic fistula, to an abdominal catastrophe. After RYGB, leaks can occur at several sites—the gastrojejunal anastomosis, gastric pouch staple lines, and the jejunojejunal anastomosis. The most common sites for leak are the gastrojejunal anastomosis followed by the jejunojejunal anastomosis, with an associated mortality of up to 18.4% and 40%, respectively [7]. Less commonly, leaks may occur at the jejunal stump, the excluded stomach, the duodenal stump (in resectional bypass), and the blind jejunal limb. In patients post-SG, leaks can occur anywhere along the long gastric staple line, although by far most leaks occur in the proximal third of the stomach near the gastroesophageal junction within 2 cm of the angle of His.

Several mechanisms for leaks after bariatric surgery have been proposed, and can be divided into technique-related and patient-related factors. Chen et al. described two main causes for staple line leaks post-SG: mechanical disruption of the staple line, and ischemic events leading to leakage from the fundus [8]. Leaks presenting within the first 2 days postoperatively are often mechanical, whereas ischemic leaks usually present 5−6 days postoperatively. Technical factors that can lead to the development of leaks include tension on the anastomosis, inappropriate staple height and stapling maneuvers, stapler misfiring, and staple line bleeding. Preservation of the pylorus and mucosal edema in the narrow sleeve following SG results in an increase in intragastric pressure. This elevated pressure is a natural sequelae of the operation, but in some instances is heightened due to a technical error, which results in a stricture or torsion of the gastric body. In addition, the esophagogastric junction represents an anatomical area of weakness. The fundic wall is thinner and vascularization is more precarious than the rest of the stomach. This combination of increased intragastric pressure and ischemia at the fundus has the propensity to result in leakage. The primary reason for the high leak rate observed at the gastrojejunal anastomosis postbypass is the presence of tension associated with this anastomosis. Additionally, tissue ischemia plays a role due to the division of the jejunal mesentery, which can compromise tissue perfusion to the antimesenteric aspect of the jejunum.

Patient-related factors contributing to the development of leaks include the presence of poor nutrition, current or recent smoking history, liver cirrhosis, and renal failure. In a study analyzing factors predictive of leaks after laparoscopic and open gastric bypass, Masoomi et al. found open gastric bypass (aOR 4.85), congestive heart failure (aOR 3.04), chronic renal failure (aOR 2.38), age >50 (aOR 1.82), Medicare payer (aOR 1.54), male gender (aOR 1.50), and chronic lung disease (aOR 1.21) to be factors associated with high risk for leaks [9]. In a meta-analysis of 4888 patients who underwent laparoscopic SG, a leak rate of 2.4% was demonstrated. BMI > 50 kg/m^2 and using a bougie smaller than 40 French were the factors found to be associated with increased leak rate [10]. In a retrospective analysis of 4444 patients put in the longitudinal assessment of bariatric surgery database, open surgery, revision surgery, and routine drain placement were associated with an increased leak rate [11].

DIAGNOSIS

The clinical presentation of an anastomotic leak can be subtle. Early leaks typically manifest constitutional symptoms within 3 days, though may still be difficult to diagnose. Gonzales et al. reported early symptoms of tachycardia, fever, and abdominal pain in most patients. Additional signs and symptoms include shoulder pain, hypoxia, leukocytosis >10,000/mm^3, and an elevated CRP of >11 mg/L [12]. The most common objective sign of leak is tachycardia, which is present in 72–92% of patients and may occur before a temperature elevation or impressive leukocytosis [12]. Tachycardia in the early postoperative period should raise suspicion for GI leak.

Chronic gastric leaks may develop with similar constitutional symptoms or as a less toxic and more insidious process with abdominal pain and fever. Chronic gastrogastric fistulas are a nonurgent complication defined as a communication between the pouch and defunctionalized stomach, and typically manifest with acid reflux, abdominal discomfort, and weight regain. Gastric acid can flow into the pouch via a gastrogastric fistula, so patients with heartburn, acid reflux, or anastomotic ulcer should be evaluated for fistula.

The diagnosis of a leak can be made radiographically or endoscopically (Fig. 29.1). Upper GI series with water-soluble contrast and computed tomography (CT) have limited sensitivity, but have a high positive predictive value. Often, an upper GI series is performed to confirm contrast extravasation, and subsequently, a CT scan to delineate the site of leak and also the presence of local or distant fluid/abscess collections. Alternatively, CT may be utilized after an inconclusive upper GI study. However, the sensitivity of an upper GI contrast study varies among reports from between 22% and 75%. In one study, CT was diagnostic in only 56% of patients, with 30% of patients having both a negative barium swallow and CT despite a leak being present [12,13]. Whether a postoperative upper GI series should be ordered routinely or selectively remains controversial. Patients with normal studies may harbor leaks, especially at the

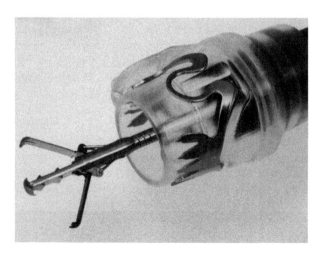

FIGURE 29.1 An Over-the-Scope clip. Over-the-Scope clips are useful for full-thickness closures of fistulas and leaks less than 2 cm.

jejunojejunal anastomosis or excluded stomach. Normal study findings should not delay therapy if clinical signs suggest a leak. Drain amylase levels have been proposed as a simple, low-cost adjunct with high sensitivity and specificity that can help to identify patients who may have a leak after gastric bypass surgery.

The most sensitive method for diagnosis of an anastomotic leak is probably a combination of maneuvers, including an endoscopic examination with careful visualization of the esophagus, stomach, or gastric pouch (including retroflexion) after suction of all fluid present, a bubble test, and injection of contrast with methylene blue while looking endoscopically and fluoroscopically. Most practitioners usually use either an endoscopic test with insufflated CO_2 or a blue dye test via an orogastric tube; both tests have rather low sensitivity but high specificity for positive results.

The high mortality associated with leaks, together with the lack of reliable clinical symptoms and poor specificity of imaging, mandates clinical vigilance. The threshold for early exploration should be low where lack of improvement usually prompts surgical reintervention. A negative exploration that does not reveal a leak should not be considered a complication, but rather evidence of good surgical judgment. Multiple studies have shown that reexploration is safe when compared with missing or delaying the treatment of a leak [11,14]. In experienced hands, surgical exploration and management of a leak is often feasible via a laparoscopic approach. In 2009, the American Society for Metabolic and Weight Loss Surgery published a position statement on prevention and detection of GI leaks after gastric bypass, including the role of imaging and surgical exploration [15]. The position statement suggested that laparoscopic or open reexploration should be an appropriate diagnostic option when a GI leak is suspected, as reliance on false-negative imaging studies may delay operative intervention.

MANAGEMENT

The surgical principles for leak management include sepsis control (including broad spectrum antibiotic coverage, identification and repair of the defect, irrigation and control of contamination, wide external drainage of the contaminated area), adequate nutrition, appropriate skin care, monitoring fistula output, and anatomical definition. Further steps to improve healing include enteral feeding, alleviation of downstream strictures, and minimizing fistula exposure to alimentary contents. The initial treatments should include NPO (i.e., not by mouth), early initiation of broad spectrum antibiotics, and fluid resuscitation. The main treatment options in managing staple line leaks vary depending on the physiological status of the patient. In cases of hemodynamic stability, nonoperative interventions may be pursued. For small, contained leaks in patients without peritoneal signs on abdominal exam, the patient should be kept NPO with total parenteral nutrition, and broad spectrum antibiotics should be administered. Accessible intraabdominal collections should be drained percutaneously. The course for nonoperative treatment should be followed with measurement of serial white blood cell (WBC) count, vitals, and a physical exam. Reports have shown 23 of 26 and 5 of 5 patients with successful conservative management when classified as "low risk" [12,16]. Patients with increasing WBC, persistent tachycardia or worsening of abdominal pain should proceed to operative management.

Surgical

Whenever hemodynamic compromise exists, or after failure of nonoperative management, immediate reoperation is the preferred course of action. Precise surgical management depends on the extent of the disruption, the extent of abdominal contamination, the site of the leak, and timing of presentation. Initial surgical intervention often takes the form of laparoscopic drainage of all fluid collections and placement of abdominal drains. An attempt may be made at suture closure of the defect if the tissue condition allows it. Sound surgical judgement is imperative in deciding whether the tissues are amenable to suturing or whether intervention will only impose further damage, as these closures tend to break down due to poor tissue integrity at the leak site. Alternatively, resection of the anastomosis may be required. Nutritional access should be considered with placement of a gastrostomy tube in the remnant stomach or a jejunostomy tube for postoperative enteral feeding.

The timing of the leak is important to the strategy of surgical repair. Early gastric sleeve leaks (<48 hours) have a better likelihood of successful immediate repair, whether proximal or distal. Furthermore, early postbypass leaks at the jejunojejunostomy or gastric remnant may be more amenable for primary closure, and revision of the anastomosis is rarely needed. Conversely, late-occurring sleeve leaks, especially if of ischemic origin, are less likely to be responsive to immediate surgical repair. Leaks from the gastric pouch or gastrojejunal anastomosis after gastric bypass occur in a lower pressure system, unless obstruction exists at the jejunal anastomosis; hence, these leaks are more likely to close with time. Conversely, the persistence of a leak after SG is perpetuated by the high pressure originated by the presence of a functioning pylorus.

An alternative surgical approach to control the leak site is placement of a t-tube directly into the defect. This technique consists of obtaining a conventional t-tube drain and placing the T part of the drain directly into the defect. The drain is then exteriorized and placed to bag drainage, converting the leak to a controlled gastrostomy. The t-tube is left in place for 4−6 weeks, and is slowly withdrawn over time (1−2 in. per week). A track is created along the drain, hence forming a controlled fistula. Upon withdrawal of the tube, the well-formed fistulous track will collapse and eventually close as long as there is no distal obstruction. Additional surgical revision options have been described and range from a Roux-en-Y pull-up of small intestine anastomosed to the leak site, conversion from sleeve to gastric bypass with resection of the leak site (if sufficiently distal), or resection of the gastric sleeve with an esophagojejunostomy. An additional novel approach has been described using a Roux limb as a serosal patch to the staple line disruption. It is cautioned that revisional surgery should not be attempted before 3−4 months postleak as this is enough time for adhesion maturation; others advocate no less than 6 months for reparative surgery. Surgical management is associated with high morbidity (up to 50%) and mortality (2%−10%), with a high conversion rate to open surgery (48%) as well as long operative time and increased intraoperative blood loss [12,17]. As such, initial management has moved toward conservative or endoscopic treatment.

Endoscopic

Surgical therapies for persistent gastric leak are technically difficult and often unsuccessful. Because of this, approaches using minimally invasive endoscopic repair techniques are increasingly being used as first-line therapies. So far, data supports that they are both safe and effective. Endoscopic techniques include endoscopic clips or suturing to close the leak, use of covered stents across the leak, sealants, plugs, or internal decompression with pigtail stents. Finally, vacuum-assisted drainage has been investigated.

Endoscopic Clips or Suturing

Clips have been used with variable efficacy for GI tract defects, although limited data are available for bariatric management of gastric leaks and fistula. Clips are used to approximate the tissue surrounding the defect to effect closure. Standard clips should be deployed perpendicular to the long axis of the defect. If needed, multiple clips can be placed sequentially, starting at either edge of a defect and meeting at the center. Currently available through-the-scope clips achieve superficial tissue apposition engaging the mucosa and submucosa (with 1.2-mm-wide and 6-mm-long arms capable of an approximately 12-mm grasp), and have been used in conjunction with thermal ablation or mechanical scraping of the tissue around the edges of the defect to achieve a more resilient seal. The Over-the-Scope Clip (Ovesco Endoscopic AG, Tubingen, Germany) is a nitinol clip placed on a cap at the endoscope tip (Fig. 29.1). Unlike clips inserted through the endoscope, the OVESCO can perform full-thickness apposition. Case series of GI tract fistula closure have shown success rates of 72−91% [18,19].

Endoscopic suturing platforms are now available and may allow intraluminal management of complications previously treated by laparoscopic or open surgery. However, there are no large series reporting the use of suturing devices in the treatment of postsurgical leaks. Device limitations combined with procedural complexity and need for specialized technical skill have limited adoption. The StomaphyX suturing system (EndoGastric Solutions, Redmond, WA), initially purposed for stomal reductions, was described for the treatment of postbariatric fistulas. Overcash et al. described successful use of the device in the management of two patients, one with a 7.6-mm fistula and the other with a 1.0-cm fistula. They used five and six polypropylene fasteners, respectively, to create pleats of gastric tissue surrounding the defect, forming a "tissue shield." Follow-up was at 6 and 3 months, respectively, which confirmed fistula healing [20]. Fernandez-Espaarach et al. examined sutured gastrogastric fistula repaired using the Bard EndoCinch (C R Bard, Murray Hill, NJ) [21]. Initial success rate was 95% and durable success rate was 35%. None of the fistulas with a diameter >20 mm remained sealed, and fistulas <10 mm had the best outcomes. There was one esophageal perforation and one occurrence of significant bleeding. It has been postulated that inadequate suture placement depth may have resulted in lack of successful closure. The Apollo OverStitch (Apollo Endosurgery, Austin, TX), which creates full-thickness plications, has shown early success (Fig. 29.2). Other endoscopic devices are under development for gastric tissue manipulation that may aid fistula closure; however, these devices are in various stages of research and require further investigation.

Clips and sutures are most effective for acute perforations or leaks. All are less effective for anastomotic leaks due to tissue thickness and anatomic position.

FIGURE 29.2   The OverStitch device. OverStitch is an endoscopic suturing device that is available in most markets.

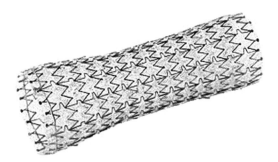

FIGURE 29.3   Endolumenal stent. Covered or partly covered endoluminal stents are frequently used to exclude acute and subacute leaks.

Stent Placement

Supported by the most substantial body of evidence, stent placement serves as a useful adjunct to drainage, creating enteral diversion and potentially allowing oral nutrition, thereby theoretically accelerating recovery and avoiding the need for parenteral nutrition (Fig. 29.3). Peritoneal contamination is ideally decreased and improvement in abdominal pain may follow. The less invasive nature of endoscopic therapy may convey a decrease in cost and possibly a mortality benefit when compared with conventional surgical management.

Several case reports and a few case series demonstrate the utility of stents for management of postbariatric surgery leaks [22–26]. Using partially covered self-expanding metal stents (pSEMS) in postbariatric surgery leak patients, Eisendrath et al. reported an overall 81% closure rate ($n = 17/21$) with a stent duration of 2 months and a median follow-up of 221 days. Complications included stent migration, bleeding, and thoracic pain [22]. Eubanks et al. used pSEMS and self-expanding plastic stents (SEPS) to heal 89% of acute leaks ($n = 11$) and 50% of chronic leaks ($n = 2$) with an average stent duration of 20 days [23]. Further studies by Blackmon et al. used pSEMS for successful treatment of 10 gastric bypass leaks, 5 of which were acute and 5 chronic [24]. An additional study by Tan et al. reported 50% resolution in 8 patients with SG leaks using pSEMS with 4 successful fistula closures after 6 weeks of stent placement [25]. A meta-analysis of stent placement for treatment of acute leaks after bariatric surgery by Puli et al. [26] demonstrated a pooled proportion for successful leak closure, defined as radiographic evidence of leak closure after stent removal, of 87.8% (95% confidence interval, 79.4–94.2%). A majority of leaks closed with one treatment, but restenting was reported in four of seven studies. Nine percent of patients had failure and required revision surgery. Stents were extracted between 4 and 8 weeks in the majority of studies. Stent migration was reported in 16.9% (95% CI, 9.3–26.3%).

Stent placement is best done with the use of fluoroscopy and a forward-viewing endoscope. Deployment precautions should include ensuring adequate distance from the upper esophageal sphincter to avoid globus sensation, and appropriate length to prevent distal enteral wall impaction because this may lead to bleeding and possibly perforation. Stent deployment may begin proximally or distally, depending on the system. Both covered SEMS and SEPS have been used. SEPS are more easily removed than SEMS, but the migration risk is higher. The use of partially covered stents can decrease migration rate because of tissue ingrowth into the stent, but subsequent extraction becomes more difficult. If tissue hyperplasia occurs, argon plasma coagulation can be used to fulgurate ingrown tissue. Alternatively, a large-diameter SEPS can be placed inside the SEMS to induce pressure necrosis of the ingrown tissue with subsequent withdrawal of the stents in tandem.

Enteral stent complications for postbariatric surgery leaks include transient chest pain radiating to the back induced by stent expansion, bleeding, nausea, and stent migration. Stents are usually left in place for 2–8 weeks because longer indwelling periods can increase extraction difficulty. One of the major difficulties with usage of stents for control of leaks is their ability to migrate, with the literature reporting an approximate 17% stent migration rate [26]. Future postbariatric anatomy-specific stent designs may help mitigate this. Several principles should be followed when an esophageal stent is considered for management of a gastric leak after SG or gastric bypass. Gastric leaks at the proximal and mid-aspect of the gastric sleeve are best suited to endoscopic treatment with a stent. A leak at the distal staple line of the gastric sleeve, near the gastric antrum, may not heal with stenting as the stent diameter would likely be too narrow and would not provide appropriate sealing of the defect. Whenever utilizing stents for leak management, the size of the endoscopic stent should be selected erring on the large size in an effort to prevent migration. Methods to prevent migration include clipping or suturing the stent in place, and anecdotal evidence suggests this decreases but doesn't eliminate migration.

Sealants and Plugs

Fibrin sealant is a biodegradable compound with a long history of surgical use. As a group, fibrin sealants are not associated with the inflammatory response and tissue necrosis inherent to other sealants such as cyanoacrylate. Currently approved for hemostasis, skin graft attachment in burn patients, and colostomy closure, fibrin glue has been used off label for esophageal leaks and GI fistulas. The endoscopic approach involves fistula identification, removal of overlying granulation tissue when possible, and subsequent excoriation or mucosal ablation prior to sealant application. The sealant is endoscopically placed using a dual lumen catheter that allows mixture of components at the distal tip, resulting in a whitish fibrin plug. Care must be taken to insert the more viscous component via the larger lumen. Multiple sessions may be required.

Several series and case reports have shown efficacy of endoscopic gastric leak management with fibrin glue. In a study reviewing 354 laparoscopic RYGB, Kowalski et al. encountered eight gastric leaks, five of which were successfully treated with fibrin glue [27]. Lippert et al. reported a series of 52 patients with GI fistula; 36.5% had sealing with fibrin alone, and 55.7% were cured with additional endoscopic therapy [28]. Patients without infection had a higher cure rate. A series of 15 patients who failed conservative therapy for fistula was reported by Rabago et al. [29]. Sealing of 86.6% of fistulas was achieved in a mean 16 days after a mean 2.5 sessions. Cure was more frequent in low-output fistulas than high-output fistulas. Wong et al. demonstrated treatment of postsurgical fistulas with insertion of fibrin glue during fistuloscopy by using a 5-mm choledochoscope [30]. Other procedures were performed concurrently, including irrigation and debridement. All nine fistulas sealed in a mean 18.7 days, without recurrence during the following year. Further work is necessary to identify characteristics that may favor fibrin glue therapy.

Biologic matrix plugs have been described for the closure of chronic fistulas, especially for enterocutaneous ones (Fig. 29.4). These are usually placed by passing a guidewire or catheter through the track attaching the plug and pulling it through the mouth and into the fistula. Fixation can be external if possible, or internal with clips or sutures. Maluf-Filho et al. demonstrated the use of Surgisis strips and fistula plugs to treat gastrocutaneous fistulas in 25 patients after RYGB [31]. The plugs were deployed in five patients, and the strips inserted via polypectomy snare into the lumenal opening of the fistulas of 20 patients. Closure was achieved in all patients with plugs, and in 75% of patients treated with strips.

Internal Drainage

More chronic leaks with walled-off adjacent abscesses are sometimes best treated by internal drainage using endoscopic pigtail catheters. This technique involves advancing the endoscope through the fistula, even if it requires dilating it, lavaging the abscess cavity and making sure any percutaneous drains are not immediately adjacent to the defect. Then one or more double pigtail stents are deployed across the defect (Fig. 29.5). External drains are removed as soon as volumes decrease, usually in a day or two. Catheters are left in place for 3–4 weeks and then removed one by one, each with an interval of 2–3 weeks. Donatelli et al. [32] described their experience with endoscopic internal drainage

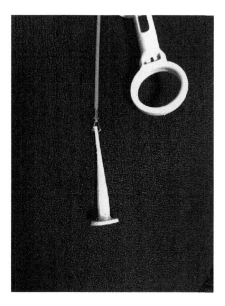

FIGURE 29.4 Biomatrix plug. Biomatrix plugs can be useful in the control and treatment of chronic fistulas.

FIGURE 29.5 Internal drainage. Double pigtail endoscopic drains are useful and effective ways to close both acute and nonacute leaks that are not free into the chest or abdomen.

for treatment of leaks following SG. Double pigtail stents were successfully delivered in 66 out of 67 patients. Fifty patients were cured by endoscopic drainage after a mean time of 57.5 days and an average of 3.14 endoscopic sessions. The technique appears to achieve drainage of perigastric collections while stimulating mucosal growth over the stent. Endoscopic internal drainage appears to be well-tolerated with few complications.

Vacuum-Assisted Sponge Closure

Endoluminal vacuum therapy is a promising new method in the treatment of upper GI leaks and perforations. The device consists of an open-cell sponge attached to external vacuum suction via tubing. The sponge induces formation of granulation tissue, while vacuum suction improves perfusion and removes secretions. A feeding tube is inserted transnasally and then orally exteriorized. A sponge, which is cut to a size smaller than the wound cavity, is fixed to the tip of the tube with suture. The sponge is grasped with endoscopic forceps and introduced into the fistula endoscopically (Fig. 29.6). The feeding tube is then attached to continuous vacuum suction. The sponge is changed two or three times weekly. The system has been used successfully to close rectal anastomotic fistulas and has also shown success in a prospective trial of treatment for intrathoracic anastomotic leaks. Ahrens et al. reported its use for gastroesophageal leak. All 5 patients had leak closure at a median 42 days after a mean 9 sponge changes. Two patients developed stenosis

FIGURE 29.6   Vacuum sponge. Catheter-based, endoscopic, negative-pressure therapy is a new therapy to promote healing of leaks.

requiring dilation, and one patient had hemorrhage after dilation [33]. Future prospective studies will be needed to validate these results.

PREVENTION

Some technical pitfalls seem to be important in preventing postsurgical leaks and fistulas. A tight sleeve is created by dividing all connective tissue and vascular attachments of the stomach except the lesser curvature vessels. Use of the appropriate staple height for the resected segment of the stomach is mandatory, and care should be taken with heat-producing instruments so as not to cause thermal injury to the created sleeve. If the dissection is too aggressive near the posterior aspect of the upper sleeve, devascularization may occur, making that area more susceptible to leakage. Bougie sizes ranging from 32 to 48 French have been used for sleeve calibration. Most advocate the use of bougies smaller than 40 French without buttressing material or oversewing of the staple line. Reports linking smaller bougie size to leak rates have been published, and a recent consensus statement by an international SG expert panel deemed the optimal bougie size to be 32–36 French [34]. In an added effort to prevent postoperative leaks, surgeons have advocated the need to strengthen the staple line. Buttressing with specific bioabsorbable materials, oversewing, or application of sealant agents have been proposed. To date, no study has unequivocally supported or obviated the use of buttressing material or suture reinforcement for leak prevention. A systematic analysis showed that oversewing or buttressing of the staple line does not have a clinically significant effect on leakage [10]. Routine or selective use of intraoperative diagnostic methods is likewise controversial. Intraoperative endoscopy, air leak testing, and transgastric dye injection have been used to detect a leak during initial surgery or in a patient returning after a suspected or proven leak. However, it is well-established that a negative intraoperative test does not totally eliminate the possibility of a leak.

CONCLUSIONS

Staple line disruption is one of the most morbid surgical complications after a stapling bariatric procedure. This can result in a lengthy recovery with multiple admissions, radiographic studies, inability to eat, malnutrition, and potential additional invasive procedures. It should also be considered that nonstaple line leaks can occur as these surgeries involve extensive manipulations and introduction of devices across the esophagus. Meticulous attention to surgical technique and a high clinical suspicion with early leak identification and management reduce the incidence of leaks and the associated morbidity and mortality. Many post-SG leaks can occur after patient discharge, and so vigilant follow-up during the first 30 days is recommended. Management depends on clinical presentation and requires treatment of sepsis, nutritional support, abscess detection, appropriate drainage, fistula monitoring, skin care, and frequently invasive procedures. Revision surgery has a high morbidity and mortality, and is usually technically difficult. In patients with clinically stable presentations, there may be a role for noninvasive endoscopic therapies with a more attractive risk profile. These include clips, fibrin glue, plugs, endoscopic suturing, stent placement, internal drainage, and vacuum-assisted sponge closure. Currently, the best evidence exists for stents, despite the relatively high migration rate. For late and chronic leaks, management may require proximal gastrectomy with esophagojejunostomy. Leaks and fistulas after bariatric surgery are likely to become an even more prominent issue with the increasing frequency of bariatric surgery being performed, and as such, will require ongoing research and attention.

MINI-DICTIONARY OF TERMS

- *Gastrointestinal Fistula*: An abnormal connection between two epithelialized surfaces that usually involves the gut and another hollow organ.
- *Upper GI Series*: Fluoroscopic and radiographic examination of the esophagus, stomach, and duodenum during and following oral ingestion of a solution of barium sulfate or a water-soluble contrast.
- *Endoscopic Clip*: A metallic mechanical device used in endoscopy in order to close two mucosal surfaces without the need for surgery and suturing. Its function is similar to a suture in gross surgical applications, as it is used to join together two disjointed surfaces, but can be applied through the channel of an endoscope under direct visualization.
- *Endoscopic Stent*: A hollow device endoscopically inserted, designed to prevent constriction or collapse of a tubular organ. Also utilized to seal and bypass GI leaks or perforations.
- *Internal Drainage*: An internal drain placed endoscopically through the staple line dehiscence that may effectively control local sepsis by draining a perigastric abscess, promote resorption of fluid, enhance healing, and avoid the formation of an external fistula.

KEY FACTS

- The incidence of a leak is approximately 2%, likely double that of RYGB.
- Most leaks occur in the proximal third of the stomach near the gastroesophageal junction.
- The development seems to be a result of local ischemia in combination with elevated intraluminal pressure, which can be exacerbated by technical aspects of the sleeve construction.
- The clinical presentation may be subtle, although one can progress to overwhelming sepsis.
- Various options in management, including nonoperative, endoscopic, and surgical, all of which typically result in a multiple month-long healing process.

SUMMARY POINTS

- Gastrointestinal leak after bariatric surgery is associated with significant morbidity and mortality.
- Early identification and management is key to minimizing further complications.
- Patients without signs of systemic toxicity may be managed nonoperatively.
- If hemodynamic compromise exists, or after failure of nonoperative management, reoperation may be required.
- Endoscopic repair techniques are increasingly being utilized as first-line therapy.

REFERENCES

[1] Fernandez AZ, DeMaria EJ, Tichansky DS, et al. Multivariate analysis of risk factors for death following gastric bypass for treatment of morbid obesity. Ann Surg 2004;239(5):698–703.
[2] Morales MP, Miedema BW, Scott JS, et al. Management of postsurgical leaks in the bariatric patient. Gastrointest Endosc Clin N Am 2011;21:295–304.
[3] Gonzalez R, Murr MM. Anastomotic leaks following gastric bypass surgery. In: Jones DB, Rosenthal R, editors. Weight loss surgery: a multidisciplinary approach. Edgemont: Matrix Medical Communications; 2008. p. 369.
[4] Lee S, Carmody B, Wolfe L, et al. Effect of location and speed of diagnosis on anastomotic leak outcomes in 3828 gastric by-pass cases. J Gastrointest Surg 2007;11:708–13.
[5] Carucci LR, Turner MA, Conklin RC, et al. Roux-en-Y gastric bypass surgery for morbid obesity: evaluation of postoperative extraluminal leaks with upper gastrointestinal series. Radiology 2006;238:119–27.
[6] Almahmeed T, Gonzalez R, Nelson LG, et al. Morbidity of anastomotic leaks in patients undergoing Roux-en-Y gastric bypass. Arch Surg 2007;142:954–7.
[7] Lee S, Carmody B, Wolfe L, et al. Effect of location and speed of diagnosis on anastomotic leak outcomes in 3828 gastric bypass cases. J Gastrointest Surg 2007;11:708–13.
[8] Chen B, Kiriakopoulos A, Tsakayannis D, et al. Reinforcement does not necessarily reduce the rate of staple line leaks after sleeve gastrectomy. A review of the literature and clinical experiences. Obes Surg 2009;19:166–72.

[9]   Masoomi H, Kim H, Reavis KM, et al. Analysis of factors predictive of gastrointestinal tract leak in laparoscopic and open gastric bypass. Ach Surg 2011;146(9):1048−51.

[10] Aurora AR, Khaitan L, Saber AA. Sleeve gastrectomy and the risk of leak: a systematic analysis of 4,888 patients. Surg Endosc 2012;26:1509.

[11] Smith MD, Adeniji A, Wahed AS, et al. Technical factors associated with anastomotic leak after Roux-en-Y gastric bypass. Surg Obes Relat Dis 2015;11:313.

[12] Gonzalez R, Sarr MG, Smith CD, et al. Diagnosis and contemporary management of anastomotic leaks after gastric bypass for obesity. J Am Coll Surg 2007;204:47−55.

[13] Lee CW, Kelly JJ, Wassef WY. Complications of bariatric surgery. Curr Opin Castroenterol 2007;23:636−43.

[14] Durak E, Inabnet WB, Schrope B, et al. Incidence and management of enteric leaks after gastric bypass for morbid obesity during a 10-year period. Surg Obes Relat Dis 2008;4:389.

[15] ASMBS Clinical Issues Committee. ASMBS guideline on the prevention and detection of gastrointestinal leak after gastric bypass including the role of imaging and surgical exploration. Sure Obes Relat Dis 2009;5(3):293−6.

[16] Csendes A, Braghetto I, León P, et al. Management of leaks after laparoscopic sleeve gastrectomy in patients with obesity. J Gastrointest Surg 2010;14:1343−8.

[17] Madan AK, Lanier B, Tichansky DS. Laparoscopic repair of gastrointestinal leaks after laparoscopic gastric bypass. Am Surg 2006;72:586−90.

[18] Surace M, Mercky P, Demarquay JF, et al. Endoscopic management of GI fistulae with the over-the-scope clip system (with video). Gastrointest Endosc 2011;74:1416−19.

[19] von Renteln D, Denzer UW, Schachschal G, et al. Endoscopic closure of GI fistulae by using an over-the-scope clip (with videos). Gastrointest Endosc 2010;72:1289−96.

[20] Overcash WT. Natural orifice surgery (NOS) using StomaphyX for repair of gastric leaks after bariatric revisions. Obes Surg 2008;18:882−5.

[21] Fernandez-Esparrach G, Lautz DB, Thompson CC. Endoscopic repair of gastrogastric fistula after Roux-en-Y gastric bypass: a less-invasive approach. Surg Obes Relat Dis 2010;6:282−8.

[22] Eisendrath P, Cremer M, Himpens J, et al. Endotherapy including temporary stenting of fistulas of the upper gastrointestinal tract after laparoscopic bariatric surgery. Endoscopy 2007;39:625−30.

[23] Eubanks S, Edwards CA, Fearing NM, et al. Use of endoscopic stents to treat anastomotic complications after bariatric surgery. J Am Coll Surg 2008;206:935−8 discussion 938-939

[24] Blackmon SH, Santora R, Schwarz P, et al. Utility of removable esophageal covered self-expanding metal stents for leak and fistula management. Ann Thorac Surg 2010;89:931−6 discussion 936-937

[25] Tan JT, Kariyawasam S, Wijeratne T, et al. Diagnosis and management of gastric leaks after laparoscopic sleeve gastrectomy for morbid obesity. Obes Surg 2010;20:403−9.

[26] Puli SR, Spofford IS, Thompson CC. Use of self-expandable stents in the treatment of bariatric surgery leaks: a systematic review and meta-analysis. Gastrointest Endosc 2012;75:287−93.

[27] Kowalski C, Kastuar S, Mehta V, et al. Endoscopic injection of fibrin sealant in repair of gastrojejunostomy leak after laparoscopic Roux- en-Y gastric bypass. Surg Obes Relat Dis 2007;3:438−42.

[28] Lippert E, Klebl FH, Schweller F, et al. Fibrin glue in the endo- scopic treatment of fistulae and anastomotic leakages of the gastrointestinal tract. Int J Colorectal Dis 2011;26:303−11.

[29] Rábago LR, Ventosa N, Castro JL, et al. Endoscopic treatment of postoperative fistulas resistant to conservative management using biological fibrin glue. Endoscopy 2002;34:632−8.

[30] Wong SK, Lam YH, Lau JY, et al. Diagnostic and therapeutic fistuloscopy: an adjuvant management in postoperative fistulas and abscesses after upper gastrointestinal surgery. Endoscopy 2000;32:311−13.

[31] Maluf-Filho F, Hondo F, Halwan B, et al. Endoscopic treatment of Roux-en-Y gastric bypass-related gastro-cutaneous fistulas using a novel biomaterial. Sure Endosc 2009;23(7):1541−5.

[32] Donatelli G, Dumon JL, Cereatti F, et al. Treatment of leaks following sleeve gastrectomy by endoscopic internal drainage. Ones Surg 2015;25 (7):1293−301.

[33] Ahrens M, Schulte T, Egberts J, et al. Drainage of esophageal leakage using endoscopic vacuum therapy: a prospective pilot study. Endoscopy 2010;42:693−8.

[34] Rosenthal RJ. International sleeve gastrectomy expert panel consensus statement: best practice guidelines based on experience of >12,000 cases. Surg Obes Relat Dis 2012;8:8−19.

Chapter 30

Gastric Leaks and Use of Endoscopic Internal Drainage With Enteral Nutrition

G. Donatelli[1] and P. Dhumane[2]

[1]Hôpital Privé des Peupliers, Paris, France, [2]Lilavati Hospital and Research Center, Mumbai, India

LIST OF ABBREVIATIONS

BS bariatric surgery
CT computed tomography
EDEN endoscopic internal drainage and enteral nutrition
GI gastrointestinal
GL gastric leak
NOTES natural orifice transluminal endoscopic surgery
SEMS self-expanding metallic stents

INTRODUCTION

The prevalence of overweight and obesity problems in multiple countries all over the world is alarmingly on the rise [1]. More than one-third (34.9% or 78.6 million) of adults and 17% of youths under the age of 20 in the United States were obese as of 2012 [2]. It has been described as a "global pandemic," and concerns about the health risks associated with rising obesity have become nearly universal [1]. This has led to the introduction of many new bariatric procedures that have quickly gained popularity among people in the medical fraternity. In 2013, 468,609 bariatric procedures were performed worldwide, 95.7% carried out laparoscopically. The highest number ($n = 154,276$) was from the United States/Canada region. The most commonly performed bariatric procedure in the world was Roux-en-Y gastric bypass at 45%; this was followed by sleeve gastrectomy (SG) at 37%. The prevalence of SG increased from 0% to 37% of the world total from 2003 to 2013, and the prevalence of Adjustable Gastric Banding (AGB) fell significantly to 10% in 2013 from its peak in 2008 of 68% [3]. Between the years 2011−2013, a total of 18,283 bariatric procedures, 16,956 primary and 1327 revisions or planned second stage procedures, were performed in the United Kingdom. A total of 52.1% (99,526/18,283) of these were gastric bypass followed by AGB and SG [4]. These surgical procedures can have some unique complications, such as malnutrition, bleeding, stenosis, and reflux, and they can also cause GLs. All the bariatric procedures are centered on stomach, the reservoir of food, and proximal jejunum; hence, gastric leaks (GLs) are one of the most common serious complications.

SG has erroneously been considered simple and easy, which has led to its adoption by a large number of surgeons. After laparoscopic sleeve gastrectomy surgery (LSGS) for morbid obesity, a considerable number of patients develop nutritional and surgical complications. Compared to gastric bypass and biliopancreatic diversion, its complications can be even more severe [5]. Incidences of bleeding, leaks, and stenosis are estimated to be between 10% and 13.2% [6], with leaks being one of the most prominent causes of morbidity and mortality. The International Sleeve Gastrectomy Expert Panel Consensus Statement 2011 (12799 LSGS) showed the leak rate was 1.06% [7], but the leak rate can vary between 1% and 3% for the primary procedure and [8] more than 10% in revision procedures.

GLs after bariatric surgery (BS) represent one of the most dreaded complications due to their associated high morbidity and mortality. According to the United Kingdom Surgical Infection Study Group [9], a GL is defined as a "leak of luminal contents from a surgical join between two hollow viscera." It can also be an effluent of gastrointestinal (GI) content through a suture line, which may collect near anastomosis, or exit through the wall or the drain [9]. GLs are most likely to occur along the proximal third of the stomach, close to the gastroesophageal junction due to creation of a stomach tube with high intraluminal pressure with impaired peristaltic activity and ischemia. No standard protocol for management of GLs exists. Surgical revision, due to surrounding inflammation and ischemic edges, is often unsuccessful and burdened with high postoperative complications [10,11]. Surgical treatment is usually reserved for patients presenting severe sepsis or Multiorgan Dysfunction Syndrome. The mainstay of nonsurgical (conservative) treatment consists of complete drainage of any fluid collection, enteral hyperalimentation, and antibiotics therapy [12].

At present, placement of self-expanding metal stents (SEMS) across a leak coupled with surgical or radiological drainage of the perigastric cavity is considered a mainstay of treatment for managing these complications. Their success rate depends largely on delay of intervention after surgery. Moreover, use of SEMS is limited by poor tolerance due to nausea, vomiting, retrosternal discomfort, and by significant morbidity and mortality. Migration of SEMS is still an open issue occurring with a frequency variable from 33% [13] to 83% [14] of cases. Widening of the leak orifice due to constant radial force exerted by SEMS and the inability of the peri-GL cavity to effectively empty its contents internally leads to refilling of the fistula tract, and this keeps the fistula patent for a long time. Occlusion of the distal end of the metallic stent due to tissue overgrowth has also been observed in our experience. This issue is in particular with the newly developed SEMS, which are designed to be longer and wider in order to reduce migration [15].

Alternative options such as glue, biodegradable plugs, Over-the-Scope clips, transorificeal plastic stenting, and also simple drainage tubes associated with enteral feeding, have been proposed, but with limited success. However, universal guidelines concerning management of these complications are lacking [16].

ENDOSCOPIC INTERNAL DRAINAGE WITH ENTERAL NUTRITION

Common surgical dictum states that if there is any leakage of GI contents from any hollow viscus in the peritoneal cavity, it needs to be drained. This drainage can be achieved by inserting wide-bore drains, which entails a surgical procedure, either a laparotomy or laparoscopy. Another option for achieving drainage of the collection is by inserting a pigtail or malecot's catheter under image guidance (ultrasonography or computed tomography (CT) scan). More recently, various techniques of endoscopic treatment of postsurgical GLs have been introduced. Pequignot et al. first reported use of pigtail drains in post-SG leaks [17]. This approach was described as more effective, better tolerated, and resulting in faster healing, as fewer procedures are required as compared to SEMS [10]. Slim et al. reported use of endobiliary stents for successful management of GLs following SG [18].

Here we present the treatment protocol of EDEN, which is an efficacious, safe, and rapid alternative for achieving early healing of GLs after BS.

The Technique

All those patients who have an established diagnosis of GL following BS are suitable candidates for the EDEN protocol, irrespective of duration of their leak. Early, acute, and late leaks can be managed with comparable success rates by this protocol. Patients are initially evaluated with a CT scan or upper GI contrast study. Patients in severe sepsis and the primary surgical team's refusal to include their patient in an EDEN protocol are the only reasons for deferring implementation of the EDEN protocol as a primary line of therapy for these leaks.

Under general anesthesia and oral endotracheal intubation, upper GI endoscopy is performed in the supine position using standard gastroscopes under cover of systematically administered antibiotics. At the beginning of all procedures, an endoscopic contrast study is done to precisely localize the peri-GL/collection (Fig. 30.1). Its exact anatomical location and its potential communication with abdominal drain, if present, are also ascertained. Having this information at hand, after reaching the gastric lumen, the tip of the endoscope is carefully maneuvered across the leak orifice (if it is of at least 1 cm diameter and permits the passage of the scope). This is to objectively gauge orifice size and to thoroughly examine the leak cavity for its size, shape, volume, contents, orientation, and inner lining. At times, the tip of the external drainage catheter can be visualized in the cavity. Then the endoscope is changed to a duodenoscope or colonoscope, depending on the size of the cavity orifice. These endoscopes have a larger working channel, which enables delivery of the appropriately sized double-pigtail plastic stent across the orifice. For the most part, the 10 French stent suffices for the purpose, but the size of the pigtail stent is dictated by the size of the collection, while the shape of the collection determines the number of stents used to ascertain complete drainage from each part of the cavity.

The steps for delivering stents are summarized in Fig. 30.2. Using an injection catheter, the leak orifice is catheterized, and contrast medium is injected into the cavity. Then, under fluoroscopic guidance, a guidewire is pushed into the peri-GL cavity, the catheter is withdrawn, and a double-pigtail plastic stent is delivered across the leak orifice so that

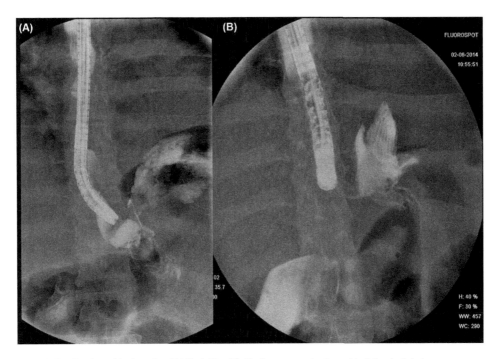

FIGURE 30.1 Endoscopic localization of leak cavity. (A) Peri-GL. (B) Cavity communicating with abdominal drain.

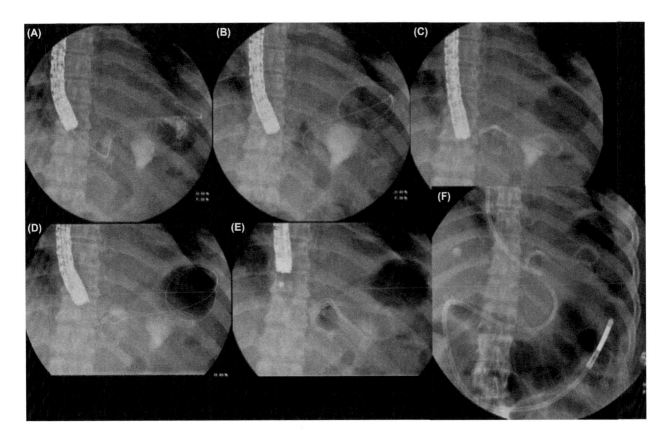

FIGURE 30.2 Steps of EDEN procedure. (A) Opacifying leak cavity. (B) Guidewire pushed into cavity. (C) Delivery of first double pigtail stent across orifice. (D, E) Delivery of second pigtail stent across orifice. (F) A nasojejunal (NJ) feeding tube is inserted up to the third part of the duodenum.

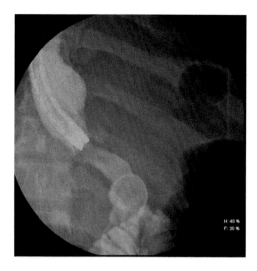

FIGURE 30.3 Good patient outcome. Absence of free passage of contrast medium from the gastric lumen to the perigastric/peritoneal cavity at 1 month of EDEN treatment.

one pigtail end remains inside the cavity and the other pigtail end rests inside the gastric lumen. More stents are delivered across the orifice if the leak orifice is large enough to accommodate two or more stents, or if the shape of the cavity is complex, making it necessary to install more stents for it to drain more effectively. Finally, a nasojejunal (NJ) feeding tube is left in place in the third part of the duodenum, under fluoroscopic control. All patients are kept nil by mouth and on exclusive enteral nutrition (101 kcal/100 mL, 1500 mL day) until the first-check endoscopy 4−6 weeks after the index endoscopic procedure. Patients are discharged between 1 and 7 days after the index endoscopic procedure, depending on their clinical condition (Fig. 30.3).

Patients are followed up on a weekly basis. During every follow-up visit, a detailed clinical evaluation is performed. A contrast CT scan is reserved for those in whom clinical evaluation indicates a persistent GL/lingering abdominal sepsis. A perigastric corrugated drain, when present, is gradually withdrawn, depending on the fluid drained, until it is completely removed over the next few weeks.

At around 4−6 weeks after the index endoscopic procedure of the EDEN protocol, patients are systematically hospitalized for stent replacement and endoscopic reevaluation. Double-pigtail stents placed during previous endoscopies are retrieved using endoscopic Foreign Body Forceps. The patient's outcome is defined as "good" if the following two criteria are met:

1. The absence of perigastric collection and/or no free abdominal fluid or pus drainage through the stent/leak orifice, and
2. The absence of free passage of contrast medium from the gastric lumen to the perigastric/peritoneal cavity and/or to the external drainage tube upon injection of the contrast medium inside the gastric lumen (Fig 30.4).

Presence of a reepithelialized, watertight, small pseudocavity or blind crossing stent at the level of leak is considered acceptable (Fig. 30.4). This sequence of regular clinical evaluation and 4−6 weekly endoscopic evaluations, coupled with stent replacement, is continued until final healing is achieved. The final decision of removing the pigtail stent, or restenting coupled with enteral nutrition, is at the discretion of the expert endoscopist. After the first-check endoscopy, the patient is gradually started with oral intake and advanced to a full oral diet over a few weeks, followed by removal of the NJ tube.

For the sake of uniformity in recording of data related to the EDEN protocol, we recommend the following nomenclature. "**Technical success**" is defined as the successful deployment of a pigtail stent across the leak. "**Clinical success**" is defined as the absence of free contrast medium extravasation in the peritoneal cavity, neither around the stomach nor through the fistula orifice. Pseudodiverticula communicating with the gastric tube with assured internal emptying will also amount to "clinical success."

Results

Experience and data related to implementation of the EDEN protocol for post-BS GLs is limited. In 2015 our team published the results of 67 patients of the EDEN protocol following GLs after BS [15]. This is probably the one of its

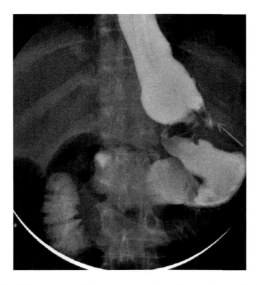

FIGURE 30.4 Pseudodiverticulum. Reepithelialized, water tight, small pseudocavity at the level of the leak (*red arrow* (gray arrow in print versions)).

kind series of successful implementations of the EDEN protocol for post-SG GLs. Technical success was achieved in 66/67 patients (98.5%). In one patient, we failed to establish internal drainage due to intraprocedural perforation related to erroneous guidewire manipulation. The patient required emergency laparotomy and recovered fully after 3 months of enteral nutrition by the NJ tube. Two patients died of pulmonary embolism at days 1 and 13 after their index endoscopic procedure. In our series, except for six patients who were started with total parenteral nutrition, all patients received exclusive enteral nutrition. In that group of patients, at the time of going to press, clinical success was achieved in 59/64 (92%) of patients after a mean stenting duration of 57.5 days (10−206) with an average of 3.14 (2−16) endoscopic sessions per patient. We registered five clinical failures (8%), i.e., the fistula persisted and needed an alternative treatment modality for final cure. Two patients with chronic fistulas were successfully cured by injection of *n*-butyl-2-cyanoacrylate glue (Glubran 2; GEM, Viareggio, Italy) after failure of endoscopic drainage (average of 368 days of treatment and 12 endoscopic sessions) due to recurrent pigtail stent migration. We could not ascertain reasons for repeated stent migration in these patients. But in both these patients, orifice size was significantly reduced. The other three patients (two late and one chronic fistula) were definitively treated by total gastrectomy for chronic sepsis.

Six of fifty patients developed stenosis after an average of 36 days (15−45) after the end of the EDEN protocol, i.e., after the last pigtail stent was removed and the patient was declared a "clinical success." All patients underwent an average of three pneumatic dilatations with an achalasia balloon up to a diameter of 40 mm. In one case, after balloon dilation failed, a fully covered SEMS was deployed for 5 weeks. All stenoses were successfully treated endoscopically. On retrospective analysis, four out of six of these patients had complete dehiscence (length >2 cm) of the staple line. Aggressive granulation tissue formation induced by the pigtails and subsequent wound contraction might have led to stenosis formation.

As of April 2016, we have implemented the EDEN protocol in more than 200 patients: overall clinical success is 90%, with average follow-up of more than 5 months (1−36 m), and an average treatment period of 2 months (10−270 days). Median delay between the index BS procedure and EDEN is 24 days (0−1430).

We also had some unique complications. One of the stents migrated to the spleen 4 weeks after the index endoscopic procedure with the fistula healed [19]. The patient was successfully managed endoscopically without surgery. In one of the patients, the gastric end of the pigtail stent caused esophageal ulceration leading to continuous pain, for which the stent needed to be removed (Fig 30.5). In another patient, an ulcer induced massive bleeding, which was treated by blood transfusion and removal of the pigtail in an emergency operation. We had a technical failure in one of these patients—we could not deliver the stent in a fistula with a crisscross trajectory despite good opacification of the tract. On injecting the contrast medium, we had a vascular perenchymography of a solid organ, mostly spleen, so we abandoned the procedure (Fig 30.6).

Five percent of our patients needed a complementary drainage tube (placed by the radiologist and/or surgeon) following EDEN because the endoscopic drainage was deemed to be not enough (persistence of inflammatory syndrome). This happened for one huge leak cavity (Fig 30.7). New devices need to be developed to achieve complete internal drainage of such difficult-to-drain leak cavities.

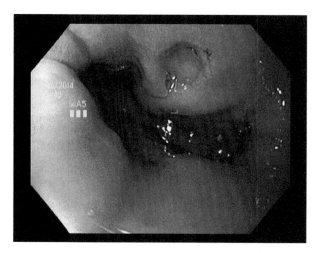

FIGURE 30.5 Complication of EDEN. Deep esophageal ulceration due to the gastric end of the pigtail.

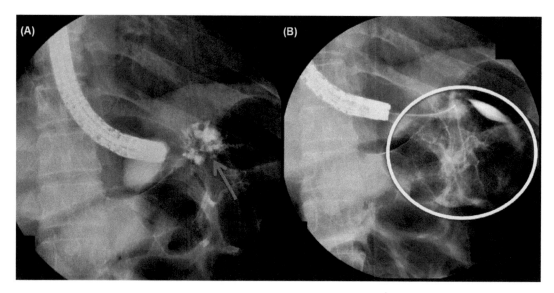

FIGURE 30.6 Technical failure of EDEN. (A) Chronic perigastric fistula (*red arrow* (gray arrow in print versions)). (B) Parenchymography (*yellow circle* (white circle in print versions)) following failure of endoscopic internal drainage.

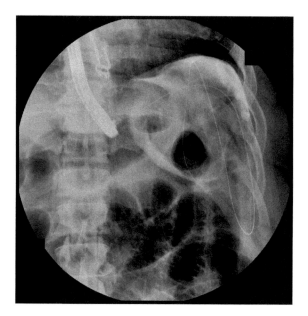

FIGURE 30.7 Need of complementary drainage tube after EDEN. Huge perigastric collection needing drainage tube placement by a radiologist following EDEN.

Philosophy Behind the Success of EDEN

We feel that the EDEN protocol is a one-step solution for management of GLs occurring after BS. Rather than just bypassing (spanning) the leak with SEMS, the key to success is to accomplish complete internal drainage of any collection with pigtail stents, which also induces microtrauma to the cavity and its orifice to promote healing. This philosophy is akin to the principles behind drainage of the pseudocyst of pancreas or walled-off pancreatic necrosis.

After encouraging results from our initial experience (since March 2013), we abandoned the use of SEMS in favor of the EDEN protocol. There are multiple aspects of the EDEN protocol that make it an efficacious, safe, and definitive treatment for these leaks.

- *It is more efficient, easier, and more physiological to drain the leak cavity internally from inside than by putting in a surgical or radiological drainage.* The post-SG leak cavity is located in anatomical proximity to the stomach sleeve and away from the skin. By draining this collection externally, a longer, nonanatomical track is created across different anatomical plains, which are otherwise not involved in the leak cavity formation. An external drainage tube takes many days to dry up, and then the track epithelializes over many days. Internal drainage puts spilled-over contents of the GI track back into its organ of origin. Double pigtail stents transport liquid contents of the cavity to the gastric lumen by capillary action. Variations in the intragastric pressure gradient with every cycle of respiration, and during acts of abdominal straining, must be helping to achieve complete drainage of the cavity via stent and orifice. So, draining the abscess cavity internally into the gastric lumen is more physiological and easier, and it significantly reduces potential complications of long-term external fistulas.

  Patients with severe sepsis, and some of the other patients with post-SG GLs, undergo initial damage control procedures in their parent surgical unit, and hence they have surgically placed or radiologically guided drainage tubes draining the leak cavity at the time of presentation to the endoscopy team. An effective EDEN protocol can successfully allow withdrawal of this external drainage tube over a couple of weeks without formation of fistulas. As the leak cavity is effectively drained in the gastric lumen, gastric contents don't spill over to the fistula track, and its epithelialization is prevented, leading to its successful closure. Better coordination between surgical and endoscopic treating teams will lead to promotion of the EDEN protocol as the first line of treatment for managing post-BS GLs.
- *A double-pigtail stent is a more physiological and efficient management alternative to achieve internal drainage and stimulate granulation tissue.* A pigtail stent, acting as a foreign body and by causing microtrauma as long as it is in situ, promotes reepithelialization. At the same time it guarantees complete internal drainage of infected fluid collection. Moreover, stents do not interfere with early oral realimentation after a short period of enteral nutrition. Pigtail stents are usually well-tolerated, and in our experience, most of our patients even tolerated a normal diet with a pigtail stent in situ. Early initiation of enteral nutrition is necessary for control of local and systemic sepsis, and for rapid progress of the healing process of the leak cavity.
- *Paramount importance needs to be given to location and orientation of the leak cavity and its internal evaluation.* We perform not only an intraprocedural contrast study, but also, whenever feasible, inspect inside the cavity by direct visualization with an endoscope (NOTES procedure). At this stage, loose necrotic tissue inside the cavity can also be removed with the help of an endoscope (Fig. 30.8).

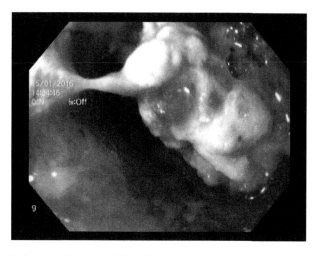

FIGURE 30.8 Necrosectomy. Removal of necrosis from the inside cavity.

- *Another important aspect is CO_2 insufflation instead of air insufflation.* It reduces risk of persistent pneumoperitoneum, air embolization, and ensures postprocedural patient discomfort.
- *Systematic endoscopic evaluation and stent replacement for all patients after every 4–6 weeks are important.* It is necessary in order to avoid stent blockage, and also because of the stent being a foreign body, that it should not remain inside the body for a longer duration. It serves one more purpose: microtrauma during the stent replacement procedure helps granulation tissue formation and eventually leads to watertight closure of the fistula tract.
- *A double-pigtail stent placed across the narrow leak orifice is less likely to migrate.* The stent seems to be well-anchored and may be expelled into the stomach tube only when the leak cavity almost completely obliterates. All the patients who encountered stent migration in our series had obliterated leak cavities. For a small cavity, it is important to deploy only a small-caliber pigtail stent. They are softer and easier to insert, thus reducing the risk of perforating the cavity.
- *Persistence of an orifice leading to a spontaneously emptying pseudodiverticulum does not have any negative impact on recovery.* This doesn't affect realimentation or gastric motility or stomach emptying. All our patients who had pigtail stents removed in spite of persistent pseudodiverticulum had uneventful recovery and didn't require any further treatment.
- *Once technical success is achieved, the EDEN protocol will almost guarantee fistula healing.* None of our patients required treatment interruption, not even in cases of complications related to stent deployment. EDEN can be safely continued until complete healing, allowing patients to continue with oral intake, without need for any external drainage.

The EDEN protocol should be considered as a primary management tool for both early and late GLs in patients after BS procedures, if no diffuse peritonitis or multiorgan failure is present. Although multiple endoscopic sessions are required, the EDEN protocol achieves the most complete drainage of perigastric collections, simultaneously stimulating mucosal growth. EDEN is better tolerated, more effective, and less expensive than SEMS, and is burdened by fewer complications. Long-term follow-up confirms good outcomes with no motility or feeding-related issues.

MINI-DICTIONARY OF TERMS

- *Gastric Leaks (GLs)*: Breaches in continuity of the gastric lining that lead to extravasation of the contents of the digestive system.
- *Fistula*: A permanent abnormal track between two hollow organs in the body or between an organ and the exterior of the body.
- *Internal drainage*: Drainage of an intraperitoneal cavity/collection into the hollow organ through the orifice of cavity without placement of external drainage tubes.
- *EDEN*: Protocol including internal drainage of a leak cavity by placing pigtail stents across the leak orifice and early enteral nutrition.
- *NOTES (natural orifice transluminal endoscopic surgery)*: A surgical approach that introduces flexible or rigid endoscopic instruments through one of the natural orifices and makes an entry into the peritoneal cavity by cutting open that visceral organ.

KEY FACTS [20]

- Soehendra and Reynders-Frederix, working in Hamburg, Germany, described the first case of insertion of a single-pigtail biliary endoprosthesis.
- They made it by using the cut end of an angiography catheter for drainage of malignant obstructive jaundice. It migrated upstream.
- Cotton, working in London, reported the first use of an endoprosthesis made with a double-pigtail design to prevent upward migration.
- They are radioopaque, and a loop at each end helps keep the stent in place.
- Apart from routine use for various biliary indications, they are used for other novel indications.

SUMMARY POINTS

- Gastric leaks (GLs) after bariatric surgery (BS) represent one of the most dreaded complications due to their associated high morbidity and mortality.
- A double-pigtail stent is a more physiological and efficient management alternative to achieve internal drainage and stimulate granulation tissue.
- Paramount importance needs to be given to location and orientation of the leak cavity and its internal evaluation.
- Systematic endoscopic evaluation and stent replacement for all patients after every 4−6 weeks are important.
- Persistence of an orifice leading to a spontaneously emptying pseudodiverticulum does not have any negative impact on recovery.
- Once technical success is achieved, the EDEN protocol will almost guarantee fistula healing.
- The EDEN protocol should be considered as a primary management tool for both early and late GLs in patients after BS procedures if no diffuse peritonitis or multiorgan failure is present; EDEN is more efficient, easier, and more physiological for draining a leak cavity internally from inside than by putting in a surgical or radiological drainage.

REFERENCES

[1] Ng M, et al. Global, regional, and national prevalence of overweight and obesity in children and adults during 1980−2013: a systematic analysis for the Global Burden of Disease Study 2013. Lancet 2014;384:766−81.

[2] Ogden CL, Carroll MD, Kit BK, et al. Prevalence of childhood and adult obesity in the United States, 2011−2012. JAMA 2014;311(8):806−14.

[3] Angrisani L, Santonicola A, Iovino P, et al. Bariatric surgery worldwide 2013. Obes Surg 2015;25(10):1822−32.

[4] The UK National Bariatric Surgery Registry of the British Obesity and Metabolic Surgery Society. Second Registry Report 2014.

[5] Fuks D, Verhaeghe P, Brehant O, et al. Results of laparoscopic sleeve gastrectomy: a prospective study in 135 patients with morbid obesity. Surgery 2009;145:106−13.

[6] Trastulli S, Desiderio J, Guarino S, et al. Laparoscopic sleeve gastrectomy compared with other bariatric surgical procedures: a systematic review of randomized trials. Surg Obes Relat Dis 2013;9(5):816−29.

[7] Rosenthal RJ, Diaz AA, Arvidsson D, et al. International Sleeve Gastrectomy Expert Panel Consensus Statement: best practice guidelines based on experience of >12,000 cases. Surg Obes Relat Dis 2012;8:8−19.

[8] Márquez MF, Ayza MF, Lozano RB, et al. Gastric leak after laparoscopic sleeve gastrectomy. Obes Surg 2010;20:1306−11.

[9] Peel AL, Taylor EW. Proposed definitions for the audit of postoperative infection: a discussion paper. Surgical Infection Study Group. Ann R Coll Surg Engl 1991;73:385−8.

[10] Dakwar Anthony, Assalia Ahmad, Khamaysi Iyad, et al. Late complication of laparoscopic sleeve gastrectomy. Case Rep Gastrointest Med 2013; Article ID 136153, 5 pages.

[11] Roller JE, Provost DA. Revision of failed gastric restrictive operations to Roux-en-Y gastric bypass: impact of multiple prior bariatric operations on outcome. Obes Surg 2006;16:865−9.

[12] Sakran N, Goitein D, et al. Gastric leaks after sleeve gastrectomy: a multicenter experience with 2,834 patients. Surg Endosc 2013;27:240−5.

[13] Csendes A, Braghetto I, León P, et al. Management of leaks after laparoscopic sleeve gastrectomy in patients with obesity. J Gastrointest Surg 2010;14(9):1343−8.

[14] Campos JM, Pereira EF, Evangelista LF, et al. Gastrobronchial fistula after sleeve gastrectomy and gastric bypass: endoscopic management and prevention. Obes Surg 2011;21(10):1520−9.

[15] Donatelli G, Dumont JL, Cereatti F, et al. Treatment of leaks following sleeve gastrectomy by endoscopic internal drainage (EID). Obes Surg 2015;25:1293−301.

[16] Donatelli G, Ferretti S, Vergeau BM, et al. Endoscopic internal drainage with enteral nutrition (EDEN) for treatment of leaks following sleeve gastrectomy. Obes Surg 2014;24:1400−7.

[17] Pequignot A1, Fuks D, Verhaeghe P, et al. Is there a place for pigtail drains in the management of gastric leaks after laparoscopic sleeve gastrectomy? Obes Surg 2012;22(5):712−20.

[18] Slim R, Smayra T, Chakhtoura G, et al. Endoscopic stenting of gastric staple line leak following sleeve gastrectomy. Obes Surg 2013;23:1942−5.

[19] Donatelli G, Airinei G, Poupardin E, et al. Double-pigtail stent migration invading the spleen: rare potentially fatal complication of endoscopic internal drainage for sleeve gastrectomy leak. Endoscopy 2016;48(Suppl 1):E74−5.

[20] Leung JW. History of bile duct stenting: rigid prostheses. Self-expandable stents in the gastrointestinal tract. New York, NY: Springer; 2013. p. 15−31. Available from: http://dx.doi.org/10.1007/978-1-4614-3746-8_2.

Chapter 31

Gastroesophageal Reflux Disease and Hiatal Hernia in Bariatric Procedures

A. Ardestani[1,2], A. Tavakkoli[1,2] and E.G. Sheu[1,2]

[1]Harvard Medical School, Boston, MA, United States, [2]Brigham and Women's Hospital, Boston, MA, United States

LIST OF ABBREVIATIONS

BE	Barrett's esophagus
BOLD	Bariatric Outcomes Longitudinal Database
GEJ	gastroesophageal junction
GERD	gastroesophageal reflux disease
GI	gastrointestinal
HH	hiatal hernia
LAGB	laparoscopic adjustable gastric band
LES	lower esophageal sphincter
LRYGB	laparoscopic Roux-en-Y gastric bypass
LSG	laparoscopic sleeve gastrectomy

INTRODUCTION

More than 70% of the adult population over 20 years old is overweight, and over one-third of US adults are obese [1,2]. Obesity is an important risk factor for the development of various comorbidities, including gastroesophageal reflux disease (GERD). Obesity increases the risk of GERD development up to twofold. While the prevalence of GERD is about 10–20% in the general population, it's substantially higher ($> 30\%$) in the obese population [3,4]. Furthermore, reflux symptoms have been reported in more than 50% of the severely obese patients who are candidates for bariatric surgery [4].

MECHANISMS

Obesity likely contributes to GERD through multiple mechanisms. Increased intraabdominal pressure leads to changes in the gastroesophageal pressure gradients. These changes facilitate retrograde flow of the gastric contents and consequently reflux. Increased intraabdominal pressure may also promote disruption of the gastroesophageal junction (GEJ), which may contribute to development of a hiatal hernia (HH). Esophageal motility disorders such as nutcracker esophagus and nonspecific motility disorder are also seen with a higher prevalence in the obese population with GERD in comparison to both nonobese and non-GERD populations.

In HH, widening of the esophageal hiatus of the diaphragm allows intermittent or persistent herniation of abdominal organs into the mediastinum [5]. HH is present in up to 40% of obese patients [6]. In a study of 345 bariatric surgery candidates, more than half of the patients had HH detected on endoscopy [7]. These patients also had a higher rate of esophagitis and abnormal pH testing. HH are classified into four types: Type I is when the GEJ alone is above the diaphragm, either constantly or intermittently (sliding HH); Type II is when the upper stomach is intrathoracic, while the GEJ remains below the diaphragm; in Type III, both the GEJ and stomach (usually, fundus and cardia) are within the

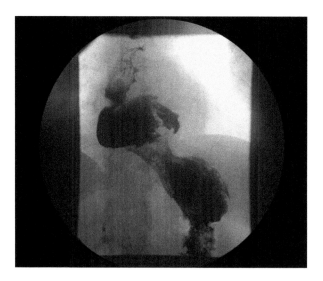

FIGURE 31.1   Hiatal hernia in an upper GI contrast study.

mediastinum (Fig. 31.1); and lastly, in Type IV, other abdominal organs (transverse colon, spleen, liver) are involved within the hernia. The majority of HH found in the preop bariatric surgery patient are small, type I hernias [6,8].

TREATMENT

Weight loss and antisecretory medications are the first line of therapy for obese patients who suffer from GERD. However, as is the case with weight loss in general, the impact of diet and lifestyle interventions for GERD are inconsistent and often transient. [4] Medical management with proton pump inhibitors (PPI) is the next line of therapy. Conflicting data exists on whether PPI effectiveness is altered in the obese. [9] However, as with the general population, a significant proportion of GERD patients have an incomplete response to PPI therapy, and side effects of medical therapy are not uncommon. Furthermore, in the presence of a significant HH, symptoms such as regurgitation and dysphagia are likely to persist even with weight loss and acid suppression.

In the nonobese patient, surgical treatment is offered for medically refractory GERD. What is currently considered the gold standard surgical treatment for GERD is Nissen fundoplication with concurrent repair of HH, which can usually be accomplished laparoscopically and transabdominally. However, significant evidence indicates the efficacy and durability of a surgical fundoplication is diminished in the obese patient. [10,11] Data has now accumulated that bariatric surgery—in particular, laparoscopic Roux-en-Y gastric bypass (LRYGB)—leads to significant improvement in GERD, and is now considered the recommended surgical therapy for GERD in the obese patient [9,10]. Small series comparing LRYGB to laparoscopic Nissen fundoplication have demonstrated improvement in objective and subjective measures of GERD, with the added benefit of weight loss and resolution of other obesity-related comorbidities [12].

PREOPERATIVE GERD AND HH EVALUATION IN THE BARIATRIC PATIENT

Currently, there are no accepted standards for the preoperative screening for GERD and HH in the bariatric surgery patient. A thorough history and physical exam to illicit symptoms suggestive of reflux or HH, such as heartburn, regurgitation, and dysphagia, is an important first step. Patients with symptoms or history suggestive of significant GERD or HH, or a family history of upper gastrointestinal (GI; esophageal, gastric) malignancy, likely merit a more intensive preoperative evaluation. The American Society of Metabolic and Bariatric Surgery recommends preoperative imaging studies such as upper GI contrast studies or endoscopy in patients with clinically significant GI symptoms [13].

However, many patients with GERD and HH are asymptomatic, so history and a physical exam are not reliable indicators to guide preoperative evaluations [14]. As a result, many practitioners advocate routine upper GI contrast studies (UGI) or upper endoscopy (esophagogastroduodenoscopy, EGD) to screen for HH or signs of significant reflux [7]. Early studies recommended esophageal manometry and even formal pH testing prior to planned laparoscopic adjustable gastric banding (LAGB) or laparoscopic sleeve gastrectomy (LSG). [15] However, a recent meta-analysis found that preoperative EGD led to results that changed or delayed surgery in less than 8% of the patients, and therefore recommended selective use of preoperative endoscopy in bariatric surgery candidates [16]. As discussed in more detail below, choice of bariatric procedure may influence aggressiveness in the preoperative evaluation—in particular, to rule out presence of Barrett's esophagus (BE).

Options for studies to screen patients preoperatively include UGI and EGD. UGI offers the advantage of being non-invasive and can be a helpful screening modality for HH, severe reflux, esophageal motility disorders, and unusual anatomic abnormalities or mass lesions, although the sensitivity and specificity varies depending on the operator. The advantages of EGD include mucosal visualization to evaluate for reflux esophagitis, BE, and unexpected tumors. As with UGI, however, the accuracy of EGD is user-dependent and does confer the additional risk of an invasive procedure and sedation in an obese patient. As detailed below, for those with severe GERD symptoms, EGD may be the preferred choice to rule out BE or other complications of GERD.

CHOICE OF BARIATRIC PROCEDURE IN PATIENT WITH GERD

Bariatric surgery can lead to significant improvement in GERD in the obese patient. The etiology is multifactorial, but significant weight loss with reduced intraabdominal pressure is likely a critical component of GERD resolution. Results from the bariatric outcomes longitudinal database (BOLD) have shown a correlation between postoperative weight loss and improved GERD symptoms [14,17]. Nevertheless, there are variations in GERD resolution between bariatric procedures. In a study of over 22,000 patients from BOLD with GERD symptoms, the highest rate of improvement was recorded in the LRYGB patients in comparison to LAGB and LSG, although overall, GERD improved in most patients (Table 31.1) [18].

It has been shown that LRYGB leads to superior improvement of GERD symptoms in comparison to other procedures in patients with similar weight loss [18]. Multiple mechanisms may contribute to the superior efficacy of LRYGB, including creation of a small gastric pouch, thus limiting potential esophageal acid exposure; decreased bile reflux with the long Roux limb and gastrojejunostomy; preservation of native GEJ anatomy and intrinsic antireflux mechanisms; and excellent weight loss. In summary, LRYGB is considered by most as the procedure of choice in obese patients suffering from GERD.

IMPACT OF LSG ON GERD

LSG has become the commonly performed bariatric operation in the United States. However, the impact of LSG on GERD remains unclear. While many patients experience resolution or improvement of GERD symptoms after LSG, a subset of patients experience worsened or de novo GERD symptoms [19]. A few studies have been published with both pre- and post-LSG physiologic testing and have shown conflicting results. One prospective of 71 patients showed that in patients with preoperative pathologic pH testing, DeMeester scores and GERD symptoms significantly improved 2 years after LSG, with resolution of pathologic acid exposure in most patients and no change in lower esophageal sphincter (LES) pressures. In patients without preoperative GERD on pH testing, the rate of de novo GERD development was 5% [20]. However, two other case series have reported reduction in LES pressures and increased esophageal acid exposure in patients after LSG [21,22].

Improvement in GERD after LSG may be due to reduction in parietal cell mass and acid production, faster gastric emptying, as well as weight loss-induced improvement in intraabdominal pressure. Conversely, other anatomic changes after LSG have been suggested to worsen GERD, including reduced distensibility and compliance of the residual stomach, leading to heightened intragastric pressure; disruption of the GEJ and angle of His impairing LES function; and division of sling fibers involved in LES function [23].

TABLE 31.1 Variation in the Percentage of Patients Experiencing Improvement or Worsening of GERD After Bariatric Surgery Based on Procedure From BOLD Database [18]

	Improvement	Worsening
LRYGB	56.6	2
LAGB	46	1.2
LSG	41	4.6

Technical factors in forming the sleeve have been argued to play a role in worsening GERD. These include creating a tighter sleeve with a smaller bougie size, leaving excessive residual gastric fundus, disrupting the angle of His, and division of sling muscle fibers involved in LES function [24]. Most surgeons feel that small to moderate HH should be diligently searched for and repaired in LSG, especially in those with preoperative GERD symptoms, as this may improve GERD outcomes postop. Some have advocated routine hiatal dissection and crural closure in all LSG patients, although the data to support this is weak [25,26].

Early on, GERD and HH were considered a relative contraindication to LSG. Data has been fairly uniform supporting the superiority of LRYGB over LSG for GERD. For example, one randomized study showed a higher incidence of GERD (12.5% vs 4%; $p = 0.12$), and decreased rate of GERD improvement or resolution (75% vs 50%; $p = 0.008$) after LSG as compared to LRYGB [27]. At the most recent International Consensus on LSG, 23% of expert surgeons, but 52% of all surveyed surgeons, considered GERD a contraindication for a sleeve, highlighting the controversy that still exists on this topic. [28]. Overall, most studies demonstrate improvement or resolution of GERD in most patients. Therefore, in patients with mild or well-controlled GERD, most feel that, with adequate patient counseling, LSG is an acceptable option, particularly if contraindications to LRYGB exist. The true incidence of worsening or de novo GERD after LSG, as well as risk factors to predict these patients ahead of time, remain to be defined.

BARRETT'S ESOPHAGUS

Persistent esophageal acid exposure leads to esophageal mucosal damage and intestinal metaplasia, termed BE. A small percentage of BE patients will progress to develop esophageal adenocarcinoma. In the obese patient with GERD and BE, most surgeons agree that LRYGB is the first option. First, LRYGB is the most effective bariatric operation for treatment of GERD, and physiologically would be predicted to have the greatest impact on BE. Small series of LRYGB patients demonstrate complete or partial regression of BE postoperatively [29]. Moreover, the gastric remnant is preserved and can be used for reconstruction should esophageal cancer develop and esophagectomy be required.

For similar reasons, 80% of expert surgeons at the most recent sleeve consensus conference consider BE a contraindication to perform the sleeve [28]. LSG is the inferior reflux operation, and moreover, eliminates the stomach as an option for esophageal reconstruction, obligating colonic or jejunal interposition down the line should esophagectomy be required. Given that the preoperative diagnosis of BE may influence bariatric surgical choice, this may be an argument in support of more aggressive screening endoscopy in the bariatric patient. Still, given the lack of good, long-term data, controversy persists on this topic. Some have argued that LSG is an acceptable option for patients with BE in the absence of dysplasia, particularly in the older patient, given the overall low rate of BE progression to cancer and effectiveness of new surveillance and endoscopic ablation techniques [30].

BARIATRIC SURGERY SELECTION IN THE PRESENCE OF HH

Similar to primary antireflux operations in the obese, while primary HH repair in the obese population is safe and feasible, durability is believed to be poor, and a concomitant operation for weight loss is recommended if indicated. Studies from two US nationwide databases have shown that simultaneous HH repair with either LAGB or LRYGB is safe and feasible [8,18,31]. More recent work has also demonstrated the safety of LSG in combination with HH repair [8,32,33].

As in reflux disease, in those with a large hiatal or paraesophageal hernia, LRYGB is the recommended procedure of choice [34]. First, it is the better antireflux procedure, which often coexists in the HH patient. Second, the anatomical changes after LRYGB may promote durability, as the Roux limb and gastrojejunostomy may function as a pexy to help provide recurrent intrathoracic herniation of the gastric pouch.

Overall, with a large HH, LAGB is not recommended. In order to attempt to prevent reflux and to restore normal GEJ for optimal band placement, most surgeons have opted to repair even small HH when placing bands. However, a study on 40,000 LAGB patients from the BOLD has shown that concurrent repair of HH with LAGB (in a subset of over 8000 patients) had minimal effect on GERD improvement after LAGB [35].

High-quality, long-term data regarding the efficacy and durability of combined LSG and HH is scarce. Only 12% of expert surgeons in the most recent international sleeve consensus survey consider HH a contraindication for LSG [28]. The consensus did not distinguish between size of the hernia, and many surgeons would likely consider large paraesophageal (type III–IV) hernias as a relative contraindication to an LSG. This aversion stems from both the more mixed outcomes of LSG on GERD, as well as the worry that the tubular stomach will be more prone to herniate intrathoracically should the crural repair break down. However, in the short-term, combined paraesophageal hernia repair with

LSG has been shown to be safe and efficacious. That being said, the durability and rate of symptomatic relapse in the medium-term follow-up may be of concern [34−37].

The more common, smaller, type 1 HH frequently found in the bariatric population are likely not a contraindication to an LSG. As indicated previously, aggressive interrogation for, and repair of, occult or small HH at the time of LSG have been advocated by many as a way to improve outcomes [36,37]. A retrospective study of 338 patients who underwent LSG showed that a combination of LSG and HH repair yields higher weight loss and GERD improvement in comparison to LSG alone [38].

Intraoperative identification of HH requires hiatal dissection, with exposure of the diaphragmatic crura, and liberal use of intraoperative endoscopy. For large HH, a complete hiatal dissection, reduction of the GEJ intraabdominally, followed by a posterior crural repair, has been linked to improved postoperative GERD outcomes [36]. However, for smaller, sliding type 1 hernias, some surgeons have advocated anterior crural repair due to technical ease and perhaps a lower rate of postoperative dysphagia [30,38].

MANAGEMENT OF GERD AND HH AFTER BARIATRIC SURGERY

Patients who present with newly developed or worsened symptoms of GERD after bariatric surgery require careful evaluation. First, anatomic issues related to the bariatric surgery, such as recurrent or new HH, stricture, or functional problems, should be ruled out. For patients who have previously undergone gastric bypass, particularly open gastric bypass, the presence of a gastrogastric fistula allowing reflux of acid from the remnant stomach should be assessed. In patients with reflux symptoms after LSG, it is important to rule out a mechanical stricture or functional obstruction within the sleeve. In general, a UGI study and/or EGD are helpful studies to evaluate for postbariatric complications. Should an anatomic cause be found that explains the GERD symptoms, consideration should be given to endoscopic or surgical correction.

In the absence of defined anatomic complications from bariatric surgery, an empiric trial of PPIs can be considered, depending on the severity of presenting symptoms. With more severe symptoms, or prior to consideration of any intervention, further testing to confirm that symptoms are attributable to pathologic GERD should be strongly considered, including ambulatory pH or impedance monitoring, esophageal manometry, barium swallow, and/or upper endoscopy.

The first line of management in patients with recurrence or development of GERD symptoms is PPIs. For those patients with GERD refractory to medical management and without structural complications, alternate therapies can be considered. A low-risk option is radiofrequency energy treatment of the LES (Stretta), which in a small study of patients post-LRYGB, showed symptomatic efficacy, and could also be considered in post-LSG patients [39]. Other surgical interventions that have been reported include magnetic sphincter augmentation with the LINX device, modified fundoplication procedures using residual fundus in post-LSG or the remnant stomach in LRYGB patients, and endoluminal fundoplication [28]. However, limited data is available to support the efficacy of these treatments. In the post-LSG patient with intractable GERD, conversion to LRYGB is considered the best option for treatment.

CONCLUSIONS

GERD and HH are common in patients undergoing bariatric surgery. In obese patients, laparoscopic gastric bypass is considered the best therapy for GERD, especially in those with severe symptoms or endoscopic evidence of esophagitis or BE. The impact of LSG on GERD is controversial, but it does appear that a subset of patients will experience persistent, worsening, or de novo GERD following LSG, although many will have improvement of GERD. More study is needed to define the preoperative and technical factors that predict GERD outcomes with LSG. For HH, a combined bariatric procedure at the time of HH repair is recommended in the obese to improve durability of hernia repair and control of GERD. For large hiatal or paraesophageal hernias, LRYGB is again the recommended bariatric procedure, although data regarding comparative efficacy of combined LSG and HH repair are lacking. De novo GERD after bariatric surgery should trigger evaluation for and correction of structural, anatomic issues that are responsible, and/or physiologic testing to confirm pathologic reflux.

MINI-DICTIONARY OF TERMS

- *Gastroesophageal reflux disease*: Pathologic esophageal acid exposure from reflux of gastric contents due to inadequate function of the lower esophageal sphincter.
- *Hiatal hernia*: Weakening of the esophageal diaphragmatic hiatus, allowing intermittent or persistent intrathoracic displacement of the stomach and, potentially, other abdominal organs.

- *Barrett's esophagus*: Intestinal metaplasia of the distal esophagus resulting from damage from pathologic acid exposure.
- *Gastroesophageal junction*: The area where esophagus and proximal stomach meet, which can be seen endoscopically as the Z-line separating squamous esophageal mucosa and columnar gastric folds.

KEY FACTS

- GERD and HH are more frequent in obese than nonobese patients.
- Durability of surgical antireflux operations is reduced in the presence of obesity.
- LRYGB is the most effective bariatric operation for treatment of GERD.
- GERD improves in some patients after LSG; however, a subset will have persistent or de novo GERD.
- Factors that predict outcomes of GERD after LSG remain undefined.

SUMMARY POINTS

- Obese patients have a high incidence of GERD and occult HH.
- Bariatric surgery alone or in combination with an HH repair (when indicated) offers a high rate of GERD resolution as well as treatment of obesity and other weight-related comorbidities.
- LRYGB is the gold standard for treatment of obese patients with GERD, particularly with coexistent BE, as well as large HH.
- LSG is a safe and effective treatment for obesity with positive impact on GERD in some patients. However, given the possibility of worsening or de novo GERD after LSG and unclear preoperative predictive factors, it should be used with caution in those with significant symptoms or complications of GERD.
- New onset GERD after bariatric surgery should prompt evaluation for anatomic causes, such as gastrogastric fistula or stricture.

REFERENCES

[1] Ogden CL, et al. Prevalence of childhood and adult obesity in the united states, 2011-2012. JAMA 2014;311(8):806–14.
[2] National Center for Health Statistics. Health, U.S., 2015: With Special Feature on Racial and Ethnic Health Disparities. Hyattsville, MD; 2016.
[3] Kindel TL, Oleynikov D. The improvement of gastroesophageal reflux disease and Barrett's after Bariatric Surgery. Obes Surg 2016;26(4):718–20.
[4] Sise A, Friedenberg FK. A comprehensive review of gastroesophageal reflux disease and obesity. Obes Rev 2008;9(3):194–203.
[5] Kahrilas PJ, Kim HC, Pandolfino JE. Approaches to the diagnosis and grading of hiatal hernia. Best Pract Res Clin Gastroenterol 2008;22(4):601–16.
[6] Che F, et al. Prevalence of hiatal hernia in the morbidly obese. Surg. Obes Relat Dis 2013;9(6):920–4.
[7] Suter M, et al. Gastro-esophageal reflux and esophageal motility disorders in morbidly obese patients. Obes Surg 2004;14(7):959–66.
[8] Boules M, et al. The incidence of hiatal hernia and technical feasibility of repair during bariatric surgery. Surgery 2015;158:911–18.
[9] Nadaleto BF, et al. Gastroesophageal reflux disease in the obese: pathophysiology and treatment. Surgery 2016;159:475–86.
[10] Varela JE, Hinojosa MW, Nguyen NT. Laparoscopic fundoplication compared with laparoscopic gastric bypass in morbidly obese patients with gastroesophageal reflux disease. Surg Obes Relat Dis 2009;5(2):139–43.
[11] Perez AR, et al. Obesity adversely affects the outcome of antireflux operations. Surg Endosc 2001;15(9):986–9.
[12] Patterson EJ, et al. Comparison of objective outcomes following laparoscopic Nissen fundoplication versus laparoscopic gastric bypass in the morbidly obese with heartburn. Surg Endosc 2003;17(10):1561–5.
[13] Mechanick JI, et al. Clinical practice guidelines for the perioperative nutritional, metabolic, and nonsurgical support of the bariatric surgery patient—2013 update: cosponsored by American Association of Clinical Endocrinologists, the Obesity Society, and American Society for Metabolic & Bariatric Surgery. Endocr Pract 2013;19(2):337–72.
[14] Carabotti M, et al. Usefulness of upper gastrointestinal symptoms as a driver to prescribe gastroscopy in obese patients candidate to bariatric surgery. A prospective study. Obes Surg 2015;1–6.
[15] Klaus A, Weiss H. Is preoperative manometry in restrictive bariatric procedures necessary? Obes Surg 2008;18(8):1039–42.
[16] Parikh M, et al. Preoperative endoscopy prior to bariatric surgery: a systematic review and meta-analysis of the literature. Obes Surg 2016;1–6.
[17] Nadaleto BF, Herbella FAM, Patti MG. Gastroesophageal reflux disease in the obese: pathophysiology and treatment. Surgery 2016;159(2):475–86.
[18] Pallati PK, et al. Improvement in gastroesophageal reflux disease symptoms after various bariatric procedures: review of the bariatric outcomes longitudinal database. Surg Obes Relat Dis 2014;10(3):502–7.
[19] Mahawar KK, et al. Sleeve gastrectomy and gastro-oesophageal reflux disease: a complex relationship. Obes Surg 2013;23(7):987–91.

[20] Rebecchi F, et al. Gastroesophageal reflux disease and laparoscopic sleeve gastrectomy: a physiopathologic evaluation. Annal Surg 2014;260:909–15.

[21] Del Genio G, et al. Sleeve gastrectomy and development of "de novo" gastroesophageal reflux. Obes Surg 2014;24:71–7.

[22] Burgerhart JS, et al. Effect of sleeve gastrectomy on gastroesophageal reflux. Obes Surg 2014;24:1436–41.

[23] Mahawar KK, et al. Simultaneous sleeve gastrectomy and hiatus hernia repair: a systematic review. Obes Surg 2014;25(1):159–66.

[24] Petersen WV, et al. Functional importance of laparoscopic sleeve gastrectomy for the lower esophageal sphincter in patients with morbid obesity. Obes Surg 2012;22(3):360–6.

[25] Hawasli A, et al. Can morbidly obese patients with reflux be offered laparoscopic sleeve gastrectomy? A case report of 40 patients. Am J Surg 2016;211(3):571–6.

[26] Samakar K, et al. The effect of laparoscopic sleeve gastrectomy with concomitant hiatal hernia repair on gastroesophageal reflux disease in the morbidly obese. Obes Surg 2016;26(1):61–6.

[27] Peterli R, et al. Early results of the Swiss Multicentre Bypass or Sleeve Study (SM-BOSS): a prospective randomized trial comparing laparoscopic sleeve gastrectomy and Roux-en-Y gastric bypass. Ann Surg 2013;258(5):690–5.

[28] Gagner M, Hutchinson C, Rosenthal R. Fifth International Consensus Conference: current status of sleeve gastrectomy. Surg Obes Relat Dis 2016;12(4):750–6.

[29] Braghetto I, Csendes A. Patients having bariatric surgery: surgical options in morbidly obese patients with Barrett's Esophagus. Obes Surg 2016;1–5.

[30] Gagner M. Is sleeve gastrectomy always an absolute contraindication in patients with Barrett's? Obes Surg 2016;26(4):715–17.

[31] al-Haddad BJS, et al. Hiatal hernia repair in laparoscopic adjustable gastric banding and laparoscopic Roux-En-Y gastric bypass: a national database analysis. Obes Surg 2013;24(3):377–84.

[32] Patel AD, et al. Combining laparoscopic giant paraesophageal hernia repair with sleeve gastrectomy in obese patients. Surg Endosc 2015;29:1115–22.

[33] Davis M, et al. Paraesophageal hernia repair with partial longitudinal gastrectomy in obese patients. JSLS 2015;19(3):e2015.00060

[34] Chaudhry UI, et al. Laparoscopic Roux-en-Y gastric bypass for treatment of symptomatic paraesophageal hernia in the morbidly obese: medium-term results. Surg Obes Relat Dis 2014;10(6):1063–7.

[35] Ardestani A, Tavakkoli A. Hiatal hernia repair and gastroesophageal reflux disease in gastric banding patients: analysis of a national database. Surg Obes Relat Dis 2014;10(3):438–43.

[36] Lyon A, et al. Gastroesophageal reflux in laparoscopic sleeve gastrectomy: hiatal findings and their management influence outcome. Surg Obes Relat Dis 2015;11(3):530–7.

[37] Soricelli E, et al. Sleeve gastrectomy and crural repair in obese patients with gastroesophageal reflux disease and/or hiatal hernia. Surg Obes Relat Dis 2013;9(3):356–61.

[38] Chaar M, et al. Short-term results of laparoscopic sleeve gastrectomy in combination with Hiatal Hernia Repair: experience in a single accredited center. Obes Surg 2015;26(1):68–76.

[39] Mattar SG, et al. Treatment of refractory gastroesophageal reflux disease with radiofrequency energy (Stretta) in patients after Roux-en-Y gastric bypass. Surg Endosc 2006;20(6):850–4.

Metabolism, Endocrinology and Organ Systems

Chapter 32

Endoscopic Treatments for Obesity-Related Metabolic Diseases

G. Lopez-Nava[1], M. Galvao Neto[2,3,4] and J.W.M. Greve[5,6]

[1]Madrid Sanchinarro University Hospital, Madrid, Spain, [2]Gastro Obeso Center and Mario Covas Hospital, São Paulo, Brazil, [3]Florida International University, Miami, FL, United States, [4]ABC Medical School, Santo Andre, Brazil, [5]Medical Director Obesity Clinics South, Heerlen, The Netherlands, [6]Zuyderland Medical Center, Heerlen, The Netherlands

LIST OF ABBREVIATIONS

%EWL	% excess weight loss
BE	bariatric endoscopy
BMI	body mass index
BS	bariatric surgery
DMR	duodenal mucosal resurfacing
ESG	endoscopic sleeve gastroplasty
GLP-1	glucagon-like polypeptide 1
IGB	intra-gastric balloon
POSE	primary obesity surgery endoscopic
PYY	peptide YY
RCT	randomized controlled trial
T2D	type 2 diabetes mellitus
TBWL	total body weight lost

INTRODUCTION

Bariatric Endoscopy (BE) is a new term created to define the interface of advanced therapeutic endoscopy with Bariatric Surgery (BS). It mainly deals with treating BS complications and primary obesity itself, even revising secondary obesity (postop weight loss failure or postop weight regain). Interest in BE among bariatric surgeons and gastroenterologists/endoscopists is growing fast, and literature and awareness are growing solidly, but are still somewhat scarce. Traditional clinical treatment of obesity is prone to failure on obese patients, and even though BS is an efficient treatment, only a few surgical candidates (<2%) undergo it. This leaves enough room for less invasive ways to treat obesity and its metabolic disorders. Traditionally BE is used to treat obesity with space-occupying temporary devices like intra-gastric balloons (IGBs), and recently it has evolved into more sophisticated and more durable devices and procedures to treat obesity by mimicking BS restrictive procedures such as bands and gastroplasty with endoscopic stapling and suturing. BE is not just restricted to obesity itself, but also reaches for imaginative solutions like altering the bowel with endoluminal sleeves, bowel diversion with magnetic anastomoses, and even duodenal mucosal resurfacing (DMR) by thermal ablation, thus expanding BE into a possible endoscopic treatment for type 2 diabetes mellitus (T2D).

INTRA-GASTRIC BALLOON

The IGB is a gastric space-occupying device that can be used as a temporary weight loss measure in patients with a body mass index (BMI) of 27 or higher. The hypothetical mechanism of action is threefold: (1) it induces early satiety

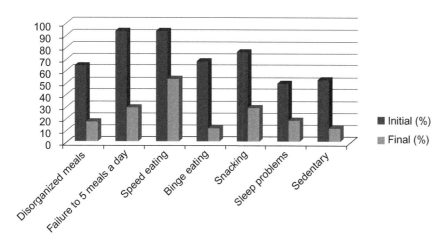

FIGURE 32.1 Change in eating, sleeping, and sedentary habits: initial and 1 year after Apollo gastroplasty.

because it occupies gastric volume; (2) it diminishes hunger due to parietal stimuli of the gastric wall; and (3) it triggers hypothalamic signals to enhance satiety.

Orbera (Apollo Endosurgery, Inc.) is the gold standard of IGB with the most experience (25 + years) and the greatest number of patients undergoing this type of treatment [1−3]. The Orbera IGB is a spherical ball of silicone with a smooth surface to reduce the risk of erosion of the mucosa. It is filled with an isotonic saline serum, and has a self-sealing radiopaque valve that can be located by simple radiological techniques. After diagnostic endoscopy, IGB placement is performed under unconscious sedation (Propofol, 2 mg/kg) or intubation (as a preanesthesia protocol). After 6 months in place, the balloon is removed. All patients are discharged after 3 hours [3]. The specific indications for the procedure are based on obesity parameters (BMI of 30−49 kg/m^2), previous failed attempts with conventional treatment, and the willingness and ability of patients to be treated by a multidisciplinary team for at least 1 year. Postprocedure care includes weekly or biweekly follow-up by a nutritionist and a psychologist, with additional emphasis on initiating an exercise program. A follow-up program is organized by the team in relation to diet, psychological support, physical activity counseling, and a visit schedule.

The objectives of BE are induction of weight loss, a change in lifestyle, and improvement of comorbidities, all with an acceptable safety profile. To define its effectiveness, weight loss following IGB intervention is determined by the following: (1) change in weight parameters, i.e., change in body weight total body weight lost (TBWL), percentage loss of initial body weight (%TBWL), and % excess weight loss (%EWL); (2) measurements of changes in comorbidities, i.e., hypertension, T2D, respiratory disorders, osteoarthropathy, or others comorbidities (comorbidity evolution can be classified as resolved, improved, or unchanged); and (3) adverse effects such as abdominal pain, gastroparesis, gastroduodenitis, esophagitis, nausea and/or vomiting, psychological intolerance, early explantation, bleeding requiring blood transfusions, gastric ulcer, gastric perforation, pneumothorax, pneumoperitoneum, liver abscess, deflation of IGB (detectable early by urine color due to methylene blue dye in the balloon) with or without intestinal obstruction, as well as death.

Weight results: Recently the American Society for Gastrointestinal Endoscopy (ASGE) Bariatric Endoscopy Task Force, comprising experts in the subject area, and the Technology Committee Chair [4] performed a meta-analysis (17 studies and 1683 patients) by using a randomized control trial to compare the efficacy and safety of IGB with conservative therapy. The results showed a %EWL with the IGB at 12 months of 25.44% with a mean difference in %EWL over controls of 26.9% in 3 randomized, controlled trials. Furthermore, the pooled %TBWL after IGB implantation was 12.3%, 13.16%, and 11.27% at 3, 6, and 12 months after implantation, respectively (Fig. 32.1).

Comorbidities: Different studies have demonstrated the benefit of a 10 kg weight loss in terms of comorbidities (T2D, blood pressure, lipids, etc.) and related mortality [5]. Besides the weight loss, only a few studies have been performed so far to assess the effect of IGB on obesity-related comorbidities. Lopez-Nava et al. reported in 2011 over a sample of 719 patients with IGB the improvement or resolution of pretreatment comorbidities in 140/162 (86.4%) patients. Table 32.1 shows the results of this study. Gottig et al. in 2009, in a study over 109 super- and super-super-obese patients with IGB, showed that in more than 95% of the patients, one or more obesity-related comorbidities were present. In 56.8% of patients at least one related disorder was improved, and 78% described an overall improvement in the quality of life [6]. Crea et al. in 2009, in a study over 138 patients with IGB, recorded results related to Metabolic Syndrome prevalence (OMS criteria). Prevalence of metabolic syndrome dropped from 34.8% (pre-IGB value) to

TABLE 32.1 Comorbidities and Their Outcome According to Resolution Criteria After Balloon Treatment [3]

Comorbidity	Prevalence n (%)	Resolution Criteria	Evolution, n (%)		
			Resolution	Improvement	No Change
Respiratory disorders	31 (19.1)	Absence of symptoms	17 (54.8)	13 (41.9)	1 (3.2)
Hypertension	22 (13.6)	Systolic <120–130 mmHg	7 (31.8)	11 (50)	4 (18.1)
Diabetes mellitus type 2	16 (9.8)	Fasting glycemia <110 mg/dL or HbA1c <6%	4 (25)	10 (62.5)	2 (12.5)
Osteo/arthropathy	14 (8.6)	Pain absence without drugs	3 (21.4)	7 (50)	4 (28.5)
Others	79 (46.9)		33 (41.7)	35 (44.3)	11 (13.9)
Total	162		64 (39.5)	76 (46.9)	12 (13.6)

14.5% at removal time, and 11.6% at the 12-month follow-up visit. Likewise, T2D, hypertriglyceridemia, hypercholesterolemia, and hypertension (HT) incidence decreased from 32.6%, 37.7%, 33.4%, and 44.9% (pre-IGB values) to 20.9%, 14.5%, 16.7%, and 30.4% at t_0 (time of removal of the balloon), and 21.3%, 17.4%, 18.9%, and 34.8% at 12 months after removal [7].

A large retrospective study by the Italian Collaborative Study Group for Lap-Band and BIB (GILB) has shown that the weight loss induced by IGB, though modest, caused most comorbidities associated with obesity to be either resolved or improved, in particular, respiratory disorders, hypertension, T2D, and osteoarthropathy [2]. They reported a resolution of comorbidities in 44.3% of subjects at the end of treatment, and an improvement of comorbidities with reduced pharmacological dosage in 44.8% of cases; no change was observed in 10.9% of patients. Similar conclusions were drawn by Mui in 2010 [8] and Ricci in 2008 [9]. Genco et al. in 2013 [10] had evaluable data at the 3-year follow-up (the longest time of follow-up) in over 121 patients. In the total sample, the rate of patients with hypertension decreased at 3 years from 29% at baseline to 16%, diabetes decreased from 15% to 10%, dyslipidemia decreased from 20% to 18%, hypercholesterolemia decreased from 32% to 21%, and osteoarthropathy decreased from 25% to 13%. In conclusion, although weight regain is commonly observed during long-term follow-up after IGB removal, improvement in comorbidities persists at a high percentage. Modifications in nutritional and physical activity habits could influence this improvement.

Safety: In general, it can be considered a safe and simple technique, with an overall average rate of complication ranging between 2.8% and 40% [11]. The aforementioned ASGE Bariatric Endoscopy Task Force meta-analysis (68 studies) [4] summarized the rates of adverse events after implantation of the IGB: pain and nausea were frequent side effects (33.7% of subjects). The early removal rate for the IGB was approximately 7%. Serious side effects with the Orbera IGB were rare, with an incidence of migration and gastric perforation of 1.4% and 0.1%, respectively. Four deaths associated with the IGB were reported in the literature, and these were either related to gastric perforation or an aspiration event [4].

ENDOSCOPIC SLEEVE GASTROPLASTY (APOLLO METHOD)

The goal of the endoscopic sleeve gastroplasty (ESG) is to reduce the gastric lumen into a tubular configuration, with the greater curvature modified by a line of sutured plication. Published clinical experiences [12,13] with ESG show that the use of ESG for the treatment of obesity is safe and feasible, and results in both significant weight loss and altered patient eating behaviors [13–15].

To perform the gastroplasty, interrupted sutures are deployed from the distal to proximal body. Each suture consists of six bites along the anterior/greater curvature/posterior gastric wall before it is cinched. This is not a continuous staple line, but rather an invagination of the greater curvature of the stomach. Reinforcing stitches are usually placed in the upper body of the stomach. The suture pattern has evolved from just a few cases addressing the fundus, to the majority of them where the fundus is not closed, so the patient can still have a pouch and some accommodation ability. Bleeding complications were excluded by blood tests at 6 hours and at 24 hours postprocedure. Postprocedure care remained unchanged with our earlier experience, and included hospitalized observation, fasting sips, a liquid diet at 8 hours post-procedure, analgesia, and 24-hour discharge [14].

The specific indications for the procedure are based on obesity parameters (BMI of 30−49 kg/m^2) with previous failed attempts with conventional treatment of obesity and the willingness and ability of patients to be treated by a multidisciplinary team for at least 1 year. The procedure is contraindicated in patients with prior gastric surgery, potentially bleeding lesions (e.g., ulcers and acute gastritis), and neoplastic findings. Individuals with psychiatric disorders (mental retardation, manic−depressive psychosis, severe depression, schizophrenia, and untreated eating behavior disorders) that interfere with their ability to actively engage with the postprocedural instruction and recommended lifestyle adjustments were excluded [14].

The gastric cavity restriction helps with caloric limitation. A multidisciplinary team (nutritionist, physiologist, and exercise physiologist) is continuously in contact to resolve all these problems and to design the best strategy to treat each patient individually. Exercise is recommended, taking into account the limitations of individual patients. The nutritional intervention changes during the course of treatment. Initially, the focus is on the transitional diet postintervention. Then, after the patient starts on solid food, the focus is on following the prescribed hypocaloric diet and discussing healthy food choices and alternatives. The psychologist coaches the patient to follow the recommended lifestyle modification program necessary to maintain the weight loss in the long term [15].

To define the effectiveness of the ESG intervention, the following parameters are recorded: (1) change in weight parameters (change in body weight (TBWL), percentage loss of initial body weight (%TBWL), and %EWL); (2) change in comorbidities (hypertension, T2D, respiratory disorders, osteoarthropathy, or other comorbidities); and (3) adverse effects (bleeding postintervention, nausea, abdominal pain, stomach perforation, pneumothorax and peri-hepatic, and/or peri-splenic abscesses).

Results at 6 months and 1 year of the first cases by ESG through transmural sutures (APOLLO Method) have been published.

Lopez-Nava et al. published results after 6 months postprocedure of 20 patients who underwent ESG, with initial BMI of 38.5 kg/m^2, showing 53.8 %EWL [14]. Sharaiha et al. published results for a group of 10 patients with an initial BMI of 45.2 kg/m^2, showing a 30 %EWL at 6 months [16]. Lopez-Nava et al. published results of 22 patients after the procedure with a 54.6 %EWL, showing that subgroups with the highest number of nutritional and psychological interactions demonstrated the most favorable weight loss (Table 32.2) [15].

Although no published articles exist that detail changes on comorbidities with ESG, in our experience with more than 150 patients with this procedure, all the related obesity comorbidities (hypertension, T2D, hypertriglyceridemia) improved in relation to weight loss and changes in physical activity. In a recent article, Lopez-Nava published initial and final values of nutritional habits [15]. Initially, the worst habits were "not eating 5 meals a day" (94.1%) and "not eating slowly" (93.3%). One year after the procedure, the most notable changes were "not eating 5 meals a day" (from 94.1% to 29.4%) and binge eating (from 68.8% to 12.5%). Among the initially sedentary patients, 55.6% began physical activity (walking or doing cardiovascular exercises in the gym), and 75% of those who were initially not sedentary improved their level of physical activity (e.g., increasing walking time or doing other activities in the gym).

The technique was applied in the first 50 patients (6) with an average BMI of 37.7 kg/m^2 (range of 30−47) with 13 having reached 1 year, and showed that all patients underwent successful gastroplasty and were discharged in less than 24 hours. There were no major intraprocedural, early, or delayed adverse events. No bleeding complications were found. Postdischarge pain was controlled with oral analgesia (2−4 days), and nausea (1 day) with antiemetics in 50% and 20% patients, respectively.

TABLE 32.2 Weight Changes With Endoscopic Sleeve Gastroplasty Procedure (Results at 1 Year)

Variable	Initial (n = 50)	1 Month (n = 45)	3 Months (n = 42)	6 Months (n = 30)	12 Months (n = 13)
Weight (kg)	107.0 ± 18.4	98.5 ± 16.5	93.5 ± 16.5	89.2 ± 17.8	88.1 ± 12.0
BMI (kg/m^2)	37.7 ± 4.6	35.2 ± 4.5	33.3 ± 4.6	31.8 ± 5.2	30.9 ± 5.1
TBWL (kg)		7.4 ± 2.7	13.5 ± 5.6	18.7 ± 8.9	21.6 ± 13.5
%TBWL (%)		6.9 ± 2.1	12.6 ± 4.3	17.2 ± 7.5	19.0 ± 10.8
%EWL (%)		22.6 ± 10.5	40.2 ± 17.3	53.5 ± 26.2	57.0 ± 33.9

The main results demonstrate that the ESG method is a technique that, used in expert hands, is safe and reproducible. In the current experience there have been no major complications. From the weight loss results, it is shown that this technique, used as an adjunct to dietary intervention and psycho—behavioral modification in obese patients, is effective in the treatment of obesity. Regarding the duration of the effect, the results at 12 months show that at least in this timeframe the technique is still effective and helpful. In this regard it should be noted that as there are no irreversible anatomical alterations in the gastric cavity, the technique is repeatable, and it can therefore be continued in the future as a means of achieving lasting results.

PRIMARY OBESITY SURGERY ENDOSCOPIC

Another technique used to perform endoscopic suturing is primary obesity surgery endoscopic (POSE). POSE as a primary treatment for obesity consists of placing sutures in the fundic area and in the antral area of the stomach. POSE is an endoscopic platform that consists of several tools to reliably place sutures with full thickness. The components are the Transport Multilumen Access Device (an endoscope-like instrument with several working channels in which an endoscope and the suturing devices are placed), the g-Lix tissue grasper, the g-Prox endoscopic grasper, and the g-Cath for delivery of the snowshoe suture anchors (Fig. 32.2) [17]. With this technique, bites of about 6 cm full thickness on the gastric wall are realized. The procedure is performed under general anesthesia, and on average 8—10 sutures are placed in the fundus and four in the antrum of the stomach. The main effect of the procedure is early satiety and delayed gastric emptying. The sutures have been shown to be durable, still being in place 1 and 2 years after the procedure [18]. In the study by Lopez-Nava and colleagues, 147 patients were treated with POSE. Baseline BMI was $38.0 +/- 4.8$ kg/m^2. Initial weight ($106.8 +/- 18.2$ kg) was significantly reduced at 3, 6, and 12 months. At 1 year, 116 patients (79% of total) who were available for follow-up had a mean TWL of $16.6 +/- 9.7$ kg, %TWL of $15.1 +/- 7.8$, and %EWL of $44.9 +/- 24.4$. Similar results were reported by Espinos and colleagues in a study with 18 patients; at 15 months the patients had a $19.1 +/- 6.6$ %TWL and a $63.7 +/- 25.1$ %EWL [19]. In this particular study several additional parameters were studied. The authors looked at intake capacity after the procedure, which was significantly reduced from 901 kcal at baseline to 473 kcal and 574 kcal at 2 and 6 months after the procedure, respectively. Glucose homeostasis was studied in the patients in a meal tolerance test, and plasma peptide YY (PYY-36), leptin, and total ghrelin were also measured. There was a significant increase of postprandial PYY 6 months after the procedure, and there was a normalization of the ghrelin response after a meal. Leptin levels decreased significantly, and there was a significantly improved glucose homeostasis (fasting glucose levels improved from 105 mg/dL to 92 at 6 months, and after a standard meal from 145 mg/dL at baseline to 126 at 6 months).

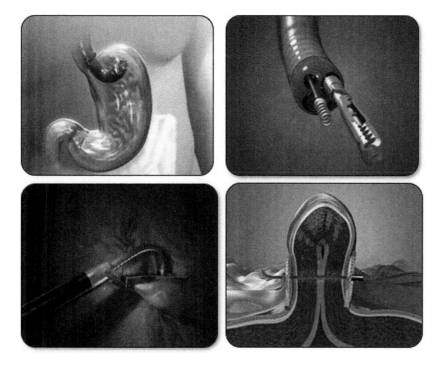

FIGURE 32.2 Transport, g-Prox, and Snowshoe suturing anchor of the Primary Obesity Surgery Endoscopic (POSE). *Courtesy USGI Medical, San Clemente, CA, USA.*

To date more than 1500 patients have been treated with the POSE procedure. The procedure is safe, with only a limited number of complications reported. Most patients report some upper gastrointestinal discomfort and a sore throat after the procedure. Major complications include postprocedure bleeding requiring transfusion, perforation of the stomach, pneumothorax, and perihepatic or perisplenic abscess [18]. Out of 1500 patients, 0.33% required an invasive procedure, and 0.67% had a prolonged hospitalization [18]. The reported outcome of metabolic disorders is limited, but there is clear evidence that POSE has a significant impact on T2D and other disorders. Like most endoscopic therapies, long-term effects are not well known, and depend on patient compliance with the required lifestyle change and improved diet. However, short-term effects are significant and even in case of failure additional therapy such as bariatric surgery can be safely performed. In particular, in the lower BMI patients, where there is no effective treatment alternative, POSE appears to be a good option.

DUODENAL MUCOSAL RESURFACING

DMR is a novel, minimally invasive, catheter-based, upper endoscopic procedure involving hydrothermal ablation of the duodenal mucosa and subsequent mucosal healing. DMR could less invasively recapitulate some of the mechanisms of action of gastrointestinal bypass surgery. It may offer a new treatment approach for T2D by altering the duodenal mucosal surface itself, thereby altering downstream signaling and eliciting metabolic improvement. Preclinical studies conducted in the Goto−Kakizaki rat, a rodent model equivalent of human T2D, support this thesis by demonstrating that selective denudation of the duodenal mucosa conducted by an abrasion device resulted in immediate lowering of glycemia during an oral glucose gavage when compared to preprocedure levels and also to a sham-treated group. Moreover, similarly conducted studies in a nondiabetic rodent model (Sprague−Dawley) showed no lowering of glycemia, indicating that this perturbation of the duodenal mucosa reduced abnormal hyperglycemia, but did not impact normoglycemia (unpublished data, Fractyl Laboratories, Inc., Waltham, MA, USA). Additional studies in a pig model also demonstrate that the DMR procedure achieves a predictable ablation of the intestinal mucosa surface without damage to the underlying muscularis mucosa or deeper structures (unpublished data, Fractyl Laboratories, Inc., Waltham, MA, USA). A first-in-human study was performed on the Clinical Center for Diabetes, Obesity and Reflux, Santiago, Chile and IRB approval and presented at the 2014 World Congress of the International Federation for the Surgery of Obesity and Metabolic Disorders in Montreal (3 month results) and on the 3rd World Congress On Interventional Therapies For T2D in London 2015 (extended results for 6 months) on 39 subjects with the following technic.

In the current study, DMR consists of intestinal luminal sizing, submucosal expansion with saline injection (designed to provide a uniform ablative surface and a thermally protective layer of saline between the ablated mucosa and deeper tissue layers), and circumferential thermal ablation along a length of the duodenum using novel polyethylene terephthalate (PET) balloon treatment catheters (Revita system, Fractyl Laboratories, Waltham, MA, USA); they were introduced into the duodenum via a transoral endoscopic approach in anesthetized patients (Fig. 32.3). The first catheter was used to determine the size of the duodenum and inject saline into the submucosal space via vacuum-assisted needle injectors around the circumference of the balloon to create a mucosal lift along the length of the postpapillary

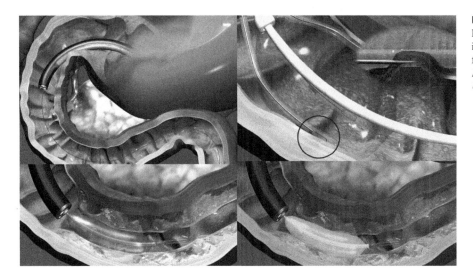

FIGURE 32.3 Duodenal Mucosal Resurfacing (DMR) procedure representing endoluminal duodenal mucosal lifting followed by thermal ablation. *Courtesy Fractyl Laboratories, Waltham, MA, USA.*

duodenum from 1 cm distal to the ampulla of Vater up to just proximal to the ligament of Treitz. After removal of the initial catheter, a second balloon catheter was introduced to perform thermal ablation on the lifted area. Under direct endoscopic visualization, discrete circumferential thermal ablations of approximately 10 seconds each were applied at temperatures of approximately 90°C to obtain up to 5 longitudinally separated ablations along the length of the postpapillary duodenum. Care was taken to avoid the ampulla of Vater to prevent damage to the biliary tree, and to avoid treatment in or beyond the ligament of Treitz. Patients were discharged within 24 hours after the procedure and prescribed a progressive diet (liquids → soft foods → pureed foods) for 2 weeks.

Among 39 subjects, 28 received DMR on a long segment ($\sim$9.3 cm; LS-DMR) and 11 received DMR on a short segment ($\sim$3.4 cm; SS-DMR). The Baseline Mean HbA1c of 9.5% (SD $\pm$ 1.3) was reduced by 2.5% (SD $\pm$ 1.3) in LS-DMR and 1.2% (SD $\pm$ 1.7) in SS-DMR at 3 months postprocedure ($p < 0.05$ for LS vs SS). In 15 LS-DMR subjects with baseline HbA1c of 7.5–10% (mean 8.7% (SD $\pm$ 0.9)), HbA1c at 6 months was reduced to 7.5% (SD $\pm$ 1.2) despite antidiabetic medication reduction in 8 of the patients. Excluding patients with medication adjustments, HbA1c at 6 months after LS-DMR was 7.0% (SD $\pm$ 0.7; $p < 0.01$), accompanied by a modest weight reduction of 2.8 kg (SD $\pm$ 3.4). There was no apparent correlation between degree of weight loss and magnitude of HbA1c improvement. Three patients experienced duodenal stenosis that required balloon dilation with good resolution. This study concludes that a single-procedure DMR substantially improves glycemic control in T2D with acceptable safety and tolerability. A new international multicenter trial is ongoing (https://clinicaltrials.gov/ct2/show/NCT02413567).

ENDOLUMINAL SLEEVES

Several endoscopic devices (noninvasively placed) have been developed to mimic the effect of bypass types of BS. Basic research in rats with diabetes has shown that exclusion of the duodenum from chyme by means of an endoluminal tube significantly improves glucose homeostasis [20]. To date two devices have been reported for use in patients. The ValenTx gastro–duodenal bypass is an endoluminal sleeve that is attached to the gastro–esophageal junction and extends far into the jejunum. The function of the ValenTx is similar to a regular gastric bypass, and initial results have been promising [21]. However, attachment to the gastro–esophageal junction in particular proved to be a difficult problem, and the device was withdrawn from clinical use; further improvement of the technique is currently under investigation. Another endoluminal sleeve is the EndoBarrier, also known as the duodeno–jejunal bypass liner [22,23]. The EndoBarrier is an impermeable fluoropolymer sleeve of 60 cm (Fig. 32.4) that is placed endoscopically and is kept in place with a nitinol anchor with barbs that is positioned in the duodenal bulb. The sleeve blocks contact of all nutrients with the duodenal wall, whereas bile and pancreatic juices pass on the outer site of the sleeve to the more distal jejunum. To date more than 3000 sleeves have been placed with acceptable-to-good weight loss results and significant improvement of metabolic disorders such as diabetes mellitus. A modification of the device included a membrane in the anchor with a narrow dilatable opening (flow restrictor) to increase restriction and thus weight loss results; this modification was not released for clinical use after an initial successful clinical study [24].

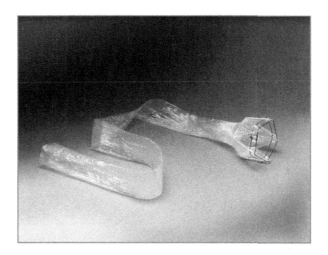

FIGURE 32.4   EndoBarrier duodenal-jejunal bypass liner. *Courtesy GI Dynamics, Lexington, MA, USA.*

The initial studies focused on weight loss, but results also showed a significant effect on T2D. The first-generation device was approved for 3-month treatment that was extended to 6 months because of the good results obtained. After modification of the device's anchor, it was proven stable for 1 year. Initial studies with the device were performed in Chile, after which CE mark was achieved in a multicenter randomized study in the Netherlands. In this study the average %EWL in a group of 26 patients was 19% after 3 months, compared to 6.9% in the control group. In a second study with the second-generation device (improved anchor), treatment was extended to 6 months and the weight loss obtained was 32 %EWL compared to 16 %EWL in the control group [25]. The second-generation device is now approved for 1-year treatment. About 70% of the treated patients will be able to complete 1 year. This results in a consistent weight loss of over 45 %EWL.

The EndoBarrier is minimally invasive and safe. To date there is no mortality due to the treatment, however, complications are relatively frequent. In particular, upper abdominal discomfort, nausea, and sometimes vomiting occur in most patients immediately after implantation of the EndoBarrier, but these complaints are usually mild and of short duration. More important are migration of the EndoBarrier (1.6%), bleeding (1.5%), intolerance (1.6%), and obstruction (0.8%), which usually result in sleeve removal (Table 32.3). Technical complications at implantation or removal are rare, but can be serious, such as esophagus perforation (0.15%). All complications were resolved without serious sequelae to the patient, only rarely requiring a surgical solution. More recently reported complications are liver abscesses (0.9%) and acute pancreatitis (0.3%); most of these patients were treated conservatively (antibiotics and sleeve removal). Overall conclusions from current clinical experience with the duodenal-jejunal bypass liner (EndoBarrier) are that it is safe and minimally invasive.

An interesting observation in the early experience of the EndoBarrier was the rapid improvement of obesity-related metabolic disorders, in particular T2D [26,27]. Not only did glucose levels improve, but most patients could reduce or even stop their antidiabetic medication. Since then, the focus of EndoBarrier treatment has been on patients with diabetes mellitus. In a multicenter randomized study in the Netherlands there was a significant improvement in glucose homeostasis with a reduction of HbA1c from 8.3% (7.7−9.0%) to 7.0% (6.4−7.5%) after 6 months of treatment, compared to 8.3% (7.7−8.9%) at baseline to 7.9% (6.6−8.3%) at 6 months in control patients [25]. In a subgroup of 17 patients, glucose homeostasis was studied in more detail [28]. A significant reduction in glucose levels after a test meal was found after just 1 week, with normalization of the glucagon response, while insulin sensitivity appeared to be unchanged. A recent meta-analysis did not confirm this positive effect, however, the studies included were diverse, with results from the first-generation device with only 3 months of treatment, preliminary studies with the second-generation device with 6 months of treatment, and nonrandomized single-center studies [29].

In a small group of patients, several potential mechanisms involved in improvement of glucose homeostasis were studied. Similar to results found in gastric bypass patients there is a significant increase in GLP-1 and PYY response

TABLE 32.3 Complications With the EndoBarrier (Second-Generation Device) Leading to Removal of the Device (Serious Adverse Events)

Complications Leading to Removal	
Complication	Incidence (%)
GI bleeding	48 (1.5%)
Liner obstruction	24 (0.8%)
Intolerance	50 (1.6%)
Migration	50 (1.6%)
Hepatic abscess	26 (0.9%)
Perforation[a]	10 (0.3%)
Pancreatitis	12 (0.3%)
Difficult removal (requiring surgery)	10 (0.3%)

[a]Five patients had a perforation during the implant or explant procedure.
Total number of devices (excluding USA) 3150.
Source: Courtesy GI Dynamics in 3150 devices from clinical trials and commercial use.

after a meal [28]. Also, a clear improvement in the response of the orexigenic hormone ghrelin was found. However, weight loss is also important in improvement in glucose metabolism. The achieved weight loss is induced by a reduced satiety (delayed gastric emptying) [30] and by early response of the satiety hormones (PYY) and increase of ghrelin with a more physiologic decrease after a meal [31]. The effect on these gut hormones was present within 1 week of the treatment and lasted until removal of the device.

Endoluminal sleeves placed by endoscopy show a significant effect on weight and obesity-related comorbidities. Clinical experience (EndoBarrier) is still accumulating, with acceptable complication rates. A major drawback is the limited duration of the treatment (maximum 1 year with the EndoBarrier); however, repeat treatment has been shown to be possible [32]. In a large patient group that has no treatment option other than diet and lifestyle change (BMI of 30–40) the duodenal-jejunal bypass liner (EndoBarrier) is certainly a valid, safe, and effective treatment option.

GENERAL CONCLUSION

With the huge number of obese and morbidly obese patients in mind, and surgery currently the only durable and long-term effective treatment option (reaching only a minority of potential patients), less invasive and effective treatments are needed. With the development of more effective endoscopic alternatives, a larger group of obese patients (BMI of 30–40) will potentially get treatment for their obesity and obesity-related metabolic disorders. For the moment, gastric balloons, endoluminal sleeves (EndoBarrier), endoscopic gastric plication (POSE), and ESG (Apollo) have a proven clinical effect with a good safety profile. New options are in development and are being extensively studied. DMR looks promising, and the still experimental endoscopic anastomosis with magnets may open an even larger number of endoscopic treatment options [33]. Because of their minimally invasive character and good safety profile, endoscopic treatment options should be a standard part of a treatment program for obesity and obesity-related metabolic disorders.

MINI-DICTIONARY OF TERMS

- *Endoluminal sleeve (EndoBarrier)*: Device that excludes the duodenum from contact with food, which results in a change in gut hormones and improvement of weight and T2D.
- *Primary obesity surgery endoscopic (POSE)*: Endoscopic full-thickness gastric plication that induces weight loss by increased satiety.
- *Endoscopic sleeve gastroplasty (Apollo, Overstitch)*: Endoscopic sutured gastric sleeve that mimics sleeve resection of the stomach.
- *Duodenal mucosal resurfacing (Revita system)*: Thermal ablation of the duodenal mucosa that affects incretin release and induces weight loss and T2D improvement.
- *Intra-gastric balloon (IGB)*: Fluid of air-filled balloon placed in the stomach to induce early satiety and reduced food intake.

KEY FACTS

- Endoscopic techniques can result in significant weight loss, and as a result, improvement of obesity-related metabolic disorders.
- Endoscopic procedures mimic parts or even entire surgical procedures such as gastric bypass, sleeve resection, and duodenal exclusion.
- The drawback of most endoscopic endoluminal procedures is the limited duration of the treatment.
- Repeat treatment is in general possible with most endoscopic techniques.
- A combination of several endoscopic procedures may increase effectiveness.

SUMMARY POINTS

- This chapter gives an overview of currently available endoscopic (minimal invasive) techniques to treat obesity and its metabolic disorders.
- The endoscopic techniques have been shown to be effective and safe.
- Currently available are intragastric balloons, endoscopic sleeve gastroplasty (Apollo), Gastric reduction with POSE, duodenal mucosal resurfacing DMR), and endoluminal sleeves (EndoBarrier, duodenal-jejnunal bypass liner).
- A number of new devices are under investigation.

- The mentioned techniques all result in significant weight loss and directly or indirectly improve type 2 DM, dyslipidemia, and other obesity-related metabolic disorders.
- Mechanisms of action are briefly discussed.

REFERENCES

[1] Sallet JA, et al. Brazilian multicenter study of the intragastric balloon. Obes Surg 2004;14(7):991–8.

[2] Genco A, et al. BioEnterics intragastric balloon: the Italian experience with 2,515 patients. Obes Surg 2005;15(8):1161–4.

[3] Lopez-Nava G, et al. BioEnterics(R) intragastric balloon (BIB(R)). Single ambulatory center Spanish experience with 714 consecutive patients treated with one or two consecutive balloons. Obes Surg 2011;21(1):5–9.

[4] Force ABET, et al. ASGE bariatric endoscopy task force systematic review and meta-analysis assessing the ASGE PIVI thresholds for adopting endoscopic bariatric therapies. Gastrointest Endosc 2015;82(3):425–38, e5.

[5] Deitel M. How much weight loss is sufficient to overcome major co-morbidities? Obes Surg 2001;11(6):659.

[6] Gottig S, et al. Analysis of safety and efficacy of intragastric balloon in extremely obese patients. Obes Surg 2009;19(6):677–83.

[7] Crea N, et al. Improvement of metabolic syndrome following intragastric balloon: 1 year follow-up analysis. Obes Surg 2009;19(8):1084–8.

[8] Ricci G, et al. Bariatric therapy with intragastric balloon improves liver dysfunction and insulin resistance in obese patients. Obes Surg 2008;18(11):1438–42.

[9] Mui WL, et al. Impact on obesity-related illnesses and quality of life following intragastric balloon. Obes Surg 2010;20(8):1128–32.

[10] Genco A, et al. Multi-centre European experience with intragastric balloon in overweight populations: 13 years of experience. Obes Surg 2013;23(4):515–21.

[11] Espinet-Coll E, et al. Current endoscopic techniques in the treatment of obesity. Rev Esp Enferm Dig 2012;104(2):72–87.

[12] Abu Dayyeh BK, Rajan E, Gostout CJ. Endoscopic sleeve gastroplasty: a potential endoscopic alternative to surgical sleeve gastrectomy for treatment of obesity. Gastrointest Endosc 2013;78(3):530–5.

[13] Lopez-Nava G, et al. Endoscopic sleeve gastroplasty: How I Do It? Obes Surg 2015;25(8):1534–8.

[14] Lopez-Nava G, et al. Endoscopic sleeve gastroplasty for the treatment of obesity. Endoscopy 2015;47(5):449–52.

[15] Lopez-Nava G, et al. Endoscopic sleeve gastroplasty with 1-year follow-up: factors predictive of success. Endosc Int Open 2016;4(2):E222–7.

[16] Sharaiha RZ, et al. Initial experience with endoscopic sleeve gastroplasty: technical success and reproducibility in the bariatric population. Endoscopy 2015;47(2):164–6.

[17] Espinos JC, et al. Early experience with the incisionless operating platform (IOP) for the treatment of obesity: the primary obesity surgery endolumenal (POSE) procedure. Obes Surg 2013;23(9):1375–83.

[18] Lopez-Nava G, et al. The primary obesity surgery endolumenal (POSE) procedure: one-year patient weight loss and safety outcomes. Surg Obes Relat Dis 2015;11(4):861–5.

[19] Espinos JC, et al. Gastrointestinal physiological changes and their relationship to weight loss following the POSE procedure. Obes Surg 2015.

[20] Aguirre V, et al. An endoluminal sleeve induces substantial weight loss and normalizes glucose homeostasis in rats with diet-induced obesity. Obesity (Silver Spring) 2008;16(12):2585–92.

[21] Sandler BJ, et al. One-year human experience with a novel endoluminal, endoscopic gastric bypass sleeve for morbid obesity. Surg Endosc 2015;29(11):3298–303.

[22] Gersin KS, et al. Duodenal-jejunal bypass sleeve: a totally endoscopic device for the treatment of morbid obesity. Surg Innov 2007;14(4):275–8.

[23] Rodriguez L, et al. Pilot clinical study of an endoscopic, removable duodenal-jejunal bypass liner for the treatment of type 2 diabetes. Diabetes Technol Ther 2009;11(11):725–32.

[24] Escalona A, et al. Initial human experience with restrictive duodenal-jejunal bypass liner for treatment of morbid obesity. Surg Obes Relat Dis 2010;6(2):126–31.

[25] Koehestanie P, et al. The effect of the endoscopic duodenal-jejunal bypass liner on obesity and type 2 diabetes mellitus, a multicenter randomized controlled trial. Ann Surg 2014;260(6):984–92.

[26] Schouten R, et al. A multicenter, randomized efficacy study of the EndoBarrier gastrointestinal liner for presurgical weight loss prior to bariatric surgery. Ann Surg 2010;251(2):236–43.

[27] Escalona A, et al. Weight loss and metabolic improvement in morbidly obese subjects implanted for 1 year with an endoscopic duodenal-jejunal bypass liner. Ann Surg 2012;255(6):1080–5.

[28] de Jonge C, et al. Endoscopic duodenal-jejunal bypass liner rapidly improves type 2 diabetes. Obes Surg 2013;23(9):1354–60.

[29] Rohde U, et al. Effect of the EndoBarrier gastrointestinal liner on obesity and type 2 diabetes: a systematic review and meta-analysis. Diabetes Obes Metab 2016;18(3):300–5.

[30] de Moura EG, et al. Effects of duodenal-jejunal bypass liner (EndoBarrier(R)) on gastric emptying in obese and type 2 diabetic patients. Obes Surg 2015;25(9):1618–25.

[31] de Jonge C, et al. Impact of duodenal-jejunal exclusion on satiety hormones. Obes Surg 2016;26(3):672–8.

[32] Koehestanie P, et al. Is reimplantation of the duodenal-jejunal bypass liner feasible? Surg Obes Relat Dis 2015;11(5):1099–104.

[33] Ryou M, Agoston AT, Thompson CC. Endoscopic intestinal bypass creation by using self-assembling magnets in a porcine model. Gastrointest Endosc 2016;83(4):821–5.

Chapter 33

Sleeve Gastrectomy: Mechanisms of Weight Loss and Diabetes Improvements

P.K. Chelikani and D. Sekhar

University of Calgary, Calgary, AB, Canada

LIST OF ABBREVIATIONS

FGF-19 fibroblast growth factor-19
FXR farnesoid X receptor
GLP-1 glucagon-like peptide 1
MC4R melanocortin-4 receptor
PYY peptide YY
RYGB Roux-en-Y gastric bypass
SG sleeve gastrectomy
TGR5 Takeda G-protein-coupled receptor

INTRODUCTION AND SIGNIFICANCE

Obesity and associated comorbidities (e.g., hypertension, dyslipidemia, and diabetes) pose serious threats to our health and the sustainability of health care systems. In obese subjects, compared to dietary and behavioral interventions, bariatric surgeries produce multiple benefits, including long-term weight loss ($\sim 57-67\%$), resolution or improvement of diabetes in 86% of patients, pronounced improvements in cardiovascular function and dyslipidemia in $70-96\%$ of individuals, and complete resolution or improvements in hypertension in 78% of patients [1]. More importantly, resolution or improvements in some comorbidities such as diabetes occur prior to substantial weight loss, suggesting that profound alterations in gut physiology have important roles in metabolic adaptations following bariatric surgery [2]. However, these surgeries are also associated with short- and long-term complications, and poor weight loss in some patients. It is noteworthy that the volume of bariatric surgeries has increased worldwide. In Canada alone, the number of operations performed increased 280% from 2007 to 2013, with the number of sleeve gastrectomy (SG) procedures rapidly increasing from 410 to 1676 [3]. However, the mechanisms of action of these surgeries are poorly understood. Understanding the mechanisms of action of bariatric surgeries is important for identifying predictors of success, and for developing novel nonsurgical approaches that could reproduce the effectiveness of surgeries without the attendant surgical risks.

BARIATRIC SURGERIES

Bariatric surgeries involve either restrictive and/or malabsorptive procedures. Purely restrictive operations include gastric banding (lap-band), SG or vertical SG, and vertical banded gastroplasty. These procedures reduce stomach capacity, resulting in earlier satiety and weight loss; however, other effects on metabolism are also important [4]. Malabsorptive procedures, such as biliopancreatic diversion and Roux-en-Y gastric bypass (RYGB), involve rearrangement of the intestines and/or stomach to reduce intake and nutrient absorption, and enhance lower gut stimulation, thereby promoting weight loss. From 2003 to 2011, the total number of bariatric procedures performed worldwide increased from 146,301 to 340,768; RYGB decreased from 65% to 47%, whereas SG procedures rapidly increased from 0% to 28% of total bariatric surgeries during the same interval [4]. In contrast to the complexity of RYGB, SG is a relatively simple procedure with fewer complications. SG involves surgical removal of ~80% of the stomach while preserving intestinal length, thereby avoiding the complications of anastomotic leaks and intestinal obstructions; it does not involve proximal intestinal bypass, thereby avoiding malabsorption of minerals and vitamins, and it circumvents the problems of band migrations, erosions, and infections that are frequently observed with gastric banding. In a recent meta-analysis of 326 studies between 2000 and 2011, SG was found to be as effective as RYGB in producing weight loss [5]. It is noteworthy that the international diabetes federation taskforce on epidemiology and prevention of diabetes recommends bariatric surgery as a treatment option for obese diabetics (body mass index (BMI) ≥ 35) and as an alternative treatment option for moderately obese subjects (BMI 30−35) with poorly controlled diabetes and concurrent major cardiovascular disease risk factors [6]. Despite such recommendations and benefits, the underlying mechanisms of weight loss and metabolic benefits of bariatric surgeries remain poorly understood. Given the increasing popularity and effectiveness of SG, this chapter focuses on the mechanisms of weight loss and diabetic improvements following SG procedure.

SLEEVE GASTRECTOMY: MECHANISMS OF WEIGHT LOSS AND DIABETIC CONTROL

There is substantial evidence that SG produces weight loss effects comparable to RYGB [5]; however, the underlying mechanisms of weight loss and metabolic benefits are poorly understood. The weight loss and metabolic effects of SG are hypothesized to be due to the following: (1) gut adaptation, including reductions in gastric volume and accelerated gastric emptying with consequent rapid delivery of nutrients to the distal intestine; (2) reduction in orexigenic (increased food intake) and enhanced secretion of anorexigenic (reduced food intake) gut hormones; (3) alterations in central neuronal circuits regulating energy balance; (4) modulation of enterohepatic bile acid cycling; (5) alterations in the gut microbiome; and (6) adaptations in glucose and lipid metabolism in peripheral tissues (Fig. 33.1). The relative importance of these mechanisms to weight loss and the metabolic benefits of SG remain largely unknown.

Gut Adaptation

Inherent to the SG procedure, a major portion of the stomach is excised, which in turn decreases gastric capacity, increases gastric pressure, and as a consequence, accelerates digestive transit to the lower tract in both rodents and humans [7,8]. Since rapid gastric emptying decreases meal size in humans [9], it is likely that part of the initial anorexic effects of SG are due to accelerated gastric emptying. In rodent models (Table 33.1), the rapid gastric transit observed with SG is resistant to the negative feedback effects of nutrient calories, muscarinic blockers, and glucagon-like peptide 1 (GLP-1) [7]. Thus, although the neurohumoral control of gastric emptying is clearly disrupted in SG, it is unknown whether this is due to loss or attenuation of neural activity in entero−vagal−brain stem circuits or higher brain areas. Little is known of the structural and functional adaptations in the intestine following SG. We [18] and others [15,24]

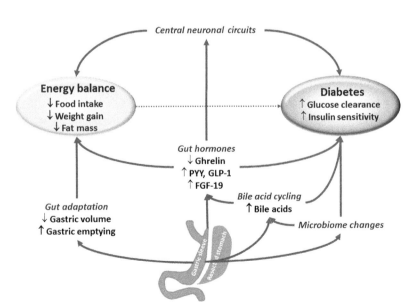

FIGURE 33.1 Hypothetical mechanisms of weight loss and diabetes improvements following sleeve gastrectomy (SG) surgery. Gastric restriction and accelerated gastric emptying likely contribute to early hypophagia. Altered neurohumoral feedback from the gut (decreased ghrelin; increased glucagon-like peptide 1, GLP-1; peptide YY, PYY; fibroblast growth factor-19, FGF-19) may engage central neuronal circuits to induce hypophagia and decrease adiposity with consequent indirect or direct improvements in glycemic control. The surgery-induced changes in gut microbiota and enhanced enterohepatic bile acid cycling may improve diabetic control directly or indirectly via modulation of gut hormone secretion. However, the relative importance of these multiple signals to the weight loss and diabetic improvements of SG remains poorly understood.

TABLE 33.1 Summary of the Effects of Sleeve Gastrectomy (SG) in Rats on Food Intake, Body Composition, Glucose and Insulin Tolerance, Circulating Hormones, and Other Metabolic Parameters

References	Model	Food Intake	Body Composition	Glucose/ Insulin Tolerance	Circulating Hormones	Other Parameters
[10]	SD rats ($n = 20$); 16 weeks postoperative follow-up; normal diet.	—	↓Weight	—	—	—
[11]	Obese Wistar rats ($n = 10$); 15 weeks postoperative follow-up; 45% high-fat diet.	—	↓Weight	—	—	—
[12]	Obese Zucker rats ($n = 12$); 2 weeks postoperative follow-up; normal diet.	↓	↓Weight	Improved	NSD	↓Cholesterol
[13]	Nonobese T2DM GK rats ($n = 10$); 13 weeks postoperative follow-up; normal diet.	—	↓Weight	Improved	↑Insulin	—
[14]	STZ treated diabetic SD rats ($n = 6$); 8 weeks postoperative follow-up; normal diet.	↓	↓Weight	Improved	↓Ghrelin	—
[15]	Long Evans rats; 28 weeks postoperative follow-up; 45% high-fat diet.	↓	↓Weight and fat mass	Improved	↑GLP-1	—
[8]	ZDF rats ($n = 15$); 6 weeks of postoperative follow-up; normal diet.	↓	↓Weight	Improved	↓Ghrelin ↑GLP-1	↓Cholesterol
[16]	GK rats ($n = 10$); 16 weeks of postoperative follow-up; 45% high-fat diet.	↓	↓Weight	Improved	↓Ghrelin ↑GLP-1 ↑PYY ↑Insulin	—
[17]	Female SD rats ($n = 18$); 4 weeks postoperative follow-up; 60% high-fat diet.	—	↓Weight	NSD	NSD	—
[18]	SD rats ($n = 9$); 8 weeks postoperative follow-up; normal diet.	↓	↓Weight	Improved	↑GLP-1 ↑PYY ↓Insulin ↓Leptin	—
[19]	GK rats; 4 weeks postoperative follow-up; normal diet.	—	↓Weight	Worsened	↓Ghrelin ↑GLP-1 ↑PYY	—
[20,21]	Wistar rats ($n = 4-5$); 2 weeks of postoperative follow-up; 45% high-fat diet.	↓	↓Weight	Improved	↑GLP-1 ↓Ghrelin	↓Cholesterol
[22]	ZDF rats ($n = 10$); 12 weeks of postoperative follow-up; 60% high-fat diet.	↓	↓Weight	Improved	↓Ghrelin (fasting) ↑GLP-1 ↓Insulin	—
[23]	UCD-T2DM rats ($n = 13$); 12 months postoperative follow-up; normal diet.	↓	↓Weight	Improved	↓Ghrelin ↑GLP-1 ↑PYY ↓Insulin ↓Leptin	↓Cholesterol ↑Bile acids

SD, Sprague Dawley; T2DM, Type-2 Diabetes Mellitus; GK, Goto-Kakizaki; STZ, Streptozotocin; ZDF, Zucker Diabetic Fatty; UCD-T2DM, University of California Davis Type-2 Diabetes Mellitus; NSD, No Significant Difference; GLP-1, Glucagon-like Peptide-1, PYY, Peptide YY. ↑ and ↓ denote an increase or decrease, and − indicates data not reported.

have shown in rat models that SG increases villus height and the number of GLP-1 immuno-positive in the jejunum or ileum at 2−8 weeks, but others [25] have reported no changes in gut adaptation at 12 weeks postsurgery. It is likely that the extent and magnitude of intestinal adaptations following SG dissipate over time. Nonetheless, during the early postoperative period, SG appears to reciprocally decrease glucose flux from the luminal to serosal side while increasing flux from the blood to the lumen, which in part may contribute to the improvements in glucose tolerance [26].

Peripheral Neurohumoral Mechanisms

Bariatric procedures often lead to enhanced secretion of anorexigenic gut hormones, while decreasing orexigenic hormones, thereby promoting earlier satiety. The L-cells of the ileum, colon, and rectum are a major source of peptide YY (PYY) and GLP-1 isoforms. There is substantial evidence that postprandial concentrations of these anorexigenic gut peptides (PYY, GLP-1) are greater in obese humans following SG [27]. Similarly, we [18] and others [7,15] have reported that SG surgeries increase plasma GLP-1 and PYY levels while decreasing leptin, insulin, and glucose-dependent insulinotropic peptide concentrations in rats, with a concomitant reduction in food intake and body weight, and improvements in glucose tolerance (Table 33.1). There is some evidence that fibroblast growth factor-19 (FGF-19), primarily of intestinal origin, is increased [28]—whereas FGF-21 is either decreased [28] or unaltered [29]—following SG in humans. However, circulating changes in these hormones are not causative for the metabolic effects of SG. The GLP-1 receptor antagonist exendin (9−39) has been shown to attenuate the improvements in glucose tolerance following SG in rats [15], but was found to be ineffective in humans [27]. Further, in mouse models of GLP-1 receptor deficiency, SG was found to produce weight loss and improve glucose clearance [30]. Thus, GLP-1 receptor signaling does not appear to be essential for the anorexic, weight loss, and glycemic improvements of SG. Circulating concentrations of the prototypical satiety hormone cholecystokinin are increased, whereas ghrelin concentrations are decreased following SG in humans [31,32]. However, SG was found to be effective in decreasing intake, body weight, and glucose excursions in ghrelin knockout mice, indicating that ghrelin may not be necessary for the metabolic effects of the surgery [31]. Together, these studies indicate that despite the postprandial increase in GLP-1 and decrease in ghrelin following SG (Tables 33.1 and 33.2), these hormones alone are unlikely to be essential for the weight loss and metabolic improvements of the surgery; however, whether these hormones interact to produce the health benefits of SG remains unknown.

The rapidity of responses with bariatric surgeries also implicates the nervous system because nerves in the gut are the major signaling pathways to other physiological systems of the body, vagal afferents are the target of the gut hormones that are altered following bariatric surgery, and ablating gastric afferents leads to reductions in visceral adiposity [41]. Though the hepatic branch of the vagus does not appear to mediate the effects of RYGB on food intake, body weight, and energy expenditure [42], RYGB decreases whereas SG increases the density of vagal afferents in the nucleus tractus solitarius [43]. In a rodent model of gastric banding, the efficacy of the band to decrease food intake and body weight was attenuated by perivagal capsaicin, indicating a role for vagal signaling in modulating the effects of gastric banding [44]. However, it is unknown whether the rapid weight loss and improvements in glucose tolerance following SG are due to surgical excision of gastric vagal terminals.

Enterohepatic Bile Acid Cycling

The normal enterohepatic cycling of bile acids entails multiple steps, including hepatic synthesis of primary bile acids from cholesterol, their storage in the gall bladder, secretion into the intestine under the control of endogenous secretagogues such as cholecystokinin, their critical role in lipid emulsification in the intestine, followed by absorption of the bile acids primarily in the terminal ileum and recycling back to the liver. It is now recognized that the metabolic effects of bariatric surgeries are in part due to alterations in enterohepatic cycling and involve at least two key bile acid receptors—Takeda G-protein-coupled receptor (TGR5) and farnesoid X receptor (FXR). Circulating bile acid concentrations were reported to be increased following SG in rodents [23,35,37,40] (Tables 33.1 and 33.2), and in some [45] (but not all [28]) studies in humans. Given that bile acid signaling via TGR5 stimulates GLP-1 secretion, it is interesting to note an association between circulating bile acids and GLP-1 in humans [45]. In support of bile acid−TGR5 signaling, improvements in glycemia and circulating bile acids following SG were attenuated in TGR5 knockout mice [40]. Similarly, the SG-induced reductions in food intake, body weight, fat mass, and improvements in glucose tolerance were attenuated in FXR knockout mice [37] (Table 33.2). Together these findings indicate that elevations in bile acids following SG in part contribute to the improvements in diabetic control via TGR5 and FXR signaling.

TABLE 33.2 Summary of the Effects of Sleeve Gastrectomy (SG) in Mice on Food Intake, Body Composition, Glucose and Insulin Tolerance, Circulating Hormones, and Other Metabolic Parameters

References	Model	Food Intake	Body Composition	Glucose /Insulin Tolerance	Circulating Hormones	Other Parameters
[33]	C57Bl/6 female mice (n = 9); 4 weeks of postoperative follow-up; normal diet.	–	↓Weight	Improved	NSD	–
[34]	C57BL/6 mice (n = 14); 8 weeks postoperative follow-up; 60% high-fat diet.	–	↓Weight and fat mass	Improved	–	–
[35]	C57Bl/6 mice; 8 weeks postoperative follow-up; 60% high-fat diet.	NSD	↓Weight	Improved	↓Ghrelin ↓Insulin	↓Cholesterol ↑Bile acids
[36]	C57BL/6 (n = 13); 3 weeks postoperative follow-up; 45% high-fat diet.	↓	↓Weight and fat	Improved	↓Ghrelin	–
[37]	FXR KO mice (n = 15); 14 weeks postoperative follow-up; 60% high-fat diet.	NSD	↓Weight	Improved with SG in wild-type but impaired in FXR KO	–	Microbes: ↓Bacteroides in wild-type SG but not FXR KO; ↑Porphyromonadaceae in FXR KO but not wild-type
[38]	Mgat2 KO mice (n = 16); 12 weeks postoperative follow-up; 60% high-fat diet.	–	↓Weight and Fat mass	Improved with SG in both wild-type and Mgat2 KO	–	–
[39]	SHP KO mice; 8 weeks postoperative follow-up; 60% high-fat diet.	↓	↓Weight	Improved with SG in both wild-type and Mgat2 KO	–	↑Bile acids
[40]	Tgr5 KO mice (n = 8); 26 weeks postoperative follow-up; 45% high-fat diet.	↓	↓Weight and Fat mass	Improved with SG in wild-type but impaired in Tgr5 KO	↑Insulin ↑GLP-1 ↓Leptin	↑Bile acids ↓Cholesterol Microbes: ↓Verrucomicrobia ↓Turicibacteraceae ↓Adlercreutzia ↑Enterococcus

FXR KO, Farnesoid X Receptor knockout; MGAT2 KO, acyl-CoA:monoacylglycerol transferase knockout; SHP KO, small heterodimer partner knockout; TGR5 KO, G-protein-coupled BA receptor 1 knockout; NSD, No Significant Difference; GLP-1, Glucagon-like Peptide-1. ↑ and ↓ denote an increase or decrease, and – indicates data not reported.

Gut Microbiota

There is increasing evidence that alterations in the gut microbial communities may play a role in the metabolic benefits of bariatric surgeries. RYGB has been associated with a reduction in the ratio of Firmicutes to Bacteroidetes phyla, increase in Gammaproteobacteria and Verrumicrobia, and an increase in bacterial genera richness in humans and rodents [46,47]. SG and RYGB were found to produce similar long-term shifts in microbial composition in women [48]. Importantly, transfer of microbiota from RYGB-treated mice (or from RYGB and SG patients) to germ-free mice decreased body weight, fat content, and respiratory quotient, and improved glucose tolerance [47,48]. These findings provide strong evidence that changes in the gut microbiome following RYGB and SG may contribute to weight loss and improve glycemic control. However, given that microbiota modulate the secretion of enteric hormones and bile acids, the causal link between the gut microbiome, gut hormones, and bile acid–TGR5–FXR signaling remains unknown.

Glucose and Lipid Metabolism in Peripheral Tissues

Despite substantial evidence that SG improves diabetic control, less is known of the downstream molecular events by which the surgery improves glucose and lipid metabolism in peripheral tissues. In humans, skeletal muscle, liver, and adipose tissues account for about 80%, 5%, and 1% of total glucose metabolized, respectively [49], with ectopic lipid deposition in muscle and liver contributing to impaired insulin signaling. There is increasing evidence that SG decreases hepatic lipidosis and improves indices of hepatic function and glycemic control in both rodent models and humans [39,50]. In rodent models, although SG produces similar reductions in weight gain in intact male and female rats [17,24], it increases severity of hepatic lipidosis in females [51] but decreases lipidosis in males [35], and improves glycemic control in males [18,24] but not females [17]. However, the mechanisms for these sex differences and the underlying molecular events are poorly understood. We have shown [18] that SG decreases the transcript abundance of the gluconeogenic enzyme phosphoenol pyruvate carboxykinase in the liver, and others have reported reduction in the lipogenic marker peroxisome proliferator-activated receptor-γ in the liver of male rats [32]. These data are suggestive of decreased hepatic lipid load and glucose output, and improved glucose tolerance. We have also demonstrated that SG increases the transcript and protein abundance of glucose transporter type 4 in muscle and adipose tissues, and decreases the insulin receptor substrate 1 (IRS-1) pS636/IRS-1 ratio in adipose tissue, which is suggestive of enhanced glucose clearance and insulin sensitivity in these tissues [18]. Further, SG has been shown to upregulate the expression of the glycerol efflux markers aquaporin-3 and -7 in white adipose tissue [32], and to increase tissue volume and oxidative metabolism in brown adipose tissue [52], which is suggestive of enhanced lipolysis. However, little is known of the relative importance of gut peptides, bile acids, and neural mechanisms involved in mediating the observed metabolic effects of SG in peripheral tissues.

Central Energy Balance Circuits

There is evidence that SG subjects show distinct shifts in eating patterns, have reduced desire to eat, sometimes show aversion to calorie-dense foods, and that those with normal eating and snacking patterns appear to have more sustained weight loss [53]. Similarly, in rodent models, SG decreases preferences for palatable high-fat diets [54,55]. The anorexic effects of SG are associated with a lower threshold for c-Fos activity in the nucleus tractus solitarius [55]. However, the effects of SG on hypothalamic neuropeptide expression are inconsistent, with some studies reporting an increase in anorexigenic (proopiomelanocortin) and decrease in orexigenic (neuropeptide Y) expression [56], whereas other studies failed to detect such differences [24]. It is well established that central melanocortins exert anorexigenic effects via melanocortin-4 receptors (MC4R), with mutations in MC4R being one of the most common causes of monogenic obesity in humans. In pediatric patients with mutations in MC4R [57], and in MC4R knockout rats [58], SG was found to be effective in promoting weight loss and improving glycemic control, indicating that MC4R is not essential for the weight loss and metabolic effects of the surgery. Despite the marked reduction in food intake, the effects of SG on energy expenditure are inconsistent, with some studies reporting increased resting energy expenditure per unit weight [40,59], and others finding stable [24] or reduced energy expenditure in rodents and humans [60]. Thus, despite the well-established anorexic effects of SG, the central neuronal homeostatic and hedonic networks that mediate improvements in energy intake and energy expenditure are poorly understood.

SUMMARY

SG surgery is becoming increasingly popular because of its relative technical simplicity, with its efficacy in weight loss and diabetic control being comparable to more complex procedures. Though initially regarded as a purely restrictive operation, it is now recognized that the health benefits realized from the metabolic component of the procedure are more important. Importantly, the improvements in glycemic control following SG appear to occur independently of changes in body weight. Some of the mechanisms that have been proposed to explain the weight loss and glycemic benefits of SG include gut adaptation, enhanced secretion and/or activity of anorexigenic lower gut peptides with concomitant reduction in orexigenic signals, alterations in enterohepatic bile acid metabolism, modulation of gut microbiota, improvements in glucose and lipid metabolism in peripheral tissues, and modulation of central circuits regulating energy balance (Fig. 33.1). An understanding of the relative importance of these mechanisms to the metabolic effects of SG surgery may lead to the development of less invasive interventions that can mimic the effects of the surgery without the attendant surgical risks and long-term complications.

ACKNOWLEDGMENTS

This work was supported in part by the Koopmans Memorial Research Fund and Natural Sciences and Engineering Research Council of Canada (NSERC) Discovery Grant to Prasanth K. Chelikani.

MINI-DICTIONARY OF TERMS

- *Sleeve gastrectomy (SG).* A relatively popular weight loss procedure in which the size of the stomach is decreased by over 80%.
- *Glucagon-like peptide 1 (GLP-1).* A peptide hormone secreted from L-cells of the lower intestine that acts to produce satiety and improve glucose control.
- *Peptide YY (PYY).* A peptide hormone secreted from L-cells of the lower intestine that acts to produce satiety and weight loss.
- *Fibroblast growth factor (FGF).* A peptide hormone produced by the intestine and liver that acts to improve energy balance and glucose tolerance.
- *Takeda G protein−coupled receptor (TGR5).* A G protein−coupled receptor for bile acids.
- *Farnesoid X receptor (FXR).* A nuclear receptor for bile acids.

KEY FACTS

- SG is a popular and relatively simple surgical weight loss procedure that also improves diabetic control.
- The procedure involves removing a substantial portion of the stomach while keeping the intestines intact.
- The decreased food consumption and weight loss from the procedure were initially believed to be due to the reduction in stomach capacity.
- We are now realizing that the health benefits of the procedure are due to more complex mechanisms that involve multiple physiological systems.

SUMMARY POINTS

This chapter deals with how SG, a popular bariatric procedure, works to produce weight loss and improve diabetic control. The proposed mechanisms include:

- Adaptations in the gastrointestinal tract, including reductions in gastric volume and rapid delivery of nutrients to the distal intestine.
- Alterations in gut hormone secretion involving reduction in orexigenic (increased food intake) and enhanced secretion of anorexigenic (reduced food intake) gut peptides.
- Changes in central brain networks that regulate energy balance.
- Modulation of gut microbiota composition and bile acid cycling between intestine and liver.
- Enhanced glucose clearance by peripheral tissue through adaptations in glucose and lipid metabolism.

REFERENCES

[1] Buchwald H, Avidor Y, Braunwald E, Jensen MD, Pories W, Fahrbach K, et al. Bariatric surgery: a systematic review and meta-analysis. JAMA 2004;292:1724−37.

[2] Couzin J. Medicine. Bypassing medicine to treat diabetes. Science 2008;320:438−40.

[3] C.I.f.H.I. (CIHI), Bariatric Surgery in Canada, 2014.

[4] Buchwald H, Oien DM. Metabolic/bariatric surgery worldwide 2011. Obes Surg 2013;23:427–36.

[5] Franco JV, Ruiz PA, Palermo M, Gagner M. A review of studies comparing three laparoscopic procedures in bariatric surgery: sleeve gastrectomy, Roux-en-Y gastric bypass and adjustable gastric banding. Obes Surg 2011;21:1458–68.

[6] Dixon JB, Zimmet P, Alberti KG, Rubino F. Bariatric surgery: an IDF statement for obese Type 2 diabetes. Surg Obes Relat Dis 2011;7:433–47.

[7] Chambers AP, Smith EP, Begg DP, Grayson BE, Sisley S, Greer T, et al. Regulation of gastric emptying rate and its role in nutrient-induced GLP-1 secretion in rats after vertical sleeve gastrectomy. Am J Physiol Endocrinol Metab 2014;306:E424–32.

[8] Masuda T, Ohta M, Hirashita T, Kawano Y, Eguchi H, Yada K, et al. A comparative study of gastric banding and sleeve gastrectomy in an obese diabetic rat model. Obes Surg 2011;21:1774–80.

[9] Torra S, Ilzarbe L, Malagelada JR, Negre M, Mestre-Fusco A, Aguade-Bruix S, et al. Meal size can be decreased in obese subjects through pharmacological acceleration of gastric emptying (The OBERYTH trial). Int J Obes 2011;35:829–37.

[10] Cai J, Zheng C, Xu L, Chen D, Li X, Wu J, et al. Therapeutic effects of sleeve gastrectomy plus gastric remnant banding on weight reduction and gastric dilatation: an animal study. Obes Surg 2008;18:1411–17.

[11] Patrikakos P, Toutouzas KG, Perrea D, Menenakos E, Pantopoulou A, Thomopoulos T, et al. A surgical rat model of sleeve gastrectomy with staple technique: long-term weight loss results. Obes Surg 2009;19:1586–90.

[12] Lopez PP, Nicholson SE, Burkhardt GE, Johnson RA, Johnson FK. Development of a sleeve gastrectomy weight loss model in obese Zucker rats. J Surg Res 2009;157:243–50.

[13] Inabnet WB, Milone L, Harris P, Durak E, Freeby MJ, Ahmed L, et al. The utility of [(11)C] dihydrotetrabenazine positron emission tomography scanning in assessing beta-cell performance after sleeve gastrectomy and duodenal-jejunal bypass. Surgery 2010;147:303–9.

[14] Wang Y, Yan L, Jin Z, Xin X. Effects of sleeve gastrectomy in neonatally streptozotocin-induced diabetic rats. PLoS One 2011;6:e16383.

[15] Chambers AP, Jessen L, Ryan KK, Sisley S, Wilson-Perez HE, Stefater MA, et al. Weight-independent changes in blood glucose homeostasis after gastric bypass or vertical sleeve gastrectomy in rats. Gastroenterology 2011;141:950–8.

[16] Donglei Z, Liesheng L, Xun J, Chenzhu Z, Weixing D. Effects and mechanism of duodenal-jejunal bypass and sleeve gastrectomy on GLUT2 and glucokinase in diabetic Goto-Kakizaki rats. Eur J Med Res 2012;17:15.

[17] Brinckerhoff TZ, Bondada S, Lewis CE, French SW, DeUgarte DA. Metabolic effects of sleeve gastrectomy in female rat model of diet-induced obesity. Surg Obes Relat Dis 2013;9:108–12.

[18] Nausheen S, Shah IH, Pezeshki A, Sigalet DL, Chelikani PK. Effects of sleeve gastrectomy and ileal transposition, alone and in combination, on food intake, body weight, gut hormones, and glucose metabolism in rats. Am J Physiol Endocrinol Metab 2013;305:E507–18.

[19] Eickhoff H, Louro TM, Matafome PN, Vasconcelos F, Seica RM, Castro ESF. Amelioration of glycemic control by sleeve gastrectomy and gastric bypass in a lean animal model of type 2 diabetes: restoration of gut hormone profile. Obes Surg 2015;25:7–18.

[20] Arapis K, Cavin JB, Gillard L, Cluzeaud F, Letteron P, Ducroc R, et al. Remodeling of the residual gastric mucosa after Roux-en-Y gastric bypass or vertical sleeve gastrectomy in diet-induced obese rats. PLoS One 2015;10:e0121414.

[21] Cavin JB, Couvelard A, Lebtahi R, Ducroc R, Arapis K, Voitellier E, et al. Differences in alimentary glucose absorption and intestinal disposal of blood glucose after Roux-en-Y gastric bypass vs sleeve gastrectomy. Gastroenterology 2016;150:454–64.

[22] Wang K, Zhou X, Quach G, Lu J, Gao W, Xu A, et al. Effect of sleeve gastrectomy plus side-to-side jejunoileal anastomosis for type 2 diabetes control in an obese rat model. Obes Surg 2016;26:797–804.

[23] Cummings BP, Bettaieb A, Graham JL, Stanhope KL, Kowala M, Haj FG, et al. Vertical sleeve gastrectomy improves glucose and lipid metabolism and delays diabetes onset in UCD-T2DM rats. Endocrinology 2012;153:3620–32.

[24] Stefater MA, Perez-Tilve D, Chambers AP, Wilson-Perez HE, Sandoval DA, Berger J, et al. Sleeve gastrectomy induces loss of weight and fat mass in obese rats, but does not affect leptin sensitivity. Gastroenterology 2010;138:2426–36.

[25] Mumphrey MB, Hao Z, Townsend RL, Patterson LM, Berthoud HR. Sleeve gastrectomy does not cause hypertrophy and reprogramming of intestinal glucose metabolism in rats. Obes Surg 2015;25:1468–73.

[26] Cavin JB, Couvelard A, Lebtahi R, Ducroc R, Arapis K, et al. Differences in alimentary glucose absorption and intestinal disposal of blood glucose after Roux-en-Y Gastric Bypass vs Sleeve Gastrectomy. Gastroenterology 2016;150:454–64.

[27] Jimenez A, Mari A, Casamitjana R, Lacy A, Ferrannini E, Vidal J. GLP-1 and glucose tolerance after sleeve gastrectomy in morbidly obese subjects with type 2 diabetes. Diabetes 2014;63:3372–7.

[28] Haluzikova D, Lacinova Z, Kavalkova P, Drapalova J, Krizova J, Bartlova M, et al. Laparoscopic sleeve gastrectomy differentially affects serum concentrations of FGF-19 and FGF-21 in morbidly obese subjects. Obesity 2013;21:1335–42.

[29] Buzga M, Holeczy P, Svagera Z, Zonca P. Laparoscopic gastric plication and its effect on saccharide and lipid metabolism: a 12-month prospective study. Wideochir Inne Tech Maloinwazyjne 2015;10:398–405.

[30] Wilson-Perez HE, Chambers AP, Ryan KK, Li B, Sandoval DA, Stoffers D, et al. Vertical sleeve gastrectomy is effective in two genetic mouse models of glucagon-like Peptide 1 receptor deficiency. Diabetes 2013;62:2380–5.

[31] Chambers AP, Kirchner H, Wilson-Perez HE, Willency JA, Hale JE, Gaylinn BD, et al. The effects of vertical sleeve gastrectomy in rodents are ghrelin independent. Gastroenterology 2013;144:50–2.

[32] Mendez-Gimenez L, Becerril S, Moncada R, Valenti V, Ramirez B, Lancha A, et al. Sleeve gastrectomy reduces hepatic steatosis by improving the coordinated regulation of aquaglyceroporins in adipose tissue and liver in obese rats. Obes Surg 2015;25:1723–34.

[33] Schlager A, Khalaileh A, Mintz Y, Abu Gazala M, Globerman A, Ilani N, et al. A mouse model for sleeve gastrectomy: applications for diabetes research. Microsurgery 2011;31:66–71.

[34]   Yin DP, Gao Q, Ma LL, Yan W, Williams PE, McGuinness OP, et al. Assessment of different bariatric surgeries in the treatment of obesity and insulin resistance in mice. Ann Surg 2011;254:73–82.

[35]   Myronovych A, Kirby M, Ryan KK, Zhang W, Jha P, Setchell KD, et al. Vertical sleeve gastrectomy reduces hepatic steatosis while increasing serum bile acids in a weight-loss-independent manner. Obesity 2014;22:390–400.

[36]   Schneck AS, Iannelli A, Patouraux S, Rousseau D, Bonnafous S, Bailly-Maitre B, et al. Effects of sleeve gastrectomy in high fat diet-induced obese mice: respective role of reduced caloric intake, white adipose tissue inflammation and changes in adipose tissue and ectopic fat depots. Surg Endosc 2014;28:592–602.

[37]   Ryan KK, Tremaroli V, Clemmensen C, Kovatcheva-Datchary P, Myronovych A, Karns R, et al. FXR is a molecular target for the effects of vertical sleeve gastrectomy. Nature 2014;509:183–8.

[38]   Mul JD, Begg DP, Haller AM, Pressler JW, Sorrell J, Woods SC, et al. MGAT2 deficiency and vertical sleeve gastrectomy have independent metabolic effects in the mouse. Am J Physiol Endocrinol Metab 2014;307:E1065–72.

[39]   Myronovych A, Salazar-Gonzalez RM, Ryan KK, Miles L, Zhang W, Jha P, et al. The role of small heterodimer partner in nonalcoholic fatty liver disease improvement after sleeve gastrectomy in mice. Obesity 2014;22:2301–11.

[40]   McGavigan AK, Garibay D, Henseler ZM, Chen J, Bettaieb A, Haj FG, et al. TGR5 contributes to glucoregulatory improvements after vertical sleeve gastrectomy in mice. Gut 2015.

[41]   Leung FW. Capsaicin as an anti-obesity drug, progress in drug research. Fortschritte der Arzneimittelforschung. Progres des recherches pharmaceutiques 2014;68:171–9.

[42]   Shin AC, Zheng H, Berthoud HR. Vagal innervation of the hepatic portal vein and liver is not necessary for Roux-en-Y gastric bypass surgery-induced hypophagia, weight loss, and hypermetabolism. Ann Surg 2012;255:294–301.

[43]   Ballsmider LA, Vaughn AC, David M, Hajnal A, Di Lorenzo PM, Czaja K. Sleeve gastrectomy and Roux-en-Y gastric bypass alter the gut-brain communication. Neural Plast 2015;2015:601985.

[44]   Kampe J, Stefanidis A, Lockie SH, Brown WA, Dixon JB, Odoi A, et al. Neural and humoral changes associated with the adjustable gastric band: insights from a rodent model. Int J Obes 2012;36:1403–11.

[45]   Steinert RE, Peterli R, Keller S, Meyer-Gerspach AC, Drewe J, Peters T, et al. Bile acids and gut peptide secretion after bariatric surgery: a 1-year prospective randomized pilot trial. Obesity 2013;21:E660–8.

[46]   Furet JP, Kong LC, Tap J, Poitou C, Basdevant A, Bouillot JL, et al. Differential adaptation of human gut microbiota to bariatric surgery-induced weight loss: links with metabolic and low-grade inflammation markers. Diabetes 2010;59:3049–57.

[47]   Liou AP, Paziuk M, Luevano Jr. JM, Machineni S, Turnbaugh PJ, Kaplan LM. Conserved shifts in the gut microbiota due to gastric bypass reduce host weight and adiposity. Sci Transl Med 2013;5:178ra141.

[48]   Tremaroli V, Karlsson F, Werling M, Stahlman M, Kovatcheva-Datchary P, Olbers T, et al. Roux-en-Y gastric bypass and vertical banded gastroplasty induce long-term changes on the human gut microbiome contributing to fat mass regulation. Cell Metab 2015;22:228–38.

[49]   DeFronzo RA, Jacot E, Jequier E, Maeder E, Wahren J, Felber JP. The effect of insulin on the disposal of intravenous glucose. Results from indirect calorimetry and hepatic and femoral venous catheterization. Diabetes 1981;30:1000–7.

[50]   Billeter AT, Senft J, Gotthardt D, Knefeli P, Nickel F, Schulte T, et al. Combined non-alcoholic fatty liver disease and type 2 diabetes mellitus: sleeve gastrectomy or gastric bypass?-a Controlled matched pair study of 34 patients. Obes Surg 2016;26(8):1867–74.

[51]   Grayson BE, Schneider KM, Woods SC, Seeley RJ. Improved rodent maternal metabolism but reduced intrauterine growth after vertical sleeve gastrectomy. Sci Transl Med 2013;5:199ra112.

[52]   Baraboi ED, Li W, Labbe SM, Roy MC, Samson P, Hould FS, et al. Metabolic changes induced by the biliopancreatic diversion in diet-induced obesity in male rats: the contributions of sleeve gastrectomy and duodenal switch. Endocrinology 2015;156:1316–29.

[53]   Sioka E, Tzovaras G, Oikonomou K, Katsogridaki G, Zachari E, Papamargaritis D, et al. Influence of eating profile on the outcome of laparoscopic sleeve gastrectomy. Obes Surg 2013;23:501–8.

[54]   Wilson-Perez HE, Chambers AP, Sandoval DA, Stefater MA, Woods SC, Benoit SC, et al. The effect of vertical sleeve gastrectomy on food choice in rats. Int J Obes (Lond) 2013;37(2):288–95.

[55]   Chambers AP, Wilson-Perez HE, McGrath S, Grayson BE, Ryan KK, D'Alessio DA, et al. Effect of vertical sleeve gastrectomy on food selection and satiation in rats. Am J Physiol Endocrinol Metab 2012;303:E1076–84.

[56]   Kawasaki T, Ohta M, Kawano Y, Masuda T, Gotoh K, Inomata M, et al. Effects of sleeve gastrectomy and gastric banding on the hypothalamic feeding center in an obese rat model. Surg Today 2015;45:1560–6.

[57]   Jelin EB, Daggag H, Speer AL, Hameed N, Lessan N, Barakat M, et al. Melanocortin-4 receptor signaling is not required for short-term weight loss after sleeve gastrectomy in pediatric patients. Int J Obes 2015;40(3):550–3.

[58]   Mul JD, Begg DP, Alsters SI, van Haaften G, Duran KJ, D'Alessio DA, et al. Effect of vertical sleeve gastrectomy in melanocortin receptor 4-deficient rats. Am J Physiol Endocrinol Metab 2012;303:E103–10.

[59]   Schneider J, Peterli R, Gass M, Slawik M, Peters T, Wolnerhanssen BK. Laparoscopic sleeve gastrectomy and Roux-en-Y gastric bypass lead to equal changes in body composition and energy metabolism 17 months postoperatively: a prospective randomized trial. Surg Obes Relat Dis 2015.

[60]   Tam CS, Rigas G, Heilbronn LK, Matisan T, Probst Y, Talbot M. Energy adaptations persist 2 years after sleeve gastrectomy and gastric bypass. Obes Surg 2016;26(2):459–63.

Chapter 34

Postprandial Hyperinsulinemic Hypoglycemia in Bariatric Surgery

L.J.M. de Heide[1], M. Emous[1] and A.P. van Beek[2]

[1]Medical Centre Leeuwarden, Leeuwarden, The Netherlands, [2]University Medical Centre Groningen, Groningen, The Netherlands

LIST OF ABBREVIATIONS

AUC	area under the curve
BCAA	branched-chain amino acids
CCK	cholecystokinin
CGM	continuous glucose monitoring
GIP	glucose-dependent insulinotropic peptide
GLP-1	glucagon-like peptide 1
GS	gastric sleeve
IGF-1	insulin-like growth factor 1
LAGB	laparoscopic gastric banding
MMTT	mixed meal tolerance test
OGTT	oral glucose tolerance test
PHH	postprandial hyperinsulinemic hypoglycemia
PYY	peptide tyrosine tyrosine
RYGB	Roux-en-Y gastric bypass
SIBO	small intestine bacterial overgrowth
SST2	somatostatin receptor subtype 2

INTRODUCTION

Postprandial hyperinsulinemic hypoglycemia (PHH) is a syndrome that occurs following ingestion of carbohydrates in patients who have undergone gastric surgery. It used to be a known complication of gastric surgery for ulcer disease and malignancies. Nowadays it is mainly seen as a consequence of bariatric surgery.

PHH is part of the so called "dumping syndrome," which can be divided into early and late dumping (see "Key Facts" section). Early dumping, occurring within 1 hour after a meal, is caused by rapid introduction of highly osmotic nutrients into the jejunum, which attracts fluid and leads to intravascular depletion followed by hypotension and vasoactive responses. Late dumping, which occurs 1–3 hours after a meal, is synonymous with PHH. Table 34.1 shows the symptoms of early and late dumping.

This chapter describes the current knowledge regarding the definition, epidemiology, pathophysiology, and treatment of PHH.

DEFINITION, EPIDEMIOLOGY, AND RISK FACTORS

Currently, no clear definition of PHH exists in the literature, and a consensus on which criteria have to be met in order to diagnose either early or late dumping is lacking. An overview of all different diagnostic modalities for PHH is present in a review by Emous et al. [1] (see Table 34.2).

TABLE 34.1 Signs and Symptoms of Early and Late Dumping

Early	Late
Vasomotor	*Adrenergic*
Palpitations, Tachycardia	Palpitations, Tachycardia
Perspiration	Perspiration
Flushing	Tremor
Fatigue, Drowsiness	Hunger
Urge to lie down	
Hypotension, Syncope	*Neuroglycopenic*
Gastrointestinal	Weakness
Abdominal pain	Confusion
Bloating	Coma
Nausea	Seizures
Diarrhea	
Borborygmi	
Abdominal distension	

Signs and symptoms of dumping divided into early dumping, with vasomotor- and gastrointestinal-related items, and late dumping, with adrenergic and neuroglycopenic items.

TABLE 34.2 Diagnostic Modalities in Postprandial Hyperinsulinemic Hypoglycemia

Study Type [Ref.]	Population, Surgery	Test Characteristic	Cutoff Value Glucose	Incidence/Prevalence of Hypoglycemia
Cohort [2]	5040, RYGB	Hospitalization	Not available	0.2%
[3]	3082, RYGB	Documented severe hypo	Not available	0.36%
[4]	257, 514, RYGB/GB/Sleeve	Self-reported	Not available	0.1%/0.01%/0.02%
[5]	450, RYGB	Validated questionnaire	Not available	34%
OGTT [6]	30, upper GI-surgery	75 g glucose	<60 mg/dL	80%
[7]	63, RYGB	100 g glucose	<60 mg/dL	68%
[8]	351, RYGB	75 g glucose	<50 mg/dL	10.4%
MMTT [9]	9, RYGB no PHH	Mixed meal	<60 mg/dL	33%
[10]	24, RYGB, 12 plus, 12 no PHH	Mixed meal	<50 mg/dL	80%, (=TP), 16%, (=FP)
[11]	16, RYGB, 9 plus, 7 no PHH	Mixed meal	<50 mg/dL	90%, (=TP), 0%, (=FP)
[12]	51, RYGB	Mixed meal	<55 mg/dL	29%
CGM [13]	8, RYGB, plus PHH	3-day CGM	<60 mg/dL	75%
[14]	30, RYGB and DS	3-day CGM	<50 mg/dL	All
[12]	40, RYGB	5-day CGM	<55 mg/dL	75%

Summary of main studies on diagnosis of PHH.
RYGB, Roux-en-Y gastric bypass; GB, gastric banding; Sleeve, sleeve gastrectomy; OGTT, oral glucose tolerance test; MMTT, mixed meal tolerance test; CGM, continuous glucose monitoring; PHH, postprandial hyperinsulinemic hypoglycemia; TP, true positive; FP, false positive.

Sigstad developed a symptom score in the area of gastric surgery for ulcer disease that can aid in discriminating dumping from other postoperative complications [15]. Arts developed a questionnaire to differentiate between early and late dumping using eight and six symptoms, respectively [6]. Neither Sigstad's score nor Arts' dumping questionnaire have been validated in bariatric surgery patients.

The incidence of hypoglycemia after bariatric surgery is highly dependent on the cutoff value of the glucose concentration that defines hypoglycemia. Most authors agree that the classical triad of Whipple has to be met [16]. This triad consists of neuroglycopenic and/or adrenergic symptoms and signs consistent with hypoglycemia (see Table 34.3), a low plasma glucose value, and resolution of the symptoms after euglycemia is restored.

In the first report of five patients with PHH, Service et al. used a postprandial value of 55 mg/dL (3.06 mmol/L) or less for the diagnosis [17]. Various other studies, however, used different cutoff values of plasma glucose after provocative testing with the oral glucose tolerance test (OGTT) or a mixed meal tolerance test (MMTT), ranging from 50 to 60 mg/dL (2.8−3.3 mmol/L). Recently, continuous subcutaneous glucose monitoring (CGM) has been introduced, which can be helpful in diagnosing PHH.

Several studies report on the incidence of severe hypoglycemia after bariatric surgery. Severe hypoglycemia is by definition a hypoglycemia requiring assistance from a third party for resolution. In the Swedish Obesity Study (SOS), the incidence of hospital admissions for hypoglycemia or symptoms suggestive of hypoglycemia in 5040 patients after Roux-en-Y gastric bypass (RYGB) was 0.2% [2]. In another study, 0.36% of 3082 RYGB patients were found to have severe hypoglycemia [3]. In the BOLD data set, comprising 275,514 patients, the incidence of self-reported hypoglycemia was 0.1% after RYGB, 0.01% after laparoscopic gastric banding (LAGB), and 0.02% with gastric sleeve (GS) procedures [4]. However, the number of self-reported hypoglycemias is probably an underestimation of the problem. Using an adapted validated hypoglycemia questionnaire in a survey in post-RYGB patients, 34% of the 450 respondents reported symptoms highly suspect of hypoglycemia [5], and 11.5% had severe or documented hypoglycemia.

Studies with an OGTT used cutoff values from 50 to 60 mg/dL (2.8−3.3 mmol/L) and glucose loads of 75 or 100 g [6−8]. Prevalence of hypoglycemia varied from 10.4% to 80%, making the diagnostic accuracy low. Furthermore, late hypoglycemia during an OGTT is often also present in nonsurgical persons without a history of hypoglycemic symptoms, making it difficult to establish cutoff values.

In comparison to an OGTT, an MMTT contains a meal with a more normal composition and is likely to trigger a more physiological response. Studies used liquid formulas with various compositions of carbohydrates (35−100 g), proteins (6−13 g), and fat (6−12 g) [1]. Most patients with postsurgical hypoglycemia developed hypoglycemia (glucose < 50 mg/dL, <2.8 mmol/L) during an MMTT. In those without neuroglycopenic symptoms, an MMTT was usually normal.

Small studies with 9−11 patients reported a true positive MMTT in 80−100% [9−11], with a false positivity of 0−33%. A larger study, however, found a positive MMTT in only 50% of symptomatic patients (12/24), with a false-positive of 10% (3/29) [18]. In an MMTT study in eight RYGB patients with well-documented neuroglycopenic episodes, no difference in glucose excursions was found in comparison to patients without hypoglycemia [19]. In summary, an MMTT seems to perform better than an OGTT, but with the current limited data, it does not prove to be discriminating enough to use as a standard diagnostic tool.

TABLE 34.3 Adrenergic and Neuroglycopenic Signs and Symptoms of Hypoglycemia

Adrenergic	Neuroglycopenic
Palpitations, Tachycardia	Visual disturbances
Shakiness, Tremor	Confusion
Perspiration	Abnormal behavior
Cold skin	Loss of consciousness
Hunger	Seizures
Anxiety	
Irritation	

Signs and symptoms of hypoglycemia divided into physiological origin.

CGM is able to continuously measure glucose in the interstitial fluid. It has the advantage of providing more information on glucose values in relation to symptoms, food intake, and activities. Due to the measurement in the interstitial fluid, a lag-time difference with blood glucose exists, varying between 5 and 15 minutes depending on different situations. Furthermore, the accuracy of the measurement in the hypoglycemic range is poor, limiting the applicability in studies on hypoglycemia [20]. CGM has been used in blinded fashion in some studies in bariatric surgery. CGM showed that hypoglycemia is often preceded by a rapid increase in glycemia after a meal, and that patients with hypoglycemia had a shortened time to peak glucose and more glycemic variability [21].

In a small study, 6 out of 8 patients with PHH had hypoglycemic episodes during a 3-day blinded CGM registration [13]. In 15 RYGB patients without known hypoglycemia, CGM recorded glucose values below 60 mg/dL (3.3 mmol/L) on average 2.9% of the time and below 50 mg/dL (2.8 mmol/L) in 1.5% [14]. Patients were aware of hypoglycemia only 1 in 5 times. This discrepancy can be explained by a phenomenon known as hypoglycemia unawareness or hypoglycemia-associated autonomic failure. It is frequently observed in type 1 diabetes on intensive insulin treatment, but we speculate that this is also part of PHH [22]. As there are no studies with systematic follow-up of glucose values of large cohorts of bariatric patients, it is currently not known how many patients develop hypoglycemia and hypoglycemia unawareness. The largest study with CGM found a prevalence of 75% of hypoglycemia in 40 post-RYGB patients without known hypoglycemia, suggesting that hypoglycemia is much more common than assumed thus far [12].

RISK FACTORS

In the survey by Lee et al., hypoglycemic symptoms were significantly more often present among women, after RYGB, with a longer interval after surgery, and in those without diabetes before surgery [5]. Furthermore, those with hypoglycemic symptoms were significantly more likely to report comparable symptoms before surgery. However, in the BOLD experience, also a study with self-reported hypoglycemias, no clear risk factors could be identified [4]. No prospective studies with well-documented hypoglycemia have been performed as yet. In the SOS follow-up the average time interval between surgery and development of symptoms was 2.7 years, comparable with the range of 1−5 years in most case studies [2].

In conclusion, the real incidence of hypoglycemia after bariatric surgery is not known. Barriers to better documentation of this complication are a lack of definition and of well-designed prospective studies. The high prevalence of asymptomatic hypoglycemic episodes documented by CGM suggests that hypoglycemia unawareness is probably often present in these patients. The clinical relevance has yet to be established.

PATHOPHYSIOLOGY

The pathophysiology of PHH after bariatric surgery is not clear. Possible mechanisms leading to hypoglycemia could be the following:

- difference in resorption of nutrients, especially carbohydrates, but possibly also amino acids;
- differences in gastrointestinal (GI) hormone responses, such as increased production of incretins and/or less production of a yet unknown anti-incretin;
- an increased insulin response to glucose or to incretins;
- a diminished contraregulation to hypoglycemia.

In patients with type 2 diabetes, an increase in insulin production is noticed within days after RYGB surgery, long before weight loss occurs [23]. This rapid improvement in insulin response to glucose is attributed to the so-called incretins. Oral administration of glucose leads to an augmented insulin response in comparison to intravenous glucose, and is called the incretin effect. This effect is mainly attributed to the release of glucagon-like peptide 1 (GLP-1) by L cells in the distal ileum in response to nutrient delivery. Glucose-dependent insulinotropic peptide (GIP), released from K cells in the duodenum and proximal jejunum, also contributes to the release of insulin. Gastric bypass surgery leads to a rapid delivery of nutrients to the L cell−rich distal part of the ileum. Within 2 weeks after RYGB, a 5−10-fold increase in GLP-1 concentration is seen after an MMTT, together with increases in the GI hormones peptide tyrosine tyrosine (PYY) and cholecystokinin (CCK), with a concomitant increase in insulin levels [24]. GIP levels remain unchanged in some studies, but increase in others (Fig. 34.1). A more rapid glucose peak and higher GLP-1 response was found during an MMTT in patients with postsurgical hypoglycemia, suggesting a role for GLP-1 in the pathophysiology [9]. Indeed, blocking the GLP-1 receptor by the GLP-1 receptor antagonist exendin 9-39 abolishes the occurrence of hypoglycemia during an MMTT in patients with post-RYGB hyperinsulinemic hypoglycemia [11].

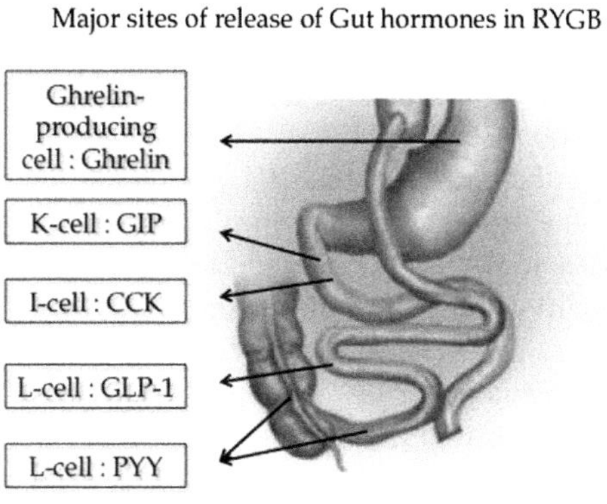

Major sites of release of Gut hormones in RYGB

FIGURE 34.1 Major sites of release of gut hormones in RYGB. *GIP*, glucose-dependent insulinotropic peptide; *CCK*, cholecystokinin; *GLP-1*, glucagon-like peptide 1; *PYY*, peptide tyrosine tyrosine.

Amino acids are also known to stimulate both GLP-1 and insulin release. Gastric bypass surgery leads to differences in resorption of amino acids, especially branched-chain amino acids (BCAA), during an MMTT [25]. Both fasting and postmeal BCAA levels were 20−30% lower after surgery. There was a positive correlation between changes in BCAA and changes in GLP-1 and insulin levels, with the hormones following the amino acids in time, suggesting a role for these amino acids in the release of these hormones. In normal physiology, the release of GLP-1 and GIP-stimulated insulin is glucose-dependent and does not lead to hypoglycemia.

The occurrence of postprandial hypoglycemia with high levels of GLP-1 suggests a different response of beta-cells to GI hormones either by beta-cell hyperplasia or by changes in function. Hypertrophic islets have been found in partial pancreatectomy specimens of patients with PHH in two studies, but were not found in a third study [17,26,27]. An increase in average diameter of nuclei in the islet cells was found in two studies, suggesting changes in properties of the cells [26,27]. In addition, we found that pancreatic ^{18}F-DOPA PET uptake was increased in a patient with postgastric bypass hypoglycemia, suggesting also an increased functionality of beta-cells [28]. Furthermore, an overexpression of the insulin-like growth factor 1 (IGF-1) receptor-alpha on the membrane and increase in cytoplasmatic IGF-2 mRNA and protein were also found; both are known to play a role in glucose metabolism [26]. Intravenous glucose loading does not lead to a difference in insulin response in patients with PHH, suggesting that there is no change in beta-cell response to glucose per se [29,30]. Oral delivery of glucose in combination with other nutrients by a mixed meal, however, leads to an exaggerated insulin response in patients with PHH [29,30]. RYGB patients with neuroglycopenic symptoms have hyperinsulinemia relative to the glucose levels during an MMTT [19]. This is attributed to both an exaggerated insulin secretion rate during the second part of the MMTT when glucose levels are declining, and a reduced insulin clearance rate. Which mechanisms underlie these observations are not known.

In addition to the role of the distal small intestine (the hindgut), bypassing of the proximal small bowel (the foregut) could also play an important contribution. Delivery of nutrients in the duodenum is postulated to induce the release of anti-incretins to balance the incretins in order to prevent hyperinsulinemia [31]. Both RYGB and GS lead to strong increases in hindgut hormones, most prominently GLP-1 and PYY. RYGB differs from GS in bypassing the duodenum and proximal jejunum (the foregut), leading to lower CCK and higher acyl−ghrelin area under the curve (AUC) during an MMTT [32]. The consequences of these differences in hormonal response caused by excluding the foregut are not clear, and other anti-incretin signals could also play a yet undefined role.

Another possible cause of postgastric surgery hyperinsulinemic hypoglycemia is small intestine bacterial overgrowth (SIBO). Postgastrectomy patients with SIBO, diagnosed with an H_2-CH_4 breath test, more often have hypoglycemia during an OGTT than patients without SIBO [33]. Whether this also holds true for postgastric bypass patients has not been studied.

Finally, there could be a predisposition to PHH in patients who have experienced symptoms, indicating postprandial hypoglycemia (not documented, however) before bariatric surgery [10].

In conclusion, the pathophysiology of PHH has not been elucidated. There is a definite role for rapid delivery of nutrients to the hindgut followed by excessive release of GI hormones, especially GLP-1. The exclusion of the foregut, leading to an imbalance between incretin and anti-incretin factors, is postulated, but has not been confirmed. Possible roles for bacterial overgrowth, altered beta-cell function, and diminished insulin clearance will need further investigation.

TREATMENT

A well-defined treatment algorithm for PHH is not available, but the following approaches are known: dietary modification, medication, and surgical (re-)intervention (see Fig. 34.2).

Modification of diet is the first step in the treatment of PHH. As rapidly absorbed carbohydrates stimulate insulin and GLP-1 release, patients are advised to eat frequent small meals with high-fiber and high-protein content, and low amounts of rapidly absorbable carbohydates [34]. In addition, intake of fluids is not advised during eating. In a substantial number of patients, these dietary modifications prove sufficient to abolish hypoglycemic symptoms. Fructose can substantially attenuate the postprandial insulin response compared to glucose, and may therefore provide protection from development of postprandial hypoglycemia in RYGB patients with recurrent postprandial hypoglycemia [35].

When dietary modifications fail, the next step in treatment is pharmacotherapy. No systematic studies have addressed which medication is the most effective or in which order medication should be started. Acarbose and somatostatin analogs are the most widely studied medications.

Acarbose, an alpha-glucosidase inhibitor, slows the digestion of carbohydrates, leading to slower resorption of glucose, which could in turn lead to less reactive hyperinsulinemia. Positive effects of acarbose have been reported in 106 RYGB patients in 10 mainly case series; questionnaires, CGM, as well as an OGTT and an MMTT were used [14,36]. A starting dose of three times 50 mg before a meal is advised, increasing to 3 dd 100 mg, depending on efficacy and side effects. Dose-dependent adverse events are diarrhea, bloating, and flatulence.

Somatostatin analogs have a beneficial effect on PHH through different mechanisms. They slow gastric emptying and intestinal motility, inhibit GLP-1 release by L cells and insulin release by beta-cells. Octreotide, a somatostatin analog with high affinity for the somatostatin receptor subtype 2 (sst2), and to a lesser extent sst5, has been best studied. In long-duration studies, short-acting subcutaneous analogs (50–100 μg t.i.d.) proved somewhat more effective than slow-release intramuscular preparations (10–20 mg/2–4 weeks) [6,37]. However, long-acting preparations were preferred by most patients. The most common side effects were diarrhea, steatorrhea, nausea, formation of gallstones, and pain on the injection side. Pasireotide is a newer somatostatin analog with high affinity for 4 of the 5 somatostatin receptor subtypes, especially sst5. In comparison to octreotide, pasireotide had a lower affinity for sst2 and a 30–40-fold higher affinity for sst5 [38]. Somatostatin inhibited glucagon release by alpha-cells via sst2 and insulin release by the beta-cells via both sst2 and sst5. Accordingly, in a single-case study using an MMTT, pasireotide inhibited glucagon less and insulin more compared to octreotide [39]. Recently, a phase II study with both short- and long-acting pasireotide for PHH was performed, but results have not been published yet.

Diazoxide, an ATP-dependent potassium channel activator, inhibits voltage-sensitive calcium channels in beta-cells, inhibiting insulin release. Successful treatment of PHH with 50–100 mg diazoxide twice daily has been published in

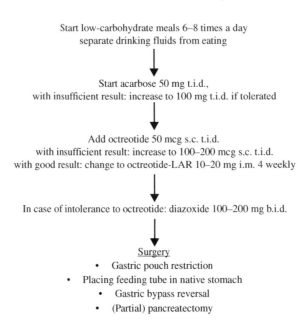

FIGURE 34.2   Treatment algorithm for postprandial hyperinsulinemic hypoglycemia. Suggested steps in treatment of PHH: *arrows* stand for next step in treatment in case of failure. Sequence of surgical options is not fixed, and must be individualized.

two case reports. However, hypotension, tachycardia, and edema often limit the use of diazoxide. The calcium channel blockers verapamil and nifedipine have been incidentally reported in the literature as a successful treatment of PHH.

GLP-1 analogs could also be useful in the treatment of PHH. Successful treatment of PHH with liraglutide (1.2 mg/ day in five patients) has been reported [40]. Liraglutide is thought to increase glucagon and decrease insulin release due to a longer stimulation of the GLP-1 receptors in alpha- and beta-cells compared to the short-acting native GLP-1.

Finally, exendin 9-39, a GLP-1 receptor antagonist, can fully abolish hypoglycemia during an MMTT in patients with PHH, but has to be administered by continuous intravenous infusion, making it not yet applicable for daily treatment [18].

We suggest the following algorithm for pharmacological treatment of PHH (see Fig. 34.2): start with acarbose, 50−100 mg t.i.d.; add octreotide s.c., 50−100 mcg t.i.d., in case of insufficient effect or intolerance; with good effect switch to octreotide−LAR, 10−20 mg i.m. 4 weekly, adjusting dose and interval. Diazoxide may be tried in case of failure. Finally, a GLP-1 analog can be tried before surgery is contemplated.

Although most patients can be treated adequately for PHH by dietary modifications with or without medication, a small group of patients might continue experiencing severe hypoglycemic episodes. In these patients, surgical reintervention must be considered. Surgery can consists of a gastric pouch restriction, placing a feeding tube into the native stomach, a gastric bypass reversal with or without sleeve resection, or a partial (80%) or total pancreas resection. Eleven cases of gastric pouch restriction, using a banding procedure, have been reported in the literature, with recurrence of hypoglycemic symptoms in two patients [41,42].

Enteral feeding via the original nutrient transit route by placement of a gastric tube in the remnant stomach may be a temporary but effective means of controlling postprandial hypoglycemia after RYGB. There is one case report in the literature, and from personal experience with two patients, this method holds promise as a new therapy [43].

Eighty-three cases of partial (80−85%), usually distal, pancreas resection have been reported in case series [27,42]. In slightly more than 50%, resolution of hypoglycemic symptoms was achieved.

Gastric bypass reversal (17 cases) was successful in all but three patients [44−47]. Some patients however, regained weight and their diabetes returned.

In some patients, more than one surgical intervention had to be performed to control the PHH. Due to the limited data in the literature and lack of studies comparing the different surgical approaches, no firm recommendation can be made as to which type of surgery is the most effective. Therefore, the type of surgery has to be individualized through good consultation with the patient.

MINI-DICTIONARY OF TERMS

- *Hyperinsulinemic hypoglycemia*: Low blood sugar caused by inadequately high insulin concentration.
- *Dumping*: Symptoms caused by rapid introduction of nutrients into the small intestine after gastric surgery.
- *Neuroglycopenia*: Symptoms of impaired brain function due to low blood sugar.
- *Incretins*: GI hormones that augment insulin response to an increase in blood glucose concentration.
- *GLP-1 analogs*: Drugs that activate the GLP-1 receptor but are not degraded by dipeptidyl peptidase 4 (DPP-4), and therefore have a longer duration of action.

KEY FACTS

- Dumping is a syndrome caused by rapid release of nutrients into the small intestine after gastric surgery.
- Dumping is divided into early and late.
- Early dumping, occurring within 1 hour after a meal, is caused by rapid introduction of highly osmotic food into the jejunum, attracting fluid and leading to intravascular depletion followed by hypotension and vasoactive responses.
- Late dumping, occurring 1−3 hours after a meal, is synonymous with PHH.

SUMMARY POINTS

- PHH is a complication of gastric surgery, nowadays mainly seen after RYGB.
- PHH is thought to be caused by rapid delivery of nutrients, especially carbohydrates, into the small intestine, thus causing release of GI hormones (incretins), which stimulate an excessive release of insulin, leading to hypoglycemia, usually after 1−1.5 hours.

- A clear definition is lacking, and the diagnosis can be made by documenting hypoglycemia during symptoms, provocating tests (OGTT; MMTT) or CGM.
- The true prevalence of PHH is not known, and it is likely that a large number of patients are unaware of their hypoglycemia.
- Treatment consists of dietary modification, medication, or if those are insufficient, surgical intervention.

REFERENCES

[1] Emous M, Ubels FL, van Beek AP. Diagnostic tools for post-gastric bypass hypoglycaemia. Obes Rev 2015;16:843.

[2] Marsk R, Jonas E, Rasmussen F, Naslund E. Nationwide cohort study of post-gastric bypass hypoglycaemia including 5,040 patients undergoing surgery for obesity in 1986-2006 in Sweden. Diabetologia 2010;53:2307.

[3] Kellogg TA, Bantle JP, Leslie DB, et al. Postgastric bypass hyperinsulinemic hypoglycaemia syndrome: characterization and response to a modified diet. Surg Obes Relat Dis 2008;4:492.

[4] Sarwar H, Chapman III WH, Pender JR. Hypoglycemia after Roux-en-Y gastric bypass: the BOLD experience. Obes Surg 2014;24:1120.

[5] Lee CJ, Clark JM, Schweitzer M, et al. Prevalence of and risk factors for hypoglycemic symptoms after gastric bypass and sleeve gastrectomy. Obesity 2015;23:1079.

[6] Arts J, Caenepeel P, Bisschops R, et al. Efficacy of the long-acting repeatable formulation of the somatostatin analogue octreotide in postoperative dumping. Clin Gastroenterol Hepatol 2009;7:432.

[7] Roslin MS, Oren JH, Polan BN, Damani T, Brauner R, Shah PC. Abnormal glucose tolerance testing after gastric bypass. Surg Obes Relat Dis 2013;9:26.

[8] Pigeyre M, Vaurs C, Raverdy V, Hanaire H, Ritz P, Pattou F. Increased risk of OGTT-induced hypoglycemia after gastric bypass in severely obese patients with normal glucose tolerance. Surg Obes Relat Dis 2015;11:573.

[9] Goldfine AB, Mun EC, Devine E, et al. Patients with neuroglycopenia after gastric bypass surgery have exaggerated incretin and insulin secretory responses to a mixed meal. J Clin Endocrinol Metab 2007;92:4678.

[10] Salehi M, Prigeon RL, D'Alessio DA. Gastric bypass surgery enhances glucagon-like peptide 1—stimulated postprandial insulin secretion in humans. Diabetes 2011;60:2308.

[11] Salehi M, Gastaldelli A, D'Alessio DA. Blockade of glucagon-like peptide 1 receptor corrects postprandial hypoglycemia after gastric bypass. Gastroenterology 2014;146:669.

[12] Kefurt R, Langer FB, Schindler K, et al. Hypoglycemia after Roux-En-Y gastric bypass: detection rates of continuous glucose monitoring (CGM) versus mixed meal test. Surg Obes Rel Dis 2015;11:564.

[13] Ritz P, Vaurs C, Bertrand M, et al. Usefulness of acarbose and dietary modifications to limit glycemic variability following roux-en-y gastric bypass as assessed by continuous glucose monitoring. Diab Technol Ther 2012;14:736.

[14] Abrahamsson N, Engström BE, Sundbom M, Karlsson FA. Hypoglycemia in everyday life after gastric bypass and duodenal switch. Eur J Endocrinol 2015;173:91.

[15] Sigstad H. A clinical diagnostic index in the diagnosis of the dumping syndrome. Changes in plasma volume and blood sugar after a test meal. Acta Med Scand 1970;188:479.

[16] Whipple AO. The surgical therapy of hyperinsulinism. J Int Chir 1938;3:237.

[17] Service GJ, Thompson GB, Service FJ, et al. Hyperinsulinemic hypoglycemia with nesidioblastosis after gastric-bypass surgery. N Engl J Med 2005;353:249.

[18] Salehi M, Gastaldelli A, D'Alessio DA. Altered islet function and insulin clearance cause hyperinsulinemia in gastric bypass patients with symptoms of postprandial hypoglycemia. J Clin Endocrinol Metab 2014;99:2008.

[19] Laurenius A, Werling M, Roux CW Le. More symptoms but similar blood glucose curve after oral carbohydrate provocation in patients with a history of hypoglycemia-like symptoms compared to asymptomatic patients after Roux-en-Y gastric bypass. Surg Obes Relat Dis 2014;10:1047.

[20] Seaquist ER, Anderson J, Childs B, et al. Hypoglycemia and diabetes: a report of a workgroup of the American Diabetes Association and the Endocrine Society. Diab Care 2013;36:1384.

[21] Hanaire H, Bertrand M, Guerci B, et al. High glycemic variability assessed by continuous glucose monitoring after surgical treatment of obesity by gastric bypass. Diab Technol Ther 2011;13:625.

[22] Cryer PE. Mechanisms of hypoglycemia-Associated Autonomic Failure in Diabetes. N Engl J Med 2013;369:362.

[23] Schauer PR, Burguera B, Ikramuddin S, et al. Effect of laparoscopic Roux-en Y gastric bypass on type 2 diabetes mellitus. Ann Surg 2003;238:467.

[24] Jacobsen SH, Olesen SC, Dirksen C. Changes in gastrointestinal hormone responses, insulin sensitivity, and beta-cell function within 2 weeks after Gastric Bypass in non-diabetic subjects. Obes Surg 2012;22:1084.

[25] Khoo CM, Muehlbauer MJ, Stevens RD, et al. Postprandial metabolite profiles reveal differential nutrient handling after bariatric surgery compared with matched caloric restriction. Ann Surg 2014;259:687.

[26] Rumilla KM, Erickson LA, Service FJ. Hyperinsulinemic hypoglycemia with nesidioblastosis: histologic features and growth factor expression. Mod Pathol 2009;22:239.

[27] Meier JJ, Butler AE, Galasso R, Butler PC. Hyperinsulinemic hypoglycemia after gastric bypass surgery is not accompanied by islet hyperplasia or increased beta-cell turnover. Diabetes Care 2006;29:1554.

[28] de Heide LJM, Glaudemans AW, Oomen PH, et al. Functional imaging in hyperinsulinemic hypoglycemia after gastric bypass surgery for morbid obesity. J Clin Endocrin Metabol 2012;97:e963.

[29] Kim SH, Abbasi F, Lamendola C, et al. Glucose-stimulated insulin secretion in Gastric Bypass patients with hypoglycemic syndrome: no evidence for inappropriate pancreatic β-cell function. Obes Surg 2010;20:1110.

[30] Patti ME, Li P, Goldfine AB. Insulin response to oral stimuli and glucose effectiveness increased in neuroglycopenia following gastric bypass. Obesity 2015;23:798.

[31] Kamvissi V, Salerno A, Bornstein SR, Mingrone G, Rubino F. Incretins or anti-incretins? A new model for the "entero-pancreatic axis". Horm Metab Res 2015;47:84.

[32] Lee WJ, Chen CY, Chong K, Lee YC, Chen SC, Lee SD. Changes in postprandial gut hormones after metabolic surgery: a comparison of gastric bypass and sleeve gastrectomy. Surg Obes Relat Dis 2011;7:683.

[33] Paik CN, Choi MG, Lim CH, et al. The role of small intestinal bacterial overgrowth in postgastrectomy patients. Neurogastroenterol Motil 2011;23:e191.

[34] Ukleja A. Dumping syndrome: pathophysiology and treatment. Nutr Clin Pract 2005;20:517.

[35] Bantle AE, Wang Q, Bantle JP. Post-Gastric Bypass hyperinsulinemic hypoglycemia: fructose is a carbohydrate which can be safely consumed. J Clin Endocrinol Metab 2015;100:3097.

[36] Valderas JP, Ahuad J, Rubio L, et al. Acarbose improves hypoglycaemia following Gastric Bypass surgery without increasing Glucagon-Like Peptide 1 levels. Obes Surg 2012;22:582.

[37] Vecht J, Lamers CB, Masclee AA. Long-term results of octreotide-therapy in severe dumping syndrome. Clin Endocrinol 1999;51:619.

[38] Bruns C, Lewis I, Briner U, et al. SOM230: a novel somatostatin peptidomimetic with broad somatotropin release inhibiting factor (SRIF) receptor binding and a unique antisecretory profile. Eur J Endocrin 2002;146:707.

[39] de Heide LJ, Laskewitz AJ, Apers JA. Treatment of severe post RYGB hyperinsulinemic hypoglycemia with pasireotide: a comparison with octreotide on insulin, glucagon, and GLP-1. Surg Obes Rel Dis 2014;10:e31.

[40] Abrahamsson N, Engström BE, Sundbom M, Karlsson FA. GLP-1 analogs as treatment of postprandial hypoglycemia following gastric bypass surgery: a potential new indication? Eur J Endocrinol 2013;21:885.

[41] Z'Graggen K, Guweidhi A, Steffen R, et al. Severe recurrent hypoglycemia after gastric bypass surgery. Obes Surg 2008;18:981.

[42] Vanderveen KA, Grant CS, Thompson GB, et al. Outcomes and quality of life after partial pancreatectomy for noninsulinoma pancreatogenous hypoglycemia from diffuse islet cell disease. Surgery 2010;148:1237.

[43] McLaughlin T, Peck M, Holst J, Deacon C. Reversible hyper-insulinemic hypoglycemia after gastric bypass: a consequence of altered nutrient delivery. J Clin Endocrinol Metab 2010;95:1851.

[44] Patti ME, McMahon G, Mun EC, et al. Severe hypoglycaemia post-gastric bypass requiring partial pancreatectomy: evidence for inappropriate insulin secretion and pancreatic islet hyperplasia. Diabetologia 2005;48:2236.

[45] Campos GM, Ziemelis M, Paparodis R, Ahmed M, Davis DB. Laparoscopic reversal of Roux-en-Y gastric bypass: technique and utility for treatment of endocrine complications. Surg Obes Rel Dis 2014;10:36.

[46] Lee CJ, Brown T, Magnuson TH, et al. Hormonal response to a mixed-meal challenge after reversal of gastric bypass for hypoglycemia. J Clin Endocrin Metabol 2013;98:e1208.

[47] Vilallonga R, van de Vrande S, Himpens J. Laparoscopic reversal of Roux-en-Y gastric bypass into normal anatomy with or without sleeve gastrectomy. Surg Endos 2013;27:4640.

Chapter 35

Bariatric Surgery Improves Type 2 Diabetes Mellitus

J.W.M. Greve

Zuyderland Medical Center, Heerlen, The Netherlands

LIST OF ABBREVIATIONS

B I, B II Billroth I or II reconstruction after gastric resection
BPD biliopancreatic diversion
CCK cholecystokinin
DAIR digestive adaptation with intestinal reserve
DJB-SG duodenojejunal bypass with sleeve gastrectomy
DS duodenal switch
GERD gastroesophageal reflux disease
GIP glucose-dependent insulinotropic polypeptide (gastric inhibitory polypeptide)
GLP-1 glucagon-like-polypeptide-1
HT hypertension
II-SG ileal interposition with a sleeve gastrectomy
MGB omegaloop or mini gastric bypass
OSAS obstructive sleep apnea syndrome
PYY polypeptide YY-36
RYGB Roux-en-Y gastric bypass
SADI-S single-anastomosis duodenoileal bypass with sleeve gastrectomy
SAGB silicone adjustable gastric band
SG sleeve gastrectomy
T2D type 2 diabetes mellitus
TB transit bipartition
VBG vertical banded gastroplasty (Mason)

INTRODUCTION

Even in the first half of the 20th century, surgeons observed a significant improvement of type 2 diabetes mellitus (T2D) after resection of the stomach and a Billroth II (BII) reconstruction [1,2]. It was not until much later, however, that it was realized that the exclusion of the duodenum was important for the immediate effect on T2D that was seen. In a recent systematic review comparing Billroth II with Billroth I (BI) reconstructions after gastric resection for cancer or ulcers, it was found that BII more effectively improved T2D [3]. However, the pivotal publication by Walter Pories' group, "Who would have thought it? An operation proves to be the most effective therapy for adult-onset diabetes mellitus," was what really opened the eyes of many surgeons involved in bariatric surgery [4]. Anecdotally, Dr. Pories went back to the lab after getting normal blood sugar levels in a severely diabetic patient immediately after the surgery because he did not believe the data! Convinced that the results were correct, he went back and studied many of his morbidly obese patients and found a consistent improvement and even amelioration of T2D in most of them [4].

The rapid improvement of T2D after bariatric surgery, in particular bypass surgery, is a result of a number of potential mechanisms. Not in the least, patients who have an immediate calorie and glucose restriction after the surgical procedure, which most certainly has an effect on the glucose homeostasis [5]. A second explanation is the weight loss: adipose tissue plays an important role in insulin sensitivity through adipokines and cytokines. Trayhurn reported on the important role of adipose tissue, and Tilg described it as the largest endocrine organ in the human body [6,7]. Adipose tissue produces a large number of adipokines and cytokines, which are directly related to the fat content of cells [7] (Fig. 35.1). Adipose tissue plays an important role in appetite regulation and energy balance, but also in acute phase response and inflammation [6]. Increasing obesity results in a chronic inflammatory state. Inflammatory mediators and adipokines are linked to insulin resistance and nonalcoholic fatty liver disease. This chronic inflammatory state was confirmed in morbidly obese patients in several studies looking at cytokines and inflammatory markers [8]. More importantly, weight loss resulted in normalization of these inflammatory mediators [9,10]. The role of weight loss was also confirmed by the metabolic effect of purely restrictive types of surgery (silicone adjustable gastric band (SAGB) and vertical banded gastroplasty (VBG/Mason)), after which a significant improvement of T2D was documented [11−13]. It was also shown by Buchwald in a literature review that resolution of T2D was directly related to the achieved weight loss [14]. Wei-J Lee compared the effect of weight loss after adjustable gastric band and gastric bypass on T2D and found that with similar weight loss, band and bypass had the same effect on T2D [15]. More direct proof of the importance of weight loss, in particular long-term weight loss, is the well-known recurrence of T2D when a patient with an initially successful bariatric procedure regains weight [16]; there is a clear relationship between T2D remission and long-term weight loss [17].

However, most attention has been focused on the role of gut hormones. The immediate and very significant effect of bypass surgery on glucose homeostasis without any weight loss has led to multiple studies exploring the role of the duodenum and small bowel. In 2004, Rubino reported a study in rats with diabetes in which duodenal exclusion resulted in a significant improvement of glucose tolerance [18]. By only making a bypass of the stomach directly to the jejunum without blocking food passage to the duodenum, this effect was not seen [19]. These results were confirmed in studies

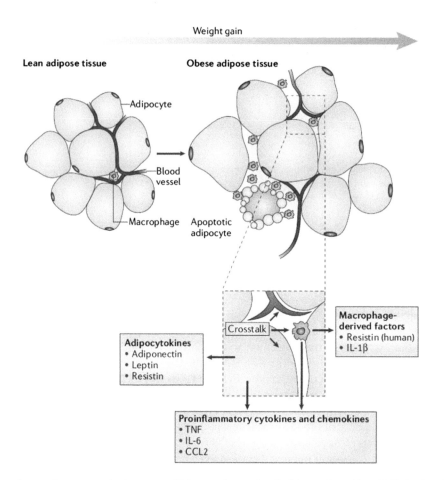

FIGURE 35.1 Adipose tissue as important endocrine organ with known changes in adipokines and cytokines. TNF, leptin, adiponectin, etc., play a role in insulin sensitivity and glucose homeostasis [7]. *Reprinted by permission from MacMillan Publishers Ltd: Nature Reviews Immunology 2006.*

in which the duodenum was excluded by placing a sleeve in the lumen, thus bypassing the duodenum without resection or transection of the gut [20]. In Rubino's theory, the foregut theory, imbalance of incretins (gut-derived hormones) results in insulin resistance [21]. By excluding the duodenum, this imbalance is restored. The exact nature of these incretins is not known, but several gut hormones have been identified, such as glucose-dependent insulinotropic poly-peptide (GIP), cholecystokinin (CCK), and ghrelin. Besides the foregut, the hindgut is also involved in release of gut hormones that are involved in insulin sensitivity and glucose homeostasis. In particular, glucagon-like-polypeptide-1 (GLP-1), which is significantly increased after bypass-type procedures, has been shown to increase insulin production and decrease glucagon [22,23]. This increase in GLP-1 is most likely a result of fast passage of undigested food into the distal small bowel. Also, the response of polypeptide YY-36 (PYY) [24,25], the so called ileal break, is significantly increased, causing early satiety that may positively affect weight loss; PYY has no effect on insulin sensitivity or secre-tion [26]. The effect of gut hormones depends on the type of bypass surgery. After a gastric bypass, a significant increase in GLP-1 is found, resulting in significantly increased insulin production [27], whereas after biliopancreatic diversion (BPD), the effect is a near-immediate normalization of insulin sensitivity [28].

CLINICAL PROOF

Restrictive Procedures (Vertical Banded Gastroplasty, Silicone Adjustable Gastric Band)

Bariatric procedures have been performed since the 50th of the 20th century. These procedures mainly focused on inducing weight loss, although Buchwald and colleagues reported in the early days of bariatric surgery that the jejuno-ileal bypass had a profound effect on dyslipidemia [29]. Realizing that weight loss resulted in a remarkable improve-ment of most obesity-related comorbidities, such as T2D, hypertension (HT), obstructive sleep apnea syndrome, and many others, the focus today is mainly on the metabolic effects. Where initially these metabolic effects were studied retrospectively, an increasing number of large randomized controlled studies are now confirming the positive effects of surgically induced weight loss.

In 2008 the group of O'Brien and Dixon reported a study comparing conservative treatment (lifestyle, diet, and med-ication) with SAGB in morbidly obese patients with T2D. In the treated group there was significantly better weight loss, but more importantly a large number of SAGB patients had complete remission of their T2D [11]. In a National Health Service cohort of T2D patients treated with SAGB, this positive effect was maintained over 5 years [30]. In the Swedish Obese Subject (SOS) study, most patients were also treated with a restrictive procedure, either SAGB or VBG [31]. This study compared surgery with conservative treatment in a large cohort of Swedish patients. Although not ran-domized (not allowed by the ethical committee at the start of the study), which may cause some bias, but now with up to 20 years of follow-up, this is a very important study. A significantly improved long-term survival rate was found in the surgically treated group, which was mainly a result of T2D remission and significantly less de novo T2D compared to the control group [32]. Purely restrictive surgical procedures thus do have a significant effect on T2D, with no or only limited effect on the incretins/gut hormones. As was shown in a study by van Dielen et al., there is an early improvement of blood glucose levels in patients with a VBG or an SAGB; however, measured with the steady state plasma glucose test, there was only a significant improvement of insulin sensitivity after more than 25% excess weight loss (%EWL) [33]. This was paralleled by an improvement in inflammatory mediators which also became obvious when the weight loss was more than 25−50 %EWL [10].

The importance of weight loss and weight maintenance for remission of T2D has been demonstrated over the years. It is clear that the more aggressive procedures with bypass of the bowel have a higher chance of reaching that goal, but every procedure that results in good weight loss will have a significant effect on T2D.

Sleeve Gastrectomy

Sleeve gastrectomy (SG) was originally part of the modified BPD, the duodenal switch (DS). After a bad experience with DS in super-superobese patients (body mass index (BMI) > 60), Gagner et al. started using the SG as a safe first step in DS. With this experience the SG was introduced as a stand-alone procedure for morbidly obese patients, and has become one of the most frequently used procedures at this time [34]. In 2013, Roux-en-Y gastric bypass (RYGB) was still in the lead with 45% of the procedures performed worldwide, but SG had reached 37% by then, and as of now (2016), is probably in the lead. The main concern is the long-term efficacy, with some studies showing significant weight regain in more than 40% of patients with follow-up of 4 years and longer [35]. In a review of SG reported in 2010, the procedure was still considered experimental [36], a view that changed rapidly, as can be concluded from the

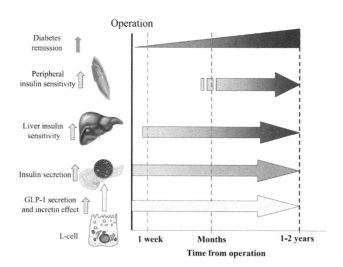

FIGURE 35.2 Chain of events leading to improved glucose homeostasis according to Holst and colleagues; shift from early gut hormone changes to longer-term effect of reduced fat content of the adipocytes (weight loss) [41]. *Reprinted by permission from the American Diabetes Association: Diabetes 2014.*

2012 consensus meeting that reported on more than 46,000 patients with low mortality (0.3%) and good results (1-year 59 %EWL, 6-year 50.5 %EWL) [37].

The effect of SG on T2D was shown to be similar to RYGB after 1 and 2 years in a meta-analysis by Li et al. [38]. In this review the RYGB was superior in terms of %EWL, resolution of HT, dyslipidemia, and GERD, but not for T2D. In the STAMPEDE trial, not included in this review, similar results were found at 3 years, with a better, but not significant, resolution of T2D after RYGB and less recurrences [39]. In general, after SG, complete remission of T2D at 1 year is typically about 53% and partial remission is over 20%, as reported by Lee and colleagues [40]. Surprisingly, despite the fact that after SG no part of the bowel is bypassed, GLP-1, which has a positive effect on insulin secretion, is significantly increased [41] (Fig. 35.2). By blocking the GLP-1 response, this early effect of SG and RYGB on T2D can be completely reversed [42]. The long-term effect of SG on T2D depends on weight loss, as do all other procedures, but Holst and colleagues suggested that an increase in gastric inhibitory peptide (GIP) may also play an important role after SG, as opposed to RYGB [41].

Gastric Bypass (Roux-en-Y Gastric Bypass)

Gastric bypass is still considered the gold standard for treatment of morbid obesity and its comorbidities. Since its introduction in the late 60th of the previous century by E.E. Mason, who performed a loop bypass to a horizontal pouch [43], the operation has been adapted and improved, with the important addition of the RYGB by Griffen in the 70th [44]. After the introduction of laparoscopic RYGB by Wittgrove [45] in 1994, laparoscopic RYGB became the most favored procedure, and because of increasing experience and the obvious benefit for patients, numbers have since rapidly increased worldwide. Important features of the currently used gastric bypass procedure are a small vertical pouch and a Roux-en-Y bypass. Many modifications are used with varying limb length (standard, long limb, distal, long biliary limb (Elegance)), type of anastomosis (linear stapler, circular stapler, hand suture), retrocolic or antecolic, and banded bypass.

Since the Annals of Surgery publication of Walter Pories' work in 1995, many groups have retrospectively and prospectively studied the effect of bypass procedures on T2D [4]. However, endocrinologists in particular were skeptical about these results and requested randomized trials similar to the large lifestyle and medical treatment studies such as the Look AHEAD, UKPDS, and others [46,47]. Since 2007, a number of randomized clinical trials have been initiated comparing optimal medical and lifestyle intervention with several surgical procedures (Table 35.1). The first study by Schauer's group, the STAMPEDE study (Surgical Treatment and Medications Potentially Eradicate Diabetes Efficiently), started in 2007 and compared intensive medical treatment with SG or RYGB [39]; it was followed by a number of other studies, including the one by Mingrone's group that compared medical treatment with RYGB or BPD [48,49]. After that, more studies followed: Courcoulas and colleagues compared intensive lifestyle and weight loss intervention with RYGB [50,51]; Halperin and colleagues looked at intensive medical treatment and weight

TABLE 35.1 Overview of Randomized Controlled Trials Comparing Medical Treatment With Bariatric Surgery (RYGB and Others)

Author	Journal	Year	Study Start	Patient Groups (N Final)	BMI (Mean)		HbA1c (Mean % Points)		%TBWL	T2D Remission
					Baseline	3 years	Baseline	3 years		
Schauer [39]	N Engl J Med	2014	2007 N=150 BMI 27–43	Medical therapy (40)	36.4	34.8	9.0	8.4	4.2 ± 8.3	5%
				SG (49)	37.1	29.2	9.5	7.0	21.1 ± 8.9	24%
				RYGB (48)	36.1	27.9	9.3	6.7	24.5 ± 9.1	38%
					Baseline	2 years	Baseline	2 years		ADA criteria
Mingrone [48]	N Engl J Med	2012	2009 N=60 BMI > 35	Medical therapy (18)	45.6	43.1	8.5	7.7	4.7 ± 6.4	0%
				RYGB (19)	44.9	29.3	8.6	6.4	33.3 ± 7.9	75%
				BPD (19)	45.1	29.2	8.9	5.0	33.8 ± 10.2	95%
					Baseline	5 years	Baseline	5 years		ADA criteria
Mingrone [49]	Lancet	2015	2009 N=60 BMI > 35	Medical therapy (15)	45.4	42.1	8.5	6.9	6.9 ± 8.4	0%
				RYGB (19)	44.0	31.3	8.7	6.7	28.4 ± 7.4	37%
				BPD (19)	44.7	30.3	8.9	6.4	31.1 ± 9.3	63%
					Baseline	1 year	Baseline	1 year		
Courcoulas [50]	JAMA Surg	2014	2009 N=69 BMI 30–40	LWLI (20)	35.7	32.1	7.0	6.9	10.2	0%
				SAGB (21)	35.5	29.3	7.9	6.8	17.3	23%
				RYGB (20)	35.5	25.8	8.7	6.4	27.0	17%
					Baseline	3 years	Baseline	3 years		
Courcoulas [51]	JAMA Surg	2015	2009 N=69 BMI 30–40	LWLI (14)	35.8	35.0	7.0	6.8	5.7	0%
				SAGB (20)	35.6	29.9	7.9	7.1	15.0	29%
				RYGB (18)	35.7	27.0	8.6	7.1	25.0	40%
					Baseline	1 year	Baseline	1 year		
Halperin [52]	JAMA Surg	2014	2010 N=38 BMI 30–42	Why WAIT (19)	36.5	34.3	8.8	8.7	7.6	5%
				RYGB (19)	36.0	25.4	8.2	6.3	27.8	58%
					Baseline	1 year	Baseline	1 year		
Cummings [53]	Diabetologia	2016	N=43 BMI 30–45	ILMI (17)	37.1		7.3	6.9	6.4	5.9%
				RYGB (15)	38.3		7.7	6.4	25.8	60%

LWLI, intensive lifestyle and weight loss intervention; Why WAIT, Weight Achievement and Intensive Treatment; ILMI, intensive lifestyle and medical intervention; RYGB, Roux-en-Y gastric bypass; SG, sleeve gastrectomy; BPD, biliopancreatic diversion; SAGB, silicone adjustable gastric band.

management with RYGB [52]; and Cummings' group performed the CROSSROADS study (Calorie Reduction or Surgery: Seeking to Reduce Obesity and Diabetes Study), in which carefully selected patients received either intensive lifestyle and medical intervention or RYGB [53]. All studies included patients in the lower BMI range (27−45) except for the Mingrone study (BMI > 35). Remission of T2D was 0−5.9% in the conservative group, and in the bariatric surgery groups this ranged from 24−40% at 3 years and 37−65% at 5 years, depending on the procedure. In the surgical groups, medication use for T2D was strongly reduced, and the effect on cardiovascular risk factors was also significantly better. An important observation is that T2D remission is high after 1 and 2 years, but with longer follow-up, T2D does recur in a number of patients. This is partly due to weight regain, and indicates that continued monitoring of glycemic control is important in these patients [48]. As weight maintenance is of great importance, this may be a plea for the banded bypass. In a review by Buchwald et al. at 5- and 10-year follow-up, data showed a sustained 72 and 70 %EWL, respectively, and a remission rate of T2D of 84.2%; the latter, however, with limited follow-up [54].

A more recent development is the Omegaloop or mini gastric bypass (MGB) developed and promoted by Rutledge. It is a single anastomosis loop gastric bypass, not dissimilar to the early gastric bypass procedures in the past (although these procedures had a horizontal pouch). The biliary loop is 200−250 cm in length. The big difference is the long and narrow vertical pouch that is supposed to prevent biliary reflux. Although bile reflux remains a concern and marginal ulcers a frequent complication, MBG has been shown to be effective in terms of weight loss at low risk [55]. In fact, due to the long pouch, it can be an escape procedure in patients with abdominal adiposity and a short small bowel mesentery. In a recent review/meta-analysis, the effect of MGB on T2D was shown to be similar to RYGB and better than SG and SAGB [56]. Compared to SG, MGB had a higher remission rate of T2D at 89% versus 76%, whereas weight loss and %EWL were similar. In one study, 5-year results were reported on SG and MGB with a remission rate that was still better in the MGB group. In this meta-analysis it was also found that MGB had a better remission rate for T2D compared to RYGB, 93.4% versus 77.6%, respectively. Apparently MBG does result in significantly better weight loss when compared to RYGB. However, with limited long-term results in mind, the effectiveness of MGB is still unclear. The short learning curve, advantages in superobese patients, and comparable results with the current gold standard (RYGB) make it a tempting option. Concerns remain however, including chronic biliary reflux, intractable biliary reflux, higher risk of severe complications after leakage (biliary peritonitis), and in some cases, serious malabsorption.

Biliopancreatic Diversion, Duodenal Switch, Single Anastomosis Duodenal Ileal Bypass

BPD, also known as the Scopinaro procedure, was first reported in 1976 by Nicola Scopinaro. This is a malabsorptive procedure that was developed to overcome the problems of the jejuno-ileal bypass. As in all other procedures, there is a restrictive component to BPD as well. Scopinaro realized that differences in diet were important in determining the success of the procedure. With a more protein-rich diet the size of the stomach needs to be larger compared to patients who have a more carbohydrate-rich diet (e.g., pasta), a feature he described as the ad hoc stomach.

In BPD the common channel is only 50 cm with an alimentary limb of 250−300 cm bypassing the remainder of the small bowel. DS is a modification of BPD that combines a sleeve resection of the stomach with a distal bypass, thus preserving the pylorus. As SG causes more restriction, the common channel in DS needs to be longer (75−100 cm).

Finally, a more recent procedure is the single-anastomosis duodenoileal bypass with SG (SADI-S) [57]. Similar to MGB, SADI-S is a single anastomosis procedure with a common channel of 250 cm; the big difference with MGB is the presence of the pylorus preventing bile reflux. The longest experience with the malabsorptive procedures has been with BPD. Weight loss results are among the best of all bariatric procedures (70.1 %EWL), and T2D remission is up to 95% [13]. After BPD or DS there is a near-immediate normalization of blood glucose levels and normalization of insulin sensitivity, which is different from RYGB. The disadvantage of these malabsorptive procedures is the higher risk of micro- and macro-nutrient malnutrition with risk of protein shortage, and if not corrected, liver cirrhosis in the long term. Despite the huge impact on gut hormones, for these malabsorptive procedures, long-term weight loss is also a key factor for maintenance of T2D remission. In a study by Pontiroli comparing long-term results of BPD with SAGB, it was shown that patients with similar weight loss after BPD or SAGB had a comparable resolution of T2D [58]. SADI-S has (comparable to DS and BPD) a significant effect on T2D, with 70−84% remission at 5 years, depending on the medication prior to treatment [57].

Duodenal Exclusion, Ileal Interposition, and Other Metabolic Procedures

Many other procedures have been proposed to treat obesity-related metabolic disorders. Duodenal exclusion was shown to be effective in treating T2D in lower-BMI patients without significant weight loss, but only a small series has been reported (by Cohen and colleagues [59]). Lee and colleagues reported a randomized study comparing SG to SG combined with duodenojejunal bypass with SG [60]. The result was significantly better weight loss and T2D remission. An even more complex procedure is the ileal interposition in combination with a sleeve gastrectomy (II-SG), as reported by de Paula et al. [61]. Midterm ($>$ 3 years) results were good, with resolution of T2D in 84% of the patients and good weight loss. An operation initially called "digestive adaptation with intestinal reserve," which included SG, omentectomy, enterectomy, and transit bipartition (TB), was developed by Santoro. After fine tuning (skipping both the omentectomy and enterectomy), this technique is now called SG with TB [62]. In 2012 results were reported on 1020 patients with a minimum of 4 months and a maximum of 5 years of follow-up. Despite the fact that in this procedure the duodenum is not excluded, the T2D remission rate was 86%. To date, the reported number of studies and patients treated is too small to consider these procedures as standard metabolic surgery.

CONCLUSION

Surgical treatment of T2D is a reality. Major diabetes associations are now advising doctors and patients to consider bariatric/metabolic surgery as a treatment option for T2D. However, it is important to realize that bariatric surgery is a tool to help the patient lose weight; the final results largely depend on patient compliance with the necessary diet. It is also important to realize that the surgical treatment does not cure, but merely ameliorates, T2D. Patients need lifelong monitoring for potential recurrence of T2D, which in general is more easy to treat than without the surgery. Many studies have looked at parameters that predict T2D remission, and it is obvious that a long duration of T2D and use of insulin result in a lower chance of complete remission. Moreover, in the SOS study, the most important factor in the increased survival of the surgically treated patients is the significantly lower risk of de novo T2D. This suggests that patients who are overweight or obese should be treated as early as possible, probably even before T2D develops.

Resolution of T2D after surgical treatment is the result of multiple mechanisms, including changes in gut hormones (GLP-1, GIP, PYY, CCK, etc.), reduced glucose (calorie) load, and weight loss. However, for long-term remission of T2D, maintenance of weight loss is the key factor. The procedure that best provides significant and lasting weight loss, irrespective of bypassing parts of the gut, will guarantee the best effect on T2D. Taking dyslipidemia and other cardiovascular risk factors into account, the bypass type of procedure is preferred.

MINI-DICTIONARY OF TERMS

- *Restriction*: An important part of all surgical procedures is reducing calorie intake mechanically.
- *Malabsorption*: Procedures that reduce utilization of ingested macronutrients, resulting in reduced calorie uptake.
- *Adipokines*: Cytokines and hormones (e.g., leptin and adiponectin) released by adipocytes that play an important role in food regulation and insulin sensitivity.
- *Incretins*: Gut hormones involved in regulation of glucose homeostasis.
- *Foregut theory*: Imbalance of incretins (e.g., GIP and CCK) causing insulin resistance.
- *Hindgut theory*: Increased release of incretins due to rapid passage of undigested food to the distal small bowel.
- *Metabolic surgery*: Surgical procedures that improve or cure metabolic disorders (T2D, dyslipidemia, etc.).

KEY FACTS

- Early resolution of T2D is caused by reduced glucose load, and depending on the type of procedure, change in incretins (e.g., GLP-1 and GIP).
- Long-term remission/improvement of T2D is only achieved with substantial and sustained weight loss.
- Reduced fat content of the adipocyte caused by weight loss induces a decrease in inflammation and normalization of adipokines, and as a result, normalization of insulin sensitivity.
- Any type of surgical procedure that induces significant weight loss will cause improvement or resolution of T2D; bypass types of procedures have the additional benefit of improving dyslipidemia.
- Surgery results in remission but not cure of T2D; lifelong monitoring of glucose homeostasis is mandatory.

SUMMARY POINTS

- Surgery is an excellent tool to treat T2D.
- The effect of surgery on T2D is multifactorial, including reduced calorie load, change in incretins, and weight loss.

- Significant and sustained weight loss is required for long-term T2D remission.
- Randomized trials have definitively shown the benefit of surgery over optimal medical and lifestyle treatments of T2D.
- Both restrictive and malabsorptive procedures can successfully improve T2D.

REFERENCES

[1] Gilbert JA, Dunlop DM. Hypoglycaemia following partial gastrectomy. Br Med J 1947;2(4521):330–2.

[2] Barnes CG. Hypoglycaemia following partial gastrectomy; report of three cases. Lancet 1947;2(6476):536–9.

[3] Kwon Y, et al. A systematic review and meta-analysis of the effect of Billroth reconstruction on type 2 diabetes: a new perspective on old surgical methods. Surg Obes Relat Dis 2015;11(6):1386–95.

[4] Pories WJ, et al. Who would have thought it? An operation proves to be the most effective therapy for adult-onset diabetes mellitus. Ann Surg 1995;222(3):339–50, discussion 350–2.

[5] Lips MA, et al. Calorie restriction is a major determinant of the short-term metabolic effects of gastric bypass surgery in obese type 2 diabetic patients. Clin Endocrinol (Oxf) 2014;80(6):834–42.

[6] Trayhurn P, Wood IS. Adipokines: inflammation and the pleiotropic role of white adipose tissue. Br J Nutr 2004;92(3):347–55.

[7] Tilg H, Moschen AR. Adipocytokines: mediators linking adipose tissue, inflammation and immunity. Nat Rev Immunol 2006;6(10):772–83.

[8] Nijhuis J, et al. Neutrophil activation in morbid obesity, chronic activation of acute inflammation. Obesity (Silver Spring) 2009;17(11):2014–18.

[9] van Dielen FM, et al. Increased leptin concentrations correlate with increased concentrations of inflammatory markers in morbidly obese individuals. Int J Obes Relat Metab Disord 2001;25(12):1759–66.

[10] van Dielen FM, et al. Macrophage inhibitory factor, plasminogen activator inhibitor-1, other acute phase proteins, and inflammatory mediators normalize as a result of weight loss in morbidly obese subjects treated with gastric restrictive surgery. J Clin Endocrinol Metab 2004;89(8):4062–8.

[11] Dixon JB, et al. Adjustable gastric banding and conventional therapy for type 2 diabetes: a randomized controlled trial. JAMA 2008;299(3):316–23.

[12] Ponce J, et al. Effect of Lap-Band-induced weight loss on type 2 diabetes mellitus and hypertension. Obes Surg 2004;14(10):1335–42.

[13] Buchwald H, et al. Bariatric surgery: a systematic review and meta-analysis. JAMA 2004;292(14):1724–37.

[14] Buchwald H, et al. Weight and type 2 diabetes after bariatric surgery: systematic review and meta-analysis. Am J Med 2009;122(3):248–256.e5.

[15] Lee WJ, et al. Improvement of insulin resistance after obesity surgery: a comparison of gastric banding and bypass procedures. Obes Surg 2008;18(9):1119–25.

[16] DiGiorgi M, et al. Re-emergence of diabetes after gastric bypass in patients with mid- to long-term follow-up. Surg Obes Relat Dis 2010;6(3):249–53.

[17] Brethauer SA, et al. Can diabetes be surgically cured? Long-term metabolic effects of bariatric surgery in obese patients with type 2 diabetes mellitus. Ann Surg 2013;258(4):628–36, discussion 636-7.

[18] Rubino F, Marescaux J. Effect of duodenal-jejunal exclusion in a non-obese animal model of type 2 diabetes: a new perspective for an old disease. Ann Surg 2004;239(1):1–11.

[19] Rubino F, et al. The mechanism of diabetes control after gastrointestinal bypass surgery reveals a role of the proximal small intestine in the pathophysiology of type 2 diabetes. Ann Surg 2006;244(5):741–9.

[20] Aguirre V, et al. An endoluminal sleeve induces substantial weight loss and normalizes glucose homeostasis in rats with diet-induced obesity. Obesity (Silver Spring) 2008;16(12):2585–92.

[21] Rubino F, et al. The early effect of the Roux-en-Y gastric bypass on hormones involved in body weight regulation and glucose metabolism. Ann Surg 2004;240(2):236–42.

[22] Valverde I, et al. Changes in glucagon-like peptide-1 (GLP-1) secretion after biliopancreatic diversion or vertical banded gastroplasty in obese subjects. Obes Surg 2005;15(3):387–97.

[23] Mingrone G, Castagneto-Gissey L. Mechanisms of early improvement/resolution of type 2 diabetes after bariatric surgery. Diabetes Metab 2009;35(6 Pt 2):518–23.

[24] Naslund E, et al. Distal small bowel hormones: correlation with fasting antroduodenal motility and gastric emptying. Dig Dis Sci 1998;43(5):945–52.

[25] Chan JL, et al. Peptide YY levels are elevated after gastric bypass surgery. Obesity (Silver Spring) 2006;14(2):194–8.

[26] Tan TM, et al. Combination of peptide YY3-36 with GLP-1(7-36) amide causes an increase in first-phase insulin secretion after IV glucose. J Clin Endocrinol Metab 2014;99(11):E2317–24.

[27] Laferrere B, et al. Effect of weight loss by gastric bypass surgery versus hypocaloric diet on glucose and incretin levels in patients with type 2 diabetes. J Clin Endocrinol Metab 2008;93(7):2479–85.

[28] Kamvissi V, et al. Incretins or anti-incretins? A new model for the "entero-pancreatic axis". Horm Metab Res 2015;47(1):84–7.

[29] Buchwald H, et al. Positive results of jejunoileal bypass surgery: emphasis on lipids with comparison to gastric bypass. Int J Obes 1981;5(4):399–404.

[30]   Egan RJ, et al. The impact of laparoscopic adjustable gastric banding on an NHS cohort of type 2 diabetics: a prospective cohort study. Obes Surg 2016;26(9):2006–13.

[31]   Sjostrom L, et al. Effects of bariatric surgery on mortality in Swedish obese subjects. N Engl J Med 2007;357(8):741–52.

[32]   Sjoholm K, et al. Incidence and remission of type 2 diabetes in relation to degree of obesity at baseline and 2 year weight change: the Swedish Obese Subjects (SOS) study. Diabetologia 2015;58(7):1448–53.

[33]   van Dielen FM, et al. Early insulin sensitivity after restrictive bariatric surgery, inconsistency between HOMA-IR and steady-state plasma glucose levels. Surg Obes Relat Dis 2010;6(4):340–4.

[34]   Angrisani L, et al. Bariatric surgery Worldwide 2013. Obes Surg 2015;25(10):1822–32.

[35]   Himpens J, Dobbeleir J, Peeters G. Long-term results of laparoscopic sleeve gastrectomy for obesity. Ann Surg 2010;252(2):319–24.

[36]   Shi X, et al. A review of laparoscopic sleeve gastrectomy for morbid obesity. Obes Surg 2010;20(8):1171–7.

[37]   Gagner M, et al. Survey on laparoscopic sleeve gastrectomy (LSG) at the Fourth International Consensus Summit on Sleeve Gastrectomy. Obes Surg 2013;23(12):2013–17.

[38]   Li J, Lai D, Wu D. Laparoscopic Roux-en-Y gastric bypass versus laparoscopic sleeve gastrectomy to treat morbid obesity-related comorbidities: a systematic review and meta-analysis. Obes Surg 2016;26(2):429–42.

[39]   Schauer PR, et al. Bariatric surgery versus intensive medical therapy for diabetes--3-year outcomes. N Engl J Med 2014;370(21):2002–13.

[40]   Lee WJ, et al. Laparoscopic sleeve gastrectomy for type 2 diabetes mellitus: predicting the success by ABCD score. Surg Obes Relat Dis 2015;11(5):991–6.

[41]   Madsbad S, Holst JJ. GLP-1 as a mediator in the remission of type 2 diabetes after gastric bypass and sleeve gastrectomy surgery. Diabetes 2014;63(10):3172–4.

[42]   Jimenez A, et al. GLP-1 and glucose tolerance after sleeve gastrectomy in morbidly obese subjects with type 2 diabetes. Diabetes 2014;63(10):3372–7.

[43]   Mason EE, Ito C. Gastric bypass in obesity. Surg Clin North Am 1967;47(6):1345–51.

[44]   Griffen Jr. WO, Young VL, Stevenson CC. A prospective comparison of gastric and jejunoileal bypass procedures for morbid obesity. Ann Surg 1977;186(4):500–9.

[45]   Wittgrove AC, Clark GW, Tremblay LJ. Laparoscopic gastric bypass, Roux-en-Y: preliminary report of five cases. Obes Surg 1994;4(4):353–7.

[46]   Intensive blood-glucose control with sulphonylureas or insulin compared with conventional treatment and risk of complications in patients with type 2 diabetes (UKPDS 33). UK Prospective Diabetes Study (UKPDS) Group. Lancet 1998;352(9131):837–853.

[47]   Look ARG, et al. Cardiovascular effects of intensive lifestyle intervention in type 2 diabetes. N Engl J Med 2013;369(2):145–54.

[48]   Mingrone G, et al. Bariatric-metabolic surgery versus conventional medical treatment in obese patients with type 2 diabetes: 5 year follow-up of an open-label, single-centre, randomised controlled trial. Lancet 2015;386(9997):964–73.

[49]   Mingrone G, et al. Bariatric surgery versus conventional medical therapy for type 2 diabetes. N Engl J Med 2012;366(17):1577–85.

[50]   Courcoulas AP, et al. Three-year outcomes of bariatric surgery vs lifestyle intervention for type 2 diabetes mellitus treatment: a randomized clinical trial. JAMA Surg 2015;150(10):931–40.

[51]   Courcoulas AP, et al. Surgical vs medical treatments for type 2 diabetes mellitus: a randomized clinical trial. JAMA Surg 2014;149(7):707–15.

[52]   Halperin F, et al. Roux-en-Y gastric bypass surgery or lifestyle with intensive medical management in patients with type 2 diabetes: feasibility and 1-year results of a randomized clinical trial. JAMA Surg 2014;149(7):716–26.

[53]   Cummings DE, et al. Gastric bypass surgery vs intensive lifestyle and medical intervention for type 2 diabetes: the CROSSROADS randomised controlled trial. Diabetologia 2016;59(5):945–53.

[54]   Buchwald H, Buchwald JN, McGlennon TW. Systematic review and meta-analysis of medium-term outcomes after banded Roux-en-Y gastric bypass. Obes Surg 2014;24(9):1536–51.

[55]   Georgiadou D, et al. Efficacy and safety of laparoscopic mini gastric bypass. A systematic review. Surg Obes Relat Dis 2014;10(5):984–91.

[56]   Quan Y, et al. Efficacy of laparoscopic mini gastric bypass for obesity and type 2 diabetes mellitus: a systematic review and meta-analysis. Gastroenterol Res Pract 2015;2015:152852.

[57]   Sanchez-Pernaute A, et al. Single-anastomosis duodenoileal bypass with sleeve gastrectomy (SADI-S) for obese diabetic patients. Surg Obes Relat Dis 2015;11(5):1092–8.

[58]   Pontiroli AE, et al. Biliary pancreatic diversion and laparoscopic adjustable gastric banding in morbid obesity: their long-term effects on metabolic syndrome and on cardiovascular parameters. Cardiovasc Diabetol 2009;8:37.

[59]   Cohen R, et al. Glycemic control after stomach-sparing duodenal-jejunal bypass surgery in diabetic patients with low body mass index. Surg Obes Relat Dis 2012;8(4):375–80.

[60]   Lee WJ, et al. Duodenal-jejunal bypass with sleeve gastrectomy versus the sleeve gastrectomy procedure alone: the role of duodenal exclusion. Surg Obes Relat Dis 2015;11(4):765–70.

[61]   DePaula AL, et al. Surgical treatment of morbid obesity: mid-term outcomes of the laparoscopic ileal interposition associated to a sleeve gastrectomy in 120 patients. Obes Surg 2011;21(5):668–75.

[62]   Santoro S, et al. Sleeve gastrectomy with transit bipartition: a potent intervention for metabolic syndrome and obesity. Ann Surg 2012;256(1):104–10.

Chapter 36

Weight Loss Surgery and the Surrogate Insulin Resistance Markers HOMA, TyG, and TG/HDL-c in Relation to Metabolic Syndrome

E. Cazzo, J.C. Pareja and E.A. Chaim
State University of Campinas (UNICAMP), Campinas, Brazil

LIST OF ABBREVIATIONS

AUROC   area under the receiver operating characteristic
HIEG   hyperinsulinemic euglycemic glucose
HOMA   homeostasis model assessment
IR   insulin resistance
IS   insulin sensitivity
IST   insulin suppression test
MetS   metabolic syndrome
SSPG   steady-state plasma glucose
T2DM   type 2 diabetes mellitus
TG/HDL-c   triglyceride to high-density lipoprotein cholesterol ratio
TyG   product of triglycerides and fasting glucose

INTRODUCTION

Insulin is a hormone produced by the β-cells of the islets of Langerhans located in the pancreas. It presents predominantly anabolic effects and is largely engaged in energy balance homeostasis, mainly regulating glucose metabolism. Its ultimate effect is to promote the influx of glucose into cells, mediated by membrane receptors that allow glucose transporters to carry glucose into the cells. Moreover, it promotes glucose storage in the form of glycogen, and inhibits lipid and protein catabolism; it is also related to several other cellular phenomena, such as cell proliferation by means of insulin-like growth factors [1,2]. The dominant hormonal control of the fasting rate of glucose release is the insulin/glucagon ratio in the prehepatic venous plasma, with insulin inhibiting, and glucagon stimulating, both glycogenolysis and gluconeogenesis [1]. The capacity of insulin to bind to its destination receptors is called insulin sensitivity (IS). Insulin resistance (IR) is defined as a condition in which there is a lack of capacity of the hormone to bind to its receptors and signal the physiological responses expected—e.g., insufficient or defective IS. Since insulin is primarily related to glucose metabolism, its malfunction readily leads to changes in glucose homeostasis, mainly hyperglycemia. Initially, pancreatic β-cells respond to hyperglycemia by producing and secreting greater amounts of insulin, leading to hyperinsulinemia [3].

The assessment of insulin activity and sensitivity can be performed by means of direct methods, which provide data as accurately as possible through simulation of situations of regular and hyperstimulated β-cells. The most accurate of these methods are clamp studies. Despite their high accuracy, clamp studies are expensive, not available at most health-providing services, and require a long time to perform. These limitations restrict this method to experimental settings, and inhibit its wide utilization. In contrast, there is the possibility of estimating IR by means of surrogate methods, which are calculated using laboratory examinations, usually available in general practice through an inexpensive, fast, and noninvasive fashion.

IMPACT OF BARIATRIC SURGERY ON INSULIN RESISTANCE

Bariatric surgery, through its several modalities, is known to promote a critical effect on IR. In particular, the surgeries that alter the normal anatomy of the gastrointestinal tract increase IS through the acute changes in glucose homeostasis that occur very early following the procedures; these changes are more highly linked to the structural changes than to weight loss itself. It is believed that duodenal exclusion and the increase in the passage of nutrients through the distal small bowel may cause the release of gastrointestinal hormones involved in glucose metabolism and immunomodulation [4−8]. They directly increase the production and secretion of insulin by the pancreas, as well as function as insulin sensitizers. The improvement in IR occurs early after the intervention, and its exact pathophysiology is not completely known, but it seems to be strongly linked to changes in gastrointestinal hormones called incretins and fat tissue-active peptides called adipokines, rather than the weight loss itself [5−8]. Immediately following surgery, there is also a decrease in the portal influx of free fatty acids, which contributes to the early amelioration of IR. Furthermore, weight loss associated with changes in liver and whole body fat content appears to have a long-term role in maintenance of the IR improvement observed [9]. Changes in gut microbiota and bile acid enteric circulation are possibly linked to some extent to the increase in IS [10,11]. It has also been shown that some subjects tend to have considerably less metabolic improvement after surgery than the majority of individuals due to mechanisms that are not completely understood and that may involve insufficient incretin response, persistent adiposopathy, and chronic inflammation [12,13]. The main gastrointestinal functions involved in glucose metabolism and satiety regulation and their respective effects are summarized in Table 36.1.

The overall effect causes an early and drastic amelioration of IR and its major clinical manifestations following surgery, mainly metabolic syndrome (MetS) and type 2 diabetes mellitus (T2DM) [14]. Biliopancreatic diversions, surgical techniques that create a long bypass of the food transit to the distal ileum, are known to cause remission of T2DM in up to 90% of the individuals who undergo the surgeries. Also, Roux-en-Y gastric bypass (RYGB), which creates a much less drastic bypass in the food transit, leads to high rates of resolution of T2DM and MetS [15,16].

ASSESSING INSULIN SENSITIVITY IN GENERAL PRACTICE

The hyperinsulinemic euglycemic glucose (HIEG) clamp technique has been described as the gold standard method for quantifying IS since it directly measures the effects of insulin in promoting glucose utilization under steady-state conditions in vivo [17].

The glucose clamp is difficult to apply in large-scale investigations because it is a chaotic procedure that involves intravenous infusion of insulin, taking frequent blood samples over a 3-hour period, and continuous adjustment of a glucose infusion [18].

TABLE 36.1 Main Gastrointestinal Hormones Enrolled in Glucose Metabolism and Satiety Regulation

Hormone	Site of Synthesis and Release	Effect
Ghrelin	Gastric fundus	• Inhibits satiety • Inhibits glucose-stimulated insulin release
Glucagon-like peptide-1 (GLP-1)	L cells in the ileum	• Promotes satiety • Increases insulin production and release
Glucagon-like peptide-2 (GLP-2)	L cells in the ileum	• Promotes nutrient absorption • Possibly influences glucose metabolism
Gastrointestinal insulinotropic polypeptide (GIP)	K cells in the ileum	• Promotes satiety • Increases insulin production and release
Peptide YY (PYY)	L cells in the distal ileum and colon	• Promotes satiety
Oxyntomodulin	L cells in the ileum	• Promotes satiety • Glucagon and GLP-1 receptor agonist

Another method of direct estimation of IR, the insulin suppression test (IST), was described by Shen et al. [19]. After an overnight fast, somatostatin (250 µg/hour) is intravenously infused to suppress the endogenous production of insulin. At the same time, glucose (6 mg/kg body weight/minute) and insulin (50 mU/minute) are infused over 150 minutes at a constant rate. Glucose and insulin determinations are performed every 30 minutes for 2.5 hours, then at 10-minute intervals from 150 to 180 minutes of the IST. The resulting steady-state plasma glucose (SSPG) concentration obtained during the last 30 minutes of infusion represents an estimation of tissue IS. The higher the SSPG concentration, the more insulin-resistant the individual is. The IST was the first test to use steady-state plasma insulin levels to promote disposal of a glucose load.

However, as with the HIEG clamp, the IST is difficult to apply in large epidemiological studies. Furthermore, there is a risk of hypoglycemia in insulin-sensitive subjects. Moreover, the IST could provoke glycosuria in some subjects, such as patients with T2DM, and could then lead to underestimation of IR by SSPG [20].

Therefore, direct estimation of IR by means of the HIEG clamp technique and IST is experimentally demanding, complicated, and impractical when large-scale epidemiological studies and daily clinical practice are involved. These methods are laborious and expensive, and are thus rarely used in larger-scale clinical research; as such, they are irrelevant for clinical practice. Consequently, over the years, a number of surrogate indices for IS or IR have been developed. Hence, surrogate tests for determining IR and cardiometabolic risk have become common and easily applied methods, and tend to get an ever-present usefulness in clinical practice, especially for epidemiological purposes [18]. Homeostasis model assessment (HOMA), the product of triglycerides and fasting glucose index (TyG), and the triglyceride to high-density lipoprotein ratio (TG/HDL-c) are the most commonly used of these tests, and have been previously validated [18,21]. The overall accuracy of all surrogate IR markers varies according to ethnicity and the cutoff values adopted, hence, widespread use requires local validation and the determination of an optimal cutoff value [18].

The most widely known and utilized is HOMA, which can be calculated by means of the formula developed by Matthews [22]:

$$\text{HOMA} - \text{IR} = \text{Fasting insulin } (\mu\text{U}/\text{L}) \times \text{Fasting glucose } (\text{nmol}/\text{L})/22.5$$

There is also the possibility of calculating HOMA by means of a computerized model available for free at the following website: https://www.dtu.ox.ac.uk/homacalculator/.

Although the computerized model is practical, HOMA is more widely calculated by means of Matthews' formula. HOMA, calculated by both the Mattews' index formula and by means of the computerized model, has been validated as a method of IS assessment comparable to clamp studies, with an 88% significance correlation [22,23].

HOMA is a model of the relationship of glucose and insulin dynamics that predicts fasting steady-state glucose and insulin concentrations for a wide range of possible combinations of IR and β-cell function. Insulin levels depend on the pancreatic β-cell effect on glucose concentrations, while glucose concentrations are regulated by insulin-mediated glucose production via the liver. IR is reflected by the decreased suppressive effect of insulin on hepatic glucose production. The HOMA model has proved to be a robust clinical and epidemiological tool for the assessment of IR [18].

The HOMA model compares favorably with other models, and presents the advantage of requiring just a single plasma sample assayed for insulin and glucose. In conclusion, the HOMA model has become a widely used clinical and epidemiological tool and, when used appropriately, it can yield valuable data. However, as with all models, the primary input data need to be robust, and the data need to be interpreted carefully [23]. Although the use of HOMA does not definitively classify a subject as insulin-resistant, it may offer information to help the clinician in deciding which first-line therapy to use. Moreover, this information may be of interest in cases of severe IR syndromes by identifying such a state [20]. The usefulness of the HOMA is limited in pathophysiological situations such as insulinoma and primary hyperaldosteronism [24].

HOMA cutoff values for IR vary according to ethnicity; therefore, they have to be locally validated. The Brazilian Metabolic Syndrome (BRAMS) study group defined the cutoff value for HOMA in the racially mixed Brazilian population as 2.7 [25]. Buccini and Wolfthal [26] identified the cutoff value of 2.6 in Argentina; Bonora et al. [27] observed the cutoff point of 2.77 in Italy; and Yeni-Komshian et al. [28] defined the cutoff value of 2.7 among North Americans.

Since HOMA calculation requires the dosage of serum insulin levels, which is not widely available in general practice, other methods were developed in order to use more generally available variables. Both TyG and TG/HDL-c were described more recently, and have the advantage of using biochemical variables more routinely measured [29–31].

The TyG index [32] is a logarithmic function calculated by means of the following formula:

$$\text{TyG} = \text{Ln} \, [\text{Fasting glucose } (\text{mg}/\text{dL}) * \text{Triglycerides } (\text{mg}/\text{dL})]/2$$

The best TyG index level for diagnosis of IR is Ln 4.65, which shows the highest sensitivity (84.0%) and specificity (45.0%) values. The positive and negative predictive values are 81.1% and 84.8%, and the probability of disease, given

TABLE 36.2 Advantages and Disadvantages of Each of the Main Surrogate Insulin Resistance Markers

Marker	Advantages	Disadvantages
HOMA	• Easy to calculate • More widely used	• Requires fasting insulin dosage • Reflects only liver insulin sensitivity
TyG	• Uses common examinations • Reflects whole body insulin sensitivity	• Requires calculation of a logarithmic function • Not widely used yet
TG/HDL-c	• Easy to calculate • Uses common examinations • Reflects whole body insulin sensitivity	• Presents ethnic limitations • Different cutoff values in men and women • Not widely used yet

TABLE 36.3 Formulas to Calculate the Main Surrogate Insulin Resistance Markers

Marker	Formula
HOMA	Fasting Insulin (μU/L) × Fasting Glucose (nmol/L)/22.5
Tyg	Ln [Fasting Glucose(mg/dL) * Triglycerides (mg/dL)]/2
TG/HDL-c	Triglycerides (mg/dL)/HDL-c (mg/dL)

a positive test, is 60.5% [32]. The major advantage of the TyG index is the use of the widely available variables of fasting glucose and triglyceride level. Vasques et al. [33] observed that the TyG index is correlated with fat distribution and fat depots, metabolic parameters, and markers of subclinical atherosclerosis related to IR, signaling a narrower correlation with whole body IS than HOMA. Once the elevation of triglyceride levels in blood and skeletal muscle interferes with glucose metabolism in muscle [34], the TyG index apparently predominantly reflects the muscle IR while HOMA mainly shows IR in the liver [35]. Therefore, it is possible that the TyG index may be superior to previously known indices for diabetes prediction, although this hypothesis needs further confirmation [36]. Triglyceride to HDL-c ratio (TG/HDL-c) is such an easily available surrogate index. It is promptly calculated by the simple division of fasting triglycerides (mg/dL) by fasting HDL-cholesterol (mg/dL), two variables that are widely available in general practice [37]. Since the HDL-c reference differs between men and women, its cutoff values differ according to gender. It has been reported that TG/HDL-c concentration ratios of ≥ 3.5 (men) and ≥ 2.5 (women) were classified as abnormal. It has been demonstrated that this marker may not be so reliable among African—American and Hispanic populations using the same cutoff points [38]. For both Hispanics and African—Americans, the area under the receiver operating characteristic (AUROC) curve was not significant, and thus did not allow the calculation of an optimal cutoff TG/HDL-c value. [39] On the other hand, Li et al. [40] showed that the optimal cutoff value of the TG/HDL-c ratio for prediction of hyperinsulinemia was 3.0 (in mg/dL units) for non-Hispanic Whites and Mexican—Americans, and 2.0 (in mg/dL units) for non-Hispanic Blacks; the corresponding sensitivity and specificity were 70.6% and 71.0% in non-Hispanic Whites, 61.6% and 77.4% in non-Hispanic Blacks, and 64.1% and 71.1% in Mexican—Americans. Applying these race/ethnicity-specific cutoff values yielded a similar magnitude of strength in the association across racial/ethnic subgroups. In White and Asian populations, it presents high sensitivity and specificity, both of which vary in men and women. In men, the overall sensitivity is 84% and specificity is 80%, whereas in women they are, respectively, 70% and 88% [41].

The main advantages and disadvantages of each surrogate IR marker cited above are listed in Table 36.2. The respective formulas to calculate the markers are summarized in Table 36.3.

THE USE OF SURROGATE INSULIN RESISTANCE MARKERS AFTER BARIATRIC SURGERY

As bariatric surgeries have become the standard treatment option for morbid obesity, many studies have shown a significant impact of these procedures on IR, thus leading to the reversal of metabolic comorbidities or prevention of their onset altogether [9,15,42].

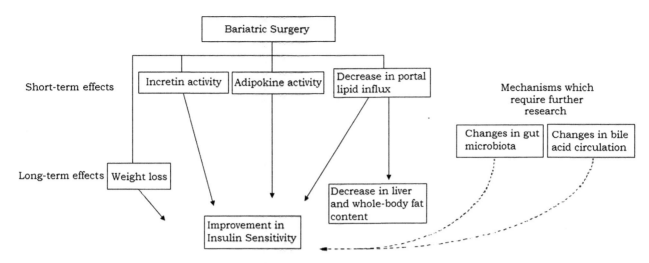

FIGURE 36.1 Mechanisms of IR improvement following bariatric surgery.

Several studies have shown a significant decrease in IR, evaluated by means of HOMA after bariatric surgery [42]. Conversely, only a few assessed postsurgical evolution of IR by means of TyG and TG/HDL. As expected, all of them observed postoperative reduction in IR prevalence, whichever marker was used. Appachi et al. [43] observed a significant decrease in TG/HDL 6 months following surgery, whereas DeMoura et al. [44] also showed a significant decrease after the use of an endoscopically placed duodenojejunal bypass liner. In a study of a sample of 96 diabetic individuals of mixed ethnicity, our group observed significant decreases in HOMA (pre: 4.4, post: 1.9; $p < 0.0001$), TyG (pre: 4.9, post: 4.5; $p < 0.0001$), and TG/HDL (pre: 5.1, post: 1.9; $p < 0.0001$). Additionally, there was a significant correlation between the post- versus preoperative ratios of Tyg and TG/HDL and the reversal of MetS [45].

CONCLUSION

Since there is a high prevalence of insulin-resistant individuals among the population who undergo bariatric surgery, assessing the postoperative evolution of IR is often needed to provide optimal assistance to these patients. Direct estimation by means of clamp studies is not possible in general practice; the use of surrogate markers is warranted in these subjects for the purpose of evaluating the surgical effects on IR, as well as observing the necessity of withdrawing medications used to treat it and to estimate prognosis of cardiometabolic risk and evolution to target-organ damage. Local and ethnic considerations must be made before widespread application of the tests. Furthermore, since there is a risk of persistent or reemergent IR within this group of individuals, the use of these markers during a carefully conducted follow-up is even more important (Fig. 36.1).

MINI-DICTIONARY OF TERMS

- *Adipokines*: Metabolically active peptides produced by fat tissue.
- *Adiposopathy*: Imbalance in production or release of adipokines.
- *Hyperaldosteronism*: Disease characterized by hyperproduction of the hormone aldosterone.
- *Incretin*: Enterohormone that directly influences glucose metabolism.
- *Insulin resistance*: Defective IS.
- *Insulin sensitivity*: Capacity of insulin to bind to its receptors and lead to its physiological effects.
- *Insulinoma*: Neoplasm that develops from the pancreatic β-cells.
- *Metabolic syndrome*: Cluster of interconnected metabolic factors that directly increase cardiometabolic risk.
- *Surrogate insulin resistance marker*: A laboratory measurement that is used as a substitute for the gold standard direct-estimation method.
- *Type 2 diabetes mellitus*: Chronic metabolic disease characterized by disturbance in glucose homeostasis, primarily caused by IR.

KEY FACTS

- Direct estimation of IS is expensive, highly unavailable, and nonpractical.
- Surrogate markers are useful tools for estimating IR by means of simple laboratory examinations.
- Structural changes caused by bariatric surgeries seem to lead to early improvement in IS.
- Incretin and adipokine activity, as well as immunomodulation, seem to be key points in metabolic improvement after bariatric surgery.

SUMMARY POINTS

- Surrogate markers of IR are useful in epidemiological studies and clinical practice.
- For correct use of surrogate IR markers, local validation and ethnicity must be considered.
- Bariatric surgery often leads to critical improvement of IR through a complex array of metabolic pathways associated with structural changes.
- Assessment of the postoperative evolution of IR is often needed to provide optimal assistance after bariatric surgery.
- There is a risk of persistence or reemergence of IR after bariatric surgery; thus, continuous follow-up is necessary.

REFERENCES

[1] Ferrannini E. Physiology of glucose homeostasis and insulin therapy in type 1 and type 2 diabetes. Endocrinol Metab Clin North Am 2012;41(1): 25–39.
[2] Cherrington AD. Banting Lecture 1997. Control of glucose uptake and release by the liver in vivo. Diabetes 1999;48(5):1198–214.
[3] Cherrington AD, Moore MC, Sindelar DK, Edgerton DS. Insulin action on the liver in vivo. Biochem Soc Trans 2007;35(Pt 5):1171–4.
[4] Campbell JE, Drucker DJ. Pharmacology, physiology, and mechanisms of incretin hormone action. Cell Metab 2013;17(6):819–37.
[5] Meek CL, Lewis HB, Reimann F, Gribble FM, Park AJ. The effect of bariatric surgery on gastrointestinal and pancreatic peptide hormones. Peptides 2016;77:28–37.
[6] Suzuki S, Ramos EJ, Gonçalves CG, Chen C, Meguid MM. Changes in GI hormones and their effect on gastric emptying and transit times after Roux-en-Y gastric *bypass* in rat model. Surgery 2005;138(2):283–90.
[7] Huda MS, Wilding JP, Pinkney JH. Gut peptides and the regulation of appetite. Obes Rev 2006;7(2):163–82.
[8] Vahl T, D'Alessio D. Enteroinsular signaling: perspectives on the role of the gastrointestinal hormones glucagon-like peptide 1 and glucose-dependent insulinotropic polypeptide in normal and abnormal glucose metabolism. Curr Opin Clin Nutr Metab Care 2003;6(4):461–8.
[9] Verna EC, Berk PD. Role of fatty acids in the pathogenesis of obesity and fatty liver: impact of bariatric surgery. Semin Liver Dis 2008;28(4): 407–26.
[10] Bell DS. Changes seen in gut bacteria content and distribution with obesity: causation or association? Postgrad Med 2015;127(8):863–8.
[11] Albaugh VL, Flynn CR, Cai S, Xiao Y, Tamboli RA, Abumrad NN. Early increases in bile acids post Roux-en-Y gastric bypass are driven by insulin-sensitizing, secondary bile acids. J Clin Endocrinol Metab 2015;100(9):E1225–33.
[12] Hirsch FF, Pareja JC, Geloneze SR, Chaim E, Cazzo E, Geloneze B. Comparison of metabolic effects of surgical-induced massive weight loss in patients with long-term remission versus non-remission of type 2 diabetes. Obes Surg 2012;22(6):910–17.
[13] DiGiorgi M, Rosen DJ, Choi JJ, Milone L, Schrope B, Olivero-Rivera L, et al. Re-emergence of diabetes after gastric bypass in patients with mid- to long-term follow-up. Surg Obes Relat Dis 2010;6(3):249–53.
[14] Pories WJ, Swanson MS, MacDonald KG, Long SB, Morris PG, Brown BM, et al. Who would have thought it? An operation proves to be the most effective therapy for adult-onset diabetes mellitus. Ann Surg 1995;222(3):339–50.
[15] Buchwald H, Avidor Y, Braunwald E, Jensen MD, Pories W, Fahrbach K, et al. Bariatric surgery: a systematic review and meta-analysis. JAMA 2004;292(14):1724–37.
[16] Buchwald H, Estok R, Fahrbach K, Banel D, Jensen MD, Pories WJ, et al. Weight and type 2 diabetes after bariatric surgery: systematic review and meta-analysis. Am J Med 2009;122(3):248–56.
[17] DeFronzo RA, Tobin JD, Andres R. Glucose clamp technique: a method for quantifying insulin secretion and resistance. Am J Physiol 1979;237(3):E214–23.
[18] Singh B, Saxena A. Surrogate markers of insulin resistance: a review. World J Diabetes 2010;1(2):36–47.
[19] Shen SW, Reaven GM, Farquhar JW. Comparison of impedance to insulin-mediated glucose uptake in normal subjects and in subjects with latent diabetes. J Clin Invest 1970;49(12):2151–60.
[20] Antuna-Puente B, Disse E, Rabasa-Lhoret R, Laville M, Capeau J, Bastard JP. How can we measure insulin sensitivity/resistance? Diabetes Metab 2011;37(3):179–88.
[21] Murguía-Romero M, Jiménez-Flores JR, Sigrist-Flores SC, Espinoza-Camacho MA, Jiménez-Morales M, Piña E, et al. Plasma triglyceride/HDL-cholesterol ratio, insulin resistance, and cardiometabolic risk in young adults. J Lipid Res 2013;54(10):2795–9.
[22] Matthews DR, Hosker JP, Rudenski AS, Naylor BA, Treacher DF, Turner RC. Homeostasis model assessment: insulin resistance and beta-cell function from fasting plasma glucose and insulin concentrations in man. Diabetologia 1985;28(7):412–19.

[23] Wallace TM, Levy JC, Matthews DR. Use and abuse of HOMA modeling. Diab Care 2004;27(6):1487−95.

[24] Skrha J, Haas T, Sindelka G, Prázný M, Widimský J, Cibula D, et al. Comparison of the insulin action parameters from hyperinsulinemic clamps with homeostasis model assessment and QUICKI indexes in subjects with different endocrine disorders. J Clin Endocrinol Metab 2004;89(1):135−41.

[25] Geloneze B, Vasques AC, Stabe CF, Pareja JC, Rosado LE, Queiroz EC, et al. HOMA1-IR and HOMA2-IR indexes in identifying insulin resistance and metabolic syndrome: Brazilian Metabolic Syndrome Study (BRAMS). Arq Bras Endocrinol Metabol 2009;53(2):281−7.

[26] Buccini GS, Wolfthal DL. Valores de corte para índices de insulinorresistencia, insulinosensibilidad e insulinosecreción derivados de la fórmula HOMA y del programa HOMA2. Interpretación de los datos. Rev Argent Endocrinol Metab 2008;45(1):3−21.

[27] Bonora E, Kiechl S, Willeit J, Oberhollenzer F, Egger G, Targher G, et al. Prevalence of insulin resistance in metabolic disorders: the Bruneck Study. Diabetes 1998;47(10):1643−9.

[28] Yeni-Komshian H, Carantoni M, Abbasi F, Reaven GM. Relationship between several surrogate estimates of insulin resistance and quantification of insulin-mediated glucose disposal in 490 healthy nondiabetic volunteers. Diabetes Care 2000;23(2):171−5.

[29] Abbasi F, Reaven GM. Comparison of two methods using plasma triglyceride concentration as a surrogate estimate of insulin action in nondiabetic subjects: triglycerides x glucose versus triglyceride/high-density lipoprotein cholesterol. Metabolism 2011;60(12):1673−6.

[30] Guerrero-Romero F, Simental-Mendía LE, González-Ortiz M, Martínez-Abundis E, Ramos-Zavala MG, Hernández-González SO, et al. The product of triglycerides and glucose, a simple measure of insulin sensitivity. Comparison with the euglycemic-hyperinsulinemic clamp. J Clin Endocrinol Metab 2010;95(7):3347−51.

[31] Malita FM, Messier V, Lavoie JM, Bastard JP, Rabasa-Lhoret R, Karelis AD. Comparison between several insulin sensitivity indices and metabolic risk factors in overweight and obese postmenopausal women: a MONET study. Nutr Metab Cardiovasc Dis 2010;20(3):173−9.

[32] Simental-Mendía LE, Rodríguez-Morán M, Guerrero-Romero F. The product of fasting glucose and triglycerides as surrogate for identifying insulin resistance in apparently healthy subjects. Metab Syndr Relat Disord 2008;6(4):299−304.

[33] Vasques AC, Novaes FS, de Oliveira Mda S, Souza JR, Yamanaka A, Pareja JC, et al. TyG index performs better than HOMA in a Brazilian population: a hyperglycemic clamp validated study. Diabetes Res Clin Pract 2011;93(3):e98−e100.

[34] Kelley DE. Skeletal muscle triglycerides: an aspect of regional adiposity and insulin resistance. Ann N Y Acad Sci 2002;967:135−45.

[35] Tripathy D, Almgren P, Tuomi T, Groop L. Contribution of insulin-stimulated glucose uptake and basal hepatic insulin sensitivity to surrogate measures of insulin sensitivity. Diab Care 2004;27(9):2204−10.

[36] Lee SH, Kwon HS, Park YM, Ha HS, Jeong SH, Yang HK, et al. Predicting the development of diabetes using the product of triglycerides and glucose: the Chungju Metabolic Disease Cohort (CMC) study. PLoS One 2014;9(2):e90430.

[37] Karelis AD, Pasternyk SM, Messier L, St-Pierre DH, Lavoie JM, Garrel D, et al. Relationship between insulin sensitivity and the triglyceride-HDL-C ratio in overweight and obese postmenopausal women: a MONET study. Appl Physiol Nutr Metab 2007;32(6):1089−96.

[38] Sumner AE, Finley KB, Genovese DJ, Criqui MH, Boston RC. Fasting triglyceride and the triglyceride-HDL cholesterol ratio are not markers of insulin resistance in African Americans. Arch Intern Med 2005;165(12):1395−400.

[39] Giannini C, Santoro N, Caprio S, Kim G, Lartaud D, Shaw M, et al. The triglyceride-to-HDL cholesterol ratio: association with insulin resistance in obese youths of different ethnic backgrounds. Diab Care 2011;34(8):1869−74.

[40] Li C, Ford ES, Meng YX, Mokdad AH, Reaven GM. Does the association of the triglyceride to high-density lipoprotein cholesterol ratio with fasting serum insulin differ by race/ethnicity? Cardiovasc Diabetol 2008;7:4.

[41] Gasevic D, Frohlich J, Mancini GJ, Lear SA. Clinical usefulness of lipid ratios to identify men and women with metabolic syndrome: a cross-sectional study. Lipids Health Dis 2014;13:159.

[42] Cazzo E, Gestic MA, Utrini MP, Machado RR, Geloneze B, Pareja JC, et al. Impact of Roux-en-Y gastric bypass on metabolic syndrome and insulin resistance parameters. Diab Technol Ther 2014;16(4):262−5.

[43] Appachi S, Kelly KR, Schauer PR, Kirwan JP, Hazen S, Gupta M, et al. Reduced cardiovascular risk following bariatric surgeries is related to a partial recovery from "adiposopathy". Obes Surg 2011;21(12):1928−36.

[44] de Moura EG, Orso IR, Martins BC, Lopes GS, de Oliveira SL, Galvão-Neto Mdos P, et al. Improvement of insulin resistance and reduction of cardiovascular risk among obese patients with type 2 diabetes with the duodenojejunal bypass liner. Obes Surg 2011;21(7):941−7.

[45] Cazzo E, Callejas-Neto F, Pareja JC, Chaim EA. Correlation between post over preoperative surrogate insulin resistance indexes' ratios and reversal of metabolic syndrome after Roux-en-Y gastric bypass. Obes Surg 2014;24(6):971−3.

Chapter 37

Cancer and Bariatric Surgery

D.S. Casagrande, M. Moehlecke, C.C. Mottin, D.D. Rosa and B.D. Schaan

Hospital de Clínicas de Porto Alegre, Porto Alegre, RS, Brazil

LIST OF ABBREVIATIONS

AMPK adenosine monophosphate−activated protein kinase
CI confidence interval
HR hazard ratio
IGF-1 insulin-like growth factor-1
IL interleukin
Jak Janus kinase
MAPK mitogen-activated protein kinase
mTOR mammalian target of rapamycin
NA data not available
OR odds ratio
PI3K phosphatidylinositol 3-kinase
RR relative risk
RYGB Roux-en-Y gastric bypass
SHBG steroid hormone−binding globulin
SIR standardized incidence ratios
SOCS suppressor of cytokine signaling proteins
SOS Swedish Obese Subjects
Stat signal transducer and activator transcription
TNF-α tumor necrosis factor-α
VBG vertical banded gastroplasty
VEGF vascular endothelial growth factor

INTRODUCTION

Obesity, conventionally defined as a body mass index (BMI) equal to or higher than 30 kg/m^2, has increased worldwide over the last several years [1]. It is a major health concern due to its epidemic presence today. More than 1.9 billion adults worldwide are overweight (BMI $\geq$ 25 kg/m^2), and more than 600 million are obese [2]. Further, obesity causes more deaths than those related to underweight conditions [2].

Obesity results from a chronic energy imbalance between food intake and energy expenditure (Fig. 37.1). In general, the obesogenic environment is associated with easy access to highly palatable and energy-dense foods, low physical activity, and an increase in sedentary behavior, causing a positive energy balance. Further, environmental and social changes are involved, as an absence of policies addressing urban planning, food processing, education, marketing, and even transportation play a role in the current epidemic [2].

Several interventions have been employed to prevent and treat obesity, including lifestyle-based behavioral approaches and pharmacotherapy. However, among severely obese patients (BMI $\geq$ 40 kg/m^2), results from therapeutic approaches are often limited. Bariatric surgery has emerged as a treatment option in these cases, particularly when

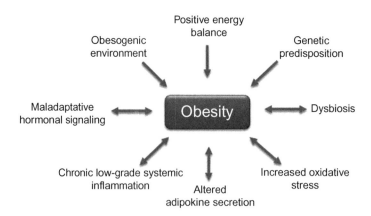

FIGURE 37.1 Disruption of homeostatic pathways associated with obesity.

comorbidities are present. Literature supports the multiple benefits of this approach, including sustained weight loss, diabetes remission, and reduced mortality [3,4]. Assuming that obesity is related to higher cancer risk [5], we would expect weight loss achieved through bariatric surgery to result in reduced risk of obesity-related cancers, which is the focus of this chapter.

OBESITY AND CANCER: OVERVIEW

Weight gain and obesity increase the risk for several diseases, including diabetes, cardiovascular diseases [6], obstructive sleep apnea [7], and cancer [5]. Poor response to nonsurgical therapies and increased cancer-related mortality have also been associated with obesity [8,9]. Obesity was first recognized as an etiological factor for cancer in 1979, when the mortality rate from cancer was estimated to be approximately 90% higher in obese patients compared to nonobese patients in a prospective study with 750,000 individuals [10].

Subsequently, several studies revealed higher cancer rates in obese patients compared to lean patients, with rates ranging from 2.12 to 5.8 cases per 1000 person-years [11−13]. A cohort study including more than 5 million UK subjects aged 16 years or older showed an association between BMI and 17 of 22 cancer types. For six of these cancers (uterus, gallbladder, kidney, cervix, thyroid, and leukemia), a linear association with increasing BMI was observed. The association between BMI and cancer occurrence remained significant, even after adjustments for multiple potential confounders such as age, smoking status, alcohol use, diabetes, and socioeconomic status. Approximately 41% of new uterine cancer cases and 20% of gallbladder cancers each year were attributable to weight gain or obesity [14].

Although most studies linking obesity to cancer risk use BMI as a parameter, some studies show an association between increased waist measurements and higher cancer risk. A systematic review from cohort and case-controlled studies shows an association between smaller waist size and a reduced risk of premenopausal breast cancer after adjusting for BMI. Interestingly, for postmenopausal women, the adjustment for BMI disproved the relationship between waist size and breast cancer risk [15]. These findings suggested that metabolic derangements associated with abdominal obesity, regardless of BMI, could be associated with cancer.

MECHANISMS INVOLVED IN THE ASSOCIATION BETWEEN CANCER AND OBESITY

The biological mechanisms that display the association between cancer and obesity are still not completely understood. This relationship involves metabolic, inflammatory, and hormonal derangements.

Obesity, especially increased visceral fat, is associated with white adipose tissue dysfunction, resulting in chronic low-grade inflammation [16]. This subclinical inflammation leads to the release of adipokines, cytokines, growth factors, and hormones that act in specific cells, promoting a protumorigenic microenvironment [17].

Adipokines are bioactive fat tissue−derived molecules that play a role in the integration of endocrine, metabolic, and inflammatory pathways that control energy homeostasis, promote insulin signaling, and stimulate angiogenesis and cellular proliferation. Leptin, an adipokine produced mainly in white adipose tissue, has concentrations proportional to the amount of adipose tissue, and has been demonstrated to activate intracellular signaling pathways, including Janus kinase (Jak)/signal transducers, transcription molecule activators (Stat), mitogen-activated protein kinase (MAPK), phosphatidylinositol 3-kinase (PI3K)/Akt, and cytokine signaling protein suppressors (SOCS), all of which promote mitogenic, antiapoptotic, and metastatic pathways [18]. Also, hyperleptinemia is associated with tumor cell proliferation and breast cancer progression in recent studies [19].

Both adipocytes and macrophages from adipose tissue produce proinflammatory cytokines such as interleukin-1β (IL-1β), IL-6, and tumor necrosis factor-α (TNF-α), which leads to increased tumor cell growth and higher estrogen bioavailability from enhanced local aromatization [20,21]. Moreover, hyperinsulinemia due to insulin resistance, usually found in patients with obesity and type 2 diabetes, inhibits hepatic production of steroid hormone—binding globulin (SHBG), leading to increased free hormone levels (estrogen and testosterone). Thus, higher breast cancer risk in post-menopausal women with higher BMI may be associated with increased estrogen levels resulting from the peripheral conversion of androgenic precursors from adipose aromatase to estradiol [22]. High estrogen levels can promote cancer cell proliferation in hormonally responsive tissues such as the breast, uterus, ovaries, and prostate [23—26]. Also, high circulating estrogen levels may promote tumor development and progression through direct effects on cellular proliferation, the inhibition of apoptosis, and the induction of high vascular endothelial growth factor (VEGF) levels that may result in angiogenesis [27]. Endometrial [24], breast [25], uterine [28], ovarian [23], and prostate cancers [26] are among the hormone-dependent cancers associated with changes in steroid hormone levels.

Most obese people have insulin resistance, and several epidemiological studies display an association of high insulin levels with some cancers. Patients with type 2 diabetes have high incidences of malignances such as breast, endometrial, colorectal, pancreatic, and hepatocellular cancers [29]. Hyperinsulinemia may also potentiate anabolic effects and promote cell proliferation and cancer progression [30]. Aside from increased free hormone levels, insulin has direct and indirect effects related to tumor development by binding to cell surface receptors and activating the PI3K/Akt/mammalian rapamycin targets (mTOR) and Ras/Raf/MAPK pathways, while also stimulating insulin-like growth factor-1 (IGF-1) synthesis in the liver [31]. Additionally, IGF-1 also binds to insulin receptors [24,31]. Insulin receptor overexpression was found in many tumor locations, including the colon, breast, lung, prostate, thyroid, and ovaries [32]. Although obesity is an established risk factor for several cancers with sex-specific differences [5], the effects of bariatric surgery on the presumably lower-incidence cancers are just recently under evaluation.

CANCER AND BARIATRIC SURGERY

Currently, bariatric surgery is the most effective treatment for sustained weight loss to improve quality of life and to reduce obesity-related morbidity and mortality among severely obese individuals who cannot achieve weight loss through lifestyle interventions and/or antiobesity medications [33]. Roux-en-Y gastric bypass (RYGB) is the most commonly performed procedure for these patients worldwide [34], as it typically leads to fewer side effects (malabsorption and malnutrition) than disabsorptive procedures (i.e., jejunoileal bypass) [35]. Increasing evidence shows that lower mortality rates are associated with bariatric surgery compared to other treatments, given reduced incidence of major cardiovascular events and cancers [37,38].

In 2010, the Swedish Obese Subjects (SOS) trial involving obese patients undergoing bariatric surgery and 2037 contemporaneously matched obese controls [37] showed cancer was the most common cause of death over 16 years of follow-up diagnoses among the surgically treated group. Exploratory analysis revealed that the number of first-time cancers was 42% lower in women undergoing surgery than the obese controls ($p = 0.0001$), whereas the surgery had seemingly no effect on the outcome of male patients. Similar results were obtained after exclusion of all cancer cases in the first 3 years following intervention [37].

In 2014, our group published the first systematic review that showed the influence of bariatric surgery on cancer outcomes [39]. Considering only controlled studies (11,087 patients in surgery groups, and 20,720 patients in control groups), bariatric surgery was clearly associated with a reduction in cancer risk (odds ratio (OR) 0.42; 95% confidence interval (CI) 0.24, 0.73) (Fig. 37.2). Considering both controlled and uncontrolled studies, we found a cancer incidence density rate of 1.06 cases per 1000 patients per year following bariatric surgery, which was similar to the nonobese subject rate (Fig. 37.3) [39].

Four systematic reviews followed our publication [39]: two considered overall cancer incidence [40,41], and the others considered colorectal [42] and endometrial cancers [43]. Yang et al. [40] reported an OR of 0.43 (95% CI 0.27—0.69) for reduced total incidence of obesity-related cancers following surgery (mostly gastric bypass) from five retrospective studies. In their meta-analysis, the entire control group was comprised of only obese subjects, though none of the studies specified the treatments for this group. Subgroup analysis of gender and varying cancer types showed a clear benefit from the surgery among women, with 61% lower risk of total cancer incidence and 24% lower colorectal

Study
ID
OR (95% CI)
Events, Bariatrics
Events, Control
% Weight
Adams 2009 (19)
0.75 (0.64, 0.88) 254/6596 477/9442 27.04
Christou 2008 (20)
0.22 (0.14, 0.35) 21/1035 487/5746 23.77
McCawley 2009 (12)
0.19 (0.12, 0.32) 17/1446 203/3495 22.90
Sjostrom 2009 (13)
0.68 (0.54, 0.87) 117/2010 169/2037 26.29
Overall (I-squared = 93.8%, p < 0.001)
0.40 (0.23, 0.72) 409/11087 1336/20720 100.00
.1
.5
1
2

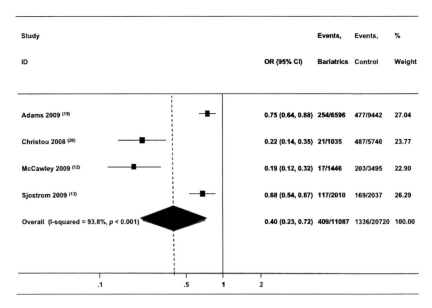

FIGURE 37.2 Association between cancer risk and bariatric surgery in controlled studies. *Reprinted from Casagrande DS et al. Incidence of cancer following bariatric surgery: systematic review and meta-analysis. Obes Surg 2014;24(9):1499–1505. Copyright (2014) by Springer Publishing Company.*

Study name
Statistics for each study
Event rate and 95% CI
Event rate
Lower limit
Upper limit
Adams 2009 19 1.67 1.48 1.89
Christou 2008 20 1.24 0.81 1.89
Forsell & Hellers 1997 22 5.00 0.70 34.61
Gagne 2009 23 1.24 0.76 2.02
Gusenoff 2009 24 0.30 0.17 0.52
Ostlund 2010 26 2.51 2.24 2.81
Sjostrom 2009 13 5.37 4.48 6.44
Srikanth 2005 28 0.20 0.08 0.48
Sugerman 2001 30 0.06 0.01 0.22
1.06 0.64 1.75
0
5 /cases per 1000
10 cases per 1000

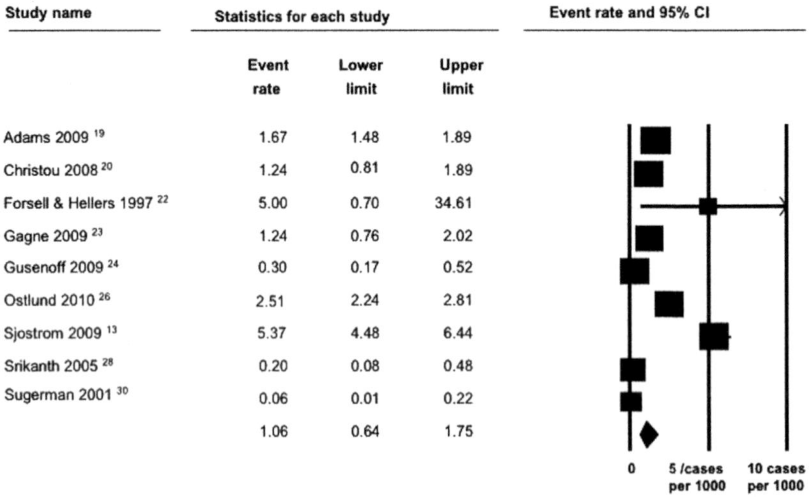

FIGURE 37.3 Studies displaying the incidence density rate of cancer following bariatric surgery. *Reprinted from Casagrande DS et al. Incidence of cancer following bariatric surgery: systematic review and meta-analysis. Obes Surg 2014;24(9):1499–1505. Copyright (2014) by Springer Publishing Company.*

cancer rates compared to the control group [40]. The main studies addressing associations between bariatric surgery and overall cancer risk are summarized in Table 37.1.

Another recent meta-analysis addressed the effects of bariatric surgery on mortality, cardiovascular events, and cancer outcomes, including three randomized clinical trials that included obesity-related cancer evaluations. However, the meta-analysis **revealed no association** between bariatric surgery and lower obesity-related cancer incidence, likely due to the small number of subjects considered and the limited follow-up periods (3-year maximum). This new information does not invalidate previous data from observational studies, due to these methodological issues [49].

In relation to the meta-analysis addressing specific cancer risk, Afshar et al. [42] described lower colorectal cancer incidence (relative risk, RR = 0.73, 95% CI 0.58–0.90) from four retrospective studies that covered a wide range of postoperative follow-up periods (1–30 years) following bariatric surgery. In this meta-analysis, only one study reported baseline BMI data for a bariatric surgery group and a control group, and only two of the other studies used age- and sex-matched controls.

TABLE 37.1 Summary of Main Studies Addressing Association Between Bariatric Surgery and Overall Cancer Risk

Author (Year)	Study	Surgical (n)	Control (n)	Age (Years)		Baseline BMI (kg/m²)		Follow-up (Years)	Type of Bariatric Surgery	Cancer Rate Following Bariatric Surgery
				Surgery	Control	Surgery	Control			
Christou et al. [44]	Retrospective	1035	5746	45 ± 12	47 ± 13	50 ± 8.2	NA	5.3	RYGB (81.3%), VBG (18.7%)	RR = 0.76 (95% CI 0.17−0.39)
Christou et al. [45]	Retrospective	1035	5746	45 ± 12	47 ± 13	50 ± 8.2	NA	5	RYGB (81.3%), VBG (18.7%)	RR = 0.22 (95% CI 0.14−0.35)
Adams et al. [46]	Retrospective	6596	9442	39 ± 10	39 ± 11	45 ± 7.6	47 ± 6.5	12.5	RYGB (100%)	HR = 0.76 (95% CI 0.65−0.89)
McCawley et al. [36]	Retrospective	1482	3495	42	47	52	NA	16	RYGB (93.5%), Gastric banding (3.8%), VBG (1.9%), Jejunal bypass (0.1%)	OR = 0.62 (p = 0.002)
Sjöström et al. [37]	Prospective	2010	2037	47 ± 6	49 ± 6	42	41	10.9	Gastric banding (18.7%), VBG (68.1%), RYGB (13.2%)	HR = 0.67 (95% CI 0.53−0.85)
Östlund et al. [47]	Population-based cohort	13,123	----	NA	NA	NA	NA	9	VBG (48.3%), Gastric banding (37.2%), RYGB (11.5%)	SIR = 1.04 (95% CI 0.93−1.17)
Derogar et al. [48]	Retrospective	15,095	62,016	39	49	NA	NA	10	RYGB (51%), VBG (25%), Gastric banding (24%)	HR = 1.60 (95% CI 1.25−2.02) for colorectal cancer

CI, confidence interval; HR, hazard ratio; OR, odds ratio; NA, data is not available; RR, relative risk; SIR, standardized incidence ratios; VBG, vertical banded gastroplasty; RYGB, Roux-en-Y gastric bypass.

Finally, Upala et al. [43] showed a 60% risk reduction for endometrial cancer following bariatric surgery compared to controls from three retrospectives studies.

METHODOLOGICAL FACTORS

When examining the relationship between bariatric surgery and cancer, it is challenging to separate the effects of the surgery itself from the multiple associated changes that it yields. In this sense, many environmental and lifestyle factors influence the risk of obesity-related cancers, and the impact of some of these variables cannot be fully determined by bariatric surgery studies [50]. Epidemiological factors that are not completely controlled following bariatric surgeries may be responsible for portions of the lower cancer incidence rates observed in several of the cited studies. The main limitations of the association between bariatric surgery and cancer incidence are summarized in Table 37.2.

Most of the cited studies did not report excess weight loss or baseline BMI data. In addition, BMI measurements are often derived from self-reported information. Hence, a relationship between weight loss magnitude and cancer risk, which could provide further evidence for a causal relationship, cannot be determined. The SOS trial displayed an excess weight loss varying from 14% to 25% in the surgical group, and a 30% decrease in the overall cancer incidence in this group, though it cited no association between cancer incidence and changes in body weight [38]. However, cancer incidence was not specified as a secondary endpoint in this trial.

With respect to selection bias, individuals who actively seek surgery as a treatment option are generally more concerned about their health than obese patients who do not pursue bariatric surgery. Moreover, socioeconomic and educational factors likely increase the number of patients with access to health care in the surgery groups. Furthermore, treatment given (if any) to participants in the control group was rarely captured in studies.

The majority of studies have not included analyses that adjust for potential confounding factors such as family history of cancer, smoking, alcohol consumption, and physical inactivity. Another environmental confounding factor with increasing importance, and which is frequently omitted from studies, is patient dietary patterns. However, the Netherlands cohort study did show a 21% higher risk of epithelial ovarian cancer in subjects who have high saturated fat intake as compared to those with low saturated fat intake [51].

Another variable that impacts cancer incidence and mortality rates in men may correspond to the low number of men participating in bariatric surgery studies. Indeed, men represent less than 20% of participants across most bariatric surgery studies [39,40].

Bariatric surgery was associated with decreased breast and endometrial cancer development in 1482 women following the procedure [36]. Similar results were observed in a follow-up mean of 7.1 years following gastric bypass. These

TABLE 37.2 Main Limitations in Interpreting Cancer Risk Following Bariatric Surgery

Bias	
Selection	• *Self-selection*: Patients seeking surgery tend to be healthier than those who do not seek surgery. • *Sample*: More women than men. Patients allocated to the bariatric surgery group are sometimes excluded from surgery when they have suffered from active or invasive cancer within 5 years preceding the study. • *Survivorship (incidence/prevalence)*: Varying follow-up periods.
Measurement	• Self-reported BMI. • Detection (diagnostic): Earlier self-detection of cancer within the surgery group.
Confounding	• Family history of cancer. • Smoking status. • Alcohol consumption. • Physical inactivity. • Dietary patterns.
Secondary treatment	• Metformin or other drugs.
Other factors	• Most studies are retrospective. • No data of excess weight loss calculated. • Most studies lack baseline BMI data for the control group.

authors hypothesized that lower cancer rates following bariatric surgery could be related to the metabolic changes associated with weight loss, or that the lower BMI following surgery resulted in earlier self-detection and diagnoses, which led to improved cancer treatment outcomes [36]. Indeed, weight loss with consequent reduction of fat mass likely improves screening accuracy.

Finally, the studies reporting cancer incidence following bariatric surgery have different follow-up periods. Given the highly variable latency period across cancer types, the effects of an intervention on subsequent risk may take decades to become apparent, and sufficient follow-up examinations over time are required to understand this relationship. Thus, since lead time for the appearance of cancer is too long to expect a sufficient number of studies to be presently available, short-term studies did not detect a positive relationship between cancer and bariatric surgery.

MECHANISMS INVOLVED IN THE ASSOCIATION BETWEEN REDUCED CANCER RISK AND BARIATRIC SURGERY

Multiple possible mechanisms may be involved in overall cancer risk reduction associated with bariatric surgery. Decreased inflammatory markers and oxidative stress, enhanced insulin sensitivity, lower steroid hormone secretion, and lower genomic damage, combined with increased antineoplastic response and improved adiponectin—leptin ratios, are some of the mechanisms that may lead to reduction in obesity-related cancer risk following bariatric surgery [50]. These mechanisms are displayed in Fig. 37.4 and are addressed in detail below.

Reduced Inflammation and Oxidative Stress

While inflammatory mediators can directly stimulate tumor development and progression through indirect and direct effects on innate and adaptive cells from the immune system [17], the reversal of these inflammatory and hormonal imbalances should be expected from weight loss following bariatric surgery. Decreases in both oxidative stress and systemic inflammatory markers were previously reported after surgery [52]. Twenty-six nondiabetic, morbidly obese women undergoing bariatric surgery showed pronounced improvement of acute phase mediators, such as C-reactive protein, and proinflammatory cytokines, such as TNF-α and IL-6, during postoperative weight loss [53]. Likewise, it has been shown that increases in cytotoxic activity of natural killer cells through production of interferon-gamma, IL-12, and IL-18 generate anticancerous effects [54]. These findings suggest that weight loss induced by RYGB improves the cytotoxic immune response against tumor cells following bariatric surgery.

Insulin Resistance

Results from a recent meta-analysis reported improvement or remission of type 2 diabetes in approximately 89% of patients, and complete remission in approximately 65% of patients, following bariatric surgery [55]. Improvement in metabolic control is often evident within days following RYGB, most likely reflecting an alteration in metabolism rather than weight loss, as weight loss is not significant within such a short time period. This metabolic control is probably influenced by improvement in insulin resistance [56], which may contribute to the late anticancerous effects noted in this population.

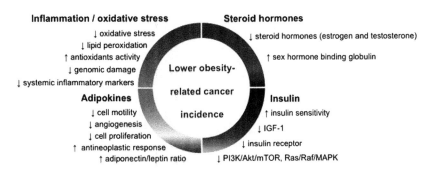

FIGURE 37.4 Biological factors potentially related to lower incidence of cancer following bariatric surgery.

Steroid Hormones

Lower breast and endometrial cancer incidence rates following bariatric surgery are particularly seen in postmenopausal women, perhaps due to a decrease in the estradiol levels in hormone-sensitive tissue, improvement in insulin sensitivity, and restoration of SHBG levels [36].

Adipokines: Leptin and Adiponectin Levels

Adiponectin seems to have a protective role against tumorigenesis through a number of mechanisms, including cell proliferation inhibition, induction of apoptosis, and activation of adenosine monophosphate (AMP)−activated protein kinase (AMPK), which leads to the upregulation of p21 and G1 cell cycle arrests [57]. Considering that obesity is characterized by hyperleptinemia and hypoadiponectinemia, sustained weight loss following bariatric surgery could increase restricted levels of tumorigenic processes, shown previously to both significantly decrease leptin levels for up to 2 years following the surgery [58] and to raise adiponectin levels through the initial year following the surgery [59].

CONCLUSION

In summary, effective weight loss through bariatric surgery seems to reduce the risk of several cancer types, particularly in women. Several mechanisms have been postulated, such as the improvement of insulin resistance, the reduction of oxidative stress and inflammation, the beneficial modulation of sex steroids, intestinal hormones, and the immune system, and the restoration of a leptin−adiponectin balance. However, most data from bariatric surgery and lower risk of obesity-related cancers are derived from observational studies, many of which lacking age-, sex-, and BMI-matched controls, and thus, are subject to bias. Moreover, we must appreciate that younger subjects account for a majority of bariatric surgeries, while cancer is more frequently observed in older individuals. Therefore, more research and better-designed trials are necessary to clarify the benefits of bariatric surgery and its exact association with cancer risk.

ACKNOWLEDGMENT

We thank Stephen Anderson for carefully reviewing the manuscript.

MINI-DICTIONARY OF TERMS

- *Confounding factors*: Factors that can cause or prevent the outcome of interest, which are not intermediate variables and are not associated with the factor(s) under investigation.
- *Jejunoileal bypass*: A procedure consisting of the surgical anastomosis of the proximal portion of the jejunum connected to the distal portion of the ileum, bypassing the nutrient-absorptive segment of the small intestine.
- *Insulin resistance*: Diminished effectiveness of insulin to lower blood glucose levels.
- *Meta-analysis*: Statistical technique for combining the findings of independent studies.
- *Oxidative stress*: A disturbance in the balance between the production of reactive oxygen species (free radicals) and antioxidant defenses.
- *Population-based cohort study*: Study in which a sample, or even the entirety of a defined population, is selected for longitudinal assessment of exposure-related outcomes.
- *Roux-en-Y gastric bypass (RYGB)*: Surgical procedure in which the stomach is transected high on the body. The resulting small proximal gastric pouch is joined to parts of the small intestine by end-to-side surgical anastomosis.
- *Selection bias*: Introduction of error due to systematic differences in the characteristics between those selected and those not selected for a given study.
- *Systematic review*: A literature review that collects and critically analyzes multiple research studies or papers.
- *Tumorigenesis*: The process involved in the production of a new tumor or tumors.

KEY FACTS

- *Obesity-related cancer*: Obesity is associated with increased risk of several cancer types, such as cancer of the esophagus, pancreas, colon and rectum, breast, endometrium, kidney, thyroid, and gallbladder.

- *Mechanisms of increased cancer risk*: Although not fully understood, the relationship between obesity and elevated cancer risk involves metabolic, inflammatory, and hormonal derangements.
- *Bariatric surgery and cancer reduction rate*: Losing weight or avoiding weight gain can reduce cancer incidence, although most data supporting this assessment comes from cohort and case-controlled studies.
- *Mechanisms of decreased tumor incidence after bariatric surgery*: Decreased inflammatory markers and oxidative stress, enhanced insulin sensitivity, lower steroid hormone secretion, lower genomic damage combined with increased antineoplastic response, and an improved adiponectin–leptin ratio are some of the mechanisms that may lead to a reduction in obesity-related cancer risk following bariatric surgery.
- *Obesity-related cancer and bariatric surgery effects*: Losing weight through bariatric surgery can reduce cancer incidence, although most data supporting this information comes from systematic reviews of cohort and case-controlled studies.

SUMMARY POINTS

- Obesity is related to several diseases, including higher cancer rates.
- Bariatric surgery is the most long-term, effective treatment for severe obesity.
- The association between obesity and cancer involves several different mechanisms.
- Inflammatory and hormonal imbalances are observed in both cancer and obesity.
- Bariatric surgery is associated with cancer reduction in several observational studies.

REFERENCES

[1] Rokholm B, Baker JL, Sorensen TI. The levelling off of the obesity epidemic since the year 1999--a review of evidence and perspectives. Obes Rev 2010;11(12):835−46.

[2] WHO. Obesity and overweight. World Health Organization; 2015. p. Fact sheet N°311.

[3] Sjostrom L, et al. Lifestyle, diabetes, and cardiovascular risk factors 10 years after bariatric surgery. N Engl J Med 2004;351(26):2683−93.

[4] Sjostrom L, et al. Effects of bariatric surgery on mortality in Swedish obese subjects. N Engl J Med 2007;357(8):741−52.

[5] Calle EE, Kaaks R. Overweight, obesity and cancer: epidemiological evidence and proposed mechanisms. Nat Rev Cancer 2004;4(8):579−91.

[6] Schmidt MI, et al. Chronic non-communicable diseases in Brazil: burden and current challenges. Lancet 2011;377(9781):1949−61.

[7] Simard B, et al. Asthma and sleep apnea in patients with morbid obesity: outcome after bariatric surgery. Obes Surg 2004;14(10):1381−8.

[8] Parekh N, Chandran U, Bandera EV. Obesity in cancer survival. Annu Rev Nutr 2012;32:311−42.

[9] Kaidar-Person O, Bar-Sela G, Person B. The two major epidemics of the twenty-first century: obesity and cancer. Obes Surg 2011;21(11):1792−7.

[10] Lew EA, Garfinkel L. Variations in mortality by weight among 750,000 men and women. J Chronic Dis 1979;32(8):563−76.

[11] Rapp K, et al. Obesity and incidence of cancer: a large cohort study of over 145,000 adults in Austria. Br J Cancer 2005;93(9):1062−7.

[12] Lukanova A, et al. Body mass index and cancer: results from the Northern Sweden Health and Disease Cohort. Int J Cancer 2006;118(2):458−66.

[13] Renehan AG, et al. Body-mass index and incidence of cancer: a systematic review and meta-analysis of prospective observational studies. Lancet 2008;371(9612):569−78.

[14] Bhaskaran K, et al. Body-mass index and risk of 22 specific cancers: a population-based cohort study of 5.24 million UK adults. Lancet 2014;384(9945):755−65.

[15] Harvie M, Hooper L, Howell AH. Central obesity and breast cancer risk: a systematic review. Obes Rev 2003;4(3):157−73.

[16] Neeland IJ, et al. Associations of visceral and abdominal subcutaneous adipose tissue with markers of cardiac and metabolic risk in obese adults. Obesity (Silver Spring) 2013;21(9):E439−47.

[17] Iyengar NM, Hudis CA, Dannenberg AJ. Obesity and cancer: local and systemic mechanisms. Annu Rev Med 2015;66:297−309.

[18] Gallagher EJ, LeRoith D. Obesity and diabetes: the increased risk of cancer and cancer-related mortality. Physiol Rev 2015;95(3):727−48.

[19] Assir AM, Kamel HF, Hassanien MF. Resistin, visfatin, adiponectin, and leptin: risk of breast cancer in pre- and postmenopausal Saudi females and their possible diagnostic and predictive implications as novel biomarkers. Dis Markers 2015;2015:253519.

[20] Zhao Y, et al. Tumor necrosis factor-alpha stimulates aromatase gene expression in human adipose stromal cells through use of an activating protein-1 binding site upstream of promoter 1.4. Mol Endocrinol 1996;10(11):1350−7.

[21] Morris PG, et al. Inflammation and increased aromatase expression occur in the breast tissue of obese women with breast cancer. Cancer Prev Res (Phila) 2011;4(7):1021−9.

[22] Judd HL, Lucas WE, Yen SS. Serum 17 beta-estradiol and estrone levels in postmenopausal women with and without endometrial cancer. J Clin Endocrinol Metab 1976;43(2):272−8.

[23] Lacey Jr. JV, et al. Menopausal hormone replacement therapy and risk of ovarian cancer. JAMA 2002;288(3):334−41.

[24] Dossus L, et al. Hormonal, metabolic, and inflammatory profiles and endometrial cancer risk within the EPIC cohort--a factor analysis. Am J Epidemiol 2013;177:787−99.

[25] Kaaks R, et al. Serum sex steroids in premenopausal women and breast cancer risk within the European Prospective Investigation into Cancer and Nutrition (EPIC). J Natl Cancer Inst 2005;97(10):755−65.

[26] Salonia A, et al. Preoperative sex steroids are significant predictors of early biochemical recurrence after radical prostatectomy. World J Urol 2013;31(2):275–80.

[27] Banerjee SN, et al. 2-Methoxyestradiol exhibits a biphasic effect on VEGF-A in tumor cells and upregulation is mediated through ER-alpha: a possible signaling pathway associated with the impact of 2-ME2 on proliferative cells. Neoplasia 2003;5(5):417–26.

[28] Zhao Z, et al. 18F-FES and 18F-FDG PET for differential diagnosis and quantitative evaluation of mesenchymal uterine tumors: correlation with immunohistochemical analysis. J Nucl Med 2013;54(4):499–506.

[29] Tsilidis KK, et al. Type 2 diabetes and cancer: umbrella review of meta-analyses of observational studies. BMJ 2015;350. g7607.

[30] Sciacca L, et al. Biological effects of insulin and its analogs on cancer cells with different insulin family receptor expression. J Cell Physiol 2014;229(11):1817–21.

[31] Wang CF, et al. Effects of insulin, insulin-like growth factor-I and -II on proliferation and intracellular signaling in endometrial carcinoma cells with different expression levels of insulin receptor isoform A. Chin Med J (Engl) 2013;126(8):1560–6.

[32] Ouban A, et al. Expression *and distribution of insulin-like growth factor-1 receptor in human carcinomas.* Hum Pathol 2003;34(8):803–8.

[33] Colquitt JL, et al. Surgery for obesity. Cochrane Database Syst Rev 2009;(2) CD003641.

[34] Angrisani L, et al. Bariatric surgery worldwide 2013. Obes Surg 2015;25(10):1822–32.

[35] Garcia OP, Long KZ, Rosado JL. Impact of micronutrient deficiencies on obesity. Nutr Rev 2009;67(10):559–72.

[36] McCawley GM, et al. Cancer in obese women: potential protective impact of bariatric surgery. J Am Coll Surg 2009;208(6):1093–8.

[37] Sjostrom L, et al. Effects of bariatric surgery on cancer incidence in obese patients in Sweden (Swedish Obese Subjects Study): a prospective, controlled intervention trial. Lancet Oncol 2009;10(7):653–62.

[38] Arterburn DE, et al. Association between bariatric surgery and long-term survival. JAMA 2015;313(1):62–70.

[39] Casagrande DS, et al. Incidence of cancer following bariatric surgery: systematic review and meta-analysis. Obes Surg 2014;24(9):1499–509.

[40] Yang XW, et al. Effects of bariatric surgery on incidence of obesity-related cancers: a meta-analysis. Med Sci Monit 2015;21:1350–7.

[41] Xu S, et al. Does bariatric surgery decrease the risk of obesity-related tumor: a meta-analysis. Zhonghua Wei Chang Wai Ke Za Zhi 2015;18(11):1144–8.

[42] Afshar S, et al. The effects of bariatric surgery on colorectal cancer risk: systematic review and meta-analysis. Obes Surg 2014;24(10):1793–9.

[43] Upala S, Anawin S. Bariatric surgery and risk of postoperative endometrial cancer: a systematic review and meta-analysis. Surg Obes Relat Dis 2015;11(4):949–55.

[44] Christou NV, et al. Surgery decreases long-term mortality, morbidity, and health care use in morbidly obese patients. Ann Surg 2004;240(3):416–23.

[45] Christou NV, et al. Bariatric surgery reduces cancer risk in morbidly obese patients. Surg Obes Relat Dis 2008;4(6):691–5.

[46] Adams TD, et al. Cancer incidence and mortality after gastric bypass surgery. Obesity (Silver Spring) 2009;17(4):796–802.

[47] Ostlund MP, et al. Risk of obesity-related câncer after obesity surgery in a population-based cohort study. Ann Surg 2010;252(6):972–6.

[48] Derogar M, et al. Increased risk of colorectal cancer after obesity surgery. Ann Surg 2013;258(6):983–8.

[49] Zhou X, et al. Effects of bariatric surgery on mortality, cardiovascular events, and cancer outcomes in obese patients: systematic review and meta-analysis. Obes Surg 2016;26:2590–601.

[50] Ashrafian H, et al. Metabolic surgery and cancer: protective effects of bariatric procedures. Cancer 2011;117(9):1788–99.

[51] Merritt MA, et al. Nutrient-wide association study of 57 foods/nutrients and epithelial ovarian cancer in the European Prospective Investigation into Cancer and Nutrition study and the Netherlands Cohort Study. Am J Clin Nutr 2015;103:161–7.

[52] Monte SV, et al. Reduction in endotoxemia, oxidative and inflammatory stress, and insulin resistance after Roux-en-Y gastric bypass surgery in patients with morbid obesity and type 2 diabetes mellitus. Surgery 2012;151(4):587–93.

[53] Vazquez LA, et al. Effects of changes in body weight and insulin resistance on inflammation and endothelial function in morbid obesity after bariatric surgery. J Clin Endocrinol Metab 2005;90(1):316–22.

[54] Moulin CM, et al. Bariatric surgery reverses natural killer (NK) cell activity and NK-related cytokine synthesis impairment induced by morbid obesity. Obes Surg 2011;21(1):112–18.

[55] Yu J, et al. The long-term effects of bariatric surgery for type 2 diabetes: systematic review and meta-analysis of randomized and non-randomized evidence. Obes Surg 2015;25(1):143–58.

[56] Rubino F, Marescaux J. Effect of duodenal-jejunal exclusion in a non-obese animal model of type 2 diabetes: a new perspective for an old disease. Ann Surg 2004;239(1):1–11.

[57] Fogarty S, Hardie DG. Development of protein kinase activators: AMPK as a target in metabolic disorders and cancer. Biochim Biophys Acta 2010;1804(3):581–91.

[58] Nijhuis J, et al. Ghrelin, leptin and insulin levels after restrictive surgery: a 2-year follow-up study. Obes Surg 2004;14(6):783–7.

[59] Pender C, et al. Muscle insulin receptor concentrations in obese patients post bariatric surgery: relationship to hyperinsulinemia. Int J Obes Relat Metab Disord 2004;28(3):363–9.

Chapter 38

Upper Gastrointestinal Diseases Before and After Bariatric Surgery

M. Carabotti and C. Severi

University "Sapienza", Rome, Italy

LIST OF ABBREVIATIONS

BMI	body mass Index
GERD	gastroesophageal reflux disease
GI	gastrointestinal
H. pylori	*Helicobacter pylori*
HH	hiatal hernia
LES	lower esophageal sphincter
LGB	laparoscopic gastric banding
LSG	laparoscopic sleeve gastrectomy
NW	normal weight
PPI	proton pump inhibitor
RYGB	Roux-en-Y gastric bypass
TLESRs	transient lower esophageal sphincter relaxations

Obesity represents an important risk factor for the development of gastrointestinal (GI) disorders, most notably gastroesophageal reflux disease (GERD), but also dyspepsia, as well as other symptoms such as nausea, vomiting, and upper abdominal pain. However, the prevalence of proximal abdominal diseases and their relative complications are difficult to assess just on clinical evaluation; the presence of motor and endoscopic alterations are often asymptomatic, probably due to a dysfunction of the autonomic nervous system, which determines an impaired visceral sensation [1,2]. The preoperative GI setting is dramatically changed after bariatric surgery, affecting both gut anatomy and physiology, and influencing the outcomes of GI diseases.

GASTROESOPHAGEAL REFLUX DISEASE

GERD is a common condition, clinically characterized by heartburn and/or regurgitation, determined by the occurrence of several pathophysiological mechanisms, all hampered by obesity, that lead an increased esophageal exposure to gastric contents (Fig. 38.1). The prevalence of GERD is higher in obese patients compared with normal weight (NW) controls, with 2.5 times the risk of developing the disease, mainly as erosive esophagitis [3]. Obesity influences the following factors:

- *Transient lower esophageal sphincter relaxations (TLESRs)*: In overweight and obese patients a substantial increase of TLESRs in the postprandial phase has been reported [4]. TLESRs are a vagal nerve−mediated phenomenon that represent the most important mechanism concurring with GERD pathogenesis in NW patients.
- *Hiatal hernia (HH)*: Obese subjects are more likely to have esophagogastric junction disruption that leads to hernia development [5]; it can be identified with high-resolution manometry [6]. HH is diagnosed in 26−52.6% of obese

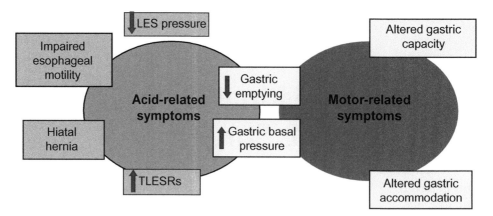

FIGURE 38.1 Pathophysiological mechanisms involved in the development of upper gastrointestinal (GI) symptoms in obese patients. The main mechanisms involved in the pathogenesis of acid-related symptoms (GERD) and motor-related diseases (dyspepsia, nausea, and vomiting) are summarized. *LES*: Lower Esophageal Sphincter; *TLESRs*: Transient Lower Esophageal Sphincter Relaxations.

candidates for bariatric surgery [7,8], and is frequently associated with more severe forms of GERD. Its prevalence is significantly higher in patients with esophagitis (68.2%) with respect to patients without (11.4%) [9].

- *Esophageal motility*: An impaired esophageal clearance can also contribute to GERD. Abnormal esophageal manometric findings have been reported in 25–51% of obese patients candidate to bariatric surgery, hypomotility of the esophageal body being the most common alteration. It has been noted, however, that esophageal dysmotility might occur in the absence of GI symptoms [1,8].

- *Basal lower esophageal sphincter (LES) pressure*: This factor likely covers a less relevant importance in the pathogenesis of GERD. Similar to NW, the contribution of LES pressure in the development of GERD in obese patients still needs to be established. Conflicting results are available, with some studies showing a decreased pressure [10], and others a normal one [11].

- *Gastric basal pressure*: The increased intraabdominal pressure, resulting from the mechanical burden of fat, promotes retrograde flow of gastric juice into the esophagus. A significant correlation between body mass index (BMI) and waist circumference with intragastric pressure and gastroesophageal pressure gradient has been observed [5]. Furthermore, an increased gastric basal pressure, associated with overeating, may contribute to the increase in TLESRs.

- *Gastric emptying*: Delayed gastric emptying could be another factor contributing to GERD. However, data show conflicting results both in NW and obese patients [12].

Postoperative Gastroesophageal Reflux Disease Outcomes

The outcomes of GERD depend on the type of surgical procedure. Nevertheless, available data are conflicting, which is probably related to the different methods used for GERD detection (questionnaire, pH-metry, manometry, endoscopy, proton pump inhibitor (PPI) treatment) and the different times of follow-up. As of now, Roux-en-Y gastric bypass (RYGB) is considered the best surgical treatment option for GERD in the morbidly obese patient, since both the small size of the gastric pouch and the diversion of bile lead to a remarkable reduction of the gastric content that can reflux into the esophagus. Several studies reported an improvement of reflux symptoms, reduction of esophageal acid exposure, erosive esophagitis, and PPI prescription after RYGB [13]. On the other hand, impact of laparoscopic sleeve gastrectomy (LSG) on GERD symptoms is controversial. The Second International Consensus Summit for Sleeve Gastrectomy reported a prevalence of postoperative GERD ranging from 0% to 83% [14]. Again, the discrepancy might arise from the different definitions and methodologies used for GERD diagnosis, but also from the different LSG surgical procedures used. In particular the different positions of the stapler at the angle of His, bougie size, and the presence of HH might influence GERD outcomes [15]. LSG-related anatomical modifications lead to a new balance between exacerbating and protective factors for GERD, the latter being mainly represented by the supposed reduced acid load and possible faster gastric emptying. Finally, the most recent data has shown notable evidence of worsening of GERD after laparoscopic gastric banding, above all regurgitation, in both previously symptomatic and asymptomatic patients [13].

DYSPEPSIA AND OTHER UPPER GASTROINTESTINAL DISORDERS

Visceral adiposity is associated with an increased risk of functional dyspepsia [16] and globus [17]. By means of a validated GI symptoms questionnaire based on the Rome Criteria, it has been reported, in a cohort of 120 consecutive patients candidate to bariatric surgery, that 89% complained of functional symptoms (31% esophageal and 38% gastroduodenal) [18]. An association has been found between increased BMI and abdominal pain associated with nausea or vomiting (OR 2.0, 95% CI 1.0−2.9; $p = 0.04$) [19]. Similarly, increased BMI categories are associated with bloating (OR = 1.3, 95% CI 1.1−1.6) [20]. The higher prevalence of upper GI symptoms has also been found in a population-based study: nausea, vomiting, early satiety, upper abdominal pain, and bloating were respectively present in 10.6%, 4.0%, 9.0%, 19.8%, and 23.4% of obese patients compared to 7.9%, 1.2%, 6.5%, 3.5%, 11.4% in the NW control subjects [21].

In contrast to GERD, whose pathophysiological mechanisms are well-known, the underlying reasons for upper functional GI disorders are poorly understood, with scarce data availability (Fig. 38.1). Different mechanisms have been proposed, such as an alteration in gastric emptying, gastric capacity, and accommodation [12,22]. Available evidence has suggested that obese patients have a higher gastric capacity or enlarged antrum, even if it remains to be established whether these are a cause or a consequence of obesity [23]. It is also possible that altered psychological behaviors and eating disorders, often present in obese patients, influence the development of GI symptoms.

Postoperative Outcomes

The prevalence of functional upper GI symptoms after bariatric surgery has been poorly analyzed, likely because upper GI symptoms (mainly dyspepsia) represent an expected outcome related to the marked anatomical manipulation intrinsic to bariatric procedures. After RYGB, the increase in gastric symptoms, such as postprandial fullness, early satiation, pain, or epigastric burning, is not observed, likely due to the exclusion of the stomach from the continuity of the digestive tract. Comparing RYGB and LSG 1 year after surgery, nausea ($p < 0.01$) and vomiting ($p < 0.001$) were significantly more frequent among LSG subjects [24]. Nevertheless, after RYGB, dysphagia can occur, probably not GERD-related, but ascribable to the impaired emptying of the gastric pouch [25,26]. When dysphagia is associated with pain, nausea, and vomiting, likely a marginal ulcer of the gastro-jejunal anastomosis occurred [27,28]. Esophageal dysmotility, with decrease in esophageal body amplitude and ineffective esophageal peristalsis, are reported in a quarter of patients after RYGB, despite absence of dysphagia or chest pain [29].

After LSG, in a prospective clinical study evaluating upper GI symptoms before and after surgery through validated questionnaires, the main postsurgical complaint was dyspepsia; this was true in almost 60% of patients, LSG being associated with de novo dyspepsia, with an OR of 7.00 (95% CI 2.9−18.3, $p < 0.0001$) [30].

HELICOBACTER PYLORI INFECTION

Helicobacter pylori infection represents one of the most common human infections, present in more than half of the world's population, with a disparity between developed and developing countries. An inverse correlation between *H. pylori* infection and BMI has been proposed based on the significant weight gain observed after *H. pylori* eradication [31]. Especially in developed areas, the reduced prevalence of infection has been proposed as a contributing factor for the increased prevalence of overweight and obese individuals [32].

The significant weight gain observed after *H. pylori* eradication might be related to different mechanisms. First, the improvement of *H. pylori*−related GI symptoms may favor dietary excess and promote weight gain. Second, the reversal of mucosal gastritis after *H. pylori* eradication can cause an increase in ghrelin, the orexigenic hormone produced by gastric mucosa relevant in the regulation of appetite and weight gain. Data on the relationship between *H. pylori* and ghrelin are however discordant. Some studies showed that *H. pylori* infection has a negative impact on density of gastric ghrelin-positive cells in obese patients [33], whereas others showed an increase in the number of immunoreactive ghrelin cells in obese patients compared to NW subjects, independent of *H. pylori* status [34]. Other studies aimed at analyzing this topic are needed.

Helicobacter pylori Infection Outcome After Bariatric Surgery

The advantages of preoperative *H. pylori* screening and eradication in reducing postsurgical complications are still controversial, mainly due to a lack of randomized controlled trials. An effective *H. pylori* eradication prior to RYGB brings a reduced incidence of viscus perforations [35] and marginal ulcers [36], whose prevalence after RYGB is about 1–16% [37]. In a retrospective study including 560 patients, the incidence of ulcers was 2.4% in patients tested and treated for *H. pylori* infection prior to surgery, compared to 6.8% in those who did not undergo such screening [36]. Nevertheless, this positive effect is not confirmed by recent data that reported a significantly lower incidence of postoperative anastomotic ulcer complications in *H. pylori*–positive versus negative patients at the time of RYGB, which does not support a routine identification and eradication of infection prior to RYGB [38].

After LSG, different authors have not reported an increase in postsurgical complications in patients with *H. pylori* who are undergoing surgery [39–41]. The gastric environment for *H. pylori* colonization may dramatically change after LSG, with possible spontaneous clearance of infection [42]. However, these data need to be confirmed in a long-term follow-up. Guidelines on the necessity of preoperative *H. pylori* screening and management are discordant. The American Association of Clinical Endocrinologists, The Obesity Society, and the American Society for Metabolic and Bariatric Surgery guidelines [43] do not provide a clear indication. *H. pylori* screening is recommended only in patients in high-prevalence areas, and upper endoscopy in selected cases. European guidelines [44] in turn recommend performing an upper GI endoscopy before bariatric surgery in all patients in order to diagnose and treat any lesions (including *H. pylori* infection) that may cause postoperative complications.

Even if controversial, there are plausible reasons to attempt eradication in *H. pylori*–positive patients (Table 38.1). First of all, a large part of the stomach is endoscopically inaccessible after RYGB. Second, it has to be taken into account that *H. pylori* is a class I carcinogen in the development of gastric cancer, with an odds ratio of 2.0–5.9 [45]. Third, its eradication might lead to a moderate benefit on symptomatic dyspepsia [46]. Nevertheless, it remains to define the correct eradication strategy, since there is a need to test new regimens. The methods available in clinical practice for *H. pylori* infection diagnosis, and the key facts of the its eradication treatment, are listed in Table 38.2 [47].

BMI represents an independent risk factor for eradication failure, with a 6% increase of risk of eradication failure for each unit of BMI [48]. With a 7-day regimen, a successful eradication was obtained in 55.0% of the overweight/obese group compared with 85.4% of the NW group [48]. The advantage of a longer treatment has been reported, with an eradication rate achieved with a 14-day triple therapy significantly higher than with a 7-day triple therapy (OR = 1.96; 95% CI 1.16–3.30; $p = 0.016$) [49]. Although the mechanisms underlying the low rates of eradication remained unclear, two hypotheses emerged: (1) the physiological changes occurring in obesity (i.e., altered gastric emptying) may lead to a

TABLE 38.1 Reasons to Attempt or Not Attempt *H. pylori* Eradication

Why Eradicate	Why Not Eradicate
H. pylori-infected persons present a higher risk for gastric cancer.	Eradication therapy is less effective in obese than NW patients.
H. pylori eradication might improve dyspepsia.	Doubts about the real benefits in terms of reduction of postsurgical complications.
Stomach may be difficult to explore after RYGB.	Spontaneous eradication has been described after LGS.

LSG, Laparoscopic Sleeve Gastrectomy; NW, Normal Weight; RYGB, Roux-en-Y Gastric Bypass.

TABLE 38.2 Diagnosis of *H. pylori* Infection

Invasive Tests (during endoscopy)	Noninvasive Tests
Histology	^{13}C Urea breath test
Rapid urease test	Stool antigen
Culture	Immunoglobin G (IgG) serology

For the diagnosis of *H. pylori*, invasive and noninvasive tests are available. All tests, except serology, require discontinuation of proton pump inhibitors and antibiotics for at least 1 month before the test. Serology is not adequate to verify the eradication treatment.

decrease in the rate of drug absorption, regardless of the characteristics of the drug and (2) the volume of distribution of drugs may be altered because tissue mass can influence medications with lipophilic properties [50,51].

In conclusion, obese patients present several proximal GI diseases whose prevalence is probably underestimated due to the impairment of visceral sensation likely occurring in this setting. To attain optimal tailored management, it is important to understand with a multidisciplinary approach the different gut perturbations consequent to the anatomical and functional modifications inherent to each procedure.

MINI-DICTIONARY OF TERMS

- *Gastroesophageal reflux disease (GERD)*: A chronic digestive disease occurring when stomach contents, mainly acid, flow back into the esophagus. The refluxate damages esophageal mucosa, causing troublesome symptoms and/ or complications.
- *Lower esophageal sphincter (LES)*: A specialized muscle segment of the distal esophagus characterized by a high-pressure zone that represents one of the mechanisms of the antireflux barrier.
- *Transient lower esophageal sphincter relaxations (TLESRs)*: LES opening not occurring after swallowing. They are mainly caused by gastric fundus relaxation.
- *Hiatal hernia (HH)*: Occurs when part of the stomach pushes upward through the diaphragm into the thorax. There are two main types: (1) sliding HH with upward shift of LES and (2) paraesophageal HH with upward shift of gastric fundus next to the esophagus without migration of LES. Only the more frequent "sliding hernia" will be discussed here.
- *Esophageal manometry*: Evaluates coordination of esophageal movements and pressures using a thin catheter with pressure sensors inserted through the esophagus. The high-resolution manometry with esophageal pressure topography plotting combines improvements in pressure-sensing technology with an increased number of pressure sensors.
- *Regurgitation*: A characteristic GERD symptom usually described as a sour taste in the mouth or a sense of fluid moving up and down in the chest.
- *Dysphagia*: Characterized by difficult swallowing that can occur both for solids and liquids. It can be a manifestation of stricture or, more commonly, of dysmotility of the esophagus.
- *Dyspepsia*: Defined by the presence of symptoms thought to originate in the gastroduodenal region. Following Roma III Criteria for functional disorders, it is further subdivided in *postprandial distress syndrome*, in the presence of postprandial fullness and/or early satiation, and *epigastric pain syndrome*, in the presence of pain or burning localized in the epigastrium and not associated with other abdominal- or chest-related symptoms.
- *Helicobacter pylori (H. pylori)*: A gram-negative, microaerophilic and spiral-shaped bacteria that is the main etiological factor of chronic gastritis, benign peptic ulcer, and gastric cancer.
- *Ghrelin*: Ghrelin is a fast-acting hormone produced mainly in the gastric fundus by neuroendocrine cells (X/A-like cells in rodents and P/D1 cells in humans), and secreted into the circulation.

KEY FACTS

- The first-line therapy is represented by the triple treatment that includes a PPI, clarithromycin, and amoxicillin or metronidazole.
- Use of clarithromycin should be avoided if the resistance rate in the region is more than 15–20%.
- Bismuth-containing quadruple therapy represents an alternative.
- Use of PPI twice a day might increase the efficacy of therapy.
- Extending the duration of PPI–clarithromycin-containing triple therapies from 7 to 10–14 days might improve the eradication success.
- Use of probiotics and prebiotics might reduce side effects.
- After the first therapy, either a bismuth-containing quadruple therapy or levofloxacin-containing triple therapy are recommended.
- The repetition of the same eradication therapy must be avoided.
- After failure of the second therapy, an antimicrobial susceptibility test might be useful.
- After at least 4 weeks from the end of treatment, a urea breath test or stool test are recommended to determine the success of the treatment.

SUMMARY POINTS

- Obesity confers an increased risk of upper GI disorders, particularly GERD.
- Different surgical approaches are associated with variable GI outcomes, but RYGB is the most efficient option for the improvement of GERD.
- Other upper GI symptoms than GERD are associated with obesity, but the impact of bariatric surgery on these disturbances is less clear.
- The reduced prevalence of *H. pylori* infection in developed areas might relate to the augmentation of overweight and obese individuals.
- In patients candidate to bariatric surgery, the need for *H. pylori* screening and eradication is still debated.

REFERENCES

[1] Côté-Daigneault J, Leclerc P, Joubert J, Bouin M. High prevalence of esophageal dysmotility in asymptomatic obese patients. Can J Gastroenterol Hepatol 2014;28:311.

[2] Carabotti M, Avallone M, Cereatti F, Paganini A, Greco F, Scirocco A, et al. Usefulness of upper gastrointestinal symptoms as a driver to prescribe gastroscopy in obese patients candidate to bariatric surgery. A prospective study. Obes Surg 2016;26:1075.

[3] El-Serag HB, Graham DY, Satia JA, Rabeneck L. Obesity is an independent risk factor for GERD symptoms and erosive esophagitis. Am J Gastroenterol 2005;100:1243.

[4] Wu JC, Mui LM, Cheung CM, Chan Y, Sung JJ. Obesity is associated with increased transient lower esophageal sphincter relaxation. Gastroenterology 2007;132:883.

[5] Pandolfino JE, El−Serag HB, Zhang Q, Shah N, Ghosh SK, Kahrilas PJ. Obesity: a challenge to esophagogastric junction integrity. Gastroenterology 2006;130:639.

[6] Tolone S, Savarino E, de Bortoli N, Frazzoni M, Furnari M, d'Alessandro A, et al. Esophagogastric junction morphology assessment by high resolution manometry in obese patients candidate to bariatric surgery. Int J Surg 2016;28(Suppl. 1):S109.

[7] Iovino P, Angrisani L, Galloro G, Consalvo D, Tremolaterra F, Pascariello A, et al. Proximal stomach function in obesity with normal or abnormal oesophageal acid exposure. Neurogastroenterol Motil 2006;18:425.

[8] Suter M, Dorta G, Giusti V, Calmes JM. Gastro-esophageal reflux and esophageal motility disorders in morbidly obese patients. Obes Surg 2004;14:959.

[9] Stene-Larsen G, Weberg R, Frøyshov Larsen I, Bjørtuft O, Hoel B, Berstad A. Relationship of overweight to hiatus hernia and reflux oesophagitis. Scand J Gastroenterol 1988;23:427.

[10] Iovino P, Angrisani L, Tremolaterra F, Nirchio E, Ciannella M, Borrelli V, et al. Abnormal esophageal acid exposure is common in morbidly obese patients and improves after a successful Lap-band system implantation. Surg Endosc 2002;16:1631.

[11] Fornari F, Callegari-Jacques SM, Dantas RO, Scarsi AL, Ruas LO, de Barros SG. Obese patients have stronger peristalsis and increased acid exposure in the esophagus. Dig Dis Sci 2011;56:1420.

[12] Park MI, Camilleri M. Gastric motor and sensory functions in obesity. Obes Res 2005;13:491.

[13] Naik RD, Choksi YA, Vaezi MF. Consequences of bariatric surgery on oesophageal function in health and disease. Nat Rev Gastroenterol Hepatol 2016;13:111.

[14] Gagner M, Deitel M, Kalberer TL, Erickson AL, Crosby RD. The Second International Consensus Summit for sleeve gastrectomy, March 19-21, 2009. Surg Obes Relat Dis 2009;5:476.

[15] Chiu S, Birch DW, Shi X, Sharma AM, Karmali S. Effect of sleeve gastrectomy on gastroesophageal reflux disease: a systematic review. Surg Obes Relat Dis 2011;7:510.

[16] Jung JG, Yang JN, Lee CG, Choi SH, Kwack WG, Lee JH, et al. Visceral adiposity is associated with an increased risk of functional dyspepsia. J Gastroenterol Hepatol 2016;31:567.

[17] Bouchoucha M, Fysekidis M, Julia C, Airinei G, Catheline JM, Cohen R, et al. Body mass index association with functional gastrointestinal disorders: differences between genders. Results from a study in a tertiary center. J Gastroenterol 2016;51:337.

[18] Fysekidis M, Bouchoucha M, Bihan H, Reach G, Benamouzig R, Catheline JM. Prevalence and co-occurrence of upper and lower functional gastrointestinal symptoms in patients eligible for bariatric surgery. Obes Surg 2012;22:403.

[19] Talley NJ, Howell S, Poulton R. Obesity and chronic gastrointestinal tract symptoms in young adults: a birth cohort study. Am J Gastroenterol 2004;99:1807.

[20] Talley NJ, Quan C, Jones MP, Horowitz M. Association of upper and lower gastrointestinal tract symptoms with body mass index in an Australian cohort. Neurogastroenterol Motil 2004;16:413.

[21] Delgado-Aros S, Locke III GR, Camilleri M, Talley NJ, Fett S, Zinsmeister AR. Obesity is associated with increased risk of gastrointestinal symptoms: a population-based study. Am J Gastroenterol 2004;99:1801.

[22] Kim DY, Camilleri M, Murray JA, Stephens DA, Levine JA, Burton DD. Is there a role for gastric accommodation and satiety in asymptomatic obese people? Obes Res 2001;9:655.

[23] Xing J, Chen JD. Alterations of gastrointestinal motility in obesity. Obes Res 2004;12:1723.

[24] Carabotti M, Silecchia G, Greco F, Leonetti F, Piretta L, Rengo M, et al. Impact of laparoscopic sleeve gastrectomy on upper gastrointestinal symptoms. Obes Surg 2013;23:1551.

[25] El Labban S, Safadi B, Olabi A. The effect of Roux-en-Y gastric bypass and sleeve gastrectomy surgery on dietary intake, Food Preferences, and Gastrointestinal Symptoms in Post-Surgical Morbidly Obese Lebanese Subjects: A Cross-Sectional Pilot Study. Obes Surg 2015;25:2393.

[26] Madalosso CA, Gurski RR, Callegari-Jacques SM, Navarini D, Thiesen V, Fornari F. The impact of gastric bypass on gastroesophageal reflux disease in patients with morbid obesity: a prospective study based on the Montreal Consensus. Ann Surg 2010;251:244.

[27] Ortega J, Escudero MD, Mora F, Sala C, Flor B, Martinez-Valls J, et al. Outcome of esophageal function and 24-hour esophageal pH monitoring after vertical banded gastroplasty and Roux-en-Y gastric bypass. Obes Surg 2004;14:1086.

[28] El-Hayek K, Timratana P, Shimizu H, Chand B. Marginal ulcer after Roux-en-Y gastric bypass: what have we really learned? Surg Endosc 2012;26:2789.

[29] Moon RC, Teixeira AF, Goldbach M, Jawad MA. Management and treatment outcomes of marginal ulcers after Roux-en-Y gastric bypass at a single high volume bariatric center. Surg Obes Relat Dis 2014;10:229.

[30] Mejía-Rivas MA, Herrera-López A, Hernández-Calleros J, Herrera MF, Valdovinos MA. Gastroesophageal reflux disease in morbid obesity: the effect of Roux-en-Y gastric bypass. Obes Surg 2008;18:1217.

[31] Danesh J, Peto R. Risk factors for coronary heart disease and infection with *Helicobacter pylori*: meta-analysis of 18 studies. BMJ 1998;316:1130.

[32] Lender N, Talley NJ, Enck P, Haag S, Zipfel S, Morrison M, et al. Review article: associations between *Helicobacter pylori* and obesity--an ecological study. Aliment Pharmacol Ther 2014;40:24.

[33] Liew PL, Lee WJ, Lee YC, Chen WY. Gastric ghrelin expression associated with *Helicobacter pylori* infection and chronic gastritis in obese patients. Obes Surg 2006;16:612.

[34] Maksud FA, Alves JS, Diniz MT, Barbosa AJ. Density of ghrelin-producing cells is higher in the gastric mucosa of morbidly obese patients. Eur J Endocrinol 2011;165:57.

[35] Hartin CW, ReMine DS, Lucktong TA. Preoperative bar- iatric screening and treatment of *Helicobacter pylori*. Surg Endosc 2009;23:2531.

[36] Schirmer B, Erenoglu C, Miller A. Flexible endoscopy in the management of patients undergoing Roux-en-Y gastric bypass. Obes Surg 2002;12:634.

[37] Rasmussen JJ, Fuller W, Ali MR. Marginal ulceration after laparoscopic gastric bypass: an analysis of predisposing factors in 260 patients. Surg Endosc 2007;21:1090.

[38] Kelly JJ, Perugini RA, Wang QL, Czerniach DR, Flahive J, Cohen PA. The presence of *Helicobacter pylori* is not associated with long-term anastomotic complications in gastric bypass patients. Surg Endosc 2015;29:2885.

[39] Brownlee AR, Bromberg E, Roslin MS. Outcomes in patients with *Helicobacter pylori* Undergoing Laparoscopic Sleeve Gastrectomy. Obes Surg 2015;25:2276.

[40] Rossetti G, Moccia F, Marra T, Buonomo M, Pascotto B, Pezzullo A, et al. Does *Helicobacter pylori* infection have influence on outcome of laparoscopic sleeve gastrectomy for morbid obesity? Int J Surg 2014;12(Suppl. 1):S68.

[41] Almazeedi S, Al-Sabah S, Alshammari D, Alqinai S, Al-Mulla A, Al-Murad A, et al. The impact of *Helicobacter pylori* on the complications of laparoscopic sleeve gastrectomy. Obes Surg 2014;24:412.

[42] Keren D, Matter I, Rainis T, Goldstein O, Stermer E, Lavy A. Sleeve gastrectomy leads to *Helicobacter pylori* eradication. Obes Surg 2009;19:751.

[43] Mechanick JI, Youdim A, Jones DB, Garvey WT, Hurley DL, McMahon MM, et al. Clinical practice guidelines for the perioperative nutritional, metabolic, and nonsurgical support of the bariatric surgery patient-2013 update: cospon- sored by American Association of Clinical Endocrinologists, The Obesity Society, and American Society for Metabolic & Bariatric Surgery. Obesity (Silver Spring) 2013;21(Suppl. 1):S1.

[44] Sauerland S, Angrisani L, Belachew M, Chevallier JM, Favretti F, Finer N, et al. Obesity surgery: evidence-based guidelines of the European Association for Endoscopic Surgery (EAES). Surg Endosc 2005;19:200.

[45] Eslick G. Gastrointestinal symptoms and obesity: a meta-analysis. Obes Rev 2012;13:469.

[46] Moayyedi P, Soo S, Deeks J, Forman D, Mason J, Innes M, et al. Systematic review and economic evaluation of *Helicobacter pylori* eradication treatment for non-ulcer dyspepsia. Dyspepsia Review Group. BMJ 2000;321:659.

[47] Malfertheiner P, Megraud F, O'Morain CA, Atherton J, Axon AT, Bazzoli F, et al. European Helicobacter Study Group. Management of *Helicobacter pylori* infection—the Maastricht IV/Florence Consensus Report. Gut 2012;61:646.

[48] Abdullahi M, Annibale B, Capoccia D, Tari R, Lahner E, Osborn J, et al. The eradication of *Helicobacter pylori* is affected by body mass index (BMI). Obes Surg 2008;18:1450.

[49] Cerqueira RM, Manso MC, Correia MR, Fernandes CD, Vilar H, Nora M, et al. *Helicobacter pylori* eradication therapy in obese patients undergoing gastric bypass surgery--fourteen days superior to seven days? Obes Surg 2011;21:1377.

[50] Cheymol G. Effects of obesity on pharmacokinetics implications for drug therapy. Clin Pharmacokinet 2000;39:215.

[51] Pai MP, Bearden DT, Antimicrobial DT. Dosing considerations in obese adult patients. Pharmacotherapy 2007;27:1081.

Chapter 39

Hematological Disorders Following Bariatric Surgery

H. Qiu, R. Green and M. Chen

Davis Medical Center, Sacramento, CA, United States

LIST OF ABBREVIATIONS

AGB adjustable gastric banding
holoTC holotranscobalamin
IFN-γ interferon-gamma
MCV mean cell volume
MDS myelodysplastic syndrome
MMA methyl malonic acid
RYGB Roux-en-Y gastric bypass
TNF-α tumor necrosis factor-alpha
VBG vertical-banded gastroplasty

INTRODUCTION

Obesity, an emergent worldwide epidemic, is associated with multiple comorbidities including hypertension, hypercholesterolemia, hypertriglyceridemia, diabetes mellitus, obstructive sleep apnea, osteoarthritis, back/extremity pain, gastroesophageal reflux disease, asthma, and depression [1]. Bariatric surgery not only results in substantial and sustained effects on weight loss, but also ameliorates obesity-attributable comorbidities in the majority of patients [2]. The two main principles of bariatric surgery that exist either alone or in combination are the following: (1) restriction, which is decreasing the size of the stomach so that there is less food intake with early satiety and (2) malabsorption, which is limiting food absorption by bypassing parts of the gastrointestinal tract [3]. Common bariatric procedures include laparoscopic vertical-banded gastroplasty, adjustable gastric banding, and Roux-en-Y gastric bypass (RYGB)—often called gastric bypass. Despite major advancements in bariatric surgery over the past two decades, RYGB, which is both a restrictive and malabsorptive operation, still remains the most frequently performed bariatric procedure worldwide. The surgery is usually performed by noninvasive laparoscopy, and is considered the "gold standard" of bariatric surgery [1].

The RYGB procedure involves creating a small gastric pouch (<30 mL), which is connected to the lower part of the small intestine, and dividing the small intestine to rearrange it into a Y-configuration. This allows outflow of food from the small stomach pouch via a "Roux limb." The gastric bypass can reduce the size of the stomach by well over 90%. The remainder of the stomach secretes gastric acid, pepsin, and intrinsic factor into a biliopancreatic limb, which combines with bile and pancreatic drainage and joins the Roux limb approximately 100 cm from the gastrojejunostomy. This procedure bypasses the duodenum along with a portion of the proximal jejunum, thereby restricting food intake and limiting absorption (Fig. 39.1) [4].

RYGB results in a substantial and durable weight loss, and significantly improves or resolves many of the debilitating obesity-related comorbidities to reduce long-term mortality [5,6]. Patients who undergo RYGB usually lose about 65−80% of their excess body weight in the first year after surgery, and sustained weight loss has been reported up to 16 years after gastric bypass [7]. RYGB has been reported to ameliorate hypertension (with normalization of blood pressure in 30−50% of patients 1−2 years after surgery) and reduce the need for antihypertensive treatment in a further

Metabolism and Pathophysiology of Bariatric Surgery.

Roux-en-Y gastric bypass (RYGB)

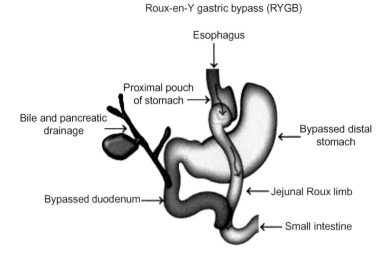

FIGURE 39.1   Gastric bypass with Roux-en-Y anastomosis. Diagrammatic representation of gastric bypass using a Roux-en-Y anastomosis. *Chen M, et al. Hematological disorders following gastric bypass surgery: emerging concepts of the interplay between nutritional deficiency and inflammation. Biomed Res Int 2013;2013:205467, with permission.*

20—30% of patients [8]. Diabetes type 2 is reversed in up to 80% of patients, usually leading to normal blood sugar levels without medication [9]. Hyperlipidemia is resolved after RYGB in 50—70% of patients [8]. In addition, the resolution of diabetes, hypertension, and hyperlipidemia postoperatively has a significant effect on the decline of the incidence of cardiovascular adverse events such as myocardial infarction [10]. Obstructive sleep apnea improves substantially in most patients [7]. Long-term improvements in pulmonary function have been documented as well after RYGB [11].

Although a highly effective therapy for morbid obesity, RYGB has predictable adverse effects and potential complications due to the alteration of normal physiology of the gastrointestinal tract [12]. Nutritional deficiencies frequently occur after RYGB despite supplementation with the standard multivitamin preparation [13,14]. Dumping syndrome develops in some patients and postprandial hypoglycemia in others following RYGB [8]. Hematological disorders may occur after gastric bypass surgery, including anemia and multilineage cytopenias that result from a complex interplay between micronutrient deficiencies, inflammation, and immune dysregulation [15]. The incidence of anemia after RYGB at different follow-up periods ranges from 5% to 64% [16]. A recent study reported that 27% of patients had anemia more than 10 years after RYGB [17]. While iron deficiency is a major cause for anemia after bariatric surgery, it does not account for all cases, and excess iron supplementation and transfusion overload can cause secondary hemosiderosis and irreversible liver injury. Therefore, postsurgical management of bariatric surgery patients requires lifelong follow-up of hematological and metabolic parameters [18]. The evaluation of hematological disorders after bariatric surgery must take into account issues unique to the postsurgery setting. This chapter focuses on the factors unique to RYGB that influence the development of anemia and other cytopenias.

NUTRITIONAL ANEMIA AFTER GASTRIC BYPASS SURGERY

RYGB can result in multifactorial nutritional anemia [19]. Various mechanisms contribute to the development of anemia at different stages following surgery [20]. Immediately after surgery, hemolysis and reduced erythrocyte survival may be important causes of anemia [15]. Furthermore, suppression of erythropoiesis and interference with the erythropoietin signaling pathway may contribute to decreased production of red blood cells. While iron deficiency is the most common cause of anemia following RYGB, deficiencies in vitamin B12, folate (folic acid), zinc, and selenium can also contribute to anemia [17,21]. In addition, copper deficiency has also been reported after RYGB, which can cause hematological abnormalities with or without associated neurological complications [22].

Iron Deficiency

Iron deficiency has been reported in 6—50% of patients who undergo gastric bypass surgery [15]. Gastric acid increases solubility of inorganic iron, and favors reduction of dietary iron to its ferrous form and enhances its absorption in the duodenum. Factors contributing to iron deficiency following RYGB include anatomic alteration of the duodenum and proximal jejunum from the path of nutrients, decreased gastric acidity, and low tolerance to red meat [23,24]. In addition, bacterial overgrowth and the induction of an iron-losing enteropathy may also be contributory factors. Notably, infection with *Helicobacter pylori* is known to alter gastric pH, in combination with hypergastrinemia, which leads to impaired iron absorption [25].

Iron deficiency is usually characterized by microcytic hypochromic anemia. The complete blood count typically shows low hematocrit, hemoglobin, mean corpuscular volume, mean corpuscular hemoglobin, and/or mean corpuscular hemoglobin concentration, as well as high red blood cell distribution width [26]. Lowered serum iron level, elevated serum transferrin, and total iron binding capacity (TIBC) indicate iron-restricted erythropoiesis. Serum ferritin is the most specific laboratory test for iron deficiency anemia. However, serum ferritin can be elevated by inflammation, and so caution should be exercised when assessing iron status if it is within normal limits [26]. Clinical symptoms of iron deficiency include fatigue, headache, exercise intolerance, pica, and also oral manifestations including stomatitis and glossitis [15].

Patients after RYGB are routinely recommended to take prophylactic oral iron supplements [27]. When there is concern over patient compliance with oral therapy, intravenous replacement can be used on an annual basis [15]. If the anemia is persistent and refractory to iron supplementation therapy, additional micronutrient deficiencies and other origins of the anemia should be considered. Special attention must be paid to deficiencies of vitamin B12, folate, copper, and zinc that may need correction [22]. Additionally, other factors such as *H. pylori* infection should be considered; studies have shown that eradication of *H. pylori* can lead to improved responsiveness to iron supplementation [18]. Interestingly, a randomized trial of iron supplementation (65 mg of elemental iron taken orally twice daily) prevented iron deficiency but did not prevent anemia, which highlights the importance of considering other micronutrient deficiencies that can contribute to anemia [28].

Vitamin B12 Deficiency

Vitamin B12 deficiency is the next most common cause of anemia following RYGB, with reported prevalences of 4–62% after 2 years and 19–35% after 5 years [13,29,30]. Absorption of food vitamin B12 requires an intact and functioning stomach, exocrine pancreas, intrinsic factor, and small bowel [31]. Hematological and neurological disorders are among the main symptoms of vitamin B12 deficiency. Megaloblastic anemia in vitamin B12 deficiency develops as a result of disrupted DNA synthesis, and in the bone marrow there is a resultant maturation disorder of the cell nuclei, whereas the cytoplasm develops normally. In the peripheral blood, macrocytic erythrocytes (MCV > 100FL) and hypersegmented neutrophils can be observed [32]. Notably, if B12 deficiency occurs concomitantly with iron deficiency, and the iron deficiency is more severe than the B12 deficiency, iron deficiency-related microcytosis can obscure or dominate over B12 deficiency-related macrocytosis [33], B12 deficiency can be masked, and the diagnosis is thus difficult [34]. Moreover, B12 deficiency can cause an additional loss of iron by means of a secondary effect on the enterocytes [32].

B12 deficiency can manifest late due to its efficient storage and low requirements; therefore, postoperative annual screening for vitamin B12 deficiency is highly recommended to allow timely intervention even before symptom development [35]. Total serum vitamin B12 measurement has limited sensitivity and specificity as a biomarker of deficiency. A lowered serum holotranscobalamin (holoTC) concentration is the earliest marker of vitamin B12 deficiency, while methyl malonic acid (MMA) is a functional B12 marker that increases when the B12 stores are depleted [36]. Isolated lowering of holoTC shows B12 depletion (negative B12 balance), while lowered holoTC combined with raised MMA and homocysteine levels are indicative of metabolically manifest vitamin B12 deficiency; however, the patients can still be asymptomatic. The diagnostic use of MMA, homocysteine, and holoTC allows treatment to be instituted before clinical manifestation of B12 deficiency, including potentially irreversible neurological damage and hematological symptoms [31,32].

Following RYGB, oral supplementation at a dosage of 1000 µg or more is recommended to maintain normal vitamin B12 levels [35]. In the case of existing vitamin B12 deficiency, parenteral administration remains the preferred route, and the given dose can be calculated to take account of the fact that clinical symptoms usually only begin to manifest when body vitamin B12 stores are depleted to as little as 5–10% [37]. The aim of therapy must therefore be to fully compensate the deficit, and choose the dosage accordingly [13].

Folate Deficiency

Folate is absorbed along the entire length of small intestine by a specific pH-dependent, proton-dependent, carrier-mediated mechanism [38]. It plays a crucial role in the synthesis of purine and thymidine as well as in the metabolism of several amino acids. It is key for the disposal of homocysteine, elevated levels of which have been associated with vascular and neurodegenerative disease. Folate deficiency leads to clinical abnormalities ranging from megaloblastic anemia to intrauterine growth retardation and congenital (neural tube) defects [13].

Since with adequate food intake, folate is well absorbed throughout the small intestine, the occurrence of folate deficiency is predominantly due to decreased dietary intake rather than malabsorption, and can be easily corrected by oral vitamin supplementation [39]. The incidence of folate deficiency following bariatric surgery has fallen dramatically in countries like the United States where folic acid fortification of the food supply is practiced and is generally low in the face of widespread vitamin supplementation and nutritional support [40]. During pregnancy, folate requirements increase several-fold because of transfer of folate to the growing fetus. Women considering pregnancy following obesity surgery should have supplementation to ensure normal serum folate levels to minimize the risk for neural tube defects [41].

It should be kept in mind that since vitamin B12 plays an important role in the conversion of inactive methyltetrahydrofolic acid to active tetrahydrofolic acid, functional folate deficiency may arise as a consequence of vitamin B12 deficiency [42].

Copper Deficiency

Copper deficiency is a potentially important, but often overlooked, cause of anemia in patients following RYGB. Copper is a trace element that acts as an essential cofactor in many enzymatic reactions vital to the normal function of the hematological, vascular, skeletal, and neurological systems [43]. Copper deficiency is associated with a variety of hematological abnormalities including microcytic, hypochromic anemia, leukopenia, neutropenia, and pancytopenia [44], as well as neurological symptoms associated with demyelinating neuropathy, which is similar to the myeloneuropathy observed in vitamin B12 deficiency [45]. Gastric surgery seems to be an increasingly common cause of acquired copper deficiency that can go unnoticed for years [46]. RYGB reduces copper absorption by decreasing food exposure to gastric acid that is required for copper bioavailability, and also by excluding the duodenum and proximal jejunum that are the main sites for copper absorption [47]. High doses of zinc supplementation may also induce copper deficiency, potentially by upregulation of enterocyte metallothionein that binds copper and inhibits its absorption [47,48].

It has been increasingly recognized that copper deficiency may be associated with anemia after RYGB [22,46], which may manifest as pancytopenia with marked changes in red blood cells (Fig. 39.2) [46]. Copper deficiency hinders iron transport and utilization at several key points and impairs heme synthesis, resulting in sideroblastic anemia [4]. The mechanism of neutropenia is unknown, although arrested granulocytic maturation has been hypothesized [49].

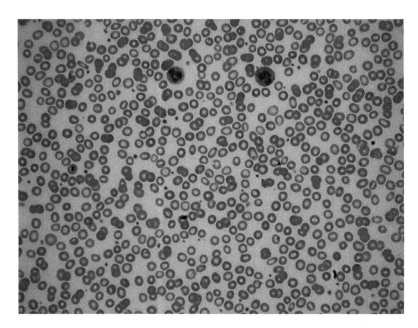

FIGURE 39.2   Microcytic hypochromic anemia in a patient with gastric bypass surgery. Peripheral blood smear of a patient with gastric bypass surgery demonstrates mild microcytic hypochromic anemia with increased anisopoikilocytosis of the red blood cells (Wright–Giemsa stain, ×400. Unpublished).

It is important to consider that copper deficiency may not occur until many years after bariatric surgery [44]. Although copper deficiency may be a relatively rare occurrence, it is also underreported in RYGB, since anemia induced by copper deficiency is often misinterpreted as iron or vitamin B12 deficiency anemia [22]. However, it should be considered in the differential diagnosis of anemia, neutropenia, and bone marrow dysplasia in patients after gastric bypass surgery, particularly when there is a concomitant neurological deficit [22]. Recent guidelines recommend that copper supplementation should be included as part of routine multivitamin and mineral supplementation following bariatric surgery [35]. Importantly, following copper supplementation, hematological abnormalities rapidly resolve, but the neurological deficits may be irreversible, making early diagnosis and prompt treatment crucial for successful outcomes [22].

OBESITY-MEDIATED INFLAMMATORY ANEMIA AFTER GASTRIC BYPASS SURGERY

Although nutrient deficiency is a prominent cause of anemia in patients after RYGB, it does not account for all cases. Many patients display persistent anemia after RYGB despite micronutrient supplementation and normal blood concentrations of iron, vitamin B12, folate, and copper, suggesting that other mechanisms may contribute to the development of anemia in these patients. Inflammation arising from various etiologies, including infection, autoimmune disorders, chronic diseases, and aging, can promote anemia [7,20]. The pathophysiology of this type of inflammatory anemia include disturbances of iron homeostasis, impaired proliferation of erythroid progenitor cells, and a blunted erythropoietin response to anemia [50].

Currently, adipose tissue is considered an endocrine and immune-modulating organ that plays important roles in the maintenance of energy homeostasis and the pathogenesis of obesity-related metabolic and inflammatory complications [51]. Obesity is characterized by chronic low-grade systemic inflammation, which is aggravated immediately after RYGB due to increased lipolysis and fat oxidation. The aggravated inflammation is associated with the expression and release of proinflammatory cytokines such as tumor necrosis factor-alpha and interferon-gamma, which have been shown to negatively affect erythropoiesis and iron homeostasis [52]. Moreover, proinflammatory cytokines may result in the release of hepcidin, a small peptide hormone acting as a key regulator of systemic iron homeostasis [53]. Hepcidin inhibits iron absorption in the enterocytes and sequesters iron in the macrophages, leading to increased iron stores in conjunction with hypoferremia, giving rise to the picture seen in the anemia of inflammation [53,54]. Hepcidin appears to block iron uptake in the duodenum and iron release from macrophages, thereby decreasing delivery of iron to erythroid precursors and diminishing response to oral iron therapy [53]. Anemia relating to the inflammatory state is also associated with alterations in erythrocyte physiology, including decreased red cell lifespan and impaired biological responsiveness to erythropoietin [55].

Changes in gut microbiota profiles may contribute to reduced host weight and adiposity after RYGB surgery, in particular switching the metabolism of microbes so that they burn fewer carbohydrates and more fat [56]. Bile acid and bacterial changes could affect the gut's communication with organs responsible for the glucose dysregulation that is associated with diabetes [57,58]. On the other hand, these changes can also induce further inflammatory dysregulation. Following RYGB, the disruption of neuronal, hormonal, and metabolic pathways can induce abnormal activation of inflammatory signaling pathways and increased cytokine production [59]. The crosstalk between chronic inflammation, metabolic disorders, and immune dysregulation on the one hand, and long-term sequelae of nutritional deficiency on the other, may lead to a complex and sustained suppression of normal hematopoiesis, which in turn leads to persistent anemia that often does not respond well to simple micronutrient supplementation and may require a distinct therapeutic approach (Fig. 39.3) [4].

In addition to anemia, RYGB is also associated with a generalized decrease of white blood cell (WBC) and platelet counts. Although these decreases of WBC and platelet counts do not appear to be clinically significant, they suggest that a generalized suppression of hematopoiesis might occur after RYGB [4,60].

BONE MARROW FINDINGS AFTER GASTRIC BYPASS SURGERY MIMICKING MYELODYSPLASTIC SYNDROME

Because the hematological complications following gastric bypass surgery can result in severe comorbidities, if anemia and single or multilineage cytopenias are present with no response to simple supplementation therapy, an early investigation by diagnostic bone marrow biopsy is warranted.

After gastric bypass surgery, the bone marrow cells exhibit dysplastic features that can be mistaken for a myelodysplastic syndrome (MDS), while no convincing evidence of clonal abnormality in marrow stem cells is detected [4]. The marrow is hypercellular in most cases with characteristic changes such as cytoplasmic vacuolization of both erythroid

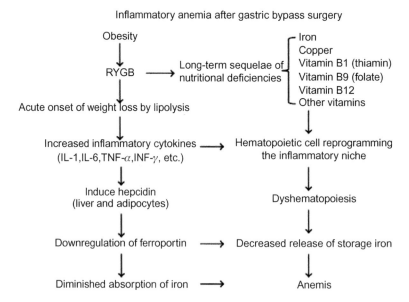

FIGURE 39.3 Mechanisms of obesity-mediated inflammatory anemia. Proposed mechanistic link between obesity and inflammatory anemia after gastric bypass surgery. *Chen M, et al. Hematological disorders following gastric bypass surgery: emerging concepts of the interplay between nutritional deficiency and inflammation. Biomed Res Int 2013;2013:205467, with permission.*

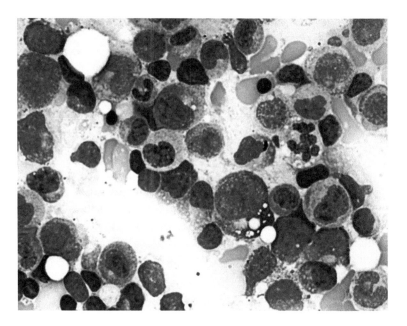

FIGURE 39.4 Bone marrow findings after gastric bypass surgery. Bone marrow aspirate smear of a patient with history of gastric bypass surgery. The marrow smear shows trilineage hematopoiesis. The erythroid precursors demonstrate megaloblastoid changes with prominent cytoplasmic vacuolization. Hypersegmented neutrophils are also present (Wright–Giemsa stain, × 1000. Unpublished).

and myeloid precursor cells and a marked left shift in granulocytic precursors with decreased numbers of late and terminally differentiated myeloid cells, giving rise to the appearance of a myeloid arrest. Iron stores are characteristically increased with prominent ringed sideroblasts (Fig. 39.4) [4,46].

There is an increasing recognition of copper deficiency associated with anemia following gastric bypass surgery, resulting in an MDS-like picture (Figs. 39.5 and 39.6) [46]. Patients often present with anemia, neutropenia, and less commonly, thrombocytopenia; and administration of copper results in the reversal of the dysplastic changes in these patients [46,61]. The pathogenesis of the copper deficiency-associated dysplastic process is complex and seemingly

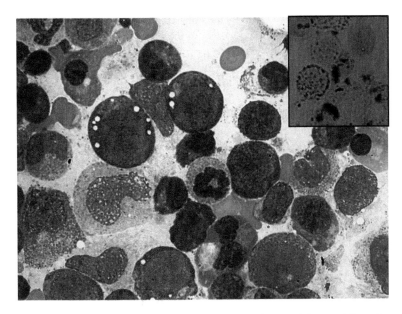

FIGURE 39.5   Bone marrow aspirate smear of a patient with gastric bypass surgery and copper deficiency. The aspirate smear demonstrates morphological changes in marrow precursors, some of which mimic a myelodysplastic syndrome. The early erythroid precursors and left-shifted immature myeloid cells have cytoplasmic vacuoles. There are scattered hypolobulated dyspoietic neutrophils and occasional reactive plasma cells. Insert shows increased cytoplasmic iron and ring sideroblasts (Wright–Giemsa stain, × 1000. Unpublished).

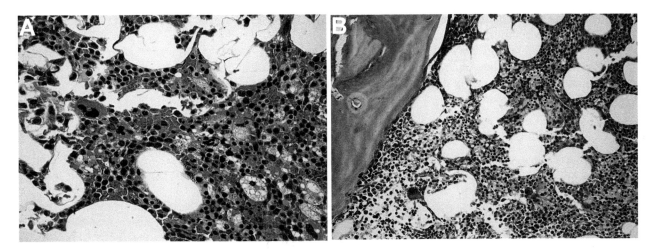

FIGURE 39.6   Bone marrow core biopsy specimen of a patient with gastric bypass surgery and copper deficiency. Stained bone marrow core biopsy specimens: (A) H&E stain shows hypercellular marrow for age with occasional dyspoietic megakaryocytes (× 200); and (B) PAS stain shows hypercellular marrow for age with occasional dyspoietic megakaryocytes (× 200) (Unpublished).

multifactorial. A decrease in the copper-dependent enzymes (ceruloplasmin and cytochromes) that aid in iron metabolism and transport has been proposed as a potential factor responsible for anemia [46]. The mechanism underlying neutropenia remains unknown. Suggested etiologies include inhibition of differentiation and self-renewal of CD34 + hematopoietic progenitor cells, destruction of myeloid progenitor cells, and impaired outflow of neutrophils from the bone marrow [4,46,62]. Early investigation of the etiology of persistent cytopenia(s) by diagnostic bone marrow biopsy is warranted to allow timely intervention before potentially irreversible clinical manifestations occur [4].

MINI-DICTIONARY OF TERMS

- *Anemia*: A quantitative deficiency of hemoglobin, often accompanied by a reduced number of red blood cells.
- *Nutritional anemia*: Types of anemia that can be directly attributed to nutritional disorders.

- *Inflammatory anemia or anemia of inflammation*: Anemia that commonly occurs with long-term inflammatory disorders.
- *Hypergastrinemia*: The presence of an excess of gastrin in the blood.
- *Myelodysplastic syndromes*: A group of diverse bone marrow disorders in which the bone marrow does not produce enough healthy blood cells.
- *Total iron-binding capacity*: A test that measures the blood's capacity to bind iron to transferrin.
- *Anisocytosis*: Red blood cells of unequal size.
- *Microcytic anemia*: Anemia characterized by small red blood cells.
- *Ring sideroblasts*: Erythroblasts with iron-loaded mitochondria visualized by Prussian blue staining.
- *Dyspoiesis*: Abnormal hematopoiesis.

KEY FACTS ABOUT SIDEROBLASTIC ANEMIA

- A group of blood disorders with failure to use iron in hemoglobin synthesis, despite adequate iron stores.
- May be hereditary or acquired.
- Normoblasts fail to use iron to synthesize hemoglobin.
- Iron is deposited in the mitochondria of normoblasts, giving a ringed appearance of iron granules around the nucleus.
- The primary damage that produces iron-laden mitochondria in sideroblastic anemia could produce a feedback loop of escalating mitochondrial injury.

SUMMARY POINTS

- Gastric bypass surgery is successful in inducing weight loss in morbidly obese individuals, but often is complicated by multifactorial nutritional anemia.
- Anemia after RYGB has been predominantly linked to iron deficiency, however, deficiencies of vitamin B12, folate, and copper also contribute.
- Obesity-mediated inflammatory anemia may occur following RYGB.
- In RYGB patients, the bone marrow cells exhibit dysplastic features that can be mistaken for a MDS.
- Long-term follow-up and early diagnosis is important to allow judicious intervention before potentially irreversible clinical manifestations occur.

REFERENCES

[1] Angrisani L, et al. Bariatric surgery Worldwide 2013. Obes Surg 2015;25(10):1822–32.
[2] Chang SH, et al. The effectiveness and risks of bariatric surgery: an updated systematic review and meta-analysis, 2003-2012. JAMA Surg 2014;149(3):275–87.
[3] Health Quality Ontario. Bariatric surgery: an evidence-based analysis. Ont Health Technol Assess Ser 2005;5(1):1–148.
[4] Chen M, et al. Hematological disorders following gastric bypass surgery: emerging concepts of the interplay between nutritional deficiency and inflammation. Biomed Res Int 2013;2013:205467.
[5] Adams TD, et al. Health benefits of gastric bypass surgery after 6 years. JAMA 2012;308(11):1122–31.
[6] Ikramuddin S, et al. Roux-en-Y gastric bypass vs intensive medical management for the control of type 2 diabetes, hypertension, and hyperlipidemia: the Diabetes Surgery Study randomized clinical trial. JAMA 2013;309(21):2240–9.
[7] Livingston EH. Bariatric surgery in the new millennium. Arch Surg 2007;142(10):919–22.
[8] Svane MS, Madsbad S. Bariatric surgery - effects on obesity and related co-morbidities. Curr Diabetes Rev 2014;10(3):208–14.
[9] Buchwald H, et al. Weight and type 2 diabetes after bariatric surgery: systematic review and meta-analysis. Am J Med 2009;122(3):248–256, e5.
[10] Sjostrom L, et al. Bariatric surgery and long-term cardiovascular events. JAMA 2012;307(1):56–65.
[11] Hewitt S, et al. Long-term improvements in pulmonary function 5 years after bariatric surgery. Obes Surg 2014;24(5):705–11.
[12] Bal B, et al. Managing medical and surgical disorders after divided Roux-en-Y gastric bypass surgery. Nat Rev Gastroenterol Hepatol 2010;7(6): 320–34.
[13] Stein J, et al. Review article: the nutritional and pharmacological consequences of obesity surgery. Aliment Pharmacol Ther 2014;40(6): 582–609.
[14] Del Villar Madrigal E, et al. Anemia after Roux-en-Y gastric bypass. How feasible to eliminate the risk by proper supplementation? Obes Surg 2015;25(1):80–4.
[15] Love AL, Billett HH. Obesity, bariatric surgery, and iron deficiency: true, true, true and related. Am J Hematol 2008;83(5):403–9.
[16] Cable CT, et al. Prevalence of anemia after Roux-en-Y gastric bypass surgery: what is the right number? Surg Obes Relat Dis 2011;7(2): 134–9.

[17] Karefylakis C, et al. Prevalence of anemia and related deficiencies 10 years after gastric bypass--a retrospective study. Obes Surg 2015;25(6): 1019−23.

[18] Poitou Bernert C, et al. Nutritional deficiency after gastric bypass: diagnosis, prevention and treatment. Diabetes Metab 2007;33(1):13−24.

[19] Bal BS, et al. Nutritional deficiencies after bariatric surgery. Nat Rev Endocrinol 2012;8(9):544−56.

[20] Green R. Anemias beyond B12 and iron deficiency: the buzz about other B's, elementary, and nonelementary problems. Hematol Am Soc Hematol Educ Program 2012;2012:492−8.

[21] Weng TC, et al. Anaemia and related nutrient deficiencies after Roux-en-Y gastric bypass surgery: a systematic review and meta-analysis. BMJ Open 2015;5(7):e006964.

[22] Griffith DP, et al. Acquired copper deficiency: a potentially serious and preventable complication following gastric bypass surgery. Obesity (Silver Spring) 2009;17(4):827−31.

[23] Ruz M, et al. Iron absorption and iron status are reduced after Roux-en-Y gastric bypass. Am J Clin Nutr 2009;90(3):527−32.

[24] Gesquiere I, et al. Iron deficiency after Roux-en-Y gastric bypass: insufficient iron absorption from oral iron supplements. Obes Surg 2014;24(1): 56−61.

[25] Hershko C, et al. Role of autoimmune gastritis, *Helicobacter pylori* and celiac disease in refractory or unexplained iron deficiency anemia. Haematologica 2005;90(5):585−95.

[26] Green R, King R. A new red cell discriminant incorporating volume dispersion for differentiating iron deficiency anemia from thalassemia minor. Blood Cells 1989;15(3):481−91, discussion 492−5.

[27] Ruz M, et al. Heme- and nonheme-iron absorption and iron status 12 mo after sleeve gastrectomy and Roux-en-Y gastric bypass in morbidly obese women. Am J Clin Nutr 2012;96(4):810−17.

[28] Brolin RE, et al. Prophylactic iron supplementation after Roux-en-Y gastric bypass: a prospective, double-blind, randomized study. Arch Surg 1998;133(7):740−4.

[29] Blume CA, et al. Nutritional profile of patients before and after Roux-en-Y gastric bypass: 3-year follow-up. Obes Surg 2012;22(11):1676−85.

[30] Dalcanale L, et al. Long-term nutritional outcome after gastric bypass. Obes Surg 2010;20(2):181−7.

[31] Stabler SP. Clinical practice. Vitamin B12 deficiency. N Engl J Med 2013;368(2):149−60.

[32] Herrmann W, Obeid R. Causes and early diagnosis of vitamin B12 deficiency. Dtsch Arztebl Int 2008;105(40):680−5.

[33] Herbert V, Das KC. The role of vitamin B12 and folic acid in hemato- and other cell-poiesis. Vitam Horm 1976;34:1−30.

[34] Obeid R, et al. The impact of vegetarianism on some haematological parameters. Eur J Haematol 2002;69(5−6):275−9.

[35] Mechanick JI, et al. Clinical practice guidelines for the perioperative nutritional, metabolic, and nonsurgical support of the bariatric surgery patient--2013 update: cosponsored by American Association of Clinical Endocrinologists, The Obesity Society, and American Society for Metabolic & Bariatric Surgery. Obesity (Silver Spring) 2013;21(Suppl. 1):S1−27.

[36] Herrmann W, et al. Vitamin B-12 status, particularly holotranscobalamin II and methylmalonic acid concentrations, and hyperhomocysteinemia in vegetarians. Am J Clin Nutr 2003;78(1):131−6.

[37] Dali-Youcef N, Andres E. An update on cobalamin deficiency in adults. QJM 2009;102(1):17−28.

[38] Said HM. Intestinal absorption of water-soluble vitamins in health and disease. Biochem J 2011;437(3):357−72.

[39] Toh SY, Zarshenas N, Jorgensen J. Prevalence of nutrient deficiencies in bariatric patients. Nutrition 2009;25(11−12):1150−6.

[40] Alvarez-Leite JI. Nutrient deficiencies secondary to bariatric surgery. Curr Opin Clin Nutr Metab Care 2004;7(5):569−75.

[41] American College of Obstetricians and Gynecologists. ACOG committee opinion no. 549: obesity in pregnancy. Obstet Gynecol 2013;121(1): 213−17.

[42] Shane B, Stokstad EL. Vitamin B12-folate interrelationships. Annu Rev Nutr 1985;5:115−41.

[43] Gabreyes AA, et al. Hypocupremia associated cytopenia and myelopathy: a national retrospective review. Eur J Haematol 2013;90(1):1−9.

[44] Lazarchick J. Update on anemia and neutropenia in copper deficiency. Curr Opin Hematol 2012;19(1):58−60.

[45] Gletsu-Miller N, et al. Incidence and prevalence of copper deficiency following Roux-en-Y gastric bypass surgery. Int J Obes (Lond) 2012;36(3): 328−35.

[46] Robinson SD, Cooper B, Leday TV. Copper deficiency (hypocupremia) and pancytopenia late after gastric bypass surgery. Proc (Bayl Univ Med Cent) 2013;26(4):382−6.

[47] Yarandi SS, et al. Optic neuropathy, myelopathy, anemia, and neutropenia caused by acquired copper deficiency after gastric bypass surgery. J Clin Gastroenterol 2014;48(10):862−5.

[48] Rowin J, Lewis SL. Copper deficiency myeloneuropathy and pancytopenia secondary to overuse of zinc supplementation. J Neurol Neurosurg Psychiatry 2005;76(5):750−1.

[49] Haddad AS, et al. Hypocupremia and bone marrow failure. Haematologica 2008;93(1):e1−5.

[50] Weiss G, Goodnough LT. Anemia of chronic disease. N Engl J Med 2005;352(10):1011−23.

[51] Wellen KE, Hotamisligil GS. Obesity-induced inflammatory changes in adipose tissue. J Clin Invest 2003;112(12):1785−8.

[52] Wessling-Resnick M. Iron homeostasis and the inflammatory response. Annu Rev Nutr 2010;30:105−22.

[53] Tussing-Humphreys L, et al. Rethinking iron regulation and assessment in iron deficiency, anemia of chronic disease, and obesity: introducing hepcidin. J Acad Nutr Diet 2012;112(3):391−400.

[54] Bekri S, et al. Increased adipose tissue expression of hepcidin in severe obesity is independent from diabetes and NASH. Gastroenterology 2006;131(3):788−96.

[55] Theurl I, et al. Regulation of iron homeostasis in anemia of chronic disease and iron deficiency anemia: diagnostic and therapeutic implications. Blood 2009;113(21):5277−86.

[56] Liou AP, et al. Conserved shifts in the gut microbiota due to gastric bypass reduce host weight and adiposity. Sci Transl Med 2013;5(178): 178ra41.

[57] Saeidi N, et al. Reprogramming of intestinal glucose metabolism and glycemic control in rats after gastric bypass. Science 2013;341(6144): 406–10.

[58] Hughes V. Weight-loss surgery: a gut-wrenching question. Nature 2014;511(7509):282–4.

[59] Faintuch J, et al. Increased gastric cytokine production after Roux-en-Y gastric bypass for morbid obesity. Arch Surg 2007;142(10):962–8.

[60] Dallal RM, Leighton J, Trang A. Analysis of leukopenia and anemia after gastric bypass surgery. Surg Obes Relat Dis 2012;8(2):164–8.

[61] Halfdanarson TR, et al. Hematological manifestations of copper deficiency: a retrospective review. Eur J Haematol 2008;80(6):523–31.

[62] Peled T, et al. Cellular copper content modulates differentiation and self-renewal in cultures of cord blood-derived CD34 + cells. Br J Haematol 2002;116(3):655–61.

Chapter 40

Enteric Hyperoxaluria, Calcium Oxalate Nephrolithiasis, and Oxalate Nephropathy After Roux-en-Y Gastric Bypass

V. Agrawal

University of Vermont College of Medicine, Burlington, VT, United States

LIST OF ABBREVIATIONS

ACEi	angiotensin-converting enzyme inhibitor
ARB	angiotensin receptor blocker
BMI	body mass index
CKD	chronic kidney disease
CaO$_x$	calcium oxalate
CaO$_x$ SS	calcium oxalate supersaturation
Cr	serum creatinine
DM	diabetes mellitus
eGFR	estimated glomerular filtration rate
ESRD	end-stage renal disease
JI bypass	jejunoileal bypass
NSAIDs	nonsteroidal antiinflammatory drugs
RYGB	Roux-en-Y gastric bypass
SG	sleeve gastrectomy

INTRODUCTION

Roux-en-Y gastric bypass (RYGB) is the most popular surgical therapy for morbid obesity (body mass index (BMI) > 40 kg/m^2) with more than 197,000 procedures performed worldwide in 2013 [1]. The profound weight loss after RYGB is associated with significant improvement in obesity-associated diabetes mellitus (DM), hypertension, hyperlipidemia, obstructive sleep apnea, and also mortality over 10-year follow-up [2]. While RYGB is a safe surgery [1], treating physicians need to be aware of uncommon adverse effects such as postoperative complications, anastomotic leak, and nutritional deficiencies, as discussed elsewhere in this textbook. In this chapter, I will review the adverse effects of RYGB on the kidney and specifically discuss the incidence, risk factors, clinical presentation, diagnosis, and management of oxalate-mediated kidney injury.

KIDNEY DISEASE DUE TO OBESITY

Obesity (BMI > 30 kg/m^2) causes injury to the glomeruli by exacerbating systemic and glomerular capillary blood pressure, hyperglycemia due to insulin resistance, an overactive renin angiotensin–aldosterone system, systemic inflammation, and oxidative stress [3]. Obesity-related glomerulopathy refers to the pathological changes in the glomeruli seen in obesity that resemble early diabetic nephropathy (glomerular enlargement, mild focal mesangial expansion, focally thickened glomerular basement membrane, and decreased podocyte density), though severe glomerular damage manifests as secondary focal segmental glomerulosclerosis [4]. The resulting glomerular damage may explain the fact that obesity is an independent, yet modifiable, risk factor for chronic kidney disease (CKD) progression to end-stage renal disease (ESRD) [5].

Medical therapies for weight loss (changes in diet, physical activity, or behavior) have not been successful in ameliorating obesity due to a high rate of nonadherence [6]. As surgical weight loss, such as that achieved by RYGB, is much greater and more durable than lifestyle changes, RYGB has a pivotal role in the treatment of obesity and its associated comorbidities [2]. In fact, numerous observational studies have noted significant improvement in albuminuria after RYGB, though its effect on long-term change in glomerular filtration rate is not evident [7]. While there is a possibility for renal benefits after RYGB, at this time there is no randomized trial comparing RYGB against medical or other surgical options on hard renal outcomes such as doubling of serum creatinine (Cr) or progression to end-stage kidney disease [8]. At the same time, these potential renal benefits need to be weighed against the risk of adverse effects of RYGB on the kidney, namely hyperoxaluria, oxalate nephropathy, and calcium oxalate (CaO_x) nephrolithiasis [9].

INCIDENCE OF RENAL COMPLICATIONS AFTER ROUX-en-Y GASTRIC BYPASS

Data on incidence of renal complications after RYGB are heterogeneous because of different study populations (stone formers versus not), time of follow-up after RYGB ($<$1 year vs later), and variable postoperative medication regimens (such as dose of calcium supplementation).

Enteric Hyperoxaluria

Enteric hyperoxaluria refers to high urine oxalate excretion ($\geq$45 mg/24 hours) as a result of increased intestinal absorption of oxalate. Enteric hyperoxaluria after RYGB is common, and is observed in 42−67% of adults at 1−3.5 years follow-up [10−12]. In one study, it was found that 90% of patients who developed hyperoxaluria after RYGB had normal urine oxalate excretion prior to the surgery [13]. While the degree of hyperoxaluria and the resulting renal complications are variable, severe hyperoxaluria ($\geq$100 mg/24 hours) has been noted among 23% of adults after RYGB, and greatly increases the risk of kidney injury [14,15].

Calcium Oxalate Nephrolithiasis

An increased risk of nephrolithiasis after RYGB has been noted with an incidence of 5.8−11.1% [10,16−18]. In a large case-control study, stone events occurred about two times more frequently among those who underwent RYGB as compared to obese controls (7.6% vs 4.6%, respectively) [19]. CaO_x is the most common type of kidney stone, and the first stone event occurs 2.2−3.6 years after RYGB [10,14]. It is important to note that among adults with a stone event after RYGB, about half of them had no prior history of nephrolithiasis—i.e., these were de novo stones [10,16,18]. Calcium oxalate supersaturation (CaO_x SS), the gold standard for assessing CaO_x stone risk in urology, [20] is found to increase significantly to 2.2−5.7 as soon as 6−12 months after RYGB [10,21]. Even higher CaO_x SS (8.8−12.1) has been reported over a long period of follow-up after RYGB [12,14].

Oxalate Nephropathy

Oxalate nephropathy refers to biopsy-proven, oxalate-mediated damage to the renal tubules. As its diagnosis hinges on performing a kidney biopsy, the incidence of oxalate nephropathy after RYGB is not clearly known. Case reports and series on oxalate nephropathy have appeared in the published literature and have been usually described in the presence of severe hyperoxaluria diagnosed at a median of 12 months after RYGB [22]. The incidence of oxalate-mediated renal tubular damage from a milder degree of hyperoxaluria among the adults who do not undertake a kidney biopsy is, thus, not known.

ETIOLOGY

Enteric Hyperoxaluria

Weight loss after RYGB is achieved by gastric restriction and intestinal malabsorption. Unfortunately, it is the fat malabsorption that plays a major role in hyperoxaluria after RYGB [23]. Food (containing fat and oxalate) that enters the gastric pouch quickly exits into the Roux limb, and then into the distal ileum. In this way, dietary fat bypasses the duodenum and jejunum, where pancreatic enzymes and bile salts normally break down fat into free fatty acids and allow its absorption into blood (Fig. 40.1A). Free fatty acids in the intestinal lumen bind to calcium (saponification), and this complex is excreted in feces. As a result, there is less calcium available to bind the oxalate in the intestine, and the unbound oxalate thus reaches the colon, where it is absorbed into blood without any hindrance (Fig. 40.1B), and eventually reaches the kidney tubules. This phenomenon is called "enteric" or "secondary" hyperoxaluria to differentiate it from "primary" hyperoxaluria, a genetic disorder of increased oxalate release by the liver. Thus, factors that enhance fat malabsorption, such as increased fat intake in diet or increased length of Roux limb [10], that permit very little binding of calcium to oxalate, lead to a greater degree of hyperoxaluria. In fact, this is the reason behind severe hyperoxaluria seen after jejunoileal (JI) bypass, an obsolete bariatric surgery, that caused intense fat malabsorption to induce weight loss.

Another factor proposed to cause hyperoxaluria after RYGB is the effect of bile salts on colonic epithelial cells. As there is a small length of terminal ileum beyond the biliopancreatic limb to allow enterohepatic circulation, the unabsorbed bile salts enter into the colon and stimulate the colonic epithelial cells to absorb oxalate by causing changes in epithelial tight junctions [24]. Decreased colonization with *Oxalobacter formigenes* in the intestine after RYGB has been proposed to be a risk factor for hyperoxaluria [25], but has not been definitively proven [26]. Because *Oxalobacter* is a member of the endogenous bacterial flora and uses intestinal oxalate for energy needs, decrease in *Oxalobacter* count due to perioperative antibiotics may increase the availability of oxalate for colonic absorption [27]. Whether alteration in gastrointestinal hormones such as peptide YY can promote oxalate absorption after RYGB by affecting intestinal transit time is not known [11].

Calcium Oxalate Nephrolithiasis

Low urine volume, increased urine oxalate, and low urine citrate after RYGB are the major factors contributing to the increased CaO_x SS and risk of CaO_x nephrolithiasis. It is to be noted that the factors contributing to increased CaO_x stone risk are different at various follow-up times after RYGB. In the early postoperative period (<6 months after RYGB), low urine volume observed as soon as 2−3 months after RYGB is the major driver of increased CaO_x SS [21], while at 1 year and beyond, hyperoxaluria increases the CaO_x stone risk [10]. Low 24-hour urine volume after RYGB is a result of decreased oral intake (because of gastric restriction) and gastrointestinal losses, but urine volume may improve over time due to gastric adaptation [21]. As citrate inhibits CaO_x crystallization in urine, low urine citrate is a risk factor for CaO_x stones. Low citrate after RYGB occurs due to low citrate intake, and also from increased citrate reabsorption by the renal tubules in order to compensate for bicarbonate losses in the form of diarrhea or steatorrhea [10]. It is to be noted that hypercalciuria is not usually seen after RYGB, likely because of vitamin D deficiency, decreased oral intake, and binding of calcium to fatty acids in the intestine, with less availability for absorption [21]. Thus, decreased urine calcium after RYGB does not contribute to increased CaO_x stone, and may even be protective of CaO_x stone formation [12].

Oxalate Nephropathy

In oxalate nephropathy, high urine oxalate injures the renal tubular epithelial cells, leading to chronic inflammatory changes in the tubules and adjoining renal interstitium. Low reserve of functioning nephrons, as seen in common causes of CKD such as DM, hypertension, or elderly age, may enhance the risk of oxalate nephropathy [22].

CLINICAL PRESENTATION AND DIAGNOSIS

Enteric Hyperoxaluria

Urine oxalate levels ranging between 67 [10] and 83 mg/24 hours [14] are reported to occur at 1 year after RYGB, though a significant rise in urine oxalate may be seen as early as 6 months after RYGB [21]. Patients with enteric hyperoxaluria are asymptomatic, but microscopic hematuria may be detected on urinalysis. Thus, unless suspected, hyperoxaluria can be missed by the treating physicians. 24-hour urine collection to measure oxalate is vital in diagnosing hyperoxaluria. As urine volume can be quite variable in the postoperative period, 24-hour urine creatinine measured simultaneous to the 24-hour urine oxalate can ensure adequate urine collection (adult 24-hour urine creatinine should be at least ≥ 800 mg in females, and ≥ 1000 mg in males) [28].

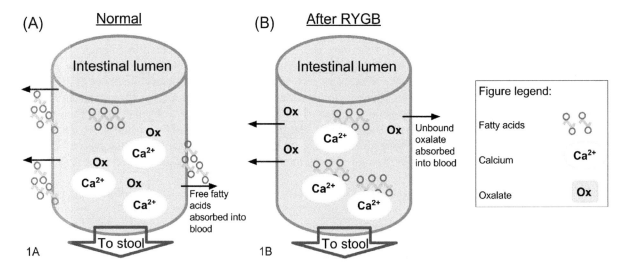

FIGURE 40.1 Schematic diagram of the mechanism of hyperoxaluria after RYGB. (A) Under normal conditions, calcium in the intestine binds to oxalate and prevents the absorption of oxalate. The calcium−oxalate complex is excreted in the stool. (B) After RYGB, excessive free fatty acids in the intestine bind to calcium and are excreted. This process allows the unbound oxalate to be absorbed unhindered into blood and leads to hyperoxaluria.

Calcium Oxalate Nephrolithiasis

CaO_x kidney stones present with flank pain, gross hematuria, urinary tract infection, or acute kidney injury (AKI) [29]. The stone composition is commonly CaO_x monohydrate or dihydrate, though <20% of stones may be CaO_x mixed with uric acid because of the hyperuricosuria seen in obesity [10]. Imaging of the kidney stones is done by spiral computerized tomography or ultrasonography. A 24-hour urine study to identify risk factors for CaO_x kidney stones usually reveals low urine volume, high oxalate, low citrate, low potassium, low magnesium, low sodium, [10,21], low urine pH (5−5.5) [14], and increased CaO_x SS. The 24-hour urine volume after RYGB is about 1.1−1.5 L/day [9,30], but may be as low as <1 L/day in 36% of adults in the first 6−12 months [13]. As alluded to earlier, hypocalciuria is usually seen after RYGB, but normal urine calcium may be seen with a moderate−high dose of calcium supplementation [23].

Oxalate Nephropathy

Oxalate nephropathy does not cause any symptoms and is only suspected when AKI (i.e., elevated Cr) is incidentally found on routine blood testing, usually at ≥6−12 months after RYGB, and that may range anywhere from 2.4 to 9.2 mg/dL [22]. It is to be noted that AKI in the early postoperative period (<6 months after RYGB) is less likely due to oxalate nephropathy, and prerenal causes such as gastrointestinal volume losses or use of nonsteroidal antiinflammatory drugs (NSAIDs), angiotensin-converting enzyme inhibitor (ACEi), or angiotensin receptor blocker (ARB) need to be considered [31]. As glomerular damage does not occur in oxalate nephropathy, blood or protein is not usually seen on urinalysis, though subnephrotic proteinuria (about 1.4 g/day) may rarely be seen, likely from tubular damage or preexisting glomerular disease [22]. Kidney ultrasound may demonstrate increased cortical echogenicity or kidney stones. A definitive diagnosis of oxalate nephropathy is made by kidney biopsy. In oxalate nephropathy, acute tubular injury and acute tubulointerstitial nephritis with lymphocytic inflammation of the renal interstitium, and occasionally foreign-body giant cell reaction, are observed (Fig. 40.2A) [22]. Oxalate is seen as crystalline material in the renal tubules and adjoining interstitium, and is highly polarizable (Fig. 40.2B). Glomeruli are not affected, but may show changes associated with obesity-related glomerulopathy or diabetic nephropathy. 24-hour urine studies always reveal hyperoxaluria [22] in oxalate nephropathy. It is important to make this diagnosis expeditiously as prolonged exposure to oxalate can cause permanent tubular damage, interstitial fibrosis, and rapid loss of kidney function [22]. In a case series of biopsy-proven oxalate nephropathy after RYGB ($n = 11$), eight patients (72.7%) quickly progressed to ESRD within 3.2 months [22].

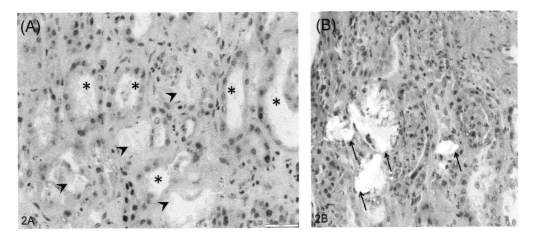

FIGURE 40.2   Kidney biopsy, light microscopy, hematoxylin, and eosin stain. (A) Standard illumination (original magnification, × 400) demonstrating tubular injury (marked with asterisks) and interstitial inflammation. Crystals were seen in lumen of many tubules (marked with arrowheads). (B) Under polarized light (original magnification, × 400), bright polarizable intratubular crystals (marked with arrows) were seen that were clear and colorless under bright light. *Courtesy of Pamela C. Gibson M.D., Department of Pathology, University of Vermont, Burlington, Vermont, USA.*

MEDICAL MANAGEMENT

Enteric Hyperoxaluria

Medical management of oxalate-mediated renal complications after RYGB is primarily directed at reducing urine oxalate (Table 40.1). Most of the following recommendations are not evidence-based, and are gathered from studies of CaO_x stone formers [29] or after JI bypass. As fat malabsorption drives the enteric hyperoxaluria, dietary changes such as low-fat and low-oxalate diet are commonly recommended to reduce urine oxalate [32]. Use of oxalate chelating agents in the intestine (such as calcium carbonate or calcium citrate) are effective in reducing hyperoxaluria, but need to be timed with every meal [15,29]. Cholestyramine, a bile acid sequestrant that can blunt the effect of bile salts on colonic oxalate transport, has been successful in reducing urine oxalate in small studies, but may need to be taken up to 4−6 times/day [33]. Probiotics such as *Oxalobacter* [34] and *Lactobacillus* sp. [27] have variable efficacy in reducing urine oxalate among CaO_x stone formers, but there is little available evidence for use in managing hyperoxaluria after RYGB, and dosage and timing of administration need to be further studied [27]. High dose of a vitamin C supplement exacerbates hyperoxaluria, and thus its intake should be limited to <100 mg/day after RYGB [35].

Calcium Oxalate Nephrolithiasis

Strategies to reduce CaO_x stone risk mainly include increasing urine volume, lowering urine oxalate, and increasing urine citrate. Patients with CaO_x stones after RYGB need to be educated regarding increasing fluid intake to achieve urine output >2−2.5 L/day. However, the restricted gastric pouch and fluid losses due to malabsorption make this a difficult task. Citrate supplementation, in the form of potassium citrate or potassium−magnesium citrate, serves to inhibit CaO_x crystallization in the urine, and acts as a CaO_x stone inhibitor. As citrate alkalinizes the urine, the urine pH needs to be carefully monitored to avoid increasing the urine to pH > 6.5, as that may precipitate calcium phosphate stones [29]. Unlike CaO_x stones in patients who have not undergone RYGB, there is limited utility of thiazide diuretics and low sodium diet (<100 mmol/day) in reducing CaO_x stone risk as hypercalciuria is not a major risk factor driving the increased CaO_x stone risk after RYGB. Furthermore, diuretics may cause volume depletion and worsen AKI. 24-hour urine studies need to be performed periodically to measure CaO_x stone risk factors and help guide institution of various preventive strategies at different time points to address these risk factors appropriately and reduce CaO_x SS. When a kidney stone event does occur, standard medical and surgical therapies need to be employed [29].

TABLE 40.1 Strategies to Reduce Urine Oxalate and Calcium Oxalate Stone Risk

Strategy	Mechanism of Action	Recommended Dose	Comments
Low fat	Reduces fat malabsorption and steatorrhea	<25% of total calories or <40 g/day	Patient may already be restricting fat intake due to limited oral intake (gastric restriction)
Low oxalate	Reduces oxalate absorption from the colon	<50 mg/day	Reducing dietary oxalate to <70–80 mg/dL is not likely to be practical in the long term
Bile acid sequestrant (cholestyramine)	Binds bile acids and salts in intestine and prevents their effect on stimulating oxalate transport in the colon	4 g 1–6 times/day with meals	May cause vitamin K deficiency and worsen diarrhea in some patients
Enteric oxalate-binding agent (calcium carbonate, calcium citrate)	Binds oxalate in intestine and prevents oxalate absorption into blood	1200–1600 mg/day elemental calcium with meals	Higher doses (2500–3000 mg/day) may help further reduce hyperoxaluria, but patients need to be monitored for hypercalciuria
Probiotics (*Oxalobacter*, *Lactobacillus* sp.)	Limits available oxalate for intestinal absorption either by using or degrading oxalate	3 doses/day	Efficacy is variable and may depend on dose and timing of administration
Increased fluid intake	Dilutes urine oxalate and factors that promote CaO$_x$ crystallization, thus reducing CaO$_x$ SS	Fluid intake to produce at least 2 L urine/day	Difficult to have high fluid intake, especially in the early postoperative period. Frequent small volumes of fluid may be better tolerated
Avoiding high animal protein intake	Reduces urine acidification by reducing phosphorus and sulfur-containing amino acids	0.8–1 g/kg/day	Patient may already be restricting protein intake due to limited oral intake (gastric restriction)
Increase citrate intake	Citrate is a stone inhibitor, alkalinizes urine, and reduces CaO$_x$ SS	Potassium citrate (40–60 mEq/day in two or three divided doses). Potassium–magnesium citrate or calcium citrate	Need to carefully monitor urine pH to avoid overalkalinization (urine pH > 6.5) that may increase risk of calcium phosphate kidney stones

Oxalate Nephropathy

Oxalate nephropathy is managed by treatment strategies aimed at reducing hyperoxaluria. Medical therapies may slow the decline in kidney function, but are unlikely to reverse the oxalate nephropathy. Physicians should ensure that the patient is not volume depleted by accounting for gastrointestinal volume losses, and should also evaluate for other causes of AKI such as exposure to nephrotoxic medications (NSAIDs, iodinated contrast) or urinary obstruction. There is no role for glucocorticoids in this form of acute interstitial nephritis.

SURGICAL MANAGEMENT

Surgical reversal of RYGB is performed for uncommon complications such as dumping syndrome or excessive weight loss. At present, there is only one case report by the author of a patient who was incidentally found to have AKI (Cr 2.1 mg/dL) 1 year after RYGB. Diagnostic workup revealed oxalate nephropathy on kidney biopsy and severe hyperoxaluria (urine oxalate 150 mg/day). Medical therapy failed to reduce urine oxalate, and progression of renal injury thus led to a consideration of reversal of RYGB. Within 1 month after surgical reversal, her hyperoxaluria resolved (urine oxalate 20 mg/day) and her oxalate nephropathy improved (Cr 1.7 mg/dL) [36]. Similar findings were noted in previous reports of reversal of JI bypass for oxalate nephropathy and CaO$_x$ stones [37]. Bariatric surgeons report that reversal of RYGB is safe and the few postoperative complications are manageable [38]. Weight gain and recurrence of obesity-associated disorders (such as diabetes) are expected consequences of reversal of

RYGB. In this situation, other surgical weight loss options such as vertical sleeve gastrectomy (SG) or gastric banding can be offered that may have less risk of causing hyperoxaluria, as these achieve weight loss through gastric restriction and not malabsorption [39].

PREVENTION

Physicians and surgeons who evaluate patients for RYGB should identify those at risk for renal complications such as prior stone formers or baseline CKD. This can be done by assessing Cr level, 24-hour urine oxalate at baseline, and renal ultrasound. In patients with risk for oxalate-mediated renal complications, vertical SG may need to be considered for surgical weight loss. In the perioperative period, nephrotoxic agents are to be avoided, while diuretics and ACEi/ARB should be held until the gastrointestinal volume losses subside. Empiric recommendations such as high fluid intake, low fat and oxalate diet, and calcium supplementation are reasonable for every patient after RYGB. Scheduled monitoring of Cr and 24-hour urine oxalate at least every 6 months for the first 2 years should be considered. If hyperoxaluria is detected, more frequent measurement of Cr and 24-hour urine oxalate may need to be done (such as every 3 months) to ensure effectiveness of the therapeutic measures.

FUTURE PERSPECTIVE

At present, Cr and prior history of kidney stones are the only pertinent data on renal health gathered at preoperative evaluation. It needs to be determined whether baseline patient factors—such as hyperoxaluria, increased CaO_x SS on 24-hour urine studies, or operative factors (such as planned length of Roux limb)—identify patients at high risk of renal complications after RYGB and influence decisions regarding the type of bariatric surgery. Urine oxalate measured in a 24-hour urine collection has variability ranging between 25% and 50%, and it is not known if two 24-hour urine studies are needed to diagnose hyperoxaluria [15]. Further studies should evaluate new potential therapies for hyperoxaluria such as preparations of oxalate decarboxylase or lyophilized *O. formigenes*. [40] Whether surgical reversal of RYGB in patients with oxalate-mediated renal complications, especially oxalate nephropathy, can thwart loss of kidney function needs to be researched. As obesity is the most important reason for patients with kidney failure being made ineligible for kidney transplantation [41], it needs to be evaluated whether RYGB or SG can be safely offered to these patients to without affecting immunosuppressant drug pharmacokinetics, metabolism, or survival of the renal allograft.

CONCLUSION

As RYGB continues to be offered to a growing number of morbidly obese adults worldwide, physicians need to weigh the risks versus benefits of this popular surgical weight loss procedure. Hyperoxaluria is a common adverse effect seen in about half of the patients after RYGB as a result of fat malabsorption. The risk of CaO_x kidney stones increases after RYGB, and is attributed mainly to low urine volume, high oxalate, and low citrate in the urine. Oxalate nephropathy is an uncommon, yet devastating, form of kidney injury after RYGB that often progresses to ESRD. Medical therapies for hyperoxaluria have limited efficacy, but nevertheless need to be promptly instituted to halt kidney damage. Surgeons, internists, and nephrologists must be aware of hyperoxaluria as a possible complication of RYGB, recognize signs of renal damage or stone disease, and take appropriate clinical measures to mitigate the renal complications of RYGB.

ACKNOWLEDGMENTS

I thank F. John Gennari, MD, Professor of Medicine Emeritus, University of Vermont College of Medicine, and Virginia L. Hood, MBBS, MPH, Professor of Medicine, University of Vermont College of Medicine, for their critical reviews of this chapter.

CONFLICTS OF INTEREST

None to declare.

MINI-DICTIONARY OF TERMS

- *Roux-en-Y Gastric Bypass (RYGB)*: Weight loss surgery that limits the size of the stomach and bypasses food absorption in the intestine.
- *Hyperoxaluria*: Increased excretion of oxalate in the urine.
- *Calcium Oxalate (CaOx) Nephrolithiasis*: Formation of CaO_x stones in the kidney.
- *Oxalate nephropathy*: Form of kidney injury due to high urine oxalate.
- *Acute Kidney Injury (AKI)*: Rapid decline in kidney function.
- *Acute tubular necrosis*: Abrupt reduction in kidney function due to damage to epithelial cells of renal tubules.
- *Acute tubulointerstitial nephritis*: Form of inflammatory injury of renal tubules and interstitium.
- *Fat malabsorption*: Failure of intestinal absorption of fat.
- *Calcium Oxalate supersaturation (CaOx SS)*: Measure of saturation of CaO_x to assess risk of forming stones.
- *End-stage renal disease (ESRD)*: Permanent and severe kidney damage.

KEY FACTS

- 24-hour urine volume decreases because of low oral intake and due to gastrointestinal losses (diarrhea or steatorrhea).
- Urine oxalate increases due to fat malabsorption and effect of bile salts on colonic epithelium. It is not clear whether decreased colonization with *Oxalobacter* contributes to hyperoxaluria.
- Urine citrate decreases due to low oral intake and increased reabsorption by the renal tubules to compensate for bicarbonate losses in stool.
- Urine calcium is either low or normal because of intraluminal binding of calcium to fatty acids in the intestine, and is due to vitamin D deficiency.
- Urine pH is low (acidic urine) due to chronic metabolic acidosis as a result of bicarbonate loss in stool.
- Urine sodium, potassium, and magnesium decrease due to low oral intake.

SUMMARY POINTS

- Enteric hyperoxaluria, calcium oxalate (CaO_x) nephrolithiasis, and oxalate nephropathy are potential renal complications of RYGB.
- Enteric hyperoxaluria is associated with fat malabsorption as a result of RYGB. Enteric hyperoxaluria affects approximately half of the adults postoperatively, and is usually evident at 1 year or more after RYGB.
- The incidence of CaO_x nephrolithiasis after RYGB is about two times higher than that of obese nonoperated controls, and may occur even among nonstone formers. Factors that increase CaO_x supersaturation (CaO_x SS) and the risk for kidney stones after RYGB include low urine volume, hyperoxaluria, and hypocitraturia.
- Oxalate nephropathy is a rare, yet fatal, form of kidney injury characterized by oxalate-mediated acute tubular injury and interstitial inflammation. Patients diagnosed with oxalate nephropathy after RYGB may experience rapid loss of kidney function.
- Dietary changes and medical therapies should be promptly initiated to reduce urine oxalate and address the lithogenic factors as appropriate.
- Physicians need to weigh the risks versus benefits of RYGB while considering this surgical procedure for the obese adult, and be cognizant of the possible renal complications.
- While not definitely proven, I recommend routine monitoring of serum creatinine (Cr) and 24-hour urine oxalate every 6 months at least for the first 2 years after RYGB. This protocol enables early diagnosis of any renal adverse effect and timely referral to a specialist (nephrologist or urologist).
- Patients at risk for oxalate complications after RYGB (such as those with prior history of kidney stones or preexisting CKD) should be considered for SG or adjustable gastric banding for surgical weight loss. From a renal perspective, these restrictive bariatric surgeries are likely safer than RYGB due to lack of fat malabsorption.

REFERENCES

[1] Angrisani L, Santonicola A, Iovino P, Formisano G, Buchwald H, Scopinaro N. Bariatric surgery worldwide 2013. Obes Surg 2015;25:1822−32.

[2] Sjöström L, Narbro K, Sjöström CD, Karason K, Larsson B, Wedel H, et al. Effects of bariatric surgery on mortality in Swedish obese subjects. N Engl J Med 2007;357:741−52.

[3] Agrawal V, Shah A, Rice C, Franklin BA, McCullough PA. Impact of treating the metabolic syndrome on chronic kidney disease. Nat Rev Nephrol 2009;5:520−8.

[4] Kambham N, Markowitz GS, Valeri AM, Lin J, D'Agati VD. Obesity-related glomerulopathy: an emerging epidemic. Kidney Int 2001;59:1498−509.

[5] Hsu C-Y, McCulloch CE, Iribarren C, Darbinian J, Go AS. Body mass index and risk for end-stage renal disease. Ann Intern Med 2006;144:21−8.

[6] Puterbaugh JS. The emperor's tailors: the failure of the medical weight loss paradigm and its causal role in the obesity of America. Diabetes Obes Metab 2009;11:557−70.

[7] Agrawal V, Khan I, Rai B, Krause KR, Chengelis DL, Zalesin KC, et al. The effect of weight loss after bariatric surgery on albuminuria. Clin Nephrol 2008;70:194−202.

[8] Friedman AN, Wolfe B. Is bariatric surgery an effective treatment for type II diabetic kidney disease? Clin J Am Soc Nephrol 2015. Available from: http://dx.doi.org/10.2215/CJN.07670715.

[9] Canales BK, Hatch M. Kidney stone incidence and metabolic urinary changes after modern bariatric surgery: review of clinical studies, experimental models, and prevention strategies. Surg Obes Relat Dis 2014;10:734−42.

[10] Sinha MK, Collazo-Clavell ML, Rule A, Milliner DS, Nelson W, Sarr MG, et al. Hyperoxaluric nephrolithiasis is a complication of Roux-en-Y gastric bypass surgery. Kidney Int 2007;72:100−7.

[11] Duffey BG, Alanee S, Pedro RN, Hinck B, Kriedberg C, Ikramuddin S, et al. Hyperoxaluria is a long-term consequence of Roux-en-Y Gastric bypass: a 2-year prospective longitudinal study. J Am Coll Surg 2010;211:8−15.

[12] Maalouf NM, Tondapu P, Guth ES, Livingston EH, Sakhaee K. Hypocitraturia and hyperoxaluria after Roux-en-Y gastric bypass surgery. J Urol 2010;183:1026−30.

[13] Park AM, Storm DW, Fulmer BR, Still CD, Wood GC, Hartle II JE. A prospective study of risk factors for nephrolithiasis after Roux-en-Y gastric bypass surgery. J Urol 2009;182:2334−9.

[14] Asplin JR, Coe FL. Hyperoxaluria in kidney stone formers treated with modern bariatric surgery. J Urol 2007;177:565−9.

[15] Patel BN, Passman CM, Fernandez A, Asplin JR, Coe FL, Kim SC, et al. Prevalence of hyperoxaluria after bariatric surgery. J Urol 2009;181:161−6.

[16] Durrani O, Morrisroe S, Jackman S, Averch T. Analysis of stone disease in morbidly obese patients undergoing gastric bypass surgery. J Endourol 2006;20:749−52.

[17] Lieske JC, Mehta RA, Milliner DS, Rule AD, Bergstralh EJ, Sarr MG. Kidney stones are common after bariatric surgery. Kidney Int 2015;87:839−45.

[18] Valezi AC, Fuganti PE, Junior JM, Delfino VD. Urinary evaluation after RYGBP: a lithogenic profile with early postoperative increase in the incidence of urolithiasis. Obes Surg 2013;23:1575−80.

[19] Matlaga BR, Shore AD, Magnuson T, Clark JM, Johns R, Makary MA. Effect of gastric bypass surgery on kidney stone disease. J Urol 2009;181:2573−7.

[20] Pak CYC, Adams-Huet B, Poindexter JR, Pearle MS, Peterson RD, Moe OW. Rapid Communication: relative effect of urinary calcium and oxalate on saturation of calcium oxalate. Kidney Int 2004;66:2032−7.

[21] Agrawal V, Liu XJ, Campfield T, Romanelli J, Enrique Silva J, Braden GL. Calcium oxalate supersaturation increases early after Roux-en-Y gastric bypass. Surg Obes Relat Dis 2014;10:88−94.

[22] Nasr SH, D'Agati VD, Said SM, Stokes MB, Largoza MV, Radhakrishnan J, et al. Oxalate nephropathy complicating Roux-en-Y Gastric Bypass: an underrecognized cause of irreversible renal failure. Clin J Am Soc Nephrol 2008;3:1676−83.

[23] Kumar R, Lieske JC, Collazo-Clavell ML, Sarr MG, Olson ER, Vrtiska TJ, et al. Fat malabsorption and increased intestinal oxalate absorption are common after Roux-en-Y gastric bypass surgery. Surgery 2011;149:654−61.

[24] Dobbins JW, Binder HJ. Effect of bile salts and fatty acids on the colonic absorption of oxalate. Gastroenterology 1976;70:1096−100.

[25] Allison MJ, Cook HM, Milne DB, Gallagher S, Clayman RV. Oxalate degradation by gastrointestinal bacteria from humans. J Nutr 1986;116:455−60.

[26] Froeder L, Arasaki CH, Malheiros CA, Baxmann AC, Heilberg IP. Response to dietary oxalate after bariatric surgery. Clin J Am Soc Nephrol 2012;7:2033−40.

[27] Lieske JC, Goldfarb DS, de Simone C, Regnier C. Use of a probiotic to decrease enteric hyperoxaluria. Kidney Int 2005;68:1244−9.

[28] Curhan GC, Willett WC, Speizer FE, Stampfer MJ. Twenty-four−hour urine chemistries and the risk of kidney stones among women and men. Kidney Int 2001;59:2290−8.

[29] Worcester EM, Coe FL. Clinical practice. Calcium kidney stones. N Engl J Med 2010;363:954–63.

[30] Wu JN, Craig J, Chamie K, Asplin J, Ali MR, Low RK. Urolithiasis risk factors in the bariatric population undergoing gastric bypass surgery. Surg Obes Relat Dis 2013;9:83–7.

[31] Thakar CV, Kharat V, Blanck S, Leonard AC. Acute kidney injury after gastric bypass surgery. Clin J Am Soc Nephrol 2007;2:426–30.

[32] Pang R, Linnes MP, O'Connor HM, Li X, Bergstralh E, Lieske JC. Controlled metabolic diet reduces calcium oxalate supersaturation but not oxalate excretion after bariatric surgery. Urology 2012;80:250–4.

[33] Smith LH, Fromm H, Hofmann AF. Acquired hyperoxaluria, nephrolithiasis, and intestinal disease. Description of a syndrome. N Engl J Med 1972;286:1371–5.

[34] Kaufman DW, Kelly JP, Curhan GC, Anderson TE, Dretler SP, Preminger GM, et al. *Oxalobacter formigenes* may reduce the risk of calcium oxalate kidney stones. J Am Soc Nephrol 2008;19:1197–203.

[35] Sunkara V, Pelkowski TD, Dreyfus D, Satoskar A. Acute kidney disease due to excessive vitamin C ingestion and remote Roux-en-Y gastric bypass surgery superimposed on CKD. Am J Kidney Dis 2015;66:721–4.

[36] Agrawal V, Wilfong JB, Rich CE, Gibson PC. Reversal of Gastric Bypass Resolves Hyperoxaluria and Improves Oxalate Nephropathy Secondary to Roux-en-Y Gastric Bypass, *Case Rep Nephrol Dial* 2016;**6**(3):114–9.

[37] Hassan I, Juncos LA, Milliner DS, Sarmiento JM, Sarr MG. Chronic renal failure secondary to oxalate nephropathy: a preventable complication after jejunoileal bypass. Mayo Clin Proc 2001;76:758–60.

[38] Dapri G, Cadière GB, Himpens J. Laparoscopic reconversion of Roux-en-Y gastric bypass to original anatomy: technique and preliminary outcomes. Obes Surg 2011;21:1289–95.

[39] Semins MJ, Asplin JR, Steele K, Assimos DG, Lingeman JE, Donahue S, et al. The effect of restrictive bariatric surgery on urinary stone risk factors. Urology 2010;76:826–9.

[40] Asplin JR. The management of patients with enteric hyperoxaluria. Urolithiasis 2016;44:33–43.

[41] Sachdeva M, Sunday S, Israel E, Varghese J, Rosen L, Bhaskaran M, et al. Obesity as a barrier to living kidney donation: a center-based analysis. Clin Transplant 2013;27:882–7.

Chapter 41

Thyroid Hormone Homeostasis in Weight Loss and Implications for Bariatric Surgery

L.B. Sweeney, G.M. Campos and F.S. Celi
Virginia Commonwealth University, Richmond, VA, United States

LIST OF ABBREVIATIONS

BAT   brown adipose tissue
FT3   free triiodothyronine
GLP-1   glucagon-like peptide 1
HPA   hypothalamus−pituitary−adrenal axis
HPT   hypothalamus−pituitary−thyroid axis
rT3   reverse T3
T3   triiodothyronine
T4   thyroxine
TH   thyroid hormone
TRH   thyrotropin-releasing hormone
TSH   thyroid stimulating hormone

INTRODUCTION

Obesity and bariatric surgery profoundly affect the thyroid hormone (TH) axis. This chapter highlights the following areas: (1) thyroid homeostasis and obesity; (2) changes in the hypothalamus−pituitary−thyroid (HPT) axis associated with bariatric surgery; (3) effects of bariatric surgery on underlying thyroid disease; (4) effects of micronutrient deficiency on thyroid function; and (5) effects of bariatric surgery on brown adipose tissue (BAT).

OBESITY, ENERGY EXPENDITURE, AND THYROID

Over time the energy balance is remarkably stable, and changes in body weight account [1] for a minimal fraction of the energy intake, indicating that several mechanisms interplay to maintain the energy balance [2]. The TH axis has a pivotal role in the energy balance, and TH action modulates energy expenditure directly (futile cycles and ion leaks) and indirectly by expanding cellular respiration capacity [3]. TH action is mediated by the interaction of triiodothyronine (T3), with its receptor isoforms and the transcription machinery. This complex system allows a fine regulation of the hormonal action at the end-organ target relatively independent of the serum TH levels. The peripheral metabolism of thyroxine (T4), the main product of the thyroid gland, into the active form T3, or into the metabolically inactive reverse T3 (rT3), represents an important prereceptor modulator of the hormonal action [4]. Indeed, T3 levels and the ratio T3/T4 are a good index of thyroid activity [5]. Fig. 41.1 and Table 41.1 illustrate the function and characteristics of the deiodinases.

While overt thyrotoxicosis is associated with weight loss, changes in weight associated with hypothyroidism are less evident, probably because of increase in peripheral conversion of TH resulting in a preservation of T3 levels [6]. The energy stores play an important role in modulating the thyroid axis via the secretion of leptin (afferent loop), which stimulates the release of thyrotropin-releasing hormone (TRH) from the hypothalamus [7]. In models of starvation, reduction in fat mass results in TH decrease, consistent with a compensatory mechanism [8]. Thus, the HPT feedback axis

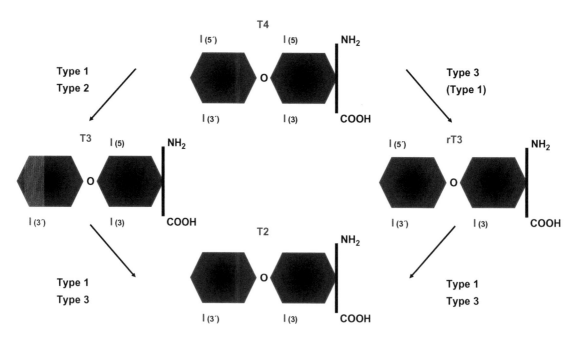

FIGURE 41.1 Thyroid hormone metabolism.T4, the main product of the thyroid, is converted into T3 by the activating deiodinases type-1 and -2, which remove an iodine from the outer ring (5′) of the molecule. The removal of an iodine from the inner ring (5) by type-3 (and partially by type-1) deiodinase catabolizes T4 into rT3.

TABLE 41.1 Deiodinase Characteristics

Deiodinase	Site of Action	Activity on Thyroid Hormone	Preferred Substrate	Tissue Localization	Effects of Weight Loss
Type 1	5′ and 5	rT3 → T2 T4 → T3	rT3 > T4 > T3	Liver, kidney, thyroid	Inhibition
Type 2	5′	T4 → T3	T4	Hypothalamus, pituitary, thyroid, BAT	Activation in hypothalamus, inhibition in thyroid
Type 3	5	T4 → rT3	T4	Placenta, hypoxic tissues	Possible activation

Biochemistry and distribution of the deiodinases.

should be expanded to the "HPT—fat" axis (Fig. 41.2). In humans the secretion of the thyroid is skewed toward T4, and a greater percentage of T3 derives from the peripheral conversion of TH, which is also directly affected by weight loss or gain [9] (Fig. 41.3).

ASPECTS OF BARIATRIC SURGERY PROCEDURES RELEVANT TO THYROID HOMEOSTASIS

The most common procedures are laparoscopic Roux-en-Y gastric bypass (LRYGB) and laparoscopic sleeve gastrectomy (LSG) [10], whose safety are similar to other common laparoscopic procedures [11]. LRYGB and LSG induce a reduction of the food bolus contact with gastric secretions, leading to decreased absorption of vitamin B12. However, the mechanisms are different [12]. After LRYGB, the small gastric pouch secretes very little acid [13], while the excluded stomach parietal cell mass may still be active, so that chyme does not mix with the food bolus until the midsmall bowel. After LSG, a large portion of the parietal cell mass is actually excised, leading to a reduction in gastric acid mixing with the food bolus. Use of acid-reducing medications (ARM) depends on procedure type, with LSG

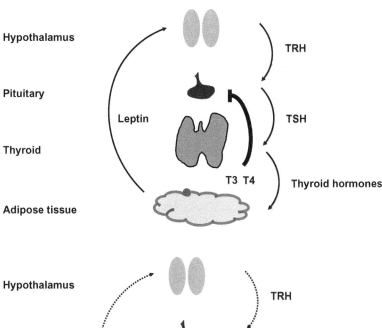

FIGURE 41.2 Energy status and hypothalamus—pituitary—thyroid (fat) axis. Obesity affects the thyroid axis via leptin, which stimulates the release of TRH, activating the secretion of thyroid hormone in a classical positive-feedback mechanism. Thyroid hormone inhibits the secretion of TSH (negative feedback).

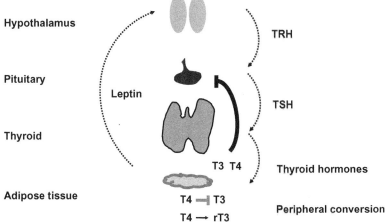

FIGURE 41.3 Weight loss and the hypothalamus—pituitary—thyroid (fat) axis. Weight loss profoundly affects thyroid homeostasis: the drop in leptin inhibits the release of TRH, depressing the entire thyroid axis. The catabolic state also inhibits the peripheral conversion of thyroid hormone, with a relative preservation of thyroxine (T4; substrate accumulation) sufficient to inhibit a compensatory rise in TSH.

subjects having a higher chance of using ARM [14]. Clinicians use ARM after LRYGB to prevent gastrojejunal ulcers, and after LSG for reflux symptoms [12,15]. Macro- and micronutrient deficiencies are common after bariatric surgery and can occur after both LSG and/or LRYGB [16]. Some studies have shown a higher rate of nutritional deficiencies with LRYGB [17], but recent publications have shown similar levels [18]. While initially it was believed that changes in gut, gastric, and pancreatic hormone levels, and in fasting bile acids, would be greater with LRYGB subjects [19], recent data suggest that these changes are similar for both procedures [20]. Collectively, both LRYGB and sleeve gastrectomy induce comparable changes in gut physiology that parallel reduced disease risk [19].

HYPOTHALAMUS—PITUITARY—THYROID AXIS IN OBESE AND IN POSTSURGICAL BARIATRIC PATIENTS

While overt hyperthyroidism causes significant weight loss, the weight gain associated with subclinical or overt hypothyroidism is minimal [21]. Recent studies suggest that thyroid dysfunction is more common among obese subjects. Michalaki et al. evaluated a cohort of 144 obese patients [22]. Of this group, 11.8% had hypothyroidism (defined as currently being treated with TH), 0.7% had subclinical hypothyroidism, and 7.7% had thyroid autoantibodies. The same author assessed thyroid function tests in a subset of 78 obese, euthyroid, and TPO-ab negative patients and matched controls. Obese subjects had increased thyroid-stimulating hormone (TSH) and TH levels (within the normal range). One should consider though, the potential of ascertainment bias—i.e., aggressive screening for perceived causes of weight gain, in this case hypothyroidism. In a cohort of healthy individuals with various degrees of adiposity, Aghniothri et al. demonstrated a direct correlation between fat mass and TSH and T3 [9], consistent with animal studies of overfeeding, which has been associated with increases in T3 and free triiodothyronine (FT3), and reductions in rT3 [23]. These changes reflect a leptin-mediated stimulation of the HPT axis, and an enhanced TH peripheral conversion.

TABLE 41.2 Remission of Subclinical Hypothyroidism After Bariatric Surgery

Authors	No. of Patients	Prevalence of Subclinical Hypothyroidism (%)	Follow-up (months)	Average BMI Reduction (kg/m^2)	Amelioration of Subclinical Hypothyroidism (%)
Chikunguwo et al.	86	10.5	12	17	100
Ruiz-Tovar et al.	60	16.7	12	15	83
Janssen et al.	71	14.1	12	14	87

Postsurgical resolution of subclinical hypothyroidism is a common occurrence.

These findings suggest a metabolic adaptation to positive energy balance as T3 augments lipogenesis and thermogenesis. Janssen et al. evaluated for TH changes in thyroxine-naïve subjects with subclinical hypothyroidism before and 12 months after RYGB [24]. While the baseline prevalence of subclinical hypothyroidism was 14.1% after RYGB, subclinical hypothyroidism resolved in 87% of subjects. Table 41.2 summarizes recent studies that have evaluated the influence of surgical weight loss on TH.

THYROID HORMONE ABSORPTION FOLLOWING BARIATRIC SURGERY

Interference with TH absorption is common following bariatric surgery. TH is absorbed across the small intestine and is excreted in the bile, generating an enterohepatic circle [25]. Factors that could influence absorption of TH after bariatric surgery include decrease in mucosal surface area, alteration in gastric pH, delayed gastric emptying, decreased gastric acid production, and bile acid modifications. Proton pump inhibitors are often employed following bariatric surgery procedures, and the reduction of the acidic environment, coupled with small frequent meals and use of supplements (iron, calcium), can contribute to a significant malabsorption of TH replacement therapy [26].

Pharmacokinetic studies to evaluate changes in TH absorption have yielded conflicting results. Rubio et al. performed TH absorption studies administering 600 mcg of oral levothyroxine in 15 patients 3 months post surgery and in 15 patients scheduled for RYGB [27]. Baseline TSH levels were higher in the presurgical group. Maximum ΔTT4 (differential between baseline and peak concentration), and area under the curve (AUC), were greater in the postsurgical group. The postsurgical group also had a significant delay in levothyroxine absorption. Gkostina et al. performed similar studies on patients before and 35 days after bariatric surgery [28]. Patients who underwent sleeve gastrectomy and biliopancreatic diversion with long limb displayed an increase in AUC for thyroxine, compared to no change in patients who underwent RYGB. The discrepancies among these studies could relate to multiple confounding factors, including differences in surgical procedures, TH status (within the normal range), length of bypassed intestine, and degree of excess weight loss achieved after surgery. Malabsorption of micronutrients can occur months to years after surgery, possibly due to bacterial overgrowth [29], suggesting the need to perform longitudinal studies. Studies in mice have demonstrated a four-times higher intestinal content of T4 than that of plasma concentration [30], and bariatric surgery is associated with changes in bile acid composition, which could affect the TH enterohepatic circulation. Finally, since the intestine participates in TH conversion of T4 to T3 [31], it is possible that intestinal TH metabolism is altered in bariatric surgery patients, secondary to the decrease in surface area or changes in the intestinal microenvironment. A practical aspect to consider is the potential for malabsorption secondary to concurrent therapy with cations such as calcium, iron, magnesium, and selenium. Patients should be encouraged to take their TH supplement first thing in the morning on an empty stomach, and delay the ingestion of the aforementioned supplements for several hours. In patients who appear to develop malabsorption of TH therapy after bariatric surgery, consideration can be given to substitution with a gel capsule or liquid TH preparation [26].

CHANGES IN ENTERO/ADIPO–INSULAR AXIS IN POSTSURGICAL PATIENTS AND THYROID FUNCTION

TH and leptin are crucial regulators of energy balance, food intake, and substrate utilization [32]. Bariatric surgery results in decreased levels of leptin, which correlate with lower insulin levels [33]. Ghrelin promotes release of growth hormone, regulates food intake, and decreases TH concentrations in humans [34]. RYGB is associated with persistently diminished ghrelin levels, which may account for the sustained weight loss observed in these patients [33]. Ghrelin modulates the HPT axis at multiple levels, particularly in the hypothalamus. Downregulation of the HPT axis may thus result from ghrelin-mediated stimulation of the hypothalamic—pituitary—adrenal axis [35]. Bariatric surgery is associated with a significant and sustained increase in glucagon-like peptide 1 (GLP-1) levels [33]. There is emerging concern that increased GLP-1 receptor activity may increase the risk for development of c-cell (medullary) thyroid cancers [36]. Additionally, GLP-1 levels have been associated with papillary thyroid cancer tumor multifocality [37].

BARIATRIC SURGERY-INDUCED NUTRIENT DEFICIENCIES AND THYROID

Bariatric surgery may result in deficiencies of important micronutrients, in particular, calcium, vitamins (A, D, and B12), folate, iron, thiamine, selenium, and zinc [38], which may go unrecognized in up to 50% of patients since symptoms may initially be subtle and progress over time. Of note, selenium, thiamine, and zinc are regulators of thyroid function. Selenium is an important modulator of thyroid function since it is a regulator of thyroid deiodinases, thioredoxin reductases, and the sodium—iodine symporter [39]. Selenium deficiency has been associated with autoimmune thyroid disease, goiter, and thyroid dysfunction [40]. Winther et al. evaluated the effects of selenium supplementation on thyroid function in a Danish population [41]. Serum TSH and T4 concentrations decreased significantly and dose-dependently. This may reflect selenium-mediated, enhanced deiodinase function. Zinc deficiency has been associated with TSH elevation [42] and impairment of TH conversion [43]. Matsuda observed an association between zinc deficiency and reduced T3 levels, which was reversed with oral zinc supplementation [44]. Conversely, excess of zinc may suppress thyroid function; hypothyroidism in rats is associated with diminished intestinal zinc absorption [45]. Thiamine (vitamin B1) deficiency has been observed in Hashimoto's and Graves' diseases [46]. This is possibly due to concurrent malabsorption. Costantini et al. administered thiamine to three patients with Hashimoto's disease who had persistent fatigue despite adequate TH supplementation. An 8-week course of thiamine resulted in resolution of fatigue among these patients [46]. Table 41.2 summarizes the micronutrient deficiencies, which may impact the thyroid.

THE INTESTINAL MICROBIOME

The intestinal neuroendocrine cells play a critical role in the regulation of nutrient absorption, gut motility, and secretion. These cells also regulate food intake through stimulation of neural and humoral pathways, in particular, the endocannabinoid system [47]. Both intestinal epithelial and neuroendocrine cells are influenced by the number, composition, and relative proportion of the gut microbiota. The human intestine harbors more than 100 trillion bacteria and archaea, composed of over 1000 different species. Current research suggests this "ecosystem" elicits variable effects on human physiological processes based on the proportion and balance of four key phyla: Proteobacteria, Firmicutes, Actinobacteria, and Bacteroidetes [48]. Graessler et al. reported a decrease in Clostridium bacteria (phylum Firmicutes) and an increase in Enterobacteriaceae (phylum Proteobacteria) in patients who had undergone RYGB as compared to controls [49]. This finding is consistent with other studies that noted a postsurgical decrease in Firmicutes and an increase in Enterobacteriaceae and Proteobacteria. In general, higher proportions of Firmicutes have been associated with increased systemic inflammation, which is common in obese individuals. It is currently unknown whether weight loss modulates thyroid autoimmunity. The gut microflora in obesity has been termed "intestinal dysbiosis" and represents a state of enhanced endocannabinoid "tone" [50]. Activation of the endocannabinoid system negatively regulates TSH secretion, and cannabinoid receptor antagonism results in increased TSH release. TH may also be an important regulator of intestinal function and microbiome composition [51,52]. TH promotes tight junctions between cells and may decrease epithelial permeability. A leaky intestinal epithelium contributes to inflammation in states of overweight or obesity. Moreover, TH enhances gut absorption of micronutrients. Therefore, the aggressive screening for and treatment of overt hypothyroidism in bariatric surgery patients appears indicated (Table 41.3).

BROWN FAT, THYROID, AND METABOLISM

BAT has become the target of intense research owing to the discovery of its presence in adult humans [53—55] and its ability to dissipate energy. Adrenergic stimulation promotes the cAMP signal transduction pathway, which drives type-2 deiodinase; this generates intracellular hyperthyroidism, promoting the transformation of substrate stores (fat) into heat [56]. Some data also suggest a direct stimulation of BAT activity by cold [57]. In humans, nonshivering thermogenesis generates an increase in energy expenditure of approximately 15% [58,59]. This moderate response does not support the use of this tissue as a treatment of obesity, but it is a promising target for improving carbohydrate

TABLE 41.3 Micronutrient Deficiencies Associated With Alterations in Thyroid Status

Micronutrient	Potential Benefits of Supplementation	ADA	Manifestations of Deficiency
Zinc	Increase in T4 to T3 conversion	8–11 mg/day	Folate deficiency, anemia, neuropsychiatric, hair loss, dermatitis
Selenium	Decrease in goiter volume, decrease in TPO-ab, increased NIS activity, increased T4 to T3 conversion	55 mcg/day	Fatigue and menstrual irregularities
Thiamine	Decrease in thyroid autoimmunity, amelioration of symptoms of hypothyroidism	1.2 mg/day	Hyperemesis, Wernike's Encephalitis

Thyroid function, aside from iodine, is heavily affected by micronutrients.

metabolism [60,61], and potentially for facilitating weight loss maintenance. BAT activity is inversely proportional to adiposity [62]. Hence, obese individuals, by virtue of their increased thermal insulation provided by the subcutaneous adipose tissue depots, tend to have lesser needs for thermogenesis [56]. Interestingly, an increase in BAT activity was observed following bariatric surgery [63,64]. A potential driver of this phenomenon is the increase in serum levels of bile acids, which promote BAT expansion and activity [65]. The actual contribution of BAT to the pathogenesis of obesity and to the protection against weight regain following bariatric surgery is under investigation.

CONCLUSIONS

Obesity results in derangements of key mediators, including gut- and adipose-derived hormones, intestinal microflora, and the endocannabinoid system, which influence the thyroid axis. The proinflammatory state of obesity may inhibit TH action and peripheral conversion, thereby resulting in a decrease in energy expenditure. Much of our understanding of these mechanisms has evolved from research on overfeeding and underfeeding. The decrease in fat mass regulates the HPT axis, favoring energy homeostasis. Dysregulation of this adaptive response could ultimately antagonize loss of excess body weight. Aside from weight loss, bariatric surgery results in anatomical and physiological changes to the intestinal microenvironment. Alterations in the intestinal surface area, motility, bacterial content, bile acid composition, and gut- and adipose-derived hormones, may further affect the HPT axis and TH action. Changes in the intestinal microenvironment may result in deficiencies of key nutrients that regulate thyroid function. Changes in gut-derived hormones may affect the HPT axis, influencing weight loss. Lastly, postsurgical changes in bile acid composition and dynamics may regulate BAT activity and energy expenditure.

MINI-DICTIONARY OF TERMS

- *C-Cells*: Neuroendocrine cells dispersed in the thyroid parenchyma.
- *Deiodinases*: Enzymes that convert the prohormone T4 into T3 or rT3.
- *Hypothalamus–pituitary–thyroid–(fat) axis*: Regulation of TH secretion and action; modulated by energy stores.
- *Multifocal papillary thyroid cancer*: This form of cancer, occasionally familiar, is not associated with a more aggressive course.
- *Nonshivering thermogenesis*: Shunting of energy from substrate (fat, glucose) to heat; driven by the uncoupling protein-1 in brown fat.
- *rT3*: Inactive metabolite of TH.
- *Selenium*: Trace element present in the catalytic domain of the deiodinase enzymes.
- *Subclinical hypothyroidism*: Elevation of TSH with TH levels within normal range.
- *Thyroid hormone absorption*: The acidic stomach environment promotes this process, and calcium and iron inhibit it.
- *Thyroid peroxidase antibodies*: Markers of thyroid autoimmunity.

KEY FACTS

- TH is a critical modulator of energy expenditure.
- The energy status affects the thyroid axis via leptin signaling.
- TH peripheral conversion is affected by the individual's energy stores.
- Subclinical and overt hypothyroidism are likely more common in obese patients.
- Subclinical hypothyroidism may resolve after bariatric surgery.
- Alterations in TH levels in obese subjects likely reflect adaptive physiological responses.
- Treatment of subclinical or overt hypothyroidism exerts moderate effects on fat mass.
- Evidence is currently insufficient to suggest a causal link between bariatric surgery and TH malabsorption.
- TH malabsorption can occur in the setting of cation supplements (calcium, iron, selenium, and magnesium).
- Bariatric surgery may induce changes in gut conversion of T4 to T3.
- Bariatric surgery may induce changes in the TH intestinal storage pool and TH enterohepatic circulation.
- Ghrelin exerts a modulatory effect on the HPT axis.
- The decrease in ghrelin levels following bariatric surgery may contribute to the normalization of TSH levels in patients with subclinical hypothyroidism.
- Bariatric surgery-induced augmentation of GLP-1 could theoretically result in a risk factor for thyroid cancer, but empirical data are lacking.
- Bariatric surgery is associated with micronutrient deficiencies that may affect the HPT axis.
- Intestinal dysbiosis is associated with obesity and may contribute to the development of autoimmune thyroid disease.
- Intestinal dysbiosis may affect the HPT axis through the endocannabinoid system.
- Alterations in TH levels after bariatric surgery may reflect surgically induced changes in the composition of gut microflora.
- TH may promote absorption of micronutrients and enhance gut epithelial barrier function.
- Brown adipose tissue activity is inversely correlated with adiposity.
- BAT stimulation results in a time- and tissue-specific intracellular increase in T3 due to activation of type-2 deiodinase.
- Bariatric surgery results in an increase in BAT, likely secondary to loss of subcutaneous fat, and possibly to changes in bile acid dynamics.

SUMMARY POINTS

- TH homeostasis reflects the energy stores.
- Weight loss induces a suppression of the HPT axis.
- TH absorption is affected by bariatric surgery.
- Medications interfere with TH absorption.
- The HPT axis and TH action are affected at multiple levels by bariatric surgery procedures.

REFERENCES

[1] Wang YC, Colditz GA, Kuntz KM. Forecasting the obesity epidemic in the aging U.S. population. Obesity (Silver Spring) 2007;15:2855.

[2] Hall KD, Sacks G, Chandramohan D, Chow CC, Wang YC, Gortmaker SL, et al. Quantification of the effect of energy imbalance on body-weight. Lancet 2011;378:826.

[3] Vaitkus JA, Farrar JS, Celi FS. Thyroid Hormone Mediated Modulation of Energy Expenditure. Int J Mol Sci 2015;16:16158.

[4] Gereben B, Zavacki AM, Ribich S, Kim BW, Huang SA, Simonides WS, et al. Cellular and molecular basis of deiodinase-regulated thyroid hormone signaling. Endocr Rev 2008;29:898.

[5] Amino N, Yabu Y, Miyai K, Fujie T, Azukizawa M, Onishi T, et al. Differentiation of thyrotoxicosis induced by thyroid destruction from Graves' disease. Lancet 1978;2:344.

[6] Abdalla SM, Bianco AC. Defending plasma T3 is a biological priority. Clin Endocrinol (Oxf) 2014;81:633.

[7] Nillni EA, Vaslet C, Harris M, Hollenberg A, Bjorbak C, Flier JS. Leptin regulates prothyrotropin-releasing hormone biosynthesis. Evidence for direct and indirect pathways. J Biol Chem 2000;275:36124.

[8] Legradi G, Emerson CH, Ahima RS, Flier JS, Lechan RM. Leptin prevents fasting-induced suppression of prothyrotropin-releasing hormone messenger ribonucleic acid in neurons of the hypothalamic paraventricular nucleus. Endocrinology 1997;138:2569.

[9] Agnihothri RV, Courville AB, Linderman JD, Smith S, Brychta R, Remaley A, et al. Moderate weight loss is sufficient to affect thyroid hormone homeostasis and inhibit its peripheral conversion. Thyroid 2014;24:19.

[10] Angrisani L, Santonicola A, Iovino P, Formisano G, Buchwald H, Scopinaro N. Bariatric Surgery Worldwide 2013. Obes Surg 2015;25:1822.

[11] Aminian A, Brethauer SA, Kirwan JP, Kashyap SR, Burguera B, Schauer PR. How safe is metabolic/diabetes surgery? Diabetes Obes Metab 2015;17:198.

[12] Ying VW, Kim SH, Khan KJ, Farrokhyar F, D'Souza J, Gmora S, et al. Prophylactic PPI help reduce marginal ulcers after gastric bypass surgery: a systematic review and meta-analysis of cohort studies. Surg Endosc 2015;29:1018.

[13] Smith CD, Herkes SB, Behrns KE, Fairbanks VF, Kelly KA, Sarr MG. Gastric acid secretion and vitamin B12 absorption after vertical Roux-en-Y gastric bypass for morbid obesity. Ann Surg 1993;218:91.

[14] Varban OA, Hawasli AA, Carlin AM, Genaw JA, English W, Dimick JB, et al. Variation in utilization of acid-reducing medication at 1 year following bariatric surgery: results from the Michigan Bariatric Surgery Collaborative. Surg Obes Relat Dis 2015;11:222.

[15] Schigt A, Coblijn U, Lagarde S, Kuiken S, Scholten P, van Wagensveld B. Is esophagogastroduodenoscopy before Roux-en-Y gastric bypass or sleeve gastrectomy mandatory? Surg Obes Relat Dis 2014;10:411.

[16] Shankar P, Boylan M, Sriram K. Micronutrient deficiencies after bariatric surgery. Nutrition 2010;26:1031.

[17] Toh SY, Zarshenas N, Jorgensen J. Prevalence of nutrient deficiencies in bariatric patients. Nutrition 2009;25:1150.

[18] Verger EO, Aron-Wisnewsky J, Dao MC, Kayser BD, Oppert JM, Bouillot JL, et al. Micronutrient and protein deficiencies after gastric bypass and sleeve gastrectomy: a 1-year follow-up. Obes Surg 2015. Available from: http://dx.doi.org/10.1007/s11695-015-1803-7.

[19] Malin SK, Kashyap SR. Differences in weight loss and gut hormones: Rouen-Y Gastric bypass and sleeve gastrectomy surgery. Curr Obes Rep 2015;4:279.

[20] Malin SK, Samat A, Wolski K, Abood B, Pothier CE, Bhatt DL, et al. Improved acylated ghrelin suppression at 2 years in obese patients with type 2 diabetes: effects of bariatric surgery vs standard medical therapy. Int J Obes (Lond) 2014;38:364.

[21] Portmann L, Giusti V. Obesity and hypothyroidism: myth or reality? Rev Med Suisse 2007;3:859.

[22] Michalaki MA, Vagenakis AG, Leonardou AS, Argentou MN, Habeos IG, Makri MG, et al. Thyroid function in humans with morbid obesity. Thyroid 2006;16:73.

[23] Danforth Jr. E, Horton ES, O'Connell M, Sims EA, Burger AG, Ingbar SH, et al. Dietary-induced alterations in thyroid hormone metabolism during overnutrition. J Clin Invest 1979;64:1336.

[24] Janssen IM, Homan J, Schijns W, Betzel B, Aarts EO, Berends FJ, et al. Subclinical hypothyroidism and its relation to obesity in patients before and after Roux-en-Y gastric bypass. Surg Obes Relat Dis 2015;11:1257.

[25] Visser TJ, Peeters RP. Metabolism of thyroid hormone. In: De Groot LJ, et al. editors. Endotext. South Dartmouth, MA: MDText.com, Inc.; 2005.

[26] Vita R, Saraceno G, Trimarchi F, Benvenga S. Switching levothyroxine from the tablet to the oral solution formulation corrects the impaired absorption of levothyroxine induced by proton-pump inhibitors. J Clin Endocrinol Metab 2014;99:4481.

[27] Rubio IG, Galrao AL, Santo MA, Zanini AC, Medeiros-Neto G. Levothyroxine absorption in morbidly obese patients before and after Roux-En-Y gastric bypass (RYGB) surgery. Obes Surg 2012;22:253.

[28] Gkotsina M, Michalaki M, Mamali I, Markantes G, Sakellaropoulos GC, Kalfarentzos F, et al. Improved levothyroxine pharmacokinetics after bariatric surgery. Thyroid 2013;23:414.

[29] Ishida RK, Faintuch J, Paula AM, Risttori CA, Silva SN, Gomes ES, et al. Microbial flora of the stomach after gastric bypass for morbid obesity. Obes Surg 2007;17:752.

[30] DiStefano 3rd JJ, Sternlicht M, Harris DR. Rat enterohepatic circulation and intestinal distribution of enterally infused thyroid hormones. Endocrinology 1988;123:2526.

[31] Hays MT. Thyroid hormone and the gut. Endocr Res 1988;14:203.

[32] Kim KJ, Kim BY, Mok JO, Kim CH, Kang SK, Jung CH. Serum concentrations of ghrelin and leptin according to thyroid hormone condition, and their correlations with insulin resistance. Endocrinol Metab (Seoul) 2015;30:318.

[33] Major P, Matlok M, Pedziwiatr M, Migaczewski M, Zub-Pokrowiecka A, Radkowiak D, et al. Changes in levels of selected incretins and appetite-controlling hormones following surgical treatment for morbid obesity. Wideochir Inne Tech Maloinwazyjne 2015;10:458.

[34] Kordi F, Khazali H. The effect of ghrelin and estradiol on mean concentration of thyroid hormones. Int J Endocrinol Metab 2015;13:e17988.

[35] Emami A, Nazem R, Hedayati M. Is association between thyroid hormones and gut peptides, ghrelin and obestatin, able to suggest new regulatory relation between the HPT axis and gut? Regul Pept 2014;189:17.

[36] Chiu WY, Shih SR, Tseng CH. A review on the association between glucagon-like peptide-1 receptor agonists and thyroid cancer. Exp Diabetes Res 2012;2012:924168.

[37] Jung MJ, Kwon SK. Expression of glucagon-like Peptide-1 receptor in papillary thyroid carcinoma and its clinicopathologic significance. Endocrinol Metab (Seoul) 2014;29:536.

[38] Stein J, Stier C, Raab H, Weiner R. Review article: the nutritional and pharmacological consequences of obesity surgery. Aliment Pharmacol Ther 2014;40:582.

[39] Leoni SG, Sastre-Perona A, De la Vieja A, Santisteban P. Selenium increases TSH-induced sodium/iodide symporter expression through Txn/Ape1-dependent regulation of Pax8 binding activity. Antioxid Redox Signal 2016;24:855.

[40] Toulis KA, Anastasilakis AD, Tzellos TG, Goulis DG, Kouvelas D. Selenium supplementation in the treatment of Hashimoto's thyroiditis: a systematic review and a meta-analysis. Thyroid 2010;20:1163.

[41] Winther KH, Bonnema SJ, Cold F, Debrabant B, Nybo M, Cold S, et al. Does selenium supplementation affect thyroid function? Results from a randomized, controlled, double-blinded trial in a Danish population. Eur J Endocrinol 2015;172:657.

[42] Alvarez-Salas E, Alcantara-Alonso V, Matamoros-Trejo G, Vargas MA, Morales-Mulia M, de Gortari P. Mediobasal hypothalamic and adenohypophyseal TRH-degrading enzyme (PPII) is down-regulated by zinc deficiency. Int J Dev Neurosci 2015;46:115.

[43] Olivieri O, Girelli D, Stanzial AM, Rossi L, Bassi A, Corrocher R. Selenium, zinc, and thyroid hormones in healthy subjects: low T3/T4 ratio in the elderly is related to impaired selenium status. Biol Trace Elem Res 1996;51:31.

[44] Nishiyama S, Futagoishi-Suginohara Y, Matsukura M, Nakamura T, Higashi A, Shinohara M, et al. Zinc supplementation alters thyroid hormone metabolism in disabled patients with zinc deficiency. J Am Coll Nutr 1994;13:62.

[45] Piao F, Yokoyama K, Ma N, Yamauchi T. Subacute toxic effects of zinc on various tissues and organs of rats. Toxicol Lett 2003;145:28.

[46] Costantini A, Pala MI. Thiamine and Hashimoto's thyroiditis: a report of three cases. J Altern Complement Med 2014;20:208.

[47] Boroni Moreira AP, Fiche Salles Teixeira T, do C Gouveia Peluzio M, de Cassia Goncalves Alfenas R. Gut microbiota and the development of obesity. Nutr Hosp 2012;27:1408.

[48] Andoh A. Physiological Role of Gut Microbiota for Maintaining Human Health. Digestion 2016;93:176.

[49] Graessler J, Qin Y, Zhong H, Zhang J, Licinio J, Wong ML, et al. Metagenomic sequencing of the human gut microbiome before and after bariatric surgery in obese patients with type 2 diabetes: correlation with inflammatory and metabolic parameters. Pharmacogenomics J 2013;13:514.

[50] Anhe FF, Varin TV, Le Barz M, Desjardins Y, Levy E, Roy D, et al. Gut microbiota dysbiosis in obesity-linked metabolic diseases and prebiotic potential of polyphenol-rich extracts. Curr Obes Rep 2015;4:389.

[51] Sirakov M, Kress E, Nadjar J, Plateroti M. Thyroid hormones and their nuclear receptors: new players in intestinal epithelium stem cell biology? Cell Mol Life Sci 2014;71:2897.

[52] Zhou L, Li X, Ahmed A, Wu D, Liu L, Qiu J, et al. Gut microbe analysis between hyperthyroid and healthy individuals. Curr Microbiol 2014;69:675.

[53] Virtanen KA, Lidell ME, Orava J, Heglind M, Westergren R, Niemi T, et al. Functional brown adipose tissue in healthy adults. N Engl J Med 2009;360:1518.

[54] van Marken Lichtenbelt WD, Vanhommerig JW, Smulders NM, Drossaerts JM, Kemerink GJ, Bouvy ND, et al. Cold-activated brown adipose tissue in healthy men. N Engl J Med 2009;360:1500.

[55] Cypess AM, Lehman S, Williams G, Tal I, Rodman D, Goldfine AB, et al. Identification and importance of brown adipose tissue in adult humans. N Engl J Med 2009;360:1509.

[56] Celi FS. Brown adipose tissue--when it pays to be inefficient. N Engl J Med 2009;360:1553.

[57] Ma S, Yu H, Zhao Z, Luo Z, Chen J, Ni Y, et al. Activation of the cold-sensing TRPM8 channel triggers UCP1-dependent thermogenesis and prevents obesity. J Mol Cell Biol 2012;4:88.

[58] Yoneshiro T, Aita S, Matsushita M, Kameya T, Nakada K, Kawai Y, et al. Brown adipose tissue, whole-body energy expenditure, and thermogenesis in healthy adult men. Obesity (Silver Spring) 2011;19:13.

[59] Chen KY, Brychta RJ, Linderman JD, Smith S, Courville A, Dieckmann W, et al. Brown fat activation mediates cold-induced thermogenesis in adult humans in response to a mild decrease in ambient temperature. J Clin Endocrinol Metab 2013;98:E1218.

[60] Celi FS, Brychta RJ, Linderman JD, Butler PW, Alberobello AT, Smith S, et al. Minimal changes in environmental temperature result in a significant increase in energy expenditure and changes in the hormonal homeostasis in healthy adults. Eur J Endocrinol 2010;163:863.

[61] Lee P, Smith S, Linderman J, Courville AB, Brychta RJ, Dieckmann W, et al. Temperature-acclimated brown adipose tissue modulates insulin sensitivity in humans. Diabetes 2014;63:3686.

[62] Saito M, Okamatsu-Ogura Y, Matsushita M, Watanabe K, Yoneshiro T, Nio-Kobayashi J, et al. High incidence of metabolically active brown adipose tissue in healthy adult humans: effects of cold exposure and adiposity. Diabetes 2009;58:1526.

[63] Vijgen GH, Bouvy ND, Teule GJ, Brans B, Hoeks J, Schrauwen P, et al. Increase in brown adipose tissue activity after weight loss in morbidly obese subjects. J Clin Endocrinol Metab 2012;97:E1229.

[64] Hankir MK, Bronisch F, Hintschich C, Krugel U, Seyfried F, Fenske WK. Differential effects of Roux-en-Y gastric bypass surgery on brown and beige adipose tissue thermogenesis. Metabolism 2015;64:1240.

[65] Broeders EP, Nascimento EB, Havekes B, Brans B, Roumans KH, Tailleux A, et al. The bile acid chenodeoxycholic acid increases human brown adipose tissue activity. Cell Metab 2015;22:418.

Chapter 42

The Ghrelin–Cannabinoid 1 Receptor Axis After Sleeve Gastrectomy

C. Fedonidis[2], D. Mangoura[2] and N. Alexakis[1]

[1]University of Athens, Athens, Greece, [2]Center for Neurosciences BRFAA, Athens, Greece

LIST OF ABBREVIATIONS

AgRP   agouti-related peptide
α-MSH   α-melanocyte-stimulating hormone
CB1R   cannabinoid 1 receptor
CNS   central nervous system
GHSR   growth hormone secretagogue receptor
GOAT   ghrelin O-acyltransferase
GPCR   G protein–coupled receptor
HFD   high-fat diet
LEPRb   leptin receptor-b
MC4R   melanocortin-4 receptor
POMC   proopiomelanocortin
SG   sleeve gastrectomy

ENERGY HOMEOSTASIS

The balance between energy gained through digestion and metabolic processes of food intake, and energy expenditure for normal metabolic processes, physiological activities, or exercise, is referred to as energy homeostasis. It results in the maintenance of constant levels of stored energy in the form of adipose tissue and a normal body weight. For most people, energy (food) intake precisely matches energy expenditure. Appetite for food intake is therefore an important factor in keeping a normal energy homeostasis. Many environmental factors, however, such as emotions, social factors, and food access and cost, interfere with eating behaviors. As a result, daily energy intake is not always correlated with energy consumption. Even worse, extended periods of fasting (diets) lead to abnormal rates of food intake (hyperphagia) and regaining of the lost body weight. These observations point out the complexity and the tight regulation of food intake and energy homeostasis.

Regulation of energy homeostasis, and thus body weight, is known to occur in the gut–brain axis. This axis is primarily governed by the hypothalamus, in cooperation with other areas of the central nervous system (CNS). More specifically, the arcuate nucleus in the hypothalamus is informed about the energy status of the body when it receives the peripheral hormonal signals of ghrelin and leptin through the growth hormone secretagogue receptor (GHSR) and leptin receptor-b (LEPRb), respectively, as well as the centrally mediated signals of endocannabinoids and α-melanocyte-stimulating hormone (α-MSH), which are detected by cannabinoid 1 receptor (CB1R) and melanocortin-4 receptors (MC4R), respectively [1–3] (Fig. 42.1A). The interplay between the responses of these receptors in response to the differential levels of their respective ligands defines in the end whether the signaling outcome, sent by the hypothalamus to the periphery, will be orexigenic or anorexigenic.

Ghrelin, mainly secreted from the stomach, primes the orexigenic pathway by activating the neuropeptide Y/agouti-related peptide (AgRP)–expressing neurons in the arcuate nucleus to secrete the orexigenic neuropeptides neuropeptide

(A) Energy homeostasis

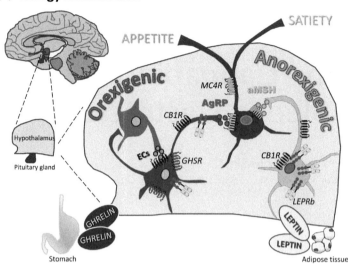

(B) Energy homeostasis after sleeve gastrectomy

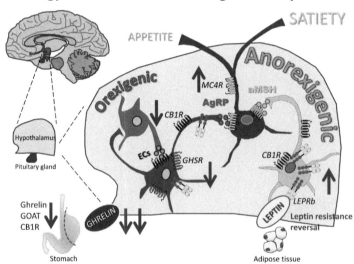

FIGURE 42.1 Graphic depiction of important energy homeostasis-related molecules along the gut−brain axis after sleeve gastrectomy. (A) Under physiological conditions energy homeostasis is maintained by a balance between orexigenic and satiety signals. In the periphery, ghrelin, mainly secreted by the stomach, and leptin, produced by the adipose tissue, reach the hypothalamus through the bloodstream to exert their actions after binding to their receptors, growth hormone secretagogue receptor (GHSR) and leptin receptor-b (LEPRb), respectively. Ghrelin primes the orexigenic pathway through the production of endocannabinoids (ECs) and activation of the cannabinoid 1 receptor (CB1R). Leptin stimulates the anorexigenic pathway through the production of α-melanocyte-stimulating hormone (α-MSH), which inhibits feeding behavior by binding to melanocortin-4 receptor (MC4R). Ghrelin may attenuate MC4R actions through the production of AgRP. These complex molecular interactions in the gut−brain axis produce the behavioral outcome that sustains physiological energy levels and body weight. (B) The energy homeostasis system balances toward the anorexigenic pathway after sleeve gastrectomy (SG). The acute loss of gastric tissue due to SG leads to downregulation of ghrelin, Ghrelin O-acyltransferase (GOAT); the enzyme that activates ghrelin, and CB1R expression in the remnant stomach resulting in less acyl−ghrelin in the periphery; low ghrelin levels impose changes into hypothalamus suppressing the orexigenic ghrelin−CB1R pathway and ablating their inhibitory effects on the anorexigenic pathway [4]. In addition, reversal of leptin resistance after SG [5] restores the control of leptin to the anorexigenic pathway. The net outcome of these molecular changes post-SG is the upregulation of the anorexigenic pathway and restoration of the satiety signal contributing to the long-term maintenance of lower body weight.

Y and AgRP [6]. At the same time, ghrelin suppresses the anorexigenic pathway through inhibition of the production and secretion of the anorexigenic peptides proopiomelanocortin (POMC) and cocaine- and amphetamine-regulated transcript by the POMC/cocaine- and amphetamine-regulated transcript neurons [7]. In contrast, leptin, produced by the adipose tissue at proportional levels to stored fat, counteracts ghrelin actions by stimulating the anorexigenic pathway to both inhibit food intake and promote lipid mobilization. The central effects of leptin are mediated through increases in the production of α-MSH [2], the agonist of MC4R that inhibits feeding behavior [2]. Moreover, leptin suppresses the orexigenic pathway by diminishing the production and secretion of AgRP, the endogenous MC4R antagonist [2].

Another G protein—coupled receptor (GPCR) involved in this network is CB1R, a pleiotropic receptor that controls many central neural functions [8] and plays an important role in appetite and metabolism regulation [9], as it mediates the effects of both ghrelin and leptin. Specifically, ghrelin elevates the production of endocannabinoids in the hypothalamus to promote appetite [10], whereas leptin reduces it [11]. CB1R regulates the anorexigenic pathway when, paradoxically, it activates the anorexigenic POMC neurons to selectively secrete the opioid peptide β-endorphin and thus promote feeding behavior [9]. In addition, orexin-A action on the same neuronal population elevates endocannabinoid production and stimulates CB1R to suppress POMC transcription, thus blunting α-MSH production and inducing hyperphagia and weight gain [12].

On the other hand, reciprocal regulation between ghrelin and CB1R is evident at two levels. First, ghrelin levels, by modulating endocannabinoid synthesis, act indirectly on the transcription of CB1R [10]. Specifically, the downregulation of CB1R mRNA is prevented by ghrelin, as for example in nodose ganglion neurons, where it counteracts the effects of postprandial cholecystokinin [13]. Secondly, CB1R regulates acyl—ghrelin secretion by the stomach, as CB1R signaling blockade correlates with a reduction in both ghrelin secretion from the stomach and ghrelin plasma levels [14,15]. On the other hand, activation of CB1R elevates ghrelin secretion into the circulation, while genetic ablation of CB1R diminishes the orexigenic effect of ghrelin [10,16]. Whether this crosstalk contributes to changes in hypothalamic receptor signaling potential remains elusive. These complex molecular interactions in the gut—brain axis produce the behavioral outcome that will sustain physiological energy levels. This chapter will focus on the ghrelin—CB1R system.

THE GHRELIN SYSTEM

Ghrelin exerts its biological actions through binding to GHSR when acylated on serine 3 by ghrelin O-acyltransferase (GOAT) before its release into the bloodstream [17]. Ghrelin acts on the anterior pituitary gland to stimulate the release of growth hormone [18]. Functional magnetic resonance imaging studies of human brain activity after ghrelin administration revealed that the main area in the CNS where ghrelin exerts its action is the hypothalamus [19]. Ghrelin is the first known peripheral hormone that has an orexigenic effect, and plain peripheral injection of ghrelin is sufficient to promote appetite in animal models or humans [20,21], all of which establishes ghrelin as a potent orexigenic signal [22].

Ghrelin has important roles in the acute as well as in the long-term regulation of energy balance. In the short term, ghrelin plasma levels rise before a meal and fall afterwards, suggesting that ghrelin acts as a signal for meal initiation in humans and animals [23]. This preprandial increase leads to voluntary food consumption, which is not related to the time of day or food type [24], while the postprandial decrease of ghrelin is proportional to the ingested calories [25]. The orexigenic action of ghrelin increases food intake while reducing metabolism by decreasing the usage of fat as a fuel for energy production, thus causing an increase in body weight. In reverse, the energy status of the organism affects ghrelin plasma levels, seen e.g., as increases of ghrelin in the bloodstream after 2 days of fasting or as decreased levels after direct application of glucose in the stomach [26]. The density of GHSR in the hypothalamus is also sensitive to energy levels, as fasting produces strong upregulation in the expression of this receptor in comparison to normally fed animals [27]. This upregulation under conditions of negative energy balance may represent restoring mechanisms for the signal strength of ghrelin in order to ensure stable energy levels and thus survival.

Ghrelin actions also affect the regulation of energy homeostasis in the long term. Concerning metabolism, ghrelin acts in peripheral tissues through the AMP-activated protein kinase as the functional opposite of insulin [28]. Specifically, ghrelin reduces insulin secretion caused by glucose ingestion, thus decreasing glucose clearance and leading to increased glucose levels in healthy people [29]. Moreover, ghrelin reverses the insulin-induced reduction of gluconeogenesis [30], while it upregulates enzymes involved in the production and storage of fatty acids to induce excessive lipid accumulation in adipose cells [26]. These peripheral metabolic actions of ghrelin are independent of its orexigenic capacity through GHSR signaling in the CNS [31], which attests to the multiplicity of energy regulation levels that ghrelin participates in.

THE ENDOCANNABINOID AND CANNABINOID 1 RECEPTOR SYSTEM

The endocannabinoid system is a complex signaling system that includes two GPCRs, namely CB1R and CB2R, their endogenous lipid ligands (endocannabinoids), and enzymes that regulate their biosynthesis and degradation. The endocannabinoids belong to the category of bioactive lipids produced on demand after cell stimulation by a signal. In most cellular types—e.g., neurons, adipose cells, or muscle cells—precursor phospholipids of the plasma membrane are activated for the synthesis of endocannabinoids after extracellular stimuli. The endocannabinoid system plays important roles in central and peripheral regulation of metabolism, hence this system is often deregulated during obesity [32].

CB1R is the main mediator of endocannabinoid action as it is widely expressed in all tissues and is the prominent type in the CNS. CB1R regulates the metabolic pathways of glucose and lipids [33], as well as food intake behavior through the hypothalamus and specific areas of the limbic system that control the reward value of food [34]. The pathways and the molecular mechanisms that CB1R control in the CNS, adipose tissue, the liver, and the gastrointestinal system to adjust energy balance are still subjects of intense research [35].

There is a direct link between the endocannabinoid system and the increase of energy availability. Central or subcutaneous application of endocannabinoids increases food intake [36], while marijuana consumption (which elicits CB1R activation through its main constituent, the agonist delta-9-tetrahydrocannabinol) increases body weight in normal subjects [37]. In obese people or animal models of obesity, an increased availability of endocannabinoids is detected in the circulation, the CNS, and in peripheral tissues, accompanied by changes in CB1R expression levels [35,38]. Ablating CB1R in the mouse (CB1R$^{-/-}$) protects the animal from the development of obesity or insulin resistance after a high-fat diet (HFD) [39]. CB1R$^{-/-}$ animals are leaner, having less adipose tissue and lower leptin levels than wild-type animals of the same age, while, when treated with leptin, they show greater loss of body weight. These events suggest a link between CB1R signaling and leptin sensitivity [39], and altogether attest to the fact that CB1R expression and activation regulate food intake and body weight.

SLEEVE GASTRECTOMY AND THE GHRELIN–CANNABINOID 1 RECEPTOR AXIS

The term "obesity" refers to the excess deposition of body fat caused by chronic disturbances in energy balance, namely, when energy intake is constantly higher than energy expenditure. As a result, during the state of obesity, the circulating levels of hormones change drastically; that is, ghrelin levels are reduced [40] and leptin's are increased [2], while the hypothalamus loses its ability to readily respond to these hormonal signals, a phenomenon characterized as resistance to leptin [2] and ghrelin [1]. Thus, an extended rewiring of the hypothalamic circuits takes place to accommodate and perpetuate the obese state. In an attempt to lose weight, obese individuals adopt diet regimens of low energy intake. The altered hormonal levels, however, contribute to a reversion of weight back to obesity levels after an initial period of weight loss [41], effectively "defending" the obese state.

A successful method to manage obesity is bariatric surgery. Sleeve gastrectomy (SG) dramatically reduces the excess fat mass, thus maintaining a lower body weight in the long term [42]. With this type of surgery, resection of the greater curvature of the stomach reduces gastric volume by almost 80%, while a unique hormonal balance between the plasma levels of ghrelin and leptin is established [5,43,44]. These changes lead to improvements in the metabolism of glucose and lipids, and to type 2 diabetes resolution [45]. Moreover, patients usually change their dietary habits, adopting the consumption of low-calorie food post-SG [43]. These effects are not fully explained by the limitation of the stomach volume after SG [5]; instead they point out an interplay between the plasma levels of hormones that regulate energy homeostasis and their receptors in the brain. Our recent work using a HFD-induced obese rat model to study the effects of SG has further elucidated the molecular events that regulate the gut–brain axis post-SG and possibly lead to the maintenance of lower body weight [4].

In accordance with previous reports [43,44], we first established that post-SG the acute loss of ghrelin-producing tissue, and decreases in the expression of both ghrelin and GOAT in the remnant stomach, result in lower circulating acyl–ghrelin levels. Lower ghrelin levels, along with higher cholecystokinin plasma levels post-SG [44], suppress the expression of CB1R in the remnant stomach. In turn, we postulate that downregulation of CB1R results in reduced ghrelin secretion in the periphery by retention of proghrelin into the A-like cells, and, along with the stable, low levels of GOAT, even lower levels of plasma acyl–ghrelin may be sustained in the long term post-SG.

Moreover, we have provided evidence that the post-SG sustained low ghrelin levels initiate changes into the expression of energy homeostasis-related receptors in a temporal fashion that hierarchically (from the stomach to the brain) modulate the orexigenic and anorexigenic pathways (Fig. 42.1B). CB1R expression is impaired after SG in the hypothalamus, simulating the decreases of CB1R mRNA during long periods of low circulating ghrelin levels like obesity, possibly due to a reduction in the hypothalamic concentration of endocannabinoids [1,10,46]. We have shown that transcriptional neuroadaptations in the hypothalamus following SG include a decrease in GHSR, as ghrelin levels may also regulate the expression of its own receptor [1,47,48]. Protein levels of GHSR increase after SG, reflecting an attenuation of the degradation of GHSR due to acyl–ghrelin scarcity in our SG model compared to sham operations. Suppression of the ghrelin system may reduce the inhibitory effect of AgRP on MC4R, and along with the constantly low impact of the leptin system after SG, leads to an increase in anorexigenic signaling through MC4R [4]. Therefore, the suppression of the ghrelin–CB1R axis primarily affects the maintenance of low endocannabinoid signaling, reprogramming the hypothalamic circuits to sustain the weight loss post-SG. These observations highlight the role of the endocannabinoid system as an effector of ghrelin actions and emphasize its importance in the success of bariatric procedures [3,4].

In accordance with our results, the concept of targeting CB1R to manage obesity began in 1994 when the substance rimonabant was introduced as a selective antagonist of CB1R with the ability to cross the blood–brain barrier. Many animal studies produced promising results, as rimonabant not only decreased food intake and body weight [32,49], but also suppressed the desire to consume palatable food [50]. Rimonabant was released in European markets as a noninvasive, pharmaceutical approach to treat obesity in 2006, yet, despite the successfully decreased appetite and body weight, it was withdrawn due to severe psychological side effects after 3 years in circulation. The most likely reason for these unwanted side effects was the high expression of CB1R in the brain [8]. Therefore, development and production of an antagonist that will specifically block CB1R in the periphery to manage obesity is necessary. We have thus suggested that peripheral intervention blocking both the CB1R and ghrelin systems, as seen after SG, should be considered as an additional therapeutic scheme for human obesity [4]. Further elucidation of the molecular mechanisms that control the maintenance of weight loss post-SG may lead to better intervention into obesity.

MINI-DICTIONARY OF TERMS

- *Energy homeostasis*: The balance between energy intake and expenditure to maintain stable energy levels in the body.
- *Gut–brain axis*: All the molecular events in the stomach and intestine that result in a signal traveling through the bloodstream or nerve fibers to the brain and vice versa.
- *Ligand*: A molecule (e.g., protein/peptide, lipid), possibly produced and secreted by a cell type to the extracellular space (e.g., hormones) or administered, that specifically binds to a membrane receptor on the same or another cell population to change the conformation of the receptor and initiate a signaling cascade.
- *Agonist*: A ligand or substance that specifically binds to a receptor and activates signaling.
- *Antagonist*: A substance that binds to a receptor and inhibits signaling.
- *Membrane receptor*: A protein, localized at the plasma membrane of a cell, that specifically binds to ligands and generates intracellular signaling to impose acute or transcriptional changes.
- *Signaling*: A phosphorylation cascade initiated at the cell membrane upon ligand–receptor binding that, through intracellular protein–protein interactions, acutely modifies existing proteins and transfers messages to the nucleus for gene expression modification.

KEY FACTS

- Energy homeostasis is the balance between energy gain, food intake, and energy expenditure for normal metabolic procedures, physiological activities, or exercise.
- The gut and brain form an interconnected physiological system, an axis that regulates energy intake so that it matches energy expenditure. This well-coordinated system maintains physiological body weight and constant levels of stored fat.
- Systematic, long-term increases in food intake and/or reduction in energy expenditure lead to obesity.
- Obesity is a "new" balance point that favors maintenance of higher body weight. This energy balance (obesity homeostasis) is resistant to diet regimens.
- Drastic weight reduction approaches, such as SG and drugs that block the action of the CB1R, appear to be necessary to upset obesity homeostasis.

SUMMARY POINTS

- Once obesity is established, the CNS "defends" the obese state through altered hormonal levels and rewiring of the hypothalamic circuits.
- SG may reset hormonal levels and thus reduce the excess fat mass in the long term.
- These metabolic improvements and healthier dietary practices by patients are not fully explained solely by stomach volume reduction after SG.

- In our model, acute loss of gastric tissue by SG decreases the expression of both ghrelin and its acylating enzyme, GOAT, in the remnant stomach, leading to reductions in acyl−ghrelin in the circulation.
- SG-dependent low levels of acyl−ghrelin suppress endogenous CB1R activation, which in turn reduce ghrelin secretion.
- More importantly, peripheral blunting of ghrelin−CB1R actions imposes reprogramming on the appetite-related receptors in the hypothalamus.
- Apparently, the endocannabinoid system is an important partner of ghrelin in the success of SG.
- Further elucidation of the molecular mechanisms that control the maintenance of weight loss post-SG may lead to better intervention into obesity.

REFERENCES

[1] Briggs DI, Enriori PJ, Lemus MB, Cowley MA, Andrews ZB. Diet-induced obesity causes ghrelin resistance in arcuate NPY/AgRP neurons. Endocrinology 2010;151:4745.

[2] Enriori PJ, Evans AE, Sinnayah P, Jobst EE, Tonelli-Lemos L, Billes SK, et al. Diet-induced obesity causes severe but reversible leptin resistance in arcuate melanocortin neurons. Cell Metab 2007;5:181.

[3] Guijarro A, Osei-Hyiaman D, Harvey-White J, Kunos G, Suzuki S, Nadtochiy S, et al. Sustained weight loss after Roux-en-Y gastric bypass is characterized by down regulation of endocannabinoids and mitochondrial function. Ann Surg 2008;247:779.

[4] Fedonidis C, Alexakis N, Koliou X, Asimaki O, Tsirimonaki E, Mangoura D. Long-term changes in the ghrelin-CB1R axis associated with the maintenance of lower body weight after sleeve gastrectomy. Nutr Diabetes 2014;4:e127.

[5] Stefater MA, Perez-Tilve D, Chambers AP, Wilson-Perez HE, Sandoval DA, Berger J, et al. Sleeve gastrectomy induces loss of weight and fat mass in obese rats, but does not affect leptin sensitivity. Gastroenterology 2010;138:2426.

[6] Seoane LM, Lopez M, Tovar S, Casanueva FF, Senaris R, Dieguez C. Agouti-related peptide, neuropeptide Y, and somatostatin-producing neurons are targets for ghrelin actions in the rat hypothalamus. Endocrinology 2003;144:544.

[7] Cowley MA, Smith RG, Diano S, Tschop M, Pronchuk N, Grove KL, et al. The distribution and mechanism of action of ghrelin in the CNS demonstrates a novel hypothalamic circuit regulating energy homeostasis. Neuron 2003;37:649.

[8] Asimaki O, Mangoura D. Cannabinoid receptor 1 induces a biphasic ERK activation via multiprotein signaling complex formation of proximal kinases PKCepsilon, Src, and Fyn in primary neurons. Neurochem Int 2011;58:135.

[9] Koch M, Varela L, Kim JG, Kim JD, Hernandez-Nuno F, Simonds SE, et al. Hypothalamic POMC neurons promote cannabinoid-induced feeding. Nature 2015;519:45.

[10] Kola B, Farkas I, Christ-Crain M, Wittmann G, Lolli F, Amin F, et al. The orexigenic effect of ghrelin is mediated through central activation of the endogenous cannabinoid system. PLoS One 2008;3:e1797.

[11] Di Marzo V, Goparaju SK, Wang L, Liu J, Batkai S, Jarai Z, et al. Leptin-regulated endocannabinoids are involved in maintaining food intake. Nature 2001;410:822.

[12] Morello G, Imperatore R, Palomba L, Finelli C, Labruna G, Pasanisi F, et al. Orexin-A represses satiety-inducing POMC neurons and contributes to obesity via stimulation of endocannabinoid signaling. Proc Natl Acad Sci USA 2016.

[13] Burdyga G, Varro A, Dimaline R, Thompson DG, Dockray GJ. Ghrelin receptors in rat and human nodose ganglia: putative role in regulating CB-1 and MCH receptor abundance. Am J Physiol Gastrointest Liver Physiol 2006;290:G1289.

[14] Senin LL, Al-Massadi O, Folgueira C, Castelao C, Pardo M, Barja-Fernandez S, et al. The gastric CB1 receptor modulates ghrelin production through the mTOR pathway to regulate food intake. PLoS One 2013;8:e80339.

[15] Cani PD, Montoya ML, Neyrinck AM, Delzenne NM, Lambert DM. Potential modulation of plasma ghrelin and glucagon-like peptide-1 by anorexigenic cannabinoid compounds, SR141716A (rimonabant) and oleoylethanolamide. Br J Nutr 2004;92:757.

[16] Zbucki RL, Sawicki B, Hryniewicz A, Winnicka MM. Cannabinoids enhance gastric X/A-like cells activity. Folia Histochem Cytobiol 2008;46:219.

[17] Yang J, Brown MS, Liang G, Grishin NV, Goldstein JL. Identification of the acyltransferase that octanoylates ghrelin, an appetite-stimulating peptide hormone. Cell 2008;132:387.

[18] Kojima M, Hosoda H, Matsuo H, Kangawa K. Ghrelin: discovery of the natural endogenous ligand for the growth hormone secretagogue receptor. Trends Endocrinol Metab 2001;12:118.

[19] Malik S, McGlone F, Bedrossian D, Dagher A. Ghrelin modulates brain activity in areas that control appetitive behavior. Cell Metab 2008;7:400.

[20] Wren AM, Seal LJ, Cohen MA, Brynes AE, Frost GS, Murphy KG, et al. Ghrelin enhances appetite and increases food intake in humans. J Clin Endocrinol Metab 2001;86:5992.

[21] Wren AM, Small CJ, Abbott CR, Dhillo WS, Seal LJ, Cohen MA, et al. Ghrelin causes hyperphagia and obesity in rats. Diabetes 2001;50:2540.

[22] Nakazato M, Murakami N, Date Y, Kojima M, Matsuo H, Kangawa K, et al. A role for ghrelin in the central regulation of feeding. Nature 2001;409:194.

[23] Cummings DE, Purnell JQ, Frayo RS, Schmidova K, Wisse BE, Weigle DS. A preprandial rise in plasma ghrelin levels suggests a role in meal initiation in humans. Diabetes 2001;50:1714.

[24] Cummings DE, Frayo RS, Marmonier C, Aubert R, Chapelot D. Plasma ghrelin levels and hunger scores in humans initiating meals voluntarily without time- and food-related cues. Am J Physiol Endocrinol Metabol 2004;287:E297.

[25] Callahan HS, Cummings DE, Pepe MS, Breen PA, Matthys CC, Weigle DS. Postprandial suppression of plasma ghrelin level is proportional to ingested caloric load but does not predict intermeal interval in humans. J Clin Endocrinol Metab 2004;89:1319.

[26] Tschop M, Smiley DL, Heiman ML. Ghrelin induces adiposity in rodents. Nature 2000;407:908.

[27] Kim MS, Yoon CY, Park KH, Shin CS, Park KS, Kim SY, et al. Changes in ghrelin and ghrelin receptor expression according to feeding status. Neuroreport 2003;14:1317.

[28] Kola B, Hubina E, Tucci SA, Kirkham TC, Garcia EA, Mitchell SE, et al. Cannabinoids and ghrelin have both central and peripheral metabolic and cardiac effects via AMP-activated protein kinase. J Biol Chem 2005;280:25196.

[29] Tong J, Prigeon RL, Davis HW, Bidlingmaier M, Kahn SE, Cummings DE, et al. Ghrelin suppresses glucose-stimulated insulin secretion and deteriorates glucose tolerance in healthy humans. Diabetes 2010;59:2145.

[30] Murata M, Okimura Y, Iida K, Matsumoto M, Sowa H, Kaji H, et al. Ghrelin modulates the downstream molecules of insulin signaling in hepatoma cells. J Biol Chem 2002;277:5667.

[31] Davies JS, Kotokorpi P, Eccles SR, Barnes SK, Tokarczuk PF, Allen SK, et al. Ghrelin induces abdominal obesity via GHS-R-dependent lipid retention. Mol Endocrinol 2009;23:914.

[32] Jbilo O, Ravinet-Trillou C, Arnone M, Buisson I, Bribes E, Peleraux A, et al. The CB1 receptor antagonist rimonabant reverses the diet-induced obesity phenotype through the regulation of lipolysis and energy balance. FASEB J Off Publ Fed Am Soc Exp Biol 2005;19:1567.

[33] Silvestri C, Ligresti A, Di Marzo V. Peripheral effects of the endocannabinoid system in energy homeostasis: adipose tissue, liver and skeletal muscle. Rev Endocr Metab Disord 2011;12:153.

[34] Bellocchio L, Lafenetre P, Cannich A, Cota D, Puente N, Grandes P, et al. Bimodal control of stimulated food intake by the endocannabinoid system. Nat Neurosci 2010;13:281.

[35] Matias I, Gonthier MP, Orlando P, Martiadis V, De Petrocellis L, Cervino C, et al. Regulation, function, and dysregulation of endocannabinoids in models of adipose and beta-pancreatic cells and in obesity and hyperglycemia. J Clin Endocrinol Metab 2006;91:3171.

[36] Kirkham TC, Williams CM. Endogenous cannabinoids and appetite. Nutr Res Rev 2001;14:65.

[37] Foltin RW, Fischman MW, Byrne MF. Effects of smoked marijuana on food intake and body weight of humans living in a residential laboratory. Appetite 1988;11:1.

[38] Cote M, Matias I, Lemieux I, Petrosino S, Almeras N, Despres JP, et al. Circulating endocannabinoid levels, abdominal adiposity and related cardiometabolic risk factors in obese men. Int J Obes (Lond) 2007;31:692.

[39] Ravinet Trillou C, Delgorge C, Menet C, Arnone M, Soubrie P. CB1 cannabinoid receptor knockout in mice leads to leanness, resistance to diet-induced obesity and enhanced leptin sensitivity. Int J Obes Relat Metab Disord J Int Assoc Study Obes 2004;28:640.

[40] Tschop M, Weyer C, Tataranni PA, Devanarayan V, Ravussin E, Heiman ML. Circulating ghrelin levels are decreased in human obesity. Diabetes 2001;50:707.

[41] Sumithran P, Prendergast LA, Delbridge E, Purcell K, Shulkes A, Kriketos A, et al. Long-term persistence of hormonal adaptations to weight loss. N Engl J Med 2011;365:1597.

[42] Eid GM, Brethauer S, Mattar SG, Titchner RL, Gourash W, Schauer PR. Laparoscopic sleeve gastrectomy for super obese patients: forty-eight percent excess weight loss after 6 to 8 years with 93% follow-up. Ann Surg 2012;256:262.

[43] Chambers AP, Kirchner H, Wilson-Perez HE, Willency JA, Hale JE, Gaylinn BD, et al. The effects of vertical sleeve gastrectomy in rodents are ghrelin independent. Gastroenterology 2013;144:50.

[44] Peterli R, Steinert RE, Woelnerhanssen B, Peters T, Christoffel-Courtin C, Gass M, et al. Metabolic and hormonal changes after laparoscopic Roux-en-Y gastric bypass and sleeve gastrectomy: a randomized, prospective trial. Obes Surg 2012;22:740.

[45] Keidar A, Hershkop KJ, Marko L, Schweiger C, Hecht L, Bartov N, et al. Roux-en-Y gastric bypass vs sleeve gastrectomy for obese patients with type 2 diabetes: a randomised trial. Diabetologia 2013;56:1914.

[46] Timofeeva E, Baraboi ED, Poulin AM, Richard D. Palatable high-energy diet decreases the expression of cannabinoid type 1 receptor messenger RNA in specific brain regions in the rat. J Neuroendocrinol 2009;21:982.

[47] Luque RM, Kineman RD, Park S, Peng XD, Gracia-Navarro F, Castano JP, et al. Homologous and heterologous regulation of pituitary receptors for ghrelin and growth hormone-releasing hormone. Endocrinology 2004;145:3182.

[48] Nogueiras R, Tovar S, Mitchell SE, Rayner DV, Archer ZA, Dieguez C, et al. Regulation of growth hormone secretagogue receptor gene expression in the arcuate nuclei of the rat by leptin and ghrelin. Diabetes 2004;53:2552.

[49] Gary-Bobo M, Elachouri G, Gallas JF, Janiak P, Marini P, Ravinet-Trillou C, et al. Rimonabant reduces obesity-associated hepatic steatosis and features of metabolic syndrome in obese Zucker fa/fa rats. Hepatology 2007;46:122.

[50] Arnone M, Maruani J, Chaperon F, Thiebot MH, Poncelet M, Soubrie P, et al. Selective inhibition of sucrose and ethanol intake by SR 141716, an antagonist of central cannabinoid (CB1) receptors. Psychopharmacology (Berl) 1997;132:104.

Chapter 43

PNPLA3 Variant p.I148M and Bariatric Surgery

R. Liebe[1], F. Lammert[1] and M. Krawczyk[1,2]

[1]Saarland University Medical Center, Homburg, Germany, [2]Laboratory of Metabolic Liver Diseases, Medical University of Warsaw, Warsaw, Poland

LIST OF ABBREVIATIONS

ALT	alanine transaminase
AST	aspartate aminotransferase
E	glutamic acid
GWAS	genome-wide association study
HCC	hepatocellular cancer
I	isoleucine
K	lysine
M	methionine
NAFLD	nonalcoholic fatty liver disease
NASH	nonalcoholic steatohepatitis
OR	odds ratio
P	*p*-value
PASH	PNPLA3-associated steatohepatitis
PNPLA3 (adiponutrin)	patatin-like phospholipase domain containing 3
TM6SF2	transmembrane 6 superfamily member 2
VLDL	very-low-density lipoprotein

INTRODUCTION

While it is entirely clear from the epidemiology and geographical distribution of affected individuals that changes in lifestyle are fueling the obesity epidemic, the last 10 years have yielded increasing evidence for a previously underestimated role of inherited predisposition in the development and progression of nonalcoholic fatty liver diseases (NAFLD) [1]. Before the advent of the human genome map and dense sets of markers for genome-wide association studies (GWAS), Schwimmer et al. [2] investigated the heritability of NAFLD based on magnetic resonance imaging to quantify the liver fat fraction. The group studied 44 probands with or without NAFLD (11/33), 41 siblings (12/29) and 74 parents (19/55) concluding that, adjusted for age, sex, race, and body mass index, the heritability of fatty liver was 1.00 and of liver fat fraction was 0.39 [2]. The investigation was not designed to differentiate between environmental and genetic contributors, but revealed that familial factors are a major determinant of NAFLD, and therefore family members of children with NAFLD should be considered at high risk for NAFLD. A cross-sectional analysis of a cohort of 60 twin pairs residing in Southern California (42 monozygotic and 18 dizygotic; average age 46 ± 22 years) estimated the heritability of hepatic steatosis to be 0.52—i.e., 52% of the risk for developing steatosis was contributed by genetic variation, while 48% was caused by environmental factors [3]. The heritability of hepatic fibrosis (based on liver stiffness) was 0.50. Hepatic steatosis was quantified noninvasively by magnetic resonance imaging based on the proton density fat fraction (MRI-PDFF); liver fibrosis was quantified by stiffness determined by magnetic resonance elastography [3]. Whereas the use of the expression "genetic predisposition" may sound deterministic, the reality could not be further

TABLE 43.1 Liver Disease Associated With the Presence of the *PNPLA3* p.I148M Genotype

Disease	Study	Year
NAFLD	Romeo et al.	2008
Liver fibrosis	Krawczyk et al.	2011
HBV steatosis	Vigano et al.	2013
HCV steatosis	Cai et al.	2011
HCV cirrhosis	Valenti et al.	2011
Alcoholic cirrhosis	Tian et al.	2010
	Stickel et al.	2011
	Salameh et al.	2015
HCC	Liu et al.	2014

Source: Adapted from Krawczyk/Lammert, Semin Liver Dis, 2013.

from it. In this chapter, we report that carriers of genetic risk factors that predispose to more severe disease when combined with excess nutrition also have the best prospect for weight loss and reconstitution of liver function. Thus, carriers of genetic risk factors might represent the most promising target group, gaining maximum benefit from dietetic or surgical intervention. The first convincing GWAS on the genetic basis of NAFLD demonstrated that carriers of the patatin-like phospholipase domain-containing 3 (*PNPLA3*) variant p.I148M (rs738409) show significantly increased risk of developing fatty liver, as well as its more severe types such as nonalcoholic steatohepatitis (NASH), cirrhosis, and hepatocellular cancer (HCC) [4–7]. As demonstrated in Table 43.1, carriers of the *PNPLA3* p.I148M variant are at risk of severe hepatic phenotypes.

Since the seminal discovery of this previously unknown and unsuspected central contributor towards liver disease in 2008, roughly 500 studies have investigated its basic biochemistry and the impact of genetic variation on its function in various ethnic groups under different environmental stressors. In this chapter we summarize selected studies that investigated the association between common genetic variants, in particular the *PNPLA3* p.I148M polymorphism, and hepatic as well as metabolic traits. We also demonstrate how these variants affect weight loss interventions.

ADIPONUTRIN (PNPLA3) FUNCTION AND BIOCHEMISTRY

The *PNPLA3* gene encodes adiponutrin, a protein involved in lipid metabolism, whose precise function in humans remains unclear. The enzyme is a 481-amino acid member of the patatin-like phospholipase domain-containing family (PNPLA). This domain was originally discovered in lipid hydrolases of potato and named after the most abundant protein of the potato tuber, patatin. In hepatoma cell lines, it demonstrates lipase activity towards triglycerides [8] and retinyl-esters [9,10]. The mechanisms behind the hepatic fat accumulation are still not completely elucidated, but according to some studies, this phenomenon might be related to a loss of the enzymatic activity, resulting in reduced very-low-density lipoprotein (VLDL) secretion [11]. Studies in mouse models suggest that carriers of the prosteatotic *PNPLA3* p.148M allele might be characterized by larger lipid droplets due to inactivation of the enzyme [12], which is in line with the idea of "loss-of-function" of the mutated protein as a major driver of the hepatic phenotypes [13].

PNPLA3 P.I148M AND THE RISK OF DEVELOPING HEPATOCELLULAR CARCINOMA

Various studies have shown a strong genetic association between the common *PNPLA3* variant p.I148M and the risk of developing HCC in patients with NAFLD [14,15]. A contribution of the rare *PNPLA3* risk allele was also shown in patients with alcoholic liver injury [16–18], rendering this variant one of the most frequent genetic risk factors for HCC development. The association study [14] comprised 100 individuals with NAFLD-associated HCC and 1476 controls, and showed that homozygous carriers of the rare p.148M variant carry a 12-fold increased HCC risk (odds ratio (OR) = 12.19, 95% CI 6.89–21.58, $p < 0.0001$) as compared to p.148I homozygotes. Lately we have detected a comparable association between the *PNPLA3* polymorphism and HCC risk in German patients [19]. The impact of the *PNPLA3* risk genotype on cancer risk is remarkable, and comparable to another inherited liver disease, hereditary hemochromatosis type 1 (OMIM #235200). This disorder is due to rare mutations in the *HFE* gene, and thus represents a monogenic ("Mendelian") liver disease [20]. A meta-analysis of hemochromatosis, encompassing a total of 66,000 cases and 226,000 controls from Europe and China [21], showed that homozygous carriers of the *HFE* mutation p. C282Y (rs1800562) have a 12-fold increased risk of developing liver disease. Considering the fact that the risk allele frequency in the European population is 21−28% for *PNPLA3* (minor allele frequency = 0.23), whereas it is a mere 2−8% for *HFE* (minor allele frequency = 0.05), the contribution of the former toward the HCC burden is hard to underestimate and should be included in future concepts for preventive strategies [22].

PNPLA3 P.I148M AND METABOLIC TRAITS

A study from Finland investigated clinical characteristics, *PNPLA3* rs738409 genotype, and serum cytokeratin 18 fragments in 296 consecutive bariatric surgery patients who underwent a liver biopsy in order to validate a novel "NASH score." The score included the *PNPLA3* genotype, serum AST activities, and fasting insulin concentrations. It was validated in an Italian cohort comprising 380 (mainly nonbariatric surgery) patients who had undergone a liver biopsy for NASH. It predicted NASH with an AUROC of 0.77 (0.71−0.84) in Finns and 0.76 (0.71−0.81) in Italians. Subsequently, the NASH score was applied to a cohort comprising 2849 random healthy subjects (ages 45−74 years) from the Finnish National Diabetes Prevention Program (FIN-D2D) to estimate the population prevalence of NASH. The prevalence of NASH in this population cohort was around 5%, and to reduce follow-up cost, bariatric surgery in patients carrying the *PNPLA3* risk genotype could be considered [23]. A recent study in 181 severely obese patients who underwent bariatric surgery confirmed the role of the variant in disease progression in an Asian cohort based on the analysis of liver histopathology, which was correlated with *PNPLA3* p.I148M genotypes [24]. The authors concluded that the homozygous *PNPLA3* p.148MM genotype increases NASH susceptibility in severely obese Asians with NAFLD and correlates with the histopathological severity of NAFLD.

PNPLA3 P.I148M AND LIVER DISEASES IN CHILDREN

Since the discovery of *PNPLA3* as a genetic risk factor for hepatic steatosis and inflammation, a series of studies have investigated its impact in cohorts of normal weight and overweight children of various ages. Valenti et al. [25] recruited 149 children and adolescents (mean age 10 years) and observed a striking association of the *PNPLA3* p.148M allele with NASH and fibrosis, but not with serum activities of AST or ALT. Giudice et al. [26] tested 1058 children between 2 and 16 years and observed an association of the *PNPLA3* prosteatotic allele with higher AST and ALT levels, which was particularly pronounced in the subgroup with a high level of abdominal fat (waist-to-height ratio > 0.62). These data were in agreement with findings by Romeo et al. [27] in 475 obese children and adolescents (mean age 10 years) who showed significantly increased ALT and AST activities in carriers of the p.148M allele. Our group studied a cohort of 142 German children aged 5−9 years (median age 7 years) and concluded that the *PNPLA3* risk variant p.148M was associated with higher serum ALT activities (i.e., subclinical liver injury, already at a very young age) [28]. This association was significant in the nonoverweight children and the total cohort, but not in the overweight subgroup. The variant was associated with increased serum glucose concentrations ($p = 0.01$) and HOMA index ($p = 0.02$) in carriers of the p.148IM genotype, but affected neither other metabolic traits nor the presence of NAFLD [28]. Interestingly, a recent analysis of children aged 6−8 years showed an association between the *PNPLA3* variant and increased liver tests in obese, but not in normal weight, children. This association became stronger during 2 years follow-up [29].

NEW CANDIDATE GENES: *TM6SF2* P.E167K AND *MBOAT7* RS641738

A GWAS for alcohol-related cirrhosis in individuals of European descent (712 cases and 1426 controls) with subsequent validation in two independent European cohorts (1148 cases and 922 controls) identified variants in the *MBOAT7* ($p = 1.03 \times 10^{-9}$) and *TM6SF2* ($p = 7.89 \times 10^{-10}$) genes as new risk factors for fatty liver disease, and confirmed rs738409 in *PNPLA3* as an important risk locus for alcohol-related cirrhosis ($p = 1.54 \times 10^{-48}$) at a genome-wide level of significance [30]. All three loci have a role in lipid processing, suggesting that lipid turnover is important in the pathogenesis of alcohol-related cirrhosis.

A subsequent study in 3854 participants from the Dallas Heart Study (a multiethnic population-based probability sample of Dallas County residents) and 1149 European individuals investigated the role of the *MBOAT7* gene in

NAFLD [31]. A total of 2736 participants from the Dallas Heart Study underwent proton magnetic resonance spectros-copy to measure hepatic triglyceride content. In the liver biopsy cross-sectional cohort, all 1149 individuals underwent liver biopsy to assess liver injury and disease severity. At the molecular level, *MBOAT7*, also called lysophosphatidyli-nositol acyltransferase, was found to be highly expressed in the liver. The *MBOAT7* rs641738 T allele was associated with lower protein expression in the liver and changes in plasma phosphatidylinositol species consistent with decreased MBOAT7 function. The group concluded that the impact of the gene variant appeared to be mediated by changes in hepatic phosphatidylinositol acyl-chain remodeling [31]. With a minor risk allele frequency of 0.42, the *MBOAT* T allele is a frequent contributor to the overall NAFLD risk in the population.

TM6SF2, like adiponutrin, appears to be involved in intracellular lipid metabolism [32]. Whereas the *TM6SF2* p. E167K variant (rs58542926, minor allele frequency 0.07) is far less frequent than the *PNPLA3* variant [33], heterozy-gous carriers are at increased risk of developing NAFLD [34], and, according to recent reports, might also be prone to developing hepatic fibrosis [35]. An additional, intriguing aspect of TM6SF2 is the differential impact of the gene vari-ant in different body compartments. Carriers of the rare *TM6SF2* p.E167K risk variant are prone to develop fatty liver as a result of impaired lipidation of VLDL. As a consequence, they have lower circulating lipids and reduced risk of myocardial infarction. A study investigating liver damage and cardiovascular outcomes in subjects at risk of NASH assessed liver injury in 1201 patients who underwent liver biopsy for suspected NASH; 427 were evaluated for carotid atherosclerosis. Among this population,188 subjects (13%) were carriers of the p.E167K risk variant. They had signifi-cantly lower serum lipid levels than noncarriers, as well as more severe steatosis, necroinflammation, ballooning, and fibrosis, and were more likely to present with NASH. As expected, carriers of the "hepatic risk" allele had a lower risk of developing carotid plaques, resulting in the intriguing conclusion that "what is bad for the liver can be good for the heart" [34].

Interestingly, although the association between both the *PNPLA3* p.I148M and the *TM6SF2* p.E167K polymorphisms and hepatic steatosis has been reproduced in different populations, their effects on metabolic traits are less apparent [36,37]. Even though there is evidence that the *TM6SF2* p.E167K variant results in lower VLDL serum levels, the exact mechanism is still not known. VLDL formation requires a multistep process where triglycerides are combined with apoli-poproteins that are then secreted, altered, and eventually taken up again by the liver as VLDL remnants. It remains to be clarified how this cycle is impacted by the *TM6SF2* p.E167K variant. Chylomicron assembly and circulation in the intes-tine parallels VLDL assembly and circulation in the liver, and *TM6SF2* is also expressed in the intestine; thus it remains to be determined how lipoprotein biology is affected by *TM6SF2* gene variation in humans and mice [32].

THE IMPACT OF *PNPLA3* GENOTYPE ON WEIGHT LOSS

Given the association between the *PNPLA3* polymorphism and progressive liver disease, therapies aimed at reducing the harmful effects of this polymorphism on liver status are urgently needed. To date, the role of genetic factors such as *PNPLA3*, *TM6SF2*, or *MBOAT7* predisposing variants on the hepatic response after bariatric surgery is still largely unknown. Interestingly, studies concerning diet and NAFLD demonstrated that carriers of the *PNPLA3* prosteatotic var-iant show improvement of liver phenotypes during healthy lifestyle modifications and might respond better to dietetic interventions than patients with the common *PNPLA3* genotype [38–40].

In a recent study, we performed a genetic analysis of a cohort of obese individuals scheduled for bariatric surgery. Changes in liver fat content were evaluated within 12 months after weight-reducing surgery using a novel (MRI)-based method to quantitatively determine hepatic triglyceride contents [41] as well as liver biopsy samples. Homozygous car-riers of the *PNPLA3* prosteatotic allele lost more weight and showed better improvement of fatty liver as compared to carriers of the common allele. Therefore, we speculated that the *PNPLA3* p.I148M risk genotype might be a good prog-nostic factor for NAFLD patients undergoing bariatric surgery.

These results are particularly interesting for several reasons. First of all, given the increased risk of fatty liver dis-ease in homozygous carriers of the *PNPLA3* risk variant [1,42], it would have been conceivable that the presence of this variant would negatively affect the improvement of hepatic steatosis. However, bariatric surgery was able to over-ride the harmful effects of the *PNPLA3* p.I148M polymorphism. These findings are in line with previous studies show-ing that weight loss [38,39] and lifestyle changes [40] can reduce hepatic steatosis in carriers of the *PNPLA3* risk variant. Interestingly, in a seminal paper published in 2008, a subgroup analysis restricted to African−Americans identi-fied a second *PNPLA3* variant (p.S453I, rs6006460) that was strongly associated with lower hepatic fat content in this ethnic group [4]. This finding supports the hypothesis that the PNPLA3 protein is not just a harbinger of liver disease, but a central node molecule in hepatic lipid and energy metabolism.

Given our results, it can be speculated that carriers of the *PNPLA3* variant do not just represent a subgroup at increased risk of progression toward the most severe phenotypes, including liver cirrhosis [43] and HCC [14,15]. They also represent a particularly promising target for intervention, because in this population long-term beneficial effects of weight reduction are likely to go beyond the sole reduction of hepatic fat content, and might also prevent progression toward chronic liver disease and cancer.

Insulin resistance is recognized as a major trigger of fatty liver [44]. An increasing body of evidence suggests, however, that the scenario of fat accumulation in livers of patients carrying the *PNPLA3* p.I148M variant might be different. Indeed, none of the previous studies provided any hints for an association between this polymorphism and insulin resistance [37]. Some studies even postulated better insulin sensitivity in obese carriers of the *PNPLA3* risk variant [45,46]. Comparable observations have been made in carriers of the *TM6SF2* risk genotype, which paradoxically also seems to be associated with an improved metabolic status regardless of the increased NAFLD risk [32,36,47]. A recent study of 511 cirrhotic patients (44% alcohol-related, 56% viral) confirmed that the risk alleles of both *PNPLA3* and *TM6SF2* were associated with the occurrence of HCC in patients with alcoholic liver disease, but *not* in patients with fatty liver disease of viral etiology [48]. Hence, patients with the genetic predisposition to develop fatty liver, especially carriers of the *PNPLA3* and most likely *TM6SF2* variants, might be characterized by a distinct pathogenesis of NAFLD and benefit from specific therapeutic approaches. Bariatric surgery, already performed at early stages of fatty liver disease, might represent one of these tailored therapies. Indeed, according to our results [41] bariatric surgery might decrease hepatic fat content regardless of the presence of the *PNPLA3* p.I148M variant, although this reduction was more pronounced in homozygous carriers of the risk allele. However, the role of bariatric surgery on liver status of patients with more advanced NAFLD needs to be investigated in future cohorts. The effects of the *TM6SF2* and *MBOAT7* risk variants on hepatic fat accumulation and liver injury appear to be less pronounced than the impact of the *PNPLA3* polymorphism. This observation, underscored by the decreasing population-attributable risk of each variant to develop fatty liver (*PNPLA3* > *TM6SF2* > *MBOAT7*) [49], is in line with the results of our transient elastography-based study [33], and suggests that *PNPLA3* p.I148M might be the most relevant prosteatotic variant.

A FUTURE CONCEPT OF *PNPLA3*-ASSOCIATED LIVER DISEASE AS AN ETIOLOGY-TRANSCENDING ENTITY: PERSONALIZED, GENE-BASED MEDICINE

For the last 50 years, NAFLD has been growing in prevalence, at the same time gathering acceptance among hepatologists in particular and physicians in general [50], as a potentially severe chronic liver disease. On a cautionary note, the diagnosis of NAFLD and its subtypes is still challenging in clinical practice. Fatty liver disease is usually diagnosed in individuals who present with a typical "bright liver" image on abdominal ultrasonography, do not consume excessive amounts of alcohol (i.e., less than 20−30 g/d according to the current European Association for the Study of the Liver guidelines [51]), and do not suffer from other specific liver diseases [52]. In addition to ultrasound, which can be used as a screening tool in risk groups, other noninvasive methods for quantifying hepatic fat content have been developed and validated [33,41,53]. In the past, liver inflammation, the critical step from mere metabolic overload to chronic injury, has been differentiated in terms of the underlying environmental etiology, such as ASH for alcoholic steatohepatitis and NASH for nonalcoholic steatohepatitis. We have lately proposed the term "PASH" for *PNPLA3*-associated steatohepatitis as a novel gene-based liver disease entity [37]. The proposal is based on the observation that in carriers of the *PNPLA3* risk allele, increased hepatic lipid content and liver inflammation can be driven predominantly by PNPLA3, but at the same time, the variant increases the likelihood of liver pathology in almost all entities of environmentally induced chronic liver injury. Thus, we speculate that patients carrying the *PNPLA3* risk allele might benefit from a more systematic, early, and careful surveillance of complications of progressive fatty liver disease [25]. The use of the *PNPLA3* genotype for construction of a "NASH score" in Finnish and Italian patients indicates its potential for disease stratification and precision medicine [23]. More studies in patients coming from different ethnic groups are required to clearly define the role of genetic predisposition as a determinant of the long-term success of bariatric surgery. This will require prospective analyses of large groups of patients with accurate estimation of liver steatosis and fibrosis both before and after intervention.

MINI-DICTIONARY OF TERMS

- *Genome-wide association study (GWAS)*: A type of genetic study based on genotyping thousands of genetic variants in cases and controls to investigate their association with the studied trait.

- *Hepatocellular cancer (HCC)*: The most frequent primary liver cancer, its frequency is increased in individuals with liver cirrhosis; this risk is further increased in the carriers of the *PNPLA3* p.I148M polymorphism.
- *Nonalcoholic fatty liver disease (NAFLD)*: One of the most frequent liver diseases. Characterized by accumulation of fat in the liver. In most cases it is not associated with liver dysfunction; however, patients with an aggressive form of NAFLD, namely NASH, are at risk of liver cirrhosis.
- *Nonalcoholic steatohepatitis (NASH)*: The malign necroinflammatory form of NAFLD, which might result in liver dysfunction. To date, liver biopsy is required to diagnose NASH.
- *PNPLA3 (Adiponutrin)*: A 481 amino acid long fat-processing enzyme member of the patatin-like phospholipase domain-containing family (PNPLA).
- *PNPLA3 p.I148M, TM6SF2 p.E167K, MBOAT rs641738*: Genetic variants associated with hepatic fat accumulation.

KEY FACTS

- Carriers of *PNPLA3* and *TM6SF2* risk variants are burdened with increased risk of progressive liver disease and hepatocellular carcinoma.
- Children bearing the *PNPLA3* risk genotype display subclinical liver injury at a young age (5−8 years).
- At the same time, risk variant bearers have a hepatic lipid metabolism that makes them benefit most from dietary or surgical intervention (i.e., bariatric surgery).
- We hypothesize that it is particularly efficient and sensible to perform bariatric surgery in carriers of these variants.
- This hypothesis needs to be tested in large cohorts and prospective studies.

SUMMARY POINTS

- Previous studies demonstrating familial clustering as well as ethnic differences indicate that genetic factors contribute to fatty liver disease.
- GWAS helped to identify the *PNPLA3* p.I148M polymorphism as a frequent risk factor for fatty liver.
- Subsequent studies demonstrated that this variant is also associated with increased risk of progressive liver damage such as liver fibrosis, cirrhosis, or HCC.
- The *TM6SF2* and *MBOAT7* variants are also associated with NAFLD, but seem to pose a weaker threat than the *PNPLA3* polymorphism.
- Bearers of the risk variants might be characterized by a fast response to dietary and surgical interventions.
- The *PNPLA3* risk allele carriers represent promising targets for interventions aimed at weight loss, such as bariatric surgery.
- Wider use of genetic testing will help to identify individuals who, due to their genetic predisposition to develop progressive liver disease, will benefit the most from weight loss.

REFERENCES

[1] Anstee QM, Day CP. The genetics of nonalcoholic fatty liver disease: spotlight on PNPLA3 and TM6SF2. Semin Liver Dis 2015;35(3):270−90.

[2] Schwimmer JB, Celedon MA, Lavine JE, Salem R, Campbell N, Schork NJ, et al. Heritability of nonalcoholic fatty liver disease. Gastroenterology 2009;136(5):1585−92.

[3] Loomba R, Schork N, Chen CH, Bettencourt R, Bhatt A, Ang B, et al. Heritability of hepatic fibrosis and steatosis based on a prospective twin study. Gastroenterology 2015;149(7):1784−93.

[4] Romeo S, Kozlitina J, Xing C, Pertsemlidis A, Cox D, Pennacchio LA, et al. Genetic variation in PNPLA3 confers susceptibility to nonalcoholic fatty liver disease. Nat Genet 2008;40(12):1461−5.

[5] Sookoian S, Pirola CJ. Meta-analysis of the influence of I148M variant of patatin-like phospholipase domain containing 3 gene (PNPLA3) on the susceptibility and histological severity of nonalcoholic fatty liver disease. Hepatology 2011;53(6):1883−94.

[6] Trepo E, Nahon P, Bontempi G, Valenti L, Falleti E, Nischalke HD, et al. Association between the PNPLA3 (rs738409 C > G) variant and hepatocellular carcinoma: evidence from a meta-analysis of individual participant data. Hepatology 2014;59(6):2170−7.

[7] Wei JL, Leung JC, Loong TC, Wong GL, Yeung DK, Chan RS, et al. Prevalence and severity of nonalcoholic fatty liver disease in non-obese patients: a population study using proton-magnetic resonance spectroscopy. Am J Gastroenterol 2015;110(9):1306−14, quiz 1315.

[8] He S, McPhaul C, Li JZ, Garuti R, Kinch L, Grishin NV, et al. A sequence variation (I148M) in PNPLA3 associated with nonalcoholic fatty liver disease disrupts triglyceride hydrolysis. J Biol Chem 2010;285(9):6706−15.

[9] Pirazzi C, Valenti L, Motta BM, Pingitore P, Hedfalk K, Mancina RM, et al. PNPLA3 has retinyl-palmitate lipase activity in human hepatic stellate cells. Hum Mol Genet 2014;23(15):4077−85.

[10] Kovarova M, Konigsrainer I, Konigsrainer A, Machicao F, Haring HU, Schleicher E, et al. The genetic variant I148M in PNPLA3 is associated with increased hepatic retinyl-palmitate storage in humans. J Clin Endocrinol Metab 2015;100(12):E1568−74.

[11] Pirazzi C, Adiels M, Burza MA, Mancina RM, Levin M, Stahlman M, et al. Patatin-like phospholipase domain-containing 3 (PNPLA3) I148M (rs738409) affects hepatic VLDL secretion in humans and in vitro. J Hepatol 2012;57(6):1276−82.

[12] Smagris E, BasuRay S, Li J, Huang Y, Lai KM, Gromada J, et al. Pnpla3I148M knockin mice accumulate PNPLA3 on lipid droplets and develop hepatic steatosis. Hepatology 2015;61(1):108−18.

[13] Pingitore P, Pirazzi C, Mancina RM, Motta BM, Indiveri C, Pujia A, et al. Recombinant PNPLA3 protein shows triglyceride hydrolase activity and its I148M mutation results in loss of function. Biochim Biophys Acta 2014;1841(4):574−80.

[14] Liu YL, Patman GL, Leathart JB, Piguet AC, Burt AD, Dufour JF, et al. Carriage of the PNPLA3 rs738409 C >G polymorphism confers an increased risk of non-alcoholic fatty liver disease associated hepatocellular carcinoma. J Hepatol 2014;61(1):75−81.

[15] Burza MA, Pirazzi C, Maglio C, Sjoholm K, Mancina RM, Svensson PA, et al. PNPLA3 I148M (rs738409) genetic variant is associated with hepatocellular carcinoma in obese individuals. Dig Liver Dis 2012;44(12):1037−41.

[16] Trepo E, Gustot T, Degre D, Lemmers A, Verset L, Demetter P, et al. Common polymorphism in the PNPLA3/adiponutrin gene confers higher risk of cirrhosis and liver damage in alcoholic liver disease. J Hepatol 2011;55(4):906−12.

[17] Trepo E, Guyot E, Ganne-Carrie N, Degre D, Gustot T, Franchimont D, et al. PNPLA3 (rs738409 C > G) is a common risk variant associated with hepatocellular carcinoma in alcoholic cirrhosis. Hepatology 2012;55(4):1307−8.

[18] Nischalke HD, Berger C, Luda C, Berg T, Muller T, Grunhage F, et al. The PNPLA3 rs738409 148M/M genotype is a risk factor for liver cancer in alcoholic cirrhosis but shows no or weak association in hepatitis C cirrhosis. PLoS One 2011;6(11):e27087.

[19] Casper M, Krawczyk M, Behrmann I, Glanemann M, Lammert F. Variant PNPLA3 increases the HCC risk: prospective study in patients treated at the Saarland University Medical Center. Z Gastroenterol 2016;54(6):585−6.

[20] Feder JN, Gnirke A, Thomas W, Tsuchihashi Z, Ruddy DA, Basava A, et al. A novel MHC class I-like gene is mutated in patients with hereditary haemochromatosis. Nat Genet 1996;13(4):399−408.

[21] Ellervik C, Birgens H, Tybjaerg-Hansen A, Nordestgaard BG. Hemochromatosis genotypes and risk of 31 disease endpoints: meta-analyses including 66,000 cases and 226,000 controls. Hepatology 2007;46(4):1071−80.

[22] Krawczyk M, Stokes CS, Romeo S, Lammert F. HCC and liver disease risks in homozygous PNPLA3 p.I148M carriers approach monogenic inheritance. J Hepatol 2015;62(4):980−1.

[23] Hyysalo J, Mannisto VT, Zhou Y, Arola J, Karja V, Leivonen M, et al. A population-based study on the prevalence of NASH using scores validated against liver histology. J Hepatol 2014;60(4):839−46.

[24] Tai CM, Huang CK, Tu HP, Hwang JC, Chang CY, Yu ML. PNPLA3 genotype increases susceptibility of nonalcoholic steatohepatitis among obese patients with nonalcoholic fatty liver disease. Surg Obes Relat Dis 2015;11(4):888−94.

[25] Valenti L, Alisi A, Galmozzi E, Bartuli A, Del Menico B, Alterio A, et al. I148M patatin-like phospholipase domain-containing 3 gene variant and severity of pediatric nonalcoholic fatty liver disease. Hepatology 2010;52(4):1274−80.

[26] Giudice EM, Grandone A, Cirillo G, Santoro N, Amato A, Brienza C, et al. The association of PNPLA3 variants with liver enzymes in childhood obesity is driven by the interaction with abdominal fat. PLoS One 2011;6(11):e27933.

[27] Romeo S, Sentinelli F, Cambuli VM, Incani M, Congiu T, Matta V, et al. The 148M allele of the PNPLA3 gene is associated with indices of liver damage early in life. J Hepatol 2010;53(2):335−8.

[28] Krawczyk M, Liebe R, Maier IB, Engstler AJ, Lammert F, Bergheim I. The frequent adiponutrin (PNPLA3) variant p.Ile148Met is associated with early liver injury: analysis of a German pediatric cohort. Gastroenterol Res Pract 2015;2015:205079.

[29] Viitasalo A, Pihlajamaki J, Lindi V, Atalay M, Kaminska D, Joro R, et al. Associations of I148M variant in PNPLA3 gene with plasma ALT levels during 2-year follow-up in normal weight and overweight children: the PANIC Study. Pediatr Obes 2015;10(2):84−90.

[30] Buch S, Stickel F, Trepo E, Way M, Herrmann A, Nischalke HD, et al. A genome-wide association study confirms PNPLA3 and identifies TM6SF2 and MBOAT7 as risk loci for alcohol-related cirrhosis. Nat Genet 2015;47(12):1443−8.

[31] Mancina RM, Dongiovanni P, Petta S, Pingitore P, Meroni M, Rametta R, et al. The MBOAT7-TMC4 variant rs641738 increases risk of nonalcoholic fatty liver disease in individuals of European descent. Gastroenterology 2016;150(5):1219−30, e6.

[32] Kahali B, Liu YL, Daly AK, Day CP, Anstee QM, Speliotes EK. TM6SF2: catch-22 in the fight against nonalcoholic fatty liver disease and cardiovascular disease?. Gastroenterology 2015;148(4):679−84.

[33] Arslanow A, Stokes CS, Weber SN, Grünhage F, Lammert F, Krawczyk M. The common PNPLA3 variant p.I148M is associated with liver fat contents as quantified by controlled attenuation parameter (CAP). Liver Int 2016;36(3):418−26.

[34] Dongiovanni P, Petta S, Maglio C, Fracanzani AL, Pipitone R, Mozzi E, et al. Transmembrane 6 superfamily member 2 gene variant disentangles nonalcoholic steatohepatitis from cardiovascular disease. Hepatology 2015;61(2):506−14.

[35] Milano M, Aghemo A, Mancina RM, Fischer J, Dongiovanni P, De Nicola S, et al. Transmembrane 6 superfamily member 2 gene E167K variant impacts on steatosis and liver damage in chronic hepatitis C patients. Hepatology 2015;62(1):111−17.

[36] Pirola CJ, Sookoian S. The dual and opposite role of the TM6SF2-rs58542926 variant in protecting against cardiovascular disease and conferring risk for nonalcoholic fatty liver: a meta-analysis. Hepatology 2015;62(6):1742−56.

[37] Krawczyk M, Portincasa P, Lammert F. PNPLA3-associated steatohepatitis: toward a gene-based classification of fatty liver disease. Semin Liver Dis 2013;33(4):369−79.

[38] Sevastianova K, Kotronen A, Gastaldelli A, Perttila J, Hakkarainen A, Lundbom J, et al. Genetic variation in PNPLA3 (adiponutrin) confers sensitivity to weight loss-induced decrease in liver fat in humans. Am J Clin Nutr 2011;94(1):104−11.

[39] Marzuillo P, Grandone A, Perrone L, del Giudice EM. Weight loss allows the dissection of the interaction between abdominal fat and PNPLA3 (adiponutrin) in the liver damage of obese children. J Hepatol 2013;59(5):1143−4.

[40] Shen J, Wong GL, Chan HL, Chan RS, Chan HY, Chu WC, et al. PNPLA3 gene polymorphism and response to lifestyle modification in patients with nonalcoholic fatty liver disease. J Gastroenterol Hepatol 2015;30(1):139–46.

[41] Krawczyk M, Jimenez-Aguero R, Alustiza JM, Emparanza JI, Perugorria MJ, Bujanda L, et al. PNPLA3 p.148M variant is associated with higher reduction of liver fat content after bariatric surgery. Surg Obes Relat Dis 2016; in press.

[42] Dongiovanni P, Donati B, Fares R, Lombardi R, Mancina RM, Romeo S, et al. PNPLA3 I148M polymorphism and progressive liver disease. World J Gastroenterol 2013;19(41):6969–78.

[43] Shen JH, Li YL, Li D, Wang NN, Jing L, Huang YH. The rs738409 (I148M) variant of the PNPLA3 gene and cirrhosis: a meta-analysis. J Lipid Res 2015;56(1):167–75.

[44] Perry RJ, Samuel VT, Petersen KF, Shulman GI. The role of hepatic lipids in hepatic insulin resistance and type 2 diabetes. Nature 2014;510 (7503):84–91.

[45] Kantartzis K, Peter A, Machicao F, Machann J, Wagner S, Konigsrainer I, et al. Dissociation between fatty liver and insulin resistance in humans carrying a variant of the patatin-like phospholipase 3 gene. Diabetes 2009;58(11):2616–23.

[46] Park JH, Cho B, Kwon H, Prilutsky D, Yun JM, Choi HC, et al. I148M variant in PNPLA3 reduces central adiposity and metabolic disease risks while increasing nonalcoholic fatty liver disease. Liver Int 2015;35(12):2537–46.

[47] Zhou Y, Llaurado G, Oresic M, Hyotylainen T, Orho-Melander M, Yki-Jarvinen H. Circulating triacylglycerol signatures and insulin sensitivity in NAFLD associated with the E167K variant in TM6SF2. J Hepatol 2015;62(3):657–63.

[48] Falleti E, Cussigh A, Cmet S, Fabris C, Toniutto P. PNPLA3 rs738409 and TM6SF2 rs58542926 variants increase the risk of hepatocellular carcinoma in alcoholic cirrhosis. Dig Liver Dis 2016;48(1):69–75.

[49] Mancina RM, Dongiovanni P, Petta S, Pingitore P, Meroni M, Rametta R, et al. The *MBOAT7-TMC4* variant rs641738 increases risk of nonalcoholic fatty liver disease in individuals of European descent. Gastroenterology 2016;150(5):1219–30.

[50] Ratziu V, Cadranel JF, Serfaty L, Denis J, Renou C, Delassalle P, et al. A survey of patterns of practice and perception of NAFLD in a large sample of practicing gastroenterologists in France. J Hepatol 2012;57(2):376–83.

[51] European Association for the Study of the Liver (EASL). EASL-EASD-EASO Clinical Practice Guidelines for the management of nonalcoholic fatty liver disease. J Hepatol 2016;64(6):1388–402.

[52] Krawczyk M, Bonfrate L, Portincasa P. Nonalcoholic fatty liver disease. Best Pract Res Clin Gastroenterol 2010;24(5):695–708.

[53] Festi D, Schiumerini R, Marzi L, Di Biase AR, Mandolesi D, Montrone L, et al. Review article: the diagnosis of non-alcoholic fatty liver disease -- availability and accuracy of non-invasive methods. Aliment Pharmacol Ther 2013;37(4):392–400.

Section V

Nutritional Aspects

Chapter 44

Dietary Reference Values

M.Y. Price[1] and V.R. Preedy[2]

[1]Royal Free London NHS Foundation Trust, London, United Kingdom, [2]King's College London, London, United Kingdom

LIST OF ABBREVIATIONS

AIs   adequate intakes
DRIs   dietary reference intakes
DRVs   dietary reference values
EAR   estimated average requirement
LRNI   lower reference nutrient intake
RDAs   recommended dietary allowances
RNI   reference nutrient intake
SACN   scientific advisory committee on nutrition
SI   safe intake
ULs   tolerable upper intake levels

INTRODUCTION

Patients who undergo bariatric surgery have specific nutritional requirements before and after surgery [1,2]. These include reducing overall caloric intake before surgery as well as addressing any micronutrient imbalance. Some micronutrient deficiencies arise before surgical procedures as a consequence of obesity per se, and also as a consequence of the presurgical weight loss procedures themselves. For example, as many as 70% of subjects who are about to undergo bariatric surgery are at risk of vitamin D deficiency [1,3]. In one study, 63% of patients were deficient in folate and 98% deficient in vitamin D, though none were deficient in vitamin E prior to surgery [4]. As a consequence of the surgery itself, micro- and macronutrient deficiencies will also arise unless treated [5,6]. For example, fat malabsorption as a consequence of bariatric weight loss procedures will have implications for the status of fat-soluble vitamins A and D, necessitating supplementation in the long term [7,8]. Supplementation with multivitamin and mineral preparations are frequently advocated for use after bariatric surgery [9–13]. An understanding of nutrient supplementations in bariatric surgery is dependent upon an understanding of the normative requirements for the general population.

MICRO- AND MACRONUTRIENTS

There are four levels at which dietary components should be considered:

Level 1. The micro- and macronutrient compositions of foods.
Level 2. The amount of ingested micro- and macronutrients.
Level 3. The requirements of micro- and macronutrients.
Level 4. The status or levels of micro- and macronutrients.

For these four levels of consideration, the amount of ingested nutrients in *Level 2* will depend on the micro- and macronutrient compositions of foods and portion size (*Level 1*). While there are numerous methods for calculating dietary intakes, e.g., with the combined use of food diaries, questionnaires, and food composition tables [14,15], a fundamental question arises as to how much of a particular macronutrient an individual needs, which is the basis of *Level 3*.

The status of micro- and macronutrients (*Level 4*) will depend on various factors such as genetic variability, the specific forms of the nutrients (e.g., heme and nonheme iron), competitive interactions, and so on (reviewed in Ref. [16]).

In the United Kingdom, the requirements of individuals and populations are termed the dietary reference values (DRVs) [16]. However in many countries, the terms DRVs, recommended values, recommended daily intakes (RDIs), or recommended daily allowance are used interchangeably, even though they mean the same thing, namely imparting advice on what people should consume in order to achieve optimal health and prevent disease. Some systems have also been updated with concomitant name changes. For example, in the United Kingdom the DRVs replaced the UK RDIs of 1979, which in turn replaced the UK recommended daily amounts (RDAs) of 1969 [16]. There may also be some confusion in that terms used in the United Kingdom (DRVs) are not used in the United States. Furthermore, terms previously used in the United Kingdom (RDAs) are presently used in the United States. To address this it is beneficial to document three systems of guidance in detail and define the terms that are often used (Table 44.1).

The UK DRVs can be applied for assessing diets of individuals, assessing diets of groups of individuals, for prescribing diets or food supplies, and finally for food labeling [16].

TABLE 44.1 Definitions of Reference Dietary Requirements for United Kingdom, United States, and Japan

Component and Abbreviation	Definition
UK Definition of Dietary Reference Values (DRVs)	
Dietary reference values (DRVs)	A term used to cover LRNI, EAR, RNI, and SI.
Estimated average requirement (EAR)	EAR of a group of people for energy, protein, vitamin, or mineral. About half will usually need more than the EAR, and half less.
Lower reference nutrient intake (LRNI)*	LRNI for protein, vitamin, or mineral. An amount of the nutrient that is enough for only the few people in a group who have low needs.
Reference nutrient intake (RNI)	RNI for protein, vitamin, or mineral. An amount of the nutrient that is enough, or more than enough, for about 97% of people in a group. If average intake of a group is at RNI, then the risk of deficiency in the group is small.
Safe intake (SI)	A term used to indicate intake or range of intakes of a nutrient for which there is not enough information to estimate RNI, EAR, or LRNI. It is an amount that is enough for almost everyone, but not so large as to cause undesirable effects.
US Definitions of Dietary Reference Intakes (DRIs)	
Adequate intakes (AIs)	A recommended average daily nutrient intake level based on observed or experimentally determined approximations or estimates of mean nutrient intake by a group (or groups) of apparently healthy people. An AI is used when the RDAs cannot be determined.
Dietary reference intakes (DRIs)	A set of nutrient-based reference values that are quantitative estimates of nutrient intakes to be used for planning and assessing diets for healthy people. DRIs expand on the periodic reports called RDAs, which were first published by the Institute of Medicine in 1941.
Estimated average requirement (EAR)	The average daily nutrient intake level estimated to meet the requirement of half the healthy individuals in a particular life stage and sex group.
Recommended dietary allowance (RDA)	The average daily dietary intake level that is sufficient to meet the nutrient requirement of nearly all (97–98%) healthy individuals in a particular life stage and sex group.
Tolerable upper intake level (UL)	The highest average daily nutrient intake level likely to pose no risk of adverse health effects for nearly all individuals in a particular life stage and sex group. As intake increases above the UL, the potential risk of adverse health effects increases.
Japanese Definition of Dietary Reference Intakes (DRIs)	
Adequate intake (AI)	A less well-defined value, generally the median of the population without evidence of deficiency.

(Continued)

TABLE 44.1 (Continued)

Component and Abbreviation	Definition
Dietary reference intakes (DRIs)	Designed to present reference values for healthy individuals and groups for intake of energy and 34 nutrients to maintain and promote health, to prevent lifestyle-related diseases due to insufficient, excess, and/or imbalanced consumption of energy and nutrients.
Estimated average requirement (EAR)	An average requirement (50%) for the nutrient for each age and sex group.
Recommended dietary allowance (RDA)	An intake that covers the needs of 97.5% of the population for each age and sex group. The RDA was calculated by EAR +2 SD EAR. However, the variance is usually unknown, so the coefficient of variation is adopted instead.
D Japan	The highest level of intake that can be tolerated without the possibility of causing adverse effects.

Source: Adapted from Department of Health. Dietary reference values for food energy and nutrients for the United Kingdom. Report of the Panel on Dietary Reference Values of the Committee on Medical Aspects of Food Policy. United Kingdom: Department of Health, 1991; ODPHP Office of Disease Prevention and Health Promotion 2015-2020 Dietary Guidelines for Americans; 2015. Available at http://health.gov/dietaryguidelines/2015/guidelines/. Accessed 14 May 2016; MHLW. Ministry of Health, Labour and Welfare. Overview of Dietary Reference Intakes for Japanese; 2015. Available at http://www.mhlw.go.jp/file/06-Seisakujouhou-10900000-Kenkoukyoku/Overview.pdf. Accessed 14 May 2016 [16–18].

The definitions for the advisory amounts for the United Kingdom, United States, and Japan are displayed in Table 44.1. The UK DRVs cover the following: estimated average requirement (EAR), lower reference nutrient intake (LRNI), reference nutrient intake (RNI), and safe intake (SI).

When specifying or describing advisory amounts, it is important to state what system is being used, such as the UK DRVs or US RDAs, or the source of updated material. Overall, however, attempts should be made to achieve the RNI (United Kingdom) or RDA (United States and Japan), which means that the requirements of the majority (97–98%) of the population are met. However, as reiterated throughout this chapter, bariatric procedures will alter normative requirements of micro- and macronutrients.

The center point for nutrient requirements is the EAR. The LRNI is set at two standard deviations below the EAR, and the RNI is set at two standard deviations above the EAR (Fig. 44.1).

If the intake of a micro- or macronutrient is above the RNI, then it means that 97.5% of the requirement is met and deemed to be adequate. However, if the intake of individuals is at or below the LRNI, then it means the requirements of only 2.5% of individuals will be met (Table 44.1). Intakes at LRNI are certainly inadequate. If the intake is between the LRNI and the RNI, then it is not certain if requirements have been met unless there are other measures of adequacy, e.g., markers of disease or other variables such as biomarkers of oxidative stress.

For some nutrients there was not sufficient information to derive an LRNI, EAR, or RNI for the UK DRVs. In these cases a SI was derived, which is an amount that is enough, but not undesirable (Table 44.1). Components within this group of SIs include pantothenic acid, biotin, vitamins E and K, manganese, molybdenum, chromium, and fluoride.

Although the UK DRV covers a considerable number of nutrients, it also documents the potential harm related to high doses. Many of the nutrients in individual chapters have sections on guidance on high intakes [16]. The United States and Japan equivalent to this is the tolerable upper intake level (UL).

An example of guidance on high intakes relates to the UK DRVs reporting that infants are particularly at risk of developing conditions related to high intakes of vitamin D, and hypercalcemia has been shown in doses of 50 μg/day, and oral doses of 15 mg/day every 3–5 months has been shown to cause hypercalcemia. However, much of the UK's guidance on high intakes is somewhat dated, and some areas are not covered. For example, the UK DRVs do not have guidance on high intakes for vitamin D in relation to adults, which is of relevance to the majority of subjects undergoing weight loss bariatric procedures.

In the United States, the system that offers guidance as to what nutrients people should consume is referred to as dietary reference intakes (DRIs), and covers terms such as adequate intakes (AIs), EARs, recommended dietary allowances (RDAs), and ULs (Table 44.1).

In Japan the system is also called DRIs, and also includes AIs, EARs, RDA, and ULs (Table 44.1). These are the same terms used in the US system, though there are slight differences in the definitions.

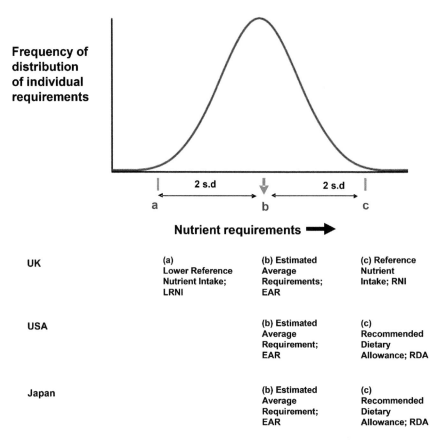

FIGURE 44.1 Schematic representation for the frequency distribution of individual requirements for the United Kingdom, United States, and Japan. Figure showing the relationship between the requirements of a nutrient and the frequency distribution of individual requirements for the United Kingdom, United States, and Japan. *SDs*, standard deviations. The points (a) and (c) are 2 SDs from the Estimated Average Requirements (EAR). *Adapted from Department of Health. Dietary reference values for food energy and nutrients for the United Kingdom. Report of the Panel on Dietary Reference Values of the Committee on Medical Aspects of Food Policy. United Kingdom: Department of Health, 1991.*

CAVEATS AND ASSUMPTIONS

To derive the UK DRVs, four working groups constituting the Committee on Medical Aspects (COMA) of Food Policy were convened to determine requirements for energy and protein; fats and carbohydrates; and vitamins and minerals. Since then, the COMA panel has been replaced with the Scientific Advisory Committee on Nutrition (SACN; see https://www.gov.uk/government/groups/scientific-advisory-committee-on-nutrition). The terms of reference for the COMA panel were to review the previously published RDAs. The COMA panel made various assumptions, namely:

- Requirements vary within a population.
- Requirements follow a normal distribution (Fig. 44.1) [16]. There are exceptions; e.g., the requirements for iron do not follow a normal distribution for women due to blood loss during the menstrual cycle.
- Requirements relate to well people, not those who are unwell or sick.
- The micro- and macronutrients are ingested or delivered by food via the oral route, not via parenteral routes.
- The requirements of one component assume that requirements of all other components are met. For example, the requirements for selenium assume that requirements for iron, magnesium, and zinc are met.

The COMA panel decided that there was no single criterion that allowed for the derivation of requirements for all nutrients, so used a variety of sources of information. Where information was limited, anecdotal and other semiquantitative information was used. In other words, there were imperfections in drawing up UK DRVs, and the information was based on the prevailing information of the day.

While there are numerous factors influencing requirements, the COMA panel assumed that factors such as gender, growth (pregnancy and lactation), physical activity, and age influence requirements. However, not all requirements are structured along these lines: physical activity will influence energy requirements, but it is not known if physical activity will influence magnesium requirements. Of course, there are numerous factors that influence the attainment of adequate nutritional status. Active metabolites or organ damage will impact nutritional status even though intakes of particular nutrients are deemed to be adequate at the RNI or above. Thus, goitrogens in food will affect iodine status [19] and alcohol will affect the absorption of a number of vitamins (e.g., vitamins A, B1, B12, C, and folate) [20]. In alcoholism, iron status may actually increase due to its enhanced absorption [21].

This aspect of considering the effects of ethanol is of particular importance as it has been suggested that alcohol misuse may be of significant importance before and after bariatric surgery [22−24].

REQUIREMENTS FOR MICRO- AND MACRONUTRIENTS

Inspection of Table 44.1 shows that there are differences in terms and definitions between the United Kingdom, United States, and Japanese systems of reference values, but essentially they aim to achieve optimal levels for health. In this review we present requirements for energy and protein (Table 44.2), carbohydrate, fat, nonstarch polysaccharides (Table 44.3), minerals (Table 44.4), and vitamins (Table 44.5).

TABLE 44.2 Energy and Protein Requirements for Adults in the United Kingdom, United States, and Japan (19−50 Years for United Kingdom and United States; 18−49 Years for Japan)

Nutrient	Unit per Day	Country	Gender	EAR	RNI/RDA
Energy	kcal	United Kingdom[a]	Male	2581−2772	−
			Female	2103−2175	−
		United States[b]	Male	2400−2800	−
			Female	2000−2200	−
		Japan[c]	Male	2650	−
			Female	1950−2000	−
Protein	g	United Kingdom	Male	44.4	55.5
			Female	36.0	45.0
		United States	Male	−	56.0
			Female	−	46.0
		Japan	Male	50.0	60.0
			Female	40.0	50.0

[a]For the United Kingdom, the breakdown for different age groups is as follows: Males 19−24 years, 2772 kcal/day; 25−34 years, 2749 kcal/day; 35−44 years, 2629 kcal/day; 45−54 years, 2581 kcal/day. Females 19−34 years, 2175 kcal/day; 35−54 years, 2103 kcal/day (note that for Japan there are four age ranges for males and two for females). For the United Kingdom, figures are based on a median physical activity level (PAL) of 1.63 [25].
[b]For the United States, the breakdown for different age groups is as follows: Males 19−25 years, 2800 kcal/day; 26−45 years, 2600 kcal/day; 46−50 years, 2400 kcal/day. Females 19−25 years, 2200 kcal/day; 26−50 years, 2000 kcal/day (note that for the United States there are three age ranges for males and two for females). Figures for the United States are based on moderate activity level, which includes physical activity equivalent to walking 1.5−3 miles/day at 3−4 miles/hour, in addition to the activities of independent living [17].
[c]For Japan, the breakdown for different age groups is as follows: Males have one value for 18−49 years. Females 18−29 years, 1950 kcal/day; 30−49 years, 2000 kcal/day. All values for females are excluding those who are lactating or pregnant. EARs for Japan are based on medium activity levels [18].
DRVs, dietary reference values; DRIs, dietary reference intakes; EAR, estimated average requirement; RNI, reference nutrient intake; RDA, recommended dietary allowance. The RDA of the United States and Japan is equivalent to the RNI of the United Kingdom. For Japan, all figures relate to requirements for 18−49-year olds.
Source: Energy EAR for United Kingdom SACN. Scientific Advisory Committee on Nutrition. 2011. Dietary Reference Values for Energy. Available at https://www.gov.uk/government/uploads/system/uploads/attachment_data/file/339317/SACN_Dietary_Reference_Values_for_Energy.pdf. Accessed 14 May 2016. Protein, EAR, and RNI for United Kingdom Department of Health. Dietary reference values for food energy and nutrients for the United Kingdom. Report of the Panel on Dietary Reference Values of the Committee on Medical Aspects of Food Policy. United Kingdom: Department of Health, 1991. Energy and protein for the United States ODPHP Office of Disease Prevention and Health Promotion 2015-2020 Dietary Guidelines for Americans. Available at http://health.gov/dietaryguidelines/2015/guidelines/. Accessed 14 May 2016. 2015. Energy and protein for Japan MHLW. Ministry of Health, Labour and Welfare. Overview of Dietary Reference Intakes for Japanese. Available at http://www.mhlw.go.jp/file/06-Seisakujouhou-10900000-Kenkoukyoku/Overview.pdf. Accessed 14 May 2016. 2015.

TABLE 44.3 Adult Carbohydrate, Fat, Nonstarch Polysaccharide Requirements for the United Kingdom, United States, and Japan (19–50 Years for United Kingdom and United States; 18–49 Years for Japan)

Nutrient	Country	Gender	Percent of Caloric Intake. Population Average (UK)/ AMDR (US)/Dietary Goal (Japan)	Energy Equivalent (kcal/day)	Amount (g/day)
Carbohydrate	United Kingdom[a]	Male	50	1291–1386	323–347
		Female	50	1052–1088	263–272
	United States[b]	Male	45–65	1080–1820	270–455
		Female	45–65	900–1430	225–356
	Japan[c]	Male	50–65	1325–1723	331–431
		Female	50–65	975–1300	244–325
Fat	United Kingdom[a]	Male	35	903–970	100–108
		Female	35	736–761	82–85
	United States[b]	Male	20–35	480–980	53–109
		Female	20–35	400–770	44–86
	Japan[c]	Male	20–30	530–795	59–88
		Female	20–30	390–600	43–67
Dietary fiber[d]	United Kingdom	Male	–	–	30
		Female	–	–	30
	United States	Male	–	–	38
		Female	–	–	25
	Japan	Male	–	–	20 or more
		Female	–	–	18 or more

[a]For the United Kingdom, the population average (%) of total dietary energy intake is used.
[b]For the United States, the AMDR (%) of total dietary energy intake is used.
[c]For Japan, the Dietary Goal (%) of total dietary energy intake is used. For Japan, the carbohydrate Dietary Goal includes alcohol as a source of energy, but it does not imply or recommend alcohol consumption. The calculation for carbohydrate amounts (g) does not take alcohol into consideration. For all countries, figures for energy equivalent (kcal/day) and amount except dietary fiber (g) are calculated on the basis of EAR of energy shown in Table 44.1. The amount of carbohydrate and fat (but not fiber) is calculated using a conversion factor of 17 kj/g (4.0 kcal/g) for carbohydrate and 37 kj (9.0 kcal/g) for fat [25].
[d]For Dietary Fiber, figures for the United Kingdom show the recommended population average; for the United States, adequate intakes; for Japan, Dietary Goal.
AMDR, acceptable macronutrient distribution range.
Source: Population average for carbohydrate and dietary fiber for the United Kingdom SACN. Scientific Advisory Committee on Nutrition. 2015. Carbohydrates and Health. Available at https://www.gov.uk/government/uploads/system/uploads/attachment_data/file/445503/SACN_Carbohydrates_and_Health.pdf. Accessed 14 May 2016 [26]. Population average for fat for the United Kingdom Department of Health. Dietary reference values for food energy and nutrients for the United Kingdom. Report of the Panel on Dietary Reference Values of the Committee on Medical Aspects of Food Policy. United Kingdom: Department of Health, 1991. AMDR and Adequate Intakes for the United States NASEM. The National Academies of Sciences Engineering Medicine. Dietary Reference Intakes Tables and Application. 2016. Available at http://www.nationalacademies.org/hmd/Activities/Nutrition/SummaryDRIs/DRI-Tables.aspx. Accessed 14 May 2016 [27]. Dietary Goal for Japan MHLW. Ministry of Health, Labour and Welfare. Overview of Dietary Reference Intakes for Japanese. Available at http://www.mhlw.go.jp/file/06-Seisakujouhou-10900000-Kenkoukyoku/Overview.pdf. Accessed 14 May 2016. 2015.

TABLE 44.4 Mineral Requirements for Adults in the United Kingdom, United States, and Japan (19−50 Years for United Kingdom and United States; 18−49 Years for Japan)

Nutrient	Unit per Day	Country	Gender	LRNI	EAR	RNI/ RDA	I/AI	UL
Calcium[a]	mg	United Kingdom	Male	400	525	700	−	−
			Female	400	525	700	−	−
		United States	Male	−	800	1000	−	2500
			Female	−	800	1000	−	2500
		Japan	Male	−	550−650	650−800	−	2500
			Female	−	550	650	−	2500
Magnesium[b]	mg	United Kingdom	Male	190	250	300	−	−
			Female	150	200	270	−	−
		United States	Male	−	330−350	400−420	−	350
			Female	−	255−265	310−320	−	350
		Japan	Male	−	280−310	340−370	−	−
			Female	−	230−240	270−290	−	−
Phosphorus	mg	United Kingdom	Male	−	−	550	−	−
			Female	−	−	550	−	−
		United States	Male	−	580	700	−	4000
			Female	−	580	700	−	4000
		Japan	Male	−	−	−	1000	3000
			Female	−	−	−	800	3000
Sodium	mg	United Kingdom	Male	575	−	1600	−	−
			Female	575	−	1600	−	−
		United States	Male	−	−	−	1500	2300
			Female	−	−	−	1500	2300
		Japan	Male	−	600	−	−	UL
			Female	−	600	−	−	−
Potassium	mg	United Kingdom	Male	2000	−	3500	−	−
			Female	2000	−	3500	−	−
		United States	Male	−	−	−	4700	−
			Female	−	−	−	4700	−
		Japan	Male	−	−	−	2500	−
			Female	−	−	−	2000	−

(Continued)

TABLE 44.4 (Continued)

Nutrient	Unit per Day	Country	Gender	LRNI	EAR	RNI/RDA	I/AI	UL
Chloride	mg	United Kingdom	Male	–	–	2500	–	–
			Female	–	–	2500	–	–
		United States	Male	–	–	–	2300	3600
			Female	–	–	–	2300	3600
		Japan	Male	–	–	–	–	–
			Female	–	–	–	–	–
Iron[c]	mg	United Kingdom	Male	4.7	6.7	8.7	–	–
			Female	8.0	11.4	14.8	–	–
		United States	Male	–	6.0	8.0	–	45
			Female	–	8.1	18.0	–	45
		Japan	Male	–	6.0–6.5	7.0–7.5	–	50–55
			Female	–	8.5–9.0	10.5	–	40
Zinc[d]	mg	United Kingdom	Male	5.5	7.3	9.5	–	–
			Female	4.0	5.5	7.0	–	–
		United States	Male	–	9.4	11.0	–	40
			Female	–	6.8	8.0	–	40
		Japan	Male	–	8.0	10.0	–	40–45
			Female	–	6.0	8.0	–	35
Copper[e]	mg	United Kingdom	Male	–	–	1.2	–	–
			Female	–	–	1.2	–	–
		United States	Male	–	0.7	0.9	–	10
			Female	–	0.7	0.9	–	10
		Japan	Male	–	0.7	0.9–1.0	–	10
			Female	–	0.6	0.8	–	10
Selenium[f]	µg	United Kingdom	Male	40	–	75	–	–
			Female	40	–	60	–	–
		United States	Male	–	45	55	–	400
			Female	–	45	55	–	400
		Japan	Male	–	25	30	–	420–460
			Female	–	20	25	–	330–350

(Continued)

TABLE 44.4 (Continued)

Nutrient	Unit per Day	Country	Gender	LRNI	EAR	RNI/ RDA	I/AI	UL
Molybdenum [g]	µg	United Kingdom	Male	–	–	–	50–400	–
			Female	–	–	–	50–400	–
		United States	Male	–	34	45	–	2000
			Female	–	34	45	–	2000
		Japan	Male	–	20–25	25–30	–	550
			Female	–	20	20–25	–	450
Manganese	mg	United Kingdom	Male	–	–	–	Above 1.4	–
			Female	–	–	–	Above 1.4	–
		United States	Male	–	–	–	2.3	11
			Female	–	–	–	1.8	11
		Japan	Male	–	–	–	4.0	11
			Female	–	–	–	3.5	11
Chromium	µg	United Kingdom	Male	–	–	–	Above 25	–
			Female	–	–	–	Above 25	–
		United States	Male	–	–	–	35	–
			Female	–	–	–	25	–
		Japan	Male	–	–	–	10	–
			Female	–	–	–	10	–
Iodine	µg	United Kingdom	Male	70	–	140	–	–
			Female	70	–	140	–	–
		United States	Male	–	95	150	–	1100
			Female	–	95	150	–	1100
		Japan	Male	–	95	130	–	3000
			Female	–	95	130	–	3000

[a]For calcium, EAR of Japan, the breakdown for different age groups is as follows: Males 18–29 years, 650 mg/day; 30–49 years, 550 mg/day. For the RDA of calcium for Japan, the values for ages are as follows: Males 18–29 years, 800 mg/day; 30–49 years, 650 mg/day.
[b]For magnesium, EAR of the United States, the breakdown for different age groups is as follows: Males 19–30 years 330 mg/day; 31–50 years, 350 mg/day. Females 19–30 years, 255 mg/day; 31–50 years, 265 mg/day. For magnesium, RDA of the United States, the breakdown for different age groups is as follows: Males 19–30 years, 400 mg/day; 31–51 years, 420 mg/day. Females 19–30 years, 310 mg/day; 31–50 years, 320 mg/day. For magnesium, EAR of Japan, the breakdown for different age groups is as follows: Males, 18–29 years, 280 mg/day; 30–49 years, 310 mg/day. Females 18–29 years, 230 mg/day; 30–49 years, 240 mg/day. For magnesium, RDA of Japan, the breakdown for different age groups is as follows: Males 18–29 years, 340 mg/day; 30–49 years, 370 mg/day. Females 18–29 years, 270 mg/day, 30–49 years, 290 mg/day.
[c]For iron, EAR of Japan, the breakdown for different age groups is as follows: Males 18–29 years, 6.0 mg/day; 30–49 years, 6.5 mg/day. Females 18–29 years, 8.5 mg/day; 30–49 years, 9.0 mg/day (excluding not-menstruating females). For iron, RDA of Japan, the breakdown for different age groups is as follows: Males 18–29 years, 7.0 mg/day; 30–49 years, 7.5 mg/day. For iron, UL of Japan, the breakdown for different age groups is as follows: Males 18–29 years, 50 mg/day; 30–49 years, 55 mg/day.
[d]For zinc, UL of Japan, the breakdown for different age groups is as follows: Males 18–29 years, 40 mg/day; 30–49 years, 45 mg/day.
[e]For copper, RDA of Japan, the breakdown for different age groups is as follows: Males 18–29 years, 0.9 mg/day; 30–49 years, 1.0 mg/day.

(Continued)

TABLE 44.4 (Continued)

[f]For selenium, UL of Japan, the breakdown for different age groups is as follows: Males 18–29 years, 420 g/day; 30–49 years, 460 g/day. Females 18–29 years, 330 g/day; 30–49 years, 350 g/day.
[g]For molybdenum, EAR of Japan, the breakdown for different age groups is as follows: Males 18–29 years, 20 g/day; 30–49 years 25 g/day. For molybdenum, RDA of Japan, the breakdown for different age groups is as follows: Males 18–29 years, 25 g/day; 30–49 years, 30 g/day. Females 18–29 years, 20 g/day; 30–49 years 25 g/day.
AI, adequate intake; EAR, estimated average requirement; LRNI, lower reference nutrient intake; RDA, recommended dietary allowance; RNI, reference nutrient intake; SI, safe intake; UL, tolerable upper intake levels. The United Kingdom uses RNI. The United States and Japan use RDA. The RDA of the United States and Japan are equivalent to the RNI of the United Kingdom. The United Kingdom uses SI. The United States and Japan use AI. AI of the United States and Japan is equivalent to SI of the United Kingdom. No LRNI values have been derived for any nutrient in the United States and Japan. UL values that indicate guidance on high intake have not been developed in the United Kingdom.
Source: All figures for the United Kingdom Department of Health. Dietary reference values for food energy and nutrients for the United Kingdom. Report of the Panel on Dietary Reference Values of the Committee on Medical Aspects of Food Policy. United Kingdom: Department of Health, 1991. All figures for Japan MHLW. Ministry of Health, Labour and Welfare. Overview of Dietary Reference Intakes for Japanese. Available at http://www.mhlw.go.jp/file/06-Seisakujouhou-10900000-Kenkoukyoku/Overview.pdf. Accessed 14 May 2016. 2015. All figures for the United States NASEM. The National Academies of Sciences Engineering Medicine. Dietary Reference Intakes Tables and Application. 2016. Available at http://www.nationalacademies.org/hmd/Activities/Nutrition/SummaryDRIs/DRI-Tables.aspx. Accessed 14 May 2016.

TABLE 44.5 Vitamin Requirements for Adults in the United Kingdom, United States, and Japan (19–50 Years for United Kingdom and United States; 18–49 Years for Japan)

Nutrient	Unit per Day	Country	Gender	LRNI	EAR	RNI/RDA	SI/AI	UL
Vitamin A [a]	μg Retinol equivalent	United Kingdom	Male	300	500	700	–	–
			Female	250	400	600	–	–
		United States	Male	–	625	900	–	3000
			Female	–	500	700	–	3000
		Japan	Male	–	600–650	850–900	–	2700
			Female	–	450–500	650–700	–	2700
Vitamin B1 [b] (Thiamin)	mg	United Kingdom	Male	0.59–0.64	0.77–0.83	1.03–1.11	–	–
			Female	0.48–0.50	0.63–0.65	0.84–0.87	–	–
		United States	Male	–	1.00	1.20	–	–
			Female	–	0.90	1.10	–	–
		Japan	Male	–	1.20	1.40	–	–
			Female	–	0.90	1.10	–	–
Vitamin B2 (Riboflavin)	mg	United Kingdom	Male	0.8	1.0	1.3	–	–
			Female	0.8	0.9	1.1	–	–
		United States	Male	–	1.1	1.3	–	–
			Female	–	0.9	1.1	–	–
		Japan	Male	–	1.3	1.6	–	–
			Female	–	1.0	1.2	–	–

(Continued)

TABLE 44.5 (Continued)

Nutrient	Unit per Day	Country	Gender	LRNI	EAR	RNI/RDA	SI/AI	UL
Vitamin B3 [c] (Niacin) [c]	mg	United Kingdom	Male	11.4–12.2	14.2–15.2	17.0–18.3	–	–
			Female	9.3–9.6	11.6–12.0	13.9–14.4	–	–
		United States	Male	–	12.0	16.0	–	35
			Female	–	11.0	14.0	–	35
		Japan	Male	–	13.0	15.0	–	300–350
			Female	–	9.0–10.0	11.0–12.0	–	250
Vitamin B6 [d]	mg	United Kingdom	Male	0.49	0.58	0.83	–	–
			Female	0.40	0.47	0.68	–	–
		United States	Male	–	1.10	1.30	–	100
			Female	–	1.10	1.30	–	100
		Japan	Male	–	1.20	1.40	–	55–60
			Female	–	1.00	1.20	–	45
Vitamin B12	μg	United Kingdom	Male	1.0	1.25	1.50	–	–
			Female	1.0	1.25	1.50	–	–
		United States	Male	–	2.00	2.40	–	–
			Female	–	2.00	2.40	–	–
		Japan	Male	–	2.00	2.40	–	–
			Female	–	2.00	2.40	–	–
Folate [e]	μg	United Kingdom	Male	100	150	200	–	–
			Female	100	150	200	–	–
		United States	Male	–	320	400	–	1000
			Female	–	320	400	–	1000
		Japan	Male	–	200	240	–	900–1000
			Female	–	200	240	–	900–1000
Pantothenic acid	mg	United Kingdom	Male	–	–	–	3–7	–
			Female	–	–	–	3–7	–
		United States	Male	–	–	–	5	–
			Female	–	–	–	5	–
		Japan	Male	–	–	–	5	–
			Female	–	–	–	4	–

(Continued)

TABLE 44.5 (Continued)

Nutrient	Unit per Day	Country	Gender	LRNI	EAR	RNI/RDA	SI/AI	UL
Biotin	µg	United Kingdom	Male	–	–	–	10–200	–
			Female	–	–	–	10–200	–
		United States	Male	–	–	–	30	–
			Female	–	–	–	30	–
		Japan	Male	–	–	–	50	–
			Female	–	–	–	50	–
Vitamin C	mg	United Kingdom	Male	10	25	40	–	–
			Female	10	25	40	–	–
		United States	Male	–	75	90	–	2000
			Female	–	60	75	–	2000
		Japan	Male	–	85	100	–	–
			Female	–	85	100	–	–
Vitamin D [f]	µg	United Kingdom	Male	–	–	0	–	–
			Female	–	–	0	–	–
		United States	Male	–	10	15	–	100
			Female	–	10	15	–	100
		Japan	Male	–	–	–	5.5	100
			Female	–	–	–	5.5	100
Vitamin E [g]	mg	United Kingdom	Male	–	–	–	Above 4.0	–
			Female	–	–	–	Above 3.0	–
		United States	Male	–	12	15	–	1000
			Female	–	12	15	–	1000
		Japan	Male	–	–	–	6.5	800–900
			Female	–	–	–	6.0	650–700
Vitamin K [h]	µg	United Kingdom	Male	–	–	–	68.8–71.5	–
			Female	–	–	–	59.0–59.9	–
		United States	Male	–	–	–	120	–
			Female	–	–	–	90	–
		Japan	Male	–	–	–	150	–
			Female	–	–	–	150	–

[a]For vitamin A, EAR of Japan, the breakdown for different age groups is as follows: Males 18–29 years, 600 µg retinol equivalents/day; 30–49 years, 650 µg retinol equivalents/day. Females 18–29 years, 450 µg retinol equivalents/day; 30–49 years, 500 µg retinol equivalents/day. For vitamin A, RDA of Japan,

(Continued)

TABLE 44.5 (Continued)

the breakdown for different age groups is as follows: Males 18–29 years, 850 µg retinol equivalents/day; 30–49 years, 900 µg retinol equivalents/day. Females 18–29 years, 650 µg retinol equivalents/day; 30–49 years, 700 µg retinol equivalents/day.
[b]For vitamin B1, all figures for the United Kingdom are calculated on the basis of mg/1000 kcal [16] using DRVs as shown in Table 44.1.
[c]For vitamin B3 (niacin), unit per day (mg) shows as niacin equivalent, except UL of Japan, which shows in nicotinamide. For vitamin B3 (niacin), all figures for the United Kingdom are based on mg niacin equivalent/1000 kcal [16] using DRVs shown in Table 44.1. For vitamin B3 (niacin), EAR of Japan, the breakdown for different age groups is as follows: Females 18–29 years, 9 mg/day; 30–49 years, 10 mg/day. For vitamin B3 (niacin), RDA of Japan, the breakdown for different age groups is as follows: Females 18–29 years, 11 mg/day; 30–49 years, 12 mg/day. For vitamin B3 (niacin), UL of Japan, the breakdown for different age groups is as follows: Males 18–29, 300 mg/day; 30–49 years 350 mg/day.
[d]For vitamin B6, all figures for the United Kingdom are calculated based on µg/g protein [16] using DRVs shown in Table 44.1. For vitamin B6, LRNI of the United Kingdom, EAR protein figures shown in Table 44.1 are used for the calculation. For vitamin B6, UL of Japan, the breakdown for different age groups is as follows: Males 18–29 years, 55.5 mg/day; 30–40 years, 60.0 mg/day.
[e]For folate, UL of Japan, the breakdown for different age groups is as follows: Males 18–29 years, 900 µg/day; 30–49 years, 1000 µg/day. Females 18–29 years, 900 µg; 30–49 years, 1000 µg/day. For folate, UL of Japan, figures such as pteroylmonoglutamic acid contained in dietary supplement and vitamin-enriched food.
[f]For vitamin D, for the United Kingdom, no DRVs are derived, though certain at-risk individuals or groups may require dietary vitamin D [16].
[g]For vitamin E, UL of Japan, the breakdown for different age groups is as follows: Males 18–29 years, 800 mg/day; 30–49 years, 900 mg/day. Females 18–29 years, 650 mg/day; 50–69 years, 700 mg/day.
[h]For Vitamin K, for the United Kingdom, figures are calculated based on g/body weight (kg)/day [16] using weight for males 19–24 years, 71.5 kg, 25–34 years, 71.0 kg; 35–44 years, 69.7 kg; 45–54 years, 68.8 kg. For Females, 19–24 years, 59.9 kg; 25–34 years, 59.7 kg; 45–54 years, 59.0 kg [25].
DRVs, dietary reference values; DRIs, dietary reference intakes; EAR, estimated average requirement; RNI, reference nutrient intake; SI, safe intake; RDA, recommended dietary allowance; AI, adequate intake; UL, tolerable upper intake level. The United Kingdom uses RNI. The United States and Japan use RDA. The RDA of the United States and Japan is equivalent to the RNI of the United Kingdom. The United Kingdom uses SI. The United States and Japan use AI. AI for the United States and Japan is equivalent of SI for the United Kingdom. UL values, which indicate guidance on high intake, have not been developed in the United Kingdom. No LRNI values have been developed in any nutrient in the United States and Japan. UL values, which indicate guidance on high intake, have not been developed in the United Kingdom.

REQUIREMENTS IN POSTSURGICAL PROCEDURES AND THE CONCEPT OF BIOAVAILABILITY

As mentioned earlier, obesity may influence micronutrient status, and the bariatric procedures themselves will also cause some malabsorption [1,28,29]. Bariatric surgery will lead to iron deficiency due to decreases in meat consumption, and the total absorptive areas for iron are also reduced. The main sites of iron absorption are the duodenum and proximal jejunum, so any bariatric procedure that affects these regions will not only affect iron, but other dietary components such as calcium and folate [28]. Calcium and fiber also affect bioavailability of iron [30]. The impact of bariatric surgery on iron, calcium, zinc, vitamins A, B12, E, D, K, and folate have been previously reviewed [28].

There are a number of guidelines relating to dietary requirements in bariatric surgery that are either general (covering a number of micro- or macronutrients) or specific (focusing on single components such as vitamin D or K) [2,31–34]. Other guidelines are covered elsewhere in this book [35], but below we have selected those from the British Obesity and Metabolic Surgery Society (see also http://www.bomss.org.uk/). Table 44.6 displays advice on micronutrient supplementation after (1) gastric bypass and sleeve gastrectomy and (2) duodenal switch [36]. The guidelines have specific mention of supplements (multiformulations) as well as specific mention of calcium, copper, iron, selenium, and zinc, and vitamins A, B1 (thiamine), B12, D, E, and K. Although these guidelines in Table 44.6 relate to gastric bypass, sleeve gastrectomy, and duodenal switch, the original guidelines also have information for gastric balloons and gastric bands [36].

We have examined these guidelines in the context of a single specific proprietary brand of multivitamins (Table 44.7). For many of the nutrients there is some agreement between supplemental amounts and RNIs. For example, the RNI for iodine (a component of thyroid hormones) is set at 140 µg/day for adults; the proprietary brand also contains this amount in a single capsule.

Two cautionary notes are required in interpreting Table 44.7. In Table 44.7 it is perceived that twice as much folic acid is in the commercial brand of nutritional supplement compared to the RNI. However, there is a difference between the naturally occurring folate and synthetic folic acid. Folic acid taken with food is about 85% bioavailable, but the naturally occurring folate is 50% bioavailable. Furthermore, it is also important to note that not all brands of multivitamin and minerals are identical in composition.

This chapter does not constitute dietary advice per se, but provides a framework of data relating to the normative requirements for general populations. There is frequent avocation of multivitamin and multimineral use after bariatric surgery. Many multivitamin and multivitamin preparations contain formulations that reflect RNIs for many of the components. However, as a consequence of bariatric surgery, subjects will need to be considered in their own right; i.e., individualized approaches should be adopted. There should be close monitoring of subjects before and after weight loss surgery [36]. Due to alterations in bioavailability, monitoring should, where possible, entail measures of circulating levels or surrogate measures in blood rather than measures of intake or prescription amounts.

TABLE 44.6 Guidelines for Micronutrient Supplementation After Gastric Bypass and Sleeve Gastrectomy and Duodenal Switch

	Gastric Bypass and Sleeve Gastrectomy	Duodenal Switch
Multivitamin and Mineral Supplement	Multivitamin and mineral supplement should include: iron, selenium, 2 mg copper (minimum), zinc (ratio of 8–15 mg zinc for each 1 mg copper); The following meet these requirements (August 2014): • One daily Forceval (soluble and capsule). • "Over the Counter" complete multivitamin and mineral supplement, two daily—e.g., Sanatogen A-Z Complete, Superdrug A-Z multivitamins and minerals, Tesco Complete multivitamins and minerals, Lloyds Pharmacy A-Z multivitamins and minerals. **For Preconception and pregnancy.** Safe to continue with Forceval as vitamin A is in beta-carotene form, or consider pregnancy multivitamin and mineral—e.g., Seven Seas Pregnancy, Pregnacare, Boots Pregnancy Support.	Multivitamin and mineral supplement should include: • Iron • Selenium • Copper (2 mg minimum) • Zinc (ratio of 8–15 mg zinc for each 1 mg copper), The following meet these requirements (August 2014): • One daily Forceval (soluble and capsule) • "Over the Counter" complete multivitamin and mineral supplement, two daily—e.g., Sanatogen A-Z Complete, Superdrug A-Z multivitamins and minerals, Tesco Complete multivitamins and minerals, Lloydspharmacy A-Z multivitamins and minerals. **Preconception and pregnancy.** Safe to continue with Forceval as vitamin A is in beta-carotene form, or consider pregnancy multivitamin and mineral—e.g., Seven Seas Pregnancy, Pregnacare, Boots Pregnancy Support.
Iron	45–60 mg daily. 200 mg ferrous sulfate, 210 mg ferrous fumarate, or 300 mg ferrous gluconate daily. 100 mg daily for menstruating women. 200 mg ferrous sulfate or 210 mg ferrous fumarate twice daily.	45–60 mg daily. 200 mg ferrous sulfate, 210 mg ferrous fumarate, or 300 mg ferrous gluconate daily. 100 mg daily for menstruating women. 200 mg ferrous sulfate or 210 mg ferrous fumarate twice daily.
Folic acid	Contained within multivitamin and mineral supplement. Encourage consumption of folate-rich foods. If deficient, check compliance with multivitamin and mineral supplement. If compliant, check for vitamin B12 deficiency before recommending additional folic acid supplements. Additional folic acid (prescribed or over the counter) if deficient. Recheck folate levels after 4 months. **Pregnancy and preconception.** Additional folic acid (5 mg) preconception and first 12 weeks of pregnancy.	Encourage consumption of folate-rich foods. Contained within multivitamin and mineral supplement. If deficient, check compliance with multivitamin and mineral supplement. If compliant, check for vitamin B12 deficiency before recommending additional folic acid supplements. Additional folic acid (prescribed or over the counter) if deficient. Recheck folate levels after 4 months. **Preconception and pregnancy.** Additional folic acid (5 mg, but see text) preconception and first 12 weeks of pregnancy.
Vitamin B12	Intramuscular injections of 1 mg vitamin B12, 3 monthly.	Intramuscular injections of 1 mg vitamin B12, 3 monthly.
Calcium and Vitamin D	Ensure good oral intake of calcium and vitamin D-rich foods. Continue with maintenance doses of calcium and vitamin D as identified preoperatively. Treat and adjust vitamin D supplementation in line with National Osteoporosis Society Guidelines. Patients are likely to be on at least 800 mg calcium and 20 mcg vitamin D—e.g., Adcal D3, Calceos, Cacit D3; however, many patients will require additional vitamin D.	Ensure good oral intake of calcium and vitamin D-rich foods. Continue with maintenance doses of calcium and vitamin D as identified preoperatively. Treat and adjust vitamin D supplementation in line with National Osteoporosis Society Guidelines. Patients are likely to be on at least 800 mg calcium and 20 mcg vitamin D—e.g., Adcal D3, Calceos, Cacit D3; however, most patients will require additional vitamin D.

(Continued)

TABLE 44.6 (Continued)

	Gastric Bypass and Sleeve Gastrectomy	Duodenal Switch
Fat-soluble Vitamins A, E, and K	Sufficient amount contained within vitamin and mineral supplement. Additional fat-soluble vitamins may be needed if patient has steatorrhea.	Additional fat-soluble vitamins are needed. AquADEKs Softgels provide additional high doses of fat-soluble vitamins A, D, E, and K, and other vitamins and minerals. Recommend one to two daily. Alternatively supplement with additional vitamins A, E, and K as required.
Zinc and Copper	Sufficient amount contained within multivitamin and mineral supplement. If additional zinc is needed, ratio of 8–15 mg zinc per 1 mg copper must be maintained.	Sufficient amount contained within multivitamin and mineral supplement. If additional zinc is needed, ratio of 8–15 mg zinc per 1 mg copper must be maintained.
Selenium	Sufficient amount contained within multivitamin and mineral supplement. If required, additional selenium may be provided by two to three Brazil nuts a day or by over-the-counter preparations, including Selenium ACE, Holland and Barrett Selenium, Boots Selenium with Vitamins A, C, and E.	Sufficient amount contained within multivitamin and mineral supplement. If required, additional selenium may be provided by two to three Brazil nuts a day or by over-the-counter preparations, including Selenium ACE, Holland and Barrett Selenium, Boots Selenium with Vitamins A, C, and E.
Thiamine	Sufficient amount contained within multivitamin and mineral supplement. If patient experiences prolonged vomiting, always prescribe additional thiamine (thiamine 200–300 mg daily, vitamin B co strong 1 or 2 tablets, three times a day) and urgent referral to bariatric center. Those patients who are symptomatic or where there is clinical suspicion of acute deficiency should be admitted immediately for administration of IV thiamine.	Sufficient amount contained within multivitamin and mineral supplement. If patient experiences prolonged vomiting, always prescribe additional thiamine (thiamine 200–300 mg daily, vitamin B co strong 1 or 2 tablets, three times a day) and urgent referral to bariatric center. Those patients who are symptomatic or where there is clinical suspicion of acute deficiency should be admitted immediately for administration of IV thiamine.

Source: Data from O'Kane M, Pinkney P, Aasheim ET, Barth JH, Batterham RL, Welbourn R. BOMSS Guidelines on peri-operative and postoperative biochemical monitoring and micronutrient replacement for patients undergoing bariatric surgery. 2014. United Kingdom, Access on 1 July 2016 BOMSS. British Obesity and Metabolic Surgery Society. Article published online http://www.bomss.org.uk/bomss-nutritional-guidance/. For other details of guidelines see BOMSS | British Obesity & Metabolic Surgery Society http://www.bomss.org.uk/

TABLE 44.7 Composition of Forceval

Micronutrient	Amount per Capsule	UK RNI for Adults 19–50 Years	Amount in Forceval as % RNI	Comment
Vitamin A	As β-Carotene, 2500 iu. As Retinol Equivalents, 750 µg	700 µg Male 600 µg Female	107% Male 125% Female	
Vitamin D2	4000 iu (10 µg)	None	NA	10 µg for pregnancy
Vitamin B1 (Thiamin)	1.2 mg	1.0 mg Male 0.8 mg Female	120% Male 150% Female	
Vitamin B2 (Riboflavin)	1.6 mg	1.3 mg Male 1.1 mg Female	123% Male 145% Female	
Vitamin B6	2.0 mg	1.4 mg Male 1.2 mg Female	143% Male 167% Female	

(Continued)

TABLE 44.7 (Continued)

Micronutrient	Amount per Capsule	UK RNI for Adults 19–50 Years	Amount in Forceval as % RNI	Comment
Vitamin B12	3.0 µg	1.5 µg Male	200% Male	
		1.5 µg Female	200% Female	
Vitamin C	60 mg	40 mg Male	150% Male	
		40 mg Female	150% Female	
Vitamin E	10 mg	No RNI	NA	There is no RNI, but there is a safe intake of above 4 mg for men and 3 mg for women.
Biotin	100 µg	No RNI	NA	There is no RNI, but there is a safe intake of 10–200 µg/day without reference to age in the DRVs.
Nicotinamide	18 mg	17 mg Male	106% Male	
		13 mg Female	138% Female	
		As nicotinic acid equivalents.		
Pantothenic Acid	4.0 mg	No RNI	NA	There is no RNI, but there is a safe intake of 10–200 µg/day.
Folic Acid	400 µg	200 µg Male	200% Male	
		200 µg Female	200% Female	
		As folate		
Calcium	108 mg	700 mg Male	17% Male	
		700 mg Female	17% Female	
Iron	12 mg	8.7 mg Male	73% Male	
		14.8 mg Female	123% Female	
Copper	2.0 mg	1.2 mg Male	167% Male	
		1.2 mg Female	167% Female	
Phosphorous	83 mg	550 mg Male	15% Male	
		550 mg Female	15% Female	
Magnesium	30 mg	300 mg Male	10% Male	
		300 mg Female	10% Female	
Potassium	4.0 mg	3500 mg Male	0.1% Male	
		3500 mg Female	0.1% Female	
Zinc	15 mg	9.5 mg Male	158% Male	
		7.0 mg Female	214% Female	
Iodine	140 µg	140 µg Male	100% Male	
		140 µg Female	100% Female	

(Continued)

TABLE 44.7 (Continued)

Micronutrient	Amount per Capsule	UK RNI for Adults 19–50 Years	Amount in Forceval as % RNI	Comment
Manganese	3.0 mg	No RNI	NA	There is no RNI, but there is a safe intake of above 1.4 mg/day for adults.
Selenium	50 µg	60 µg Male	120% Male	
		75 µg Female	150% Female	
Chromium	200 µg	No RNI	NA	There is no RNI, but there is a safe intake of above 25 µg/day for adults.
Molybdenum	250 µg	No RNI	NA	There is no RNI, but there is a safe intake of 50–400 µg/day.

Source: Data on Forceval from Alliance Pharmaceuticals Limited, United Kingdom, with permission.

KEY FACTS

- The majority of the population in many countries consume ethanol (i.e., alcohol) in various types of beverages.
- It has been reported that Roux-en-Y gastric bypass (RYGB) increases the risk of alcohol misuse, and this has been supported by work on experimental animals.
- In chronic alcohol misuse there is damage to virtually all regions of the gastrointestinal tract. Alcohol misuse damages microvilli, and disrupts cell membranes and active transport processes.
- Pancreatic damage due to alcohol will cause fat malabsorption, which also contributes to impaired absorption of fat-soluble vitamins.
- Alcohol misuse perturbs the absorption, storage, metabolism, or nutritional status of many micro- and macronutrients.
- Iron absorption, in contrast, may increase due to the effects of alcohol.
- There is little information as to how low-to-moderate alcohol consumption interferes with the metabolism or absorption of micro- and macronutrients.

SUMMARY POINTS

- There are various international guidelines that give advice on the amounts of micro- and macronutrients to be consumed in order to maintain optimal health. These guidelines are designed for the general population in good health.
- When specifying or describing advisory amounts, it is important to state what system is being used, such as the UK DRVs or the source of updated material, such as the UK's SACN.
- Micro- and macronutrients can be considered at four levels, namely, the composition of foods, the amount ingested, requirements, and the actual status of the individual.
- DRVs can be used for providing benchmarks for the provision of micro- and macronutrients.
- The amounts of micronutrients in many vitamin and mineral supplements match, in broad terms, the amounts suggested in DRVs.
- DRVs do not take into account illness or interventional procedures such as bariatric surgery, which also alter the bioavailability of micro- and macronutrients.

ACKNOWLEDGMENTS

We wish to thank Paul Sharp (King's College London) for providing the framework for some of this chapter, encouragement, and advice.

REFERENCES

[1] Mechanick JI. Clinical practice guidelines for the perioperative nutritional, metabolic, and nonsurgical support of the bariatric surgery patient-2013 update: cosponsored by American association of clinical endocrinologists, the obesity society, and American society for metabolic & bariatric surgery. Obesity 2013;21:S1—27.

[2] Bosnic G. Nutritional requirements after bariatric surgery. Crit Care Nurs Clin N Am 2014;26:255—62.

[3] Peterson LA, Peterson LA. Treatment for vitamin D deficiency prior to bariatric surgery: a prospective cohort study. Obes Surg 2016;26:1146—9.

[4] Krzizek EC. Prevalence of micronutrient deficiency in patients with morbid obesity before bariatric surgery. Obesity Facts 2015; Conference: 22nd Congress of the European Congress on Obesity, ECO 2015 Prague Czech Republic.

[5] Slater GH. Serum fat-soluble vitamin deficiency and abnormal calcium metabolism after malabsorptive bariatric surgery. J Gastrointes Surg 2004;8:48—55.

[6] Aasheim ET. Vitamin status after bariatric surgery: a randomized study of gastric bypass and duodenal switch. Am J Clin Nutr 2009;90:15—22.

[7] Billeter AT. Malabsorption as a therapeutic approach in bariatric surgery. Viszeralmed Gastrointes Med Surg 2014;30:198—204.

[8] Lanzarini E. High-dose vitamin D supplementation is necessary after bariatric surgery: a prospective 2-year follow-up study. Obes Surg 2015;25:1633—8.

[9] Xanthakos SA. Nutritional deficiencies in obesity and after bariatric surgery. Pediat Clin N Am 2009;56:1105—21.

[10] Devlieger R. Micronutrient levels and supplement intake in pregnancy after bariatric surgery: a prospective cohort study. PLoS One 2014;9: e114192.

[11] Goodman JC. Neurological complications of bariatric surgery. Curr Neurol Neurosci Rep 2015;15.

[12] Papamargaritis D. Copper, selenium and zinc levels after bariatric surgery in patients recommended to take multivitamin-mineral supplementation. J Trace Elem Med Biol Organ Soc Miner Trace Elem (GMS) 2015;31:172

[13] Dunstan MJ. Variations in oral vitamin and mineral supplementation following bariatric gastric bypass surgery: a national survey. Obes Surg 2015;25:648—55.

[14] Klassen K. Dietary intakes of HIV-infected adults in urban UK. Eur J Clin Nutr 2013;67:890—3.

[15] Livingstone KM. Fat mass- and obesity-associated genotype, dietary intakes and anthropometric measures in European adults: the Food4Me study. Br J Nutr 2016;115:440—8.

[16] Department of Health. Dietary reference values for food energy and nutrients for the United Kingdom. Report of the Panel on Dietary Reference Values of the Committee on Medical Aspects of Food Policy. United Kingdom: Department of Health; 1991.

[17] ODPHP. Office of Disease Prevention and Health Promotion 2015-2020 Dietary Guidelines for Americans, 2015. Available at http://health.gov/dietaryguidelines/2015/guidelines/. Accessed 14 May 2016.

[18] MHLW. Ministry of Health, Labour and Welfare. Overview of Dietary Reference Intakes for Japanese, 2015. Available at http://www.mhlw.go.jp/file/06-Seisakujouhou-10900000-Kenkoukyoku/Overview.pdf. Accessed 14 May 2016.

[19] Mondal C. Studies on goitrogenic/ antithyroidal potentiality of thiocyanate, catechin and after concomitant exposure of thiocyanate-catechin. Inter J Pharm Clin Res 2016;8:108—16.

[20] World MJ. Alcoholic malnutrition and the small intestine. Alcohol Alcoholism 1985;20:89—124.

[21] Harrison-Findik DD. Alcohol metabolism-mediated oxidative stress down-regulates hepcidin transcription and leads to increased duodenal iron transporter expression. J Biol Chem 2006;281:22974—82.

[22] Cuellar-Barboza AB. Change in consumption patterns for treatment-seeking patients with alcohol use disorder post-bariatric surgery. J Psych Res 2015;78:199—204.

[23] de Araujo Burgos MGP. Prevalence of alcohol abuse before and after bariatric surgery associated with nutritional and lifestyle factors: A study involving a Portuguese population. Obes Surg 2015;25:1716—22.

[24] Ferrario C. Increased risk of alcohol use disorders after bariatric surgery. Revue Medicale Suisse 2016;12:602—5.

[25] SACN. Scientific Advisory Committee on Nutrition. 2011. Dietary Reference Values for Energy. Available at https://www.gov.uk/government/uploads/system/uploads/attachment_data/file/339317/SACN_Dietary_Reference_Values_for_Energy.pdf. Accessed 14 May 2016.

[26] SACN. Scientific Advisory Committee on Nutrition. 2015. Carbohydrates and Health. Available at https://www.gov.uk/government/uploads/system/uploads/attachment_data/file/445503/SACN_Carbohydrates_and_Health.pdf. Accessed 14 May 2016.

[27] NASEM. The National Academies of Sciences Engineering Medicine. Dietary Reference Intakes Tables and Application. 2016. Available at http://www.nationalacademies.org/hmd/Activities/Nutrition/SummaryDRIs/DRI-Tables.aspx. Accessed 14 May 2016.

[28] Sawaya RA. Vitamin, mineral, and drug absorption following bariatric surgery. Curr Drug Metabol 2012;13:1345—55.

[29] Zhou J-C. Oral vitamin D supplementation has a lower bioavailability and reduces hypersecretion of parathyroid hormone and insulin resistance in obese Chinese males. Public Health Nutr 2015;18:2211—19.

[30] Bueno L. Iron (FeSo4) bioavailability in obese subjects submitted to bariatric surgery. Nutr Hospit 2013;28:100—4.

[31] Jastrzebska-Mierzynska M. Dietetic recommendations after bariatric procedures in the light of the new guidelines regarding metabolic and bariatric surgery. Roczniki Panstwowego Zakladu Higieny 2015;66:13—19.

[32] Ewang-Emukowhate M. Vitamin K and other markers of micronutrient status in morbidly obese patients before bariatric surgery. Inter J Clin Pract 2015;69:638—42.

[33] Chakhtoura MT, Chakhtoura MT. ORCID: http:/. guidelines on vitamin D replacement in bariatric surgery: identification and systematic appraisal. Metabol Clin Exp 2016;65:586—97.

[34] O'Kane M. Guidelines for the follow-up of patients undergoing bariatric surgery. Clin Obes 2016;6:210−24.

[35] Rajendram R, Martin C, Preedy VR. Metabolism and pathophysiology of bariatric surgery: nutrition, procedures, outcomes and adverse effects. Cambridge, MA: Academic Press; 2016.

[36] O'Kane M, Pinkney P, Aasheim ET, Barth JH, Batterham RL, Welbourn R. BOMSS Guidelines on peri-operative and postoperative biochemical monitoring and micronutrient replacement for patients undergoing bariatric surgery, UK, 2014. Access on 1 July 2016 BOMSS. British Obesity and Metabolic Surgery Society. Article published online http://www.bomss.org.uk/bomss-nutritional-guidance/.

Chapter 45

Meal Disposal After Bariatric Surgery

S. Camastra and B. Astiarraga
University of Pisa, Pisa, Italy

LIST OF ABBREVIATIONS

ApoB48 apolipoprotein B48
BPD biliopancreatic diversion
CCK cholecystokinin
EGP endogenous glucose production
FFA free fatty acid
GIP gastric inhibitory polypeptide/glucose-dependent insulinotropic polypeptide
GLP-1 glucagon-like peptide-1
GLUT-4 glucose transporter-4
NGT normal glucose tolerance
OGTT oral glucose tolerance test
RYGB Roux-en-Y gastric bypass
SG sleeve gastrectomy
T2D type 2 diabetes mellitus

INTRODUCTION

The metabolic changes after bariatric surgery have attracted considerable attention over the last several years; however, the mechanisms of action of surgical procedures are not completely understood. The gastrointestinal tract is the direct target of most bariatric procedures, so it is reasonable to suppose that the intestinal rearrangement, its functional adaptation to the new anatomical condition, as well as the changes in the kinetics of nutrient absorption, could play an important role in the metabolic changes obtained. The available different surgical procedures can be categorized depending on whether they are predominantly restrictive of meal assumption, predominantly malabsorptive of nutrients, or a combination of both.

Macronutrient disposal after a meal can be different depending on the type of surgery, but many aspects still remain poorly understood. Fig. 45.1 describes the main physiological events following meal ingestion and the principal roles of the different gastrointestinal tracts. Table 45.1 summarizes the principal gastrointestinal changes that occur in most practiced surgical techniques representative of the three categories of bariatric surgery: sleeve gastrectomy (SG) for restrictive surgery, biliopancreatic diversion (BPD) for malabsorptive surgery, and Roux-en-Y gastric bypass (RYGB) for mixed surgery. This chapter describes the current knowledge regarding this topic, with particular focus on the effects of RYGB and BPD.

ROUX-ᴇɴ-Y GASTRIC BYPASS

RYGB determines extensive rearrangements of the upper gastrointestinal tract. The anatomical changes of the gastrointestinal apparatus induce essential modifications in nutrient disposal after a meal as a consequence of a number of mechanisms; among these are the changes in gastrointestinal transit [1,2], nutrient absorption [3], enterohormonal release [4], intestinal microbiome [5], and gut adaptation [6].

Metabolism and Pathophysiology of Bariatric Surgery.

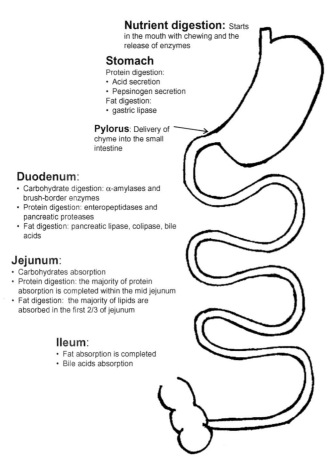

Nutrient digestion: Starts in the mouth with chewing and the release of enzymes

Stomach
Protein digestion:
• Acid secretion
• Pepsinogen secretion
Fat digestion:
• gastric lipase

Pylorus: Delivery of chyme into the small intestine

Duodenum:
• Carbohydrate digestion: α-amylases and brush-border enzymes
• Protein digestion: enteropeptidases and pancreatic proteases
• Fat digestion: pancreatic lipase, colipase, bile acids

Jejunum:
• Carbohydrates absorption
• Protein digestion: the majority of protein absorption is completed within the mid jejunum
• Fat digestion: the majority of lipids are absorbed in the first 2/3 of jejunum

Ileum:
• Fat absorption is completed
• Bile acids absorption

FIGURE 45.1 Nutrient digestion and absorption in normal condition.

TABLE 45.1 Gastrointestinal Changes Produced by Bariatric Surgery

	SG	RYGB	BPD
Gastric emptying	↑	↑↑ for liquid ↓↑ for solid	n/a
Gastric pouch (mL)	<100	30−50	200−500
Acid secretion	↓	↓	↓
Pylorus	Yes	No	No
Biliopancreatic limb	No	Yes	Yes
Intestinal transit time	↔	↓↑	n/a

The table summarizes anatomic and functional changes in the gastrointestinal tract after bariatric surgery: increase (↑); decrease (↓); unchanged (↔); information unavailable or not found (n/a).

The meal ingested passes freely from the stomach into the jejunum. The studies that have evaluated gastric emptying using the scintigraphy technique [7,8] or the indirect method by acetaminophen [9] show a rapid gastric emptying after liquid meal ingestion in patients who have undergone RYGB compared to control subjects. Contrasting results are reported after ingestion of a solid meal, gastric emptying has been shown to be both more rapid [8] and slower [7]. Regarding intestinal transit in patients who underwent RYGB, after the arrival of food into the jejunum, the transit in the small bowel has been found to be prolonged both for liquid and solid meals in some studies [8], while in others more rapid for liquid [10] or unchanged [11].

Carbohydrate Handling

Studies directed at evaluating nutrient absorption in the intestinal tract after RYGB failed to find an impairment of carbohydrate absorption [3,12].

The impact of RYGB on carbohydrate disposal of a mixed meal was evaluated using a meal test combined with a tracer technique in which the carbohydrate component of the meal consisted of a glucose solution [13,14]. The results of these studies indicated that both early [14] and long after surgery [13], the postmeal plasma glucose and insulin profiles were extremely altered by the surgery. Thus, there was a large excursion of glucose peaking at 60 minutes after the meal ingestion followed by a sharp drop to basal levels or below (Fig. 45.2). The time course of insulin secretion ran parallel to that of plasma glucose. Both in diabetic and in nondiabetic patients, the glycemic and insulinemic pattern observed after surgery matched a similar pattern of oral glucose appearance, both very different compared to presurgery. Most of the glucose was absorbed in the first hour after meal ingestion, approximately 35 of the 75 g of the glucose ingested, corresponding to ≈45% of the total amount ingested compared to ≈30% of the presurgery condition (Fig. 45.3). This amount represented 70% of the glucose appearing in 5 hours after the meal for nondiabetic patients and 60% for diabetic patients, with an increase of 30% in nondiabetic and 20% in diabetic patients compared to presurgery [13,14]. Other studies using similar techniques confirmed these proportions [15,16].

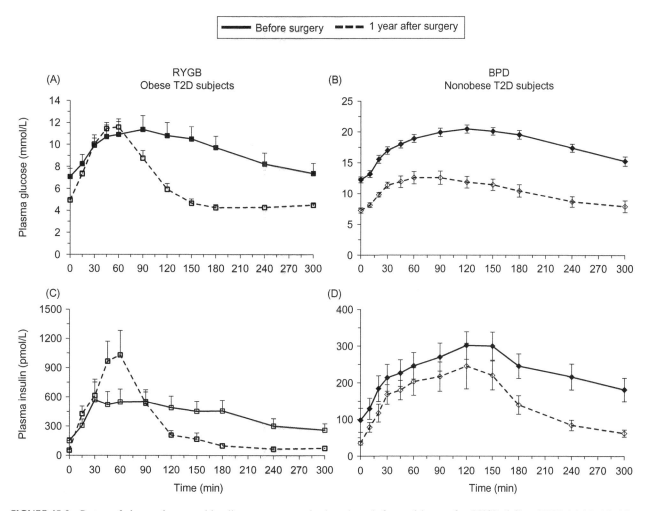

FIGURE 45.2 Pattern of plasma glucose and insulin response to a mixed meal test before and 1 year after RYGB (*left*) and BPD (*right*). (A) After RYGB there is a faster increase in plasma glucose followed by rapid return to basal values or lower. (B) After BPD plasma glucose shows a large improvement, although the curve pattern is similar to presurgery. (C) After RYGB plasma insulin ran in parallel with plasma glucose. (D) After BPD plasma insulin shows a similar pattern of plasma glucose. *Adapted from Camastra S, Muscelli E, Gastaldelli A, Holst JJ, Astiarraga B, Baldi S, et al. Long-term effects of bariatric surgery on meal disposal and β-cell function in diabetic and nondiabetic patients, Diabetes 2013;62:3709—17; Astiarraga B, Gastaldelli A, Muscelli E, Baldi S, Camastra S, Mari A, et al. Biliopancreatic diversion in non-obese patients with type 2 diabetes: impact and mechanisms, J Clin Endocrinol Metab 2013;98:2765—73.*

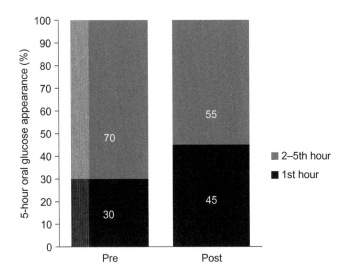

FIGURE 45.3 Percentage of the appearance of oral glucose during the 5-hour period after mixed meal (75 g glucose) in normal glucose-tolerant subjects before and after RYGB. In *blue* (black in print versions) the glucose appeared in the first hour and in *green* (gray in print versions) the glucose appeared from the second to the fifth hour. During the first hour (*blue* (black in print versions)), 30% and 45% of the total oral glucose appearance was recovered before and 1 year after RYGB, respectively. *Data calculated from Camastra S, Muscelli E, Gastaldelli A, Holst JJ, Astiarraga B, Baldi S, et al. Long-term effects of bariatric surgery on meal disposal and β-cell function in diabetic and nondiabetic patients, Diabetes. 2013;62:3709—17.*

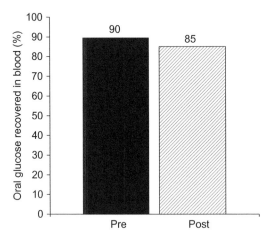

FIGURE 45.4 Percentage of oral glucose recovery during a 6-hour mixed meal test (75 g glucose) in normal glucose-tolerant subjects before and after surgery. The figure shows that the total percentage of ingested glucose that appeared in the circulation was slightly reduced (about 4%) after surgery. *Calculated from data from Magkos F, Bradley D, Eagon JC, Patterson BW, Klein S. Effect of Roux-en-Y gastric bypass and laparoscopic adjustable gastric banding on gastrointestinal metabolism of ingested glucose. Am J Clin Nutr. 2016;103:61—5.*

However, if the time course of oral glucose appearance was changed after surgery, the total amount of oral glucose appearing in plasma in the 5-hour postmeal period was only slightly reduced (about 6% and 3% in nondiabetic and diabetic patients, respectively) [13]. Other studies confirmed a similar slight reduction (about 4%) of the total oral glucose appearance after a mixed meal compared to presurgery [16] (Fig. 45.4).

Thus the observed pattern of plasma glucose after a meal is mainly the consequence of the altered glucose delivery of the ingested glucose in the systemic circulation, in turn a result of accelerated gastric emptying in accord with the observation that plasma glucose concentration increased with the increase in the rate of gastric emptying [17].

The accelerated gastric emptying after RYGB determines an increase of GLP-1 release in response to a meal [4,13], which contributes to the increased insulinemic response to the meal and to the consequent tendency to hypoglycemia observed 2 hours after glucose ingestion in patients who underwent RYGB [13].

In synergy with glucose absorption, the hepatic glucose production after meal ingestion is characterized by an initial suppression followed by a recovery in glucose production mainly from the second hour after the meal when plasma glucose concentrations are back to basal levels or below [13]. The time course of the endogenous glucose production (EGP) after the meal is critical to preventing excessive postprandial plasma glucose concentrations early after meal ingestion and to preventing hypoglycemia later when plasma glucose declines. Plasma glucagon rose after the meal,

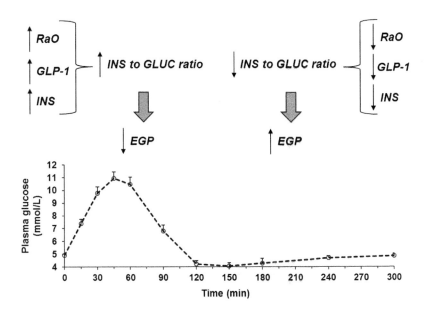

FIGURE 45.5 Determinants of postmeal plasma glucose profile after RYGB. *RaO*, rate of oral glucose appearance; *GLP-1*, glucagon like peptide-1; *INS*, plasma insulin; *INS-to-GLUC ratio*, molar ratio from insulin secretion values and plasma glucagon during the entire postmeal period (5 hours); *EGP*, endogenous glucose production.

and consequently, the prehepatic insulin-to-glucagon ratio droppes rapidly after the first hour peak, which coincides with the lower plasma glucose levels [13]. The postprandial EGP is probably a consequence of the integrated liver response to insulin, glucagon, and glucose levels (Fig. 45.5).

An important consequence of weight loss after RYGB is the improvement of tissue insulin sensitivity [18], with a more efficient peripheral utilization of glucose that contributes to the improvement of glycemic control, especially in diabetic patients.

Studies in rats demonstrate that the Roux limb displays hyperplasia and hypertrophy [19] and that exposing the Roux limb to undigested nutrients determines an adaptation of intestinal glucose metabolism such as change in glucose transporter and glucose uptake [20]. The reprogramming of intestinal glucose metabolism could make the gut an important organ for glucose disposal, contributing to the improvement in glycemic control after RYGB.

Protein Handling

Although 50% of protein absorption occurs in the duodenum, the entire small bowel can perform this function through the action of the enzymes from exocrine pancreas and the proteases located in the brush border membrane of the enterocytes. Despite exclusion of a major part of the stomach and the mixing of gastropancreatic secretions in the distal part of the jejunum, RYGB does not appear to impair protein digestion. Using a semiliquid mixed meal containing intrinsically labeled calcium caseinate as a protein source combined with labeled leucine infusion [21] 3 months after RYGB, an accelerated appearance of amino acids, derived from the protein ingested, was observed. Thus, while the alteration of gastric acid secretion was expected not to influence protein absorption, as demonstrated from a study on total gastrectomy [22], it was less obvious that the delay of the encounter of the pancreatic juice with the meal would not influence protein digestion [23], also considering the possibility of protein malnutrition in patients undergoing RYGB [24].

Since pancreatic enzymes are crucial for protein absorption, the effect of RYGB is likely to be a reflection of a rapid delivery of nutrients from the upper anastomosis to the common limb, as observed after a liquid meal [10].

The small influence of RYGB on protein absorption has also been shown in a study that found a nonsignificant reduction of protein absorption after long-limb RYGB (a variant of RYGB in which the excluded limb is lengthened). Protein absorption was estimated by the coefficient of absorption of nutrient calculated from the nutrient intake and fecal output [3].

Accelerated protein digestion and amino acid absorption can influence the metabolic and hormonal changes that occur after RYGB. Amino acids, such as leucine and phenylalanine, are able to stimulate insulin secretion from pancreatic beta-cell [25] and GLP-1 secretion from intestine L-cells [26]. After RYGB, the accelerated protein digestion that follows a mixed meal may enhance postprandial insulin and GLP-1 release [15]. Moreover, high concentrations of amino acids could contribute to the paradoxical increase in postprandial glucagon release [27] observed after RYGB in response to mixed meals [13,28].

Lipid Handling

Under physiological conditions, lipids enter into the duodenum, eliciting the secretion of CCK, which stimulates the release of bile and pancreatic enzymes. Following RYGB, lipids do not pass through the duodenum, and consequently the secretion of bile and lipolytic enzymes is reduced. The delayed breakdown of dietary fats and the delayed formation of micelles should limit the amount of fat available for absorption. From a study conducted in patients who underwent long-limb RYGB [3], it has been calculated that the coefficients of fat absorption averaged 92.1% before surgery and decreased an average of 71.9% at 5 months and 68.1% at 14 months after surgery.

The serum concentrations of triglycerides after a mixed meal result decreased after RYGB compared to presurgery [29], suggesting a reduced lipid absorption.

However, from a recent study [30] conducted on a small group of patients, it emerged that after a mixed meal ingestion, RYGB patients compared to a nonoperated control group had earlier but lower peak triglyceride response, a higher peak of apoliprotein B48 (ApoB48) response, and an earlier peak and higher values of bile acid and CCK concentration. ApoB48, the primary proteic component of chylomicrones, can be considered a biomarker for postmeal lipid metabolism. Since intestinal lipoprotein production is determined primarily by the amount of fat digested and absorbed, the authors interpreted that the combination of a lower postprandial triglycerides peak with the early postprandial ApoB48 and bile acid response reflected an accelerated delivery of lipids to the distal small intestine with a consequent enhanced fat absorption and, at the same time, an increased metabolic clearance of plasma triglycerides.

BILIOPANCREATIC DIVERSION

BPD is a malabsorptive bariatric surgery designed to promote high malabsorption, mainly for fats, with moderate restriction of food intake. The metabolic effects of BPD go beyond the simple weight loss from energy restriction [31]. BPD resolves several obesity-associated comorbidities such as type 2 diabete mellitus (T2D), hyperlipidemia, and low-grade inflammation, and it also reduces cardiovascular risk.

Carbohydrate Handling

Data concerning glucose metabolism after meal ingestion in BPD derive mainly from oral glucose tolerance tests (OGTT) and mixed meals. To our knowledge there are no studies using a tracer to evaluate the rate of appearance of carbohydrates deriving from a meal ingestion after BPD. However, analyzing the glycemic curves after OGTT or after carbohydrate ingestion of a mixed meal, we can observe that the mean glucose levels or the area under the glycemic curve are reduced after surgery both in diabetic and in nondiabetic patients; however, the glycemic time course does not change after surgery compared to before surgery [32−34]. The latter is a distinctive characteristic compared to RYGB, and it suggests that in BPD there is no effect of accelerated gastric emptying observed in others surgeries (Fig. 45.2).

The reduction of plasma glucose is followed by a reduction in insulin secretion, which confirms the evidence that after BPD there is a drastic amelioration in insulin sensitivity [35] and, in diabetic patients, in beta cell function [36] already evident early after surgery [32]. After BPD, an increase in gene expression of GLUT-4 and hexokinase in skeletal muscle was observed, as well as a higher carbohydrate oxidation as assessed by indirect calorimetry [37]. Others have shown early recovery of hepatic insulin resistance in nonobese T2D [33] and recovery in incretin effect after BPD [38].

Protein Handling

BPD delays the mixing of biliopancreatic juice and nutrients, provoking maldigestion, and therefore results in malabsorption [39,40]. Protein deficiency is one of the most serious negative impacts of BPD. Analyzing the rate of surgical revision in patients who have undergone BPD, it was found that 3−18% of patients were reoperated to extend the common limb to avoid severe protein malnutrition [41].

A study conducted in a group of 95 subjects who underwent BPD showed levels of serum albumin ≤ 3 mg/dL in 2.8% and 3% of the population 1 and 2 years postsurgery, respectively [42]. Studies directed at investigating diet protein absorption in humans after BPD found that about 30% of the protein intake was not absorbed [39,40]. In these studies, since the amount of nitrogen found in urine and stool exceeded the amount explained from malabsorption, the authors hypothesized a loss of endogenous proteins as a contributor to protein deficiency that can occur after BPD [39]. Similar results were obtained when BPD was applied to rats [43].

In contrast to the findings for fat and starch, for which an increase of intake corresponded to an increase in the amount lost in stool, for proteins an increase of dietary intake does not correlate with urine and fecal loss. This indicated that for proteins, the increase of consumption correlates to major absorption [40]. From the data available, protein malnutrition is quite rare in BPD, although the rates are higher than in RYGB. However, contrary to what happens for fats and carbohydrates, an increase of protein intake should overcome the protein loss.

Lipid Handling

In BPD, fat malabsorption can be explained by a small absorptive area in the common limb in which biliopancreatic juice meets ingested fat. In patients who underwent BPD and maintained an isoenergetic and isonitrogenic diet for 3 days, the analysis of stool and urine collected found that subjects absorbed about 32%, ranging from 12% to 59% of fat. The rest was lost in stool [39].

Lipid malabsorption has been documented by measuring postprandial lipemia. From the analysis of the 24-hour profile of serum triglycerides, it is possible to note that after BPD patients do not show the postprandial triglycerides increase observed before surgery [35]. Similar results were obtained with other malabsorptive surgeries such as BPD with duodenal switch [44]. In patients who underwent BPD, lipid withdrawal was associated with an important reduction in muscle fat, as well as an increase in the skeletal muscle expression of molecular markers of insulin sensitivity (GLUT-4) [35]. Finally, in a group of 15 nonobese diabetic subjects studied with a 5-hour meal test 2 months and 1 year after BPD, the data showed a gradual recovery in plasma free fatty acid (FFA) to ranges no longer different from the control group [33]. The general metabolic improvement was also observed in a fasting lipid profile, in insulin sensitivity, and in beta-cell function of T2D patients. Several studies, using the euglycemic hyperinsulinemic clamp technique, found a complete recovery of insulin sensitivity after BPD [35,45,46]. In some cases, nondiabetic morbidly obese subjects showed supranormal insulin sensitivity value during the clamp [35]. From this information, it could be assumed that the improvement in lipid profile after BPD could be a result of the combination of reduced fat absorption and amelioration of fat metabolism, mainly regarding insulin action in skeletal muscle, liver, and adipose tissue.

MINI-DICTIONARY OF TERMS

- *Endogenous glucose production (EGP)*: Production of glucose by the kidney and liver from compounds such as amino acids, lactate, and glycerol. EGP ensures constant plasma glucose concentrations to glucose-dependent tissues.
- *Insulin-to-glucagon molar ratio*: The molar ratio from insulin secretion values and plasma glucose concentration. Insulin and glucagon are regulators of glucose metabolism with opposite functions.
- *Insulin resistance*: The resistance of the tissues to the biological effects of insulin.
- *Rate of oral glucose appearance*: The rate of appearance in the circulation of ingested glucose, measured by the glucose tracer method.
- *Chylomicrons*: Lipoproteins that consist of phospholipids, triglycerides, cholesterol, and proteins (apolipoproteins). They transport dietary lipids from the intestine to the tissues.
- *Apolipoprotein B48 (ApoB48)*: An apolipoprotein produced exclusively by the intestine and the main protein component of chylomicrons.
- *Bile acids*: Steroid acids synthesized by the liver and released from the gall bladder into the duodenum. They are responsible for the emulsification and solubilization of lipids and liposoluble vitamins, facilitating their digestion and absorption. They are reabsorbed by the ileus and return to the liver through the enterohepatic circulation.

KEY FACTS

- The digestion and absorption of carbohydrates, proteins, and lipids are influenced by the type of surgical procedure.
- The accelerated gastric delivery after RYGB influences the glucose rate of appearance of glucose in the circulation and determines hormonal changes.
- Hormonal changes and gut adaptation influence nutrient disposal.
- Protein and lipid absorption are reduced after malabsorptive procedures.
- Bariatric surgery improves insulin sensitivity.

SUMMARY POINTS

- The gastrointestinal rearrangement after bariatric surgery plays a role in the metabolic changes observed after surgery.
- The rapid delivery of a meal into the jejunum after Roux-en-Y gastric bypass (RYGB) contributes to the rapid appearance of glucose in peripheral circulation; the majority of the glucose appears in the first hour after meal ingestion.
- After RYGB, the rapid delivery of the meal into the jejunum and the rapid oral glucose appearance determine the postprandial increase of insulin secretion, increased glucagon-like peptide-1 and gastric inhibitory polypeptide, and paradoxical increase of glucagon and reduced suppression of endogenous glucose production. These changes contribute to the postprandial glycemic pattern observed and characterized by the initial increase in plasma glucose levels followed by a sharp drop.
- Protein digestion and amino acid absorption appear to be accelerated after RYGB.
- The reduction of serum triglyceride concentrations after meal ingestion suggests a reduced lipid absorption or an increased clearance after RYGB.
- Biliopancreatic diversion (BPD) is characterized by a significant improvement of insulin sensitivity and beta-cell function, and a reduction of plasma glucose, but the postmeal glycemic pattern is not influenced. Protein, and especially lipid absorption, is reduced.

REFERENCES

[1] Marathe CS, Rayner CK, Jones KL, Horowitz M. Relationships between gastric emptying, postprandial glycemia, and incretin hormones. Diabetes Care 2013;36:1396–405.
[2] Chambers AP, Smith EP, Begg DP, et al. Regulation of gastric emptying rate and its role in nutrient-induced GLP-1 secretion in rats after vertical sleeve gastrectomy. Am J Physiol Endocrinol Metab 2014;306:E424–32.
[3] Odstrcil EA, Martinez JG, Santa Ana CA, Xue B, Schneider RE, Steffer KJ, et al. The contribution of malabsorption to the reduction in net energy absorption after long-limb Roux-en-Y gastric bypass. Am J Clin Nutr 2010;92:704–13.
[4] Holst JJ. Enteroendocrine secretion of gut hormones in diabetes, obesity and after bariatric surgery. Curr Opin Pharmacol 2013;13:983–8.
[5] Aron-Wisnewsky J, Clement K. The effects of gastrointestinal surgery on gut microbiota: potential contribution to improved insulin sensitivity. Curr Atheroscler Rep 2014;16:454.
[6] Seeley RJ, Chambers AP, Sandoval DA. The role of gut adaptation in the potent effects of multiple bariatric surgeries on obesity and diabetes. Cell Metab 2015;21:369–78.
[7] Horowitz M, Cook DJ, Collins PJ, Harding PE, Hooper MJ, Walsh JF, et al. Measurement of gastric emptying after gastric bypass surgery using radionuclides. Br J Surg 1982;69:655–7.
[8] Dirksen C, Damgaard M, Bojsen-Møller KN, Jørgensen NB, Kielgast U, Jacobsen SH, et al. Fast pouch emptying, delayed small intestinal transit, and exaggerated gut hormone responses after Roux-en-Y gastric bypass. Neurogastroenterol Motil 2013;25 346-e255
[9] Falkén Y, Hellström PM, Holst JJ, Näslund E. Changes in glucose homeostasis after Roux-en-Y gastric bypass surgery for obesity at day three, two months, and one year after surgery: role of gut peptides. J Clin Endocrinol Metab 2011;96:2227–35.
[10] Nguyen NQ, Debreceni TL, Bambrick JE, Bellon M, Wishart J, Standfield S, et al. Rapid gastric and intestinal transit is a major determinant of changes in blood glucose, intestinal hormones, glucose absorption and postprandial symptoms after gastric bypass. Obesity 2014;22:2003–9.
[11] Carswell KA, Vincent RP, Belgaumkar AP, Sherwood RA, Amiel SA, Patel AG, et al. The effect of bariatric surgery on intestinal absorption and transit time. Obes Surg 2014;24:796–805.
[12] Wang G, Agenor K, Pizot J, Kotler DP, Harel Y, Van Der Schueren BJ, et al. Accelerated gastric emptying but no carbohydrate malabsorption 1 year after gastric bypass surgery (GBP). Obes Surg 2012;22:1263–7.
[13] Camastra S, Muscelli E, Gastaldelli A, Holst JJ, Astiarraga B, Baldi S, et al. Long-term effects of bariatric surgery on meal disposal and β-cell function in diabetic and nondiabetic patients. Diabetes 2013;62:3709–17.
[14] Bradley D, Conte C, Mittendorfer B, Eagon JC, Varela JE, Fabbrini E, et al. Gastric bypass and banding equally improve insulin sensitivity and β cell function. J Clin Invest 2012;122:4667–74.
[15] Jacobsen SH, Olesen SC, Dirksen C, Jørgensen NB, Bojsen-Møller KN, Kielgast U, et al. Changes in gastrointestinal hormone responses, insulin sensitivity, and beta-cell function within 2 weeks after gastric bypass in non-diabetic subjects. Obes Surg 2012;22:1084–96.
[16] Magkos F, Bradley D, Eagon JC, Patterson BW, Klein S. Effect of Roux-en-Y gastric bypass and laparoscopic adjustable gastric banding on gastrointestinal metabolism of ingested glucose. Am J Clin Nutr 2016;103:61–5.
[17] Horowitz M, Edelbroek MA, Wishart JM, Straathof JW. Relationship between oral glucose tolerance and gastric emptying in normal healthy subjects. Diabetologia 1993;36:857–62.
[18] Camastra S, Gastaldelli A, Mari A, Bonuccelli S, Scartabelli G, Frascerra S, et al. Early and longer term effects of gastric bypass surgery on tissue-specific insulin sensitivity and beta cell function in morbidly obese patients with and without type 2 diabetes. Diabetologia 2011;54:2093–102.

[19] Mumphrey MB, Patterson LM, Zheng H, Berthoud HR. Roux-en-Y gastric bypass surgery increases number but not density of CCK-, GLP-1, 5-HT-, and neurotensin-expressing enteroendocrine cells in rats. Neurogastroenterol Motil 2013;25:e70−9.

[20] Saeidi N, Meoli L, Nestoridi E, Gupta NK, Kvas S, Kucharczyk J, et al. Reprogramming of intestinal glucose metabolism and glycemic control in rats after gastric bypass. Science 2013;341:406−10.

[21] Bojsen-Møller KN, Jacobsen SH, Dirksen C, Jørgensen NB, Reitelseder S, Jensen JE, et al. Accelerated protein digestion and amino acid absorption after Roux-en-Y gastric bypass. Am J Clin Nutr 2015;102:600−7.

[22] Bradley EL, Isaacs J, Hersh T, Davidson ED, Millikan W. Nutritional consequences of total gastrectomy. Ann Surg 1975;182:415−29.

[23] Goodman BE. Insights into digestion and absorption of major nutrients in humans. Adv Physiol Educ 2010;34:44−53.

[24] Bal BS, Finelli FC, Shope TR, Koch TR. Nutritional deficiencies after bariatric surgery. Nat Rev Endocrinol 2012;8:544−56.

[25] van Loon LJ, Kruijshoop M, Menheere PP, Wagenmakers AJ, Saris WH, Keizer HA. Amino acid ingestion strongly enhances insulin secretion in patients with long-term type 2 diabetes. Diabetes Care 2003;26:625−30.

[26] Calbet JA, Holst JJ. Gastric emptying, gastric secretion and enterogastrone response after administration of milk proteins or their peptide hydrolysates in humans. Eur J Nutr 2004;43:127−39.

[27] Lindgren O, Pacini G, Tura A, Holst JJ, Deacon CF, Ahrén B. Incretin effect after oral amino acid ingestion in humans. J Clin Endocrinol Metab 2015;100:1172−6.

[28] Dirksen C, Jørgensen NB, Bojsen-Møller KN, Jacobsen SH, Hansen DL, Worm D, et al. Mechanisms of improved glycemic control after Roux-en-Y gastric bypass. Diabetologia 2012;55:1890−901.

[29] Griffo E, Cotugno M, Nosso G, Saldalamacchia G, Mangione A, Angrisani L, et al. Effects of sleeve gastrectomy and gastric bypass on postprandial lipid profile in obese type 2 diabetic patients: a 2-year follow-up. Obes Surg 2016;26:1247−53.

[30] De Giorgi S, Campos V, Egli L, Toepel U, Carrel G, Cariou B, et al. Long-term effects of Roux-en-Y gastric bypass on postprandial plasma lipid and bile acids kinetics in female non-diabetic subjects: a cross-sectional pilot study. Clin Nutr 2015;34:911−17.

[31] Muscelli E, Mingrone G, Camastra S, Manco M, Pereira JA, Pareja JC, et al. Differential effect of weight loss on insulin resistance in surgically treated obese patients. Am J Med 2005;118:51−7.

[32] Mari A, Manco M, Guidone C, Nanni G, Castagneto M, Mingrone G, et al. Restoration of normal glucose tolerance in severely obese patients after bilio-pancreatic diversion: role of insulin sensitivity and beta cell function. Diabetologia 2006;49:2136.

[33] Astiarraga B, Gastaldelli A, Muscelli E, Baldi S, Camastra S, Mari A, et al. Biliopancreatic diversion in non-obese patients with type 2 diabetes: impact and mechanisms. J Clin Endocrinol Metab 2013;98:2765−73.

[34] Camastra S, Manco M, Mari A, Baldi S, Gastaldelli A, Greco AV, et al. beta-cell function in morbidly obese subjects during free living: long-term effects of weight loss. Diabetes 2005;54:2382−9.

[35] Greco AV, Mingrone G, Giancaterini A, Manco M, Morroni M, Cinti S, et al. Insulin resistance in morbid obesity: reversal with intramyocellular fat depletion. Diabetes 2002;51:144−51.

[36] Camastra S, Manco M, Mari A, Greco AV, Frascerra S, Mingrone G, et al. Beta-cell function in severely obese type 2 diabetic patients-Long-term effects of bariatric surgery. Diabetes Care 2007;30:1002−4.

[37] Iesari S, le Roux CW, De Gaetano A, Manco M, Nanni G, Mingrone G. Twenty-four hour energy expenditure and skeletal muscle gene expression changes after bariatric surgery. J Clin Endocrinol Metab 2013;98:E321−7.

[38] Novaes FS, Vasques AC, Pareja JC, Knop FK, Tura A, Chaim ÉA, et al. Recovery of the incretin effect in type 2 diabetic patients after biliopancreatic diversion. J Clin Endocrinol Metab 2015;100:1984−8.

[39] Scopinaro N, Marinari GM, Pretolesi F, Papadia F, Murelli F, Marini P, et al. Energy and nitrogen absorption after biliopancreatic diversion. Obes Surg 2000;10:436−41.

[40] Scopinaro N, Adami GF, Marinari GM, Gianetta E, Traverso E, Friedman D, et al. Biliopancreatic diversion. World J Surg 1998;22:936−46.

[41] Topart PA, Becouarn G. Revision and reversal after biliopancreatic diversion for excessive side effects or ineffective weight loss: a review of the current literature on indications and procedures. Surg Obes Relat Dis 2015;11:965−72.

[42] Skroubis G, Sakellaropoulos G, Pouggouras K, Mead N, Nikiforidis G, Kalfarentzos F. Comparison of nutritional deficiencies after Roux-en-Y gastric bypass and after biliopancreatic diversion with Roux-en-Y gastric bypass. Obes Surg 2002;12:551−8.

[43] Evrard S, Aprahamian M, Loza E, Guerrico M, Marescaux J, Damgé C. Malnutrition and body weight loss after biliopancreatic bypass in the rat. Int J Obes 1991;15:51−8.

[44] Johansson HE, Haenni A, Karlsson FA, Edén-Engström B, Ohrvall M, Sundbom M, et al. Bileopancreatic diversion with duodenal switch lowers both early and late phases of glucose, insulin and proinsulin responses after meal. Obes Surg 2010;20:549−58.

[45] García-Díaz, Jde D, Lozano O, Ramos JC, Gaspar MJ, Keller J, Duce AM. Changes in lipid profile after biliopancreatic diversion. Obes Surg 2003;13:756−60.

[46] Mingrone G, DeGaetano A, Greco AV, Capristo E, Benedetti G, Castagneto M, et al. Reversibility of insulin resistance in obese diabetic patients: role of plasma lipids. Diabetologia 1997;40:599−605.

Chapter 46

Underreporting of Energy Intake and Bariatric Surgery

M.N. Ravelli[1], Y.P.G. Ramirez[1], K. Quesada[2], I. Rasera, Jr[3] and M.R.M. de Oliveira[4]

[1]School of Pharmacological Sciences of the University Júlio de Mesquita Filho, UNESP, Araraquara, SP, Brazil, [2]University of Marília, Marília, SP, Brazil, [3]Center of Excellence in Bariatric Surgery of Piracicaba, Piracicaba, SP, Brazil, [4]Bioscience Institute of the University Júlio de Mesquita Filho, UNESP, Botucatu, SP, Brazil

LIST OF ABBREVIATIONS

‰   per mille
δ   delta
$^{13}C/^{12}C$ isotopic ratio of carbon atoms
$^{15}N/^{14}N$ isotopic ratio of nitrogen atoms
BMR   basal metabolic rate
IRMS   isotope ratio mass spectrometer

INTRODUCTION

Bariatric surgery is a clinical intervention that results in significant modifications to food intake behaviors. The effects are not only from anatomic modifications, as it may seem at first. The decision to undergo bariatric surgery, and the surgery's effects on food intake, are experiences unique to the individuals who choose to undergo the procedure. In this context, the obtainment of accurate data about food intake is a challenging task, but indispensable for decision making in the health care process and for reliability of the epidemiological data.

Eating practices respond to a complex set of determinations that reconstruct sensations and motivations [1]. Eating is a social behavior interweaved with representations, practices, and relationships that lead to preferences and aversions to certain foods [1]. Current dietary practices reflect the economic and social transformations of the last century, informing choices that have been contributing to the increase in obesity [2], a problem with multiple causes [3]. To consider food consumption as a cause for these disorders, and to propose interventions, requires important information about the amount and diversity of food consumption. The collection of information regarding dietary practices allows researchers to assess food and nutrient consumption, and to get direct indicators about the nutritional status of individuals and population groups. Dietary surveys are important tools for the detection of deficiencies or excesses in food intake, and are used in nutritional recommendations and in redefining health care actions and nutritional education [4,5]. Such instruments are also used to study the relationship between health and diseases in epidemiological studies [6]. However, the methods available for food intake evaluation are vulnerable to the subjectivity of the report, which distorts food intake information and even the perception of what is consumed [7].

METHODS OF ASSESSMENT OF FOOD CONSUMPTION

Since eating habits are steeped in the conventions of social life, it is difficult to find accurate methods to evaluate the consumption of food [8]. The known methods include, on one hand, the objectivity of observation, and on the other hand, the subjectivity of story. Few methods include biological indicators that correlate with consumed nutrients. An example of a direct method, one with a great degree of subjectivity, is the observation of consumption by trained researchers, or even the evaluation of food and nutrients that duplicate what is found on a served plate of food. Indirect methods of food consumption evaluation include the 24-hour recall, the food record, dietary history, and food frequency questionnaires. Data can be collected by a trained researcher or by the person under observation.

Measurement of sodium urinary excretion is an example of biological evaluation of nutrient consumption. In addition, various food consumption evaluation methods have been developed that use computerized tools [9].

The reliability of the information collected in dietary assessments can be affected by several factors. Intrapersonal (due to the variability of the food consumption patterns) or interpersonal sources of error have great impact on the reliability of food intake data analysis [10]. Thus, food intake assessment studies often report the apparent consumption of an individual rather than the real consumption [11]. Through statistical strategies it is possible to evaluate the information reported by the interviewee with regards to the real consumption of energy and nutrients [12].

An important source of error is misreporting, comprising the sub- and supernotification of food consumption, which affects the estimate of energy and nutrient consumption [13,14], and is a serious obstacle for effective research in this area [9].

Despite the method chosen, the obtainment of valid data is not easy; there is no gold standard, and there are often measurement variations and errors, including with subnotification [15,16]. Thus, it is always necessary to look at the pros and cons of each method, and to consider the target population and the purpose of the study [17].

ISOTOPIC TRACERS AS AN ALTERNATIVE TO FOOD INTAKE ASSESSMENT

There is great future potential in the use of biological markers of food intake, which have the potential to become the gold standard for objective assessment of food consumption. They are not dependent on personal accounts of consumption, and therefore they are not affected by the bias of over- or underreporting, which is frequently seen on food surveys [18]. In this sense, the use of stable isotopes as dietary biomarkers has strong potential for nutritional epidemiology. The stable carbon ($^{13}C/^{12}C$) and nitrogen isotopes ($^{15}N/^{14}N$) are present in food items, and are incorporated into the tissues of those who consume them; integrating the isotopic signal of that tissue, they can be measured in an isotope ratio mass spectrometer [18,19].

Foods show different isotope values, allowing us to track them throughout the metabolic process. In the case of carbon, the plants of the photosynthetic cycle C_3 (wheat, rice, beans, and most fruits and vegetables) have an isotopic signal of about $-28‰$ because the enzyme rubisco (ribulose bisphosphate carboxylase—oxygenase 1.5), a carbon fixer, has a preference for light carbon ($\delta^{12}C$) at the time of fixing that stems from atmospheric CO_2 [18]. The cycle C_4 photosynthetic plants (maize, sugarcane, and sorghum) are less active than the rubisco enzyme, favoring the fixing of heavy carbon ($\delta^{13}C$) with isotope at approximately -12 to $-13‰$ [18]. Nitrogen is a potential biomarker of protein consumption, especially for marine animal sources, since the value of $\delta^{15}N$ is high compared to the amount of this isotope in plants [18]. During the nitrogen excretion process, metabolic processes occur, preferably eliminating ^{14}N and favoring the retention of ^{15}N, thus allowing for an isotopic sign from $3‰$ to $4‰$ higher than from the diet—or even higher, depending on the trophic level of the animal [18].

Nardoto et al. [20] worked with isotopic biomarkers in human nail samples, and showed different values of $\delta^{13}C$ and $\delta^{15}N$ between omnivores and individual resident vegetarians in Brazil and the United States; results were consistent with the isotopic values of local foods. The intake of additional sugar and sweetened beverages can also be detected in $\delta^{13}C$ blood analysis [21], just as with hair samples in the evaluation of animal protein consumption using the $\delta^{13}C$ and $\delta^{15}N$ analysis [19]. Even though data regarding stable isotopes as food consumption biomarkers is consistent, more research is needed to improve the specificity of these analyses [18].

METHODOLOGIES FOR ASSESSING THE ACCURACY OF INFORMATION ON FOOD CONSUMPTION

Dietary surveys have been used in the absence of a validated technique for accurate assessment of food consumption. These are based on the assumption that personal accounts of consumption reflect typical food intake. However, the reported information can be biased, affecting the reliability. Studies [22,23] have shown the distortions, which can lead to incorrect interpretations of energy and nutrient intake. Underreporting is the most widely recognized limitation, and is influenced by factors such as age, sex, and body mass index [24,25]. Among the obese population, underreporting of food intake is an even greater challenge [26].

The use of dietary biomarkers is potentially more objective with regards to consumption as compared with traditional methods [27]; they can be used as reference measurements to assess the validity of measures of food consumption [22]. Several biomarkers have been used to check the results of dietary assessment, including protein intake (nitrogen in the urine), salt (sodium) levels, and total energy expenditure. The reference method for the evaluation of total energy

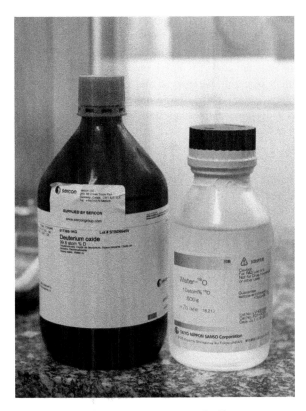

FIGURE 46.1 Doubly labeled water using isotopes of deuterium and oxygen-18 ($^2H_2^{18}O$). Enriched water used in the total energy expenditure assessment with the individual in free life.

expenditure is doubly labeled water (Fig. 46.1) [13], and it can be highly accurate for identifying implausible reports of consumption. This is done by comparing reported energy consumption with the objective estimation of energy consumption by doubly labeled water. However, it is not feasible in large-scale studies because it is relatively expensive [23,28]. Alternative methods suggested for identifying individuals who underestimate food intake [29,30] are more indirect, since estimation of energy requirements is also indirect [31].

In the method described by Goldberg, energy consumption is a multiple of the basal metabolic rate (BMR), and uses this index (consumption of energy/BMR) compared with estimated energy expenditure. The Goldberg equation calculates confidence limits (cutoffs) to assess whether the reported food intake and energy is plausible [29]. The sensitivity of the Goldberg cutoff improved when subjects were classified according to physical activity levels (low, medium, or high) at different cutoff values [30]. The resting metabolic rate used to calculate the Goldberg cutoff can be measured or estimated from specific prediction equations for age and sex. In the study of obese individuals on the waiting list for bariatric surgery, the percentage of undernotification did not vary significantly according to the estimated metabolic rate; however, indirect calorimetry helped to identify individuals with a low resting metabolic rate [31].

It has been shown that when individuals who underreport are excluded, it can influence the associations between diet and obesity, strengthening associations with consumption of sugar and fat, for example [32].

Research results have indicated that to better understand the influence of diet on health, especially on obesity, it is crucial to reduce dietary measurement error by improving data collection, evaluation, and analysis [26].

OBESITY AND UNDERREPORTING

The accuracy of information obtained from food consumption surveys is affected by intraindividual and seasonal variability, and errors in the reporting of consumption. Underreporting is influenced by excess weight, age, and gender [13,14,33,34]. Other factors that lead to underreporting include cigarette smoking, lack of education, low social level, lack of motivation for recording consumption, the nature of the test applied, and difficulties in describing portion size and recalling food items consumed (especially snacks between meals) [34,35].

One indication of the need for a surgical procedure for obesity treatment is failure in weight loss efforts via clinical treatment. Also, patients on the bariatric surgery waiting list may underreport their food consumption to a great extent because they are unaware of the necessity for surgery [31].

Underreporting of food consumption can be selective. Studies show that there is not only underreporting of food groups designated as "less healthy," but there is also overreporting of food groups known as "healthy" [20,25].

Food intake involves elements that go beyond just biochemical aspects [36]. This introduces both conceptual and methodological limitations to making an analysis based on a single or only a few nutrients [37]. One alternative approach is based on the assessment of eating patterns; a multivariate statistical technique is used to evaluate information from dietary surveys to identify factors or underlying patterns common to consumption [28,29]. Cluster analysis is another multivariate method that "clusters" individuals into relatively homogeneous subgroups based on similar diet characteristics [37,38].

Our research group conducted a study using macronutrients as a reference in order to set food standards. In this chapter, we use some of this unpublished data as an example of underreporting of consumption results for different food intake patterns. Through factor analysis, the dietary patterns of obese women (no = 412) who underreported (no = 255) or did not underreport (no = 157) energy consumption (classified by the Goldberg equation) were evaluated in terms of food subgroups, and in a second evaluation, by cluster analysis, in order to discriminate food consumption characterized into subgroups. Information was used from three records for 24 hours of consumption and three records of physical activities. Overall, the breakdown of dietary patterns in terms of macronutrients (factor analysis) was equivalent for individuals who reported correctly and those who underreported; fats and saccharides were highlighted in the consumption of both groups of women (Fig. 46.2).

Reporters

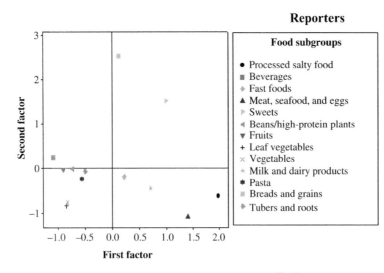

Food subgroups
- ● Processed salty food
- ■ Beverages
- ◆ Fast foods
- ▲ Meat, seafood, and eggs
- ▷ Sweets
- ◁ Beans/high-protein plants
- ▼ Fruits
- + Leaf vegetables
- × Vegetables
- ✳ Milk and dairy products
- ✴ Pasta
- ▨ Breads and grains
- ◈ Tubers and roots

Variables	1° factor	2° factor
	Fat	Saccharides
Protein	0.532	−0.044
Total fat	0.939	0.045
Carbohydrates	0.092	0.917
Fat acids and Saturated fat	0.940	0.118
Added sugar	0.077	0.448
Variance (%)*	**52.8**	**35.9**
Auto-values*	**2.64**	**1.80**

Factor charges using VARIMAX rotation.

Underreporters

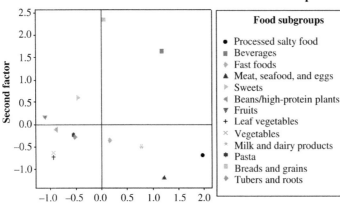

Food subgroups
- ● Processed salty food
- ■ Beverages
- ◆ Fast foods
- ▲ Meat, seafood, and eggs
- ▷ Sweets
- ◁ Beans/high-protein plants
- ▼ Fruits
- + Leaf vegetables
- × Vegetables
- ✳ Milk and dairy products
- ✴ Pasta
- ▨ Breads and grains
- ◈ Tubers and roots

Variables	1° Factor	2° Factor
	Fat	Saccharides
Protein	0.474	-0.047
Total fat	0.879	0.219
Carbohydrates	0.201	0.938
Fat acids and Saturated fat	0.877	0.228
Added sugar	0.275	0.549
Variance (%)*	**61.5**	**29.8**
Auto-values*	**3.08**	**1.49**

Factor charges using VARIMAX rotation.

* Variance (%) and auto-values: Calculated by the facto analysis (Principal component analysis) before the orthogonal VARIMAX rotation.

FIGURE 46.2 Food pattern of women who correctly reported energy intake and those who underreported according to macronutrients. The figure shows the same factor analysis (fat and saccharides) for the correctly reporting and the underreporting individuals, but different food groups were highlighted between the analyses. For reporters, sweets were highlighted, while for underreporters beverages were highlighted. *Data unpublished from our research group.*

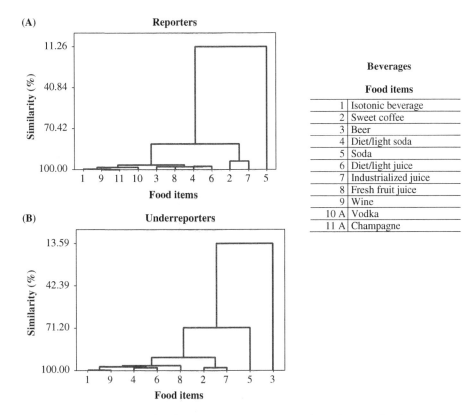

		Beverages
		Food items
1	Isotonic beverage	
2	Sweet coffee	
3	Beer	
4	Diet/light soda	
5	Soda	
6	Diet/light juice	
7	Industrialized juice	
8	Fresh fruit juice	
9	Wine	
10 A	Vodka	
11 A	Champagne	

FIGURE 46.3 Similarities between food consumption for beverages among women who underreported and those who reported correctly. The figure shows different foods highlighted in the beverage group by the cluster analysis comparing reporters and underreporters. The most-consumed beverage to reporters was soda, while to underreporters it was beer. *Data unpublished from our research group.*

In the breakdown of dietary patterns by factor analysis, the first factor (Fig. 46.2) for both groups of women was represented by macronutrients, total fat, saturated fatty acid, and protein. Fig. 46.2 shows that the food intake subgroups representative of these macronutrients were processed salty foods; meat, fish, and eggs; and dairy products, which are simultaneous sources of fat and protein. The second factor (Fig. 46.2) for both groups of women was represented by additional carbohydrates and sugar. Breads and cereals were highlighted for both groups of women as one of the representatives of the consumption of macronutrients for this factor. With cluster analysis we sought to deepen the study of the food patterns identified in the factor analysis; assessment of the similarities between the food that made up each food subgroup accounted for the second factor differently for the underreporting patients versus those who were plausibly reporting. Cluster analysis indicated a difference between underreporting patients and those who correctly reported the intake of foods from the subgroups of beverages and sweets (Figs. 46.2 and 46.3).

Underreporting of energy consumption is typical for food items whose main calorie sources are fats and sugars [33,39,40]. In our study, sweets were reported more by the patients who reported correctly, and beverages showed up in the food patterns of those who underreported. Fig. 46.3 shows that underreporting of food intake can be selective. In the cluster analysis of candy consumption, for both groups of women (Fig. 46.4) in general, it can be seen that patients who report correctly describe a greater number of food items, like energy quantities. In this study, fat consumption was the main discriminating factor in the diet of obese women, followed by the consumption of carbohydrates, regardless of underreporting. However, it was shown that in the underreporting of food consumption, beyond the quantitative component was a qualitative component; the consumption of sweets was the main qualitative discrimination factor for underreporting women on the waiting list for bariatric surgery.

UNDERREPORTING AFTER BARIATRIC SURGERY

A growing number of studies indicate that weight regain after surgery is very likely, and it often requires an intervention to contain or reverse the process [41]. People who regain weight after surgery are most certainly consuming beyond their weight-maintenance needs. The need for energy after surgery can be reduced by a metabolic adaptation triggered by lower food consumption (because of surgery), but this is less common than one might expect [42]. In our studies we

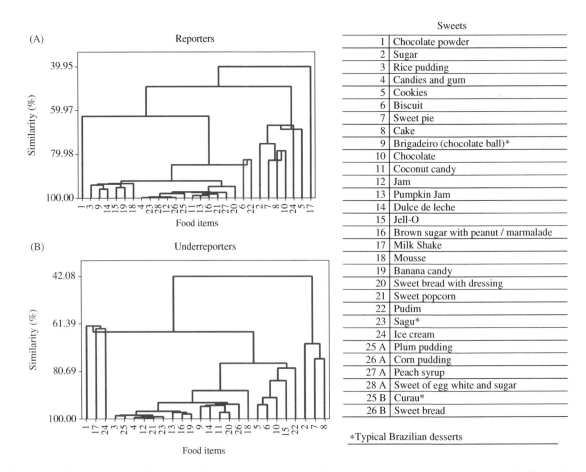

	Sweets
1	Chocolate powder
2	Sugar
3	Rice pudding
4	Candies and gum
5	Cookies
6	Biscuit
7	Sweet pie
8	Cake
9	Brigadeiro (chocolate ball)*
10	Chocolate
11	Coconut candy
12	Jam
13	Pumpkin Jam
14	Dulce de leche
15	Jell-O
16	Brown sugar with peanut / marmalade
17	Milk Shake
18	Mousse
19	Banana candy
20	Sweet bread with dressing
21	Sweet popcorn
22	Pudim
23	Sagu*
24	Ice cream
25 A	Plum pudding
26 A	Corn pudding
27 A	Peach syrup
28 A	Sweet of egg white and sugar
25 B	Curau*
26 B	Sweet bread

*Typical Brazilian desserts

FIGURE 46.4 Differences in the intake of sweets for women who reported and for those who underreported energy intake. The figure shows different foods highlighted in the sweets group by the cluster analysis comparing reporters and underreporters. In general, patients who reported correctly described a greater number of food items, like energy quantities. *Data unpublished from our research group.*

have sought answers regarding weight regain after surgery, and have found no difference between women who achieve or do not achieve the expected body weight reduction; this includes factors such as food consumption, and biochemical [42,43] and genetic variables [43]. In evaluations of consumption, it is possible that women who regain weight after 2 or more years of surgery are underreporting their food consumption, overreporting their physical activity, or, they may in fact represent the small portion of this population who experience metabolic adaptation.

It is difficult to assess dietary intake by conventional methods since these techniques are based on the assumption that they are valid only for a stable weight. Through the use of isotopic techniques and statistical schemes, we hope to develop future tools to more accurately assess underreporting of consumption after surgery. For now, however, what we have are the equations used for the general population, which have wide variations in results, depending on the parameters used in the equations [31]. Therefore, underreporting of consumption can only be considered in clinical and epidemiological studies, and food consumption data remains essential in the management of bariatric surgery.

MINI-DICTIONARY OF TERMS

- *Goldberg equation*: Predictive equation for evaluating whether the reported food intake energy is plausible.
- *Total energy expenditure*: Energy needed to maintain the balance between energy intake and expenditure.
- *Reported energy intake*: Energy consumed as estimated from dietary surveys.
- *Dietary surveys*: Questionnaires and forms for collecting food intake information.
- *Stable isotopes*: Atoms that do not emit radiation.
- *Physical activity level*: Ratio of total energy expenditure and basal energy expenditure.

- *Food frequency questionnaire*: Consists of a predefined list of food products and a section asking about frequency of quantitative and qualitative consumption in a given time period.
- *24-Hour recall*: Type of dietary survey that reports the consumption of all foods and beverages consumed over a period of 24 hours.
- *Underreporting of food consumption*: Omission, lack of perception, and intentional or unintentional forgetting of the quantity of consumption of specific foods.
- *Basal metabolic rate*: The amount of energy necessary for the maintenance of the vital functions of the body, as measured under standard conditions of fasting, and physical and mental rest in a quiet environment with controlled temperature, lighting, and noise.
- *Resting metabolic rate*: Minimum rate of energy consumed to maintain physiological functions in standby mode.

KEY FACTS

- *Underreporting*: Misreporting can affect not only the estimation of the intake of energy, but also of other nutrients. It may lead to bias in the results of food consumption associated with health/disease.
- *Weight Regain*: It is observed in 10–20% of bariatric patients. The reasons for weight regain after bariatric surgery are not clear, which may be related to inadequate follow-up of nutritional guidance, physical inactivity, endocrine and metabolic changes, and other factors.

SUMMARY POINTS

- Food consumption in the context of bariatric surgery includes social, psychological, and cultural variables.
- Underreporting of food intake is a confounding factor in studies on diet and health. It is more prevalent among obese people and, among these individuals, it is more prevalent among those in treatment for weight loss.
- The available methodologies are based on energy balance equations and are not susceptible to instability of body weight.
- Stable isotopes associated with alternative statistical approaches have potential for this type of evaluation.
- Dietary studies show that underreporting is selective, and that higher-energy foods are underreported in the obese population.
- In our studies, we found no differences in the food intake of people who respond differently to changes in body weight after surgery, indicating the need for more accurate approaches to the assessment of food consumption.
- We do not have a method to accurately assess the food intake of people who have undergone bariatric surgery. Underreporting will always have to be considered, and for now all we have are generic equations for this type of evaluation.

REFERENCES

[1] De oliveira SP. Estudo do consumo alimentar: em busca de uma abordagem multidisciplinar. Rev Saúde Públ 1997;31(2):201–8.
[2] Popkin BM. The nutrition transition and its health implications in lower income countries. Pub Health Nutr 1998;1(1):5–21.
[3] World Health Organization. Obesity: preventing and managing the global epidemic: report of a WHO consultation. Geneva: World Health Organization. WHO Technical Report 894;2000.
[4] Hirvonen T, Mannisto S, Ross E, Pietinen P. Increasing prevalence of underreporting does not necessarily distort dietary surveys. Eur J Clin Nutr 1997;51(5):297–301.
[5] Cintra IP, Von Der Heyde MED, Schimitz BAS, Franceschini SCC, Taddei JA, Sigulem DM. Métodos de inquéritos dietéticos. Cad Nutr 1997;13:11–23.
[6] Willett WC, editor. Nutritional epidemiology. 2nd ed. New York, NY: Oxford University Press; 1998.
[7] Murtagh L, Ludwig DS. State intervention in life-threatening childhood obesity. JAMA 2011;306(2):206–7. Available from: http://dx.doi.org/10.1001/jama.2011.903.
[8] Garcia RWD. Representações sobre consumo alimentar e suas implicações em inquéritos alimentares: estudo qualitativo em sujeitos submetidos à prescrição dietética. Rev Nutrição Campinas 2004;17(1):15–28.
[9] Shim JS, Oh K, Kim HC. Dietary assessment methods in epidemiologic studies. Epidemiol Health 2014;36:e2014009. Available from: http://dx.doi.org/10.4178/epih/e2014009. eCollection 2014.
[10] Nusser SM, Carriquiry AL, Dood KW, Fuller WA. A semiparametric transformation approach to estimating usual daily intake distribuitions. JASA 1996;91(436):1440–9.

[11] Institute of Medicine - IOM. Dietary reference intakes (DRIs): applications in dietary assessment. Washington, D.C.: National Academy Press; 2001.

[12] Slater B, Marchioni DL, Fisberg RM. Estimando a prevalência da ingestão inadequada de nutrientes. Rev Saúde Públ 2004;38(4):599−605.

[13] Castro-Quezada I, Ruano-Rodríguez C, Ribas-Barba L, Serra-Majem L. Misreporting in nutritional surveys: methodological implications. Nutr Hosp 2015;31(Suppl. 3):119−27. Available from: http://dx.doi.org/10.3305/nh.2015.31.sup3.8760.

[14] Murakami K, Livingstone MB. Prevalence and characteristics of misreporting of energy intake in US adults: NHANES 2003-2012. Br J Nutr 2015;114(8):1294−303. Available from: http://dx.doi.org/10.1017/S0007114515002706. Epub 2015 Aug 24.

[15] Lopes ACS, Caiaffa WT, Mingoti AS, Lima-Costa MFF. Ingestão alimentar em estudos epidemiológicos. Rev Bras de Epidemiol 2003;6(3): 209−19.

[16] Poslusna K, Ruprich J, de Vries JH, Jakubikova M, van't Veer P. Misreporting of energy and micronutrient intake estimated by food records and 24 hour recalls, control and adjustment methods in practice. Br J Nutr 2009;101(Suppl. 2):S73−85.

[17] Innovative Dietary Assessment Methods in Epidemiological Studies and Public Health I. WebHome 2015 [cited 2015]. Available from: <http://nugo.dife.de/wiki/bin/view/IDAMES/>.

[18] O'Brien D. Stable isotope ratio as biomarkers od diet for health research. Annu Rev Nutr 2015;35:565−94. Available from: http://dx.doi.org/10.1146/annurev-nutr-071714-034511.

[19] Petzke KJ, Boeing H, Klaus S, Metges CC. Carbon and nitrogen stable isotopic composition of hair protein and amino acids can be used as biomarkers for animal-derived dietary protein intake in humans. J Nutr 2005;135(6):1515−20.

[20] Nardoto GB, Silva S, Kendall C, Ehleringer JR, Chesson LA, Ferraz ES, et al. Geographical patterns of human diet derived from stable-isotope analysis of fingernails. Am J Phys Anthropol 2006;131(1):137−46.

[21] Hedrick VE, Davy BM, Wilburn GA, Jahren AH, Zoellner JM. Evaluation of a novel biomarker of added sugar intake ($\delta^{13}C$) compared with self-reported added sugar intake and the Healthy Eating Index-2010 in a community-based, rural US sample. Pub Health Nutr 2016;19(3): 429−36. Available from: http://dx.doi.org/10.1017/S136898001500107X.

[22] Pfrimer K, Vilela M, Resende CM, Scagliusi FB, Marchini JS, Lima NK, et al. Under-reporting of food intake and body fatness in independent older people: a doubly labelled water study. Age Ageing 2015;44(1):103−8. Available from: http://dx.doi.org/10.1093/ageing/afu142. Epub 2014 Oct 22.

[23] Rhee JJ, Sampson L, Cho E, Hughes MD, Hu FB, Willett WC. Comparison of methods to account for implausible reporting of energy intake in epidemiologic studies. Am J Epidemiol 2015;181(4):225−33. Available from: http://dx.doi.org/10.1093/aje/kwu308. Epub 2015 Feb 5.

[24] Stice E, Palmrose CA, Burger KS. Elevated BMI and male sex are associated with greater underreporting of caloric intake as assessed by doubly labeled water. J Nutr 2015;145(10):2412−18. Available from: http://dx.doi.org/10.3945/jn.115.216366. Epub 2015 Sep 2.

[25] Souza DR, Anjos LA, Wahrlich V, Vasconcellos MT. Energy intake underreporting of adults in a household survey: the impact of using a population specific basal metabolic rate equation. Cad Saúde Pública 2015;31(4):777−86.

[26] Mendez MA. Invited commentary: dietary misreporting as a potential source of bias in diet-disease associations: future directions in nutritional epidemiology research. Am J Epidemiol 2015;181(4):234−6. Available from: http://dx.doi.org/10.1093/aje/kwu306. Epub 2015 Feb 5.

[27] O'Gorman A, Gibbons H, Brennan L. Metabolomics in the identification of biomarkers of dietary intake. Comput Struct Biotechnol J 2013;4: e201301004. Available from: http://dx.doi.org/10.5936/csbj.201301004.

[28] Livingstone MB, Black AE. Markers of the validity of reported energy intake. J Nutr 2003;133(Suppl. 3):895S−920S.

[29] Goldberg GR, Black AE, Jebb SA, Cole TJ, Murgatroyd PR, Coward WA, et al. Critical evaluation of energy intake data using fundamental principles of energy physiology: 1. Derivation of cut-off limits to identify under-recording. Eur J Clin Nutr 1991;45(12):569−81.

[30] Black AE. Critical evaluation of energy intake using the Goldberg cut-off for energy intake: basal metabolic rate. A practical guide to its calculation, use and limitations. Int J Obes Relat Metab Disord 2000;24(9):1119−30.

[31] Quesada KR, Novais PFS, Detregiachi CRP, Barbalho SM, Rasera Jr I, Oliveira MRM. Comparative analysis of approaches for assessing energy intake underreporting by female bariatric surgery candidates. J Am Coll Nutr 2014;33(2):155−62. Available from: http://dx.doi.org/10.1080/07315724.2013.874893.

[32] Mendez MA, Popkin BM, Buckland G, Schroder H, Amiano P, Barricarte A, et al. Alternative methods of accounting for underreporting and overreporting when measuring dietary intake-obesity relations. Am J Epidemiol 2011;173(4):448−58.

[33] Berta Vanrullen I, Volatier JL, Bertaut A, Dufour A, Dallongeville J. Characteristics of energy intake under-reporting in French adults. Br J Nutr 2014;111(7):1292−302. Available from: http://dx.doi.org/10.1017/S0007114513003759. Epub 2013 Dec 13.

[34] Abbot JM, et al. Psychosocial and behavior profile and predictors of self-reported energy underreporting in obese middle-aged women. J Am Diet Assoc 2008;108(1):114−19.

[35] Scagliusi FB, Polacow VO, Artioli GG, Benatti FB, Lancha Jr AH. Selective underreporting of energy intake in women: magnitude, determinants, and effect of training. J Am Diet Assoc 2003;103(10):1306−13.

[36] Pedraza DF. Padrões alimentares: da teoria à prática − o caso do Brasil. Mneme − Rev Virtual de Humanidades 2004;9(3):1−10.

[37] Hu FB. Dietary pattern analysis: a new direction in nutritional epidemiology. Curr Opin Lipidol 2002;13(1):3−9.

[38] Moeller SM, et al. Dietary patterns: challenges and opportunities in dietary patterns research. An experimental biology workshop, April 1, 2006. J Am Diet Assoc 2007;107(7):1233−9.

[39] Lafay L, Menne L, Basdevant A, Charles MA, Borys JM, Eschewege E, et al. Does energy intake underreporting involve all kinds of food or only specific food items? Results from the Fleurbaix Laventie Ville Santé (FLVS) study. Inter J Obes Relat Metab Disord 2000;24(11):1500−6.

[40] Pryer JA, Vrijheid M, Nichols R, Kiggins M, Elliott P. Who are the 'low energy reporters' in the dietary and nutrition Survey of British Adults? Inter J Epidemiol 1997;26(1):146–54.

[41] Karmali S, Brar B, Shi X, Sharma AM, Gara C, Birch DW. Weight recidivism post-bariatric surgery: a systematic review. Obes Surg 2013;23(11): 1922–33. Available from: http://dx.doi.org/10.1007/S11695-013-1070-4.

[42] Fogaça KCP. Investigação de fatores envolvidos na recuperação de peso após derivação gástrica. [doctoral thesis]. Post-Graduate Program in Food and Nutrition. UNESP – Fcfar 2009.

[43] Novais PFS. Polimorfismos genéticos como moduladores do consumo alimentar, peso corporal e comorbidades após um ano de cirurgia bariátrica. [doctoral thesis]. Post-Graduate Program in Food and Nutrition. UNESP - Fcfar 2015.

[44] Edwards SL, Slattery ML, Murtaugh MA, Edwards RL, Bryner J, Pearson M, et al. Development and use of touch-screen audio computer-assisted self-interviewing in a study of American Indians. Am J Epidemiol 2007;165(11):1336–42.

Chapter 47

Control Eating Following Bariatric Surgery

E.M. Conceição

Universidade do Minho, Braga, Portugal

LIST OF ABBREVIATIONS

BE	binge eating
BED	binge eating disorder
BES	Binge Eating Scale
DEBQ	Dutch Eating Behavior Questionnaire
DSM-5	Diagnostic and Statistical Manual of Mental Disorders, 5th Edition
EDE	Eating Disorder Examination
EDE-Q	Eating Disorder Examination Questionnaire
EDI	Eating Disorder Inventory
ELOCS	Eating Loss of Control Scale
LOC	loss of control
OBE	objective binge eating
QEWP-5	Questionnaire on Eating and Weight Patterns
Rep(eat)	Repetitive Eating Interview
Rep(eat)-Q	Repetitive Eating Questionnaire
SBE	subjective binge eating

INTRODUCTION

Variability of weight loss outcomes following bariatric surgery, particularly in the long-term [1], has encouraged researchers and clinicians to investigate factors associated with poorer weight-related outcomes. Psychosocial and behavioral aspects have received great attention as they represent potential modifiable factors and require clinical attention both at pre- and during postoperative periods [2].

After surgery, an improvement in social relationships, quality of life, depressive symptoms, self-esteem, as well as a normalization of eating behavior and eating patterns, is generally observed [3–5]. Yet, as time elapses after surgery, a subset of patients experience deterioration of their psychological status and eating patterns, with significant impact on weight-related outcomes [6].

Currently, there is a strong body of evidence suggesting that the presence of problematic eating behaviors before surgery is not associated with postoperative weight outcomes [7,8]. In fact, it is the onset of problematic eating behaviors in the postoperative period that has consistently been associated with less weight loss and weight regain [9,10], and thus the interest in properly identifying and assessing such behaviors seems to play a central role for maintenance of long-term outcomes.

Yet, assessing problematic eating behaviors in the bariatric surgery population has been a challenge [11]. Surgery induces an extreme modification of the gastrointestinal track, which in turn necessarily alters eating behaviors and eating patterns. One of the most dramatic changes experienced by patients is the amount of food that patients can accommodate in their new stomach, which poses important considerations when assessing problematic eating in postoperative patients, as detailed later in this chapter [10].

BINGE EATING DISORDER, BINGE EATING, AND LOSS OF CONTROL EATING

Binge eating disorder (BED) is currently a Diagnostic and Statistical Manual of Mental Disorders (5th edition; DSM-5) [12] diagnosis under the Feeding and Eating Disorders section. Table 47.1 presents DSM-5 diagnostic criteria for BED. Binge eating (BE) episodes are the core feature for BED and bulimia nervosa (BN), and are sometimes reported by patients with a compulsive variant of anorexia nervosa (AN). In the interest of capturing the loss of control (LOC) nature of eating episodes even when an unambiguously large amount of food is not eaten [13], some authors in the eating disorders field make a distinction between objective binge eating (OBE), corresponding to BE as defined in the DSM-5, and subjective binge eating (SBE), characterized by the hallmark feature of LOC overeating an amount of food that is not unambiguously large, but is considered excessive by the respondent [14,15].

Given the alterations induced by surgery in the gastric track, ingestion of large amounts of food is not physically possible postoperatively, and full criteria for BED diagnosis are usually not met in the postoperative time, resulting in rather lower rates of BED ranging from 0% to 10.3% [16,17], compared to preoperative time (up to 49%; [7]). To date, no specific guidelines have been proposed for assessing BED in postbariatric patients [10], and current DSM-5 diagnostic criteria for BED do not capture the atypical presentations seen postoperatively. Thus the distinction between OBE episodes and SBE episodes, as well as the focus on the experience of LOC eating, is particularly relevant when assessing postbariatric patients.

Table 47.1 describes specific considerations of assessing BED and BE in postbariatric patients.

Consistent with the above, studies report a significant decrease of OBE from pre- to postsurgery, but a rebound is observed in longer-term follow-up as patients begin to be able to accommodate increasingly larger amounts of food [6,17–19]. However, a higher percentage of patients (up to 39%; [10]) report engaging in SBE or LOC eating postoperatively, suggesting that the experience of LOC overeating is maintained or initiated after surgery even when patients cannot ingest large amounts of food.

TABLE 47.1 DSM-5 Criteria for Binge Eating Disorder and Binge Eating Episodes and Summary of Considerations for Their Assessment Postbariatric Surgery

A. Recurrent episodes of *binge eating*	
a. Eating an *unambiguous and objectively large* amount of food compared to what most people would eat under comparable circumstances, and in a discrete period of time (e.g., within 2 h).	1. Physical restriction induced by surgery, which greatly limits the total gastric capacity. 2. Consider episodes of binge eating regardless of the amount of food ingested.
b. Such episodes are accompanied by a *sense of loss of control (LOC)* over eating, i.e., feeling that one cannot stop eating once started or cannot control what or how much is eaten.	1. Loss of control eating may be the hallmark feature of binge eating in postoperative patients. 2. Loss of control eating May be present in other maladaptive eating such as grazing behavior.
B. Three or more of the following: 　1. Eating much *more rapidly* than normal 　2. Eating until feeling *uncomfortably full* 　3. Eating *large* amounts *when not feeling physically hungry* 　4. *Eating alone* because of feeling embarrassed by how much 　5. *Feeling disgusted with oneself, depressed, or very guilty after overeating.*	1. Speed of ingestion may be hard to assess when eating small amounts of food. 2. Feeling uncomfortably full is frequent among post-bariatric patients. 3. Large amounts of food are not tolerated post operatively.
C. **Marked distress** regarding binge eating is present.	No special consideration needed.
D. The binge eating occurs, on average, at least once a week for 3 months.	1. Symptoms might fluctuate depending on the surgical procedure and surgical aspects that need attention from surgeon.
E. No use of inappropriate compensatory behavior. Not better explained by bulimia nervosa or anorexia nervosa diagnoses.	1. It might be challenging to differentiate self-induced vomiting to compensate previous overeating or self-induced vomiting which is often experienced by bariatric patients to avoid physical discomfort after eating foods not well tolerated by the new stomach.

This table presents the Diagnostic and Statistical Manual of Mental Disorders, 5th edition (DSM-5) [12], criteria for binge eating disorder diagnosis. For each criterion, special considerations for assessing postbariatric patients are described.

BED, BE, or LOC Eating and Weight- and Psychological-Related Outcomes

Different studies sought to investigate the role of preoperative BED, BE, or LOC eating as predictors of poor weight loss postoperatively. However, results appear to be rather mixed. Only a few studies reported association with worst weight-related outcomes [16,20], and interestingly, other authors found associations with greatest weight loss for preoperative binge eaters [7]. Several other papers report BE, BED, and LOC eating as nonsignificant predictors of postoperative results (e.g., Refs. [7,17]). More consistent are findings regarding the role of postoperative BED, BE, or LOC eating on weight outcomes (see Meany and collaborators [7] for a review). SBE episodes, uncontrolled eating, or just LOC eating are concepts used in the literature that refer to the experience of LOC eating regardless of the amount of food eaten. Despite different terminologies, studies brought evidence for their association with less weight loss and/or weight regain [7,10]. The prospective effects of eating behaviors have also been proposed by White and collaborators [21], who suggested that the presence of LOC eating postoperatively predicts weight loss in later assessment times.

The Clinical Significance of LOC/SBE Versus OBE

Such studies support the experience of LOC eating as clinically relevant, warranting therapeutic attention as OBE does. The interest in capturing LOC eating regardless of the amount of food ingested is supported by studies showing a significant association with weight outcomes and psychological status. Such studies suggest that LOC may be the hallmark indicator of BE [14,15].

Research on LOC eating conducted with several populations showed that people reporting LOC (over amounts of food not unambiguously large) experience high levels of psychosocial distress, psychiatric comorbidities [15], eating disorder psychopathology, as well as poor quality of life [22]—comparable to those experiencing only OBE.

Compared to objective binge eaters, people engaging in SBE present similar levels of food avoidance, restraint eating, hunger, maturity fears, perfectionism, and other eating-related psychopathological features [22]. Antecedents of BE (e.g., boredom, cravings) and the use of compensatory behaviors (to compensate the amount of food ingested) such as self-induced vomiting, are also similarly associated with both OBE and SBE [22,23]. Additionally, Palavras and collaborators [15] found that there are no differences in binge eating severity nor in the number of LOC items endorsed between people reporting only OBE or only SBE. Similarly, the degree of obesity did not differ between OBE and SBE individuals [15].

Psychological Assessment Measures of BED, BE, or LOC Eating

Literature highlights the importance of systematically and adequately assessing subsyndromal presentations of problematic eating behaviors as an early prevention strategy to optimize postoperative weight outcomes. For this reason, assessment of eating behavior in postbariatric surgery should include measures targeting a broad conceptualization of disordered or problematic eating behaviors.

Assessment methods for BED and BE include self-report measures and clinical interviews that were traditionally developed for eating disorder patients. Although such measures should be used with caution, particularly with postoperative bariatric patients, most of them have been systematically employed with this population, producing important and reliable information.

Another consideration in assessing LOC eating is its measurement as a dichotomous variable (present/absent). Only recently a few have authors proposed a continuous assessment rating scheme to assess LOC in association with a variety of eating behaviors. For example, Conceição and colleagues [24] suggested that such a continuous scale would enable professionals in the field to capture different degrees of LOC in association with a variety of problematic eating behaviors (from grazing to OBE).

Table 47.2 summarizes the self-report measures most administered to assess BED, BE, LOC eating, and disordered eating among obese and bariatric samples.

Regarding clinical interviews, the Eating Disorder Examination (EDE) [25] is a semistructured, interviewer-based clinical interview that allows a methodological identification of both SBE and OBE, as well as the diagnosis of BED or any other DSM-5 eating disorder. EDE takes about 45–75 minutes to administer, and assesses current eating disorder diagnoses. Besides diagnostic items to identify overeating and BE episodes, it also generates a global score and four subscales: restraint, eating concern, shape concern, and weight concern. EDE has been adapted by de Zwaan et al. [17] to include items for postbariatric patients.

TABLE 47.2 Self-Report Measures for Eating Disorders Including Binge Eating Disorder, Binge Eating Episodes, Loss of Control Eating, and Grazing Before and After Surgery

Measure	Authors/ Year	Validated for Bariatric Surgery	Used in Obese or Bariatric Samples	Description
Eating Disorder Examination Questionnaire (EDE-Q)	Fairburn and Beglin (2008) [40]	✓	✓	This 28-item measure assesses eating disorder behaviors and eating disorder symptomatology. Generates global score and 4 subscales: restraint, eating concern, shape concern, and weight concern.
Binge Eating Scale (BES)	Gormall et al. (1982) [41]	✓	✓	A 16-item questionnaire with a total score reflecting severity of binge eating behaviors.
Dutch Eating Behavior Questionnaire (DEBQ)	van Strien et al. (1986) [42]	✓	✓	A 33-item questionnaire assessing 3 patterns of eating resulting in 3 subscales: restrained eating, emotional eating, and external eating.
Questionnaire on Eating and Weight Patterns (QEWP-5)	Spitzer et al. (1993) [43]		✓	This 28-item questionnaire provides information on frequency and duration of BE to draw a BED diagnosis.
Eating Disorder Inventory (EDI)	Garner et al. (1983) [44]		✓	This is a 91-item questionnaire that generates 11 scales: drive for thinness, bulimia, body dissatisfaction, ineffectiveness, perfectionism, interpersonal distrust, interoceptive awareness, maturity fears, asceticism, impulse regulation, and social insecurity.
Eating Loss of Control Scale (ELOCS)	Bloomquist et al. (2014) [45]		✓	This is an 18-item self-report measure to assess loss of control overeating.
Repetitive Eating Questionnaire (Rep (eat)-Q)	Conceição et al. (2014) [24]	*Work in progress*		A 15-item questionnaire that assesses grazing and generates a total score reflecting levels of associated grazing eating patterns.

This table presents self-report measures that have been used in the literature to assess different eating behaviors of relevance in the bariatric surgery population.
Adapted from the comprehensive reviews of Parker K, Brennan L. Measurement of disordered eating in bariatric surgery candidates: a systematic review of the literature. Obes Res Clin Pract 2015;9(1):12–25 and Parker K, O'Brien P, Brennan L. Measurement of disordered eating following Bariatric surgery: a systematic review of the literature. Obes Surg, 20104;24(6):945–53 on assessment measures for bariatric patients [28,29].

The Structured Clinical Interview for DSM (SCID-I) [26] is a semistructured interview that has been validated for the bariatric population. It assesses current and lifetime eating disorders, and it takes about 60 minutes to conduct.

For a detailed review on measures of disordered eating in bariatric surgery candidates and postoperative samples, please refer to the reviews of Parker and colleagues [27,28].

OTHER PROBLEMATIC EATING BEHAVIORS AND LOC

Some authors have suggested that other forms of problematic eating associated with LOC may emerge in the postoperative time [2,18]. A growing number of studies have provided support for the association between grazing, an eating behavior characterized by the repetitive ingestion of small to modest amounts of food, and poor weight loss or weight regain in the long-term after surgery [19,39,30]. However, different terminologies, such as picking or nibbling, and snacking or nibbling, have been used in the literature to identify similar eating behaviors, producing rather inconsistent data [24]. The major difference in the criteria employed in each study is in its association with LOC, as some authors completely exclude the sense of LOC eating from grazing behavior [25]. This debate was previously expanded by Conceição and colleagues [24], who proposed a unifying definition for the concept. Their proposal includes two distinct types of grazing, putting great attention on the unplanned nature of repeatedly eating small/modest amounts of food when not hungry: (1) compulsive grazing, reflecting an association with the sense of LOC, is usually characterized by willingness to resist but returning to repeatedly eat and (2) noncompulsive grazing, which precludes the sense of LOC, characterizing a mindless eating behavior. However, the authors note that despite being noncompulsive and supposedly with no sense of LOC eating, noncompulsive grazing is markedly different from an eating pattern wherein the person plans or anticipates repeatedly eating small-size meals, suggesting this is not an eating behavior under control, and some degree of uncontrolled eating must be associated with it.

Conceição and colleagues [10,24] proposed a categorization system for distinguishing different eating behaviors (OBE, SBE, grazing, overeating, and normal eating; Table 47.3) in reference to a continuous scale representing different degrees of LOC eating.

Grazing, Weight, and Psychological-Related Outcomes

Although the variability of definitions enable us to draw firm conclusions, there is increasing evidence for the association of grazing with less weight loss and weight regain [19,29,31,32]. If findings regarding preoperative grazing and postoperative weight loss are mixed, more consistent evidence has been published on postoperative grazing. The association with weight management is hypothesized to be related to its unplanned nature, which would result in an excessive caloric intake repetitively throughout the day, poor dietary adherence, and poor compliance with follow-up appointments [18,32,33].

Research has produced mixed results for its association with psychopathology. A few studies failed to find association with frequency of meals, BE, overeating, dietary restraint, or shape, eating, and weight concerns, or compensatory behaviors in BED, BN, AN, or college samples [34–37]. Other studies presented support for its association with negative affect [38], depression, and poor mental health [19,29]. In fact, studies show patients presenting BE or LOC eating are more likely to engage in grazing [6], and that grazing may be a postoperative variant of BE. These mixed results may be hindered by the variety of criteria used across studies.

TABLE 47.3 Eating Behaviors in Relation to the Degree of LOC

Eating Episode	Degree LOC	Example of Eating Behavior
None	0	Plan to repeatedly have small amounts of food throughout the day.
Deliberate overeating	0	Plan to repeatedly have small amounts of food to be able to accommodate large amounts of food in total.
Grazing, noncompulsive subtype	1	"Mindless" eating; eating in distracted way repetitively; eating whatever is available "on the spur of the moment." Not planned/anticipated.
Grazing, compulsive subtype	2	Trying to resist but repetitively going back and eating small/modest amounts of food. Not planned/anticipated.
(Subjective) Binge eating episode	3	Feeling that one cannot stop eating after starting, or cannot control the amount eaten. Eating episode occurs in a circumscribed period of time rather than repeatedly over time.
(Objective) Binge eating episode	4	Feeling that one cannot stop eating after starting, or cannot control the amount being eaten. Eating extremely large amounts of food in a short period of time.

Eating behaviors are listed in relation to their supposed degree of associated LOC eating. Higher degrees correspond to increased levels of LOC (uncontrolled) eating.
LOC, loss of control.
Retrieved from Conceição EM, Utzinger LM, Pisetsky EM. Eating disorders and problematic eating behaviours before and after bariatric surgery: characterization, assessment and association with treatment outcomes. Eur Eat Disord Rev 2015;35:417–25. Copyright © 1999–2016 John Wiley & Sons, Inc. All Rights Reserved [10].

Assessment of Grazing

Only recently have a few assessment instruments been proposed to assess grazing and picking or nibbling. Conceição and colleagues [24] developed a self-report measure (Rep(eat)-Q; see Table 47.2) based on their proposed definition of grazing, designed to capture both compulsive and noncompulsive grazing.

A clinical interview [Rep(eat)] was also developed by Conceição and collaborators [24] to assess both types of grazing. This is an investigator-based semistructured interview using a series of probe questions to identify each of the indicators of grazing. Interviewers are asked to rate the number of episodes in the previous month.

The current version of the EDE [25] also includes diagnostic items to assess the number of days with picking or nibbling in the past month, which are rated from 0 (no picking) to 6 (every day).

It should be noted, however, that to date there is only preliminary evidence on the use of these measures among bariatric patients, and further research is still required to provide a solid ground for such concepts and instruments.

CONCLUSION

Assessment of LOC eating-related behaviors has been in the spotlight for its supposed association with both weight- and psychological-related outcomes after surgery. With a growing body of research supporting the importance of assessing a variety of eating behaviors, measures that capture the experience of LOC eating regardless of the amount ingested, and other forms of noncontrolled eating behaviors such as grazing, should be used in a more consistent and systematic manner.

Of note, a few studies showed that deterioration of weight-related outcomes is more notorious in longer assessment times, particularly after 2 years of surgery, when the apparent normalization of eating behaviors immediately after surgery supposedly erodes, and a stronger association with poor weight outcomes is observed [20,39]. Thus, assessing long-term eating behaviors seems to represent a reasonable strategy to prevent weight regain and optimize weight and psychological-related outcomes among postbariatric patients.

MINI-DICTIONARY OF TERMS

Definitions of the different eating episodes: loss of control (LOC) eating; binge eating (BE); subjective binge eating (SBE); and objective binge eating (OBE); grazing; compulsive grazing and noncompulsive grazing; picking or nibbling.

LOC	1. Sense of LOC overeating accompanied by the feeling that one cannot stop eating or cannot control how much (or what) one is eating. 2. It is considered independently of what and how much is ingested. 3. Experienced in different eating behaviors.
BE	1. Eating in a circumscribed period of time independently of how much is eaten. 2. Associated experience of LOC (regardless of the amount of food eaten). 3. The DSM-5 considers BE episodes associated with eating an unambiguous amount of food.
SBE	1. Ingestion of amounts of food not objectively large, but experienced as excessive by the respondent. 2. Associated experience of LOC. 3. Currently not included as criterion for BED in the DSM-5.
OBE	1. Ingestion of objectively large amounts of food. 2. Objectively large amounts of food are twice or more compared to what a person would eat under similar circumstances. 3. Associated experience of LOC. 4. These are BE episodes as defined in the DSM-5. 5. Criterion for both BED and bulimia nervosa DSM-5 diagnoses.
Grazing	1. Defined in the Rep(eat) interview (Conceição et al.) [24]. 2. Repetitive eating of small/modest amounts of food in an unplanned manner. 3. Eating is not in response to hunger/satiety sensations. 4. "Repetitive" is eating twice or more in the same period of time (e.g., morning, afternoon, evening, or consecutively throughout the day).
Compulsive grazing	1. Associated with the sense that one cannot resist eating, returning to snack on food even when trying to avoid it. 2. Associated with the experience of some degree of LOC.
Noncompulsive grazing	1. Characterized by eating in a distracted way over a long period. 2. Not associated with experience of LOC, but an uncontrolled behavior.

Picking or nibbling	1. Defined in the Eating Disorders Examination (EDE; Fairburn & Cooper) [25].
	2. Characterized by repetitively picking or nibbling on food in an unplanned manner.
	3. It is distinguished from meals and snacks, and it should happen in between meals or snacks.
	4. The amount of food should be uncertain at the beginning of the episode and should not be trivial.
	5. LOC eating is absent.

KEY FACTS

- BED was only recently (2013) included as a diagnostic entity in DSM-5.
- BED is integrated in the Feeding and Eating Disorders category.
- Other disorders under the same Feeding and Eating Disorders Category are Anorexia Nervosa (AN); Bulimia Nervosa (BN); Pica; Rumination; avoidant/restrictive food intake disorder; otherwise specified feeding and eating disorders; and Unspecified Feeding and Eating Disorder.
- Episodes of binge eating (BE) are also diagnostic criteria for BN. In BN, they are followed by compensatory behaviors such as vomiting, misuse of laxatives/diuretics, excessive exercise, or intense caloric restriction.
- BE episodes can be engaged in by AN patients in a compulsive variant of the disorder.
- At least three of the following LOC items must be endorsed in BE episodes:
  - The speed of eating is much higher than in normal eating;
  - Eating until feeling uncomfortably full;
  - Eating large amounts of food despite not feeling physically hungry;
  - Eating alone or hidden because of embarrassment about how much is ingested;
  - Followed by feeling disgusted with oneself, depressed, or very guilty.
- Traditionally, LOC was considered only in association with BE episodes (both SBE and OBE).
- The original dichotomous (present/absent) classification ruled out LOC from other eating behaviors such as grazing or picking or nibbling.
- New assessments of LOC using continuous rating schemes allow the association of LOC with other eating behaviors.
- Psychological assessment of eating disorders and interpretation of results must be conducted by a qualified clinician or mental health professional trained in the DSM-5 classification.
- The most common and easy to administer instruments are self-report measures.
- Self-report measures ARE NOT diagnostic measures. Usually, they are good screening measures for patients scoring high in related psychopathology, but final diagnosis MUST be confirmed by interview.
- Structured clinical interviews are usually composed of a set of questions rigorously asked, and there is usually not room for additional questions from the interviewer.
- Semistructured clinical interviews consist of a set of compulsory questions, but include a variety of probe questions and allow new questions from the interviewer to be brought up for clarification. In investigator-based interviews, the interviewer decides what constitutes "large amounts," "excessive," or "loss of control"; their opinions should not be shared with the respondent. For example, the respondent should be asked to describe episodes where they felt as if they had overeaten, and the interviewer should judge whether the amount of food was objectively large or not.

SUMMARY POINTS

- This chapter describes different eating behaviors associated with the experience of loss of control (LOC) overeating in the postoperative bariatric population.
- Such eating behaviors are relevant in the postoperative time for their presumed association with weight- and psychological-related outcomes of bariatric surgery.
- Considering the limited gastric capacity induced by bariatric surgery, diagnostic criteria for binge eating disorder (BED) is rare in the postoperative time.
- Despite lower rates of BED, postbariatric patients still engage is binge eating (BE) episodes over amounts of food that are not unambiguously large, but are experienced as excessive.
- The experience of LOC eating, regardless of how much and what is eaten, has been considered the hallmark feature of BE.
- Authors have distinguished objective binge eating (OBE) episodes (LOC over an unambiguously and objectively large amount of food) from subjective binge eating (SBE), which is LOC over amounts of food that are not objectively large, but are experienced as excessive by the respondent.

- SBE episodes have been associated with poor weight loss and high food avoidance, restraint eating, hunger, maturity fears, perfectionism, and other eating-related psychopathological features.
- Grazing, generally characterized by repetitive eating of small/modest amounts of food in an unplanned manner, is another eating behavior that has been associated with some degree of uncontrolled eating.
- Grazing has been consistently associated with poor weight loss and weight regain, but inconsistently related to increased levels of psychological distress.
- Patients presenting BE or LOC eating are more likely to engage in grazing.
- Assessment measures for the postoperative samples should capture subsyndromal presentations of BED and the experience of LOC regardless of the amount ingested.
- A variety of self-report measures and semistructured interviews have been used and validated for the postoperative bariatric population.

REFERENCES

[1] Courcoulas AP, Nicholas C, Belle S, Berk PD, Flum D, Garcia L, et al. Weight change and health outcomes at 3 years after bariatric surgery among individuals with severe obesity. J Am Med Assoc Dec. 2013;310(22):2416−25.

[2] Mitchell JE, Christian NJ, Flum DR, Pomp A, Pories WJ, Wolfe BM, et al. Postoperative behavioral variables and weight change 3 years after bariatric surgery. J Am Med Assoc Surg 2016;58107:1−6.

[3] Bocchieri LE, Meana M, Fisher BL. A review of psychosocial outcomes of surgery for morbid obesity. J Psychosom Res 2002;52:155−65.

[4] Sarwer DB, Lavery M, Spitzer JC. A review of the relationships between extreme obesity, quality of life, and sexual function. Obes Surg 2012;22(4):668−76.

[5] de Zwaan M, Enderle J, Wagner S, Mühlhans B, Ditzen B, Gefeller O, et al. Anxiety and depression in bariatric surgery patients: a prospective, follow-up study using structured clinical interviews. J Affect Disord 2011;133:61−8.

[6] Conceição E, Mitchell JE, Vaz A, Bastos AP, Ramalho S, Silva C, et al. The presence of maladaptive eating behaviors after bariatric surgery in a cross sectional study: importance of picking or nibbling on weight regain. Eat Behav 2014;15(4):558−62.

[7] Meany G, Conceição E, Mitchell JE. Binge eating, binge eating disorder and loss of control eating: effects on weight outcomes after bariatric surgery. Eur Eat Disord Rev 2014;22(2):87−91.

[8] Livhits M, Mercado C, Yermilov I, Parikh JA, Dutson E, Mehran A, et al. Preoperative predictors of weight loss following bariatric surgery: systematic review. Obes Surg 2012;22:70−89.

[9] Sheets CS, Peat CM, Berg KC, White EK, Bocchieri-Ricciardi L, Chen EY, et al. Post-operative psychosocial predictors of outcome in bariatric surgery. Obes Surg 2014;25(2):330−45.

[10] Conceição EM, Utzinger LM, Pisetsky EM. Eating disorders and problematic eating behaviours before and after bariatric surgery: characterization, assessment and association with treatment outcomes. Eur Eat Disord Rev 2015;35:417−25.

[11] Conceição E, Mitchell JE. Assessment of eating disorders and problematic eating behavior in bariatric surgery patients. In: Walsh TB, Attia E, Glasofer DR, Sysko R, editors. Handbook of assessment and treatment of eating disorders. Arlington: American Psychiatric Association Publishing; 2016. p. 83−104.

[12] American Psychiatric Association. Diagnostic and statistical manual of mental disorders. 5th ed. Washington, DC: American Psychaitric Association; 2013.

[13] Wolfe BE, Baker CW, Smith AT, Kelly-Weeder S. Validity and utility of the current definition of binge eating. Int J Eat Disord 2009;42(8):674−85.

[14] Mond JM, Latner JD, Hay PH, Owen C, Rodgers B. Objective and subjective bulimic episodes in the classification of bulimic-type eating disorders: another nail in the coffin of a problematic distinction. Behav Res Ther 2010;48(7):661−9.

[15] Palavras MA, Morgan CM, Borges FMB, Claudino AM, Hay PJ. An investigation of objective and subjective types of binge eating episodes in a clinical sample of people with co-morbid obesity. J Eat Disord 2013;1:26.

[16] Scholtz S, Bidlake L, Morgan J, Fiennes A, El-Etar A, Lacey JH, et al. Long-term outcomes following laparoscopic adjustable gastric banding: postoperative psychological sequelae predict outcome at 5-year follow-up. Obes Surg 2007;17:1220−5.

[17] de Zwaan M, Hilbert A, Swan-Kremeier L, Simonich H, Lancaster K, Howell LM, et al. Comprehensive interview assessment of eating behavior 18−35 months after gastric bypass surgery for morbid obesity. Surg Obes Relat Dis 2010;6(1):79−85.

[18] Sarwer DB, Wadden TA, Moore RH, Baker AW, Gibbons LM, Raper SE, et al. Preoperative eating behavior, postoperative dietary adherence and weight loss following gastric bypass surgery. Surg Obes Relat Dis 2008;4(5):640−6.

[19] Colles SL, Dixon JB, O'Brien PE. Grazing and loss of control related to eating: two high-risk factors following bariatric surgery. Obesity 2008;16:615−22.

[20] Sallet PC, Sallet JA, Dixon JB, Collis E, Pisani CE, Levy A, et al. Eating behavior as a prognostic factor for weight loss after gastric bypass. Obes Surg 2007;17:445−51.

[21] White MA, Kalarchian MA, Masheb RM, Marcus, Marsha D, Grilo CM. Loss of control over eating predicts outcomes in bariatric surgery: a prospective 24-month follow-up study. J Clin Psychiatry 2010;71(2):175−84.

[22] Latner JD, Hildebrandt T, Rosewall JK, Chisholm AM, Hayashi K. Loss of control over eating reflects eating disturbances and general psychopathology. Behav Res Ther 2007;45(9):2203–11.

[23] Griez E, Jansen A, van den Hout M. Clinical and non-clinical binges. Behav Res Ther 1990;28(5):439–44.

[24] Conceição E, Mitchell J, Engle S, Machado P, Lancaster K, Wonderlich S. What is 'grazing'? Reviewing its definition, frequency, clinical characteristics and impact on bariatric surgery outcomes, and proposing a standardized definition. Surg Obes Relat Dis 2014;10(5):973–82.

[25] Fairburn Z, Cooper CG, O'Connor M. The eating disorder examination (Edition 17.0). [Online]. Available: <http://www.credo-oxford.com/pdfs/EDE_17.0D.pdf>; 2014.

[26] First MB, Spitzer RL, Gibbon M, Williams JB. User's guide for the structured clinical interview for DSM-IV axis I disorders SCID-I: clinician version. Am Psychiatr Pub 1997.

[27] Parker K, Brennan L. Measurement of disordered eating in bariatric surgery candidates: a systematic review of the literature. Obes Res Clin Pract 2015;9(1):12–25.

[28] Parker K, O'Brien P, Brennan L. Measurement of disordered eating following bariatric surgery: a systematic review of the literature. Obes Surg 2014;24(6):945–53.

[29] Kofman MD, Lent MR, Swencionis C. Maladaptive eating patterns, quality of life, and weight outcomes following gastric bypass: results of an Internet survey. Obesity 2010;18(10):1938–43.

[30] Nicolau J, Ayala L, Rivera R, Speranskaya A, Sanchís P, Julian X, et al. Postoperative grazing as a risk factor for negative outcomes after bariatric surgery. Eat Behav 2015;18:147–50.

[31] Faria SL, de E, Kelly O, Lins RD, Faria OP. Nutritional management of weight regain after bariatric surgery. Obes Surg 2010;20(2):135–9.

[32] Robinson AH, Adler S, Stevens HB, Darcy AM, Morton JM, Safer DL. What variables are associated with successful weight loss outcomes for bariatric surgery after 1 year? Surg Obes Relat Dis 2014;10(4):697–704.

[33] Sarwer D, Dilks R, West-Smith L. Dietary intake and eating behavior after bariatric surgery: threats to weight loss maintenance and strategies for success. Surg Obes Relat Dis 2011;7(5):644–51.

[34] Conceição EM, Crosby R, Mitchell JE, Engel SG, Wonderlich SA, Simonich HK, et al. Picking or nibbling: Frequency and associated clinical features in bulimia nervosa, anorexia nervosa, and binge eating disorder. Int J Eat Disord 2013;46(8):815–18.

[35] Masheb RM, Grilo CM, White MA. An examination of eating patterns in community women with bulimia nervosa and binge eating disorder. Int J Eat Disord 2011;44:618–24.

[36] Masheb RMB, Roberto CA, White MA. Nibbling and picking in obese patients with binge eating disorder. Eat Behav 2013;14(4):424–7.

[37] Reas DL, Wisting L, Kapstad H, Lask B. Nibbling: Frequency and relationship to BMI, pattern of eating, and shape, weight, and eating concerns among university women. Eat Behav 2012;13:65–6.

[38] Poole NA, Al Atar A, Kuhanendran D, Bidlake L, Fiennes A, McCluskey S, et al. Compliance with surgical after-care following bariatric surgery for morbid obesity: a retrospective study. Obes Surg 2005;15:261–5.

[39] Hsu LKG, Sullivan SP, Benotti PN. Eating disturbances and outcome of gastric bypass surgery: a pilot study. Int J Eat Disord 1997;21:385–90.

[40] Fairburn CG, Beglin SJ. Eating Disorder Examination Questionnaire (6.0). In: Fairburn CG, editor. Cognitive Behavior Therapy and Eating Disorders. New York: NY: Guilford Press; 2008.

[41] Gormally J, Black S, Daston S, Rardin D. The assessment of binge eating severity among obese persons. Addict Behav 1982;7(1):47–55.

[42] van Strien T, Frijters JER, Bergers GPA, Defares PB. The Dutch eating behavior questionnaire (DEBQ) for assessment of restrained, emotional, and external eating behavior. Int J Eat Disord 1986;5(2):295–315.

[43] Spitzer RL, Yanovski S, Wadden T, Wing R, Marcus MD, Stunkard A, et al. Binge eating disorder: its further validation in a multisite study. Int J Eat Disord Mar. 1993;13(2):137–53.

[44] Garner DM, Olmstead M, Polivy J. Development and validation of a multidimensional eating disorder inventory for anorexia nervosa and bulimia. Int J Eat Disord 1983;2(2):15–34.

[45] Blomquist KK, Roberto CA, Barnes RD, White MA, Masheb RM, Grilo CM. Development and validation of the eating loss of control scale. Psychol Assess 2014;26(1):77–89.

Chapter 48

Dietary Planning in Bariatric Surgery Postoperative

L. Crovesy, F.C.C.M. Magno and E.L. Rosado
Federal University of Rio de Janeiro, Rio de Janeiro, RJ, Brazil

LIST OF ABBREVIATIONS

25(OH)D 25-hydroxy vitamin D
PTH parathyroid hormone
RYGB Roux-en-Y gastric bypass
TEE total energy expenditure
USA United States of America
VG vertical gastrectomy
WHO World Health Organization

INTRODUCTION

Obesity is characterized by abnormal or excessive accumulation of adipose tissue that has harmful consequences for one's health, including cardiovascular disease, diabetes, musculoskeletal disorders, and some types of cancer [1]. Worldwide, approximately 39% of adults are overweight, and 13% are obese [1]. More than 64% of obesity cases occur in developed countries. The five countries with the highest number of obese individuals are the United States, China, India, Russia, and Brazil [2].

Bariatric bypass surgery is an alternative to other obesity treatments to promote weight loss and long-term maintenance of the weight loss [3]. There are several different types of techniques for bariatric bypass surgery, but only a few are permitted in Brazil and the rest of the world: gastric banding, Roux-en-Y gastric bypass (RYGB), sleeve gastrectomy, and biliopancreatic diversion with duodenal switch [4,5]. Performance of these surgeries is increasing worldwide. In the United States, in 1998, 12,775 of these surgical procedures were carried out; however, this number increased to 179,000 in 2013 [5,6]. In Brazil, in 2012 alone, approximately 72,000 bariatric and metabolic surgeries were performed [4].

Despite the promising results of surgery to control obesity and comorbidities, some adverse effects and complications can occur in the immediate and late postoperative periods; these adverse effects and complications are often associated with inadequate dietary planning or low compliance with the proposed plan. Therefore, follow-ups with nutritionists are essential in the bariatric surgery pre- and postoperative periods [7].

NUTRITIONAL RECOMMENDATIONS IN THE IMMEDIATE POSTOPERATIVE PERIOD

The transitions and durations of the diet phases depend on which surgical procedure was used and on the tolerance of each patient. Nutritional supplements and specific food products may be required to ensure adequate nutrition. Table 48.1 shows the phases of the suggested diet following bariatric surgery, and Table 48.2 presents foods (along with their quantities) suggested for each phase of the diet [8,9].

It is not uncommon for patients to experience some food intolerance at this point, therefore, it may be necessary to provide some guidelines regarding foods that may be difficult for patients to consume, such as rice, milk, and red meat. In such cases, substitution of these foods is important so that the patient will still eat nutrients from these food groups,

TABLE 48.1 Diet Phases in the Immediate Postoperative Period Following Bariatric Surgery

Diet Phase	Length	First Postoperative Day
Clear liquids	1–2 days	1–2 days
Full liquid or semiliquid	10–14 days	2–16 days
Pureed diet	10–14 days	16–30 days
Soft	≤ 14 days	30–60 days

Source: Adapted from Allied Health Sciences Section Asd Hoc Nutrition Committee, Aills L, Blankenship J, Buffington C, Furtado M, Parrott J. ASMBS Allied Health Nutrition guidelines for the surgical weight loss patients. Surg Obes Relat Dis 2008;4: S73–S108; Mechanick JI, Youdim A, Jones DB, Garvey WT, Hurley DL, McMahon MM, et al. Clinical practice guidelines for the perioperative nutritional, metabolic, and nonsurgical support of the bariatric surgery patient – 2013 update: cosponsored by American Association of Clinical Endocrinologists, The Obesity Society, and American Society for Metabolic & Bariatric Surgery. Endocr Pract 2013;19:337–372.

TABLE 48.2 Suggested Foods and Quantities for Each Phase of the Diet

Diet Phase	Foods	Amount/Frequency
Clear liquids	Water Coconut water Clear broth (meat and vegetable) Herbal tea	• 30–50 mL • 20–30 min
Full liquid or semiliquid	Water Coconut water Blended and strained soup Skim milk or soybean beverages Fruit juice diluted in water	• 80–100 mL • Every 1 h or • Every hour and a half
Pureed	Water Skim milk blended with fruit Bread and biscuits soaked in milk Mashed food or porridge (rice, vegetables, meat) Blended grains Mashed and cooked fruit	• 100–150 mL • Every 2 h
Soft	Water Cooked vegetables Boiled eggs and cooked meat Soft bread Milk and skim dairy Fruit	• 150 mL • Every 2 h

Source: Adapted from Allied Health Sciences Section Asd Hoc Nutrition Committee, Aills L, Blankenship J, Buffington C, Furtado M, Parrott J. ASMBS Allied Health Nutrition guidelines for the surgical weight loss patients. Surg Obes Relat Dis 2008;4: S73–S108; Mechanick JI, Youdim A, Jones DB, Garvey WT, Hurley DL, McMahon MM, et al. Clinical practice guidelines for the perioperative nutritional, metabolic, and nonsurgical support of the bariatric surgery patient – 2013 update: cosponsored by American Association of Clinical Endocrinologists, The Obesity Society, and American Society for Metabolic & Bariatric Surgery. Endocr Pract 2013;19:337–372.

such as the following: rice can be replaced with potatoes or pasta, milk with yogurt, and red meat with more easily digestible white meat. It is important that the nutritionist direct the patient to consume solid foods and chew efficiently because the patient may experience fear of consuming solid food. Food intolerances and aversions have been reduced through preoperative guidance and with the gradual introduction of solid foods [10].

NUTRITIONAL RECOMMENDATIONS IN THE LATE POSTOPERATIVE PERIOD

The nutritional pyramid proposed by Moizé et al. [11] aims to assist the dietary treatment of individuals in the late postoperative period following bariatric surgery. The base of the pyramid consists of physical activity, vitamin and mineral supplements, and sugar-free, noncaffeine, or noncarbonated liquids. The second level includes protein-rich foods such as lean meats, fish, low-fat or fat-free dairy products, legumes, and eggs. The third level includes fruits, vegetables, and vegetable oils. Emphasis should be on consumption of foods from the second and third levels of the pyramid. Foods that make up the fourth level of the pyramid, such as cereals and tubers, should be sparingly consumed. Foods high in saturated and trans fatty acids, cholesterol, sugar, and alcoholic or sweetened drinks are at the top of the pyramid and should be avoided [11].

Patients should have a hypocaloric diet, which is characterized by reduced caloric intake. However, an energy requirement prediction formula has not been validated for postbariatric surgery patients [11]. The protein recommendation for these individuals is 0.8−2.1 g/kg of ideal weight/day, while a daily intake of 60−80 g is recommended to prevent protein malnutrition [8,9,11]. However, many patients are intolerant to dry and hard protein food sources such as red meat, pork, and poultry, which makes it difficult for them to consume the ideal recommendations of protein and prevent protein malnutrition. To increase the consumption of proteins in cases of intolerance, other protein sources such as fish, eggs, legumes with cereals, cheese, and tofu should be included in the diet [11].

The use of protein supplements can help to achieve the consumption of this macronutrient and avoid the excessive loss of lean body mass. There are different types of protein supplements on the market, including concentrated complete proteins, which are derived from egg whites, soybeans, or milk; frozen collagen-based proteins; combinations of casein and other complete proteins; amino acids; hybrid proteins plus concentrated amino acids, a complete protein, or a collagen base associated with dispensable amino acids (alanine, aspartic acid, asparagine, arginine, cysteine, glutamic acid, glycine, proline, serine, and tyrosine) [9].

Lipids should be monitored and reduced if their intake is high in order to decrease calorie consumption. An intake of 30−35% of the total energy expenditure (TEE) is recommended for lipids. The amount and quality of consumed lipids is important, with priority given to foods containing unsaturated fatty acids. Extra virgin olive oil and soybean, canola, and flaxseed oils have higher levels of fatty acids of the n-3 and n-6 series, so n-3 supplementation via fish oil could be necessary [11].

The recommended carbohydrate intake is 40−45% of the TEE, and simple sugars should be avoided, while consumption of whole grains and fiber should be prioritized; the recommended consumption of whole grains and fiber is 14 g/1000 kcal [11].

Nutritional guidelines during the late postoperative period are essential to enable adherence to the diet plan and to promote healthy eating and prevent complications. Nutritional guidelines include the following: avoid drinking liquids during meals; consume liquids at least 30 minutes after meals; avoid high-calorie and/or carbonated liquids; eat many small portions (six meals per day, up to 300 g/meal); respect the amount of food; and chew the food well [8,11].

MICRONUTRIENT DEFICIENCIES: PREVALENCE, ETIOLOGY, AND CONSEQUENCES

Micronutrient deficiencies observed in patients who undergo bariatric surgery primarily occur because of anatomical and physiological alterations that impair the absorption of nutrients and/or food intake [12] (Fig. 48.1).

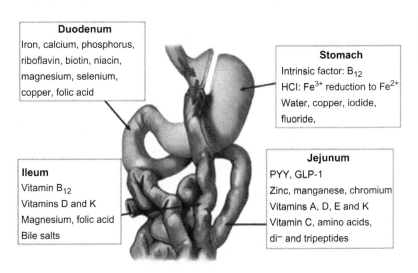

FIGURE 48.1 RYGB technique and its main metabolic changes. Smaller rectangles show the nutrients and/or hormone production [12].

The most common types of bariatric surgery currently performed are vertical gastrectomy (VG), which restricts the size of the gastric cavity and reduces the amount of ingested food; and RYGB, which also reduces stomach size and even excludes the duodenum and the first portion of the jejunum from the absorption process, causing malabsorption of some nutrients [4].

Thiamine or Vitamin B1

The prevalence of a deficiency in this vitamin is 18% within 1 year after and 11% within 2 years after RYGB, while the prevalence is between 0% and 11% within 1 year after and 31% within 5 years after VG. Reserves of thiamine are low, approximately 30 mg, and are rapidly depleted, within an average of 20 days. The causes of this deficiency are the low intake of whole carbohydrates, malabsorption, excessive intake of alcohol, and excessive vomiting. The consequences usually include cardiovascular issues and disorders in the central nervous system, with 13 cases of Wernicke encephalopathy after VG being reported in the literature [13−15].

Folic Acid or Vitamin B9

The prevalence of a deficiency in this vitamin is 18.4% within 4 years after and 12% within 10 years after RYGB, and 16.7% within 1 year after, 21.4% within 2 years after, and 20% within 4 years after VG. Folic acid is absorbed via transport mechanisms through the intestinal wall and low pH. Folic acid deficiency is associated with low intake and preoperative deficiency. Past research points to the relationship between folate metabolism and lipid metabolism in obesity and that serum concentrations of folic acid are significantly lower during obesity, regardless of food intake. The consequences include anemia, memory loss, fatigue, and drowsiness [16−18].

Cyanocobalamin or Vitamin B12

The prevalence of a deficiency in this vitamin is 42.1% within 4 years after and 35% within 5 years after RYGB, and 13% within 2 years after and 5% within 4 years after VG. However, the prevalence of this deficiency may range from 4% to 62%, depending on the surgical technique and the length of time after surgery. This deficiency occurs because of decreased production of gastric acid (hydrochloric acid and pepsin), a decline in the bioavailability of intrinsic factor (IF), and the consequent formation of an IF−B12 complex. Furthermore, cyanocobalamin deficiency may also occur because of a low intake of source foods and inadequate supplementation. The consequences usually include anemia, hyperhomocysteinemia, and neurological disorders [15−17,19].

Vitamin A

The prevalence of a vitamin A deficiency is 35.9% within the 30 days after and 18−27.5% within 1 year after RYGB, and 0% within 1 year after VG. Bariatric surgery as a treatment for obesity might affect the biomarkers of oxidative stress. In addition, the decreased intake and absorption of antioxidant vitamins may be exacerbating factors that increase systemic oxidative stress. Other causes for this deficiency are the malabsorption of lipids, low intake of certain nutrients, and the presence of nonalcoholic, common fatty liver disease in obese patients with high body mass indexes (BMIs). The consequences usually include Bitot spot, xerosis, hyperkeratinization of the skin, night blindness, and poor wound healing [20−24].

Vitamin D

The prevalence of a vitamin D deficiency is 21% within 1 year after and 39.6% within 4 years following RYGB, and 36−93% within 1 year after, 42% within 2 years after and 56.3% within 4 years after VG. There is consensus in the literature that vitamin D deficiency is high in obese people due to the excess of adipose tissue, but other common factors that contribute to the occurrence of this deficiency in the postoperative period are low sun exposure and malabsorption of lipids. The consequences include calcium deficiency, increased fractures, secondary hyperparathyroidism, and immune system dysfunction. Depending on the surgical technique and on the malabsorption induced, difficulties can occur in the normalization of concentrations of 25-hydroxy vitamin D (25(OH)D) and parathyroid hormone (PTH), which requires lifelong surveillance [17,18,25,26].

Vitamin E

The prevalence of a vitamin E deficiency after bariatric surgery is 0% within 1 year, 14% after 2 years, 3% after 3 years and 4% after 4 years, and it is most prevalent after disabsortive surgeries, but unusual after RYGB or VG. The most common causes for this deficiency are oxidative stress and malabsorption of lipids. The consequences include asthenia, an association with the loss of lean mass, retinopathy, ataxia, impaired immunity, and atherosclerosis [8,21,27].

Vitamin K

The prevalence of a vitamin K deficiency following bariatric surgery is 14% after 1 year, 21% after 2 years, 13% after 3 years and 42% after 4 years, and it is most prevalent after disabsortive surgeries, but unusual after RYGB or VG. According to the literature, its deficiency occurs because of the malabsorption of lipids. The consequences include blood clotting issues and the loss of bone quality [8,27].

Iron

The prevalence of an iron deficiency is 13.4% 1 year after, 23.0% 2 years after and 47% 5 years after RYGB, and 28% 1 year after, 38% 2 years after and 30% 4 years after VG. In addition, the prevalence of an iron deficiency may increase up to 52% in fertile women who menstruate and are extremely obese. Studies indicate that with an excess of adipose tissue, there is an increase in inflammatory adipokines (tumor necrosis factor α, resistin, and interleukins-1 and -6), and that this inflammation impairs regulation of the physiological homeostasis of iron by the systemic iron regulatory protein (hepcidin). The most common causes of iron deficiency in the postoperative period are decreased hydrochloric acid, which impairs the conversion of ferric iron to ferrous iron, the most absorbable form; low intake of certain nutrients, which has been reported to be more frequent with intolerance to meat; and intestinal bypass. The consequences usually include anemia, dysphagia, increased heart rate, and fatigue [8,15−18,28,29].

Calcium

The prevalence of a calcium deficiency after bariatric surgery is 15% after 1 year, 27% after 2 years, 31% after 3 years and 48% after 4 years. After RYGB, it occurs in 10.8−14.7% of patients after 1 year and 6.4−8.2% after 2 years, depending on the size of the common strap, and following VG, calcium deficiency occurs in 4−67% of patients after 1 year. The most common causes for this deficiency are reduced hydrochloric acid, low intake of certain nutrients, and intestinal bypass. The consequences usually include loss of bone health (osteopenia, osteoporosis, evidence of increased bone remodeling, and a steady increase in parathyroid hormone (PTH)), brittle nails, susceptibility to cavities, hypertension, insomnia, and irritability [18,22,27,30].

Zinc

The prevalence of a zinc deficiency is 4.8−7.4% after 1 year, 5.4−38.1% after 2 years, 15.4% after 4 years and 21.2% after 5 years of RYGB, and 30% after 6 months, 5% after 1 year and 22.2% after 2 years of VG. However, this prevalence can reach 50% in 1 year after surgeries that cause more malabsorption. There is a high level of intolerance to meat and low intake of other sources of zinc, which may be key factors for zinc deficiency. Other causes cited in the literature include fat malabsorption and intestinal bypass. The consequences usually include an increase in susceptibility to hair loss, infection, and damage to the gastrointestinal, musculoskeletal, immune, and reproductive systems [12,27,31,32].

Copper

Copper deficiencies are uncommon and rarely reported in the literature. They may occur in 9.6−14% of patients 2 years after RYGB. Studies show that an excess of zinc in the diet or in a supplemental form may cause copper deficiency. However, factors such as low oral intake and intestinal bypass may also contribute to this deficiency after surgery. The consequences usually include anemia, leukopenia, several neuromuscular abnormalities, and impaired immune function [31−35].

Selenium

Selenium deficiencies are rarely reported in the literature. They may occur in 14.3% of patients 3 years after bariatric surgery and in 14−22% of patients after RYGB; there have been no reports on evaluation time. Published studies have shown that in obese patients there is an increase in metabolic consumption of antioxidants, which results in their depletion in the body. Therefore, low serum concentrations of selenium could be related to deficiencies that occur even before surgery, but could also be due to intestinal bypasses that occur during mixed and disabsortive surgeries like RYGB. The consequences usually include cardiovascular problems, loss of lean body mass, impaired immunity, low thyroid function, loss of skin and hair pigmentation, whitened nails, decreased fertility in males, and progressive encephalopathy [33,36].

VITAMIN AND MINERAL SUPPLEMENTATION

The most appropriate way to replenish and maintain the storages of micronutrients in the body is through a balanced and varied diet. However, patients undergoing bariatric surgery may be unable to fully restore the necessary vitamins and minerals because their intake and absorption are impaired from their reduced gastric capacity and intestinal bypass, respectively. Therefore, the daily use of multivitamins/minerals is an important strategy to ensure that patients receive the required amount of micronutrients for their bodies to function properly [8,12,37].

To demonstrate the importance of supplemental vitamins and minerals, the nutritional pyramid for the postoperative treatment of bariatric surgery patients, developed by Moizé and collaborators, takes into account the specifics of this population. It is a tool built to help professionals and patients in nutritional counseling; it covers food and portion choices to provide adequate food and nutrition, and also details the importance of supplemental multivitamins/minerals [11].

Multivitamin/mineral supplementation involves 1−2 tablets per day, plus additional supplements involving 20−30 mg/day (oral) or 50−100 mg/day (intravenous or muscle) of thiamine; 400 mcg/day of folic acid; 350 mcg/day (prevention/oral) or 500 mcg/day (treatment/oral/day) or 500 mcg/day (sublingual/week) or 1000 mcg (intramuscular) of cyanocobalamin; 10,000−1,000,000 IU/day of vitamin A; 400 IU/day of vitamin D; 100−400 IU/day of vitamin E; 25 mcg/day of vitamin K; 40−65 mg/day of iron (ferrous fumarate); 1200−2000 mg/day calcium (calcium citrate); 6.5 mcg/day of zinc; 50−900 mcg/day of copper; and 50 mcg/day of selenium [8,12,37].

There are several multivitamin options on the market that are used by the bariatric population. In recent years, specific formulations directed at these patients in particular have been produced to facilitate adherence to the supplements, as these supplements are chewable or are very small pills and are manufactured in different types of capsules (so that there is no nutrient/nutrient interaction).

MINI-DICTIONARY OF TERMS

- *Diet phases* are different consistency of diet in which the postbariatric patients must pass until to be able to consume solid food.
- *Food intolerance* is a physiological condition in which the body does not have some enzyme able to digest a specific nutrient ingested.
- *Food aversions* repulse for certain food due to association with uneasiness.
- *Food groups* is a food set with similar nutritional composition.
- *Nutritional pyramid* is a graphic illustration with the amount of food that must be consumed in a day according to a balanced diet.
- *Macronutrients* compose the foods and provide energy necessary to the functioning of the body. They are: carbohydrate, protein, and lipid.

KEY FACTS

- Obesity is difficult to control and is increasing worldwide.
- Bariatric surgery is an alternative that has been successful in treating obesity.
- Bariatric surgery can lead to nutritional deficiencies.
- Nutritional counseling is important postoperatively, both in the short-term and during follow-up.
- The dietary plan for postbariatric surgery patients may require more protein and some micronutrients.

SUMMARY POINTS

- This chapter focuses on short-term and follow-up dietary plans after bariatric surgery.
- In the short-term after bariatric surgery, the diet is liquid and small in volume because of the reduction in stomach size.
- Quick weight loss typically promotes free fat mass loss because of low protein intake.
- Some types of surgery promote deficiencies in micronutrients.
- The follow-up after bariatric surgery focuses on having a diet that provides adequate intake of energy, protein, and micronutrients.

REFERENCES

[1] World Health Organization (WHO). Fact sheet: obesity and overweight, n. 311 Available from: <http://www.who.int/mediacentre/factsheets/fs311/en/index.html>; 2015 [accessed 11.03.16].

[2] Europe PMC. Funders Group, Global, regional and national prevalence of overweight and obesity in children and adults 1980–2013: a systematic analysis for the Global Burden of Disease Study 2013. Lancet 2014;385:117–71.

[3] International Federation for the Surgery of Obesity and Metabolic Disorders – IFSO, Available from: <http://www.ifso.com/about-ifso/>; [accessed 11.03.16].

[4] Sociedade Brasileira de Cirurgia Bariátrica e Metabólica, Available from: <http://www.sbcbm.org.br/wordpress/>; [accessed 11.03.16].

[5] Connect the official new magazine of ASMBS. Available from: <http://connect.asmbs.org/may-2014-bariatric-surgery-growth.html>; [accessed 11.03.16].

[6] Nguyen NT, Masoomi H, Magno CP, Nguyen XM, Laugenour K, Lane J. Trends in use o bariatric surgery, 2003–2008. J Am Coll Surg 2011;213:261–6.

[7] Costa LD, Valezi AC, Matsu T, Dichi I, Dichi JB. Repercussão da perda de peso sobre parâmetros nutricionais e metabólicos de pacientes obesos graves após um ano de gastroplastia em Y-de-Roux. Rev Col Bras Cir 2010;37:96–101.

[8] Allied Health Sciences Section Asd Hoc Nutrition Committee, Aills L, Blankenship J, Buffington C, Furtado M, Parrott J. ASMBS Allied Health Nutrition guidelines for the surgical weight loss patients. Surg Obes Relat Dis 2008;4:S73–S108.

[9] Mechanick JI, Youdim A, Jones DB, Garvey WT, Hurley DL, McMahon MM, et al. Clinical practice guidelines for the perioperative nutritional, metabolic, and nonsurgical support of the bariatric surgery patient – 2013 update: cosponsored by American Association of Clinical Endocrinologists, The Obesity Society, and American Society for Metabolic & Bariatric Surgery. Endocr Pract 2013;19:337–72.

[10] França DLM, Nascimento EA, Gravena AAF. Aspectos gastrointestinais, perda de peso e uso de suplementos vitamínicos em pacientes pós-operatório de cirurgia bariátrica. Rev Saude Pesq 2011;4:23–8.

[11] Moizé VL, Pi-Sunyer X, Mochari H, Vidal J. Nutritional pyramid for post-gastric bypass patients. Obes Surg 2010;20:1133–41.

[12] Bordalo LA, Teixeira TFS, Bressan J, Mourão DM. Bariatric surgery: how and why to supplement. Rev Assoc Med Bras 2011;57:113–20.

[13] Kröll D, Laimer M, Borbély YM, Laederach K, Candinas D, Nett PC. Wernicke Encephalopathy: a future problem even after sleeve gastrectomy? A Systematic Literature Review. Obes Surg 2015; [Epub ahead of print]

[14] Stroh C, Meyer F, Manger T. Beriberi, a severe complication after metabolic surgery – review of the literature. Obes Facts 2014;7:246–52.

[15] Saltzman E, Karl JP. Nutrient deficiencies after gastric bypass surgery. Annu Rev Nutr 2013;33:183–203.

[16] Ben-Porat T, Elazary R, Yuval JB, Wieder A, Khalaileh A, Weiss R. Nutritional deficiencies after sleeve gastrectomy: can they be predicted preoperatively? Surg Obes Relat Dis 2015;S1550–7289 [Epub ahead of print]

[17] Alexandrou A, Armeni E, Kouskouni E, Tsoka E, Diamantis T, Lambrinoudaki I. Cross-sectional long-term micronutrient deficiencies after sleeve gastrectomy versus Roux-en-Y gastric bypass: a pilot study. Surg Obes Relat Dis 2014;10:262–8.

[18] Alvarez V, Cuevas A, Olivos C, Berry M, Farías MM. Micronutrient deficiencies one year after sleeve gastrectomy. Nutr Hosp 2014;29:73–9.

[19] da Silva RP, Kelly KB, Rajabi A, Jacobs RL. Novel insights on interactions between folate and lipid metabolism. Biofactors 2014;40:277–83.

[20] Belfiore A, Cataldi M, Minichini L, Aiello ML, Trio R, Rossetti G, et al. Short-term changes in body composition and response to micronutrient supplementation after laparoscopic sleeve gastrectomy. Obes Surg 2015;25:2344–51.

[21] Cuesta M, Pelaz L, Pérez C, Torrejón MJ, Cabrerizo L, Matía P, et al. Fat-soluble vitamin deficiencies after bariatric surgery could be misleading if they are not appropriately adjusted. Nutr Hosp 2014;30:118–23.

[22] Van Rutte PW, Aarts EO, Smulders JF, Nienhuijs SW. Nutrient deficiencies before and after sleeve gastrectomy. Obes Surg 2014;24:1639–46.

[23] Dadalt C, Fagundes RL, Moreira EA, Wilhelm-Filho D, de Freitas MB, Jordão Jr. AA, et al. Marcadores de estresse oxidativo emadultos 2 ano-sapós a Roux-en-Y bypass gástrico. Eur J Gastroenterol Hepatol 2013;25:580–6.

[24] Zalesin KC, Miller WM, Franklin B, Mudugal D, Rao Buragadda A, Boura J, et al. Vitamin a deficiency after gastric bypass surgery: an under-reported postoperative complication. J Obes 2011;760695.

[25] Maeda SS, Borba VZ, Camargo MB, Silva DM, Borges JL, Bandeira F, et al. Brazilian Society of Endocrinology and Metabology (SBEM). Recommendations of the Brazilian Society of Endocrinology and Metabology (SBEM) for the diagnosis and treatment of hypovitaminosis D. Arq Bras Endocrinol Metabol 2014;58:411–33.

[26] Obinwanne KM, Riess KP, Kallies KJ, Mathiason MA, Manske BR, Kothari SN. Effects of laparoscopic Roux-en-Y gastric bypass on bone mineral density and markers of bone turnover. Surg Obes Relat Dis 2014;10:1056–62.

[27] Slater GH, Ren CJ, Siegel N, Williams T, Barr D, Wolfe B, et al. Serum fat-soluble vitamin deficiency and abnormal calcium metabolism after malabsorptive bariatric surgery. J Gastrointest Surg 2004;8:48–55.

[28] Weng TC, Chang CH, Dong YH, Chang YC, Chuang LM. Anaemia and related nutrient deficiencies after Roux-en-Y gastric bypass surgery: a systematic review and meta-analysis. BMJ Open 2015;5:e006964.

[29] Aigner E, Feldman A, Datz C. Obesity as an emerging risk factor for iron deficiency. Nutrients 2014;6:3587–600.

[30] Abellan I, Luján J, Frutos MD, Abrisqueta J, Hernández Q, López V, et al. The influence of the percentage of the common limb in weight loss and nutritional alterations after laparoscopic gastric bypass. Surg Obes Relat Dis 2014;10:829–33.

[31] Basfi-Fer K, Rojas P, Carrasco F, Valencia A, Inostroza J, Codoceo J, et al. Evolution of the intake and nutritional status of zinc, iron and copper in women undergoing bariatric surgery until the second year after surgery. Nutr Hosp 2012;27:1527–35.

[32] Balsa JA, Botella-Carretero JI, Gómez-Martín JM, Peromingo R, Arrieta F, Santiuste C, et al. Copper and zinc serum levels after derivative bariatric surgery: differences between Roux-en-Y Gastric bypass and biliopancreatic diversion. Obes Surg 2011;21:744–50.

[33] Papamargaritis D, Aasheim ET, Sampson B, le Roux CW. Copper, selenium and zinc levels after bariatric surgery in patients recommended to take multivitamin-mineral supplementation. J Trace Elem Med Biol 2015;31:167–72.

[34] Robinson SD, Cooper B, Leday TV. Copper deficiency (hypocupremia) and pancytopenia late after gastric bypass surgery. Proc (Bayl Univ Med Cent). 2013;26:382–6.

[35] Gletsu-Miller N, Broderius M, Frediani JK, Zhao VM, Griffith DP, Davis Jr. SS, et al. Incidence and prevalence of copper deficiency following Roux-en-Y gastric bypass surgery. Int J Obes (Lond) 2012;36:328–35.

[36] Freeth A, Prajuabpansri P, Victory JM, Jenkins P. Assessment of selenium in Roux-en-Y gastric bypass and gastric banding surgery. Obes Surg 2012;22:1660–5.

[37] Torezan EFG. Revisão das principais deficiências de micronutrientes no pós-operatório do Bypass Gástrico em Y de Roux. Int J Nutrol 2013;6:37–42.

Chapter 49

Protein Nutrition and Status and Bariatric Surgery

V. Moizé[1], B. Laferrère[2] and J. Vidal[1]

[1]Hospital Clinic, Barcelona, Spain, [2]Columbia University College of Physicians and Surgeons, New York, NY, United States

LIST OF ABBREVIATIONS

AA	amino acid
BC	body composition
BCAA	branched-chain amino acid
BS	bariatric surgery
EE	energy expenditure
FFM	fat-free mass
GBP	gastric bypass
IBW	ideal body weight
LBM	lean body mass
NB	nitrogen balance
PI	protein intake
PS	Protein status
RDA	recommended dietary allowance
REE	resting energy expenditure
SM	skeletal muscle
T2D	type 2 diabetes
VLCD	very-low-calorie diet

INTRODUCTION

It is established that bariatric surgery (BS) is the most effective long-term therapy for the treatment of obesity. Several studies have demonstrated that BS is associated with favorable impact on hard endpoints such as overall and cardiovascular mortality, incidence of first occurrence of fatal or nonfatal cardiovascular events, as well as prevention and remission of type 2 diabetes (T2D) and quality of life [1]. As described in previous chapters of this book, BS techniques include restrictive, malabsorptive, and mixed procedures (which combine restrictive and malabsorptive components) [2]. The mechanism of improved comorbidities is associated, but not limited to, calorie restriction and weight loss. Clinical and translational studies over the last decade have shown that a number of gastrointestinal mechanisms, including changes in gut hormones [3], neural signaling [4], intestinal flora [5], bile acids [6], and lipid metabolism [7], can play a significant role in the effects of this procedure on energy homeostasis, independent of weight loss. Sustained negative energy balance is needed to promote weight loss, and dietary changes after BS entail calorie restriction. The protein content of diet is potentially important during surgical weight loss since protein intake (PI) has been associated with lean body mass (LBM) retention during weight loss and weight maintenance [8,9], satiety [10], thermogenesis [11], glucose homeostasis (specifically circulating levels of branched-chain amino acids (BCAAs) after BS) [12] and, if insufficient, with malnutrition [13]. It is well known that protein is one of the main nutrients that can be affected by BS [14]. A series of successive steps can compromise nutritional status with detrimental consequences for the overall health of BS individuals (Fig. 49.1).

Factors associated with the development of a compromised protein status (PS) after BS

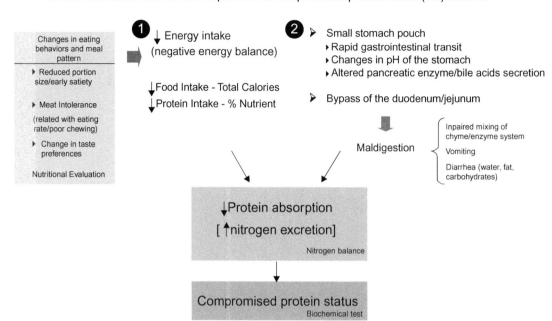

FIGURE 49.1   Mechanism associated with the development of a compromised protein status after BS.

DECREASE IN TOTAL ENERGY AND PROTEIN INTAKE

After BS, the development of new eating behaviors and the changes in anatomic conditions limit total energy and PI, which create an energy deficit situation. The smaller size of the gastric pouch, along with gut hormonal changes, may contribute to this phenomenon [15]. Specifically, changes in food behaviors and meal patterns, including reduced portion size, early satiety experienced after the meal, changes in taste preferences, and the potential for the development of persistent vomits, can contribute to a negative energy balance and limited PI [16,17]. It is generally accepted that not only the absolute amount of protein, but a dietary content of all of the essential amino acids (AAs), are required for optimal protein synthesis and balance [18,19]. Based on limited evidence, current BS guidelines suggest that a minimal PI of 60 g/day and up to 1.5 g/kg ideal body weight (IBW) per day could be adequate after BS as an achievable and meaningful target to minimize postsurgical complications. However, the limited volume of the stomach and the higher satiety experienced after a meal following BS could make this recommendation unrealistic; consequently, dietary protein supplementation has been postulated as a useful tool to achieve daily protein needs [14].

GASTROINTESTINAL CHANGES

Protein digestion is initiated in the stomach by the primary proteolytic enzyme *pepsin*, a nonspecific protease that is maximally active at pH2; therefore, the acidic environment favors protein denaturation [20]. Subsequently, protein degrades in the lumen of the intestine due to the activities of proteolytic enzymes secreted by the pancreas. The end products of protein digestion—free AAs and small peptides—are absorbed in the mucosal cells of the small intestine. After protein is absorbed and enters the blood stream, the free AAs are taken up mainly by the liver, and one-third are used for protein synthesis, energy metabolism, or gluconeogenesis. Skeletal muscle (SM) is the main site of metabolism of BCAAs (leucine, isoleucine, and valine) [21].

After BS, exclusion of the duodenum and proximal part of the jejunum where protein is mainly absorbed, together with the anatomical changes in the gastric pouch (affecting gastric acid's secretion and pepsin), modify the enabling environment for optimal protein uptake [13]. Procedures with a high malabsortive component (jejunoileal bypass; biliopancreatic diversion (BPD) with or without duodenal switch; and distal gastric bypass, GBP) normally cause diarrhea, especially of dietary fats and complex carbohydrates [22,23]. Of note, when malabsorption occurs, fecal losses of

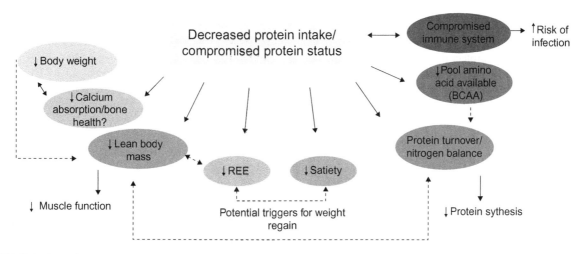

FIGURE 49.2 Potential metabolic consequences of varying PI in individuals after BS. *REE*, resting energy expenditure; *BCAA*, branched-chain amino acids.

nitrogen may be as high as 3.5 g/day [21]. This is much larger when compared to the estimated amount of 0.4 g/day that is normal [24]. A potential negative nitrogen balance (NB) status can then be developed and sustained until the body reaches a new equilibrium by mobilizing the body's protein storage (visceral tissues), affecting the protein turnover system. Finally, other gastrointestinal symptoms can be developed after BS, such as dumping syndrome and gastroesophageal reflux, and can contribute to gastrointestinal changes [25].

Protein in the body is widely distributed throughout the different tissues, and is a major component of all living cells. Protein represents 17% of the weight of an average healthy adult, making it the second largest component after water [26]. SM represents about 43% of total protein, followed by the visceral protein pool, and together they make up the metabolically available protein known as "body cell mass" [21]. Protein performs vital functions. It is composed of macromolecules made up of linear chains of AAs that contain carbon, hydrogen, oxygen, nitrogen, sulfur, and in some cases, selenium. Unlike fat and carbohydrates, there are no dispensable stores in humans. Hence, loss of body protein in response to a reduced or inadequate supply of dietary nitrogen and/or of specific indispensable AAs results in the loss of essential structural elements as well as impaired functions. The four major biochemical systems responsible for maintaining body protein and AA homeostasis are the following: (1) AA transport and uptake, (2) AA oxidation and catabolism, (3) protein synthesis, and (4) protein breakdown [27].

The first response to a reduced protein and AA intake is the reduction in the rate of AA oxidation, which is associated with a decline in the rate of specific organ and tissue protein synthesis. Protein and AA metabolism in muscle and liver is greatly affected when dietary PI is restricted, with reduced rates of muscle protein synthesis and the synthesis of export proteins from the liver, which occurs at an earlier period [28], resulting in a significant and negative effect on the SM mass [29].

As shown in Fig. 49.2, due to the behavioral and anatomical changes and consequences of BS, there is an increased risk of compromised PS. Consequently, the equilibrium and status quo of various key systems of the body can be altered and need to be carefully monitored in order to avoid complications. The actual knowledge of the potential complications associated with changes in the PS after BS will be discussed below. In that context, existing gaps in the literature will also be addressed further in the chapter.

Weight Loss

The consequences of negative energy and protein balance on total body and SM mass are well established [30]. It has been demonstrated during weight loss that there is close to 75% reduction of adipose tissue and 25% of fat-free mass (FFM) [31]. Apart from the beneficial reduction on adipose tissue in obese individuals during weight loss, the decrease in LBM may downregulate metabolic processes, such as protein turnover and basal metabolic rate, thus compromising long-term healthy weight management [32]. Weight loss induced by BS can reach about 60% excess weight loss (EWL), or about 30% total weight loss at 1 year (30–50 kg), with changes in body composition being of comparable proportions to conventional treatment [33].

Nitrogen Balance and Skeletal Mass Loss: Effect of Protein Intake

Long-term negative NB can be associated with loss of lean and fat tissue [34]. An increase in absolute grams of PI would improve NB and thus protect the amount of FFM loss during weight loss. Some studies have investigated the effect of different levels of PI on body composition changes during weight loss. Uncontrolled studies carried out by Bistrian et al. [35] showed that under severe caloric restriction, PI at recommended dietary allowance (RDA) levels (0.8 g/IBW/day) does not allow for nitrogen equilibrium. Rather, increasing PI up to 1.5 g/IBW/day would result in mean nitrogen equilibrium. In an attempt to establish the adequate amount of dietary protein needed for FFM preservation, Krieger performed a meta-analysis of the literature available in the field [36]. From this meta-analysis it was suggested that in the short term (less than 12 weeks), PI > 1.05 g/kg/day was associated with an extra 0.60 kg of FFM retention when compared with intakes of less than 1.05 g/kg/day. In studies lasting more than 12 weeks, FFM retention was increased up to 1.22 kg when PI was >1.05 g/kg/day. Studies performed in obese individuals undergoing very-low-calorie diet (VLCD; <800 kcal/day) can also help us understand the effect of PI on protein balance while losing weight in the BS patient. As a matter of fact, a similar situation (regarding kcal intake) occurs during the first trimester of BS when energy intake is drastically and spontaneously reduced (400−800 kcal/day) and the rate of weight loss is significant [37]. To understand PI adequacy under this condition, Hoffer et al. compared the metabolic effect of two levels of PI during VLCD: 0.8 g protein/kg IBW/day versus 1.5 g/kg IBW/day during 8 weeks intervention [38]. They observed that in the higher PI group, NB was closer to zero, reaching NB after 3 weeks of dietary intervention, while the low-protein group remained in negative NB during the entire 8 weeks. A more recent random controlled trial (RCT) [39] studied the effect of high-protein, low-calorie diets on FFM and muscle protein synthesis during a 31-day weight loss intervention. Participants were allocated to different groups provided protein at three levels: 0.82 g/kg/day (RDA), 1.63 g/kg/day (2 × RDA), and 2.43 g/kg/day (3 × RDA). Changes in body mass (kg) were significantly different between the RDA and 3 × RDA groups. The RDA group lost more weight (0.8), but had a significantly higher FFM% compared with the other groups. During the first 2 weeks of intervention, all groups were in negative NB, but by the end of the study (day 30) the RDA group remained in negative balance, while the 2 × RDA group reached NB, and the 3 × RDA group improved their NB and was not significantly different from the 2 × RDA group. In summary, based on the studies discussed above, it was concluded that during active weight loss FFM was better preserved with higher levels of PI.

Effect of Protein Intake on Resting Energy Expenditure After Weight Loss

LBM is the main determinant of resting energy expenditure (REE), explaining 75% of the REE variance [40], with REE being the largest component of 24-hour energy expenditure (EE). The impact of daily PI not only on changes in body composition, but also on REE, was studied in a 6-month randomized parallel study that compared two low-energy diets containing the required protein level (0.8 g protein/kg BW/day) and a level above (1.2 g protein/kg BW/day). During the 6 months of energy restriction, sustaining PI at the required level appeared to be sufficient to induce body weight loss while preserving FFM. However, above-PI requirements resulted in a greater decrease in fat mass and greater preservation of FFM and REE with similar effects on body weight loss [41]. Of note, reduced REE may trigger weight regain in the BS population [42]. A diet high in protein may increase REE while preventing LBM loss [40] during weight loss. It has also been suggested that increased EE from dietary protein is attributable to an enhanced thermic effect ($15 \pm 4\%$) compared to carbohydrates ($6 \pm 2\%$) or lipids ($7 \pm 3\%$) [43].

Dietary Protein Intake and Satiety

A reduction in PI can also affect satiety. A high-protein diet has been shown to increase satiety even in the context of energy restriction [44]. Underlying proposed mechanisms involve the following: a ketogenic state, relatively elevated plasma AA levels, an increase of anorexigenic hormones such as neuropeptide YY, glucagon-like peptide 1, and cholecystokinin, and a decrease of the orexigenic hormone ghrelin produced in response to peripheral and central detection of AAs and feedback to the central nervous system to prolong the duration of satiation [45,46]. Support has been found for a high PI promoting a negative energy balance by enhancing satiety, decreasing hunger, and decreasing energy intake [46].

Roles of Dietary Protein Intake on Circulating Levels of Branched-chain Amino Acids

It has long been recognized that circulating levels of AAs, including BCAAs, are elevated in persons with obesity, insulin resistance, or T2D, as compared to healthy controls [47]. More recently, using targeted metabolomic analysis,

BCAAs, including leucine, phenylalanine, tyrosine, and products of BCAA catabolism, were shown to be associated with insulin resistance [12]. Infusion of AAs in humans resulted in insulin resistance [48]. A recent epidemiological study reported that elevations in plasma levels of essential AAs, including BCAAs, phenylalanine, and tyrosine in healthy individuals, predicted a fivefold increase in the risk of developing T2D [49]. This indicated that elevated BCAAs may be a marker of the disease process, or may contribute to the development of insulin resistance, and are sensitive to therapeutic interventions; this includes BS, which is associated with reduced concentrations of plasma BCAAs [50], and therefore improved insulin sensitivity [51,52].

Protein Turnover

Altered intakes of protein and AAs modulate the rates of the major systems (synthesis, degradation, and oxidation) responsible for the maintenance of organ and whole-body protein and AA homeostasis [28]. The body tends to adapt to inadequate nutritional conditions by reducing whole-body protein turnover ($\downarrow$synthesis and $\uparrow$breakdown) [53], resulting in SM loss that is simultaneously impaired by surgery-induced weight loss. This adaptation can compromise the host's capacity to resist stressful stimuli, and can have a deleterious effect on overall health. For a more detailed cellular and mechanistic explanation of the integrative aspects of whole-body AAs and protein metabolism, see Young and Marchini [28].

Bone Health

Protein malnutrition or insufficiency compromise bone health. There is evidence that increased essential AAs or protein availability can enhance muscle protein synthesis and anabolism, as well as improve bone homeostasis [54]. Hence, optimal PI (higher than recommended levels) is recommended for bone health, particularly in the elderly [55]. It was suggested that a low-protein diet (0.7 g/kg of body weight) compared to a high-protein diet (2.1 g/kg) induced lower calcium intestinal absorption [56]. Also, calcium and vitamin D deficiencies, and hyperparathyroidism, are prevalent after BS [37].

Immune System

The immune system can be impaired with protein malnutrition [57].

METHODS TO ASSESS PROTEIN STATUS AFTER BARIATRIC SURGERY

Multiple related factors could potentially contribute to the development of protein deficiencies after BS. Therefore, the approach to assess PS after BS should also be multifactorial. Since any assessment method has its limitations, a combination of different approaches helps us to better appraise PS in the BS patient. Unfortunately, only a few of these methods are applicable to everyday clinical practice, whereas the vast majority should be seen mainly as research tools. The available methods are summarized in Table 49.1.

Nutritional Assessment

The evaluation of dietary intake is considered an indirect measure of nutritional status. Since clinical signs can appear late in the development of the deficiency, this approach could serve as an orientation to the patient's situation. A variety of dietary assessment methods could be used to estimate the new dietary habits, energy, and PI after BS. A brief description of the most frequent dietary assessment methods successfully implemented as part of the nutritional interview is given in Table 49.1. Its application can be either in clinical practice or in research [58]. The dietary assessment allows one to obtain qualitative and quantitative information about PI, and thereby identify targets of dietetic intervention. Over the years, it has been observed that protein-containing foods are poorly tolerated after BS [15,37]. Meat intolerance was observed in over 50% of a cohort of 200 patients who were followed up to 8 years after surgery [59]. To determine adequacy of dietary protein, a careful and detailed dietary interview should be performed, including questions about the patient's/participant's food preferences, intolerances, rate of eating, number of meals day, and presence of vomiting; this is based on the development of new eating behaviors identified post-BS [16].

TABLE 49.1 Methods to Assess Protein Status After Bariatric Surgery

1. Nutritional assessment methods:
   Dietary history (3−7 days)
   Dietary recalls (24 hours)
   Weighing method
   Food frequency questionnaire

2. Laboratory assessment:
   Albumin
   Transferrin, retinol-binding protein, prealbumin
   Urinary creatinine excretion
   Total lymphocyte count

3. In vivo measurement of muscle mass:
   Anthropometry
   Bioelectrical impedance analysis
   Dual-energy X-ray absorptiometry
   Magnetic resonance imaging
   Computed tomography
   Isotopic methods (^{15}N-creatinine)
   In vivo neuron activation analysis
   Nuclear magnetic resonance

4. Nitrogen balance

5. Protein turnover:
   Isotope kinetic study (L-[^{13}C] leucine as a tracer)

6. Functional Test:
   Handgrip dynamometer
   Functional strength (isometric knee extension and sit-to-stand test)

Clinical and research resources available in the assessment of protein status.

Clinical Assessment: Serum Protein and Amino Acid Concentrations

Clinical assessment of PS after BS might include, but is not limited to, the determination of the serum levels of proteins of hepatic origin listed in Table 49.1. To further assist the interpretation of blood protein measurements, it is necessary to include a measure of an acute-phase reactant C−reactive protein; this is because serum proteins are affected by inflammation and infection, in addition to a nutritional status per se [60]. Plasma free AAs can be affected after acute alterations in dietary protein and AA intake. This determination is frequently limited to research studies.

Physical Measurements

Evaluations of muscle protein mass have been approached using several methods. These methods are helpful for estimation of the protein content of specific tissues and body regions [61]. However, because these methods are expensive, time-consuming, and/or require radiation (and may have limited availability), they are unrealistic for use in clinical settings, and their use is limited to large research studies.

Nitrogen Balance

NB methodology is a measure of the net status of protein metabolism. It does not provide information on the size of protein stores or nutritional status [21]. However, it provides a holistic assessment of protein balance, allowing insight into the relationship between energy status, dietary protein, and FFM. Several factors may precipitate a negative NB in the BS setting. These include inadequate protein or energy intakes, imbalance in the nonessential/essential AA ratio, accelerated protein catabolism, and excessive diarrhea. A negative energy balance may have a negative and direct effect on the protein synthesis rate and, consequently, on SM mass.

Whole-Body Muscle Protein Turnover

Decreases in SM mass in response to negative energy balance are due to imbalanced rates of muscle protein synthesis and degradation [54]. It has been shown that PI stimulates muscle protein synthesis [62]. Limited to the area of metabolism and nutritional research, tracer AA techniques make it possible to directly measure both whole-body and specific-tissue protein turnover [63]. Leucine is the most frequent isotope used. It is predominantly metabolized in muscle, and acts as an intracellular marker. Stable isotopes can be used to measure flux across a tissue (i.e., SM), and the fractional synthesis rate of individual proteins (such as albumin) in interpreting whether AA intake is adequate [64]. Research in this specific area is needed to further our understanding of muscle protein turnover in BS patients.

Muscle Function Test

An easy and available tool that is useful for assessment of muscle function is the determination of voluntary handgrip strength. It can be measured with a handgrip dynamometer, which tests grip strength up to 90 kg. Other tests of muscle function include a pulmonary function test and electrical stimulation, which are normally used in critically ill patients. Other tests of function include evaluations of isometric knee extension, sit-to-stand capability, and in older individuals, the senior fitness test.

Although post-BS changes in FFM have been reported in several studies, reports on changes in muscle strength and physical performance after BS are scarce. To the best of our knowledge, evaluation of handgrip strength in BS subjects has been reported in only two studies. In a series of 25 middle-aged subjects (mean age of 37 years), Otto et al. did not find significant changes in handgrip strength in the short term after GBP [65], but it was observed in the long term by Cole et al. [66]. Interestingly, in both studies, an association between FFM and handgrip strength was found. On the other hand, Handrigan evaluated isometric knee extension in a series of 10 middle-aged subjects (mean age of 46 years) who underwent duodenal switch [67]. Lower limb maximum force at 12 months decreased approximately 33% relative to baseline. A series of handgrip measurements combined with the isometric knee extension test or sit-to-stand test would successfully assess changes in muscle function after BS.

ACTUAL KNOWLEDGE, GAPS IN THE LITERATURE, AND FUTURE DIRECTIONS IN THE FIELD

Observational studies have shown that caloric intake dramatically decreases during the first 3−6 months after surgery, and is frequently associated with long-term vitamin, mineral [14,15], and protein deficiencies [13−16,45]. As described, following PI recommendations can be challenging, and patients often do not reach their PI goal [8,15]. Our group and others have shown that consuming 60 g of protein a day can be challenging during the first months after surgery, even when the recommended protein supplements are supplied free of charge [68]. Also, 1.5 g protein/kg IBW/day would represent (when considering IBW for a BMI = 25 kg/m^2) a PI of 105 g/day for a woman with an IBW of 70 kg; this amount is higher than the PI reported. From our point of view, the adequate amount of dietary protein required after BS still needs to be defined.

The impact of PI on body composition after BS has not been deeply investigated, and the available results are not in agreement. Uncontrolled observational studies carried out by our group did not find a significant correlation between absolute PI and FFM loss relative to total weight loss at 4, 9, and 12 months after surgery. However, PI ≥ 60 g/day was associated with better maintenance of LBM after 12 months compared to PI < 60 g/day, meaning that PI was an independent predictor of decreased %loss of LBM at 12 months [8]. However, in a recent randomized, double-blind pilot study, 60 g/day led to higher fat mass loss with no significant difference in LBM [41] at 6 months.

Prospective studies observed that low albumin levels, a clinical marker of protein deficiency, can occur up to 2 years after GBP [13,45], with a prevalence ranging from 3% to 18% [13,45]; however, it is more commonly seen in malabsorptive procedures, such as BPD [69]. To our knowledge, only one study has been done to assess nutrient absorption after a long-limb GBP. Malabsorption accounted for 13% of the total reduction in protein absorption [70]. Since this study did not measure serum albumin, we cannot definitively state that malabsortion is the cause of low albumin levels.

Serum BCAA levels were shown to decrease after BS [51]. Although this can be seen as a positive effect in relation to glucose homeostasis, it could have a negative impact on SM synthesis. On the other hand, it has been shown that adequate nitrogen intake (about 10 g/day) of a balanced AA mixture, as well as a solution enriched with BCAAs, can maintain protein synthesis in the immediate postoperative period (3 days), and can reduce protein loss equivalent to complete AA formulas for 5 days after GBP surgery [71]; this mitigates the effects of dietary protein restriction on SM. Whether this protein sparing continues in the longer term has not been demonstrated.

CONCLUSION

Although we know that PI is dramatically reduced after BS, and may be mediated by gastrointestinal changes (i.e., hormones levels, stomach size, gastrointestinal transit time, pancreatic and biliary secretions), the consequences of compromised PS on NB, SM mass, and protein turnover still need to be addressed in this population. For a complete understanding we need to study all the interrelated systems addressed in this chapter since they are all implicated in PS.

A well-designed RCT that evaluates the effect of different levels of PI will add to this knowledge. The following are suggested areas of study: (1) nitrogen balance, (2) changes in body composition, (3) REE, (4) muscle protein kinetics, (5) serum levels of BCAAs, (6) satiety, (7) dietary evaluation, and (8) muscle function. The adequate amount of dietary protein necessary to ensure an optimal LBM retention, along with maintenance on REE and NB, and prevention of reduction in muscle protein breakdown after BS, are yet to be determined.

MINI-DICTIONARY OF TERMS

- *Protein turnover*: Term used to describe the process of synthesis and breakdown that occurs in the body. Protein synthesis is always taking place, even in nongrowing adults. AAs are made into new protein molecules, a process that is matched by an equal amount of protein being broken down into AAs; the whole process is known as protein turnover. This can be demonstrated by feeding patients isotopically labeled AAs (kinetics study) and observing that some of the label is retained in body proteins even though the amount of nitrogen going in as food protein is exactly matched by the amount of nitrogen excreted as urea [28].
- *Energy balance*: It is assumed that a healthy individual who eats an adequate diet can maintain equilibrium or balance for essential nutrients. This means that over a certain period of time, the amount of each of these elements that enters the body (i.e., energy, protein), is equaled by the amount that leaves the body. This balance is calculated as the difference between the amount consumed and the amount excreted.
  Energy balance is represented by the following equation: [Energy (Kcal) In] = [Energy (Kcal) Out].
  - A negative energy balance equation is represented as: [Energy In] < [Energy Out].
  - A positive energy balance equation is represented as: [Energy In] > [Energy Out].
- *Nitrogen balance (NB)*: NB methodology is classically used to determine adequacy of PI and to measure whole-body protein balance in response to nutritional interventions. NB is a measure of the net status of protein metabolism and is based on the assumption that nearly all of total body nitrogen is incorporated into a protein. The expression for NB is the following: NB = IN−(UN + FN + MN). Factors in that equation are defined as: IN = Daily Dietary Nitrogen Intake (protein/6.25); UN = Daily Urinary Nitrogen Excretion; FN = Daily Fecal Nitrogen Excretion (as unabsorbed protein); MN = Daily Miscellaneous Nitrogen Losses.
- *Resting energy expenditure (REE)*: This is the amount of energy needed to maintain energy balance in a nonactive period. Prediction equations exist to estimate REE for adults. REE can also be determined directly, by measuring the heat production with the individual inside a metabolic chamber, or indirectly with a calorimeter. In that case, the REE is determined by measuring the flow gas exchange during breathing.
- *Satiety*: Satiety is a central concept in the understanding of appetite control, and is associated with the inhibition of eating. The feeling of satiety starts after the end of eating, and prevents further eating before the return of hunger. Enhancing satiety can facilitate weight control.

KEY FACTS

- Bariatric surgery (BS) can compromise protein balance.
- Compromised protein status (PS) has negative effects on SM mass, protein turnover, REE, bone density, circulating BCAA levels, the immune system, and satiety.
- Prevalence of protein deficiency is more common after malabsorptive procedures such as distal bypass, duodenal switch, or BPD.
- Protein supplements are recommended after BS, at least during the first year.
- Nutritional evaluation and a muscle function (MF) test are key parameters that help monitor PS.

- This table lists the key facts that compromise PS after BS, including the consequences on other health-related systems of the body, and evaluation and treatment approaches.

SUMMARY POINTS

- Despite the beneficial effect of BS on overall health, this therapeutic approach may compromise PS.
- A change in PS may have a negative impact on several systems of the body, such as SM mass, protein turnover, REE, bone density, circulating BCAA levels, the immune system, and satiety.
- Mechanisms by which BS alters PS include: (1) decrease of food and PI and (2) changes in the gastrointestinal tract, which may result in maldigestion and malabsorption.
- As a consequence of these changes, PI is drastically reduced and protein absorption could be altered.
- Quantitative and qualitative measures of PS can be performed using a variety of assessments, including nutritional, laboratory, physical, NB, protein turnover, and functional tests.
- The long-term impact of BS over the systems addressed in this chapter need to be explored further since they all appear to impact PS.

REFERENCES

[1] Sjöström L. Review of the key results from the Swedish Obese Subjects (SOS) trial — a prospective controlled intervention study of bariatric surgery. J Intern Med 2013;273(3):219—34.

[2] Demaria E, Jamal M. Surgical options for obesity. Gastroenterol Clin N Am 2005;34:127—42.

[3] Yousseif A, Emmanuel J, Karra E, Millet Q, Elkalaawy M, Jenkinson AD, et al. Differential effects of laparoscopic sleeve gastrectomy and laparoscopic gastric bypass on appetite, circulating acyl-ghrelin, peptide YY3-36 and active GLP-1 levels in non-diabetic humans. Obes Surg 2014;24(2):241—52.

[4] Björklund P, Laurenius A, Een E, Olbers T, Lönroth H, Fändriks L. Is the Roux limb a determinant for meal size after gastric bypass surgery? Obes Surg 2010;20(10):1408—14.

[5] Furet JP, Kong LC, Tap J, Poitou C, Basdevant A, Bouillot JL, et al. Differential adaptation of human gut microbiota to bariatric surgery-induced weight loss: links with metabolic and low-grade inflammation markers. Diabetes 2010;59(12):3049—57.

[6] Kohli R, Kohli R, Bradley D, Setchell KD, Eagon JC, Abumrad N, et al. Weight loss induced by Roux-en-Y gastric bypass but not laparoscopic adjustable gastric banding increases circulating bile acids. J Clin Endocrinol Metab 2013;98(4):E708—12.

[7] Watanabe M, Houten SM, Mataki C, Christoffolete MA, Kim BW, Sato H, et al. Bile acids induce energy expenditure by promoting intracellular thyroid hormone activation. Nature 2006;439(7075):484—9.

[8] Moizé V, Andreu A, Rodríguez L, Flores L, Ibarzabal A, Lacy A, et al. Protein intake and lean tissue mass retention following bariatric surgery. Clin Nutr 2013;32(4):550—5.

[9] Abdeen G, le Roux CW. Mechanism underlying the weight loss and complications of roux-en-y gastric bypass. Review. Obes Surg 2016;26(2):410—21.

[10] Morínigo R, Moizé V, Musri M, Lacy AM, Navarro S, Marín JL, et al. Glucagon-like peptide-1, peptide YY, hunger, and satiety after gastric bypass surgery in morbidly obese subjects. J Clin Endocrinol Metab 2006;91:1735—40.

[11] Faria SL, Faria OP, Buffington C, de Almeida Cardeal M, Rodrigues de Gouvêa H. Energy expenditure before and after roux-en-y gastric bypass. Obes Surg 2012;22:1450—5.

[12] Newgard CB. Interplay between lipids and branched-chain amino acids in development of insulin resistance. Cell Metab 2012;2(5):606—14 15

[13] Brolin RE, LaMarca LB, Kenler HA, Cody RP. Malabsorptive gastric bypass in patients with superobesity. J Gastrointest Surg 2002;6(2):195—203.

[14] Mechanick JI, Youdim A, Jones DB, et al. Clinical practice guidelines for the perioperative nutritional, metabolic, and nonsurgical support of the bariatric surgery patient 2013 update: cosponsored by American association of clinical endocrinologists, the obesity society, and American society for metabolic & bariatric surgery. Obesity (Silver Spring) 2013;21(suppl. 1):S1—27.

[15] Moize V, Geliebter A, Gluck ME, Yahav E, Lorence M, Colarusso T, et al. Obese patients have inadequate protein intake related to protein intolerance up to 1 year following roux-en-Y gastric bypass. Obes Surg 2003;13(1):23—8.

[16] Laurenius A, Larsson I, Bueter M, Melanson KJ, Bosaeus I, Forslund HB, et al. Changes in eating behaviour and meal pattern following Roux-en-Y gastric bypass. Int J Obes (Lond) 2012;36(3):348—55.

[17] Coluzzi I, Raparelli L, Guarnacci L, Paone E, Del Genio G, le Roux CW, et al. Food Intake and Changes in Eating Behavior After Laparoscopic Sleeve Gastrectomy. Obes Surg 2016; [Epub ahead of print].

[18] Rose WC, Wixom RL. The AA requirements in man: XVI. The role of Nitrogen intake. J Biol Chem 1955;217:997—1004.

[19] Powanda MC. Changes in body balances of nitrogen and other key nutrients: description and underlying mechanisms. Am J Clin Nutr 1977;30(8):1254—68. PubMed PMID: 70166.

[20] Biochemistry, 5th edition. J.M. Berg, John L Tymoczko, and Lubert Stryer. New York: W H Freeman; 2002. ISBN-10: 0-7167-3051-0.

[21] Gibson RS. Assessment of protein status. Principles for Nutrition Assessment 422-430. New York, NY: Oxford University Press; 2005.

[22] Brolin RE, Kenler HA, Gorman JH, Cody RP. Long-limb gastric bypass in the superobese. A prospective randomized study. Ann Surg 1992;215(4):387−95.

[23] Buchwald H. Overview of bariatric surgery. J Am Coll Surg 2002;194:367−75.

[24] FAO/WHO/UNU. Energy and protein requirements. Report of a Joint Expert. Consultation. World Health Organ Tech Rep Ser 1985;724:1−206.

[25] Tack J, Deloose E. Complications of bariatric surgery: dumping syndrome, reflux and vitamin deficiencies. Best Pract Res Clin Gastroenterol 2014;28(4):741−9.

[26] Emery PW. Basic metabolism: protein. Surgery (Oxford) 2012;30(5):209−13.

[27] Young V, Marchini S, Cortiella J. Assessment of protein status. J of Nutr 1990;120:1496−502.

[28] Young VR, Marchini JS. Mechanism and nutritional significance of metabolic responses to altered intakes of protein and AA, with reference to nutritional adaptation in humans. Am J Clin Nutr 1990;51(270):289.

[29] Waterlow JC, Garlic PJ, Millward DJ. Protein turnover in mammalian tissues and in the whole body. New York: North-Holland; 1978.

[30] Waterlow JC. Metabolic adaptation to low intakes of energy and protein. Annu Rev Nutr 1986;6:495−526.

[31] Weinheimer EM, Sands LP, Campbell WW. A systematic review of the separate and combined effects of energy restriction and exercise on fat-free mass in middle-aged and older adults: implications for sarcopenic obesity. Nutr Rev 2010;68(7):375−88.

[32] Ravussin E, Lillioja S, Knowler WC, Christin L, Freymond D, Abbott WG, et al. Reduced rate of energy expenditure as a risk factor for body-weight gain. N Engl J Med 1988;318(8):467−72, 25.

[33] Carey DG, Pliego GJ, Raymond RL. Body composition and metabolic changes following bariatric surgery: effects on fat mass, lean mass and basal metabolic rate: six months to one-year follow-up. Obes Surg 2006;16(12):1602−8.

[34] Carbone JW, McClung JP, Pasiakos SM. Skeletal muscle responses to negative energy balance: effects of dietary protein. Adv Nutr 2012;3(2): 119−26, 1.

[35] Bistrian DR, Winterer J, Blackburn GL, Young V, Sherman M. Effect of a protein-sparing diet and Brief fast on nitrogen metabolism in mildly obese subjects. J Lab Clin Med 1977;89(5):1030−5.

[36] Krieger JW, Sitren HS, Daniels MJ, Langkamp-Henken B. Effects of variation in protein and carbohydrate intake on body mass and composition during energy restriction: a meta-regression. Am J Clin Nutr 2006;83(2):260−74.

[37] Moizé V, Andreu A, Flores L, Torres F, Ibarzabal A, Delgado S, et al. Long-term dietary intake and nutritional deficiencies following sleeve gastrectomy or Roux-En-Y gastric bypass in a mediterranean population. J Acad Nutr Diet 2013;113(3):400−10.

[38] Hoffer LJ, Bistrian BR, Young VR, Blackburn GL, Matthews DE. Metabolic effects of very low calorie weight reduction diets. J Clin Invest 1984;73(3):750−8.

[39] Pasiakos SM, Cao JJ, Margolis LM, Sauter ER, Whigham LD, McClung JP, et al. Effects of high-protein diets on fat-free mass and muscle protein synthesis following weight loss: a randomized controlled trial. FASEB J 2013;27(9):3837−47.

[40] Cunningham JJ. Body composition as a determinant of energy expenditure: a synthetic review and a proposed general prediction equation. Am J Clin Nutr 1991;54(6):963−9.

[41] Soenen S, Martens EA, Hochstenbach-Waelen A, Lemmens SG, Westerterp-Plantenga MS. Normal protein intake is required for body weight loss and weight maintenance, and elevated protein intake for additional preservation of resting energy expenditure and fat free mass. J Nutr 2013;143(5):591−6.

[42] Astrup A, Gøtzsche PC, van de Werken K, Ranneries C, Toubro S, Raben A, et al. Meta-analysis of resting metabolic rate in formerly obese subjects. Am J Clin Nutr 1999;69(6):1117−22.

[43] Nair KS, Halliday D, Garrow JS. Thermic response to isoenergetic protein, carbohydrate or fat meals in lean and obese subjects. Clin Sci (Lond) 1983;65(3):307−12.

[44] Martens EA, Westerterp-Plantenga MS. Protein diets, body weight loss and weight maintenance. Curr Opin Clin Nutr Metab Care 2014;17(1): 75−9.

[45] Westerterp-Plantenga M, Lemmens SG, Westerterp KR. Dietary protein - its role in satiety, energetics, weight loss and health. Br J Nutr 2012;108(suppl. 2):S105−12.

[46] Leidy HJ, Carnell NS, Mattes RD, Campbell WW. Higher protein intake preserves lean mass and satiety with weight loss in pre-obese and obese women. Obesity (Silver Spring) 2007;15(2):421−9.

[47] Felig P, Marliss E, Cahill Jr. GF. Plasma amino acid levels and insulin secretion in obesity. N Engl J Med 1969;281(15):811−16, 9.

[48] Tremblay F, Jacques H, Marette A. Modulation of insulin action by dietary proteins and amino acids: role of the mammalian target of rapamycin nutrient sensing pathway. Curr Opin Clin Nutr Metab Care 2005;8(4):457−62.

[49] Wang TJ, Larson MG, Vasan RS, Cheng S, Rhee EP, McCabe E, et al. Metabolite profiles and the risk of developing diabetes. Nat Med 2011;17(4):448−53.

[50] Tan HC, Khoo CM, Tan MZ, Kovalik JP, Ng AC, Eng AK, et al. The effects of sleeve gastrectomy and gastric bypass on branched-chain amino acid metabolism 1 year after bariatric surgery. Obes Surg 2016;4 [Epub ahead of print].

[51] Magkos F, Bradley D, Schweitzer GG, Finck BN, Eagon JC, Ilkayeva O, et al. Effect of Roux-en-Y gastric bypass and laparoscopic adjustable gastric banding on branched-chain amino acid metabolism. Diabetes 2013;62(8):2757−61.

[52] Laferrère B, Reilly D, Arias S, Swerdlow N, Gorroochurn P, Bawa B, et al. Differential metabolic impact of gastric bypass surgery versus dietary intervention in obese diabetic subjects despite identical weight loss. Sci Transl Med 2011;3(80):80re 27.

[53] Waterlow JC. Protein turnover with special reference to man. Q J Exp Physiol 1984;69(3):409−38.

[54] Genaro Pde S, Martini LA. The effect of protein intake on bone and muscle mass in the elderly. Nutr Rev 2010;68(10):616−23.

[55] Heaney RP, Layman DK. Amount and type of protein influences bone health. Am J Clin Nutr 2008;87(5):1567S−70S.

[56] Kerstetter JE, O'Brien KO, Insogna KL. Dietary protein affects intestinal calcium absorption. Am J Clin Nutr 1998;68(4):859−65.

[57] Calder PC, Jackson AA. Undernutrition, infection and immune function. Nutr Res Rev 2000;13(1):3−29.

[58] Marr JW. Individual dietary surveys: purposes and methods. World Rev Nutr Diet 1971;13:105−64.

[59] Avinoah E, Ovnat A, Charuzi I. Nutritional status seven years after Roux-en-Y gastric bypass surgery. Surgery 1992;111(2):137−42.

[60] Benjamin DR. Laboratory tests and nutritional assessment. Protein-energy status. Pediatr Clin North Am 1989;36(1):139−61.

[61] Coward WA, Parkinson SA, Murgatroyd PR. Body composition measurements for nutrition research. Nutr Res Rev 1988;1(1):115−24.

[62] Bauer J, Biolo G, Cederholm T, Cesari M, Cruz-Jentoft AJ, Morley JE, et al. Evidence-based recommendations for optimal dietary protein intake in older people: a position paper from the PROT-AGE Study Group. J Am Med Dir Assoc 2013;14(8):542−59.

[63] Guggenheim KY. Rudolf Schoenheimer and the concept of the dynamic state of body constituents. J Nutr 1991;121(11):1701−4.

[64] Guillet C, Boirie Y, Walrand S. An integrative approach to in-vivo protein synthesis measurement: from whole tissue to specific proteins. Curr Opin Clin Nutr Metab Care 2004;7(5):531−8.

[65] Otto M, Kautt S, Kremer M, Kienle P, Post S, Hasenberg T. Handgrip strength as a predictor for post bariatric body composition. Obes Surg 2014;24(12):2082−8.

[66] Cole AJ, Kuchnia AJ, Beckman LM, Jahansouz C, Mager JR, Sibley SD, et al. Long-term body composition changes in women following Roux-en-Y gastric bypass surgery. JPEN J Parenter Enteral Nutr 2016;2 [Epub ahead of print].

[67] Handrigan G, Hue O, Simoneau M, Corbeil P, Marceau P, Marceau S, et al. Weight loss and muscular strength affect static balance control. Int J Obes (Lond) 2010;34(5):936−42.

[68] Andreu A, Moize V, Rodriguez L, Flores L, Vidal J. Protein intake, body composition, and protein status following bariatric surgery. Obes Surg 2010;20(11):1509−15.

[69] Dolan K, Hatzifotis M, Newbury L, Lowe N, Fielding G. Newbury. A clinical and nutritional comparison of biliopancreatic diversion with and without duodenal switch. Ann Surg 2004;240:51−6.

[70] Schollenberger AE, Karschin J, Meile T, Küper MA, Königsrainer A, Bischoff SC. Impact of protein supplementation after bariatric surgery: A randomized controlled double-blind pilot study. Nutrition 2016;32(2):186−92.

[71] Desai SP, Bistrian BR, Moldawer LL, Blackburn GL. Whole-body nitrogen and tyrosine metabolism in surgical patients receiving branched-chain AA solutions. Arch Surg 1985;120(12):1345−50.

Chapter 50

Micronutrient Deficiencies and Sleeve Gastrectomy for Weight Reduction

L. Schiavo, G. Scalera and A. Barbarisi

Second University of Naples, Naples, Italy

LIST OF ABBREVIATIONS

AACE   American Association of Clinical Endocrinologists
ASMBS American Society for Metabolic and Bariatric Surgery
BMI   body mass index
BPD   biliopancreatic diversion
BS   bariatric surgery
LRYGB laparoscopic Roux-en-Y gastric bypass
SG   sleeve gastrectomy
TOS   the obesity surgery

INTRODUCTION

Obesity is a well-recognized global health crisis, and weight reduction has been demonstrated to improve survival and obesity-related conditions [1]. Contrary to lifestyle modifications and dietary programs, bariatric surgery (BS) has been shown to be most effective in terms of significant and durable weight loss [2]. In this context, sleeve gastrectomy (SG) represents one of the most performed BS procedures for the long-time treatment of morbid obesity and its associated comorbidities [3]. There are many recognized complications of SG [4]. Early post-SG complications include bleeding, staple-line leak, and development of intraabdominal abscesses. Delayed post-SG complications include stricture, gastroesophageal reflux, and nutritional and micronutrient deficiencies.

SG patients are at risk of vitamin and mineral deficiencies due to numerous mechanisms. These include reduced food intake, decreased hydrochloric acid and intrinsic factor secretion, postsurgical vomiting and nausea, poor food choice, food avoidance (probably due to intolerance), and last but not least, the patient's poor adherence to the prescribed post-SG diet and micronutrient supplement intake guidelines [5,6].

Concerning the correlation between SG and micronutrient deficiencies, this chapter focuses on the following topics:

- Pre-SG micronutrient deficiencies.
- Post-SG micronutrient deficiencies.
- Micronutrient-monitoring frequency after SG.
- Micronutrient supplementation after SG.

PRESLEEVE GASTRECTOMY MICRONUTRIENT DEFICIENCIES

The presence of vitamin and mineral deficiencies in obese patients scheduled for SG may appear contradictory in light of the evident and excessive caloric intake. However, several authors have demonstrated that vitamin and mineral deficiencies may be at a higher prevalence in obese adults, particularly in those with high BMI ($>40 \, kg/m^2$), with a reported prevalence in low concentration of vitamin B12, folic acid, 25-vitamin D, vitamin A, vitamin E, vitamin B1, iron, zinc, and selenium [5–14].

Metabolism and Pathophysiology of Bariatric Surgery.

The cause and the mechanisms contributing to micronutrient deficiencies in obesity are not fully known, but poorly balanced diet (high intake of high-calorie foods with poor nutritional quality and low intake of vegetables, cereals, legumes, and fruit) and lifestyle habits (sedentary, lack of physical activity) may represent the most important contributors [15]. Other factors, such as alcohol consumption [16], chronic subclinical inflammation [17−21], several diseases (such as diabetes, hypertension, hypothyroidism, and renal and liver disease), and medications (such as corticosteroids, bisphosphonates, etc.) can affect micronutrient levels in the blood [15]. In addition, body composition may also impact vitamin status; e.g., low plasma vitamin C has been associated (independently of BMI) with central fat distribution [22].

In addition, morbidly obese patients are characterized by a high level of total body water, and the extracellular compartment is more expanded than the intracellular compartment [23]. This possibly influences vitamin concentration (dilution effect).

The most common vitamin deficiency associated with obesity seems to be low concentration of 25-vitamin D [5−14]. The low concentration of 25-vitamin D in obesity may partially be a result of its sequestration in adipose tissue [24−25]. In addition to 25-vitamin D deficiency, mineral malnutrition is common in obese patients admitted for SG, with a reported prevalence of low levels of zinc, iron, copper, calcium, and selenium [5−14]. However, there is little consideration paid to those nutrients, which if left untreated may have several clinical postoperative consequences (i.e., hair loss, anemia), and more frequent postoperative infections caused by poor immunity associated with macro- and micronutrient deficiencies [26−28].

An analysis of the relevant literature reveals that in the last 10 years several authors have started to study the correlation between micronutrient deficiencies and obesity in patients who are candidates for SG, or BS general.

Recently, Ben-Porat et al. [10] analyzed the preoperative micronutrient deficiencies in 192 patients who were candidates for laparoscopic SG. Prior to surgery they found the following deficiencies: 25-vitamin D (99.4%), iron (47.1%), vitamin B12 (13.1%), and folate (32%).

In 2013, Nicoletti et al. [29] characterized through a retrospective study the anthropometric, dietary, and biochemical profiles of morbidly obese adult patients who were candidates for BS. The occurrence of micronutrient deficiencies was high for vitamin B12 (3%), iron (9%), vitamin A (15%), magnesium (19%), and vitamin C (16%).

Also, in 2013, de Luis et al. [30] described the vitamin and mineral status in 115 obese women prior to BS. Deficiencies were found in 6.1% of the subjects for albumin, 21.7% for prealbumin, 2.6% for hemoglobin, and 5.2% for ferritin. In the vitamin analysis, no deficiencies were found in the patients for vitamins A, E, or K, but 71.3% had a moderate deficiency of 25-vitamin D, and 26.1% a severe deficiency of 25-vitamin D (<15 ng/mL). Deficiencies of 9.5% were found for vitamin B12, 25.2% for folic acid, 67.8% for copper, and 73.9% for zinc.

In 2012, Damms-Machado et al. [31] assessed obese SG candidates for both pre- and postsurgical micronutrient deficiencies. In 51% of the tested patients they found at least one vitamin or mineral deficiency prior to surgery. In particular, pre-SG micronutrient concentrations were below normal for iron (29%), vitamin B12 (9%), 25-vitamin D (27%), folate (6%), potassium (7%), and vitamin B6 (11%).

In 2010, Schweiger et al. [32], with a similar aim, collected blood samples from 83 women and 31 men, and found that the frequency of preoperative micronutrient deficiencies were as follows: iron (35%), folate (24%), vitamin B12 (3.6%), calcium (0.9%), and phosphorous (2%).

In 2009, Ernst et al. [7] assessed more than 200 patients with morbid obesity prior to BS. They looked at levels of serum vitamin B12, calcium, iron, magnesium, phosphate, zinc, folate, vitamin D, and other vitamins and minerals. Deficiencies were found in 12.5% of the subjects for albumin, 8.0% for phosphate, 4.7% for magnesium, 6.9% for ferritin, 6.9% for hemoglobin, 24.6% for zinc, 3.4% for folate, and 18.1% for vitamin B12. In addition, 25.4% showed a severe 25-vitamin D deficiency. In the subsample, subjects showed deficiencies of 32.6% for selenium, 5.6% for vitamin B3, 2.2% for vitamin B6, and 2.2% for vitamin E.

In 2006, Flancbaum et al. [33] retrospectively analyzed the preoperative macro- and micronutrient status of patients scheduled for laparoscopic Roux-en-Y gastric bypass (LRYGB) for morbid obesity. Deficiencies in iron (43.9%), ferritin (8.4%), hemoglobin (22%), thiamine (29%), and vitamin D (68.1%) were observed. Preoperatively, women had more significant deficits in many micronutrients, including lower levels of iron and 25-vitamin D, compared to men [10]. In both male and female obese patients, the occurrence of pre-BS micronutrient deficiencies was the strongest predictor of their presence in the postsurgical period [10].

Although all nutritional guidelines for obesity surgery patients [5,6,12] suggest that a comprehensive preoperative macro- and micronutrient assessment and correction of deficiencies is advised (ideally with sufficient time before surgery), data concerning the correction of preoperative micronutrient deficiencies are scarce. Recently, Schiavo et al. [14] reported that 10 weeks of preoperative micronutrient supplementation is capable of effectively treating micronutrient deficiencies in morbidly obese candidates for BS. Clearly, more research is needed to determine whether the correction of micronutrient deficiencies prior to BS will have a positive impact on micronutrient deficiency prevention after surgery.

POSTSLEEVE GASTRECTOMY MICRONUTRIENT DEFICIENCIES

Micronutrients play a vital role as factors and cofactors in numerous physiological processes of the human body (i.e., hunger, resting metabolic rate, nutrient digestion and adsorption, fatty acid and carbohydrate metabolism, gland function, nervous central system activities, and others) [34]. Taking daily vitamin and mineral supplements and, most importantly, eating foods rich in such micronutrients, are key factors in any successful weight loss strategy, even obesity surgery (such as SG) [5,6,12]. Because SG is a restrictive procedure (surgical details are beyond the scope of this chapter, but the main classical steps are shown in Fig. 50.1), and therefore lacks the malabsorptive component, the risk for developing micronutrient deficiencies after SG is often considered low, and is therefore frequently not tested.

However, resection of the stomach fundus with a consequent decrease in hydrochloric acid and intrinsic factor secretion (which plays a key role in vitamin B12 adsorption; Fig. 50.2) is associated with reduced intake, postoperative nausea and vomiting, poor food choice, and food avoidance. This is due to intolerance and may render the patient prone to several vitamin and mineral deficiencies, which in turn increases susceptibility to the development of several disorders, especially anemia, hair loss, and metabolic and neurological disorders.

Therefore, following SG, multivitamin and multimineral supplementation is mandatory not only for health, but also for weight loss and long-term weight maintenance. Because many micronutrient deficiencies progress with time, patients who undergo SG must be monitored regularly to avoid micronutrient deficiencies [5,6,12].

Although it is universally accepted that the best way to introduce vitamins and minerals into our bodies is through food sources, after SG many patients unfortunately do not adhere to their post-SG prescribed diet, and consequently often fail to consume an appropriate amount of foods rich in vitamins and minerals (i.e., fruits, vegetables, legumes, and cereals) [35−37]. As a consequence, a daily multivitamin and multimineral supplement is often essential after all bariatric procedures, even for purely restrictive procedures such as SG. A number of authors agree that it is necessary to improve adherence to a post-SG diet in order to maintain weight loss [5,6,12].

An analysis of the relevant literature reveals that in the past few years several authors have begun to study the prevalence of micronutrient deficiencies in morbidly obese patients who undergo SG for weight reduction.

Although SG tends to cause fewer micronutrient deficiencies than malabsorptive procedures such as LRYGB and biliopancreatic diversion (BPD), it is well established that patients who undergo SG are at risk of vitamin B12, iron, 25-vitamin D, and zinc deficiencies [38−44]. The key physiological roles, sources, and deficiency symptoms of these micronutrients are reported in Table 50.1.

In this context, a low serum concentration of vitamin B12 is often evident among SG patients; this effect could exacerbate pre-SG deficiencies. The incidence of post-SG B12 deficiency ranges from 0% to 20% up to 36 months after SG [45]. Consequently, vitamin B12 monitoring and supplementation should be standard care both pre- and post-SG.

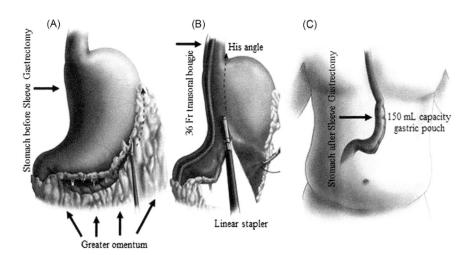

FIGURE 50.1 **Main steps in the sleeve gastrectomy procedure in our bariatric surgery unit.** (A) The skeletization of the great curvature of the stomach started at 6 cm from the pylorus up to the angle of His. (B) A 36 French transoral bougie was positioned from the hiatus to the duodenum along the lesser curvature, and the SG was performed using a linear stapler with different cartridges (*green*, *blue*, and *gold*). (C) A 150 mL-capacity gastric pouch was finally obtained.

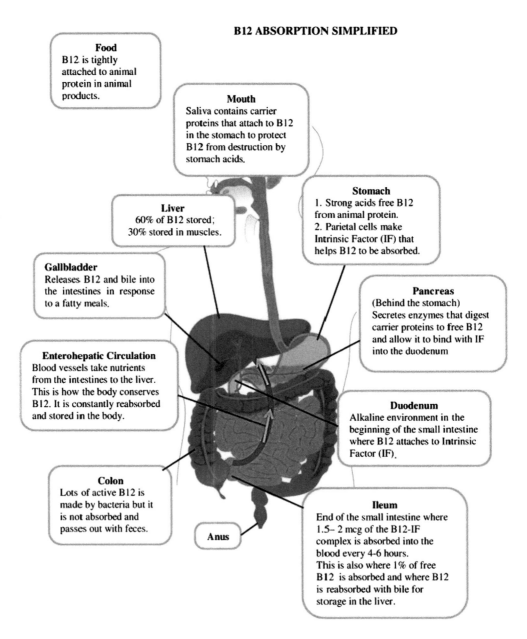

B12 ABSORPTION SIMPLIFIED

Food
B12 is tightly attached to animal protein in animal products.

Mouth
Saliva contains carrier proteins that attach to B12 in the stomach to protect B12 from destruction by stomach acids.

Stomach
1. Strong acids free B12 from animal protein.
2. Parietal cells make Intrinsic Factor (IF) that helps B12 to be absorbed.

Liver
60% of B12 stored; 30% stored in muscles.

Gallbladder
Releases B12 and bile into the intestines in response to a fatty meals.

Pancreas
(Behind the stomach) Secretes enzymes that digest carrier proteins to free B12 and allow it to bind with IF into the duodenum

Enterohepatic Circulation
Blood vessels take nutrients from the intestines to the liver. This is how the body conserves B12. It is constantly reabsorbed and stored in the body.

Duodenum
Alkaline environment in the beginning of the small intestine where B12 attaches to Intrinsic Factor (IF).

Colon
Lots of active B12 is made by bacteria but it is not absorbed and passes out with feces.

Anus

Ileum
End of the small intestine where 1.5– 2 mcg of the B12-IF complex is absorbed into the blood every 4-6 hours. This is also where 1% of free B12 is absorbed and where B12 is reabsorbed with bile for storage in the liver.

FIGURE 50.2   General aspects of vitamin B12 absorption. Diagram kindly provided by Dr. Jennifer Rooke, MD, MPH. Used with permission.

Iron deficiencies after SG occur in 20−50% of patients, and are mainly due to the following factors: (1) reduced dietary foods rich in iron (i.e., red meat, legumes, etc.) and (2) reduced gastric acid secretion (which is necessary for iron absorption). For these reasons, numerous authors agree that iron monitoring in patients undergoing SG should continue forever, even after initial iron repletion [27].

Calcium and 25-vitamin D deficiencies seen post-SG should not be due to the surgery because the duodenum, which is the main calcium absorption point, is not bypassed during SG. Therefore, calcium and 25-vitamin D deficiencies following SG are probably due to both a post-SG decreased intake in amounts of calcium- and 25-vitamin D-containing foods, and the preoperative calcium and 25-vitamin D deficiencies that are frequently observed in obese patients who are candidates for SG. Lastly, with regards to zinc deficiency: a common complication after SG is hair loss, which is related not only to rapid weight loss, but also to zinc and other micronutrient deficiencies.

TABLE 50.1 Main Micronutrient Deficiencies Observed After Sleeve Gastrectomy: Key Physiological Roles, Food-Rich Sources, and Symptoms of Deficiency

Micronutrient	Main Physiological Role	Main Source	Main Symptoms of Deficiency
Vitamin B12	Water-soluble vitamin that plays a key role in the production of energy from fats and proteins; maintains healthy nerve cells; supports red blood cell production; supports healthy immune responses.	Eggs, milk, cheese, milk products, fortified cereals, fortified soy products, red meat, fish, shellfish, crustaceans, poultry, and liver beef.	Low energy, fatigue, muscle pain, depression, anxiety, irritability, vision problems, hearing problems, mood disorders, and memory loss.
Iron	Plays a key role in adequate erythropoietic function, oxidative metabolism, and cellular immune responses.	Legumes, lentils, soy beans, whole grains, green leafy vegetables, cereals, breads, spinach, turnip, fish, meat, sprouts, broccoli, and dried fruits.	Anemia, pallor, fatigue, lethargy, headache, muscle pain, difficulty concentrating, impaired immune system, and shortness of breath.
25-Vitamin D	Fat-soluble vitamin that plays a key role in gene transcription, vision, optimal immune function, and skin health.	The body makes vitamin D when it is exposed to ultraviolet rays from the sun. Foods high in vitamin D include cheese, margarine, butter, fortified milk, cereals, fatty fish, beef liver, and mushrooms.	Muscle pain, weakness, osteomalacia, osteoporosis, insomnia, myopia, and depression.
Zinc	Participates in several enzymatic reactions; plays an important role in cell division and growth, and in supporting wound healing and immune responses.	Sea food, beef, lamb, oysters, pumpkin seeds, beef, pork, chicken, nuts, seeds, almonds, peas, and beans.	Compromised immune system, diarrhea, hair loss, acne, eczema, dermatitis, white spots on fingernail beds, and low energy.

MICRONUTRIENT-MONITORING FREQUENCY AFTER SLEEVE GASTRECTOMY

Another important point regards the frequency of the commonly suggested practice of micronutrient monitoring after SG. The recommendations stated by the American Society for Metabolic and Bariatric Surgery (ASMBS), the American Association of Clinical Endocrinologists (AACE), and The Obesity Surgery (TOS) for macro- and micronutrient-monitoring frequency after SG are shown in Table 50.2. Although it is typically advised to follow these recommendations until the fifth year post-SG, many experts recommend lifelong monitoring after all bariatric procedures, even restrictive procedures such as SG.

MICRONUTRIENT SUPPLEMENTATION AFTER SLEEVE GASTRECTOMY

It is important to emphasize that there is no unanimity with regards to the appropriate type and amount of multivitamin and multimineral supplementation that should be prescribed after SG, and therefore supplementation practices vary widely [46–50]. Post-SG micronutrient prescriptions are often based on expert opinion and personal experience rather than on the results of randomized-controlled trials. With regards to post-SG supplementation, the first recommendation is typically oral supplementation, but unfortunately not all patients perfectly adhere to the prescribed postsurgical multi-vitamin and multimineral regimen. In most cases, the main reasons for nonadherence are difficulty in swallowing pills or capsules, and costs. Therefore, in some cases it is necessary to find a different approach in order to enhance a patient's compliance, such as supplement administration in the form of liquid or drops.

CONCLUSIONS

As reported in the ASMBS guidelines [5], macro- and micronutrient assessment and dietary management are crucial components of the BS process. The main goal should be to create a plan for postsurgical dietary intake that will

TABLE 50.2 Common Suggested Micronutrient Mandatory Monitoring Frequency After Sleeve Gastrectomy

Micronutrient	Pre-SG	Post-SG (month 3)	Post-SG (month 6)	Post-SG (month 12)	Post-SG (month 24)	Post-SG (month 36)	Post-SG (month 48)	Post-SG (month 60)
Calcium	X							
Iron	X		X	X	X	X	X	X
Zinc	X	X	X					
Vitamin A	X							
25-Vitamin D	X	X	X	X	X	X	X	X
Vitamin B1	X							
Vitamin B2	X							
Vitamin B6	X							
Vitamin B9	X		X	X	X	X	X	X
Vitamin B12	X		X	X	X	X	X	X

X, assessment of serum levels is mandatory.
Recommendations of the American Society for Metabolic and Bariatric Surgery (ASMBS), the American Association of Clinical Endocrinologists (AACE), and The Obesity Surgery (TOS).

enhance the likelihood of success. The above-mentioned document also highlights the importance for bariatric patients to regularly take vitamin and mineral supplements, not only to prevent nutrient deficiencies that can worsen after surgery, but because some micronutrients such as calcium can enhance weight loss and help avoid weight regain.

In conclusion, SG is extensively performed as a BS procedure, but can lead to micronutrient (vitamin and mineral) deficiencies. Therefore, preoperative lifestyle modifications [51], micronutrient assessment, and a rigorous postoperative follow-up plan with administration of multivitamin and multimineral supplements, and a regular assessment of micronutrient serum levels, are lifelong recommendations after all bariatric procedures, even for restrictive procedures such as SG.

MINI-DICTIONARY OF TERMS

- *Sleeve gastrectomy*: A surgical weight loss procedure that significantly reduces the size of the stomach, and consequently limits the amount of food that can be eaten.
- *Micronutrients*: A chemical element or substance required in trace amounts for the normal growth and development of living organisms. Micronutrients are subdivided into vitamins and minerals, and play a vital role as factors and cofactors in numerous physiological processes of the human body.
- *Vitamins*: Organic substances necessary (even if in small quantities) for general metabolic reactions and for the growth and normal level of efficiency of an organism.
- *Minerals*: Micronutrients that do not provide energy (unlike carbohydrates, lipids, and proteins); their presence is necessary because metabolic reactions can occur with the release of energy. The body is unable to synthesize any mineral, so it is therefore necessary to introduce them via food and drinks.
- *Micronutrient deficiency*: A deficiency of essential vitamins and minerals that are required in small amounts by the body for proper growth and development.
- *Intrinsic factor*: A protein mostly produced by the stomach's parietal cells. Intrinsic factor plays a crucial role in the absorption mechanism of vitamin B12.
- *Stomach fundus*: An anatomical portion of the stomach involved in undigested food storage.

KEY FACTS

- Micronutrient deficiencies are frequent in morbidly obese patients scheduled for SG.
- In morbidly obese SG candidates, micronutrient deficiencies are particularly frequent in those with a high BMI.

- Micronutrient deficiencies in morbidly obese patients are mostly due to a poorly balanced diet that lacks micronutrients.
- Micronutrient deficiencies in morbidly obese SG candidates are often left untreated.
- Identifying micronutrient deficiencies preoperatively and correcting them before SG may be useful in preventing micronutrient deficiencies after surgery.
- SG may exacerbate preexisting micronutrient deficiencies.
- Micronutrient deficiencies are well-known, long-term complications of SG.
- Deficiencies in micronutrients after SG are frequent, despite routine supplementation.
- Although SG does not cause malabsorption, it seems to alter micronutrient status.
- Several studies have demonstrated that micronutrient (vitamins and minerals) deficiencies seen after SG may be a result of stomach surgery-induced changes (decreased synthesis and release of both hydrochloric acid and intrinsic factor) and a substantial decrease in daily food intake.
- Micronutrient deficiencies observed after SG are also due to patients' poor adherence to both postoperative dietary guidelines and prescribed intake of vitamin and mineral supplementation.

SUMMARY POINTS

- This chapter offers information on the prevalence and frequency of micronutrient deficiencies in morbidly obese patients prior to and following SG.
- Micronutrient deficiency is common in morbidly obese patients admitted for SG, with a reported prevalence of low levels of 25-vitamin D, folate, vitamin B1, vitamin B12, vitamin A, vitamin E, zinc, iron, and selenium.
- SG may exacerbate preexisting vitamin and mineral deficiencies or produce new ones, depending on dietary intake, and adherence to recommended postoperative diet and supplementation.
- It is imperative to screen for micronutrient deficiencies preoperatively in these patients in order to intervene in a timely manner and prevent more serious complications after SG.
- Pre-SG micronutrient evaluation and a rigorous post-SG follow-up plan with administration of multivitamin and mineral supplements and assessment of serum levels is strongly suggested after all bariatric procedures, even purely restrictive procedures such as SG.
- Little is known about optimal nutritional care during the preoperative period, but it could be used to help correct deficiencies in sufficient time prior to surgery.
- There is scarce information and/or agreement on which micronutrient supplementations should be used (and how much) prior to and after SG.
- There is scarce information and/or agreement on micronutrient-monitoring frequency after SG.
- Up-to-date data on patient adherence to a prescribed diet and micronutrient supplementation program after SG, and its impact on postoperative weight loss and micronutrient status, are scarce.
- Further studies are necessary to determine whether or not the correction of micronutrient deficiencies prior to SG really has a positive impact on the prevention of micronutrient deficiencies after surgery.

REFERENCES

[1] Bolen SD, Chang HY, Weiner JP, Richards TM, Shore AD, Goodwin SM, et al. Clinical outcomes after bariatric surgery: a five-year matched cohort analysis in seven US states. Obes Surg 2012;22:749–63.

[2] Colquitt J, Clegg AJ, Loveman E, Royle P, Sidhu MK. Surgery for morbid obesity. Cochrane Database Syst Rev 2005;4 Art. No.: CD003641.

[3] Schiavo L, Scalera G, Barbarisi A. Sleeve gastrectomy to treat concomitant polycystic ovary syndrome, insulin and leptin resistance in a 27-years morbidly obese woman unresponsive to insulin-sensitizing drugs: A 3-year follow-up. Int J Surg Case Rep 2015;17:36–8.

[4] Sarkhosh K, Birch DW, Sharma A, Karmali S. Complications associated with laparoscopic sleeve gastrectomy for morbid obesity: a surgeon's guide. Can J Surg 2013;56:347–52.

[5] Aills L, Blankenship J, Buffington C, Furtado M, Parrott J. ASMBS Allied health nutritional guidelines for the surgical weight loss patient. Surg Obes Relat Dis 2008;4:73–108.

[6] Thibault R, Olivier H, Azagury DE, Pichard C. Twelve key nutritional issues in bariatric surgery. Clin Nutr 2015. Available from: http://dx.doi.org/10.1016/j.clnu.2015.02.012, [Epub ahead of print].

[7] Ernst B, Thurnheer M, Schmid SM, Schultes B. Evidence for the necessity to systematically assess micronutrient status prior to bariatric surgery. Obes Surg 2009;19:66–73.

[8] de Lima KV, Costa MJ, Gonçalves MD, Sousa BS. Micronutrient deficiencies in the pre-bariatric surgery, ABCD ARQ. Bras Cir Dig 2013;26:63–6.

[9] Wolf E, Utech M, Stehle P, Büsing M, Stoffel-Wagner B, Ellinger S. Preoperative micronutrient status in morbidly obese patients before undergoing bariatric surgery: results of a cross-sectional study. Surg Obes Rel Dis 2015;11:1157−63.

[10] Ben-Porat T, Elazary R, Yuval JB, Wieder A, Khalaileh A, Weiss R. Nutritional deficiencies following sleeve gastrectomy − can they be predicted pre-operatively? Surg Obes Relat Dis 2015;11:1029−36.

[11] Peterson LA, Cheskin LJ, Furtado M, Papas K, Schweitzer MA, Magnuson TH, et al. Malnutrition in bariatric surgery candidates: multiple micronutrient deficiencies prior to surgery. Obes Surg 2016;26(4):833−8. [Epub ahead of print].

[12] Snyder-Marlow G, Taylor D, Lenhard J. Nutrition care for patients undergoing laparoscopic sleeve gastrectomy for weight loss. J Am Diet Assoc 2010;110:600−7.

[13] Sanchez A, Rojas P, Basfi-fer K, Carrasco F, Inostroza J, Codoceo J, et al., Micronutrient deficiencies in morbidly obese women prior to bariatric surgery, Obes Surg 2016;26(2):361−8.

[14] Schiavo L, Scalera G, Pilone V, De Sena G, Capuozzo V, Barbarisi A. Micronutrient deficiencies in obese patients candidate for bariatric surgery: a prospective, preoperative trial of screening, diagnosis, and treatment J Vit Nutr Res 2016;10:1−8. [Epub ahead of print]. Available from: http://dx.doi.org/10.1024/0300-9831/a000282.

[15] Aasheim ET, Hofsø D, Hjelmesæth J, Birkeland KI, Bøhmer T. Vitamin status in morbidly obese patients: a cross-sectional study. Am J Clin Nutr 2008;87:362−9.

[16] Lissner L, Lindroos AK, Sjostrom L. Swedish obese subject (SOS): an obesity intervention study with a nutritional perspective. Eur J Clin Nutr 1998;52:316−22.

[17] Stephensen CB, Gildengorin G. Serum retinol, the acute phase response, and the apparent misclassification of vitamin A status in the third National Health and Nutrition Examination Survey. Am J Clin Nutr 2000;72:1170−8.

[18] Gray A, McMillan DC, Wilson C, Williamson C, O'Reilly DS, Talwar D. The relationship between plasma and red cell concentrations of vitamins thiamine diphosphate, flavin adenine dinucleotide and pyridoxal 5-phosphate following elective knee arthroplasty. Clin Nutr 2004;23:1080−3.

[19] Friso S, Jacques PF, Wilson PW, Rosenberg IH, Selhub J. Low circulating vitamin B(6) is associated with elevation of the inflammation marker C-reactive protein independently of plasma homocysteine levels. Circulation 2001;103:2788−91.

[20] Wannamethee SG, Lowe GD, Rumley A, Bruckdorfer KR, Whincup PH. Associations of vitamin C status, fruit and vegetable intakes, and markers of inflammation and hemostasis. Am J Clin Nutr 2006;83:567−74.

[21] Wellen KE, Hotamisligil GS. Inflammation, stress, and diabetes. J Clin Invest 2005;115:1111−19.

[22] Canoy D, Wareham N, Welch A, Bingham S, Luben R, Day N, et al. Plasma ascorbic acid concentrations and fat distribution in 19,068 British men and women in the European Prospective Investigation into Cancer and Nutrition Norfolk cohort study. Am J Clin Nutr 2005;82:1203−9.

[23] Waki M, Kral JG, Mazariegos M, Wang J, Pierson Jr. RN, Heymsfield SB. Relative expansion of extracellular fluid in obese vs. nonobese women. Am J Physiol 1991;261:199−203.

[24] Harris SS, Dawson-Hughes B. Reduced sun exposure does not explain the inverse association of 25-hydroxy25-Vitamin D with percent body fat in older adults. J Clin Endocrinol Metab 2007;92:3155−7.

[25] Wortsman J, Matsuoka LY, Chen TC, Lu Z, Holick MF. Decreased bioavailability of 25-Vitamin D in obesity. Am J Clin Nutr 2000;72:690−3.

[26] Ruiz-Tovar JR, Oller I, Llavero C, Zubiaga L, Diez M, Arroyo A, et al. Hair loss in females after sleeve gastrectomy: predictive value of serum zinc and iron levels. Am Surg 2014;80:466−71.

[27] Jàuregui-Lobera I. Iron deficiency after bariatric surgery. Nutrients 2013;5:1595−608.

[28] Schiavo L, Scalera G, De Sena G, Ciorra FR, Pagliano P, Barbarisi A. Nonsurgical management of multiple splenic abscesses in an obese patient that underwent laparoscopic sleeve gastrectomy: case report and review of literature. Clin Case Rep 2015;3:870−4.

[29] Nicoletti CF, Lima TP, Donadelli SP, Salgado Jr. W, Marchini JS, Nonino CB. New look at nutritional care for obese patient candidates for bariatric surgery. Surg Obes Relat Dis 2013;9:520−5.

[30] de Luis DA, Pacheco D, Izaola O, Terroba MC, Cuellar L, Cabezas G. Micronutrient status in morbidly obese women before bariatric surgery. Surg Obes Relat Dis 2013;9:323−7.

[31] Damms-Machado A, Friedrich A, Kramer KM, Stingel K, Meile T, Küper MA, et al. Pre- and postoperative nutritional deficiencies in obese patients undergoing laparoscopic sleeve gastrectomy. Obes Surg 2012;22:881−9.

[32] Schweiger C, Weiss R, Berry E, Keidar A. Nutritional deficiencies in bariatric surgery candidates. Obes Surg 2010;20:193−7.

[33] Flancbaum L, Belsley S, Drake V, Colarusso T, Tayler E. Preoperative nutritional status of patients undergoing Roux-en-Y gastric bypass for morbid obesity. J Gastrointest Surg 2006;10:1033−7.

[34] Eden T, Rajput-Ray M, Ray S. Micronutrient and vitamin physiology and requirements in critically ill patients. In: Faber P, Siervo M, editors. Nutrition in critical care. Cambridge: Cambridge University Press; 2014. p. 33−42.

[35] Gjessing HR, Nielsen HJ, Mellgren G, Gudbrandsen OA. Energy intake, nutritional status and weight reduction in patients one year after laparoscopic sleeve gastrectomy. Springerplus 2013;2:352.

[36] Soares FL, Bissoni de Sousa L, Corradi-Perini C, Ramos da Cruz MR, Nunes MG, Branco-Filho AJ. Food quality in the late postoperative period of bariatric surgery: an evaluation using the bariatric food pyramid. Obes Surg 2014;24:1481−6.

[37] Ammon BS, Bellanger DE, Geiselman PJ, Primeaux SD, Yu Y, Greenway FL. Short-term pilot study of the effect of sleeve gastrectomy on food preference. Obes Surg 2015;25:1094−7.

[38] Verger EO, Aron-Wisnewsky J, Dao MC, Kayser BD, Oppert JM, Bouillot JL, et al. Micronutrient and Protein Deficiencies After Gastric Bypass and Sleeve Gastrectomy: a 1-year Follow-up. Obes Surg 2015; [Epub ahead of print].

[39] van Rutte PW, Aarts EO, Smulders JF, Nienhuijs SW. Nutrient deficiencies before and after sleeve gastrectomy. Obes Surg 2014;24:1639−46.

[40] Alvarez V, Cuevas A, Olivos C, Berry M, Farías MM. Micronutrient deficiencies one year after sleeve gastrectomy. Nutr Hosp 2014;29:73−9.

[41] Alexandrou A, Armeni E, Kouskouni E, Tsoka E, Diamantis T, Lambrinoudaki I. Cross-sectional long-term micronutrient deficiencies after sleeve gastrectomy versus Roux-en-Y gastric bypass: a pilot study. Surg Obes Relat Dis 2014;10:262−8.

[42] Moizé V, Andreu A, Flores L, Torres F, Ibarzabal A, Delgado S, et al. Long-term dietary intake and nutritional deficiencies following sleeve gastrectomy or Roux-En-Y gastric bypass in a Mediterranean population. J Acad Nutr Diet 2013;113:400−10.

[43] Aarts EO, Janssen IM, Berends FJ. The gastric sleeve: losing weight as fast as micronutrients? Obes Surg 2011;21:207−11.

[44] Gehrer S, Kern B, Peters T, Christoffel-Courtin C, Peterli R. Fewer nutrient deficiencies after laparoscopic sleeve gastrectomy (LSG) than after laparoscopic Roux-Y-gastric bypass (LRYGB) − a prospective study. Obes Surg 2010;20:447−53.

[45] Eltweri AM, Bowrey DJ, Sutton CD, Graham L, Williams RN. An audit to determine if vitamin b12 supplementation is necessary after sleeve gastrectomy. Springerplus 2013;2:218.

[46] Moore CE, Sherman V. 25-Vitamin D supplementation efficacy: sleeve gastrectomy versus gastric bypass surgery. Obes Surg 2014;24:2055−60.

[47] Belfiore A, Cataldi M, Minichini L, Aiello ML, Trio R, Rossetti G, et al. Short-Term changes in body composition and response to micronutrient supplementation after laparoscopic sleeve gastrectomy. Obes Surg 2015;25:2344−51.

[48] Wolf E, Utech M, Stehle P, Büsing M, Helfrich HP, Stoffel-Wagner B, et al. Oral high-dose 25-vitamin D dissolved in oil raised serum 25-hydroxy-vitamin D to physiological levels in obese patients after sleeve gastrectomy-A double-blind, randomized, and placebo-controlled trial. Obes Surg 2015; [Epub ahead of print].

[49] Donadelli SP, Junqueira-Franco MV, de Mattos Donadelli CA, Salgado Jr. W, Ceneviva R, Marchini JS, et al. Daily vitamin supplementation and hypovitaminosis after obesity surgery. Nutrition 2012;28:391−6.

[50] Shannon C, Gervasoni A, Williams T. The bariatric surgery patient--nutrition considerations. Aust Fam Physician 2013;42:547−52.

[51] Schiavo L, Scalera G, Sergio R, De Sena G, Pilone V, Barbarisi A. Clinical impact of Mediterranean-enriched-protein diet on liver size, visceral fat, fat mass, and fat-free mass in patients undergoing sleeve gastrectomy. Surg Obes Relat Dis 2015;11:1164−70.

Chapter 51

Thiamine (Vitamin B1) After Weight Loss Bariatric Surgery

A. Nath[1], T.R. Shope[1,2,3] and T.R. Koch[1,2,3]

[1]MedStar-Washington Hospital Center, Washington, DC, United States, [2]Georgetown University School of Medicine, Washington, DC, United States, [3]Center for Advanced Laparoscopic General & Bariatric Surgery, Washington, DC, United States

LIST OF ABBREVIATIONS

ATP	adenosine triphosphate
Cardio	cardiovascular
DS	duodenal switch
GI	gastrointestinal
Neuro-Psych	neuropsychiatric
RYGB	Roux-en-Y gastric bypass
TPP	thiamine pyrophosphate
VSG	vertical sleeve gastrectomy

SURGICAL MANAGEMENT OF OBESITY

Bariatric surgery is a major tool for treating medically complicated obesity. Multiple studies have shown that bariatric surgery leads to more sustained weight loss when compared to nonsurgical interventions [1]. By 2013, the most commonly performed procedures included Roux-en-Y gastric bypass (RYGB) at 45%, and vertical sleeve gastrectomy (VSG), which had risen to 37%; adjustable gastric banding had declined to 10% [2]. This chapter describes mechanisms by which these procedures may lead to deficiency of vitamin B1 or thiamine.

As shown in Fig. 51.1, bariatric surgeries involving the stomach are generally restrictive in nature. These bariatric procedures include adjustable gastric banding and VSG. Adjustable gastric banding does not bypass any small intestine. VSG induces an immediate restriction of caloric intake with meal sizes of approximately 250 mL. By contrast, surgery that involves shortening of the small intestine may also induce additional malabsorption. Currently performed surgical procedures that may promote malabsorption include RYGB, biliopancreatic diversion, and sleeve gastrectomy with duodenal switch (DS). A short channel (generally <150 cm in length) of small intestine distal to the jejuno-jejunal (JJ) anastomosis (in gastric bypass surgery and biliopancreatic diversion) or distal to the jejuno-ileal anastomosis (in sleeve gastrectomy with DS) may interfere with complete mixing of nutrients with bile and pancreatic secretions, and thus could promote malabsorption. As we discuss later in this chapter, the major absorption of thiamine occurs in the proximal jejunum.

OVERVIEW OF THE BIOCHEMISTRY AND PHYSIOLOGY OF THIAMINE

Thiamine is a water-soluble vitamin, and as the first B vitamin discovered, it was termed vitamin B1. Thiamine is involved in a wide variety of the intricate biochemical pathways necessary for proper tissue and organ function (see Fig. 51.2). Thiamine plays a pivotal role in the pentose phosphate pathway, which is not only an alternate glucose metabolism pathway, but also a major route for the synthesis of several neurotransmitters, nucleic acids, lipids, amino acids, steroids, and glutathione. Thiamine, as thiamine pyrophosphate (TPP), is an important coenzyme for essential

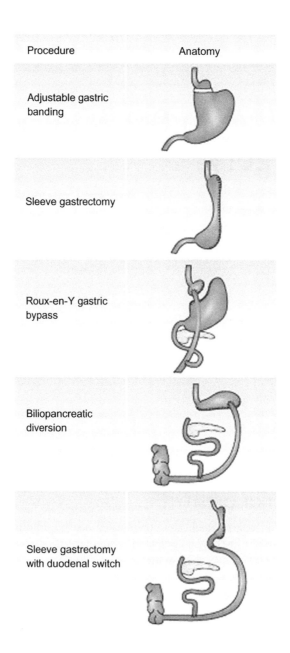

	Procedure	Anatomy
Adjustable gastric banding		
Sleeve gastrectomy		
Roux-en-Y gastric bypass		
Biliopancreatic diversion		
Sleeve gastrectomy with duodenal switch		

FIGURE 51.1 Bariatric Surgical Procedures. These procedures are mainly restrictive (adjustable gastric banding and sleeve gastrectomy) in nature or are partially restrictive and partially malabsorptive in nature (Roux-en-Y gastric bypass, biliopancreatic diversion, and sleeve gastrectomy with duodenal switch). Small intestine distal to the JJ anastomosis (in gastric bypass surgery and biliopancreatic diversion) or distal to the jejuno-ileal anastomosis (in sleeve gastrectomy with duodenal switch) is termed the common channel. A short common channel (generally <150 cm in length) does not allow complete mixing of nutrients with bile and pancreatic secretions. *Reproduced with the permission of Nature Publishing Group from Bal BS, et al. Nature Rev Endocrinol 2012;8:544−56.*

steps in the Krebs cycle, including decarboxylation of pyruvate and oxidation of alpha-ketoglutamic acid; thiamine therefore provides an important link between the glycolytic and citric acid cycles.

The Recommended Daily Allowance (RDA) for thiamine is 1.2 mg/day in males age 14 and older, 1.1 mg/day in females age 19 and older, and 1.4 mg/day in pregnant and lactating mothers. Thiamine is present in lean pork and other meats, wheat germ, liver and other organ meats, poultry, eggs, fish, beans and peas, nuts, and enriched, fortified, or whole-grain products; refined carbohydrates, fruits, and vegetables have low thiamine content [3]. Not all dietary thiamine is bioavailable; e.g., polyphenolic compounds in coffee and tea may inactivate thiamine, while bacterial thiaminases, which can be present in contaminated meat, raw fish, and shellfish, can also degrade thiamine.

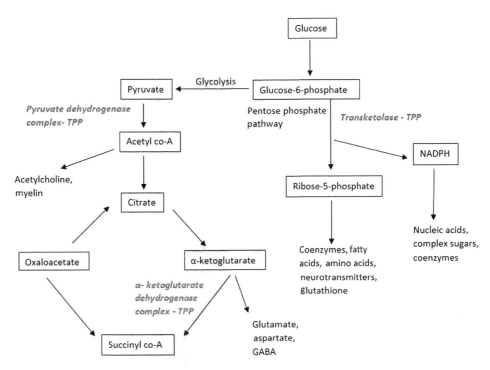

FIGURE 51.2 Summary of Major Biochemical Pathways Involving Thiamine. The involvement of vitamin B1 or thiamine as thiamine pyrophosphate (TPP) is shown in the Pentose Phosphate Pathway (Right Side of Panel) and in the Krebs Cycle (Left Side of Panel).

Thiamine is absorbed primarily in the proximal part of the small intestine. Dietary thiamine exists mainly in the phosphorylated forms, which are hydrolyzed to free thiamine in the intestinal lumen before absorption through the brush border membrane. The uptake of thiamine into enterocytes takes place through specific Na^+-independent, pH-dependent (uptake enhanced by outward pH gradient), electroneutral, carrier-mediated mechanisms; however, diffusion can occur at higher concentrations [4,5]. The uptake of thiamine is enhanced by thiamine deficiency, but can be reduced by the presence of diabetes mellitus. Additionally, thiamine absorption is inhibited by thiamine analogs (such as metronidazole), ethanol, and diuretics (such as amiloride).

Thiamine undergoes phosphorylation in enterocytes via the action of the cytoplasmic thiamine pyrophosphokinase, and then undergoes some degree of intracellular metabolism. The exit of thiamine from the enterocyte across the basolateral membrane is Na^+-dependent and directly coupled to adenosine triphosphate (ATP) hydrolysis by Na^+-K^+-ATPase. In erythrocytes, thiamine transport is saturable at low thiamine concentrations and diffusive at higher ones. Thiamine is mainly excreted in urine. Transport of thiamine by renal brush border membrane is similar to mechanisms involved in intestinal absorption.

Since thiamine is a water-soluble vitamin, there are no substantial stores of thiamine in the body, and reserves can be depleted in as little as 2−3 weeks. Indeed, biopsy studies have shown that thiamine content in human tissues is lower than in animal species [6]. It has been suggested that the combination of low circulating thiamine concentrations and low thiamine tissue content may make humans highly sensitive to thiamine deficiency.

CLINICAL MANIFESTATIONS OF THIAMINE DEFICIENCY

As shown in Table 51.1, the clinical manifestations of thiamine deficiency can be divided into four major subtypes: cardiovascular (wet beriberi), neuropsychiatric (neuro-psych), peripheral neurologic (dry beriberi), and gastrointestinal (GI). The classic symptoms of beriberi were described in the 1890s by a Dutch physician, Dr. Christian Eijkman. Dr. Eijkman famously declared that "white rice is poisonous," he described the antiberiberi factor in rice skin, and he was awarded the Nobel Prize in Physiology or Medicine in 1929.

In the 1940s, Williams and associates studied 14 young, healthy women at the Mayo Clinic in Rochester, MN in order to better understand the effects of thiamine deficiency [7]. These young women were maintained on a thiamine-restricted diet for 20 days. They developed symptoms of anorexia, nausea, and vomiting, in addition to other vague

TABLE 51.1 Symptoms and Findings in Subsets of Thiamine Deficiency

Cardiovascular	Neuropsychiatric	Peripheral Neurologic	Gastrointestinal
Tachycardia	Blurred vision	Tingling	Nausea
Bradycardia	Nystagmus	Burning pain	Vomiting
Cardiomegaly	Ophthalmoplegia	Numbness	Dysphagia
Pulmonary edema	Acute visual loss	Decreased strength of extremities	Constipation
Lower leg edema	Ataxia		Anorexia
Dyspnea on exertion	Confusion		
L-Lactic acidosis	Mood Swings		
	Memory Loss		
	Confabulation		

symptoms including depression, paresthesias, malaise, giddiness, palpitations, and pseudoangina. Serial interval roentgenograms after ingestion of a barium sulfate meal demonstrated impairment of GI motility. With the restoration of a thiamine-sufficient diet, these vague symptoms, which an inattentive physician might consider functional in nature, disappeared quickly.

In clinical practice, symptoms and signs can be identified in individual patients and then categorized into one or more of the four major subtypes of thiamine deficiency. These potential signs and symptoms (see Table 51.1) are consistent with cardiovascular thiamine deficiency, neuro-psych thiamine deficiency, peripheral neurologic thiamine deficiency, and/or GI thiamine deficiency. In addition to its well-recognized neuropathic features, thiamine deficiency has also been reported to induce a myopathic disorder [14].

The feared deficiency syndrome of thiamin, Wernicke–Korsakoff Syndrome [15], is seen relatively rarely in Western countries. The classic triad is a severe cognitive or psychotic disorder in an individual who has cranial nerve dysfunction and ataxia. By contrast, multiple authors have suggested that marginal thiamine deficiency may be common in at-risk patients and may give rise to symptoms that are vague and likely to be overlooked. Physicians may not consider thiamine deficiency in their differential diagnosis for obscure or functional symptoms, perhaps because of their trust in vitamin fortification of food products in the Western diet.

In evaluation of this concern, the frequency of Wernicke's encephalopathy in series of consecutive autopsies ranges from 0.4% to 2.8%. In one series of Wernicke's encephalopathy diagnosed at the autopsy stage, thiamine deficiency was suspected during life in only about 32% of alcoholic and 6% of nonalcoholic patients [16]. These autopsy studies support the concern that thiamine deficiency is underdiagnosed in life, since a classical clinical presentation is either uncommon or unrecognized. A more recent study by O'Keefe and associates [17] examined thiamine levels in 36 consecutive nondemented, community-dwelling patients admitted to an acute geriatric unit. Marginal deficiency was found in 11 (31%) patients, and definite deficiency in 6 (17%). Delirium was reported to be more common in thiamine-deficient patients (76%) compared to patients with normal thiamine status (32%).

The importance of considering a clinical diagnosis of thiamine deficiency in at-risk individuals has also been described by the World Health Organization [18]. In their summary of thiamine deficiency, a clinical case in the field is described as consistent with an individual having at least two of the following signs: bilateral edema of the lower limbs, dyspnea with exertion or at rest, paresthesias of the extremities (hands or feet), a symmetrical drop in muscular strength or motor deficiencies, or loss of balance.

CONFIRMATION OF A DIAGNOSIS OF THIAMINE DEFICIENCY

It is presently unclear whether routine symptom screening can prevent the development of thiamine deficiency in a highly susceptible population, such as patients who have undergone bariatric surgery. However, due to the major end-organ risks present in individuals with thiamine deficiency, physicians should understand the elements for a clinical diagnosis of thiamine deficiency. The presence of thiamine deficiency should be strongly considered in patients with rapid weight loss, protracted vomiting, parenteral nutrition, excessive alcohol use, neuropathy or encephalopathy, or heart failure after bariatric surgery. Using the classic description of thiamine deficiency, a symptomatic improvement in individuals given thiamine supplementation supports a clinical diagnosis of thiamine deficiency [8].

Due to its serious nature, if a diagnosis of Wernicke's encephalopathy is being considered, a brain image should be obtained [19].

By contrast, the biochemical assessment of thiamine status poses many challenges because, despite the many methods available, the available methods have specific problems. Whole blood or plasma thiamine levels and urinary thiamine excretion are the most widely used methods, but these determinations are not reliable for the assessment of thiamine status. Since thiamine activity is mostly intracellular, these tests may not always reflect marginal deficiencies [20]. During interpretation of these tests, the physician should keep in mind the potential relevance of the clinical symptoms of the patient. Functional assays such as direct measurement of thiamine concentration by High Performance Liquid Chromatography in plasma [21], or an indirect assay of thiamine status by determination of transketolase (a thiamine-dependent enzyme) activity in erythrocytes [22], are more precise [23]. However, specific assays are technically demanding, are not routinely available, and can be expensive. Activities of thiamine-dependent enzymes have not been well-validated in patients who have undergone bariatric surgery. Clearly, there is a need to develop more reliable, economical, and widely available biomarkers for thiamine status.

THIAMINE DEFICIENCY IN OBESITY

A potential cause for a micronutrient deficiency after bariatric surgery is exacerbation of a preexisting deficiency. There is indeed evidence that obese individuals are at-risk for thiamine deficiency.

Obesity is often viewed as a disease of "overnutrition" because it can be the result of excessive caloric intake. However, dietary factors could be a potential root cause for both obesity as well as for the development of micronutrient deficiencies [11,24–27]. Diets consumed by obese patients can include foods rich in carbohydrates and fat, but low in thiamine. Furthermore, eating disorders such as loss of control eating, night eating syndrome, binge eating disorder, bulimia nervosa, etc. are more common in the obese population. In a large multicenter trial performed using 2266 obese patients, loss of control eating was present in 43.4% of patients, night eating syndrome in 17.7% of patients, binge eating disorder in 15.7% of patients, and bulimia nervosa in 2% of patients [28]. The role of thiamine is quintessential in carbohydrate metabolism. The increased intake of the carbohydrate results in an increased requirement for vitamin B1, which accelerates its depletion [29–31]. In a study described by Elmadfa and associates, 12 healthy volunteers were evaluated to investigate the influence of stepwise increases of carbohydrate intake on the status of thiamine; an increase in dietary carbohydrate intake from 55% to 75% of total caloric intake at isocaloric conditions caused a decrease of plasma and urine levels of thiamine [31].

In addition, obese patients are at increased risk for the development of metabolic syndrome. Increased urinary excretion of thiamine has been reported in patients with diabetes mellitus [32–34]. Decreased plasma thiamine concentration in 76% of type 1 diabetic patients and 75% of type 2 diabetic patients was reported by Thornalley and associates [32]. An increase in renal clearance of thiamine by 24-fold in type 1 diabetic patients and 16-fold in type 2 diabetic patients when compared to the normal volunteers was also reported [32]. Another potential problem in obese individuals could be an increased renal clearance of thiamine induced by the use of diuretics. Obese individuals sometimes use diuretics either for treatment of associated comorbidities or for attempted weight loss.

And finally, we could have the wrong daily target when suggesting dietary thiamine intake for obese individuals. RDAs are reference values intended for the general population. Obese patients may have increased physiological requirements due to increased body size. In a study of an obese population by Damms-Machado and associates, the RDAs for micronutrients did not appear to be adequate to meet the metabolic demands of these individuals [35].

This concern of potential thiamine deficiency in obese individuals is supported by a number of studies that examine the prevalence of biochemical thiamine deficiency in preoperative morbidly obese individuals. Carrodeguas and associates reported low preoperative thiamine levels in 47 out of the 303 patients (15.5%) presenting to their preoperative clinic prior to RYGB [36]. In another study, Flancbaum and associates studied 379 consecutive patients presenting to their bariatric clinic for preoperative evaluation and identified low thiamine levels in 29% of these individuals [37]. These observational studies suggest that thiamine deficiency may be present even before bariatric surgery has been performed in morbidly obese patients.

THIAMINE DEFICIENCY AFTER BARIATRIC SURGERY

Thiamine deficiency as a complication of bariatric surgery has been well described (see Table 51.2). There are wide variations reported in the postoperative prevalence of thiamine deficiency, likely due to the different patient care settings, the length of follow-up, socioeconomic factors, and the use of different methods for defining thiamine deficiency. Bariatric procedures involving alteration of the anatomy of the small intestine induce an additional component of malabsorption. Several authors have suggested that bariatric procedures involving alteration of the anatomy of the small intestine produce greater nutritional deficiencies than purely restrictive surgical procedures [38].

We have reported a prospective study of clinical vitamin B1 (thiamine) deficiency that was diagnosed in 27 out of 151 patients (or 18%) who had undergone RYGB and were seen in a bariatric GI clinic in a large, urban community hospital [8]. As shown in Fig. 51.3, in this study multiple subtypes of thiamine deficiency were simultaneously identified in individual patients. The most commonly identified subtypes of thiamine deficiency were cardiovascular manifestations (68% of individuals with thiamine deficiency) and neuro-psych manifestations (52% of individuals with thiamine deficiency).

In a study reported from a University Hospital [10], 318 individuals had vitamin B1 levels examined at 1 year after RYGB. The levels of thiamine were found to be low in 18.3% of these patients. In a more recent study [9], 22 individuals had blood thiamine levels measured after RYGB. The prevalence of thiamine deficiency in these patients was only 5%, but this was with 6 months of follow-up.

Biliopancreatic diversion may result in more weight loss when compared to other bariatric procedures. However, this more extensive surgery is usually performed in supermorbidly obese patients (e.g., a body mass index of $\geq 50\,kg/m^2$),

TABLE 51.2 Prevalence of Thiamine Deficiency After Bariatric Surgery

Surgery	Postsurgery	No. of Patients	Thiamine Deficiency	References
RYGB	12 months	151	18%	Shah et al. [8]
RYGB	6 months	22	5%	Verger et al. [9]
RYGB	12 months	318	18.3%	Clements et al. [10]
VSG	12 months	200	5.5%	Van Rutte et al. [11]
VSG	60 months	35	30.8%	Saif et al. [12]
DS	18–30 months	128	14.1%	Botella et al. [13]

RYGB, Roux-en-Y gastric bypass; VSG, vertical sleeve gastrectomy; DS, duodenal switch.

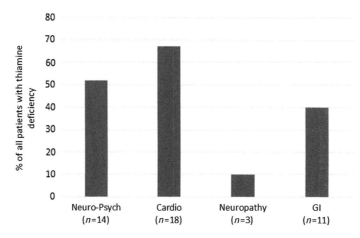

FIGURE 51.3 Distribution of the Percentages of Patients with Subtypes of Thiamine Deficiency. After Roux-en-Y gastric bypass surgery, the four major subtypes depicted include neuropsychiatric (Neuro-Psych), cardiovascular (Cardio), peripheral neuropathic (Neuropathy), and gastrointestinal (GI) manifestations of thiamine deficiency. *Adapted from Shah H, et al. Digestion 2013;88:119–24.*

and the increased malabsorption may create more severe nutritional deficiencies [39]. Aasheim and associates reported an increased risk of thiamine deficiency in individuals after DS [40]. These authors reported results from patients in whom thiamine concentrations declined steeply after DS when compared to patients who had undergone gastric bypass during a 1-year follow-up period. In another study of 128 patients after DS reported by Botella and associates, 14.1% of the patients were found to have low levels of water-soluble vitamins (other than vitamin B12) at 30 months of follow-up [13].

Despite the restrictive nature of VSG, there is emerging evidence that thiamine deficiency is not an uncommon complication after this bariatric procedure. This is important due to the dramatic rise in the performance of VSG surgeries over the last 12 years. Both short-term and long-term studies of the prevalence of thiamine deficiency after VSG have been reported. In a 1-year follow-up of 200 individuals after VSG, van Rutte and associates reported a 5.5% prevalence of thiamine deficiency [11]. In a second prospective study, Saif et al. reported biochemical thiamine deficiency in 30.8% at 5 years of follow-up in 35 individuals after VSG [12]. Indeed, a recent systematic literature review has warned that individuals can progress to Wernicke's encephalopathy after VSG [41].

In contrast to VSG, the worldwide performance of adjustable gastric banding has declined rapidly in recent years. Systematic studies evaluating the prevalence of thiamine deficiency after adjustable gastric banding are sparse, although cases of thiamine deficiency have been reported [42,43].

MECHANISMS OF THIAMINE DEFICIENCY AFTER BARIATRIC SURGERY

Despite multiple studies describing thiamine deficiency after bariatric surgery, the underlying mechanisms involved in the postoperative development of thiamine deficiency are not fully understood. Pathophysiological factors that may alter thiamine status after bariatric surgery are mainly thought to include reduced intake and impaired absorption, as shown in Table 51.3.

As described above in "Thiamine Deficiency in Obesity," biochemical thiamine deficiency is commonly found in individuals before bariatric surgery and may persist or worsen after bariatric surgery. Other factors associated with early thiamine deficiency after bariatric surgery include a reduction in dietary nutrient intake due to decreased gastric capacity, and reduced absorptive capacity due to performance of a bariatric surgery that bypasses the proximal jejunum. Studies evaluating food records found that dietary intake was inadequate in almost all patients presenting for follow-up after bariatric surgery [44]. Vomiting can be an early GI disorder after bariatric surgery, and this problem can limit both oral intake of food as well as multivitamin supplements [45]. Many patients may experience episodes of vomiting as they develop a new sense of upper abdominal fullness following restrictive bariatric surgery, and this problem may be exacerbated by overeating. There is extensive literature on the risks of using alcohol products after bariatric surgery. Use of dietary supplements, which is common in obese individuals, is also associated with the promotion of thiamine deficiency [8].

Small intestinal bacterial overgrowth can result in micronutrient deficiencies as a result of alterations in the upper gut flora. Colonization of bacteria in the small intestinal tract is present in low concentrations in normal healthy anatomy. The loss of gastric acid secretion after bariatric surgery appears to impair a defense mechanism. The flora of the upper small intestine can change to higher concentrations of fermentation-inducing anaerobic bacteria. These bacteria may secrete the enzyme thiaminase, thus altering the metabolism and absorption of thiamine.

We have reported an association between thiamine deficiency and small intestinal bacterial overgrowth after RYGB [46]. Our studies have expanded upon prior reports of increased serum levels of folate (reported specificity of 79%)

TABLE 51.3 Proposed Mechanisms for the Development of Thiamine Deficiency After Bariatric Surgery

- Preexisting subclinical thiamine deficiency.
- Consumption of alcohol-containing products.
- Use of dietary supplements.
- Inadequate nutritional supplements.
- Medication side effects.
- Reduction in total food intake due to decreased gastric capacity.
- Postoperative vomiting.
- Reduction in intestinal absorption surface.
- Dumping syndrome with rapid oral–cecal transit.
- Small intestinal bacterial overgrowth.

produced by bacteria in individuals with small intestinal bacterial overgrowth. Oral thiamine was not well absorbed in patients after gastric bypass surgery. Postoperative patients who were then given oral antibiotics to treat small intestinal bacterial overgrowth and oral thiamine did increase their whole blood thiamine levels.

MANAGEMENT OF THIAMINE DEFICIENCY AFTER BARIATRIC SURGERY

Prevention would certainly be the best management option for this potentially serious micronutrient deficiency. Given the high prevalence of micronutrient deficiencies, it is now recommended that all individuals receive daily micronutrient supplementation consisting of multivitamin supplements (to include iron, folic acid, thiamine, and vitamin B12), elemental calcium, and vitamin D for at least the first 3−6 months after bariatric surgery [47]. European guidelines also advise routine supplementation of thiamine in the immediate postoperative period in individuals who have undergone gastric bypass surgery [16].

In postoperative follow-up visits, patients who have a high clinical suspicion of thiamine deficiency should be screened for compatible clinical symptoms, and empiric thiamine supplementation should be initiated without waiting for the biochemical laboratory results.

For minor, chronic, or subclinical symptoms of thiamine deficiency, the standard oral dosing of thiamine is 100 mg by mouth taken with meals twice daily. The benefit of thiamine supplementation in these patients is believed to be nutritional rather than pharmacological. Nutritional benefit is present when the change caused by low levels of the nutrient has detrimental consequences, and health restorative effects (supplementation) are a response to an apparent deficiency. Pharmacological benefit, in contrast, is when supplementation alleviates a condition other than the nutritional deficiency of the particular nutrient, or alters biochemical or biological structures in a therapeutic way [48].

If an oral dose of thiamine is not effective for resolution of a patient's symptoms, a course of oral antibiotic for treatment of potential small intestinal bacterial overgrowth is presently recommended [47]. The antibiotic chosen for a 7- to 10-day oral course remains empiric due to the paucity of controlled trials, and treatment is usually focused on the use of metronidazole, doxycycline, or cefaclor (taken one to three times daily). Reviews of the treatment of small intestinal bacterial overgrowth is focused on treatment of this disorder with rifaximin (200 mg or 550 mg) [49]. However, this nonabsorbable antibiotic may not be the best option for postbariatric surgery patients who have bypassed intestinal loops (and thus may have blind loop syndrome).

For treatment of serious eye, cardiac, or neurological symptoms, parenteral thiamine should be given to the patient. Although the doses of thiamine have not been well studied, we did not identify a consistent rise in whole blood thiamine levels in patients given 100 mg thiamine-HCl intramuscularly after gastric bypass surgery [8]. Vials of commercially available thiamine-HCl do contain 200 mg in a 2 mL volume. Delivery of this 2 mL volume is by deep intramuscular injection, generally into the gluteus muscle. When considering redosing, as mentioned above, there are no substantial stores of thiamine in the human body, and so reserves of thiamine can be depleted in as little as 2−3 weeks.

By contrast, a sudden onset of vision loss, psychosis, encephalopathy, ataxia, or suspected Wernicke's disease should be treated as a medical emergency [41]. Patients with these sudden, severe symptoms should be hospitalized for treatment with intravenous thiamine-HCl, using a minimum of 500 to 1000 mg of thiamine-HCl given daily, diluted in 250 to 500 mL of 0.9% NaCl, and then given by slow intravenous infusion over 10−20 hours; alternatively, split doses of up to three times daily with each intravenous infusion given over a minimum of 4 hours (to reduce the risk of anaphylaxis) have also been described. Intravenous treatment should be given for a minimum of 3−5 days, or until resolution of the patient's ongoing symptoms. For these more severe symptoms of thiamine deficiency, we empirically favor maintenance therapy with intramuscular thiamine-HCl rather than immediate conversion to maintenance with oral thiamine.

MINI-DICTIONARY OF TERMS

- *Thiamine*: Vitamin B1 with a chemical formula of 3-((4-amino-2-methyl-5-pyrimidinyl)methyl)-5-(2-hydroxyethyl)-4-methylthiazolium chloride. It is an essential nutrient in humans, and is present in cells as TPP.
- *Recommended daily allowance*: The dietary intake level of a nutrient considered sufficient by the Institute of Medicine's Food and Nutrition Board to meet the daily requirements of 97.5% of healthy individuals grouped by sex and age categories.
- *Wernicke's encephalopathy*: A manifestation of thiamine deficiency that may include the classic triad of encephalopathy or confusion, nystagmus, and ataxia.

- *Transketolase*: An enzyme of the pentose phosphate pathway, which in the nonoxidative pathway, catalyzes transfer of a 2-carbon fragment from a 5-carbon ketose to thiamine diphosphate.
- *Small intestinal bacterial overgrowth*: In the classic definition, the flora of the proximal small intestine begins to resemble the flora of a healthy subject's colon; there are $\geq 10^5$ colony-forming units/mL of bacteria identified in cultures from fluid aspirated from the proximal small intestine.
- *Thiaminase*: Enzymes that metabolize thiamine into two molecules and are categorized as thiaminase I (EC 2.5.1.2) and thiaminase II (EC 3.5.99.2).
- *Blind loop syndrome*: Also termed "stagnant loop syndrome" or "stasis syndrome," this disorder occurs when a loop of small intestine is formed at surgery that will be bypassed by food during normal digestion of a meal. Complications of this syndrome are often thought to be caused by small intestinal bacterial overgrowth.

KEY FACTS

- Thiamine is absorbed primarily in the proximal part of the small intestine.
- Thiamine plays a pivotal role in the pentose phosphate pathway and in the Krebs cycle.
- In humans, thiamine reserves can be depleted in as little as 2–3 weeks.
- Subclinical thiamine deficiency may be present in morbidly obese individuals prior to bariatric surgery.
- Thiamine deficiency has been identified after all bariatric surgical procedures, including VSG.
- Thiamine absorption may be improved in individuals after bariatric surgery by using oral antibiotics to treat concomitant small intestinal bacterial overgrowth.
- Wernicke's encephalopathy requires treatment with high doses of intravenous thiamine-HCl.

SUMMARY POINTS

- Bariatric surgery is a major tool for treating medically complicated obesity.
- Subclinical thiamine deficiency is common in obese individuals.
- In humans, thiamine stores can be depleted in as little as 2–3 weeks.
- Thiamine deficiency has been reported both after restrictive bariatric procedures as well as after malabsorptive bariatric procedures, including VSG.
- The most common clinical subtypes of thiamine deficiency after RYGB are cardiovascular and neuropsychiatric (neuro-psych) manifestations.
- Small intestinal bacterial overgrowth is a major mechanism for development of symptomatic thiamine deficiency after bariatric surgery.
- Treatment of bacterial overgrowth with an oral antibiotic can improve oral absorption of thiamine after bariatric surgery.
- Wernicke's disease is a complication of thiamine deficiency that should be managed with immediate intravenous infusions of high doses of thiamine-HCl.

REFERENCES

[1] Adams TD, Pendleton RC, Strong MB, et al. Health outcomes of gastric bypass patients compared to nonsurgical, nonintervened severely obese. Obesity (Silver Spring) 2010;18(1):121.

[2] Angrisani L, Santonicola A, Iovino P, Formisano G, Buchwald H, Scopinaro N. Bariatric surgery worldwide 2013. Obes Surg 2015;25(10):1822.

[3] US Department of Health and Human Services and US Department of Agriculture. Dietary guidelines for Americans. 7th ed. Washington, DC: US Government Printing Office; 2010.

[4] Rindi G, Laforenza U. Thiamine intestinal transport and related issues: recent aspects. Exp Biol Med 2000;224(4):246.

[5] Dudeja PK, Tyagi S, Kavilaveettil RJ, Gill R, Said HM. Mechanism of thiamine uptake by human jejunal brush-border membrane vesicles. Am J Physiol Cell Physiol 2001;281(3):C786.

[6] Gangolf M, Czerniecki J, Radermecker M, Detry O, Nisolle M, Jouan C, et al. Thiamine status in humans and content of phosphorylated thiamine derivatives in biopsies and cultured cells. PLoS One 2010;5(10):e13616.

[7] Williams RD, Mason HI, Power MH, et al. Induced thiamine (vitamin B1) deficiency in man: relation of depletion of thiamine to development of biochemical defect and of polyneuropathy. Arch Int Med 1943;71:38.

[8] Shah HN, Bal B, Finelli FC, Koch TR. Constipation in patients with thiamine deficiency after Roux-en-Y gastric bypass surgery. Digestion 2013;88:119.

[9] Verger EO, Aron-Wisnewsky J, Dao MC, Kayser BD, Oppert JM, Bouillot JL, et al. Micronutrient and protein deficiencies after gastric bypass and sleeve gastrectomy: a 1-year follow-up. Obes Surg 2016;26:785.

[10] Clements RH, Katasani VG, Palepu R, Leeth RR, Leath TD, Roy BP, et al. Incidence of vitamin deficiency after laparoscopic Roux-en-Y gastric bypass in a university hospital setting. Am Surg 2006;72:1196.

[11] van Rutte PW, Aarts EO, Smulders JF, Nienhuijs SW. Nutrient deficiencies before and after sleeve gastrectomy. Obes Surg 2014;24:1639.

[12] Saif T, Strain GW, Dakin G, Gagner M, Costa R, Pomp A. Evaluation of nutrient status after laparoscopic sleeve gastrectomy 1, 3, and 5 years after surgery. Surg Obes Relat Dis 2012;8:542.

[13] Botella Romero F, Milla Tobarra M, Alfaro Martínez JJ, García Arce L, García Gómez A, Salas Sáiz MA, et al. Bariatric surgery in duodenal switch procedure: weight changes and associated nutritional deficiencies. Endocrinol Nutr 2011;58:214.

[14] Koike H, Watanabe H, Inukai A, Iijima M, Mori K, Hattori N, et al. Myopathy in thiamine deficiency: analysis of a case. J Neuro Sci 2006;249:175.

[15] Platt BS. Thiamine deficiency in human beriberi and in 'Wernicke's encephalopathy. In: Wolstenholme GEW, O'Connor M, editors. Thiamine deficiency. Boston: Little, Brown and Company; 1967. p. 135−43.

[16] Galvin R, Bråthen G, Ivashynk A, Hillbom M, Tanasescu R, Leone MA. EFNS, EFNS guidelines for diagnosis, therapy and prevention of Wernicke encephalopathy. Eur J Neurol 2010;17(12):1408.

[17] O'Keefe ST, Tormey WP, Glasgow R, Lavan JN. Thiamine deficiency in hospitalized elderly patients. Gerontology 1994;40:18.

[18] Prinzo Z.W., World Health Organization, Department of Nutrition for Health and Development, WHO, Thiamine deficiency and its prevention and control in major emergencies, Report no: WHO/NHD/99.13, Geneva, 1999, p. 1−52, Available at: <http://www.who.int/nutrition/publications/emergencies/WHO_NHD_99.13/en/> [accessed 22.02.16].

[19] Beh SC, Frohman TC, Frohman EM. Isolated mammillary body involvement on MRI in Wernicke's encephalopathy. J Neurol Sci 2013;334:172.

[20] Butterworth R. Thiamin. In: Shils ME, Shike M, Ross AC, et al., editors. Modern nutrition in health and disease. New York, NY: Lipincott Williams and Wilkins; 2006. p. 426−33.

[21] Brunnekreeft JW, Eidhof H, Gerrits J. Optimized determination of thiochrome derivatives of thiamine and thiamine phosphates in whole blood by reversed-phase liquid chromatography with precolumn derivatization. J Chromatogr 1989;491(1):89.

[22] Dreyfus PM. Clinical application of blood transketolase determinations. N Engl J Med 1962;267:596.

[23] Talwar D, Davidson H, Cooney J, O'Reilly D. Vitamin B1 status assessed by direct measurement of thiamin pyrophosphate in erythrocytes or whole blood by HPLC: comparison with erythrocyte transketolase activation assay. Clin Chem 2000;46:704.

[24] Kaidar-Person O, Person B, Szomstein S, Rosenthal RJ. Nutritional deficiencies in morbidly obese patients: a new form of malnutrition? Part A: vitamins. Obes Surg 2008;18:870.

[25] Kaidar-Person O, Person B, Szomstein S, Rosenthal RJ. Nutritional deficiencies in morbidly obese patients: a new form of malnutrition? Part B: minerals. Obes Surg 2008;18:1028.

[26] Damms-Machado A, Friedrich A, Kramer KM, Stingel K, Meile T, Küper MA, et al. Pre- and postoperative nutritional deficiencies in obese patients undergoing laparoscopic sleeve gastrectomy. Obes Surg 2012;22:881.

[27] Schweiger C, Weiss R, Berry E, Keidar A. Nutritional deficiencies in bariatric surgery candidates. Obes Surg 2010;20:193.

[28] Mitchell JE, King WC, Courcoulas A, Dakin G, Elder K, Engel S, et al. Eating behavior and eating disorders in adults before bariatric surgery. Int J Eat Disord 2015;48:215.

[29] Lonsdale D. A review of the biochemistry, metabolism and clinical benefits of thiamin(e) and its derivatives. Evid Based Complement Alternat Med 2006;3:49.

[30] Via M. The malnutrition of obesity: micronutrient deficiencies that promote diabetes. ISRN Endocrinol 2012;2012:103472.

[31] Elmadfa I, Majchrzak D, Rust P, Genser D. The thiamine status of adult humans depends on carbohydrate intake. Int J Vitam Nutr Res 2001;71:217.

[32] Thornalley PJ, Babaei-Jadidi R, Al Ali H, Rabbani N, Antonysunil A, Larkin J, et al. High prevalence of low plasma thiamine concentration in diabetes linked to a marker of vascular disease. Diabetologia 2007;50:2164.

[33] Saito N, Kimura M, Kuchiba A, Itokawa Y. Blood thiamine levels in outpatients with diabetes mellitus. J Nutr Sci Vitaminol (Tokyo) 1987;33:421.

[34] Jermendy G. Evaluating thiamine deficiency in patients with diabetes. Diab Vasc Dis Res 2006;3:120.

[35] Damms-Machado A, Weser G, Bischoff SC. Micronutrient deficiency in obese subjects undergoing low calorie diet. Nutr J 2012;11:34.

[36] Carrodeguas L, Kaidar-Person O, Szomstein S, Antozzi P, Rosenthal R. Preoperative thiamine deficiency in obese population undergoing laparoscopic bariatric surgery. Surg Obes Relat Dis 2005;1:517.

[37] Flancbaum L, Belsley S, Drake V, Colarusso T, Tayler E. Preoperative nutritional status of patients undergoing Roux-en-Y gastric bypass for morbid obesity. J Gastrointest Surg 2006;10:1033.

[38] Davies DJ, Baxter JM, Baxter JN. Nutritional deficiencies after bariatric surgery. Obes Surg 2007;17:1150.

[39] Dolan K, Hatzifotis M, Newbury L, et al. A clinical and nutritional comparison of biliopancreatic diversion with and without duodenal switch. Ann Surg 2004;240:51.

[40] Aasheim ET, Björkman S, Søvik TT, Engström M, Hanvold SE, Mala T, et al. Vitamin status after bariatric surgery: a randomized study of gastric bypass and duodenal switch. Am J Clin Nutr 2009;90:15.

[41] Kröll D, Laimer M, Borbély YM, Laederach K, Candinas D, Nett PC. Wernicke encephalopathy: A future problem even after sleeve gastrectomy? A systematic literature review. Obes Surg 2016;26:205.

[42] Aron-Wisnewsky J, Verger EO, Bounaix C, Dao MC, Oppert JM, Bouillot JL, et al. Nutritional and protein deficiencies in the short term following both gastric bypass and gastric banding. PLoS One 2016;11:e0149588.

[43] Bozbora A, Coskun H, Ozarmagan S, Erbil Y, Ozbey N, Orham Y. A rare complication of adjustable gastric banding: Wernicke's encephalopathy. Obes Surg 2000;10:274.

[44] Andersen T, Larsen U. Dietary outcome in obese patients treated with a gastroplasty program. Am J Clin Nutr 1989;50:1328.

[45] Frank P, Crookes PF. Short- and long-term surgical follow-up of the postbariatric surgery patient. Gastroenterol Clin North Am 2010;39:135.

[46] Lakhani SV, Shah HN, Alexander K, Finelli FC, Kirkpatrick JR, Koch TR. Small intestinal bacterial overgrowth and thiamine deficiency after Roux-en-Y gastric bypass surgery in obese patients. Nutr Res 2008;28:293.

[47] Mechanick JI, Youdim A, Jones DB, et al. Clinical practice guidelines for the perioperative nutritional, metabolic, and nonsurgical support of the bariatric surgery patients-2013 update: cosponsored by American Association of Clinical Endocrinologists, the Obesity Society, and American Society for Metabolic & Bariatric Surgery. Obesity 2013;21:S1.

[48] Nielsen F. Importance of making dietary recommendations for elements designated as nutritionally beneficial, pharmacologically beneficial, or conditionally essential. J Trace Element Exp Med 2000;13:113.

[49] Shah SC, Day LW, Somsouk M, Sewell JL. Meta-analysis: antibiotic therapy for small intestinal bacterial overgrowth. Aliment Pharmacol Ther 2013;38:925.

Chapter 52

Vitamin A and Roux-EN-Y Gastric Bypass

G.V. Chaves[1] and J.S. Silva[2]

[1]Brazilian National Cancer Institute, Rio de Janeiro, Brazil, [2]Federal University of Rio de Janeiro, Rio de Janeiro, Brazil

LIST OF ABBREVIATIONS

BMI	body mass index
CVD	cardiovascular disease
HDL-c	high-density lipoprotein cholesterol
HOMA-IR	homeostatic model assessment of insulin resistance
IR	insulin resistance
LDL-c	low-density lipoprotein cholesterol
NEFAs	nonesterified fatty acids
OS	oxidative stress
RBP	retinol-binding protein
RYGBP	Roux-en-Y gastric bypass
VAD	vitamin A deficiency

INTRODUCTION

Obesity is a widespread disease of rising prevalence that is reaching epidemic proportions. It is one of the principal public health issues in modern society [1]. Between 1995 and 2011−12, the prevalence of obesity (all classes) increased from 19.1% to 27.2%. During this 17-year period, the relative increase in class III obesity (body mass index (BMI) $\geq 40.0\ kg/m^2$) was 2.2-fold. In 2011−12, the prevalence of class III obesity was 2.0% in men and 4.2% in women. There was an increase in severe obesity from 1 in 20 people (1995) to 1 in 10 people (2011−12); women were disproportionally represented in this population. Over the last two decades, there have been substantial increases in the prevalence of obesity, particularly at the more severe levels [2].

Morbid obesity is a disease that presents one of the highest mortality rates in the world. Compared with normal weight adults, obese adults have at least a 20% higher rate of dying of all-cause or cardiovascular disease (CVD). These rates advance death by 5.0 years (class III obesity) for CVD-specific mortality [3].

A clinical approach is generally ineffective in these cases. Bariatric operations are the most effective treatment, and lead to significant improvement in comorbidities [4]. However, because the operations create a restriction of food intake and a partial bypass of the upper intestinal tract, the risk of nutritional deficiencies is increased. Macronutrient deficiency is frequently associated with that of micronutrients, and can lead to the development of anemia, bone demineralization, and several hypovitaminoses [5].

Vitamin A deficiency (VAD) is one of the most prevalent public health issues in the world. It results in damage to the health of individuals affected by it, and sometimes death [6]. Vitamin A, besides participating in several primordial functions in the human organism—such as visual acuity, immunological activity, and cellular proliferation and differentiation [7]—has recently received attention due to its action against free radicals, i.e., protection of the organism against oxidative stress (OS). Consequently, vitamin A prevents tissue injury related to several nontransmissible chronic diseases often attributed to the effects of obesity [8].

In this chapter, we aim to identify the principal factors that may contribute to the appearance and/or aggravation of VAD in individuals with morbid obesity in pre- and postoperative stages of Roux-en-Y gastric bypass (RYGBP). Emphasis is on intervention strategies.

Metabolism and Pathophysiology of Bariatric Surgery.

ROUX-ᴇɴ-Y GASTRIC BYPASS AND NUTRITIONAL DEFICIENCIES

The advance of scientific knowledge with regards to the rise of morbidity—mortality rates emphasizes the necessity of medical intervention against obesity. Operations for morbid obesity are acknowledged as the only methods that effectively control obesity both in the medium- and long-term [4].

RYGBP has the potential to generate metabolic disorders, taking into account the reduction of the gastric reservoir with consequent hypochlorhydria and enteric deviation (total exclusion of the duodenum and of the proximal jejunum with significant reduction of digestive enzymes). This leads to macro- and micronutrient deficiencies [9].

According to Bavaresco [10], patients maintain food consumption between 700 and 900 kcal. This can cause serious nutritional deficiencies if not well monitored, especially during the first 6 months after the procedure. These deficiencies occur mainly due to malabsorption secondary to the decrease of gastrointestinal tract segments where nutrients are absorbed. It is also possible that some nutrient deficiencies may occur due to decreased intake of food postoperatively, which can be caused by the tendency to avoid foods linked to gastrointestinal intolerances (e.g., dumping syndrome).

The prophylactic use of nutritional supplements may avoid these deficiencies [5]. While the magnitude of these deficiencies is not certain, due to the lack of specific recommendations for supplementation for this group, they are frequently not diagnosed [5].

Dunstan et al. [11] examined multivitamin supplementation prescription following bariatric gastric bypass in 37 hospitals in England. There was no standard practice regarding postoperative supplementation among the 35 responding hospitals. Only 7 out of 35 (20%) hospitals followed the guidance for two-tablet doses of multivitamins. Immediately postoperatively, 25 out of 35 (71%) hospitals administered calcium carbonate and vitamin D (but no calcium citrate, as recommended by the American Association of Clinical Endocrinologists to improve absorption), and only 9 out of 35 (26%) recommended iron to all patients.

The most frequent deficiencies were hypoproteinemia, iron deficiency anemia, hypocalcemia, and liposoluble vitamins [12,13]. The reports focused on a limited number of micronutrients, such as calcium, vitamin D [12,13], iron, folic acid, and vitamin B [14].

Few studies have reported VAD after RYGBP. Pereira et al. [15] assessed vitamin A and beta-carotene levels preoperatively and 30 and 180 days after surgery in patients who underwent gastric bypass and detected below-normal vitamin A levels in 14%, 50.8%, and 52.9% of patients, respectively. Although there is not much knowledge regarding clinical consequences of VAD after bariatric surgery, some case reports have indicated the occurrence of ophthalmological complications such as night blindness [15,16].

CONTRIBUTING FACTORS TO VITAMIN A DEFICIENCY IN MORBID OBESITY, BEFORE AND AFTER ROUX-ᴇɴ-Y GASTRIC BYPASS

Table 52.1 summarizes the main factors related to VAD before and after RYGBP, which will be discussed below.

Oxidative Stress

OS may be defined as the adverse effects of oxidative reactions induced by free radicals inside biological systems that cause lesions in relevant molecules such as deoxyribonucleic acid, proteins, carbohydrates, and lipids. It is a condition where oxidant metabolites exert their toxic effects either due to increased production or from disorders in cellular protective mechanisms [17].

Obesity and related metabolic diseases are associated with chronic low-grade inflammation, leading to immune cell infiltration into adipose tissue and subsequent activation of tumor necrosis factor-α and interleukin-6 production and release. In addition, hormone-sensitive lipase activity can be enhanced in adipose tissue by tumor necrosis factor-α, which further increases the release of nonesterified fatty acids (NEFAs) into the blood, and concomitantly reduces insulin-stimulated glucose uptake via impaired insulin signaling [18,19].

In turn, insulin resistance (IR) is implicated in the increase in OS [20]. Paolisso et al. [21] found a positive correlation between plasma levels of reactive oxygen species and insulin. Chaves et al. [22] found a significant negative correlation between the homeostatic model assessment of insulin resistance (HOMA-IR) index and beta-carotene serum levels in morbidly obese patients undergoing RYGBP. In addition, 97.6% of the individuals with low levels of serum beta-carotene had IR. As regards retinol, an inadequacy of 100% was observed in individuals resistant to insulin, and individuals with adequate retinol also presented lower averages of basal insulin. These

TABLE 52.1 Main Factors Related to Vitamin A Deficiency Before and After Roux-en-Y Gastric Bypass

Causes	Mechanism	Relation to Vitamin A Status
Oxidative stress	• Chronic low-grade inflammation. • Dyslipidemias. • Insulin resistance. • Obesity per se.	Retinol and carotenoids stand out as substances that act against the oxidative attack of oxygen.
Deficiency of other micronutrients (zinc, iron)	• Before surgery: Obese individuals consume lesser quantities of fruits and vegetables. • After surgery: Meats and meat products, leafy vegetables (sources of zinc and iron) may be rejected because they require prolonged mastication.	Zinc and iron influence vitamin A metabolism in many aspects, including its absorption, transport, and utilization.
Malabsorption and fewer food sources of vitamin A after RYGBP	• Exclusion of the duodenum and of the first 100 cm of jejunum from food traffic. • Lipid restriction in the postoperative stage. • Milk and its derivatives may be rejected or not tolerated, and can contribute to a low intake of vitamin A.	Decreased area of the sites of vitamin A absorption; reduction of vitamin A sources in alimentary diet; and lower absorption because of lipid restriction, since vitamin A is a liposoluble nutrient.

results support the hypothesis that carotenoids may have a protective effect in the pathogenesis of IR, probably due to their role as protective agents in OS; it has been suggested that the increase of OS is implicated in a decrease of insulin action [23].

Dyslipidemias are also implicated in the increase of OS. Individuals with dyslipidemia have higher levels of lipid peroxidation. With the aim of evaluating the hypothesis that low concentrations of high-density lipoprotein cholesterol (HDL-c) interfere with lipoprotein oxidation, Toikka et al. [24] found significantly lower levels of oxidation of low-density lipoprotein cholesterol (LDL-c) in the group with high HDL-c levels when compared to the one with low HDL levels. They concluded that constant low HDL-c levels were related to an increase in OS.

Serban et al. [25] observed that individuals with dyslipidemia and glycemic levels higher than 200 mg/dL presented higher levels of lipid peroxidation. They concluded that the association between dyslipidemia and hyperglycemia, and high levels of lipid peroxides and a decrease in antioxidant capacity, were implicated in the increased risk for development of cardiovascular attacks. Therefore, both hyperglycemia and dyslipidemias, factors commonly present in morbidly obese individuals, tend to increase lipid peroxides; this indicates a decrease in serum antioxidant levels, such as beta-carotene.

The effect of dietary VAD on plasma lipid profiles has been investigated. A vitamin A-deficient diet has been associated with decreased HDL-c cholesterol and/or triglycerides [26,27]. More recently, Relevy et al. [28], using an atherosclerotic mouse model, showed that dietary VAD significantly accelerated atherogenesis despite unaffected plasma retinol and retinoic acid concentrations. They also found that a diet enriched with beta-carotene can compensate for the effects of dietary VAD with regards to atherosclerosis development.

Obesity per se has been related to impaired vitamin A nutritional status. The literature has shown the association between BMI and serum levels of retinol and carotenoids. Viroonudomphol et al. [29] found a negative correlation between BMI levels and those of serum retinol in a study performed with overweight and obese patients. A statistically significant decrease in the serum levels of carotenoids accompanying an increase in BMI has been verified, with obesity being the nutritional condition most associated with low levels of carotenoids [15]. On the other hand, weight reduction was shown to have a significant role in increasing antioxidant enzyme activities, especially glutathione reductase, and catalase enzymes in obese women [30].

Retinol and carotenoids stand out as substances that act against the oxidative attack of oxygen. Studies have demonstrated that carotenoids effectively participate in antioxidant defense, which promotes the elimination of the singlet oxygen involved in the oxidative attack on nucleic acids, amino acids, and polyunsaturated fatty acids. Therefore, retinol and carotenoids reduce nitric oxide synthesis through action on the inducible nitric oxide synthase, and decrease the production of oxygen-reactive species. Hence, adequate intake of vitamin A (particularly carotenoids) is important for protection against the free radical oxidative attack on cell membranes, thus reducing oxidative damage [31].

Deficiency of Other Micronutrients

Compared to nonobese individuals, obese individuals consume larger amounts of energy-dense foods, and relatively lesser quantities of foods of low energetic density (such as fruits and vegetables) [32]. Thus, it is possible to verify the presence of micronutrient deficiencies, even if subclinical, that characterize the "hidden hunger" in obese individuals.

Frequently, deficiencies of vitamins and minerals occur in a combined manner because of the association between food sources, metabolic routes, and physiological functions. Some nutrients are notable for their relationship with vitamin A, such as zinc and iron [33].

Zinc nutritional status influences vitamin A metabolism in many aspects, including its absorption, transport, and utilization. On the other hand, there is also evidence that vitamin A has influence on the absorption and utilization of zinc [33]. Two mechanisms are accepted to explain the interrelationship between those two micronutrients. One is related to the regulatory role of zinc in the transport of vitamin A mediated by the synthesis of proteins. Zinc deficiency may decrease the synthesis of retinol-binding protein (RBP) in liver, and may lead to low concentrations of RBP in plasma. Another proposed mechanism is the conversion of retinol into retinaldehyde, which requires the action of zinc-dependent retinol dehydrogenase enzymes [34]. Zinc deficiency can also affect the absorption of vitamin A by decreasing the lymphatic absorption of retinol; this may be attributed to the lower exit of lymphatic phospholipids, which results from the decrease in bile secretion in the intestinal lumen [33].

Cominetti et al. [35] observed zinc deficiency in 71% and 68% of RYGBP patients pre- and postoperatively. The percentage of zinc intake 2 months after RYGBP was 31% of the adequacy of zinc ingestion, which showed that with reduced food intake it was not possible to guarantee adequate intake of this nutrient. Neve et al. [36], in spite of not having determined serum levels of zinc, observed a prevalence of ~36% alopecia 6 months after vertical banded gastroplasty. Thus, zinc deficiency may be one more of the many contributing factors to VAD in bariatric patients.

A possible relationship also exists between VAD and anemia. Vitamin A participates in iron metabolism at several stages, including iron intestinal absorption, transport in serum, release of iron in liver stores, mobilization of iron to bone marrow, and hemoglobin synthesis. The availability of iron for hematopoietic tissue synthesis is inhibited during VAD. Conversely, iron influences the bioavailability of vitamin A. The function of the intestinal mucosa is compromised by iron deficiency, which can impair gut vitamin A absorption [37].

Jang et al. [37] demonstrated that iron deficiency caused a reduction in plasma retinol levels due to lower mobilization of vitamin A by the liver, besides damaging the absorption and utilization of this vitamin. The lower mobilization of vitamin A by the liver may lead to a reduction in the plasma pool of retinol. Likewise, plasma levels of RBP are low in iron deficiency.

The presence of iron deficiency has been demonstrated in 20–49% of patients who have undergone RYGBP, with a higher incidence in females. Patients submitted to RYGBP are particularly vulnerable to iron malabsorption because the duodenum and proximal jejunum, which are the largest sites of absorption of this nutrient, are excluded from the normal digestive traffic. Also, iron absorption is severely limited by achlorhydria, which results from the decrease in stomach size [38].

Besides the significant micronutrient deficiency observed prior to surgery [39], reduced intake of several micronutrients has been described despite the use of vitamin–mineral supplements [40,41]. This reinforces the fact that patients need continuous follow-up and individualized prescription of supplementation after surgery to prevent and treat vitamin deficiencies.

Malabsorption and Reduced Food Sources of Vitamin A After Roux-en-Y Gastric Bypass

In RYGBP, exclusion of the duodenum and of the first 100 cm of jejunum from food traffic decreases the area of the sites of vitamin A absorption. The preformed vitamin A absorption occurs in the intestinal lumen where retinol esters from the diet (mainly retinyl palmitate) are emulsified with biliary salts and hydrolyzed to retinol by several pancreatic enzymes and retinyl ester hydrolases; this occurs prior to its absorption [42].

Furthermore, the reduction in alimentary intake with frequent lipid restriction in the postoperative stage, including a decrease in consumption of food sources of vitamin A, can also be a contributing factor to VAD in this group.

The stomach remaining after RYGBP has a volume of 30–50 mL. Thus, restricted liquid intake (evolving to a complete liquid diet) in the immediate postoperative period is the preferred diet until discharge from the hospital. Intake in the first 15 days consists of a liquid diet of small volumes (~50 mL at each meal). Its main objective is gastric rest, adaptation to small volumes, and hydration. The following stage involves the introduction of pasty foods to avoid gastric distension, ultimately progressing to solid alimentation 12 weeks after surgery [43].

TABLE 52.2 Main Sources of Vitamin A (Retinol and Carotenoids)

Form of Vitamin A	Food Sources of Vitamin A
Retinol (animal sources)	Whole milk Butter Eggs Tuna fish Salmon Eel Cheese Liver (chicken, lamb, pork)
Carotenoids (plant sources)	Sweet potatoes Carrots Dark leafy greens Squash (butternut, cooked) Cos or romaine lettuce Apricots Cantaloupe melon Sweet red peppers Mango Broccoli Tomatoes

TABLE 52.3 Daily Vitamin A and Other Micronutrients Related to Vitamin A Metabolism Needs

Age in Years	Vitamin A (mcg/day)[a]	Zinc (mg/day)	Iron (mg/day)
Men (>19)	900	11	8
Women (19–50)	700	8	18
Women (>50)	700	8	8

[a]As retinol activity equivalents (RAEs). Institute of Medicine. Food and Nutrition Board. Dietary Reference Intakes. Washington, DC: National Academic Press, 1999–2001.

Not until the third month post-RYGBP does the gradual evolution to a consistency close to the ideal for satisfactory nutrition occur; this is when almost all foods are introduced into the diet. In this phase, foods with high-fiber consistency are not tolerated. Generally, a diet with a normal consistency that contains all foods occurs around the fourth month after surgery [43].

Meats and meat products, leafy vegetables, bread, and other foods may be rejected because they require prolonged mastication. Milk and its derivatives are also frequently rejected. Although easy to swallow, they are not well-tolerated by these patients [44]. This can contribute to a low intake of vitamin A because preformed vitamin A (or retinol) is found in animal sources like liver, milk and its derivatives, and in a lower proportion, yolk. On the other hand, sources of provitamin A (i.e., carotenoids) are dark−green and yellow−orange vegetables [42]. Sources of vitamin A, and daily needs of vitamin A and related micronutrients, are summarized in Tables 52.2 and 52.3.

Also, the lower fat absorption after RYGBP makes restriction of fats in the diet necessary, which may also compromise vitamin A intake and absorption. Lipid intake, in parallel with vitamin A intake, is of extreme importance for this absorption [45].

MANAGING VITAMIN A DEFICIENCY AFTER ROUX-ᴇɴ-Y GASTRIC BYPASS

It has been reported that routine supplementation with 5000 IU of retinol acetate is not enough to reverse VAD statuses over 6 and 24 months after surgery [15,46]. Pereira et al. [47], comparing patients who received a daily oral dose of

5000 IU or 10,000 IU, or a monthly intramuscular dose of 50,000 IU, of retinol acetate, suggested that a daily dose of 10,000 IU should be considered for some patients. They emphasized the importance of preoperative screening for VAD. Clearly, a stronger supplementation may be required in order to prevent visual and other consequences of VAD after an RYGBP procedure, but no definitive data is currently available on optimal supplementation.

Currently, the American Society for Metabolic and Bariatric Surgery guidelines do not recommend an additional prophylactic vitamin A post-RYGBP [48]. An updated guideline recommends that routine screening for VAD (which may present as ocular complications) be performed after purely malabsorptive bariatric procedures such as biliopancreatic diversion with duodenal switch—supplementation alone or in combination with other fat-soluble vitamins may be indicated in this situation. The guideline also mentions that the minimum daily nutritional supplementation for patients with RYGBP (all in chewable form initially, i.e., 3−6 months) should include two adult multivitamins plus minerals; there is no mention of specific recommendations for long-term vitamin A supplementation [49]. Therefore, studies are needed to evaluate the most effective dose.

MINI-DICTIONARY OF TERMS

- *Vitamin A*: A fat-soluble vitamin necessary for vision, reproduction, growth, immune function, epithelial tissue differentiation, skeletal tissue maintenance, spermatogenesis, placenta generation, and maintenance.
- *Beta-carotene*: The most important carotenoid precursor of retinol. Effectively participates in antioxidant defense, promoting the elimination of the singlet oxygen involved in the oxidative attack on nucleic acids, amino acids, and polyunsaturated fatty acids.
- *Vitamin A deficiency*: One of the most prevalent public health issues in the world, resulting in damage to the health of individuals affected by it, even causing death. VAD leads to a variety of ocular manifestations, including corneal and conjunctival xerosis, keratinization of the conjunctiva, keratomalacia, and potentially corneal perforation, retinopathy, visual loss, and nyctalopia.
- *Night blindness*: Impaired dark adaptation. A subclinical symptom of VAD that occurs before the decrease in serum retinol levels.
- *Roux-en-Y gastric bypass*: Currently the preferred bariatric operation, it involves two surgical alterations: restriction of gastric volume and diversion of ingested nutrients away from the proximal small intestine.
- *Oxidative stress*: Adverse effects of oxidative reactions induced by free radicals inside biological systems. It causes lesions in relevant molecules such as deoxyribonucleic acid, proteins, carbohydrates, and lipids.
- *Morbid obesity or class III obesity*: Individuals classified according to BMI ≥ 40.0 kg/m^2.
- *Supplementation routine*: Daily supplementation of micronutrients before and/or after RYGBP surgery to prevent clinical and functional alterations related to micronutrient deficiency.

KEY FACTS

- A multidisciplinary approach is important to guarantee adequate management of vitamin A status before and after surgery.
- While there is no consensus on the need to change the route of administration and vitamin dose after surgery, monitoring of subclinical signs of VAD and periodic assessment of serum retinol concentrations are recommended.

SUMMARY POINTS

- Morbid obesity per se appears to be a risk factor for VAD, and can be associated with chronic inflammation and OS.
- VAD after RYGBP surgery can be exacerbated by the deficiency of other micronutrients and malabsorption due to exclusion of the duodenum and the first 100 cm of jejunum.
- Lower fat absorption after RYGBP makes restriction of fats in the diet necessary, which may compromise vitamin A intake and absorption.
- The current supplementation routine provided to patients who have undergone RYGBP seems to be insufficient for meeting the high vitamin A requirements.
- It is not known whether changing the means of administration or dose size for vitamin A supplementation will improve vitamin A levels after surgery.

REFERENCES

[1] Ng M, Fleming T, Robinson M, et al. Global, regional, and national prevalence of overweight and obesity in children and adults during 1980–2013: a systematic analysis for the Global Burden of Disease Study 2013. Lancet 2014;384(9945):766–81.

[2] Keating C, Backholer K, Gearon E, Stevenson C, Swinburn B, Moodie M, et al. Prevalence of class-I, class-II and class-III obesity in Australian adults between 1995 and 2011-12. Obes Res Clin Pract 2015;9(6):553–62.

[3] Borrell LN, Samuel L. Body mass index categories and mortality risk in US adults: the effect of overweight and obesity on advancing death. Am J Public Health 2014;104(3):512–19.

[4] Schroeder R, Harrison TD, McGraw SL. Treatment of adult obesity with bariatric surgery. Am Fam Physician 2016;93(1):31–7.

[5] Dogan K, Aarts EO, Koehestanie P, Betzel B, Ploeger N, de Boer H, et al. Optimization of vitamin suppletion after Roux-en-Y gastric bypass surgery can lower postoperative deficiencies: a randomized controlled trial. Medicine (Baltimore) 2014;93(25):e169.

[6] Akhtar S, Ahmed A, Randhawa MA, Atukorala S, Arlappa N, Ismail T, et al. Prevalence of vitamin A deficiency in South Asia: causes, outcomes, and possible remedies. J Health Popul Nutr 2013;31(4):413–23.

[7] D'Ambrosio DN, Clugston RD, Blaner WS. Vitamin A metabolism: an update. Nutrients 2011;3:63–103.

[8] Iqbal S, Naseem I. Role of vitamin A in type 2 diabetes mellitus biology: effects of intervention therapy in a deficient state. Nutrition 2015;31(7–8):901–7.

[9] Knop FK, Taylor R. Mechanism of metabolic advantages after bariatric surgery: it's all gastrointestinal factors versus it's all food restriction. Diabetes Care 2013;36(suppl. 2):S287–91.

[10] Bavaresco M, Paganini S, Lima TP, Salgado Jr. W, Ceneviva R, Dos Santos JE, et al. Nutritional course of patients submitted to bariatric surgery. Obes Surg 2010;20(6):716–21.

[11] Dunstan MJ, Molena EJ, Ratnasingham K, Kamocka A, Smith NC, Humadi S, et al. Variations in oral vitamin and mineral supplementation following bariatric gastric bypass surgery: a national survey. Obes Surg 2015;25(4):648–55.

[12] Moizé V, Andreu A, Flores L, Torres F, Ibarzabal A, Delgado S, et al. Long-term dietary intake and nutritional deficiencies following sleeve gastrectomy or Roux-En-Y gastric bypass in a Mediterranean population. J Acad Nutr Diet 2013;113(3):400–10.

[13] Rosa CLR, Saubermann AP, Silva JS, Pereira SE, Saboya CJ, Ramalho A. Routine supplementation does not warrant the nutritional status of vitamin D adequate after gastric bypass Roux-en-Y. Nutr Hosp 2013;28(1):169–72.

[14] Aarts EO, van Wageningen B, Janssen IM, Berends FJ. Prevalence of anemia and related deficiencies in the first year following laparoscopic gastric bypass for morbid obesity. J Obes 2012;2012:193705.

[15] Pereira S, Saboya C, Chaves G, Ramalho A. Class III obesity and its relationship with the nutritional status of vitamin A in pre- and postoperative gastric bypass. Obes Surg 2009;19:738–44.

[16] Chagas CB, Saunders C, Pereira S, Silva SJ, Saboya C, Ramalho A. Vitamin A deficiency in pregnancy: perspectives after bariatric surgery. Obes Surg 2013;23(2):249–54.

[17] Ferrari R, Guardigli G, Mele D, Percoco GF, Ceconi C, Curello S. Oxidative stress during myocardial ischaemia and heart failure. Curr Pharm Des 2004;10:1699.

[18] Tateya S, Kim F, Tamori Y. Recent advances in obesity-induced inflammation and insulin resistance. Front Endocrinol 2013;4:93.

[19] Matsuda M, Shimomura I. Increased oxidative stress in obesity: implications for metabolic syndrome, diabetes, hypertension, dyslipidemia, atherosclerosis, and cancer. Obes Res Clin Pract 2013;7:330–41.

[20] Orzechowski A, Jank M, Gajkowska B, Sadkowski T, Godlewski MM, Ostaszewski P. Delineation of signalling pathway leading to antioxidant-dependent inhibition of dexamethasone-mediated muscle cell death. J Muscle Res Cell Motil 2003;24:33–53.

[21] Paolisso G, D'amore A, Volpe C, Balbi V, Saccomanno F, Galzeran D, et al. Evidence for a relationship between oxidative stress and insulin action in non-insulin-dependent (type II) diabetic patients. Metabolism 1994;43:1426–9.

[22] Chaves GV, Pereira SE, Saboya CJ, Ramalho A. Non-alcoholic fatty liver disease and its relationship with the nutritional status of vitamin A in individuals with class III obesity. Obes Surg 2008;18:378–85.

[23] Sigiura M, Nakamura M, Ikoma Y, Yano M, Ogawa K, Matsumoto H, et al. The homeostasis model assessment-insulin resistance index is inversely associated with serum carotenoids in non-diabetic subjects. J Epidemiol 2006;16:71–8.

[24] Toikka JO, Ahotupa M, Viikari JS, Niinikoski H, Taskinen M, Irjala K, et al. Constantly low HDL-cholesterol concentration relates to endothelial dysfunction and increased in vivo LDL oxidation in healthy young men. Atherosclerosis 1999;147:133–8.

[25] Serban MG, Negru T. Lipoproteins, lipidic peroxidation and total antioxidant capacity in serum of aged subjects suffering from hyperglicemia. Rom J Intern Med 1998;36:65–70.

[26] Gatica LV, Vega VA, Zirulnik F, Oliveros LB, Gimenez MS. Alterations in the lipid metabolism of rat aorta: effects of vitamin A deficiency. J Vasc Res 2006;43:602–10.

[27] Oliveros LB, Domeniconi MA, Vega VA, Gatica LV, Brigada AM, Gimenez MS. Vitamin A deficiency modifies lipid metabolism in rat liver. Br J Nutr 2007;97:263–72.

[28] Relevy NZ, Harats D, Harari A, Ben-Amotz A, Bitzur R, Ruhl R, et al. Vitamin A-deficient diet accelerated atherogenesis in apolipoprotein E (-/-) mice and dietary β-carotene prevents this consequence. Biomed Res Int 2015;75–87.

[29] Viroonudomphol D, Pongpaew P, Tungtrongchitr R, Changbumrung S, Tungtrongchitz A, Phonrat B, et al. The relationships between anthropometric measurements, serum vitamin A and E concentrations and lipid profiles in overweight and obese subjects. Asia Pac J Clin Nutr 2003;12:73–9.

[30] Ramezanipour M, Jalali M, Sadrzade-Yeganeh H, Keshavarz SA, Eshraghian MR, Bagheri M, et al. The effect of weight reduction on antioxidant enzymes and their association with dietary intake of vitamins A, C and E. Arq Bras Endocrinol Metabol 2014;58:744—9.

[31] Palace VP, Khaper N, Qin Q, Singal PK. Antioxidant potentials of vitamin A and carotenoids and their relevance to heart disease. Free Radic Biol Med 1999;26:746—61.

[32] Crino M, Sacks G, Vandevijvere S, Swwinburn B, Neal B. The influence on population weight gain and obesity of the macronutrient composition and energy density of the food supply. Curr Obes Rep 2015;4:1—10.

[33] Christian P, West Jr. KP. Interactions between zinc and vitamin A: an update. Am J Clin Nutr 1998;68:435S—41S.

[34] Smith JC, Brown ED, McDaniel EG, Chan W. Alterations in vitamin A metabolism during zinc deficiency and food and growth restriction. J Nutr 1976;106:569—74.

[35] Cominetti C, Garrido JR, Cozzolino SMF. Zinc nutritional status of morbidly obese patients before and after Roux-en-Y gastric bypass: a preliminary report. Obes Surg 2005;16:448—53.

[36] Neve HJ, Bhatti WA, Soulsby C, Kincey J, Taylor TV. Reversal of hair loss following vertical gastroplasty when treated with zinc sulphate. Obes Surg 1996;6:63—5.

[37] Jang JT, Green JB, Beard JL, Green MH. Kinetic analysis shows that iron deficiency decreases liver vitamin A mobilization in rats. J Nutr 2000;130:1291—6.

[38] Bloomberg RD, Fleishman A, Nalle JE, Herron DM, Kini S. Nutritional deficiencies following bariatric surgery: what have we learned? Obes Surg 2005;15:145—54.

[39] Peterson LA, Cheskin LJ, Furtado M, Papas K, Scweitzer MA, Magnuson TH, et al. Malnutrition in bariatric surgery candidates: multiple micronutrient deficiencies prior to surgery. Obes Surg 2015;22:1—6.

[40] Mercachita T, Santos Z, Limão J, Carolino E, Mendes L. Anthropometric evaluation and micronutrients intake in patients submitted to laparoscopic Roux-en-Y gastric bypass with a postoperative period of ≥ 1 year. Obes Surg 2014;24:102—8.

[41] Donadelli SP, Junqueira-Franco MV, de Mattos Donadelli CA, Sangalo Jr W, Ceneviva R, Marchini JS, et al. Daily vitamin supplementation and hypovitaminosis after obesity surgery. Nutrition 2012;28:391—6.

[42] Blomhoff R. Transport and metabolism of vitamin A. Nutr Rev 1994;52:13—23.

[43] Nelson DK, Pieramico O, Malfertheiner P. Gastrointestinal motility and hormones in obesity. Am J Gastroenterol 1994;89:1120—1.

[44] Pereira S, Saboya C, Ramalho A. Impact of different protocols of nutritional supplements on the status of vitamin A in class III obese patients after Roux-en-Y gastric bypass. Obes Surg 2013;23:1244—51.

[45] Van het Hof KH, West CE, Weststrate JA, Hautvast JG. Dietary factors that affect the bioavailability of carotenoids. J Nutr 2000;130:503—6.

[46] Brolin RE, Leung M. Survey of vitamin and mineral supplementation after gastric bypass and biliopancreatic diversion for morbid obesity. Obes Surg 1999;9:150—4.

[47] Pereira S, Saboya C, Ramalho A. Impact of different protocols of nutritional supplements on the status of vitamin A in class III obese patients after Roux-en-Y gastric bypass. Obes Surg 2013;23:1244—51.

[48] Aills L, Blankenship J, Buffington C. ASMBS Allied Health Nutritional Guidelines for the surgical weight loss patient. Surg Obes Relat Dis 2008;4:S73—108.

[49] Mechanick J, Youdim A, Jones DB, Garvey WT, Hurley DL, McMahon MM, et al. Clinical Practice Guidelines for the perioperative nutritional, metabolic, and nonsurgical support of the bariatric surgery patient - 2013 update: cosponsored by American Association of Clinical Endocrinologists, the Obesity Society, and American Society for Metabolic & Bariatric Surgery. Endocr Pract 2013;19:337—72.

Chapter 53

Iron and Bariatric Surgery

I. Gesquiere, C. Matthys and B. Van der Schueren

Clinical and Experimental Endocrinology, University Hospitals Leuven/KU Leuven, Leuven, Belgium

LIST OF ABBREVIATIONS

AUC area under the curve
BPD biliopancreatic diversion
CRP C-reactive protein
DcytB duodenal cytochrome B
DMT1 divalent metal transporter 1
EAR estimated average requirements
FPN ferroportin
HCP-1 heme carrier protein-1
HEPH hephaestin
HO-1 heme oxygenase-1
MCV mean corpuscular volume
RDA recommended dietary allowances
RYGB Roux-en-Y gastric bypass
SG sleeve gastrectomy
sTfR soluble transferrin receptor
Tf transferrin
TSAT transferrin saturation

INTRODUCTION

Iron is included in a lot of mechanisms in the body, such as oxygen transport, mitochondrial function, muscle activity, and DNA synthesis and repair [1]. The human body contains approximately 3.5−5 g of iron, mostly in hemoglobin [2]. Daily, we lose 1−2 mg of iron by means of epithelial cells of the gastrointestinal tract, urine, sweat, and feces. To compensate for this loss, iron intake from food is necessary [3]. If iron intake is insufficient to meet iron needs or to compensate for iron loss, an iron deficiency will develop—e.g., iron need increases in women during pregnancy and lactation [4].

In food, iron is available as heme iron (in animal products) and especially as nonheme iron (in plant and animal products). The bioavailability of heme iron is higher and more uniform compared to nonheme iron, but heme iron constitutes the minor fraction of dietary iron [3,5]. Both forms of dietary iron are absorbed through different mechanisms, as shown in Fig. 53.1.

The pathway for heme iron has not been fully elucidated, but the heme carrier protein-1 (HCP-1) has been proposed as the responsible transporter for the uptake of heme iron from the intestinal lumen into the cytoplasm of the enterocytes [6]. Once absorbed in the enterocyte, iron is extracted from the protoporphyrin ring by heme oxygenase-1 (HO-1) [7]. The absorption of dietary nonheme iron is regulated by a separate process and depends on the iron status and on the presence of absorption inhibitors (i.e., phytate, polyphenol, oxalate, and calcium) or enhancers (i.e., ascorbic acid and dietary proteins) present in the diet [8]. Furthermore, an excess zinc intake can compete with iron absorption, and copper is needed for iron transport by transferrin through the circulation [9]. The ingested iron is mainly absorbed at

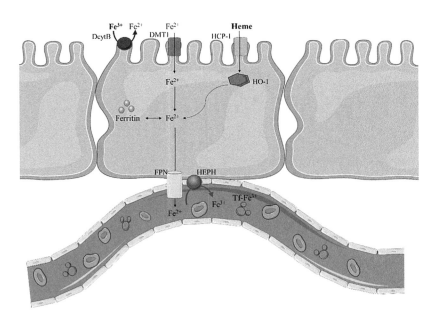

FIGURE 53.1   Absorption pathways of heme and nonheme iron. Heme iron is absorbed by heme carrier protein-1 (HCP-1) and in the enterocyte, iron is extracted from the protoporphyrin ring by heme oxygenase-1 (HO-1). For the absorption of nonheme iron by the divalent metal transporter 1 (DMT1), Fe^{3+} must be reduced to Fe^{2+} by gastric acid and duodenal cytochrome B (DcytB). In the enterocyte, iron can be stored inside ferritin or it can be transferred across the basolateral membrane of the enterocyte into the blood circulation by ferroportin (FPN). Hephaestin (HEPH) converts Fe^{2+} into Fe^{3+} so that binding to transferrin for transportation is possible.

the level of the duodenum and proximal jejunum. Furthermore, to absorb nonheme iron, the ferric form (Fe^{3+}) must be reduced to the ferrous form (Fe^{2+}), which can be achieved by gastric acid and by duodenal cytochrome B (DcytB), a reductase enzyme. Fe^{2+} can be absorbed by the divalent metal transporter 1 (DMT1) in the brush border membrane of duodenal enterocytes [2]. After absorption of iron in the enterocyte, it can be stored inside the cytoplasmic protein ferritin or can be transferred across the basolateral membrane of the enterocyte into the blood circulation. The transport protein ferroportin (FPN) is responsible for the transfer of Fe^{2+} from the enterocyte into the blood circulation. Additionally, hephaestin converts Fe^{2+} into Fe^{3+}, which is crucial for the association of iron with the most important iron transport protein, transferrin (Tf) in the blood [1]. Transferrin is distributed in the body and delivered to the iron-dependent tissues. More than two-thirds of total body iron is incorporated in hemoglobin in developing erythroid precursors and mature circulating erythrocytes. Most of the remaining part is found in the iron storage protein ferritin in the hepatocytes and the reticuloendothelial macrophages, which both serve as major iron storage pools, and in the oxygen storage protein myoglobin in the muscles (see Fig. 53.2) [10].

Iron absorption is strictly regulated, as iron deficiency and overload are harmful. Iron absorption and transport are stimulated by low iron stores in the body, erythropoiesis, hypoxia, and pregnancy [3]. In contrast, when body iron stores are high or when there is inflammation, iron absorption is reduced to protect against iron overload. Hepcidin has a central role in iron homeostasis. It is a negative regulator of iron metabolism, and functions by inhibiting intestinal iron absorption, iron recycling by macrophages, and iron mobilization from hepatic stores by internalization and degradation of ferroportin, as shown in Fig. 53.3 [11]. However, when iron homeostasis is deregulated, iron deficiency can develop. It is a common problem worldwide, and obese patients before and after bariatric surgery have an increased risk for the development of iron deficiency.

IRON DEFICIENCY IN BARIATRIC SURGERY CANDIDATES

Iron deficiency is extremely common in bariatric surgery candidates; an impaired iron status is observed in about 20% of the candidates for bariatric surgery [12]. This can be explained by low iron intake due to the Western affluent diet, increased iron requirements due to higher blood volume, and the obesity-associated, chronic, low-grade, inflammation-inducing hepcidin production and subsequent inhibition of iron absorption and release that lead to iron deficiency, as shown in Fig. 53.3 [13,14]. This chronic inflammation is caused by the secretion of proinflammatory adipokines; IL-6 is the most important for iron metabolism because it stimulates hepcidin synthesis [15].

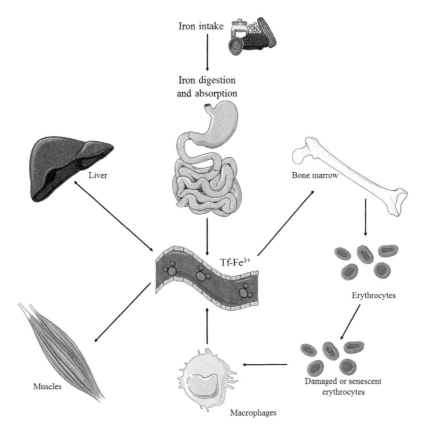

FIGURE 53.2 Iron absorption, transport, recycling, and storage. After intake, iron needs to be digested and absorbed before it can be distributed through the body. Transferrin will distribute iron to the iron-dependent tissues, especially to the bone marrow for incorporation in hemoglobin in erythrocytes. Macrophages phagocytize and degrade damaged and senescent erythrocytes to recover iron. Most of the remaining part of iron is found in the iron storage protein ferritin in the hepatocytes and the reticuloendothelial macrophages, and to a lesser extent in the oxygen storage protein myoglobin in the muscles.

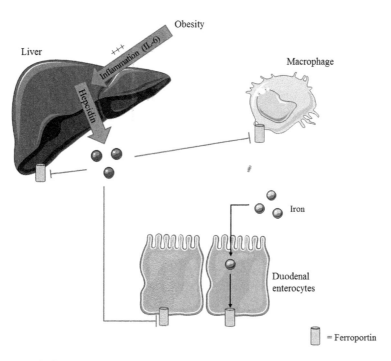

FIGURE 53.3 Hepcidin has a central role in iron homeostasis. Iron homeostasis is strictly regulated by hepcidin, a negative regulator of iron metabolism as it inhibits iron absorption, recycling by macrophages, and mobilization from hepatic stores by internalization and degradation of ferroportin. Obesity is associated with a chronic low-grade inflammation resulting in the secretion of IL-6, which stimulates the synthesis of hepcidin.

Thus, patients are at risk for nutritional deficiencies even before bariatric surgery. The evidence therefore supports extensive preoperative nutritional screening to allow for treatment of deficiencies presurgery [16]. This is important because preoperative micronutrient deficiencies persist postoperatively, and many patients end up developing new deficiencies, including iron deficiency [17].

IMPACT OF BARIATRIC SURGERY ON IRON INTAKE

Patients who have undergone bariatric surgery consume less food postoperatively because of increased satiety and reduced hunger, which result in reduced intake of macro- and micronutrients. Furthermore, food preferences can change, and some patients avoid certain foods because of the development of food intolerances. Intolerance for red meat, a major source of highly bioavailable heme iron, is one example. This may further contribute to the reduced intake of some micronutrients postoperatively [18,19]. Also, insufficient supplementation due to nonadherence to the prescribed diet, or inadequate amounts of iron or inadequate nutritional support, contribute to the development of iron deficiency after bariatric surgery. Table 53.1 shows an overview of studies about the impact of bariatric surgery on iron intake.

Because of reduced iron intake, it is important that diet is optimized to a healthy pattern after bariatric surgery and that the intake of nutrient-dense food is encouraged. Overall, it is important that all the nutrition advice is given to patients in a convenient way, taking into account personal and cultural preferences to increase adherence to the proposed diet.

IMPACT OF BARIATRIC SURGERY ON IRON ABSORPTION

The anatomical and physiological changes associated with bariatric surgery may alter iron absorption.

The reduced gastric acid secretion, which is required for the absorption of iron as it transforms the ferric form (Fe^{3+}) to the ferrous form (Fe^{2+}), has an impact on absorption. Furthermore, after undergoing procedures that bypass parts of the intestinal tract, the intestinal absorption surface is diminished; this is especially true when the duodenum and proximal jejunum are bypassed as these are the main absorption sites for iron. Several studies have shown the impact of bariatric surgery on iron absorption [28–32].

Ruz et al. [28] performed iron absorption tests in RYGB patients with ferrous ascorbate using radioactively labeled iron, and showed in women that the absorption from a standard diet and a standard dose of ferrous ascorbate was significantly reduced 6 months post-RYGB; the absorption values post-RYGB were approximately 30% of the observed iron absorption before surgery. No further significant changes of the absorption were observed 12 and 18 months post-RYGB. Furthermore, in another study performed by Ruz et al. ($n = 32 ♀$) [29], the absorption of both heme- and nonheme iron was reduced 1 year after RYGB. The extent of the reduced absorption was greater for heme iron than for nonheme iron [29]. To date, this is the only iron absorption study that also included patients who had undergone sleeve gastrectomy (SG); it was shown that both heme and nonheme iron absorption were markedly reduced after SG [29].

Rosa et al. [30] performed iron tolerance tests ($n = 9 ♀$) with iron sulfate (15 mg elemental iron) before and 3 months post-RYGB, and demonstrated delayed iron absorption post-RYGB. During 4 hours after administration, no statistically significant differences in total iron plasma response were shown, only a delayed response, even though six of the nine patients presented a mean decrease in area under the curve (AUC) of 51%. The first hour after administration, the plasma concentration of iron was significantly lower after surgery compared to the situation before surgery [30]. We also performed oral iron absorption tests with 100 mg of iron sulfate in patients post-RYGB, and only 1 out of 23 patients showed sufficient iron absorption [31].

Previously, Rhode et al. [32] performed iron absorption tests with iron gluconate in patients after gastric bypass. Normal absorption, which was defined as more than 100% change in serum iron concentration over 3 hours after administration, was observed in 36 of the 55 patients. The patients with normal absorption had a higher incidence of anemia and had lower serum ferritin concentrations compared to those with low absorption.

We need to take into account that all iron absorption studies had other designs and that the formulations of the supplements were also different. This can have an impact on the results of the study and makes it difficult to compare the effectiveness of the different iron supplements. Moreover, changes in gut microbiota and intestinal adaptation can occur postoperatively, and can subsequently play a role in nutrient absorption. Thus, further research regarding nutrient absorption is needed.

TABLE 53.1 Overview of Studies on Influence of Bariatric Surgery on Iron Intake

Reference	Methodology	Number of Subjects	Time Points/Surgery Studied	Effect on Intake
Colossi et al. [20]	24-hour food recall	210 patients (147♀, 63♂)	1, 3, 6, 9, 12, 18, and 24 months post-RYGB	The intake of micronutrients increased with time after surgery. The minimal requirements for iron were not attained at the different time points post-RYGB.
de Torres Rossi et al. [21]	4-day food record	44 patients post-RYGB (♀); 38 age and economic condition–matched controls (♀)	At least 1 year post-RYGB	Iron intake was significantly lower post-RYGB.
Gesquiere et al. [22]	2-day food record	54 patients (33♀, 21♂)	Before RYGB 1, 3, 6 and 12 months post-RYGB	The usual dietary intake of iron enormously decreased one month post-RYGB, but gradually increased until one year post-RYGB. However, one year post-RYGB still 14.3% had a dietary iron intake below the corresponding age- and gender-specific EAR; this percentage was 7.1 when also including the iron intake from nutritional supplements.
McGrice et al. [23]	Food frequency questionnaire	52 patients (38♀, 14♂)	1 year post-LAGB	In premenopausal women, iron intake from food and fluids was insufficient post-LAGB compared to the Recommended Dietary Allowances (RDA).
Mercachita et al. [24]	24-hour food recall	60 patients (39♀, 21♂)	Before RYGB 1 year post-RYGB 2 years post-RYGB	Significant reduction of iron intake 1 year post-RYGB in comparison with intake before RYGB; intake increased again 2 years post-RYGB (NS). The percentage of patients not meeting the RDA increased after RYGB; 1 and 2 years after RYGB, more than 65% of the patients had an intake below RDA for iron.
Miller et al. [25]	4-day food record	17 patients (16♀, 1♂)	Before RYGB 3 weeks post-RYGB 3, 6, and 12 months post-RYGB	Intake of micronutrients was the highest before RYGB, the lowest 3 weeks post-RYGB, and increased gradually from 3 weeks to 12 months post-RYGB, some returning to the values before RYGB. The percentage of patients not meeting the Estimated Average Requirements (EAR) increased after RYGB.
Moizé et al. [26]	3-day food record	294 patients (226♀, 68♂)	Before RYGB 3, 6, 12, 18, 24, 30, 36, 48, and 60 months post-RYGB	The mean dietary intake of the studied micronutrients was below RDA.
Wardé-Kamar et al. [27]	24-hour food recall	69 patients (61♀, 8♂)	At least 18 months post-RYGB	Dietary intake of iron was at or above RDA.

The methodology and results of these studies are shown.

IRON DEFICIENCY AFTER BARIATRIC SURGERY

Because of the impact of bariatric surgery on iron intake and absorption (see Fig. 53.4), iron deficiency is one of the most common nutrient deficiencies postsurgery. The prevalence of iron deficiency after RYGB ranges from 20% to 49%

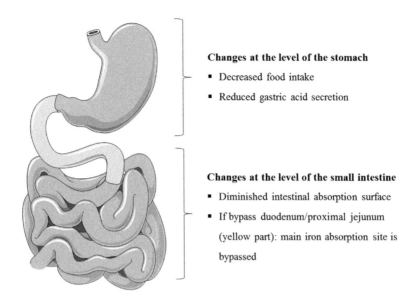

Changes at the level of the stomach

- Decreased food intake
- Reduced gastric acid secretion

Changes at the level of the small intestine

- Diminished intestinal absorption surface
- If bypass duodenum/proximal jejunum (yellow part): main iron absorption site is bypassed

FIGURE 53.4   Impact of bariatric surgery on iron intake and absorption. Bariatric surgery can be associated with changes at the level of the stomach and/or the small intestine and subsequently can influence iron intake and/or absorption.

[33]. Nevertheless, iron deficiency in bariatric surgery patients is often underdiagnosed as the assessment of the iron status is difficult and iron deficiency can occur without symptoms. When it is symptomatic, clinical symptoms will arise depending on the severity of the deficiency and the rapidity of anemia development. These patients typically complain about fatigue, weakness, lack of energy, and pica [34]. When iron deficiency is untreated, its severity progresses and can lead to anemia, specifically microcytic, hypochromic anemia. However, anemia can also be caused by other nutrient deficiencies, especially those of vitamin B12 and folic acid, but also vitamin B6, copper, zinc, and selenium [16,35].

From a theoretical point of view, SG is believed to cause less malabsorption of micronutrients, and subsequently leads to fewer deficiencies postoperatively. However, the same frequencies of iron deficiency are observed after RYGB and SG [26,36,37]. Patients who have undergone SG develop iron deficiency by low nutrient-dense food choices, food intolerances, poor eating behavior, reduced portion size, and decreased gastric acid secretion [38].

We need to consider that there are risk factors for the development of iron deficiency other than just the type of bariatric surgery. Avinoah et al. showed that in patients 7 years post-RYGB who ate red meat less than once a week, the incidence of iron deficiency was significantly higher compared to patients who ate red meat more than once a week [39]. Additionally, females have a higher risk for the development of iron deficiency compared to males as a result of menstrual blood loss. Furthermore young age, vitamin B12 deficiency, and a poor preoperative iron status predispose one to iron deficiency [31]. The risk for the development of iron deficiency after RYGB and biliopancreatic diversion (BPD) increases with time after surgery as body iron stores gradually decrease [31,40].

COMPREHENSIVE ASSESSMENT OF IRON STATUS IS REQUIRED TO DIAGNOSE IRON DEFICIENCY

There are three stages of iron deficiency that take it from iron depletion to iron deficiency anemia, as shown in Fig. 53.5 [4,41].

In all bariatric patients, iron status should be monitored [16]. An overall approach to diagnose iron deficiency in obese patients before and after bariatric surgery is important, as a proper assessment of iron status is difficult in this patient population. The diagnosis of iron deficiency is often only based on serum ferritin concentrations, but in obese patients the concentration of serum ferritin as an acute phase reactant might be increased despite an iron deficiency. This can be explained by the chronic, low-grade inflammation associated with obesity, which results in increased ferritin [14]. When ferritin concentrations are elevated, an adequate body iron store is apparent, but it is possible that there is insufficient incorporation of iron into erythroid precursors. This is defined as functional iron deficiency [42]. Including C-reactive protein (CRP) can help to indicate the presence of underlying inflammation. However, it is important to include more parameters than only ferritin and CRP when determining iron deficiency. Not doing so can lead to underdiagnosis and lack of treatment in this patient population, which was recently confirmed by Careaga et al. [43].

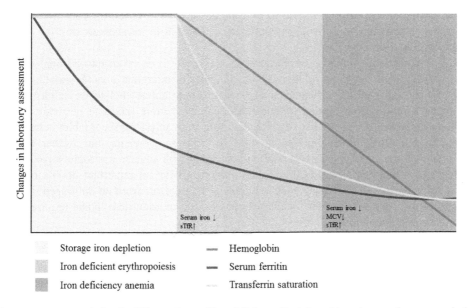

Storage iron depletion   — Hemoglobin

Iron deficient erythropoiesis   — Serum ferritin

Iron deficiency anemia   — Transferrin saturation

FIGURE 53.5 Laboratory assessment during the different stages of iron deficiency. Depletion of iron stores evolves progressively over time, leading to iron-deficient erythropoiesis; the final and most severe stage is iron deficiency anemia. *MCV*: mean corpuscular volume; *sTfR*: soluble transferrin receptor.

Furthermore, the determination of transferrin saturation (TSAT) is important in this patient population, especially in those individuals with elevated CRP levels. When TSAT is below 20%, these patients suffer from significant impairment of iron metabolism [43].

A marker that is not influenced by inflammation is soluble transferrin receptor (sTfR), but its use in clinical practice is limited because of cost, availability, and limited external quality assessment issues [42].

Furthermore, to assess whether patients have already developed iron deficiency anemia, the hemoglobin concentration is measured. However, this measurement has low specificity and low sensitivity. Therefore, the mean corpuscular volume (MCV) of the erythrocytes can be measured to differentiate between various causes of anemia.

Further, hepcidin, which has a central role in iron homeostasis, can help with an appropriate assessment of iron status. Until now it has been the only index that integrated sensitivity to iron levels, as well as inflammation and erythropoietic demand for iron, and subsequently gave direct information on the availability of iron for erythropoiesis [44–46]. However, the serum concentration of hepcidin is currently not determined in clinical practice. In a study in patients who had undergone restrictive bariatric surgery (gastric banding or SG), a significant decrease in hepcidin was shown 6 months postoperatively [47]. This decrease can be explained by the reduced inflammation associated with weight loss, which allows for enhanced iron absorption [47].

As mentioned before, iron absorption and metabolism might be affected by the status of other nutrients, which emphasizes the need to also check the status of other micronutrients.

PREVENTION AND TREATMENT OF IRON DEFICIENCY AFTER BARIATRIC SURGERY

The long-term use of multivitamin/mineral supplements should be recommended to all patients after bariatric surgery to prevent the development of nutritional deficiencies. In some cases, additional supplementation may be required [16,48]. Therefore, monitoring of iron status is necessary in all bariatric surgery patients, even in those taking preventive multivitamin/mineral supplements. Furthermore, preoperative screening and treatment of deficiencies is important, otherwise these deficiencies persist after bariatric surgery.

For patients with RYGB or SG, the American Association of Clinical Endocrinologists (AACE), The Obesity Society (TOS), and the American Society for Metabolic & Bariatric Surgery (ASMBS) suggest a minimum use of 45–60 mg of iron daily via multivitamins and additional supplements, initially (the first 3–6 months) in chewable form to prevent iron deficiency [16]. When patients develop iron deficiency anemia, a daily treatment regimen with oral ferrous sulfate, fumarate, or gluconate containing 150–200 mg of elemental iron is recommended. Further, the concomitant intake of vitamin C with iron should be considered as vitamin C can increase iron absorption [16]. If oral iron supplementation is insufficient to correct iron deficiency, a switch to intravenous iron (i.e., ferric gluconate, glucose, or carboxymaltose) is warranted [16].

Anatomical and physiological changes after bariatric surgery need to be taken into account when choosing an oral supplement, as bariatric surgery can result in altered bioavailability of micronutrients (as mentioned before). Therefore, changing the type of a particular micronutrient, such as the salt form or amino chelate, as well as the supplement formulation, can contribute to improved absorption. For instance, calcium citrate seems to be better absorbed after RYGB than calcium carbonate, and the bioavailability of calcium citrate from an effervescent tablet is higher than from a regular tablet formulation [49,50]. This might also be true for iron supplements, but further research is necessary. Furthermore, the intake of oral iron supplements is often associated with adverse gastrointestinal effects, which contribute to low adherence to a regimen. Thus, healthcare professionals have an important double role, first to encourage these patients, and secondly to stimulate adherence to supplement use apart from an optimized diet. Stimulating adherence to supplement use is necessary in order for the patient to meet micronutrient intake requirements and prevent the development of micronutrient deficiencies [16].

CONCLUSION

Nutritional guidance is essential following bariatric surgery, including stimulation of adjustments in eating behavior, encouraging the intake of nutrient-dense food, and supporting lifelong follow-up and adherence.

MINI-DICTIONARY OF TERMS

- *DMT1*: Divalent metal transporter 1. This transporter is located in the brush border membrane of duodenal enterocytes and is responsible for apical Fe^{2+} absorption.
- *DcytB*: A reductase enzyme that catalyzes the reduction of Fe^{3+} to Fe^{2+} in the duodenum.
- *Ferritin*: An iron storage protein, but also an acute phase reactant.
- *Ferroportin*: Ferroportin is responsible for the export of Fe^{2+} from enterocytes, hepatocytes, and macrophages into the blood circulation.
- *Heme iron*: Present in animal products like red meat, fish, and poultry. It is better absorbed than nonheme iron.
- *Hepcidin*: Hepcidin has a central role in iron homeostasis. It is a negative iron regulator by inhibiting iron absorption, recycling, and mobilization by internalization and degradation of ferroportin.
- *Hephaestin*: A protein, especially present in the small intestine, that converts Fe^{2+} into Fe^{3+} and mediates iron efflux.
- *Iron deficiency anemia*: The most severe stage of iron deficiency, characterized by microcytic, hypochromic erythrocytes.
- *Nonheme iron*: Present in plant foods and animal products. It represents the majority of iron derived from food, but it has a lower bioavailability than heme iron.
- *Transferrin*: The most important iron transporter. It transports iron through the blood to iron-dependent tissues.
- *Transferrin saturation (TSAT)*: A useful parameter to determine iron status, determined by the amount of iron that is bound to transferrin circulating in the blood.

KEY FACTS

- It is the most prevalent and widespread mineral deficiency disorder worldwide, and affects billions of people.
- It has a high prevalence in both developing and industrialized countries.
- Many causes of iron deficiency have been identified, including *Helicobacter pylori* infection or chronic treatment with proton pump inhibitors, antacids, and histamine (H_2)-receptor antagonists.
- A healthy diet with an adequate iron intake does not guarantee good levels of iron as a lot of dietary factors can inhibit iron absorption, i.e., polyphenols in tea, coffee, and wine, or phytates in grains and nuts.
- The first link between obesity and iron status was reported in the beginning of 1960 when lower serum iron concentrations were observed in obese adolescents compared to adolescents of normal weight.

SUMMARY POINTS

- This chapter focuses on iron deficiency, which is a common problem after bariatric surgery.
- Bariatric surgery candidates have an increased risk for iron deficiency due to the unbalanced Western affluent diet in combination with obesity-associated, chronic, low-grade inflammation.
- Preoperative screening and treatment of deficiencies is important.
- Multiple factors contribute to the development of iron deficiency after bariatric surgery: decreased food intake (including iron-rich foods), reduction of gastric acid secretion, and diminished intestinal absorption surface leading to reduced iron absorption.
- Lifelong monitoring of iron status in bariatric surgery patients is essential, and an overall approach to diagnose iron deficiency is necessary, as a proper assessment of iron status is difficult in this patient population.
- Long-term use of multivitamin/mineral supplements should be recommended to all patients after bariatric surgery to prevent the development of nutritional deficiencies.
- If an iron deficiency is developed, appropriate treatment is necessary.
- Nutritional guidance is essential following bariatric surgery, including stimulation of adjustments in eating behavior, encouragement of the intake of nutrient-dense food, and support of lifelong follow-up and adherence to multivitamin/mineral supplementation.

REFERENCES

[1] Evstatiev R, Gasche C. Iron sensing and signalling. Gut 2012;61(6):933–52.
[2] Stein J, Hartmann F, Dignass AU. Diagnosis and management of iron deficiency anemia in patients with IBD. Nat Rev Gastroenterol Hepatol 2010;7(11):599–610.
[3] Miret S, Simpson RJ, McKie AT. Physiology and molecular biology of dietary iron absorption. Annu Rev Nutr 2003;23:283–301.
[4] Clark SF. Iron deficiency anemia. Nutr Clin Pract 2008;23(2):128–41.
[5] Carpenter C, Mahoney A. Contributions of heme and nonheme iron to human nutrition. Crit Rev Food Sci Nutr 1992;31:333–67.
[6] Shayeghi M, Latunde-Dada GO, Oakhill JS, Laftah AH, Takeuchi K, Halliday N, et al. Identification of an intestinal heme transporter. Cell 2005;122(5):789–801.
[7] Andrews NC. Understanding heme transport. N Engl J Med 2005;353(23):2508–9.
[8] Hurrell R, Egli I. Iron bioavailability and dietary reference values. Am J Clin Nutr 2010;91(5):1461S–7S.
[9] Olivares M, Pizarro F, Ruz M, Lo D. Acute inhibition of iron bioavailability by zinc: studies in humans. Biometals 2012;657–64.
[10] Andrews NC. Disorders of iron metabolism. N Engl J Med 1999;341(26):1986–95.
[11] Nemeth E, Ganz T. Regulation of iron metabolism by hepcidin. Annu Rev Nutr 2006;26(1):323–42.
[12] Salgado W, Modotti C, Nonino CB, Ceneviva R. Anemia and iron deficiency before and after bariatric surgery. Surg Obes Relat Dis 2014;10(1):49–54.
[13] Moizé V, Deulofeu R, Torres F, de Osaba JM, Vidal J. Nutritional intake and prevalence of nutritional deficiencies prior to surgery in a Spanish morbidly obese population. Obes Surg 2011;21(9):1382–8.
[14] Zimmermann MB, Zeder C, Muthayya S, Winichagoon P, Chaouki N, Aeberli I, et al. Adiposity in women and children from transition countries predicts decreased iron absorption, iron deficiency and a reduced response to iron fortification. Int J Obes (Lond) 2008;32(7):1098–104.
[15] Nemeth E, Rivera S, Gabayan V, Keller C, Taudorf S, Pedersen BK, et al. IL-6 mediates hypoferremia of inflammation by inducing the synthesis of the iron regulatory hormone hepcidin. J Clin Investig 2004;113(9):1271–6.
[16] Mechanick JI, Youdim A, Jones DB, Timothy W, Hurley DL, Mcmahon MM, et al. Clinical practice guidelines for the perioperative the bariatric surgery patient—2013 update. Endocr Pract 2013;19(2):337–72.
[17] Toh SY, Zarshenas N, Jorgensen J. Prevalence of nutrient deficiencies in bariatric patients. Nutrition 2009;25(11–12):1150–6.
[18] Ortega J, Ortega-Evangelio G, Cassinello N, Sebastia V. What are obese patients able to eat after Roux-en-Y gastric bypass? Obes Facts 2012;5(3):339–48.
[19] Godoy CMDA, Caetano AL, Viana KRS, de Godoy EP, Barbosa ALC, Ferraz EM. Food tolerance in patients submitted to gastric bypass: the importance of using an integrated and interdisciplinary approach. Obes Surg 2012;22(1):124–30.
[20] Colossi FG, Casagrande DS, Chatkin R, Moretto M, Barhouch AS, Repetto G, et al. Need for multivitamin use in the postoperative period of gastric bypass. Obes Surg 2008;18:187–91.
[21] de Torres Rossi RG, Dos Santos MTA, de Souza FIS, de Cássia de Aquino R, Sarni ROS. Nutrient intake of women 3 years after Roux-en-Y Gastric bypass surgery. Obes Surg 2012;22(10):1548–53.
[22] Gesquiere I, Foulon V, Augustijns P, Gils A, Lannoo M, Van der Schueren B, et al. Micronutrient intake, from diet and supplements, and association with status markers in pre- and post-RYGB patients, Clin Nutr 2016. Available from: http://dx.doi.org/10.1016/j.clnu.2016.08.009.
[23] McGrice MA, Porter JA. The micronutrient intake profile of a multicentre cohort of Australian LAGB patients. Obes Surg 2014;24(3):400–4.
[24] Mercachita T, Santos Z, Limão J, Carolino E, Mendes L. Anthropometric evaluation and micronutrients intake in patients submitted to laparoscopic Roux-en-Y gastric bypass with a postoperative period of ≥1 year. Obes Surg 2014;24(1):102–8.

[25] Miller GD, Norris A, Fernandez A. Changes in nutrients and food groups intake following laparoscopic Roux-en-Y gastric bypass (RYGB). Obes Surg 2014;24(11):1926–32.

[26] Moizé V, Andreu A, Flores L, Torres F, Ibarzabal A, Delgado S, et al. Long-term dietary intake and nutritional deficiencies following sleeve gastrectomy or Roux-En-Y gastric bypass in a Mediterranean population. J Acad Nutr Diet 2013;113(3):400–10.

[27] Wardé-Kamar J, Rogers M, Flancbaum L, Laferrère B. Calorie intake and meal patterns up to 4 years after Roux-en-Y gastric bypass surgery. Obes Surg 2004;14(8):1070–9.

[28] Ruz M, Carrasco F, Rojas P, Codoceo J, Inostroza J, Rebolledo A, et al. Iron absorption and iron status are reduced after Roux-en-Y gastric bypass 1–3. Am J Clin Nutr 2009;90:527–32.

[29] Ruz M, Carrasco F, Rojas P, Codoceo J, Inostroza J, Basfi-Fer K, et al. Heme- and nonheme-iron absorption and iron status 12 mo after sleeve gastrectomy and Roux-en-Y gastric bypass in morbidly obese women 1–3. Am J Clin Nutr 2012;96:810–17.

[30] Rosa FT, De Oliveira-Penaforte FR, De Arruda Leme I, Padovan GJ, Ceneviva R, Marchini JS. Altered plasma response to zinc and iron tolerance test after Roux-en-Y gastric bypass. Surg Obes Relat Dis 2011;7(3):309–14.

[31] Gesquiere I, Lannoo M, Augustijns P, Matthys C, Van der Schueren B, Foulon V. Iron deficiency after Roux-en-Y gastric bypass: insufficient iron absorption from oral iron supplements. Obes Surg 2014;24(1):56–61.

[32] Rhode BM, Shustik C, Christou NV, MacLean LD. Iron absorption and therapy after gastric bypass. Obes Surg 1999;9(1):17–21.

[33] Bordalo LA, Teixeira TF, Bressan MDM J, Bordalo LA, Teixeira TF, Bressan J, et al. Bariatric surgery: how and why to supplement. Rev Assoc Med Bras 2009;57(1):113–20.

[34] Ross C, Caballero B, Cousins R, Tucker K, Ziegler T. Modern nutrition in health and disease 2014;1648.

[35] Marinella M. Anemia following Roux-en-Y suryger for morbid obesity: a review. South Med J 2008;101(10):1024–31.

[36] Alexandrou A, Armeni E, Kouskouni E, Tsoka E, Diamantis T, Lambrinoudaki I. Cross-sectional long-term micronutrient deficiencies after sleeve gastrectomy versus Roux-en-Y gastric bypass: A pilot study. Surg Obes Relat Dis 2014;10(2):262–8.

[37] Gehrer S, Kern B, Peters T, Christofiel-Courtin C, Peterli R. Fewer nutrient deficiencies after laparoscopic sleeve gastrectomy (LSG) than after laparoscopic Roux-Y-gastric bypass (LRYGB)—a prospective study. Obes Surg 2010;20(4):447–53.

[38] Snyder-Marlow G, Taylor D, Lenhard MJ. Nutrition care for patients undergoing laparoscopic sleeve gastrectomy for weight loss. J Am Diet Assoc 2010;110(4):600–7.

[39] Avinoah E, Ovnat A, Charuzi I. Nutritional status seven years after Roux-en-Y gastric bypass surgery. Surgery 1992;111(2):137–42.

[40] Skroubis G, Sakellaropoulos G, Pouggouras K, Mead N, Nikiforidis G, Kalfarentzos F. Comparison of nutritional deficiencies after Roux-en-Y gastric bypass and after biliopancreatic diversion with Roux-en-Y gastric bypass. Obes Surg 2002;12(4):551–8.

[41] Cook JD, Finch CA. Assessing iron status of a population. Am J Clin Nutr 1979;32(10):2115–19.

[42] Thomas DW, Hinchliffe RF, Briggs C, Macdougall IC, Littlewood T, Cavill I. Guideline for the laboratory diagnosis of functional iron deficiency. Br J Haematol 2013;161(5):639–48.

[43] Careaga M, Moizé V, Flores L, Deulofeu R, Andreu A, Vidal J. Inflammation and iron status in bariatric surgery candidates. Surg Obes Relat Dis 2015;11(4):906–11.

[44] Drakesmith H. Next-generation biomarkers for iron status. Nestle Nutr Inst Workshop Ser 2016;84:59–69.

[45] Kroot JJC, Tjalsma H, Fleming RE, Swinkels DW. Hepcidin in human iron disorders: diagnostic implications. Clin Chem 2011;57(12): 1650–69.

[46] Theurl I, Aigner E, Theurl M, Nairz M, Seifert M, Schroll A, et al. Regulation of iron homeostasis in anemia of chronic disease and iron deficiency anemia: diagnostic and therapeutic implications. Blood 2009;113(21):5277–86.

[47] Tussing-Humphreys LM, Nemeth E, Fantuzzi G, Freels S, Holterman AL, Galvani C, et al. Decreased serum hepcidin and improved functional iron status 6 months after restrictive bariatric surgery. Obesity (Silver Spring) 2010;18(10):2010–16.

[48] Heber D, Greenway FL, Kaplan LM, Livingston E, Salvador J, Still C. Endocrine and nutritional management of the post-bariatric surgery patient: an Endocrine Society clinical practice guideline. J Clin Endocrinol Metab 2010;95(11):4823–43.

[49] Tondapu P, Provost D, Adams-Huet B, Sims T, Chang C, Sakhaee K. Comparison of the absorption of calcium carbonate and calcium citrate after Roux-en-Y gastric bypass. Obes Surg 2009;19(9):1256–61.

[50] Sakhaee K, Pak C. Superior calcium bioavailability of effervescent potassium calcium citrate over tablet formulation of calcium citrate after Roux-en-Y gastric bypass. Surg Obes Relat Dis 2013;9(5):743–8.

Section VI

Cardiovascular, Body Composition, and Physiological Aspects

Chapter 54

Long-Term Cardiovascular Risks in Bariatric Surgery

L. Busetto, R. Fabris, R. Serra and R. Vettor

Padova University Hospital, Padova, Italy

LIST OF ABBREVIATIONS

BMI   body mass index
CAC   Coronary Artery Calcium
FRS   Framingham Risk Score
NNT   number needed to treat
RYGB   Roux-en-Y gastric bypass
SOS   Swedish Obese Subjects
WHR   waist-to-hip ratio

INTRODUCTION

At present, bariatric surgery is the most effective method for weight loss and weight maintenance in patients with morbid obesity. The sustained and stable weight loss achieved in the majority of morbidly obese patients treated with bariatric surgery has been found to be associated with a general improvement of obesity-related comorbidities and a longer life expectancy, as compared to morbidly obese patients treated with medical therapy. This chapter reviews the effects of bariatric surgery on atherosclerosis and cardiovascular diseases. Atherosclerosis can be seen as a chronic disease process with a long preclinical phase characterized by the action of cardiovascular risk factors and the accumulation of lipids in the arterial wall, followed by a subclinical or clinical chronic phase, with the formation of substenotic or stenotic plaques, and finally by the abrupt occurrence of major cardiovascular events related to plaque rupture and thrombosis with acute vascular occlusion (Fig. 54.1). We can therefore examine the preventive effects of bariatric surgery at three distinct levels: first, at the level of risk factor control; second, at the level of a possible regression of early markers of atherosclerosis (like intima-media thickness, coronary artery calcium (CAC) score, or others), and finally at the level of prevention of fatal and nonfatal cardiovascular events.

CARDIOVASCULAR RISK IN MORBIDLY OBESE CANDIDATES TO BARIATRIC SURGERY

Obesity is a complex and heterogeneous disease with a wide spectrum of clinical manifestations and comorbidities. One of the most important clinical consequences of obesity is an increased incidence of cardiovascular disease. A collaborative analysis of 57 prospective epidemiological studies, including 894,576 participants, demonstrated both in males and females a strict association between baseline body mass index (BMI) and total mortality [1]. Minimum mortality was observed in the BMI range $22.5-25.0$ kg/m^2. The progressive excess mortality above this range, with each 5 kg/m^2 higher BMI on average associated with about 30% higher overall mortality, was mainly due to cardiovascular diseases [1]. Therefore, we can conclude that obesity represents a relevant cardiovascular risk factor and that most of the excess mortality observed in obese patients is cardiovascular in nature.

Cardiovascular risk, however, is not the same in all obese patents, and can vary greatly even in patients who have the same BMI levels. Cardiovascular risk and the incidence of cardiovascular events and death differ within the obese

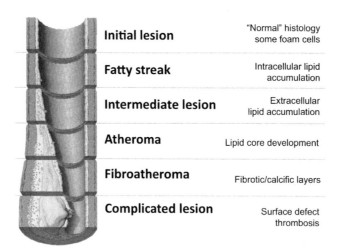

Initial lesion	"Normal" histology some foam cells
Fatty streak	Intracellular lipid accumulation
Intermediate lesion	Extracellular lipid accumulation
Atheroma	Lipid core development
Fibroatheroma	Fibrotic/calcific layers
Complicated lesion	Surface defect thrombosis

FIGURE 54.1 Progression of atherosclerosis. Atherosclerosis can be seen as a chronic disease process with a long preclinical phase characterized by the action of cardiovascular risk factors and the accumulation of lipids in the arterial wall, followed by a subclinical or clinical chronic phase, with the formation of substenotic or stenotic plaques, and finally by the abrupt occurrence of major cardiovascular events related to plaque rupture and thrombosis with acute vascular occlusion.

population according to age, sex, ethnicity, familial history, and lifestyle behaviors, but probably the most important factor in determining cardiovascular risk in obesity, whatever the sex and the age of the patient, is fat distribution. Patients with central or abdominal or visceral obesity frequently have multiple metabolic abnormalities and a rather high cardiovascular risk, whereas patients with peripheral or subcutaneous obesity may have normal metabolic profiles, and usually have a low probability of cardiovascular events. The importance of fat distribution in determining cardiovascular risk was demonstrated in an analysis of the INTERHEART database, a large case-control study on myocardial infarction (MI) that included 27,098 participants (12,461 cases and 14,637 controls) enrolled in 52 different countries with different ethnic backgrounds [2]. In this study, adjustment for other cardiovascular risk factors greatly attenuated the association between BMI levels and the risk of a first MI. On the contrary, the association between risk of a first MI and wait to hip circumfernece ratio (WHR), an anthropometric index for fat distribution, was not affected by such an adjustment, thus demonstrating that fat distribution is a more important determinant of cardiovascular risk than the level of obesity per se [2]. More interestingly, when the two anthropometric components of WHR were considered separately, only waist circumference was strongly related to MI risk, with a continuous relation persistent even after adjustment for BMI and height. In contrast, a trend toward lower risk of MI was noted as hip circumference increased. This trend was highly significant after adjustment for BMI and height [2]. The independent protective effect against metabolic and cardiovascular diseases of a large hip circumference has been confirmed in other epidemiological studies [3]. The difference in cardiovascular risk and disease between visceral and subcutaneous obesity deserves great attention in explaining the relationships between obesity and atherosclerosis. This dichotomy may have relevant clinical implications, considering that the effects of weight loss on cardiovascular events and death are obviously dependent on the baseline cardiovascular risk.

In the end, what level of baseline cardiovascular risk should be expected in a clinical sample of morbidly obese candidates to obesity surgery? Taking into account the epidemiological observations previously presented, a substantial variability in mean cardiovascular risk levels between different bariatric series should be expected, according to the composition of the series in terms of age, sex, fat distribution, and obesity-related comorbidities. This variability was confirmed by Batsis et al. in an interesting analysis of the cardiovascular risk levels observed before and after surgery in several studies [4]. In these studies [5–10], the baseline 10-year cardiovascular risk was calculated before surgery both using the Framingham Risk Score (FRS) and the Prospective Cardiovascular Munster Heart Study (PROCAM) risk score (Table 54.1). Mean baseline 10-year cardiovascular risk ranged between studies from a moderate level of risk (PROCAM risk score <2%) to higher risk levels (PROCAM risk score >5%), mainly depending on the mean age and the prevalence of type 2 diabetes mellitus (T2DM) in the series (Table 54.1) [4]. Therefore, a substantial variability in baseline cardiovascular risk is confirmed, even within series all formed by morbidly obese patients who had the same high BMI levels.

Besides the large variability in cardiovascular risk observed between different groups of morbidly obese patients, a substantial variability can also be observed within the groups. In an analysis of long-term cardiovascular risk

TABLE 54.1 Baseline Mean Cardiovascular Risk Levels in Several Published Series of Morbidly Obese Patient Candidates to Bariatric Surgery

Surgical Series	Age	Female (%)	T2DM (%)	FRS	PROCAM
Sjostrom et al. [5]	47	71	74	12.2	7.1
Stoopen-Margain et al. [6]	31	63	24	4.4	<1.0
He and Stubbs [7]	42	77	17	7.3	5.1
Batsis et al. [8]	44	73	24	3.8	1.8
Pontiroli et al. [9]	43	81	46	4.7	2.2
Busetto et al. [10]	38	76	11	3.4	1.9

Ten-year cardiovascular risk was calculated using both the Framingham Risk Score (FRS) and the Prospective Cardiovascular Munster Heart Study (PROCAM) risk score. T2DM, type 2 diabetes mellitus.
Source: Data were extracted from Batsis JA, Sarr MG, Collazo-Clavell ML, et al. Cardiovascular risk after bariatric surgery for obesity. Am J Cardiol. 2008;102:930.

(> 10 years) and coronary events in morbidly obese patients treated with gastric banding, Busetto et al. calculated baseline 10-year probability of experiencing an MI according to the PROCAM algorithm [11]. Mean baseline cardiovascular risk in this series was quite low (2.0 ± 4.9%), as expected in a sample composed mostly of female patients with a relatively young age and a low prevalence of T2DM. Nevertheless, a substantial variability in cardiovascular risk was observed within the sample. The majority of the patients (67.1%) had a low 10-year cardiovascular risk (<1%), 22.7% of the patients had a 10-year cardiovascular risk in the range of 1–5%, 5.8% of the patients had a 5–10% risk, and 4.4% of the patients had a high cardiovascular risk (> 10%) [11]. As expected, 10-year cardiovascular risk was higher in males than in females, higher in patients age > 50 years than in younger patients, and higher in patients with T2DM than in patients without diabetes [11]. This analysis confirmed that a proportion of patients with high cardiovascular risk can be observed even in bariatric series with a low mean cardiovascular risk. Therefore, any analysis of the effects of bariatric surgery on cardiovascular risk and/or cardiovascular events should be focused not only on the study of the mean variations, but more specifically on the effects on the subgroup of patients who have the higher baseline cardiovascular risk, in which a higher number of events are expected and the benefits of weight loss on cardiovascular risk may be more important.

EFFECTS OF BARIATRIC SURGERY ON CARDIOVASCULAR RISK FACTORS

Several studies analyzed the changes in cardiovascular risk factors after bariatric surgery. In 2012, Vest et al. published a large systematic review including 73 studies with a total of 19,543 patients [12]. The review confirmed that weight loss induced by bariatric surgery is associated with a significant reduction in the prevalence of hypertension, T2DM, and dyslipidemia. These results have been confirmed in a more recent and rigorous meta-analysis including 22 studies with 4160 subjects [13]. However, the evidence level of these reviews is reduced by the fact that most of the studies are uncontrolled nonrandomized studies with a relatively short follow-up. A publication bias favoring studies with positive results should also be taken into account.

The Swedish Obese Subjects (SOS) study is considered the landmark study on the long-term effects of bariatric surgery. SOS is a prospective nonrandomized controlled study investigating the long-term outcome in 2010 morbidly obese patients treated by various bariatric procedures (mostly vertical banded gastroplasty) in the last decade of the 20th century (surgical group) and in 2037 matched obese patients not surgically treated (control group). Matching was performed according to a large number of variables (gender, age, weight, height, waist and hip circumferences, systolic blood pressure, cholesterol, triglycerides, smoking habits, diabetes, menopausal status, and psychosocial and personality status) [14]. Ten-year results of the SOS demonstrated a 16% reduction of body weight in the surgical group and a slight increase in weight in the control group. This important difference in weight loss was associated with significant differences in the rates of normalization of cardiovascular risk factor levels in the surgical group: glucose levels (36% vs 13%; odds ratio (OR): 3.45; 95% CI 1.64–7.28), triglyceride levels (46% vs 24%; OR: 2.57; 95% CI 1.85–3.57), high-density lipoprotein cholesterol (HDL-c) levels (73% vs 53%; OR: 2.35; 95% CI 1.44–3.84), and blood pressure

(49% vs 11%; OR: 1.68; 95% CI 1.09−2.58) [5]. An additional important point for cardiovascular prevention in the SOS study was the reduction in the new cases of T2DM observed in the surgical group. The incidence of new cases of T2DM at the 15-year follow-up was 40% in the control group, and it was greatly reduced in the surgical group, with a relative risk of 0.12 (95% CI 0.05−0.27; $p < 0.001$) for gastric bypass, 0.20 (95% CI 0.13−0.32; $p < 0.001$) for gastric banding, and 0.25 (95% CI 0.19−0.31; $p < 0.001$) for vertical banded gastroplasty [14]. The number needed to treat (NNT) by surgery in order to prevent one new case of T2DM was only 7 in patients with a normal glucose tolerance at baseline, and was further lowered to 1.3 in patients with impaired fasting glucose at baseline [14].

The improvements in multiple cardiovascular risk factors observed in the nonrandomized SOS study were confirmed (with shorter follow-up times) by several randomized controlled trials comparing the effects of bariatric surgery with those of medical therapy, both in nondiabetic [15] and diabetic [16−19] obese patients. In particular, Ikramuddin et al. randomized 120 obese diabetic patients to receive an intensive lifestyle−medical management protocol alone or with the same protocol with RYGB surgery in a study with a composite goal of hemoglobin A1c (HbA1c) less than 7.0%, low-density lipoprotein cholesterol (LDL-c) less than 100 mg/dL, and systolic blood pressure less than 130 mm Hg. After 12-month, 49% of the participants (95% CI 36−63%) in the RYGB group and 19% (95% CI 10−32%) in the life-style−medical management group achieved the primary composite end point (OR: 4.8; 95% CI 1.9−11.7) [17].

The changes in cardiovascular risk after bariatric surgery can be studied not only by examining the modifications of any single cardiovascular risk factor, but also by analyzing the variations in the global cardiovascular risk, calculated as the future probability of suffering a cardiovascular event, with the application of one of the algorithms that takes into account several cardiovascular risk factors simultaneously (FRS or PROCAM score). This type of analysis was performed by Batsis et al. in their review of the cardiovascular risk levels observed before and after surgery in several studies [4]. The variations in the FRS observed in the surgical series analyzed in the review [5−10] and in the nonsurgically treated control groups included in some of the studies [5,8] are reported in Table 54.2. Global cardiovascular risk calculated with the FRS tended to increase over time in all the control groups. On the contrary, a substantial and significant reduction in the 10-year probability of suffering a fatal or nonfatal MI was observed in the surgical series, with the only exception being represented by the 10-year follow-up data from the SOS study, where the aging of the population probably played a more important role [4]. A similar analysis was also performed in five studies included in the large systematic review published by Vest et al. in 2012 [12]. The analysis confirmed a significant reduction in FRS after surgery in all five studies. Reduction in risk was more relevant in those subgroups of patients who had higher levels of

TABLE 54.2 Variation in the Framingham Risk Score (FRS) in Several Published Series of Obese Patients Treated With Bariatric Surgery and in Control Groups Treated Nonsurgically

	Baseline FRS	Final FRS	Absolute Change	Relative Change (%)
Surgical Series				
Batsis et al. [8]	3.8	2.7	− 1.1	− 29
Pontiroli et al. [9]	4.7	3.5	− 1.2	− 26
Stoopen-Margain et al. [6]	4.4	1.7	− 2.7	− 61
Busetto et al. [10]	3.4	2.6	− 0.8	− 34
He and Stubbs [7]	7.3	3.3	− 4.0	− 55
SOS 2-year [5]	12.2	5.4	− 6.8	− 56
SOS 10-year [5]	7.7	11.8	+ 4.1	+ 54
Control Groups				
Batsis et al. [8]	4.3	4.8	+ 0.5	+ 12
SOS 2-year [5]	5.9	8.4	+ 2.5	+ 42
SOS 10-year [5]	5.9	12.3	+ 6.4	+ 108

Ten-year cardiovascular risk was calculated using the Framingham Risk Score (FRS).
Source: Data were extracted from Batsis JA, Sarr MG, Collazo-Clavell ML, et al. Cardiovascular risk after bariatric surgery for obesity. Am J Cardiol. 2008;102:930.

cardiovascular risk before surgery and who showed better metabolic improvements after surgery, such as patients with T2DM and hypertension at baseline, which demonstrated a remission of both diabetes and hypertension after weight loss [12].

A possible criticism of these analyses based on FRS variation derives from the fact that algorithms invented for the calculation of the probability of future cardiovascular events have never been tested in a longitudinal way. This means that even if patients obtain a better cardiovascular risk profile after surgery, this could not be automatically translated into a corresponding reduction of the risk of cardiovascular events because the same patients have been exposed to a high level of risk before surgery. A study design invented in order to overcome this criticism was proposed by Torquati et al. in an interesting study in 500 patients treated with RYGB. Torquati et al. not only calculated the 10-year probability of cardiovascular events before and 1 year after the procedure, but also compared these probabilities with the real number of events that occurred in the same patients in the following 5 years [20]. First, authors confirmed a significant improvement in FRS-calculated cardiovascular risk after surgery. Improvements were more relevant in patients with higher cardiovascular risk levels at baseline, such as in patients with diabetes compared with nondiabetic subjects, older patients compared to younger ones, and patients with high baseline 10-year event probability compared to patients with lower baseline risk. Second, the actual number of cardiovascular events observed in the first 5 years after surgery was similar to the number of events predicted according to the cardiovascular risk calculated 1 year after surgery, and substantially lower than the number of events predicted by the baseline levels of cardiovascular risk [20].

Busetto el al. analyzed the long-term effects on cardiovascular risk in patients treated with gastric banding and compared the real number of coronary events that occurred in the first 10 years after surgery with the number of events expected according to the cardiovascular risk calculated at baseline with the PROCAM risk score. The real 10-year incidence of events (1.6%) was lower than the 10-year probability of events calculated before surgery (2.0%), with a difference between observed and predicted events that tended to be more relevant in the groups of patients with the higher baseline event probability. For instance, observed incidence of events in patients with diabetes was 2.6% with a predicted value of 4.7%, whereas in nondiabetic patients observed incidence was 1.0% and predicted incidence was 1.2% [11]. These comparisons between predicted and observed events are suggestive, but it remains obvious that they cannot produce levels of evidence comparable to those of prospective controlled studies.

EFFECTS OF BARIATRIC SURGERY ON EARLY MARKERS OF ATHEROSCLEROSIS

As schematized in Fig. 54.1, chronic exposure to cardiovascular risk factors may lead to the deposition of lipids in the arterial wall and to the formation of atherosclerotic plaques that grow progressively and cause subclinical or clinical arterial stenosis. This process can be detected in the preclinical phase by analyzing with noninvasive methodologies the modification of arterial wall morphology and function. Typical early markers of subclinical atherosclerosis that can be detected and measured in clinical studies include arterial intima-media thickness and flow-mediated or nitrate-mediated arterial dilatation. These markers are considered independent predictors of cardiovascular events and provide important prognostic data beyond the traditional cardiovascular risk factors examined so far.

The effects of bariatric surgery on these early markers of atherosclerosis were recently reviewed and meta-analyzed by Lupoli et al. [21]. Ten articles (including a total of 314 obese patients treated with different bariatric procedures) were included in the analysis [22−31]. The changes in carotid intima-media thickness after surgery were analyzed in six of these studies, which included 206 patients [24,25,27−29,31]. Meta-analysis of their results showed a significant reduction of intima-media thickness after bariatric surgery (−0.17 mm; 95% CI −0.290, −0.049; $p = 0.006$). Flow-mediated arterial dilatation was evaluated in eight studies [22,23,25−30] with 269 subjects, and a significant improvement was observed after surgery (+5.65%; 95% CI 2.87, 8.03; $p < 0.001$). Finally, nitrate-mediated flow dilatation was evaluated in four studies [26,27,29,30] (149 patients). In this case, only a nonsignificant increase was observed after surgery (+2.17%; 95% CI −0.796, 5.142; $p = 0.151$). On the basis of these results, the authors concluded that despite heterogeneity among studies, bariatric surgery is associated with improvement of subclinical atherosclerosis and endothelial function, and that these effects may significantly contribute to the reduction of cardiovascular risk after bariatric surgery [21]. However, it should be noted that the studies published so far, albeit consistent in their results, are generally small, uncontrolled, and have a short follow-up range of 3−24 months. One single-study [28] presented a prolonged follow-up (5 years) in a set of patients previously analyzed at 18 months after surgery [29] and confirmed that the improvements in intima-media thickness and flow-mediated arterial dilatation observed in the early phase could be maintained in the long term.

An emerging clinical early marker of coronary atherosclerosis is represented by the detection of areas of calcification at the level of cardiac atherosclerotic plaques. Calcified plaques can be readily detected and quantified using computed tomography. CAC scores are related to the total atherosclerotic burden in the coronary arterial tree and can predict coronary events with greater accuracy than FRS. Priester et al. recently measured CAC scores in 149 consecutive patients returning for 6-year follow-up in a prospective, longitudinal study comparing a variety of endpoints in severely obese subjects undergoing RYGB versus severely obese subjects treated without bariatric surgery. The CAC scores were significantly lower in RYGB patients than in nonsurgical controls ($p < 0.01$), and RYGB subjects had a lower likelihood of measureable levels of coronary calcium (OR: 0.39 for CAC > 0; 95% CI 0.17–0.90) [32].

EFFECTS OF BARIATRIC SURGERY ON CARDIOVASCULAR EVENTS

Analysis of the number of acute fatal or nonfatal cardiovascular events observed after bariatric surgery continues to provide the most important data on the effects that surgically induced weight loss has on cardiovascular risk. In 2012, Sjöström et al. reported the results of the SOS study on this specific outcome (MI and stroke), and in the surgical group demonstrated a clear and highly significant reduction of both fatal (28 events in 2010 patients in the surgical group vs 49 events in 2037 patients in the control group; adjusted relative risk: 0.47; 95% CI 0.29–0.76; $p = 0.0020$) and total first cardiovascular events (199 events in 2010 patients in the surgical group vs 234 events in 2037 patients in the control group; adjusted relative risk: 0.67; 95% CI 0.54–0.83; $p < 0.001$) [33]. Reduction of cardiovascular events in the surgical group observed in the general SOS population was also confirmed specifically in the SOS patients affected by T2DM, where a reduction in the incidence of both macrovascular and microvascular complications was reported [34]. An interesting aspect that emerged from the analysis of the incidence of cardiovascular events in the SOS study was that both in the surgical and in the control group there was no relationship between presurgical baseline BMI levels and event rates in the follow-up. On the contrary, baseline fasting insulin levels were significantly related to event rates in both groups, and the reduction of events observed in the surgical group was more evident in the patients with higher insulin levels at enrollment [33]. This last observation seems to confirm that cardiovascular risk in obese patients is not dependent on the degree of obesity, but on the severity of obesity-associated metabolic abnormalities; also, the weight loss–induced benefits on cardiovascular risk are more important in patients with the highest levels of risk before surgery.

The reduction in cardiovascular events observed in SOS diabetic patients enrolled in the surgical groups was recently confirmed in a nationwide Swedish observational cohort study. It matched data (i.e., sex, age, BMI, and calendar time) for 6132 patients registered in the Scandinavian Obesity Surgery Registry who had undergone RYGB, and 6132 controls registered in the National Diabetes Registry who had not undergone bariatric surgery [35]. Median follow-up was 3.5 years. Authors reported a 58% relative risk reduction (hazard ratio: 0.42, 95% CI 0.30–0.57; $p < 0.0001$) in overall mortality in the RYGB group compared with the controls. The risk of fatal or nonfatal MI was 49% lower (hazard ratio: 0.51, 95% CI 0.29–0.91; $p = 0.021$) and that of cardiovascular death was 59% lower (0.41, 95% CI 0.19–0.90; $p = 0.026$) in the RYGB group than in the control group [35].

In addition, the rates of macrovascular and microvascular events observed in diabetic patients who had undergone (2580 cases) or not undergone (13,371 cases) bariatric surgery were retrospectively analyzed by using data derived from the administrative health database of South Carolina [36]. The primary outcome of this study was a composite of any first major macrovascular or microvascular complication. Macrovascular complications were defined as acute MI, stroke, or all-cause death, and microvascular complications were defined as a new diagnosis of blindness in at least one eye, laser eye or retinal surgery, nontraumatic amputation, or creation of permanent arteriovenous access for dialysis. Secondary outcomes included macrovascular and microvascular complications considered separately, as well as other vascular complications, including revascularization of coronary, carotid, or lower extremity arteries, or a new diagnosis of congestive heart failure or angina pectoris. After adjusting for baseline differences between groups, bariatric surgery remained significantly associated with a reduced risk of any major macro- or microvascular event combined (hazard ratio: 0.36, 95% CI 0.27–0.47). A similar relationship was observed for each of the secondary outcomes (macrovascular hazard ratio: 0.39, 95% CI 0.29–0.51; microvascular hazard ratio: 0.22, 95% CI 0.09–0.49; and other vascular hazard ratio: 0.25, 95% CI 0.19–0.32) [36]. The authors concluded that bariatric surgery was associated with a 65% reduction in major macrovascular and microvascular events in obese patients with T2DM.

The impact of bariatric surgery on cardiovascular disease observed in all parallel group studies that evaluated the clinical outcomes associated with bariatric surgery as compared to nonsurgical treatment was recently reviewed and meta-analyzed by Kwok et al. [37]. A total of 14 studies including 29,208 patients who underwent bariatric surgery and 166,200 nonsurgical controls were considered. Compared to nonsurgical controls, there was more than 50% reduction in mortality among patients who had bariatric surgery (OR: 0.48; 95% CI 0.35–0.64). In a more specific pooled analysis of four studies [33,36,38,39], bariatric surgery was associated with a significantly reduced risk of composite cardiovascular adverse events (OR: 0.54; 95% CI 0.41−0.70), and with a significant decrease in specific endpoints of MI (OR: 0.46; 95% CI 0.30−0.69) and stroke (OR: 0.49; 95% CI 0.32−0.75).

In summary, prospective and retrospective studies analyzing the relationship between bariatric surgery and the incidence of new cardiovascular events showed quite consistently a roughly 50% reduction in cardiovascular events after surgery, both in diabetic and nondiabetic patients. At the moment, however, no randomized control studies with sufficient size and duration to explore this specific outcome have been published.

CONCLUSION

Weight loss induced by modern bariatric procedures is associated with a stable and clinically significant improvement of cardiovascular risk, an improvement in structural and functional markers of atherosclerosis, and a reduction in the incidence of new cardiovascular events. The reduction observed in risk and events is more important in those obese patients with the worst cardiovascular profiles and the highest probability of events before surgery.

MINI-DICTIONARY OF TERMS

- *Cardiovascular risk*: The probability of suffering in the future from a clinical cardiovascular event. It depends on the presence and severity of cardiovascular risk factors, and can be calculated by cardiovascular risk assessment tools.
- *Cardiovascular risk assessment tools*: Electronic algorithms derived from analysis of the events observed in large epidemiological prospective studies; they permit the calculation of probability of future coronary or cardiovascular events on the basis of predefined risk factor levels. The most commonly used instruments are the FRS in the United States and the PROCAM risk score in Europe.
- *Coronary artery calcium score*: A quantitative measure of the deposits of calcium in the walls of the coronary artery, as visualized by computed tomography scanning of the heart. Calcifications in the coronary arteries are an early sign of coronary heart disease.
- *Early markers of atherosclerosis*: The atherosclerotic process may be detected in a preclinical phase by analyzing with noninvasive methodologies the modification of the morphology and the function of the arterial walls. Typical early markers of subclinical atherosclerosis that can be detected and measured in clinical studies include arterial intima-media thickness, flow-mediated or nitrate-mediated arterial dilatation, and CAC scores. These markers are considered independent predictors of cardiovascular events and provide important prognostic data beyond the traditional cardiovascular risk factors.
- *Flow-mediated or nitrate-mediated arterial dilatation*: The diameter of a target artery, usually the brachial artery, is measured by ultrasound in response to an increase in blood flow during reactive hyperemia (induced by cuff inflation and then deflation). This leads to endothelium-dependent dilatation. Impairment of flow-mediated dilatation is considered a marker of endothelial dysfunction or early atherosclerosis.
- *Intima-media thickness*: A measurement of the thickness of the tunica intima and tunica media, the two innermost layers of the wall of an artery. The measurement is usually made by external ultrasound at the level of the carotid artery, and is used to detect the presence of atherosclerotic disease.
- *Nitrate-mediated arterial dilatation*: The diameter of a target artery, usually the brachial artery, is measured by ultrasound before and after administration of sublingual nitroglycerin, which causes endothelium-independent smooth muscle−mediated vasodilatation. Impairment of nitrate-mediated dilatation is considered a marker of early atherosclerosis.

KEY FACTS

- Cardiovascular risk is defined as the probability of suffering in the future from a clinical cardiovascular event. It depends on the presence and severity of cardiovascular risk factors.
- Cardiovascular risk can be more precisely estimated by using cardiovascular risk assessment tools. These are electronic algorithms, derived by the analysis of the events observed in large epidemiological prospective studies, that permit the calculation of the future probability (usually 10 years) of coronary or cardiovascular events on the basis of the levels of multiple predefined risk factors. The most commonly used risk assessment tools are the FRS in the United States and the PROCAM risk score in Europe.

- The FRS was first developed based on data obtained from the Framingham Heart Study. In this algorithm, the 10-year risk of coronary events was estimated based on the following parameters: age, gender, total cholesterol, HDL-c, current treatment for hypertension, systolic blood pressure, and current smoking habits. The algorithm is available online at the National Heart, Lung, and Blood Institute website (http://cvdrisk.nhlbi.nih.gov/).
- PROCAM Health Check Algorithm is based on 462 coronary events occurring among 18,460 men and 49 events occurring among 8518 women aged 20−78 years enrolled in the PROCAM study. In this algorithm, 10-year risk of coronary events was estimated with the use of the following parameters: age, gender, known diabetes mellitus or fasting blood glucose levels ≥120 mg/dL, current nicotine consumption, positive family history of coronary heart disease (a first-degree relative suffering a heart attack before the age of 60), systolic blood pressure, low-density lipoprotein cholesterol (LDL-c) levels, HDL-c levels, and triglyceride levels. The algorithm is available online at the International Task Force for Prevention of Coronary Heart Disease website (http://www.chd-taskforce.com/procam_interactive.html).

SUMMARY POINTS

- Cardiovascular risk and the incidence of cardiovascular events and death vary within the obese population according to age, sex, ethnicity, familial history, lifestyle, and fat distribution.
- Weight loss induced by bariatric surgery is associated with a significant reduction in the prevalence of hypertension, type 2 diabetes mellitus (T2DM), and dyslipidemia.
- Reduction in cardiovascular risk after bariatric surgery is more relevant in those subgroups of patients who have higher levels of cardiovascular risk before surgery and show better metabolic improvements after surgery.
- Improvements in some early markers of atherosclerosis (e.g., carotid intima-media thickness, flow-mediated arterial dilatation, CAC score) have been demonstrated after bariatric surgery in small and uncontrolled studies.
- Prospective and retrospective studies analyzing the relationship between bariatric surgery and the incidence of new cardiovascular events showed quite consistently a 50% reduction in cardiovascular events after surgery, both in diabetic and nondiabetic patients.

REFERENCES

[1] Prospective Studies Collaboration. Body-mass index and cause-specific mortality in 900 000 adults: collaborative analyses of 57 prospective studies. Lancet 2009;373:1083.

[2] Yusuf S, Hawken S, Ôunpuu S, On behalf of the INTERHEART Study Investigators, et al. . Obesity and the risk of myocardial infarction in 27 000 participants from 52 countries: a case-control study. Lancet 2005;366:1640.

[3] Heitmann BL, Lissner L. Hip Hip Hurrah! Hip size inversely related to heart disease and total mortality. Obes Rev 2011;12:478.

[4] Batsis JA, Sarr MG, Collazo-Clavell ML, et al. Cardiovascular risk after bariatric surgery for obesity. Am J Cardiol 2008;102:930.

[5] Sjöström L, Lindroos A-K, Peltonen M, et al. Lifestyle, diabetes, and cardiovascular risk factors 10 years after bariatry surgery. N Engl J Med 2004;351:2683.

[6] Stoopen-Margain E, Fajardo R, Espana N, et al. Laparoscopic Roux-en-Y gastric bypass for morbid obesity: results of our learning curve in 100 consecutive patients. Obes Surg 2004;14:201.

[7] He M, Stubbs R. Gastric bypass surgery for severe obesity: what can be achieved? N Z Med J 2004;117:U1207.

[8] Batsis JA, Romero-Corral A, Collazo-Clavell ML, et al. Effect of weight loss on predicted cardiovascular risk: change in cardiac risk after bariatric surgery. Obesity (Silver Spring) 2007;15:772.

[9] Pontiroli AE, Pizzocri P, Librenti MC, et al. Laparoscopic adjustable gastric banding for the treatment of morbid (grade 3) obesity and its metabolic complications: a three-year study. J Clin Endocrinol Metab 2002;87:3555.

[10] Busetto L, Sergi G, Enzi G, et al. Short-term effects of weight loss on the cardiovascular risk factors in morbidly obese patients. Obes Res 2004;12:1256.

[11] Busetto L, De Stefano F, Pigozzo S, et al. Long-term cardiovascular risk and coronary events in morbidly obese patients treated with laparoscopic gastric banding. Surg Obes Relat Dis 2014;10:112.

[12] Vest AR, Heneghan HM, Agarwal S, Schauer PR, Young JB. Bariatric surgery and cardiovascular outcomes: a systematic review. Heart 2012;98:1763.

[13] Ricci C, Gaeta M, Rausa E, et al. Long-term effects of bariatric surgery on type II diabetes, hypertension and hyperlipidemia: a meta-analysis and meta-regression study with 5-year follow-up. Obes Surg 2015;25:397.

[14] Sjöström L. Review of the key results from the Swedish Obese Subjects (SOS) trial − a prospective controlled intervention study of bariatric surgery. J Intern Med 2013;273:219.

[15] O'Brien PE, Dixon JB, Laurie C, et al. Treatment of mild to moderate obesity with laparoscopic adjustable gastric banding or an intensive medical program: a randomized trial. Ann Intern Med 2006;144:625.

[16] Dixon JB, O'Brien PE, Playfair J, et al. Adjustable gastric banding and conventional therapy for type 2 diabetes: a randomized controlled trial. JAMA 2008;299:316.

[17] Ikramuddin S, Korner J, Lee WJ, et al. Roux-en-Y gastric bypass vs intensive medical management for the control of type 2 diabetes, hypertension, and hyperlipidemia. The Diabetes Surgery Study randomized clinical trial. JAMA 2013;309:2240.

[18] Schauer PR, Bhatt DL, Kirwan JP, et al. Bariatric surgery versus intensive medical therapy for diabetes--3-year outcomes. N Engl J Med 2014;370:2002.

[19] Mingrone G, Panunzi S, De Gaetano A, et al. Bariatric-metabolic surgery versus conventional medical treatment in obese patients with type 2 diabetes: 5 year follow-up of an open-label, single-centre, randomised controlled trial. Lancet 2015;386:964.

[20] Torquati A, Wright K, Melvin W, Richards W. Effect of gastric bypass operation on Framingham and actual risk of cardiovascular events in class II to III obesity. J Am Coll Surg 2007;204:776.

[21] Lupoli R, Di Minno MND, Guidone C, et al. Effects of bariatric surgery on markers of subclinical atherosclerosis and endothelial function: a meta-analysis of literature studies. Int J Obes 2015;40(3):395−402 [Epub ahead of print].

[22] Gokce N, Vita JA, McDonnell M, et al. Effect of medical and surgical weight loss on endothelial vasomotor function in obese patients. Am J Cardiol 2005;95:266.

[23] Lind L, Zethelius B, Sundbom M, Edén Engström B, Karlsson FA. Vasoreactivity is rapidly improved in obese subjects after gastric bypass surgery. Int J Obes 2009;33:1390.

[24] Sarmento PL, Plavnik FL, Zanella MT, Pinto PE, Miranda RB, Ajzen SA. Association of carotid intima-media thickness and cardiovascular risk factors in women pre- and post-bariatric surgery. Obes Surg 2009;19:339.

[25] Habib P, Scrocco JD, Terek M, Vanek V, Mikolich JR. Effects of bariatric surgery on inflammatory, functional and structural markers of coronary atherosclerosis. Am J Cardiol 2009;104:1251.

[26] Brethauer SA, Heneghan HM, Eldar S, et al. Early effects of gastric bypass on endothelial function, inflammation, and cardiovascular risk in obese patients. Surg Endosc 2011;25:2650.

[27] Saleh MH, Bertolami MC, Assef JE, et al. Improvement of atherosclerotic markers in non-diabetic patients after bariatric surgery. Obes Surg 2012;22:1701.

[28] Tschoner A, Sturm W, Gelsinger C, et al. Long-term effects of weight loss after bariatric surgery on functional and structural markers of atherosclerosis. Obesity (Silver Spring) 2013;21:1960.

[29] Sturm W, Tschoner A, Engl J, et al. Effect of bariatric surgery on both functional and structural measures of premature atherosclerosis. Eur Heart J 2009;30:2038.

[30] Nerla R, Tarzia P, Sestito A, et al. Effect of bariatric surgery on peripheral flow-mediated dilation and coronary microvascular function. Nutr Metab Cardiovasc Dis 2012;22:626.

[31] García G, Bunout D, Mella J, et al. Bariatric surgery decreases carotid intima-media thickness in obese subjects. Nutr Hosp 2013;28:1102.

[32] Priester T, Ault TG, Davidson L, et al. Coronary Calcium Scores 6 years after bariatric surgery. Obes Surg 2015;25:90.

[33] Sjöström L, Peltonen M, Jacobson P, et al. Bariatric surgery and long-term cardiovascular events. JAMA 2012;307:56.

[34] Sjöström L, Peltonen M, Jacobson P, et al. Association of bariatric surgery with long-term remission of type 2 diabetes and with microvascular and macrovascular complications. JAMA 2014;311:2297.

[35] Eliasson B, Liakopoulos V, Franzén S, et al. Cardiovascular disease and mortality in patients with type 2 diabetes after bariatric surgery in Sweden: a nationwide, matched, observational cohort study. Lancet. Diabetes Endocrinol 2015;3:847.

[36] Johnson BL, Blackhurst DW, Latham BB, et al. Bariatric surgery is associated with a reduction in major macrovascular and microvascular complications in moderately to severely obese patients with type 2 diabetes mellitus. J Am Coll Surg 2013;216:545.

[37] Kwok CS, Pradhan A, Khanc MA, et al. Bariatric surgery and its impact on cardiovascular disease and mortality: a systematic review and meta-analysis. Int J Cardiol 2014;173:20.

[38] Adams TD, Gress RE, Smith SC, et al. Long-term mortality after gastric bypass surgery. N Engl J Med 2007;357:753.

[39] Scott JD, Johnson BL, Blackhurst DW, Bour ES. Does bariatric surgery reduce the risk of major cardiovascular events? A retrospective cohort study of morbidly obese surgical patients. Surg Obes Relat Dis 2013;9:32.

Chapter 55

QT Interval After Bariatric Surgery

A. Al-Salameh[1] and M. Fysekidis[2]

[1]Hôpital Bicêtre, Le Kremlin-Bicêtre, France, [2]Hôpital Avicenne, Bobigny, France

LIST OF ABBREVIATIONS

BMI	body mass index
BPD	biliopancreatic diversion
CREB	cyclic AMP response element binding protein
ECG	electrocardiogram
FFAs	free fatty acids
LAGB	laparoscopic adjustable gastric banding
PKD	protein kinase D
PPAR-γ	peroxisome proliferator-activated receptor gamma
QTc	QT corrected for heart rate
QTd	QT dispersion
VBG	vertical banded gastroplasty

INTRODUCTION

Obesity is closely associated with cardiovascular disease and represents an independent risk factor for coronary heart disease [1], heart failure [2], stroke [3], atrial fibrillation [4], and sudden cardiac death [5]. Obesity also contributes to subclinical cardiovascular alterations like coronary artery calcifications [6], left ventricular hypertrophy [7], increased carotid artery intima-media thickness [8], and increased arterial stiffness [9]. Finally, obesity is consistently and strongly associated with cardiovascular risk factors like hypertension [10,11], type 2 diabetes mellitus, and obstructive sleep apnea [12].

The impact of obesity on cardiovascular morbidity and mortality goes beyond coronary artery disease. Sudden cardiac death occurs more frequently in obese patients compared to lean individuals [13]. In fact, obesity is one of the main predictors of sudden cardiac death in middle-aged adults [14]. Another main predictor is obstructive sleep apnea, which commonly coexists and interacts with obesity. The presence of cardiomyopathy of the obese is the leading cause of nonischemic sudden cardiac death, accounting for 23.7% of cases [15]. Therefore, sudden cardiac death is expected to become a challenging health problem because of the obesity pandemic [16].

Ventricular tachyarrhythmias (ventricular tachycardia, torsade de pointes, and ventricular fibrillation) are the most common mechanisms leading to sudden cardiac death [17]. QT interval represents the time from the onset of ventricular depolarization to the end of ventricular repolarization, and its prolongation is known to be proarrhythmic. Prolonged QT interval is associated with an increased risk of fatal ventricular arrhythmias, especially torsade de pointes. Furthermore, QT interval prolongation is an independent risk factor for sudden cardiac death [18]. Thus, measurement of the QT interval from a standard electrocardiogram (ECG) represents an easy and efficient noninvasive way to assess the risk of sudden cardiac death in obese patients.

QT INTERVAL IN OBESITY

The QT interval is usually evaluated after correction for heart rate (QTc) to enable the comparison of QT interval at different heart rates. Many studies have reported QTc prolongation in obese patients even after adjustments for age, sex, and blood pressure [19]. Obesity-related left ventricular hypertrophy is a major determinant of QTc interval in severely obese patients [20]. Interestingly, QTc prolongation seems to be more significant with increasing intraabdominal fat distribution in women [21,22]. QT interval prolongation in obesity does not always meet the criteria of the long QT syndrome, but this prolongation may have a synergistic effect in the presence of other factors causing QT prolongation like drugs, hypertension, and/or cardiac failure. QT dispersion (QTd), represents heterogeneity in myocardial repolarization, and despite methodological issues that make its use complicated, is also used to evaluate QT interval and is increased in obese patients [23]. This increase in QTd is positively correlated with visceral obesity in women [24]. Most studies have shown an association between obesity and QT interval (QTc prolongation and/or increased QTd) in adults [25] or in children [26].

EFFECT OF WEIGHT LOSS ON QT INTERVAL

Early studies of obese patients treated with an unsupervised very-low-calorie diet reported significant prolongation of QTc interval [27], torsades de pointes [28], and sudden cardiac death [29]. This deleterious effect was probably related to severe caloric restriction and a low-protein diet rather than to the weight loss itself. Inversely, studies on medically supervised diets have confirmed a beneficial effect of weight loss by shortening (decreasing) QTc interval [30,31] and/ or by improving QTd during their follow-up [32–34]. This beneficial effect was attributed to the decrease in plasma free fatty acids (FFAs) [35] or the improvement in insulin resistance [36].

QT INTERVAL AFTER BARIATRIC SURGERY

QT Interval After Restrictive Procedures

Following earlier reports of QT interval prolongation after weight loss, Rasmussen et al. [37] prospectively followed 22 female patients for 6 months after gastroplasty. An ECG was obtained monthly, and food intake was recorded at the end of follow-up period. QTc interval was 0.41 ± 0.03 seconds before gastroplasty, 0.40 ± 0.03 seconds 3 months after surgery, and 0.40 ± 0.02 seconds 6 months after surgery, but the difference was not statistically significant. However, seven patients (32%) developed QTc interval superior to 0.44 seconds at some time during the follow-up period. The presence of QTc prolongation was associated with low protein intake and lower plasma prealbumin concentrations, while serum levels of sodium, potassium, calcium, and magnesium were not different. Weight loss did not affect the length of QTc. Moreover the administration of vitamins, minerals, and trace elements did not prevent QTc prolongation. The authors of this study concluded that adequate protein intake was mandatory during weight loss in order to prevent QTc prolongation.

Pontiroli et al. [38] studied the QT interval in 116 obese patients who underwent laparoscopic adjustable gastric banding (LAGB). Forty-one patients were hypertensive at baseline, while the others were normotensive. QTc interval decreased significantly at 1 year in obese patients whether they had hypertension or not, 15.4 versus 16.5 milliseconds (ms), respectively (Fig. 55.1). The authors proposed that better control of sympathetic overactivity contributed to the decrease in QT interval after weight loss. This study demonstrated that the effect of weight loss on QTc interval was not related to its effect on blood pressure.

The same group [39] measured QT interval on 24-hour Holter monitoring in 12 obese patients before and 6 months after LAGB (Fig. 55.1). The reduction in QT interval was not significant after weight loss (438.7 ± 19.5 before surgery and 424.9 ± 47.89 ms 6 months later). However, this study was not specifically designed to evaluate bariatric surgery impact on QT interval. Moreover, the number of subjects might have been too small to detect a decrease in QT interval.

Four studies reported results from vertical banded gastroplasty (VBG)–induced weight loss on QT interval (Fig. 55.2). In the first one, Papaioannou et al. [40] reported 17 patients with morbid obesity who underwent VBG. Two patients were excluded from final analysis because they hadn't achieved the predefined weight loss. The QTc interval decreased from 428 ± 14 ms before surgery to 393 ± 26 ms 10 months postsurgery.

The second study was with 39 normotensive obese patients (body mass index (BMI) ≥ 40 kg/m^2) who were evaluated before VBG and at the nadir of postoperative weight loss. QTc interval decreased from 428.7 ± 18.5 ms before surgery to 410.3 ± 11.9 ms after weight loss. QTc dispersion also decreased from 44.1 ± 11.2 ms before surgery to 32.2 ± 3.3 ms after weight loss (Fig. 55.4). However, significant reduction in QTc interval and QTc dispersion was observed only in patients with left ventricular hypertrophy at baseline (before surgery). The authors identified left ventricular mass/height$^{2.7}$ as the most important predictor of QTc interval and dispersion before surgery. They also

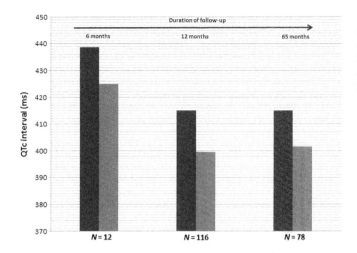

FIGURE 55.1 Effect of laparoscopic adjustable gastric banding (LAGB) on QTc interval. Studies are shown according to follow-up duration. Blue bars (black bars in print version) represent mean QTc interval before surgery while red bars (gray bars in print version) represent mean QTc interval after surgery. The number of study participants is noted below each couple of bars.

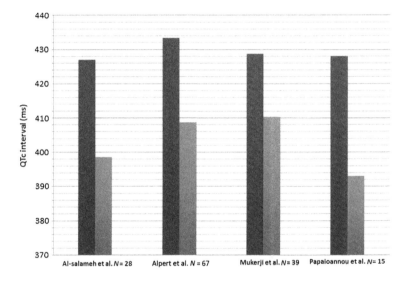

FIGURE 55.2 Effect of vertical banded gastroplasty (VBG) on QTc interval. Studies reporting the effect of VBG on QTc interval are shown. Blue bars (black bars in print version) represent mean QTc interval before surgery while red bars (gray bars in print version) represent mean QTc interval after surgery. The number of study participants is noted below each couple of bars.

identified that weight loss—related decrease in left ventricular mass was a key predictor of the reduction in QTc interval and dispersion [41].

In the third study, 67 normotensive obese patients who underwent VBG were included and followed up prospectively. Twenty-eight patients had heart failure while the remaining 39 patients had normal cardiac function. All patients were followed until the nadir of postoperative weight loss (5.0 ± 0.6 months after surgery). QTc interval and QTc dispersion decreased in all patients and in the subgroups of patients with or without heart failure. The decrease of QTc interval and dispersion were more significant in patients with heart failure. These decreases were strongly correlated with reduction in left ventricular mass [42].

Finally, a recent brief report by Al-Salameh et al. [43] found a significant decrease in QTc interval in 28 patients who underwent sleeve gastroplasty (28.5 ± 15.61 ms). This decrease was observed 3 months after surgery and was statistically significant. No significant correlation was found between weight loss and the decrease in QTc.

QT Interval After Restrictive and Malabsorptive Procedures

Bezante et al. [44] retrospectively analyzed data from 85 obese patients; among them, 55 underwent biliopancreatic diversion (BPD). The maximum and minimum QTc intervals decreased rapidly from 446 ± 27.9 and 394 ± 23.1 ms before surgery to 420 and 387 ± 22 ms 1 month after surgery, respectively (Fig. 55.3). QTd also decreased from

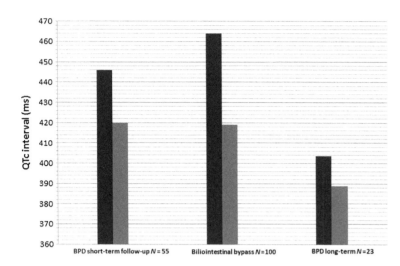

FIGURE 55.3 QT interval after restrictive and malabsorptive procedures. Studies reporting the effect of restrictive and malabsorptive procedures on QTc interval are shown. Violet bars (black bars in print version) represent mean QTc interval before surgery while brown bars (gray bars in print version) represent mean QTc interval after surgery. The number of study participants is noted below each couple of bars.

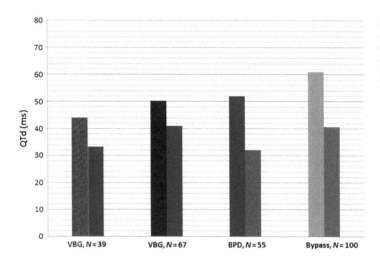

FIGURE 55.4 Effect of different techniques of bariatric surgery on QT dispersion (QTd). Green bars represent mean QTd before surgery while brown—yellow bars represent mean QTd after surgery. The technique and number of each study participant is noted below each couple of bars.

52 ± 20.3 ms before surgery to 32 ± 14 ms 1 month after surgery (Fig. 55.4). At the end of the follow-up period (1 year), the maximum, the minimum QTc intervals and QTd remained stable. The longer the preoperative maximum QTc interval was, the more it decreased postoperatively. The authors suggested that elevated sympathetic tone due to hyperinsulinemia might be responsible for prolonged preoperative QTc and increased QTd in obese patients.

Russo et al. [45] studied indexes of ventricular repolarization heterogeneity in 100 patients who underwent biliointestinal bypass. They found that QTc interval had decreased from 464 ± 29.9 ms before surgery to 419.4 ± 21.6 ms 1 year after (Fig. 55.3). Similarly, QTc dispersion decreased from 60.9 ± 23 to 40.5 ± 12.7 ms; this decrease was correlated with weight loss (Fig. 55.4).

Long-term Effect of Bariatric Surgery on QT Interval

All the above-mentioned studies were of short duration (≤ 12 months). Only one study looked for long-term effects of bariatric surgery on QT interval. Pontiroli et al. [46] recalled 101 patients who had undergone bariatric surgery (78 LAGB and 23 BPD) between 1989 and 2001. Mean interval since surgery was 65.4 months. In every patient, they recorded an ECG and compared it to the presurgery ECG. QTc interval was 412.4 ± 2.99 ms before surgery and 398.7 ± 2.81 65 months later ($p < 0.05$). QTc interval decreased from 403.5 ± 5.44 ms to 388.9 ± 6.27 ms in BPD patients (Fig. 55.3) and from 415.1 ± 3.48 to 401.6 ± 3.09 ms in LAGB patients (Fig. 55.1). This study was the only one that demonstrated a long-term effect of bariatric surgery on QT interval.

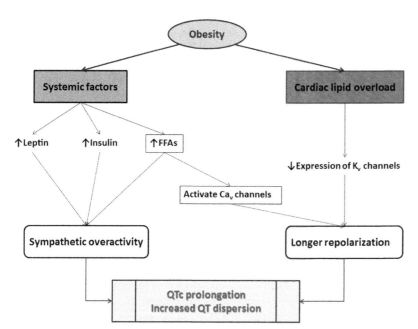

FIGURE 55.5 Schematic representation of proposed mechanisms to explain QTc prolongation and increased QT dispersion (QTd) in obese patients. Obesity increases systemic factors such as free fatty acids (FFAs), leptin, and insulin that enhance sympathetic activity. In addition, intracardiac lipid overload affects ion transport through the plasma membrane, leading to longer repolarization and consequently to QTc prolongation. K_v channels: Voltage-dependent potassium channels; Ca_v channels: Voltage-dependent calcium channels.

MECHANISMS OF QT INTERVAL REDUCTION AFTER BARIATRIC SURGERY

In order to understand how bariatric surgery improves QT prolongation it is necessary to know the mechanisms of QT prolongation in obese patients. However, these mechanisms are poorly understood. Some of them are related to systemic effects of obesity, while others are related to local effects of lipid accumulation in the heart.

Insulin is one of the systemic factors implicated in obesity-related QT prolongation, as it was suggested that hyperinsulinemia prolongs QT interval by direct insulin effect on cardiomyocyte plasma membrane potential and/or indirectly by enhancing central sympathetic activity [47]. Sympathetic overactivity leads to QT prolongation and increased QT variability (dispersion) [48]. Secondly, leptin may play a role in obesity-related QT prolongation as it enhances sympathetic activity via its multiple central interactions [49]. Finally, several lines of evidence suggest a role for elevated FFA concentrations, but the precise way leading to prolonged QT interval and/or increased QTd is still to be elucidated. Elevated FFA concentrations might enhance sympathetic tone, and also might affect (directly or indirectly) ion transport through the plasma membrane [50]. In a study on obese women, baseline QTc interval and QTc dispersion correlated well with plasma FFAs, epinephrine, and norepinephrine concentrations. Moreover, the change in QTc interval after weight loss was significantly and independently correlated with the change in plasma FFAs concentrations and the plasma levels of catecholamines [35]. It has been shown that long-chain fatty acids directly activate voltage-dependent calcium channels (essentially L-type calcium channels) in ventricular cardiomyocytes [51].

Transgenic mice with cardiac-specific overexpression of peroxisome proliferator-activated receptor gamma (PPAR-γ) represent a model for intracellular lipid accumulation in the heart. Transgenic PPAR-γ mice present significant QT prolongation, and more frequent ventricular arrhythmias when compared to age-matched, wild-type littermates. They also have reduced expression of several voltage-gated potassium channels, which are responsible for cardiomyocyte repolarization. The model of transgenic PPAR-γ mice gives some insight into the links between lipid accumulation and QT prolongation [52]. However, it is worth mentioning that ion channels responsible for cardiomyocyte repolarization are not the same in mice and humans. Huang et al. [53] studied QT interval in diet-induced obese mice. They found that QT interval was increased by 15% in diet-induced obese mice when compared with controls. This increase was related to reduced transcription of several voltage-gated potassium channels. The authors concluded that activation of protein kinase D during intracardiac lipid accumulation (overload) led to increased degradation of cyclic AMP response element binding protein (CREB). CREB binds to the promoters of these voltage-gated potassium channels and increases their transcription.

Overall, no single mechanism fully explains QT prolongation in obese patients. It seems that many systemic factors work together to enhance sympathetic activity, and besides the fact that they share a synergistic effect, every factor may have an independent action. In addition, cardiomyocyte lipid accumulation as it has been shown in animal models may contribute to the repolarization abnormality, and consequently to QT prolongation (Fig. 55.5).

CONCLUSION AND PERSPECTIVES

Most published studies support a positive effect of bariatric surgery on QT interval. The decrease in QT interval seems to correct obesity-related QT prolongation. This decrease is consistent and reproducible in both restrictive and malabsorptive techniques of bariatric surgery. This effect has been observed in a short period after surgery and seems to persist in the long term. The postoperative decrease in QT seems to be independent of the effect on blood pressure, but depends on the existence of preoperative left ventricular hypertrophy. Contradictory results were reported concerning the correlation between weight loss and QT decrease.

However, some important points need to be mentioned. First of all, the presence of publication bias—studies showing a significant effect of bariatric surgery on QT interval—are usually preferred by medical journals and editors. Secondly, published studies had small numbers of patients and limited follow-up, making it difficult to have good quality data from those studies. The lack of sufficient information about concomitant cardiovascular disease and medications that can modify QT interval duration provides additional bias in published studies. Thirdly and most importantly, QT decrease per se is a surrogate marker for the risk of sudden cardiac death. None of the above-mentioned studies presented data for the positive effect on QT interval resulting in a reduction in sudden cardiac deaths. Actually, all studies except one were of short duration, too short to detect a decrease in sudden cardiac death. Finally, the pathophysiology of QT prolongation in obesity and QT decrease after bariatric surgery need to be better illustrated.

Based on the limitations of existing studies, a well-conducted study with a sufficient number of patients and a longer follow-up period is needed. This study should be extended beyond QT interval measurements to evaluate postoperative sudden cardiac death rates. More studies are necessary to fully understand and analyze the systemic and local mechanisms of QT prolongation in obesity.

MINI-DICTIONARY OF TERMS

- *Sudden cardiac death*: Unexpected death of cardiovascular cause that occurs rapidly outside of the hospital or in the emergency room, usually in less than 24 hours.
- *Cardiomyopathy of the obese*: Structural and functional myocardial modifications induced by obesity independently of other comorbidities; these modifications usually lead to cardiomyopathy.

KEY FACTS

- QT interval is measured by standard ECG from the beginning of the QRS complex to the end of the T wave.
- It is usually measured in lead II, but leads V5 or V6 can be used to measure QT interval.
- Many formulas are used to correct QT interval for heart rate. The Bazett formula ($QTc = QT/RR^{\frac{1}{2}}$) is one of them.
- Normal QTc interval is 350−450 ms in males and 360−460 ms in females.
- QTd is the difference between the longest and shortest QT interval on standard ECG.

SUMMARY POINTS

- QTc prolongation is an independent risk factor for sudden cardiac death.
- The obese patient has an increased rate of sudden cardiac death.
- Most studies show that QTc interval is prolonged and QTd is increased in the obese patient.
- Bariatric surgery is associated with a reduction in QTc interval and/or decrease in QTd.
- This effect of bariatric surgery on QT interval seems to be reproducible with different surgical techniques.
- The effect of bariatric surgery on QT interval seems to be of rapid onset and to persist in the long term.
- The effect of bariatric surgery on QT interval is mediated by a decrease in systemic factors such as insulin and FFAs, as well as improvement in cardiac lipid accumulation.

REFERENCES

[1] Bogers RP, et al. Association of overweight with increased risk of coronary heart disease partly independent of blood pressure and cholesterol levels: a meta-analysis of 21 cohort studies including more than 300,000 persons. Arch Intern Med 2007;167(16):1720−8.
[2] Kenchaiah S, et al. Obesity and the risk of heart failure. N Engl J Med 2002;347(5):305−13.
[3] Guh DP, et al. The incidence of co-morbidities related to obesity and overweight: a systematic review and meta-analysis. BMC Public Health 2009;9:88.

[4]   Wanahita N, et al. Atrial fibrillation and obesity - results of a meta-analysis. Am Heart J 2008;155(2):310—15.

[5]   Jouven X, et al. Predicting sudden death in the population: the Paris Prospective Study I. Circulation 1999;99(15):1978—83.

[6]   Cassidy AE, et al. Progression of subclinical coronary atherosclerosis: does obesity make a difference? Circulation 2005;111(15):1877—82.

[7]   Bombelli M, et al. Impact of body mass index and waist circumference on the long-term risk of diabetes mellitus, hypertension, and cardiac organ damage. Hypertension 2011;58(6):1029—35.

[8]   Ingelsson E, et al. Burden and prognostic importance of subclinical cardiovascular disease in overweight and obese individuals. Circulation 2007;116(4):375—84.

[9]   Zebekakis PE, et al. Obesity is associated with increased arterial stiffness from adolescence until old age. J Hypertens 2005;23(10):1839—46.

[10] Brown CD, et al. Body mass index and the prevalence of hypertension and dyslipidemia. Obes Res 2000;8(9):605—19.

[11] Gupta AK, et al. Baseline predictors of resistant hypertension in the Anglo-Scandinavian Cardiac Outcome Trial (ASCOT): a risk score to identify those at high-risk. J Hypertens 2011;29(10):2004—13.

[12] Chrostowska M, et al. Impact of obesity on cardiovascular health. Best Pract Res Clin Endocrinol Metab 2013;27(2):147—56.

[13] Hubert HB, et al. Obesity as an independent risk factor for cardiovascular disease: a 26-year follow-up of participants in the Framingham Heart Study. Circulation 1983;67(5):968—77.

[14] Noheria A, et al. Distinctive profile of sudden cardiac arrest in middle-aged vs. older adults: a community-based study. Int J Cardiol 2013;168 (4):3495—9.

[15] Hookana E, et al. Causes of nonischemic sudden cardiac death in the current era. Heart Rhythm 2011;8(10):1570—5.

[16] Plourde B, et al. Sudden cardiac death and obesity. Expert Rev Cardiovasc Ther 2014;12(9):1099—110.

[17] Huikuri HV, Castellanos A, Myerburg RJ. Sudden death due to cardiac arrhythmias. N Engl J Med 2001;345(20):1473—82.

[18] Algra A, et al. QTc prolongation measured by standard 12-lead electrocardiography is an independent risk factor for sudden death due to cardiac arrest. Circulation 1991;83(6):1888—94.

[19] Frank S, Colliver JA, Frank A. The electrocardiogram in obesity: statistical analysis of 1,029 patients. J Am Coll Cardiol 1986;7(2):295—9.

[20] Mukerji R, et al. Relation of left ventricular mass to QTc in normotensive severely obese patients. Obesity (Silver Spring) 2012;20(9):1950—4.

[21] Peiris AN, et al. Relationship of regional fat distribution and obesity to electrocardiographic parameters in healthy premenopausal women. South Med J 1991;84(8):961—5.

[22] Park JJ, Swan PD. Effect of obesity and regional adiposity on the QTc interval in women. Int J Obes Relat Metab Disord 1997;21 (12):1104—10.

[23] Seyfeli E, et al. Effect of obesity on P-wave dispersion and QT dispersion in women. Int J Obes (Lond) 2006;30(6):957—61.

[24] Esposito K, et al. Autonomic dysfunction associates with prolongation of QT intervals and blunted night BP in obese women with visceral obesity. J Endocrinol Invest 2002;25(11):RC32—5.

[25] Ziegler D, et al. Selective contribution of diabetes and other cardiovascular risk factors to cardiac autonomic dysfunction in the general population. Exp Clin Endocrinol Diabetes 2006;114(4):153—9.

[26] Nigro G, et al. Increased heterogenity of ventricular repolarization in obese nonhypertensive children. Pacing Clin Electrophysiol 2010;33 (12):1533—9.

[27] Pringle TH, et al. Prolongation of the QT interval during therapeutic starvation: a substrate for malignant arrhythmias. Int J Obes 1983;7 (3):253—61.

[28] Thwaites BC, Bose M. Very low calorie diets and pre-fasting prolonged QT interval. A hidden potential danger. West Indian Med J 1992;41 (4):169—71.

[29] Spencer IO. Death during therapeutic starvation for obesity. Lancet 1968;1(7555):1288—90.

[30] Carella MJ, et al. Obesity, adiposity, and lengthening of the QT interval: improvement after weight loss. Int J Obes Relat Metab Disord 1996;20(10):938—42.

[31] Pietrobelli A, et al. Electrocardiographic QTC interval: short-term weight loss effects. Int J Obes Relat Metab Disord 1997;21(2):110—14.

[32] Mshui ME, et al. QT interval and QT dispersion before and after diet therapy in patients with simple obesity. Proc Soc Exp Biol Med 1999;220 (3):133—8.

[33] Esposito K, et al. Sympathovagal balance, nighttime blood pressure, and QT intervals in normotensive obese women. Obes Res 2003;11 (5):653—9.

[34] Seyfeli E, et al. Effect of weight loss on QTc dispersion in obese subjects. Anadolu Kardiyol Derg 2006;6(2):126—9.

[35] Corbi GM, et al. FFAs and QT intervals in obese women with visceral adiposity: effects of sustained weight loss over 1 year. J Clin Endocrinol Metab 2002;87(5):2080—3.

[36] Seshadri P, et al. Free fatty acids, insulin resistance, and corrected qt intervals in morbid obesity: effect of weight loss during 6 months with differing dietary interventions. Endocr Pract 2005;11(4):234—9.

[37] Rasmussen LH, Andersen T. The relationship between QTc changes and nutrition during weight loss after gastroplasty. Acta Med Scand 2009;217(3):271—5.

[38] Pontiroli AE, et al. Left ventricular hypertrophy and QT interval in obesity and in hypertension: effects of weight loss and of normalisation of blood pressure. Int J Obes Relat Metab Disord 2004;28(9):1118—23.

[39] Pontiroli AE, et al. Effect of weight loss on sympatho-vagal balance in subjects with grade-3 obesity: restrictive surgery versus hypocaloric diet. Acta Diabetol 2013;50(6):843—50.

[40] Papaioannou A, et al. Effects of weight loss on QT interval in morbidly obese patients. Obes Surg 2003;13(6):869—73.

[41] Mukerji R, et al. Effect of weight loss after bariatric surgery on left ventricular mass and ventricular repolarization in normotensive morbidly obese patients. Am J Cardiol 2012;110(3):415–19.

[42] Alpert MA, et al. Effect of weight loss on ventricular repolarization in normotensive severely obese patients with and without heart failure. Am J Med Sci 2015;349(1):17–23.

[43] Al-Salameh A, et al. Shortening of the QT interval is observed soon after sleeve gastrectomy in morbidly obese patients. Obes Surg 2013;24(1):167–70.

[44] Bezante GP, et al. Biliopancreatic diversion reduces QT interval and dispersion in severely obese patients*. Obesity 2007;15(6):1448–54.

[45] Russo V, et al. Effect of weight loss following bariatric surgery on myocardial dispersion of repolarization in morbidly obese patients. Obes Surg 2007;17(7):857–65.

[46] Pontiroli AE, et al. Biliary pancreatic diversion and laparoscopic adjustable gastric banding in morbid obesity: their long-term effects on metabolic syndrome and on cardiovascular parameters. Cardiovasc Diabetol 2009;8(1):37.

[47] Gastaldelli A, et al. Insulin prolongs the QTc interval in humans. Am J Physiol Regul Integr Comp Physiol 2000;279(6):R2022–5.

[48] Piccirillo G, et al. Autonomic nervous system activity measured directly and QT interval variability in normal and pacing-induced tachycardia heart failure dogs. J Am Coll Cardiol 2009;54(9):840–50.

[49] Hall JE, Hildebrandt DA, Kuo J. Obesity hypertension: role of leptin and sympathetic nervous system. Am J Hypertens 2001;14(6 Pt 2):103S–15S.

[50] Marfella R, et al. Elevated plasma fatty acid concentrations prolong cardiac repolarization in healthy subjects. Am J Clin Nutr 2001;73(1):27–30.

[51] Huang JM, Xian H, Bacaner M. Long-chain fatty acids activate calcium channels in ventricular myocytes. Proc Natl Acad Sci USA 1992;89(14):6452–6.

[52] Morrow JP, et al. Mice with cardiac overexpression of peroxisome proliferator-activated receptor have impaired repolarization and spontaneous fatal ventricular arrhythmias. Circulation 2011;124(25):2812–21.

[53] Huang H, et al. Diet-induced obesity causes long QT and reduces transcription of voltage-gated potassium channels. J Mol Cell Cardiol 2013;59:151–8.

Chapter 56

Plasma Polyunsaturated Fatty Acids After Weight Loss Surgery

I. Aslan[1] and M. Aslan[2]

[1]Research and Education Hospital, Antalya, Turkey, [2]Akdeniz University Medical Faculty, Antalya, Turkey.

LIST OF ABBREVIATIONS

AA arachidonic acid
ALA alpha-linolenic acid
COX cyclooxygenase
DGLA dihomo-gamma-linolenic acid
DHA docosahexaenoic acid
EPA eicosapentaenoic acid
HDL high-density lipoprotein
hs-CRP high sensitive C-reactive protein
LA linoleic acid
LOX lipoxygenase
LSG laparoscopic sleeve gastrectomy
MUFA monounsaturated fatty acids
n-3 omega-3
n-6 omega-6
NAFLD nonalcoholic fatty liver disease
PGE2 prostaglandin E2
PUFAs polyunsaturated fatty acids
RYGBP Roux-en-Y gastric bypass
T2DM type 2 diabetes mellitus

INTRODUCTION

Polyunsaturated fatty acids (PUFAs) are mainly localized in cell membranes. They play an important role in membrane fluidity and many physiological functions, some of which include inflammation, blood clotting, regulation of blood pressure, and cell signaling [1]. PUFAs are categorized as omega-3 (*n*-3) and omega-6 (*n*-6) depending on the location of the last double bond with reference to the terminal methyl end of the molecule [2]. Humans can synthesize all fatty acids utilized by the body except for the two essential PUFAs, linoleic acid (LA, C18:2*n*-6) and alpha-linolenic acid (ALA, C18:3*n*-3) [3]. Major dietary sources for these two essential PUFAs are plant oils such as corn and sunflower oil [4]. LA is the precursor of *n*-6 while ALA is the precursor of the *n*-3 series of PUFAs [2]. Even though LA and ALA cannot be synthesized in humans, they can be metabolized to other PUFAs by the addition of double bonds and acyl chains via desaturases and elongases, respectively [2,5,6] (Fig. 56.1).

Arachidonic acid (AA, C20:4n6) and eicosapentaenoic acid (EPA, C20:5n3) can be metabolized to eicosanoids via cyclooxygenase (COX) and lipoxygenase (LOX) pathways (Fig. 56.1). The metabolism of docosahexaenoic acid (DHA, C22:6n3) via LOX can also yield resolvins and protectins that display potent antiinflammatory properties and are recognized in the resolution of inflammation [7]. It is known that eicosanoids derived from *n*-6 PUFAs have proinflammatory and immunoactive functions, whereas eicosanoids derived from *n*-3 PUFAs have antiinflammatory properties,

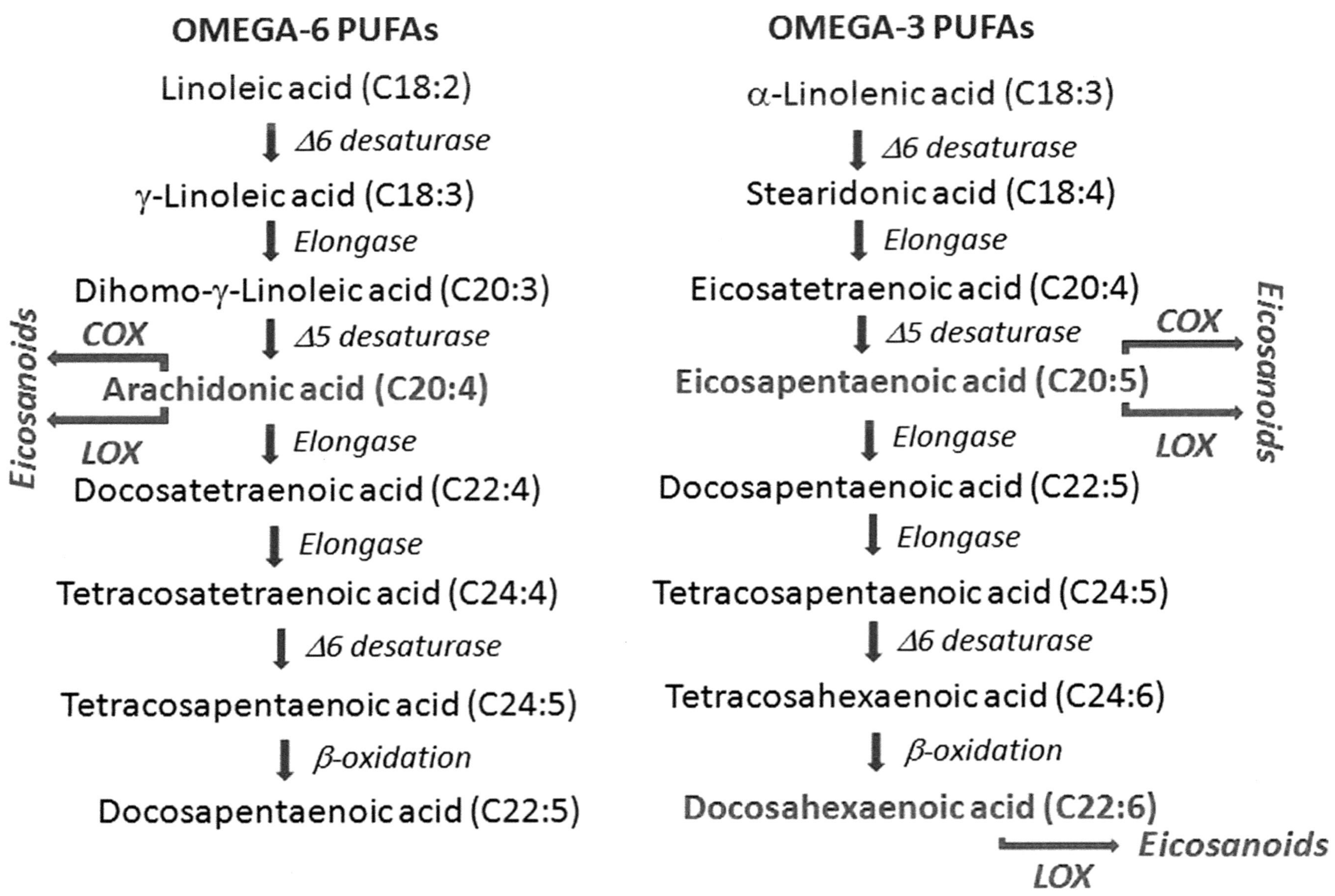

OMEGA-3 PUFAs

α-Linolenic acid (C18:3)
↓ Δ6 desaturase
Stearidonic acid (C18:4)
↓ Elongase
Eicosatetraenoic acid (C20:4)
↓ Δ5 desaturase
Eicosapentaenoic acid (C20:5) → COX → Eicosanoids / LOX → Eicosanoids
↓ Elongase
Docosapentaenoic acid (C22:5)
↓ Elongase
Tetracosapentaenoic acid (C24:5)
↓ Δ6 desaturase
Tetracosahexaenoic acid (C24:6)
↓ β-oxidation
Docosahexaenoic acid (C22:6) → LOX → Eicosanoids

OMEGA-6 PUFAs

Linoleic acid (C18:2)
↓ Δ6 desaturase
γ-Linoleic acid (C18:3)
↓ Elongase
Dihomo-γ-Linoleic acid (C20:3)
↓ Δ5 desaturase
Arachidonic acid (C20:4) → COX → Eicosanoids / LOX → Eicosanoids
↓ Elongase
Docosatetraenoic acid (C22:4)
↓ Elongase
Tetracosatetraenoic acid (C24:4)
↓ Δ6 desaturase
Tetracosapentaenoic acid (C24:5)
↓ β-oxidation
Docosapentaenoic acid (C22:5)

FIGURE 56.1 Synthesis of polyunsaturated fatty acids. Scheme depicting polyunsaturated fatty acid production [2.5.6]. *COX*, cyclooxygenase; *LOX*, lipoxygenase.

attributable to their ability to inhibit the formation of *n*-6 PUFA-derived eicosanoids [2]. Recent studies have shown a low serum EPA/AA ratio in male subjects with visceral obesity [8].

Bariatric surgery is considered to be the treatment with the best long-term results for severe obesity [9]. However, impaired absorption of important nutrients and food restriction in the postoperative period after bariatric surgery can cause changes in circulating and tissue levels of PUFAs [10]. Bariatric surgery is also associated with a high rate of resolution of type 2 diabetes mellitus (T2DM) and other obesity-associated comorbidities such as hyperlipidemia [11]. The improvement of insulin action occurs very early following bariatric surgery, with a significant reduction in insulin resistance [12]. Insulin stimulates the conversion of essential fatty acids (LA and ALA) to longer-chain PUFAs [13]. This review focuses on the postoperative effects of bariatric surgery on plasma *n*-6 and *n*-3 PUFA levels, and summarizes studies that have investigated changes in PUFA levels following weight loss surgery.

ALTERATIONS OF POLYUNSATURATED FATTY ACIDS AFTER WEIGHT LOSS SURGERY

The restrictive and malabsorptive effects of bariatric surgery have been reported to cause a reduction in circulating levels of PUFAs [10,14], which could have detrimental effects on vitamin absorption, metabolism, and cardiovascular function. Thus, recognition of PUFA deficiencies before and after bariatric surgery is critical in order to minimize the effects of the surgery and to reduce the incidence of metabolic and/or cardiovascular complications.

A recent study assessed nutritional status in 25 premenopausal women who had undergone Roux-en-Y gastric bypass (RYGBP), and 33 age- and body mass index (BMI)—matched women who had not undergone surgery [10]. A semiquantitative food frequency questionnaire was used for dietary assessment, and it was recorded that premenopausal women who had undergone RYGBP had significantly lower intake of PUFAs.

Plasma polyunsaturated *n*-6 and *n*-3 fatty acids were also measured in 38 morbidly obese patients before RYGBP, and 28 of them were reexamined 6 months postoperatively [14]. Measured *n*-6 PUFAs, including LA, dihomo-gamma-linolenic acid (DGLA, C20:3n6), and *n*-3 PUFAs, including ALA, DHA, and EPA, were significantly decreased in patients 6 months postoperatively. On the contrary, plasma AA levels were significantly increased in RYGBP patients 6 months postoperation. This study also reported that EPA was negatively related to both circulating fasting insulin and high-sensitivity C-reactive protein (hs-CRP), and positively to high-density lipoprotein cholesterol (HDL-c). Other studies have also shown an association with *n*-3 PUFA and metabolic syndrome in obese patients [15]. In fact, the beneficial effect of EPA on insulin resistance has been attributed to its antiinflammatory action [16].

We recently evaluated early postoperative effects of bariatric surgery on plasma *n*-6 and *n*-3 PUFA levels by measuring AA, DGLA, EPA, and DHA in 10 obese patients who had undergone laparoscopic sleeve gastrectomy (LSG) and 11 normal weight control patients who had undergone laparoscopic abdominal surgery [17]. We observed a significant decrease in insulin levels and insulin resistance in sleeve gastrectomy patients after postoperation oral feeding compared to preoperation levels. Plasma AA levels and AA/EPA ratio were significantly increased in sleeve gastrectomy patients after postoperation oral feeding compared to postoperation day 1. Serum prostaglandin E2 (PGE2) levels and AA/DHA ratio were significantly higher in sleeve gastrectomy patients when compared to control group patients. Although our study determined early postoperative effects of bariatric surgery, the results were in agreement with Chalut-Carpentier et al. [14], who reported that AA levels were significantly increased in RYGBP patients 6 months postoperation. Increased peripheral insulin sensitivity associated with bariatric surgery may play a role in the significant increase of plasma AA levels in sleeve gastrectomy patients following surgery. The significant increase in PGE2 levels and AA/DHA ratio in morbidly obese patients also confirmed the presence of a proinflammatory state in obesity.

As stated in the introduction, insulin stimulates the conversion of essential fatty acids (LA and ALA) to longer-chain PUFAs. Improved insulin sensitivity in diabetic obese patients after bariatric surgery, before any weight loss, seems to be related to hormonal changes of possible gastric origin [18]. Improvements in glucose metabolism and insulin resistance in the long term following bariatric surgery are due to decreased fat mass and resulting changes in the release of adipocytokines [19]. Disturbed fatty acid metabolism is an important feature of the insulin-resistant state [20]. The hepatic microsomal delta-6-desaturation of LA and ALA was found to be depressed in alloxan-induced diabetic rats [21]. The observed enzymatic defect was corrected by insulin injection in 2 days [22]. It was later shown that delta-6- desaturase mRNA was seven fold lower in the streptozotocin—diabetic rat than in the control [22], and the administration of insulin induced the enzyme mRNA eight fold within 24 hours [23]. In vivo experiments also showed similar effects of diabetes on rat liver delta-5 desaturation and the correcting effect of insulin [24]. We showed that insulin analog initiation therapy in T2DM patients increased long-chain PUFAs in human plasma and resulted in a significant decrease in AA/EPA ratio [25]. These alterations may be of importance to understand the role of improved insulin sensitivity in increasing AA levels.

An ^{1}H NMR−based metabolomics assay was used to study hepatic levels of PUFA, DHA, AA, EPA, and the ratio between poly- and monounsaturated fatty acids (PUFA/MUFA) in patients undergoing bariatric surgery [26]. Data were obtained immediately before and after a 12-month period following surgery. There was a significant depletion of PUFA in moderate to severe nonalcoholic fatty liver disease (NAFLD). Conversely, there were increased hepatic levels of PUFAs, DHA, AA, EPA, and PUFA/MUFA in patients with no or very moderate NAFLD. NAFLD decreased in all patients and was nearly absent (0−7%) 1 year after surgery. Severe NAFLD is known to be accompanied by mitochondrial dysfunction and progressive inhibition of fatty acid oxidation, which increases glucose oxidation and metabolites that cause defects in insulin signaling [27]. Increased peripheral insulin sensitivity associated with bariatric surgery [12,19] may also play a role in the significant decrease of NAFLD postoperation.

A recent randomized, open-label controlled clinical trial investigated whether *n*-3 PUFA reduced adipose tissue inflammation in severely obese nondiabetic patients. Fifty-five severely obese nondiabetic patients scheduled to undergo elective bariatric surgery were treated with long-chain *n*-3 PUFAs, including EPA and DHA or an equivalent amount of butter fat as control, for 8 weeks. Inflammatory gene expression was measured in visceral and subcutaneous adipose tissue samples collected during surgery. Adipose tissue production of *n*-3 PUFA−derived eicosanoids and plasma concentrations of inflammatory markers were also determined. Treatment with *n*-3 PUFAs significantly decreased gene expression of most analyzed inflammatory genes in subcutaneous adipose tissue and significantly increased production of antiinflammatory eicosanoids in visceral adipose tissue and subcutaneous adipose tissue compared with controls [28].

The constructive effects of *n*-3 PUFA have also been shown in obese diabetic mice [29]. An *n*-3 PUFA−enriched diet increased expression of genes involved in glucose transport and insulin signaling, as well as genes involved in insulin sensitivity such as peroxisome proliferator−activated receptor gamma (PPARγ) [29]. There is also evidence showing that use of *n*-3 PUFA supplementation improves clinical outcomes of patients with diabetes. In diabetic and insulin-resistant patients, *n*-3 PUFAs bind to G protein−coupled receptor 120 (GPR120), which leads to decreased cytokine synthesis from inflammatory macrophages and enhanced signaling in adipocytes, and causes a reduction in insulin resistance [30].

In conclusion, restrictive and malabsorptive effects of bariatric surgery have been reported to cause a reduction in circulating levels of PUFAs, which could have detrimental effects on vitamin absorption, metabolism, and cardiovascular function. Thus, recognition of PUFA deficiencies before and after bariatric surgery is critical in order to minimize the effects of the surgery and to reduce the incidence of metabolic and/or cardiovascular complications. The improvement of insulin action following bariatric surgery, and a significant reduction in insulin resistance, can also lead to conversion of essential fatty acids to longer-chain PUFAs. Indeed, increased peripheral insulin sensitivity associated with bariatric surgery may play a role in the significant increase of plasma AA levels in sleeve gastrectomy patients following surgery. Decreased NAFLD in patients after bariatric surgery and increased hepatic levels of PUFAs in patients with no or very moderate NAFLD suggest that increased peripheral insulin sensitivity associated with bariatric surgery may also play a role in the significant decrease of NAFLD postoperation.

MINI-DICTIONARY OF TERMS

- *Cell signaling*: A multipart organization of interaction that regulates basic cellular activities and harmonizes cellular events.
- *Clotting*: A physiological and complex process intended to stop bleeding in response to an injury or cut.
- *Cyclooxygenase*: Also known as prostaglandin−endoperoxide synthase (PTGS), it is an enzyme that leads to the formation of prostanoids, including thromboxane and prostaglandins.
- *Desaturases*: An enzyme that extracts two hydrogen atoms from a fatty acid, creating a carbon−carbon double bond.
- *Diabetes mellitus*: A metabolic disease in which there are high blood sugar levels over a prolonged period.
- *Elongase*: An enzyme that generates many of the long-chain mono- and PUFAs.
- *Hyperlipidemia*: Increased levels of lipids in the blood, including cholesterol and triglycerides.
- *Inflammation*: A localized bodily situation in which a part of the tissue becomes swollen, reddened, hot, and frequently painful. Takes place as a reaction to infection or tissue injury.
- *Lipoxygenase*: Iron-containing enzyme that catalyzes the dioxygenation of PUFAs in lipids.
- *Malabsorption*: A disorder in which the intestine cannot effectively absorb specific nutrients into the bloodstream. It can impede the absorption of fats, proteins, carbohydrates, vitamins, and minerals.
- *Steatosis*: A term that describes the accumulation of fat in the liver.

KEY FACTS

- Fatty acids are carboxylic acids with a long saturated or unsaturated aliphatic chain.
- Many naturally occurring fatty acids have 4–28 carbon atoms.
- Fatty acids are derived from triglycerides or phospholipids.
- Fatty acids are important sources of energy, generating ATP when metabolized.
- Many cell types can use fatty acids to generate energy.
- Long-chain fatty acids cannot cross the blood–brain barrier, and therefore cannot be an energy source for cells of the central nervous system.
- Medium-chain fatty acids can be used as an energy source by the central nervous system.

SUMMARY POINTS

- This chapter focuses on the postoperative effects of bariatric surgery on plasma omega-3 (*n*-3) and omega-6 (*n*-6) PUFA levels.
- Restrictive and malabsorptive effects of bariatric surgery have been reported to cause alterations in circulating levels of PUFAs.
- Recognition of PUFA alterations before and after bariatric surgery is critical in order to minimize the effects of the surgery and to reduce the incidence of metabolic and/or cardiovascular complications.
- Increased peripheral insulin sensitivity associated with bariatric surgery may play a role in the significant change of PUFA levels in sleeve gastrectomy patients following surgery.
- Reduced hepatic steatosis in patients after bariatric surgery suggests that decreased insulin levels after weight loss surgery may also play a role in eliminating NAFLD.

REFERENCES

[1] Calder PC. n-3 polyunsaturated fatty acids, inflammation, and inflammatory diseases. Am J Clin Nutr 2006;83:1505S–19S.
[2] Wall R, Ross RP, Fitzgerald GF, Stanton C. Fatty acids from fish: the anti-inflammatory potential of long-chain omega-3 fatty acids. Nutr Rev 2010;68:280–9.
[3] Patterson E, Wall R, Fitzgerald GF, Ross RP, Stanton C. Health implications of high dietary omega-6 polyunsaturated fatty acids. J Nutr Metab 2012;2012:539426.
[4] Spector AA, Kim HY. Discovery of essential fatty acids. J Lipid Res 2015;56:11–21.
[5] Chilton FH, Murphy RC, Wilson BA, Sergeant S, Ainsworth H, Seeds MC, et al. Diet-gene interactions and PUFA metabolism: a potential contributor to health disparities and human diseases. Nutrients 2014;6:1993–2022.
[6] Lenihan-Geels G, Bishop KS, Ferguson LR. Alternative sources of omega-3 fats: can we find a sustainable substitute for fish? Nutrients 2013;5:1301–15.
[7] Serhan CN, Chiang N, Van Dyke TE. Resolving inflammation: dual anti-inflammatory and pro-resolution lipid mediators. Nat Rev Immunol 2008;8:349–61.
[8] Inoue K, Kishida K, Hirata A, Funahashi T, Shimomura I. Low serum eicosapentaenoic acid/arachidonic acid ratio in male subjects with visceral obesity. Nutr Metab (Lond) 2013;10:25.
[9] Bult MJ, van Dalen T, Muller AF. Surgical treatment of obesity. Eur J Endocrinol 2008;158:135–45.
[10] Menegati GC, de Oliveira LC, Santos AL, Cohen L, Mattos F, Mendonça LM, et al. Nutritional status, body composition, and bone health in women after bariatric surgery at a University Hospital in Rio de Janeiro. Obes Surg 2015. Available from: http://dx.doi.org/10.1007/s11695-015-1910-5.
[11] Silecchia G, Boru C, Pecchia A, Rizzello M, Casella G, Leonetti F, et al. Effectiveness of laparoscopic sleeve gastrectomy (first stage of biliopancreatic diversion with duodenal switch) on co-morbidities in super-obese high-risk patients. Obes Surg 2006;16:1138–44.
[12] Rizzello M, Abbatini F, Casella G, Alessandri G, Fantini A, Leonetti F, et al. Early postoperative insulin-resistance changes after sleeve gastrectomy. Obes Surg 2010;20:50–5.
[13] Brenner RR. Hormonal modulation of delta6 and delta5 desaturases: case of diabetes. Prostaglandins Leukot Essent Fatty Acids 2003;68:151–62.
[14] Chalut-Carpentier A, Pataky Z, Golay A, Bobbioni-Harsch E. Involvement of dietary fatty acids in multiple biological and psychological functions, in morbidly obese subjects. Obes Surg 2015;25:1031–8.
[15] Robinson LE, Buchholz AC, Mazurak VC. Inflammation, obesity, and fatty acid metabolism: influence of n-3 polyunsaturated fatty acids on factors contributing to metabolic syndrome. Appl Physiol Nutr Metab 2007;32:1008–24.
[16] Kalupahana NS, Claycombe K, Newman SJ, Stewart T, Siriwardhana N, Matthan N, et al. Eicosapentaenoic acid prevents and reverses insulin resistance in high-fat diet-induced obese mice via modulation of adipose tissue inflammation. J Nutr 2010;140:1915–22.

[17] Aslan M, Aslan I, Özcan F, Eryılmaz R, Ensari CO, Bilecik T. A pilot study investigating early postoperative changes of plasma polyunsaturated fatty acids after laparoscopic sleeve gastrectomy. Lipids Health Dis 2014;13:62.

[18] Basso N, Capoccia D, Rizzello M, Abbatini F, Mariani P, Maglio C, et al. First-phase insulin secretion, insulin sensitivity, ghrelin, GLP-1, and PYY changes 72 h after sleeve gastrectomy in obese diabetic patients: the gastric hypothesis. Surg Endosc 2011;25:3540–50.

[19] Gumbs AA, Modlin IM, Ballantyne GH. Changes in insulin resistance following bariatric surgery: role of caloric restriction and weight loss. Obes Surg 2005;15:462–73.

[20] Sjögren P, Sierra-Johnson J, Gertow K, Rosell M, Vessby B, de Faire U, et al. Fatty acid desaturases in human adipose tissue: relationships between gene expression, desaturation indexes and insulin resistance. Diabetologia 2008;51:328–35.

[21] Mercuri O, Peluffo RO, Brenner RR. Depression of microsomal desaturation of linoleic to gamma-linolenic acid in the alloxan-diabetic rat. Biochim Biophys Acta 1966;116:409–11.

[22] Mercuri O, Peluffo RO, Brenner RR. Effect of insulin on the oxidative desaturation ofa-linolenic, oleic and palmitic acids. Lipids 1967;2:284–5.

[23] Rimoldi OJ, Finarelli GS, Brenner RR. Effects of diabetes and insulin on hepatic delta6 desaturase gene expression. Biochem Biophys Res Commun 2001;283:323–6.

[24] Poisson JP. Comparative in vivo and in vitro study of the influence of experimental diabetes on rat liver linoleic acid delta 6- and delta 5-desaturation. Enzyme 1985;34:1–14.

[25] Aslan M, Özcan F, Aslan I, Yücel G. LC-MS/MS analysis of plasma polyunsaturated fatty acids in type 2 diabetic patients after insulin analog initiation therapy. Lipids Health Dis 2013;12:169.

[26] Calvo N, Beltrán-Debón R, Rodríguez-Gallego E, Hernández-Aguilera A, Guirro M, Mariné-Casadó R, et al. Liver fat deposition and mitochondrial dysfunction in morbid obesity: an approach combining metabolomics with liver imaging and histology. World J Gastroenterol 2015;21:7529–44.

[27] Yki-Järvinen H. Nutritional modulation of non-alcoholic fatty liver disease and insulin resistance. Nutrients 2015;7:9127–38.

[28] Itariu BK, Zeyda M, Hochbrugger EE, Neuhofer A, Prager G, Schindler, et al. Long-chain n-3 PUFAs reduce adipose tissue and systemic inflammation in severely obese nondiabetic patients: a randomized controlled trial. Am J Clin Nutr 2012;96:1137–49.

[29] González-Périz A, Horrillo R, Ferré N, Gronert K, Dong B, Morán-Salvador E, et al. Obesity-induced insulin resistance and hepatic steatosis are alleviated by omega-3 fatty acids: a role for resolvins and protectins. FASEB J 2009;23:1946–57.

[30] Kazemian P, Kazemi-Bajestani SM, Alherbish A, Steed J, Oudit GY. The use of ω-3 poly-unsaturated fatty acids in heart failure: a preferential role in patients with diabetes. Cardiovasc Drugs Ther 2012;26:311–20.

Chapter 57

Bariatric Procedures and Dual Energy X-Ray Absorptiometry

S.L. Faria[1], M.F. de Novais[2], O.P. Faria[2] and M. de Almeida Cardeal[2]

[1]University of Brasilia, Brasilia, DF, Brazil, [2]Gastrocirurgia de Brasília, Brasilia, DF, Brazil

LIST OF ABBREVIATIONS

BMI body mass index
CV coefficient of variation
DXA dual energy X-ray absorptiometry
FFM fat-free mass
SAD sagittal abdominal diameter

In recent decades, obesity has grown at alarming rates worldwide. In Brazil, 51% of the population is overweight, 17% of whom are obese. In developing countries, the number of overweight adults has increased by four times in the last three decades, and in the United States the prevalence of obesity is 34% among men and 55% among women between 20 and 64 years of age [1].

Obesity is characterized by an elevated amount of adipose tissue mass and the total of extracellular water, considering the individual's body composition. Despite the measurement of total body weight and body mass index (BMI) being well-known indices and widely used in nutritional assessment, they are extremely inaccurate and understate body fat values in the sedentary, obese, elderly, and in certain clinical conditions [2–4].

Bariatric interventions are the best treatment for grave obesity, being essential in promoting health, reducing mortality, and offering long-term weight loss, while avoiding a risk factor for other diseases and the reduction of one's quality of life [2]. Thus, to better assess the quality of weight loss of these procedures, it is crucial that an accurate assessment of body composition be routinely included as part of the clinical evaluation of these patients. The assessment of body composition among obese and postobese patients may be difficult due to the limitations of the equipment, the characteristics of the methods used, and the intrinsic characteristics of the patient's own nutritional state [5]. Thus, the present chapter aims to present the features, advantages, and limitations of using one of the body composition assessment methods on obese adults and postoperative bariatric patients: dual energy X-ray absorptiometry (DXA), known as DXA.

Proper body composition assessment requires studies that address varied issues, such as the quantification of the main structural components of the human body, namely, the specific and differentiated tissues that make up the total body mass. Quantification of body components can be done using different methods, which are classified as either direct or indirect [2]. Using these methods, it is possible to quantify the major components of the body, obtaining important information regarding the size, shape, and constitution, characteristics influenced by genetic and environmental factors [5]. In summary, there are two major body components in terms of tissues, which are described as fat-free mass (FFM) (water, bones, and muscles) and fat (Fig. 57.1). The variations and differences in the quantities of these tissues are influential in forming the wide variety of body mass found among individuals, given the particularities of gender, age, and lifestyles [5]. Human body composition is divided into organs and metabolically active tissues. In the figure, one can identify that the muscle and adipose tissue are the most metabolically active tissues. Thus, the variation of body composition is directly related to the basal metabolic rate of the individual, whether obese or not.

Body compartments

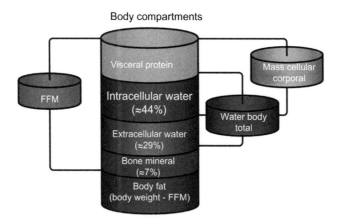

FIGURE 57.1   Body compartments. *Kyle UG et al. Bioelectrical impedance analysis--part I: review of principles and methods. Clin nutr 2004;23(5): 1226–43.*

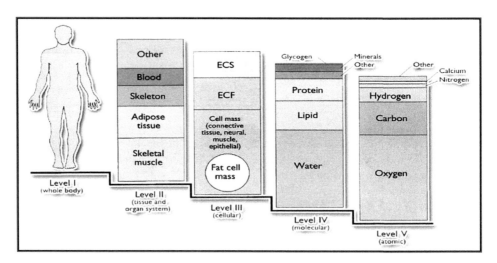

FIGURE 57.2   The five levels of body composition: whole body, tissue and organ system, cellular, molecular, and atomic levels. *ECS*: extracellular solids; *ECF*: extracellular fluid. *Adapted from Wang ZM, Pierson Jr RN, Heymsfield SB. The five-level model: a new approach to organizing body composition research. Am J Clin Nutr 1992;56:19–28.*

In postoperative bariatric patients, protein deficiency is common, which interferes in body composition and subsequently in basal metabolic rate. The body cell mass is the part of the body rich in protein, corresponding to the visceral proteins and intracellular water, which is the compartment affected in catabolic states. In the case of superhydrated individuals, as with patients who are or were obese (e.g., bariatric patients), a difficulty arises in detecting FFM, which in turn makes it difficult to identify any protein malnutrition due to the increase of extracellular water in these individuals (Fig. 57.2).

The various methods of assessing body composition already used and studied have been widely discussed. This is especially true for the assessment of body composition among the obese population, given that the assessment of these individuals is hampered by a number of limitations, such as the equipment and body composition protocols employed.

DXA results and measurements are regarded as the "gold standard" in validation studies of methods and equations used for the assessment of body composition. DXA is one of the densitometry techniques most frequently used to determine bone density in healthy populations [6,7]. Despite these advantages, its use is limited due to factors such as the high cost of equipment, the need for specialized training, and the construction of appropriate facilities [7].

The confirmation of the exam's excellent results, with only minor deviations of precision, as seen in comparative studies with chemical analysis of animal carcasses, has made this method a reference for the study of body composition in human beings [8].

The densitometry of the whole body as presented by DXA fully assesses every body compartment in a direct manner, whether it be bone, muscle, fat, or water, and does not infer data by measuring only one compartment. In DXA body composition results, body water is included and incorporated into the compartment of muscle mass (lean mass), and as such, body water does not affect the measurement of the content of adipose or bone tissue. In addition to the advantages of DXA, when compared to other noninvasive methods for assessing body composition (analysis of neutron activation, submersion in water, labeled water, total body potassium, among others), there are advantages because these other methods require the performance of more than one exam to complete data collection, since results are based on assessment of just one body compartment and inference is applied for the others. For these and other reasons, such methods require more work and effort and are not available in clinical practice—they are restricted to research laboratories [4,8,9].

Body composition assessment and measurement of body fat and lean mass, as well as evaluating the loss of muscle mass and bone mass among the obese, are of great interest because they are parameters widely used in diagnosing sarcopenic obesity [7,8]. Bariatric patients lose a considerable amount of weight after surgery, so regular evaluation of body composition (from the preoperative phase) is an important means for diagnosing and preventing excessive loss of lean mass.

Another advantage of the DXA densitometry method is the determination of regional composition, which permits the individualized study of the composition of the arms, legs, trunk, and abdomen. In obesity, abdominal and central adiposity (related to insulin resistance, dyslipidemia, and arterial hypertension) are regarded as an additional risk for cardiovascular disease and diabetes [3,4,7]. The examination, which is noninvasive, is simple for the patient, requires no preparation, and lasts only 10–15 minutes. Radiation is extremely low (similar to environmental radiation), and can be repeated as often as necessary for a complete evaluation of the patient [7].

The coefficients of variation (CV%) of parameters assessed in body composition by DXA are likely to be compared to results obtained in densitometry examinations of the lumbar column and femur (approximately 1–2%), thus providing reliable and reproducible results. Only fat mass presents a somewhat higher CV% (around 3–4%), probably due to the heterogeneous soft tissue in obese patients [4,7].

Calibration of the densitometer allows for the assessment of individuals with a fat percentage between 4% and 50%, and up to 135 kg, with a sagittal abdominal diameter (SAD) of up to 30 cm; considerable precision is lost due to these limits. Such factors hamper the assessment of "the superobese." Moreover, it is important to remember that, as stated earlier, body water is already incorporated into the lean mass compartment, and therefore hydration disorders that are observed in cardiac, renal, and hepatic insufficiencies, and in nephrotic syndrome, diabetes, and aging (higher muscle hydration and fibrosis) can alter the values obtained and should not be interpreted as alterations of muscle mass [10,11].

True effectiveness of the DXA method requires a minimum of cooperation between the patient and an examiner with technical skill. However, for more precise and accurate results, the information and instructions on how to use the scanner must be clearly stated by the manufacturer. In addition, some locations require a qualified radiology technician to perform the scan.

Of the guidelines and information provided for users of this method, the following are especially important:

1. Before performing the test, calibrate the DXA scanner using a well-known calibration marker provided by the manufacturer.
2. Request that the patients use as little clothing as possible and that they remove their shoes when measuring their height and weight.
3. Carefully position the patients in order to maintain them in the supine position on the scanner table for full front and back views, from head to toe.
4. The thickness of the patient's body must be accurately determined.
5. Set the scanner at medium speed to perform the exam of the entire body, which normally takes 20 minutes. For patients with a SAD exceeding 27 cm, set the scan at a low speed, which typically lasts 40 minutes.

Fig. 57.3 shows an example of the scanning result from a DXA exam of body composition after a nutritional monitoring of weight loss of a patient (male). The equipment used in clinical practice is also shown (Fig. 57.4).

Besides these steps, the DXA method should not be used to assess the body composition of larger patients, i.e., individuals whose physical dimensions (height or width) exceed the length or width of the scanner table [12].

There are several methodologies for measuring the percentage of total body fat, and the consensus is that DXA is the gold standard. DXA uses two different low-energy X-rays with a whole body scanner to simultaneously determine global and local fat content, bone mineral content (and therefore bone density), and muscle mass with excellent precision. Scanning times are 5–7 minutes, and radiation doses are low. The desirable body fat for fit males should be

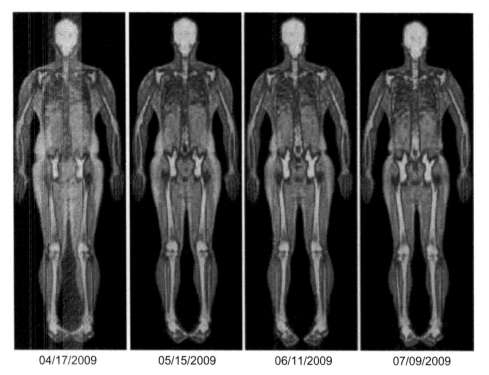

04/17/2009   05/15/2009   06/11/2009   07/09/2009

FIGURE 57.3 Results from a DXA exam showing bone distribution, muscle mass, and fat mass. *Faria SL. The effect of a very low calorie diet on visceral fat and anthropometric parameters of patients with morbid obesity, PhD Thesis, University of Brasilia, Brazil; 2014.*

FIGURE 57.4 A GE equipment image. *Faria SL. The effect of a very low calorie diet on visceral fat and anthropometric parameters of patients with morbid obesity, PhD Thesis, University of Brasilia, Brazil; 2014.*

14−17% of total weight. Males with values in excess of 26% fat, and females in excess of 32% are generally considered overweight or obese. The next figures show two examples of test results, one of a nonobese patient (Fig. 57.5), and the other of an obese patient (Fig. 57.6). Fig. 57.6 shows a follow-up case of an obese patient. From the first figure (January 2013) to the second figure (August 2013), the patient (female) lost 19.4 kg (16.9 kg of fat mass). For a better understanding, the figure is color-coded: yellow (light gray in print versions) is fat mass, red (dark gray in print versions) is lean muscle, and blue (white in print versions) represents bone mass. This patient had 52% fat mass in the first scan and 38% in the second scan. This was an example of a good opportunity to use DXA in obese patients, specifically bariatric patients.

In comparing DXA with other reference methods, there is less need for involvement and compliance on the part of the individual with DXA. The individual does not need to perform the breathing maneuvers required by other methods [12,13]. Furthermore, it is not necessary to collect body fluid samples as in other evaluation methods. Some manufacturers note in the scanner protocol that calcium supplements should not be used on the day of the exam. There are no restrictions, however, for maintaining the patient in a full or partial fasting state, nor is there any restriction on physical activity prior to the exam [12].

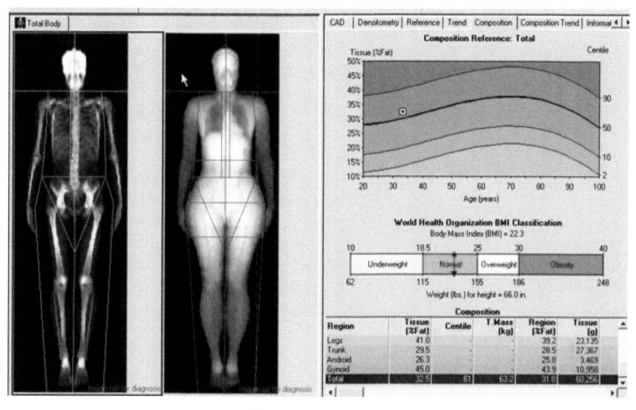

FIGURE 57.5   Results of body composition in a nonobese patient. *Faria SL, The effect of a very low calorie diet on visceral fat and anthropometric parameters of patients with morbid obesity, PhD Thesis, University of Brasilia, Brazil; 2014.*

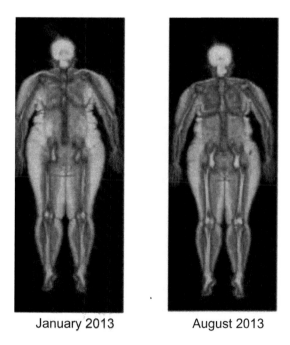

January 2013   August 2013

FIGURE 57.6   Results and a comparison of body composition in an obese patient. *Faria SL. The effect of a very low calorie diet on visceral fat and anthropometric parameters of patients with morbid obesity, PhD Thesis, University of Brasilia, Brazil; 2014.*

TABLE 57.1 Articles That Used the DXA Method to Assess Body Composition

Author	Year	Intervention	%FFM Loss	Weight Loss (kg)	n	Sex	Method
Tacchino et al. [13]	2003	BPD	19.2	31.7	101	F	DXA
Mingrone et al. [14]	2002	BPD	25.1	35.1	31	F	DXA
Mingrone et al. [14]	2002	BPD	27.8	52.1	15	M	DXA
Benedetti et al. [15]	2000	BPD	31.5	60.4	130	–	DXA
Greco et al. [16]	2002	BPD	48.5	33.0	8	–	DXA
Coupaye et al. [17]	2005	LAGB	12.7	23.7	36	F	DXA
Pugnale et al. [18]	2003	LAGB	14.1	27.7	31	F	DXA
Sergi et al. [19]	2003	LAGB	14.1	17.0	6	F	DXA
Gasteyger et al. [20]	2006	LAGB	16.4	39.6	36	F	DXA
Giusti et al. [21]	2005	LAGB	16.9	29.5	37	F	DXA
Garrapa et al. [22]	2005	LAGB	26	23.6	15	–	DXA

BPD, biliopancreatic diversion; DXA, dual energy X-ray absorptiometry; LAGB, laparoscopic adjustable gastric banding.

Table 57.1 shows several studies using the DXA method to assess the body composition of patients undergoing surgical procedures for weight loss.

As we can see, several studies on bariatric patients used the DXA method in the evaluation of body composition. Assessments of FFM loss as a proportion of total weight loss could provide an important measure of the safety of individual weight loss methods, especially in bariatric surgery, which provides the greatest sustained weight loss, and is arguably of greatest concern. In addition to significant caloric restriction, bariatric surgical methods can cause malabsorption, malnutrition, and changes to gastrointestinal hormone levels that may influence %FFM.

A recent study suggests that DXA may underestimate body fat loss occurring 12 months after Roux-en-Y gastric bypass (RYGB). However, this misconception seems to stem from errors of preoperative measurements when the amount of fat seems to be underestimated by 7% [23].

The International Society for Clinical Densitometry stated its position regarding the use of DXA to evaluate body composition of bariatric patients: "DXA total body composition with regional analysis can be used in the following conditions: in obese patients undergoing bariatric surgery (or medical, diet, or weight loss regimens with anticipated large weight loss) to assess fat and lean mass changes when weight loss exceeds approximately 10%." However, it states that the impact on clinical results remains uncertain, and assigns a weak grade to it [24].

Besides, for the post-RYGB measurement of bone mineral density (BMD), studies have shown certain limitations on using DXA where it seems to overestimate bone loss when compared to computerized tomography (CT), especially when reporting femoral and phalange readings [25,26]. There is no gold standard for measuring BMD in obese patients or after massive weight loss. The authors claim that several theoretical and experimental models suggest that bone measurements by DXA may be more susceptible than the measurements by CT to the presence of excess soft tissue and changes in body composition. Studies claim that the accuracy of DXA for this purpose decreases with increasing BMI [27], which may affect the ability to detect significant changes over time in the obese population. Also, weight loss increases the distance between the source X-ray and the beam fan detector (fan-beam detector), which leads to image magnifications that may artificially increase the bone area designated and BMD [28].

Another suggestion for lack of concordance between the methods would be that DXA technology is inherently limited by relying on only two different energies of X-ray beam, which are not able to identify the three types of tissue: bone, fat, and lean tissue. To compensate for this "limitation bicomponent," DXA technology assumes that the ratio of fat to lean tissue within the region of verifying DXA interest (ROI) is uniform, an assumption that may not be true in severely obese individuals or those undergoing substantial weight loss [25]. The International Society of Clinical Densitometry says more studies are needed to better define how the changes of fat and lean mass can affect the reliability of DXA in body composition measurements before and after intense weight loss [24].

However, a recent study from Cole et al. [29] evaluated five women from a larger cohort (pre-RYGB; and 1.5 months, 1 year, and 9 years post-RYGB) using different methods for body composition evaluation (DXA, air displacement plethysmography, and multiple dilution of deuterium oxide). They found that the three methods were highly correlated ($r > 0.98$, $p < 0.001$). Among the three methods, DXA was the least expensive and most accessible; it could be used even with some limitations inherent to body composition analysis. Therefore, DXA was determined to be a safe and useful method to evaluate the obese population during weight loss treatment.

MINI-DICTIONARY OF TERMS

- *Body composition*: The proportion of fat, muscle, and bone of an individual's body. It is expressed as percentage of body fat and percentage of lean body mass.
- *Dual energy X-ray absorptiometry*: A means of measuring BMD. Two X-ray beams, with different energy levels, are aimed at the patient's bones. When soft tissue absorption is subtracted out, the BMD can be determined from the absorption of each beam by bone. DXA is the most widely used and most thoroughly studied bone density measurement technology.
- *Sarcopenic obesity*: Most of the literature focuses on the obesity/low muscle mass combination.
- *Sagittal abdominal diameter*: A measure of visceral obesity, which is the amount of fat in the gut region. SAD is the distance from the small of the back to the upper abdomen. SAD can be measured in a standing or supine position. It can be measured from the narrowest point between the last rib and the iliac crests to the midpoint of the iliac crests. SAD is a strong predictor of coronary disease, with higher values indicating increased risk, independent of BMI [14].

KEY FACTS

- The body is composed of water, protein, minerals, and fat. A two-component model of body composition divides the body into a fat component and fat-free component. Body fat is the most variable constituent of the body.
- Storage fat is located around internal organs (internal storage fat) and directly beneath the skin (subcutaneous storage fat). It provides bodily protection and serves as an insulator to conserve body heat. The relationship between subcutaneous fat and internal fat may not be the same for all individuals and may fluctuate during one's life cycle.
- Factors that influence body composition: heredity, growth and development, activity levels, and diet.
- Excess weight around the abdominal area is associated with increased risk of cardiovascular disease.
- Lean body mass represents the weight of one's muscles, bones, ligaments, tendons, and internal organs. Lean body mass differs from FFM. Since there is some essential fat in the marrow of bones and internal organs, the lean body mass includes a small percentage of essential fat.

SUMMARY POINTS

- The DXA scanner is a device that can be used not only in the investigation of bone degeneration, but also as a method of assessing body composition.
- The use of the densitometry method for DXA assessment of body composition allows the measurement of both bone mass and volume of body fat and lean body mass.
- The DXA bone densitometry of the whole body is the only method that directly measures all body compartments (bone, muscle and water mass, fat mass) without inferring from the measurement data of only one compartment.
- An additional advantage of densitometry by DXA is the determination of regional composition, allowing detailed study of arms, legs, and abdomen. Abdominal fat, which is related to insulin resistance, dyslipidemia, and hypertension, is an additional risk for cardiovascular disease and diabetes, especially in obese patients.
- The exam is simple for the patient, requires no preparation, and lasts 10—15 minutes. Radiation is extremely low and can be repeated as often as necessary.
- Among the bariatric population, the evaluation of body composition in the preoperative phase and at regular postoperative moments is a key point for preventing sarcopenia.
- DXA can be safely used among bariatric patients. Even with some limitation, which is inherent to body composition evaluation, DXA is a precise and less invasive exam.

REFERENCES

[1] World Health Organization. World health statistics: progress on the health-related millennium development goals. Geneva: WHO; 2012. [cited 2013 Nov 30]. Available from: <http://www.who.int/gho/publications/world_health_statistics/2012/en/index.html>.

[2] Heymsfield SB, et al. Human body composition. Champaign: Human Kinetics; 2005, 524p.

[3] Sites CK, Calles-Escandon J, Brochu M, Butterfield M, Ashiga T, Poehlman ET. Relation of regional fat distribution to insulin sensitivity in post-menopausal women. Fertil Steril 2000;73:61–5.

[4] Heymsfield SB, Wang Z, Baumgartner RN, Ross R. Human body composition: Advances in models and methods. Annu Rev Nutr 1997;17(1): 527–58.

[5] Petroski EL, editor. Antropometria: técnicas e padronizações. Blumenau: Nova Letra; 2007, 182p.

[6] Khan K, et al. Physical activity and bone health. Champaign: Human Kinetics; 2001, 276p.

[7] Silva LK. Avaliação tecnológica em saúde: densitometria óssea e terapêuticas alternativas na osteoporose pós-menopausa. Cad Saude Publica 2003;19(4):987–1003.

[8] Ellis KJ. Human body composition: in vivo methods. Physiol Rev 2000;80:650–71.

[9] Kiebzak GM, Leamyb LJ, Pierson LM, Nord RH, Zhang ZY. Measurement precision of body composition variables using the Lunar DPX-L densitometer. J Clin Densitometry 2000;3:35–41.

[10] Bolanowski M, Nilsson BE. Assessment of human body composition using dual-energy X-ray absorptiometry and bioelectrical impedance analysis. Med Sci Monit 2001;7:1029–33.

[11] Ziai S, et al. Agreement of bioelectric impedance analysis and dual-energy X-ray absorptiometry for body composition evaluation in adults with cystic fibrosis. J Cyst Fibros 2014;13:585–8.

[12] Heyward VH, Wagner DR. Applied body composition assessment. 2nd ed. United States of America: Human Kinetics; 2004.

[13] Tacchino RM, Mancini A, Perrelli M, Bianchi A, Giampietro A, Milardi D, et al. Body composition and energy expenditure: relationship and changes in obese subjects before and after iliopancreatic diversion. Metabolism 2003;52:552–8.

[14] Mingrone G, Greco AV, Giancaterini A, Scarfone A, Castagneto M, Pugeat M. Sex hormone-binding globulin levels and cardiovascularrisk factors in morbidly obese subjects before and afterweight reduction induced by diet or malabsorptive surgery. Atherosclerosis 2002;161:455–62.

[15] Benedetti G, Mingrone G, Marcoccia S, Benedetti M, Giancaterini A, Greco AV, et al. Body composition and energy expenditure afterweight loss following bariatric surgery. J Am Coll Nutr 2000;19:270–4.

[16] Greco AV, Mingrone G, Giancaterini A, Manco M, Morroni M, Cinti S, et al. Insulin resistance in morbid obesity: reversal withintramyocellular fat depletion. Diabetes 2002;51:144–51.

[17] Coupaye M, Bouillot JL, Coussieu C, Guy-Grand B, Basdevant A, Oppert JM. One-year changes in energy expenditure and serumleptin following adjustable gastric banding in obese women. Obes Surg 2005;15:827–33.

[18] Pugnale N, Giusti V, Suter M, Zysset E, Heraief E, Gaillard RC, et al. Bone metabolism and risk of secondary hyperparathyroidism 12 onths after gastric banding in obese pre-menopausal women. Int J Obes Relat Metab Disord 2003;27:110–16.

[19] Sergi G, Lupoli L, Busetto L, Volpato S, Coin A, Bertani R, et al. Changes in fluid compartments and body composition in obesewomen after weight loss induced by gastric banding. Ann Nutr Metab 2003;47:152–7.

[20] Gasteyger C, Suter M, Calmes JM, Gaillard RC, Giusti V. Changesin body composition, metabolic profile and nutritional status 24 months after gastric banding. Obes Surg 2006;16:243–50.

[21] Giusti V, Suter M, Heraief E, Gaillard RC, Burckhardt P. Effects oflaparoscopic gastric banding on body composition, metabolicprofile and nutritional status of obese women: 12-months followup. Obes Surg 2004;14:239–45.

[22] Garrapa GGM, Canibus P, Gatti C, Santangelo M, Frezza F, eliciotti F, et al. Changes in body composition and insulinsensitivity in severely obese subjects after laparoscopic adjustablesilicone gastric banding (LASGB). Med Sci Monitor 2005;11:CR522–8.

[23] Levitt DG, Beckman LM, Mager JR, Valentine B, Sibley SD, Beckman TR, et al. Comparison of DXA and water measurements of body fat following gastric bypass surgery and a physiological model of body water, fat, and muscle composition. J Appl Physiol (1985) 2010;109 (3):786–95.

[24] Kendler DL, Borges JL, Fielding RA, Itabashi A, Krueger D, Mulligan K, et al. The Official Positions of the International Society for Clinical Densitometry: indications of use and reporting of DXA for body composition. J Clin Densitom 2013;16(4):496–507.

[25] Yu EW, Bouxsein ML, Roy AE, Baldwin C, Cange A, Neer RM, et al. Bone loss after bariatric surgery: discordant results between DXA and QCT bone density. J Bone Miner Res 2014;29(3):542–50.

[26] Lima TP, Nicoletti CF, Marchini JS, Junior WS, Nonino CB. Effect of weight loss on bone mineral density determined by ultrasound of phalanges in obese women afterRoux-en-y gastric bypass: conflicting results with dual-energy X-ray absorptiometry. J Clin Densitom 2014 Oct-Dec;17(4):473–8.

[27] Knapp KM, Welsman JR, Hopkins SJ, Fogelman I, Blake GM. Obesity increases precision erros in dual-energy X-ray absorptiometry measurements. J Clin Densitom 2012;15(3):315–19.

[28] Blake GM, Parker JC, Buxton FM, Fogelman I. Dual X-ray absorptiometry: a comparison between fan beam and pencil beam scans. Br J Radiol 1993;66(790):902–6.

[29] Cole AJ, Kuchnia AJ, Beckman LM, Jahansouz C, Mager JR, Sibley SD, et al. Long-term body composition changes in women following Roux-en-Y gastric bypass surgery. JPEN J Parenter Enteral Nutr 2016.

Chapter 58

Factors Associated with Metabolic Bone Disorders and Its Complications After Bariatric Surgery

L.J. Hintze[1] and N. Nardo Jr.[2]

[1]University of Ottawa, Ottawa, ON, Canada, [2]State University of Maringá, Maringá, Brazil

LIST OF ABBREVIATIONS

BMD bone mineral density
PHT parathyroid hormone
SAS Private Health System
SUS Unified Health System of Brazilian government

INTRODUCTION

As the obesity pandemic has become a real public health problem around the world over the last 20 years, the prevalence of severe obesity (BMI > 40 kg/m^2) has grown quickly, and treatment options (focused on efficacy) have become crucial [1].

Several guidelines have been published since the late 1990s that clearly state that an algorithm must be followed with a multidisciplinary intervention based on lifestyle changes as the cornerstone. Despite these guidelines, several issues still remain regarding the way these algorithms are followed. The rationale is to prevent a high-cost, high-risk intervention from being applied before a less aggressive treatment can solve the problem; e.g., an intensive lifestyle intervention that includes a physical activity component along with dietary behavior alterations under the supervision of kinesiology, nutrition, and psychology specialists. If the subject does not improve as expected after that comprehensive intervention, then medication or surgery must be considered [2,3].

Bariatric surgery is a last-case-scenario method of treating obesity that has gained popularity over the years because of two main factors (as revealed by long-term studies): (1) expressive and sustained weight loss and (2) resolution of the many comorbidities associated with obesity [4,5]. Despite all the advantages cited by reviews on bariatric surgery, a high number of them fail to mention metabolic bone disorders and their complications among patients undergoing surgery [6]. Other important information is related to the surgical procedures that used to be classified as "malabsortive," which are related to alterations to the alimentary tract that are restrictive and/or limit the absorption of nutrients; these occur when the stomach capacity is reduced, thus limiting the amount of food that can be consumed at each meal. There are still some techniques in which these principles (malabsortive and restrictive) are applied at the same time and promote more pronounced results. Depending on the category of surgical procedure, there will be greater or smaller risks of nutrient deficiency and associated health problems [2,7].

One of the potential consequences of these procedures is a change in bone metabolism that leads to decreased bone mineral density (BMD), and in the long term, higher risks of osteoporosis. Osteoporosis and low BMD are important health issues that lead to an increased risk of fracture, poor quality of life, and functional impairment. Many factors, including age, sex, race, body weight, diet, physical activity level, hormonal status, and body composition, can affect BMD [8,9]. Recent evidence has also associated BMD (potentially leading to osteoporosis) with bariatric surgery

[10,11]. This association is due to several reasons: (1) structural changes to the digestive system in malabsortive techniques [12−14]; (2) severe weight loss promoted by the surgery [8,15]; (3) high prevalence of type II diabetes in the surgery candidates [6,16]; and (4) profiles of most of the candidates seeking surgical intervention. All those variables combined highlight the importance of discussing bone metabolism and its complications after gastric surgery. The objective of this chapter is to discuss factors associated with metabolic bone disorders and their complications after bariatric surgery.

MINERAL DEFICIENCIES AFTER BARIATRIC SURGERY

As mentioned before, there is evidence associating bariatric surgery to decreased BMD and to bone metabolism abnormalities. One of the reasons is that calcium (Ca) is mainly absorbed in the duodenum and proximal jejunum, while vitamin D is mainly absorbed in the jejunum and ileum. The reduced calcium serum levels lead to an increased production of parathyroid hormone (PTH), which causes increased production of 1,25-dihydroxyvitamin D [17]. This condition, known as secondary hyperparathyroidism, promotes a higher release of calcium from bone, which decreases BMD and bone mineral content. In this way, in patients who have undergone surgical procedures where the duodenum and proximal jejunum were excluded or bypassed, the risk of development of bone mass abnormalities becomes higher [12−14].

In a prospective cohort of 33 obese adults undergoing Roux-en-Y gastric bypass (RYGB), Schafer et al. [18] verified a decrease of $32.7\% \pm 14.0\%$ preoperatively to $6.9\% \pm 3.8\%$ postoperatively for intestinal fractional Ca absorption. Consistent with the intestinal fractional Ca absorption decline, 24-hour urinary Ca decreased, PTH increased, and vitamin D increased. Bone turnover markers also increased markedly, areal BMD decreased at the proximal femur, and volumetric BMD decreased at the spine, even with most vitamin D levels ≥ 30 ng/mL and with recommended Ca intake.

Similarly, Slater et al. [19] found abnormal levels of calcium and vitamin D in 57% and 63% of patients 1 and 4 years after biliopancreatic diversion, respectively. Hypocalcemia was also present in 15% and 48% of patients 1 and 4 years after biliopancreatic diversion, respectively. Secondary hyperparathyroidism was present in 69% of patients after 4 years, and 27% of the cases were clinically significant.

Costa et al. [10] verified that 60.4% of the operated group presented vitamin D deficiency versus 16.6% in the control group. PTH levels were significantly higher in the operated group, and 41.7% of the operated subjects presented secondary hyperparathyroidism. Menegati et al. [11] showed higher mean PTH in the operated group and vitamin D plasma insufficiency (especially in women with longer time since surgery), even when no reductions in BMD were verified in the subjects.

Yu et al. [20] assessed the rate of bone loss in 30 patients in the 24 months after RYGB and compared the results to 20 nonoperated controls. Even though the mean serum calcium, 25(OH)-vitamin D, and PTH were maintained within the normal range in both groups, the microarchitecture at the distal radius and tibia in the operated group presented a deterioration over 24 months, and BMD was 5−7% lower at the spine and 6−10% lower at the hip in subjects who underwent gastric bypass as compared with nonsurgical controls (Fig. 58.1). The effect of all those changes promoted by gastric surgeries on bone metabolism was reinforced in a recent meta-analysis [21]. Results showed a significant decrease in BMD, and an increase in bone turnover, which led to accelerated bone remodeling following bariatric surgery.

We are currently analyzing the bone metabolism markers from 330 participants enrolled in a project conducted in our lab form 2010−2015 (manuscript in preparation). Participants were divided into three groups: SUS (individuals undergoing bariatric surgery in the Unified Health System of the Brazilian government), $n = 99$; SAS (individuals undergoing bariatric surgery in the Private Health System), $n = 136$; and a control group, $n = 95$. The bone metabolism marker analysis has shown higher urinary calcium in the control group compared to the two operated groups ($F_{2,218} = 14.83$, $p < 0.001$); also, osteocalcin concentrations were lower in the control group compared to the operated groups ($F_{2,260} = 26.91$, $p < 0.001$). No differences were found in the PTH concentrations among the three experimental groups (Table 58.1).

Another recent academic contribution from our research group has shown some interesting results regarding BMD and bone metabolism activity in individuals who have undergone bariatric surgery compared to a nonoperated group of obese individuals. This study included data for 58 subjects (29 operated, 29 nonoperated), and analyzed the BMD and bone metabolism activity from both groups. Our results have shown lower BMD in total body, total hip, neck, and total femur BMD in the operated group compared to the obese nonoperated group. Prevalence of low femur bone mass was 44.8% in the operated group and 10.3% in the nonoperated group, while the prevalence of low bone mass in the spine was 21.7% for the operated group and 13.8% for the nonoperated group. Regarding osteoporosis prevalence—it was verified as 6.9% and 17.2% in the nonoperated and operated groups, respectively. Our data regarding the biochemical

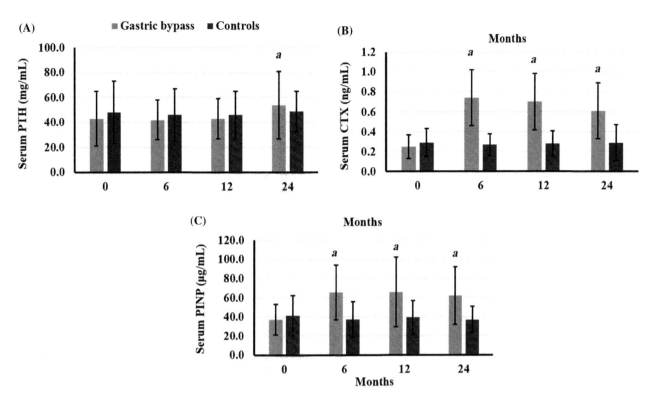

FIGURE 58.1   Responses to bone markers through time after bariatric surgery: (A) changes in *PTH*; (B) serum *CTX*; and (C) serum *PINP*. Adapted from Yu EW, Bouxsein ML, Putman MS, Monis EL, Roy AE, Pratt JSA, et al. Two-year changes in bone density after Roux-en-Y gastric bypass surgery. J Clin Endocrinol Metab 2015;100(4):1452–9. Note: Data are expressed as mean and standard deviation; *PTH*, parathormone; *CTX*, C-telopeptide; *PINP*, amino-terminal propeptide; *a* = difference in percentage change between gastric bypass and control groups in longitudinal mixed models ($p < 0.05$).

TABLE 58.1 Bone Metabolism Activity in Three Different Experimental Groups

	Calcium (mg/dL)	Urinary Calcium (mg/24 h)	Osteocalcin (ng/mL)	Parathormone (pg/mL)
SUS ($n = 99$)	9.6 (0.6)	90.7 (64.4)	29.9 (13.8)	61.1 (29.9)
SAS ($n = 136$)	9.4 (0.5)[a]	114.3 (78.6)	27.3 (13.6)	69.7 (31.7)
Control Group ($n = 95$)	9.7 (0.5)	163.3 (106.6)[a]	17.4 (7.1)[a]	55.2 (53.3)

Note: Data presented as mean and standard deviation.
[a]Significant difference ($p < 0.05$); for calcium concentrations, there was a difference between the experimental groups ($F_{2,259} = 6.87$, $p < 0.01$), with lower values than the SAS group when compared to SUS, and the control group ($p < 0.01$) (for both comparisons).
Abbreviations: SUS, Unified Health System of Brazilian government; SAS, Private Health System.

variables showed the following: higher values of PTH were observed in individuals who underwent bariatric surgery compared to the nonoperated group. Serum and urinary calcium were found to be lower in the operated group [22].

As a potential way to attenuate those negative side effects on the bone metabolism caused by bariatric surgery, some authors defend the idea of adequate use of supplements [23]. However, much evidence has shown that surgically induced weight loss can be associated with reductions in bone mineral content and BMD, despite calcium and vitamin D supplementation [24]. This is mainly because of the structural changes in the gastric system promoted by the surgery. A systematic review conducted by Chakhtoura et al. [25], which included 51 observational studies assessing vitamin D status pre- and/or postbariatric surgery, has shown that despite the various vitamin D supplementation regimens, hypovita-minosisis D persists in obese patients undergoing bariatric surgery. In postmenopausal women it seems that all those effects on bone metabolism can be even more severe, independent of calcium and vitamin D supplement intake [26].

SEVERE WEIGHT LOSS PROMOTED BY SURGERY

There are a few studies suggesting that obesity can be a protective factor against osteoporosis [27,28], since there is a significant relationship between body weight and BMD [8]. The basic principle states that the increase in bone mass can be attributed to the greater mechanical impact of excessive body weight on bone structures [29]. Consequently, a severe weight loss leads to a smaller mechanical effect that can promote a decrease in bone mass [8,15]. Previous studies state that in parallel with weight reduction, in approximately 10% of obese individuals, a reduction from 15% to 2% bone mass can occur at various skeletal sites. The bone loss during the weight reduction process may be associated with loss of lean body mass and with inadequate Ca and protein intake [30]. Considering that the individuals undergoing bariatric surgery go through massive weight loss, they are probably even more exposed to bone mass losses.

There is evidence stating that the greater the weight loss, the greater the reduction in BMD at all skeletal sites. In a study conducted by Abbasi et al. [31] involving 136 patients who had undergone a malabsorptive bariatric surgery 5 years prior, the BMD at all sites was reduced and correlated with the lowest lean body mass and weight loss. Similarly, Carrasco et al. [32] found a positive correlation between BMD in spine and femur at the initial weight and the percentage of weight loss in 42 premenopausal women 12 months after RYGB. Costa et al. [10] also found an inverse correlation between BMD and percentage of excess weight loss. Individuals who dropped a larger amount of weight also presented higher changes in BMD.

An especially low body weight can be considered a risk factor for osteoporotic fracture; however, an adequate amount of muscle mass or lean body mass seems to be helpful in maintaining adequate bone density. Recent studies have shown a positive and significant correlation between lower BMD and lower lean body mass [9,10,33].

A study conducted by Blain et al. [34] has shown that appendicular lean mass explained 35% variance of femoral neck BMD after adjusting for age, lifestyles, and serum hormones.

Wu et al. [9] demonstrated that subjects with sarcopenia were more likely to have low BMD in the full body, and in the femoral neck and lumbar spine compared to the nonsarcopenia group. Even in different gender groups with age categorized in the study, sarcopenia was shown to be an important independent factor in the female group.

Although the mechanism is not clear, these results reinforce the idea that the prevention of bone loss should be initiated even before surgery by increasing physical activity levels and promoting an increase in lean body mass, and potentially in BMD, among surgery candidates. This prevention should be continued in the first months after surgery, which is a period associated with expressive reduction in the lean body mass and increased bone turnover.

HIGH PREVALENCE OF TYPE II DIABETES AMONG SURGERY CANDIDATES

Even considering the evidence suggesting that obesity has a protective effect against osteoporosis, the evidence is not unanimous when the comorbidities related to excessive body weight are considered [6].

Evidence has shown that type 1 and type 2 diabetes mellitus are known to increase fracture risk. Type 1 diabetes mellitus is associated with lower BMD, while type 2 diabetes mellitus is associated with decreased bone strength [16,35]. Considering that type 2 diabetes is a common condition among obese individuals, and consequently there is a high prevalence of type 2 diabetes in candidates for gastric surgeries, it becomes clear that there is a higher risk in this population to present bone metabolism complications.

In a trial involving 54 subjects with type 2 diabetes who had undergone RYGB ($n = 18$) or sleeve gastrectomy ($n = 19$), or intensive medical therapy ($n = 16$), Maghrabi et al. [33] showed that a reduction in 8% and 9% were verified in the whole body bone mineral content and hip bone mineral content in both surgical groups at 2 years (Fig. 58.2). Those reductions were significantly greater in the operated patients compared to the patients who had undergone intensive medical therapy. The results occurred despite the use of vitamin D supplements in all groups. Similarly, Billeter et al. [36] compared the BMD and micronutrient deficiencies within 24 months after RYGB in obese patients living with type 2 diabetes. The authors verified a decrease in BMD in the hip, resulting in a significant increase in osteopenia rates (18−50%) among the subjects enrolled in the study, despite the use of multivitamin and micronutrient supplements. All this evidence suggests the importance of balancing the metabolic benefits of surgery against unintended consequences, including the impact on bone health; this is important to consider in defining the role of bariatric surgery in the treatment of obesity and type 2 diabetes.

PROFILE OF CANDIDATES SEEKING BARIATRIC SURGERY

One of the common things about the profiles of candidates to bariatric surgery is the predominance of middle-aged women. The age factor becomes even more important because the majority of patients seeking bariatric surgery are in

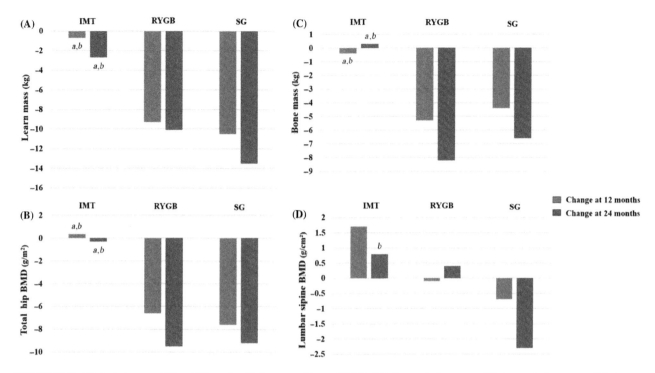

FIGURE 58.2 Clinical changes at 12 and 24 months: Body composition and BMD; (A) changes in lean mass; (B) changes in bone mass; (C) changes in hip *BMD*; and (D) changes in lumbar spine *BMD*. Adapted from Maghrabi AH, Wolski K, Abood B, Licata A, Pothier C, Bhatt DL, et al. Two-year outcomes on bone density and fracture incidence in patients with T2DM randomized to bariatric surgery versus intensive medical therapy. Obesity (Silver Spring) 2015;23(12):2344–8. Note: Data as expressed by median; *IMT*, intensive medical therapy; *RYGB*, Roux-en-Y gastric bypass; *SG*, sleeve gastrectomy; *BMD*, bone mineral density; a = IMT difference of RYGB ($p < 0.001$); b = IMT difference of SG ($p < 0.001$).

their 40s and female—i.e., the vulnerability to osteoporosis is increased, especially after menopause. The prevalence of osteoporosis reportedly ranges from 0.4% in premenopausal to 40% in postmenopausal women who have not undergone bariatric surgery [37].

Other previous publications have presented evidence that women (> 50 years old) have greater chances of lower BMD in the lumbar spine, the whole body, and the femur and femoral neck when compared to a younger group (< 50 years old) [38].

Other recent data of BMD in 293 women enrolled in a project conducted in our laboratory from 2010 to 2015 (manuscript in preparation) have shown a significant difference in the pelvis mineral density comparing the experimental groups ($F_{2,292} = 4.840$, $p = 0.008$) with the control group, which presented the highest value ($p < 0.01$) (Fig. 58.3). Similarly, the female femoral neck Z-score was different among the experimental groups ($F_{2,290} = 3.240$, $p = 0.040$). The control group presented the highest values compared to the SUS and the SAS groups ($p < 0.001$) (Fig. 58.4).

The biggest issue when BMD decreases is increased fracture risk, which heavily affects not only the individual, but puts an economic burden on society as well, due to the associated medical problems. A recent study conducted by Nakamura et al. [39] evaluated 258 patients who underwent bariatric surgery (82% females); it verified an increased relative risk for any fracture was 2.3-fold, wherein the risk for a first fracture at the hip, spine, wrist, or humerus was elevated by 1.9-fold. On the other hand, better preoperative activity status was associated with a lower age-adjusted risk (HR, 0.4). Along the same lines, Maghrabi et al. [33] reported that at the 2-year mark after surgery, 10 individuals reported spontaneous fractures, all of those in the surgery groups versus only one spontaneous fracture reported in the control group.

Final Considerations

Bariatric surgery is currently the most successful weight loss technique for patients who require massive weight loss. However, the potential for adverse effects on skeletal integrity remains an important concern, especially among postmenopausal women and/or individuals living with type 2 diabetes. All those factors such as age, sex, and glycemic levels point to the necessity of reviewing a patient's follow-up, treatment, and referral protocols, and to balancing the metabolic benefits of surgery against unintended consequences, including the impact on bone health.

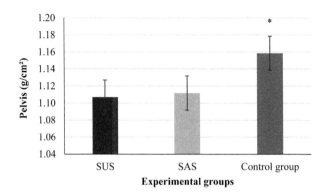

FIGURE 58.3 Bone mineral density of pelvis in the three experimental groups. Note: Data presented as mean and standard deviation; *SUS*, Unified Health System of Brazilian government; *SAS*, Private Health System; *indicates significant differences ($p < 0.05$).

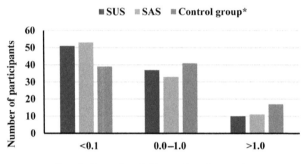

FIGURE 58.4 The female femoral neck T-score of the three experimental groups. Note: Data distributed into three classifications, i.e., it comprises negative values ($<0.1 =$ from -4.0 to 0.0 for a standard deviation of population average); *SUS*, Unified Health System of Brazilian government; *SAS*, Private Health System; *indicates significant differences ($p < 0.001$).

MINI-DICTIONARY OF TERMS

Osteoporosis: Disease characterized by low bone mass and deterioration of bone tissue. This leads to increased bone fragility and risk of fracture.

Bone mineral density (BMD): Measurement of how dense and strong the bones are. The most widely used technique to measure BMD is called dual-energy X-ray absorptiometry.

Parathyroid hormone (PTH): Secreted by the chief cells of the parathyroid glands. It acts to increase the concentration of ionic calcium and calcitonin in the blood and to decrease ionic calcium concentration.

1,25-Dihydroxyvitamin D: Hormonally active metabolite of vitamin D with three hydroxyl groups. It increases the level of calcium in the blood by increasing the uptake of calcium from the gut into the blood, and possibly increasing the release of calcium into the blood from bone.

Secondary hyperparathyroidism: Refers to the excessive secretion of PTH in response to hypocalcemia (low blood calcium levels).

Bone turnover: Refers to the total volume of bone that is both resorbed and formed over a period of time.

Roux-en-Y gastric bypass (RYGB): Gastric bypass (GBP) surgery refers to a surgical procedure in which the stomach is divided into a small upper pouch and a much larger lower "remnant" pouch; the small intestine is then rearranged to connect to both. Surgeons have developed several different ways to reconnect the intestine, thus leading to several different GBP procedures.

Sarcopenia: The degenerative loss of skeletal muscle mass, quality, and strength associated with aging.

Bone mineral content: Sum of all skeletal tissue within the body measured by dual-energy X-ray absorptiometry.

Osteopenia: Refers to bone density that is lower than normal peak density; however, it is not low enough to be classified as osteoporosis.

KEY FACTS

Osteoporosis: definition, causes consequences, and treatment

- Osteoporosis is a disease characterized by low bone mass and deterioration of bone tissue.
- The bones become thin and porous, decreasing BMD and bone strength, which leads to increased risk of fractures (particularly of the hip, spine, wrist, and shoulder).
- Osteoporosis also can cause a decrease in quality of life and functional impairment.
- No single cause for osteoporosis has been identified, however, there are several factors that contribute to the development of this condition, such as race, age, gender, body weight, diet, exercise, hormonal status, and body composition.
- Women are more likely to develop osteoporosis compared to men, especially due to hormonal changes at menopause, such as reduction of estrogen levels.
- The elderly, Caucasians, and people of Asian descent present a greater risk of osteoporosis.
- Dieting factors, such as eating disorders and gastrointestinal surgery, that severely restrict calcium intake affect the development of osteoporosis.
- Treatment involves lifestyle changes, and often the use of prescription medication. Increased calcium and vitamin D intake and regular physical activity are some of the changes necessary to treat and/or control this condition.
- Strength training exercises can be helpful for improving the strength and mobility of bones and muscles, which in turn helps prevent falls.

SUMMARY POINTS

- One of the potential consequences of malabsortive and restrictive procedures is a change in bone metabolism leading to a decreased BMD, and in the long-term, a higher risk of osteoporosis.
- This association is due to several reasons: (1) structural changes to the digestive system; (2) severe weight loss; (3) high prevalence of type II diabetes among surgery candidates; and (4) profiles of most of the candidates.
- The reduced calcium serum levels found in patients who have undergone bariatric surgery lead to an increased production of PTH, which causes increased production of 1,25-dihydroxyvitamin D. This condition promotes a higher release of calcium from bone, which decreases mineral density and mineral content.
- There is a significant relationship between body weight and BMD. Its basic principle states that the increase in bone mass can be attributed to the greater mechanical impact of excessive body weight on bone structures. Consequently, severe weight loss leads to a smaller mechanical effect, which can promote a decrease in bone mass.
- Type 1 and type 2 diabetes mellitus are known to increase fracture risk. Type 1 diabetes mellitus is associated with lower BMD, while type 2 diabetes mellitus is associated with a decrease in bone strength.
- The majority of patients seeking bariatric surgery are in their 40s and female—i.e., their vulnerability to osteoporosis is increased, especially after menopause.
- The potential for adverse effects on skeletal integrity remains an important concern, especially among postmenopausal women and/or individuals living with type 2 diabetes. All those factors, such as age, sex, and glycemic levels, point to the necessity of reviewing the patient's follow-up, treatment, and referral protocols.

REFERENCES

[1] Strain GW, Gagner M, Pomp A, Dakin G, Inabnet WB, Hsieh J, et al. Comparison of weight loss and body composition changes with four surgical procedures. Surg Obes Relat Dis 2009;5(5):582–7.

[2] Fried M, Yumuk V, Oppert J, Scopinaro N, Torres AJ, Weiner R, et al. Interdisciplinary European guidelines on metabolic and bariatric surgery. Obes Facts 2013;6:449–68.

[3] Lau DCW, Douketis JD, Morrison KM, Hramiak IM. 2006 Canadian clinical practice guidelines on the management and prevention of obesity in adults and children. CMAJ 2007;176(8):1–117.

[4] Picot J, Jones J, Colquitt JL, Gospodarevskaya E, Loveman E, Baxter L, et al. The clinical effectiveness and cost-effectiveness of bariatric (weight loss) surgery for obesity: a systematic review and economic evaluation. Health Technol Assess 2009;13(41):1–190, 215–357.

[5] Sjöström L, Lindroos A-K, Peltonen M, Torgerson J, Bouchard C, Carlsson B, et al. Lifestyle, diabetes, and cardiovascular risk factors 10 years after bariatric surgery. N Engl J Med 2004;351(26):2683–93.

[6] Laurent MR. Bariatric surgery: give more weight to bone loss. BMJ 2014;349:g6189.

[7] Mechanick JI, Kushner RF, Sugerman HJ, Gonzalez-Campoy JM, Collazo-Clavell ML, Guven S, et al. American Association of Clinical Endocrinologists, The Obesity Society, and American Society for Metabolic & Bariatric Surgery Medical guidelines for clinical practice for the perioperative nutritional, metabolic, and nonsurgical support of the bariatric. Endocr Pract 2008;14(Suppl 1):1−83.

[8] Reid IR. Relationships among body mass, its components, and bone. Bone 2002;31(5):547−55.

[9] Wu C-H, Yang K-C, Chang H-H, Yen J-F, Tsai K-S, Huang K-C. Sarcopenia is related to increased risk for low bone mineral density. J Clin Densitom 2013;16(1):98−103.

[10] Costa TL, Paganotto M, Radominski RB, Kulak CM, Borba VC. Calcium metabolism, vitamin D and bone mineral density after bariatric surgery. Osteoporos Int 2015;26(2):757−64.

[11] Menegati GC, de Oliveira LC, Santos ALA, Cohen L, Mattos F, Mendonça LMC, et al. Nutritional status, body composition, and bone health in women after bariatric surgery at a University Hospital in Rio de Janeiro. Obes Surg 2015;26(7):1517−24.

[12] Ducloux R, Nobécourt E, Chevallier J-M, Ducloux H, Elian N, Altman J-J. Vitamin D deficiency before bariatric surgery: should supplement intake be routinely prescribed? Obes Surg 2011;21(5):556−60.

[13] Signori C, Zalesin KC, Franklin B, Miller WL, McCullough PA. Effect of gastric bypass on vitamin D and secondary hyperparathyroidism. Obes Surg 2010;20(7):949−52.

[14] Valderas JP, Padilla O, Solari S, Escalona M, González G. Feeding and bone turnover in gastric bypass. J Clin Endocrinol Metab 2014;99(2):491−7.

[15] Fleischer J, Stein EM, Bessler M, Della Badia M, Restuccia N, Olivero-Rivera L, et al. The decline in hip bone density after gastric bypass surgery is associated with extent of weight loss. J Clin Endocrinol Metab 2008;93(10):3735−40.

[16] Piscitelli P, Neglia C, Vigilanza A, Colao A. Diabetes and bone: biological and environmental factors. Curr Opin Endocrinol Diabetes Obes 2015;22(6):439−45.

[17] Alvarez-Leite JI. Nutrient deficiencies secondary to bariatric surgery. Curr Opin Clin Nutr Metab Care 2004;7(5):569−75.

[18] Schafer AL, Weaver CM, Black DM, Wheeler AL, Chang H, Szefc GV, et al. Intestinal calcium absorption decreases dramatically after gastric bypass surgery despite optimization of vitamin D status. J Bone Miner Res 2015;30(8):1377−85.

[19] Slater GH, Ren CJ, Siegel N, Williams T, Barr D, Wolfe B, et al. Serum fat-soluble vitamin deficiency and abnormal calcium metabolism after malabsorptive bariatric surgery. J Gastrointest Surg 2004;8:48−55.

[20] Yu EW, Bouxsein ML, Putman MS, Monis EL, Roy AE, Pratt JSA, et al. Two-year changes in bone density after Roux-en-Y gastric bypass surgery. J Clin Endocrinol Metab 2015;100(4):1452−9.

[21] Liu C, Wu D, Zhang J-F, Xu D, Xu W-F, Chen Y, et al. Changes in bone metabolism in morbidly obese patients after bariatric surgery: a meta-analysis. Obes Surg 2016;26(1):91−7.

[22] Cremon AS, Moreira VM, Terra CM, Pagan DC, Hintze LJ, Dada RP, et al. Bone mineral density in patients who underwent bariatric surgery and in a non-operated equivalent group: a comparative analysis of serum parameters, and urinary biochemical markers of bone metabolism. Can J Diabetes 2015;39(2015):S50.

[23] Shea MK, Booth SL, Gundberg CM, Peterson JW, Waddell C, Dawson-Hughes B, et al. Adulthood obesity is positively associated with adipose tissue concentrations of vitamin K and inversely associated with circulating indicators of vitamin K status in men and women. J Nutr 2010;140(5):1029−34.

[24] Gasteyger C, Suter M, Calmes JM, Gaillard RC, Giusti V. Changes in body composition, metabolic profile and nutritional status 24 months after gastric banding. Obes Surg 2006;16(3):243−50.

[25] Chakhtoura MT, Nakhoul NN, Shawwa K, Mantzoros C, El Hajj Fuleihan G. Hypovitaminosis D in bariatric surgery: a systematic review of observational studies. Metabolism 2016;65(4):574−85.

[26] Valderas JP, Velasco S, Solari S, Liberona Y, Viviani P, Maiz A, et al. Increase of bone resorption and the parathyroid hormone in postmenopausal women in the long-term after Roux-en-Y gastric bypass. Obes Surg 2009;19(8):1132−8.

[27] Barrera G, Bunout D, Gattás V, de la Maza MP, Leiva L, Hirsch S. A high body mass index protects against femoral neck osteoporosis in healthy elderly subjects. Nutrition 2004;20(9):769−71.

[28] Puzziferri N, Roshek TB, Mayo HG, Gallagher R, Belle SH, Livingston EH. Long-term follow-up after bariatric surgery: a systematic review. JAMA 2014;312(9):934−42.

[29] Olmos JM, Vázquez LA, Amado JA, Hernández JL, González Macías J. Mineral metabolism in obese patients following vertical banded gastroplasty. Obes Surg 2008;18(2):197−203.

[30] Riedt CS, Brolin RE, Sherrell RM, Field MP, Shapses SA. True fractional calcium absorption is decreased after Roux-en-Y gastric bypass surgery. Obesity (Silver Spring) 2006;14(11):1940−8.

[31] Abbasi AA, Amin M, Smiertka JK, Grunberger G, MacPherson B, Hares M, et al. Abnormalities of vitamin D and calcium metabolism after surgical treatment of morbid obesity: a study of 136 patients. Endocr Pract 2007;13(2):131−6.

[32] Carrasco F, Ruz M, Rojas P, Csendes A, Rebolledo A, Codoceo J, et al. Changes in bone mineral density, body composition and adiponectin levels in morbidly obese patients after bariatric surgery. Obes Surg 2009;19(1):41−6.

[33] Maghrabi AH, Wolski K, Abood B, Licata A, Pothier C, Bhatt DL, et al. Two-year outcomes on bone density and fracture incidence in patients with T2DM randomized to bariatric surgery versus intensive medical therapy. Obesity (Silver Spring) 2015;23(12):2344−8.

[34] Blain H, Jaussent A, Thomas E, Micallef J-P, Dupuy A-M, Bernard PL, et al. Appendicular skeletal muscle mass is the strongest independent factor associated with femoral neck bone mineral density in adult and older men. Exp Gerontol 2010;45(9):679−84.

[35] Ikramuddin S. Bariatric surgery and bone health. Obesity (Silver Spring) 2015;23(12):2323.

[36] Billeter AT, Probst P, Fischer L, Senft J, Kenngott HG, Schulte T, et al. Risk of Malnutrition, Trace Metal, and Vitamin Deficiency Post Roux-en-Y Gastric Bypass—a Prospective Study of 20 Patients with BMI < 35 kg/m2. Obes Surg 2015;25(11):2125—34.

[37] Frazão P, Naveira M. Prevalence of osteoporosis: a critical review. Rev Bras Epidemiol 2006;9(2):206—14.

[38] Hintze LJ, Cremon ADS, Bevilaqua CA, Bianchini JAA, Nardo Junior N. Factors associated with bone mineral density in women who underwent bariatric surgery. Acta Sci 2014;36(1):105.

[39] Nakamura KM, Haglind EGC, Clowes JA, Achenbach SJ, Atkinson EJ, Melton LJ, et al. Fracture risk following bariatric surgery: a population-based study. Osteoporos Int 2014;25(1):151—8.

Chapter 59

Gait Patterns After Bariatric Surgery

A.W. Froehle[1], R.T. Laughlin[1], R.J. Sherwood[2] and D.L. Duren[2]

[1]Wright State University Boonshoft School of Medicine, Dayton, OH, United States, [2]University of Missouri, Columbia, MO, United States

LIST OF ABBREVIATIONS

EBW excess body weight
eKAM external knee adduction moment
KOA knee osteoarthritis
PA physical activity
ROM range of motion
vGRF vertical ground reaction force vector

INTRODUCTION

Walking is the most fundamental human movement, and abnormal gait is linked to significant loss of mobility, function, and health-related quality-of-life [1]. Bariatric surgery candidates carry a great deal of excess body weight (EBW), which alters gait biomechanics [2], including the inertial properties of the entire body and individual limb segments. These alterations challenge whole body balance and stability [3] and make the limbs more resistant to acceleration [4]. Gait in bariatric surgery candidates is generally slower, stiffer, and more careful than normal gait, resulting in substantially reduced mobility.

Increased joint loading with EBW can also lead to knee joint cartilage degeneration and the onset and progression of knee osteoarthritis (KOA) [5,6], creating additional challenges to mobility in bariatric surgery candidates [7]. The same abnormal gait biomechanics that characterize obesity also characterize KOA as adaptations to pain and other symptoms [8]. Individuals suffering from obesity and KOA are therefore prone to exacerbated gait deviations and limitations on activity and mobility relative to people with only one of these conditions. Accordingly, obesity and KOA are now leading causes of general disability worldwide [2,9].

Interactions between obesity, KOA, abnormal gait, and reduced mobility create an "obesity—immobility" feedback loop (Fig. 59.1), in which greater obesity leads to poorer mobility, and vice versa. By reducing mobility, excess weight increases the difficulty of performing the simple activities of daily living, let alone the higher-intensity physical activity (PA) required for massive weight loss [10]. Reduced mobility and self-perceived disability may also increase the psychological burden common among bariatric surgery candidates [11], further discouraging them from engaging in weight loss—oriented PA [10,12]. These factors combine to increase sedentary behavior among already highly sedentary individuals, raising risk for additional weight gain and cardiometabolic disease [13]. As more weight is gained, mobility is further compromised, KOA risk rises, and the feedback loop continues until surgical intervention becomes the only viable solution.

Given the centrality of immobility to obesity-related disability, recovery of normal walking gait should be among the most important postbariatric surgical outcomes. Postsurgical gait restoration is critical for patients in several ways. First is reduction of the burden of obesity-related disability, which is linked to increasing the ability to engage in activities of daily living and boosting self-esteem and independence [14]. Second, the ability to engage in elevated levels of PA is crucial to maintaining postsurgical weight loss [15,16] and recovering the healthy muscle mass wasted as part of surgical weight loss [17–20]. Walking as PA has substantial health benefits in and of itself [21], and can be effective as the core of a low-impact PA program for reducing KOA symptoms and progression [22]. By reducing disability,

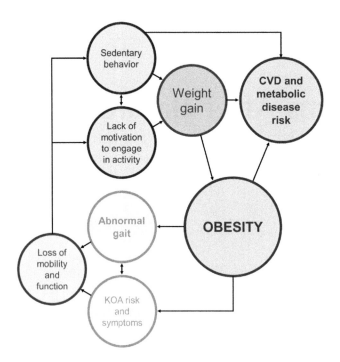

FIGURE 59.1 The obesity–immobility feedback loop. Obesity, immobility, and several related factors interact to create a feedback loop whereby greater obesity increases immobility, and vice versa.

helping with weight and body composition maintenance, and alleviating KOA symptoms, restoring normal walking gait can serve broadly as the basis for increasing other, higher-intensity PA, thereby improving health-related quality-of-life.

Below, we review the available data on postsurgical recovery of walking gait. We have constrained the scope of our review to quantitative studies of walking gait biomechanics, excluding studies of other functional tests or movements more easily conducted in the typical clinical setting. We have made this choice for several reasons. First, most clinical tests (e.g., 6-minute walk test, timed up-and-go) rely at least in part on walking to evaluate function, and poor performance is linked to underlying gait deficiencies that must ultimately be the targets of rehabilitation. Second, recovery of normal gait should be considered a precursor to engagement in other forms of PA. Finally, walking gait is one of the least-studied postsurgical outcomes despite its importance to functional restoration and maintenance of postsurgical health. We take this opportunity to highlight existing data and call attention to multiple outstanding issues that require resolution through further research. Prior to discussing postsurgical gait, we present fundamental concepts in gait analysis and a summary of obese gait.

OBESE GAIT

See "Key Facts" and Fig. 59.2 for basic information about gait. Obesity has broad effects on gait, largely attributable to the fundamental mechanical principle that greater mass is more resistant to change in motion. Substantial excess fat mass with obesity requires more force to put into motion in the first place, and is more difficult to control once in motion. Greater obesity is thus associated with increasingly poorer walking performance [23]. Excess mass also increases joint loads and can raise risk of tissue damage in the absence of compensatory adjustments in loading velocity, time, or frequency. Patterns of obese gait can therefore be broadly explained as adaptations to maintain balance and stability, minimize energy expenditure, and reduce joint loading while safely moving large quantities of excess mass [24–27].

These adaptations are readily manifest in spatiotemporal aspects of gait, showing that obese people adopt a slower, more careful gait than nonobese peers. The most apparent characteristics of obese gait are lower self-selected walking speed [24,26,28–36] and reductions in stride length and cadence subcomponents [24,26,28,29,32,34,36]. Reduced walking speed reduces energy expenditure, since greater muscle force would be required to more rapidly accelerate the body and limbs. Slower speed also reduces whole body and segmental momentum, requiring less muscle force to alter motion and prevent falling if stability is compromised. Stability is also related to the tendency of obese individuals to

Percent of gait cycle

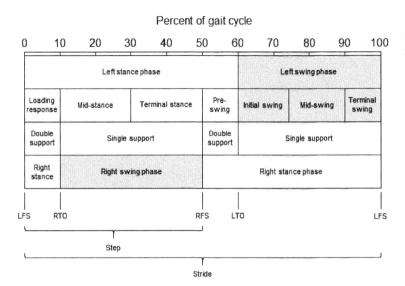

FIGURE 59.2 The normal gait cycle. Phases of the normal gait cycle. *LFS*: left foot-strike; *LTO*: left toe-off; *RFS*: right foot-strike; *RTO*: right toe-off. See Mini-Dictionary of Terms for further information.

walk with wider steps [24,28,31,37], likely as compensation for a wider range of motion (ROM) of the center of gravity due to exaggerated trunk oscillation [38,39]. Finally, stability is also maintained by spending more of the gait cycle in stance versus swing [24–26,29,31,37], especially in the percentage of time spent in double support [26,37]. When taken together, these traits mean that in a practical sense obese people need more time to walk to a destination. If hurried, obese walkers can become rapidly fatigued and may have greater difficulty controlling their bodies, thus raising risk for falls or other injuries.

Compared to spatiotemporal parameters, evidence for obesity-related alterations in gait kinematics is less consistent. *Hip:* Reduced flexion has been observed in some obese samples [25], but not in others [24,26,37], and two studies of frontal plane angles have generated contradictory results [24,26]. *Knee:* Two studies found a reduction in flexion during stance in obese gait [25,40], but others showed no differences between obese and nonobese subjects [24,35,37]. Frontal plane data on stance are also contradictory, with one study finding a tendency for greater adduction in the obese [26] and another finding greater abduction [35]. *Ankle:* Obese walkers exhibit greater frontal plane ROM and tend to evert the foot during stance compared to nonobese subjects [26,41]. The data for sagittal plane ankle kinematics remain contradictory [24–26,37]. The overall picture of kinematic alterations due to obesity is therefore quite unclear, and much additional research is needed to clarify these issues. Where kinematic differences between obese and nonobese subjects do exist, they may simply be due to slower walking speed, or may compensate for muscle weakness relative to excess weight [40].

Data on kinetic aspects of gait overwhelmingly find few, if any, differences between obese and nonobese walkers in terms of joint moments. The best explanation for these findings is that spatiotemporal changes in obese gait compensate for higher vertical ground reaction force vectors (vGRF) to minimize joint torques. Surprisingly few studies report vGRF data, but in absolute terms these forces are 130–160% higher in obese versus nonobese subjects [26,37]. Expressed per unit body mass, however, relative vGRF vectors are smaller in obese versus nonobese subjects, supporting the concept that obese gait is partly geared toward minimizing forces, joint moments, and attendant risk for joint damage [26,37]. Data on internal and external moments at all lower limb joints are broadly inconsistent [25,26,31,37]. Multiple studies have, however, found a relationship between obesity and high peak external knee adduction moment (eKAM) in stance [30,37,42], which results in excessive medial knee compartment compression and elevates risk for KOA [43,44].

GAIT RECOVERY FOLLOWING BARIATRIC SURGERY

As of this writing, we are aware of only six peer-reviewed, quantitative studies of walking gait following surgical weight loss [38,45–49]. Samples and methodological details are presented in Table 59.1, showing wide variation in length of follow-up and study design. Despite their differences, however, the studies' results are remarkably consistent, reporting substantial recovery of normal gait with surgical weight loss.

TABLE 59.1 Studies of Postsurgical Gait: Sample Characteristics

Authors	Year	Ref. No.	Study Type[a]	Time Point(s) Sampled	Surgical Cohort				Control or Reference Cohort			Walking Speed(s) Analyzed[c]	Gait Analysis Method(s)[d]	Gait Variables Analyzed
					N	M,F	Average Age (y)	Weight Loss[b]	N	M,F	Average Age (y)			
Hortobágyi et al.	2011	[45]	PC	Presurgery, 7.0 & 12.8 months postsurgery	10	Not reported	42.8	33.6%	Not reported	Not reported	43.6	SS, SZ	3DQ/FP	Spatiotemporal, kinematic, kinetic
Vartiainen et al.	2012	[46]	PC	Presurgery, 8.8 months postsurgery	13	3,10	45.5	21.5%		N/A		SZ	3DQ/FP	Spatiotemporal, kinematic, kinetic
Vincent et al.	2012	[47]	PC	Presurgery, 3 months postsurgery	25	5,20	41.0	19.4%	20	3,17	50.0	SS, FC	GM	Spatiotemporal, kinematic
Bragge et al.	2014	[49]	PC	Presurgery, 8.8 months postsurgery	15	Unclear, mostly F	45.7	21.8%		N/A		SZ	FP/SMA	Kinematic, kinetic
Froehle et al.	2014	[48]	RC	56.4 months postsurgery	9	0,9	48.7	31.8%	132	0,132	49.5	SS	3DQ	Spatiotemporal
Ponta et al.	2014	[38]	PC	Presurgery, 11–14 months postsurgery	10	0,10	36.7	33.2%	10	0,10	34.0	SS	DV	Spatiotemporal

[a]PC: prospective cohort; RC: retrospective cohort.
[b]Calculated as (|presurgical weight − weight at gait measurement|/presurgical weight · 100%).
[c]SS: self-selected walking speed; SZ: standardized walking speed; FC: fastest comfortable speed.
[d]3DQ: three-dimensional quantitative gait analysis; FP: force plate(s); GM: gait mat; SMA: skin-mounted accelerometer; DV: digital video.

TABLE 59.2 Postsurgical Spatiotemporal Gait Parameters Relative to Presurgical Conditions

Variable[a]	Hortobágyi et al.[b]	Vartiainen et al.	Vincent et al.	Froehle et al.[c]	Ponta et al.
Speed[d]	Increased	—	Increased	Typical for current BMI	—
Cadence	Decreased (SZ only)	—	No change	Typical for current BMI	—
Step/stride length	Increased (SS and SZ)	No change	Increased	Typical for current BMI	—
Base of support	—	Decreased	Decreased	Typical for current BMI	Decreased
Double support time	—	No change	—	Typical for current BMI	—
Single support/swing time	Increased (SS and SZ)	No change	Increased	Typical for current BMI	—

[a]An entry of "—" indicates no data for the specific variable reported.
[b]SS: self-selected walking speed; SZ, standardized walking speed.
[c]Presurgical gait analysis not performed. Postsurgical data compared to a reference sample matched for current age and BMI.
[d]Refers to walking speed during trials at self-selected speed.

Five studies examined postsurgical spatiotemporal aspects (Table 59.2) of gait at self-selected walking speed. With weight loss, speed, stride length, and single support/swing time all increased, while step width and double support time decreased [38,45,47]. Five years after surgery, all of these parameters and cadence were typical for current body mass index (BMI) when compared to a nonsurgical reference sample with normal gait [48]. Step width also decreased postsurgically at standardized speeds [46]. These results are consistent with a transition to normal gait related to more typical, nonobese (or less obese) weight-bearing conditions following surgery.

A spatiotemporal discrepancy between studies occurred at standardized speeds, where one study [45] found reduced cadence, longer strides, and increased swing time, in contrast to another [46] that found no such changes. The difference likely relates to greater joint disease and pain burdens among the latter study's subjects versus the healthier subjects of the former. In the healthy subjects, self-selected speed increased postsurgically to almost reach the tested standardized speed of 1.5 m/s, suggesting this was a speed at which the subjects could walk comfortably. In contrast, the subjects with more pain and joint disease showed no improvement other than step width at either 1.2 m/s or 1.5 m/s, indicating that although they made some gains in stability, their gait remained uncomfortable and abnormal even at slow speeds. This comparison highlights the need to consider underlying joint disease when evaluating postsurgical gait.

There is very little overlap between studies in kinematic variables studied, but significant pre-to-post-surgical changes were nonetheless observed (Table 59.3). *Hip*: Postsurgical weight loss was related to decreased flexion at foot-strike [46] and greater sagittal plane ROM during the swing phase [45], but no overall change in average sagittal plane angle during stance [45]. *Knee*: At self-selected speeds, maximum flexion angle in early stance increased [45], but did not change at standardized speeds [45,46]. Peak vertical and shear limb segment accelerations decreased after surgery, pointing to smoother, less abrupt foot-strikes and reduced impulsive loading of the knee joint [49]. *Ankle*: increased plantar flexion during stance was observed postsurgically [45], but toe-out angle, a risk factor for KOA progression [50], did not change [47]. *Trunk*: Oscillation ROM decreased following surgery, indicating improved control of postural stability [38].

Despite their importance to joint disease risk with obesity, kinetic variables are the least-studied aspect of postsurgical gait (Table 59.4). We focus on changes in *absolute* magnitudes of forces and moments, because joint tissue dimensions and properties do not adapt proportionally to massive EBW. Higher absolute joint loads can damage joints and lead to degenerative disease [37,51], meaning that reducing absolute loading via surgical weight loss can theoretically reduce risk for KOA onset or progression. It is noteworthy, however, that normalization for body weight generally shows that pre-to-post-surgical changes in loading are directly proportional to weight loss [45,46,49].

With surgical weight loss, absolute magnitudes of ground reaction forces decreased as expected, including peak vGRF during braking and propulsion, maximum vertical loading rate, peak mediolateral GRF, and peak anteroposterior braking and propulsion forces [45,49]. In terms of internal joint moments, Hortobágyi and colleagues [45] found no changes in hip or knee extension angular impulses or knee extension moment in early stance, but did report decreased ankle plantar flexion impulse and lower internal knee abduction moment in early stance. If the internal knee abduction moment reflects the muscular response to stabilize the knee against an eKAM, then this decrease may indicate a corresponding reduction in eKAM and thus less risk for KOA. It is unclear whether Varitainen and colleagues [46] also

TABLE 59.3 Postsurgical Gait Kinematics Relative to Presurgical Conditions

Variable[a,b]		Hortobágyi et al.[c]	Vartiainen et al.	Vincent et al.	Bragge et al.
Hip	Flexion angle at FS	—	Decreased	—	—
	Average angle in ST	No change	—	—	—
	Range of motion in SW	Increased (SS only)		—	—
Knee	Flexion angle at FS	—	No change	—	—
	Maximum flexion angle in ES	Increased (SS only)	No change	—	—
	Minimum flexion angle in ST	—	No change	—	—
	Average angle in ST	No change	—	—	—
	Flexion angle at TO	—	No change	—	—
	Initial peak vertical acceleration	—	—	—	Decreased
	Initial peak shear acceleration	—	—	—	Decreased
	Peak-to-peak vertical acceleration	—	—	—	Decreased
	Peak-to-peak shear acceleration	—	—	—	Decreased
Ankle/Foot	Average plantar flexion angle in ST	Increased (SS and SZ)	—	—	—
	Average toe-out angle	—	—	No change	—
Trunk	Oscillation	—	—	—	—

[a]An entry of "—" indicates no data for the specific variable reported.
[b]FS: foot-strike; ST: stance phase; SW: swing phase; ES: early stance phase (loading response); TO: toe-off.
[c]SS: self-selected walking speed; SZ: standardized walking speed.

reported internal moments (see Table 59.4 footnotes), but if so, they found the same decrease in knee abduction moment, further suggesting that weight loss has an important impact on kinetic risk factors for KOA. Finally, the finding of reduced peak hip extension and knee flexion moments during stance [46] likely reflects the effects of reduced body weight on overall joint angular loading.

CONCLUSIONS

Despite relatively few data on gait following surgical weight loss, gait and mobility are generally shown to improve rapidly after surgery [47], with sustained improvement over several years with weight loss maintenance [48]. Gait recovery is also linked to broader functional gains and reduced disability, as shown by concomitant improvements in physical function and pain scores on self-assessment questionnaires [46–48]. These results are promising, suggesting that gait recovery is an important, achievable surgical outcome, and that surgical weight loss can break the obesity–immobility feedback loop.

There are, however, several limitations and areas for improvement that can be addressed through future research. At this early stage of inquiry, the primary limitation is simply insufficient data on both obese and postsurgical gait. Although spatiotemporal assessments are consistent, substantial variation in kinematic and kinetic results makes clear that there are multiple ways to "walk obese." This variation is likely related to the myriad and variable underlying causes of obesity, which may very well have different impacts on movement. Variation in muscle mass and strength, e.g., could underlie variability in gait among obese people, and the age of onset of obesity relative to the timing of biomechanical maturation could play a role in variability of adult obese gait. There is a basic need for more research on why the kinematics and kinetics of obese gait vary, so that variation in postsurgical outcomes can be better evaluated.

Another important question concerns the generalizability of the collective results on fewer than 100 patients. Existing study cohorts were, outside of obesity, largely healthy, and most were female. Most studies also had relatively

TABLE 59.4 Postsurgical Gait Kinetics Relative to Presurgical Conditions

Variable[a]			Hortobágyi et al.[b]	Vartiainen et al.	Bragge et al.
Forces[c]		Peak braking force (anteroposterior)	Decreased (SS and SZ)	—	Decreased
		Peak vertical GRF during braking	Decreased (SS and SZ)	—	Decreased
		Peak propulsion force (anteroposterior)	—	—	Decreased
		Peak vertical GRF during propulsion	—	—	Decreased
		Peak mediolateral GRF	—	—	Decreased
		Maximum vertical loading rate	Decreased (SS and SZ)	—	Decreased
Joint moments[d]	Hip	Average extension impulse	No change (SS and SZ)	—	—
		Peak extension moment (full cycle)	—	Decreased	—
		Peak flexion moment (full cycle)	—	No change	—
	Knee	Extension angular impulse in ES	No change (SS and SZ)	—	—
		Extension moment in ES	No change (SS and SZ)	—	—
		Peak extension moment (full cycle)	—	No change	—
		Peak flexion moment (full cycle)	—	Decreased	—
		Peak abduction moment in ES	Decreased (SS and SZ)	Decreased	—
		Peak abduction moment in LS	—	Decreased[e]	—
	Ankle	Average plantar flexion impulse	Decreased (SS and SZ)	—	—

[a]An entry of "—" indicates no data for the specific variable reported.
[b]SS: self-selected walking speed; SZ: standardized walking speed.
[c]Only absolute values (i.e., not normalized for body mass or height) are considered here.
[d]Hortobágyi et al. reported internal (i.e., muscle-generated) joint moments. Vartiainen et al. did not clarify whether their reported joint moments were internal or external (i.e., due to GRF).
[e]Significant decrease only occurred at the 1.2 m/s walking speed; no significant change occurred at 1.5 m/s.

short follow-up times, leaving open the question of the sustainability of normal gait. Our own long-term follow-up study suggested that BMI-typical gait was maintained for 4–5 years, but we were only able to rerecruit 9 of 47 women from an original presurgical cohort. This loss-to-follow-up over longer study periods is common [52], but raises the question as to whether people who stay in such studies tend to be healthier, with better surgical outcomes, than those who drop out. More studies are thus needed on patients with significant comorbidities, on men, and over longer time periods with frequent contact to minimize loss to follow-up.

There are also other critical areas that future research should address, related to designing and implementing postsurgical functional recovery plans. One such area is the relationship between function and loss of muscle and bone tissues with surgical weight loss. Musculoskeletal tissue loss affects function and the ability to engage in PA, and we suggest that monitoring and rehabilitation of muscle mass, strength, and bone quality should be central to any functional recovery program. We also recommend that other barriers to increased postsurgical PA should be evaluated, including KOA symptoms and the psychological and motivational factors related to engaging in PA. Finally, biomechanical analyses need to be better adapted to clinical settings with simpler and less expensive technologies. Researchers have already made important strides in this direction using skin-mounted accelerometers [49,53] and simple digital video setups [38], and we encourage further development of these techniques. Clinical adaptation of these methods would allow for ongoing monitoring and evaluation of gait recovery during postsurgical rehabilitation.

Clarification of these outstanding issues through continued research on gait patterns following bariatric surgery holds great promise for improving the quality of functional recovery in formerly obese patients. We suggest that the overarching goal of such research should be the improvement of postsurgical therapeutic and PA interventions that are better-tailored to individual limitations and barriers to function [10]. This type of individualized approach to rehabilitation can maximize mobility and independence for the large majority of bariatric surgery patients, helping to significantly alleviate the burden of obesity-related disability.

MINI-DICTIONARY OF TERMS

- *Double support*: The first and final 10% of stance, with both limbs in ground contact (one in foot-strike, the other in toe-off). Double support time increases with obesity.
- *Foot-strike*: The point of initial contact with the ground, foot-strike defines the beginning of the gait cycle and the start and end points of a stride.
- *Kinematics*: The area of biomechanics concerned with angular positions, velocities, and accelerations of limb segments. Kinematic data provide information on joint ROM, can identify compensatory strategies for dealing with excess weight or joint disease, and affect the lengths of the moment arms of muscle and ground reaction force vectors.
- *Kinetics*: The area of biomechanics concerned with the forces and moments of movement, and the resulting mechanical loads on joints and related tissues.
- *Moment*: Also referred to as torque, moments describe the tendency of eccentric forces to cause rotation of limb segments around joint axes of rotation.
- *Spatiotemporal gait parameters*: Broad descriptors of movement in time and space, including speed, cadence, stride length, and time spent in gait cycle phases.
- *Stance phase*: The period of gait between foot-strike and toe-off, during which time body weight is accepted and controlled by the stance limb and then shifted for toe-off.
- *Stride*: The period between two sequential foot-strikes of the same (ipsilateral) limb. One stride delimits one full gait cycle.
- *Swing phase*: The period of gait between toe-off and the next foot-strike, critical to advancing the limb to its next stance phase position.
- *Toe-off*: The final event of stance phase that also initiates swing phase. Toe-off is crucial to providing the propulsive force used to make forward progress.

KEY FACTS

- Gait is a cyclical movement whereby one limb is used to push the body forward off the ground while the other limb swings forward to break the body's fall and initiate the next cycle. One gait cycle is defined as the period between two consecutive, ipsilateral episodes of foot contact with the ground (foot-strike), termed a stride. Each cycle is divided into two main phases (see definitions in mini-dictionary): stance, which consists of foot-strike through toe-off and normally accounts for the first 60% of the gait cycle; and swing, which follows toe-off and occupies the remaining 40% of the gait cycle. The cycles of the two limbs overlap so that during the first and final 10% of stance both limbs are in contact with the ground, referred to as double support. Biomechanical measurements of gait fall into three general categories: spatiotemporal, kinematic, and kinetic (see "Mini-Dictionary of Terms"). There are multiple well-established methods for measuring these parameters, but a detailed discussion of methods is beyond the scope of this review (for more information, see Ref. [54], for example).

SUMMARY POINTS

- Obesity is associated with slower, stiffer, more careful gait, leading to loss of mobility and physical function.
- Obese gait also contributes to knee osteoarthritis (KOA) onset and progression.
- Obese gait and KOA diminish health-related quality-of-life and cause disability.
- Thus, restoration of normal gait is an important outcome of bariatric surgery.
- Six studies show general improvement in postsurgical walking gait.
- Questions about kinematic and kinetic changes postsurgery remain.
- Additional research is needed to clarify several issues related to postsurgical recovery of gait and physical function.

REFERENCES

[1] Larsson U, Karlsson J, Sullivan M. Impact of overweight and obesity on health-related quality of life--a Swedish population study. Int J Obes Relat Metab Disord 2002;26:417.

[2] Vincent HK, Vincent KR, Lamb KM. Obesity and mobility disability in the older adult. Obes Rev 2010;11:568.

[3] Blaszczyk JW, Cieslinska-Swider J, Plewa M, et al. Effects of excessive body weight on postural control. J Biomech 2009;42:1295.

[4] Matrangola SL, Madigan ML, Nussbaum MA, et al. Changes in body segment inertial parameters of obese individuals with weight loss. J Biomech 2008;41:3278.

[5] Felson DT. Does excess weight cause osteoarthritis and, if so, why? Ann Rheum Dis 1996;55:668.

[6] Runhaar J, Koes BW, Clockaerts S, et al. A systematic review on changed biomechanics of lower extremities in obese individuals: a possible role in development of osteoarthritis. Obes Rev 2011;12:1071.

[7] Jordan JM, Helmick CG, Renner JB, et al. Prevalence of knee symptoms and radiographic and symptomatic knee osteoarthritis in African Americans and Caucasians: the Johnston County Osteoarthritis Project. J Rheumatol 2007;34:172.

[8] Russell E. Knee OA and obesity: a cyclical clinical challenge. Low Extrem Rev 2010; July.

[9] Woolf AD, Pfleger B. Burden of major musculoskeletal conditions. Bull World Health Organization 2003;81:646.

[10] Waldburger R, Schultes B, Zazai R, et al. Comprehensive assessment of physical functioning in bariatric surgery candidates compared with subjects without obesity. Surg Obes Relat Dis 2016; In Press.

[11] Burgmer R, Legenbauer T, Muller A, et al. Psychological outcome 4 years after restrictive bariatric surgery. Obes Surg 2014;24:1670.

[12] Elfhag K, Rossner S. Who succeeds in maintaining weight loss? A conceptual review of factors associated with weight loss maintenance and weight regain. Obes Rev 2005;6:67.

[13] Thorp AA, Owen N, Neuhaus M, et al. Sedentary behaviors and subsequent health outcomes in adults a systematic review of longitudinal studies, 1996–2011. Am J Prev Med 2011;41:207.

[14] Forhan M, Gill SV. Obesity, functional mobility and quality of life. Best Pract Res Clin Endocrinol Metab 2013;27:129.

[15] Josbeno DA, Kalarchian M, Sparto PJ, et al. Physical activity and physical function in individuals post-bariatric surgery. Obes Surg 2011;21:1243.

[16] Bond DS, Phelan S, Leahey TM, et al. Weight-loss maintenance in successful weight losers: surgical vs non-surgical methods. Int J Obes (Lond) 2009;33:173.

[17] Carey DG, Pliego GJ, Raymond RL. Body composition and metabolic changes following bariatric surgery: effects on fat mass, lean mass and basal metabolic rate: six months to one-year follow-up. Obes Surg 2006;16:1602.

[18] Metcalf B, Rabkin RA, Rabkin JM, et al. Weight loss composition: the effects of exercise following obesity surgery as measured by bioelectrical impedance analysis. Obes Surg 2005;15:183.

[19] Frimel TN, Sinacore DR, Villareal DT. Exercise attenuates the weight-loss-induced reduction in muscle mass in frail obese older adults. Med Sci Sports Exerc 2008;40:1213.

[20] Henriksen M, Christensen R, Danneskiold-Samsoe B, et al. Changes in lower extremity muscle mass and muscle strength after weight loss in obese patients with knee osteoarthritis: a prospective cohort study. Arth Rheum 2012;64:438.

[21] Chodzko-Zajko WJ, Proctor DN, Singh MAF, et al. Exercise and physical activity for older Adults. Med Sci Sports Exerc 2009;41:1510.

[22] Bliddal H, Leeds AR, Christensen R. Osteoarthritis, obesity and weight loss: evidence, hypotheses and horizons – a scoping review. Obes Rev 2014;15:578.

[23] Santarem GCDS, de Cleva R, Santo MA, et al. Correlation between body composition and walking capacity in severe obesity. PLoS One 2015;10.

[24] Spyropoulos P, Pisciotta JC, Pavlou KN, et al. Biomechanical gait analysis in obese men. Arch Phys Med Rehabil 1991;72:1065.

[25] DeVita P, Hortobagyi T. Obesity is not associated with increased knee joint torque and power during level walking. J Biomech 2003;36:1355.

[26] Lai PP, Leung AK, Li AN, et al. Three-dimensional gait analysis of obese adults. Clin Biomech (Bristol Avon) 2008;23(Suppl. 1):S2.

[27] Lyytinen T, Bragge T, Hakkarainen M, et al. Repeatability of knee impulsive loading measurements with skin-mounted accelerometers and lower limb surface electromyographic recordings during gait in knee osteoarthritic and asymptomatic individuals. J Musculoskelet Neuronal Interact 2016;16:63.

[28] de Souza SAF, Faintuch J, Valezi AC, et al. Gait cinematic analysis in morbidly obese patients. Obes Surg 2005;15:1238.

[29] Vismara L, Romei M, Galli M, et al. Clinical implications of gait analysis in the rehabilitation of adult patients with "Prader-Willi" Syndrome: a cross-sectional comparative study ("Prader-Willi" Syndrome vs matched obese patients and healthy subjects). J Neuroeng Rehabil 2007;4:14.

[30] Segal NA, Yack HJ, Khole P. Weight, rather than obesity distribution, explains peak external knee adduction moment during level gait. Am J Phys Med Rehabil 2009;88:180.

[31] Ko S, Stenholm S, Ferrucci L. Characteristic gait patterns in older adults with obesity--results from the Baltimore Longitudinal Study of Aging. J Biomech 2010;43:1104.

[32] Russell EM, Braun B, Hamill J. Does stride length influence metabolic cost and biomechanical risk factors for knee osteoarthritis in obese women? Clin Biomech (Bristol Avon) 2010;25:438.

[33] Hergenroeder AL, Wert DM, Hile ES, et al. Association of body mass index with self-report and performance-based measures of balance and mobility. Phys Ther 2011;91:1223.

[34] da Silva-Hamu TC, Formiga CK, Gervasio FM, et al. The impact of obesity in the kinematic parameters of gait in young women. Int J Gen Med 2013;6:507.

[35] Freedman Silvernail J, Milner CE, Thompson D, et al. The influence of body mass index and velocity on knee biomechanics during walking. Gait Posture 2013;37:575.

[36] Pataky Z, Armand S, Muller-Pinget S, et al. Effects of obesity on functional capacity. Obesity (Silver Spring) 2014;22:56.

[37] Browning RC, Kram R. Effects of obesity on the biomechanics of walking at different speeds. Med Sci Sports Exerc. 2007;39:1632.

[38] Ponta ML, Gozza M, Giacinto J, et al. Effects of obesity on posture and walking: study prior to and following surgically induced weight loss. Obes Surg 2014;24:1915.

[39] Ledin T, Odkvist LM. Effects of increased inertial load in dynamic and randomized perturbed posturography. Acta Otolaryngol 1993;113:249.

[40] Lerner ZF, Board WJ, Browning RC. Effects of obesity on lower extremity muscle function during walking at two speeds. Gait Posture 2014;39:978.

[41] Messier SP, Davies AB, Moore DT, et al. Severe obesity: effects on foot mechanics during walking. Foot Ankle Int 1994;15:29.

[42] Blazek K, Asay JL, Erhart-Hledik J, et al. Adduction moment increases with age in healthy obese individuals. J Orthop Res 2013;31:1414.

[43] Davies-Tuck ML, Wluka AE, Teichtahl AJ, et al. Association between meniscal tears and the peak external knee adduction moment and foot rotation during level walking in postmenopausal women without knee osteoarthritis: a cross-sectional study. Arth Res Ther 2008;10.

[44] Chang AH, Moisio KC, Chmiel JS, et al. External knee adduction and flexion moments during gait and medial tibiofemoral disease progression in knee osteoarthritis. Osteoarth Cartilage 2015;23:1099.

[45] Hortobagyi T, Herring C, Pories WJ, et al. Massive weight loss-induced mechanical plasticity in obese gait. J Appl Physiol 2011;111:1391.

[46] Vartiainen P, Bragge T, Lyytinen T, et al. Kinematic and kinetic changes in obese gait in bariatric surgery-induced weight loss. J Biomech 2012;45:1769.

[47] Vincent HK, Ben-David K, Conrad BP, et al. Rapid changes in gait, musculoskeletal pain, and quality of life after bariatric surgery. Surg Obes Relat Dis 2012;8:346.

[48] Froehle AW, Laughlin RT, Teel DD, et al. Excess body weight loss is associated with nonpathological gait patterns in women 4 to 5 years after bariatric surgery. Obes Surg 2014;24:253.

[49] Bragge T, Lyytinen T, Hakkarainen M, et al. Lower impulsive loadings following intensive weight loss after bariatric surgery in level and stair walking: a preliminary study. Knee 2014;21:534.

[50] Chang A, Hurwitz D, Dunlop D, et al. The relationship between toe-out angle during gait and progression of medial tibiofemoral osteoarthritis. Ann Rheum Dis 2007;66:1271.

[51] Mueller MJ, Maluf KS. Tissue adaptation to physical stress: a proposed "Physical Stress Theory" to guide physical therapist practice, education, and research. Phys Ther 2002;82:383.

[52] Higa K, Ho TC, Tercero F, et al. Laparoscopic Roux-en-Y gastric bypass: 10-year follow-up. Surg Obes Relat Dis 2011;7:516.

[53] Liikavainio T, Bragge T, Hakkarainen M, et al. Reproducibility of loading measurements with skin-mounted accelerometers during walking. Arch Phys Med Rehabil 2007;88:907.

[54] Perry J, Burnfield JM. Gait analysis: normal and pathological function. 2nd ed. New Jersey: SLACK Inc; 2010.

Section VII

Psychological and Behavioral Aspects

Chapter 60

Preoperative Psychosocial Assessment for the Bariatric Patient

S. Edwards-Hampton[1] and S. Wedin[2]

[1]Wake Forest Baptist Medical Center, Winston-Salem, NC, United States, [2]Medical University of South Carolina, Charleston, SC, United States

LIST OF ABBREVIATIONS

AMA American Medical Association
APA American Psychiatric Association
DSM-5 Diagnostic and Statistical Manual of Mental Disorders
TOS The Obesity Society
WHO World Health Organization
WLS weight loss surgery

INTRODUCTION

Since the inception of operations for weight loss in the 1950s, there has been increasing emphasis placed on the importance of preoperative psychosocial assessment and treatment [1,2]. Within the United States (US), preoperative psychosocial assessment is required by surgical centers for national accreditation by leading weight loss surgery (WLS) societies, and is required by all managed care providers for financial coverage of WLS [2]. Ninety percent of surgical centers within the US require preoperative psychosocial assessment [3,4]. Increased recognition of the value of preoperative psychosocial assessment is not surprising, particularly as the World Health Organization (WHO), American Medical Association, and The Obesity Society have recognized obesity as a chronic progressive *disease whose etiology is a myriad of interacting biological and environmental variables* [1,5,6]. Accordingly, obesity requires life-long treatment and symptom management, often via identification and modification of environmental contributors [5]. Preoperative psychosocial assessment not only promotes positive surgical outcomes, but also provides a wealth of information that can be utilized in service of individualized treatment to manage and prevent "flares" in the symptoms of obesity (i.e., weight gain) [7–10]. The purpose of this chapter is to provide an overview of the importance, purpose, and benefits of preoperative psychosocial assessment.

IMPORTANT CONSIDERATIONS

It is beyond the scope of this chapter to provide an exhaustive review of the literature related to the psychosocial functioning of pre- and postoperative weight loss patients. Further, there are few well-controlled studies that comprehensively explore the impact of psychosocial functioning in WLS patients. Among the studies that do exist, findings conflict, and suggest a complex interplay between psychosocial variables [11–15]. However, it is well documented that patients seeking WLS have high rates of psychosocial difficulties/diagnoses [13,16–20]. Thus, it is valuable to discuss areas of psychosocial functioning and patterns that are likely to impact pre- and postoperative treatment and outcomes [7].

Maladaptive Eating

With the release of the 5th edition of the Diagnostic and Statistical Manual of Mental Disorders (DSM-5), the American Psychiatric Association (APA) officially recognized binge eating as an autonomous psychiatric disorder [21]. Diagnostic criteria characterize binge eating as recurrent eating episodes characterized by a large amount of food intake within a discrete period of time (<2 hours), during which one feels a sense of being "out-of-control" [22]. A recent metaanalysis highlights binge eating disorder as the second most common (17−25%) psychiatric diagnosis among patients seeking WLS [12,16]. Importantly, measures intended to identify binge eating behaviors have been shown to lack diagnostic clarity and are less accurate than assessment via clinical interview [23]. Findings related to the impact of presurgical binge eating behaviors on postsurgical weight loss outcomes are mixed [15]. However, postoperative binge eating behaviors have largely been associated with poor weight loss outcomes [24].

While binge eating is the most commonly cited maladaptive eating behavior, other maladaptive eating behaviors, including irregular timing of eating, middle-of-the-night eating, grazing, and emotional eating behaviors, may be present in patients seeking WLS [25]. Importantly, these behaviors are not only associated with obesity, but are also incompatible with recommended postsurgery dietary behaviors. They are associated with poorer surgical outcomes, such as failure to meet recommended weight loss goals and weight regain. Further, these behaviors can limit quality-of-life (QoL) following WLS due to associated gastrointestinal symptoms [15,26,27]. Behaviors that allow patients to eat around the surgical restriction that is in place, such as emotional eating, grazing, binge eating, night eating, impulsive and loss of control eating, and high intake of calorie-dense liquid foods, can be particularly problematic [28]. Given that these behaviors are extremely common among the WLS-seeking population, excluding patients from WLS based on these criteria is counterproductive, particularly as these behaviors have not been consistently identified as predictors of poor weight loss outcomes [15,28−30]. Rather, the prevalence and possible negative impact of these behaviors indicate the potential value of identifying and treating maladaptive behaviors pre- and postoperatively.

Mood Disorders

Mood disorders are not uncommon among individuals seeking WLS [31]. A recent study indicated that 34% of patients seeking WLS present with an active Axis I disorder, 70% have at least one lifetime disorder, and 40% have a lifetime history of major depressive disorder. These statistics are significantly higher than population-based prevalence rates for these conditions [32]. Anxiety diagnoses have been documented in up to 48% of patients seeking WLS and identified as the most common psychiatric condition present at the time of the presurgical psychosocial evaluation [14,33−35].

Patients often attribute mood symptoms to their current weight status, and believe that mood symptoms will remit with weight loss. Research findings related to the short- and long-term resolution of depressive symptoms following WLS are mixed; however, overall symptoms seem to improve initially (1 year postoperation), but may gradually increase over time from the 1-year levels [31,36]. Importantly, it is the severity of mood symptoms and the presence of multiple psychiatric comorbidities preoperatively that place patients at higher risk for suboptimal postoperative outcomes [17,37]. Findings related to the negative impact of postsurgical mood symptoms on weight loss outcomes are less equivocal; the presence of a postoperative depressive disorder and disordered eating behaviors are negative predictors of weight loss outcomes [24]. While there is less evidence related to the relationships between postoperative anxiety symptoms and weight loss outcomes, these symptoms may be more resilient following WLS compared to other mood symptoms [30]. While the specific relationship between mood symptoms and weight loss and weight maintenance is unknown, it is not difficult to conceive how mood symptoms such as increased appetite, low energy, low motivation, limited physical activity, social isolation, and sleep disturbance are likely to negatively impact QoL and weight-related goals. Comprehensive psychosocial assessment and treatment of mood symptoms are likely to improve compliance and QoL. However, assessment and treatment becomes invaluable when it comes to managing the impact of severe mood symptoms and multiple psychiatric comorbidities on surgical weight loss outcomes [17,39].

Personality Disorders

Personality Disorders (PD) are a class of psychiatric disorders characterized by a pervasive, lifelong pattern of interpersonal and behavioral dysfunction, resulting in significant distress and impairment in functioning [40]. Due to the persistent nature of PDs, they are extremely difficult to treat and tend to take a significant financial and emotional toll on the identified patient, their interpersonal system, and health care providers [40,41]. The proportion of obese individuals with PDs significantly exceeds that of the general population, and a positive correlation between body weight and PD has been identified [18,42−44].

Similarly, bariatric patients also present with high rates of PDs. As high as 38% of bariatric patients may present with a PD; Cluster C disorders (avoidant, dependent, and obsessive—compulsive PD) represent the most common diagnoses (24%) [45]. Given the pervasive, challenging nature of PDs, patients who fit criteria are likely to present with multiple psychosocial liabilities, such as limited social support, high rates of interpersonal conflict/distress, limited insight and judgment, poor compliance, and difficulties with emotional identification and regulation [41]. Further, patients with PDs may be at higher risk for interacting with health care providers in a counterproductive manner, thus affecting their overall treatment. For example, patients with PDs are often high-utilizers of care, report frequent dissatisfaction with care, have unrealistic expectations of health care providers, distrust providers, are noncompliant with recommendations, and can be litigious [44]. A recent systematic review highlighted PDs as a variable that appears to be negatively associated with weight loss following bariatric surgery. Specific personality traits associated with poor outcomes were low patient self-directedness and high grievance rates [15].

Substance Use

Leading American medical societies have identified current alcohol or substance abuse/dependence as a contraindication to WLS [1]. However, findings related to the prevalence of substance abuse among WLS patients are nuanced. Current substance abuse tends to have a lower prevalence rate in a bariatric surgery—seeking population then in the general population, possibly because food is the substance/addiction of choice [46]. However, individuals who present for bariatric surgery with a prior history of substance abuse are at higher risk for postoperative substance abuse [47,48]. To make matters even more complex, two studies demonstrated continued remission and improved weight loss outcomes among a sample of patients who were recovered from substance addiction. The duration of abstinence among that population is unknown [49,50].

Of greater concern are the risks associated with postoperative substance use, particularly alcohol [51,52]. A significant minority of post-WLS patients develop new alcohol-use disorders after surgery [47,48]. Alcohol is metabolized differently following WLS. Time to maximum concentration decreases, while the length of time to eliminate alcohol increases. This results in a combined effect of individuals experiencing the intoxicating effects of alcohol more quickly, and for a prolonged duration of intoxication [53—55]. Altered alcohol metabolism may place patients at higher risk for alcohol abuse or addiction [52]. Other associated risks include additional nonnutritive calories, disinhibited eating behaviors, alcohol sugars that may produce dumping syndrome, and in the case of beer or sparkling wine, the presence of carbonation, which is discouraged postsurgery. Other possible ramifications of alcohol use post-WLS include vitamin deficiencies, liver problems, and dehydration [30,53]. Comprehensive, preoperative assessment of substance use via objective and subjective means is supported by literature and offers providers the opportunity not only to identify postoperative risks, but also strengths, such as how addiction recovery could be associated with greater postoperative success [7,48,53].

Economic Implications

A patient's access to resources is an important consideration for psychosocial assessment. Limited availability and affordability of nutritious food in one's environment is likely to negatively impact a patient's ability to comply with pre- and postoperative dietary recommendations [56]. Similarly, socioeconomic disparities may interfere with a patient's access to pre- and postoperative pharmacological supplements, negatively impacting weight loss and increasing risk for nutritional deficiencies [57]. Long-term medications and supplements for obesity are often costly if not covered by insurance, and racial/ethnic minorities are less likely to have health insurance coverage than whites [58]. Further, regular engagement in follow-up care with bariatric providers and support groups is a positive predictor of weight loss and weight maintenance [59,60]. Difficulty accessing transportation and inability to afford copayments are likely to inhibit patients from participating in valuable postoperative opportunities. Participation in physical activity-related interventions may be impeded by unsafe living environments and geographical conditions [61]. Importantly, while a causal relationship between environment and obesity has not been established, higher socioeconomic status, female gender, and Caucasian race are all significantly associated with positive postoperative weight loss outcomes [62,63].

ASSESSMENT PROCEDURES

While the need for presurgical psychosocial assessment is well-supported, consensus guidelines regarding the structure and content of psychosocial evaluations have not been established, and existing published recommendations are vague [10—12]. The absence of consensus guidelines has led to substantial differences in psychosocial assessment practices across centers. This inconsistency impedes the provision of a standardized continuum of care for patients and sends a divisive message regarding the significant role of psychosocial variables in obesity treatment [2]. Additionally, divergent motivations for the preoperative assessment may exist among treatment providers and patients. Patients are often motivated to undergo WLS as soon as possible, and thus present themselves in an overly favorable light, minimizing symptoms and difficulties [64]. While strong clinical acumen and assessment skills are likely to reveal a patient's attempts to present well, personality measures and validity scales are valuable in corroborating a patient's self-report during the clinical interview [65].

Despite inconsistencies across centers, comprehensive preoperative psychosocial evaluations typically gather information via clinical interviews and administration of testing measures [1,8,13−15]. Commonly assessed domains include: psychiatric adjustment, neurocognitive functioning, attitudes and surgical knowledge, expectations for outcomes, health behaviors and adherence, level of social support, coping and stress management, and postsurgical planning [6,11,16]. At the very minimum, preoperative psychosocial assessments should be used to identify possible contraindications for surgery, such as uncontrolled substance abuse, suicidality, or inability to consent to care [4]. However, the value of more detailed assessment across domains is well-documented, and there is movement away from use of psychological evaluations for mere screening purposes [7,24].

THE PURPOSE OF ASSESSMENT: BENEFITS AND OPPORTUNITIES

Unlike many other surgical procedures, WLS outcomes are inextricably tied to psychological and behavioral variables. A "successful" operation is only the first, and perhaps simplest, step in the journey to sustained weight loss. Lifelong food- and activity-related changes in behavior are essential [24]. Similarly, few surgeries have such a progressive, profound impact on an individual's psychosocial environment, particularly an individual's physical appearance and social functioning [66]. Thus, it is not surprising that there is a trend away from use of preoperative psychosocial screens for risk management purposes alone. Rather, comprehensive evaluations with feedback and treatment recommendations are becoming the gold standard of patient care [1]. Further, there is a historical bias of only examining weight loss when evaluating postoperative outcomes [67]. However, other variables such as resolution of medical comorbities, improved physical and psychological functioning, body image satisfaction, and QoL are also significant health variables to consider when evaluating what constitutes successful WLS. Thorough psychosocial assessment and treatment is a critical step in achieving positive outcomes among these variables. Finally, a comprehensive psychosocial assessment and intervention model is consistent with addressing obesity as a chronic disease state that requires lifelong medical management and follow-up, presumably even following WLS [68].

Comprehensive psychosocial assessment that extends beyond mere gatekeeping holds significant potential for minimizing postoperative challenges via preventative behavioral health care, identification of future possible postsurgical barriers to success, and an established plan for overcoming obstacles, should they occur [30]. In a recent review, Sogg and Friedman [7] described the multiple unique benefits of preoperative psychosocial assessments for bariatric patients and clinics. These included setting expectations related to the importance of behavioral health in achieving and maintaining weight loss goals, and establishing a strong therapeutic rapport that increased the likelihood that patients would seek behavioral support in the future. In addition, identification of patients' individual strengths and limitations could be utilized to help interdisciplinary team members maximize patient strengths and minimize areas of difficulty. Patients who presented as at-risk for high utilization of care or interpersonal conflict could be identified early on in the preoperative process. Behavioral health specialists could assist interdisciplinary team members in obtaining realistic expectations and clear boundaries with challenging patients, increasing provider support and reducing burn-out [7].

EMERGING CONSIDERATIONS

Over the past two decades there have been notable reductions in post-WLS morbidity and mortality, reductions in out-of-pocket costs associated with WLS, and increased diversity in the type of WLS available. These variables, among others, have resulted in increased access to, and utilization of, surgical obesity treatment [69−71]. Naturally, this growth has also resulted in greater diversity in the presentation of patients seeking WLS. Increased diversification in the WLS-seeking population continues to affirm the importance of comprehensive psychosocial assessments that take into consideration patients' unique cultural context. Further, there is a growing need for creative, diversified treatment interventions [30].

Surgical Selection

Historically, the specific type of surgical intervention utilized for a given patient has been limited by managed care reimbursement and relative medical contraindications [1]. However, within the past decade managed care providers have reimbursed for several additional procedures, and as mentioned above, there has been increased depth and breadth in the definition of "positive outcomes" following WLS [7]. Given these developments, there may be real benefit in considering psychosocial variables when making recommendations for specific WLS procedures. For example, given increased risk for nutritional deficiencies and gastrointestinal distress following the biliopancreatic diversion duodenal switch procedure, this intervention may be best suited for an individual who presents with a strong history of adherence to dietary and medical recommendations. Similarly, an individual who reports a strong affinity for sweets or fried foods may benefit from the powerful gastrointestinal distress that typically occurs when gastric bypass patients ingest these foods. While determination of surgical procedure based on psychosocial presentation might be rash, individualized and focused psychosocial assessment and treatment recommendations based on the surgical procedure that the patient plans to undergo are certainly indicated.

Supplemental Treatments

Pharmacological interventions are also emerging treatment opportunities to improve pre- and postoperative outcomes. The Food and Drug Administration recently approved lisdexamfetamine dimesylate for the treatment of binge eating disorder; this medication may also reduce impulsive eating behaviors. Bupropion/naltrexone can be effective in minimizing distressing, intrusive food-related thoughts, as are phentermine and topiramate, which can be used in combination or separately to suppress appetites and cravings. Lorcaserin can facilitate portion-controlled eating via early satiation. Pre- and postoperative patients who report relevant symptoms during the psychosocial assessment should be referred for consideration of these treatment options to medical providers [30].

Age

Historically, upper (> 65 years) and lower (< 18 years) age limits have been championed as age-related contraindications for undergoing WLS. This practice has been supported by claims that age is a risk factor for increased postoperative morbidity and mortality [1]. However, obesity is not a problem limited to middle-aged adults; it is a problem that also places adolescents and aging adults at significant medical risk. Importantly, there is limited empirical evidence that supports the historical bias against surgical treatment of obesity in these populations [72–74]. Findings among the adolescent population in particular suggest that WLS is an effective, well-tolerated obesity treatment [75]. Research findings simply do not support rigid implementation of firm, relatively arbitrary age cutoffs for surgical candidacy. Accordingly, there has been an increase in the surgical treatment of obesity in younger and older patients [75,76]. However, there are little-to-no published guidelines related to the unique pre- and postoperative psychosocial and multidisciplinary care that these patients require [30,75]. A protocol that considers patients' overall health status, including physiological, psychosocial, and cognitive functioning is prudent [76]. Increased participation of family members and assessment of patients' comprehension of the short- and long-term ramifications of WLS may be particularly important among these populations.

CONCLUSION

Preoperative psychosocial assessment of patients seeking WLS surgery is a well-established practice. While there has been limited consensus and consistency among professionals conducting these assessments, there appears to be an evolution in the use of WLS psychosocial assessments. Evaluations that were historically used for mere identification of psychosocial contraindications have developed into a means of providing comprehensive, multifaceted psychosocial assessments and interventions for the purpose of achieving immediate and long-term positive outcomes for patients and interdisciplinary teams. This requires a specialized mental health provider who is well-versed in the complexities of treating this disease state and patient population. This evolution seems natural and timely, as it parallels the emersion of various obesity surgical treatments and growth in understanding of obesity as a chronic disease state.

MINI-DICTIONARY OF TERMS

- *Binge eating disorder*: A psychiatric disorder characterized by recurrent eating episodes of an objectively large quantity of food within a discrete period of time (< 2 hours), accompanied by a sense of loss of control.
- *Maladaptive eating behaviors*: A variety of eating behaviors that are associated with difficulty maintaining proper nutrition and/or healthy weight.
- *Personality disorders*: A class of psychiatric disorders characterized by a pervasive, lifelong pattern of interpersonal and behavioral dysfunction, resulting in significant distress and impairment in functioning.

- *Alcohol use disorder*: A psychiatric disorder that consists of a variety of specific criterion, including difficulties controlling intake of alcohol, continued use despite negative consequences from drinking, development of tolerance, drinking that leads to risky situations, and development of withdrawal symptoms.
- *Psychosocial assessment*: A comprehensive evaluation conducted by a mental health provider that evaluates a wide range of psychological domains within an individual, including their mental health, substance use, lifestyle behaviors, cognitive functional capacity, and personality and coping style.

KEY FACTS

- Presurgical psychosocial assessments are a standard of care for WLS patients.
- There is no "universal" protocol for assessing patients prior to WLS.
- A variety of domains are evaluated during the psychosocial assessment, including cognitive functioning, knowledge of the procedure, psychiatric symptoms and functioning, coping skills, support system, health behaviors, and plans and preparation for surgery.
- Presurgical psychosocial assessments can be most beneficial if they are utilized to help assess strengths and potential challenges for success postsurgery.
- Psychosocial assessments can help the multidisciplinary treatment team anticipate challenges and target interventions based on individual strengths and vulnerabilities.
- Mental health providers with expertise in obesity and WLS are best-prepared to perform these assessments.

SUMMARY POINTS

- Preoperative psychosocial assessment of patients seeking WLS is a well-established practice.
- Many preoperative bariatric patients present with psychosocial challenges, such as maladaptive eating behaviors, mood symptoms, interpersonal distress, and low socioeconomic status.
- Comprehensive assessment of preoperative psychosocial functioning is likely to minimize pre- and postoperative barriers to successful postoperative outcomes for patients and bariatric clinics.
- With increased diversity of available weight loss procedures and types of patients seeking WLS, psychosocial assessment and treatment can be used to accommodate patients' unique cultural context and inform surgical decision-making.

REFERENCES

[1] Mechanick JI, Youdim A, Jones DB, et al. Clinical practice guidelines for the perioperative nutritional, metabolic, and nonsurgical support of the bariatric surgery patient--2013 update: cosponsored by American Association of Clinical Endocrinologists, The Obesity Society, and American Society for Metabolic & Bariatric Surgery. Obesity (Silver Spring) 2013;21(Suppl. 1):S1−27.

[2] Resources for optimal care of the metabolic and bariatric surgery patient. In: Surgery ASfmab, surgeons Aco, eds. Metabolic and bariatric surgery accreditation and quality improvement program; 2014.

[3] Bauchowitz AU, Gonder-Frederick LA, Olbrisch ME, et al. Psychosocial evaluation of bariatric surgery candidates: a survey of present practices. Psychosom Med 2005;67(5):825−32.

[4] Peacock JC, Zizzi SJ. An assessment of patient behavioral requirements pre- and post-surgery at accredited weight loss surgical centers. Obes Surg 2011;21(12):1950−7.

[5] Apovian CM, Aronne LJ. The 2013 American Heart Association/American College of Cardiology/The Obesity Society Guideline for the Management of Overweight and Obesity in Adults: What Is New About Diet, Drugs, and Surgery for Obesity? Circulation 2015;132(16): 1586−91.

[6] Obesity: preventing and managing the global epidemic. Report of a WHO consultation. World Health Organ Tech Rep Ser; 2000;894:i−xii, 1−253.

[7] Sogg S, Friedman KE. Getting off on the right foot: the many roles of the psychosocial evaluation in the bariatric surgery practice. Eur Eat Disord Rev 2015;23(6):451−6.

[8] Greenberg I, Sogg S, Perna FM. Behavioral and psychological care in weight loss surgery: best practice update. Obesity (Silver Spring) 2009;17(5):880−4.

[9] Wadden TA, Sarwer DB. Behavioral assessment of candidates for bariatric surgery: a patient-oriented approach. Surg Obes Relat Dis 2006;2(2): 171−9.

[10] Dziurowicz-Kozlowska AH, Wierzbicki Z, Lisik W, Wasiak D, Kosieradzki M. The objective of psychological evaluation in the process of qualifying candidates for bariatric surgery. Obes Surg 2006;16(2):196−202.

[11] Edwards-Hampton SA, Madan A, Wedin S, Borckardt JJ, Crowley N, Byrne KT. A closer look at the nature of anxiety in weight loss surgery candidates. Int J Psychiatry Med 2014;47(2):105−13.

[12]   Dawes AJ, Maggard-Gibbons M, Maher AR, et al. Mental health conditions among patients seeking and undergoing bariatric surgery: a meta-analysis. JAMA 2016;315(2):150−63.

[13]   Abilés V, Rodríguez-Ruiz S, Abilés J, et al. Psychological characteristics of morbidly obese candidates for bariatric surgery. Obes Surg 2010;20(2):161−7.

[14]   de Zwaan M, Enderle J, Wagner S, et al. Anxiety and depression in bariatric surgery patients: a prospective, follow-up study using structured clinical interviews. J Affect Disord 2011;133(1−2):61−8.

[15]   Livhits M, Mercado C, Yermilov I, et al. Preoperative predictors of weight loss following bariatric surgery: systematic review. Obes Surg 2012;22(1):70−89.

[16]   Wadden TA, Sarwer DB, Fabricatore AN, Jones L, Stack R, Williams NS. Psychosocial and behavioral status of patients undergoing bariatric surgery: what to expect before and after surgery. Med Clin North Am 2007;91(3):451−69, xi-xii

[17]   Pull CB. Current psychological assessment practices in obesity surgery programs: what to assess and why. Curr Opin Psychiatry 2010;23(1):30−6.

[18]   Petry NM, Barry D, Pietrzak RH, Wagner JA. Overweight and obesity are associated with psychiatric disorders: results from the National Epidemiologic Survey on Alcohol and Related Conditions. Psychosom Med 2008;70(3):288−97.

[19]   Barry D, Pietrzak RH, Petry NM. Gender differences in associations between body mass index and DSM-IV mood and anxiety disorders: results from the National Epidemiologic Survey on Alcohol and Related Conditions. Ann Epidemiol 2008;18(6):458−66.

[20]   Mauri M, Rucci P, Calderone A, et al. Axis I and II disorders and quality of life in bariatric surgery candidates. J Clin Psychiatry 2008;69(2):295−301.

[21]   Amianto F, Ottone L, Abbate Daga G, Fassino S. Binge-eating disorder diagnosis and treatment: a recap in front of DSM-5. BMC Psychiatry 2015;15:70.

[22]   American Psychiatric Association. Diagnostic and statistical manual of mental disorders. 5th ed. Washington, DC: American Psychiatric Publishing; 2013.

[23]   Allison KC, Wadden TA, Sarwer DB, et al. Night eating syndrome and binge eating disorder among persons seeking bariatric surgery: prevalence and related features. Surg Obes Relat Dis 2006;2(2):153−8.

[24]   Sheets CS, Peat CM, Berg KC, et al. Post-operative psychosocial predictors of outcome in bariatric surgery. Obes Surg 2015;25(2):330−45.

[25]   Berg C, Lappas G, Wolk A, et al. Eating patterns and portion size associated with obesity in a Swedish population. Appetite 2009;52(1):21−6.

[26]   Livhits M, Mercado C, Yermilov I, et al. Behavioral factors associated with successful weight loss after gastric bypass. Am Surg 2010;76(10):1139−42.

[27]   Colles SL, Dixon JB, O'Brien PE. Grazing and loss of control related to eating: two high-risk factors following bariatric surgery. Obesity (Silver Spring) 2008;16(3):615−22.

[28]   Rusch MD, Andris D. Maladaptive eating patterns after weight-loss surgery. Nutr Clin Pract 2007;22(1):41−9.

[29]   White MA, Masheb RM, Rothschild BS, Burke-Martindale CH, Grilo CM. The prognostic significance of regular binge eating in extremely obese gastric bypass patients: 12-month postoperative outcomes. J Clin Psychiatry 2006;67(12):1928−35.

[30]   Edwards-Hampton SA, Wedin S. Preoperative psychological assessment of patients seeking weight-loss surgery: identifying challenges and solutions. Psychol Res Behav Manag 2015;8:263−72.

[31]   Sarwer DB, Wadden TA, Fabricatore AN. Psychosocial and behavioral aspects of bariatric surgery. Obes Res 2005;13(4):639−48.

[32]   Mitchell JE, Selzer F, Kalarchian MA, et al. Psychopathology before surgery in the longitudinal assessment of bariatric surgery-3 (LABS-3) psychosocial study. Surg Obes Relat Dis 2012;8(5):533−41.

[33]   Kinzl JF, Schrattenecker M, Traweger C, Mattesich M, Fiala M, Biebl W. Psychosocial predictors of weight loss after bariatric surgery. Obes Surg 2006;16(12):1609−14.

[34]   Andersen JR, Aasprang A, Bergsholm P, Sletteskog N, Våge V, Natvig GK. Anxiety and depression in association with morbid obesity: changes with improved physical health after duodenal switch. Health Qual Life Outcomes 2010;8:52.

[35]   Rojas C, Brante M, Miranda E, Pérez-Luco R. Anxiety, depression and self-concept among morbid obese patients before and after bariatric surgery. Rev Med Chil 2011;139(5):571−8.

[36]   Sullivan M, Karlsson J, Sjostrom L, Taft C, editors. Why quality of life measures should be used in the treatment of patients with obesity. New York, NY: Wiley and Sons; 2001B.jorntrop P, ed. International textbook of obesity.

[37]   Herpertz S, Kielmann R, Wolf AM, Hebebrand J, Senf W. Do psychosocial variables predict weight loss or mental health after obesity surgery? A systematic review. Obes Res 2004;12(10):1554−69.

[38]   Leahy CR, Luning A. Review of nutritional guidelines for patients undergoing bariatric surgery. AORN J 2015;102(2):153−60.

[39]   Legenbauer T, De Zwaan M, Benecke A, Muhlhans B, Petrak F, Herpertz S. Depression and anxiety: their predictive function for weight loss in obese individuals. Obes Facts 2009;2(4):227−34.

[40]   Diagnostic and Statistical Manual of Mental Health Disorders, Text Revision. Washington DC: American Psychiatric Association; 2000.

[41]   Gerlach G, Herpertz S, Loeber S. Personality traits and obesity: a systematic review. Obes Rev 2015;16(1):32−63.

[42]   Stanley SH, Laugharne JD, Addis S, Sherwood D. Assessing overweight and obesity across mental disorders: personality disorders at high risk. Soc Psychiatry Psychiatr Epidemiol 2013;48(3):487−92.

[43]   Mather AA, Cox BJ, Enns MW, Sareen J. Associations between body weight and personality disorders in a nationally representative sample. Psychosom Med 2008;70(9):1012−19.

[44]   Maclean JC, Xu H, French MT, Ettner SL. Personality disorders and body weight. Econ Hum Biol 2014;12:153−71.

[45]   Capoccia D, Monaco V, Coccia F, Leonetti F, Cavaggioni G. Axis II disorders, body image and childhood abuse in bariatric surgery candidates. Clin Ter 2015;166(4):e248−53.

[46] Kalarchian MA, Marcus MD, Levine MD, et al. Psychiatric disorders among bariatric surgery candidates: relationship to obesity and functional health status. Am J Psychiatry 2007;164(2):328−34, quiz 374.

[47] King WC, Chen JY, Mitchell JE, et al. Prevalence of alcohol use disorders before and after bariatric surgery. JAMA 2012;307(23):2516−25.

[48] Mitchell JE, Steffen K, Engel S, et al. Addictive disorders after Roux-en-Y gastric bypass. Surg Obes Relat Dis 2015;11(4):897−905.

[49] Clark MM, Balsiger BM, Sletten CD, et al. Psychosocial factors and 2-year outcome following bariatric surgery for weight loss. Obes Surg 2003;13(5):739−45.

[50] Heinberg LJ, Ashton K. History of substance abuse relates to improved postbariatric body mass index outcomes. Surg Obes Relat Dis 2010;6(4): 417−21.

[51] Mitchell JE, Lancaster KL, Burgard MA, et al. Long-term follow-up of patients' status after gastric bypass. Obes Surg 2001;11(4):464−8.

[52] Ertelt TW, Mitchell JE, Lancaster K, Crosby RD, Steffen KJ, Marino JM. Alcohol abuse and dependence before and after bariatric surgery: a review of the literature and report of a new data set. Surg Obes Relat Dis 2008;4(5):647−50.

[53] Woodard GA, Downey J, Hernandez-Boussard T, Morton JM. Impaired alcohol metabolism after gastric bypass surgery: a case-crossover trial. J Am Coll Surg 2011;212(2):209−14.

[54] Maluenda F, Csendes A, De Aretxabala X, et al. Alcohol absorption modification after a laparoscopic sleeve gastrectomy due to obesity. Obes Surg 2010;20(6):744−8.

[55] Hagedorn JC, Encarnacion B, Brat GA, Morton JM. Does gastric bypass alter alcohol metabolism? Surg Obes Relat Dis 2007;3(5):543−8, discussion 548.

[56] Larson NI, Story MT, Nelson MC. Neighborhood environments: disparities in access to healthy foods in the U.S. Am J Prev Med 2009;36(1): 74−81.

[57] Lynch CS, Chang JC, Ford AF, Ibrahim SA. Obese African-American women's perspectives on weight loss and bariatric surgery. J Gen Intern Med 2007;22(7):908−14.

[58] Golden SH, Brown A, Cauley JA, et al. Health disparities in endocrine disorders: biological, clinical, and nonclinical factors--an Endocrine Society scientific statement. J Clin Endocrinol Metab 2012;97(9):E1579−639.

[59] Song Z, Reinhardt K, Buzdon M, Liao P. Association between support group attendance and weight loss after Roux-en-Y gastric bypass. Surg Obes Relat Dis 2008;4(2):100−3.

[60] Hildebrandt SE. Effects of participation in bariatric support group after Roux-en-Y gastric bypass. Obes Surg 1998;8(5):535−42.

[61] Gordon-Larsen P, Nelson MC, Page P, Popkin BM. Inequality in the built environment underlies key health disparities in physical activity and obesity. Pediatrics 2006;117(2):417−24.

[62] Sarwer DB, Wadden TA, Moore RH, et al. Preoperative eating behavior, postoperative dietary adherence, and weight loss after gastric bypass surgery. Surg Obes Relat Dis 2008;4(5):640−6.

[63] Toussi R, Fujioka K, Coleman KJ. Pre- and postsurgery behavioral compliance, patient health, and postbariatric surgical weight loss. Obesity (Silver Spring) 2009;17(5):996−1002.

[64] Ambwani S, Boeka AG, Brown JD, et al. Socially desirable responding by bariatric surgery candidates during psychological assessment. Surg Obes Relat Dis 2013;9(2):300−5.

[65] Belanger SB, Wechsler FS, Nademin ME, Virden TB. Predicting outcome of gastric bypass surgery utilizing personality scale elevations, psychosocial factors, and diagnostic group membership. Obes Surg 2010;20(10):1361−71.

[66] Ogden J, Clementi C, Aylwin S, Patel A. Exploring the impact of obesity surgery on patients' health status: a quantitative and qualitative study. Obes Surg 2005;15(2):266−72.

[67] Rigby A, Rogers AM. Response to commentary: "A comparative study of three-year weight loss and outcomes after laparoscopic gastric bypass in patients with 'yellow light' psychological clearance". Obes Surg 2015;25(3):541−2.

[68] Garvey WT, Garber AJ, Mechanick JI, et al. American association of clinical endocrinologists and american college of endocrinology position statement on the 2014 advanced framework for a new diagnosis of obesity as a chronic disease. Endocr Pract 2014;20(9):977−89.

[69] Nguyen NT, Higa K, Wilson SE. Improving the quality of care in bariatric surgery: the volume and outcome relationship. Adv Surg 2005;39:181−91.

[70] Surgery CICotASfMaB. Updated position statement on sleeve gastrectomy as a bariatric procedure. Surg Obes Relat Dis 2010;6(1):1−5.

[71] Kuo LE, Simmons KD, Kelz RR. Bariatric centers of excellence: effect of centralization on access to care. J Am Coll Surg 2015;221(5): 914−22.

[72] Dorman RB, Abraham AA, Al-Refaie WB, Parsons HM, Ikramuddin S, Habermann EB. Bariatric surgery outcomes in the elderly: an ACS NSQIP study. J Gastrointest Surg 2012;16(1):35−44, discussion 44.

[73] Ard J, Cook M, Frain A, et al. Effect of high intensity medical weight loss treatment in severely obese older adults on physical function. Paper presented at: Obesity Week 2015. Los Angeles, California.

[74] Fotenos AF, Snyder AZ, Girton LE, Morris JC, Buckner RL. Normative estimates of cross-sectional and longitudinal brain volume decline in aging and AD. Neurology 2005;64(6):1032−9.

[75] Thakkar RK, Michalsky MP. Update on bariatric surgery in adolescence. Curr Opin Pediatr 2015;27(3):370−6.

[76] Caceres BA, Moskowitz D, O'Connell T. A review of the safety and efficacy of bariatric surgery in adults over the age of 60: 2002−2013. J Am Assoc Nurse Pract 2015;27(7):403−10.

Chapter 61

Neurocognitive Factors Associated With Obesity, Obesity-Related Disorders, and Bariatric Surgery

K.L. Votruba and N. Dasher

University of Michigan, Ann Arbor, MI, United States

LIST OF ABBREVIATIONS

Aβ	amyloid beta
ACC	anterior cingulate cortex
AUDIT	alcohol use disorders identification test
BBB	blood−brain barrier
BMI	body mass index
BVMT-R	brief visual memory test−Revised
CPAP	continuous positive airway pressure
CRP	C-reactive protein
CVLT-II	California verbal learning test−2nd Edition
CVLT-SF	California verbal learning test−2nd Edition−Short Form
FFA	free fatty acids
IR	insulin resistance
LABS	longitudinal assessment of Bariatric surgery
MMPI2-RF	Minnesota multiphasic personality inventory−2nd Edition−Restructured Form
MCI	mild cognitive impairment
OSA	obstructive sleep apnea
PFC	prefrontal cortex
QEWP-R	Questionnaire on eating and weight patterns−Revised
TNF	tumor necrosis factor
WASI-II	Wechsler Abbreviated scale of intelligence−2nd Edition
WAIS-IV	Wechsler adult intelligence scale−4th Edition
WMS-IV	Wechsler memory scale−4th Edition
WRAT-4	Wide range achievement test−4th Edition

INTRODUCTION

The consequences of excessive adiposity include adverse neurological changes [1], increased risk of several comorbid conditions including hypertension [2], diabetes [3], and obstructive sleep apnea (OSA) [4], and cognitive changes. Bariatric surgery is utilized as an effective intervention in reducing the consequences of obesity and the prevalence of these weight-related comorbidities. However, to achieve positive outcomes following surgery requires adherence to complex pre- and postoperative instructions and maintenance of strict dietary and lifestyle changes. Given the importance of following these demands, patients often undergo preoperative neuropsychological evaluations assessing for cognitive risk factors that may undermine their success. The following chapter discusses the impact of excessive adiposity and some of its common comorbidities on neurological health and cognitive functioning. We then report possible cognitive benefits or negative consequences associated with weight loss and the reduction of medical comorbidities through bariatric surgery.

RELATIONSHIP BETWEEN NEUROLOGICAL FUNCTIONING, OBESITY, AND OBESITY-RELATED CONDITIONS

Obesity has been shown to have direct, adverse impacts on neurological integrity. Longitudinal data suggests that increased body mass index (BMI), particularly in mid-life, is associated with reduced gray and white matter volume [5], particularly in the prefrontal cortices and basal ganglia [1], even after controlling for relevant demographic and clinical factors. However, comorbid illness is highly prevalent in obese individuals, suggesting that obesity both contributes directly to poor neurological functioning and works in a synergistic fashion with other risk factors to exacerbate neuropsychological impairment (see Fig. 61.1).

One prominent comorbidity is hypertension, which afflicts approximately 75% of obese individuals [2]. The relationship between hypertension and neurological impairment has been validated in obese and nonobese populations [6]. In a recent meta-analysis, Beauchet et al. [7] evaluated 28 studies, finding that hypertension was associated with global and regional brain volume reductions. While the mechanism behind hypertension's impact on neurological function is multifaceted, a prominent etiology is atherosclerosis resulting in discrete lacunar infarcts and diffuse ischemic changes [8], particularly in the prefrontal and medial temporal regions [9]. These changes increase the risk of stroke and cognitive decline, particularly with mid-life hypertension.

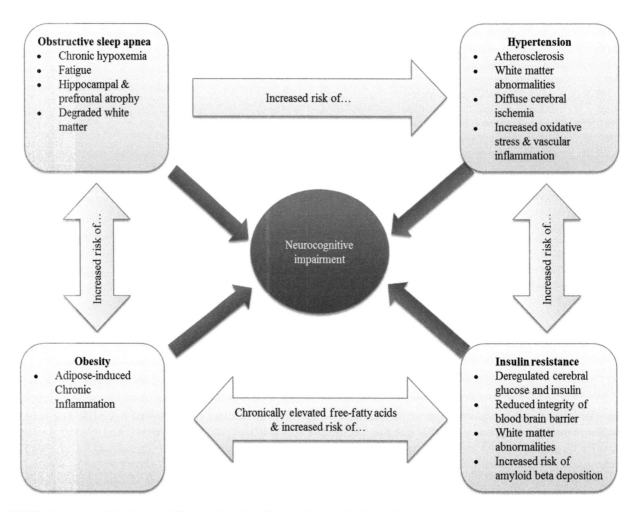

FIGURE 61.1 Potential Pathways of Obesity-related Cognitive Impairment. The figure shows the neurological changes associated with obesity, hypertension, insulin resistance, and obstructive sleep apnea that can have negative effects on cognitive functioning.

Metabolic impairment, such as diabetes, is also prevalent with obesity. It is estimated that 80% of obese individuals are insulin resistant (IR) [3], and there is evidence that insulin abnormalities can have an insidious, adverse impact on neural health and may contribute to a cascade of neuropathology consistent with Alzheimer's disease [10]. One mechanism involved is chronic inflammation, as the combination of increased adiposity and IR has been associated with chronically elevated levels of free fatty acids (FFAs), particularly tumor necrosis factor alpha (TNF-α) and C-reactive protein (CRP) [3]. Chronically elevated FFAs have been shown to deregulate normal insulin signaling in the brain, promoting deposition of amyloid beta (Aβ), increasing microstructural white matter damage, and reducing the stability of the blood—brain barrier (BBB), placing the cerebrum at increased risk of neurotoxic infection [10]. Furthermore, markers of inflammation in those with poor metabolic function have been found in otherwise cognitively healthy, middle-aged adults, suggesting early precursors to compromised cortical and subcortical integrity [11].

OSA is another common comorbidity with obesity, with 25—45% of obese individuals being afflicted [4]. Individuals with moderate—severe OSA have demonstrated smaller volumes of the hippocampus and caudate nucleus [12], reduced activation of the prefrontal cortex (PFC) and anterior cingulate cortex (ACC) [13], and degraded white matter integrity [14]. The frontal lobes appear especially vulnerable to the effects of intermittent hypoxia and fragmented sleep [15].

RELATIONSHIP BETWEEN COGNITION, OBESITY, AND OBESITY-RELATED CONDITIONS

Cognition and Obesity

The neurological changes described above have been reported to translate into varied cognitive difficulties. These deficits may begin in childhood, as evidenced by reduced inhibitory control [16], and accelerate cognitive decline associated with aging [17]. Individuals who are overweight or obese, particularly in mid-life, are at increased risk of cognitive decline in their older years [18]. After controlling for medical risk factors associated with obesity, there is evidence of an independent relationship between obesity and psychomotor performance, visual construction, verbal memory, and most prominently, aspects of executive functioning [19], reflecting the prominence of frontostriatal pathways. Even in otherwise healthy middle-aged obese adults, Gunstad and colleagues have found deficits in cognitive functioning, including memory deficits in up to 22.9% of severely obese patients [20], and significant difficulties with executive functioning [21]. In obese patients pursuing bariatric surgery, Alosco and colleagues [22] found high rates of preoperative cognitive impairments in attention, executive functioning, memory, and language. These deficits could compromise an individual's ability to plan, organize, and remember information relevant to surgery (see Table 61.1). As recent studies indicate that cognitive performance is strongly associated with adherence to postoperative guidelines after bariatric surgery [23], thorough evaluation of cognitive abilities prior to this procedure appears warranted in order to foster a positive surgical outcome.

TABLE 61.1 Potential Influences of Cognitive Impairment on Real World Behaviors

Neurocognitive Domain	Ecological Impact of Proper Function
Attention	• Effectively attend to information on proper nutrition provided by health care provider. • Appropriately focus attention in order to compare nutritional labels, cook healthy meals, and avoid distractions.
Memory	• Recall pre- and postoperative instructions. • Retain proper dietary information while grocery shopping, cooking, or dining out. • Remember to take vitamins and supplements.
Executive functioning	• Plan & organize meals ahead of time. • Organize & implement effective exercise regimen. • Inhibit prior unhealthy eating habits. • Successfully manage interpersonal relationships.
Language	• Read and comprehend written instructions/consent forms. • Understand written nutritional information on package labels, supplements, and medications.

This table describes how various cognitive deficits may be manifested in bariatric surgery candidates and how these deficits can interfere with surgical success.

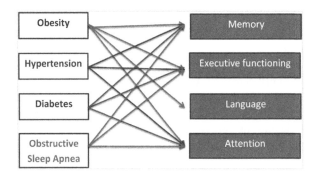

FIGURE 61.2 Influence of Obesity and Obesity-related Disorders on Cognitive Domains. The figure shows how obesity and obesity-related disorders influence cognitive functioning. Specifically, obesity has been shown to affect virtually all cognitive domains, while hypertension, diabetes, and obstructive sleep apnea are reported to affect memory, executive functioning, and attention.

Cognition and Obesity-related Disorders

In addition to the cognitive effects of obesity, the relationship between cognition and vascular disease is well established [24], accelerating cognitive decline in the aging brain [25] and increasing the risk for dementia of the Alzheimer's type [26]. This relationship appears particularly strong if hypertension emerges in mid-life [27]. As the atherosclerosis evident in hypertension has affinity for the frontal and temporal lobes, skills requiring memory and executive functioning appear prominently affected. Somewhat consistent with these findings, Waldstein and colleagues [28] found that individuals with hypertension perform worse than controls on tests of memory, attention, and abstract reasoning, and less consistently on tests of perception, constructional ability, and psychomotor speed.

Diabetes has also been shown to impair several cognitive functions, including psychomotor speed [29], executive function [29,30,31], memory [32], complex motor functioning [29], working memory [30,31], and verbal fluency [29]. Cognitive deficits in patients with diabetes are apparent early in the disease course [33], with an increased risk of cognitive impairment with earlier onset of diabetes [34].

The potential neurological changes associated with OSA also may work synergistically with sleepiness to produce cognitive difficulties. Specifically, impaired executive functioning is a consistent finding in patients with OSA, most prominently on tests of set-shifting, mental flexibility, and planning [35]. This highlights evidence presented earlier suggesting that the frontal lobes are particularly vulnerable to the effects of intermittent hypoxia and fragmented sleep. Memory deficits have also been reported in patients with OSA [36]. A systematic review regarding OSA and cognition found that the majority of evidence supports deficits in attention/vigilance, delayed long-term memory, visuospatial/constructional abilities, and executive function. This review also reported that language ability and psychomotor function are relatively unaffected by OSA. Data were reported to be equivocal for the effects of OSA on working memory and global cognitive functioning [37].

In summary, a review of the literature reveals consistent findings that obesity and its related comorbidities can have a direct and synergistic adverse impact on neuroanatomical integrity and cognitive function (see Fig. 61.2), which makes preoperative identification of neurocognitive impairment vital in the context of developing a comprehensive treatment plan that is idiosyncratic to the patient's strengths and weaknesses. While the course of obesity-related neuropathology can vary as a function of illness severity, the sum of the evidence implicates the frontostriatal and medial temporal structures as being especially vulnerable [5], which can particularly impact memory and executive abilities. The result can manifest in real-life difficulties in behaviors essential to sustaining long-term change after surgery, including learning and retaining new information, meal planning, craving inhibition, and emotional regulation. Therefore, identification of these cognitive deficits is vital for increasing prognostic success, as prior research has shown that poor neurocognitive performance is linked to poor medical adherence in the bariatric surgery population [38].

PREOPERATIVE NEUROPSYCHOLOGICAL BATTERY

Given the degree of cognitive impairment that can result as a consequence of obesity-related neuropathology, a presurgical neuropsychological battery maximizing sensitivity and specificity for detecting subtle frontostriatal compromise is important. Table 61.2 includes examples of a recommended comprehensive battery and an abbreviated battery used in case of time and resource limitations. An ideal neuropsychological battery should assess multiple cognitive domains

TABLE 61.2 Recommended Comprehensive and Abbreviated Neuropsychological Test Batteries

Cognitive Domain	Comprehensive Battery	Abbreviated Battery
General intellectual functioning	WASI-II (4-subtest)	WASI-II (2-subtest)
Attention/executive functioning	WAIS-IV digit span	Trail making test A & B
	WAIS-IV letter—number sequencing	–
	Trail making test A & B	–
	Stroop color—word test	–
	Digit vigilance test	–
	Short category test	–
Memory	CVLT-II	CVLT-II-SF
	BVMT-R	–
	WMS-IV logical memory	–
Language	Letter & semantic fluency	–
	Boston naming test	–
Visuospatial skills	Judgment of line orientation (JOLO)	–
	Rey—osterrieth complex figure test	–
Motor abilities	Grooved pegboard	–
	Finger tapping test	–
Academic/reading abilities	WRAT-4 word reading	WRAT-4 Word Reading
	WRAT-4 sentence comprehension	–
Personality/emotional functioning	MMPI-2-RF	MMPI-2-RF
	AUDIT	AUDIT
	QEWP-R	QEWP-R

AUDIT, Alcohol Use Disorders Identification Test; *BVMT-R*, Brief Visual Memory Test—Revised; *CVLT-II*, California Verbal Learning Test—2nd Edition; *CVLT-SF*, California Verbal Learning Test—2nd Edition—Short Form; *MMPI2-RF*, Minnesota Multiphasic Personality Inventory—2nd Edition—Restructured Form; *QEWP-R*, Questionnaire on Eating and Weight Patterns—Revised; *WASI-II*, Wechsler Abbreviated Scale of Intelligence—2nd Edition; *WAIS-IV*, Wechsler Adult Intelligence Scale—4th Edition; *WMS-IV*, Wechsler Memory Scale—4th Edition; *WRAT-4*, Wide Range Achievement Test—4th Edition. This table provides examples of both comprehensive and abbreviated neuropsychological pre-surgical test batteries to be used with a bariatric surgery sample.

including attention, executive functions, memory, language, visuospatial skills, and motor ability. If cognitive impairments are found, recommendations can be implemented to reduce the potential negative consequences of such impairment (see Table 61.3). In addition to a thorough cognitive assessment, psychological evaluation is important in this population, given the high prevalence of psychological disorders [39]. Though discussion of the importance of a psychosocial assessment prior to bariatric surgery is discussed elsewhere in this book, we emphasize the importance of including an objective mood inventory, as the motivation to appear in a positive light in this population is strong, and evaluation through solely a clinical interview may make interpretation vulnerable to this positive impression management.

IMPROVEMENT IN COGNITION WITH WEIGHT LOSS AND BARIATRIC SURGERY

Although a thorough discussion of improvements in obesity-related comorbidities following bariatric surgery is beyond the scope of this chapter, meta-analysis has shown that a substantial majority of patients with hypertension, diabetes, and OSA experience complete resolution or improvement in these conditions. Specifically, hypertension is resolved or improved in 78.5% of cases, diabetes is resolved or improved in 86.0% of cases, and OSA is resolved or improved in 83.6% of patients [40].

TABLE 61.3 Recommendations to Reduce Complications Associated With Cognitive Impairments in Bariatric Surgery Candidates

Neurocognitive Domain	Recommendation to Minimize Consequence of Impairment
Attention	• Eliminate external distractions (television, internet). • Patient can paraphrase what s/he was told by medical providers to assure proper attention.
Memory	• Medical providers can provide written instructions in addition to verbal directions. • Write down important dietary information to reference when patient is shopping or cooking. • Use a pillbox and set alarms to remember to take appropriate vitamins and supplements.
Executive functioning	• Focus on one task at a time to reduce mindless eating. • Utilize a smart phone calendar to assist with meal planning and exercising. • Enlist a support figure with whom you can prepare meals in advance. • Write down consequences of dietary nonadherence.
Language	• Medical providers can verbalize all important written information. • Bring a support person to pre- and postbariatric surgery appointments to assist with comprehension. • Utilize visual stimuli (take pictures) of healthy/unhealthy food choices.

This table provides recommendations for bariatric surgery patients that may help to minimize potential negative consequences of cognitive impairments.

With reduced weight frequently comes improvement in cognitive abilities associated with obesity. Siervo and colleagues [41] found in a sample of middle-aged to older adults that a weight loss of 8−12% of initial body weight resulted in a beneficial effect on cognitive function. Similarly, Napoli and colleagues [42] found that both weight loss and exercise improve cognition and health-related quality of life. Even in older adults with a diagnosis of mild cognitive impairment (MCI), weight loss through dieting was associated with cognitive improvements. Specifically, decreases in BMI were associated with improvements in verbal memory, verbal fluency, executive function, and global cognition, after adjustment for education, gender, physical activity, and baseline cognitive skills [43]. Furthermore, though the impact of hypertensive treatment on cognitive functioning in younger adults varies, treatment has been reported to improve cognition in older patients (particularly on tests of memory [44]) and reduce the risk of Alzheimer's disease or all-cause dementia based on brief cognitive screening [45].

Evidence consistently documents improvements in cognition following treatment of diabetes. Kawamura, Umemura, and Hotta [46] report that the cognitive effects of diabetes are often attenuated by improved glycemic control. Particularly in the elderly, studies show improvement in cognitive skills as diabetes is better controlled. Meneily and colleagues [47] reported that after 6 months of treatment, patients improved on tests of psychomotor processing speed, speeded word naming, cued recall, and planning.

Though some studies have found that treatment with continuous positive airway pressure (CPAP) to treat OSA did not translate into cognitive improvement, other studies have documented significant improvement in attention [48], executive functioning [49], memory [50], and global cognitive functioning [37,48,49,50]. The potential effects of stimulant medication on cognitive functioning in OSA have also been investigated, showing positive effects on reaction time [51], memory [52], and sustained attention [51]. Ancoli-Israel and colleagues [53] suggested that OSA may be a reversible cause of cognitive loss in patients with Alzheimer's disease, with treatment of OSA resulting in improvement of some of the cognitive dysfunction experienced by these patients. They suggest improvements in episodic verbal learning, verbal memory, and cognitive flexibility.

POTENTIAL COGNITIVE CHANGES FOLLOWING BARIATRIC SURGERY

Perhaps reflecting the improvement in the medical conditions discussed above, a review of the literature cited improvements in physical, cognitive, and emotional functioning following bariatric surgery [54]. In likely the largest evaluation of patients progressing through bariatric surgery, the Longitudinal Assessment of Bariatric Surgery (LABS) project documented improvement in cognitive performance that generally mapped onto weight loss. Specifically, the main effects were evident for attention, executive functioning, and memory up to 3 years after bariatric surgery [55]. Following this initial improvement, memory and executive functioning remained at this improved level, but scores on tests of attention declined if persons regained substantial weight [56]. Though these cognitive benefits were not precluded by older age [57], the type of bariatric surgery may have impacted cognitive functioning. Specifically, Grayson and colleagues [58] found that gastric bypass surgery resulted in improvement in cognition, whereas sleeve gastrectomy did not, despite relatively equivalent improvements in BMI and glucose regulation with these two procedures.

Despite evidence of cognitive improvements after bariatric procedures, surgery also poses potential cognitive risks. Several neurological complications may follow bariatric surgery, including behavioral abnormalities, seizures, and nutritional deficiencies that can impair cognitive functioning [59]. Prominent among nutritional deficiencies is vitamin B1 deficiency, which can result in mild memory problems, with the most serious cases of malnutrition after bariatric surgery potentially resulting in Wernicke−Korsakoff syndrome (WKS) [60]. Fortunately, by the combination of a thorough clinical interview by a qualified clinician and objective cognitive and psychological measures prior to surgery, risk factors in surgical candidates can be identified, idiosyncratic treatment plans can be developed, and surgical outcomes can be maximized.

MINI-DICTIONARY OF TERMS

- *Neuropsychology*: The study of the relationship between neurological function and behavior.
- *Neurocognitive*: Cognitive skills needed to complete various activities of daily living.
- *Executive functioning*: Complex cognitive functions involving the frontal lobe that require significant resources. These skills include tasks such as initiating actions, alternating attention, and inhibiting behaviors.
- *Mild cognitive impairment*: A term used to describe cognitive impairment that can be objectively measured on neuropsychological tests, but does not significantly interfere with the ability to complete activities of daily living.
- *Wernicke−Korsakoff syndrome (WKS)*: A neurological complication associated with vitamin B1 (thiamine) deficiency that often results in an amnestic syndrome and hallucinations.

KEY FACTS

- Neuropsychological evaluations can range in terms of time, intensity, and cost, depending on the reason for referral.
- In the case of bariatric surgery, a neuropsychological evaluation can be helpful in determining a potential patient's reading ability, overall intellect, and cognitive skills required for planning and executing various daily activities.
- A typical presurgical evaluation in our clinic takes approximately 3 hours and includes a clinical interview, objective cognitive testing, and administration of a structured mood inventory and other self-report questionnaires pertaining to alcohol use and eating behaviors.
- Executive functioning is a term used to describe cognitive skills typically associated with the frontal lobe. These skills include the ability to initiate, execute, and terminate specific behaviors. This broad term also reflects an individual's ability to plan actions and reflect on self-behaviors. Executive functioning typically refers to complex processes requiring significant cognitive resources.

SUMMARY POINTS

- Obesity is associated with cognitive dysfunction.
- Obesity is related to medical comorbidities that are also associated with cognitive deficits.
- There is often improved cognitive abilities following bariatric surgery.
- Complications of surgery such as vitamin deficiencies can impair cognition.
- Pre- and postsurgical neuropsychological evaluations are important to maximize surgical outcome.

REFERENCES

[1] Pannacciulli N, Parigi AD, Chen K, Le DS, Reiman EM, Tataranni PA. Brain abnormalities in human obesity: a voxel-based morphometric study. NeuroImage 2006;31:1419−25.
[2] Landsberg L, Aronne LJ, Beilin LJ, Burke V, Igel LI, Lloyd-Jones D, et al. Obesity-related hypertension: pathogenesis, cardiovascular risk, and treatment. J Clin Hypertension 2013;15:14−33.
[3] Connier MA, Dabelea D, Hernandez TL, Lindstrom AJ, Steig J, Stob NR, et al. The metabolic syndrome. Endocr Rev 2008;29:777−822.
[4] Romero-Corral A, Caples SM, Lopez-Jimenez F, Somers VK. Interactions between obesity and obstructive sleep apnea. Chest 2010;137:711−19.
[5] Raji CA, Ho AJ, Parikshak NN, Becker JT, Lopez OL, Kuller LH, et al. Brain structure and obesity. Hum Brain Map 2010;31:353−64.
[6] Beauchet O, Herrmann FR, Annweiler C, Kerlerouch C, Gosse P, Pichot V, et al. Association between ambulatory 24-hour blood pressure levels and cognitive performance: a cross-sectional elderly population-based study. Rejuvenation Res 2010;13:39−46.

[7] Beauchet O, Celle S, Roche F, Bartha R, Montero-Odasso M, Allali G, et al. Blood pressure levels and brain volume reduction: a systematic review and meta-analysis. J Hypertension 2013;31:1502–16.

[8] Veglio F, Paglieri C, Rabbia F, Bisbocci D, Bergui M, Cerrato P. Hypertension and cerebrovascular damage. Atherosclerosis 2009;205:331–41.

[9] Gonzalez CE, Pacheco J, Beason-Held LL, Resnick SM. Longitudinal changes in cortical thinning associated with hypertension. J Hypertension 2015;33:1242–8.

[10] Cholerton B, Baker LD, Craft S. Insulin resistance and pathological brain aging. Diabetic Med 2011;12:1463–75.

[11] Haley AP, Gonzolez MM, Tarumi T, Miles SC, Goudarzi K, Tanaka H. Elevated cerebral glutamate and myo-inositol levels in cognitively normal middle-aged adults with metabolic syndrome. Metab Brain Dis 2010;25:397–405.

[12] Torelli F, Moscufo N, Garreffa G, Placidi F, Romigi A, Zannino S, et al. Cognitive profile and brain morphological changes in obstructive sleep apnea. NeuroImage 2011;54:787–93.

[13] Xi Z, Lin M, Shunwei L, Yuping W, Luning W. A functional MRI evaluation of frontal dysfunction in patients with severe obstructive sleep apnea. Sleep Med 2011;12:335–40.

[14] Macey PM, Kumar R, Woo MA, Valladares EM, Yan-Go FL, Harper RM. Brain structural changes in obstructive sleep apnea. Sleep 2008;31:967–77.

[15] Alchanatis M, Deligiorgis N, Zias N, Amfilochiou A, Gotsis E, Karakatsani A, et al. Frontal brain lobe impairment in obstructive sleep apnoea: a proton MR spectroscopy study. Eur Respir J 2004;24:980–6.

[16] Blanco-Gomez A, Ferre N, Lugue V, Cardona M, Gispert-Llaurado M, Escribano J, et al. Being overweight or obese is associated with inhibition control in children from six to ten years of age. Acta Paediatr 2015;104(6):619–25.

[17] Gunstad J, Lhotsky A, Wendell CR, Ferrucci L, Zonderman AB. Longitudinal examination of obesity and cognitive function: results from the Baltimore Longitudinal Study of Aging. Neuroepidemiology 2010;34:222–9.

[18] Dahl AK, Hassing LB, Fransson EI, Gatz M, Reynolds CA, Pedersen NL. Body mass index across midlife and cognitive change in late life. Int J Obes 2013;37(2):296–302.

[19] Prickett C, Brennan L, Stolwyk R. Examining the relationship between obesity and cognitive function: a systematic literature review. Obes Res Clin Pract 2014;9:93–113.

[20] Gunstad J, Strain G, Devlin MJ, Wing R, Cohen RA, Paul RH, et al. Improved memory function 12 weeks after bariatric surgery. Surg Obes Relat Dis 2011;7:465–72.

[21] Gunstad J, Paul RH, Cohen RA, Tate DF, Spitznagel MB, Gordon E. Elevated body mass index is associated with executive dysfunction in otherwise healthy adults. Compr Psychiatry 2007;48:57–61.

[22] Alosco ML, Galioto R, Spitznagel MB, Strain G, Devlin M, Cohen R, et al. Cognitive function after bariatric surgery: evidence of improvement 3 years after surgery. Am J Surg 2014;207(6):870–6.

[23] Spitznagel MB, Garcia S, Miller LA, Strain G, Devlin M, Wing R, et al. Cognitive function predicts weight loss following bariatric surgery. Surg Obes Relat Dis 2013;9:453–9 #0.

[24] Waldstein SR, Brown JR, Maier KJ, et al. Diagnosis of hypertension and high blood pressure levels negatively affects cognitive function in older adults. Ann Behav Med 2005;29:174–80.

[25] Elias MF, Goodell AL, Dore GA. Hypertension and cognitive functioning: a perspective in historical context. Hypertension 2012;60:260–8.

[26] Purnell C, Gao S, Csllahan CM, Hendrie HC. Cardiovascular risk factors and incident Alzheimer's disease: a systematic review of the literature. Alzheimer Dis Assoc Disord 2009;23:1–10.

[27] Barnes DE, Yaffe K. The projected impact of risk factor reduction on Alzheimer's disease prevalence. Lancet 2011;10(9):819–28.

[28] Waldstein SR, Manuck SB, Ryan CM, Muldoon MF. Neuropsychological correlates of hypertension: review and methodological considerations. Psychol Bull 1991;110:451–68.

[29] Reaven GM, Thompson LW, Nahum D, Haskins E. Relationship between hyperglycemia and cognitive function in older NIDDM patients. Diabetes Care 1990;13(1):16–21.

[30] Munshi M, Grande L, Hayes M, et al. Cognitive dysfunction is associated with poor diabetes control in older adults. Diabetes Care 2006;29:1794–9.

[31] Perlmuter LC, Hakami MK, Hodgson-Harrington C, Ginsberg J, Katz L, Singer DE, et al. Decreased cognitive function in aging non-insulin dependent diabetic patients. Am J Med 1984;77(6):1043–8.

[32] Messier C. Impact of impaired glucose tolerance and type 2 diabetes on cognitive aging. Neurobiol Aging 2005;1S:26–30.

[33] Ruis C, Biessels GJ, Gorter KJ, van den Donk M, Jaap Kappelle L, Rutten GEH. Cognition in the early stages of type 2 diabetes. Diabetes Care 2009;32(7):1261–5.

[34] Awad N, Gagnon M, Messier C. The relationship between impaired glucose tolerance, type 2 diabetes, and cognitive function. J Clin Exp Neuropsychol 2004;26(8):1044–80.

[35] Saunamaki T, Himanen SL, Polo O, Jehkonen M. Executive dysfunction in patients with obstructive sleep apnea syndrome. Eur Neurol 2009;62:237–42.

[36] Salorio CF, White DA, Piccirillo J, Duntley SP, Uhles ML. Learning, memory, and executive control in individuals with obstructive sleep apnea syndrome. J Clin Exp Neuropsychol 2002;24:93–100.

[37] Bucks RS, Olaithe M, Eastwood P. Neurocognitive function in obstructive sleep apnoea: a meta-review. Respirology 2013;18(1):61–70.

[38] Spitznagel MB, Galioto R, Limbach K, Gunstad J, Heinberg L. Cognitive function is linked to adherence to bariatric postoperative guidelines. Surg Obes Relat Dis 2013;9:580–5.

[39] Malik S, Mitchell JE, Engel S, Crosby R, Wonderlich S. Psychopathology in bariatric surgery candidates: a review of studies using structured diagnostic interviews. Compr Psychiatry 2014;55:248−59.

[40] Buchwald H, Avidor Y, Braunwald E, Jensen M, Pories W, Fahrbach K, et al. Bariatric surgery: a systematic review and meta-analysis. J Am Med Assoc 2004;292(14):1724−37.

[41] Siervo M, Nasti G, Stephan BC, Papa A, Muscariello E, Wells JC, et al. Effects of intentional weight loss on physical and cognitive function in middle-aged and older obese participants: a pilot study. J Am Coll Nutr 2012;31(2):79−86.

[42] Napoli N, Shah K, Waters DL, Sinacore DR, Qualls C, Villareal DT. Effect of weight loss, exercise, or both on cognition and quality of life in obese older adults. Am J Clin Nutr 2014;100(1):189−98.

[43] Horie NC, Serrao VT, Sanz Simon S, Polo Gascon MR, dos Santos AX, Zambone MA, et al. Cognitive effects of intentional weight loss in elderly obese individuals with mild cognitive impairment. J Clin Endocrnol Metab 2015;101(3):1104−12.

[44] Jaiswal A, Bhavsar V, Javkaran, Kantharia ND. Effect of antihypertensive therapy on cognitive functions of patients of hypertension. Ann Indian Acad Neurol 2010;13(3):180−3.

[45] Lighart SA, Moll van Charante EP, Van Gool WA, Richard E. Treatment of cardiovascular risk factors to prevent cognitive decline and dementia: a systematic review. Vasc Health Risk Manage 2010;6:775−85.

[46] Kawamura T, Umemura T, Hotta N. Cognitive impairment in diabetic patients: can diabetic control prevent cognitive decline? J Diabetes Investig 2012;3(5):413−23.

[47] Meneilly GS, Cheung E, Tessier D, Yakura C, Tuokko H. The effect of improved glycemic control on cognitive functions in the elderly patient with diabetes. J Gerontol 1993;48:M117−21.

[48] Lim W, Bardwell WA, Loredo JS, Kim EJ, Ancoli-Israel S, Morgan EE, et al. Neuropsychological effects of two-week continuous Positive airway pressure treatment and supplemental oxygen in patients with obstructive sleep apnea: a randomized placebo-controlled study. J Clin Sleep Med 2007;3(4):380−6.

[49] Gagnadoux F, Fleury B, Vielle B, Petelle B, Meslier N, N'Guyen XL, et al. Titrated mandibular advancement versus positive airway pressure in sleep apnoea. Eur Respir J 2009;34:914−20.

[50] Joseph S, Zuriqat M, Husari A. Sustained improvement in cognitive and emotional status of apneic patients after prolonged treatment with positive airway pressure. South Med J 2009;102:589−94.

[51] Dinges DF, Weaver TE. Effects of modafinil on sustained attention performance and quality of life in OSA patients with residual sleepiness while being treated with nCPAP. Sleep Med 2003;4(5):393−402.

[52] Arnulf I, Homeyer P, Garma L, Whitelaw WA, Derenne J. Modafinil in obstructive sleep apnea-hypopnea syndrome: a pilot study in 6 patients. Respiration 1997;64:159−61.

[53] Ancoli-Israel S, Palmer BW, Cooke JR, Corey-Bloom J, Fiorenrino L, Natarajan L, et al. Cognitive effects of treating obstructive sleep apnea in Alzheimer's disease: a randomized controlled study. J Am Geriatr Soc 2008;56(11):2076−81.

[54] Votruba KL, Marshall D, Giordani B, Finks J. Neuropsychological factors and bariatric surgery: a review. Curr Psychiatry Rep 2014;16:448.

[55] Alosco ML, Cohen R, Spitznagel MB, Strain G, Devlin M, Crosby RD, et al. Older age does not limit postbariatric surgery cognitive benefits: a preliminary investigation. Surg Obes Relat Dis 2014;10(6):1196−201.

[56] Spitznagel MB, Hawkins M, Alosco M, Galioto R, Garcia S, Miller L, et al. Neurocognitive effects of obesity and bariatric surgery. Eur Eat Disorders Rev 2015;23:488−95.

[57] Alosco ML, Spitznagel MB, Strain G, Devlin M, Cohen R, Paul R, et al. Improved memory function two years after bariatric surgery. Obesity (Silver Spring) 2014;22(1):32−8.

[58] Grayson BE, Fitzgerald MF, Hakala-Finch AP, Ferris VM, Begg DP, Tong J, et al. Improvements in hippocampal-dependent memory and microglial infiltration with calorie restriction and gastric bypass surgery, but not with vertical sleeve gastrectomy. Int J Obes 2014;38:349−56.

[59] Berger JR. The neurological complications of bariatric surgery. Arch Neurol 2004;61:1185−9.

[60] Bozbora A, Coskun H, Ozarmagan S, Erbil Y, Ozbey N, Orham Y. A rare complication of adjustable gastric banding: Wernicke's encephalopathy. Obes Surg 2000;10(3):274−5.

Chapter 62

Temperament and Outcome of Bariatric Surgery for Severe Obesity

C. De Panfilis and I. Generali

University of Parma, Parma, Italy

LIST OF ABBREVIATIONS

BAS   behavioral activation system
BIS   behavioral inhibition system
BMI body mass index
BS   bariatric surgery
EA   executive attention
EF   executive function
EC   effortful control
HA   harm avoidance
LAGB laparoscopic adjustable gastric banding
NS   novelty seeking
RD   reward dependence
RYGB Roux-en-Y gastric bypass
STR   sensitivity to reward
TT   temperament traits
WL   weight loss

INTRODUCTION

Successful and sustained weight loss (WL) after bariatric surgery (BS) for severe obesity is not merely a function of the surgical operation. Nontechnical factors, such as patients' psychosocial status before surgery, can influence patients' abilities to implement permanent lifestyle changes after surgery, thus accounting for the high variability in terms of WL after BS [1]. Identifying preoperative psychological factors that correlate with nonoptimal WL may help to guide patient selection and assist in developing follow-up plans for those at risk of losing less weight. Individual personality traits may play a central role in this respect [2]. In fact, personality can be defined as the "organization of the systems that determine the person's unique adjustment to his environment" [3]; as such, personality factors may influence how individuals adapt and adjust to the demands imposed by BS. Temperament traits (TT) are a subset of personality traits that are apparent since childhood and tend to show consistency across situations and stability over time. This chapter will focus on the relationship between TT and BS outcome.

WHAT IS TEMPERAMENT?

TT reflect individual differences in reactivity to environmental stimuli and self-regulation [4]. These differences are biologically based, are linked to an individual's genetic endowment, and result from the activity of underlying neural systems related to motivation and attention [4,5]. Specifically, reactivity refers to the motor, affective, and attentional tendencies, whereas self-regulation refers to the processes that modulate such reactivity [4]. Thus, personality differences arise over time from the interaction between reactive (i.e., motivational) and regulatory aspects of temperament

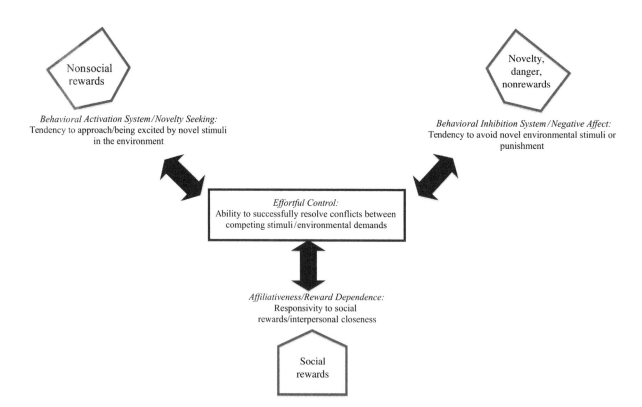

FIGURE 62.1   Interactions between reactive and self-regulatory aspects of temperament. Red arrows (dark gray in print versions) represent environmental inputs for the various motivational systems.

(Fig. 62.1). All of these temperament features can be reliably assessed by means of various questionnaires and are related to general dimensions of adult personality, as described in five-factor models (Table 62.1).

"Reactive" TT represent relatively stable biases toward automatic reactions in response to definite environmental stimuli (i.e., novelty, danger, or reward), and are underlaid by specific motivational systems that serve appetitive, defensive, and nurturant needs, and shape individual dispositions toward, respectively, positive emotions, negative emotions, and affiliative social behaviors [6] (Fig. 62.1).

The appetitive or behavioral activation motivational system (BAS) mobilizes approach behavior to stimuli that predict positive events. It mediates sensitivity to rewarding stimuli (STR), facilitates the tendency to being excited by novel events as well as approach responses directed toward the rewarding input, but also promotes irritative/appetitive aggression when goals are blocked. Individual differences in this system are related to a general personality dimension of Extraversion/Positive Affect/Impulsivity/Novelty Seeking (NS), which has been found to be stable from infancy to adulthood [6–9].

The *defensive or behavioral inhibition motivational system* (BIS) is associated with fear-related behavioral inhibition. This system responds to novel signals and signals predicting punishment or nonreward. It directs attention toward relevant and potentially dangerous information in the environment, promotes passive avoidance, and inhibits the appetitive system. In emotional terms, these multiple outputs set up a state of "anxiety" and may motivate some forms of defensive aggression. Individual differences in fearful motivation have been found to be stable from infancy to early adulthood, and relate to a general dimension of Neuroticism/Negative Emotionality/Harm Avoidance (HA) [6–9].

The *affiliative motivational system* is related to social reward motivation. It responds to the loss of social support and promotes social bonding by facilitating friendly, trusting, and helpful behaviors, as well as feelings of warmth and empathy. These types of affectionate and nurturant behaviors relate to a general personality dimension of Affiliativeness/Agreeableness/Reward–Dependence (RD) [6–8].

The regulatory or "voluntary" aspect of temperament, or Effortful Control (EC), is related to the ability to strategically regulate motivational tendencies as needed for adapting to circumstances even if one does not feel like doing so (Fig. 62.1). It allows individuals to successfully resolve conflicts between immediate reactive tendencies and long-term demands, and to overcome a prepotent, impulsive response in order to produce a more appropriate and goal-directed

TABLE 62.1 Temperament Scales and Related Dimensions of Personality

Instruments	Scales of Temperament/ Personality Traits	High Scores	Low Scores
Adult Temperament Questionnaire (ATQ) Evans and Rothbart (2007)	Positive Affect/ Extraversion (E)	High level of enthusiasm; subjects are more likely to develop externalizing problems like acting out.	Fear, tendency toward withdrawal/avoidance.
	Negative Affect (NA)	Unpleasant effects such as lowered mood, anxiety, fearfulness, frustration, and anticipation of danger.	Underestimation of danger.
	Affiliativeness (A)	Emotional dependence, empathy, warmth, and sensitivity to social cues.	Detachment, coldness.
	Effortful Control (EC)	Capacity to overcome dominant, impulsive responses in order to execute a nondominant, goal-directed response.	Distractibility and poor self-determination.
Temperament and Character Inventory (TCI) Cloninger et al. (1994)	Novelty Seeking (NS)	Enthusiasm and quick engagement with whatever is new that leads to exploration of potential rewards.	Indifference, lack of enthusiasm, tolerance of monotony.
	Harm Avoidance (HA)	Apprehension, insecurity, pessimism in situations that do not normally worry other people.	Light-heartedness, courage, and optimism in situations that worry most people.
	Reward Dependence (RD)	Kindness and gentleness, dependence, and sociability.	Coldness and callousness.
	Persistence (P)	Industriousness, hard work, perseverance, and stability of behavior despite frustrations and fatigue.	Indolence, inactivity, unreliability, and instability.
Behavioral Activation System/Behavioral Inhibition System (BAS/ BIS) Gray (1970)	Behavioral Activation System (BAS)	High levels of positive emotions such as elation, happiness, and hope in response to environmental cues consistent with nonpunishment and reward.	Anxiety and fear.
	Behavioral Inhibition System (BIS)	Heightened sensitivity to nonreward, punishment, and novel experiences.	Extraversion and openness.
NEO Five-Factor Inventory (NEO-FFI) McCrae and Costa (2004)	Extraversion (E)	Positive emotions, surgency, assertiveness, sociability.	Reserved and reflective personality.
	Neuroticism (N)	Tendency to easily experience unpleasant emotions, such as anger, anxiety, and depression.	Low perception of risk.
	Agreeableness (A)	Interpersonal warmth, kindness, and trust.	Tendency toward social withdrawal.
	Conscientiousness (C)	Self-discipline and aim for achievement.	Immaturity.

response. Thus, EC reflects the efficiency of executive attention (EA)—i.e., the ability to inhibit a dominant response and to activate a subdominant response for the sake of valued goals [5,6,10]. Dispositional differences in EC, including executive functioning (EF) skills, develop in the first and second year of life, improve greatly in the third year of life and through childhood, and then seem to remain stable from late childhood [5]. In fact, the maturation of EC in childhood parallels performance on EA tasks [11−13], and predicts successful self-control abilities in adulthood [14,15]. Among adults, EC encompasses the capacity to inhibit inappropriate approach behavior, to perform an action when there is a strong tendency to avoid it, and to focus/shift attention when needed [8]. In this way, EC decreases the risks associated with excessive emotional reactivity. Not surprisingly, limitations in EC are associated with several negative psychological and interpersonal consequences in children, adolescents, and adults [16—19]. EC variability in adulthood taps into several personality dimensions (i.e., high Consciousness, Persistence, and Self-directedness; and low Neuroticism) [8].

WHAT ARE THE RELATIONSHIPS BETWEEN TEMPERAMENT AND OBESITY?

Given the central role of TT in shaping individuals' responses and adjustments to environmental cues, research has been directed toward investigations of whether temperament (1) may play a role as a risk or protective factor in the development of obesity and (2) could be associated with treatment-seeking status or success of medical/behavioral treatments for obesity.

In general, however, empirical evidence regarding the association between TT and weight control over time in the population and in clinical medical samples yields mixed results. These often contrasting findings are concisely summarized and referenced in Tables 62.2 and 62.3, respectively.

With respect to the *motivational (i.e., "reactive") aspects of temperament*, *BAS features* such as Impulsivity/ Extraversion/NS have been associated with childhood overweight status, and greater weight gain and overweight status over time (Table 62.2). NS was found to be increased in obese subjects in the community and to be associated with lower success of medical treatment (Tables 62.2 and 62.3). However, other data suggest a more complex association between the BAS system and weight control; for instance, STR (i.e., the ability to derive pleasure or reward from natural reinforcers like food) exhibits a curvilinear relationship with body weight in children and adults, with a positive association in normal weight and overweight individuals, and a negative association in individuals with obesity (Table 62.2). In keeping with this complex pattern of findings, overall population and clinical studies failed to detect any firm association between "extraverted" TT and overweight status, obesity, and success of medical/behavioral treatments for obesity [20].

Similarly, data concerning the relationship between the *BIS system* and overweight/obesity suggest that temperament-based negative emotionality might be associated with either higher or lower likelihood of being overweight (Table 62.2). In general, neither population-based nor clinical studies found consistent associations between TT of negative emotionality/HA and obesity or treatment-seeking status [20].

Affiliative aspects of temperament such as decreased RD have been associated with higher BMI in women (Table 62.2) and increased risk for dropout from behavioral treatments for obesity (Table 62.3), but the majority of studies could not find any association between affiliativeness, overweight/obesity, and response to behavioral treatments for obesity [20].

Available evidence on the association between *self-regulatory aspects of temperament* and obesity is more consistent. For instance, among children and adolescents, poor EC and EC-related executive dysfunction are associated with obesity-related behaviors such as increased food intake, disinhibited eating, and less physical activity, as well as excess adiposity (Table 62.2). In addition, a decreased ability to delay gratification as preschoolers (an aspect of EC) is associated with increased BMI three decades later; obese subjects in the community reported lower persistence than lean controls (Table 62.2). Overall, features of "conscientiousness" (i.e., self-control, orderliness) emerge as protective factors against obesity in some population studies, and seem to be associated with treatment-seeking behavior and greater WL in patients with obesity [20]. Importantly, however, one study suggested that too much "self-control" might be detrimental, as it might increase the risk of cognitive dietary restraint and subsequent loss of control over eating, and thus lead to a higher BMI [21].

A sensible approach toward interpreting this body of results would recommend caution in drawing firm conclusions about any association between definite TT and obesity. In fact, to date, research has provided no evidence of a general "personality or temperament profile" associated with obesity, such that limited and inconsistent differences are reported compared to control populations [22,23]. Rather, the various findings in both population-based and clinical samples seem to suggest that patients with severe obesity, including BS candidates, are highly heterogeneous in terms of TT.

In this regard, temperament has proven useful to differentiate clinically meaningful phenotypes within the large group of treatment-seeking individuals with obesity. In fact, latent profile analyses revealed a "resilient/high-functioning" cluster exhibiting high EC and low BIS/BAS scores, and an "emotionally dysregulated/undercontrolled" cluster exhibiting low EC and high BIS/BAS scores as well as greater eating and general psychopathology [23,24]. Importantly, these results point out the need to carefully evaluate temperament styles in treatment-seeking patients with obesity (including BS candidates) in order to provide clinicians with a deeper understanding of patients' psychopathology statuses and problem behaviors. Specifically, these findings suggest that a subgroup of patients with obesity may display a marked imbalance between motivational and regulatory TT, such that their reactive tendencies (i.e., tendency toward experiencing negative emotions and impulsivity) are not buffered by a corresponding capacity to restrain or

TABLE 62.2 Contrasting Findings on the Association Between Temperament Traits (TT) and Overweight/Obesity in Nonclinical Samples

Specific Findings	Studies that Favor the Hypothesis	Studies that Contrast the Hypothesis
Reactive TT and Obesity		
Extraversion/Novelty-Seeking (NS) is positively associated with increased risk of overweight status and obesity in children and adults.	Burton et al. Association between infant correlates of impulsivity–surgency (extraversion) and early infant growth. Appetite 2011;57:504–9	Bradley et al. The relationship between body mass index and behaviour in children. J Peiatr 2008;153:629–34
	Davis et al. Sensitivity to reward: implications for overeating and overweight. Appetite 2004;42:131–8	Gong et al. Longitudinal study on infants' temperament and physical development in Beijing, China. Int J Nurs Pract 2013;19:487–97
	Davis Fox. Sensitivity to reward and body mass index (BMI): evidence for a nonlinear relationship. Appetite 2008;50:43–9	Hampson et al. Personality and overweight in 6–12 year-old children. Pediat Obes 2015;10:e5–7
	de Zwaan et al. Temperamental factors in severe Weight cycling. A cross-sectional study. Appetite 2015;91:336–42	Niegel et al. Is difficult temperament related to overweight and rapid early weight gain in infants? A prospective cohort study. J Dev Behav Pediatr 2007;28:462–6
	Hintsanen et al. Temperament and character predict body-mass index: A population-based prospective cohort study. J Psychosom Res 2012;73:391–397	Provencher et al. Personality traits in overweight and obese women: associations with BMI and eating behaviours. Eat Behav 2008;9:294–302
	Sullivan et al. Personality characteristics in obesity and relationship with successful weight loss. Int J Obes 2007;31:669–74	
Negative Emotionality/Harm Avoidance is positively associated with overweight/obesity in children and adults.	Anzman-Frasca et al. Infant temperament and maternal parenting self-efficacy predict child weight outcomes. Inf Behav Develop 2013;36:494–7	Gong et al. Longitudinal study on infants' temperament and physical development in Beijing, China. Int J Nurs Pract 2013;19:487–97
	Bradley et al. The relationship between body mass index and behaviour in children. J Pediatr 2008;153:629–34	Niegel et al. Is difficult temperament related to overweight and rapid early weight gain in infants? A prospective cohort study. J Dev Behav Pediatr 2007;28:462–6
	Hampson et al. Personality and overweight in 6–12 year-old children. Pediat Obes 2015;10:e5–7	
	Slining et al. Infant temperament contributes to early infant growth: a prospective cohort of African American infants. International Journal of Behavioral Nutrition and Physical Activity. Int J Behav Nutr Phys 2009;6;51	Provencher et al. Personality traits in overweight and obese women: associations with BMI and eating behaviours. Eat Behav 2008;9:294–302
	Wu et al. Joint effects of child temperament and maternal sensitivity on the development of childhood obesity. Mat Child Health J 2011;15:469–77	Wells et al. Investigation of the relationship between infant temperament and later body composition. Int J Obes 1997;21:400–6
Reward Dependence is inversely associated with overweight/obesity.	Hintsanen et al. Temperament and character predict body-mass index: a population-based prospective cohort study. J Psychosom Res 2012;73:391–7	–
Regulatory TT and Obesity		
Effortful Control/Conscientiousness is inversely associated with problem eating behavior, overweight status, and obesity in children and adults.	Anzman-Frasca et al. Infant temperament and maternal parenting self-efficacy predict child weight outcomes. Inf Behav Develop 2013;36:494–7	de Zwaan et al. Temperamental factors in severe weight cycling. A cross-sectional study. Appetite 2015;91:336–42
	Faith and Hittner. Infant temperament and eating style predict change in standardized weight status and obesity risk at 6 years of age. Int J Obes 2010;34:1515–23	Provencher et al. C. Personality traits in overweight and obese women: associations with BMI and eating behaviours. Eat Behav 2008;9:294–302
	Hampson et al. Personality and overweight in 6 12 year-old children. Pediat A Obes 2015;10:e5–7	
	Schlam et al. Preschoolers' delay of gratification predicts their body mass 30 years later. J Pediat 2013;162:90–3	
	Godefroy et al. Modelling the effect of temperament on BMI through appetite reactivity and self-regulation in eating: a structural equation modelling approach in young adolescents. Int J Obes 2012;573–80; doi:10,1038/ijo.2016.6	
	Sullivan et al. Personality characteristics in obesity and relationship with successful weight loss. Int J Obes 2007;31:669–74	

TABLE 62.3 Contrasting Findings on the Association Between Temperament Traits (TT) and Outcome of Medical Treatment for Obesity

Specific Findings	Studies That Favor the Hypothesis	Studies That Contrast the Hypothesis
Reactive Temperament Traits and Obesity		
Extraversion/Novelty-Seeking (NS) is negatively associated with success of medical treatments for obesity.	Dalle Grave et al. Personality, attrition and weight loss in treatment seeking women with obesity. Clinic Obes 2015;5: 266–72	De Panfilis et al. Personality and attrition from behavioral weight-loss treatment for obesity. Gen Hosp Psychiat 2008;30:515–20
		Elfhag. Personality correlates of obese eating behaviour: Swedish universities Scales of Personality and the Three Factor Eating Questionnaire. Eating and Weight Disorders-Studies on Anorexia, Bulimia and Obesity 2005;10:210–15
		Sullivan et al. Personality characteristics in obesity and relationship with successful weight loss. Int J Obes 2007;31:669–74
Negative Emotionality/Harm Avoidance is negatively associated with success of medical treatments for obesity.	Canetti et al. Psychosocial predictors of weight loss and psychological adjustment following bariatric surgery and a weight-loss program: the mediating role of emotional eating. Int J Eat Disord 2009;42:109–17	Dalle Grave et al. Personality, attrition and weight loss in treatment seeking women with obesity. Clinic Obes 2015;5:266–72
	Elfhag and Morey. Personality traits and eating behavior in the obese: poor self-control in emotional and external eating but personality assets in restrained eating. Eat Behav 2008;9:285–93	Munro et al. Using personality as a predictor of diet induced weight loss and weight management. Int J Behav Nutr Phys Act 2011;8:129
		Poston et al. Personality and the prediction of weight loss and relapse in the treatment of obesity. Int J Eat Disord 1999;25:301–9
Reward Dependence is positively associated with success of medical treatments for obesity.	De Panfilis et al. Personality and attrition from behavioural weight-loss treatment for obesity. Gen Hosp Psychiat 2008;30:515–20	Dalle Grave et al. Personality, attrition and weight loss in treatment seeking women with obesity. Clinic Obes 2015;5:266–72
		Sullivan et al. Personality characteristics in obesity and relationship with successful weight loss. Int J Obes 2007;31:669–74
Regulatory TT and Obesity		
Effortful Control (EC)/ Conscientiousness is inversely associated with success of medical treatments for obesity.	Carels et al. The early identification of poor treatment outcome in a women's weight loss program. Eat Behav 2003;4:265–82	Dalle Grave et al. Personality, attrition and weight loss in treatment seeking women with obesity. Clinic Obes 2015;5:266–72
	Crescioni et al. High trait self control predicts positive Health behaviours and success in weight loss. J Health Psychol 2011;16:750–9	Munro et al. Using personality as a predictor of diet induced weight loss and weight management. Int J Behav Nutr Phys Act 2011;8:129
	Elthag and Morey. Personality traits and eating behavior in the obese: poor self-control in emotional and external eating but personality assets in restrained eating. Eat Behav 2008;9:285–93	

modulate their own behavior for the sake of weight and eating control. This failure in self-regulation could also influence the success of BS.

DOES TEMPERAMENT INFLUENCE THE OUTCOME OF BARIATRIC SURGERY?

Research on the influence of motivational/reactive and regulatory TT on BS outcome is still in its infancy. However, available data support a role for self-regulation failures (i.e., excessive temperamental reactivity coupled with insufficient regulatory skills) in preventing optimal WL following BS.

With respect to motivational facets of temperament, high NS/Impulsivity is associated with reduced weight loss 2 years after Roux-en-Y gastric bypass (RYGB) [25] and decreased weight maintenance 9–15 years after RYGB [26]. Similarly, negative emotionality is associated with a worse outcome following BS through increased emotional eating [27]. However, some studies show that negative emotionality decreases at least temporarily after BS [28–30], whereas impulsivity does not differ between presurgery and postsurgery patients [29,31]. The fact that impulsivity does not seem to change after BS may make these patients more vulnerable to loss of control overeating and unsatisfactory BS outcome.

Conversely, self-regulatory features (EC; either assessed through self-reports or EA/EF tasks) seem to be positively associated with WL following BS. Self-directedness, a personality dimension associated with EC, predicted greater 6-month WL after laparoscopic vertical banded gastroplasty [32]. Notably, Persistence was inversely related with WL 1 year after laparoscopic adjustable gastric banding (LAGB) as well as 1–2 years following RYGB [25,33,34]. These results held true even after controlling for other preoperative psychological, psychiatric, and medical variables that can potentially influence the amount of postoperative WL. Furthermore, preoperative impairments in EF predicted greater BMI reductions and percentage of excess WL 1 year after BS, and postoperative (i.e., 12 weeks) EF predicted 36-month WL [35,36].

Overall, these findings suggest that a good capacity to maintain a hard-working behavior and resist distracting stimuli for the sake of valued long-term goals (as indicated by higher scores on measures of persistence, self-directedness, or EA/EF) may be adaptive after BS. In fact, such ability could enable BS patients to maintain control over their eating behavior in spite of frustrating stimuli (e.g., uneven WL rate) (Fig. 62.2).

Emerging evidence that good cognitive control in the context of food-related stimuli is associated with successful weight regulation and BS outcome further supports this hypothesis. In fact, successful dieters more readily activate brain regions devoted to cognitive control when processing food cues, and report less desire for food and greater positive mood [37]. Most importantly, the ability to exercise cognitive control over food-related cues differentiates between poor and good responders approximately 12 years after RYGB. Poor responders (i.e., patients with an excessive weight and BMI loss <50% and BMI > 30 kg/m^2 9–15 years after BS) performed worse than good responders (i.e., patients with an excessive weight and BMI loss >75% and BMI < 30 kg/m^2) on food-modified cognitive control tasks, which evaluated the extent to which an individual was "distracted" by food cues [26]. Finally, presurgical brain activity in regions associated with cognitive and behavioral regulation during a functional magnetic resonance imaging (fMRI) food-motivation paradigm predicted 6-month WL following LAGB [38].

At the behavioral level, these findings suggest a potential way through which EC and EC-related EF influence BS outcome. One intriguing possibility is that good EC/EF fosters adherence to postoperative lifestyle changes in diet and physical activity by facilitating control over eating and exercise behavior. In fact, preliminary evidence indicates that

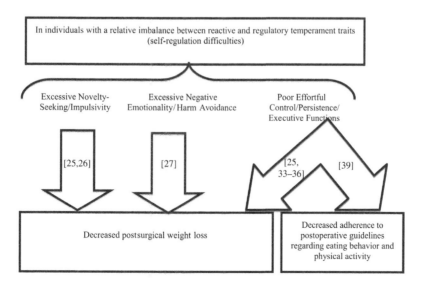

FIGURE 62.2   Available evidence on the relationship between temperament traits and outcome of bariatric surgery. Numbers in brackets refer to the corresponding references.

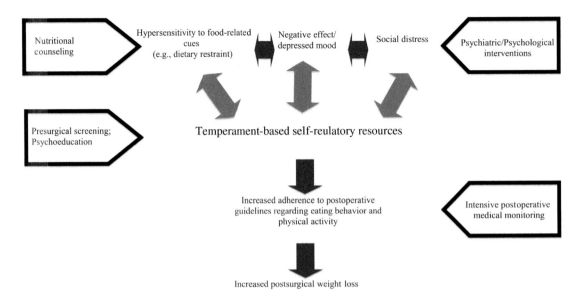

FIGURE 62.3   Clinical implications of the findings regarding temperament and bariatric surgery outcome. Blue arrows (light gray in print versions) indicate a diminishing effect. Red arrows (dark gray in print versions) indicate an augmenting effect. Purple arrows (black in print versions) indicate potential foci of intervention.

preoperative impairments in EF are associated with decreased adherence to postoperative guidelines 1 month after LAGB [39]. Future studies will clarify whether this represents the mediating mechanisms in the observed relationship between poor self-regulatory TT and decreased WL following BS.

The clinical implications of these findings are twofold. First, it is important to include TT in the preoperative psychosocial evaluation of patients undergoing BS, as this could help identify individuals at risk for suboptimal outcomes. In this respect, patients low in EC and/or high in NS could benefit from presurgical psychoeducation about potential barriers to postsurgical persistence in healthy eating and physical activity habits. These at-risk individuals may also be assisted in developing plans to overcome such difficulties in order to maximize their likelihood of WL [40].

Secondly, given the central role of self-regulation in weight control, patients who have undergone BS, and especially those low in temperamental EC and high in NS, could benefit from close monitoring of self-regulation failures in the postsurgical period. Social cognitive neuroscience of self-regulation shows that several stimuli can trigger "lapses" in self-regulation (e.g., insufficient adherence to dietary prescriptions). For instance, among chronic dieters, negative affect (i.e., low mood) or social distress (i.e., interpersonal rejection) can foster disinhibited eating. In fact, experiencing emotional/social distress powerfully engages self-regulatory resources, resulting in decreased availability of the self-regulatory skills needed to deal with impulsive eating. In this way, the brain's reward system (e.g., BAS) gets "sensitized" to appetitive cues (e.g., food), leading to a critical imbalance between the strength of the impulse toward food and the capacity to regulate this desired behavior [41]. Therefore, it is important to closely monitor BS patients' psychosocial functioning and actively detect and treat potential mood/interpersonal disturbances after surgery. Doing so could improve patients' postsurgical self-regulatory abilities and, thus, their compliance with postoperative guidelines about diet, medical monitoring, and regular exercise, thereby ultimately promoting optimal WL after BS (Fig. 62.3).

CONCLUSIONS AND FUTURE DIRECTIONS

Temperament-based personality traits enabling one to successfully solve conflicts between immediate emotional demands and the long-term goal of WL seem to favor BS success. These findings have important clinical implications for patients undergoing BS. In fact, although TT are constitutionally based, they are not unmodifiable; clarifying the processes and behaviors through which TT contribute to satisfactory postsurgical WL will allow researchers to identify potential areas of intervention. In this way, multiprofessional treatments that also address psychological issues could improve BS patients' self-regulation capacities arising from the interplay of their underlying TT. This will ultimately optimize patient care and enhance surgical success.

MINI-DICTIONARY OF TERMS

- *Affiliativeness*: Personality trait denoting sensitivity to social reward, and promoting empathy toward others, interpersonal trust, and attachment.
- *Behavioral activation system (BAS)*: Basic brain system that, according to Gray's Reinforcement Sensitivity Theory, gets activated by cues signaling rewards. It fosters approach-oriented behaviors.
- *Behavioral inhibition system (BIS)*: Basic brain system that, according to Gray's Reinforcement Sensitivity Theory, gets activated by cues of punishment/danger. It promotes avoidance-oriented behaviors.
- *Effortful control (EC)*: The self-regulatory or "voluntary" aspect of temperament that, according to Rothbart's model, allows individuals to successfully solve conflicts between prepotent impulses/urges and one's own goals. It reflects the efficiency of EA.
- *Executive attention (EA)*: The capacity to inhibit a dominant response in order to execute/activate a nondominant but goal-directed response, to engage in planning, and to detect errors. It is essential for the cognitive control of behavior.
- *Extraversion/novelty seeking (NS)*: Temperament dimension that reflects BAS activity. It promotes sensitivity to nonsocial reward, positive anticipation, increased energy, approach behavior, and eventually poor tolerance of boredom and impulsivity.
- *Negative affect/harm avoidance (HA)*: Temperament dimension that reflects BIS activity. It represents the tendency to inhibit behavior and experience anxiety in response to novel stimuli, eventually leading to negative emotions such as fear/anxiety, excessive worrying/pessimism, frustration, and sadness.
- *Persistence*: In Cloninger's psychobiological model, it is the temperament-based ability to maintain a behavior in spite of frustrating or nonrewarding stimuli for the sake of long-term goals. It partially overlaps with EC.
- *Reward dependence (RD)*: In Cloninger's psychobiological model, it is the temperament disposition toward responding to signals of social reward. It partially overlaps with affiliativeness.
- *Self-directedness*: In Cloninger's psychobiological model, it is the personality trait that allows one to flexibly regulate and adapt behavior to situational demands according to personally chosen goals and values. It partially overlaps with EC.
- *Self-regulation*: The capacity to make plans, choose between diverse options, inhibit undesired thoughts, and regulate impulses and appetitive behaviors. It is underlaid by the activity of various brain regions devoted to executive/cognitive control.
- *Temperament*: Biologically based personality traits that are apparent since childhood and reflect individual differences in emotional reactivity and self-regulatory abilities.

KEY FACTS

- Temperament traits represent constitutionally based personality features that reflect individual differences in emotional reactivity and self-regulation.
- There is no evidence of any specific personality/temperament style associated with obesity.
- Rather, individuals with (severe) obesity are highly heterogeneous in terms of personality/temperament features.
- Evaluating temperament traits helps to identify clinically meaningful subgroups of patients with obesity.
- Obese patients who display the greatest difficulties in eating behavior and general psychopathology show temperament styles denoting a relative imbalance between reactive and regulatory tendencies.
- Successful weight loss following bariatric surgery is largely dependent on patients' abilities to implement permanent lifestyle changes after surgery.
- Temperament traits (TT) may theoretically influence patients' capacities to adhere to postsurgical demands.
- Self-regulatory TT have been associated with greater weight loss following bariatric surgery.
- Future research needs to clarify the behavioral mechanisms that mediate this association.

SUMMARY POINTS

- TTs shape individuals' responses to environmental stimuli.
- The temperamental ability to effortfully regulate behavior is rooted in the EA system.
- A relative difficulty in this temperament-based capacity to modulate impulsive reactions in favor of long-term goals may influence how individuals cope with the demands of BS.

- Good self-regulatory skills and EF predict increased WL following BS.
- Good EF enhances adherence to postoperative medical guidelines.
- Assessing the TT of BS patients allows the implementation of specific interventions for those at risk of suboptimal WL.

REFERENCES

[1] van Hout GCM, van Heck G. Bariatric psychology, psychological aspects of weight loss surgery. Obes Facts 2009;2:10–15.

[2] Livhits M, Mercado C, Yermilov I, Parikh JA, Dutson E, Mehran A, et al. Preoperative predictors of weight loss following bariatric surgery: systematic review. Obes Surg 2012;22:70–89.

[3] Allport GW. Personality: a psychological interpretation. New York: Henry Holt and Company; 1937.

[4] Rothbart MK, Zentner M, Shiner RL. Advances in temperament. Handbook of Temperament. New York: Guilford Press; 2012. p. 3–20, Chapter 1.

[5] Posner MI, Rothbart MK. Toward a physical basis of attention and self-regulation. Phys Rev 2009;6:103–20.

[6] Derryberry D, Rothbart MK. Reactive and effortful processes in the organization of temperament. Dev Psychopathol 1997;9:633–52.

[7] Cloninger CR, Przybeck TR, Svravick DM. A psychobiological model of temperament and character. Arch Gen Psychiatry 1993;50:975–90.

[8] Evans DE, Rothbart MK. Developing a model for adult temperament. J Res Pers 2007;41:868–88.

[9] Gray JA. The psychophysiological basis of introversion-extraversion. Behav Res Therapy 1970;8:249–66.

[10] Casey BJ, Tottenham N, Fossella J. Clinical, imaging, lesion, and genetic approaches toward a model of cognitive control. Develop Psychobiol 2002;40:237–54.

[11] Gerardi-Caulton G. Sensitivity to spatial conflict and the development of self-regulation in children 24–36 months of age. Develop Science 2000;3:397–404.

[12] Chang F, Burns BM. Attention in preschoolers: associations with effortful control and motivation. Child Develop 2005;76:247–63.

[13] Rothbart MK, Ellis LK, Rosario Rueda M, Posner MI. Developing mechanisms of temperamental effortful control. J Pers 2003;71:1113–44.

[14] Casey BJ, Somerville LH, Gotlib IH, Ayduk O, Franklin NT, Askren MK, et al. Behavioral and neural correlates of delay of gratification 40 years later. Proc Natl Acad Sci 2011;108:14998–5003.

[15] Eigsti IM, Zayas V, Mischel W, Shoda Y, Ayduk O, Dadlani MB, et al. Predicting cognitive control from preschool to late adolescence and young adulthood. Psychol Sci 2006;17:478–84.

[16] Eisenberg N, Smith CL, Spinrad TL. Effortful control: relations with emotion regulation, adjustment, and socialization in childhood. Handbook of self-regulation. New York: Guilford Press; 2011. p. 263–84, Chapter 14.

[17] Cain NM, De Panfilis C, Meehan KB, Clarkin JF. Assessing interpersonal profiles associated with varying levels of effortful control. J Pers Asses 2013;95:640–4.

[18] De Panfilis C, Meehan KB, Cain NM, Clarkin JF. The relationship between effortful control, current psychopathology and interpersonal difficulties in adulthood. Compr Psych 2013;54:454–61.

[19] Meehan KB, De Panfilis C, Cain NM, Clarkin JF. Effortful control and externalizing problems in young adults. PAID 2013;55:553–8.

[20] Gerlach G, Herpertz S, Löber S. Personality traits and obesity: a systematic review. Obes Rev 2015;16:32–63.

[21] Provencher V, Bégin C, Gagnon-Girouard MP, Tremblay A, Boivin S, Lemieux S. Personality traits in overweight and obese women: associations with BMI and eating behaviors. Eat Behav 2008;9:294–302.

[22] Rydén A, Sullivan M, Torgerson JS, Karlsson J, Lindroos AK, Taft C. Severe obesity and personality: a comparative controlled study of personality traits. Int J Obes 2003;27:1534–40.

[23] Claes L, Müller A. Temperament and Personality in Bariatric Surgery Resisting Temptations? Eur Eat Disord Rev 2015;23:435–41.

[24] Müller A, Claes L, Wilderjans TF, Zwaan M. Temperament subtypes in treatment seeking obese individuals: A latent profile analysis. Eur Eat Disord Rev 2014;22:260–6.

[25] Gordon PC, Sallet JA, Sallet PC. The impact of temperament and character inventory personality traits on long-term outcome of Roux-en-Y gastric bypass. Obes Surg 2014;24:1647–55.

[26] Hogenkamp PS, Sundbom M, Nilsson VC, Benedict C, Schiöth HB. Patients lacking sustainable long-term weight loss after gastric bypass surgery show signs of decreased inhibitory control of prepotent responses. PLoS One 2015;10:e0119896.

[27] Canetti L, Berry EM, Elizur Y. Psychosocial predictors of weight loss and psychological adjustment following bariatric surgery and a weight-loss program: the mediating role of emotional eating. Int J Eat Disord 2009;42:109–17.

[28] Maddi SR, Fox SR, Harvey RH, Lu JL, Khoshaba DM, Persico M. Reduction in psychopathology following bariatric surgery for morbid obesity. Obes Surg 2001;11:680–5.

[29] Rydén A, Sullivan M, Torgerson JS, Karlsson J, Lindroos AK, Taft C. A comparative controlled study of personality in severe obesity: a 2-y follow-up after intervention. Int J Obes 2004;28:1485–93.

[30] van Hout GCM, Fortuin FA, Pelle AJ, van Heck GL. Psychosocial functioning, personality, and body image following vertical banded gastroplasty. Obes Surg 2008;18:115–20.

[31] Georgiadou E, Gruner-Labitzke K, Köhler H, de Zwaan M, Müller A. Cognitive function and nonfood-related impulsivity in post-bariatric surgery patients. Front Psychol 2014;5:1502.

[32] Leombruni P, Pierò A, Dosio D, Novelli A, Abbate-Daga G, Morino M, et al. Psychological predictors of outcome in vertical banded gastroplasty: a 6 months prospective pilot study. Obes Surg 2007;17:941–8.

[33] De Panfilis C, Cero S, Torre M, Salvatore P, Dall'Aglio E, Adorni A, et al. Utility of the temperament and character inventory (TCI) in outcome prediction of laparoscopic adjustable gastric banding: preliminary report. Obes Surg 2006;16:842–7.

[34] De Panfilis C, Generali I, Dall'Aglio E, Marchesi F, Ossola P, Marchesi C. Temperament and one-year outcome of gastric bypass for severe obesity. Surg Obes Relat Dis 2014;10:144–8.

[35] Spitznagel MB, Garcia S, Miller LA, Strain G, Devlin M, Wing R, et al. Cognitive function predicts weight loss after bariatric surgery. Surg Obes Relat Dis 2013;9:453–9.

[36] Spitznagel MB, Alosco M, Galioto R, Strain G, Devlin M, Sysko R, et al. The role of cognitive function in postoperative weight loss outcomes: 36-month follow-up. Obes Surg 2014;24:1078–84.

[37] Lopez RB, Milyavskaya M, Hofmann W, Heatherton TF. Motivational and neural correlates of self-control of eating: a combined neuroimaging and experience sampling study in dieting female college students. Appetite 2016. Available from: http://dx.doi.org/10.1016/j.appet.2016.03.027.

[38] Bruce AS, Bruce JM, Ness AR, Lepping RJ, Malley S, Hancock L, et al. A comparison of functional brain changes associated with surgical versus behavioral weight loss. Obesity 2014;22:337–43.

[39] Spitznagel MB, Galioto R, Limbach K, Gunstad PDJ, Heinberg PDL. Cognitive function is linked to adherence to bariatric postoperative guidelines. Surg Obes Relat Dis 2013;9:580–5.

[40] Sogg S. Comment on: temperament and one-year outcome of gastric bypass for severe obesity. Surg Obes Relat Dis 2014;10:148–50.

[41] Kelley WM, Wagner DD, Heatherton TF. In search of a human self-regulation system. Annu Rev Neurosci 2015;8:389–411.

Chapter 63

Cognitive Behavioral Therapy for Bariatric Surgery Patients

M. Atwood, L. David and S. Cassin
Ryerson University, Toronto, ON, Canada

LIST OF ABBREVIATIONS

BED binge eating disorder
BMI body mass index (kg/m^2)
CBT cognitive behavioral therapy
LOC loss of control
TAU treatment as usual
WLC waitlist control

INTRODUCTION

Bariatric surgery is associated with a wide range of health benefits, including substantial weight loss and resolution of many comorbid medical conditions [1]; however, there exists significant variability in long-term outcomes, and weight regain remains a pressing issue for many patients. Up to half of patients begin to regain their weight within 2 years following surgery [2]. In almost one quarter of bariatric patients, weight regain during this postoperative period is defined as considerable relative to their overall weight loss [3]. As a result of weight regain, patients also experience a relapse of obesity-related medical conditions, such as type 2 diabetes mellitus [4]. The variability in long-term outcomes may be attributed in part to behavioral and psychological factors implicated in the development and maintenance of obesity that surgical interventions alone do not directly address. This chapter begins with an overview of relevant psychological and behavioral factors that have been shown to attenuate postoperative weight loss, including general psychopathology, eating pathology, and nonadherence to postoperative dietary guidelines. We then present an overview of cognitive behavioral therapy (CBT), a psychosocial intervention that has shown promise in improving some of the psychosocial factors that influence bariatric surgery outcomes.

PSYCHOPATHOLOGY IN BARIATRIC SURGERY PATIENTS

Severe obesity is often associated with significant psychiatric comorbidity [5]. According to recent prevalence estimates, 34% of bariatric surgery candidates have a current psychiatric disorder, and 68% have a lifetime history of psychiatric disorders [6]. The most common current diagnoses are anxiety disorders (18%) and mood disorders (12%) [6]. It has been noted that these prevalence rates may be an underestimate due to bariatric surgery candidates' desires to appear psychologically fit for surgery during presurgical evaluations with the medical team [7].

The presence of psychopathology pre- and post-operatively has been associated with attenuated weight loss after surgery [8–10]. One mechanism that might explain this relationship is that people eat as a means of coping with psychological difficulties [8]. Indeed, in addition to biological factors, psychological factors are important determinants of weight because many individuals overeat in an attempt to regulate their emotions [9]. Preoperative depression has been shown to predict postoperative depressive symptoms, which in turn predict less weight loss up to 5 years postsurgery [10]. Lifetime rates of both affective and anxiety disorders in bariatric candidates are associated with less weight loss following surgery [7]. Thus, psychopathology both pre- and postoperatively may contribute to challenges with weight loss and weight maintenance following surgery.

EATING PATHOLOGY IN BARIATRIC SURGERY PATIENTS

In addition to general psychopathology, some bariatric patients experience clinically severe eating disorders, and many others experience maladaptive eating more broadly. The prevalence rate of current binge eating disorder (BED) has been estimated at 10% in bariatric surgery candidates [6]. The prevalence rate decreases significantly immediately following surgery, in part because the procedure imposes restrictions that physically make it difficult to consume large quantities of food during a discrete period of time (i.e., objective eating binges) [11]. However, patients remain susceptible to other maladaptive eating behaviors. For example, while only a small percentage of postoperative patients are diagnosed with BED, upwards of 60% display high-risk eating behaviors, such as loss of control (LOC) eating and grazing patterns [11]. It appears that for many patients, preoperative binge eating is replaced with LOC eating and grazing following surgery [11].

Patients who report disordered eating behaviors prior to surgery are at an increased risk of returning to old patterns postoperatively [12], and the presence of binge eating, LOC eating, and grazing in the postoperative period contributes to poorer weight loss and greater weight regain following bariatric surgery [11−14]. For example, preoperative LOC eating has been shown to predict postoperative LOC eating, which, in turn, predicts poorer weight loss at 12 and 24 months postsurgery [14]. Moreover, maladaptive eating patterns that reemerge postsurgery have been shown to worsen over time [11], and are strongly associated with depressive symptoms and poorer quality of life [11,12]. Thus, although disordered eating resolves for some patients following bariatric surgery [15], for a subset of patients, these dietary patterns persist and are associated with suboptimal weight loss and psychosocial outcomes.

NONADHERENCE TO POSTOPERATIVE DIETARY GUIDELINES

Bariatric surgery patients are strongly advised to follow postoperative dietary guidelines, including slow consumption of portioned, protein-dense meals and snacks at preplanned time intervals. However, over half (57%) of bariatric patients do not adequately follow these guidelines after surgery [16]. The adverse impact of nonadherence to dietary guidelines on postoperative weight loss has been observed at follow-up appointments as long as 5 years postsurgery [17]. Given the importance of initiating and maintaining these long-term behavioral changes for optimal weight loss, it is important that patients develop coping skills to overcome barriers that interfere with adherence to postoperative guidelines.

CBT INTERVENTIONS FOR BARIATRIC SURGERY PATIENTS

Untreated psychopathology and nonadherence to postoperative guidelines can contribute to weight regain in bariatric surgery patients, underscoring the need for adjunct psychosocial interventions pre- and/or postbariatric surgery. CBT is a short-term, skills-based psychosocial intervention that is based on the premise that thoughts (cognitions), behaviors, and emotions are all interconnected [18]. That is, the way an individual thinks about, or interprets, a situation affects how he/she behaves and feels in that situation, and conversely, the way an individual behaves affects his/her thoughts and emotions (see Fig. 63.1). According to cognitive behavioral theory, distorted thoughts ("cognitive distortions") are at the root of psychopathology and emotional difficulties [18]. Bariatric surgery patients often report cognitive distortions that have the potential to lead to negative emotions and maladaptive behaviors, which in turn can reinforce the cognitive distortion, similar to a self-fulfilling prophecy [19,20]. Some examples of cognitive distortions include, "I will never lose weight" (distortion: overgeneralization), "I should have lost more weight by now" (distortion: should statement), "I've totally blown it by eating this one 'forbidden' food" (distortion: all-or-nothing thinking), or "I will regain all of my weight" (distortion: fortune-telling error). If a bariatric patient has a thought such as, "I will regain all of my weight," he/she will likely feel frustrated, sad, pessimistic, and hopeless, and if the individual believes that he/she will regain weight regardless of his/her efforts, these feelings can lead to maladaptive eating behaviors (e.g., emotional eating, grazing, and nonadherence to dietary guidelines). Over time, these maladaptive eating behaviors actually provide evidence to support the cognitive distortion because they can lead to weight gain, thus confirming the thought, "I will regain all of my weight," and creating a vicious feedback cycle [20].

CBT seeks to disrupt this feedback cycle by identifying and altering maladaptive cognitions and behaviors that currently maintain emotional difficulties (see Fig. 63.2). As applied to bariatric surgery patients, CBT techniques focus on

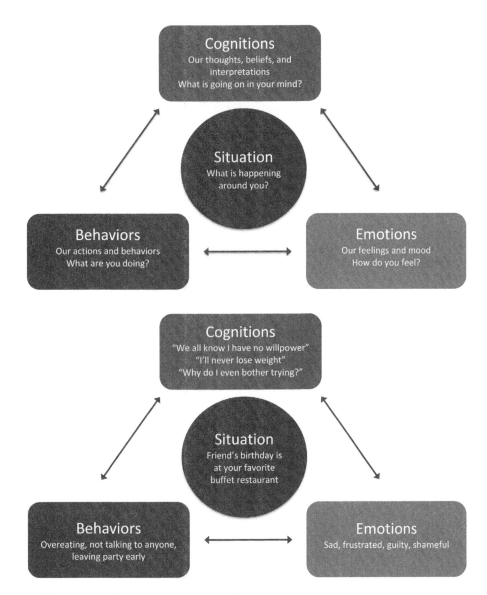

FIGURE 63.1 General CBT model and CBT model applied to overeating.

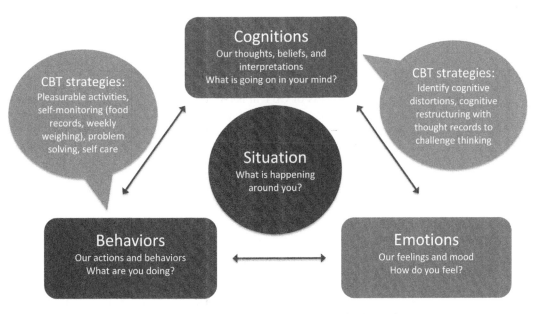

FIGURE 63.2 CBT model of overeating and corresponding intervening cognitive and behavioral strategies.

Time	Place and people you were with	Description of food and/or liquid and amount consumed	Meal or snack? Graze or binge?	Loss of control?	Situation, thoughts, feelings, other behaviors
9am	Alone, at home	2 cups of cereal with milk 1 banana	Breakfast	No	Tired, long day ahead of me
11am	Alone, in the car, running errands	1 granola bar	Snack	No	Long to-do list, I hate going shopping
1pm	Alone, in the mall looking for friend's birthday present	6 inch sub sandwich (cold cut meats with cheese) 1 large soft drink 1 cookie	Lunch	No	Rushed, no time to sit down and eat, feeling guilty I didn't need to make it a combo again.
4pm	With friend, coffee shop	1 cup of coffee with cream	Snack	No	Nice to sit down and chat with old friend, coffee was comforting after a hectic day
9pm	At friend's birthday with wife and 10 of our friends	5 pieces of fried chicken A cup of salad 2 large scoops of pasta 3 spring rolls	Dinner	Yes	Really hungry, dinner is later than usual, it's at a buffet and everything looks so good
11pm	Still at birthday dinner	3 pieces of birthday cake Specialty coffee with whipped cream	Binge	Yes	I've already blown my diet, might as well have dessert, embarrassed for my wife, why can't I just stop eating!!

FIGURE 63.3   Daily food record example.

interrupting the maladaptive cognitions and behaviors that perpetuate the cycle of problematic eating patterns, and teaching adaptive coping skills. In order to maintain weight loss and optimize psychosocial functioning, it is important that patients learn coping skills such as preparing and eating healthy meals and snacks at regular time intervals, scheduling pleasurable activities as alternatives to eating, planning for difficult eating situations, and reducing susceptibility to emotional overeating by solving problems and by challenging negative, unhelpful thinking patterns [19]. Components of CBT protocols for bariatric surgery patients frequently include topics such as the cognitive behavioral model of disordered eating (e.g., binge eating and overeating), the importance of mechanical eating and using food records (see Fig. 63.3), planning pleasurable activities, and reducing vulnerability to overeating through stimulus control, problem solving (see Fig. 63.4), and cognitive restructuring (i.e., identifying, challenging, and altering maladaptive thoughts; see Fig. 63.5) [21–24]. For preoperative patients, the sessions may also include discussion of surgery preparation and expectations of surgery [19,20,25]. CBT can be delivered as an individual or group therapy, and CBT "homework" exercises are typically assigned so that patients have an opportunity to practice the skills between sessions and generalize the skills to their home environment.

EMPIRICAL SUPPORT FOR CBT IN BARIATRIC SURGERY PATIENTS

The number of studies examining the effectiveness of CBT in bariatric patients has grown in recent years. This section reviews empirical evidence with a focus on interventions targeting eating pathology and other psychological factors, as well as adherence to dietary and physical activity guidelines. Interventions delivered in the preoperative and postoperative periods are reviewed separately, and recommendations are provided for further research on cognitive behavioral interventions aimed at improving the long-term outcomes of bariatric surgery.

PSYCHOPATHOLOGY AND EATING PATHOLOGY

To date, several studies have examined the effectiveness of CBT delivered preoperatively in improving psychological factors implicated in poor postoperative outcomes. In an uncontrolled trial, Ashton et al. [25] demonstrated that a 4-session CBT group intervention delivered to 243 bariatric surgery candidates significantly reduced the frequency of binge eating episodes in a given week from pre- ($M = 2.84$, $SD = 2.03$) to immediate posttreatment ($M = 1.18$, $SD = 1.34$). In addition, self-reported binge eating symptomatology (i.e., cognitions, emotions, and behaviors assessed via the Binge Eating Scale (BES)) also significantly decreased following treatment. Patients who exhibited a positive response to CBT (i.e., 0 binge eating episodes at posttreatment and a BES score in the minimal range, $n = 67$)

Step 1. Define the problem in simple terms:
My boss has assigned me too many projects and doesn't realize how much work they are

Step 2. Brainstorm possible solutions (without blocking out ideas):
I could ask a co-worker to help me with a few of the smaller tasks on the projects
I could talk to my boss and tell her that I am having trouble keeping up with the workload
I could quit
I could look for a new job
I could stay late each evening to keep up with the workload

Step 3. Evaluate how practical and effective each solution seems (with '+' or '-'):
I could ask a co-worker to help me with a few of the smaller tasks on the projects   +
I could talk to my boss and tell her that I am having trouble keeping up with the workload   +
I could quit   -
I could look for a new job   -
I could stay late each evening to keep up with the workload   -

Step 4. Choose one or a combination of solutions:
I will talk to my boss and tell her that while I am grateful for the opportunities, I will need to take on fewer projects in the future. In the meantime, I will ask a co-worker if they can help me with aspects of the projects I am currently working on

Step 5. Commit to following through and putting your solution into action:
I will request a meeting with my boss after the staff meeting on Friday. I am having lunch with Cindy tomorrow and will ask for her help then

Step 6. Evaluate the entire problem-solving method:
My boss was very receptive to my request, and told me to let her know if my workload becomes unmanageable again in the future. Cindy also helped out with a few administrative tasks

FIGURE 63.4 Problem-solving worksheet example.

Situation	Emotion	Automatic Thought (Cognitive Distortion)	Evidence That Supports the Thought	Evidence Against The Thought	Alternative Thought	Emotion
Weight increased by 2 pounds this week	Sad (60%) Frustrated (80%) Disappointed (90%)	I was good all week; I should have lost weight. (Should Statement) I might as well give up trying I will gain all of the weight back (Fortune Telling Error)	This is not the first time my weight has gone up I know other people from the program who have gained most of their weight back	Weight fluctuates for lots of reasons There have been many more weeks where my weight has gone down or stayed the same One week is not a good measure of whether or not the changes I've made will help me to keep the weight off Change is a process. I've made progress since the surgery.	My weight will fluctuate from time to time; all I can do is continue making healthy choices	Sad (20%) Frustrated (20%) Disappointed (40%)

FIGURE 63.5 Thought record example.

demonstrated significantly greater excess weight loss at both 6 and 12 months postsurgery relative to patients who did not exhibit a positive response to the intervention [26]. This study suggests that a relatively brief CBT intervention can reduce binge eating behavior prior to surgery, which in turn appears to positively impact postsurgical weight loss. However, participants in the study who exhibited a suboptimal response to CBT were characterized by more severe binge eating symptoms prior to the intervention, suggesting that a subset of patients might require more prolonged intervention [26].

In another uncontrolled trial, Abiles et al. [21] demonstrated that a 12-session CBT group intervention (followed by weekly individual sessions for 1 year) delivered to 110 bariatric candidates reduced binge eating in 36% of patients and eliminated it in 8% of patients at posttreatment. Further, patients reported improvements in anxiety, depression, and self-esteem, and in shape, weight, and eating concerns, particularly the subset of patients with BED. Of the patients who exhibited a "successful" response to surgery (i.e., $\geq 50\%$ excess weight loss) at 2 years postsurgery, 88% had received CBT [27].

While encouraging, the above-reviewed studies are limited by lack of a control group. In the only published controlled trial of preoperative CBT to date, 102 bariatric candidates were randomized to a 10-session group CBT intervention, or a treatment as usual (TAU) control group that received nutritional support and education [22]. Patients in the CBT group reported significant reduction in disordered eating behaviors (i.e., emotional eating and uncontrolled eating) and body mass index ($BMI = kg/m^2$), as well as depression and anxiety symptoms immediately posttreatment, relative to controls. However, at 1 year postoperative follow-up (with 80 patients retained), CBT was no longer superior to TAU [28]. Thus, while CBT demonstrated superior beneficial effects presurgery relative to TAU, these improvements did not necessarily translate to improved outcomes in the postoperative period.

Several studies have also examined the effectiveness of CBT delivered postoperatively. Leahey et al. [23] examined the effects of a weekly 10-session CBT group intervention (which incorporated elements of mindfulness-based practice) conducted with seven postoperative bariatric patients who reported engaging in binge eating between 2 and 11 months postsurgery. Participants reported a reduction in binge eating and depressive symptoms, as well as an increase in eating-related self-efficacy and emotional regulation following the intervention. Although weight loss outcomes varied across patients, overall there was a small improvement in the percentage of weight loss patients achieved from pre- to postintervention.

In a more recent uncontrolled study, Cassin et al. [20] developed a 6-session CBT protocol suitable for delivery to patients in both face-to-face and telephone formats. The authors reported on outcome data for eight patients; two received the intervention presurgery, and six received the intervention postsurgery. Results indicated that those who received CBT prior to surgery reported reductions in the frequency of binge eating behavior, while postoperative patients who reported LOC eating prior to the intervention reported reduction in this eating behavior at posttreatment. Patients also reported reductions in binge eating symptomatology (i.e., cognitions, emotions, and behaviors), emotional eating, and depressive symptoms.

Similar promising results were reported by Beaulac and Sandre [29], who found that an 8-session CBT group intervention delivered to bariatric patients ($n = 17$) who were experiencing psychosocial difficulties (e.g., poor mood) at least 6 weeks postsurgery significantly reduced psychological and weight-related distress and perceived life difficulties from pre- to posttreatment. These improvements were maintained at the 3-month follow-up. Although not statistically significant, patients also reported improvement in emotional overeating, and anxiety and avoidance in close relationships.

Together, the above-reviewed studies suggest that CBT holds promise for improving eating pathology and psychosocial functioning in bariatric surgery patients. However, the absence of control groups and small sample sizes are limitations, and the impact of treatment on short- and longer-term weight loss remains relatively unknown.

ADHERENCE TO PHYSICAL ACTIVITY AND DIETARY GUIDELINES

Few studies have examined the effectiveness of cognitive and/or behavioral modification interventions delivered preoperatively in optimizing adherence to dietary and physical activity guidelines, with the aim of promoting longer-term weight loss and positive postoperative adjustment. One study examined the effects of a 6-month behavioral intervention, consisting of 12 weekly face-to-face group meetings, followed by 5 biweekly telephone sessions [30]. The intervention targeted increased physical activity and decreased caloric intake through self-monitoring and goal-setting strategies, in addition to addressing disordered eating behaviors (e.g., binge eating, LOC eating, grazing) associated with poor weight loss. The intervention was superior to TAU in facilitating significant weight loss prior to surgery (8.3 ± 7.8 kg vs 3.3 ± 5.5 kg); however, this difference was not maintained at 2 years postsurgery [31].

Similar results were obtained by Lier and colleagues [24], who compared 6 weekly sessions of a preoperative CBT group intervention followed by three postoperative sessions (at 6, 12, and 24 months) to TAU. Although the authors did not report on outcomes at immediate posttreatment, there was no difference between groups at 1 year postsurgery follow-up with respect to weight loss, physical activity levels, or adherence to recommended dietary guidelines. Thus, the limited available evidence to date does not support the notion that targeting adherence to lifestyle guidelines prior to surgery results in beneficial effects following surgery.

Cognitive behavioral interventions targeting dietary and physical activity habits may be helpful when delivered postoperatively in order to aid patients in enhancing and/or maintaining weight loss following surgery. Similar to the preoperative literature, studies are sparse and have utilized predominantly behavioral modification as opposed to cognitive change strategies. Kalarchian and colleagues [32] recruited 36 bariatric patients who had undergone surgery more than 3 years prior ($M = 6.6$ years), had a BMI greater than 30 kg/m^2, and who had lost less than 50% excess body weight. Patients were randomly assigned to either a weekly 12-session group behavioral intervention combined with 5 biweekly telephone sessions, or to a waitlist control (WLC) group. The behavioral intervention consisted of setting incremental goals for dietary (i.e., daily caloric intake of 1200−1400 calories) and physical activity change, as well as addressing psychological factors related to postoperative weight control such as disordered eating behaviors (e.g., LOC eating, grazing, and overconsumption of high-calorie liquids). Results indicated that participants in the intervention group exhibited greater excess weight loss at both 6 (6.6% vs 3.4%, respectively) and 12 months (5.8% and 0.9%) posttreatment, compared to WLC. Patients with the least amount of weight regain before the intervention experienced superior weight loss outcomes. In addition, greater baseline levels of depressive symptoms were associated with greater excess weight loss in the intervention group only, suggesting that the strategies utilized in this study might be particularly beneficial for patients with depressive symptoms.

Papalazarou and colleagues [33] reported similar findings in a sample of 37 bariatric patients who were randomized to either a long-term behavioral intervention or TAU. Both groups received consultations providing general information on adopting healthier eating and physical activity habits weekly for the first 3 months postsurgery, biweekly for the subsequent 3 months, monthly for the following 6 months, and quarterly for the second postoperative year. In addition, the intervention group received individual sessions at each meeting, during which behavioral modification strategies such as self-monitoring, goal setting, reinforcement, and stimulus control were utilized to help patients overcome barriers to maintaining a healthy lifestyle. While weight loss remained equivalent across groups at 3, 6, and 9 months postsurgery, at 3 years of follow-up the intervention group had maintained their weight loss, whereas the TAU group had begun to regain weight. The intervention group had also increased physical activity levels and improved dietary habits relative to the TAU group. It should be noted that participants in this study were all females who had undergone vertical banded gastroplasty; thus, it is not certain whether the findings generalize to a broader population of bariatric surgery patients.

CONCLUSIONS

The empirical literature on CBT in bariatric surgery patients is still in its infancy; however, the field continues to grow as the potential role of adjunct psychosocial interventions in enhancing bariatric surgery outcomes is increasingly acknowledged. It is still premature to draw firm conclusions regarding the effectiveness of CBT for bariatric surgery patients given the relatively small literature base and heterogeneity in study design, treatment protocols (e.g., treatment strategies, and length and intensity of intervention), and outcome measures. To date, the literature suggests positive short-term effects of both preoperative CBT [20,22,25,27] and postoperative CBT [20,23] on eating pathology, as well as psychosocial functioning [29]. Although no studies have directly compared the effectiveness of a preoperative and postoperative CBT intervention, a recent review concluded that the evidence is currently strongest for postoperative interventions delivered before significant weight regain occurs [34]. It will be important for future studies to examine the durability of treatment improvements and the impact of CBT on long-term weight loss outcomes. Given the promising findings from uncontrolled studies and pilot studies, larger randomized controlled trials are warranted to examine the efficacy of CBT versus TAU in improving eating pathology, psychosocial functioning, and weight loss over the long term, and to determine the optimal time for the intervention.

MINI-DICTIONARY OF TERMS

- *Behavior modification*: CBT technique used to decrease maladaptive behaviors and increase adaptive behaviors.
- *Cognitive behavioral therapy*: A short-term, skills-based, psychosocial intervention aimed at improving negative thinking patterns and maladaptive behaviors.

- *Cognitive distortions*: Biased or inaccurate ways of thinking about oneself and the world that often reinforce negative thoughts, emotions, and behaviors.
- *Cognitive restructuring*: CBT technique used to identify, challenge, and alter negative thinking patterns.
- *Emotional eating*: Overeating in response to emotions.
- *Grazing*: Consuming a larger number of small servings in the place of meals, regardless of appetite or hunger.
- *Loss of control eating*: Experiencing LOC over eating in the absence of objective eating binges.
- *Objective eating binge*: Consuming an unusually large amount of food during a discrete period of time and experiencing a LOC over eating.

KEY FACTS

- CBT improves negative emotions by changing maladaptive thoughts and behaviors.
- CBT was originally developed as a treatment for depression.
- CBT has been shown to be effective in treating a number of mental health issues.
- CBT is considered an effective treatment for disordered eating.
- CBT is used to help bariatric surgery patients improve eating behaviors such as emotional eating and LOC eating.

SUMMARY OF POINTS

- Bariatric surgery is effective at producing large amounts of weight loss initially; however, long-term results show that premature weight loss plateaus and weight regain are pressing issues for patients.
- Poor weight loss outcomes have been attributed, in part, to psychological and behavioral factors that bariatric surgery alone does not directly address.
- Psychiatric disorders and maladaptive eating behaviors can persist and/or reemerge after surgery, making it difficult to initiate and maintain healthy eating habits after surgery.
- The empirical literature on CBT in bariatric patients suggests positive short-term effects of both pre- and postoperative CBT on eating pathology and psychosocial functioning.
- Future studies employing larger randomized controlled trials are needed to determine the durability of reported treatment gains and the impact of CBT versus TAU on long-term weight loss outcomes.

REFERENCES

[1]   Buchwald H, Avidor Y, Braunwald R, Jensen M, Pories W, Fahrbach K, et al. Bariatric surgery: a systematic review and meta-analysis. JAMA 2004;292:1724–37.

[2]   Shah M, Simha V, Garg A. Review: long term impact of bariatric surgery on body weight, comorbidities, and nutritional status. J Clin Endocrinol Metab 2006;91:4223–31.

[3]   Coucoulas AP, Christian NJ, Belle SH, Berk PD, Flum DR, Garcia L, et al. Weight change and health outcomes at 3 years after surgery among individuals with severe obesity. JAMA 2013;310:2416–25.

[4]   DiGiorgi M, Rosen DJ, Choi JJ, Milone L, Schrope B, Olivero-Rivera L, et al. Re-emergence of diabetes after gastric bypass in patients with mid- to long-term follow-up. Surg Obes Relat Dis 2010;6:249–53.

[5]   Taylor VH, McIntyre RS, Remington G, Levitan RD, Stonehocker B, Sharma A. Beyond pharmacotherapy: understanding the links between obesity and chronic mental illness. Can J Psych 2012;57:5–12.

[6]   Mitchell JE, Selzer F, Kalarchian MA, Devlin MJ, Strain GW, Elder KA, et al. Psychopathology before surgery in the longitudinal assessment of bariatric surgery-3 (LABS-3) psychosocial study. Surg Obes Relat Dis 2012;8:533–41.

[7]   Malik S, Mitchell JE, Engel S, Crosby R, Wonderlich S. Psychopathology in bariatric surgery candidates: a review of studies using structured diagnostic interviews. Compr Psych 2014;55:248–99.

[8]   Kral J. Evaluation of bariatric surgery. Surg Obes Relat Dis 2005;1:35–63.

[9]   Whiteside U, Chen E, Neighbors C, Hunter D, Lo T, Larimer M. Difficulties regulating emotions: do binge eaters have fewer strategies to modulate and tolerate negative affect? Eat Behav 2007;8:162–9.

[10]   Sheets CS, Peat CM, Berg KC, White EK, Bocchieri-Ricciardi L, Chen EY, et al. Post-operative psychosocial predictors of outcome in bariatric surgery. Obes Surg 2015;25:330–45.

[11]   Saunders R. "Grazing": a high-risk behavior. Obes Surg 2004;14:98–102.

[12]   Kalarchian MA, Marcus MD, Wilson TG, Labouvie EW, Brolin RE, LaMarca LB. Binge eating among gastric bypass patients at long-term follow-up. Obes Surg 2002;12:270–5.

[13]   Sallet PC, Sallet JA, Dixon JB, Collis E, Pisani CE, Levy A, et al. Eating behavior as a prognostic factor for weight loss after gastric bypass. Obes Surg 2007;17:445–51.

[14] White MA, Kalarchian MA, Masheb RM, Marcus MD, Grilo CM. Loss of control over eating predicts outcomes in bariatric surgery patients: a prospective, 24-month follow-up study. J Clin Psychol 2010;71:175–84.

[15] van Hout GC, Boekestein P, Foutin FA, Pelle AJ, van Heck GL. Psychosocial functioning following bariatric surgery. Obes Surg 2006;16:787–94.

[16] Toussi R, Fujioka K, Coleman KJ. Pre- and post-surgery behavioral compliance, patient health, and postbariatric surgical weight loss. Obesity 2009;17:996–1002.

[17] Wolnerhanssen BK, Peters T, Kern B, Schotzau A, Ackermann C, von Flue M, et al. Predictors of outcome in treatment of morbid obesity by laparoscopic adjustable gastric banding: results of a prospective study of 380 patients. Surg Obes Relat Dis 2008;4:500–6.

[18] Beck JS. Cognitive behavior therapy: basics and beyond. second ed. New York: Guilford; 2011.

[19] Apple RF, Lock J, Peebles R. Preparing for weight loss surgery: therapist guide. New York: Oxford University Press; 2006.

[20] Cassin SE, Sockalingham S, Wnuk S, Strimas R, Royal S, Hawa R, et al. Cognitive behavioral therapy for bariatric surgery patients: preliminary evidence for feasibility, acceptability, and effectiveness. Cog Behav Pract 2013;20:529–43.

[21] Abiles V, Rodriguez-Ruis S, Abiles J, Obispo A, Gandara N, Luna V, et al. Effectiveness of cognitive-behavioral therapy in morbidity obese candidates for bariatric surgery with and without binge eating disorder. Nutr Hosp 2013;28:1523–9, behavior.

[22] Gade H, Hjelmesæth J, Rosenvinge JH, Friborg O. Effectiveness of a cognitive behavioral therapy for dysfunctional eating among patients admitted for bariatric surgery: a randomized controlled trial. J Obes 2014;21:127936.

[23] Leahey TM, Crowther JH, Irwin SR. A cognitive-behavioral mindfulness group therapy intervention for the treatment of binge eating in bariatric surgery patients. Cog Behav Pract 2008;15:364–75.

[24] Lier HO, Biringer E, Stubhaug B, Tangen T. The impact of preoperative counseling on postoperative adherence in bariatric surgery patients: a randomized controlled trial. Patient Educ Couns 2012;87:336–42.

[25] Ashton K, Drerup M, Windover A, Heinber L. Brief, four-session group CBT reduces binge eating behaviors among bariatric surgery candidates. Surg Obes Relat Dis 2009;5:257–62.

[26] Ashton K, Heinberg L, Windover A, Merrell J. Positive response to binge eating intervention enhances postoperative weight loss. Surg Obes Relat Dis 2011;7:315–20.

[27] Abiles V, Abiles J, Rodrigues-Ruiz S, et al. Effectiveness of cognitive-behavioral weight loss therapy after two years of surgery in patients with morbid obesity. Nutr Hosp 2013;28:1109–14.

[28] Gade H, Friborg O, Rosenvinge JH, Smastuen MC, Hjelmesæth J. The impact of a preoperative cognitive behavioral therapy (CBT) on dysfunctional eating behaviors, affective symptoms and body weight 1 year after bariatric surgery: a randomized controlled trial. Obes Surg 2015;25:2112–19.

[29] Beaulac J, Sandre D. The impact of a CBT psychotherapy group on post-operative bariatric patients. Springer Plus 2015;4:764.

[30] Kalarchian MA, Marcus MD, Courcoulas AP, Cheng Y, Levine MD. Preoperative lifestyle intervention in bariatric surgery: initial results from a randomized, controlled trial. Obesity 2013;21:254–60.

[31] Kalarchian MA, Marcus MD, Courcoulas AP, Cheng Y, Levine MD. Preoperative lifestyle intervention in bariatric surgery: a randomized clinical trial. Surg Obes Relat Dis 2015. Available from: http://dx.doi.org/10.1016/j.soard.2015.05.004. [Epub ahead of print].

[32] Kalarchian MA, Marcus MD, Courcoulas AP, Cheng Y, Levine MD, Josbeno D. Optimizing long-term weight control after bariatric surgery: a pilot study. Surg Obes Relat Dis 2012;8:710–16.

[33] Papalazarou A, Yannakoulia M, Kavouras SA, Komesidou V, Dimitriadis G, Papakonstantinou A, et al. Lifestyle intervention favorably affects weight loss and maintenance following obesity surgery. Obesity 2010;18:1348–53.

[34] Kalarchian MA, Marcus MD. Psychosocial interventions pre and post bariatric surgery. Eur Eat Disord Rev 2015;23:457–62.

.

Chapter 64

Does Body Dysmorphic Disorder Have Implications for Bariatric Surgery?

D. Larkin[1] and C.R. Martin[2]

[1]Edge Hill University, Ormskirk, United Kingdom, [2]Buckinghamshire New University, Middlesex, United Kingdom

LIST OF ABBREVIATIONS

BD   body dysmorphia
BDD   body dysmorphic disorder
CBT   cognitive behavioral therapy
DSM   Diagnostic and Statistical Manual of Mental Disorders
OCD   obsessive−compulsive disorder
QoL   quality-of-life
SSRIs selective serotonin reuptake inhibitors

INTRODUCTION

There has been a significant increase internationally in the rate of obesity, which has fueled public health concerns (see 1−3). Obesity has a number of potential implications for individuals, such as an increased risk of cancer [4], cardiovascular disease, and diabetes [5], and has a direct impact on mortality. Obesity costs the world economy around $2 trillion annually, but more importantly, results in 5% of all deaths worldwide [6]. Melissas et al. [6] suggested that 2.1 billion people, equal to 30% of the world's population, are either overweight or obese, with an increasing likelihood that obesity will affect 50% of the world's population by 2030. It is not surprising, therefore, that an increasing number of individuals turn to surgical intervention as a solution to their weight issues.

Bariatric surgery has been described as the most effective treatment for extreme obesity [7]. Once the initial surgery has taken place, and the weight is lost, the once-obese individuals are frequently faced with loose or hanging skin due to the dramatic weight loss. The weight loss can be as much as 40−50% of preoperative weight [8]. Following the weight loss, many of the comorbidities associated with obesity often significantly improve. There is, however, debate with regards to particular psychopathologies such as BDD; a few studies suggest improvements, whereas others indicate little change [8].

OVERVIEW OF BODY DYSMORPHIA

The Diagnostic and Statistical Manual of Mental Disorders (DSM-5) [9] characterizes body dysmorphic disorder (BDD) as a subcategory of obsessive−compulsive disorder (OCD), under the heading of Obsessive−Compulsive and Related Disorders. Individuals with BDD report looking ugly, unattractive, or deformed (Criterion A). The perceived flaws are not observable, or appear only slight to other individuals. Those with BDD may describe their flaws as unattractive or not right, hideous, and monster-like. According to the DSM-5, preoccupied individuals may focus on one or many body areas, including the skin, hair, and facial features, but may include most parts of the body. The preoccupations are intrusive, unwanted, and time-consuming. Prevalence among those seeking cosmetic surgery is between 3% and 16%, with an overall medium onset age of 15. Rates of suicidal ideation and suicide attempts are high in both adults and children/adolescents with BDD. According to the DSM-5, nearly all individuals with BDD experience impaired psychosocial functioning because of the concerns surrounding their appearance. Impairment can range from moderate (e.g., avoidance of some social situations) too extreme and incapacitating (e.g., unable to leave the family home), with obvious consequences on work/school and social life. There are notably no criteria for BDD that include obesity or distress concerning weight-related issues, outside those associated with eating disorders.

BODY DYSMORPHIC DISORDER, CLINICAL FEATURES

BDD is characterized by a distressing or impairing preoccupation with an imagined or slight defect in appearance [10]. The contemporary psychiatric classification hierarchy and main self-perceptions associated with the presentation are shown in Fig. 64.1. In the past, patients with BDD have been diagnosed with various conditions such as dysmorphophobia, dermatologic hypochondriasis, beauty hypochondria, Hasslichkeitskummerer (one who is worried about being ugly), dermatologic nondisease, and primary monosymptomatic hypochondriacal psychosis [11]. The face (including nose, mouth, and hair), as well as the breasts, genitals, thighs, stomach, and waist, are areas commonly reported to be the focus and preoccupation in people with BDD (Fig. 64.2) [11,12]. Patients notice small imperceptible blemishes that become the epicenter for obsessional repetitive ritualistic behaviors that can force them to spend many hours in front of mirrors examining and reexamining the same area. This obsession may prevent them from leading a normal life, and may leave them unable to work, leave the house, or socialize with others [11].

The prevalence of BDD is hard to establish, but Koran et al. [13] put the figure at 2.7% of the population, which would mean BDD is more common than schizophrenia or bipolar I disorder. Bjornsson et al. [14], however, put the

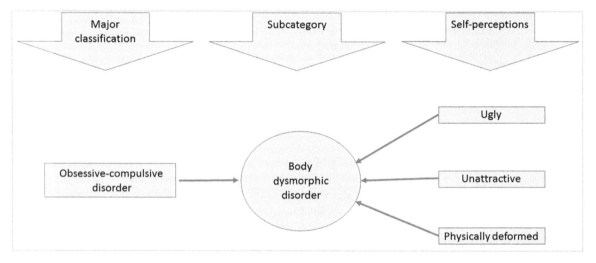

FIGURE 64.1 Psychiatric classification of body dysmorphic disorder and associated typical self-perceptions. Shows the relationship of body dysmorphic disorder as a subcategory of obsessive–compulsive disorder. The dominant perceptual anomalies typically associated with body dysmorphic disorder are highlighted as are the negative attributional biases associated with such perceptions.

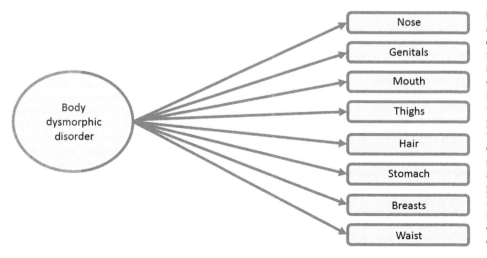

FIGURE 64.2 Preoccupation focus areas typical of patients with body dysmorphic disorder. Patients may focus on any one of these physical attributes or a combination. Direction of arrows indicates the direction of self-perceptual anomaly. Emphasizes the spectrum of physical attribute preoccupations typically associated with patients with body dysmorphic disorder. It is important to note that a single attribute may be the focus, or, alternatively, a broad range of physical attributes may be the focus. Irrespective of the breadth of individual attribute focus, the impact on psychosocial function can be debilitating.

figure more conservatively at 0.07−2.4%; they also suggested that the figure could be as high as 13% in nonclinical student populations. There appears to be a relatively equal division between males and females; males are more likely to fixate on the genital areas, and females are more likely to become preoccupied with their hair, face, breasts, and stomach [11]. BDD has been reported in children as young as 4 and adults as old as 80 [14−16], and generally involves exploring the problem area for up to 8 hours/day. Even though individuals may become preoccupied with just one area or aspect of their body, some individuals may also become dissatisfied with their overall appearance; this can manifest as muscular dysmorphia, in which they believe they have underdeveloped muscular form [14].

There is a strong association between BDD and other psychiatric disorders, in particular, major depression, which can be found in 59% of individuals with a lifetime prevalence of 83% [17]. It would appear that BDD precedes the depressive symptoms by around 12 months [11], which would therefore indicate that depression is not the predominant condition. There is a significant association between OCD and BDD. OCD is found in almost 37% of individuals with BDD [17]. Phillips et al. [18] stated that in many respects, BDD is on the OCD spectrum, which includes disorders that share features with OCD. According to Cotterill [11], the diagnosis of BDD is relatively easy, but treatment is far more problematic. Cotterill made the observation that patients with this disorder are often angry, and can turn on themselves or the consulting physician. There have been reported cases of suicide and death threats against dermatologists and plastic surgeons.

Individuals with BDD can be treated in a number of ways, primarily using drug therapy, but with mixed outcomes. According to Sarwer [19], 90% of those who underwent aesthetic medical treatment for BDD showed no improvement or a worsening of their BDD following treatment. This therefore suggests that surgical intervention for BDD only proves to exacerbate the underlying condition. Antidepressants, including monoamine oxidase inhibitors (MAOIs), along with tricyclic antidepressants, have been used to treat BDD, but with unpromising results [11]. Selective serotonin reuptake inhibitors (SSRIs) showed a much-improved profile, but in much higher doses than generally prescribed for depression [14]. Cognitive behavioral therapy (CBT) has been reported as having some success, but its efficaciousness is compromised by the relatively high attrition rate. Treatment approaches are summarized in Fig. 64.3.

BODY DYSMORPHIA BEFORE BARIATRIC SURGERY

According to Sarwer et al. [12] the vast majority of people report at sometime in their life dissatisfaction with some aspect of their appearance. Sarwer et al. [12] also suggested that body image dissatisfaction may be the motivational catalyst for surgical intervention.

A small number of studies have found a relationship between body image dissatisfaction, increased levels of depression, and low levels of self-esteem among obese women who were seeking weight loss treatment [20], but the level of dissatisfaction was not related to the level of obesity. In general terms, it would seem that obese women reported far greater levels of body dissatisfaction compared to their slimmer counterparts, but did not report greater levels of depressive symptoms or lower self-esteem [20]. Sarwer et al. [20] reported that about 8% of obese women met the criteria for a diagnosis of BDD. The problem, however, is that the DSM does not have categories or diagnoses for individuals who

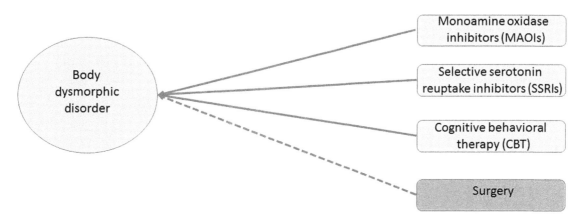

FIGURE 64.3 Treatment options for body dysmorphic disorder. Surgery may have either little or indeed a negative effect on body dysmorphic disorder. Highlights the currently limited range of treatment options available for body dysmorphic disorder and the distribution between pharmacological interventions, antidepressants such as MAOIs or SSRIs, or psychological therapies such as cognitive behavioral therapy. Body dysmorphic disorder appears resistant to surgical interventions, and indeed may, in some instances, impact deleteriously on body dysmorphic disorder.

have marked distress due to medical conditions that affect physical appearance, such as obesity. Instead, research has explored body image in general, and quality-of-life (QoL) specifically, with regards to weight-related issues.

According to Schwartz and Brownell [21], the psychological impact of obesity may be influenced by a small number of factors, one of which is social processes. Social processes may include negative messages, which according to Schwartz and Brownell [21] are omnipresent and reflect a strong antifat bias in the media, schools, and business, as well as in everyday interactions. Schwartz et al. [22] also reported an institutional weight-related bias among health care professionals, such as nurses, doctors, and medical students, leading to outright discrimination [21]. Cash and Roy [23] suggested that simply being female is a risk factor in forming a poor body image. It would seem that women who labeled themselves as overweight, and were in fact overweight, were found to be significantly less happy than women who labeled themselves as overweight but were not [21]. It could be argued that women might be more prone to self-deprecation, and as such, are at a higher risk of a poor body image, regardless of their actual weight; the same was not seen in men. Schwartz and Brownell [21] argued that overweight males may see themselves as "big and strong" rather than "big and fat." This process does, however, lead to a more stable body image.

Many individuals seeking weight loss surgery report a number of psychological and physiological reasons. Sarwer et al. [12] suggested that body image dissatisfaction was a major contributing factor in the motivation of those who elected for surgical intervention. Sarwer et al. [8] reported that 20–60% of individuals seeking bariatric surgery have been characterized as suffering from Axis I psychiatric disorders, the most common of which are mood and anxiety disorders. Sarwer [19] also reported that candidates for bariatric surgery generally scored significantly lower on QoL scales in comparison to their slimmer counterparts. In addition, Sarwer et al. [12] suggested that the relationship between obesity and psychopathology is complicated. They argued that obesity may be a consequence of an individual's psychopathology, and that an individual may use food to excess as a maladaptive coping mechanism for the underlying psychological problems. The social stigma related to obesity may contribute to mood and anxiety disorders and other psychopathologies, such as BDD. A model of interacting factors impacting on the expression of BDD is summarized in Fig. 64.4.

BODY DYSMORPHIA AFTER BARIATRIC SURGERY

For obese individuals who have tried the established weight loss programs of diet and exercise but have failed to lose substantial and sustainable weight, bariatric surgery remains the ultimate treatment [24]. The events that occur after bariatric surgery with regards to BDD, body image, and QoL remain largely unexplored. However, the goal of the initial surgery is to assist the obese individual to lose substantial amounts of weight, and in most cases this is achieved; the once-obese individual can lose as much as 40–50% of preoperative weight [8]. Sarwer et al. [12] suggested that around 20% failed to lose a significant amount of weight as a consequence of poor adherence to a diet, which was linked to disordered eating behaviors, such as vomiting or gastric dumping (food lost through rapid transportation from the stomach to small intestine, voided as diarrhea).

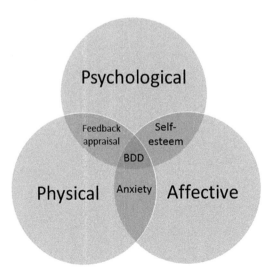

FIGURE 64.4   Interactional model of body dysmorphic disorder. The attributes within the overarching domains of psychological, physical, and affective dimensions are illustrative examples and not exhaustive. Emphasizes the notion of the gestalt nature of body dysmorphic disorder as a complex constellation of psychological, psychosocial, physical, and mood-related factors associated with this distressing clinical presentation.

Sarwer et al. [12] reported that most, but not all, studies suggest that QoL improves dramatically after weight loss surgery. Individuals reported improvements in marital satisfaction and sexual functioning, and, importantly, body image. In general, changes in QoL and body image were positively associated with weight loss following bariatric surgery [25]. The antecedents to BDD, anxiety and depression, were reported to be much improved in a dose-dependent fashion—i.e., the greater the weight loss, the less the anxiety and depression symptoms [26].

More than two-thirds of bariatric surgery patients considered the development of excess skin to be a negative consequence of surgery [27]. The skin folds were predominately found on the upper arms, breasts, abdomen, and thighs. Those who lost relatively minor amounts of weight reported greater levels of body satisfaction than those who reported larger weight loss, presumably because postoperative skin folds were less noticeable. The vast majority of postoperative individuals opted for reconstructive cosmetic surgery to remove skin folds. The skin folds sometimes became a source of frustration and anxiety, and even prevented further weight loss [28].

Song et al. [29] reported that obese individuals "suffered" on two fronts: (1) the psychological trauma of obesity and (2) the consequences of medicosurgical slimming methods, where patients were often faced with traumatic morphologic repercussions following their weight loss that they may have found difficult to accept. Lazar et al. [24] suggested that an individual's new body image might become a source of deep frustration, shame, and even humiliation. In spite of these negative outcomes, Lazar et al. [24] found that the vast majority of their participants would undergo the abdominoplasty procedure again. Pecori et al. [30] found that as a rule, following stable weight loss, the postsurgical individuals were highly satisfied with their new, lean body shape, and their QoL remarkably improved. However, they [30] also found that postoperative patients who required cosmetic or body contouring procedures still showed high degrees of body dissatisfaction compared to individuals who did not require these procedures. In many ways, postoperative obese individuals who required body-contouring procedures displayed a pattern of behavior that was clinically similar to BDD. Pecori et al. [30] ended their study by suggesting that postoperative individuals who had lost weight after bariatric surgery and esthetic operations appeared to have complex (and still unknown) psychological issues. This finding could be extended, and it could be suggested that preoperative individuals displayed the symptoms of BDD, and that regardless of the surgical procedures, still had clinically significant body image difficulties postoperatively.

MINI-DICTIONARY OF TERMS

- *Diagnostic and statistical manual of mental disorders*: Edited by the American Psychiatric Association; designed to augment the diagnosis and treatment of mental health disorders.
- *Muscular dysmorphia*: Individuals (most commonly males) generally feel they are too small in physique, insufficiently muscular, or insufficiently lean.
- *Gastric dumping*: Food lost through rapid transportation from stomach to the small intestine, voided as diarrhea.
- *Psychosocial*: Psychological development interacting with social environment.
- *Axis I psychiatric disorders*: Axis I is part of the DSM "multiaxial" system for assessment. Axis I conditions represent acute symptoms that need treatment.
- *Monoamine oxidase inhibitors (MAOIs)*: Antidepressants that prevent the breakdown of neurotransmitters such as noradrenaline (norepinephrine) and serotonin.
- *Selective serotonin reuptake inhibitors*: Work as antidepressants by inhibiting reuptake of serotonin, thus making more serotonin available as a neurotransmitter.
- *Tricyclic antidepressants*: Cyclic antidepressants that are designed to block the activity of the neurotransmitters serotonin and noradrenaline (norepinephrine) in the brain.
- *Cognitive behavioral therapy (CBT)*: Talking therapy designed to explore current behaviors as opposed to seeking causes of psychological stress related to the past.
- *Obsessive–compulsive disorder (OCD)*: Obsessive thoughts and compulsive activity that cause feelings of anxiety, disgust, or unease.

KEY FACTS

- BDD is characterized by a preoccupation with one or more perceived defects or flaws in one's appearance, which are unnoticeable to others or only slightly noticeable.
- The prevalence of BDD is roughly equal in males and females, but with a slightly different emphasis on parts of the body.

- BDD has a strong association with OCD, is found in 37% of those with the condition, and has a lifetime prevalence of 83%.
- It is thought that the prevalence among those seeking cosmetic surgery is between 3% and 16%, with an overall medium onset age of 15.
- Treatment for BDD is generally in the form of talking therapies and antidepressants.
- The DSM does not have a category or diagnosis for individuals who have a marked distress due to medical conditions that affect physical appearance, such as obesity. Instead, research has explored body image in general terms, and QoL specifically, as they pertain to weight-related issues.
- Twenty to sixty percent of individuals seeking bariatric surgery have been characterized as suffering from Axis I psychiatric disorders, the most common of which are mood and anxiety disorders.
- Weight loss following bariatric surgery may result in loose skin folds, which can become a source of frustration and anxiety, and can even prevent further weight loss.

SUMMARY POINTS

- BDD is a condition in which an individual perceives an area of their body to be fundamentally flawed.
- Small, imperceptible blemishes become the epicenter for obsessional repetitive ritualistic behaviors that can force individuals to spend many hours in front of mirrors examining and reexamining the same area.
- Prevalence of BDD among those seeking cosmetic surgery is between 3% and 16%, with an overall medium onset age of 15.
- Body image dissatisfaction is a major contributing factor in the motivation of elected surgical intervention for obesity.
- There are no criteria in the Diagnostic and Statistical Manual for Mental Disorders (DSM) for BDD that include obesity or distress-concerning weight-related issues, outside those associated with eating disorders.
- The events that occur after bariatric surgery with regards to BDD, body image, and QoL remain largely unexplored.

REFERENCES

[1] Wang Y, Beydoun MA. The obesity epidemic in the United States--gender, age, socioeconomic, racial/ethnic, and geographic characteristics: a systematic review and meta-regression analysis. Epidemiol Rev 2007;29:6−28. Available from: http://dx.doi.org/10.1093/epirev/mxm007.

[2] Rokholm B, Baker JL, Sørensen TIA. The levelling off of the obesity epidemic since the year 1999 − a review of evidence and perspectives. Obes Rev 2010;11(12):835−46. Available from: http://dx.doi.org/10.1111/j.1467-789X.2010.00810.x.

[3] Bodenlos JS, Wormuth BM. Watching a food-related television show and caloric intake. A laboratory study. Appetite 2013;61(1):8−12. Available from: http://dx.doi.org/10.1016/j.appet.2012.10.027.

[4] Louie SM, Roberts LS, Nomura DK. Review: mechanisms linking obesity and cancer. BBA 2013;1831:1499−508. Available from: http://dx.doi.org/10.1016/j.bbalip.2013.02.008.

[5] Kopelman PG. Obesity as a medical problem. Nature 2000;404(6778):635−43.

[6] Melissas J, Torres AJ, Yashkov YI, Lemmens LAG, Weiner RA. International Federation for Surgical Obesity (IFSO) center of excellence program for bariatric surgery. In: Agrawal S, editor. Obesity, bariatric and metabolic surgery: a practical guide. Cham: Springer International Publishing; 2016. p. 575−80.

[7] Maggard MA, et al. Meta-analysis: surgical treatment of obesity. Ann Inter Med 2005;142(7):547−59.

[8] Sarwer DB, Thompson JK, Mitchell JE, Rubin JP. Psychological considerations of the bariatric surgery patient undergoing body contouring surgery. Plastic Reconstr Surg 2008;121(6):423e−34e.

[9] APA. Diagnostic and statistical manual of mental disorders. *5th ed.* Washington, DC: American Psychiatric Association; 2013.

[10] Phillips KA, et al. Body dysmorphic disorder: some key issues for DSM-V. Depres Anxiety 2010;27(6):573−91.

[11] Cotterill JA. Body dysmorphic disorder. Dermatol Clin 1996;14(3):457−64.

[12] Sarwer DB, Pruzinsky T, Cash TF, Goldwyn RM, Persing JA, Whitaker LA. Psychological aspects of reconstructive and cosmetic plastic surgery: clinical, empirical, and ethical perspectives. Philadelphia: Lippincott Williams & Wilkins; 2006.

[13] Koran LM, Abujaoude E, Large MD, Serpe RT. The prevalence of body dysmorphic disorder in the United States adult population. CNS Spectrums 2008;13(04):316−22.

[14] Bjornsson AS, Didie ER, Phillips KA. Body dysmorphic disorder. Dialog Clin Neurosci 2010;12(2):221.

[15] Albertini RS, Phillips KA, Guevremont D. Body dysmorphic disorder. J Am Acad Child Adolescent Psych 1996;35(11):1425−6.

[16] Phillips KA. The broken mirror: understanding and treating body dysmorphic disorder. USA: Oxford University Press; 2005.

[17] Phillips KA. Body dysmorphic disorder. CNS Drugs 1995;3(1):30−40.

[18] Phillips KA, McElroy SL, Hudson JI, Pope HG. Body dysmorphic disorder: an obsessive-compulsive spectrum disorder, a form of affective spectrum disorder, or both? J Clin Psych 1995;56:41−51.

[19] Sarwer DB. Commentary on: prevalence of body dysmorphic disorder symptoms and body weight concerns in patients seeking abdominoplasty. Aesth Surg J 2016;36(3):333−4.

[20] Sarwer DB, Thompson JK, Cash TF. Body image and obesity in adulthood. Psych Clin 2005;28(1):69−87.

[21] Schwartz MB, Brownell KD. Obesity and body image. Body Image 2004;1(1):43−56. Available from: http://dx.doi.org/10.1016/S1740-1445(03)00007-X.

[22] Schwartz MB, Chambliss HON, Brownell KD, Blair SN, Billington C. Weight bias among health professionals specializing in obesity. Obes Res 2003;11(9):1033−9. Available from: http://dx.doi.org/10.1038/oby.2003.142.

[23] Cash TF, Roy RE. Pounds of flesh: weight, gender, and body images. Interpreting weight: the social management of fatness and thinness. Hawthorne, NY: Aldine de Gruyter; 1999. p. 209−28.

[24] Lazar CC, Clerc I, Deneuve S, Auquit-Auckbur I, Milliez P. Abdominoplasty after major weight loss: improvement of quality of life and psychological status. Obes Surg 2009;19(8):1170−5.

[25] Kolotkin RL, Crosby RD, Gress RE, Hunt SC, Adams TD. Two-year changes in health-related quality of life in gastric bypass patients compared with severely obese controls. Surg Obes Rel Dis 2009;5(2):250−6. Available from: http://dx.doi.org/10.1016/j.soard.2009.01.009.

[26] Larsen F. Psychosocial function before and after gastric banding surgery for morbid obesity. A prospective psychiatric study. Acta Psychiatrica Scandinavica 1989;359:1−57.

[27] Kitzinger HB, et al. The prevalence of body contouring surgery after gastric bypass surgery. Obes Surg 2012;22(1):8−12. Available from: http://dx.doi.org/10.1007/s11695-011-0459-1.

[28] Sarwer DB, Crerand CE. Body dysmorphic disorder and appearance enhancing medical treatments. Body Image 2008;5(1):50−8. Available from: http://dx.doi.org/10.1016/j.bodyim.2007.08.003.

[29] Song AY, Rubin JP, Thomas V, Dudas JR, Marra KG, Fernstrom MH. Body image and quality of life in post massive weight loss body contouring patients. Obesity 2006;14(9):1626−36. Available from: http://dx.doi.org/10.1038/oby.2006.187.

[30] Pecori L, Cervetti GGS, Marinari GM, Migliori F, Adami GF. Attitudes of morbidly obese patients to weight loss and body image following bariatric surgery and body contouring. Obes Surg 2007;17(1):68−73.

Chapter 65

Issues Surrounding the Relationship Between Sexual Function and Bariatric Surgery

D. Larkin[1] and C.R. Martin[2]

[1]Edge Hill University, Ormskirk, United Kingdom, [2]Buckinghamshire New University, Middlesex, United Kingdom

LIST OF ABBREVIATIONS

BS bariatric surgery
DSM Diagnostic and Statistical Manual of Mental Disorders
ED erectile dysfunction
FSD female sexual dysfunction
FSF female sexual function
MSD male sexual dysfunction
MSF male sexual function
PE premature ejaculation
SD sexual dysfunction
SF sexual function

INTRODUCTION

Epidemiological accounts of significant increases in obesity internationally are instrumental in causing express public health concerns [1−3]. Obesity represents one of the major health issues facing many societies, with significant, direct impact on mortality, and fundamentally, comorbidities [4], including cancer [5], cardiovascular disease, and diabetes [6]. Obesity also represents an important aspect in health-related quality-of-life (QoL) in relation to sexual dysfunction (SD) in both males [7] and females [8]. The picture of sexual function (SF) is complicated by the different ways in which it can be defined and measured. Kolotkin et al. [9], e.g., suggested that instruments designed to assess SF should be relevant to either gender, or be gender-specific, study-specific, or even clinically driven.

SEXUAL FUNCTION AND DYSFUNCTION

The Diagnostic and Statistical Manual of Mental Disorders (DSM) [10] has characterized SD as a disturbance in the process that embodies the sexual response cycle, or pain associated with sexual intercourse. The DSM divides the sexual response cycle into a small number of phases: (1) *Desire*, which is the desire to engage in sexual activity; (2) *Excitement*, which is the subjective sense of sexual pleasure culminating in an erection in males, and vaginal lubrication and swelling of external genitalia in females; (3) *Orgasm*, which culminates in the ejaculation of semen in males, and the contraction of the walls of the vagina in females; and (4) *Resolution*, which is muscular relaxation and a subjective sense of well-being.

SEXUAL FUNCTION: CLINICAL FEATURES

Females: Masters and Johnson [11] first characterized the female sexual response in 1966 as consisting of four successive phases: excitement, plateau, orgasm, and resolution. Kaplan [12] proposed the aspect of "desire" and the three-phase model (desire, arousal, and orgasm), desire being the factor that directs the overall response cycle. It was Kaplan's [12] three-phase model that informed the DSM's definition of female sexual dysfunction (FSD). FSD is a multicausal and multidimensional medical problem that has biological and psychosocial components [13]. According to Mimoun and Wylie [14], female sexuality is mutable and prone to change as a woman ages, and as such, the process of defining FSD is multifarious. Nevertheless, regardless of its complexity, FSD is highly prevalent and invariably misunderstood among the general populous [15]. According to Jha and Thakar [16] and Laumann and Paik [17], FSD is relatively common, with 43% of women (a representative cross-sectional sample of 18- to 59-year-olds) reporting SD, which only becomes dysfunctional when it causes marked distress, as opposed to a normal physiological response to difficult circumstances. FSD is associated with a number of interdependent factors such as age, educational level, race, and poor physical and emotional health, and is correlated with poor experiences in sexual relationships [17]. Women who have a history of sexual abuse also have an increased risk of SD. The etiology of FSD is complex and multilayered, and can be linked to normative changes in relation to aging, pregnancy, and parturition and breastfeeding [16]. Studies [18] have shown that menopause need not automatically lead to low sexual desire and to FSD, particularly in relation to new sexual partners and new relationships. The characteristics of the female sexual response are shown in Table 65.1. The relationships between key factors and FSD are shown in Fig. 65.1.

The psychological component of FSD is also complex and multilayered, but can include the emotional relationship with a partner; a good relationship can lead to good sexual health and vice versa—i.e., a bad relationship can lead to poor sexual health and even FSD [16]. This, however, does not presuppose that sexual functionality is sufficient to create a sexually satisfying experience [19], good sexual health, or good emotional relationships.

Particular medical conditions are known to cause FSD, such as genital tract atrophy, stoke, Parkinsonism, and diabetes, to name but a few. The treatments for these and other conditions are also associated with FSD—e.g., antiestrogens (Tamoxifen and Reloxifen), antihistamines, anticonvulsants, drugs of abuse, alcohol, and sedatives. Hormonal balance is an essential component to maintaining SF. Testosterone is the predominant androgen in women [16]. Testosterone is predominantly excreted from the adrenal glands and ovaries; circulating androstenedione is also a major source of testosterone, and is also secreted from the adrenal glands and ovaries. A decrease in the levels of testosterone is correlated with a decreased sexual response cycle, which is already known to decrease after menopause. Long-term use of

TABLE 65.1 Female Sexual Response Cycle

Desire	Consists of sexual fantasies and the desire to have sex.
Excitement	Consists of the subjective sense of sexual pleasure accompanied by vaginal lubrication, possible flushing of the skin, nipples becoming hard or erect, breasts becoming fuller, and the vaginal wall swelling. The clitoris may become highly sensitive, and the labia minora may engorge.
Orgasm	This is the climax of the sexual response cycle, generally resulting in muscular contraction within the vagina and uterus.
Resolution	Marked by a general sense of well-being. Some women are able to reach orgasm almost immediately if stimulated.

Key dimensions of the female sexual response cycle.

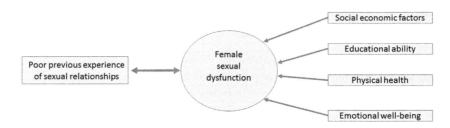

FIGURE 65.1 Factors contributing to female sexual dysfunction. Single arrows are predictive, and double arrows are correlational. Shows the factors that impact and influence female sexual dysfunction. A number of these factors are predictive, as illustrated by the unidirectional arrows. By contrast, the relationship between sexual dysfunction and poor previous experience of sexual relationships is bidirectional; thus, one may impact the other and vice versa.

TABLE 65.2 Male Sexual Response Cycle

Desire	Consists of sexual fantasies and the desire to have sex.
Excitement	The penis may becoming engorged to reach an erection.
Orgasm	In men, rhythmic contractions of the muscles at the base of the penis result in the ejaculation of semen.
Resolution	Generally marked by a refractory period in which men are unable to reach orgasm for a short time; this period is extremely subjective, and may lengthen as the individual ages.

Key dimensions of the male sexual response cycle.

testosterone supplementation, however, is questionable for safety reasons. High concentrations of endogenous testosterone are associated with a high risk of ischemic heart disease and even death [20].

Males: Male sexual dysfunction (MSD) is categorized according to its occurrence in the sexual response cycle: desire, arousal, orgasm, and resolution. The characteristics of the male sexual response are summarized in Table 65.2.

Dysfunction with arousal may manifest as erectile dysfunction (ED), whereas disorders with orgasm may manifest as premature ejaculation (PE) or delayed ejaculation [19]. Laumann and Paik [17] found in their cross-sectional representative sample of 18- to 59-year-olds that 31% of males reported SD. Dysfunctions in relation to desire are often presented alongside ED as the primary complaint, sometimes masking other issues such as relationship concerns, exhaustion, and, not uncommonly, sexual preference issues [19]. ED is a condition in which there is a persistent inability to attain and/or maintain an erection sufficient to permit satisfactory sexual performance [21]. ED and premature rejection (PE minus ejaculation with minimal stimulation) are the two most prevalent complaints in male sexual medicine [22]. Hatzimouratidis et al. [22] reported that approximately 5–20% of men have moderate to severe ED. The risk factors of ED can include cardiovascular disease, diabetes mellitus, hypertension, poor physical fitness, and obesity. The risk factors that include poor physical fitness and obesity can be mitigated with weight loss and exercise. Further risk factors may comprise radical prostatectomy, because of the risk of cavernosal nerve injury, poor oxygenation of the corpora cavernosa, and vascular insufficiency [22]. ED reported in men over the age of 40 can be an early indicator of ischemic heart disease. PE is the single most prevalent SD in men. Rösing and Klebingat [19] reported that 20–25% of adult men surveyed reported PE with distress. However, this method of reporting has a number of limitations; e.g., the normal interval between penetration and ejaculation is largely subjective, and subject to a wide range of individual differences based on age, sexual experience, and cultural variations [23].

If the etiology of the ED is judged to be psychogenic, then psychosexual therapy may be used as a course of treatment. Surgical treatments are also available, such as penile revascularization to improve blood flow following trauma to the groin area, resulting in ED [22]. If hormones have been judged to be the cause of the ED, testosterone replacement therapy has proven to be an effective treatment. However, as with hormone replacement therapy in women, testosterone administration has been known to increase the risk of cancer; in men, endogenous testosterone administration may increase the risk of prostate cancer. Another line of therapy is oral pharmacotherapy, such as Sildenafil (Viagra), Tadalafil (Cialis), and Vardenafil (Levitra), which do not achieve, but help maintain, an erect penis, thus allowing vaginal penetration. However, these pharmacotherapies have potent adverse events, which may include headaches, flushing, dyspepsia, and nasal congestion, as well as dizziness. A second line of treatment with potentially fewer potent side effects includes intracavernous injections of Alprostadil administered 5–15 minutes before sexual intercourse; these can last 4 hours, but may result in priapism (prolonged and painful erection) and fibrosis (formation of fibrous connective tissue). A third line of therapy includes penile prostheses, which are semiridged or inflatable devices implanted inside the penis to form a rigid penis that allows for vaginal penetration [22].

SEXUAL FUNCTION BEFORE AND AFTER BARIATRIC SURGERY

Females: There is an increasing trend toward obesity, which is causing concern among health authorities worldwide in terms of its comorbidities, including type 2 diabetes, heart disease, hypertension, and in particular, cancers. There are also less-reported conditions such as reduced life expectancy, infertility, sleep apnea, and (rarely reported) SD. SF has been shown to be impaired in women who are overweight and obese, particularly those seeking bariatric surgery [24]. Kolotkin and Binks [25] reported that women more than men have a greater impairment in sexual QoL, and greater sexual difficulties. Assimakopoulos and Panayiotopoulos [26] found that obese women reported significant impairment on most domains of SF, including sexual desire, sexual arousal, lubrication, orgasm, and sexual satisfaction. Janik and

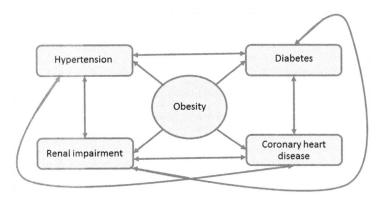

FIGURE 65.2 The relationship of obesity to major pathology, and the potential relationships between significant pathologies associated with obesity. Single arrows are predictive, and double arrows are relational. Demonstrates the relationship of obesity to major pathologies, and highlights the complexity of such relationships across pathologies as an indicator of comorbidities. The example relationships highlighted are not exhaustive, and other important physiological change states that are influenced by obesity may also exist, and indeed, coexist, with those represented in the figure (e.g., metabolic syndrome).

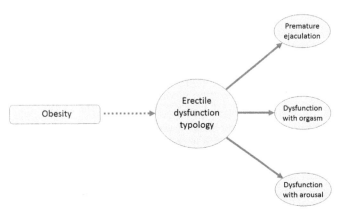

FIGURE 65.3 Obesity as a contributory and/or exacerbating factor for erectile dysfunction. Single arrows relate to typologies, and dashed arrows refers to a potential contributory/exacerbating factor. Highlights the distinct erectile dysfunction categories and how these may be influenced by obesity. An important central tenet emphasized within the figure is that although erectile dysfunction represents a global classification of distinct typologies, obesity may influence all types of erectile dysfunction.

Bielecka [8] suggested that bariatric surgery was the most effective treatment for obesity and its comorbidities, including FSD. Women who lost weight after bariatric surgery reported feeling more attractive and more comfortable with themselves. Their self-esteem had increased, anxiety around sex had diminished, and they had more intense feelings of desire and arousal. Efthymiou and Hyphantis [27] argued that the sudden and dramatic weight loss following bariatric surgery significantly improved health-related QoL, whereas more conservative interventions such as diet, exercise, and medication proved to be relatively ineffective. Dixon and Dixon [28] found that health-related QoL has been observed in both physical and mental subdomains following bariatric surgery. There have also been reports of significant positive impact following bariatric surgery on patients' sexual health, including sexual satisfaction [24]. Efthymiou and Hyphantis [27] found that bariatric surgery could lead to significant improvements not only in health-related QoL, but also many aspects of the sexual response cycle, including arousal, lubrication, and satisfaction, 1 year postoperatively. However, approaches of this kind carry with them various well-known limitations; body mass index (BMI) and QoL improvements are not linear [29], and weight loss has poor predictive power on QoL [28]. There is, however, consensus that sexual satisfaction significantly improves in most aspects 1 year postoperatively, and that weight loss has a major influence on total sexual function. It is in combination with other factors though, such as hormonal regulation [30]. Sarwer and Spitzer [30] reported that among their female patients seeking bariatric surgery, 51% reported SD accompanied by psychosocial distress, which is associated not only with BMI, but sex hormone levels (namely free testosterone, sex hormone—binding globulin, and estradiol). They found a significant improvement in testosterone and estradiol levels 2 years following bariatric surgery, after a weight loss of 32.7%. Improvements both in weight and sexual function following bariatric surgery for women might therefore be explained as a function of weight loss and improved hormonal regulation.

Males: Obesity is a contributing factor to diabetes, coronary heart disease, hypertension, and sleep apnea, and all of these individually may negatively impact sexual and reproductive health in males and females. The relationships between significant pathologies that are influenced, and in many instances, precipitated by, obesity are shown in Fig. 65.2.

Rosenblatt et al. [31] suggested that young males represent one form of the major groups at risk of obesity. These young males are at significant risk of poor sexual health and infertility. ED is commonly reported in severely obese males [31]. The potential contribution of obesity to ED is shown in Fig. 65.3. The treatment options for ED are shown in Fig. 65.4 as a function of etiological orientation.

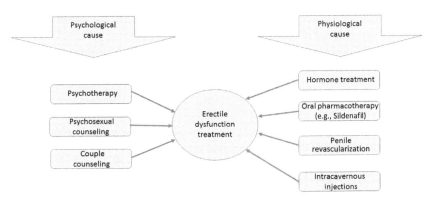

FIGURE 65.4 Treatment intervention approaches and options for erectile dysfunction as indicated by etiological orientation. Showcases the broad range of treatment interventions for erect dysfunction. The suitability and appropriateness of a particular treatment modality is circumscribed by the prevailing causative factors; however, it is possible for interventions to be combinatorial in certain circumstances, and thus a combination of psychological and pharmacological approaches may be taken.

There are surprisingly few studies exploring SD in males alone pre- and postbariatric surgery; the small number of studies generally focus on male fertility in the form of ED, and other areas of the sexual response cycle such as desire presumably remain intact. One study [32] showed that ED occurred in 79% of males who were obese. According to Reis and Favaro [33], elevated proinflammatory cytokines, which induced endothelial dysfunction through the nitric oxide pathway, could explain the obesity—ED relationship, quite apart from the hypergonadotropic—hypoestrogenic—hyperandrogenemia association. This association was a result of decreased total and free testosterone, and decreased gonadotropins, with an increased level of circulating estrogen [33]. Effectively, ED was a consequence of the deregulation of male hormones, and obesity-associated systemic inflammation. Depending on the nature of the surgical intervention, following bariatric surgery, ED, sexual QoL, and hormonal levels tended to normalize once the weight was lost [31]. Conversely, sperm production and quality of sperm may still be compromised. Nonetheless, there is an inconsistency with this argument; research [33] has found no significant differences in sperm variables pre- and postoperatively. The problem is confounded, however, by a small number of specific studies with small sample sizes. The vast majority of research has found that following bariatric surgery, ED returned to normative levels.

MINI-DICTIONARY OF TERMS

- *Diagnostic and statistical manual of mental disorders*: Edited by the American Psychiatric Association, and designed to augment the diagnosis and treatment of mental health disorders.
- *Sexual function*: Process that embodies the sexual response cycle of desire, excitement, orgasm, and resolution.
- *Sexual dysfunction*: Disturbance in the process that embodies the sexual response cycle, which may occur in any one of the phases.
- *Erectile dysfunction*: An inability to attain or retain an erection that would allow for penetrative sex.
- *Premature ejaculation*: Male ejaculation resulting from no or minimal stimulation of the penis.
- *Delayed ejaculation*: Sometimes referred to impaired ejaculation in which the male requires an extended period of sexual stimulation to reach orgasm.
- *Testosterone*: Steroid hormone from the androgen group of hormones. Secreted by the testes in males, and ovaries in females, and adrenal glands.
- *Psychosocial*: Psychological development interacting with social environments.
- *Psychogenic*: A disorder in which there is a psychological origin or cause rather than a physical one.

KEY FACTS

- FSD can manifest in any one the phases of the sexual cycle.
- Forty-three percent of premenopausal women have reported subclinical instances of SD.
- SDs have interdependent factors such as educational attainment, poor physical health, and even race.
- SD has been linked to normative changes in aging, pregnancy, parturition (giving birth), and breastfeeding.
- Many of the features associated with obesity and SD are alleviated following bariatric surgery.
- MSD manifests primarily in the excitement phase of the sexual response cycle.
- Thirty-one percent of men reported instances of SD.

- Risk factors include cardiovascular disease, diabetes, colitis, hypertension, poor physical health, and obesity.
- The two prevalent complaints of SD in males is ED and PE.
- Many of the features related to SD are alleviated following bariatric surgery with regards to sperm production and quality.

SUMMARY POINTS

- Sexual function disorder is a common condition among those seeking surgical intervention for excessive weight.
- Symptoms of SD can manifest in any one of the phases of the sexual response cycle.
- Prior to bariatric surgery, females may report impairment in sexual QoL and greater sexual difficulties.
- Females who lost weight following bariatric surgery reported feeling more attractive, more comfortable with themselves, and had increased feelings of desire and arousal.
- SD in males manifests primarily as ED or PE.
- Following bariatric surgery, male ED and sexual QoL tend to normalize once weight is lost; a few studies, however, have reported variations in sperm production and quality, but more research is required.

REFERENCES

[1] Wang Y, Beydoun MA. The obesity epidemic in the United States--gender, age, socioeconomic, racial/ethnic, and geographic characteristics: a systematic review and meta-regression analysis. Epidemiol Rev 2007;29:6–28.
[2] Rokholm B, Baker JL, Sørensen TIA. The levelling off of the obesity epidemic since the year 1999 – a review of evidence and perspectives. Obes Rev 2010;11(12):835–46.
[3] Bodenlos JS, Wormuth BM. Watching a food-related television show and caloric intake. A laboratory study. Appetite 2013;61(1):8–12.
[4] Vetter ML, et al. Relation of health-related quality of life to metabolic syndrome, obesity, depression and comorbid illnesses. Inter J Obes 2011;35(8):1087–94.
[5] Louie SM, Roberts LS, Nomura DK. Review: mechanisms linking obesity and cancer. BBA 2013;1831:1499–508.
[6] Kopelman PG. Obesity as a medical problem. Nature 2000;404(6778):635–43.
[7] Larsen SH, Wagner G, Heitmann BL. Sexual function and obesity. Inter J Obes 2007;31(8):1189–98.
[8] Janik MR, et al. Female sexual function before and after bariatric surgery: a cross-sectional study and review of literature. Obes Surg 2015;25(8):1511–17.
[9] Kolotkin RL, Zunker C, Ostbye T. Sexual functioning and obesity: a review. Obesity 2012;20(12):2325–33.
[10] APA. Diagnostic and statistical manual of mental disorders. 5th ed. Washington, DC: American Psychiatric Association; 2013.
[11] Masters WH, Johnson VE. Human sexual response, Vol. 185. Boston: Little, Brown & Co; 1966.
[12] Kaplan HS. The new sex therapy. Volume II Disorders of sexual desire and other new concepts and techniques in sex therapy. New York, NY: Brunner/Mazel; 1979.
[13] Berman JR, Goldstein I. Female sexual dysfunction. Urol Clin N Am 2001;28(2):405–16.
[14] Mimoun S, Wylie K. Female sexual dysfunctions: definitions and classification. Maturitas 2009;63(2):116–18.
[15] Raina R, et al. Female sexual dysfunction: classification, pathophysiology, and management. Fertil Steril 2007;88(5):1273–84.
[16] Jha S, Thakar R. Female sexual dysfunction. Eur J Obs Gynecol Reprod Biol 2010;153(2):117–23.
[17] Laumann EO, Paik A, Rosen RC. Sexual dysfunction in the United States: prevalence and predictors. JAMA 1999;281(6):537–44.
[18] Dennerstein L, Dudley E, Burger H. Are changes in sexual functioning during midlife due to aging or menopause? Fertil Steril 2001;76(3):456–60.
[19] Rösing D, et al. Male sexual dysfunction: diagnosis and treatment from a sexological interdisciplinary perspective. Dtsch Arztebl Int 2009;106(50):821–8.
[20] Benn M, et al. Extreme concentrations of endogenous sex hormones, ischemic heart disease, and death in women. Arterios Thromb Vasc Biol 2015;35(2):471–7.
[21] Montorsi F, et al. Summary of the recommendations on sexual dysfunctions in men. J Sexual Med 2010;7(11):3572–88.
[22] Hatzimouratidis K, et al. Guidelines on male sexual dysfunction: erectile dysfunction and premature ejaculation. Eur Urol 2010;57(5):804–14.
[23] Althof SE. Prevalence, characteristics and implications of premature ejaculation/rapid ejaculation. J Urol 2006;175(3):842–8.
[24] Bond DS, et al. Prevalence and degree of sexual dysfunction in a sample of women seeking bariatric surgery. Surg Obes Rel Dis 2009;5(6):698–704.
[25] Kolotkin RL, et al. Obesity and sexual quality of life. Obesity 2006;14(3):472–9.
[26] Assimakopoulos K, et al. Assessing sexual function in obese women preparing for bariatric surgery. Obes Surg 2006;16(8):1087–91.
[27] Efthymiou V, et al. The effect of bariatric surgery on patient HRQOL and sexual health during a 1-year postoperative period. Obes Surg 2015;25(2):310–18.
[28] Dixon JB, Dixon ME, O'brien PE. Quality of life after lap-band placement: influence of time, weight loss, and comorbidities. Obes Res 2001;9(11):713–21.

[29] Mar J, et al. Two-year changes in generic and obesity-specific quality of life after gastric bypass. Eating Weight Dis-Stud Anorexia Bulimia Obes 2013;18(3):305−10.

[30] Sarwer DB, et al. Changes in sexual functioning and sex hormone levels in women following bariatric surgery. JAMA Surg 2014;149 (1):26−33.

[31] Rosenblatt A, Faintuch J, Cecconello I. Abnormalities of reproductive function in male obesity before and after bariatric surgery—a comprehensive review. Obes Surg 2015;25(7):1281−92.

[32] Cabler S, et al. Obesity: modern man's fertility nemesis. Asian J Androl 2010;12(4):480−9.

[33] Reis L, et al. Erectile dysfunction and hormonal imbalance in morbidly obese male is reversed after gastric bypass surgery: a prospective randomized controlled trial. Inter J Androl 2010;33(5):736−44.

Chapter 66

Depression and Intragastric Balloon Treatment

K. Kotzampassi, A.D. Shrewsbury and G. Stavrou
Aristotle's University of Thessaloniki School of Medicine, Thessaloniki, Greece

LIST OF ABBREVIATIONS

BIB   bioenterics intragastric balloon
BMI   body mass index
%EWL   percentage of excess weight loss
HRQoL   health-related quality-of-life

INTRODUCTION

The reciprocal relationship between obesity and depression is well recognized. The intragastric balloon as a method for weight reduction owes much of its efficacy to the personal support program that incorporates it. In this chapter we examine whether for an obese, depressed individual this combination of weight loss and support for personal lifestyle change might kill two birds with one stone and achieve both weight reduction and alleviate depression.

OBESITY AND DEPRESSION

The prevalence of obesity in modern society has increased dramatically over the past 30 years, and is now considered the most common chronic physical illness by far. However, apart from the obesity itself, it is often associated with debilitating physical ailments, including hypertension, cardiovascular diseases, type 2 diabetes, sleep apnea, and loss of libido. Obesity thus becomes either a risk factor or a comorbidity that relates to a reduction in life expectancy [1].

The deeper link to debilitating psychosocial conditions such as depression, anxiety, and low self-esteem means that depression and obesity share common health complications, but there is also evidence of synergistic negative effects on health and treatment response when the two disorders coexist [2]. Depression is thus considered another comorbidity of obesity, and insurance companies have started to require presurgical psychological evaluations for obesity surgery candidates [3,4]. It is estimated that up to 70% of prebariatric surgery patients present with high rates of current and lifetime Axis I disorders [5], and that approximately 66% of bariatric patients have a lifetime history of psychiatric diagnoses, with 38% meeting the criteria for an Axis I diagnosis [6]. In addition, approximately 25% of female candidates have Beck's Depression Inventory scores, which is indicative of a mood disorder.

Although the precise obesity/depression relationship remains obscure, it appears to be reciprocal. A meta-analysis found a 55% increased risk of an obese individual becoming depressed and a 58% risk of a depressed individual becoming obese [7]. Psychological distress in reaction to personal social negative judgments, failure to achieve a socially idealized standard of leanness, and obesity-related comorbidities have also been identified as triggers for depression. Obesity has also been related to a variety of sexual problems, particularly for females [5,8,9].

Our cultural idealization of body shape, thinness, and fitness is generally accepted as a major cause of high levels of body image dissatisfaction [10–12], even for merely overweight or abdominally obese individuals. The stigmatization and prejudice in most areas of social functioning [12,13] result in impairment in quality of life, which is defined by the

World Health Organization (WHO) as "an individual's perception of their position in life in the context of the culture and value systems in which they live and in relation to their goals, expectations, standards and concerns" [14].

The National Institute of Mental Health has found higher levels of obesity in those with schizophrenia and depression [15]. The increased excretion of cortisol observed during depression due to hypothalamic—pituitary—adrenal axis dysregulation is considered a risk factor for obesity [7,8]. Depressed individuals are also prone to consume alcohol and carbohydrates, particularly in the form of junk food, thus exacerbating abdominal obesity, while further affecting the vegetative symptoms of depression via central serontogic activity [16,17]. Further links between depression and obesity are (1) insufficient physical exercise, which is useful for increasing endorphin levels, improving fitness, and enhancing self-esteem [18] and (2) antidepressant medication, particularly the second-generation antipsychotics, which are proven to contribute to increased visceral adiposity [15,19]. Finally, cognitive processes, family and social support, and financial and social resources are additional factors that deserve consideration in the reciprocal association between mental disorders and obesity [15].

THE INTRAGASTRIC BALLOON

Endoscopically inserted intragastric balloons—particularly the BioEnterics Intragastric Balloon (BIB), which is the most widely studied—are used worldwide as short-term treatment options in obese categories analyzed elsewhere [20—23]. As a medical procedure, it is minimally invasive, of minimal operational risk, generally safe, effective, and low cost. Moreover, it is a transient intervention, completely reversible, does not impose permanent alterations to the digestive tract anatomy, and is repeatable at any time [22,24]. The possible disadvantages are that it is not a permanent treatment, and success in terms of permanent weight reduction requires a permanent change in lifestyle and diet; it may also present an ethical dilemma for the surgeon if requested for other than strictly medical reasons [25].

The creation of a continuous sensation of satiety by mechanical gastric volume restriction is a mechanism similar to that of gastric restrictive operations, and leads to ingestion of smaller portions of food. When this is associated with a low-calorie diet, its effectiveness for weight loss and reduction in comorbidities is greater than for diet alone [20]. However, the exact mechanisms by which the balloons achieve weight loss are not completely understood, and are a point of debate with regards to their real effectiveness. The prevailing notion is that weight loss results from increased satiety and a delay in gastric emptying; however, controversy still exists concerning their role in plasma ghrelin levels and the best placement site (fundus or antrum) [26,27].

A recent meta-analysis revealed a significant reduction in body mass index (BMI), as well as a significant increase in weight lost and in percentage of excess weight loss (%EWL) in balloon versus sham-treatment-plus-diet groups [23]. Previously published data emphasize the safety and effectiveness of the balloon [20,28,29]; a single-center study from our institution on a total of 500 obese subjects showed a BMI reduction of 18.82% (8.34 ± 3.14 kg/m^2), and %EWL of 42.34 ± 19.07. However, the striking finding is that individuals who achieved 80% of total weight loss during the first 3 months were found to have much better results in the maintenance of the weight lost over the 5-year follow-up [22].

Generally speaking, analysis of data for 583 obese individuals identified advanced age, female gender, basic educational level, and single or divorced marital status as the most powerful determinants of %EWL > 50%. Attendance at more than 4 monthly interviews and adherence to a strict exercise program contributed significantly to a favorable outcome [30]. However, even without lifestyle modification, balloon placement was considered a serious psychological means of supporting weight loss in the obese. This is further backed up by randomized control trial data, where a statistically significant improvement in weight loss parameters was prominent in the study group versus the placebo balloon—treated groups or the diet-only groups [20,26]. Furthermore, the latest Position Statement issued by the American Society for Metabolic and Bariatric Surgery on the role of balloons in the treatment of obesity recommends their use "since they result in notable weight loss." However, separating the effects on individuals treated with the balloon alone from those also supervised for diet and lifestyle, the changes seemed to be challenging, and a multidisciplinary team experienced in providing medical, nutritional, psychological, and exercise counseling was considered mandatory [21].

The influence of preexisting mood disorders (especially depression) on weight loss after various bariatric interventions, although well studied, is still controversial. Kalarchian et al. [6,31] reported only 25.1% weight loss 6 months after gastric bypass surgery for patients with lifetime mood disorders (in relation to 29.5% for patients without), while the NIH Practical Guide [32] considered the evidence of detrimental psychiatric problems a reason for failure. Palmeira et al. [33], however, claimed the existence of reciprocal effects: weight loss—mediated improvement in body image, which in turn reduced concerns about body image, improved the chances of weight loss.

In our study [24], focusing on depression reduction in obese individuals treated with a balloon, there were some interesting findings. First, obesity parameters presented a comparable reduction (from 0 to 6 months) in the depressed and not-depressed, although the reduction was statistically significant within each group. Second, by splitting depressed subjects using the cutoff point of 40 into BMI $\leq$ 40 and BMI $>$ 40 groups, there was no clear difference in the depression score. Third, when individuals of each group were assessed for depression severity, no statistically significant difference was found. And finally, by splitting depressed women according to %EWL into $\leq$30% and $>$30% groups, there was a significant weight loss difference in favor of those who were less severely depressed initially.

Depressed subjects, by identifying and understanding their own gradual psychological improvement in parallel with weight loss, gained a greater understanding of how weight loss may be increased and, more importantly, maintained [34]. However, for changes in the various domains of health-related quality-of-life (HRQoL) to become significant, a minimum of 5% weight loss was required [35]. Since weight loss started early after balloon implantation, early development of autonomy and self-efficacy by the client was found to positively affect long-term weight reduction and behavioral changes [36].

Finally, it must be noted that the encouraging results of balloon treatment appear to be somewhat negatively affected in cases of binging and night eating, which are associated with more severe mood problems and a greater incidence of comorbid psychopathology. However, in a series of balloon-treated patients, a significant reduction in BMI against baseline was prominent, although obese controls exhibited a significantly higher weight loss [37].

INTRAGASTRIC BALLOON AND DEPRESSION IMPROVEMENT

According to the theory indicating hypothalamic—pituitary—adrenal axis dysregulation in depression progress and the consequently uncontrolled body weight gain (or vice versa), a significant weight loss (by any means) appeared to work positively by deactivating the inflammatory cascades driven by the obesity-related metabolic syndrome and by reversal of insulin resistance [5]. In addition, strikingly visible weight loss in the first month (often 5—8 kg, mostly fat), achieved by less effort and fewer difficulties than the individual had expected, produced significant mood improvement. The internal urge to gorge was frustrated, there was an initial, temporary increase in pre- and postprandial satiety, and side effects became increasingly tolerable and decreased [24]. Together these initiated a belief in the achievability of success and the first steps in the significant reduction of depressive symptoms.

Although today psychiatric evaluation before bariatric surgery is a prerequisite, there is inadequate literature in the field of psychiatric disorder evolution after intervention [3,31]. Furthermore, in relation to the use of an endoscopically placed balloon, the only data clearly focused on the impact of achieved weight loss on the modification of mood status are those presented in a prospective study: the 6-month outcome of depression status was recorded and then correlated with the weight loss findings [24].

One hundred obese females initially completed the Beck Depression Inventory BDI-II [38] for blind discrimination between not-depressed ($n = 35$) and depressed ($n = 65$), as well as for the severity of depression. This well-established psychological screening tool enabled assessment of the depression status of an individual into one of four categories: severe, moderate, mild, or not depressed [38]. According to these categories, the 65 individuals found to be depressed were subdivided into those with mild ($n = 26$), moderate ($n = 21$), and severe depression ($n = 18$). The depression statuses of these 65 people were then reevaluated at 3 and 6 months after balloon insertion. Regarding both obesity parameters and age, there were no significant differences between the two groups at time 0.

At 6 months (time of balloon removal) there was no difference in weight loss in the two groups. However, there was a highly statistically significant difference with respect to depression severity: 46 of the 65 depressed individuals (70.8%) were found totally not-depressed, 13 (20%) of 26 had mild depression, 5 (7.7%) of 21 were moderate, and 1 (1.5%) of 18 suffered from severe depression. This reduction, when expressed as the number of depressed subjects, represented a 70% recovery to "no depression" status after weight loss [24].

Improvements in depression status as early as 2—4 weeks after gastric bypass have been reported [39], despite the fact that no changes in self-reported physical health had occurred. Significantly improved body satisfaction as early as 6 months posttreatment was also demonstrated [40], possibly attributable to subjects taking an active role in changing their lives, despite still being overweight [41]. However, Andersen et al. [42] argued that early mood improvement was probably related to hope for long-term relief, and hypothesized that sustained relief of depression was definitely associated with (and dependent on) permanent improvement in self-reported physical health. Mood deterioration, even after the first posttreatment year, could thus be attributed to disappointment from unrealistic expectations, weight gain or mere stabilization, and/or the reoccurrence of comorbidities [13,41,43]. A residual body image dissatisfaction may also be caused by sagging skin following successful weight loss [41,44]. A significant improvement in quality-of-life (QoL) as early as 3 months after a BIB insertion was reported using a disease-specific HRQoL questionnaire in 40 patients with a BMI of 46.5 kg/m^2 at time 0 who exhibited a reduction of 4.7 kg/m^2 [45]. In addition, there was a positive correlation between the decrease in depression and anxiety scores and the amount of weight lost over a 10-year period following both surgical and conventional treatment of the severely obese [46].

Based on all of the above, we can conclusively suggest that the significant body weight reduction achieved in depressed individuals—independent of the severity of depression and the degree of obesity—worked in a positive manner to significantly reduce the depression status. However, it is beyond any doubt that long-term weight reduction cannot be achieved by balloon treatment alone.

Since there appears to be a tendency for some patients to see balloon insertion as a magic bullet to effortlessly improve their lives [13], we requested a brief statement of precise motivation prior to treatment [22,30]. The literature on coping with addictive behavior suggests that problem-focused coping leads to better obesity treatment outcomes than achieved by emotion-focused or avoidant-focused coping [13]. In a total of 583 obese balloon candidates, the most common motive for losing weight was to improve body image (55.1%), either for self-esteem or increased attractiveness (emotion-focused management). This was followed by the perceived necessity to alleviate comorbidities (32.6%) (problem-focused coping), and less commonly, to increase the chances of pregnancy (12.3%). At the conclusion of treatment, when patients were divided into successful-responders (achieving a %EWL > 50%) and poor-responders (%EWL < 20%), we surprisingly found that the most successful motivation in the successful-responders group was the necessity to alleviate comorbidities, and in the poor-responders group, the desire to improve body image and self-esteem [30].

It must, however, be mentioned that depression improvement through obesity relief alone may be insufficient for, and no indicator of, longer-term depression relief. A multidisciplinary approach seems to be essential. In addition to the intragastric balloon, this typically consists of diet, physical activity, and at least monthly interviews with a dietician, a psychologist, and the attending physician [13,22].

Regarding diet, patients were put on a well-balanced diet and advised to keep detailed records of food consumed for purposes of assessment and discussion during the monthly interviews. Avoidance of carbohydrates/sugars affecting the vegetative symptoms of depression, and consumption of fresh fruit and vegetables, were suggested [16]. The holy trinity of sufficient protein, water, and physical activity was essential for muscle mass maintenance. Finally, the importance of environmental control during the action stage of weight loss was taught. The "stages of change" model taught patients to reduce the probability of unhealthy behaviors by decreasing "opportunities" in their environment—e.g., junk food or chocolate should not be kept in the house for other family members. Healthy eating was thus encouraged for all cohabitants [47].

Physical activity, besides using energy, can also improve well-being, which can in turn facilitate other positive behaviors. Progressive amounts of physical activity increase endorphin release, improve norepinephrine regulation, and improve fitness, all of which work synergistically towards enhancing self-esteem [18].

All balloon-treated patients, irrespective of current mood status, were invited for monthly personal interviews to assess weight, review and revise their monthly dietary regimens, and to work on reeducation of eating habits to reduce total caloric intake, portion sizes, and the frequency of snacks, sweets, and dietary fat [30]. Perri [48] suggested the benefit of a health care professional to support individuals posttreatment and to teach them not only to forgive themselves for "slips and lapses," but to view those incidences as opportunities for learning.

"Adherence," the modern term for "compliance," describes the extent to which a patient follows through with previously agreed-upon actions, such as keeping appointments or attending support groups [13]. Adherence to a schedule of personal support interviews (at least 70% attendance) was found to be of great significance to overall success [30]. The maintenance of necessary behaviors to achieve and stabilize weight loss seemed to be directly related to the maintenance of contact with treatment providers over an extended period [48]. This personal contact also strengthened self-confidence and boosted mood and motivation, particularly when treatment was initiated by an individual's own volition and for specific positive reasons. According to self-determination theory, the probability of a person persisting with a behavior depends on the extent to which he or she believes that the idea for initiating and subsequently continuing to regulate said behavior comes from within [30,49].

PERSPECTIVES

Intragastric balloon treatment in depressed individuals appears to be effective, not only in relation to progressive weight reduction, but also in the significant alleviation of depression by means of increased self-esteem, autonomy, and self-reported improvement in physical health. However, it is exactly this relationship that may present the surgeon with an ethical dilemma if it is requested for other than strictly medical reasons, e.g., by a nonobese individual suffering depression due to low self-esteem based on their weight/physical appearance. For the surgeon, the problem is exacerbated if the client is not even overweight because while balloon placement for weight loss is completely out of the question on ethical grounds, there is still the subjectivity of body image and its relationship to depression to consider.

The traditional boundaries of medicine have always seen the profession as dedicated to saving lives, healing, and promoting health [50]. The definition of "cosmetic surgery," however, appears to lie outside these boundaries and to release the physician from the responsibility or obligation to refuse such treatment: "cosmetic surgery is any cosmetic procedure performed to reshape normal structures of the body or to adorn part of the body, with the aim of improving the consumer's appearance and self-esteem...being initiated by the consumer, not medical need" [51].

This appears to directly link medical intervention with individual well-being and QoL. It might reasonably be claimed (and by a petitioning client almost certainly would be) that in our western culture appearance and size are known to directly affect employment, earnings, social life, marriage opportunities, and to influence the instant evaluations of a first impression [52]. Leaving aside the problem of the client so determined to get treatment that they argue its necessity as a preventative measure against future depression, the depressed status of a prospective client itself presents the physician with a problem and raises its own set of questions: To what degree is the "right to self-determination" applicable? Is a depressed client capable of making decisions with sufficient clarity? If the client refuses diet and lifestyle change, stating that only balloon-induced satiety will work, would the doctor be justified in refusing to carry out a procedure doomed to failure? Would refusal to carry out the procedure worsen the client's mental state, or would the failure to achieve weight loss in such a case worsen the client's mental health and put the doctor in direct contravention of the Hippocratic Oath "to help or at least not to do harm?" And last but not least, would a physician be covered ethically and legally after having informed the client (regardless of his assessment of the client's ability to think rationally) of the possible consequences of his/her decision?

Who can answer these questions and how, and also how responsibility is allocated, are opening a new chapter in bariatric surgery.

MINI-DICTIONARY OF TERMS

- *Depression*: A mental disorder consisting of a constant feeling of low mood and self-esteem, loss of interest in everyday pleasures, reduced physical activity, sleep disturbances, or eating disorders.
- *Intragastric balloon*: A balloon-shaped device endoscopically placed in the stomach and filled with saline to reduce food intake, promote satiety, and thus reduce weight.

KEY FACTS

- A serious mental disorder.
- Constant feelings of low mood, low self-esteem, inactivity, and sleeping and eating disturbances.
- Commonly associated with obesity; up to 70% of obese individuals have been shown to demonstrate Axis I disorders.
- When combined with obesity, tends to have a synergistic negative effect on a person's health and to worsen response to treatment.
- A balloon made of silicone or polyurethane and inflated with saline.
- Endoscopically placed in the stomach.
- Promotes weight loss by reducing gastric volume and causing early satiety.
- May remain in the stomach for a maximum of 6 months.
- Must be used in combination with diet and exercise.

SUMMARY POINTS

- Obesity is the most common chronic physical illness.
- Obesity is reciprocally linked to depression.
- An intragastric balloon is an established device for noninvasive weight loss.
- Depressed obese subjects, by losing weight with an intragastric balloon, improve their self-esteem and thus depression severity.
- A multidisciplinary approach with diet, physical activity, and compliance with treatment providers is mandatory.

REFERENCES

[1] Fontaine KR, Redden DT, Wang C, Westfall AO, Allison DB. Years of life lost due to obesity. JAMA 2003;289:187.

[2] Singh G, Jackson CA, Dobson A, Mishra GD. Bidirectional association between weight change and depression in mid-aged women: a population-based longitudinal study. Int J Obes (Lond) 2014;38:591.

[3] Dixon JB, Dixon ME, O'Brien PE. Depression in association with severe obesity: changes with weight loss. Arch Intern Med 2003;163:2058.

[4] Walfish S, Vance D, Fabricatore AN. Psychological evaluation of bariatric surgery applicants: procedures and reasons for delay or denial of surgery. Obes Surg 2007;17:1578.

[5] van Hout GC, Hagendoren CA, Verschure SK, van Heck GL. Psychosocial predictors of success after vertical banded gastroplasty. Obes Surg 2009;19:701.

[6] Kalarchian MA, Marcus MD, Levine MD, Courcoulas AP, Pilkonis PA, Ringham RM, et al. Psychiatric disorders among bariatric surgery candidates: relationship to obesity and functional health status. Am J Psychiatry 2007;164:328.

[7] Luppino FS, de Wit LM, Bouvy PF, Stijnen T, Cuijpers P, Penninx BW, et al. Overweight, obesity, and depression: a systematic review and meta-analysis of longitudinal studies. Arch Gen Psychiatry 2010;67:220.

[8] Katz JR, Taylor NF, Goodrick S, Perry L, Yudkin JS, Coppack SW. Central obesity, depression and the hypothalamo-pituitary—adrenal axis in men and postmenopausal women. Int J Obes Relat Metab Disord 2000;24:246.

[9] Carrilho PJ, Vivacqua CA, Godoy EP, Bruno SS, Brígido AR, Barros FC, et al. Sexual dysfunction in obese women is more affected by psychological domains than that of non-obese. Rev Bras Ginecol Obstet 2015;37:552.

[10] Hogan MJ, Strasburger VC. Body image, eating disorders, and the media. Adolesc Med State Art Rev 2008;19:521.

[11] Nouri M, Hill LG, Orrell-Valente JK. Media exposure, internalization of the thin ideal, and body dissatisfaction: comparing Asian American and European American college females. Body Image 2011;8:366.

[12] Grilo CM, Wilfley DE, Brownell KD, Rodin J. Teasing, body image, and self-esteem in a clinical sample of obese women. Addict Behav 1994;19:443.

[13] Ayad FM, Martin LF. The evolving role of the psychologist. In: Schauer PR, Schirmer BD, Brethauer S, editors. Minimally invasive bariatric surgery. New York: Springer Verlag; 2007. p. 65—86.

[14] The WHOQOL Group. The development of the World Health Organization Quality of Life Assessment Instrument (WHOQOL). In: Orley J, Kuyken W, editors. Quality of life assessment: international perspectives. New York: Springer Berlin Heidelberg; 1994. p. 41—57.

[15] Allison DB, Newcomer JW, Dunn AL, Blumenthal JA, Fabricatore AN, Daumit GL, et al. Obesity among those with mental disorders: a National Institute of Mental Health meeting report. Am J Prev Med 2009;36:341.

[16] Wurtman RJ, Wurtman JJ. Carbohydrates and depression. Sci Am 1989;260:68.

[17] Greenfield TK, Rehm J, Rogers JD. Effects of depression and social integration on the relationship between alcohol consumption and all-cause mortality. Addiction 2002;97:29.

[18] Ross CE, Hayes D. Exercise and psychologic well-being in the community. Am J Epidemiol 1988;127:762.

[19] Aronne LJ, Segal KR. Weight gain in the treatment of mood disorders. J Clin Psychiatry 2003;64:22.

[20] Genco A, Balducci S, Bacci V, Materia A, Cipriano M, Baglio G, et al. Intragastric balloon or diet alone? A retrospective evaluation. Obes Surg 2008;18:989.

[21] Ali MR, Moustarah F, Kim JJ, Kothari SN, on behalf of the American Society for Metabolic and Bariatric Surgery Clinical Issues Committee. American Society for Metabolic and Bariatric Surgery Position Statement on Intra-Gastric Balloon Therapy, Surg. Obes. Relat. Dis. In press. Available from: http://dx.doi.org/10.1016/j.soard.2015.12.026.

[22] Kotzampassi K, Grosomanidis V, Papakostas P, Penna S, Eleftheriadis E. 500 intragastric balloons: what happens 5 years thereafter? Obes Surg 2012;22:896.

[23] Moura D, Oliveira J, De Moura EGH, Wanderlei B, Neto MG, Campos J, Popov VB, Thompson C. Effectiveness of intragastric balloon for obesity: a systematic review and meta-analysis based on randomized control trials, Surg. Obes. Relat. Dis. In press. Available from: http://dx.doi.org/10.1016/j.soard.2015.10.077.

[24] Deliopoulou K, Konsta A, Penna S, Papakostas P, Kotzampassi K. The impact of weight loss on depression status in obese individuals subjected to intragastric balloon treatment. Obes Surg 2013;23:669.

[25] Kotzampassi K, Shrewsbury AD. Intragastric balloon: ethics, medical need and cosmetics. Dig Dis 2008;26:45.

[26] Mathus-Vliegen EM, Eichenberger RI. Fasting and meal-suppressed ghrelin levels before and after intragastric balloons and balloon-induced weight loss. Obes Surg 2014;24:85.

[27] Papavramidis TS, Grosomanidis V, Papakostas P, Penna S, Kotzampassi K. Intragastric balloon fundal or antral position affects weight loss and tolerability. Obes Surg 2012;22:904.

[28] Imaz I, Martínez-Cervell C, García-Alvarez EE, Sendra-Gutiérrez JM, González-Enríquez J. Safety and effectiveness of the intragastric balloon for obesity. A meta-analysis. Obes Surg 2008;18:841.

[29] Dumonceau JM. Evidence-based review of the Bioenterics intragastric balloon for weight loss. Obes Surg 2008;18:1611.

[30] Kotzampassi K, Shrewsbury AD, Papakostas P, Penna S, Tsaousi GG, Grosomanidis V. Looking into the profile of those who succeed in losing weight with an intragastric balloon. J Laparoendosc Adv Surg Tech A 2014;24:295.

[31] Kalarchian MA, Marcus MD, Levine MD, Soulakova JN, Courcoulas AP, Wisinski MS. Relationship of psychiatric disorders to 6-month outcomes after gastric bypass. Surg Obes Relat Dis 2008;4:544.

[32] National Institutes of Health. National Heart, Lung, and Blood Institute, and North American Association for the Study of Obesity, The Practical Guide: Identification, Evaluation, and Treatment of Overweight and Obesity in Adults. Washington DC: N.I.H; 2002.

[33] Palmeira AL, Markland DA, Silva MN, Branco TL, Martins SC, Minderico CS, et al. Reciprocal effects among changes in weight, body image, and other psychological factors during behavioral obesity treatment: a mediation analysis. Int J Behav Nutr Phys Act 2009;6:9.

[34] Lasikiewicz N, Myrissa K, Hoyland A, Lawton CL. Psychological benefits of weight loss following behavioural and/or dietary weight loss interventions. A systematic research review. Appetite 2014;72:123.

[35] Ross R, Bradshaw AJ. The future of obesity reduction: beyond weight loss. Nat Rev Endocrinol 2009;5:319.

[36] Michie S, Abraham C, Whittington C, McAteer J, Gupta S. Effective techniques in healthy eating and physical activity interventions: a meta-regression. Health Psychol 2009;28:690.

[37] Puglisi F, Antonucci N, Capuano P, Zavoianni L, Lobascio P, Martines G, et al. Intragastric balloon and binge eating. Obes Surg 2007;17:504.

[38] Beck AT, Steer RA, Garbin MG. Psychometric properties of the Beck Depression Inventory: Twenty-five years of evaluation. Clin Psychol Rev 1988;8:77.

[39] Dymek MP, Le Grange D, Neven K, Alverdy J. Quality of life after gastric bypass surgery: a cross-sectional study. Obes Res 2002;10:1135.

[40] Hrabosky JI, Masheb RM, White MA, Rothschild BS, Burke-Martindale CH, Grilo CM. A prospective study of body dissatisfaction and concerns in extremely obese gastric bypass patients: 6- and 12-month postoperative outcomes. Obes Surg 2006;16:1615.

[41] Kubik JF, Gill RS, Laffin M, Karmali S. The impact of bariatric surgery on psychological health. J Obes 2013;2013:837989.

[42] Andersen JR, Aasprang A, Bergsholm P, Sletteskog N, Våge V, Natvig GK. Anxiety and depression in association with morbid obesity: changes with improved physical health after duodenal switch. Health Qual Life Outcomes 2010;8:52.

[43] Wadden TA, Sarwer DB, Womble LG, Foster GD, McGuckin BG, Schimmel A. Psychosocial aspects of obesity and obesity surgery. Surg Clin North Am 2001;81:1001.

[44] Kinzl JF, Traweger C, Trefalt E, Biebl W. Psychosocial consequences of weight loss following gastric banding for morbid obesity. Obes Surg 2003;13:105.

[45] Rutten SJ, de Goederen-van der Meij S, Pierik RG, Mathus-Vliegen EM. Changes in quality of life after balloon treatment followed by gastric banding in severely obese patients - the use of two different quality of life questionnaires. Obes Surg 2009;19:1124.

[46] Karlsson J, Taft C, Rydén A, Sjöström L, Sullivan M. Ten-year trends in health-related quality of life after surgical and conventional treatment for severe obesity: the SOS intervention study. Int J Obes (Lond) 2007;31:1248.

[47] Prochaska JO, DiClemente CC, Norcross JC. In search of how people change. Applications to addictive behaviors. Am Psychol 1992;47:1102.

[48] Perri M. Improving maintenance of weight lost in behavioral treatment of obesity. In: Wadden TA, Stunkard AJ, editors. Handbook of Obesity Treatment. New York: Guilford Press; 2002. p. 357–79.

[49] Elfhag K, Rössner S. Who succeeds in maintaining weight loss? A conceptual review of factors associated with weight loss maintenance and weight regain. Obes Rev 2005;6:67.

[50] World Medical Association. World medical association declaration of Helsinki. Ethical principles for medical research involving human subjects. Bull World Health Org 2001;79:373.

[51] Committee of Inquiry into Cosmetic Surgery (CICS). The cosmetic surgery report to the NSW Minister of Health, Health Care Complaints Commission, Sydney, 1999.

[52] Abraham SF. Dieting, body weight, body image and self-esteem in young women: doctors' dilemmas. Med J Aust 2003;178:607.

Section VIII

Resources

Chapter 67

Recommended Resources on Metabolism and Physiology of Bariatric Surgery

R. Rajendram[1,2,3], C.R. Martin[4] and V.R. Preedy[2]

[1]Stoke Mandeville Hospital, Aylesbury, United Kingdom, [2]King's College London, London, United Kingdom, [3]King Abdulaziz Medical City, Ministry of National Guard Hospital Affairs, Riyadh, Saudi Arabia, [4]Buckinghamshire New University, Middlesex, United Kingdom

INTRODUCTION

There is an ongoing epidemic of obesity. Patients undergoing bariatric surgery are a major subgroup of the morbidly obese. The focus on developing and refining surgical techniques for bariatric surgery has driven down perioperative morbidity and mortality. As a result, despite the significant comorbidities in this cohort, the 30-day mortality rate after elective bariatric surgery is surprisingly low (0.3%) [1].

Weight loss in this cohort is associated with a significant reduction in morbidity and mortality. Therefore, the perceived safety of bariatric surgery has led to a massive increase in uptake of these techniques. Alongside the development of novel surgical techniques, recent studies have identified a major effect of psychosocial factors on long-term outcomes after bariatric surgery [2]. These include emotional adjustment, compliance with the recommended

TABLE 67.1 Regulatory Bodies and Organizations

Australian Government Department of Health
www.health.gov.au/internet/main/publishing.nsf/Content/obesityguidelines-index.htm

Centers for Disease Control and Prevention (CDC)
www.cdc.gov/globalhealth/countries/egypt

European Medicines Agency (EMA)
ema.europa.eu/ema/index.jsp?curl = pages/special_topics/general/general_content_000349.jsp

International Federation of Clinical Chemistry and Laboratory Medicine (IFCC)
www.ifcc.org

Medicines and Healthcare Products Regulatory Agency (MHRA)
mhra.gov.uk

National Institutes of Health (NIH)
www.nlm.nih.gov/medlineplus/ency/anatomyvideos/000023.htm

National Institute for Health and Care Excellence (NICE)
www.nice.org.uk

National Institutes of Health (NIH) Obesity Research
www.obesityresearch.nih.gov

This table lists the regulatory bodies and organizations involved with various aspects of the metabolism and physiology of bariatric surgery.

631

TABLE 67.2 Professional Societies

American College of Gastroenterology (ACG)
gi.org

American Society for Metabolic and Bariatric Surgery (ASMBS)
asmbs.org

American Society for Parenteral and Enteral Nutrition (ASPEN)
www.nutritioncare.org

Associação Brasileira para o Estudo da Obesidade e da Síndrome Metabólica
(Brazilian Association for the Study of Obesity and Metabolic Syndrome; ABESO)
www.abeso.org.br

Association for the Study of Obesity (ASO)
www.aso.org.uk

Belgian Association for the Study of Obesity (BASO)
belgique.easo.org

British Obesity and Metabolic Surgery Society (BOMSS)
www.bomss.org.uk

Canadian Association of Bariatric Physicians and Surgeons (CABPS)
cabps.ca/news/index.php

Deutsche Adipositas Gesellschaft (German association for the study of obesity; DAG)
www.adipositas-gesellschaft.de

Deutsche Gesellschaft für Allgemein- und Viszeralchirurgie (German society for general and visceral surgery; DGAV)
www.dgav.de

Deutsche Gesellschaft für Allgemein- und Viszeralchirurgie, Chirurgische Arbeitsgemeinschaft Adipositastherapie und metabolische
Chirurgie (CAADIP)
www.dgav.de/arbeitsgemeinschaften/caadip.html

EuropeanAssociation for the Study of Obesity (EASO)
easo.org

International Federation for the Surgery of Obesity & Metabolic Disorders (IFSO)
www.ifso.com

The Obesity Society (TOS)
www.obesity.org

Sociedade Brasileira de Cirurgia Bariátrica e Metabólica (Brazilian Society for Bariatric and Metabolic Surgery; SBCBM)
www.sbcbm.org.br

World Obesity Federation (WOF)
www.worldobesity.org

This table lists the professional societies involved with the metabolism and physiology of bariatric surgery.

postoperative regimen, favorable weight loss outcomes, and postoperative improvement of comorbidities. It is now recommended that specialists in the health behaviors of bariatric patients be involved in the perioperative evaluation of obese patients [2].

While the involvement of specialists in health behaviors is likely to improve outcomes after bariatric surgery, this is only one of many recent developments in the field. It is now nearly impossible even for experienced scientists and clinicians to easily remain up-to-date. For those new to the field it is difficult to know which of the myriad of available resources are reliable. To assist colleagues who are interested in understanding more about the metabolism and physiology of bariatric surgery, we have compiled tables containing reliable, up-to-date resources in this chapter. The experts who assisted with the compilation of these tables are acknowledged below.

Tables 67.1−67.7 list the most up-to-date information on the regulatory bodies (Table 67.1), professional bodies (Table 67.2), journals (Tables 67.3 and 67.4), books (Table 67.5), emerging techniques and platforms (Table 67.6), and

TABLE 67.3 Journals Covering Bariatric Surgery

Obesity Surgery
Surgery for Obesity and Related Diseases
Surgical Endoscopy and Other Interventional Techniques
Obesity
Nutricion Hospitalaria (Hospital Nutrition)
Plos One
International Journal of Obesity
Annals of Surgery
Journal of Clinical Endocrinology and Metabolism
Bariatric Surgical Practice and Patient Care
Obesity Reviews
World Journal of Gastroenterology
Journal of Gastrointestinal Surgery
Journal of Obesity
Diabetes Care
Journal of the American College of Surgeons
Arquivos Brasileiros De Cirurgia Digestiva Obesite (Brazilian Archives of Digestive Surgery)
American Surgeon
Journal of Laparoendoscopic and Advanced Surgical Techniques
Obesity Facts
Surgical Laparoscopy Endoscopy and Percutaneous Techniques
Wideochirurgia I Inne Techniki Maloinwazyjne (Videosurgery and Other Minimally Invasive Techniques)
Current Atherosclerosis Reports
JAMA Surgery

This table lists the top 25 journals publishing original research and review articles related to bariatric surgery. The list was generated from SCOPUS (www.scopus.com) using the general descriptors "bariatric surgery." The journals are listed in descending order of the total number of articles published in the past 5 years. Of course, different indexing terms or different databases will produce different lists, so this is a general guide only. For example, journals associated with obesity will produce a different list and include a greater proportion of preclinical studies (see Table 67.4).

websites (Table 67.7) that are most relevant to an evidence-based study of the metabolism and physiology of bariatric surgery.

ACKNOWLEDGMENTS

We would like to thank the following authors (in alphabetical order) for contributing to the development of this resource: V. Agrawal, K. Apostolou, M. Aslan, S.M. Bruschi Kelles, E. Cazzo, F.S. Celi, L. Crovesey, I. Gesquiere, T. Koch, and K. Kotzampassi.

KEY FACTS

- There is an ongoing epidemic of obesity.
- Patients having bariatric surgery are a major subgroup of the morbidly obese.

TABLE 67.4 Journals Covering Obesity

Plos One
Obesity
Obesity Surgery
International Journal of Obesity
BMC Public Health
Surgery for Obesity and Related Diseases
Journal of Clinical Endocrinology and Metabolism
Nutricion Hospitalaria
Obesity Reviews
Appetite
Diabetes
British Journal of Nutrition
Public Health Nutrition
American Journal of Clinical Nutrition
Diabetes Care
Endocrinology
Diabetologia
Metabolism: Clinical and Experimental
American Journal of Physiology: Endocrinology and Metabolism
Journal of Obesity
Nutrients
Preventive Medicine
European Journal of Clinical Nutrition
Cell Metabolism
Pediatric Obesity

This table lists the top 25 journals publishing original research and review articles related to obesity. The list was generated from SCOPUS (www.scopus.com) using the general descriptor "obesity." The journals are listed in descending order of the total number of articles published in the past 5 years. Of course, different indexing terms or different databases will produce different lists, so this is a general guide only. For example, journals associated with bariatric surgery will produce a different list (see Table 67.3). In this table we do not identify those that pertain to preclinical studies (animals, cells, and other systems) or epidemiology, etc.

TABLE 67.5 Relevant Books and Other Publications

S. Agrawal, Obesity, Bariatric and Metabolic Surgery: A Practical Guide, Springer, Switzerland, 2016.

G.A. Bray, C. Bouchard, Handbook of Obesity, Volumes 1 and 2, CRC press, USA, 2014.

H. Buchwald, Buchwald's Atlas of Metabolic & Bariatric Surgical Techniques and Procedures: Expert Consult, Saunders, Philadelphia, 2012.

H. Buchwald, G.S.M. Cowan Jr., W.J. Pories, Surgical Management of Obesity, Saunders, Philadelphia, 2006.

A. El Solh, Critical Care Management of the Obese Patient, John Wiley & Sons, Hoboken, 2012.

(Continued)

TABLE 67.5 (Continued)

S. Hakim, F. Favretti, G. Segato, B. Dillemans, Bariatric Surgery, Imperial College Press, UK, 2011.

A. Hochstrasser, Patient's Guide to Weight Loss Surgery: Everything You Need To Know About Gastric Bypass and Bariatric Surgery, revised ed., Hatherleigh Press, USA, 2009.

D.A. Johnson, Gastroenterological Issues in the Obese Patient, Gastroenterology Clinics of North America, Elsevier, Atlanta, 2010.

J.I. Mechanick, A. Youdim, D.B. Jones, W.T. Garvey, D.L. Hurley, M.M. McMahon, et al., American Association of Clinical Endocrinologists; Obesity Society; American Society for Metabolic & Bariatric Surgery. Clinical Practice Guidelines for the Perioperative Nutritional, Metabolic, and Nonsurgical Support of the Bariatric Surgery Patient—2013 Update: Co-sponsored by American Association of Clinical Endocrinologists, The Obesity Society, and American Society for Metabolic & Bariatric Surgery. Obesity. 21 Suppl 1 (2013) S1.

N.T. Nguyen, R.P. Blackstone, J.M. Morton, J. Ponce, R.J. Rosenthal, The ASMBS Textbook of Bariatric Surgery Volume 1: Bariatric Surgery, Springer, New York, 2015.

C. Pitombo, K. Jones, K. Higa, J. Pareja, Obesity Surgery: Principles and Practice, McGraw-Hill Professional, New York, 2008.

C. Still, D.B. Sarwer, J. Blankenship, The ASMBS Textbook of Bariatric Surgery Volume 2: Integrated Health, Springer, New York, 2014.

This table lists books on the metabolism and physiology of bariatric surgery.

TABLE 67.6 Sources and Resources for Emerging Techniques and Platforms

American Society for Gastrointestinal Endoscopy (ASGE)
www.asge.org

American Society for Metabolic and Bariatric Surgery (ASMBS)—Emerging Technologies and Procedures Inventory
asmbs.org/resources/emerging-technologies-and-procedures-inventory

Bariatric and Metabolic Center of Colorado—Emerging Bariatric Technology
bariatricsurgeryco.org/bariatric-surgery/emerging-bariatric-technology

Glucagon.com
www.glucagon.com

Robotic Bariatric Surgery
link.springer.com/article/10.1007/s00268-013-2125-3/fulltext.html

Vibration Therapy After Bariatric Surgery
www.vibrationexercise.com/vibration-therapy-after-bariatric-surgery

This table lists some emerging resources, and resource platforms, on the metabolism and physiology of bariatric surgery.

TABLE 67.7 Relevant Internet Resources

BariataricPal
www.bariatricpal.com/page/articles.html

Bariatric Skinny
www.bariatricskinny.com/weight_loss_surgery_journal

Bariatric Surgery Journal Articles – Index – Medscape
www.medscape.com/index/list_6950_0

Canadian Institute for Health Information
https://secure.cihi.ca/estore/productSeries.htm?locale=en&pc=PCC1075

(Continued)

TABLE 67.7 (Continued)

Eetexpert
www.eetexpert.be

Mayo Clinic
http://www.mayoclinic.org/tests-procedures/bariatric-surgery/basics/definition/prc-20019138

MedHelp Bariatric Forum
www.medhelp.org/forums/Bariatric---Weight-Loss-Surgery/show/213

Medscape
www.medscape.com

National Institutes of Health: Office of Dietary Supplements
ods.od.nih.gov

News, Medical
www.news-medical.net

Obesity Action Coalition
www.obesityaction.org

Obesity Help
www.obesityhelp.com

Suggested Reading Before and After Bariatric Surgery
www.obesitycoverage.com/review/suggested-reading

WebMD
www.webmd.com/diet/obesity/surgery-for-you

This table lists some Internet resources on the metabolism and physiology of bariatric surgery.

- It is difficult to keep up-to-date with developments in bariatric surgery.
- This chapter lists the most up-to-date resources on the regulatory bodies, journals, books, professional bodies, and websites that are relevant to an evidence-based approach to bariatric surgery.

SUMMARY POINTS

- There is an ongoing epidemic of obesity.
- Weight loss in this cohort is associated with a significant reduction in morbidity and mortality.
- The involvement of specialists in health behaviors is likely to improve outcomes after bariatric surgery.
- It difficult even for specialists to remain up-to-date and determine which of the many available resources are reliable.
- This chapter includes lists of the most up-to-date information on the regulatory bodies, professional bodies, journals, books, emerging techniques and platforms, and websites that are relevant to bariatric surgery.

REFERENCES

[1] Flum DR, Belle SH, King WC, et al. Perioperative safety in the longitudinal assessment of bariatric surgery. New Engl J Med 2009;361:445–54.
[2] Sogg S, Lauretti J, West-Smith L. Recommendations for the presurgical psychosocial evaluation of bariatric surgery patients. Surg Obes Relat Dis 2016;12:731–49.

Index

Note: Page numbers followed by "*f*" and "*t*" refer to figures and tables, respectively.

Apolipoprotein B48 (ApoB48), 425
  response, 424
Apollo method, 287–289, 293
Apollo OverStitch for endoscopic sleeve
    gastroplasty, 130
Appetite, 47–48
  evolution
    of dietary compliance and, 43f
    of weight and, 42f
  PENS of dermatome T6 to reducing, 39–40
Appetitive activation motivational system,
    584
APTT. *See* Activated partial thromboplastin
    time (APTT)
Arachidonic acid (AA), 529–531
ARB. *See* Angiotensin receptor blocker (ARB)
Arbeitsgemeinschaft Adipositas im Kindes-und
    Jugendalter (AGA), 202
Area under curve (AUC), 309, 374, 502
Area under receiver operating characteristic
    curve (AUROC curve), 327–328
ARM. *See* Acid-reducing medications
    (ARM)
ASGE. *See* American Society for
    Gastrointestinal Endoscopy (ASGE)
ASMBS. *See* American Society for Metabolic
    and Bariatric Surgery (ASMBS)
Aspartate aminotransferase (AST), 99
Aspiration therapy, 132
AspireAssist Aspiration Therapy System, 51
AST. *See* Aspartate aminotransferase (AST)
Asthma
  antiinflammatory action of insulin, 34
  asthma-related genes in obesity, 34
  insulin resistance, 34
    effect of intravenous insulin infusion on
      inflammation, 35
    obesity, and bariatric surgery, 33–34
    proinflammatory state of obesity, 34
    reversal of proinflammatory state, 34–35
Atherosclerosis, 13, 511, 517
  bariatric surgery effects on early markers of,
    515–516
  progression of, 512f
Attrition rate, 245
AUC. *See* Area under curve (AUC)
AUROC curve. *See* Area under receiver
    operating characteristic curve
    (AUROC curve)
Axis I psychiatric disorders, 608–609

B

Balloon, 286, 287t
  devices, 53
Band technology modification (EASYBAND),
    192
Bariatric
  bypass surgery, 449
  interventions, 54, 535
  operations, 491
Bariatric endoscopy (BE), 285
Bariatric outcomes longitudinal database
    (BOLD), 277

Bariatric patients
  endoluminal procedures for, 128
    intraoperative endoscopy, 128–129
    obesity, 127–128
    postoperative evaluation, 129
    preoperative evaluation, 128
    primary endoscopic procedures,
      129–135
    revisional endoscopic procedures, 135
    preoperative GERD and HH evaluation in,
      276–277
Bariatric procedures
  accreditation, 72–73
  bariatric procedure in patient with GERD,
    277
  bariatric surgery, 71
    laparoscopy in, 71–72
  bariatric surgery selection in presence of
    HH, 278–279
  Barrett's esophagus, 278
  effect of fellowship training, 72
  impact of LSG on GERD, 277–278
  management of GERD and HH after
    bariatric surgery, 279
  mechanisms, 275–276
  preoperative GERD and HH evaluation in
    bariatric patient, 276–277
  standardization, 72–73
  treatment, 72
Bariatric surgery (BS), 9, 19, 30, 39, 50–54,
    61, 71, 103, 120–122, 139, 155,
    157–158, 171, 199, 217, 265–266,
    275, 285, 295–296, 305, 315–316,
    321, 333–338, 336f, 338t, 354, 371,
    399, 429, 457, 469, 479, 500, 511,
    531–532, 540, 543, 545f, 567, 583,
    595, 605
  absorption pathways of heme and nonheme
    iron, 500f
  addressing association, 337t
  American society for metabolic and, 51t
  assessment of iron status, 504–505, 505f
  asthma, obesity and, 33–34
  bariatric surgical procedures, 480f
  body dysmorphia after, 608–609
  body dysmorphia before, 607–608
  BS outcome, temperament influencing,
    588–590, 589f, 590f
  cardiovascular risk in morbidly obese
    candidates to, 511–513, 513t
  cognition improvement with, 577–578, 578t
  critical care after
    comorbidities, 182–183
    ICU admission, 182–183
    ICU postoperative management, 183–186
    resources by obese patients after bariatric
      surgery, 182
  demographic of physiological differences,
    62t
  devices
    altering gastric emptying or capacity,
      53–54
    for nutrients, 51–52
    for stomach space, 53

effects
  on cardiovascular events, 516–517
  on cardiovascular risk factors, 513–515
  on early markers of atherosclerosis,
    515–516
fetal outcomes, 165
gait recovery following, 555–558
H. pylori infection outcome after, 345–347
impact
  on insulin resistance, 326
  on iron absorption, 502
  on iron intake, 502, 503t
implications
  defining success, 66
  healthcare, 66–67
  medical outcomes of surgical procedure,
    66
interventions, 135
intragastric balloon on subsequent,
    142–143
iron absorption, 500
iron deficiency after, 503–504
  impact of bariatric surgery on iron intake,
    504f
iron deficiency in candidates, 500–502
laparoscopy in, 71–72
management of GERD and HH, 279
management of thiamine deficiency after,
    486
mechanisms involved in association,
    339–340
  reduced inflammation and oxidative stress,
    339
mechanisms of thiamine deficiency after,
    485–486, 485t
mineral deficiencies after, 544–545
mortality
  long-term mortality after bariatric surgery,
    210–213, 211t, 214f
  perioperative mortality, 207–210
  mortality rate and long-term outcomes after,
    171–174
  hospital admissions after, 174–175
  long-term mortality of obese individuals
    and operated subjects, 172–173
  mid-and long-term mortality due to
    specific causes, 174
  mid-term mortality of obese individuals
    and operated subjects, 172–173, 173t
  QoL, 175–176
  weight regain trends after, 176–177, 177f
obesity treatment, 51
obstetric outcomes, 164
patients
  appearance and body image, 65–66
  CBT interventions for, 596–598
  choosing surgery, 63–66
  eating pathology in, 596
  gender differences, 65f
  health and medical issues, 64–65
  motivations, 64
  obesity as chronic disease, 66
  patient experiences, 63
  primary reasons for choosing, 64f